FUNDAMENTALS OF
HEAT AND MASS TRANSFER

FUNDAMENTALS OF HEAT AND MASS TRANSFER

FOURTH EDITION

(In two colours)

C P KOTHANDARAMAN

Former Head
Department of Mechanical Engineering
PSG College of Technology, Coimbatore
Tamil Nadu

An Imprint of

NEW AGE INTERNATIONAL (P) LIMITED, PUBLISHERS
LONDON • NEW DELHI • NAIROBI
Bangalore • Chennai • Cochin • Guwahati • Hyderabad • Kolkata • Lucknow • Mumbai
Visit us at **www.newagepublishers.com**

Published by New Age International (P) Ltd., Publishers
First Edition: 1994
Fourth Edition: 2012
Reprint: 2020

GLOBAL OFFICES

- **New Delhi** — NEW AGE INTERNATIONAL (P) LIMITED, PUBLISHERS
  7/30 A, Daryaganj, New Delhi-110002, (INDIA)
  Tel.: (011) 23253771, 23253472, Telefax: 23267437, 43551305
  E-mail: contactus@newagepublishers.com • Visit us at www.newagepublishers.com
- **London** — NEW AGE INTERNATIONAL (UK) LTD.
  27 Old Gloucester Street, London, WC1N 3AX, UK
  E-mail: info@newacademicscience.co.uk • Visit us at www.newacademicscience.co.uk
- **Nairobi** — NEW AGE GOLDEN (EAST AFRICA) LTD.
  Ground Floor, Westlands Arcade, Chiromo Road (Next to Naivas Supermarket)
  Westlands, Nairobi, KENYA, Tel.: 00-254-713848772, 00-254-725700286
  E-mail: kenya@newagepublishers.com

BRANCHES

- **Bangalore** 37/10, 8th Cross (Near Hanuman Temple), Azad Nagar, Chamarajpet, Bangalore- 560 018
  Tel.: (080) 26756823, Telefax: 26756820, E-mail: bangalore@newagepublishers.com
- **Chennai** 26, Damodaran Street, T. Nagar, Chennai-600 017, Tel.: (044) 24353401, Telefax: 24351463
  E-mail: chennai@newagepublishers.com
- **Cochin** CC-39/1016, Carrier Station Road, Ernakulam South, Cochin-682 016, Tel.: (0484) 2377303
  Telefax: 4051304
  E-mail: cochin@newagepublishers.com
- **Guwahati** Hemsen Complex, Mohd. Shah Road, Paltan Bazar, Near Starline Hotel, Guwahati-781 008,
  Tel.: (0361) 2513881, Telefax: 2543669, E-mail: guwahati@newagepublishers.com
- **Hyderabad** 105, 1st Floor, Madhiray Kaveri Tower, 3-2-19, Azam Jahi Road, Near Kumar Theater, Nimboliadda
  Kachiguda, Hyderabad-500 027, Tel.: (040) 24652456, Telefax: 24652457
  E-mail: hyderabad@newagepublishers.com
- **Kolkata** RDB Chambers (Formerly Lotus Cinema) 106A, 1st Floor, S N Banerjee Road, Kolkata-700 014
  Tel.: (033) 22273773, Telefax: 22275247, E-mail: kolkata@newagepublishers.com
- **Lucknow** 16-A, Jopling Road, Lucknow-226 001, Tel.: (0522) 2209578, 4045297, Telefax: 2204098
  E-mail: lucknow@newagepublishers.com
- **Mumbai** 142C, Victor House, Ground Floor, N.M. Joshi Marg, Lower Parel, Mumbai-400 013
  Tel.: (022) 24927869, Telefax: 24915415, E-mail: mumbai@newagepublishers.com
- **New Delhi** 22, Golden House, Daryaganj, New Delhi-110 002, Tel.: (011) 23262368, 23262370, Telefax: 43551305
  E-mail: sales@newagepublishers.com

ISBN: 978-81-224-3364-7
C-19-05-12102

Printed in India at Paramanand Offset, Delhi.
Typeset at Goswami Associates, Delhi.

NEW AGE INTERNATIONAL (P) LIMITED, PUBLISHERS
7/30 A, Daryaganj, New Delhi-110002
Visit us at www.newagepublishers.com
(CIN: U74899DL1966PTC004618)

Preface to the Fourth Edition

In this edition two new chapters are added. One deals with various methods of cooling adopted for electronic devices. Design method for simple cases is indicated. The other chapter explains the principle operation of heat pipes and its applications. Heat pipes with very high conductivity are found highly suitable for cooling systems with high rates of heat generation. These are flexible and can transfer heat over long distances. Heat pipes are now used extensively in temperature control in satellites and chemical processes.

The chapter organization is maintained the same as in the previous editions except for the addition of a summary and review questions at the end of text material.

The author expresses his sincere thanks to all those who have suggested additions and modifications for the improvement of the book. Sincere thanks are also due to New Age International (P) Ltd., Publishers and members of the editorial and production department.

Author

Preface to the Third Edition

Professor Obert has observed in his famous treatise on Thermodynamics that concepts are better understood by their repeated applications to real life situations. A firm conviction of this principle has prompted the author to arrange the text material in each chapter in the following order.

In the first section after enunciating the basic concepts and laws mathematical models are developed leading to rate equations for heat transfer and determination of temperature field, simple and direct numerical examples are included to illustrate the basic laws. More stress is on the model development as compared to numerical problems.

A section titled "SOLVED PROBLEMS" comes next. In this section more involved derivations and numerical problems of practical interest are solved. The investigation of the effect of influencing parameters for the complete spectrum of values is attempted here. Problems involving complex situations are shown solved in this section. Two important ideas are stressed in this section. These are checking of dimensional homogeneity in the case of all equations derived and the validation of numerical answers by cross checking. This concept of validation in professional practice is a must in all design situations.

In the next section objective type questions are given. These are very useful for understanding the basis and resolving misunderstandings.

In the final section a large number of graded exercise problems involving simple to complex situations are included.

In the first of the 14 chapters the basic laws for the three modes of heat transfer are introduced and the corresponding rate equations are developed. The use of electrical analogy is introduced and applied to single and multimode heat transfer situations. The need for iterative working is stressed in the solved problems.

The second chapter deals with one dimensional steady state conduction. Mathematical models are developed by the three geometries namely Plate, Hollow Cylinder and Hollow Sphere. Multilayer insulation is also discussed. The effect of variation of thermal conductivity on heat transfer and temperature field is clearly brought out. Parallel flow systems are discussed. Examples on variation of area along the heat flow direction are included. The use of electrical analogy is included in all the worked examples. The importance of calculating the temperature gradient is stressed in many of the problems.

In the third chapter models for conduction with heat generation are developed for three geometric configurations namely plate, cylinder and sphere. The effect of volume to surface area and the convection coefficient at the surface in maintaining lower material temperature is illustrated. Hollow cylindrical shape with different boundary conditions is discussed. Conduction with variable heat generation rate is also modelled.

Fins/extended surface or conduction-convection situation is discussed in the fourth chapter. Models for heat transfer and temperature variation are developed for four different boundary conditions. Optimisation of the shape of the fin of specified volume for maximum heat flow is discussed. Circumferential fins and variable area fins are analysed. The use of numerical method is illustrated. Error in measurement of temperature using thermometer is well discussed. The possibility of measurement of thermal conductivity and convective heat transfer coefficient using fins is illustrated.

Two dimensional steady state conduction is discussed in the fifth chapter. Exact analysis is first developed for two types of boundary conditions. The use of numerical method is illustrated by developing nodal equations. The concept and use of conduction shape factor is illustrated for some practical situations.

One dimensional transient (unsteady) heat conduction is discussed in Chapter 6. Three types of models arise in this case namely lumped heat capacity system, semi-infinite solid and infinite solid. Lumped heat capacity model for which there are a number of industrial applications is analysed in great detail and problems of practical interest are shown solved. The condition under which semi-infinite solid model is applicable as compared to infinite solid model is clearly explained. Three types of boundary conditions are analysed. Infinite solid model for three geometric shapes is analysed next. The complexity of the analytical solution is indicated. Solution using charts is illustrated in great detail. Real solids are of limited dimensions and these models cannot be applied directly in these cases. In these cases product solution is applicable. A number of problems of practical interest for these types of solids are worked out in this section. In both cases a number of problems are solved using numerical methods. Periodic heat flow problems are also discussed.

Concepts and mechanism of convection are discussed in the seventh chapter. After discussing the boundary layer theory continuity, momentum and energy equations are derived. Next the different methods of solving these equations are discussed. In addition to the exact analysis approximate integral method, analogy method and dimensional analysis are also discussed and their applicability is indicated. General correlations for convective heat transfer coefficient in terms of dimensionless numbers are arrived at in this chapter.

In Chapter 8, in addition to the correlations derived in the previous chapter, empirical correlations arrived at from experimental results are listed and applied to flow over surfaces like flat plate, cylinder, sphere and banks of tubes. Both laminar and turbulent flows situation are discussed.

Flow through ducts is discussed in Chapter 9. Empirical correlations for various situations are listed. Flow developing region, fully developed flow conditions, constant wall temperature and constant wall heat flux are some of the conditions analysed. Flow through non-circular pipes and annular flow are also discussed in this chapter.

Natural convection is dealt with in Chapter 10. Various geometries including enclosed space are discussed. The choice of the appropriate correlation is illustrated through a number of problems. Combined natural and forced convection is also discussed.

Chapter 11 deals with phase change processes. Boiling, condensation, freezing and melting are discussed. Basic equations are derived in the case of freezing and melting and condensation. The applicable correlations in boiling are listed and their applicability is illustrated through numerical examples.

Chapter 12 deals with heat exchangers, both recuperative and regenerative types. The LMTD and NTU-effectiveness methods are discussed in detail and the applicability of these methods is illustrated. Various types of heat exchangers are compared for optimising the size.

Thermal radiation is dealt with in Chapter 13. The convenience of the use of electrical analogy for heat exchange among radiating surfaces is discussed in detail and is applied in almost all the solved problems. Gas radiation and multi-body enclosures are also discussed.

Chapter 14 deals with basic ideas of mass transfer in both diffusion and convection modes. A large number of problems with different fluid combinations are worked out in this chapter.

A large number of **short problems** and **fill in the blank type** and **true or false type** questions are provided to test the understanding of the basic principles.

Author

UNIT CONVERSION CONSTANTS

Quantity	*S.I. to English*	*English to S.I.*
Length	1 m = 3.2808 ft	1 ft = 0.3048 m
Area	1 m^2 = 10.7639 ft^2	1 ft^2 = 0.0929 m^2
Volume	1 m^3 = 35.3134 ft^3	1 ft^3 = 0.02832 m^3
Mass	1 kg = 2.20462 lb	1 lb = 0.4536 kg
Density	1 kg/m^3 = 0.06243 lb/ft^3	1 lb/ft^3 = 16.018 kg/m^3
Force	1 N = 0.2248 lb_f	1 lb_f = 4.4482 N
Pressure	1 N/m^2 = 1.4504 × 10^{-4} lb_f/in^2	1 lb_f/in^2 = 6894.8 N/m^2
Pressure	1 bar = 14.504 lb_f/in^2	1 lb_f/in^2 = 0.06895 bar
Energy	1 kJ = 0.94783 Btu	1 Btu = 1.0551 kJ
(heat, work)	1 kW hr = 1.341 hp hr	1 hp hr = 0.7457 kW hr
Power	1 W = 1.341 × 10^{-3} hp	1 hp = 745.7 W
Heat flow	1 W = 3.4121 Btu/hr	1 Btu/hr = 0.29307 W
Specific heat	1 kJ/kg°C = 0.23884 Btu/lb°F	1 Btu/lb°F = 4.1869 kJ/kg°C
Surface tension	1 N/m = 0.068522 lb_f/ft	1 lb_f/ft = 14.5939 N/m
Thermal conductivity	1 W/m°C = 0.5778 Btu/hr ft°F	1 Btu/hrft°F = 1.7307 W/m°C
Convection coefficient	1 W/m^2°C = 0.1761 Btu/hrft^2°F	1 Btu/hr ft^2°F = 5.6783 W/m^2°C
Dynamic viscosity	1 kg/ms = 0.672 lb/fts = 2419.2 lb/ft hr	1 lb/fts = 1.4881 kg/ms or Ns/m^2
Kinematic viscosity	1 m^2/s = 10.7639 ft^2/s	1 ft^2/s = 0.092903 m^2/s
Universal gas const.	8314.41 J/kg mol K = 1545 ft lb_f /mol R = 1.986 B tu/lb mol R	
Stefan Boltzmann const.	5.67 W/m^2K^4 = 0.174 Btu/hr ft^2 R^4	

Quantity	*S.I. to Metric*	*Metric to S.I.*
Force	1 N = 0.1019 kg_f	1 kg_f = 9.81 N
Pressure	1 N/m^2 = 10.19 × 10^{-6} kg_f/cm^2	1 kg_r/cm^2 = 98135 N/m^2
Pressure	1 bar = 1.0194 kg_f/cm^2	1 kg_f/cm^2 = 0.9814 bar
Energy	1 kJ = 0.2389 kcal	1 kcal = 4.186 kJ
(heat, work)	1 Nm (= 1 J) = 0.1019 kg_f m	1 kg_f m = 9.81 Nm (J)
Energy (heat, work)	1 kWhr = 1.36 hp hr	1 hp hr = 0.736 kW hr
Power (metric)	1 W = 1.36 × 10^{-3} hp	1 hp = 736 W
Heat flow	1 W = 0.86 kcal/hr	1 kcal/hr = 1.163 W
Specific heat	1 kJ/kg°C = 0.2389 kcal/kg°C	1 kcal/kg°C = 4.186 kJ/kg°C
Surface tension	1 N/m = 0.1019 kg_f/m	1 kg_f/m = 9.81 N/m
Thermal conductivity	1 W/m°C = 0.86 kcal/hrm°C	1 kcal/hrm°C = 1.163 W/m°C
Convection coefficient	1 W/m^2°C = 0.86 kcal/hrm^2°C	1 kcal/hrm^2°C = 1.163 W/m^2°C
Dynamic viscosity	1 kg/ms (Ns/m^2) = 0.1 Poise	1 poise = 10 kg/ms (Ns/m^2)
Kinematic viscosity	1 m^2/s = 3600 m^2/hr 1 Stoke = cm^2/s = 0.36 m^2/hr = 10^{-4} m^2/s	1 m^2/hr = 2.778 × 10^{-4} m^2/s
Universal gas const.	8314.41 J/kg mol K = 847.54 m kg_f/kg mol K = 1.986 kcal/kg mol K	
Gas constant in air (SI)	= 287 J/kg K	
Stefan Boltzmann const.	5.67 × 10^{-8} W/m^2 K^4 = 4.876 × 10^{-8} kcal/hr m^2 K^4	

Contents

1 CHAPTER

An Overview of Heat Transfer

1.0 INTRODUCTION

The present standard of living is made possible by the energy available in the form of heat from various sources like fuels. The process by which this energy is converted for everyday use is studied under thermodynamics. Thermodymics describes the end state of processes involving heat/work transfer. But the rate at which heat is transferred is left out in the study. In all applications, the rate at which energy is transferred as heat, plays an important role. The design of all equipments involving heat transfer requires the estimation of the rate of heat transfer. There is no need to list the various equipments where heat transfer rate influences their operation.

The driving potential or the force which causes the transfer of energy as heat is the difference in temperature between systems. Other such transport processes are the transfer of momentum, mass and electrical energy. In addition to the temperature difference, physical parameters like geometry, material properties like conductivity, emissivity flow parameters like flow velocity also influence the rate of heat transfer.

The aim of this text is to introduce the various rate equations and methods of determination of the rate of heat transfer across system boundaries under different situations.

1.1 HEAT TRANSFER

The study of heat transfer is directed to (*i*) the estimation of rate of flow of energy as heat through the boundary of a system both under steady and transient conditions, and (*ii*) the determination of temperature field under steady and transient conditions, which also will provide the information about the gradient and time rate of change of temperature at various locations and time. *i.e.* $T(x, y, z, \tau)$ and dT/dx, dT/dy, dT/dz, $dT/d\tau$ etc. These two are interrelated, one being dependent on the other. However explicit solutions may be generally required for one or the other.

The basic laws governing heat transfer and their application are as below:

1. **First law of thermodynamics** postulating the energy conservation principle: This law provides the relation between the heat flow, energy stored and energy generated in a given system. The relationship for a closed system is: **The net heat flow across the system boundary + heat generated inside the system = change in the internal energy, of the system**. This will also apply for an open system with slight modifications.

The change in internal energy in a given volume in a short time dT is equal to the product of volume density and specific heat ρcV and dT where the group ρcV is called the heat capacity of the system. The basic analysis in heat transfer always has to start with one of these relations.

2. **The second law of thermodynamics** establishing the direction of energy transport as heat. The law postulates that the flow of energy as heat through a system boundary will always be in the direction of lower temperature or along the negative temperature gradient which is the driving force for transfer of heat.

3. **Newtons laws of motion** used in the determination of fluid flow parameters connected with heat transfer by convection.

4. **Law of conservation of mass,** used in the determination of flow parameters.

5. **The rate equations** as applicable to the particular mode of heat transfer.

1.2 MODES OF HEAT TRANSFER

1.2.1. Conduction: This is the mode of energy transfer as heat due to temperature difference within a body or between bodies in thermal contact without the involvement of mass flow and mixing

This is the mode of heat transfer through solid barriers and is encountered extensively in heat transfer equipment design as well as in heating and cooling of various materials as in the case of heat treatment. Conduction mode of heat transfer applies to liquids and gases also. The rate equation in this mode is based on **Fourier's law of heat conduction which states that the heat flow by conduction in any direction is proportional to the temperature gradient and area perpendicular to the flow direction and is in the direction of the negative gradient.** The proportionality constant obtained in the relation is known as thermal conductivity, k, of the material. Thermal conductivity is a transport property which directly controls the heat transfer by conduction. Higher the value of 'k' higher the heat flow. The mathematical formulation is given in equation (1.1). Assuming the direction of heat flow is along x and area A is measured perpendicular to the flow direction x.

Heat flow, $$Q = -kA\, dT/dx \quad ...(1.1)$$

The units used in the text for various parameters are:

Q – W, (Watt), A – m^2, dT – °C or K (as this is only temperature interval, °C and K can be used without any difficulty). x – m, k – W/mK.

Instead of (dT/dx), the expression (dT/dn), where dn is the distance along the normal to the area A is more general.

Conduction, Plane Wall (Fig. 1.1), the integration of the equation (1.1) for a plane wall of thickness, L between the two surfaces at T_1 and T_2 under steady condition leads to equation (1.2). The equation can be considered as the mathematical model for this problem.

$$Q = \frac{T_1 - T_2}{(L/kA)} \quad ...(1.2)$$

Rate equations for other shapes are derived in Chapter 2. Equation (1.2) is applicable only if the thermal conductivity does not vary with temperature or location.

Example 1.1: *Determine the heat flow across a plane wall of 10 cm thickness with a constant thermal conductivity of 8.5 W/mK when the surface temperatures are steady at 100°C and 30°C. The wall area is 3m^2. Also find the temperature gradient in the flow direction.*

Solution: Refer to Fig. 1.1 and equation (1.2):

Known: T_1 = 100°C, T_2 = 30°C,

L = 10 cm = 0.1 m,

k = 8.5 W/mK, A = 3 m^2.

Find: (1) Heat flow (2) Temperature gradient.

Assumptions: (1) Thermal conductivity is constant (2) Steady flow of heat.

Therefore, heat flow, **Q** = (100 – 30) / (0.1/(8.5 × 3))

= **17850 W or 17.85 kW.**

Referring to equation (1.1)

$$Q = -kA\, dT/dx$$

$$17850 \text{ W} = -8.5 \times 3\, dT/dx.$$

Therefore, $dT/dx = -17850/(8.5 \times 3)$

= **– 700°C/m**

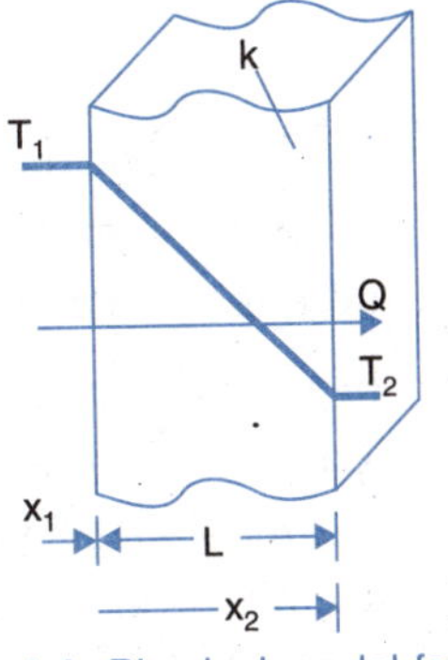

Fig. 1.1. Physical model for example 1.1

This is also equal to – (100 – 30)/0.1 = – 700°C/m, as the gradient is constant all through the thickness.

Electrical analogy is very useful in visualising heat transfer situations. The concept of driving potential as temperature and the factors affecting the heat flow as thermal resistance is useful in solving more complex problems. The same is illustrated in Fig. 1.2.

(a) $Q = \dfrac{(T_1 - T_2)}{(L/kA)}$ (b) $I = \dfrac{V_1 - V_2}{R}$

Fig. 1.2. Electrical analogy (a) Conduction circuit (b) Electrical circuit.

The denominator in equation (1.2), namely L/kA can be considered as **thermal resistance for conduction.**

COMMENTS

1. Higher the value of thermal conductivity higher will be heat flow for the system considered. For example, if the value of k is doubled, the heat flow will be also doubled.
2. Higher the value of thermal conductivity for the same heat flow lower will be temperature gradient.
3. Two distinct terms are used in connection with heat flow.

   (*a*) Heat flux q, which gives the flow in W/m^2

   and (*b*) Heat rate Q which gives the flow for the given area, denoted as W.

   Heat rate = heat flux × area.

1.2.2. Thermal Conductivity

It is the constant of proportionality in Fourier's equation and plays an important role in heat transfer. The unit in SI system for conductivity is W/mK. It is a material property. Its value is higher for good electrical conductors and single crystals like diamond. Next in order are alloys of metals and non metals. Liquids have conductivity less than these materials. Gases have the least value for thermal conductivity.

In solids heat is conducted in two modes. 1. The flow of thermally activated electrons and 2. Lattice waves generated by thermally induced atomic activity. In good conductors the predominant mode is by electron flow. In alloys it is equal between the two modes. In insulators, the lattice wave mode is the main one. In liquids, conduction is by atomic or molecular diffusion. In gases conduction is by diffusion of molecules from higher energy level to the lower level.

Thermal conductivity is found to vary with temperature. **In good conductors, thermal conductivity decreases with temperature** due to impedance to electron flow at higher electron densities. **In insulators,** as temperature increases, thermal atomic activity also increases and **hence thermal conductivity increases with temperature.** In the **case of gases,** thermal conductivity **increases with temperature** due to increased random activity of atoms and molecules. Thermal conductivity of some materials is given in Table 1.1.

Table 1.1. Thermal conductivity of some materials at 293 K

Material	***Thermal conductivity, W/mK***
Copper	386.0
Aluminium	204.2
Carbon Steel 1% C	43.3
Chrome Steel 20% Cr	22.5
Chrome Nickel Steel	12.8
Concrete	1.13
Glass	0.67
Water	0.60
Asbestos	0.11
Air	0.026

The variation of thermal conductivity of various materials with temperature is shown in Fig. 1.3.

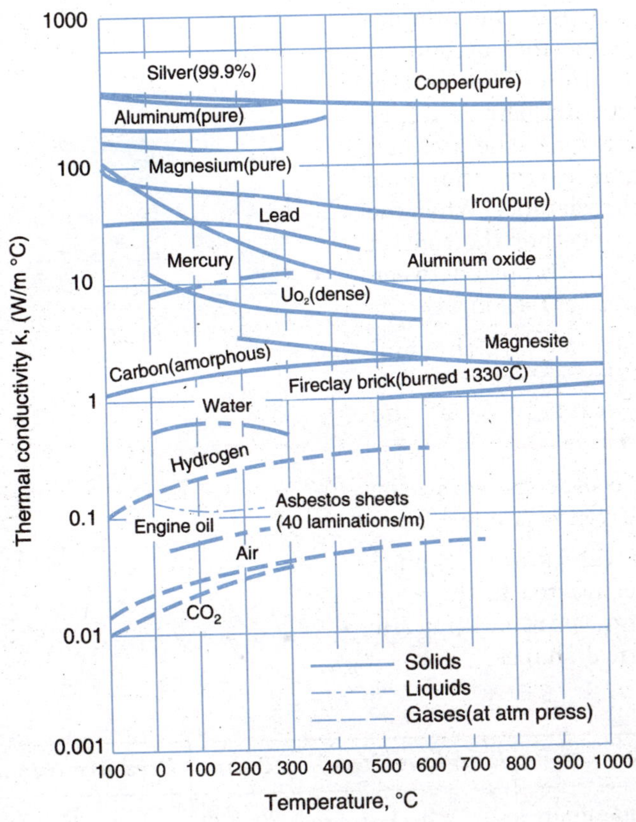

Fig. 1.3. Effect of temperature on thermal conductivity of materials

1.2.3. Thermal Insulation

In many situations to reduce the heat loss, equipments have to be insulated. Thermal insulation materials should have a low thermal conductivity. This is achieved in solids by trapping air or a gas in small cavities inside the material. It may also be achieved by loose filling of solid particles. The insulating property depends on the material as well as transport property of the gases filling the void spaces. There are essentially three types of insulating materials:

1. **Fibrous:** Small diameter particles or filaments are loosely filled in the gap between surfaces to be insulated. Mineral wool is one such material, for temperatures below 700°C. Fibre glass insulation is used below 200°C. For higher temperatures refractory fibres like Alumina (Al_2O_3) or silica (S_1O_2) are useful.

2. **Cellular:** These are available in the form of boards or formed parts. These contain voids with air trapped in them. Examples are polyurethane and expanded polystyrene foams.

3. **Granular:** These are of small grains or flakes of inorganic materials and used in preformed shapes or as powders.

4. Multilayered parallel thin sheets or foils with high reflectivity are used reflect the energy back into the source and thus reduce heat loss.The effective thermal conductivity of these materials is in the range of 0.02 to 0.04 W/mK.

1.2.4. Contact Resistance

When two different layers of conducting materials are placed in thermal contact, a thermal resistance develops at the interface. This is termed as contact resistance. A significant temperature drop develops at the interface and this has to be taken into account in heat transfer calculation. The contact resistance depends on the surface roughness to a great extent. The pressure holding the two surfaces together also influences the contact resistance. When the surfaces are brought together the contact is partial and air may be trapped between the voids as shown in Fig. 1.4. (inset)

In the equation for thermal resistance (*L*/*kA*), as the contact area, *A* is reduced at the contact surface, the resistance value increases.

Some values of contact resistance for different surfaces is given in Table 1.2.

The tabulated values are for 1 m^2 area and should be converted to the area under consideration, particularly in cylindrical and spherical shapes.

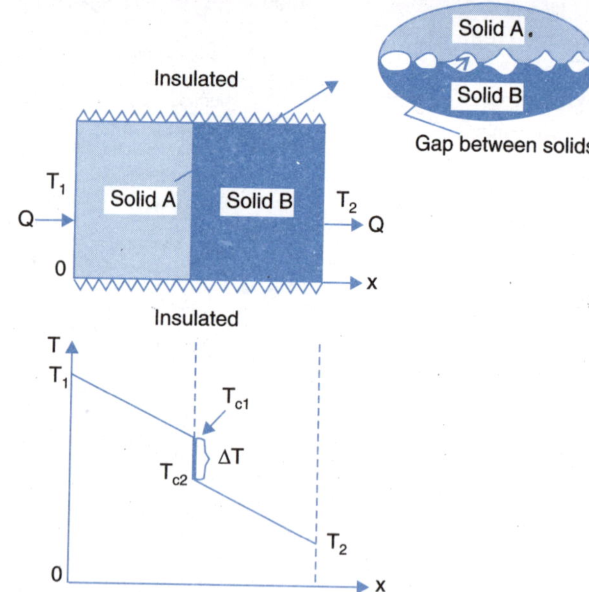

Fig. 1.4. Contact resistance temperature drop

Table 1.2

Surface type	*Roughness (μm)*	*Temp.*	*Pressure (atm.)*	*R, m^2°C/W × 10^4*
Stainless Steel ground in air	2.54	20–200	3–25	2.64
Stainless Steel ground in air	1.14	20	40–70	5.28
Aluminium ground in air	2.54	150	12–25	0.88
Aluminium ground in air	0.25	150	12–25	0.18

1.2.5. Convection

This mode of heat transfer is met with in situations where energy is transferred as heat to or from a flowing fluid at the surface over which the flow occurs. This mode is basically conduction in a very thin fluid layer at the surface and then mixing caused by the flow. The energy transfer is by combined molecular diffusion and bulk flow and mixing. The heat flow is independent of the properties of the material of the surface and depends only on the fluid properties. However the shape and nature of the surface will influence the flow and hence the heat transfer. Convection is not a pure mode as conduction or radiation and hence involves several parameters. If the flow is caused by **external means** like a fan or pump, then the mode is known as **forced convection.** If the flow is due to the **buoyant forces** caused by temperature difference in the fluid body, then the mode is known as **free or natural convection.** In most applications heat is transferred from one fluid to another separated by a solid surface. So heat is transferred from the hot fluid to the surface and then from the surface to the cold fluid by convection. In the design process thus convection mode becomes the most important one in the point of view of application. The rate equation is due to Newton who clubbed all the parameters into a single one called convective heat transfer coefficient (*h*) as given in equation (1.3). The physical configuration is shown in Fig. 1.5 (*a*).

Heat flow,
$$Q = hA\,(T_1 - T_2) = \frac{T_1 - T_2}{1/hA} \quad ...(1.3)$$

where, $Q \rightarrow W. A \rightarrow m^2$, $T_1, T_2 \rightarrow °C$ or K, $\therefore$ $h \rightarrow W/m^2K$.

The quantity $1/hA$ is called convection resistance to heat flow. The equivalent circuit is given in Fig. 1.5(*b*).

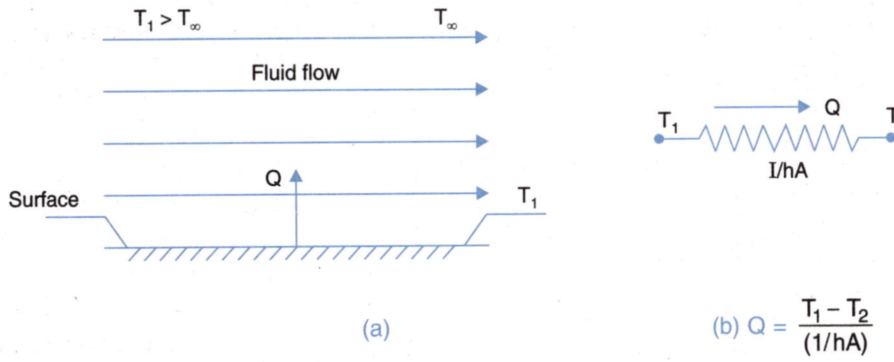

Fig. 1.5. Electrical analogy for convection heat transfer

Boiling, condensation, melting and freezing are special cases of heat transfer situation. These are discussed in later sections.

Example 1.2: *Determine the heat transfer by convection over a surface of 0.5 m² area if the surface is at 160°C and fluid is at 40°C. The value of convective heat transfer coefficient is 25 W/m²K. Also estimate the temperature gradient at the surface given k = 1 W/mK.*

Solution: Refer to Fig. 1.5 (*a*) and equation (1.3)

Known: $A = 0.5\ m^2$, $T_1 = 160°C$, $T_\infty = 40°C$

$h = 25\ W/m^2K$, $k = 1\ W/mK$.

Find: (1) Heat rate (2) Temperature gradient at the surface.

$$Q = hA\,(T_1 - T_\infty) = 25 \times 0.5 \times (160 - 40)\ W = \mathbf{1500\ W\ or\ 1.5\ kW}$$

The resistance $= 1/hA = 1/25 \times 0.5 = 0.08°C/W$.

The fluid has a conductivity of 1 W/mK, then the **temperature gradient at the surface** where y tends to zero is

$$Q = -\,kA\ dT/dy$$

Therefore, $$\mathbf{dT/dy = -\,Q/kA}$$
$$\mathbf{= -\,1500/1.0 \times 0.5 = -\,3000°C/m.}$$

COMMENTS

1. The fluid temperature is often referred as T_∞ for indicating that it is the fluid temperature well removed from the surface.
2. The convective heat transfer coefficient is dependent on several parameters and the determination of the value of this quantity is rather complex, and is discussed in later chapters.
3. The gradient at the surface depends on the heat flux and thermal conductivity.

1.2.6. Radiation

Electromagnetic radiation carries energy and the radiation is emitted by various means. Such radiation due to temperature, called thermal radiation is part of the electromagnetic spectrum in the limited wave length range of 0.1 to 10 μm and is emitted at all surfaces, irrespective of the temperature. Such radiation when incident on surfaces is absorbed and thus radiation heat transfer takes place between surfaces at different temperatures. No medium is required for radiative transfer but the surfaces should be in visual contact for direct radiation transfer. The rate equation is due to **Stefan-Boltzmann law** which states that heat radiated is proportional to the fourth power of the absolute temperature of the surface and heat transfer rate between surfaces is given in equation 1.4. The situation is represented in Fig. 1.6 (*a*).

$$Q = F\,\sigma\,A\,(T_1^4 - T_2^4) \qquad \text{...(1.4)}$$

where, F—a factor depending on relative geometric positioning of the surfaces and surface properties,

σ—Stefan Boltzmann constant 5.67×10^{-8} W/m^{2}K^4 (SI units)

A—m^2, T_1, $T_2 \rightarrow$ K (only absolute unit of temperature to be used).

This equation can also be rewritten as:

$$Q = \frac{(T_1 - T_2)}{1/\{F\sigma\,A\,(T_1 + T_2)\,(T_1^2 + T_2^2)\}} \qquad \text{...(1.5)}$$

where the denominator is referred to as radiation resistance and $F\,\sigma\,A\,(T_1 + T_2)\,(T_1^2 + T_2^2)$ is termed as h_r radiation coefficient or conductance. (Fig. 1.6)

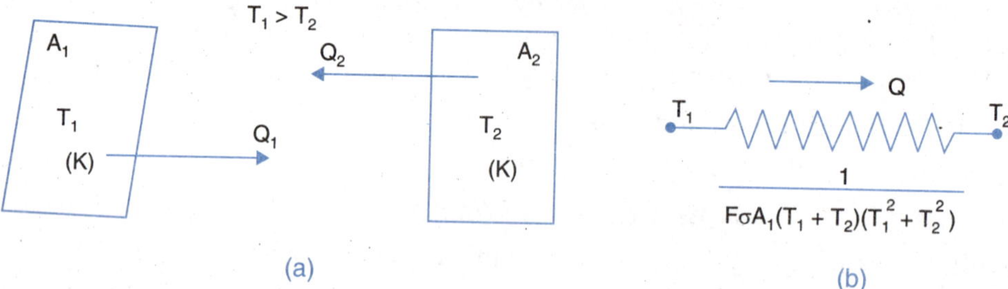

Fig. 1.6. Electrical analogy-radiation heat transfer

Radiation is a surface phenomenon.

The radiation emitted by a body due to its temperature is termed as emissive power of the body E.

The radiation from an ideal surface (black surface) is termed as E_b.

The ratio E/E_b is called emissivity of the surface. All radiant energy received by the surface is called irradiation, G,

All radiant energy leaving the surface is called radosity, J.

Example 1.3: *A surface is at 200°C and has an area of 2m². It exchanges heat with another surface B at 30°C by radiation. The value of factor due to the geometric location and emissivity is 0.46. Determine the heat exchange. Also find the value of thermal resistance and equivalent convection coefficient.*

Solution: Refer to equation (1.4) and (1.5) and Fig. 1.6.

Known: T_1 = 200°C = 200 + 273 = 473 K, T_2 = 30°C = 30 + 273 = 303 K.

(This conversion of temperature unit is very important)

$\sigma = 5.67 \times 10^{-8}$, $A = 2\text{m}^2$, $F = 0.46$.

Find: (1) Heat rate, Q

(2) Radiation resistance R,

(3) h_r

Therefore, $Q = 0.46 \times 5.67 \times 10^{-8} \times 2[473^4 - 303^4]$

$= 0.46 \times 5.67 \times 2\ [(473/100)^4 - (303/100)^4]$

(This step is also useful for calculation and will be followed in all radiation problems-taking 10^{-8} inside the bracket).

Therefore, **Q = 2171.4 W**

Resistance can be found as

$Q = \Delta T/R$, $R = \Delta T/Q = (200 - 30)/2171.4$

Therefore, **R = 0.07829°C/W or K/W**

Resistance is also given by $1/h_rA$.

Therefore, $h_r = 6.3865$ W/m²K

Check $Q = h_rA\Delta T = 6.3865 \times 2 \times (200 - 30) = 2171.4$ W

COMMENTS

1. The denominator in the resistance term is also denoted as h_rA. where $h_r = F\sigma\ (T_1 + T_2)(T_1^2 + T_2^2)$ and is often used due to convenience.
2. The determination of F is rather involved and values are available for simple configurations in the form of charts and tables. For problems which involve radiation all temperatures should be expressed in the absolute units.

In this chapter only simple cases will be dealt with and the determination of F will be taken up in the chapter on radiation. The concept of h_r is convenient, though difficult to arrive at if temperature is not specified. The value also increases rapidly with temperature.

1.3 COMBINED MODES OF HEAT TRANSFER

Previous sections treated each mode of heat transfer separately. But in practice all the three modes of heat transfer can occur simultaneously. Additionally heat generation within the solid may also be involved. Most of the time conduction and convection modes occur simultaneously when heat from a hot fluid is transferred to a cold fluid through an intervening barrier. Consider the following example. A wall receives heat by convection and radiation on one side. After conduction to the next surface heat is transferred to the surroundings by convection and radiation. This situation with equivalent circuit is shown in Fig. 1.7.

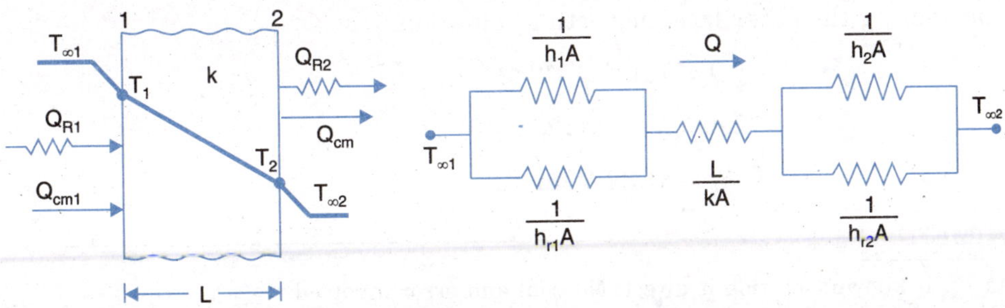

Fig. 1.7. Combined modes of heat transfer

The heat flow is given by equation (1.6).

$$\frac{Q}{A} = \frac{T_{\infty_1} - T_{\infty_2}}{\dfrac{1}{h_{r_1} + h_1} + \dfrac{L}{k} + \dfrac{1}{h_{r_2} + h_2}} \quad ...(1.6)$$

where h_{r_1} and h_{r_2} are radiation coefficients and h_1 and h_2 are convection coefficients.

Example 1.4: *A slab 250 mm thick having a thermal conductivity of 25 W/mK receives heat by convection from hot gases at 320°C. The slab surface is at 200°C. The convection coefficient has a value of 60 W/m²K. On the other surface of the slab, air at 40°C flows. At steady state conditions, determine the value of convective heat transfer coefficient on this side.*

Solution: Consider area as 1 m².

All known data are indicated in Fig. Ex. 1.4.

Find: h_2

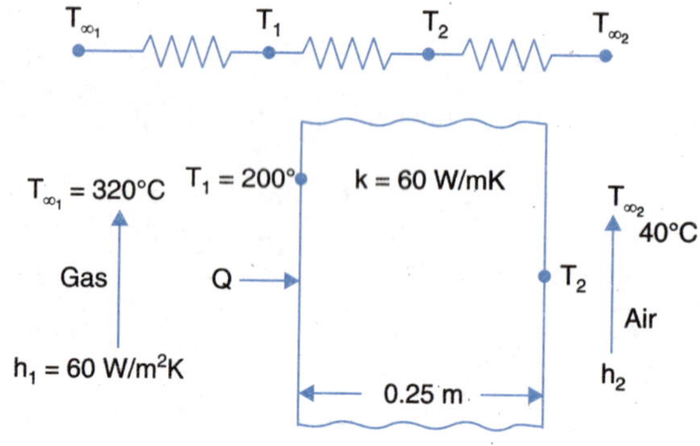

Fig. Ex. 1.4

$$Q = h_1 A\,(T_{\infty 1} - T_1)$$

$$= 60 \times 1\,(320 - 200) = 7200 \text{ W}.$$

Considering conduction through the slab

$$Q = \frac{T_1 - T_2}{L/kA} \quad \therefore \quad 7200 = (320 - T_2)/(0.25/25)$$

Solving, $T_2 = 128°\text{C}$

Considering the other face, neglecting radiation,

$$Q = h_2 A\,(T_2 - T_{\infty 2})$$

$$7200 = h_2 \times 1\,(128 - 40)$$

Solving, $h_2 =$ **81.82 W/m²K.**

COMMENT

As $T_{\infty 2}$ is known, solution is direct. No trial and error involved.

Example 1.5: *A slab 30 cm thick, having a thermal conductivity of 28 W/mK is exposed to radiation from a source at 530 K. The surface temperature is to be maintained at 470 K. The other face is exposed to radiation to space. Determine the space temperature if steady state should be maintained (without heating or cooling of slab).*

T_1 = 470 °K   k = 28 W/mK
T_{∞_1} = 530 °K T_2 T_{∞_2}
0.3 m

Fig. Ex. 1.5

Solution: All known data are shown in Fig. Ex. 1.5

Find : $T_{\infty 2}$

Considering face 1, assuming unit area and $F = 1$,

$$Q = \sigma A\,(T_{\infty 1}{}^4 - T_1{}^4)$$
$$= 5.67 \times 10^{-8} \times 1\,(530^4 - 470^4)$$
$$= 1707.12 \text{ W.}$$

Considering conduction through the slab

$$T_2 = T_1 - Q \times L/kA = 470 - 1707.12 \times 0.3/28 = 451.7 \text{ K}$$

Considering radiation on the other side,

$$1707.12 = 5.67 \times 10^{-8}\,(451.7^4 - T_{\infty 2}{}^4),$$

Solving, $\boldsymbol{T_{\infty 2} = 327.6 \text{ K}}$

COMMENT

As heat rate is known, direct solution is possible. No trial and error involved.

Example 1.6: *A very thin metal sheet is exposed on one side to convection and radiation from a source at 500°C. On the otherside the sheet is exposed to convection and radiation to surroundings at 40°C. The convective heat transfer coefficient on both sides is 52 W/m²K. At steady condition determine the temperature of the sheet.*

Solution: All known data are shown in Fig. Ex. 1.6.

Find : Temperature of the sheet

Consider 1 m² area.

Let the sheet temperature be T

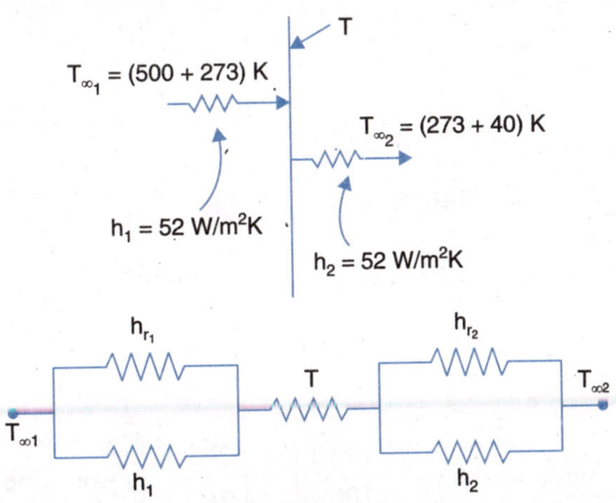

Fig. Ex. 1.6

Heat received by the sheet

$Q = 1 \times 5.67 \times 10^{-8} (773^4 - T^4) + 52 \times 1 \times (773 - T)$ under steady condition,

Heat loss from the sheet = heat received

Heat loss $\quad Q = 1 \times 5.67 \times 10^{-8} (T^4 - 313^4) + 1 \times 52 \times (T - 313)$

Equating and rearranging,

$$5.67 \times 10^{-8} (773^4 + 313^4) + 52 (773 + 313)$$
$$= 2 \times 5.67 \times 10^{-8} (T^4) + 2 \times 52T$$

This reduces to

$$11.34 \times 10^{-8} T^4 + 104T - 77260.4 = 0.$$

Solving by trial, $\quad T = 601$ K or 328°C.

COMMENTS

1. When convection and radiation are involved, an equation of 4th power is obtained. This needs solution by trial and error.
2. When a temperature value involving both convection and radiation is not known, trial solution will be necessary.
3. When both convection and radiation is involved all temperatures should be expressed in absolute units.

Example 1.7: *A slab 0.2 m thick with thermal conductivity of 45 W/mK receives heat from a furnace at 500 K both by convection and radiation. The convection coefficient has a value of 50 W/m²K. The surface temperature is 400 K on this side. The heat is transferred to surroundings at* T_{∞_2} *both by convection and radiation. The convection coefficient on this side being 60 W/m²K. Determine the surrounding temperature. Assume F = 1 for radiation.*

Solution: Refer Fig. 1.7. Consider 1 m^2 area. Steady state condition.

Known: $\quad L = 0.2$ m, $k = 45$ W/mK

$T_{\infty 1} = 500$ K, $h_1 = 50$ W/m²K, $T_2 = 400$ K

$h_2 = 60$ W/m²K,

Find: $T_{\infty 1}$

Heat received $\quad = \sigma (T_{\infty 1}^4 - T_1^4) + h (T_{\infty 1} - T_1)$

$$= 5.67 \left\{ \left(\frac{500}{100}\right)^4 - \left(\frac{400}{100}\right)^4 \right\} + 50 (500 - 400)$$

$$= \mathbf{7092.2\ W.}$$

To determine T_2, $\quad Q = \dfrac{\Delta T}{R}$ or $\Delta T = QR = 7092.2 \times \dfrac{0.2}{45} = 31.57$ K.

$\therefore \quad T_2 = 400 - 31.57 = 368.43$ K.

On the other side,

$$7092.2 = 5.67 \left\{ \left(\frac{368.43}{100}\right)^4 - \left(\frac{T_{\infty 2}}{100}\right)^4 \right\} + 60 (368.43 - T_{\infty 2})$$

or $$5.67\left(\frac{T_{\infty 2}}{100}\right)^4 + 60\,T_{\infty 2} = 16056. \text{ Solving by trial}$$

$$\mathbf{T_{\infty 2} = 263.3\ K.}$$

Check: $$5.67\left(\frac{263.3}{100}\right)^4 + 60 \times 263.3$$

$$= 16070.$$

Difference of 14 is due to truncation.

1.3.1. Overall Heat Transfer Coefficient

Often when several resistances for heat flow is involved, it is found convenient to express the heat flow as

$$Q = U\,A\,\Delta T, \qquad \text{...(1.7)}$$

where **U is termed as overall heat transfer coefficient** having the same unit as convective heat transfer coefficient, h. The value of U can be obtained for a given area A by equation 1.8.

$$\frac{1}{UA} = \frac{1}{R_1} + \frac{1}{R_2} + \frac{1}{R_3} + \ldots\ldots \qquad \text{...(1.8)}$$

where R_1, R_2, R_3, ... are the resistances in series calculated based on the areas A_1, A_2, A_3 etc.

1.3.2. Energy Balance with Heat Transfer

There are situations when a body receives heat by convection and radiation and transfer part of it to the surroundings and stores the remaining in the body by means of increase in temperature. In such a situation, the time rate of temperature change can be obtained by the equation (1.9). Heat generation may be involved in cases like nuclear fuel rod and this also should be included.

$$d\tau\,(Q_{in} - Q_{out}) + d\tau\,\dot{q} = \rho V c\, dT.$$

or $$\frac{dT}{d\tau} = \frac{Q_{in} - Q_{out} + \dot{q}}{\rho V c} \qquad \text{...(1.9)}$$

where $\dot{q}$ is the heat generation rate per unit volume and ρ, V and c are the density, Volume and specific heat of the body and τ is time.

When equilibrium is reached, $\dfrac{dT}{d\tau} = 0$, So

$$Q_{in} = Q_{out} \qquad \text{...(1.10)}$$

Example 1.8: *In a cylindrical shaped body of 30 cm diameter and 30 cm length heat is generated at a rate 1.5 × 10⁶ W/m³. The surface temperature is 400°C. The convection coefficient is 200 W/m²K. Heat is convected and radiated to the surroundings at 100°C. The radiation factor is one. The solid has a density of 19000 kg/m³ and a specific heat of 0.118 kJ/kgK. Determine the rate of change of temperature of the body at that instant in °C/s.*

Solution: **Known:** $\dot{q} = 1.5 \times 10^6$ W/m³, $T_1 = 400$°C, $T_2 = 100$°C, $h = 200$ W/m²K.

$\dot{q} = 19000$ kg/m³, $c = 0.118$ kJ/kgK.

Find: Time rate of temperature change.

Refer equation (1.9) and Fig. Ex. 1.8

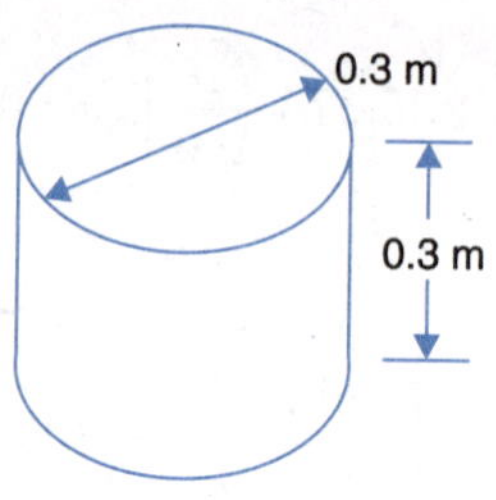

Fig. Ex. 1.8

$$(\dot{q} - Q_R - Q_C)\, d\tau = \rho V c\, dT$$

$$\therefore \quad \frac{dT}{d\tau} = \frac{\dot{q} - Q_R - Q_C}{\rho V c}$$

The surface area $= 2\pi r^2 + 2\pi r h$

$= 2\pi \times 0.15^2 + 2\pi \times 0.15 \times 0.3$

$= 0.4241 \text{m}^2$

Heat capacity = Volume × density × sp. heat

$= \pi r^2 h \times 19000 \times 118$

$= \pi \times 0.15^2 \times 0.3 \times 19000 \times 118$

= 47543 J/°C

Heat generated = Volume × $q = \pi r^2 h \times 1.5 \times 10^6$ W = 31809 W or 31809 J/s

Heat convected = $hA\, \Delta T = 200 \times 0.4241 \times (400 - 100)$ = 25446 W or 25446 J/s

Heat radiated $= \sigma A\, (T_1^4 - T_2^4)$

$= 5.67 \times 10^{-8} \times 0.4241\, [(400 + 273)^4 - (100 + 273)^4]$

= 4468 W or 4468 J/s

Therefore, Heat generated – Heat convected – Heat radiated

= 31809 – 25446 – 4468 = 1895 W or 1895 J/s

$\rho c V$ = 47543 J/°C

Therefore, $\dfrac{dT}{d\tau} = \dfrac{1895}{47543} = 0.03985°\text{C/s}$

COMMENTS

1. This problem involves energy balance. Energy balance can be done in two ways.
   (*a*) Over a time period in which case energy rate multiplied by time is used.
   (*b*) Energy balance at an instant. In this case energy rates have to be balanced. This problem is of the second type.

   Possible simplifications are:

   (*i*) no heat generation, (*ii*) no radiation or, (*iii*) steady state etc., which will reduce one of the terms to be zero.

1.4 DIMENSIONS AND UNITS

For numerical estimation of heat transfer rate units of various parameters become necessary. All equations should be dimensionally homogeneous. Dimensions like length, time and mass are universal and there is no difference from country to country. But the systems of unit varies from country to country. Three popular systems are (1) *FPS* (foot, pound, second, °F) (2) *MKS* (metre, kilogram, second °C) and (3) *SI* (metre, kilogram, second, K) system of units.

In this text *SI* system of units is adopted. The units used for various quantities is listed in Table 1.3 and conversion factors are tabulated at the beginning of the book.

Table 1.3. Units adopted for various quantities

Parameter	*Unit and symbol*	*Unit multiples*
Mass	kilogram, kg,	Ton = 1000 kg
Length	metre, m	cm, mm, km
Time	Seconds, s,	minute, hour
Force	Newton, N, (kg m/s^2)	kN, MN
Energy, (heat)	Joule, J ≡ Nm	kJ, MJ
Power	Watt, W (J/s)	kW, MW
Temperature	kelvin, K,	also °C
Dynamic viscosity	μ, Nm/s^2	Poise = 0.1 Nm/s^2
Kinematic viscosity	υ, m^2/s	Stoke = 10^{-4} m^2/s
Specific heat	c, J/kg K	kJ/kg K

The units for other parameters will be defined as and when these are used.

In solving numerical problems, consistent set of units should be used. Otherwise the answer will be meaningless.

Example 1.9: *Convert the following units into their equivalent SI units:*

(i) BTU/hr ft°F, (ii) BTU / hr ft^2 °F. From published tables the following conversion factors are read. 1J = 9.4787 × 10^{-4} BTU, 1m = 39.370 inches, kg = 2.2046 lb, °C = 9/5°F.

Solution: (*i*) Therefore, 1 *BTU* = 1/9.4787 × 10^{-4} J = 1054.997 J, *ft* = (12/39.37) m

Therefore, *BTU/hr ft* °F = 1054.997J/3600*s* (12/39.37) m. (5/9)°C

= 1.7306 J/s m°C or 1.7306 W/m°C

or **1 W/mK = 0.5778 BTU/hr ft°F.**

(*ii*) **BTU/hr ft^2 °F** = 1054.997J/3600*s* (12/39.37)^2 m^2 (5/9)°C

= **5.67792 W/m^2°C** or 1W/m^2 °C = 0.1761 *BTU/hr ft*^{2}°F.

COMMENTS

1. Note that the value of °C cannot be converted to °F by this factor 9/5. As the quantities involved are per degree interval. Hence in this conversion the constant can be used.
2. Some truncation is involved in the answers.

1.5 CLOSURE

An overview of the field of heat transfer is presented in this chapter. Each mode of heat transfer will be discussed in greater detail in the following chapters.

The series of steps listed below will be useful in analysing and estimating heat transfer.

1. List the available data for the problem situation. Then look for additional data from other sources, like property listings.

2. Sketch a schematic diagram for the system involved and identify the basic processes involved. (Physical model) Invariably this will make the problem clear.

3. List the simplifying assumptions that are reasonable. This should be checked later.

4. Apply the rate equations and conservation laws to the situation. (Mathematical model).

5. Try to validate the results obtained by back work or by other methods. This is an important step, which is often overlooked with disastrous results.

ADVANCED EXAMPLES

Combined Convection and Radiation

Example A-1.1: *A surface is at 200°C and is exposed to surroundings at 60°C and convects and radiates heat to the surroundings. The convection coefficient is 80 W/m²K. The radiation factor is one. If the heat is conducted to the surface through a solid of conductivity 12 W/mK, determine the temperature gradient at the surface in the solid.*

Solution: In this case both convection and radiation are involved.

Known: Surface temp T_s = 200°C, Surrounding temperature T_∞ = 60°C

h = 80 W/m²K, F = 1, k = 12 W/mK

Find: Temperature gradient at surface

Consider unit area.

Heat conducted into the solid = heat given out by both processes

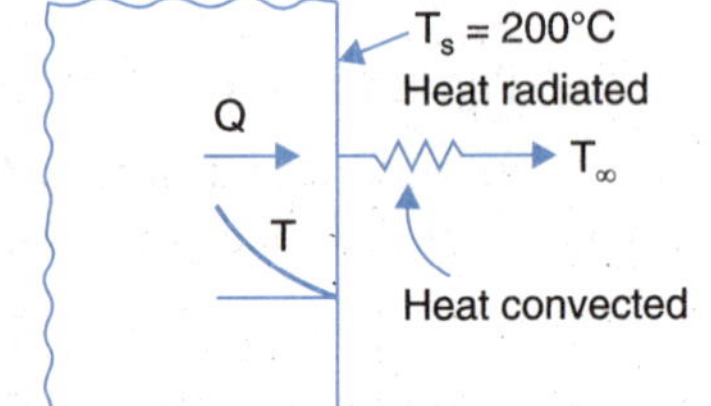

Fig. AE-1.1

$$-kA\frac{dT}{dx} = hA\,(T_s - T_\infty) + \sigma A\,[T_s^{\,4} - T_\infty^{\,4}]$$

Therefore, $80\,[(200 + 273) - (60 + 273)] + 5.67\,\{[(200 + 273)/100]^4 - [(60 + 273)/100]^4\} = -12\,\frac{dT}{dx}$

Solving $\frac{dT}{dx} = -(11200 + 2140.9)/12 = -$ **1111.7°C/m.**

COMMENTS

1. This problem involves energy balance at an instant. So heat rates are balanced.
2. When convection and radiation are involved simultaneously, all temperatures should be in absolute units.

Example A-1.2: *Heat is conducted through a material with a temperature gradient of – 9000 °C/m. The conductivity of the material is 25W/mK. If this heat is convected to surroundings at 30°C with a convection coefficient of 345W/m²K, (i) Determine the surface temperature.*

(ii) If the heat is radiated to the surroundings at 30°C determine the surface temperature.

Solution: (*i*) In this case only convection and conduction are involved. Temperatures can be in °C.

Known: $\frac{dT}{dx} = -9000$°C/m

k = 25 W/mK, T_2 = 30°C

h = 345 W/m²K

Find: T_1 $\quad - kAdT/dx = hA(T_1 - T_2)$. Considering unit area,

$$- 25 \times 1 \times (- 9000) = 345 \times 1\ (T_1 - 30)$$

Therefore, $\mathbf{T_1 = 682.17°C}$

(*ii*) In this case conduction and radiation are involved.

Heat conducted = Heat radiated

$$- 25 \times 1 \times (- 9000) = 5.67\ [(T_1/100)^4 - (303/100)^4]$$

Solving, $\mathbf{T_1 = 1412.14K = 1139°C.}$

As only radiation is involved direct solution is possible.

Example A-1.3: *There is a heat flux through a wall of 2250 W/m². The same is dissipated to the surroundings by convection and radiation. The surroundings is at 30°C. The convection coefficient has a value of 75 W/m²K. For radiation F = 1. Determine the wall surface temperature.*

Solution: Known: Heat flux = 2250 W/m², h = 45 W/m²K, $F = 1$ T_s = 30 °C = 303 K

Find: T_2

For the specified condition, Consider unit area.

The heat conducted = heat convected + heat radiated

Using the rate equations, with absolute temperature

$$2250 = 75 \times 1\ (T_2 - 303) + 5.67 \times 1[(T_2/100)^4 - (303/100)^4]$$

$$= 75T_2 - 22725 + 5.67(T_2/100)^4 - 477.92$$

or $$(T_2/100)^4 + 13.2275T_2 - 4489.05 = 0.$$

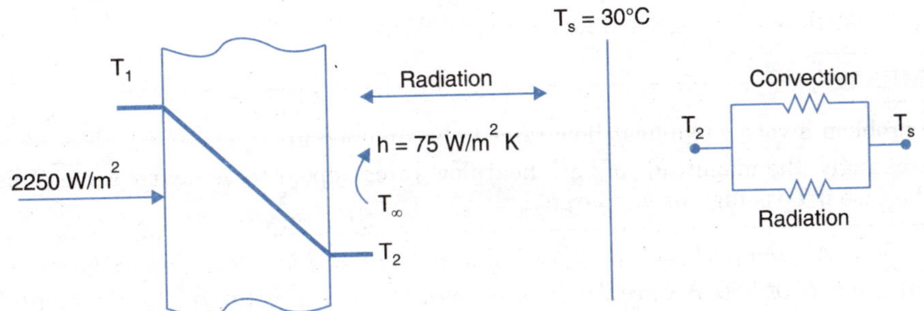

Fig. AE-1.3

This equation can be solved only by trial. It may be noted that the contribution of $(T_2/100)^4$ is small and so the first choice of T_2 can be a little less than 4489/13.227 = 340 K. The values of the reminder for T_2 = 300, 310, 320, 330 are given below:

Assumed value of T_2	300	310	320	330	330.4	330.3
Remainder	– 439.80	– 296.2	– 15.1	– 5.38	0.484	– 0.98

So, the temperature T_2 is near 330 K. By one more trial T_2 is obtained **as 330.4 K or 57.4°C.**

COMMENTS

1. The problem involves energy balance of an instant and so energy flux is balanced.
2. As both convection and radiation are involved trial solution is necessary.

Check: $$Q = 75(330.4 - 303) + 5.69(3.304^4 - 3.03^4)$$

$$= 2047.5 + 206 = 2253.5\ \text{W}.$$

Example A-1.4: *Heat leaks both by convection and radiation from the surroundings at 5°C into a cylindrical cryogenic container of outside radius 8 cm. The container outside surface is at – 10°C. If the convective heat transfer coefficient is 4.35 W/m²K determine the heat leaking into the container per meter length.*

Solution: Known: T_∞ = 5°C = 278 K, T_s = – 10°C = 263 K

Find: Heat rate, Q h = 4.35 W/m²K

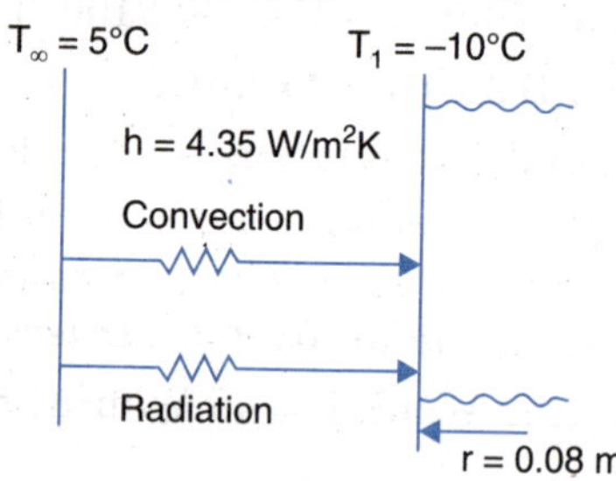

Fig. AE-1.4

Outside area per meter length = 2 × π × 0.08 × 1 = 0.50265 m².

Heat convected $= hA\,(T_\infty - T_s)$

$= 4.35 \times 0.50265 \times [5 - (-10)] = 32.8$ W/m

Heat radiated $= 5.67 \times 10^{-8} \times 0.50265\,[278^4 - 263^4]$

$= 33.87$ W/m

Total heat flow in = 32.8 + 33.87 = 66.67 W/m

COMMENTS

1. The problem involves two heat flow rates to be summed up.
2. Interestingly, the magnitude of both heat flow rates appear to be nearly same. This may not be the case if T_∞ is high or h is large.

Example A-1.5: *A chemical reactor vessel of spherical shape of outside radius of 0.5 m has to loose heat at the rate of 650 W in order to maintain the temperature of the chemical. The surface temperature of the vessel is 125°C. The surrounding is at 113°C. If heat loss is by both convection and radiation, determine the value of convective heat transfer coefficient required.*

Solution: Known: Heat rate: 650 W, T_2 = 125°C, T_∞ = 113°C, R = 0.5 m

Assume: Steady state.

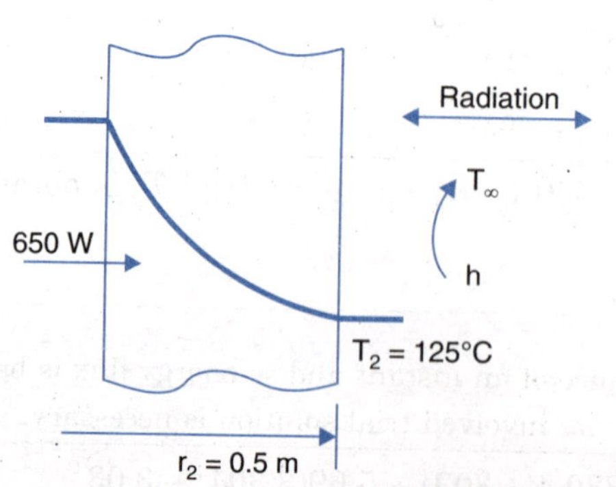

Fig. AE-1.5

Find: h

Surface area = $4\pi r^2 = 4 \times \pi \times 0.5^2 = 3.1416$ m^2.

Heat radiated: $T_s = 273 + 125 = 398$ K, $T_\infty = 273 + 113 = 386$ K

$= 3.1416 \times 5.67 \times (3.98^4 - 3.86^4)$

$= 515.15$ W.

Heat to be convected $= 650 - 515.15 = 134.85$ W

$= h \times 3.1416\,(398 - 386)$

Solving, $h =$ **3.58 W/m²K**

COMMENT

This problem involves what is known as "surface heat balance".

Example A-1.6: *A solid receives heat by radiation over its surfaces at 4 kW and the heat convection rate over the surface of the solid to the surroundings is 5.2 kW, and heat is generated at a rate of 1.7 kW over the volume of the solid, determine the heat capacity of the solid if the time rate of change of the average temperature of the solid is 0.5°C/s.*

Solution: Refer equation 1.9. The energy balance yields: Heat received by radiation – heat convected + heat generated = heat stored.

Known: $Q_r = 4000$ W, $Q_c = 5200$ W, $\dot{q} = 1700$ W, $\dfrac{dT}{d\tau} = 0.5$°C/s

Find: Heat capacity = (ρVc)

But, heat stored = heat capacity × change in temperature.

$$Q_r d\tau - Q_c d\tau + \dot{q} d\tau = \rho Vc\, dT$$

$$\frac{dT}{d\tau} = \frac{Q_r - Q_c + \dot{q}}{\rho Vc} \quad \text{or} \quad \rho VC = \frac{Q_r - Q_c + \dot{q}}{\dfrac{dT}{d\tau}}$$

$$\frac{dT}{d\tau} = 0.5\text{°C/s},\ Q_r = 4000 \text{ J/s},\ Q_c = 5200 \text{ J/s}$$

$$\dot{q} = 1700 \text{ J/s}.$$

$$\therefore \quad \rho Vc = \frac{4000 - 5200 + 1700}{0.5} = \mathbf{1000\ J/°C}.$$

The problem involves heat balance at an instant. Heat rates are used to obtain balance.

Example A-1.7: *A cube shaped solid 20 cm side having a density of 2500 kg/m³ and specific heat of 0.52 kJ/kgK has a uniform heat generation rate of 100 kJ/m³/s. If heat is received over its surfaces at 240 W, determine the time rate of temperature change of the solid.*

Solution: Known: density, $\rho = 2500$ kg/m^3

Specific heat, $c = 0.52$ kJ/kgK or 520 J/kgK.

$\dot{q} = 100000$ J/m^3/s. $Q = 240$ J/s.

Find: Time rate of temperature change, $dT/d\tau$.

During a small time interval $d\tau$, the energy equation yields:

Heat received + heat generated = heat stored

Heat stored = Volume × density × specific heat × temperature rise.

$$Q\,d\tau + \dot{q}V\,d\tau = \rho Vc\,dT.$$

$$\frac{dT}{d\tau} = \frac{Q + \dot{q}V}{\rho Vc},$$

$$V = 0.2 \times 0.2 \times 0.2 \text{ m}^3,$$

$$\therefore \quad \frac{dT}{d\tau} = \frac{240 + 100000 \times 0.2^3}{2500 \times 0.2^3 \times 520} = 0.1°\text{C/s}$$

Time rate of temperature change = 0.1°C/s.

Example A-1.8: *A spherical mass 1 m diameter receives heat from a source at 160°C by radiation and convects heat to the surroundings at 30°C, the convection coefficient being 45 W/m²K. Determine the steady state temperature of the solid.*

Assume F = 1 for radiation.

Solution: All known data are shown in Fig. AE-1.8. Using energy balance,

As $$\frac{dT}{d\tau} = 0,$$

heat received by radiation = heat convected.

$$\sigma A\,(T_s^{\,4} - T^4) = hA(T - T_\infty)$$

It is to be noted that the temperature values should be in absolute units. cancelling A on both sides and substituting the values.

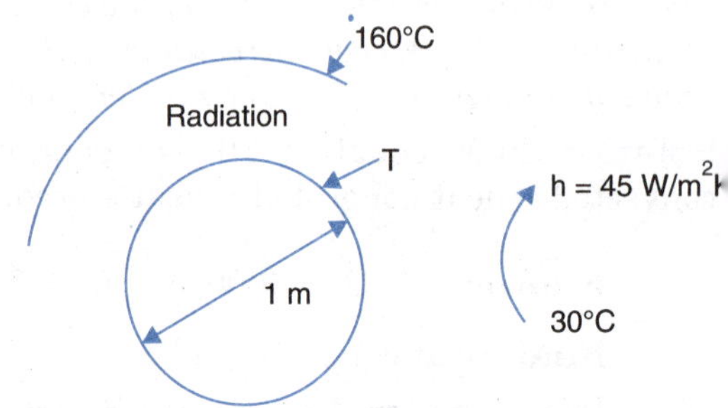

Fig. AE-1.8

$$5.67\left\{\left[\frac{160 + 273}{100}\right]^4 - (T/100)^4\right\} = 45\,[T - (273 + 30)]$$

Rearranging: $1993.13 - 5.67(T/100)^4 - 45\,T + 13635 = 0$

or $$(T/100)^4 + 7.9365\,T - 2756.30 = 0$$

Solving by trial and first taking value near (2756.30/7.9365) approximately, 330

Temp. K	330	331	332	332.1
Residue	– 18.67	– 9.6	– 0.11	+ 0.72

Therefore, the equilibrium temperature is **332 K or 59°C.**

Check: heat convected 45 [332 – (273 + 30)] = 1305 W/m²

heat received = $5.67\,(4.33^4 - 3.32^4) = 1304.3$ W/m²

Checks within reasonable limits. As both convection and radiation are involved together a fourth order equation results, necessitating trial solution.

Example A-1.9: *A person sits in a room with surrounding air at 26°C and convection coefficient over the body surface is 6 W/m²K. The walls in the room are at 5°C as the outside temperature is below freezing. If the body temperature is 37°C, determine the heat losses by convection and radiation. Assume F = 1.0 for radiation. Consider a surface area of 0.6 m².*

Solution: **Known:** $T_1 = 37°C = 310\text{ K}$, $T_\infty = 26°C$

$T_2 = 5°C = 278\text{ K}$, $h = 6\text{ W/m}^2\text{K}$, $F = 1$

Heat loss by convection: $hA(T_1 - T_\infty) = 6 \times 0.6\,(37 - 26)\text{ W} = 39.6\text{ W}$

Heat loss by radiation: $\sigma A\,(T_1^4 - T_2^4)$. Note that T should be in K.

$$= 5.67 \times 0.6\left[\left(\frac{273+37}{100}\right)^4 - \left(\frac{273+5}{100}\right)^4\right] = 110.99\text{ W}$$

Total = **150.59 W**

The direct heat loss by radiation makes one feel cooler though the surrounding temperature is not that low. Calculate the same when the wall temperature is also 26°C in summer.

Convection loss = 39.6 W (already calculated and unchanged)

Radiation loss = Now T_2 becomes 26°C or 299 K

$$= 5.67 \times 0.6\left[\left(\frac{273+37}{100}\right)^4 - \left(\frac{273+26}{100}\right)^4\right] = 42.28\text{ W}$$

Total heat loss = **81.88 W.**

COMMENTS

1. The loss from the body is reduced. But to keep the wall at 26°C, the heat required will be more.
2. This and the next problems bring out the conditions for comfort.

Example A-1.10: *A person stands in front of a fire at 650 C in a room where air is at 5°C. Assuming the body temperature to be 37°C and a convection coefficient of 6 W/m²K, the area exposed to convection as 0.6 m², determine the net heat flow from the body. The fraction of radiation from the fire of 1 m² are reaching the person is 0.01.*

Solution: Heat loss by convection = $hA(T_1 - T_2) = 6 \times 0.6(37 - 5) = 115.2\text{ W}$

Substituting the values, heat gain by radiation = $\sigma AF(T_1^4 - T_2^4)$

$$= 5.67 \times 0.01 \times 0.6\left[\left(\frac{650+273}{100}\right)^4 - \left(\frac{273+37}{100}\right)^4\right] = 243.8\text{ W}$$

Net heat gain = 243.8 – 115.2 = **128.6 W.**

COMMENTS

1. The body actually gains heat as radiation temperature is high.
2. In case the flame temperature was 495°C there will be no gain or loss. (Better comfort)
3. This shows that sudden exposure to the high temperature warms up a person quickly.

Example A-1.11: *A electric room heater (radiator) element is 25 cm long and 4 cm in diameter. The element dissipates heat to the surroundings at 1500 W mainly by radiation, the surrounding temperature being 15°C. Determine the equilibrium temperature of the element surface.*

Solution: Known data are shown in Fig. AE-1.11. At equilibrium, neglecting convection,

$$Q = \sigma A(T_1^4 - T_2^4)$$

Using absolute units of temperature,

$$1500 = 5.67 \times \pi \times 0.04 \times 0.25\,[(T_1/100)^4 - (288/100)^4]$$

Solving, $\mathbf{T_1 = 959.9\ K}$ **or** $\mathbf{686.9°C}$

Check: $Q = 5.67 \times \pi \times 0.04 \times 0.25\,[9.599^4 - 2.88^4] = 1500\ \text{W}$

Fig. AE-1.11

The requirement for steady state condition is solved in this problem.

Example A-1.12: *A steel plate is exposed to solar heat flux of 800 W/m² on one side. The plate is exposed to air at 30°C on both sides. The convection coefficients are 10 W/m²K on the back side and 15 W/m²K on the front. Determine the equilibrium temperature. Neglect radiation loss.*

Solution: All known data are indicated in Fig. AE-1.12. The energy balance yields.

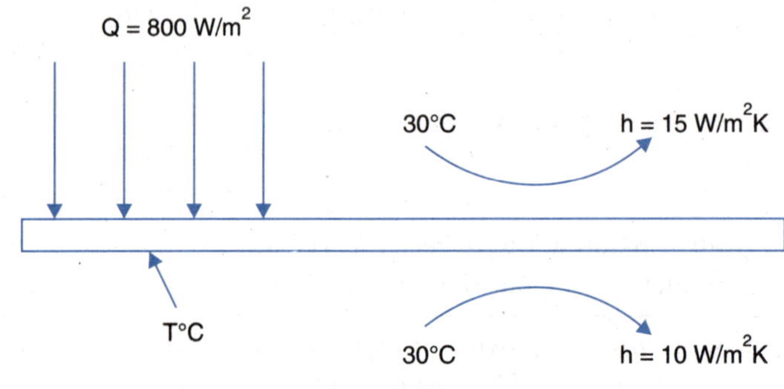

Fig. AE-1.12

The incident heat rate = convection on the front side + convection on the back side

Substituting the values, and considering 1 m²

$$800 = 15\,(T - 30) + 10(T - 30)$$

Therefore, $\mathbf{T = 62°C}$.

Check: $15(62 - 30) + 10(62 - 30) = 800$ W.

If 500 W is usefully collected by water circulation, determine the surface temperature.

The loss now equals $800 - 500 = 300$ W

$$\therefore \quad 300 = 15\,(T - 30) + 10\,(T - 30), \quad \therefore \quad T = 42°C$$

This gives an efficiency of 62.5%

For various efficiencies of collection, the collection temperature is tabulated below.

Efficiency %	30	40	50	60	62.5	0
Collection temperature°C	55.4	49.2	46	42.8	42	62

Higher the efficiency of collection lower will be the collection temperature.

Example A-1.13: *A thin plate receives radiation on one side from a source at 650°C and radiates on the other face to a surface at 150°C. Determine the temperature of the plate. Take F = 1. Neglect convection heat flow.*

Solution: The known data are indicated in Fig. AE-1.13. The energy conservation leads to radiation received by the surface = radiation from the surface

$$\sigma A[T_1^4 - T^4] = \sigma A[T^4 - T_2^4]$$

Remembering to use Kelvin scale,

$$\left[\left(\frac{650+273}{100}\right)^4 - (T/100)^4\right] = \left[(T/100)^4 - \left(\frac{150+273}{100}\right)^4\right]$$

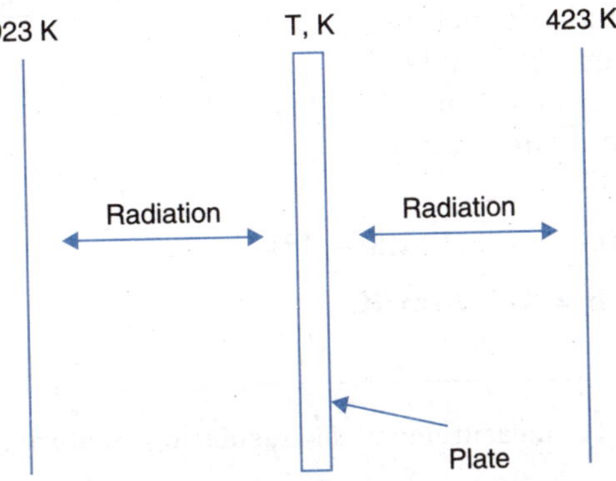

Fig. AE-1.13

Therefore, **T = 784.6 K or 511.6°C**

Later, this concept will be called as radiation shielding. The calculation of the heat flow with and without the intervening sheet will show that the heat flow is reduced by half.

With sheet, $$Q_1 = 5.67\left[\left(\frac{650+273}{100}\right)^4 - (784.6/100)^4\right] = \mathbf{19664.9\ W}$$

Without the sheet, $$Q_2 = 5.67\left[\left(\frac{650+273}{100}\right)^4 - \left(\frac{150+273}{100}\right)^4\right]$$

= 39336.66 which is 2 × Q_1.

COMMENT

In the article 1.2.3 on thermal insulation mention has been made about multilayer reflecting sheets to reduce heat loss. It is based on the principle illustrated in this problem.

Example A-1.14: *Air at 120°C flows over a plate 20 mm thick and the temperatures in the middle 10 mm layer of the plate was measured using thermo couples and were found to be 42°C and 30°C. The thermal conductivity of the material is known to be 22.5 W/mK. Determine the average convection coefficient over the plate.*

Solution: The surface temperature T_s and heat flow Q can provide the means for the determination of the convection coefficient.

Known data are indicated in Fig. AE-1.14.

Using the rate equation,

$$Q = hA(T_s - T_\infty).$$

Using the temperature drop and the thermal conductivity of the wall material, Q can be determined using

$$\mathbf{Q} = \frac{\Delta T}{L/kA} = \frac{(42-30)}{(0.01/22.5 \times 1)}$$

$$= \mathbf{27000\ W/m^2}$$

The **surface temperature** can be found assuming the material to be isotropic and having constant thermal conductivity. The drop in temperature over a 10 mm layer is, 42 – 30 = 12°C. Hence, over 5 mm, the drop will be 6°C. Hence the surface temperature = 42 + 6 = **48°C.**

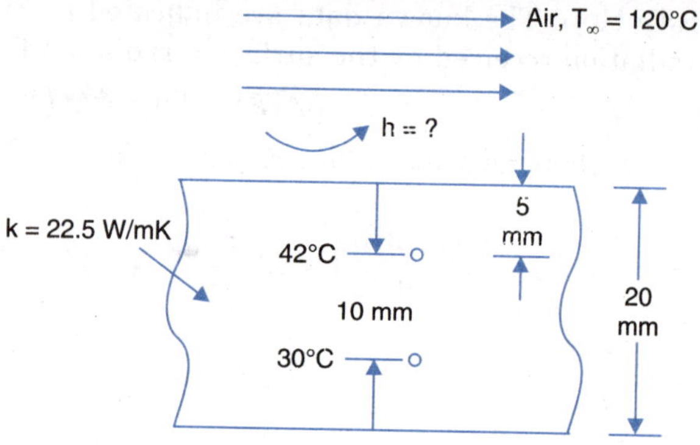

Air, T_∞ = 120°C
h = ?
k = 22.5 W/mK
42°C
30°C
5 mm
10 mm
20 mm

Fig. AE-1.14

Substituting, $27000 = h \times 1\,(120 - 48)$

Therefore, $\mathbf{h = 375\ W/m^2K.}$

COMMENT

In thermal conductivity measurement of insulating materials, the set up shown in Fig. AE-1.14 is used.

Example A-1.15: *In a solar flat plate heater some of the heat is absorbed by a fluid while the remaining heat is lost over the surface by convection the bottom being well insulated. The fraction absorbed is known as the efficiency of the collector. If the flux incident has a value of 800 W/m² and if the collection temperature is 60°C while the outside air is at 32°C with a convection coefficient of 15 W/m²K, determine the collection efficiency. Also find the collection efficiency if collection temperature is 45°C.*

Solution: The known data are indicated in Fig. AE-1.15. The heat lost by convection

$$= Q = hA(T_1 - T_2)$$

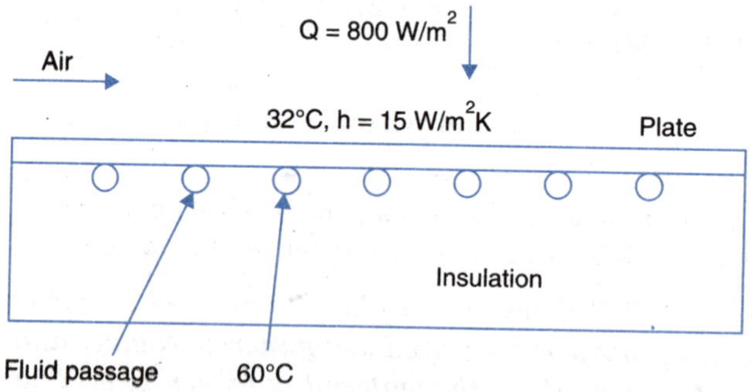

Q = 800 W/m²
Air
32°C, h = 15 W/m²K
Plate
Insulation
Fluid passage
60°C

Fig. AE-1.15

Heat loss, Assuming unit area , $Q = 15 \times 1(60 - 32) = 420$ W

Therefore efficiency of the collector $= \dfrac{800 - 420}{800} = \mathbf{0.475\ or\ 47.5\%.}$

If collection temperature is 45°C,

Heat lost by convection $= 15 \times 1(45 - 32) = 195$ W

Collection **efficiency** $= \frac{800 - 195}{800}$ = **0.75625 or 75.625%**

The efficiency improves with lower collection temperature and also with lower convection heat transfer coefficient over the surface. The efficiency at various collection temperatures are tabulated. At the same time the usefulness of the heat collected decreases.

Solar heat flux: 800 W/m², h = 15 W/m²K. Ambient temperature = 32°C

Collection temp.	°C	40	50	60	70	80	85.34
Efficiency	%	85	66.25	47.5	28.75	10	0.0

Example A-1.16: *A glass plate at 40°C is heated by passing hot air over it with a convection coefficient of 18 W/m²K. If the temperature change over 1 mm thickness is not to exceed 5°C to avoid distortion damage, determine the maximum allowable temperature of the air. Thermal conductivity of the plate material is 1.4 W/mK.*

Solution: Known data are indicated in Fig. AE-1.16. The heat flow by conduction = heat flow by convection.

The conduction heat flow is found using the allowable temperature drop over 1 mm thickness.

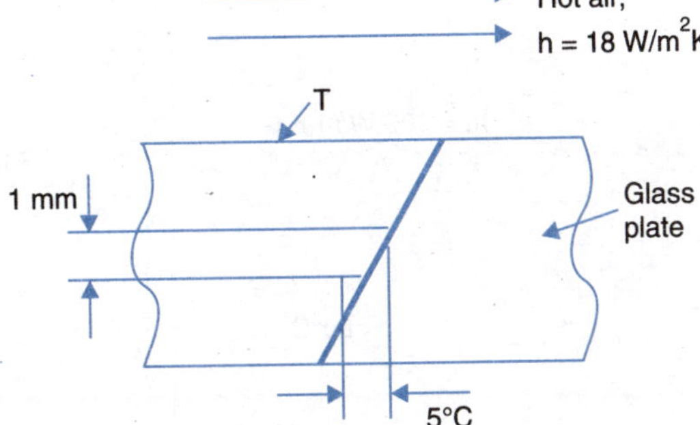

Fig. AE-1.16

$$Q = \frac{T_1 - T_2}{L / kA}$$

Assuming unit area,

$$Q = \frac{5}{0.001/(1.4 \times 1)} = 7000 \text{ W}$$

Therefore,

$$7000 = hA(T_{air} - 40)$$
$$= 18(T_{air} - 40)$$

Therefore, $\mathbf{T_{air}}$ **= 428.9°C.**

Similar problems are met within production shops.

Example A-1.17: *A surface at 260°C convects heat at steady state to Air at 60°C with a convection coefficient of 30 W/m²K. If this heat has to be conducted through wall with thermal conductivity of 9.5 W / mK, determine the temperature gradient in the solid.*

Solution: Energy balance yields the relation, heat conducted = heat convected

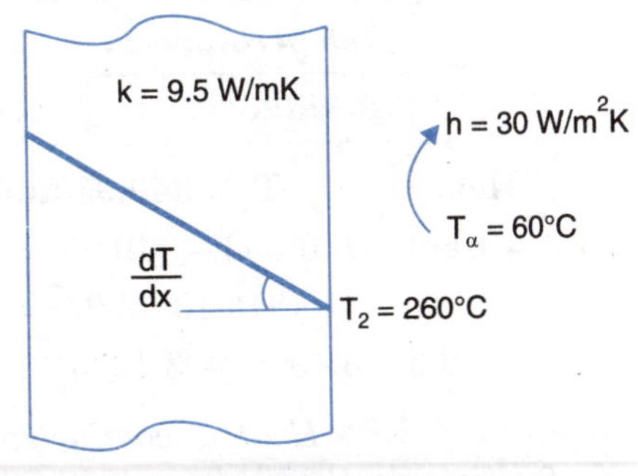

Fig. AE-1.17

Assuming unit area

$$= - kA(dT/dx) = hA(T_2 - T_\infty)$$

Therefore $dT/dx = (-h/k)\ (T_2 - T_\infty)$

$= (30/9.5)\ (260 - 60) = - 631.5$°C/m or, **– 6.315°C/cm.**

Example A-1.18: *A thin metal sheet receives heat on one side from a fluid at 80°C with a convection coefficient of 100 W/m²K while on the other side it radiates to another metal sheet parallel to it. The second sheet loses heat on its other side by convection to a fluid at 20°C with a convection coefficient of 15 W/m²K. Determine the steady state temperature of the sheets. The two sheets exchange heat only by radiation and may be considered to be black and fairly large in size.*

Solution: Known data are indicated in Fig. AE-1.18.

The energy balance provides heat received by convection by sheet 1

= Radiation heat exchange between sheet 1 and 2.

= Heat convected by sheet 2.

or
$$h_1 A(T_{\infty 1} - T_1) = \sigma A(T_1^4 - T_2^4) = h_2 A(T_2 - T_{\infty 2})$$

Substituting the values: considering unit area

$$100 \times 1(353 - T_1) = 15 \times 1(T_2 - 293)$$

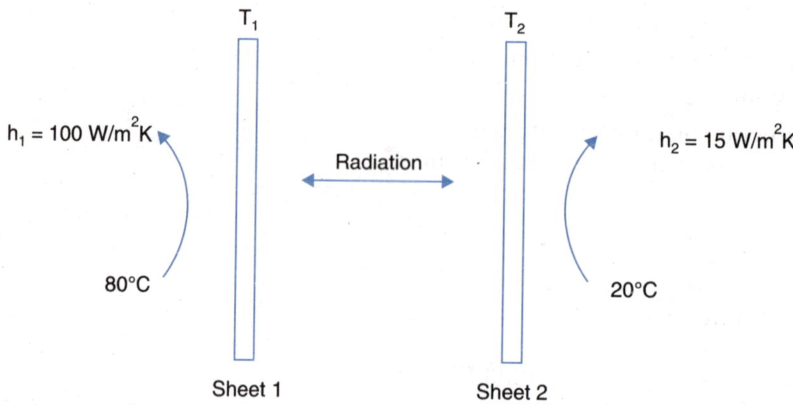

Fig. AE-1.18

Therefore, $T_2 = (100/15)\ (353 - T_1) + 293$. Considering radiation,

$$100 \times 1(353 - T_1) = 5.67 \times 1[(T_1/100)^4 - (T_2/100)^4]$$

Combining, $17.64\ (353 - T_1) + (26.46 - T_1/15)^4 - (T_1/100)^4 = 0$

This is solved by trial for T_1.

Temperature T_1	345	349	350	346.9	349.95
Residue	142.76	26.19	– 1.57	1.18	– 0.20

Hence, $\mathbf{T_1 = 349.95}$ **and** $\mathbf{T_2 = 313.33\ K}$

Check: 100(353 – 349.95) = 305 W

15(313.13 – 293) = 305 W

$5.67\ (3.4995^4 - 3.1333^4) = 303.81$ W hence checks.

Example A-1.19: *Heat is conducted at steady state through a solid with temperature gradient of – 5°C/cm, the thermal conductivity of the solid being 22.5 W/mK. If the heat is exchanged by radiation from the surface to the surroundings at 30°C, determine the surface temperature.*

Solution: Known data are indicated in Fig. AE-1.19. Energy balance yields the relation.

Heat conducted = heat radiated

$$-kA.dT/dx = \sigma A(T_2^4 - T_s^4)$$

Considering unit area and substituting the values

$$-22.5 \times -5 \times 100 = 5.67\,[(T_2/100)^4 - (303/100)^4]$$

(The gradient should be converted to °C/m by multiplying by 100)

Therefore, $\mathbf{T_2}$ **= 674.4 K** or **401.4°C.**

Fig. AE-1.19

Example A-1.20: *A satellite in space is of 2 m dia and internal heat generation is 2000 W. If it is protected from direct solar radiation by earths shadow determine its surface temperature.*

Solution: In the absence of atmosphere the only possible way heat is dissipated is by radiation. The temperature of the space may taken as 0 K.

Hence heat generated = heat radiated.

$$2000 = 4\pi \times 1^2 \times 5.67 \left\{\left(\frac{T}{100}\right)^4 - 0\right\}$$

$\therefore$ **T = 230.18 K.**

Example A-1.21: *A heat flux meter on the outside surface of a wall shows 10 W/m². The wall is 0.2 m thick and conductivity is 1.5 W/mK. Determine the temperature drop through the wall.*

Solution: $Q = \frac{\Delta T}{R}$ or $\Delta T = QR$, $R = \frac{L}{kA}$, A = 1, Q = 10 J/s.

$\therefore$ **ΔT = 10 × 0.2/(1.5 × 1) = 1.33°C.**

REVIEW QUESTIONS

1.1 (*a*) Explain the physical mechanism of heat transfer by (*i*) conduction (*ii*) convection and (*iii*) Radiation.

(*b*) Compare heat transfer and current flow in a conductor in terms of potential, flow and resistance.

(*c*) Distinguish between heat rate and heat flux. Write down the units of each.

(*d*) What is the unit for temperature gradient? How does it relate to heat transfer?

(*e*) What is the unit of thermal conductivity? Explain its effect on heat transfer.

(*f*) Distinguish between steady and unsteady state of heat transfer.

(*g*) Distinguish between natural or free convection and forced convection.

(*h*) What is the unit of convection coefficient?

(*i*) State Newton's law of cooling. How does it lead to the definition of convection coefficient?

(*j*) State Stefan-Boltzmann law.

(*k*) What is the unit of radiation heat transfer coefficient.

(*l*) Define emissivity and irradiation.

(*m*) Explain the difference between application of conservation of energy over (*i*) a time interval and (*ii*) at an instant of time.

OBJECTIVE QUESTIONS

1.1 Choose the correct statement from each question.

(*i*) A pipe carrying steam at about 300°C traverses a room, the air being still at 30°C. The major fraction of the heat loss will be by

(*a*) conduction to the still air (*b*) convection to the air

(*c*) radiation to the surroundings (*d*) conduction and convection put together.

(*ii*) A satellite in space exchanges heat with its surroundings by

(*a*) conduction (*b*) convection

(*c*) radiation (*d*) conduction as well as convection.

(*iii*) For the same temperature drop in the temperature ranges of 300–400°C the heat flow rate will be highest by

(*a*) conduction process

(*b*) convection process

(*c*) radiation process

(*d*) other factors should be known before any conclusion.

(*iv*) In the cold season a person would prefer to be near a fire because

(*a*) the conduction from the fire will be better

(*b*) the convection will be better if he is near the fire

(*c*) direct unimpeded radiation will provide quick warmth

(*d*) combined conduction and convection will be better.

(*v*) A finned tube hot water radiator with a fan blowing air over it is kept in rooms during winter. The major portion of the heat transfer from the radiator to air is due to

(*a*) radiation (*b*) convection

(*c*) conduction (*d*) combined conduction and radiation.

(*vi*) For a specified heat input and a given volume which material will have the smallest temperature rise (Use data book if necessary)

(*a*) steel (*b*) aluminium

(*c*) water (*d*) copper.

(*vii*) When a hot metal piece is left to cool in air the time rate of cooling of the outer layer will be

(*a*) slower at start and faster near the end

(*b*) faster at start and slower near the end

(*c*) both rates will be the same

(*d*) this will depend on the material.

(*viii*) A thin black plate at temperature T receives radiation from a surface at temperature T_1 and radiates to a surface at T_2. If all surfaces are black at steady state

(*a*) $(T_1 - T) > (T - T_2)$ (*b*) $(T_1 - T) < (T - T_2)$

(*c*) $(T_1 - T) = (T - T_2)$ (*d*) can be anyone of *a*, *b* or *c*.

(*ix*) The temperature profile (in) a slab initially at a constant temperature and then allowed to cool by convection for a short time will be as shown in Fig. 1.1.

(*a*) E1.1*a* (*b*) E1.1*b*

(*c*) E1.1*c* (*d*) E1.1*d*.

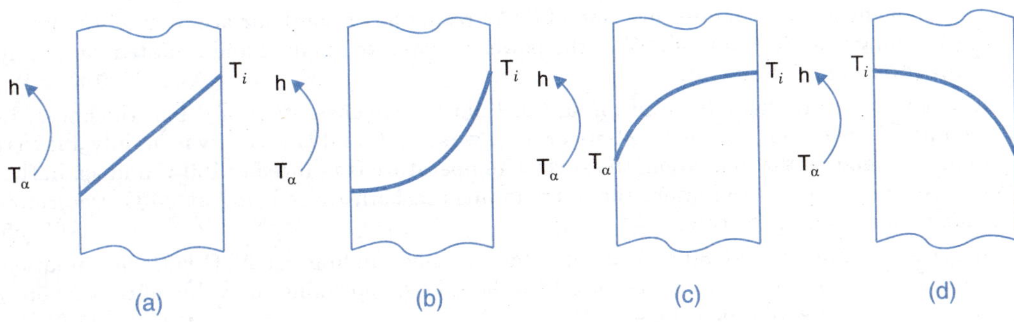

Fig. 1.1

Answers

1.1 (*i*) (*c*), (*ii*) (*c*), (*iii*) (*d*), (*iv*) (*c*), (*v*) (*b*), (*vi*) (*c*), (*vii*) (*b*), (*viii*) (*b*), (*ix*) (*c*).

EXERCISE PROBLEMS

1.1 Model the following heat transfer situations. Specify heat flows and storages. Try to write down the mathematical expressions.

(*i*) Solar heating of the road surface

(*ii*) A steam pipe passing through an open space between two buildings

(*iii*) Heat transfer from a person in a warm room in the cold season

(*iv*) Pressure cooker-warming up-cooling down

(*v*) Pressure cooker-steady conditions

(*vi*) A rod with one end in a furnace and the remaining surface in atmosphere

(*vii*) A wire carrying current, exposed to air

(*viii*) A water heater (electrical) with hot water being drawn out with cold water admission.

(*ix*) Cake being baked in an oven or a fruit placed in a refrigeration

(*x*) A frying pan placed on a stove.

1.2 A wall is exposed on one side to a heat flux of 1.5 kW/m^2 which is conducted through the wall. For the following combinations, determine the temperature drop through the wall (*a*) thickness 0.16 m and k = 1.4, 15, 25, 45, 210 and 340 W/mK. (*ii*) Thickness 0.25 m and k as above. Plot the temperature drop against the radio (L/k) and also (k/L).

1.3 The heat flux through a layer of material 40 mm thick conducting heat under steady state with a temperature drop of 40°C, was measured as 106 W/m^2. Determine the thermal conductivity of the material. [**Ans.** 0.106 W/mK]

1.4 A glass pane is 8 mm thick and the inside surface temperature was 25°C and outside surface temperature was 33°C. If k = 1.4 W/mK. Determine the heat flow through an area of 0.8 m × 1 m size pane.

1.5 The surface temperature of a plate over which air flows was measured as 80°C. The air temperature was 40°C. In order to maintain the surface temperature over an area of 0.1m^2, the heater rating required was found to be 1.5 kW. Determine the value of convection coefficient. [**Ans.** 375 W/m^{2}K]

1.6 A strip heater of area 0.2 m^2 and rating of 1200 W is fixed on a vertical wall and mostly convects the heat into the room air at 20°C. Determine the value of convective heat transfer coefficient if the surface temperature of the heater is not to exceed 60°C. Indicate whether such a value can be achieved by natural convection. [**Ans.** 150 W/m^{2}K, Not possible]

1.7 A strip heater with an area of 0.05 m^2 has to radiate at 600°C to surroundings at 30°C. Determine the rating assuming that convection is negligible. [**Ans.** 1622.8 W]

1.8 The filament of an incandescent lamp of 60 W rating has a total surface area of 40 mm^2. If the surrounding is at 30°C and if 90% of the power is converted to heat and radiated, determine the temperature of the filament. [**Ans.** 2209 K or 1936°C]

1.9 A sphere of 0.5 m diameter containing hot fluid has an insulation of 0.1 m thickness. Before adding the insulation, the surface temperature was 260°C and heat loss was mainly due to radiation. After adding the insulation, the surface temperature is reduced to 160°C and again the heat loss is mainly due to radiation, the surroundings in both cases being at 25°C. Determine the change in heat transfer rate. [**Ans.** 558 W]

1.10 A surface maintained at 80°C dissipates heat to surroundings at 25°C both by radiation and convection. If convection accounts for 60% of heat loss, determine the value of convection coefficient and also the total heat loss. [**Ans.** 11.81 W/m^2K]

1.11 A pipe of internal diameter 0.2 m and wall thickness 0.008 m carries steam at 220°C, the thermal conductivity of the material is 12.5 W/mK. Heat is lost from the outside surface by convection to surroundings at 30°C. Determine the outside surface temperature if the convection coefficient has a value of 38.5 W/m^2. Also find the heat loss /m length. Consider the pipe wall as a flat plate of thickness 0.008 m and width as $\pi \times 0.02$ m and length 1 m. [**Ans.** 215.4°C]

1.12 Considering the problem 1.11, if the heat loss from the surface is only by radiation, determine the surface temperature.

1.13 If in problem 1.11, the loss from the surface is both by radiation and convection, determine the surface temperature.

1.14 A flat plate solar collector absorbs 80% of radiation of 820 W/m^2 received. The top loss coefficient is 12 W/m^2K. Determine the temperature of heat collection if the efficiency of collection is 0.6. Atmospheric temperature is 35°C. [**Ans.** 48.67°C]

1.15 Heat is generated in a solid having a total surface area of 0.56 m^2 and a volume of 0.26 m^3 at a rate of 1 MW/m^3. Determine the surface temperature if the heat is convected as well as radiated to the surroundings at 27°C. The convective heat transfer coefficient has a value of 245 W/m^2K.

1.16 A thermocouple junction of 2 mm dia spherical shape receives heat from a flowing fluid at 800°C with a convection coefficient of 250 W/m^2K and radiates to the surroundings at 450°C. Determine the temperature of the surface at equilibrium. [**Ans.** 950 K or 677°C]

1.17 Repeat the problem 1.16 for a cylinder of 2 mm diameter and 8 mm length with convection coefficient of 125 W/m^2K. [**Ans.** 900 K or 627°C]

1.18 One side of an insulation layer of 25 mm thickness is maintained at 300°C, while the other side is exposed to convection at 30°C with a convective heat transfer coefficient of 400 W/m^2K. Determine the surface temperature neglecting radiation. Also find the heat flow. Conductivity of the material is 40 W/mK.

1.19 Repeat problem 1.18, if radiation is also present on the other side to the same surrounding temperature.

1.20 A steel billet of 0.2 m × 0.2 m × 0.4 m lies on its 0.2 m × 0.4 m face on the floor of a furnace and is exposed to radiation on all the other faces from a source at 1500°C. The density of the material is 7830 kg/m^3 and the specific heat is 960 J/kgK. Determine the rate of rise in the average temperature of the billet when its surface temperature is 500°C. Will the heating rate increase or decrease with rise in temperature of the surface?

1.21 For an air conditioned space, a double glass window pane with a thin air gap between the plates the preferred when compared to a single glass pane. Discuss the reasons for the above.

1.22 Heat is convected and also radiated from the surface of a solid at 160°C to the surroundings at 30°C. The convection coefficient is 45 W/m^2K. Assuming the solid to be isotropic and to have constant properties and the heat is conducted through the solid at steady state, determine the temperature gradient in the solid for $k = 15$, $k = 45$, $k = 210$ and $k = 340$ W/mK.

1.23 Heat is conducted through a solid layer of 0.3m and is then convected at the surface. The temperature drop in the solid and the temperature drop in the convective layer are equal. The conductivity of the material is 12.5 W/mK. Determine the value of convection coefficient. Also evaluate the conduction and convection resistances. [**Ans.** 41.67 W/m^2K]

1.24 The temperature gradient in a solid of thermal conductivity k = 12.5 W/mK, conducting heat under steady conditions is – 6°C/cm. If the heat is radiated at the surface, determine the temperature at the surface. The surroundings are at 30°C.

1.25 In a solid conducting heat at steady state, sketch the variation of the temperature gradient if (*i*) the thermal conductivity increases along the thickness (*ii*) the thermal conductivity decreases along the thickness.

1.26 The temperature variation along a slab is shown in Fig. 1.26. Determine the heat flow directions at locations x_1, x_2, x_3 and x_4.

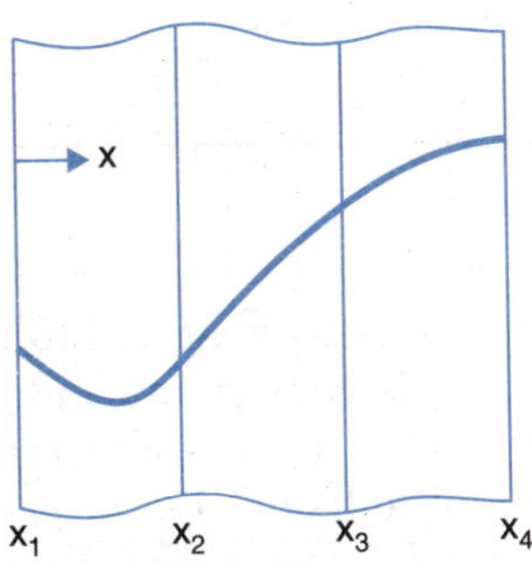

Fig. 1.26

1.27 Discuss the conditions at which (under thermal conduction) the following temperature variations as in Fig. 1.27 in a slab is possible.

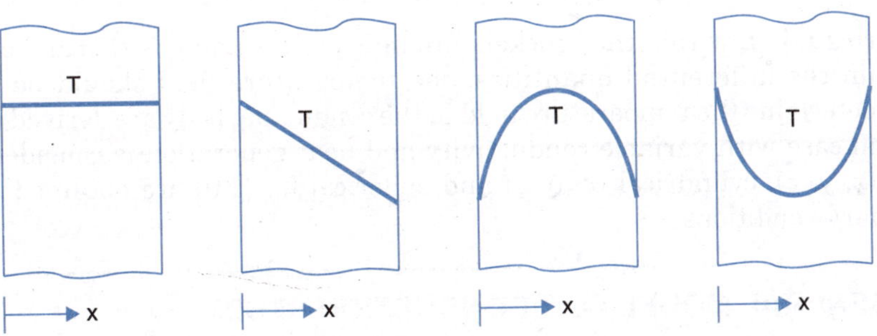

Fig. 1.27

1.28 A surface at 200°C receives a radiation flux of 800 W/m^2 and reflects 40% of the same. It also radiates with an emissivity of 0.6. Determine the total radiation flux coming out of the surface (This is also known as radiosity). [**Ans.** 2022 W]

1.29 Calculate the value of h_r between two surfaces at 100°C and 200°C. Also investigate the percentage variation in h_r as the surface temperature of the second surface increases to 300°C and 400°C.

1.30 A small thin metal plate absorbs solar energy at the rate 500 W/m^2 and dissipates it by convection into ambient air at 300 K. The convection coefficient is 20 W/m^{2}K and by radiation to the surroundings it at 280 K. The surface emissivity is 0.9. The backside of the plate is insulated. Determine the temperature of the plate. [**Ans.** 315.5 K]

1.31 A thin metal sheet is placed between two large parallel surfaces. The surface 1 is at 1000 K. The other surface 2 is at 400 K. Assume black body condition and determine the temperature of the thin metal sheet. Also calculate the percentage reduction in heat flow due to the placement of the sheet. [**Ans.** 846.2 K and 50%]

Steady State Conduction

2

CHAPTER

2.0 CONDUCTION

Conduction is the mode of energy transfer as heat due to temperature difference in a solid or any phase of material where the mass is contiguous and in thermal contact. Microscopically this mode of energy transfer is attributed to free electron flow from higher to lower energy levels, lattice vibration and molecular collision. However no macroscopic mass movement is involved. The applications have been indicated earlier. Heat flux a directional (vector) quantity is determined by an equation of the form (in cartesian coordinates)

$$\mathbf{q} = -k\left(\mathbf{i}\frac{\partial T}{\partial x} + \mathbf{j}\frac{\partial T}{\partial y} + \mathbf{k}\frac{\partial \pi}{\partial z}\right).$$

The terms **i**, **j**, **k**, within the brackets are unit vectors and k is thermal conductivity. In order to obtain the differential quantities, the temperature field should be first obtained. Equations to determine the temperature field in the conducting body are derived in this chapter. A very general case with variable conductivity and heat generation is considered. Equations in cartesian (x, y, z), cylindrical (r, ϕ, z) and spherical (r, ϕ, θ) are obtained and solved for simple boundary conditions.

2.1 THE GENERAL MODEL FOR CONDUCTION STUDY

The general model for conduction study is a body in which the following conditions are assumed

(*i*) heat flows in all the coordinate directions

(*ii*) heat is generated uniformly all over the volume and

(*iii*) the temperature at any location in the body changes with time. In addition, the properties also vary with the coordinate directions.

The energy equation for the model is given below for a time period considered.

The net heat flow through the body surface in all the coordinate directions considered + heat generated in the volume = Change in the energy content of the body.

This can be reduced to $(Q_k + qV)\, d\tau = \rho c V\, dT$ ...(2.1)

where Q_k — Net heat conducted over all surfaces, W

q — Heat generation for unit volume, W/m^3

V — Volume, m^3

ρ — Density kg/m^3

c — Specific heat J/kg K

dT — Change in temperature during time period considered °C

$d\tau$ — Time period considered, s

The temperature in the body will be a function of location as well as time. In the popular cartesian coordinates

$$T = T(x, y, z, \tau)$$

The temperature field is obtained by deriving and solving the differential equation based on energy balance relations for the volume.

2.1.1. Considering the Subvolume—Cartesian Coordinate System

With dimensions dx, dy and dz in the cartesian coordinate system and taking a time interval $d\tau$, and noting that temperature is a function of all the four, the following steps lead to the general differential equation in cartesian coordinates [Ref. Fig. 2.1(*a*)].

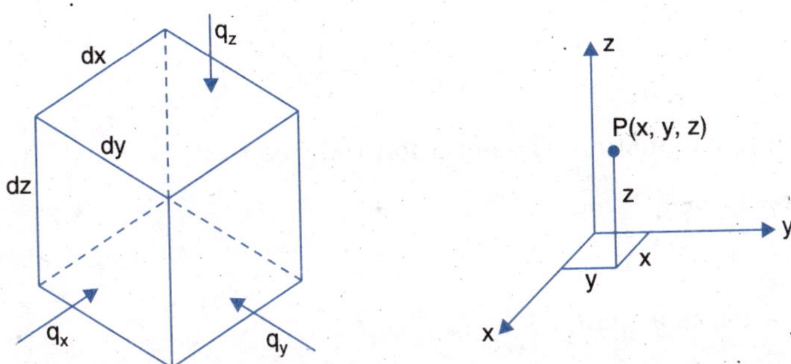

Fig. 2.1. (a) An elemental volume considered in cartesian coordinates

The heat flow by conduction through the boundary is obtained by considering the conduction over all six faces. The area perpendicular to x direction has a value $dydz$. Using Fourier's law, the heat flow at the $dydz$ face at x over time period $d\tau$ is $= -k\, dydz \,.\, \dfrac{\partial T}{\partial x}\, d\tau$

The flow at $x + dx$ is $= -k\, dy\, dz\, d\tau \dfrac{\partial T}{\partial x} + \dfrac{\partial}{\partial x}\left(-k\, dy\, dz\, d\tau \dfrac{\partial T}{\partial x}\right) dx$

The net flow in the x direction is obtained by taking the sum of these two which yields

$$\frac{\partial}{\partial x}\left(k \,.\, \frac{\partial T}{\partial x} \,.\, dy\, dz\, d\tau\right) dx = \frac{\partial}{\partial x}\left(k \,.\, \frac{\partial T}{\partial x} .\right) dx\, dy\, dz\, d\tau$$

Similar consideration in y and z directions give

$$\frac{\partial}{\partial y}\left(k \frac{\partial T}{\partial y} .\right) dx\, d\tau\, dz\, dy \quad \text{and} \quad \frac{\partial}{\partial z}\left(k \frac{\partial T}{\partial z} .\right) dx\, dy\, d\tau\, dz$$

Heat generated in the volume during $d\tau$ is given by $\dot{q}\; dx\, dy\, dz\, d\tau$.

The change in the energy of this volume over the time $d\tau$ when temperature changes is $\rho c\; dx\, dy\, dz\, \partial T$.

The sum is simplified by assuming dx, dy and dz do not change along the coordinate directions. This leads to the equation (2.2).

$$\frac{\partial}{\partial x}\left(k\frac{\partial T}{\partial x}\right)+\frac{\partial}{\partial y}\left(k\frac{\partial T}{\partial y}\right)+\frac{\partial}{\partial z}\left(k\frac{\partial T}{\partial z}\right)+q=\rho\frac{\partial T}{\partial \tau} \quad ...(2.2)$$

If thermal conductivity k is assumed to be constant with location equation [2.2(a)] is obtained from the above.

$$\frac{\partial^2 T}{\partial x^2}+\frac{\partial^2 T}{\partial y^2}+\frac{\partial^2 T}{\partial z^2}+\frac{q}{k}=\frac{\rho c}{k}\cdot\frac{\partial T}{\partial \tau} \quad ...[2.2(a)]$$

$\dfrac{k}{\rho c}$ is defined as thermal diffusivity denoted by α.

Thermal diffusivity is the ratio of heat conducted to heat stored. This becomes very important in unsteady heat conduction.

$$\therefore \quad \frac{\partial^2 T}{\partial x^2}+\frac{\partial^2 T}{\partial y^2}+\frac{\partial^2 T}{\partial z^2}+\frac{\dot{q}}{k}=\frac{1}{\alpha}\cdot\frac{\partial T}{\partial \tau} \quad ...(2.3)$$

For steady state conduction, the equation reduces to

$$\frac{\partial^2 T}{\partial x^2}+\frac{\partial^2 T}{\partial y^2}+\frac{\partial^2 T}{\partial z^2}+\frac{q}{k}=0 \quad ...[2.3(a)]$$

If heat generation is absent,

$$\frac{\partial^2 T}{\partial x^2}+\frac{\partial^2 T}{\partial y^2}+\frac{\partial^2 T}{\partial z^2}=0 \quad ...[2.3(b)]$$

If say k varies with location it should be within the differential like $\dfrac{\partial}{\partial x}\left(k\dfrac{\partial T}{\partial x}\right)$ and the integration of the equation has to consider the functionality of k with direction.

2.1.2. Cylindrical and Spherical Coordinate System

Similar steps may be followed to obtain the following differential equations in the cylindrical and spherical coordinate systems. Here it has to be noted that the area value changes in the coordinate directions

In cylindrical coordinate system (r, ϕ, z), Fig. 2.1(b), we get

$$\frac{1}{r}\frac{\partial}{\partial r}\left(kr\frac{\partial T}{\partial r}\right)+\frac{1}{r^2}\frac{\partial}{\partial \phi}\left(k\frac{\partial T}{\partial \phi}\right)+\frac{\partial}{\partial z}\left(k\frac{\partial T}{\partial z}\right)+q=\rho c\frac{\partial T}{\partial \tau} \quad ...(2.4)$$

With k constant equation (2.4) reduces to

$$\frac{1}{r}\frac{\partial}{\partial r}\left(r\frac{\partial T}{\partial r}\right)+\frac{1}{r^2}\frac{\partial^2 T}{\partial \phi^2}+\frac{\partial^2 T}{\partial z^2}+\frac{q}{k}=\frac{1}{\alpha}\cdot\frac{\partial T}{\partial \tau} \quad ...[2.4(a)]$$

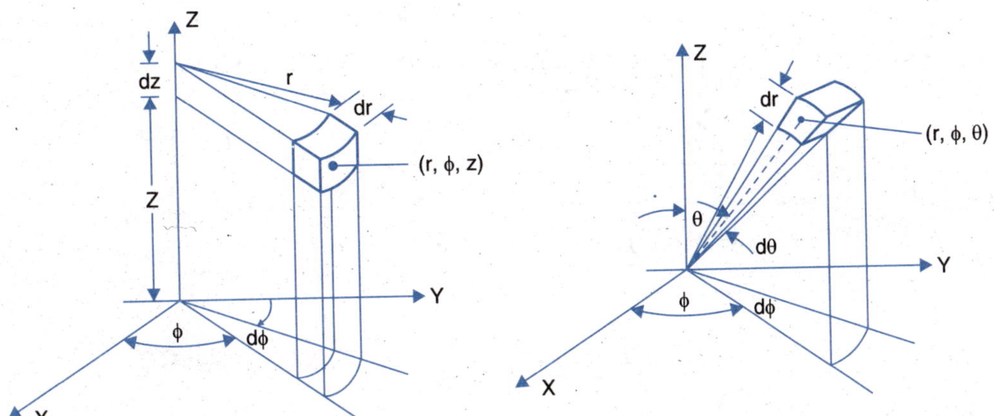

Fig. 2.1. (b) Elemental volume in cylindrical coordinates

Fig. 2.1. (c) Elemental volume in spherical coordinates

In spherical coordinate system (r, ϕ, θ), Fig. 2.1(c), we get

$$\frac{1}{r^2}\frac{\partial}{\partial r}\left(kr^2\frac{\partial T}{\partial r}\right)+\frac{1}{r^2\sin^2\theta}\cdot\frac{\partial}{\partial \phi}\left(k\frac{\partial T}{\partial \phi}\right)+\frac{1}{r^2\sin\theta}\cdot\frac{\partial}{\partial \theta}\left(k\sin\theta\frac{\partial T}{\partial \theta}\right)+q=\rho c\left(\frac{\partial T}{\partial \tau}\right) \quad ...(2.5)$$

With k constant equation (2.5) reduces to

$$\frac{1}{r^2}\frac{\partial}{\partial r}\left(r^2\frac{\partial T}{\partial r}\right)+\frac{1}{r^2\sin^2\theta}\cdot\left(\frac{\partial^2 T}{\partial \phi^2}\right)+\frac{1}{r^2\sin\theta}\cdot\frac{\partial}{\partial \theta}\left(\sin\theta\frac{\partial T}{\partial \theta}\right)+\frac{q}{k}=\frac{1}{\alpha}\left(\frac{\partial T}{\partial \tau}\right) \quad ...[2.5(a)]$$

The detailed derivation is given in the appendix.

2.1.3. The Complete Solutions to the General Model

It is rather complex. Some of the simplified models for which solutions are attempted are listed below:

1. One dimensional steady flow (x or r directions) with constant or variable properties, without heat generation.
2. Same as above but with heat generation
3. Two dimensional steady flow (with constant properties, without heat generation) and
4. One dimensional unsteady state without heat generation.

The simplified expressions in these cases in the various coordinate systems for above conditions 1, 2, 3 and 4 are:

Cartesian

$$\frac{\partial}{\partial x}\left[k\frac{\partial T}{\partial x}\right]=0 \text{ and } \frac{\partial^2 T}{\partial x^2}=0$$

$$\frac{\partial^2 T}{\partial x^2}+\frac{q}{k}=0$$

$$\frac{\partial^2 T}{\partial x^2}+\frac{\partial^2 T}{\partial y^2}=0$$

$$\frac{\partial^2 T}{\partial x^2}=\frac{1}{\alpha}.\frac{\partial T}{\partial \tau} \qquad ...[2.6(a, b, c, d)]$$

Cylindrical coordinates (for conditions 1, 2 and 4)

$$\frac{1}{r}\frac{\partial}{\partial r}\left[kr\frac{\partial T}{\partial r}\right]=0 \text{ and } \frac{1}{r}\frac{\partial}{\partial r}\left[r\frac{\partial T}{\partial r}\right]=0$$

$$\frac{1}{r}\frac{\partial}{\partial r}\left[r\frac{\partial T}{\partial r}\right]+\frac{q}{k}=0$$

$$\frac{1}{r}\frac{\partial}{\partial r}\left[r\frac{\partial T}{\partial r}\right]=\frac{1}{\alpha}\frac{\partial T}{\partial \tau} \qquad ...[2.7(a, b, c)]$$

Spherical coordinates (for conditions 1, 2 and 4)

$$\frac{1}{r^2}\frac{\partial}{\partial r}\left[kr^2\frac{\partial T}{\partial r}\right]=0 \text{ and } \frac{1}{r^2}\frac{\partial}{\partial r}\left[r^2\frac{\partial T}{\partial r}\right]=0$$

$$\frac{1}{r^2}\frac{\partial}{\partial r}\left[r^2\frac{\partial T}{\partial r}\right]+\frac{q}{k}=0$$

$$\frac{1}{r^2}\frac{\partial}{\partial r}\left[r^2\frac{\partial T}{\partial r}\right]=\frac{1}{\alpha}\frac{\partial T}{\partial \tau} \qquad ...[2.8(a, b, c)]$$

These sets of equations (2.6), (2.7) and (2.8) can be integrated for specified boundary and initial conditions and these solutions follow in this and subsequent chapters.

Presently softwares are available to model and solve conduction problems for various types of boundary and initial conditions. These softwares can be effectively used only if one has a basic understanding of the theory. Two such softwares are

1. Interactive Heat Transfer (*IHT*) developed by Inteli Pro. Inc, New Brunswick, New Jersey.
2. Finite Element Heat Transfer (*FEHT*) developed by F-chart Software of Middleton, Wisconsin.

2.2 STEADY CONDUCTION IN ONE DIRECTION (ONE DIMENSIONAL)

2.2.1. Plane Wall: Steady Flow in the x Direction, (Fig. 2.2)

Assuming constant properties, the problem can be solved by applying the basic law due to Fourier or by integrating the simplified differential equation [2.6(a)]

From Fourier's law

$$Q = -kA\frac{dT}{\partial x}$$

As k and A are constants, the equation can be integrated between the limits x_1 and x_2 to obtain

$$Q = kA\frac{T_1 - T_2}{x_2 - x_1} \qquad ...(2.9)$$

Taking $x_2 - x_1 = L$

$$Q = \frac{T_1 - T_2}{(L/kA)} \qquad ...(2.10)$$

The heat flow up to a thickness $(x - x_1)$ is the same as total heat flow

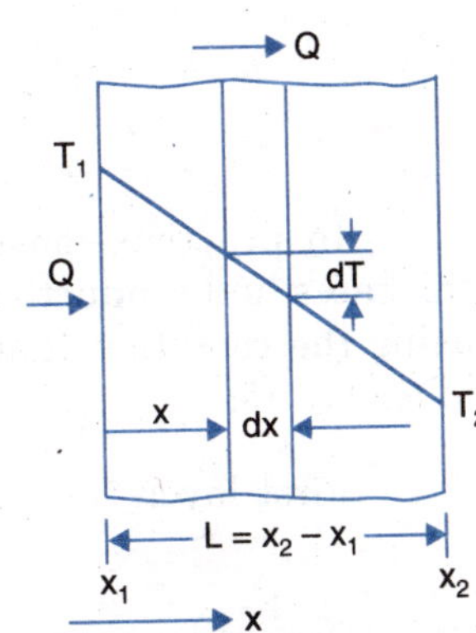

Fig. 2.2. Conduction–Plane wall

As $$Q = \frac{2\pi k(T_1 - T_2)}{x_2 - x_1} = \frac{2\pi k(T_1 - T)}{x - x_1}$$

$$\therefore \quad \frac{T_1 - T}{T_1 - T_2} = \frac{x - x_1}{x_2 - x_1}$$

same as equation (2.11) derived from differential equation.

where L/kA is called thermal resistance

Integrating equation [2.6(a)]

$$\frac{\partial^2 T}{\partial x^2} = 0, \frac{\partial T}{\partial x} = c_1, T = c_1 x + c_2.$$

$$\text{At } x = x_1, \quad T = T_1 \quad \therefore \quad T_1 = c_1 x_1 + c_2 \qquad ...(A)$$

$$\text{At } x = x_2, \quad T = T_2 \quad \therefore \quad T_2 = c_1 x_2 + c_2 \qquad ...(B)$$

(A – B) leads to $$c_1 = \frac{T_1 - T_2}{x_1 - x_2}$$

From A $$c_2 = T_1 - c_1 x_1 = T_1 - \frac{T_1 - T_2}{x_1 - x_2}\cdot x_1$$

Substituting in the general solution

$$T = \frac{T_1 - T_2}{x_1 - x_2}\cdot x + T_1 - \frac{T_1 - T_2}{x_1 - x_2} x_1$$

$$\therefore \quad T - T_1 = \frac{T_1 - T_2}{x_1 - x_2}\cdot (x - x_1)$$

or $$\frac{T - T_1}{T_1 - T_2} = \frac{x - x_1}{x_1 - x_2} \quad \text{or} \quad \frac{T_1 - T}{T_1 - T_2} = \frac{x - x_1}{x_2 - x_1} \qquad ...(2.11)$$

Equation (2.11) shows that the temperature varies linearly with x or $\frac{dT}{dx}$ is constant. To find heat flow, when k is constant.

$$Q = -kA\frac{dT}{dx}, \text{ but } \frac{dT}{dx} = c_1 = \frac{T_1 - T_2}{x_1 - x_2}$$

$$\therefore \quad Q = -kA \cdot \frac{T_1 - T_2}{x_1 - x_2} = kA \cdot \frac{T_1 - T_2}{x_2 - x_1}. \text{ same as equation (2.9)}$$

In practice, most heat transfer situations involve multilayer barriers of different thickness and conductivity. For a wall made of several layers of different materials (Fig. 2.3), using the condition that heat flow through one layer is equal to the heat flow through the other layers,

First layer $$Q = \frac{T_1 - T_2}{L_1 / k_1 A} \quad \text{or} \quad T_1 - T_2 = Q\frac{L_1}{k_1 A}$$

Second layer $$Q = \frac{T_2 - T_3}{L_2 / k_2 A} \quad \text{or} \quad T_2 - T_3 = Q \cdot \frac{L_2}{k_2 A}$$

Third layer $$Q = \frac{T_3 - T_4}{L_3 / k_3 A} \quad \text{or} \quad T_3 - T_4 = Q\frac{L_3}{k_3 A}$$

Adding $$T_1 - T_4 = Q \cdot \left\{\frac{L_1}{k_1 A} + \frac{L_2}{k_2 A} + \frac{L_3}{k_3 A}\right\} \quad \text{or} \quad Q = \frac{(T_1 - T_4)}{\frac{L_1}{k_1 A} + \frac{L_2}{k_2 A} + \frac{L_3}{k_3 A}},$$

In general $$Q = \frac{\Delta T}{\frac{L_1}{k_1 A} + \frac{L_2}{k_2 A} + \frac{L_3}{k_3 A} + ...} \qquad ...(2.12)$$

As flow area is constant along heat flow direction, adding layers of insulation will always reduce the heat flow in plane walls. Note that the temperature gradients are different in different layers.

Any interface temperature can be obtained using

$$Q = \frac{\Delta T_i}{R_i} \qquad ...(2.13)$$

where ΔT_i is the temperature drop in layer i and $R_i = L_i/(k_i A)$.

In case convection is involved on either side with $T_{\infty 1}$, h_1 and $T_{\infty 2}$ and h_2

$$Q = \frac{T_{\infty 1} - T_{\infty 2}}{R_{c1} + R_1 + ... + R_{c2}} \qquad ...(2.14)$$

where $R_{c1} = \frac{1}{h_1 A}$ and $R_{c2} = \frac{1}{h_2 A}$ are convection resistances.

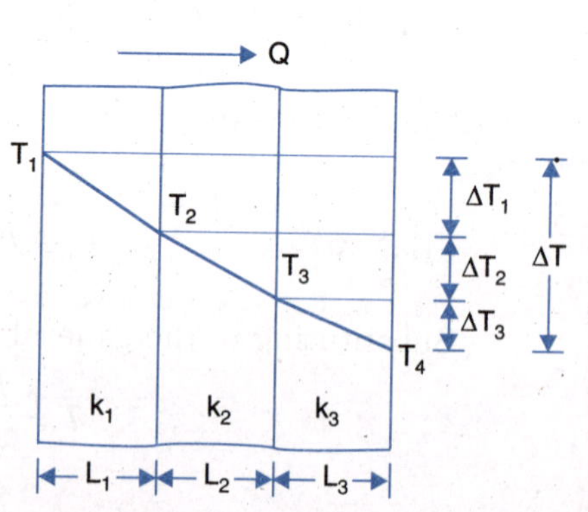

Fig. 2.3. Conduction–Composite wall

In certain cases contact between two layers may not be perfect, introducing contact resistance. Some typical values can be found in data books. This resistance is due to the surface roughness leading to material contact at high spots only. A temperature drop will occur at the interface

$$Q = \frac{\Delta T}{R_{c1} + R_1 + R_{s1} + ... + R_{c2}} \quad ...[2.15(a)]$$

where R_{s1} is the contact resistance between the first and second layers. The temperature variation in such a case is shown in Fig. Ex. 2.1.

The contact resistance will depend on the material, surface roughness, the fluid between surfaces and the pressure applied. However, contact resistance may not be considered at the first instance in many situations. Refer Fig. Ex. 2.1.

A overall heat transfer coefficient is also used for convenience where $Q = UA\,\Delta T$ where

$$UA = \frac{1}{R_1 + R_2 + R_3 + ...} \quad ...[2.15(b)]$$

Example 2.1: *A furnace wall is of three layers, first layer of insulation brick of 12 cm thickness of conductivity 0.6 W/mK. The face is exposed to gases at 870°C with a convection coefficient of 110 W/m²K. This layer is backed by a 10 cm layer of firebrick of conductivity 0.8 W/mK. There is a contact resistance between the layers of 2.6 × 10⁻⁴ m² °C/W. The third layer is the plate backing of 10 mm thickness of conductivity 49 W/mK. The contact resistance between the second and third layers is 1.5 × 10⁻⁴ m² °C/W. The plate is exposed to air at 30°C with a convection coefficient of 15 W/m²K. Determine the heat flow, the surface temperatures and the overall heat transfer coefficient.*

Solution: The data and equivalent circuit are shown in Fig. Ex. 2.1. The quantities to be found are Q, T_1, T_2^1, T_2^2, T_3^1, T_3^2 and T_4 are also indicated in the figure.

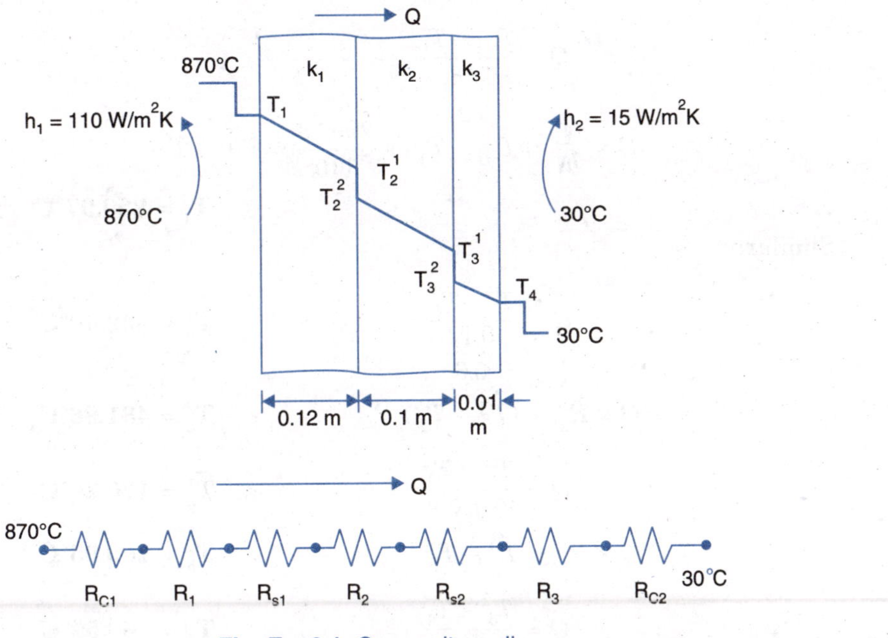

Fig. Ex. 2.1. Composite wall

Using equation (2.5)

$$Q = \frac{\Delta T}{\Sigma R}$$

The resistances (°C/W) are: Taking A = 1.0 m^2,

$$\Delta T = T_{\infty 1} - T_{\infty 2} = (870 - 30) = 840°C$$

$$Rc_1 = \frac{1}{h_1 A} = \frac{1}{110}$$

$$R_1 = \frac{L_1}{k_1 A} = (0.12/0.6) \quad R_{s1} = 2.6 \times 10^{-4}$$

$$R_2 = \frac{L_2}{k_2 A} = (0.1/0.8) \quad R_{s2} = 1.5 \times 10^{-4}$$

$$R_3 = \frac{L_3}{k_3 A} = \frac{0.01}{49} \quad R_{C2} = \frac{1}{15}$$

$$\therefore \quad \mathbf{Q} = \frac{840}{\frac{1}{110} + \frac{0.12}{0.6} + 2.6 \times 10^{-4} + \frac{0.1}{0.8} + 1.5 \times 10^{-4} + \frac{0.01}{49} + \frac{1}{15}}$$

$= \mathbf{2092.8\ W/m^2}$

$Q = UA\ \Delta T$. Where U is the overall heat transfer coefficient.

As $A = 1$ $\therefore$ $\mathbf{U} = \frac{Q}{\Delta T} = \frac{2092.8}{840} = \mathbf{2.491\ W/m^2°C}$

Surface temperatures:

$$Q = \frac{T_{\infty 1} - T_1}{1/h_1}$$

$$\therefore \quad Q \times \frac{1}{h_1} = T_{\infty 1} - T_1 \text{ or } \frac{2092.8}{110} = 870 - T_1$$

$$\therefore \quad \mathbf{T_1 = 850.97°C}$$

Similarly

$$Q = \frac{T_1 - T_2^1}{\frac{0.12}{0.6}} \qquad \therefore \quad \mathbf{T_2^1 = 432.40°C}$$

$$Q \times R_{s1} = (T_2^1 - T_2^2) \qquad \therefore \quad \mathbf{T_2^2 = 431.86°C}$$

$$Q = \frac{T_2^2 - T_3^1}{0.1/0.8} \qquad \therefore \quad \mathbf{T_3^1 = 170.26°C}$$

$$Q \times R_{s2} = T_3^1 - T_3^2 \qquad \therefore \quad \mathbf{T_3^2 = 169.95°C}$$

$$Q = \frac{T_3^2 - T_4}{0.01/49} \qquad \therefore \quad \mathbf{T_4 = 169.52°C}$$

COMMENT

For same layer thickness and heat flux the temperature drop will depend on the conductivity of the material. Higher the conductivity lower will be the drop. Note the drops due to contact resistance and the metal layer.

2.2.2. Steady Radial Heat Flow Through Hollow Cylinder

As mentioned in the previous article, the solution can be obtained either using Fourier's law or integrating equation [2.7(*a*)] directly.

A quarter of circle is shown due to symmetry. The integration of (with constant k)

$$\frac{1}{r}\frac{\partial}{\partial r}\left(kr\frac{\partial T}{\partial r}\right) = 0 \text{ yields}$$

$$\frac{dT}{dr} = \frac{C_1}{kr} \qquad \text{...[2.16(a)]}$$

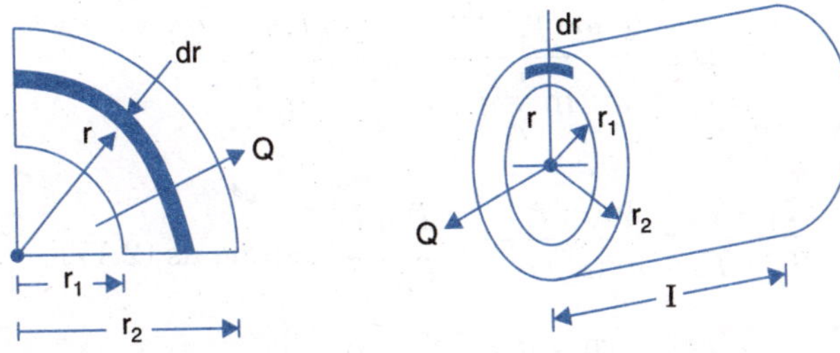

Fig. 2.4. Heat conduction–Hollow cylinder

Integrating again $\quad kT = C_1 \ln r + C_2 \qquad$ ...[2.16(*b*)]

Using the boundary conditions, that at $r = r_1$, $T = T_1$ and at $r = r_2$, $T = T_2$,

$$kT_1 = C_1 \ln r_1 + C_2 \qquad \text{...(A)}$$

$$kT_2 = C_1 \ln r_2 + C_2 \qquad \text{...(B)}$$

Subtracting and solving

$$C_1 = \frac{k(T_2 - T_1)}{\ln \frac{r_2}{r_1}} \qquad \text{...(C)}$$

From A

$$C_2 = kT_1 - C_1 \ln r_1$$

$$C_2 = kT_1 - \frac{k(T_2 - T_1)}{\ln \frac{r_2}{r_1}} . \ln r_1 \qquad \text{...(D)}$$

∴ From [2.16(*b*)]

$$kT = \frac{k(T_2 - T_1)}{\ln \frac{r_2}{r_1}} . \ln r + kT_1 - \frac{k(T_2 - T_1)}{\ln \frac{r_2}{r_1}} \ln r_1$$

∴

$$T - T_1 = \frac{(T_2 - T_1)}{\ln \frac{r_2}{r_1}} [\ln r - \ln r_1]$$

$$\frac{T - T_1}{T_2 - T_1} = \frac{\ln (r / r_1)}{\ln (r_2 / r_1)} \quad ...(2.17)$$

Note that in the case of hollow cylindrical shapes the temperature variation along the radii is logarithmic.

$$A = 2\pi rl, \quad \frac{dT}{dr} = \frac{C_1}{kr} = \frac{T_2 - T_1}{r \ln \frac{r_2}{r_1}}$$

Using $Q = - kA \dfrac{dT}{dr}$, and substituting for $\dfrac{dT}{dr}$ and A

$$Q = \frac{T_1 - T_2}{\ln \frac{r_2}{r_1} / 2\pi kl} \quad ...(2.18)$$

As the heat flow up to the radius r is the same as the heat flow up to r_2

$$Q = \frac{2\ kl\ (T_1 - T_2)}{\ln \frac{r_2}{r_1}} = \frac{2\ kl\ (T_1 - T)}{\ln \frac{r_2}{r_1}}$$

$$\therefore \quad \frac{T_1 - T}{T_1 - T_2} = \frac{\ln(r / r_1)}{\ln (r_2 / r_1)} = \frac{T - T_1}{T_2 - T_1}, \text{ same as (2.17)}$$

$$Q = \frac{T_1 - T_2}{\ln \frac{r_2}{r_1} / 2\ kl} = \frac{T_1 - T}{\ln \frac{r}{r_1} / 2\ kl}$$

where $\ln \frac{r_2}{r_1} / 2\pi kl$ **is the thermal conduction resistance for 1 m length.**

Equation (2.18) can also be derived from basic treating layers of thickness dr as a slab.

At radius r the flow area is $2\pi rl$,

Thickness is dr.

Using Fourier's equation, heat flow in the layer is given by

$$Q = - k\ 2\pi rl \cdot \frac{dT}{dr}$$

$$Q \frac{dr}{r} = - 2\pi kl \cdot dT.$$

Integrating between limits r_1 and r_2 equation (2.18) is obtained.

When a number of layers of different materials are involved with convection on both sides, (Refer Fig. Ex. 2.2).

$$Q = \frac{T_{\infty 1} - T_{\infty 2}}{\frac{1}{h_1 2\pi r_1 l} + \frac{\ln r_2 / r_1}{2\pi k_1 l} + \frac{\ln r_3 / r_2}{2\pi k_2 l} + \frac{\ln r_4 / r_3}{2\pi k_3 l} + \frac{1}{h_2 2\pi r_4 l}} \quad ...(2.19)$$

$$= \frac{T_{\infty 1} - T_{\infty 1}}{\Sigma R_1}$$

It will be noted under discussion on critical thickness of insulation, that there are some situations where adding insulation may increase the heat flow instead of reducing the same.

The interface temperature can be obtained using

$$Q = \frac{\Delta T_i}{R_i} \quad \text{...(2.20)}$$

where ΔT_i is the temperature drop in that layer and R_i is the thermal resistance of that layer. The equation will automatically indicate whether heat flow is inwards or outwards.

In the case of cylindrical shape, overall heat transfer coefficient has to be referred to the inside or outside as the occasion demands.

$$Q = U_i A_i \, \Delta T = U_o A_o \, \Delta T$$

where A_i and A_o are the inside and outside areas and U_i and U_o are the corresponding overall heat transfer coefficients.

Most of the time, the insulation over cylindrical pipes are of several layers of different materials. Example 2.2 involves such a situation.

Example 2.2: *A pipe carrying steam at 230°C has an internal diameter of 12 cm and the pipe thickness is 7.5 mm. The conductivity of the pipe material is 49 W/mK the convective heat transfer coefficient on the inside is 85 W/m²K. The pipe is insulated by two layers of insulation one of 5 cm thickness of conductivity 0.15 W/mK and over it another 5 cm thickness of conductivity 0.48 W/mK. The outside is exposed to air at 35°C with a convection coefficient of 18 W/m²K. Determine the heat loss for 5 m length. Also determine the interface temperatures and the overall heat transfer coefficient based on inside and outside areas.*

Solution: Data are shown in Fig. Ex. 2.2(*a*). The equivalent circuit is shown in Fig. Ex. 2.2(*b*). A quarter of the section is shown due to symmetry. All known data are shown in the figure.

Find: Q, T_1, T_2, T_3, T_4 and U.

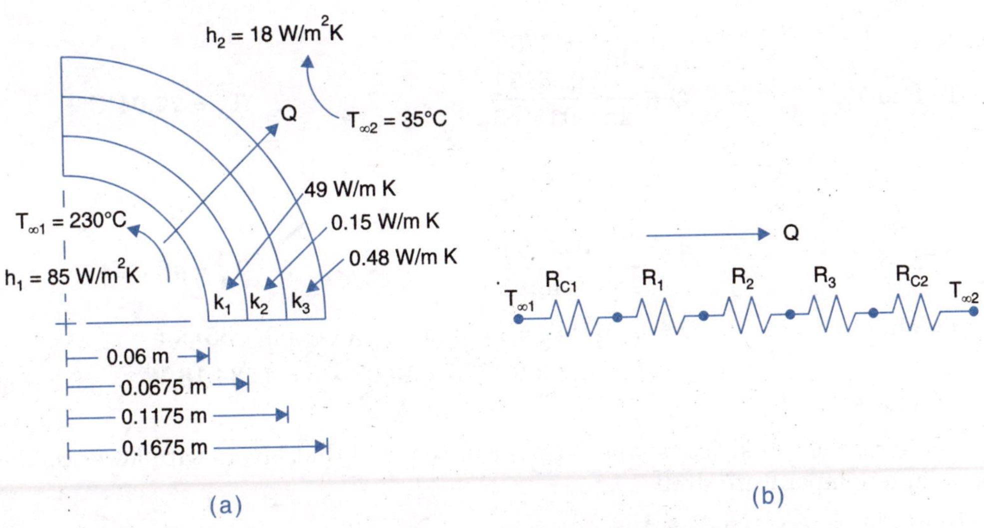

Fig. Ex. 2.2. Composite cylinder

Using equation (2.19)

$$Q = \frac{T_{\infty 1} - T_{\infty 2}}{\dfrac{1}{2\pi r_1 h_1 l} + \dfrac{\ln r_2/r_1}{2\pi k_1 l} + \dfrac{\ln r_3/r_2}{2\pi k_2 l} + \dfrac{\ln r_4/r_3}{2\pi k_3 l} + \dfrac{1}{2\pi r_4 h_2 l}}$$

Substituting

$$Q = \frac{230 - 35}{\dfrac{1}{2\pi \times 0.06 \times 85 \times 5} + \dfrac{\ln \dfrac{0.0675}{0.06}}{2\pi \times 49 \times 5} + \dfrac{\ln \dfrac{0.1175}{0.0675}}{2\pi \times 0.15 \times 5} + \dfrac{\ln \dfrac{0.1675}{0.1175}}{2\pi \times 0.48 \times 5} + \dfrac{1}{2\pi \times 0.1675 \times 18 \times 5}}$$

$= \dfrac{195}{0.1580} =$ **1234.06 W** for 5 m length.

Overall heat transfer coefficient.

Based on inside area: $Q = UA_i \, \Delta T$

$\therefore$ $1234.06 = U_i \times 2\pi r_i l \times (230 - 35)$

$\therefore$ $\mathbf{U_i = 1234.06/2\pi \times 0.06 \times 5 \times 195 = 3.3574\ W/m^2K}$

Using the same procedure, $\mathbf{U_o}$ **based on outside area = 1.203 W/m²K**

Interface Temperatures:

To find T_1, $Q = \dfrac{T_{\infty 1} - T_1}{\dfrac{1}{2\pi r_1 h_1 l}} = \dfrac{230 - T_1}{\dfrac{1}{2\pi \times 0.06 \times 85 \times 5}}$ $\therefore$ $\mathbf{T_1 = 222.3°C}$

To find T_2, $Q = \dfrac{T_1 - T_2}{\dfrac{\ln \left[\dfrac{0.0675}{0.06}\right]}{2\pi \times 49 \times 5}}$ $\therefore$ $\mathbf{T_2 = 222.2°C}$

To find T_3, $Q = \dfrac{T_2 - T_3}{\dfrac{\ln \left[\dfrac{0.1175}{0.0675}\right]}{2\pi \times 0.15 \times 5}}$ $\therefore$ $\mathbf{T_3 = 77.04°C}$

To find T_4, $Q = \dfrac{T_3 - T_4}{\dfrac{\ln \left[\dfrac{0.1675}{0.1175}\right]}{2\pi \times 0.48 \times 5}}$ $\therefore$ $\mathbf{T_4 = 48.03°C}$

To check: $Q = h_2 A_2 \, \Delta T_2 = 18 \times 2\pi \times r_o \times (48.03 - 35) \times 5$

$= 18 \times 2\pi \times 0.1675 \times 13.03 \times 5 =$ **1234.18 W**

$\therefore$ checks.

In the case of cylindrical shapes also it can be noted that layers with lower conductivity undergo large temperature drop.

Note that $U_i > U_o$ as $A_i < A_o$.

2.2.3. Contact Resistance

It can also be added taking care to use the proper value of area. Contact resistance is left out ordinarily due to the difficulty in the estimation. The overall heat transfer coefficient concept is also useful, except that it has to be based on either outside or inside area,

$$U_o A_o = U_i A_i = 1/(R_1 + R_2 \ldots) = \frac{1}{\Sigma R} \quad \ldots(2.21)$$

and

$$Q = U_o A_o \Delta T \text{ or } U_i A_i \Delta T \quad \ldots(2.22)$$

The temperature variation along the radius will not be linear and is logarithmic as shown by equation (2.17).

In handbooks contact resistance value is based on unit area. In the case of plane wall, the value may be used as such. For the cylindrical and spherical shapes, the value should be corrected and should be based on the actual area like $2\pi rl$ or $4\pi r^2$. For the cylinder and sphere.

Example 2.3: *A composite cylinder is made of 6 mm thick layers each of two materials of thermal conductivities of 30 W/m°C and 45 W/m°C. The inside is exposed to a fluid at 500°C with a convection coefficient of 40 W/m² °C and the outside is exposed to air at 35°C with a convection coefficient of 25 W/m²K. There is a contact resistance of 1 × 10⁻³ m² °C/W between the layers. Determine the heat loss for a length of 2 m and the surface temperatures. Inside dia = 20 mm.*

The data and equivalent circuit are shown in Fig. Ex. 2.3. Find: Q, T_1, T_2, T_2^1, T_3, U_i, U_o as, indicated in the figure.

Solution: Using the equation (2.19) and adding the contact resistances, contact resistance has to be for the area encountered. The area here is $2\pi \times 0.016 \times 2$ m²

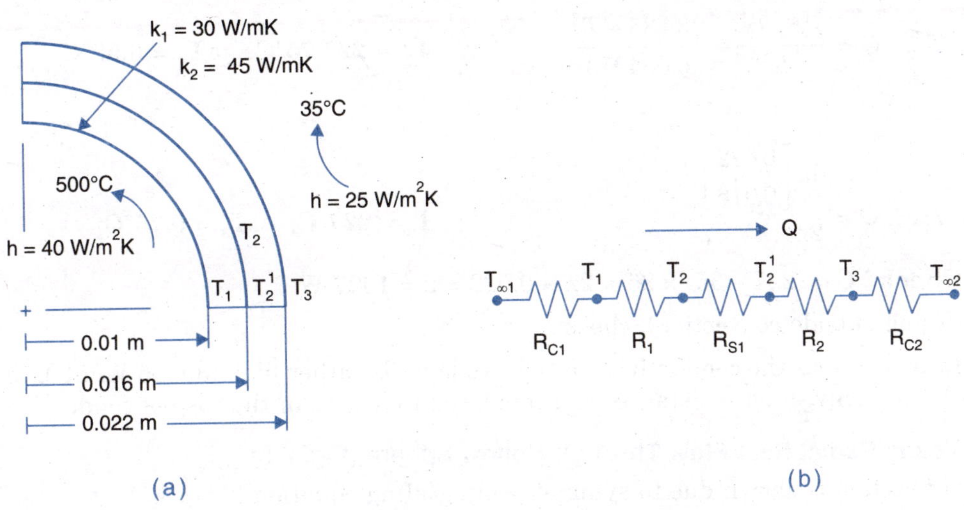

Fig. Ex. 2.3. Composite cylinder with contact resistance

$$\text{Contact resistance} = 1 \times 10^{-3} \text{ m}^2 \text{ °C/W} \cdot \frac{1}{2\pi \times 0.016 \times 2 \text{ m}^2}$$

$$= 0.004973592\text{°C/W}$$

$$Q = \frac{T_{\infty 1} - T_{\infty 2}}{\dfrac{1}{2\pi r_1 h_1 l} + \dfrac{\ln r_2/r_1}{2\pi k_1 l} + \dfrac{\ln r_3/r_2}{2\pi k_2 l} + \dfrac{1}{2\pi r_3 h_2 l} + \text{contact resistance}}$$

$$\mathbf{Q} = \frac{500 - 35}{\dfrac{1}{2\pi \times 0.01 \times 40 \times 2} + \dfrac{\ln \dfrac{0.016}{0.010}}{2\pi \times 30 \times 2} + \dfrac{\ln \dfrac{0.022}{0.016}}{2\pi \times 45 \times 2} + \dfrac{1}{2\pi \times 0.022 \times 25 \times 2} + 0.004974}$$

$$= \mathbf{1327\ W}$$

Overall heat transfer coefficient:

Based on inside area, $U_i = Q/\Delta T A_i = 22.7\ \text{W/m}^2\text{K}$

Based on outside area, $U_o = Q/\Delta T A_o = 16.52\ \text{W/m}^2\text{K}$

Note that $U_i > U_o$ as $A_i < A_o$

Interface Temperatures and temperature drops:

$$Q = \frac{T_{\infty 1} - T_1}{\dfrac{1}{2\pi \times 0.01 \times 40 \times 2}} \qquad \therefore\ \mathbf{T_1 = 236°C,\ \Delta T_1 = 264°C}$$

$$T_2,\quad Q = \frac{T_1 - T_2}{\dfrac{\ln\left[\dfrac{0.016}{0.01}\right]}{2\pi \times 30 \times 2}} \qquad \therefore\ \mathbf{T_2 = 234.35°C,\ \Delta T_2 = 1.65°C}$$

$$T_2^1,\quad Q = \frac{T_2 - T_2^1}{R} = \frac{234.35 - T_2^1}{0.004974} \qquad \therefore\ \mathbf{T_2^1 = 227.75°C,\quad \Delta T_3 = 6.60°C}$$

$$T_3,\quad Q = \frac{T_2^{\,1} - T_3}{\dfrac{\ln\left[\dfrac{0.022}{0.016}\right]}{2\pi \times 45 \times 2}} \qquad \therefore\ \mathbf{T_3 = 227°C,\quad \Delta T_4 = 0.75°C}$$

Check $Q = (227 - 35) \times 25 \times 2\pi \times 0.022 \times 2 = \mathbf{1327\ W}$

(Using outside convection), checks.

In this case as the conductivity of the two layer is rather high (30, 45 W/mK), the main drop is by the convection resistance. (0.2 compared to 0.0012 in the case of conduction)

2.2.4. Steady Radial Heat Flow Through Hollow Sphere (Fig. 2.5)

A quarter section is shown due to symmetry integrating equation [2.8(*a*)].

$$\frac{1}{r^2}\frac{\partial}{\partial r}\left(r^2 \frac{\partial T}{\partial r}\right) = 0$$

$$r^2 \frac{\partial T}{\partial r} = C_1 \quad \text{or} \quad \frac{\partial T}{\partial r} = \frac{C_1}{r^2}$$

$$\therefore \qquad T = -\frac{C_1}{r} + C_2 \qquad \text{...(A)}$$

At $r = r_1, T = T_1$ and $r = r_2, T = T_2$

$$T_1 = -\frac{C_1}{r_1} + C_2 \text{ and } T_2 = -\frac{C_1}{r_2} + C_2 \qquad \text{...(B)}$$

$$T_1 - T_2 = C_1\left(\frac{1}{r_2} - \frac{1}{r_1}\right) \text{ or } C_1 = \frac{T_1 - T_2}{\frac{1}{r_2} - \frac{1}{r_1}} \qquad \text{...(C)}$$

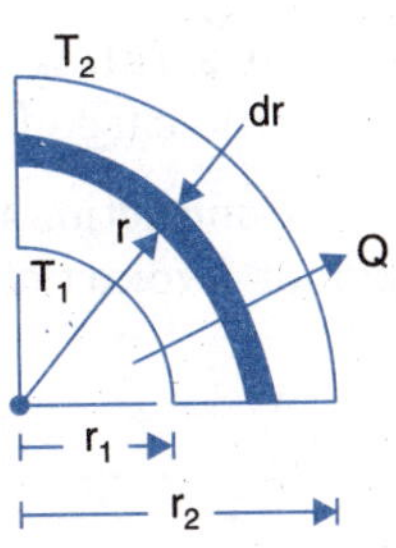

Fig. 2.5. Radial heat conduction–Hollow sphere

From B, $C_2 = T_1 + \frac{C_1}{r_1} = T_1 + \frac{T_1 - T_2}{\frac{1}{r_2} - \frac{1}{r_1}} \cdot \frac{1}{r_1}$ ...(D)

Substituting in A

$$T = -\frac{T_1 - T_2}{\frac{1}{r_2} - \frac{1}{r_1}} \cdot \frac{1}{r} + T_1 + \frac{T_1 - T_2}{\frac{1}{r_2} - \frac{1}{r_1}} \frac{1}{r_1}$$

$$T - T_1 = \frac{T_1 - T_2}{\frac{1}{r_2} - \frac{1}{r_1}} \left\{\frac{1}{r_1} - \frac{1}{r}\right\} \qquad \text{...(2.23)}$$

$$\frac{T - T_1}{T_1 - T_2} = \frac{\frac{1}{r_1} - \frac{1}{r}}{\frac{1}{r_2} - \frac{1}{r_1}} \text{ or } \frac{T_1 - T}{T_1 - T_2} = \frac{\frac{1}{r_1} - \frac{1}{r}}{\frac{1}{r_1} - \frac{1}{r_2}} \qquad \text{...(2.24)}$$

$$Q = -kA\frac{dT}{dr} = -k4\pi r^2 \frac{T_1 - T_2}{\frac{1}{r_2} - \frac{1}{r_1}} \cdot \frac{1}{r_2} = 4\pi k \frac{T_1 - T_2}{\frac{1}{r_1} - \frac{1}{r_2}} \qquad \text{...(2.25)}$$

The term $\frac{1}{4\pi k}\left(\frac{1}{r_1} - \frac{1}{r_2}\right)$ is called **conduction resistance in this case.**

The temperature variation along radius is non-linear and follows second degree curve.

This expression can also be derived from basics. Consider a thin section of thickness dr at radius r. The area of the section is $4\pi r^2$.

Considering the section as plane wall (as the thickness is small and applying Fourier equation,

$$Q = -k4\pi r^2 \frac{dT}{dr}$$

$$Q\left(\frac{dr}{r^2}\right) = 4\pi k\, dT \qquad \text{...(2.26)}$$

Integrating, between limits r_1 and r_2 the equation (2.25) can be obtained. This is possible only in the case of one dimensional flow of heat.

Conduction through composite sphere can be solved by using the same method as in the previous two articles. (Refer Fig. Ex. 2.4).

$$Q = \frac{\Delta T}{R_{c1} + R_1 + \ldots + R_{c2}}$$

$$= \frac{T_{\infty 1} - T_{\infty 2}}{\frac{1}{h_1 4\pi r_1^2} + \frac{1}{4\pi k}\left[\frac{1}{r_1} - \frac{1}{r_2}\right] + \ldots + \frac{1}{h_2 4\pi r_2^2}} \quad \ldots(2.27)$$

The temperature at any radius can also be obtained by the same method adopted in the case of plane wall and hollow cylinder.

In the case of hollow sphere also the overall heat transfer coefficient has to be based on either inside or the outside area being $4\pi r_1^2$ or $4\pi r_2^2$

$$Q = U_i A_i \, \Delta T = U_o A_o \, \Delta T \quad \ldots(2.28)$$

Example 2.4: *A spherical vessel of ID 0.3 m and thickness of 20 mm is made of steel with conductivity of 40 W/mK. The vessel is insulated with two layers of 60 mm thickness of conductivity 0.05 and 0.15 W/mK. The inside surface is at –196°C. The outside is exposed to air at 30°C with convection coefficient of 35 W/m²K. There is a contact resistance of 1 × 10⁻³ m²°C/W between the two insulations. Determine the heat gain and also the surface temperatures and the overall heat transfer coefficient based on the outside surface area of the metallic vessel.*

Solution: The data and equivalent circuit is shown in Fig. Ex. 2.4.

A quarter section is shown due to symmetry. The given data are shown in the figure. Also T_1, T_3, T_4, T_4^1, T_2 are indicated.

Using the equation (2.26) and adding the contact resistance,

$$Q = \frac{T_1 - T_{\infty 2}}{\frac{1}{4\pi k_1}\left(\frac{1}{r_1} - \frac{1}{r_2}\right) + \frac{1}{4\pi k_2}\left(\frac{1}{r_2} - \frac{1}{r_3}\right) + R_s + \frac{1}{4\pi k_3}\left(\frac{1}{r_3} - \frac{1}{r_4}\right) + \frac{1}{4\pi r_4^2 h_2}}$$

Here the inside surface temperature is specified. Hence there is no convective resistance on the inside.

$$Q = \frac{-196 - 30}{\frac{2}{4\pi \times 40}\left(\frac{1}{0.15} - \frac{1}{0.17}\right) + \frac{1}{4\pi \times 0.05}\left(\frac{1}{0.17} - \frac{1}{0.23}\right) + \frac{1}{4\pi \times 0.15}\left(\frac{1}{0.23} - \frac{1}{0.29}\right) + \frac{1}{4\pi \times 0.29^2 \times 35} + R_s}$$

$R_s = 1 \times 10^{-3}$ m² °C/W is based on 1 m²

The area in this problem is $4\pi \times 0.23^2$ m²

$\therefore$ R_s for the area $= 1 \times 10^{-3}/(4\pi \times 0.23^2) = 1.5 \times 10^{-3}$ °C/W

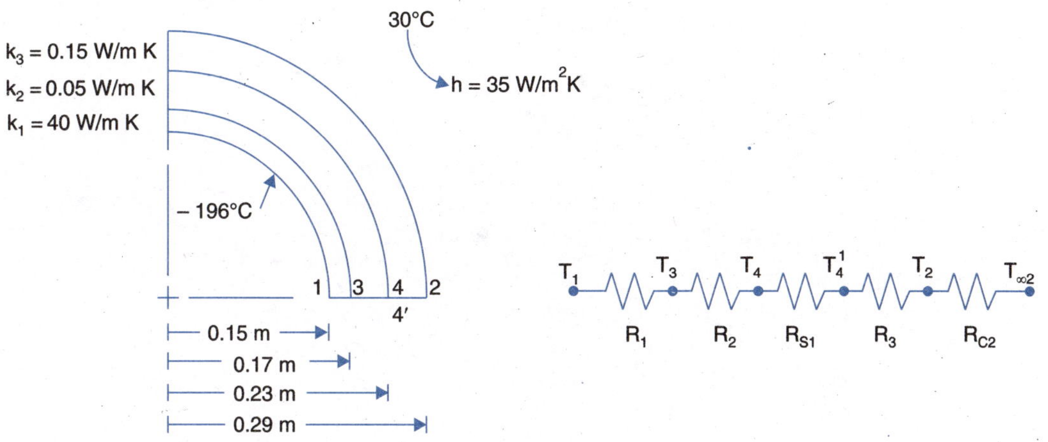

Fig. Ex. 2.4. Composite sphere

Substituting $\mathbf{Q} = \dfrac{-226}{2.95} = \mathbf{-76.6\ W}$ (heat flows in)

overall heat transfer coefficient based on the outside area of the vessel

To calculate U_o and U_i,

$$76.6 = U_i \times 4\pi \times 0.15^2 \times 226, \qquad \therefore\ \mathbf{U_i = 1.199\ W/m^2{}^\circ C}$$

$$76.6 = U_o \times 4\pi \times 0.29^2 \times 226 \qquad \therefore\ \mathbf{U_o = 0.321\ W/m^2{}^\circ C}$$

Note: $\mathbf{U_i > U_o}$ **as** $\mathbf{A_i < A_o}$

Interface temperatures:

To find T_3: $$Q = -76.6 = \frac{-196 - T_3}{\dfrac{1}{4\pi \times 10}\left(\dfrac{1}{0.15} - \dfrac{1}{0.17}\right)},\ \mathbf{T_3 = -195.88^\circ C}$$

Similarly $\mathbf{T_4 = -8.75^\circ C.}$

After contact drop $T_4^1 = -8.64$°C and $\mathbf{T_2 = 27.93^\circ C}$

Check: Using outside convection,

$$Q = (-30 + 27.93)\, 4\pi \times 0.29^2 \times 35 = -76.7\ \text{W}$$

As in the previous case contact resistance can also be taken into account by adding the value in the equation (2.27).

2.3 CONDUCTION IN OTHER SHAPES

The analysis is more complex. Correlation equations are available in handbooks and references. Two such shapes are considered in examples 2.5 and 2.6.

Example 2.5: *A circular pipe of OD 20 cm is enclosed centrally in a square section insulation of 36 cm side. The thermal conductivity of the material is 8.5 W/mK. The inside surface is at 200°C. The outside is exposed to convection at 30°C with h = 35 W/m²K. Determine the heat flow per a length of 5 m.*

Solution: The data are shown in Fig. Ex. 2.5.

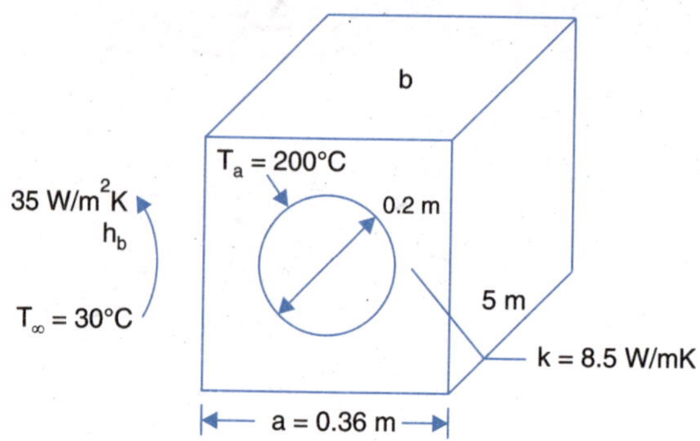

Fig. Ex. 2.5. Problem model

Using handbook (a = side length)

The resistance is given by

$$R = \frac{1}{2\pi l}\left[\frac{1}{k}\ln\frac{1.08a}{2r} + \frac{\pi}{2h_b a}\right]$$

(Internal convection is absent)

$$R = \frac{1}{2\pi \times 5}\left[\frac{1}{8.5}\ln\frac{1.08 \times 0.36}{0.2} + \frac{\pi}{2 \times 35 \times 0.36}\right] = 0.002529°\text{C/W}$$

$$Q = \frac{\Delta T}{R} = \frac{200 - 30}{0.002529} = 67219 \text{ W.}$$

Example 2.6: *A pipe of 30 cm OD is insulated by a material of thermal conductivity of 0.45 W/mK. Due to space restriction the insulation of outside diameter is placed with an eccentricity of 5 cm (Refer Fig. Ex. 2.6). The inner surface is at 250°C. The outer surface is at 60°C. Determine the loss for a length of 5 m. Outside diameter = 0.6 m.*

Solution: From handbook the thermal resistance is obtained by

$$\frac{1}{2\pi kl}\ln\frac{\sqrt{[(r_2 + r_1)^2 - e^2]} + \sqrt{[(r_2 - r_1)^2 - e^2]}}{\sqrt{[(r_2 + r_1)^2 - e^2]} - \sqrt{[(r_2 - r_1)^2 - e^2]}}$$

$$R = \frac{1}{2\pi \times 0.45 \times 5}\ln\frac{\sqrt{[(0.15 + 0.3)^2 - 0.05^2]} + \sqrt{[(0.3 - 0.15)^2 - 0.05^2]}}{\sqrt{[(0.15 + 0.3)^2 - 0.05^2]} - \sqrt{[(0.3 - 0.15)^2 - 0.05^2]}}$$

$$= \frac{1}{2\pi \times 0.45 \times 5}\ln\frac{0.4472 + 0.14142}{0.4472 - 0.14142} = 0.046325°\text{C/W}$$

$$Q = \frac{\Delta T}{R} = \frac{250 - 60}{0.046326} = \mathbf{4101\ W.}$$

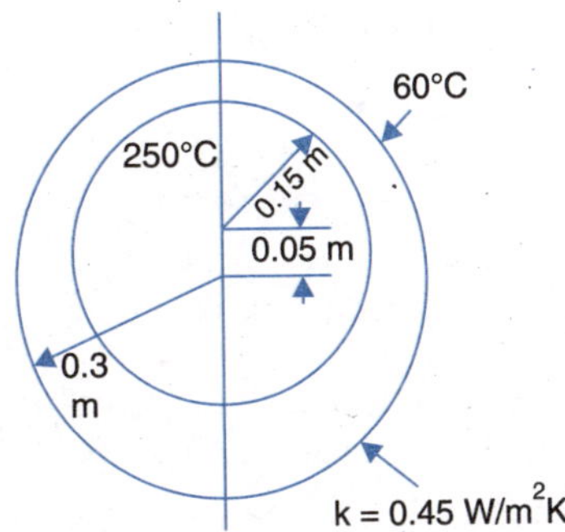

Fig. Ex. 2.6. Eccentric insulation

The situations shown in section 2.3 occurs often in hot oil transportation and similar transports.

2.4 ONE DIMENSIONAL STEADY STATE HEAT CONDUCTION WITH VARIABLE THERMAL CONDUCTIVITY OR VARIABLE AREA ALONG THE SECTION

A glance at the graphs in handbooks showing the variation of thermal conductivity with temperature will convince the learner that the assumption of constant thermal conductivity is valid only for heat flow with small temperature differences. Even then the conductivity value for the corresponding temperature range should be used. Generally a linear relationship as in equation (2.29) below can be used without sacrificing accuracy

$$k = k_o (1 \pm \beta T) \qquad \text{...(2.29)}$$

where k_o is conductivity at some reference temperature and T is the temperature above the reference and β is a constant and may be positive or negative. Substituting this in the Fourier's equation, yields **for slab**

$$Q = -k_o (1 \pm \beta T) \frac{dT}{dx},$$

$$\int_1^2 Q\, dx = -k_o A \int_1^2 (1 + \beta T)\, dT$$

$$Q(x_2 - x_1) = -A k_o \left[T + \frac{\beta T^2}{2}\right]_1^2 = -A k_o \left[T_2 + \frac{\beta T_2^2}{2} - T_1 - \frac{\beta T_1^2}{2}\right]$$

$$= -A k_o \left[(T_2 - T_1) + \frac{\beta}{2}(T_2^2 - T_1^2)\right]$$

$$= +A (T_1 - T_2) \left[k_o \left(1 + \beta \frac{(T_1 + T_2)}{2}\right)\right]$$

$$= A (T_1 - T_2) k_m, \text{ as } k_o \left(1 + \beta \frac{T_1 + T_2}{2}\right) = k_m$$

$\therefore \qquad (x_2 - x_1) = L,$

$$\therefore \qquad Q = \frac{(T_1 - T_2)}{L / k_m A} \qquad ...(2.30)$$

This holds good in the case $k = k_o\,(1 - \beta T)$ also.

where
$$k_m = k_o \left\{1 - \beta\left(\frac{T_1 + T_2}{2}\right)\right\} \qquad ...(2.31)$$

Hence if linear variation is assumed, the average value of thermal conductivity between these temperatures can be used in the equation (2.10). It is found that for other shapes also, this result $k = k_m$ is found valid.

(Note that the R.H.S. in the derivations for hollow cylinder and hollow sphere will be the same as per plane wall and hence the use of k_m as in equation 2.31)

Thus for a cylinder $Q = \dfrac{\Delta T}{\ln (r_1/r_2) / 2\pi k_m l}$ ...(2.32)

For a sphere $Q = \dfrac{\Delta T}{\dfrac{1}{4\pi k_m}\left(\dfrac{1}{r_1} - \dfrac{1}{r_2}\right)}$ ...(2.33)

Example 2.7: *An insulating wall 16 cm thick has one face at 600°C while the other is at 100°C. The thermal conductivity of the material is given by k = 0.078 (1 + 17.95 × 10⁻⁴ T) W/mK and T is in °C. Determine the heat loss per unit area and the mid plane temperature. Repeat the calculations if k = 0.078 (1 – 17.95 × 10⁻⁴ T)*

Solution: Using equations (2.30) and (2.31).

Case 1: $Q = \dfrac{\Delta T}{L / k_m A}$ and $k_m = k_o \left(1 + \beta \dfrac{T_1 + T_2}{2}\right)$

$$\mathbf{k_m} = 0.078\,[1 + 17.95 \times 10^{-4}\,(600 + 100)/2] = \mathbf{0.127\ W/mK}$$

$$\mathbf{Q} = \frac{(600 - 100)}{0.16 / (0.127 \times 1)} = \mathbf{396.9\ W/m^2}$$

To find the mid plane temperature, the heat flow for thickness upto mid plane is equated to the calculated heat flow. Taking T as the mid plane temperature,

$$\therefore \qquad 396.9 = \frac{T_1 - T}{(L_{mid} / k_{mid} A)}$$

$$k_{mid} = 0.078\,[1 + 17.95 \times 10^{-4}\,(600 + T)/2]$$

$$\therefore \qquad 396.9 = (600 - T)/(0.08/0.078\,\{1 + 17.95 \times 10^{-4}\,(600 + T)/2\})$$

$$\therefore \qquad 17.95 \times 10^{-4}\,T^2 + 2T - 1032.05 = 0$$

Solving $\qquad \mathbf{T = 383.8°C}$

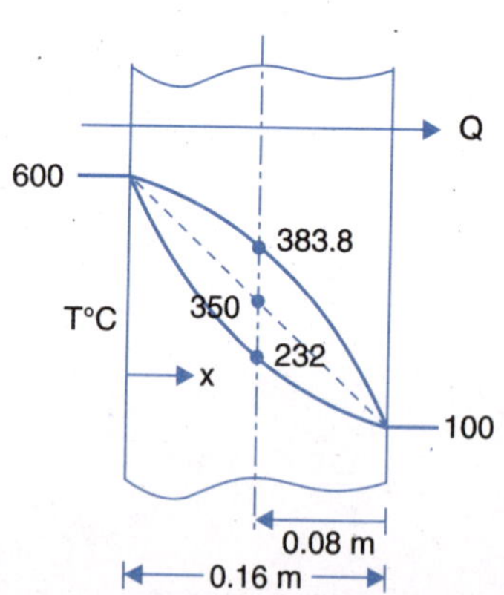

Fig. Ex. 2.7. Temperature distribution in slab with variable thermal conductivity

Case 2: $k_m = 0.078\ (1 - 17.95 \times 10^{-4} \times (600 + 100)/2) = 0.029$

$\therefore$ $Q = 500/(0.16/0.029) = \mathbf{90.625\ W/m^2}$

To find mid plane temperature:

$$90.625 = \frac{(600 - T)}{0.08} \times 0.078\ (1 - 17.95 \times 10^{-4} \times 300 - 17.95 \times 10^{-4}\ T/2)$$

Simplifying: $$\frac{90.625 \times 0.08}{0.078} = (600 - T)\ (0.4615 - 8.975 \times 10^{-4}\ T)$$

This leads to $8.975 \times 10^{-4}\ T^2 - T + 183.95 = 0$

Solving the quadratic, $T =$ **232°C**

These data are shown plotted in Fig. Ex. 2.7 (check using equation (2.34) page 55.

Note the shape of the temperature profile. When plus sign in equation (2.29) is used, the temperature variation curve will be above the linear variation curve. The slope of the curve will increase along x as the conductivity decreases along x. If k decreases, for the same heat flow, (dT/dx) should increase. In the case of the sign in the above equation is negative the temperature curve will be below the linear variation curve. The slope will decrease along x as the value of k increases along x.

Example 2.8: *The thermal conductivity of an insulating material used over a hot pipe varies as $k = 0.0545\ (1 + 28.4 \times 10^{-4}\ T)$ where T is in °C and k is in W/mK. This insulation is used for a thickness of 12 cm over a pipe of diameter 0.6 m. The pipe surface is at 300°C and the outside insulation temperature is 60°C. Determine the heat flow for a length of 5 m. Also find the mid layer temperature.*

Solution: The heat flow is calculated using equation (2.32). The data are shown in Fig. Ex. 2.8. Quarter section is shown due to symmetry. Find Q and mid plane temperature T

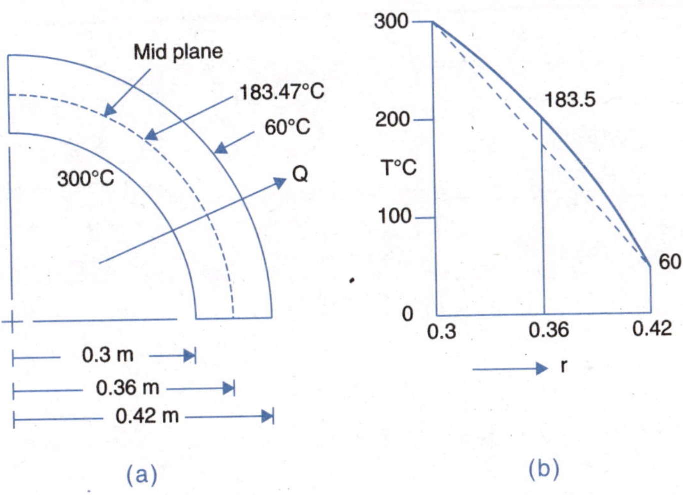

Fig. Ex. 2.8. Temperature variation in hollow cylinder with variable thermal conductivity

$$Q = \frac{\Delta T}{\ln (r_2 / r_1)} \times 2\pi\ k_m\ l$$

$$k_m = 0.0545 \left[1 + 28.4 \times 10^{-4} \left(\frac{300 + 60}{2}\right)\right] = \mathbf{0.08236\ W/mK}$$

$$Q = \frac{(300 - 60)}{\ln \dfrac{0.42}{0.3}} 2\pi \times 0.08736 \times 5 = \mathbf{1845.6\ W}$$

To find the mid plane temperature, equate this heat flow for half thickness of the cylinder

$$1845.6 = \frac{300 - T}{\ln \dfrac{0.36}{0.3}} 2\pi \times 5 \times 0.0545 \left[1 + 28.4 \times 10^{-4} \left(\frac{300 + T}{2}\right)\right]$$

Simplifying, $14.2 \times 10^{-4}\, T^2 + T - 231.27 = 0$

Solving for **T = 183.47°C. Check using equation (2.35) page 55.**

The temperature variation along the radius is shown plotted to scale in Fig. Ex. 2.8.

Example 2.9: *The thermal conductivity of an insulating material used to reduce heat gain into a cryogenic spherical shaped container varies as k = 0.028 (1 + 50 × 10⁻⁴ T). Where T is in degree centigrade and k is in W/mK. The inner radius is 16 cm and the insulation thickness is 12 cm. The inner surface is at –190°C while the outer surface is at 10°C. Determine the heat flow, the temperature at mid radius and the radius at which the temperature is –40°C.*

Solution: Known data are indicated in Fig. Ex. 2.9. To find heat flow Q and mid plane temperature T. Using equation (2.33). One quarter view is shown for clarity.

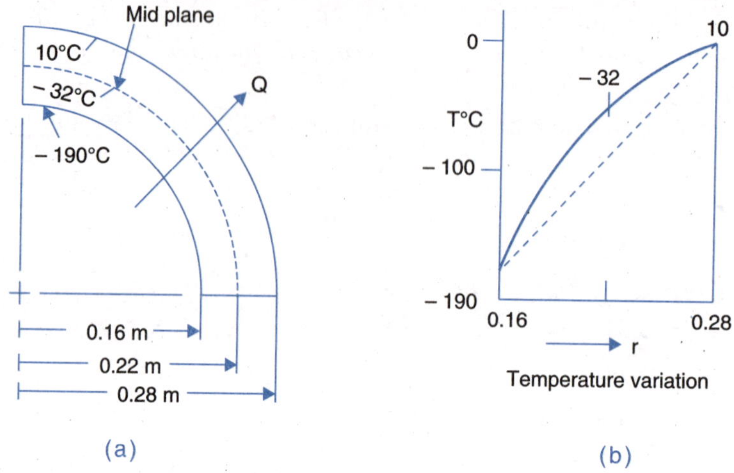

Fig. Ex. 2.9. Model

$$Q = \frac{T}{\dfrac{1}{4\pi k_m}\left(\dfrac{1}{r_1} - \dfrac{1}{r_2}\right)}$$

as $$k_m = 0.028 \left[1 + 50 \times 10^{-4} \frac{10 + (-190)}{2}\right] = \mathbf{0.0154\ W/mK}$$

Substituting $$Q = \frac{-190 - 10}{\dfrac{1}{4\pi \times 0.0154}\left[\dfrac{1}{0.16} - \dfrac{1}{0.28}\right]} = \mathbf{-14.45\ W, heat\ flow\ is\ inwards}$$

Mid radius temperature

$$-14.45 = \frac{-190 - T}{\dfrac{1}{4\pi \times 0.028\left[1 + 50 \times 10^{-4}\left(\dfrac{-190 + T}{2}\right)\right]}\left[\dfrac{1}{0.16} - \dfrac{1}{0.22}\right]}$$

Simplifying $25 \times 10^{-4}\, T^2 + T + 29.506 = 0$

Solving **T = – 32.37°C** (the other solutions – 367.63°C being not acceptable

To locate the radius at which $T = -40°C$

$$Q = \frac{\Delta T \times 4\pi\, k_m}{\left[\dfrac{1}{r_1} - \dfrac{1}{r}\right]}$$

$$k_m = 0.028\left[1 + 50 \times 10^{-4}\left(\frac{-190 - 40}{2}\right)\right] = 0.0119 \text{ W/mK}$$

$$14.45 = \frac{\{-190 - (-40)\} \times 4\pi \times 0.0119}{\left[\dfrac{1}{0.16} - \dfrac{1}{r}\right]}$$

Solving $\boldsymbol{r}$ **= 0.21287 m.**

(Check: This is less than 0.22 where the temperature is – 32.08°C).

COMMENT

Location where a specified temperature occurs is required particularly in cryogenic applications. This is required for selection of material.

2.4.1. Determination of the Temperature Variation

For a slab equating the heat flow in section L_1 and L_2.

$$\frac{\dfrac{T_1 - T_2}{\dfrac{x_2 - x_1}{k_{m1}A}}}{} = \frac{T_1 - T}{\dfrac{x - x_1}{k_{m2}A}}$$

$$\frac{T_1 - T}{T_1 - T_2} = \frac{(x - x_1)\, k_{m1}}{(x_2 - x_1)\, k_r} = \frac{(x - x_1)[1 + \beta(T_2 + T_1)/2]}{(x_2 - x_1)[1 + \beta(T + T_1)/2]}$$

$$\frac{(T_1 - T)[1 + \beta(T_1 + T)/2]}{(T_1 - T_2)[1 + \beta(T_1 + T_2)/2]} = \frac{x - x_1}{x_2 - x_1} \quad ...(2.34)$$

For cylinder

$$\frac{(T_1 - T)[1 + \beta(T_1 + T)/2]}{(T_1 - T_2)[1 + \beta(T_1 + T_2)/2]} = \frac{\ln \dfrac{r}{r_1}}{\ln \dfrac{r_2}{r_1}} \quad ...(2.35)$$

For sphere
$$\frac{(T_1 - T)[1 + \beta(T_1 + T)/2]}{(T_1 - T_2)[1 + \beta(T_1 + T_2)/2]} = \frac{\dfrac{1}{r_1} - \dfrac{1}{r}}{\dfrac{1}{r_1} - \dfrac{1}{r_2}} \quad ...(2.36)$$

The sign for β is as in the equation which specifies conductivity variation.

Given the temperature, T, the location at which this value occurs are the distances x or r.

It is suggested to solve for the temperature from basics, *i.e.,* using the relation, heat flow over the full thickness = heat flow over the given thickness.

2.4.2. Variation of Area in the Direction of Heat Flow

Some possible relationships are

$$A = A_o\,(1 + Cx),\ A = A_oCx,\ A = A_oCx^2,\ A = A_o\,(1 + Cx^2)$$

These are for truncated wedge, developed hollow cylinder, developed hollow sphere and truncated cone (Fig. 2.6*a, b, c, d*)

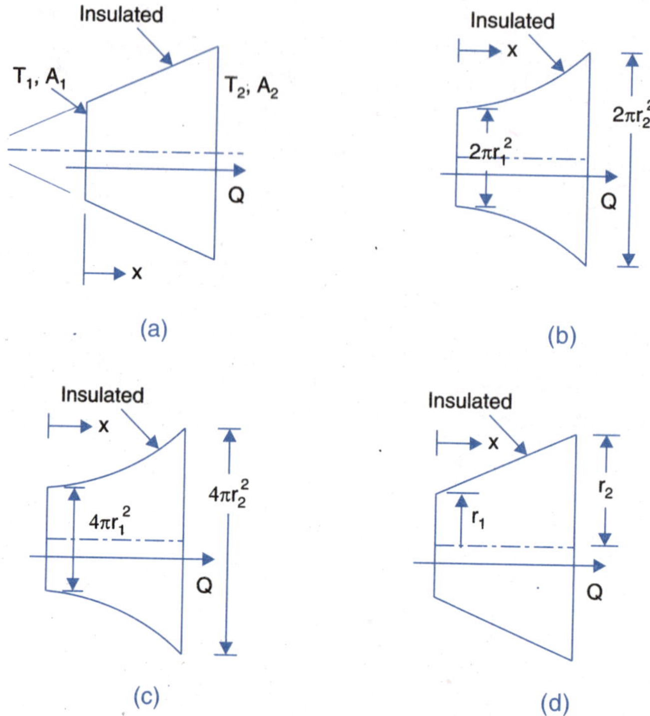

Fig. 2.6. Variable area sections

The heat flow can be found by substituting the relation in the Fourier's equation and integrating the same between the two surfaces.

The method is illustrated in example 2.10.

Example 2.10: *A truncated conelike solid has its circumferential surface insulated. The base is at 300°C and the area along the flow direction at x is given by A = 1.3 (1 – 1.5x). Where x is measured from the base in the direction of flow in m and A is in m². If the thermal conductivity* *is 2.6 W/mK and the plane at x = 0.2 m is maintained at 100°C, determine the heat flow and also the temperature at x = 0.1 m. Calculate the temperature gradients at the three sections.*

Solution: The known data are indicated in Fig. Ex. 2.10.

Find: $Q, T_{0.1}, \left.\frac{dT}{dx}\right|_0, \left.\frac{dT}{dx}\right|_{0.1}, \left.\frac{dT}{dx}\right|_{0.2}$

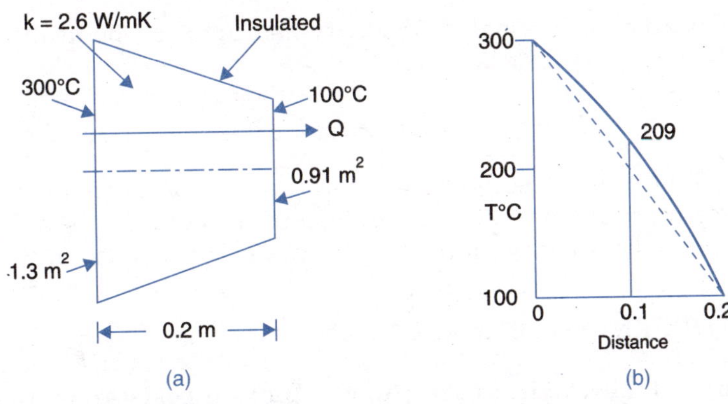

Fig. Ex. 2.10. Heat transfer in variable area section

Using the Fourier's equation

$$Q = -kA\frac{dT}{dx}$$

and substituting the relation for A and integrating after separating variables.

$$Q = -k \times 1.3\,(1 - 1.5x)\,.\,\frac{dT}{dx}$$

$$Q\int_0^{0.2}\frac{dx}{(1-1.5x)} = -1.3k\int_{T_1}^{T_2} dT$$

$$Q\left[\frac{1}{-1.5}\ln(1-1.5x)\right]_0^{0.2} = -1.3k\,(T_2 - T_1)$$

substituting the values at $x = 0.2$ and $x = 0$, $\left[-\frac{1}{1.5}\ln(0.7)\right] = 1.3\,(T_1 - T_2)$

$$\therefore \qquad Q = \frac{1.3\times k(T_1 - T_2)\times 1.5}{-\ln(0.70)}$$

$$= \frac{1.3\times 2.6\times 1.5(300-100)}{-\ln(0.70)} = \mathbf{2842.9\ W}$$

At 0.1 m $\qquad 2842.9 = \dfrac{1.3\times 1.5\times 2.6(300-T)}{-\ln(1-1.5\times 0.1)}$

$$\therefore \qquad T = \mathbf{208.87°C}$$

Temperature gradients:

$$Q = -kA\frac{dT}{dx} \quad \therefore \quad \frac{dT}{dx} = -\frac{Q}{kA}$$

Section $x = 0, \frac{dT}{dx} = -2842.9/2.6 \times 1.3 = -841.09°C/m$

$$x = 0.1, A = 1.3(1 - 1.5 \times 0.1) = 1.105, \frac{dT}{dx} = -989.52°C/m$$

$$x = 0.2, \frac{dT}{dx} = -1201.6°C/m$$

Combined variation of thermal conductivity and area can be dealt with in a similar way. For example,

$$Q = -k_o (1 + \beta T) A_o (1 + cx) \frac{dT}{dx} \quad ...(2.37)$$

Equation (2.37) can be integrated by separating the variables to obtain the value of heat flow. The temperature at any location can be found by equating the heat flow with thickness up to that section to the heat flow with the full thickness, as was done in the earlier sections.

2.5 CRITICAL THICKNESS OF INSULATION

Adding insulation will always **increase the conduction resistance**. But when the total resistance is made up of both conduction resistance and convection resistance, **the addition of insulation in some cases may reduce the convection resistance due to the increase in surface area (as in the case of a cylinder and sphere)** and the total resistance may actually decrease resulting in increased heat flow. It may be shown that the resistance actually decreases and then increases in certain cases. **The thickness upto which heat flow increases and after which heat flow decreases is termed as critical thickness.** In the case of cylinders and spheres it is called critical radius. An expression for the same can be obtained by finding the rate of change of the total resistance with change in radius and equating the same to zero *i.e.* zero slope condition.

Cylinder:

Total resistance for radius r, $R = \frac{\ln\left(\frac{r}{r_1}\right)}{2\pi kl} + \frac{1}{h2\pi rl}$

cancelling the common factors $R = \left(\frac{1}{k}\ln\frac{r}{r_1} + \frac{1}{h}\frac{1}{r}\right) \times \text{constant}$

$$\frac{dR}{dr} = \frac{1}{k} \cdot \frac{1}{r} - \frac{1}{h}\frac{1}{r^2}, \text{ equating this to zero}$$

$$\mathbf{r_{cr} = \frac{k}{h}} \quad ...(2.38)$$

For spheres the following relationship can be derived, using similar method

$$R = \frac{1}{4\pi k}\left[\frac{1}{r_1} - \frac{1}{r}\right] + \frac{1}{4\pi r^2 h}$$

$$\frac{dR}{dr} = \frac{1}{4\pi k}[1] - \frac{2}{4\pi rh}, \text{ equating to zero}$$

$$\mathbf{r_{cr}} = \frac{\mathbf{2k}}{\mathbf{h}} \quad ...(2.39)$$

Example 2.11: *A copper pipe carrying refrigerant at – 20°C is 10 mm in OD and is exposed to convection at 50 W/m²K to air at 25°C. It is proposed to apply insulation of conductivity 0.5 W/mK. Determine the thickness beyond which the heat gain will be reduced. Calculate the heat gains for 2.5 mm, 5.0 mm and 7.5 mm thicknesses for 1 m length. The convection coefficient remains constant.*

Solution: Using equation (2.38), the critical thickness up to which the heat flow will increase is calculated as below

$$\mathbf{r_c} = \frac{k}{h} = 0.5/50 = \mathbf{0.01\ m}$$

$\mathbf{r_c}$ this means that up to a thickness of **5 mm**, the heat flow will increase:

The heat flow at various configurations is given below. **Bare pipe:** unit length

$$Q = hA\,(\Delta T) = 50 \times 2\pi \times 0.005 \times 1\,\{25 - (-20)\} = \mathbf{70.69\ W/m}$$

For **2.5 mm thickness, r = 0.0075 m**

$$Q = \frac{45}{\dfrac{1}{50 \times 2\pi \times 0.0075 \times 1} + \dfrac{\ln \dfrac{0.0075}{0.005}}{2\pi \times 0.5}} = \mathbf{81.3\ W/m}$$

For **5 mm thickness, r = 0.010 m**

$$Q = \frac{45}{\dfrac{1}{50 \times 2\pi \times 0.01 \times 1} + \dfrac{\ln \dfrac{0.010}{0.005}}{2\pi \times 0.5}} = \mathbf{83.49\ W/m}$$

For **7.5 mm thickness, r = 0.0125 m**

$$Q = \frac{45}{\dfrac{1}{50 \times 2\pi \times 0.0125 \times 1} + \dfrac{\ln \dfrac{0.0125}{0.005}}{2\pi \times 0.5}} = \mathbf{82.37\ W/m}$$

The heat flow for various thickness is shown in Fig. Ex. 2.11

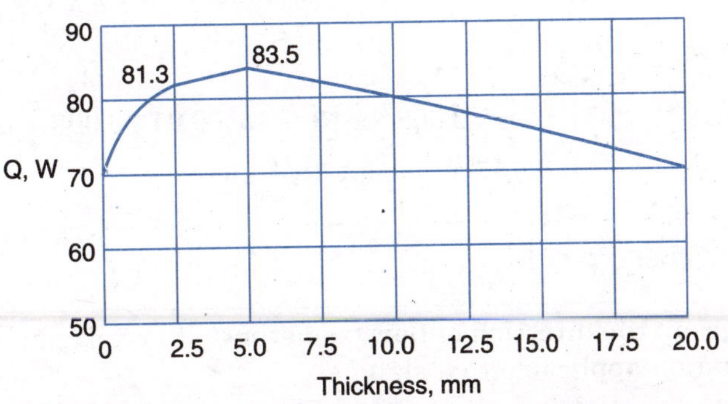

Fig. Ex. 2.11. Heat flow with thickness of insulation

Chapter 2

It will be seen that a thickness of **20 mm** and $r = 0.025$ m insulation only will reduce the heat flow to the level of bare pipe.

$$Q = \frac{45}{\dfrac{1}{50 \times 2\pi \times 0.025 \times 1} + \dfrac{\ln \dfrac{0.025}{0.005}}{2\pi \times 0.5}} = \mathbf{70.35\ W/m}$$

Only after 20 mm thickness of insulation the heat flow will decrease due to insulation.

This principles applies generally to small diameter pipes. Also insulation thickness for electrical wire has to be designed on this basis. Adding the electrical insulation may increase the heat flow and reduce the temperature of the wire.

The variation of resistances with addition of insulation is shown in Table 2.1.

Table 2.1. Variation of resistances to heat flow with increase in insulation. Base radius: 5 mm.

Insulation thickness, mm	*Conduction resistance*	*Convection resistance*	*Total resistance*
2.5	0.129	0.424	0.553
5.0	0.221	0.318	0.539
7.0	0.292	0.255	0.547
20	0.512	0.127	0.1039

Note that at 5 mm insulation, the resistance is minimum and then increases. The heat flow is maximum at the same radius because of the above.

Example 2.12: *It is desired to increase the heat dissipated over the surface of an electronic device of spherical shape of 5 mm radius exposed to convection with $h = 10\ W/m^2K$ by encasing it in a transparent spherical sheath of conductivity 0.04 W/mK. Determine the diameter of the sheath for maximum heat flow. For a temperature drop of 120°C from device surface determine the heat flow for bare and sheathed device.*

Solution: The requirement is the calculation of critical radius

For a sphere equation (2.39) gives

$$\mathbf{r_c} = \frac{2k}{h} = 2 \times 0.04/10 = 0.008 \text{ or } \mathbf{8\ mm}$$

So the diameter is 16 mm.

Bare device $Q = 4\pi r_1^2\, h(\Delta T) = 4 \times \pi \times 0.005^2 \times 10 \times 120 = \mathbf{0.377\ W}$

Sheathed device
$$Q = \frac{120}{\dfrac{1}{4\pi \times 0.008^2 \times 10} + \dfrac{1}{4\pi \times 0.04}\left(\dfrac{1}{0.005} - \dfrac{1}{0.008}\right)}$$

$= \mathbf{0.439\ W}$, about 16% increase.

2.6 MEAN AREA CONCEPT

When the area changes in the direction of flow, sometimes it is convenient to use a mean area and use the heat equation applicable to a slab *i.e.*,

$$Q = \frac{(T_1 - T_2)}{L / kA}$$

when this concept is applied to a cylinder

$$Q = \frac{(T_1 - T_2)\, kA_m}{L} = \frac{(T_1 - T_2) \times 2\pi kl}{\ln \frac{r_2}{r_1}}$$

Slab area, $$A_m = \frac{2\pi l(r_2 - r_1)}{\ln \frac{r_2}{r_1}} = \frac{(A_2 - A_1)}{\ln \frac{A_2}{A_1}} \quad \text{...(2.40)}$$

∵ Surface area outside = $2\pi r_2 l$,

Surface area inside = $2\pi r_1 l$

as $$\frac{r_2}{r_1} = \frac{A_2}{A_1},\ L = r_2 - r_1$$

A_m is also called log mean area.

Similarly for a sphere $A_m = 4\pi \sqrt{(r_1 r_2)}$ ...(2.41)

or the area at the geometric mean of the radii.

Example 2.13: *A hollow cylinder of inner radius 0.16 m and thickness 8 cm conducts heat radially. Determine the mean area and check for the heat flow.*

Solution: Using equation 2.40

$$A_m = \frac{A_2 - A_1}{\ln \frac{A_2}{A_1}} = \frac{2\pi(0.24 - 0.16)\, l}{\ln \frac{2\pi \times 0.24 \times l}{2\pi \times 0.16 \times l}}$$

Considering as a slab of A_m area and thickness $(r_2 - r_1)$ and conductivity k

$$Q = k \times \frac{2\pi(0.24 - 0.16)\, l}{\ln \frac{2\pi \times 0.24 \times l}{2\pi \times 0.16 \times l}} \frac{k\Delta T}{(0.24 - 0.16)} = \frac{2\pi k\, l\, \Delta T}{\ln \frac{0.24}{0.16}} \quad \text{same as equation 2.18.}$$

2.7 PARALLEL FLOW

There are occasions where the barrier to heat flow is made up of several materials in the same thickness as in a brick wall where bricks and mortar form the same layer. In this case the heat flow is along parallel paths [Fig. 2.7(*a*)].

The assumption made here is that temperature drop in each layer is the same. The drop in layer 1 is the same as drop in layer 2 and so on. The problem is solved by using electrical analogy as shown in Fig. 2.7(*b*), (*c*).

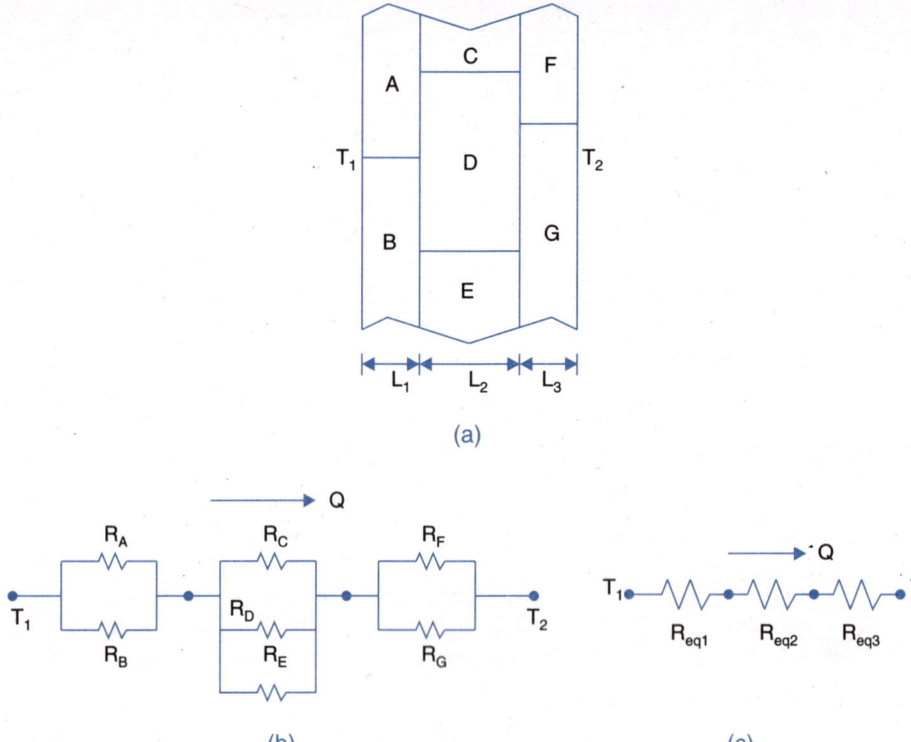

Fig. 2.7

It is necessary that the thickness, area occupied and thermal conductivity of each material be specified.

$$Q = \frac{T_1 - T_2}{\text{Total resistance}}$$

$$R_A = \frac{L_1}{k_A\, A_A},\ R_B = \frac{L_1}{k_B\, A_B},\ R_C = \frac{L_2}{k_C\, A_C},\ R_D = \frac{L_2}{k_D\, A_D}$$

$$R_E = \frac{L_2}{k_E\, A_E},\ R_F = \frac{L_3}{k_F\, A_F} \text{ and } R_G = \frac{L_3}{k_G\, A_G}$$

The equivalent resistance for a parallel circuit is given by

$$R_{eq1} = \frac{1}{\dfrac{1}{R_A} + \dfrac{1}{R_B}}$$

$$R_{eq2} = \frac{1}{\dfrac{1}{R_C} + \dfrac{1}{R_D} + \dfrac{1}{R_E}}$$

$$R_{eq3} = \frac{1}{\dfrac{1}{R_F} + \dfrac{1}{R_G}}$$

$$Q = \frac{T_1 - T_2}{\Sigma R_{EQ}} \quad ...(2.42)$$

In case interface temperature is to be determined, the method is similar to composite slab analysis *i.e.,*

$$\Delta T_i = Q/R_i \text{ and so on.}$$

Example 2.14: *Two slabs are placed in contact, but due to roughness, only 40% of area is in contact and the gap in the remaining area is 0.02 mm thick and is filled with air. The slabs are 10 cm thick each and their conductivities are 15.5 W/mK and 200 W/mK. The temperature of the face of the hot surface is at 250°C and the outside surface of the other slab is at 35°C. Determine the heat flow and the contact resistance. The conductivity of air is 0.0321 W/mK.*

Solution: Of the contact area it is assumed that half of the contact is due to either metal. Total area is taken as 1m².

The equivalent circuit and schematic diagram with all known data are given below (suffixes refer to the circuit)

$$\frac{1}{R_{eq}} = \frac{1}{R_2} + \frac{1}{R_3} + \frac{1}{R_4}$$

$$R_1 = \frac{0.1}{15.5 \times 1} = 0.0065°\text{C/W}, \; R_2 = \frac{0.00002}{15.5 \times 0.2}, \; R_3 = \frac{0.00002}{200 \times 0.2}$$

$$R_4 = \frac{0.00002}{0.0321 \times 0.6}, \; R_5 = \frac{0.1}{200 \times 1} = 0.0005°\text{C/W}$$

$$\frac{1}{R_{eq}} = \frac{15.5 \times 0.2}{0.00002} + \frac{0.0321 \times 0.6}{0.00002} + \frac{200 \times 0.2}{0.00002}$$

$$\mathbf{R_{eq} = 0.464 \times 10^{-6}}$$

$$\mathbf{R} = 0.0065 + 0.464 \times 10^{-6} + 0.0005 = \mathbf{0.00695°C/W}$$

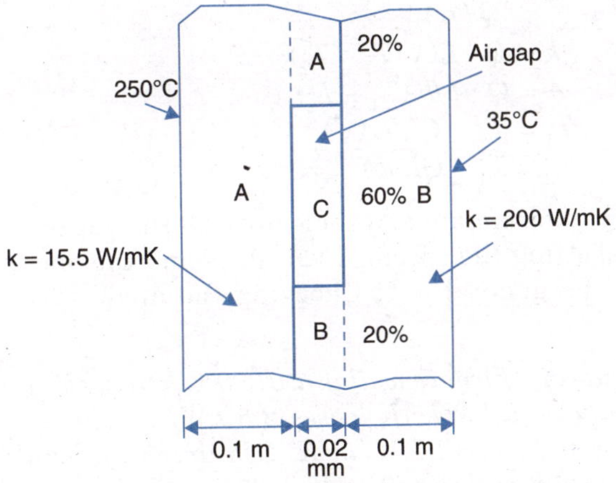

250°C
A
20%
Air gap
35°C
A
C
60% B
k = 200 W/mK
k = 15.5 W/mK
B
20%
0.1 m
0.02 mm
0.1 m

(a)

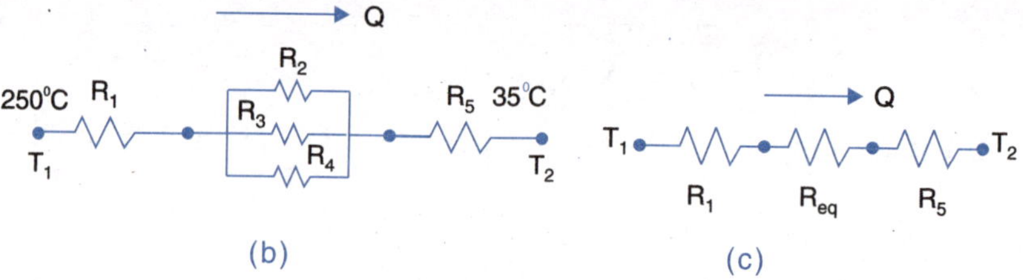

Fig. Ex. 2.14. Model and equivalent circuit

$$Q = \frac{\Delta T}{R} = \frac{250 - 35}{0.00695} = 30926 \text{ W}$$

Contact resistance, $\boldsymbol{R_{eq} = 0.464 \times 10^{-6} \,°C/W}$

Temperature drop at contact $= Q \cdot R_{eq} = \mathbf{0.0143°C.}$

ADVANCED EXAMPLES

After the derivation of each equation one or two examples directly involving the equation are worked out.

Problems requiring more reasoning and also involving more than one of the equations derived are given under this section. Only the study of this section will complete the learning under this chapter. The problems in this section are on steady state heat conduction involving the various situations discussed in this chapter. Most of the problems are closed end type and have a unique solution. A situation involving n variables may give rise to n type of problems at least. As an example, the equation for heat flow through a plane wall, is given below in six forms.

$$Q = \frac{T_1 - T_2}{L/kA} \quad \text{...(1)}$$

$$L = \frac{T_1 - T_2}{Q/kA} \quad \text{...(2)}$$

$$k = Q \cdot L/A\,(T_1 - T_2) \quad \text{...(3)}$$

$$A = Q \cdot L/k\,(T_1 - T_2) \quad \text{...(4)}$$

$$T_1 = T_2 + QL/kA \quad \text{...(5)}$$

$$T_2 = T_1 - QL/kA \quad \text{...(6)}$$

It is not necessary to formulate and remember all the various explicit forms. However, one has to be adept in spotting the specified variables and the one to be determined and get the explicit form from the general one. Checking the answer is also facilitated by such formulations.

Example A-2.1: *The inside of a furnace wall is at 1000°C and with the existing wall of material with thermal conductivity of 1.7 W/mK looses 2.5 kW/m² when the outside is exposed to convection to air at 30°C with h = 27 W/m²K. Determine the wall thickness. Additionally 75 mm thickness of insulation with k = 0.15 W/mK is added on the inside. The furnace wall temperature and the surrounding temperature and convection coefficient remain unchanged. Determine the reduction in heat flow and reduction in the outside surface temperature. Also find the overall heat transfer coefficient and the temperature gradient in each layer.*

Solution:

Case 1: Fig. AE-2.1(*a*). The schematic diagrams in Fig. AE-2.1 shows all the known data both before and after the modification. The quantities required to be found are also indicated. These are L, T_2, T_3, $\frac{dT}{dx}$ etc.

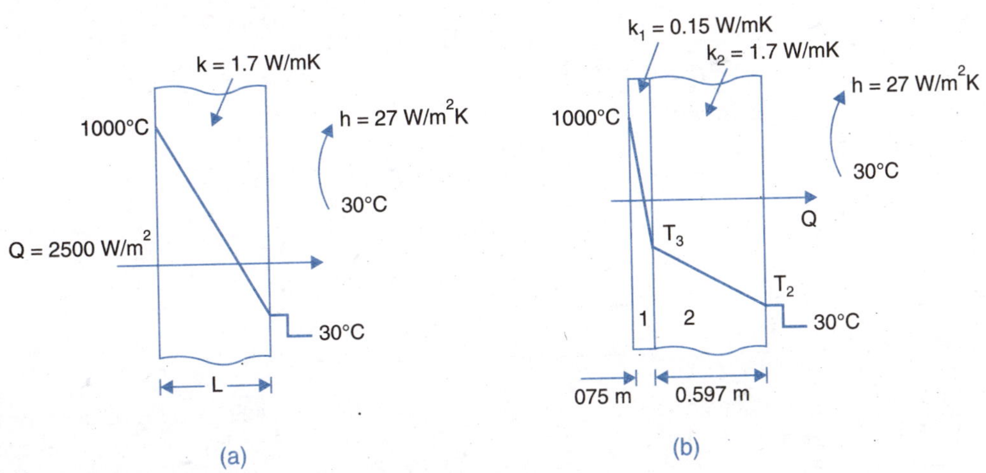

Fig. AE-2.1. Problem model

To find the outside wall temperature: Consider unit area and convection on the outside $Q = hA\Delta T$, $A = 1$

$$2500 = 27(T_2 - 30) \qquad \therefore \qquad \mathbf{T_2 = 122.6°C}$$

To find the thickness: $Q = \Delta T/(L/kA)$, $\quad A = 1$

$$2500 = \frac{1000 - 122.60}{L/1.7} \qquad \therefore \qquad \mathbf{L = 0.597\ m}$$

The temperature gradient, $(dT/dx) = -\dfrac{1000 - 122.60}{0.597} = \mathbf{-\ 1469.7°C/m}$

Case 2: Fig. AE-2.1(*b*)

$$\text{Heat flow} = \frac{\Delta T}{R_1 + R_2 + R_3} = \frac{1000 - 30}{\dfrac{0.075}{0.15} + \dfrac{0.597}{1.7} + \dfrac{1}{27}}$$

$$= \frac{970}{0.888} = \mathbf{1092.1\ W}$$

$$\%\text{ reduction is } 100\left[\frac{2500 - 1092.1}{2500}\right] = 56.32\%$$

The outside wall temperature is found using the convection layer

$$Q = h(T_2 - 30) = 1092.1 \quad h = 27\ \text{W/m}^2\text{K}$$

$$\therefore \qquad \mathbf{T_2 = 70.45°C},\ 1092.1 = \frac{(T_3 - 70.45)\,1.7}{0.597} \qquad \therefore \quad \mathbf{T_3 = 453.95°C}$$

check using second layer: $\quad Q = \dfrac{453.95 - 70.45}{(0.597 / 1.7)} = 1092$ W, checks.

In the case of the plane wall there is only one area to be considered. In this problem the area is 1m^2

$$\therefore \quad U.1 = \frac{1}{R_1 + R_2 + R_3} = \frac{1}{\frac{0.075}{0.15} + \frac{0.597}{1.7} + \frac{1}{27}}$$

$$U = \frac{1}{0.5 + 0.351 + 0.037} = \frac{1}{0.8882} = \mathbf{1.126\ W/m^2 °C}$$

Check: $Q = UA(\Delta T) = 1.126 \times 1(1000 - 30) = 1092.2$ W

The temperature gradient in the first layer is

$$= -\frac{1000 - 453.95}{0.075} = \mathbf{-\ 7280.67°C/m}$$

The temperature gradient in the second layer is

$$= -\frac{453.95 - 70.45}{0.597} = \mathbf{-\ 642.38°C/m.}$$

COMMENT

Will the addition of the layer on the outside change the heat flow? Obviously not. Because the temperature drop and resistances remain unchanged. The addition of the layer with lower thermal conductivity will cause larger temperature drop so that a cheaper material which can withstand the lower temperature can be used by such arrangement. It may be also noted that materials with lower conductivity are insulating materials that can withstand higher temperature.

Example A-2.2: *A composite slab is made of 75 mm thick layer of material with thermal conductivity 0.15 W/mK and 0.597 m thick layer of material of thermal conductivity 1.7 W/mK. The inner surface is maintained at 1000°C while the outer surface was exposed to convection to air at 30°C with convection coefficient of 27 W/m²K. The heat flow was measured as 1080 W as against the calculated value of 1.092 kW. It is presumed that this may be due to contact resistance. Determine the contact resistance and the temperature drop at the interface.*

Solution: The actual resistance is calculated from heat flow and temperature drop. This should consist of the two convection resistances and conduction resistances and contact resistance. By difference the contact resistance is calculated.

$$Q = \frac{\Delta T}{R} = \frac{1000 - 30}{R} = 1080 \text{ W} \quad \therefore \quad R = 0.898°\text{C/W}$$

$$R = 0.898 = \frac{0.075}{0.15} + \frac{0.597}{1.7} + \frac{1}{27} + R_{contact}$$

$$\therefore \quad R_{contact} = \mathbf{0.00995°C/W}$$

$$\Delta T_{contact} = 0.00995 \times 1080 = \mathbf{10.75°C.}$$

Example A-2.3: *A double glazed window is made of 2 glass panes of 6 mm thick each with an air gap of 6 mm between them. Assuming that the air layer is stagnant and only conduction is involved, determine the thermal resistance and overall heat transfer coefficient. The inside is exposed to convection with h = 1.5 W/m²K and the outside to 9 W/m²K. Compare the values with that of a single glass of 12 mm thickness. The conductivity of the glass = 1.4 W/mK and that for air is 0.025 W/mK.*

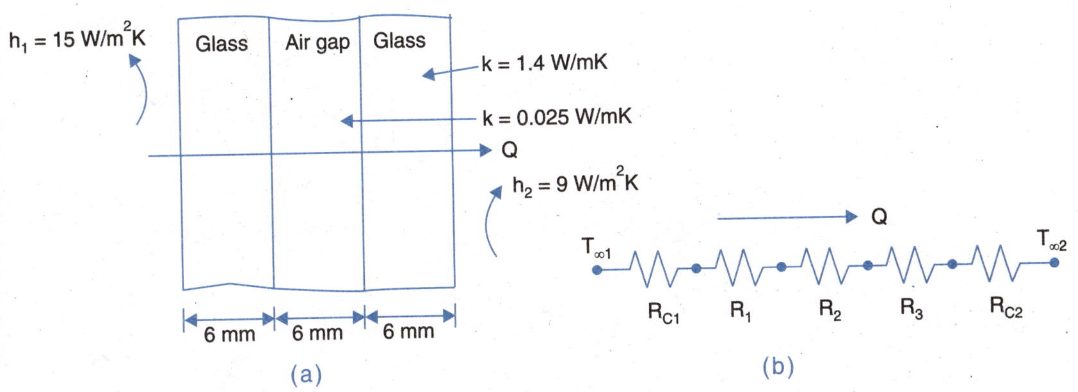

Fig. AE-2.3. Problem model

Solution: Considering unit area. The schematic diagram indicates all the known data.

Double glazing: (Fig. AE-2.3)

Total resistance $= \frac{1}{15} + \frac{0.006}{1.4} + \frac{0.006}{0.025} + \frac{0.006}{1.4} + \frac{1}{9}$

$= 0.067 + 0.0043 + 0.24 + 0.0043 + 0.11 = 0.426°C/W$

$$UA = \frac{1}{\Sigma R}, A = 1 \text{ here}$$

$$U = \frac{1}{\Sigma R \times A} = \frac{1}{0.426} = \mathbf{2.35\ W/m^2°C}$$

Single glass, $U = 1/\left[\frac{1}{15} + \frac{0.012}{1.4} + \frac{1}{9}\right] = \mathbf{5.37\ W/m^2°C}$

COMMENTS

1. Air is a good insulator. Air pockets in insulation materials and air gap between layers reduce the thermal conductivity of the system. But the thickness of the air gap should be small to avoid convection.
2. The heat flow will be almost doubled for the same temperature drop when single glass is used.
3. The resistance for some unit of thickness like inch or cm is often used to compare insulating materials in air-conditioning applications.

Example A-2.4: *A composite wall is made up of 3 layers of thicknesses 25 cm, 10 cm and 15 cm with thermal conductivities of 1.7, k_B and 9.5 W/mK. The outside surface is exposed to air at 20°C with convection coefficient of 15 W/m²K and the inside is exposed to gases at 1200°C with a convection coefficient of 28 W/m² K and the inside surface is at 1080°C. Determine the unknown thermal conductivity, all surface temperatures, resistances of each layer and the over all heat transfer coefficient. Compare the temperature gradients in the three layers.*

Solution: The schematic diagram (Fig. AE-2.4) indicates all known data and the equivalent circuit. As the gas temperature, inside surface temperature and convection coefficients are known, the heat flow can be found. Assuming unit area,

$$Q = 28\,(1200 - 1080) = \mathbf{3360\ W/m^2}$$

The heat flow is the same all through and is equal to Total temperature deep/Total resistance.

Assuming unit area,

$$\therefore \qquad 3360 = \frac{1200-20}{\frac{1}{28}+\frac{0.25}{1.7}+\frac{0.1}{k_B}+\frac{0.15}{9.5}+\frac{1}{15}} = \frac{1180}{0.2652+\frac{0.1}{k_B}}$$

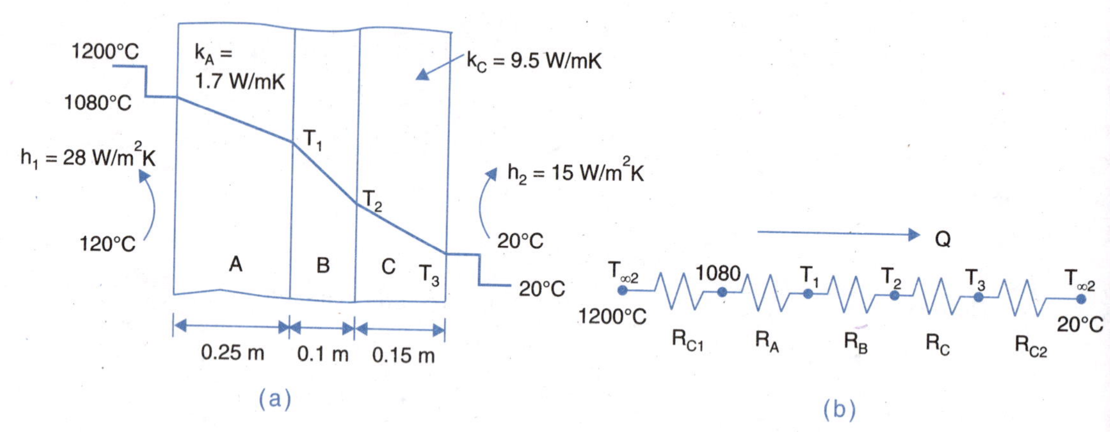

Fig. AE-2.4

$$891.07 + \frac{336}{k_B} = 1180 \quad \therefore \quad \mathbf{k_B = 1.163\ W/mK}$$

Total resistance $= \frac{1}{28}+\frac{0.25}{1.7}+\frac{0.1}{1.163}+\frac{0.15}{9.5}+\frac{1}{15} = 0.3512°C/W$

Overall heat transfer coefficient $= U = \frac{1}{R \times A} = \mathbf{2.85\ W/m^2K}$

To determine the surface temperatures, the heat flow is equated to the (temperature drop/resistance) of each layer

$$3360 = \frac{1080-T_1}{0.25/1.7} \qquad \therefore \quad \mathbf{T_1 = 585.9°C}$$

$$3360 = \frac{T_1-T_2}{0.1/1.163} = \frac{585.9-T_2}{0.1/1.163} \qquad \mathbf{T_2 = 297°C}$$

$$3360 = \frac{T_2-T_3}{0.15/9.5} = \frac{297-T_3}{0.15/9.5} \qquad \therefore \quad \mathbf{T_3 = 243.95°C}$$

Check: Using outside convection

$$Q = \frac{243.95-20}{1/15} = \mathbf{3359.25\ W}$$

Resistance of layers including convection are: 0.0357, 0.147, 0.086, 0.0158, 0.067°C/W.

Temperature gradient $= \frac{T_2 - T_1}{x_2 - x_1}$

1st layer, ΔT = 585.9 – 1080 = – 494.1°C, gradient $= \frac{\Delta T}{L} = \frac{-494.1}{0.25}$ = **– 1976.4°C/m**

2nd layer, $\Delta T = 297 - 585.9 = -288.9°C$, gradient = **– 2889°C/m**

3rd layer, $\Delta T = 243.95 - 297 = -53.05°C$, gradient = **– 353.7°C/m**

Higher the resistance larger the gradient. Also lower the conductivity larger the gradient.

Example A-2.5: *A composite wall is made of 12 mm and 18 mm layers of materials of thermal conductivity 12.5 and 22.5 W/mK. The contact resistance between surfaces is 5×10^{-4} m²°C/W. The hot side is exposed to fluid at 400°C with h = 75 W/m²K and the cold side is exposed to fluids at 60°C with h = 400 W/m²K. Determine the heat flow, temperature drop at various resistances and overall heat transfer coefficient. Comment on the contribution of contact resistance.*

Solution: The specified data are shown in Fig. AE-2.5. Required quantities like ΔT_s, are also indicated in the sketch. Assuming unit area

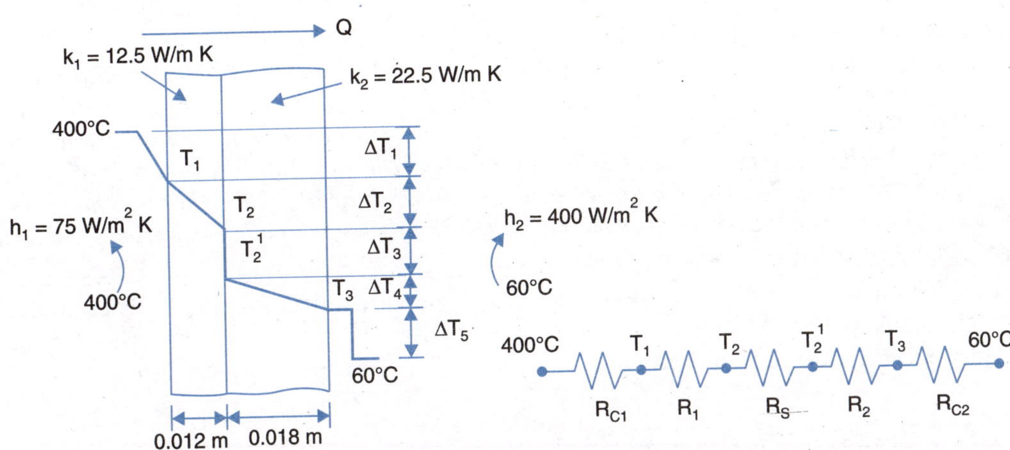

Fig. AE-2.5. Problem model

$$\text{Total resistance} = \frac{1}{h_1} + \frac{L_1}{k_1} + R_{\text{contact}} + \frac{L_2}{k_2} + \frac{1}{h_2}$$

$$= \frac{1}{75} + \frac{0.012}{12.5} + 5 \times 10^{-4} + \frac{0.018}{22.5} + \frac{1}{400}$$

$$= 0.018093333 \text{ K/W}$$

$$U = \frac{1}{R \times A} = 55.27 \text{ W/m}^2\text{K}$$

$$Q = \frac{\Delta T}{R} = \frac{400 - 60}{0.018093333} = 18791.45 \text{ W}$$

Temperature drop:

1st convection layer: $Q = h_1(T_\infty - T_1)$

$$18791.45 = 75(400 - T_1) \quad \mathbf{\Delta T_1 = 250.553°C} \quad \therefore \quad T_1 = 149.447°C$$

1st conduction layer $\quad Q = \dfrac{\Delta T}{L/k} = \dfrac{149.447 - T_2}{(0.012/12.5)}$

$$\mathbf{\Delta T_2 = 18.04°C} \quad \therefore \quad T_2 = 131.407°C$$

Contact surface drop: $\mathbf{\Delta T_3} = Q \times R = 5 \times 10^{-4} \times 18791.45 = \mathbf{9.4°C}$

$\therefore \qquad T_2^1 = 122°C$

2nd conduction layer

$$Q = \frac{\Delta T}{L_2/K_2} \quad \Delta T_4 = Q \times \frac{L_2}{K_2} = \frac{18791.45 \times 0.018}{22.5}$$

$$\Delta \mathbf{T}_4 = 15.03°\text{C} \quad T_3 = 122 - 15.03 = \mathbf{106.97°C}$$

Check: Using last convection layer

$$Q = h(\Delta T_5) = 400(106.97 - 60) = 18788 \text{ W. checks}$$

The contribution of contact resistance is small as this involves a temperature drop of 9.4°C out of 340°C. This is the general order of contact resistance. The heat flow calculated neglecting contact resistance is 19325.50 W which is higher by 2.76%. This is less than errors due to uncertainty in values of k. However, whenever possible the contact resistance should be taken into account. Note that because of the higher values of conductivity the main temperature drop is in the convection layer.

Example A-2.6: *A composite slab is made of three layers 15 cm, 10 cm and 12 cm thickness. The first layer is of material with thermal conductivity 1.45 for 60% of the area and the rest is of material with conductivity of 2.5 W/mK. The second layer is made of material with conductivity of 12.5 W/mK for 50% area and of material with conductivity 18.5 W/mK is used for the other 50%. The third layer is of single material of thermal conductivity 0.76 W/mK. The slab is exposed on one side to warm air at 26°C and to cold air at – 20°C on the other side. The convection coefficients are 15 and 20 W/m²K on the inside and outside respectively. Determine the heat flow and interface temperatures.*

Solution: The schematic diagram [Fig. AE-2.6(*a*)] indicates all known data. The equivalent circuit is shown in Fig. AE-2.6(*b*). The thermal resistances of the five material section are found assuming a total area of 1 m².

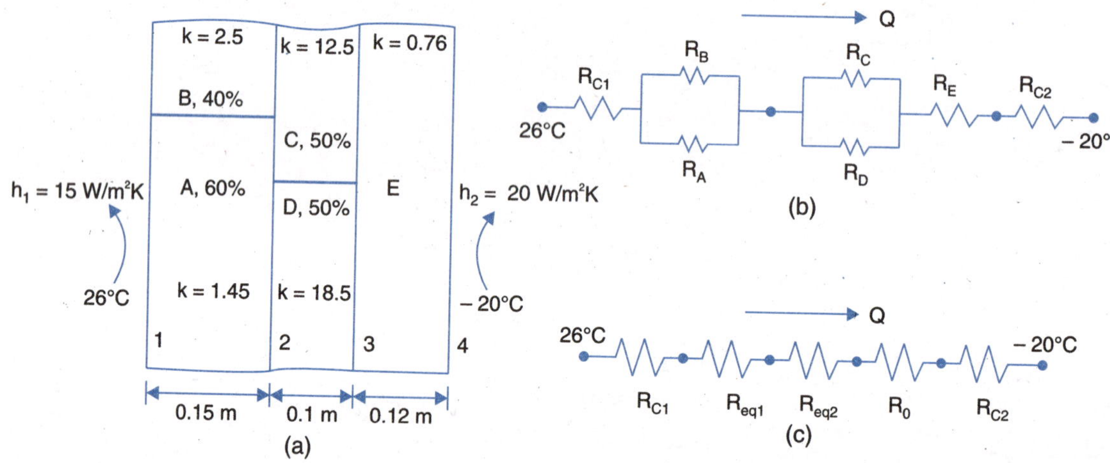

Fig. AE-2.6. Problem model

$$\therefore \quad R_{c1} = \frac{1}{1 \times 15} = 0.06667, R_A = \frac{0.15}{1.45 \times 0.6}, R_B = \frac{0.15}{0.4 \times 2.5}$$

$$\therefore \quad R_C = \frac{0.1}{12.5 \times 0.5}, R_D = \frac{0.1}{18.5 \times 0.5}, R_E = \frac{0.12}{0.76 \times 1} = 0.1579$$

$$R_{c2} = \frac{1}{1 \times 20} = 0.05$$

The equivalent resistance for the parallel paths are found using

$$1/R_{eq} = \frac{1}{R_1} + \frac{1}{R_2} \text{ where } R_1 \text{ and } R_2 \text{ are the resistances in parallel}$$

$$\therefore \quad 1/R_{eq1} = [1/(0.15/1.45 \times 0.6)] + [1/(0.15/0.4 \times 2.5)]$$

$$\therefore \quad R_{eq1} = 0.0802 \text{ m}^2\text{C/W}$$

$$\therefore \quad 1/R_{eq2} = [1/(0.1/18.5 \times 0.5)] + [1/(0.1/12.5 \times 0.5)]$$

$$\therefore \quad R_{eq2} = 0.006452 \text{ m}^2\text{C/W}$$

$$\Sigma \mathbf{R} = 0.06667 + 0.0802 + 0.006452 + 0.1579 + 0.05 = \mathbf{0.3612}$$

$$\therefore \quad \mathbf{Q} = \frac{\Delta T}{\Sigma R} = \frac{26 - (-20)}{0.3612} = \mathbf{127.35 \ W/m^2}$$

Interface temperatures:

face 1: $$Q = \frac{\Delta T_1}{R_1} = \frac{26 - T_1}{1/15} = 127.35$$

$$\Delta T_1 = 8.49 \qquad \therefore \ \mathbf{T_1 = 17.51°C}$$

face 2: $$Q = \frac{\Delta T_2}{R_{eq1}} = \frac{\Delta T_2}{0.0802} = 127.35$$

$$\Delta T_2 = 10.21°C \qquad \therefore \ \mathbf{T_2 = 7.3°C}$$

face 3: $$Q = \frac{\Delta T_3}{R_{eq2}} = \frac{\Delta T_3}{0.006452} = 127.35$$

$$\Delta T_3 = 0.82°C \qquad \therefore \ \mathbf{T_3 = 6.48°C}$$

face 4: $$Q = \frac{\Delta T_4}{R_4} = \frac{\Delta T_4}{0.12/0.76} = 127.35$$

$$\Delta T_4 = 20.11°C \qquad \therefore \ \mathbf{T_4 = -13.63°C}$$

Check: heat flow by convection on the cooler side:

$h\{-13.63 - (-20)\} = 20 \times 6.372 = 127.44$ W

COMMENT

It may be noted that some heat flow occurs between A and B as well as between C and D at their interface so that the surface temperatures to be equal and hence the temperature variation in individual materials cannot be established by this analysis. This is the situation of heat leakage into a cold storage chamber.

Example A-2.7: *A heat flux of 1500 W/m² is incident on the surface of a slab 10 cm thick with thermal conductivity of 7.5 W/mK. The hot side is found to be at 120°C. On the otherside, the heat is passed on to the surroundings at 30°C both by convection and radiation. It radiation is ideal, determine the convection coefficient and also the share of heat flow between the two processes.*

Solution: The specified data are shown in Fig. AE-2.7. This is one type of boundary condition, where heat flux at the surface is specified, as different from convection boundary. Looking at the equivalent circuit, Fig. AE-2.7(*b*), the value of h is identified as the unknown. The value of h can be found if the surface temperature is determined. Using the conduction layer only,

Assuming unit area, $Q = 1500 = \dfrac{120 - T_2}{0.1/7.5}$

$\therefore$ $\mathbf{T_2 = 100°C}$

To find h_r, temperatures should be in absolute units

$$T_2 = 273 + 100 = 373 \text{ K}, T_s = 273 + 30 = 303 \text{ K}$$

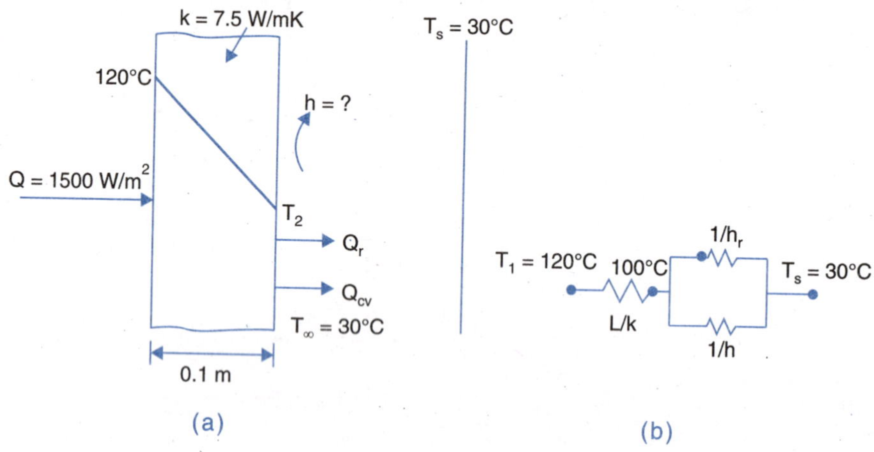

Fig. AE-2.7. Problem model

$$1/h_r = \frac{1}{\sigma(T_2 + T_s)(T_2^{\,2} + T_s^{\,2})} = \frac{1}{5.67 \times 10^{-8}\,(373 + 303) \times (373^2 + 303^2)}$$

$= 0.112973$ C/W $\qquad \therefore\ h_r = 8.8516$ W/m² °C

$Q = \dfrac{100 - 30}{R} = 1500 \qquad \therefore\ R = 0.0466$°C/W

Using the parallel circuit [Fig. AE-2.7 (*b*)]

$\dfrac{1}{R} = h_r + h \qquad \therefore\ h = (1/0.0466) - 8.8516$

$\therefore \qquad h = 12.58$ W/m²°C

$Q_r = \sigma\,(T_2^4 - T_s^4) = 5.67\,(3.73^4 - 3.03^4) =$ **619.62 W, 41.3%**

$Q_{cv} = 12.58\,(100 - 30) =$ **880.6 W. 58.7%** Total: 1500.72 W checks.

COMMENT

When convection coefficient has a low value as in natural convection, it will be desirable to consider radiation loss. Otherwise the heat loss will be underestimated and the design will be in error.

Example A-2.8: *A 2 kW heater element of area 0.04 m² is protected on the backside with insulation 50 mm thick of k = 1.4 W/mK and on the front side by a plate 10 mm thick with thermal conductivity of 45 W/mK. The backside is exposed to air at 5°C with convection coefficient of 10W/m² K and the front is exposed to air at 15°C with convection coefficient including radiation of 250 W/m²K. Determine the heater element temperature and the heat flow into the room under steady conditions.*

Solution: All known data are indicated in Fig. AE-2.8(a). The equivalent circuit can be drawn as in Fig. AE-2.8(b). k_1 = 45 W/mK, A = 0.04 m^2

$$Q_1 + Q_2 = 2000 \text{ W} \quad Q_1 = \frac{T_o - 15}{\dfrac{0.01}{45 \times 0.04} + \dfrac{1}{250 \times 0.04}} = 9.4737\,(T_o - 15)$$

$$Q_2 = \frac{T_o - 5}{\dfrac{0.05}{1.4 \times 0.04} + \dfrac{1}{10 \times 0.04}} = 0.29473\,(T_o - 5)$$

$$Q_1 + Q_2 = 9.4737\,(T_o - 15) + 0.29473\,(T_o - 5) = 2000$$

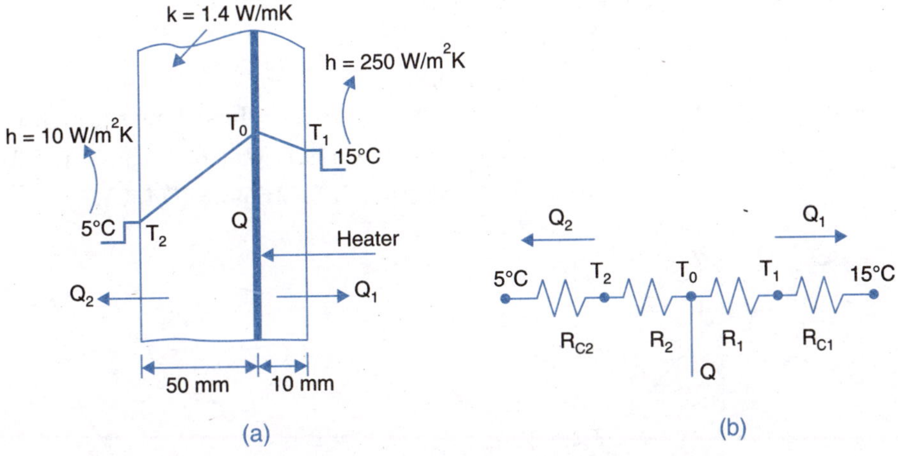

Fig. AE-2.8. Problem model

$$9.7684\,T_o - 143.58 = 2000 \quad \therefore \quad \mathbf{T_o = 219.44°C}$$

Room side
$$Q_1 = \frac{(219.44 - 15)}{\dfrac{0.01}{45 \times 0.04} + \dfrac{1}{250 \times 0.04}} = \mathbf{1936.8\ W,\ 96.84\%}$$

Back side
$$Q_2 = \frac{(219.44 - 5)}{\dfrac{1}{10 \times 0.04} + \dfrac{0.05}{1.4 \times 0.04}} = \mathbf{63.2\ W,\ 3.16\%.}$$

Surface temperature $\therefore$ Room side

$$\Delta T_1 = Q_1 \times \frac{1}{h_1 A} = \frac{1936.8}{250 \times 0.04} = \mathbf{193.68°C}$$

$$\mathbf{T_1 = 193.68 + 15 = 208.68°C}$$

Back side
$$\Delta T_2 = Q_2 \times \frac{1}{h_2 A} = 63.2 \times \frac{1}{10 \times 0.04} = 158°C$$

$$\mathbf{T_2 = 158 + 5 = 163°C.}$$

As the backside temperature is rather high, radiation loss should also be taken into account. However as the fraction loss is low in this case, it may be ignored.

Example A-2.9: *To reduce frosting it is desired to keep the outside surface of a glazed window at 4°C. The outside is at – 10°C and the convection coefficient is 60 W/m²K. In order to maintain the conditions a uniform heat flux is provided at the inner surface which is in contact with room air at 22°C with a convection coefficient of 12 W/m²K. The glass is 7 mm thick and has a thermal conductivity of 1.4 W/mK. Determine the heating required per m² area.*

Solution: The known data are shown is Fig. AE-2.9. Assume unit area.

The heat flow on the outside = heat convected on the outside

$= h\,(T - T_{\infty 1})$

$= 60\,(4 - (-10)) = 840\ \text{W/m}^2$

The heat flow through the glass is the same

$$840 = \frac{\Delta T}{R} = \frac{T_1 - 4}{0.007/1.4} \quad \therefore \quad \mathbf{T_1 = 8.2°C}$$

Heater input + heat received by convection from room = heat flow through barrier

Heater input = heat flow through glass barrier – heat convected from inside

$= 840 - 12\,(22 - 8.2) = 840 - 165.6 = \mathbf{674.4\ W}$

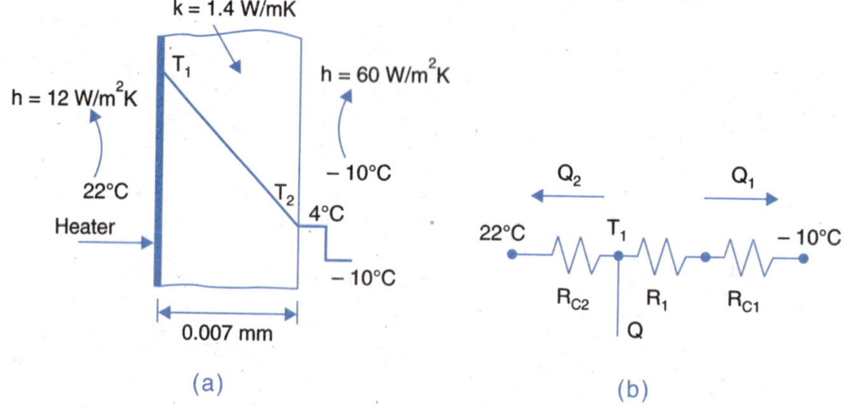

Fig. AE-2.9. Problem model

If it is desired that the inside wall temperature and room temperatures should be equal for comfort, determine the heat flux. In this case $T_1 = 22°C$ and T_2 is not known

But heat conducted = heat convected

$$\frac{22 - T_2}{0.007/1.4} = 60\,(T_2 - (-10))$$

Solving $\quad 22 - T_2 = 0.3\,T_2 + 3, \quad \mathbf{T_2 = 14.62°C}$

$$Q = \frac{22 - 14.62}{0.007/1.4} = \mathbf{1477\ W}$$

Check: $\quad Q = h(T - T_\infty) = 60 \times 24.62 = \mathbf{1477.2\ W}$

This is almost double. There is no convection input as temperatures are the same.

COMMENT

For comfort it is desirable that the room wall temperature is high, the reason being that radiation loss from the body will be less. But the power required will be high.

Example A-2.10: *In a slab of material 0.25 m thick and having a thermal conductivity of 45 W/mK, the temperature °C at x under steady state is given by T = 100 + 200x – 400x² when x is measured from one face in m. Determine the heat flow at x = 0, x = 0.125 and x = 0.25 m and also the temperatures and temperature gradients at these planes. If the difference in heat flow at these sections is due to the heat generation, determine the heat generation rate per unit volume.*

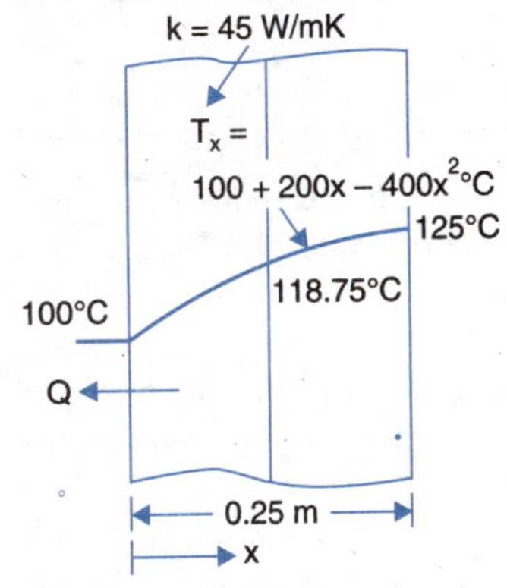

Fig. AE-2.10. Problem model

Solution: The known data are shown in Fig. AE-2.10. The temperatures are determined from the equation

$$T = 100 + 200x - 400x^2$$

at $x = 0.$ $T = 100°C$

at $x = 0.125$ $T = 100 + 200 \times 0.125 - 400 \times 0.125^2 = 118.75°C$

at $x = 0.25$ $T = 100 + 200 \times 0.25 - 400 \times 0.25^2 = 125°C$

Differentiating the above equation w.r.t. x

$$\frac{dT}{dx} = 200 - 800x \qquad \textbf{Considering unit area}$$

at $x = 0,$ $\frac{dT}{dx} = 200$ heat flow $= - kA \frac{dT}{dx}$

at $x = 0.125,$ $\frac{dT}{dx} = 100$

at $x = 0.25,$ $\frac{dT}{dx} = 0,$

at $x = 0$ $Q = - 45 \times 1 \times 200 = - 9000 \text{ W/m}^2$

at $x = 0.125,$ $Q = - 45 \times 1 \times 100 =$ **– 4500 W/m²**

at $x = 0.25$ $Q = - 45 \times 0 = 0$

Heat is flowing out at $x = 0$ at the rate of 9000 W/m². There is no flow at $x = 0.25$ m. So this heat should have been generated in the slab of thickness 0.25 m and area 1 m². **So the generation for m³ is 4 × 4000 = 36000 W/m³.**

The flow out at the middle section is 4500 W/m² which is what is generated in the slab between 0.125 m and 0.25 m.

This problem illustrates the use of the temperature gradient in the determination of heat flow. The general one dimensional heat flow equation with heat generation can also be obtained from the above as detailed. Differentiating the given equation twice,

$$\frac{d^2T}{dx^2} = -800 \quad \text{or} \quad \frac{d^2T}{dx^2} + 800 = 0$$

or
$$\frac{d^2T}{dx^2} + \frac{36000}{45} = 0 \quad \text{or} \quad \frac{d^2T}{dx^2} + \frac{q}{k} = 0$$

COMMENT

A non linear temperature distribution in a slab can be due to variable conductivity. If conductivity is constant, then such distribution indicates either heat generation or heat absorption or cooling.

Example A-2.11: *A composite cylinder consists of 10 cm radius steel pipe of 25 mm thickness over which two layers of insulation 30 mm and 35 mm are laid. The conductivities are 25 W/mK, 0.25 W/mK and 0.65 W/mK. The inside is exposed to convection at 300°C with h = 65 W/m²K. The outside is exposed to air at 30°C with h = 15 W/m²K. Determine the heat loss/m. Also find the interface temperatures.*

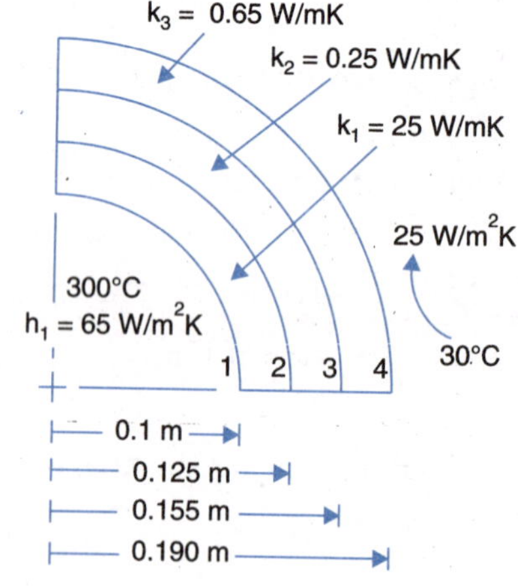

Fig. AE-2.11. Problem model

Solution: All known data are shown in Fig. AE-2.11. The heat flow is found by,

$$Q = \frac{\text{Overall temperature drop}}{\text{Total thermal resistance}}$$

considering 1 m length:

Inside convection:

$$R_1 = \frac{1}{h_1 A} = \frac{1}{65 \times 2\pi \times 0.1 \times 1}$$
$$= 0.0245°\text{C/W}$$

Conduction layer 1: $$R_2 = \frac{\ln \frac{r_2}{r_1}}{2\pi k_1 l} = \frac{\ln(0.125/0.1)}{2\pi \times 25 \times 1} = 0.0014206°\text{C/W}$$

Conduction layer 2: $$R_3 = \frac{\ln(0.155/0.125)}{2\pi \times 0.25 \times 1} = 0.13694°\text{C/W}$$

Conduction layer 3: $$R_4 = \frac{\ln(0.190/0.155)}{2\pi \times 0.65 \times 1} = 0.04985°\text{C/W}$$

Convection on the outside $$R_5 = \frac{1}{h_2 A_2} = \frac{1}{15 \times 2\pi \times 0.190 \times 1} = 0.05584°\text{C/W}$$

As total $R = 0.2685459$, $$Q = \frac{300 - 30}{0.2685459} = \mathbf{1005.42\ W/m}$$

To find interface temperatures:

$$Q = \frac{\Delta T_1}{R_1}, \Delta T_1 = Q \times R_1 = 1005.42 \times 0.0245$$

	$\Delta T_1 = 24.61$	$\mathbf{T_1 = 275.39°C}$
Similarly	$\Delta T_2 = 1.43$	$\mathbf{T_2 = 273.96°C}$
	$\Delta T_3 = 137.69$	$\mathbf{T_3 = 136.27°C}$
	$\Delta T_4 = 50.12$	$\mathbf{T_4 = 86.15°C}$
	$\Delta T_5 = 56.15$	
	Total = 270.00 checks	

Check: $Q = hA\,\Delta T = 15 \times 2\pi \times 0.19 \times 1\,(86.15 - 30)$

$= 1005.48$ W (using outside convection)

COMMENTS

1. The temperature drop is highest for the layer with low value of conductivity.
2. The drop in the metal wall is very small.
3. Lower value of k leads to larger drop.

Example A-2.12: *A 6 cm thick insulation is laid on a steam pipe with surface temperatures of 240°C and diameter of 30 cm with a contact resistance of 0.02 m²°C/W. The conductivity of the material is 0.4 W/mK. Under natural convection conditions to air at 30°C the convective heat transfer coefficient has a value of 15 W/m²K. When winds blow (the pipe runs outside) the coefficient of convective heat transfer reaches a value of 75 W/m²K. Determine the heat loss/m length in these cases. Also find the temperatures at interfaces, the temperature gradient at the two surfaces under free convection conditions. Plot the variation of temperature along the radius.*

Solution: The specified data are presented in Fig. AE-2.12(a)

$$Q = \frac{\Delta T}{\Sigma R}$$

There are three resistances:

1. Contact resistance, 2. Conduction resistance and 3. Convection resistance.

Considering 1 m length.

1. Contact resistance = 0.02 m^2°C/W. This is for 1 m^2. The contact area is $2\pi rl = 2 \times \pi \times 0.15 \times 1$ m^2. Contact resistance for the area considered

$$= \frac{0.02}{2 \times \pi \times 0.15 \times 1} = 0.02122066°\text{C/W}$$

2. Conduction resistance $= \ln\left(\frac{r_2}{r_1}\right)/2\pi kl = \ln\left(\frac{0.21}{0.15}\right)/\,2 \times \pi \times 0.4 \times 1$

$= 0.1338781°$C/W

3. Convection resistance on the outside $= \dfrac{1}{hA_o}$

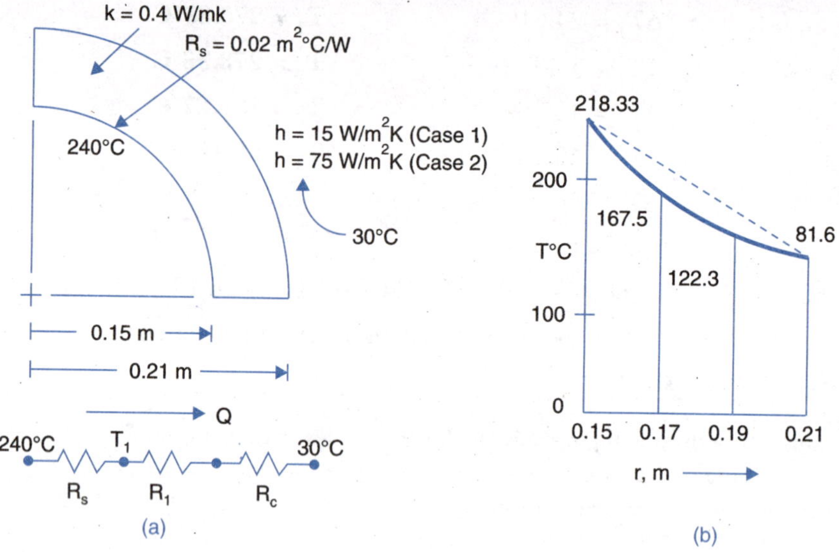

Fig. AE-2.12. Problem model

Case 1: $\dfrac{1}{15 \times 2\pi r_0 l} = \dfrac{1}{15 \times 2\pi \times 0.21 \times 1} = 0.05052538$°C/W

Case 2: $\dfrac{1}{75 \times 2\pi \times 0.21 \times 1} = 0.010105075$°C/W

Case 1: Total resistance $(R_1) = 0.02122066 + 0.1338781 + 0.05052538 = 0.205624$

Heat flow $= \dfrac{\Delta T}{R} = \dfrac{240 - 30}{0.205624} =$ **1021.28 W/m** length

Case 2: Total resistance

$$R_2 = 0.02122066 + 0.1338781 + 0.010105075 = 0.01652$$

$$Q = \frac{\Delta T}{R_2} = \frac{240 - 30}{0.1652} = \textbf{1271.15 W}, \text{ an increase of } \textbf{24.5\%}$$

The increase is due to the reduced convection resistance caused by higher value of convection coefficient.

The temperatures at various locations are calculated under free convection conditions

The pipe surface temperature: 240°C

Temperature drop in the contact resistance

$$= Q \times R_{contact} = 1021.28 \times 0.02122066 = 21.67°C$$

Insulation inside surface temperature

$$= 240 - 21.67 = \textbf{218.33°C}$$

Temperature drop in the insulation

$$= Q \times R_{insulation} = 1021.28 \times 0.1338781 = 136.73°C$$

Outside surface temperature

$$= 218.33 - 136.73 = \textbf{81.6°C}$$

Check: Convection heat flow on the outside

$$= 15 \times 2\pi \times 0.21 \times 1\,(81.6 - 30) = 1021.27 \text{ W}$$

To find temperature gradient at the inner surface:

$$Q = -kA \left.\frac{dT}{dr}\right|_{0.15 \text{ m}}$$

$$1021.28 = 0.4 \times 2\pi \times 0.15 \times 1 \frac{dT}{dr} \quad \therefore \quad \left.\frac{dT}{dr}\right|_{\text{inner}} = -\mathbf{2709°C/m}$$

Outside:

$$1021.28 = 0.4 \times 2\pi \times 0.21 \times 1 \frac{dT}{dr} \quad \therefore \quad \left.\frac{dT}{dr}\right|_{\text{outer}} = -\mathbf{1935°C/m}$$

The reduced temperature gradient is due to larger outside area,

To find and plot temperature along the radius, a radius of 0.17, 0.19 are chosen

$$Q = 1021.28 = \frac{\Delta T}{\left(\ln \dfrac{0.17}{0.15}\right) / 2\pi \times 0.4 \times 1}$$

$$\therefore \quad \Delta T = 50.86°\text{C}$$

$$T_{0.17} = 218.33 - 50.86 = \mathbf{167.47°C}$$

$$1021.28 = \frac{\Delta T}{\left(\ln \dfrac{0.19}{0.15}\right) / 2\pi \times 0.4 \times 1}, \ \Delta T = \mathbf{96.06°Cs}$$

$$T_{0.19} = 218.33 - 96.06 = \mathbf{122.27°C}$$

These are plotted in Fig. AE-2.12(*b*). The variation is logarithmic.

Example A-2.13: *Insulation is added in 3 cm layers over a steel pipe of 30 cm dia. The convection on the outside is 25 W/m²K. The conductivity of the material is 0.47 W/mK. Determine the total thermal resistance for the addition of 5 such layers. Compare the % increase in resistance and % increase in the volume of material over the first layer.*

Solution: Calculations are based on:1 m length

$$R = \frac{1}{hA_o} + \frac{\ln \dfrac{r_2}{r_1}}{2\pi k} \ ; \text{and } V = \pi\,(r_2^2 - r_1^2) \times 1$$

$$A = 2\pi r l \quad l = 1 \text{ m}, k = 0.47 \text{ W/mK} \quad h = 25 \text{ W/m}^2\text{K}$$

Insulation thickness	r_1	r_2	*Thermal resistance*	*% increase in R over the previous*	*Volume*	*% increase in V over the previous*	*Q for 100°C drop*
0	0.15	0.15	0.0424	—	—	—	2358.5
0.03	0.15	0.18	0.0971	129.0	0.0311	—	1029.9
0.06	0.15	0.21	0.1443	48.6	0.0679	118.2	693.0
0.09	0.15	0.24	0.1857	28.7	0.1103	62.4	538.5
0.12	0.15	0.27	0.2226	19.9	0.1583	43.6	449.2
0.15	0.15	0.30	0.2559	15.0	0.2121	34.0	390.8

Addition of insulation beyond a certain point is only marginally effective as seen from the tabulation. Note the change in the value of Q from 3rd to 4th layer and onwards. This is due to the reduction in convective resistance. Another point to be noted is that resistance increases continuously but at a lower rate. Critical radius $R_c = \frac{k}{h} = \frac{0.47}{25} = 0.0188$, much smaller than the base radius. **For larger diameters, the addition of insulation will always reduce the heat flow.**

Example A-2.14: *A copper wire 5 mm dia. carrying current generates 66.76 W/m length*

(i) If the surface is exposed to air at 30°C with h = 25 W/m²K. Determine the surface temperature.

(ii) If a very thin coating of varnish like insulation with a contact resistance of 0.02 m²°C/W is added, determine the surface temperature.

Solution: Case 1: heat generated = heat convected over the surface

$$66.76 = 25 \times \pi \times 0.005 \times 1(T - 30)$$

$$\therefore \qquad \mathbf{T = 200°C}$$

Case 2: heat generated = heat conducted = heat convected

$$\text{or} \qquad Q = \frac{\Delta T}{R}.$$

The resistance now is made up of contact resistance and convection resistance.

$$\text{Contact resistance} = \frac{0.02}{\pi \times d \times 1} = \frac{0.02}{\pi \times 0.005 \times 1} = 1.273°\text{C/W}$$

$$\text{Convection resistance} = \frac{1}{hA} = \frac{1}{25 \times \pi \times 0.005 \times 1} = 2.5465°\text{C/W}$$

$$Q = \frac{T - 30}{(2.5465 + 1.273)} = 66.76 \text{ W}, \ \mathbf{T = 255 + 30 = 285°C.}$$

COMMENT

In this case resistance is added by the insulation without any change in the outside area. As the convection resistance is the same in both cases. This is not a critical radius situation.

Example A-2.15: *A copper cable of 5 mm dia. carrying current generates 294 W/m length. It is exposed to convection at 30°C with h = 25 W/m²K. Investigate the effect of adding an insulation with k = 0.5 W/mK in steps of 2.5 mm thicknesses, on the cable.*

Solution: The cable temperature is found using $\frac{\Delta T}{R} = 294$ where

$$R = \frac{1}{hA} + \frac{\ln \frac{r_2}{r_1}}{2\pi k}$$

Inner radius = 0.0025 m, 1 m length. The results are tabulated.

Outer radius m	*Conduction resistance °C/W*	*Convection resistance °C/W*	*Total resistance °C/W*	*Wire surface temperature °C*
Bare wire	—	2.5465	2.5465	778.67
0.0050	0.2206	1.2732	1.4938	469.2
0.0075	0.3497	0.8488	1.1985	382.2
0.0100	0.4413	0.6366	1.0779	346.9
0.0125	0.5123	0.5093	1.0216	330.3
0.0150	0.5703	0.4244	0.9947	322.5
0.0175	0.6194	0.3638	0.9832	319.1
0.0200	**0.6619**	**0.3183**	**0.9802**	**318.2**
0.0225	0.6994	0.2829	0.9823	318.8
0.0250	0.7329	0.2546	0.9875	320.33

Chapter 2

COMMENTS

1. The wire surface temperature actually decreases by the addition of insulation. After a certain thickness it again increases. The radius for this condition is called the critical radius and this is given by

$$r_c = \frac{k}{h} = \frac{0.5}{25} = 0.02 \text{ m or } 20 \text{ mm.}$$

2. The problem shows that in small diameter wires or pipes, a critical radius exists.

Example A-2.16: *A 10 mm OD pipe has cold fluid flowing inside which keeps the surface temperature at 3°C. The pipe is exposed to air at 30°C with a convection coefficient of 10 W/m²K. Insulation with thermal conductivity of 0.15 W/mK is added in 5 mm layers. Investigate the effect of this addition on the heat flow in. The pipe surface temperature, surrounding temperature and convection coefficients remain unchanged.*

Solution: Heat flow $= \frac{\Delta T}{R_c + R_k} = \frac{30-3}{R}$

$$R_C = \text{convection resistance} = \frac{1}{hA}$$

$$R_k = \frac{\ln\left(\frac{r_2}{r_1}\right)}{2\pi k}, \; l = 1 \text{ m}, \; r_1 = 5 \text{ mm.}$$

The values calculated are tabulated:

Insulation thickness mm	*Outer radius m*	*Resistance °C/W*			*heat flow/W/m*
		R_k	R_c	*Total R*	
0	0.005	—	3.1831	3.1831	8.48
5	0.010	0.7355	1.5915	2.3270	11.60
10	**0.015**	1.1657	1.0610	**2.2267**	**12.13**
15	0.020	1.4709	0.7958	2.2667	11.91
20	0.025	1.7077	0.6366	2.3443	11.52
25	0.030	1.9011	0.5305	2.4316	11.10

The addition of insulation has actually increased the heat leakage. This is due to the large reduction in convection resistance compared to the increased conduction resistance. It may be noted that the resistance decreased and then increased. $r_c = \frac{k}{h} = \frac{0.15}{10}$ m = 15 mm. If this is a case where the heat gain is to be reduced, then no insulation should be added.

Example A-2.17: *A steel pipe of 0.4 m dia. carrying oil in the cold region is proposed to be protected by insulations A and B of 8 cm and 10 cm thickness with conductivities of 0.03 and 0.3 W/mK. These are purchased in required volumes in powder form. During the execution, by mistake the material B with conductivity 0.3 W/mK was applied first and then the other material. Investigate the heat transfer rate in the two situations.*

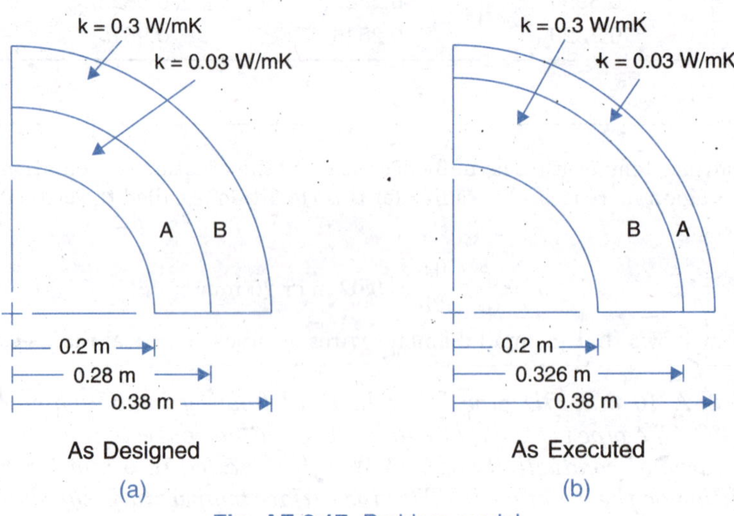

Fig. AE-2.17. Problem model

Solution: Known data are shown in Fig. AE-2.17 (*a*, *b*). The volume of materials applied are the same, considering 1 m length

Volume of $A = \pi\,(0.28^2 - 0.2^2) \times 1 = 0.12064\ \text{m}^3$

Volume of $B = \pi\,(0.38^2 - 0.28^2) \times 1 = 0.2073\ \text{m}^3$

After the mistaken application the outside radius of B is

$$0.2073 = \pi\,(r^2 - 0.2^2) \times 1 \qquad r_2 = 0.326\ \text{m}$$

The outer radius has to be 0.38 m as the total volume remains unchanged.

Check: $V = \pi\,(0.38^2 - 0.326^2) \times 1 = 0.12064\ \text{m}^3$

Case 1: Total resistance (1 m length)

$$\frac{\ln\frac{r_2}{r_1}}{2\pi k_1} + \frac{\ln\frac{r_3}{r_2}}{2\pi k_2} = \frac{\ln\frac{0.28}{0.2}}{2\pi \times 0.03} + \frac{\ln\frac{0.38}{0.28}}{2\pi \times 0.3}$$

$$= 1.785 + 0.162 = \mathbf{1.947°C/W}$$

Case 2: $\dfrac{\ln\frac{0.326}{0.2}}{2\pi \times 0.3} + \dfrac{\ln\frac{0.38}{0.326}}{2\pi \times 0.03} = 0.2592 + 0.8131 = \mathbf{1.0723°C/W}$

The resistance is reduced to 55% of the value and heat flow will increase. For insulation this is not desirable. Thus the order of application becomes important. **In the case of cylindrical and spherical shapes materials with lower thermal conductivity should be applied near the hot surface.**

Example A-2.18: *Orders were placed to supply formed layers of insulation A and B of thickness 5 cm each to be applied over a pipe of 0.3 m dia., the insulation A to be of 0.3 m ID and 0.4 m OD and insulation B is to be of 0.4 m ID and 0.5 m OD. The conductivities of A and B are to be 0.04 and 0.08 W/mK. However the supplier by mistake has supplied insulation A of size 0.4 m ID and OD of 0.5 m and insulation B m of 0.3 m ID and 0.4 m OD. Check whether there will be any change in heat flow.*

Solution: The given data are indicated in Fig. AE-2.18 (*a*, *b*).

Case 1: As proposed: (1 m length)

$$\text{Resistance} = \frac{1}{2\pi k_A} \ln \frac{r_2}{r_1} + \frac{1}{2\pi k_B} \ln \frac{r_3}{r_2}$$

$$= \frac{1}{2\pi \times 0.04} \ln \frac{0.2}{0.15} + \frac{1}{2\pi \times 0.08} \ln \frac{0.25}{0.2}$$

$$= 1.145 + 0.444 = \mathbf{1.589°C/W}$$

Case 2: As supplied:

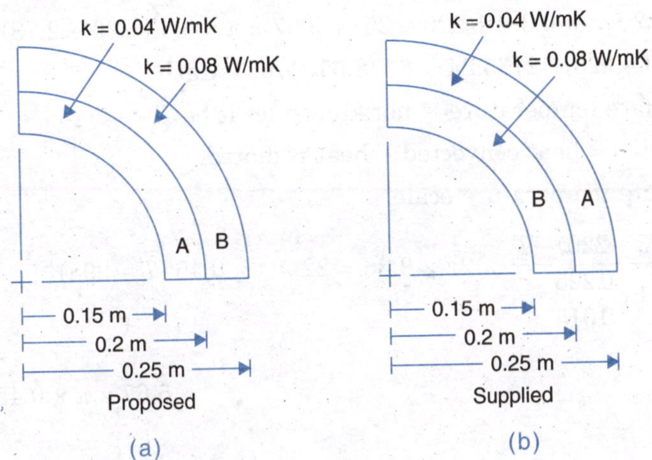

Fig. AE-2.18. Problem model

$$\text{Resistance} = \frac{1}{2\pi \times 0.08} \ln \frac{0.2}{0.15} + \frac{1}{2\pi \times 0.04} \ln \frac{0.25}{0.2}$$

$$= 0.572 + 0.888 = \mathbf{1.46°C/W}$$

There is a **reduction of 8.1% in the resistance.**

For a given temperature drop, the heat flow will increase as reciprocal of the resistance *i.e.*, as

$$\frac{1}{1.584} : \frac{1}{1.46} \quad \textit{i.e.,}\ 0.63 : 0.685 \ \textit{i.e.,}\ 1 : 1.087 \quad \text{or} \quad 8.7\%.$$

The results of this problem reinforces the comment in the previous problem.

Chapter 2

Example A-2.19: *A steel pipe of outside diameter 30 cm carries steam and its surface temperature is 220°C. It is exposed to surroundings at 25°C. Heat is lost both by convection and radiation. The convective heat transfer coefficient has a value of 22 W/m²K. Determine the heat loss per 1 m length. Check the economical merits of adding insulation pads of 7.5 cm thickness with thermal conductivity of 0.36 W/mK. The cost of heat is ₹200/- per 10^6 kJ. The cost of insulation is ₹8000/m length. The unit is in operation for 200 hr/year. The capital should be recovered in 2 years. After adding the insulation also the same convection and radiation prevail over the surface.*

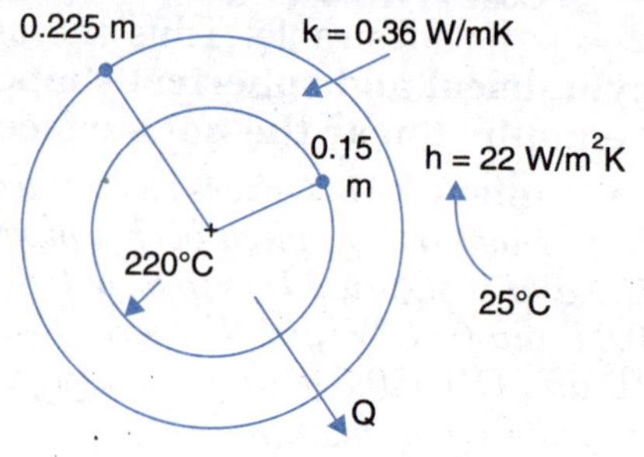

Fig. AE-2.19. Problem model

Solution: The data specified are shown in Fig. AE-2.19.

Case 1: Bare pipe. As both the pipe outside temperature and the surrounding temperature are known,

Heat lost over the surface Q = heat convected + heat radiated

$$Q = h\,A\,(\Delta T) + \sigma A(T_1^4 - T_2^4)$$

considering unit length

$$Q = 22 \times \pi \times 0.3 \times 1(220 - 25) + 5.67 \times \pi \times 0.3\,\{(220 + 2.73)^4 - (25 + 2.73)^4\}$$

$$= 4043.23 + 2735.34 = \mathbf{6778.57\ W/m\ length.}$$

Case 2: Surface temperature is not known let it be T.

Heat conducted = heat convected + heat radiated.

Using absolute temperature scale.

$$\frac{(273 + 220) - T}{\ln \dfrac{0.225}{0.015}} \times 2\pi \times 0.36 = 22 \times \pi \times 0.45\,(T - 298)$$

$$+ 5.67 \times \pi \times 0.45 \left[\left(\frac{T}{100}\right)^4 - \left(\frac{298}{100}\right)^4\right]$$

This reduces to $\left(\dfrac{T}{100}\right)^4 + 4.58T - 1578.2 = 0.$

Solving by trial, T = 321.3 K or 48.3°C.

Heat loss:

$$\frac{220 - 48.3}{\ln \dfrac{0.225}{0.15} / 2\pi \times 0.36} + 2\pi \times 0.225 \times 1 \times 5.67\,(3.213^4 - 2.98^4)$$

$$= 1179.9\ \text{W}$$

Saving = 6778.57 – 1179.9 = 5598.67 W

Cost of heat saved per year in ₹

$$= \frac{5\,598.67 \times 200 \times 3600}{10^9} \times 2000 = 8062$$

As the cost of insulation is ₹ 8000/m, the saving of 8062 per year makes it economical.

Example A-2.20: *A hollow cylinder has in internal diameter of 20 mm and thickness of 10 mm. The inner surface is at 500°C and the outside is at 100°C. Determine the temperature at 2 mm intervals and plot to scale.*

Solution: The data are shown in Fig. AE-2.20 (*a*).

Using equation 2.17

$$\frac{T_3 - T_1}{T_2 - T_1} = \frac{\ln \frac{0.012}{0.01}}{\ln \frac{0.02}{0.01}} \quad \therefore \quad \frac{T_3 - 500}{100 - 500} = 0.263 \quad \therefore \quad \mathbf{T_3 = 394.8°C}$$

$$\frac{T_4 - T_1}{T_2 - T_1} = \frac{\ln \frac{0.014}{0.01}}{\ln \frac{0.02}{0.01}} \quad \therefore \quad \frac{T_4 - 500}{100 - 500} = 0.485 \quad \therefore \quad \mathbf{T_4 = 305.8°C}$$

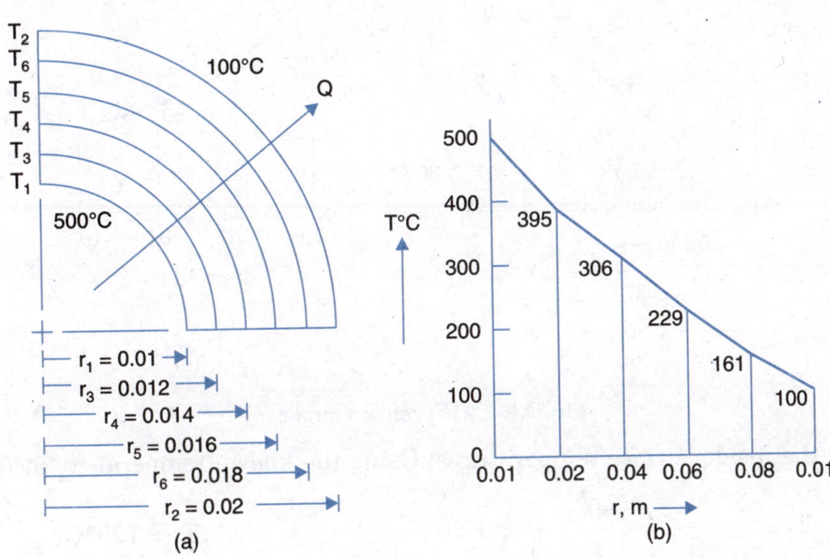

Fig. AE-2.20. Variation of temperature along the radius in hollow cylinder

$$\frac{T_5 - T_1}{T_2 - T_1} = \frac{\ln \frac{0.016}{0.01}}{\ln \frac{0.02}{0.01}} \quad \therefore \quad \frac{T_5 - 500}{100 - 500} = 0.678 \quad \therefore \quad \mathbf{T_5 = 228.8°C}$$

$$\frac{T_6 - T_1}{T_2 - T_1} = \frac{\ln \frac{0.018}{0.01}}{\ln \frac{0.02}{0.01}} \quad \therefore \quad \frac{T_6 - 500}{100 - 500} = 0.848 \quad \therefore \quad \mathbf{T_6 = 160.8°C}$$

The result is shown in Fig. AE-2.20(*b*). The variation is logarithmic.

Example A-2.21: *A hollow spherical form is used to determine the conductivity of materials. The inner diameter is 20 cm and the outer diameter is 50 cm. A 30 W heater is placed inside and under steady conditions, the temperatures at 15 and 20 cm radii were found to be 80 and 60°C. Determine the thermal conductivity of the material. Also find the outside temperature. If the surrounding is at 30°C, determine the convection heat transfer coefficient over the surface. Plot the temperature along the radius.*

Solution: The data are presented in Fig. AE-2.21(*a*).

Under steady conduction, the heat input by the heater has to pass through this layer between r = 0.15 m and 0.2 m

$$Q = \frac{\Delta T}{\frac{1}{4\pi k}\left(\frac{1}{r_1} - \frac{1}{r_2}\right)} = \frac{80 - 60}{\frac{1}{4 \times \pi \times k}\left(\frac{1}{0.15} - \frac{1}{0.20}\right)} = 30$$

$$\frac{20 \times \pi \times k \times 4}{\frac{1}{0.15} - \frac{1}{0.20}} = 30 \qquad \therefore \quad \mathbf{k = 0.199\ W/mK}$$

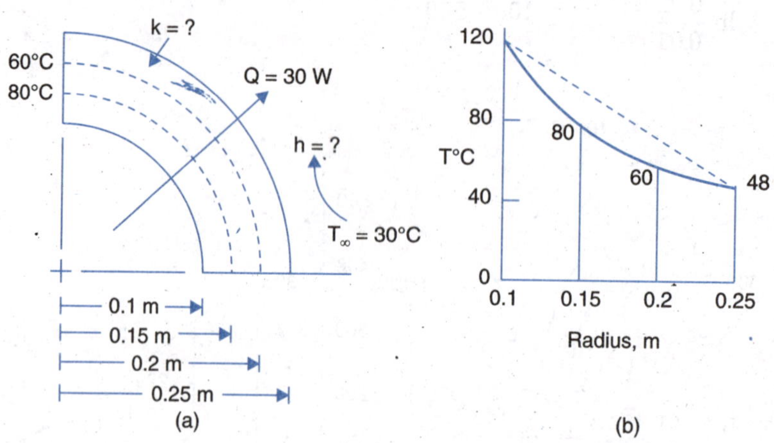

Fig. AE-2.21. Problem model

To find the inside surface temperatures: Using the known temperature at 0.2 m

$$Q = \frac{T_i - 60}{\frac{1}{4\pi \times 0.199}\left[\frac{1}{0.1} - \frac{1}{0.2}\right]} = 30 \qquad \therefore \quad \mathbf{T_i = 120°C}$$

For the outer surface Temperature

$$Q = \frac{120 - T_o}{\frac{1}{4\pi \times 0.199}\left[\frac{1}{0.1} - \frac{1}{0.25}\right]} = 30 \qquad \therefore \quad \mathbf{T_o = 48°C}$$

For the convection coefficient: $h \times 4\pi r_o^2 (48 - 30) = 30$

$r_o = 0.25$ $\qquad \therefore \quad \mathbf{h = 2.12\ W/m^2K}$

Check: $Q = \dfrac{60-48}{\dfrac{1}{4\pi \times 0.199}\left[\dfrac{1}{0.2} - \dfrac{1}{0.25}\right]}$ = 30 W checks.

The temperature plot is shown in Fig. AE-2.21(*b*). The slope is higher at the inside surface and lower at the outside surface. As the radius increases, heat flow area also increases. The thermal conductivity and heat flow are constant. So the slope (dT/dr) decreases.

Example A-2.22: *A spherical container holding a cryogenic fluid at – 140°C and having an outer diameter of 0.4 m is insulated with three layers each of 50 mm thick insulations of* $k_1 = 0.02$, $k_2 = 0.06$ *and* $k_3 = 0.16$ *W/mK (starting from inside). The outside is exposed to air at 30°C with* $h = 15$ *W/m²K. Determine the heat gain and the various surface temperatures.*

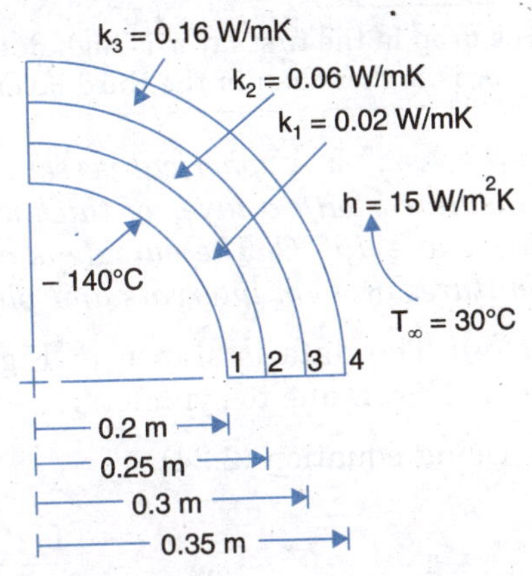

Fig. AE-2.22 Problem model

Solution: The data available are shown in Fig. AE-2.22.

$$\text{Heat flow} = \frac{\text{Temp. drop}}{\text{Thermal resistance}}$$

The resistance consist of convection resistance at the surface and conduction resistances of three layers

These are:

1st layer $= \dfrac{1}{4\pi \times 0.02}\left(\dfrac{1}{0.2} - \dfrac{1}{0.25}\right) = 3.979°\text{C/W}$

2nd layer $= \dfrac{1}{4\pi \times 0.06}\left(\dfrac{1}{0.25} - \dfrac{1}{0.3}\right) = 0.884°\text{C/W}$

3rd layer $= \dfrac{1}{4\pi \times 0.16}\left(\dfrac{1}{0.3} - \dfrac{1}{0.35}\right) = 0.237°\text{C/W}$

convection $= \dfrac{1}{4\pi \times 0.35^2 \times 15} = 0.043°\text{C/W}$

Total resistance $= 5.1432°\text{C/W}$

$$Q = \frac{30-(-140)}{R} = \frac{170}{5.143} = \mathbf{33\ W}\ (33.053)$$

To find the interface temperatures

1st layer $= \dfrac{T_2 - (-140)}{\dfrac{1}{4\pi \times 0.02}\left(\dfrac{1}{0.2} - \dfrac{1}{0.25}\right)} = 33.053$ $\quad$ $\mathbf{T_2 = -8.5°C}$

2nd layer $= \dfrac{T_3 - (-8.5)}{\dfrac{1}{4\pi \times 0.06}\left(\dfrac{1}{0.25} - \dfrac{1}{0.3}\right)} = 33.053$ $\quad$ $\mathbf{T_3 = 20.7°C}$

$$\textbf{3rd layer} = \frac{T_4 - 20.7}{\dfrac{1}{4\pi \times 0.16}\left(\dfrac{1}{0.3} - \dfrac{1}{0.35}\right)} = 33.053 \qquad \mathbf{T_2 = 28.53°C}$$

Check: $Q = 15 \times 4\pi \times 0.35^2\,(30 - 28.53) = 33.94$ W (using outside convection)

COMMENT

The drop in the first layer is high due to the lower thermal conductivity. The drop in the second layer is higher than in the third layer as its conductivity is lower than that of the third layer.

Example A-2.23: *A spherical vessel of outside diameter 0.6 m is insulated with a layer of thickness 0.16 m. The inside surface is at – 190°C. The outside is at 20°C. Determine the temperatures at 4 cm intervals and plot to scale.*

Solution: The data is shown in Fig. AE-2.23. A quarter section is shown due to symmetry.

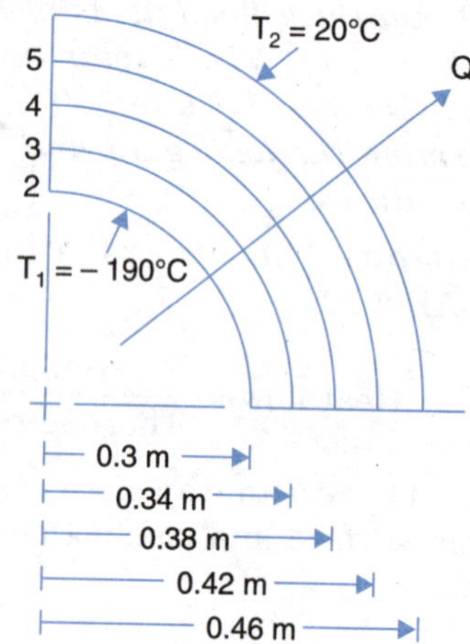

Fig. AE-2.23. (a) Problem model

Using equation (2.24)

$$\frac{T_1 - T}{T_1 - T_2} = \frac{\dfrac{1}{r_1} - \dfrac{1}{r}}{\dfrac{1}{r_1} - \dfrac{1}{r_2}}$$

∴ T at 0.34 radius, *i.e.*, T_3

$$\frac{T_1 - T_3}{T_1 - T_2} = \frac{-190 - T_3}{-190 - 20}$$

$$= \frac{-190 - T_3}{-210} = \frac{\dfrac{1}{0.3} - \dfrac{1}{0.34}}{\dfrac{1}{0.3} - \dfrac{1}{0.46}}$$

∴ $\mathbf{T_3 = -118.9°C}$

$$\frac{T_1 - T_4}{T_1 - T_2} = \frac{-190 - T_4}{-190 - 20}$$

$$= \frac{-190 - T_4}{-210} = \frac{\dfrac{1}{0.3} - \dfrac{1}{0.38}}{\dfrac{1}{0.3} - \dfrac{1}{0.46}}$$

∴ $\mathbf{T_4 = -62.89°C}$

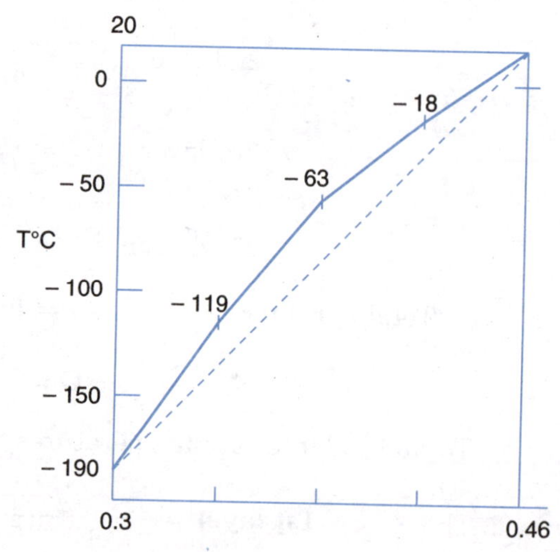

Fig. AE-2.23. (b) Radial temperature variation-sphere

$$\frac{T_1 - T_5}{T_1 - T_2} = \frac{-190 - T_4}{-190 - 20} = \frac{-190 - T_4}{-210} = \frac{\dfrac{1}{0.3} - \dfrac{1}{0.42}}{\dfrac{1}{0.3} - \dfrac{1}{0.46}}$$

∴ $T_5 = -17.5°C$

The plot is shown in Fig. AE-2.23(*b*). The remark of Example A-2.21 about the temperature gradient applies in this case also. The slope decreases as radius increases, increasing the flow area.

Example A-2.24: *A spherical vessel of 1 m outer dia. contains hot fluid at 240°C investigate the effect of adding 5 cm layers of insulation with k = 0.26 W/mK. The outside is exposed to air at 30°C with a convection coefficient of 18 W/m²K (including radiation). Compare the percentage change in heat flow and material volume added (m³).*

Solution: The data is presented in Fig. AE-2.24.

The tabulated quantities are calculated using:

$$\text{Conduction resistance} = \frac{1}{4\pi k}\left(\frac{1}{r_1} - \frac{1}{r_2}\right)$$

$$\text{Convection resistance} = \frac{1}{4\pi r_2^2 h}$$

$$Q = \frac{\Delta T}{R_1 + R_2},$$

$$\text{Volume added} = \frac{4}{3}\pi\,(r_2^3 - r_1^3)$$

$r_1 = 0.50$ m, $k = 0.26$ W/mK,

$h = 18$ W/m²K $\Delta T = 210°C$

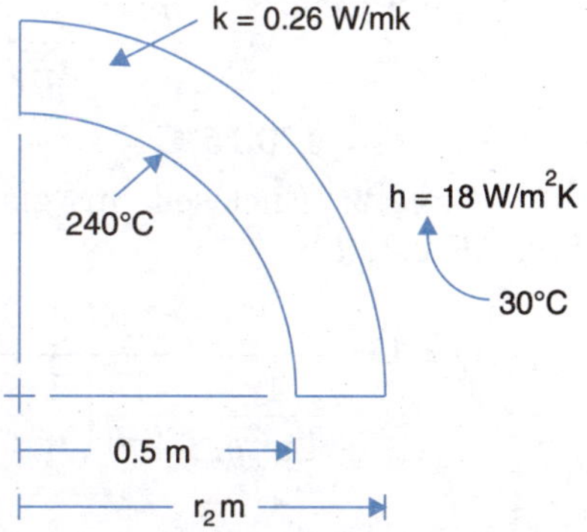

Fig. AE-2.24. Problem model

The results are tabulated below:

Outer radius (m)	*Conduction resistance °C/W*	*Convection resistance °C/W*	*Total resistance °C/W*	*Heat flow W*	*Total volume added m³*	*ΔQ/ΔV*
0.50	—	0.017684	0.017684	11875		
0.55	0.05563	0.014615	0.070265	2989	0.1733	51275
0.60	0.10202	0.012280	0.11430	1837	0.3812	5541
0.65	0.14126	0.010463	0.15172	1384	0.6268	1845
0.70	0.17490	0.009022	0.18392	1142	0.9132	845
0.75	0.20405	0.007860	0.21191	991	1.2435	457
0.80	0.22955	0.006908	0.23646	888	1.6211	273

This type of calculations and comparisons can be done effortlessly if computers are used. The results can be also graphically presented.

COMMENT

The values of ΔQ/ΔV provides a good guidance for selection of material thickness. In this case it can be observed that the first and second layers are more effective than others. So it will be economical to stop with two layers. This ratio can also be converted to return/investment ratio if costs of energy and material are known.

Chapter 2

Example A-2.25: *A spherical electronic device of 10 mm dia. generates 1 W. It is exposed to air at 20°C with a convection coefficient of 20 W/m²K. Find the surface temperature. The heat transfer consultant advices to enclose it in a glass like material of k = 1.4 W/mK, to a thickness of 5 mm all around to reduce the temperature. Investigate the problem and also find the thickness to obtain 50°C surface temperature.*

Solution. Case 1: Bare device:

The available information is presented in [Fig. AE-2.25(*a*)]

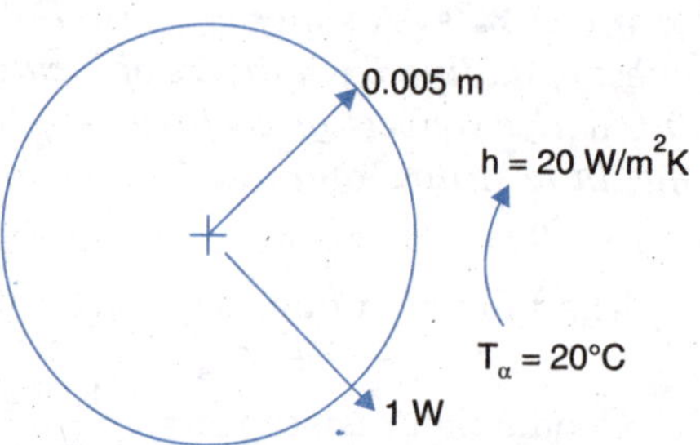

Fig. AE-2.25(a)

$$Q = 1 \quad Q = \frac{\Delta T}{R} = \frac{T_s - 20}{\dfrac{1}{4\pi \times 0.005^2 \times 20}} = 1$$

$\therefore$ $\mathbf{T_s = 179.15°C}$

Case 2: Enclosed in glass like material [Fig. AE-2.25(*b*)].

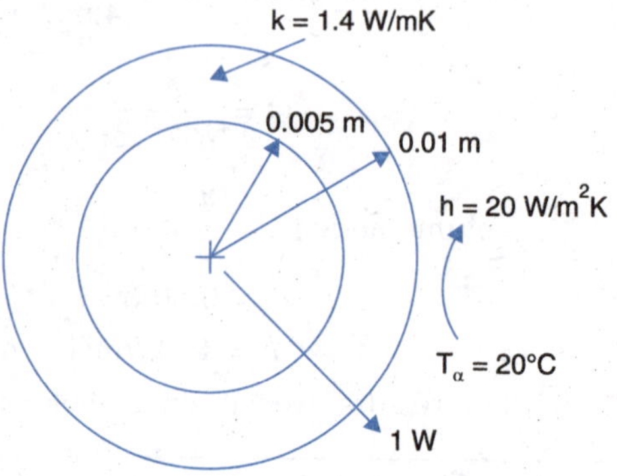

Fig. AE-2.25(b)

$$Q = 1 = \frac{\Delta T}{\dfrac{1}{4\pi r_2^2 h} + \dfrac{1}{4\pi k}\left(\dfrac{1}{r_1} - \dfrac{1}{r_2}\right)}$$

$$\Delta T = \left\{\frac{1}{4\pi r_2^2 h} + \frac{1}{4\pi k}\left(\frac{1}{r_1} - \frac{1}{r_2}\right)\right\} \times Q$$

$$= \left\{\frac{1}{4\pi \times 0.01^2 \times 20} + \frac{1}{4\pi \times 1.4}\left(\frac{1}{0.05} - \frac{1}{0.1}\right)\right\} \times 1$$

$= 40.36$

$\therefore$ $T_s = (20 + 40.36) = \mathbf{60.36°C}$. The temperature is reduced considerably.

Case 3: The problem reduces to finding r_2 such that $\Delta T = 30$°C

$$\Delta \text{T} = 30 = -\frac{1}{4\pi r_2^2 h} + \frac{1}{4\pi k}\left(\frac{1}{0.005} - \frac{1}{r_2}\right) \text{ as } Q = 1$$

$$307 = \frac{1}{20 r_2^2} + \frac{1}{1.4} \times \frac{1}{0.005} - \frac{1}{1.4 r_2}$$

$$\frac{1}{20r^2} - \frac{1}{1.4r} - 234.14 = 0$$

this can be solved exactly or by trial: $r_2 = 0.01315$ or a thickness of 8.15 mm. Such a reduction will go on up to $\mathbf{r_2} = \frac{2k}{h} = \frac{2.8}{20} = \mathbf{0.14\ m}$ or 140 mm and the temperature corresponding to this is 31.2°C. The problem comes under the topic critical radius of insulation in the case of spheres.

Note: Usual method to reduce temperature is to increase *h* or increase surface area.

Example A-2.26: *The surface of a spherical container with 0.4 m outer diameter is at –195°C. Two layers of insulation each of 2.5 cm thickness is added. The thermal conductivities of the materials are 0.004 and 0.03 W/mK. The contact resistances are each 5×10^{-4} m² °CW. The outside is exposed to air at 30°C with a convection coefficient of 16 W/m²K. Determine the heat gain and the temperatures at various surfaces and also the drops due to contact resistance.*

Solution: The data are presented in Fig. AE-2.26. A quarter section is shown due to symmetry.

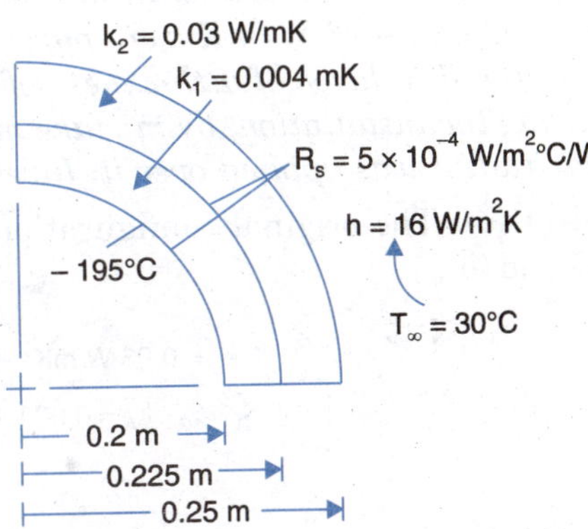

Fig. AE-2.26. Problem model

$$\text{Heat flow} = \frac{\text{Temp. drop}}{\text{Total thermal resistance}}$$

The resistances are:

1. Contact resistance (to be calculated for the area)

$$R_1 = \frac{5\times10^{-4}}{4\pi r_1^2} = \frac{5\times10^{-4}}{4\pi\times0.2^2}\ ^\circ\text{C/W} = 9.95\times10^{-4}\ ^\circ\text{C/W}$$

2. Conduction resistance of first layer

$$R_2 = \frac{1}{4\pi k}\left(\frac{1}{r_1}-\frac{1}{r_2}\right) = \frac{1}{4\pi\times0.04}\left(\frac{1}{0.2}-\frac{1}{0.225}\right) = 11.05^\circ\text{C/W}$$

3. Contact resistance $= R_3 = \dfrac{5\times10^{-4}}{4\pi\times0.225^2} = 7.86\times10^{-4}\ ^\circ\text{C/W}$

4. Conduction resistance of second layer $= R_4$

$$= \frac{1}{4\pi\times0.03}\left(\frac{1}{0.225}-\frac{1}{0.25}\right) = 1.179^\circ\text{C/W}$$

5. Convection resistance $= R_5 = \dfrac{1}{4\pi r_0^2 h} = \dfrac{1}{4\pi\, 0.25^2\times16} = 0.0796^\circ\text{C/W}$

Total resistance $= 12.3127^\circ$C/W

$$Q = \frac{\Delta T}{R} = \frac{30-(-195)}{12.3127} = \mathbf{18.27\ W}$$

To find temperatures:

Drop in the first contact	$= Q{\cdot}R_1 = 0.0182^\circ$C	∴ **T_1 = – 194.98°C**
Drop in the first layer	$= Q{\cdot}R_2 = 201.97^\circ$C	∴ **T_2 = 6.99°C**
Drop in the contact	$= Q{\cdot}R_3 = 0.0144^\circ$C	∴ **T_3 = 7.00°C**
Drop in the second layer	$= Q{\cdot}R_4 = 21.55^\circ$C	∴ **T_4 = 28.55°C**
Drop in the convection	$= Q{\cdot}R_5 = 1.46^\circ$C	∴ **T_5 = 30.01°C**

Total = 225.01°C checks

COMMENT

When other resistances are large, the effect of contact resistances become negligible. Neglecting contact resistances in this case, Q = 18.28 W

Example A-2.27: *Two insulating materials A and B in powder form with thermal conductivities of 0.004 and 0.03 W/mK were purchased for use over a sphere of 0.4 m dia. Material A is to form the first layer of a thickness of 4 cm and B is to form the next layer to 5 cm thickness. During the installation, by mistake material B was applied first using up all the material and material. A was applied over it. Investigate whether the thermal resistance will change.*

Solution: The original configuration and modified configuration are shown in Fig. AE-2.27 (*a*) and (*b*).

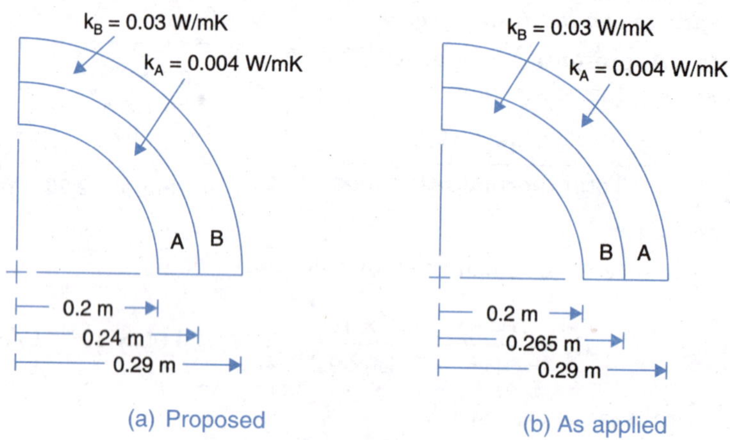

Fig. AE-2.27. Problem model

When the materials are interchanged, there is a change in the radius:

$$\text{Volume of } A = \frac{4\pi}{3}(0.24^3 - 0.2^3) = 0.0243955 \text{ m}^3$$

$$\text{Volume of } B = \frac{4\pi}{3}(0.29^3 - 0.24^3) = 0.044255 \text{ m}^3$$

the new radius is found by, $0.044255 = \frac{4}{3}\pi(r_2^3 - 0.2^3)$ $\therefore$ $r_2 = 0.2648$ m, $r_3 = 0.29$ m

Case 1: $$\text{Resistance} = \frac{1}{4\pi \times 0.004}\left[\frac{1}{0.2} - \frac{1}{0.24}\right] + \frac{1}{4\pi \times 0.08}\left[\frac{1}{0.24} - \frac{1}{0.29}\right]$$

$$= 16.57 + 1.91 = \mathbf{18.48°C/W}$$

Case 2: $$\text{Resistance} = \frac{1}{4\pi \times 0.03}\left[\frac{1}{0.2} - \frac{1}{0.2648}\right] + \frac{1}{4\pi \times 0.004}\left[\frac{1}{0.2648} - \frac{1}{0.29}\right]$$

$$= 3.25 + 6.53 = \mathbf{9.78°C/W}$$

Heat flow will almost double.

COMMENT

The remarks of Example A-2.17 applies here also. The material with lower conductivity should be applied near the hot surface.

Example A-2.28: *The thermal conductivities of some materials at 127°C and 527°C are tabulated. Assuming* ***linear variation of thermal conductivity*** *with temperature. Work out the values of k_o and β, constants for the material.* **k = k$_o$ (1 + βT) is assumed with T in °C.**

Material	*k at 127°C, W/mK*	*k at 527°C, W/mK*
Aluminium	240.0	218.0
Chromium	90.9	71.3
Copper	393.0	366.0
Carbon Steel	56.7	39.2
Cr. Steel	42.0	34.5
Stainless Steel	16.6	22.6
Uranium	29.6	38.8
Carbon	1.89	2.37

Solution: The values are worked out using (taking 0°C as base)

$$k_o = k_{127} - \frac{127}{(527-127)}(k_{527} - k_{127})$$

$$\beta = \frac{1}{(527-127)} \cdot \frac{1}{k_o}(k_{527} - k_{127})$$

For Aluminium: $\mathbf{k_o} = 240 - \frac{127}{400}(218 - 240) = \mathbf{246.985\ W/mK}$

$$\beta = \frac{1}{400} \times \frac{1}{246.985}(218 - 240) = -\mathbf{2.227 \times 10^{-4}}$$

The values of k_o and β are shown in the tabulation.

Material	*k_o, W/mK*	*β × 10⁴ (1/K)*
Aluminium	246.985	– 2.227
Chromium	97.123	– 5.045
Copper	401.5725	– 1.681
Carbon Steel	62.256	– 7.027
Cr. Steel	44.380	– 4.225
Stainless Steel	14.695	+ 10.208
Uranium	26.679	+ 8.621
Carbon	1.7376	+ 6.906

Check: For Aluminium $k_{527} = 246.985\,(1 - 2.227 \times 10^{-4} \times 527)$

$= 217.998$ or 218 checks

For Uranium $k_{527} = 26.679\,(1 + 8.621 \times 10^{-4} \times 527)$

$= 38.799998$ or 38.8 checks.

Chapter 2

Example A-2.29: *The values of thermal conductivities of some insulating materials at 300°C 100°C are tabulated. Determine the values of k_o and β in the equation $k = k_o (1 + \beta T)$ for the variation of thermal conductivity. Where k is the thermal conductivity at T°C and k_o is the thermal conductivity at 0°C and β, is a material constant.*

Material	*k at 100°C, W/mK*	*k at 300°C, W/mK*
Fire clay brick	0.76	0.895
Slag brick	0.68	0.77
Red brick	0.56	0.66
Diatomaceous earth brick	0.138	0.176
Sovelite	0.092	0.12
85% Magnesia	0.08	0.101
Slag wool	0.07	0.101
Mineral wool	0.042	0.070

Solution: The value of k_o is obtained using the given data. The results are shown in the tabulation below.

$$k_o = k_{100} - \left[\frac{k_{300} - k_{100}}{300 - 100}\right] 100$$

$$\beta = \frac{1}{200} \cdot \frac{1}{k_o} (k_{300} - k_{100})$$

For example, For fire clay brick:

$$k_o = 0.76 - \left[\frac{0.895 - 0.76}{200}\right] \times 100 = 0.6925$$

$$\beta = 9.747 \times 10^{-4}.$$

Material	*k_o, W/mK*	*β × 10^4 (1/K)*
Fire clay brick	0.6925	9.747
Slag brick	0.635	7.087
Red brick	0.51	9.804
Diatomaceous earth brick	0.119	15.966
Sovelite	0.078	17.959
85% Magnesia	0.0695	15.108
Slag wool	0.0545	28.440
Mineral wool	0.028	50.000

Check: 85% Magnesia: $k_{300} = 0.0695 (1 + 15.108 \times 10^{-4} \times 300) = 0.101.$

Example A-2.30: *A furnace wall insulation is of fireclay brick with thermal conductivity $k = 0.6925 (1 + 9.747 \times 10^{-4} T)$ where T is in °C and k is in W/mK. The wall is 30 cm thick. The inside surface is at 500°C while the outside surface is at 70°C. Determine the heat flow and also the temperatures at the mid section and at 7.5 cm from the surfaces. Also find the value of the slopes at the surfaces.*

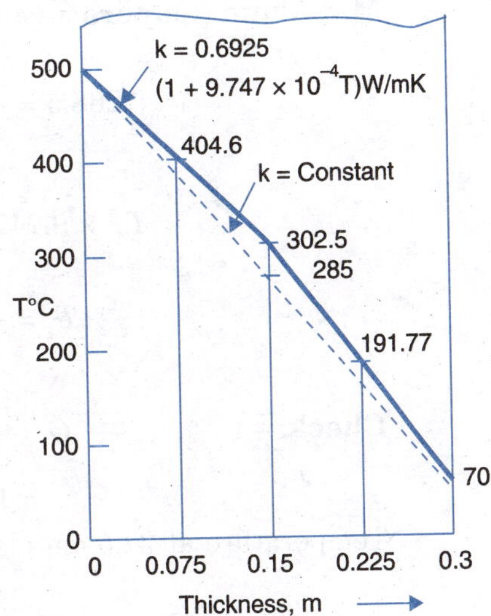

Fig. AE-2.30. Temperature variation in wall

Solution: The results are presented in Fig. AE-2.30 considering unit area: Heat flow

$$Q = \frac{\Delta T}{L / k_m}$$

$$k_m = 0.6925 \left[1 + 9.747 \times 10^{-4} \left(\frac{500 + 70}{2}\right)\right]$$

$$= \mathbf{0.885\ W/mK.}$$

$$Q = \frac{500 - 70}{0.3 / 0.885} = \mathbf{1268.5\ W/m^2}$$

The slope at the hot side:

$$Q = -kA\frac{dT}{dx},$$

$$k \text{ at } 500°\text{C} = 0.6925\ [1 + 9.747 \times 10^{-4} \times 500)$$

$$= 1.03 \text{ W/mK}$$

$$\frac{dT}{dx} = -\frac{Q}{1.03} = \frac{-1268.5}{1.03}$$

$$= \mathbf{-\ 1231.6°C/m}$$

Slope at the cold side: $k = 0.6925\ (1 + 9.747 \times 10^{-4} \times 70) = 0.74$ W/mK

Slope $$\frac{dT}{dx} = \frac{-1268.5}{0.74} = \mathbf{-\ 1714.2°C/m\ (steeper)}$$

Temperature at 7.5 cm:

$$Q = \frac{(500 - T_3)}{0.075} \times k_m = 1268.5 \text{ W}$$

$$= \frac{(500 - T_3) \times 0.6925 \left[1 + 9.747 \times 10^{-4}\, \frac{500 + T_3}{2}\right]}{0.075} = 1268.5$$

$$0.6925\ (500 - T_3) \left\{1 + 9.747 \times 10^{-4} \left(\frac{500 + T_3}{2}\right)\right\} = 1268.5 \times 0.075$$

$$(500 - T_3) \left\{1 + 9.747 \times 10^{-4} \left(\frac{500 + T_3}{2}\right)\right\} = 137.4$$

This can solved as a quadratic equation or by trial.

Simplifying $$2 \times 484.44 - 2T_3 - 9.747 \times 10^{-4}\, T_3^2 = 0$$

$$T_3 = \frac{-2 + \sqrt{(2^2 + 4 \times 2 \times 484.44 \times 9.747 \times 10^{-4})}}{2 \times 9.747 \times 10^{-4}} = \mathbf{404.64°C}$$

Check: $$Q = \frac{500 - 404.64}{0.075} \times 0.6925 \left\{1 + \frac{500 + 404.61}{2} \times 9.747 \times 10^{-4}\right\}$$

$$= \mathbf{1267.8\ W}$$

Mid plane temperature: (at 15 cm)

$$1268.5 = \left[\frac{500 - T_4}{0.15}\right] \times 0.6925 \left\{1 + \left(\frac{500 + T_4}{2}\right) 9.747 \times 10^{-4}\right\}$$

or,

$$\therefore \quad T_4^2 \times 9.747 \times 10^{-4} + 2T_4 - 694.14 = 0$$

$$\therefore \quad T_4 = \frac{-2 + \sqrt{(4 + 4 \times 694.14 \times 9.747 \times 10^{-4})}}{2 \times 9.747 \times 10^{-4}} = \mathbf{302.48°C.}$$

Check:

$$Q = \left[\frac{500 - 302.48}{0.15}\right] \times 0.6925 \left\{1 + \frac{500 + 302.48}{2} \times 9.747 \times 10^{-4}\right\}$$

$$= \mathbf{1268.51\ W}$$

Temperature at 22.5 cm plane:

$$1268.5 = \left[\frac{500 - T_5}{0.225}\right] \times 0.6925 \left\{1 + \left(\frac{500 + T_5}{2}\right) 9.747 \times 10^{-4}\right\}$$

This reduces to $T_5^2 \times 9.747 \times 10^{-4} + 2T_5 - 419.38 = 0$

$$\therefore \quad T_5 = \frac{-2 + \sqrt{(4 + 4 \times 419.38 \times 9.747 \times 10)}}{2 \times 9.747 \times 10^{-4}} = \mathbf{191.77°C}$$

Check:

$$Q = \left[\frac{500 - 191.77}{0.225}\right] . 0.6925 \left\{1 + \frac{500 + 191.77}{2} \times 9.747 \times 10^{-4}\right\}$$

$$= \mathbf{1268.49\ W}$$

slope at the other planes: ($x_3 = 0.075$, $x_4 = 0.15$, $x_5 = 0.225$). Using the temperature calculated, are – 1313°C/m, – 1414°C/m, – 1543.2°C/m. It can be seen that the slopes increase with distance.

COMMENT

The temperature curve is above the straight line. This is because higher the temperature higher the value of k and smaller the gradient for the same heat flow.

Example A-2.31: *A steel slab 16 cm thickness, has thermal conductivity $k = 51\ (1 - 8.6 \times 10^{-4}\ T)$ where T is in °C and k is the thermal conductivity at T°C in W/mK. The slab surfaces are maintained at 400 and 180°C. Determine the heat flow, the temperatures at 4, 8 and 12 cm distances and also the temperature gradients at these locations.*

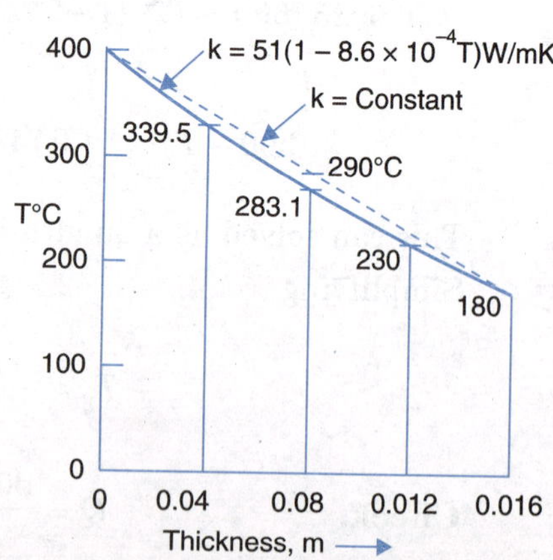

Fig. AE-2.31. Problem model

Solution: The given data are presented in Fig. AE-2.31.

Assuming unit area

$$q = \frac{\Delta T}{L} k_m$$

$$k_m = 51\left(1 - 8.6 \times 10^{-4} \times \frac{400 + 180}{2}\right)$$

$$= 38.28 \text{ W/mK}$$

$$\boldsymbol{Q} = \frac{400 - 180}{0.16} \times 38.28 = \mathbf{52635\ W/m^2}$$

Temperatures : Plane 3 : 0.04 m

$$Q = 52635 = \left[\frac{400 - T_3}{0.04}\right].\ 51\left\{1 - 8.6 \times 10^{-4}\left(\frac{400 + T_3}{2}\right)\right\}$$

This reduces to $8.6 \times 10^{-4}\, T_3^2 - 2T_3 + 579.84 = 0$

$$T_3 = \frac{+2 - \sqrt{(4 - 4 \times 8.6 \times 10^{-4} \times 579.84)}}{2 \times 8.6 \times 10^{-4}} = \mathbf{339.5°C}$$

Check: $Q = \frac{400 - 339.5}{0.04} \times 51\left\{1 - 8.6 \times 10^{-4} \times \frac{400 + 339.5}{2}\right\}$ = 52609 W checks.

plane 4 : 0.08 m

$$Q = 52635 = \frac{400 - T_4}{0.08} \times 51\left\{1 - 8.6 \times 10^{-4}\,\frac{400}{2} - 8.6 \times 10^{-4}\,\frac{T_4}{2}\right\}$$

This reduces to $8.6 \times 10^{-4}\, T_4^2 - 2T_4 + 497.28 = 0$

$$T_4 = \frac{+2 - \sqrt{(4 - 4 \times 8.6 \times 10^{-4} \times 497.28)}}{2 \times 8.6 \times 10^{-4}} = \mathbf{283.1°C}$$

Check: $Q = \frac{400 - 283.1}{0.08} \times 51\left\{1 - 8.6 \times 10^{-4} \times \frac{400 + 283.1}{2}\right\}$ = 52634 checks

plane 5 : 0.12 m $Q = 52635 = \frac{400 - T_5}{0.12}.\,51\left\{1 - 8.6 \times 10^{-4} \times \frac{400}{2} - 8.6 \times 10^{-4}\,\frac{T_5}{2}\right\}$

This reduces to $8.6 \times 10^{-4}\, T_5^2 - 2T_5 + 414.7 = 0$

$$T_5 = \frac{+2 - \sqrt{(4 - 4 \times 8.6 \times 10^{-4} \times 414.7)}}{2 \times 8.6 \times 10^{-4}} = \mathbf{230.12°C}$$

Slopes are found using $\frac{dT}{dx} = -\frac{Q}{k}$ as $A = 1$

The mean values of k are listed below for various sections.

$k_1 = 33.456$, $k_3 = 36.1$, $k_4 = 38.6$, $k_5 = 40.9$, $k_2 = 43.1$.

Slopes = – 1573.3°C/m, – 1458°C/m, – 1364.2°C/m, – 1286°C/m, – 1221°C/m

The slope is higher at higher temperature levels. This is the case where k decreases along x.

Chapter 2

Example A-2.32: *A steam pipe of 20 cm OD carrying steam at 260°C is insulated with a material having thermal conductivity $k = 0.07\,(1 + 15 \times 10^{-4}\,T)$, where k is the thermal conductivity in W/mK at temperature T°C. The outer surface is at 60°C. Determine the heat flow and also the temperature at mid thickness. The insulation thickness is 6 cm. Also find the slopes.*

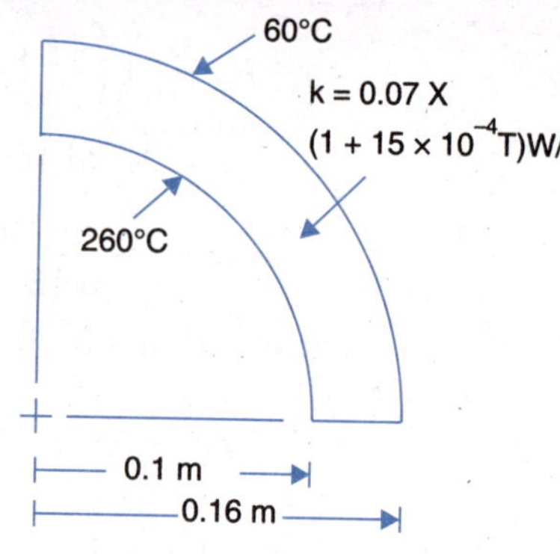

Fig. AE-2.32. Problem model

Solution: The data are presented in Fig. AE-2.32.

$$Q = \frac{\Delta T}{\ln \frac{r_2}{r_1} / 2\pi k_m l} \quad \text{let } l = 1$$

$$k_m = 0.07 \left\{1 + 15 \times 10^{-4} \left(\frac{260 + 60}{2}\right)\right\} = 0.0868$$

$$Q = \frac{260 - 60}{\ln \frac{0.16}{0.1} / 2\pi \times 0.0868} = \mathbf{232.08\ W/m}$$

To find mid plane Temperatures:

$$232.08 = \frac{260 - T}{\ln \frac{0.13}{0.1} / 2\pi \times 0.07 \left[1 + 15 \times 10^{-4} \left(\frac{260 + T}{2}\right)\right]}$$

$$138.44 = (260 - T)\left(1 + 15 \times 10^{-4} \frac{260}{2} + 15 \times 10^{-4} \frac{T}{2}\right)$$

$$= 260 + \frac{260^2 \times 15 \times 10^{-4}}{2} + 260 \times 15 \times 10^{-4} \frac{T}{2} - T$$

$$- 15 \times 10^{-4} \times \frac{260}{2} T - 15 \times 10^{-4} \frac{T^2}{2}$$

or $$15 \times 10^{-4}\, T^2 + 2T - 344.52 = 0$$

$$T = \frac{-2 + \sqrt{(4 + 4 \times 15 \times 10^{-4} \times 344.52)}}{2 \times 15 \times 10^{-4}} = \mathbf{154.38°C}$$

Check:

$$Q = \left[\frac{260 - 154.38}{\ln \frac{0.13}{0.1}}\right] \times 2\pi \times 0.07 \left(1 + 15 \times 10^{-4} \times \frac{260 - 154.38}{2}\right)$$

$= 232.08$ W, checks

Slopes $$\frac{dT}{dr} = -\frac{Q}{kA}$$

Inside surface: $k = 0.07\,(1 + 15 \times 10^{-4} \times 260) = 0.0973$ W/mK

$A = 2\pi r_1 = 2\pi \times 0.1$

$$\therefore \quad \left.\frac{dT}{dr}\right|_{\text{inside}} = \frac{-232.08}{0.0973 \times 2\pi \times 0.1} = -\mathbf{3796.2°C/m}$$

Mid plane: $k = 0.07\,(1 + 15 \times 10^{-4} \times 154.38) = 0.08621$ W/mK

$$\therefore \quad \frac{dT}{dr} = \frac{-232.08}{0.08621 \times 2\pi \times 0.13} = -\mathbf{3295.8°C/m}$$

Outer surface: $k = 0.07\,(1 + 15 \times 10^{-4} \times 60) = 0.0763$ W/mK

$$\therefore \quad \frac{dT}{dr} = \frac{-232.08}{0.0763 \times 2\pi \times 0.16} = -\mathbf{3025.6°C/m.}$$

COMMENT

Along the radius the thermal conductivity decreases, but area increases. The net effect is that slope decreases. In case heat flow is calculated on area basis, then the slope should increase.

Example A-2.33: *A spherical container with OD 0.4 m and surface temperature of –180°C is insulated by 8 cm thick layer of material with thermal conductivity k = 0.028 (1 + 5 × 10⁻³ T) W/mK where T is in °C. If the outside surface is at 15°C, determine the heat flow in.*

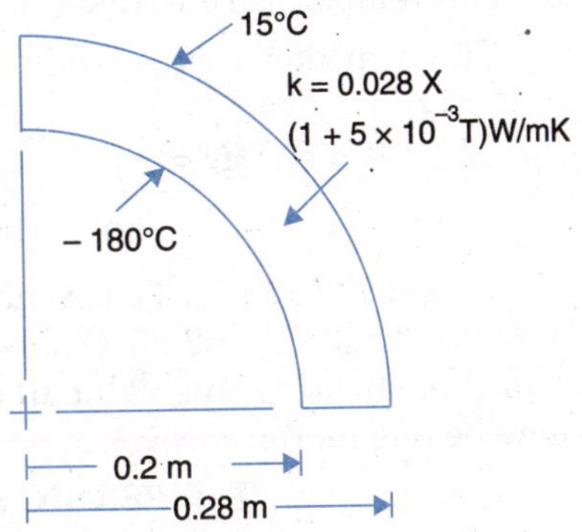

Fig. AE-2.33. Problem model

Solution: The data given are presented in Fig. AE-2.33.

$$Q = \frac{\Delta T}{\dfrac{1}{4\pi k_m}\left(\dfrac{1}{r_1} - \dfrac{1}{r_2}\right)}$$

$$k_m = 0.028\left[1 + 5 \times 10^{-3}\left(\frac{-180 + 15}{2}\right)\right]$$

$$= 0.01645 \text{ W/mK}$$

$$Q = \frac{15 - (-180)}{\dfrac{1}{4\pi \times 0.01645}\left(\dfrac{1}{0.2} - \dfrac{1}{0.28}\right)} = \mathbf{28.22\ W}$$

COMMENT

If the temperature effect on thermal conductivity is not considered, the value of *k* will be 0.028 W/mK. In that case, the heat flow will be 48 W, higher by 70%. Hence the necessity to consider variation of thermal conductivity with temperature.

Example A-2.34: *A wall of 20 cm thickness put up as insulation in an oven has its thermal conductivity given by $k = 0.5\ (1 + 10^{-3}\ T)$ where k is the thermal conductivity at T°C in W/mK. One side of the slab is exposed to convection at 300°C with h = 40 W/m²K and the otherside is exposed to air at 30°C with h = 20 W/m²K. Determine the heat flow through the wall per unit area.*

Solution: The data given are presented in Fig. AE-2.34.

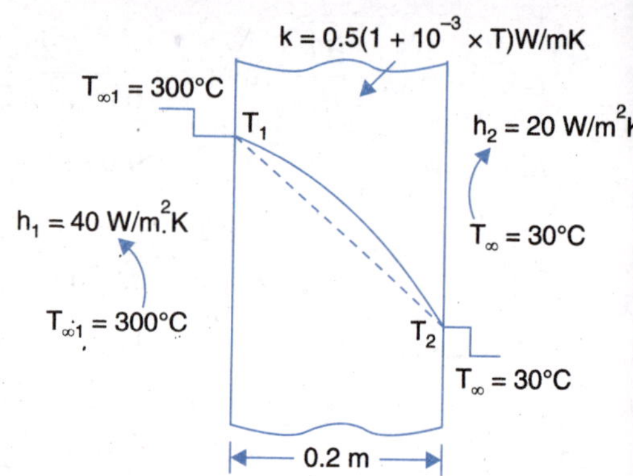

Fig. AE-2.34. Problem model

$$Q = \frac{\Delta T}{\dfrac{1}{h_1} + \dfrac{L}{k_m} + \dfrac{1}{h_2}}$$

$k_m = 0.5\ \{1 + 10^{-3}\ (T_1 + T_2)/2\}$ as T_1 and T_2 are unknown, a trial solution may be attempted. The mean temperature is taken as (330/2), and k_m = 0.5825 W/mK.

The heat flow corresponding to this value is

$$Q = \frac{300 - 30}{\dfrac{1}{40} + \dfrac{0.2}{0.5825} + \dfrac{1}{20}} = \mathbf{645\ W/m^2.}$$

T_1 and T_2 can now be estimated using $Q = h_1\ (300 - T_1)$ and $Q = h_2\ (T_2 - 30)$. This gives T_1 = 283.9°C and T_2 = 62.25°C now estimating $k_m = 0.5\ \{1 + 10^{-3}\ (283.9 + 62.25)/2\} = 0.5865$ W/mK and substituting this value in the expression for Q, Q = 649.06 W/m². Using a second trial or may be one more

T_1 = 283.8°C and T_2 = 62.4°C and Q = 649 W/m²

In this case the first guess is very near the true value, but it may not be always so.

COMMENT

For multilayer or composite walls with convection boundary with variable thermal conductivity, direct solution for heat flow will not be possible. Trial solution becomes necessary. This is not difficult with the use of calculators and is very simple with the use of computers.

Example A-2.35: *The sectional area of cone like solid varies as per the law, $A(m^2) = \dfrac{\pi \times 0.5^2}{4}$ x (m²). The solid between sections $x_1 = 0.025$ and $x_2 = 0.125$ conducts heat along the x direction having its outside surface well insulated. The surface at x_1 is maintained at 600°C while the surface at x_2 is at 400°C, k = 210 W/mK. Determine the heat flow, the temperatures and temperature gradients along the length and plot the same to scale.*

Solution: In the integration of Fourier's equation, the variation of the area should be taken into account. In this case assume

$A = Cx$ where $C = 0.19635$ *i.e.*, $(\pi \times 0.5^2/4)$

For derivation C can be used in place of the numerical value

$$Q = -kA\frac{dT}{dx} = -k \cdot Cx\frac{dT}{dx}$$

$$Q \cdot \frac{dx}{x} = -kC \cdot dT$$

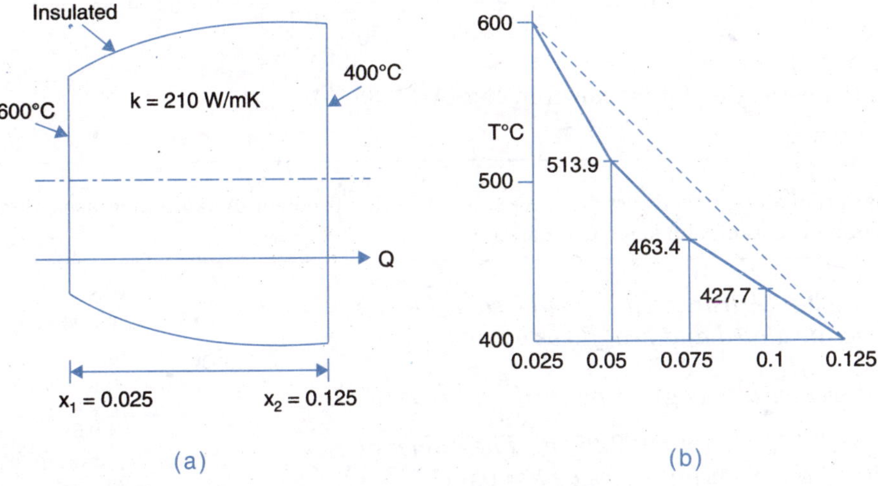

Fig. AE-2.35. Problem model

Integrating and simplifying

$$Q = \frac{kC(T_1 - T_2)}{\ln \frac{x_2}{x_1}} = \frac{210\pi \times 0.5^2}{4} \frac{(600 - 400)}{\ln \frac{0.125}{0.025}} = \mathbf{5124\ W}$$

$$\left.\frac{dT}{dx}\right| x = 0.025 = -\frac{Q}{kA} = \frac{-5124 \times 4}{210 \times \pi \times 0.5^2 \times 0.025} = -4970.72°\text{C/m}$$

$$\left.\frac{dT}{dx}\right| x = 0.125 = \frac{-5124 \times 4}{210 \times \pi \times 0.5^2 \times 0.125} = -994.15°\text{C/m}$$

for other values of x similar calculation is made

$$\left.\frac{dT}{dx}\right| x = 0.05,\ 0.075,\ 0.10 \text{ are: } -2485.36°\text{C/m},\ -1656.9°\text{C/m and } -1242.68°\text{C/m}$$

To determine the temperatures, use

$$Q = \frac{kC(T_1 - T_2)}{\ln \frac{x_2}{x_1}}, \text{ at } 0.05,\ 5124 = \frac{210 \times \pi \times 0.5^2}{4} \times \frac{(600 - T_2)}{\ln \frac{0.05}{0.025}}$$

∴ $\mathbf{T_2 = 513.86°C}$

at 0.075 $\mathbf{T_3 = 463.48°C}$

at 0.10 $\mathbf{T_4 = 427.73°C}$

Chapter 2

To check: Find Q between 0.075 and 0.10

$$Q = \frac{210 \times \pi \times 0.5^2}{4}\left[\frac{463.48 - 427.73}{\ln\left[\frac{0.1}{0.075}\right]}\right] = 5124 \text{ W}$$

The temperature plot is shown in Fig. AE-2.35 (*b*).

COMMENT

Area variation along flow direction is considered in this problem. As area increases along x, with k remaining constant, the slope decreases along x.

Example A-2.36: *A truncated cone like solid has its circumference insulated and heat flows along the axis.*

The area of section at x is given by $A = \frac{\pi}{4} x^3$ *and the faces are at 0.075 m and 0.225 m. The thermal conductivity of the material varies as* $k = 0.5\ (1 + 5 \times 10^{-3}\ T)$ *W/mK. The surface at x = 0.075 m is at 300°C and the surface at x = 0.225 is at 50°C. Determine the heat flow.*

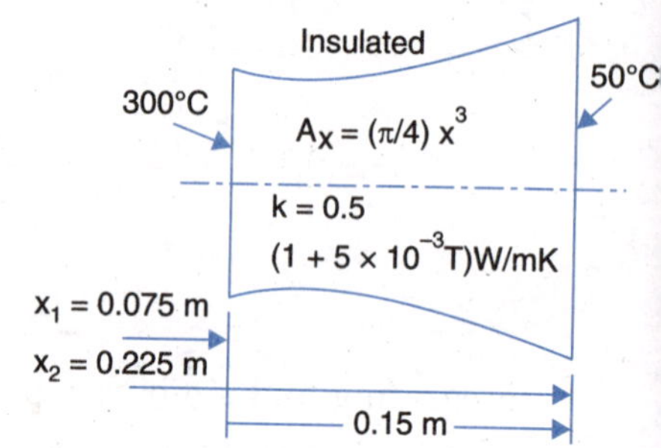

Fig. AE-2.36. Problem model

Solution: The data given are presented in Fig. AE-2.36.

$$Q = -kA\frac{dT}{dx}$$

$$= -0.5\ (1 + 5 \times 10^{-3}\ T)\left(\frac{\pi}{4}x^3\right)\frac{dT}{dx}$$

separating variables

$$Q \times \frac{dx}{x^3} = \frac{-0.5 \times \pi}{4}(1 + 5 \times 10^{-3}\ T)\ dT$$

Integrating between the limits

$$Q\left[-\frac{1}{3x^2}\right]_{0.075}^{0.225} = -\frac{\pi}{8}\left[T + \frac{5 \times 10^{-3}}{2}T^2\right]_{300}^{50}$$

$$Q\ [52.68] = -\frac{\pi}{8}\left[(50 - 300) + \frac{5 \times 10^{-3}}{2}(50^2 - 300^2)\right] = 184.07$$

$\therefore$ $\qquad$ **Q = 3.5 W.**

COMMENT

In this problem the variation of area along the flow direction as well as variation of conductivity are considered. To obtain the temperature at other sections will involve more work.

Example A-2.37: *The thermal conductivity of a slab varies as $k = k_o (1 + \alpha T^2)$. Determine the expression for heat flow through a slab of thickness L m and surface temperatures variation along x.*

Solution:
$$Q = -kA\frac{dT}{dx} = -k_o(1+\alpha T^2)A\frac{dT}{dx}$$

$$\frac{Q}{Ak_o}dx = -(1+\alpha T^2)\,dT$$

Integrating: $$\frac{Q}{Ak_o}(x_2 - x_1) = \left[-T - \frac{\alpha T^3}{3}\right]_{T_1}^{T_2} = (T_1 - T_2) + \frac{\alpha}{3}(T_1^3 - T_2^3)$$

$$Q = \frac{Ak_o}{x_2 - x_1}[T_1 - T_2]\left[1 + \frac{\alpha}{3}(T_1^2 + T_1T_2 + T_2^2)\right] \quad ...(p1)$$

$$= \frac{Ak_o\left[1 + \frac{\alpha}{3}(T_1^2 + T_1T_2 + T_2^2)\right]}{L}(T_1 - T_2)$$

The heat flow upto x is the same as full heat flow.

$$\frac{Ak_o}{x_2 - x_1}[T_1 - T_2]\left[1 + \frac{\alpha}{3}(T_1^2 + T_1T_2 + T_2^2)\right]$$

$$= \frac{Ak_o}{x - x_1}[T_1 - T]\left[\left(1 + \frac{\alpha}{3}\right)(T_1^{\,2} + T_1T + T^2)\right]$$

$$\therefore \quad \left[\frac{T_1 - T}{T_1 - T_2}\right] = \frac{\left[1 + \frac{\alpha}{3}(T_1^2 + T_1T_2 + T_2^2)\right]}{\left[1 + \frac{\alpha}{3}(T_1^2 + T_1T + T^2)\right]} \times \frac{x - x_1}{x_2 - x_1} \quad ...(p2)$$

This equation gives the temperature distribution. Of course T is involved on both sides and the solution has to be by trial.

In the case of copper the conductivity is found to vary roughly as $k = 408.21\ (1 - 3.75 \times 10^{-7}\ T^2)$. Determine the heat flow through a plate of 55 cm thickness when faces are held at 100°C and 50°C. Using the equation $p1$ of this problem, substituting the values with $A = 1\ m^2$.

$$Q = \frac{408.21}{0.55}[100 - 50]\ [1 - (375/3) \times 10^{-7}(100^2 + 100 \times 50 + 50^2)]$$

$$= \mathbf{37029\ W/m^2}.$$

COMMENT

The heat flow calculation appears straight forward. But it is rather difficult to determine the temperature at intermediate locations.

Example A-2.38: *A through which is in the form of one half of a hollow cylinder of thickness t has heat flow only in the circumferential direction. Formulate the general differential equation.*

Solve the same for 180° with the temperature $T = T_1$ at 0° and $T = T_2$ at 180°, for steady conditions and without heat generation

Solution: The configuration is shown in Fig. AE-2.38. Assume unit depth ⊥ r to the paper.

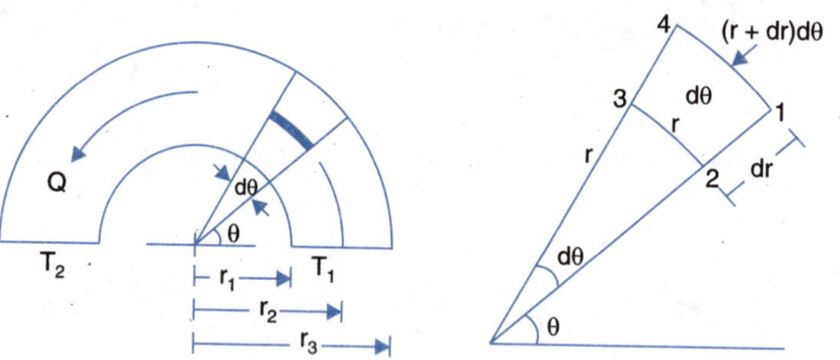

Fig. AE-2.38. Problem model

Considering the elemental volume at radius r and thickness dr, and taking heat flow in the θ direction (circumferential)

$$\begin{bmatrix}\text{heat conducted} \\ \text{through barriers}\end{bmatrix} + \begin{matrix}\text{heat generated} \\ \text{in the volume}\end{bmatrix} \begin{matrix}\text{time} \\ \text{interval}\end{matrix} = \begin{matrix}\text{heat stored in the} \\ \text{volume during the time}\end{matrix}$$

Heat is conducted only through face 1-2 and 3-4 ($rd\theta$ is the distance between planes 1-2 and 3-4)

$$\text{Heat conducted through 1-2} = -k\,dr \times \frac{1}{r}\frac{\partial T}{\partial \theta}.\partial\tau = \frac{-kdr}{r}.\frac{\partial T}{\partial \theta}.\partial\tau$$

Heat conducted through 3-4

= heat conducted through 1-2 + rate of change of heat conduction with distance × distance

$$= \frac{-kdr}{r}.\frac{\partial T}{\partial \theta}.\partial\tau + \frac{\partial}{\partial(r\partial\theta)}\left[-k\frac{\partial r}{r}.\frac{\partial T}{\partial \theta}.\partial\tau\right]rd\theta$$

Taking the difference, the net heat conducted is

$$\frac{1}{r}\frac{\partial}{\partial\theta}\left(\frac{k}{r}\frac{\partial T}{\partial\theta}.\partial\tau\,\partial r\right)rd\theta$$

Heat generated $= q\,rd\theta\,dr\,\partial\tau$

Heat stored $= \rho\,.\,c\,r\,d\theta\,1dr\,dT$. Summing up

$$\frac{1}{r^2}\frac{\partial^2 T}{\partial\theta^2} + \frac{q}{k} = \frac{1}{\alpha}\frac{\partial T}{\partial\tau}$$

which is also obtainable from the general equation in cylindrical coordinates.

For steady conduction without heat generation, the equation reduces to $\frac{d^2T}{d\theta^2} = 0$. The general solution is $T = C_1\theta + C_2$

at $\theta = 0, \quad T = T_1 \quad \therefore \quad C_2 = T_1$

at $\theta = \pi, \quad T = T_2 \quad T_2 = \pi \cdot C_1 + T_1 \quad \therefore \quad C_1 = \frac{T_2 - T_1}{\pi}$

$\therefore$ Substituting in the general solution.

$$T = \frac{T_2 - T_1}{\pi}\,\theta + T_1 \quad \text{or} \quad \frac{\mathbf{T} - \mathbf{T_1}}{\mathbf{T_2} - \mathbf{T_1}} = \frac{\theta}{\pi}$$

Heat flow can be obtained as

$$Q = \frac{k}{\pi} \ln \frac{r_2}{r_1} \cdot [T_1 - T_2] \times \text{depth.}$$

Note: Refer exercise Example A-2.11 for a similar situation. Instead of part of cylinder, part of sphere is considered in the problem.

Example A-2.39: *The temperature of air outside is found to vary at a location as per the relation $T°C = 25 + 20 \sin\left(\frac{2\pi}{12}t\right)$, (angle in radians.) where t is time in hours starting from 9 am as zero time and going up to 3 pm, the time t having value of 6 at this point. An air conditioned space is protected from this atmosphere by a wall having a total resistance 2.8449 m^2K/W. The inside is maintained at 20°C. Considering the thermal capacity of the wall as low, determine the maximum, minimum and average load on the AC. The total barrier area is 400 m^2.*

Solution: The outside temperature is calculated using the given equation for T°C

$$= 25 + 20 \sin\left(\frac{2\pi}{12}t\right) \text{ (the angle should be in radians)}$$

Time of day	9	10	11	12	1	2	3
t (hrs)	0	1	2	3	4	5	6
T °C	25	35	42.32	45	42.32	35	25

The maximum flux is at 12 noon:

$$\mathbf{Q_{max}} = \frac{(45 - 20) \times 400}{2.8449} = \mathbf{3515\ W}$$

$$\mathbf{Q_{min}} = \frac{(25 - 20) \times 400}{2.8449} = \mathbf{703\ W}$$

Average: using arithmetic average of temperatures at middle of each interval

Average temperature = 37.44°C

Average flux $= \frac{(37.44 - 20)}{2.8449} \times 400 = \mathbf{2452\ W}$

Calculating the average temperature by integration:

The angle $\theta = \left(\frac{2\pi}{12}t\right)$ varies from $0 - \pi$, by letting $\theta = \left(\frac{2\pi}{12}t\right)$ and t taking values from

Chapter 2

Average excess over 25°C $= \frac{20}{\pi}\int_0^{\pi} \sin\theta\, d\theta = \frac{20}{\pi} \times 2 = \frac{40}{\pi} = 12.732$

Average temperature $= \mathbf{37.732}$

Average flux $= \frac{(37.732 - 20)\,400}{2.8449} = \mathbf{2493.16\ W/m^2}$.

COMMENT

This is a practical situation in cooling load calculation for airconditioning of buildings.

REVIEW QUESTIONS

2.1 Is the heat flux independent of radius in one dimensional conduction without heat generation in the case of hollow cylinders and spheres?

2.2 Will the heat rate be independent of radius in one dimensional conduction without heat generation in the case of hollow cylinders and spheres?

2.3 Indicate the shape of temperature distribution in one dimensional conduction without heat generation in the case of (*a*) plane wall (*b*) hollow cylinder and (*c*) hollow sphere.

2.4 Define and state the unit of thermal resistance.

2.5 Explain the factors that control critical insulation radius.

2.6 Draw the equivalent circuit for combined convection and radiation from a surface to the surroundings.

2.7 Explain what is contact resistance in heat conduction. Indicate the factors affecting its value.

2.8 Define overall heat transfer coefficient. Indicate its units.

OBJECTIVE QUESTIONS

2.1 Choose the correct answer from each question.

(*i*) In a slab under steady conduction if the thermal conductivity increases along the thickness, the temperatures gradient along the direction will become

(*a*) Steeper (*b*) flatter

(*c*) will depend upon the heat flow (*d*) will remain constant.

(*ii*) In steady state heat conduction in the x direction, the sectional area increases along the flow direction. Then the temperature gradient in the x direction will

(*a*) Remain constant (*b*) will become flatter

(*c*) will become steeper (*d*) either b or c depending on the heat flow rate.

(*iii*) In steady state conduction with variable thermal conductivity if the conductivity decreases along the flow direction, then the temperature gradient along the flow direction will become

(*a*) steeper (*b*) flatter

(*c*) remain constant (*d*) either of the three depending on heat flow rate (*iv*) In steady state conduction with thermal conductivity given by $k = k_o (1 + \beta T)$ where β, is +ve, a slab of given thickness and given temperature drop will conduct

(*a*) more heat at lower temperature levels

(*b*) more heat at higher temperature levels

(*c*) will be the same as flow depends on the temperature drop

(*d*) will be the same as flow depends on the thickness only.

(*v*) Choose the correct statement or statements

(*a*) the thermal conductivity of gases decreases with temperature

(*b*) the thermal conductivity of insulating solids increase with temperature

(*c*) The thermal conductivity of good electrical conductors are generally low

(*d*) The thermal conductivity variation is of low percentage in gases as compared to solids.

Answers

2.1 (*i*) (*b*), (*ii*) (*b*), (*iii*) (*a*), (*iv*) (*b*), (*v*) (*b*).

EXERCISE PROBLEMS

2.1 A furnace operating at 900°C is to be insulated. The outside is to be exposed to air at 30°C with h = 15 W/m^{2}K. The convection coefficient on the inside (including radiation) is 85 W/m^{2}K. The maximum space available is 0.25 m. The heat loss should not exceed 300 W/m^2. Determine the thermal conductivity of the material to be chosen for the insulation. [**Ans.** 0.0886 W]

2.2 A composite wall is to be used to insulate a freezer chamber at – 35°C. Two insulating materials are to be used with conductivities of 0.04 W/mK and 0.1 W/mK. If the outside surface temperature of the inner layer (0.04 W/mK) should not go below zero and if the exposed surface temperature should not go below (the expected wet bulb temperature) 22°C determine the insulation thicknesses. The heat gain is to be limited to 10 W/m^2. Also estimate the value of convection coefficient at such a situation. Outside is at 25°C. [**Ans.** 0.14 m, 0.22 m, 3.33 W/m^{2}K]

2.3 A composite wall consists of 20 mm thick steel plate backed by insulation brick (k = 0.39 W/mK) of 50 cm thickness and overlaid by mineral wool of 20 cm thickness (k = 0.05 W/mK) and 70 cm layer of brick of (k = 0.39 W/mK). The inside is exposed to convection at 650°C with h = 65 W/m^{2}K. The outside is exposed to air at 35°C with a convection coefficient of 15 W/m^{2}K. Determine the heat loss per unit area, interface temperatures and temperature gradients in each materials.

2.4 A solar collector receives 880 W/m^2. Its surface temperature is 60°C. The back side is to be insulated so that back losses are limited to 15%. Insulating material with a thermal conductivity of 0.5 W/mK is available. The atmospheric temperature is 30°C and the convection coefficient on the back side is 5 W/m^{2}K. Determine the insulation thickness. [**Ans.** 13.64 mm]

2.5 A composite plate is made up of stainless steel sheet of 25 mm thickness backed by 30 mm carbon steel plate. The thermal conductivities are 19.1 W/mK and 39.2 W/mK. A contact resistance of 5.28×10^{-4} m^2°C/W exists between the sheets. If the total temperature drop in the composite wall is 18°C, determine the heat flow. If convection on the stainless steel side is from fluid at 160°C with h = 45 W/m^{2}K find the surface temperatures of the plates. If the outside is exposed to air at 35°C, determine the convection coefficient on the outside.

2.6 A membrane type electrical heater of 20,000 W/m^2 capacity is sandwiched between an insulation of 25 mm thickness with thermal conductivity of 0.029 W/mK and a metal plate with k = 12.6 W/mK of thickness 15 mm. The convection coefficient is 150 W/m^{2}K. The surroundings are at 5°C. Determine the surface temperature of the heater and the flow on either side. [**Ans.** 155.7°C]

2.7 A composite wall is made of two layers of 0.3 m and 0.15 m thickness with surfaces held at 600°C and 20°C respectively. If the conductivities are 20 and 50 W/mK, determine the heat conducted. In order to restrict the heat loss to 5 kW/m^2 another layer of 0.15 m thickness is proposed. Determine the thermal conductivity of the material required. [**Ans.** 32.22 kW, 1.53 W/mK]

2.8 A three layer insulation is proposed for a computer room to be maintained at 20°C. The materials chosen are 10 mm layer finishing material inside with k = 0.12 W/mK and a 20 mm layer of weathering material on the outside with k = 0.25 W/mK. The heat gain is limited to 7 W/m^2. Determine the conductivity of the material required for the middle layer if the thickness is restricted to 100 mm. The convection coefficient on the inside is 30 W/m^{2}K. The outside air is at 39°C with a convection coefficient of 60 W/m^{2}K **[Ans.** 0.04 W/mK]

2.9 Determine the heat flow and interface temperatures for the composite wall section shown in Fig. 2.9.

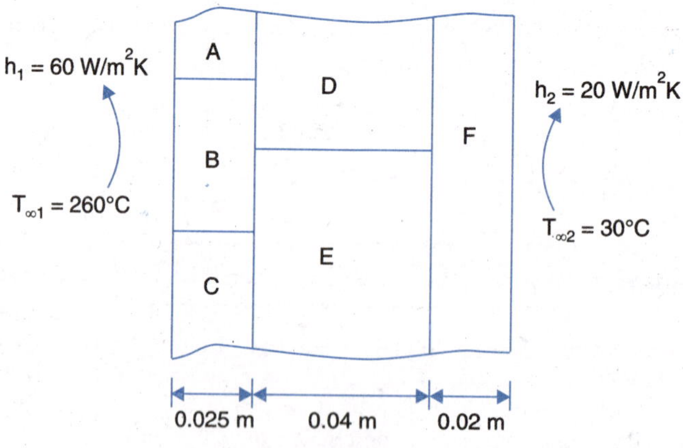

Fig. 2.9

A-1/3 area	k_A = 0.08 W/mK	B-1/3 area	k_B = 0.12 W/mK
C-1/3 area	k_C = 0.21 W/mK	D-40% area	k_D = 0.6 W/mK
E-60% area	k_E = 0.82 W/mK	F-Full area	k_F = 1.3 W/mK.

2.10 Determine the temperatures at all faces for the arrangement shown in Fig. 2.10.

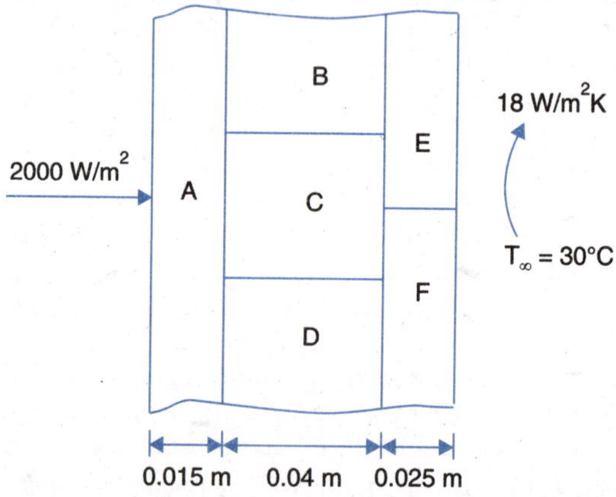

Fig. 2.10

A-full area	k_A = 0.82 W/mK	B-20% area	k_B = 0.3 W/mK
C-50% area	k_C = 0.08 W/mK	D-30% area	k_D = 0.14 W/mK
E-40% area	k_E = 1.2 W/mK	F-60% area	k_F = 2.3 W/mK.

2.11 A part of a sphere of ID $D1$ and OD $D2$ and cut as shown in Fig. 2.11 conducts heat along the direction shown. Derive the general conduction equation and integrate the same for steady conditions without heat generation. Assume that the curved surfaces are insulated.

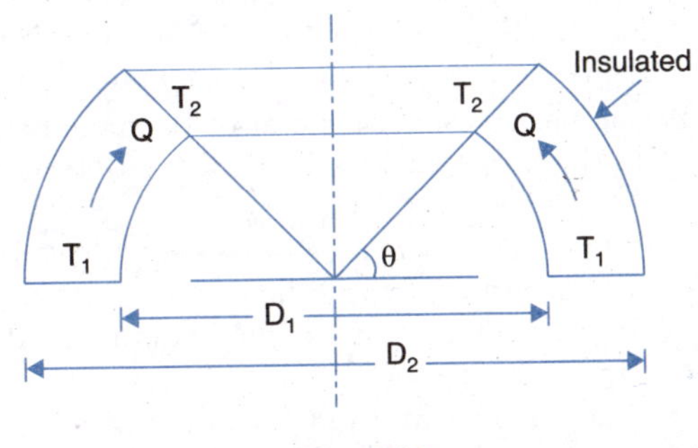

Fig. 2.11

2.12 A pipe carrying brine at – 8°C passes through a room at 30°C. The outside diameter of the pipe is 15 cm and the convection coefficient has a value of 18 W/m^{2}K. Determine the heat gain for 10 m length. In order to reduce the heat gain moulded insulations of 2.5 cm, 4 cm, 6 cm and 8 cm thicknesses are available with conductivities of 0.06 W/mK. Determine the percentage reduction due to each of these. If the cost of heat is ₹ 3/1000 kJ determine in each case the break even cost per m^3 of insulation material if the investment is to be realized in one year of operation. The unit is in operation for 7000 hr/year.

2.13 A hot water pipe of outside diameter 2.5 cm and surface temperature 85°C passes from the boiler room to the heater through a room at 20°C. The convective heat transfer coefficient is 15 W/m^{2}K. An insulation with conductivity of 0.11 W/mK is recommended. Determine the thickness of insulation to reduce the heat loss by 60% of that of the bare pipe. Also find the surface temperature of the insulation. [**Ans.** 3 cm, 12.35°C]

2.14 An electronic device in the form of cylinder of 5 mm dia. generates 25 W/m length. The heat is convected to air at 25°C with a convection coefficient of 10 W/m^{2}K. Determine the surface temperature of the device. An insulation of 4 mm thickness with thermal conductivity of 1.4 W/mK is applied over the device and is exposed to the same convective conditions. Determine the surface temperature of the device. If the thermal conductivity of the material chosen is 0.065 W/mK, determine the surface temperature. What will happen if the thickness is increased to 6 mm in the later case?

2.15 A hollow cylindrical insulation has an internal diameter of 16 cm thickness of 8 cm. The inner surface is at – 10°C while the outer surface is at 35°C. Determine the radius at which the temperature is 0°C. [**Ans.** 0.1372 m]

2.16 A pipe carrying steam at 220°C has an internal diameter of 15 cm. The convection coefficient on the inside is 60 W/m^{2}K. The pipe wall thickness is 15 mm and the thermal conductivity is 35 W/mK. The outside is exposed to a chemical at 130°C with a convection coefficient of 15 W/m^{2}K. Determine the overall heat transfer coefficient, based on (*i*) inner and (*ii*) outer area. If the pipe wall is covered with two insulation layers, the first 3 cm thickness with thermal conductivity of 0.12 W/mK and the second 4 cm thickness with a thermal conductivity of 0.35 W/mK. A contact resistance of 6×10^{-4} m^2 °C/W also is there between the two layers. Determine the heat flow and interface temperatures. Also calculate the overall heat transfer coefficient based on outside area.

2.17 A steam pipe carrying steam at 260°C is of 15 cm dia. and 20 mm thickness. The convection coefficient on the inside is 45 W/m^{2}K. The conductivity of the material is 35 W/mK. An insulation of 4 cm thickness with a conductivity of 0.08 W/mK is installed over the pipe with a contact resistance of 6×10^{-4} °C m^2/W. The outside is exposed to surroundings at 30°C with a convection coefficient of 18 W/m^{2}K and to also radiation to the same surroundings. Determine the heat flow, overall heat transfer coefficient based on outside area, and the interface temperatures.

2.18 A spherical container filled with a fluid of good thermal conductivity which is constantly being stirred is to be insulated. The inner diameter of the container is 0.4 m. The heat capacity of the fluid is 70 kJ/°C. The wall thickness is 15 mm and the conductivity of the material is 15 W/mK. The fluid is at 160°C. The convection coefficient on the inside is 45 W/m^{2}K. The outside is exposed to air at 30°C with a convection coefficient of 25 W/m^{2}K. Determine the overall heat transfer coefficient based on outside area, heat loss and the instantaneous rate of cooling of the fluid.

2.19 A hollow spherical insulation has internal diameter of 16 cm and a thickness of 8 cm. The inner surface is at – 10°C while the outer surface is at 35°C. Determine the radius at which the temperature is 0°C. [**Ans.** 0.131 m]

2.20 A spherical vessel 10 cm OD at – 24°C is exposed to air at 20°C with convection coefficient of 10 W/m^{2}K. Determine the heat gain rate. If the sphere is insulated with 5 cm thick insulation of thermal conductivity of 0.5 W/mK, determine the heat gain rate. Also investigate thicknesses of 2.5 cm and 7.5 cm of insulation.

2.21 A spherical vessel containing hot fluid at 160°C (in a chemical process) is of 0.4 m OD and is made of Titanium of 25 mm thickness. The thermal conductivity is 20 W/mK. The vessel is insulated with two layers of 5 cm thick insulations of thermal conductivities 0.06 and 0.12 W/mK. There is a contact resistance of 6×10^{-4} and 5×10^{-4} m^2 °C/W between the metal and first insulation and between the insulating layers. The outside is exposed to surrounding at 30°C with a convection coefficient of 15 W/m^{2}K. Determine the rate of heat loss, the interface temperatures and the overall heat transfer coefficient based on the metal surface area.

2.22 An insulating wall of a furnace has the following relationship for its thermal conductivity $k = 0.7 (1 + 15 \times 10^{-4} T)$ W/mK where T is in °C. The wall is 0.25 m thick. The inside surface is at 760°C and the outside surface is at 60°C. Determine the heat loss per unit area and also the temperature at the mid section.

2.23 The wall in problem 2.22 has the inside surface exposed to gases at 760°C with a convection coefficient of 40 W/m^{2}K and the outer surface is exposed to air at 35°C with a convection coefficient of 20 W/m^{2}K. Determine the heat flow rate, the surface temperatures and the mid plane temperature.

2.24 A pipe of 0.3 m outer diameter at a temperature of 160°C is insulated with a material having a thermal conductivity of $k = 0.055 (1 + 2.8 \times 10^{-3}T)$ W/mK where T is in °C. The outside surface temperature is 40°C. Determine the heat flow/m length and the temperature at the mid radius.

2.25 A cylindrical pressure vessel of ID 2 m made of material with thermal conductivity $k = 14.7 (1 + 10^{-3} T)$ W/mK where T is in °C and has a wall thickness of 20 cm. Plot the temperature along the radius if the inside surface is at 400°C and the outside is 300°C.

2.26 A cylindrical pressure vessel of 1 m inner diameter and wall thickness 15 cm is made of material with conductivity $k = 44 (1 - 4.2 \times 10^{-4} T)$ W/mK where T is in °C. If the inside surface temperature is 420°C and the outer surface temperature is 300°C, plot the temperature variation along the radius.

2.27 A spherical vessel of 0.4 m inside diameter an 15 cm thickness is made of material with thermal conductivity of $k = 51 (1 - 8 \times 10^{-4} T)$ W/mK where T is in °C. The inside surface is at – 190°C while the outside is at – 90°C. Plot the temperature variation with radius. Also find the heat gain.

2.28 A spherical vessel of 2 m outer dia and a surface temperature of 80°C is insulated with a material having its thermal conductivity given by $k = 0.12 (1 + 1.6 \times 10^{-3} T)$ W/mK, where T is in °C. The thickness of insulation is 18 cm and the outer surface is at 20°C. Determine the heat loss. Also find the temperature at mid thickness.

2.29 The temperature variation in a slab is given by $t(x) = 40 - 200 x^2$ where x is in m and x coordinate being zero at the left face of the slab and 0.1 m at the right face. Determine the heat flow at the faces $x = 0$ and $x = 0.1$ m. The conductivity of the material is 45 W/mK. If the conditions are steady, determine the heat generation rate in the slab.

2.30 The temperature distribution in a body is given by the equation

$$T(x, y, z, \tau) = y^2 - 2x^2 + z^2 - xy + 2xz.$$

Examine whether it represents steady state conduction.

2.31 A truncated cone of height L has heat conduction along the axis as detailed in Fig. 2.31. Derive an expression for the variation of thermal conductivity along x, given

$$A = A_o\left(1 - \frac{x}{L}\right),\ T(x) = 300\left\{1 - 2\left(\frac{x}{L}\right) - \left(\frac{x}{L}\right)^3\right\}$$

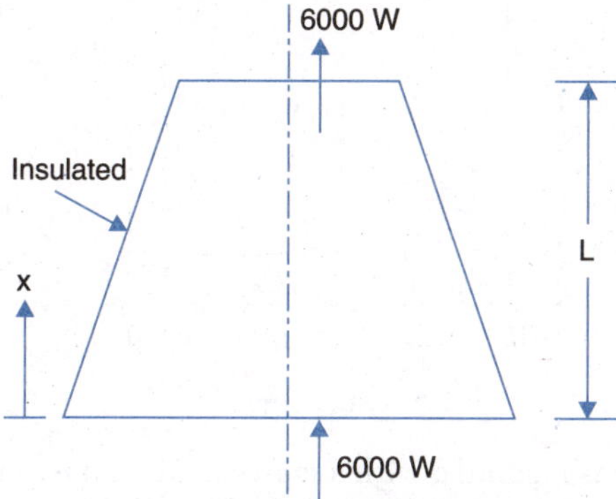

Fig. 2.31

2.32 In an aluminium surface with a roughness depth (after applying pressure) of 0.01 mm, the area of crests is 50% of the total area. The outer plate is in contact with stainless steel with roughness depth of 0.01 mm with the area of contact of 50%. Determine the contact resistance (*i*) If the inter space is filled with air (*ii*) If it is filled with oil of $k = 0.15$ W/mK.

2.33 A furnace with an operating temperature of 900°C is insulated by a wall 0.25 m thick with convection coefficient of 85 W/m²K on the inside and exposed on the outside to air at 30°C with a convection coefficient of 15 W/m²K. The supplier reported that the conductivity of the material was 0.08 W/m²K. After installation, a note was received that the material thermal conductivity varies with temperature and it is given by $k = 0.08\ (1 + 9 \times 10^{-4}\ T)$ W/mK. Calculate the actual heat flow and the original designed flow T is in °C.

2.34 A wall 2 m thick has the following temperatures distribution: $T(x) = 60 + 18x - 6x^3$ where x is in m and $T(x)$ is in °C. Determine the location of maximum temperature and the heat flow per m² area at both faces, $k = 30$ W/mK. [**Ans.** max. temp. at 1 m, 540 W at $x = 0$ 1620 W at $x = 2$ m]

2.35 A hollow cylinder of inner radius R_1 and outer radius R_o has the following variation of temperature along the radius. $T(r) = 300 - 300 \ln (r/R_1)$. The conductivity of the material is 45 W/mK. If the inner radius is 6 cm and the outer radius is 9 cm determine the direction (towards inner or outer side) and rate of flow of heat at the two surfaces for 1 m length of pipe.

2.36 A hollow sphere of inner radius R_1 and outer radius R_o has the following variation of temperature along the radius. $T(r) = 300 + 300\ 1\text{n}\ (r/R_o)$. The conductivity of the material is 45 W/mK. If the inner radius is 6 cm and the outer radius is 9 cm determine the direction and rate of flow of heat at the two surfaces.

2.37 A cross section of a wall is shown in Fig. 2.37. Determine the heat flow for a 0.25 m height for unit depth. The surface temperatures are – 15°C and 20°C. The conductivities of the materials are $k_A = 0.65$ W/mK, $k_B = 0.08$ W/mK, $k_C = 0.16$ W/mK, $k_D = 2.10$ W/mK, $k_E = 3.5$ W/mK. Material A is exposed to – 15°C. Also find interface temperatures.

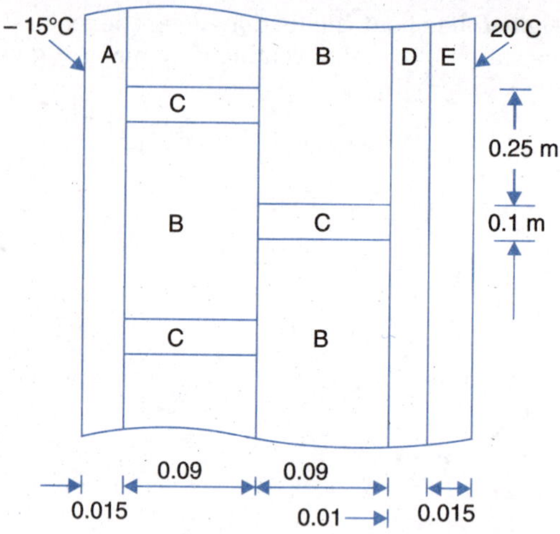

Fig. 2.37

2.38 A 5 mm dia. copper cable is insulated with a material of conductivity of 0.16 W/mK and is exposed to air at 30°C with a convection coefficient of 20 W/m^{2}K. If the surface temperature of the wire can be 120°C, determine the insulation thickness for maximum heat flow and the heat dissipated per m length. [**Ans.** 8 mm, 41.8 W/m]

2.39 An aircraft canopy in the form of a part of a sphere of diameter of 1.2 m is made of a transparent material 15 mm thickness having a conductivity of 2.5 W/mK is exposed on the outside to – 60°C and a convection coefficient of 180 W/m^{2}K. On the inside the temperature is 25°C and the convection coefficient is 10 W/m^{2}K. If the area exposed to the surroundings is a quarter of a sphere, determine the heat loss through the canopy.

2.40 To determine the contact resistance between two surfaces, stainless steel rods of 2.5 cm diameter with the ends machined to the required surface roughness and held together at the required pressure. One of the rod is heated at the free end, while the remaining portion is insulated. The temperatures are measured at points as shown in Fig. 2.40. Determine the contact resistance, k = 40 W/mK. The temperatures at A, B, C and D are 110°C, 105°C, 100°C and 95°C respectively.

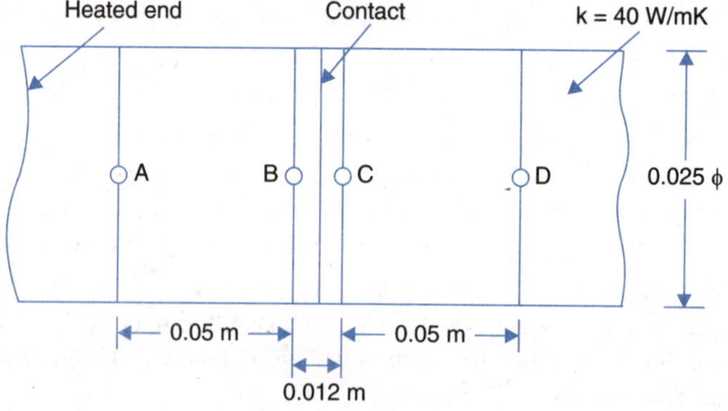

Fig. 2.40

3 CHAPTER

Conduction with Heat Generation

3.0 INTRODUCTION

In the previous chapter, steady state heat conduction problems were discussed. In these cases the boundary conditions governed the temperature distribution. In this chapter the additional factor of heat generation within the body is to be considered. Heat generation is met with in electrical conductors carrying current. The heat generated by the resistance has to be dissipated in such a way that the conductor does not fail. The maximum temperature in the wire has to be limited. A more serious case is the heat generation in nuclear reactor fuel rods. Here the generation rate is very high and accidents will be catastrophic. The heat generated has to be collected at the required rates at the boundary to avoid failure. Microwave heating is another example. Exothermic reaction in chemical processes also may cause problems if proper care is not taken to maintain the temperature at or below the allowable level. The heat flow under steady conditions is easily estimated as the product of the volume of the body and heat generation rate. The maximum temperature attained by heat generating components will depend mainly on (1) The heat generation rate, (2) Thermal conductivity of the material (3) Physical dimension like thickness or diameter and (4) convection coefficient over the surface. As in the previous chapter three geometric shapes, namely plane, cylinder and sphere, are analysed. Hollow cylinder with internal and external heat collection is also discussed. Variable heat generation rate is also considered.

3.1 STEADY STATE ONE DIMENSIONAL CONDUCTION IN A SLAB WITH UNIFORM HEAT GENERATION

The physical model is shown in Fig. 3.1(*a*) and 3.1(*b*). Two types of boundary specifications are possible as shown in these figures. Another model is half of a slab with one side insulated. The temperature distribution, heat flow at the surface etc., will be the same as the model shown in Fig. 3.1(*b*).

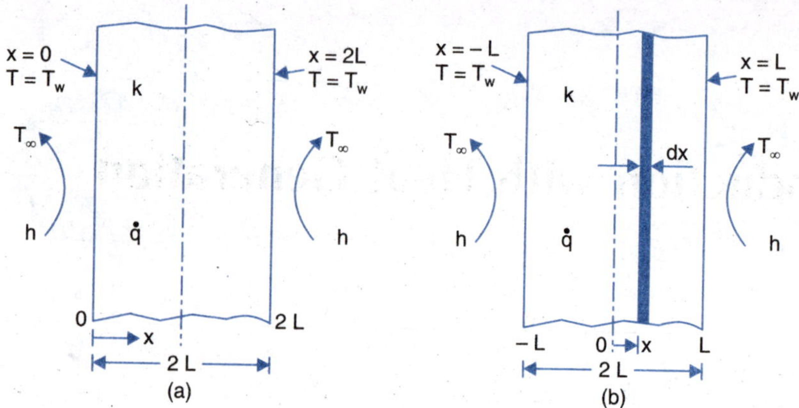

Fig. 3.1. Slab with heat generation

The results of analysis are the same in both cases.

In the following discussions the model in 3.1(*b*) is adopted.

A large slab of thickness 2L with uniform heat generation rate of $\dot{q}$ (W/m^3) is considered. There is equal convection at T_∞ on both sides such that the surface temperature is T_w on both sides. The centre plane is taken as the origin for x and the slab extends to $+L$ on the right and $-L$ on the left. The conductivity of the materials is k. A thin section of dx is considered at a distance x from the origin for heat balance.

Calculations are made on the basis of unit area. The differential equation applicable for this case has been derived in chapter 2. The equation [2.6(*b*)] is

$$\frac{d^2T}{dx^2} + \frac{\dot{q}}{k} = 0 \qquad ...(3.1)$$

Considering the thin section of thickness dx at x, the energy balance **at steady state conditions** can be written as

heat conducted in section x + heat generated in the dx thickness – heat flowing out at x + dx = 0.

or $$-kA\frac{dT}{dx} + \dot{q}\,dx\,A - \left[-kA\frac{dT}{dx} + \frac{d}{dx}\left(-kA\frac{dT}{dx}\right)dx\right] = 0$$

Simplifying $\dfrac{d^2T}{dx^2} + \dot{q}/k = 0.$

Writing the equation as below and integrating

$$d\left(\frac{dT}{dx}\right) = -\frac{\dot{q}}{k}\,dx$$

$$\frac{dT}{dx} = -\frac{\dot{q}}{k}\,x + C_1 \qquad ...(3.2)$$

$$\therefore \qquad T = -\frac{\dot{q}x^2}{2k} + C_1x + C_2 \qquad ...(3.3)$$

At the mid section, at $x = 0$, there is no heat flow across the section or $\frac{dT}{dx} = 0$. From equation (3.2), $C_1 = 0$.

At $x = L, T = T_w$, $\therefore$ $T_w = -\left(\frac{\dot{q}}{2k}\right) L^2 + C_2$

$$\therefore \quad C_2 = T_w + \frac{\dot{q}}{2k} L^2 \quad ...(3.4)$$

Substituting in equation (3.3),

$$T = -\left(\frac{\dot{q}}{2k}\right) x^2 + T_w + \frac{\dot{q}}{2k} L^2$$

or
$$T - T_w = \frac{\dot{q}}{2k} (L^2 - x^2) \quad ...[3.5(a)]$$

Sometimes only T_∞ and h will be known. In such cases the equation [3.5(*a*)] can be modified.

At the boundary $\quad AL\dot{q} = hA\,(T_w - T_\infty) \;\therefore\; T_w = T_\infty + \frac{\dot{q}L}{h}$

Equation [3.5(*a*)] can be written as

$$T - T_\infty = \frac{\dot{q}}{2k} (L^2 - x^2) + \frac{\dot{q}L}{h} \quad ...[3.5(b)]$$

For these equations to be applicable model [3.1(*b*)] should be used.

The temperature at $x = 0$ is obtained from [3.5(*a*)] and [3.5(*b*)] as

$$T_o = T_w + \frac{\dot{q}}{2k} L^2 \quad ...[3.6(a)]$$

$$T_o = T_\infty + \frac{\dot{q}}{2k} L^2 + \frac{\dot{q}L}{h} \quad ...[3.6(b)]$$

Equation [3.5(*a*)] can also be modified as below. Using [3.5(*a*)] and [3.6(*a*)].

$$T - T_o = \frac{\dot{q}}{2k}(L^2 - x^2) + T_w - T_w - \frac{\dot{q}}{2k} L^2 = -\frac{\dot{q}}{2k} x^2$$

From [3.6(*a*)] $\quad T_w - T_o = -\frac{\dot{q}}{2k} L^2.$

$$\therefore \quad \frac{T - T_o}{T_w - T_o} = \frac{x^2}{L^2} = \left(\frac{x}{L}\right)^2 \quad ...(3.7)$$

This shows that the temperature variation is parabolic. It may be also seen that the temperature gradient becomes steeper along x as more heat is to be conducted as x increases.

Solution with Boundary condition of Fig. 3.1 (*a*):

The general solution for equation (3.1) is given by

$$T = -\frac{\dot{q}x^2}{2k} + C_1 x + C_2. \quad ...(A)$$

Boundary conditions are:

(1) At $x = 0$, $T = T_w$, (2) At $x = 2L$, $T = T_w$.

Using the first boundary condition,

$$T_w = C_2 \qquad \text{...(B)}$$

Using the second boundary condition

$$T_w = \frac{-\dot{q}\,4L^2}{2k} + C_1 \times 2L + T_w.$$

$$\therefore \qquad C_1 = (\dot{q}L/k) \qquad \text{...(C)}$$

Substituting in the equation (A),

$$T = \frac{-\dot{q}x^2}{2k} + \frac{\dot{q}L}{k}.x + T_w$$

$$\therefore \qquad T - T_w = \frac{-\dot{q}x^2}{2k} + \frac{\dot{q}Lx}{k}$$

Multiplying and dividing by L^2

$$T - T_w = \frac{-\dot{q}L^2}{2k}\left[2\left(\frac{x}{L}\right) - \left(\frac{x}{L}\right)^2\right] \qquad \text{...[3.5(c)]}$$

To locate the position of T_{max},

$$\frac{dT}{dx} = \text{constant}\left[\frac{2}{L} - \frac{2x}{L^2}\right]$$

Equating to zero, $x_{max} = L$

$$T_{max} - T_w = \frac{\dot{q}L^2}{2k} \qquad \text{...[3.5(d)]}$$

$$\therefore \qquad \frac{T - T_w}{T_{max} - T_w} = 2\left(\frac{x}{L}\right) - \left(\frac{x}{L}\right)^2, \qquad \text{...[3.7(b)]}$$

When convection is considered,

$$\dot{q}L = h\,(T_w - T_\infty) \quad \therefore \quad T_w = T_\infty + \frac{\dot{q}L}{h}$$

$$\therefore \qquad T - T_\infty = \frac{\dot{q}L^2}{2k}\left[2\left(\frac{x}{L}\right) - \left(\frac{x}{L}\right)^2 + \frac{\dot{q}L}{h}\right] \qquad \text{...[3.6(c)]}$$

Example 3.1: *Heat is generated in a slab of 120 mm thickness with a conductivity of 200 W/mK at a rate of 10^6 W/m^3. Determine the temperature at the mid and quarter planes if the surface of the solid on both sides are exposed to convection at 30°C with a convection coefficient of 500 W/m^{2}K. Also find the heat flow rate at these planes and the temperature gradients at these planes.*

Solution: The data are shown in Fig. Ex. 3.1.

Using equation (3.5): at quarter plane, (L = 0.12/2 = 0.06 m)

Considering unit area heat flow on the right side = 0.06 × 1 × 1 × 10⁶. This equals the convection heat flow.

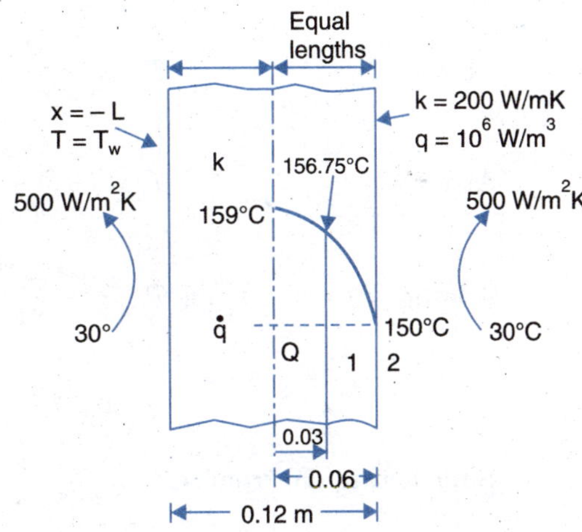

Fig. Ex. 3.1. Problem model

$\therefore$ $0.06 \times 1 \times 10^6 = 1 \times 500\,(T_w - 30)$

Solving $\mathbf{T_w = 150°C}$.

$$T - T_w = (\dot{q}/2k)(L^2 - x^2), \text{ at } \boldsymbol{x = 0.03\text{ m}}$$

$$T_1 - 150 = (10^6/2 \times 200)\,(0.06^2 - 0.03^2)$$

Therefore $\mathbf{T_1 = 156.75°C}$.

at x = 0

$$T_o - 150 = (10^6/2 \times 200) \times (0.06^2 - 0)$$

Therefore $\mathbf{T_o = 159°C}$

Heat flow at x = 0.03 m. Heat generated from mid plane to this section is conducted at this section. So for unit area, *i.e.* 1 m²

$$Q = 10^6 \times 1 \times 0.03 = 30{,}000 \text{ W/m}^2.$$

Therefore Temperature gradient = – Q/kA = – 30000/200 × 1 = **– 150°C/m.**

At x = 0.06 m as above $Q = 10^6 \times 1 \times 0.06 = 60000$ W/m².

Therefore temperature gradient = – 60000/200 × 1 = **– 300°C/m.**

In case the equation [3.5 (*d*)] is used,

$$\mathbf{T_{max} - T_w} = \frac{\dot{q}L^2}{2k} = \frac{10^6 \times 0.06^2}{2 \times 200} = \mathbf{9°C}$$

$$\therefore \qquad T_{max} = \mathbf{159°C}$$

Using equation [3.5 (*c*)], Temperature at x = 0.09

$$T - T_w = \frac{\dot{q}L^2}{2k}\left[2\left(\frac{x}{L}\right) - \left(\frac{x}{L}\right)^2\right]$$

$$= \frac{10^6 \times 0.06^2}{2 \times 200}\left[2\left(\frac{0.09}{0.06}\right) - \left(\frac{0.09}{0.06}\right)^2\right] = \mathbf{6.75}$$

At 0.09, $\therefore$ $\mathbf{T = 156.75°C}$

The results are the same by both analysis.

3.1.1. The Other Boundary of Interest

The Other Boundary of Interest is when the two surfaces are maintained at different specified temperatures as in Fig. 3.2.

The solution for equation (3.1) has the form

$$T = \frac{-\dot{q}x^2}{2k} + C_1x + C_2 \qquad \text{...(A)}$$

Chapter 3

At $x = -L$, $$T_{w1} = \frac{-\dot{q}L^2}{2k} - C_1L + C_2 \quad ...(B)$$

At $x = L$, $$T_{w2} = \frac{-\dot{q}L^2}{2k} + C_1L + C_2 \quad ...(C)$$

Adding (B) + (C), $$T_{w1} + T_{w2} = \frac{-2\dot{q}L^2}{2k} + 2C_2$$

or $$C_2 = \frac{T_{w1} + T_{w2}}{2} + \frac{\dot{q}L^2}{2k}$$

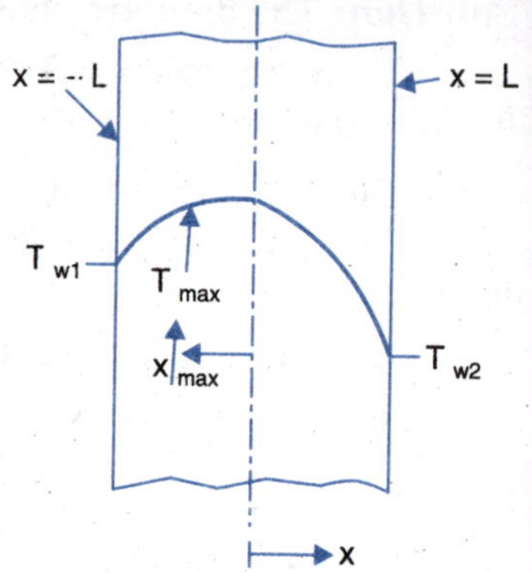

Subtracting (B) from (C),

Fig. 3.2. Problem model

$$2C_1 = T_{w2} - T_{w1} \quad \text{or} \quad C_1 = \frac{T_{w2} - T_{w1}}{2L}$$

Subtituting in (A), the temperature at any x value is

$$T_x = \frac{-\dot{q}x^2}{2k} + \frac{T_{w2} - T_{w1}}{2L}x + \frac{\dot{q}L^2}{2k} + \frac{T_{w1} + T_{w2}}{2}$$

$$= \frac{\dot{q}}{2k}(L^2 - x^2) + \frac{T_{w2} - T_{w1}}{2L}x + \frac{T_{w1} + T_{w2}}{2} \quad ...(3.8)$$

Note that if $T_{w1} = T_{w2}$, the equation reduces to 3.5 (a)

$$T_x = \frac{T_{w1} + T_{w2}}{2} + (\dot{q}/2k)(L^2 - x^2) + \left(\frac{T_{w2} - T_{w1}}{2L}\right)x$$

The location for the maximum temperature is obtained by calculating (dT/dx) and equating the value to zero. This procedure gives

$$x_{max} = (k/2\dot{q}L)(T_{w2} - T_{w1}) \quad ...(3.9)$$

and $$T_{max} = \frac{\dot{q}L^2}{2k} + (k/8\dot{q}L^2)(T_{w2} - T_{w1})^2 + \frac{T_{w2} + T_{w1}}{2} \quad ...(3.10)$$

If $T_{w2} = T_w$, $T_{max} - T_w = (\dot{q}L^2/2k)$, [same as 3.6($a$)]

Example 3.2: *In example 3.1, if the temperatures at the surfaces are maintained at 130°C and 150°C on the left and right sides determine the location and value of the maximum temperature. Also find the heat flow on each side and the temperature at the centre plane.* $\dot{q} = 10^6$ W/m^3, $L = 0.06$ m, $k = 200$ W/mK.

Solution: Using equation (3.9) the location of the maximum temperature is found as below:

$$x_{max} = (k/2\dot{q}L)(T_{w2} - T_{w1})$$

$$= (200/2 \times 10^6 \times 0.06)(150 - 130) = \mathbf{0.0333\ m}$$

This moves to right.

Using equation (3.10), the maximum temperature is found

$$\mathbf{T}_{\mathbf{max}} = qL^2/2k + (k/8qL^2)\,(T_{w2} - T_{w1})^2 + \frac{T_{w2} - T_{w1}}{2}$$

$$= \frac{10^6 \times 0.06^2}{2 \times 200} + \frac{200}{8 \times 10^6 \times 0.06^2}(150 - 130)^2 + \frac{150 + 130}{2}$$

$$= 9 + 2.778 + 140 = \mathbf{151.778°C}$$

The heat flow on either side is the heat generated from the maximum temperature plane to the surface.

On the 130°C side:

$$Q = 10^6 \times 0.09333 = 93333.3 \text{ W/m}^2$$

On the 150°C side:

$$Q = 10^6 \times (0.06 - 0.03333) = 26666.7 \text{ W/m}^2$$

Total = 120,000 W

Center plane temperature is found using equation (3.8)

$$T_x = \frac{T_{w1} - T_{w2}}{2} + \frac{T_{w2} - T_{w1}}{2}(x/L) + \frac{q}{2k}(L^2 - x^2)$$

and $x = 0$

$$T_o = \frac{130 + 150}{2} + \frac{150 - 130}{2} \times (0.0/L) + \frac{10^6}{2 \times 200}$$

$$\times (0.06^2 - 0) = 140 + 0 + 9 = \mathbf{149°C}$$

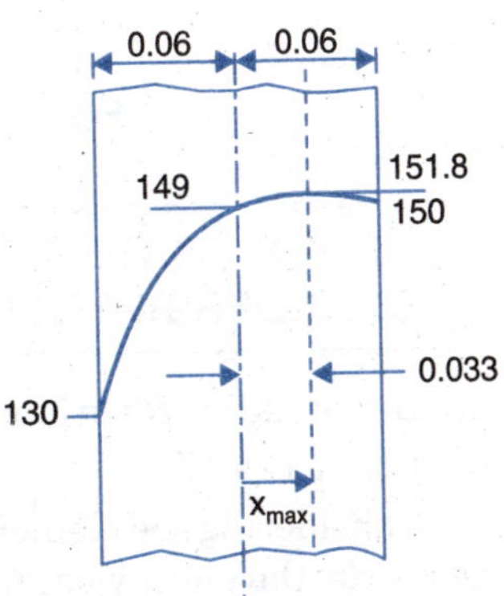

Fig. Ex. 3.2. Problem model

Also refer Advanced example 3.10.

Example 3.3: *In the case of a slab of thickness 0.12 m generating heat at the rate of 10^6 W/m³, investigate the effect of convection coefficient on the maximum temperature. Take values from 100 W/m²K to 1000 W/m²K. Thermal conductivity of the materail is 200 W/mK. Surrounding temperature is 30°C.*

Solution: Use equation [3.6 (*b*)].

$$T_o = T_\infty + \frac{\dot{q}L^2}{2k} + \frac{\dot{q}L}{h} = 30 + \frac{10^6 \times 0.06^2}{2 \times 200} + \frac{10^6 \times 0.06}{h}$$

$$= \mathbf{30 + 9 + \frac{60000}{h}}$$

The results are tabulated below:

h, W/m²K	100	250	500	750	1000
T_o°C	**639**	**279**	**159**	**119**	**99**

COMMENT

The convection coefficient greatly influences the maximum temperature in the slab, with other factors remaining the same. The variation of T_o with h is shown in Fig. Ex. 3.3.

Chapter 3

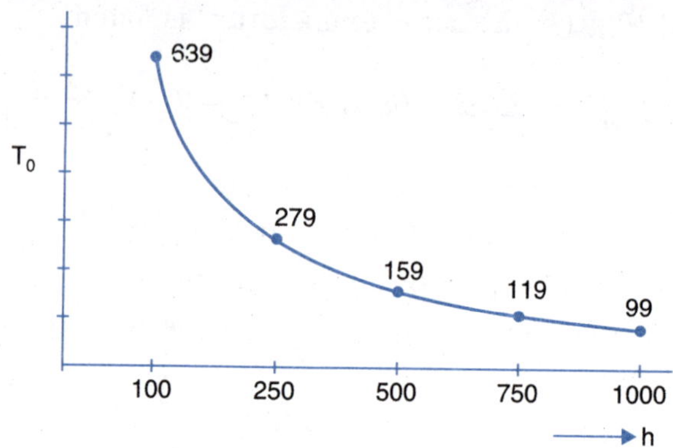

Fig. Ex. 3.3. Variation of T_o with convection coefficient, h

3.2 STEADY STATE RADIAL HEAT CONDUCTION IN SOLID/HOLLOW CYLINDERS WITH UNIFORM HEAT GENERATION. (FIG. 3.3)

A cylinder of radius R with uniform heat generation $\dot{q}$ and conductivity k is considered. Outside convection h is at T_∞

Considering the elemental annular volume between r and $r + dr$, the energy balance gives (taking 1 m length)

heat conducted in + heat generated – heat conducted out = 0.

$$- k\ 2\pi r.l\ (dT/dr) + \dot{q}2\pi r dr.l - (- k2\pi r.l dT/dr)$$
$$- (d/dr)\ (- k2\pi r.l.dT/dr)dr = 0$$

Therefore $\quad (d/dr)\ (r\,dT/dr) + (qr/k) = 0 \quad$ ...(3.11)

This equation is the same as the simplified equation [2.7(*b*)] of Chapter 2.

Fig. 3.3. Model

Integrating after separating variables

$$dT/dr = - (\dot{q}/2k)r + C_1/r \qquad ...(3.12)$$

$$T = - (\dot{q}/4k)r^2 + C_1 \ln r + C_2 \qquad ...(3.13)$$

This is the general solution and the constants C_1 and C_2 are evaluated using the boundary conditions. The four boundary conditions of interest are discussed in the following paras.

3.2.1. Solid Cylinder

The boundary conditions are:

(*i*) at $r = 0$, $dT/dr = 0$, $\qquad$ (*ii*) at $r = R$, $T = T_w$.

From equation (3.12), $r\dfrac{dT}{dr} = \dfrac{\dot{q}r^2}{2k} + C_1$

At $\quad r = 0, \quad \dfrac{dT}{dr} = 0, \quad \therefore \quad C_1 = 0.$ $\qquad$ (eqn. 3.12)

From equation (3.13), condition (*ii*) yields $C_2 = T_w + (\dot{q}/4k)\ (R^2)$

Therefore $T - T_w = (\dot{q}/4k)(R^2 - r^2)$ ...[3.14(a)]

The maximum temperature, T_o is at the axis and is $(r = 0)$

$$T_o - T_w = (\dot{q}/4k)(R^2) \quad \text{...[3.15(a)]}$$

Therefore $$\frac{T - T_w}{T_o - T_w} = 1 - (r/R)^2 \quad \text{...(3.16)}$$

This can also be rearranged as

$$(r/R)^2 = 1 - \left[\frac{T - T_w}{T_o - T_w}\right] = \frac{T_o - T}{T_o - T_w} \quad \text{...[3.16(a)]}$$

The temperature variation will be parabolic with radius.

Taking convection into account, heat generation for unit length = $\pi R^2 \times 1 \times \dot{q}$

This is absorbed by the fluid in the outside area $2\pi R \times 1$

$$\therefore \quad \pi R^2 \dot{q} = 2\pi R h (T_w - T_\infty) \quad \therefore \quad T_w = T_\infty + \frac{R\dot{q}}{2h}.$$

Equation [3.14(a)] reduces to $$T - T_\infty = \frac{\dot{q}}{4k}(R^2 - r^2) + \frac{R\dot{q}}{2h} \quad \text{...(3.14)}$$

Equation [3.15(a)] reduces to $$T_o = T_\infty + \frac{\dot{q}}{4k}R^2 + \frac{R\dot{q}}{2h} \quad \text{...(3.15)}$$

Example 3.4: *A solid cylinder of 12 cm diameter has a heat generation rate 10^6 W/m^3. The conductivity of the material is 200 W/mK. The surface is exposed to a fluid at 30°C. The convection coefficient is 500 W/m^{2}K. Determine the temperatures at the center and also at mid radius. Also determine the heat flow at the surface on unit area basis. Calculate the temperature gradients at the mid radius and surface.*

Solution: Surface temperature, $$T_w = T_\infty + \frac{R\dot{q}}{2h} = 30 + \frac{0.06 \times 10^6}{2 \times 500} = 90°\text{C}$$

Using equation [3.14(a)]

$$T - T_w = (\dot{q}/4k)[R^2 - r^2]$$

Temperature at radius $r = 0$ is

Therefore centre line temperature,

$$T = 90 + \frac{10^6}{4 \times 200}(0.06^2 - 0) = \mathbf{94.5°C.}$$

At mid radius $r = 0.03$

$$\therefore \quad T = 90 + \frac{10^6}{4 \times 200}(0.06^2 - 0.03^2) = \mathbf{93.375°C}$$

Heat flow upto mid radius equals the heat generated in the cylinder.

Therefore $Q = \dot{q} \times \pi r^2 \times 1$ W/m $= 10^6 \times \pi \times 0.03^2 =$ **2827.4 W/m.**

Same on area basis: $Q/A = 2827.4/2\pi r \times 1 = 2827.4/2\pi \times 0.03 =$ **15,000 W/m²**

Heat flow at the surface $Q = \dot{q} \times \pi R^2 \times 1 = 10^6 \times \pi \times 0.06^2 =$ **11309.7 W/m.**

Same on area basis $Q/A = 11309.7/\pi \times 0.12 \times 1 =$ **30,000 W/m²**

Temperature gradient

$Q = -kAdT/dr$. Using the heat flow on area basis.

at **mid radius** $15{,}000 = -200 \times 1\, dT/dr$. Therefore $dT/dr = -$ **75°C/m.**

at **surface** $30{,}000 = -200 \times 1 \times dT/dr$. Therefore $dT/dr = -$ **150°C/m.**

Example 3.5: *With the data of example 3.4, investigate the variation of centre temperature, T_o with h in the range 100 to 1000.*

Solution: Using equation (3.15),

$$T_o = T_\infty + \frac{\dot{q}R^2}{4k} + \frac{R\dot{q}}{2h}$$

$R = 0.06$ m, $T_w = 30$°C, $k = 200$ W/mK, $\dot{q} = 10^6$ W/m³.

$$\therefore \quad T_o = 30 + \frac{10^6 \times 0.06^2}{4 \times 200} + \frac{0.06 \times 10^6}{2h}$$

$$\therefore \quad T_o = 30 + 4.5 + 60000/2h.$$

The values of T_o calculated with different h is tabulated below:

h, W/m^2K	100	250	500	750	1000
T_o,°C	334.5	154.5	94.5	74.5	64.5

COMMENTS

1. The value of convection coefficient plays an important role in deciding the maximum temperature in the cylinder.
2. Compared to plane wall of same thickness and properties, the values of T_o are lower in the case of the cylinder. This is because the heat generated per unit length is lower in the case of the cylinder.

Example 3.6: *A 3 mm diameter wire of thermal conductivity of 20 W/mK carries 200 Amps. The resistivity ρ of the material is 75 μΩ cm. The heat generated is absorbed by a fluid at 120°C. With convective heat transfer coefficient of 4000 W/m²K. Determine the maximum temperature of in the wire.*

Solution: Heat generated = I^2R, Resistance $R = \rho \dfrac{L}{A}$. Assume 1 m length where ρ is the resistivity, L is the length and A is sectional area.

Resistance $$R = \frac{75 \times 10^{-6} \times 1}{100 \times \pi \times 0.0015^2} = 0.1061\ \Omega$$

Heat generated in 1 m wire = $200^2 \times 0.1061 = 4244.3$ W

$$\dot{q} = \frac{\text{heat generated}}{\text{volume}} = \frac{4244.3}{\pi \times 0.0015^2 \times 1} = 600.41 \times 10^6 \text{ W/m}^3.$$

Using equation (3.15), R-radius

$$T_o = T_\infty + \frac{\dot{q}R^2}{4k} + \frac{\dot{q}R}{2h}$$

$$= 120 + \frac{600.41 \times 10^6 \times 0.0015^2}{4 \times 20} + \frac{600.41 \times 10^6 \times 0.0015}{2 \times 4000}$$

$$= \mathbf{249.5°C}$$

The **maximum temperature** in the wire is **249.5°C.**

3.2.2. Hollow Cylinder of Radius R_i, R_o with T_i and T_o Specified

The Boundary Conditions are:

(i) $r = R_i$, $T = T_i$ (ii) $r = R_o$, $T = T_o$

After a somewhat long arithmetic work the equation that can be obtained is

$$T_r - T_o = (\dot{q}/4k)(R_o^2 - r^2) + [(T_o - T_i) + (\dot{q}/4k)(R_o^2 - R_i^2)] \; [\ln(r/R_o)/\ln(R_o/R_i)] \quad \text{...[3.17(a)]}$$

or

$$T_r - T_i = (\dot{q}/4k)(R_i^2 - r^2) + [(T_o - T_i) + (\dot{q}/4k)(R_o^2 - R_i^2)] \; [\ln(r/R_i)/\ln(R_o/R_i)] \quad \text{...[3.17(b)]}$$

The radius at which the maximum temperature occurs is obtained by differentiating the expression (3.17) and equating to zero.

$$R^2_{\max} = \frac{2k}{\dot{q}\ln(R_o/R_i)}\left[(T_o - T_i) + \frac{\dot{q}}{4k}(R_o^2 - R_i^2)\right]. \quad \text{...(3.18)}$$

Check weather the equation is dimensionally homogeneous.

Example 3.7: *A hollow cylinder 6 cm ID and 12 cm OD has a heat generation rate of 5×10^6 W/m^3. The inner surface is maintained at 380°C and the outer surface at 360°C by circulation of a fluid to extract the heat generated. The conductivity of the material is 30 W/mK. Determine (i) the location and value of the maximum temperature (ii) the fraction heat generated going to the inner surface.*

Solution: The location of the maximum temperature is determined using equation (3.18).

$$R_i = 0.03 \text{ m}, R_o = 0.06 \text{ m}, T_i = 380°C, T_o = 360°C$$

$$R^2_{\max} = \frac{2 \times 30}{5 \times 10^6 \ln\left(\frac{0.06}{0.03}\right)}\left[(360 - 380) + \frac{5 \times 10^6}{4 \times 30}(0.06^2 - 0.03^2)\right]$$

$$= 0.0016.$$

$$\therefore \quad \mathbf{R_{max} = 0.04\ m.}$$

To determine T_{max}, equation [3.17(b)] is used.

$$T_r - 380 = \frac{5 \times 10^6}{4 \times 30}(0.03^2 - 0.04^2) + \frac{\ln\left(\frac{0.04}{0.03}\right)}{\ln\left(\frac{0.06}{0.03}\right)}\left[(360 - 380) + \frac{5 \times 10^6}{4 \times 30}(0.06^2 - 0.03^2)\right]$$

$$= 9.22°C$$

$$\therefore \quad \mathbf{T_{max} = 389.22°C.}$$

Heat flow to the inside is the heat generated upto R_{max} per m length

$\therefore$ $$Q_{in} = 5 \times 10^6 \times \pi\,(0.04^2 - 0.03^2) = \mathbf{10995.6\ W/m.}$$

Total heat generated $= \dfrac{\pi}{4} \times (0.12^2 - 0.06^2) \times 1 \times 5 \times 10^6 = 42411.5$ W/m

Fraction going to inner surface

$$= 10995.6/42411.5 = \mathbf{0.2593}$$

3.2.3. Hollow Cylinder with Inner Surface Adiabatic

The boundary conditions are:

(*i*) $dT/dr = 0$ at $r = R_i$ (*ii*) $T = T_o$ at $r = R_o$. This gives

$$T - T_o = (\dot{q}/4k)\,(R_o^{\,2} - r^2) + (\dot{q}/2k)\,R_i^{\,2}\,.\,\ln(r/R_o) \qquad ...(3.19)$$

The maximum temperature occurs at the inner surface.

$$T_{max} = T_i = T_o + (\dot{q}/4k)\,(R_o^{\,2} - R_i^{\,2}) + (\dot{q}/2k)\,.\,R_i^{\,2}\,.\,\ln(R_i/R_o) \qquad ...(3.20)$$

$$T_{max} - T_o = (\dot{q}/4k)\,(R_o^{\,2} - R_i^{\,2}) + (\dot{q}/2k)\,.\,R_i^{\,2}\,.\,\ln(R_i/R_o) \qquad ...[3.20(a)]$$

COMMENT

For the same volume of material hollow cylinder provides larger outside area compared to solid cylinder. As outside area and convection coefficient influence the maximum temperature, this configuration is found to be preferable. In this case heat is removed from the outside surface.

Example 3.8: *The heat generation rate in a hollow cylinder of ID 0.14082 m and OD 0.018 m is 5 × 10⁶ W/m³. The conductivity of the material is 30 W/mK. If the inside surface is insulated, determine the temperature at the inside surface. The outside surface is exposed to a fluid at 100°C with a convection coefficient of 335.73 W/m²K.*

Solution: As the inside surface is insulated, the heat generated is given to the fluid flowing over the outside surface, by convection process. The outside surface temperature is determined using (considering 1 m length)

$$\dot{q} \times \pi\,(R_o^2 - R_i^2) \times 1 = h \times 2\pi \times R_o \times 1 \times (T_o - T_\infty)$$

$$5 \times 10^6 \times (0.09^2 - 0.07041^2) \times 1 = 335.73 \times 2 \times 0.09 \times 1\,(T_o - 100)$$

Therefore $\mathbf{T_o = 360°C.}$

The temperature at the inner radius is determined using equation (3.19) and substituting R_i for r

$$T_i - T_o = (q/4k)\,(R_o^{\,2} - R_i^{\,2}) + (q/2k)\,R_i^{\,2}\,.\,\ln(R_i/R_o)$$

$$T_i - 360 = \frac{5 \times 10^6}{4 \times 30}(0.09^2 - 0.07041^2) + \frac{5 \times 10^6}{2 \times 30} \times 0.07041^2\,.\,\ln\left[\frac{0.07041}{0.09}\right]$$

$$T_i - 360 = 130.93 - 101.41 = 29.52°\text{C}$$

Therefore $T_i = \mathbf{389.52°C.}$

3.2.4. Hollow Cylinder with Outside Surface Adiabatic

The boundary conditions are:

(*i*) $dT/dr = 0$ at $r = R_o$

(*ii*) $T = T_i$ at $r = R_i$.

This gives

$$T_r - T_i = (\dot{q}/4k)(R_i^2 - r^2) + (\dot{q}/2k)\, R_o^2 \,.\, \ln (r/R_i) \qquad ...(3.21)$$

The maximum temperature is at the outer surface:

$$T_{max} = T_o = T_i + (\dot{q}/2k)\, R_o^2 \,.\, \ln(R_o/R_i) - (\dot{q}/4k)(R_o^2 - R_i^2) \qquad ...(3.22)$$

In all the cases the total heat flow is the heat generated in the volume between the immediate adiabatic section and the section considered. The fluid extracting the heat from the hollow cylinder can flow either outside or inside of the cylinder. In cases where the flow inside the cylinder is advantageous, this type of configuration is used (tubular flow).

Example 3.9: *A hollow cylinder of ID 0.12 m and OD 0.14082 has a heat generation rate of 5×10^6 W/m^3. The conductivity of the material is 30 W/mK. The outside surface is insulated and heat is removed by convection over the inside surface by a fluid at 130°C with a convection coefficient of 226.26 W/m^{2}K. Determine the temperature on the outside surface.*

Solution: The heat generation is convected at the inner surface.

Therefore $\quad \dot{q} \times \pi(R_o^2 - R_i^2) \times 1 = h2\pi R_i \times 1(T_i - T_\infty)$

$$5 \times 10^6 \times (0.07041^2 - 0.06^2) \times 1 = 226.26 \times 2 \times 0.06 \times (T_i - 130)$$

Therefore $\quad T_i = 380°C$

To determine the outside surface temperature, equation (3.22) is used.

$$T_{max} = T_i + (\dot{q}/2k)(R_o^2) \,.\, \ln(R_o/R_i) - (\dot{q}/4k)(R_o^2 - R_i^2)$$

$$\mathbf{T_{max}} = 380 + \frac{5\times10^6}{2\times30} \times 0.070412 \,.\, \ln\left[\frac{0.07041}{0.06}\right] - \frac{5\times10^6}{4\times30}(0.070412 - 0.062)$$

$$= 380 + 66.1 - 56.57 = \mathbf{389.53°C.}$$

Example 3.10: *In the case of nuclear reactors, the fuel rod is clad with a stronger material for various reasons. A fuel rod of 20 mm diameter is clad with zirconium for a thickness of 5 mm. The heat generation in the rod is 2×10^6 W/m^3. The properties of fuel material and cladding are: k_f = 27.4 W/mK, k_c = 22.7 W/mK. The outside fluid temperature is 130°C. The convection coefficient is 500 W/m^{2}K. Determine the fuel rod surface and maximum temperature.*

Solution: Consider 1 m length $\dot{q} = 2 \times 10^6$ W/m^3.

Other known data are shown in Fig. Ex. 3.10.

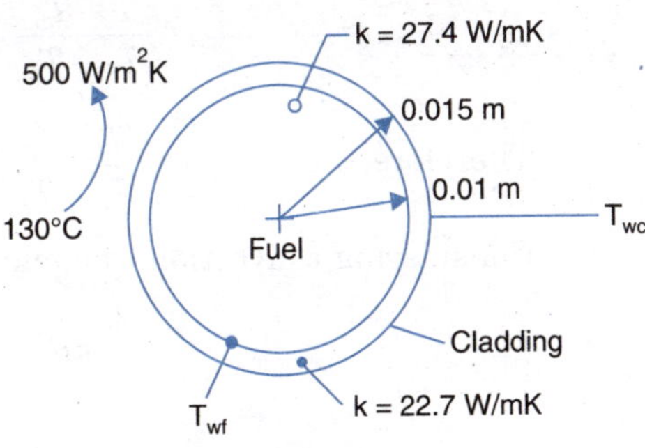

Fig. Ex. 3.10

All the heat generated is convected on the outside.

$$\pi \times 0.01^2 \times 1 \times 2 \times 10^6 = 500 \times \pi \times 0.03 \times 1 \times (T_{wc} - 130)$$

Solving $\quad \mathbf{T_{wc} = 143.3°C}$

The same heat is conducted by the cladding

$$\pi \times 0.01^2 \times 1 \times 2 \times 10^6 = 2\pi \times 22.7 \times 1\, (T_{wf} - 143.3)/\ln \frac{0.015}{0.01}$$

Solving $\quad \mathbf{T_{wf} = 147.7°C}$

$$T_o - T_w = \dot{q}k^2/4k = \frac{2 \times 10^6 \times 0.01^2}{4 \times 27.4} = 1.82$$

$\therefore$ $$T_o = 149.52°C$$

In the absence of cladding

$$T_o = 151.8°C.$$

3.3 RADIAL CONDUCTION IN SOLID SPHERE WITH UNIFORM HEAT GENERATION (FIG. 3.4)

Considering a thin layer of thickness dr at r, the energy equation gives:

$$- k4\pi r^2 dT/dr + 4\pi r^2 \dot{q} dr + k4\pi r^2 dt/dr$$
$$+ (d/dr)(k4\pi r^2 dT/dr)\, dr = 0.$$

$$(d/dr)(r^2 dT/dr) + (\dot{q}/k)r^2 = 0 \quad \text{...(3.23)}$$

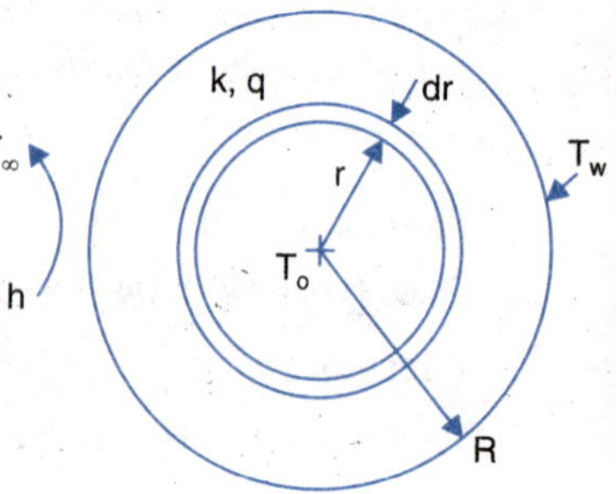

Fig. 3.4. Model

This equation is the same as [2.8(b)] of Chapter 2.

Integrating $$dT/dr = -(\dot{q}/3k)r + C_1/r^2$$

$$dT/dr = 0 \text{ at } r = 0$$

Therefore $$C_1 = (\dot{q}/3k)r^3 = 0$$

Therefore $$T = \frac{-\dot{q}r^2}{6k} + C_2$$

at $r = R, \quad T = T_w$

Therefore $$C_2 = T_w + \frac{\dot{q}R^2}{6k}$$

Therefore $$T - T_w = (\dot{q}/6k)(R^2 - r^2) \quad \text{...[3.24(a)]}$$

The maximum temperature is at $r = 0$.

$$T_o - T_w = (\dot{q}/6k)R^2 \quad \text{...(3.25)}$$

$\therefore$ $$\frac{T - T_w}{T_o - T_w} = 1 - (r/R)^2 \quad \text{...(3.26)}$$

Therefore, $$1 - \frac{T - T_w}{T_o - T_w} = \frac{T_o - T}{T_o - T_w} = (r/R)^2 \quad \text{...[3.26(a)]}$$

Considering convection, the energy balance at the outside is

$$\frac{4}{3}\pi R^3 \dot{q} = 4\pi R^2 h\,(T_w - T_\infty)$$

$\therefore$ $$T_w = T_\infty + \frac{R\dot{q}}{3h}$$

The equation (3.24) can be written as

$$T - T_\infty = \frac{\dot{q}}{6k}(R^2 - r^2) + \frac{R\dot{q}}{3h} \quad \text{...[3.24(b)]}$$

The equation (3.25) can be written as

$$T_o - T_\infty = \left(\frac{\dot{q}}{6k}\right)R^2 + \frac{R\dot{q}}{3h} \qquad ...[3.24(c)]$$

The temperature distribution is parabolic as seen from equation [3.26(*a*)]. The temperature gradient at any section can be found using $-kA\frac{dT}{dr} = Q$. And Q is the heat generated up to the section.

Example 3.11: *A solid sphere of 0.09 m radius generates heat at 5 × 10⁶ W/m³. The conductivity of the material is 30 W/mK. The heat generated is convected over the outer surface to a fluid at 160°C, with a convective heat transfer coefficient of 750 W/m²K. Determine the maximum temperature in the material and the temperature at radius = 0.06 m.*

Solution: To find surface temperature: $(4/3)\,\pi R^3\dot{q} = 4\pi R^2 h(T_w - T_\infty)$

$$(4/3)\,\pi \times 0.09^3 \times 5 \times 10^6 = 4 \times \pi \times 0.09^2 \times 750\,(T_w - 160)$$

Therefore $\mathbf{T_w = 360°C}$

To determine the maximum temperature equation (3.25) is used.

$$\mathbf{T_o} = T_w + \dot{q}R^2/6k = 360 + \frac{5\times10^6\times0.09^2}{6\times30} = \mathbf{585°C}$$

To calculate the temperature at a radius of 0.06 m, equation (3.24) is used.

$$T - T_w = (\dot{q}/6k)\,(R^2 - r^2) = \frac{5\times10^6}{6\times30}\,(0.09^2 - 0.06^2)$$

$$T - T_w = 125°C$$

Therefore $\mathbf{T} = 360 + 125 = \mathbf{485°C.}$

COMMENT

In the case of spheres also, convection coefficient plays an important role in deciding about the maximum temperature.

Example 3.12: *A hollow sphere of ID 12 cm and OD 21 cm made of material with conductivity of 30 W/mK generates heat at a rate of 5 × 10⁶ W/m³. The inside is closed and so can be taken as insulated. Determine the maximum temperature in the solid. The outside temperature is 360°C.*

Solution: Considering the differential equation (3.23) and integrating and applying the boundary conditions of $dT/dr = 0$ at $r = R_i$ and $T = T_o$ at $r = R_o$

$$(d/dr)\,(r^2 dT/dr) + (\dot{q}/k)r^2 = 0,$$

Therefore $dT/dr = (-\dot{q}/3k)r + C_1/r^2$

Therefore $C_1 = +(\dot{q}/3k)\,(R_i^3)$, Substituting

$$dT/dr = -(\dot{q}/3k)\,r + \frac{\dot{q}R_i^3}{3k}\,(1/r^2),$$ Integrating the equation.

$$T = -(\frac{\dot{q}R_i^3}{3k}/6k)r^2 - \frac{\dot{q}R_i^3}{3k}\frac{1}{r} + C_2$$ Using the second boundary

$$T_o = (-\dot{q}/6k)\,(R_o^2) - \frac{\dot{q}}{3k}\frac{R_i^3}{R_o} + C_2$$

Therefore $C_2 = T_o + (\dot{q}/6k)\, R_o^{\,2} + (\dot{q}/3k)\, R_i^{\,3}/R_o$ Substituting and rearranging

$$\mathbf{T - T_o = (\dot{q}/6k)\,(R_o^2 - r^2) - (\dot{q}\,R_i^3/3k)\,(1/r - 1/R_o)} \quad ...(1)$$

Substituting for $r = R_i$ to get the maximum temperature

$$\mathbf{T} = 360 + \frac{5\times 10^6}{6\times 30}(0.105^2 - 0.06^2) - \frac{5\times 10^6 \times 0.06^3}{3\times 30}\ \mathbf{(1/0.06 - 1/0.105)}$$

$= \mathbf{480.53°C}$ (compare with example 3.9, 585°C)

COMMENT

The volume of sphere in Example 3.7 is nearly the same as in this problem. But the maximum temperature is reduced considerably because of the increase in the surface area for heat flow.

Example 3.13: *Heat is generated in a sphere of radius R_o at a rate depending on the radius given by $q_r = q_o\,(r/R_o)$ where q_r is the generation rate in W/m^3 at any radius r. The material thermal conductivity is k. Derive an expression for the temperature variation along the radius. Also obtain the value of total heat generated in the sphere. Check the value using the heat conducted at the surface. Considering $q_o = 5 \times 10^6\ W/m^3$, $R_o = 0.09$ m, $k = 30$ W/mK and $T_o = 360°C$, determine the maximum temperature in the sphere.*

Solution: Using the general differential equation (3.23).

$(d/dr)\,(r^2 dT/dr) + (\dot{q}/k)r^2 = 0$, substituting for $\dot{q} = q_o \times r/R_o$

$(d/dr)\,(r^2 dT/dr) + (q_o/kR_o)\,(r^3) = 0.$

Integrating $r^2 dT/dr = -(q_o/4kR_o)r^4 + C_1$ and C_1 is zero as $dT/dr = 0$ at $r = 0$.

Therefore $$dT/dr = -\frac{q_o}{4kR_o}\,.\,r^2 \quad ...(1)$$

Therefore $$T = -\frac{q_o}{12kR_o}\,.\,r^3 + C_2 \quad ...(2)$$

Using the condition that $T = T_o$ at $R = R_o$.

$$C_2 = T_o + \left[\frac{q_o}{12k}\,.\,R_o^2\right]$$

Substituting the value the temperature variation with radius is given by

$$T - T_o = \left[\frac{q_o}{12kR_o}(R_o^3 - r^3)\right] \quad ...(3)$$

Total heat generation can be found by summing up the heat generated in layers of dr thickness at various radii.

$$\mathbf{Q} = q_o \int_0^{R_o} 4\pi r^2\,.\,(r/R_o)\,dr = \mathbf{q_o \pi R_o^3}$$

$Q = (-kAdT/dr)$ at $r = R_o$ from equation (E10.1) (dT/dr) at $r = R_o = -q_o/4k.R_o$

$A = 4\pi R_o^2$ Therefore $Q = -k \times 4\pi R_o^2 \times (-q_o R_o/4k)$

$= q_o \pi R_o^{\,3}$. Thus checked.

Using equation (3) and substituting the specified values,

Maximum temperature $= T_o + \left[\frac{q_o}{12kR_o} \cdot R_o^3\right] = 360 + \left[\frac{5 \times 10^6 \times 0.09^3}{12 \times 30 \times 0.09}\right]$

$=$ **472.5°C** (compare with Example 3.9)

3.4 CONCLUSION

The heat generated in a solid is conducted through the solid and is then convected at the surface. The maximum temperature in the solid is often the limiting factor in design. Once the conductivity, heat generation rate and size are specified, the convective heat transfer coefficient has to be increased to the limiting value. In case the convection coefficient has limitations, then the size or the heat generation rate has to be reduced to obtain the limiting maximum temperature. In such a case cylinder of same linear dimension is better compared with slab configuration. In case the heat generation rate increases with temperature as in the case of electrical conductors, instability or continued increase in temperature may result.

ADVANCED EXAMPLES

Example A-3.1: *A plate shaped nuclear fuel element of 24 mm thickness exposed on the sides to convection at 200°C with a convective heat transfer coefficient of 900 W/m²K generates heat at 20 MW/m³. Determine (i) the surface temperature, (ii) the maximum temperature in the plate and (iii) the temperature gradient at the surface. Thermal conductivity of the material is 25 W/mK.*

Solution: The data given are shown in Fig. AE-3.1.

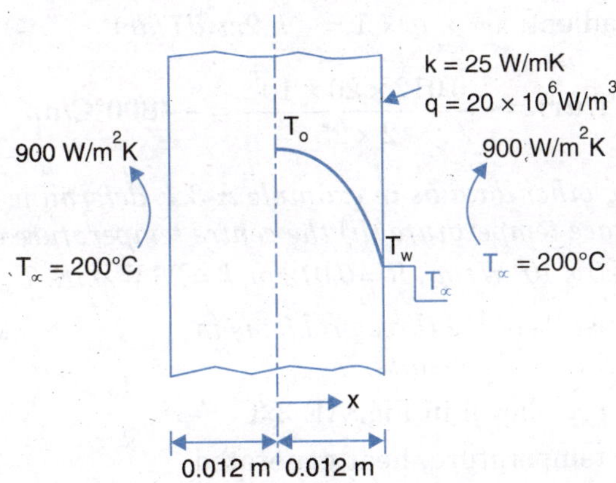

Fig. AE-3.1. Problem model

(*i*) To find the surface temperatures, heat generated = heat convected

$$\dot{q}L \times 1 = h \times 1(T_w - T_\infty)$$

$20 \times 10^6 \times 0.012 = 900(T_w - 200)$. Therefore $\mathbf{T_w = 466.7°C}$

(*ii*) $$T_o - T_w = (\dot{q}/2k)\,(L^2) = \frac{20 \times 10^6 \times 0.012^2}{2 \times 25} = 57.6°C$$

Therefore T_o = Maximum temperature = 466.7 + 57.6 = **524.3°C**

(*iii*) The temperature gradient at the surface is found by using

$$Q = -kA\, dT/dx/x = L,\ Q = \dot{q} \times L \times 1$$

Therefore $$\textbf{(dT/dx)} = \frac{-q \times L \times 1}{k \times 1} = \frac{20 \times 10^6 \times 0.012}{25} = \mathbf{-9600°C/m.}$$

Example A-3.2: *Determine in the case of a cylindrical fuel rod of radius 0.012 m, considering other details as in example A-3.1 (i) the surface temperature (ii) the centre temperature and (iii) the temperature gradient at the surface. The data given are shown in Fig. AE-3.2.*

$k = 25\ W/mK,\ Q = 20 \times 10^6\ W/m^3$

$T_\infty = 200°C,\ h = 900\ W/m^2K.$

Solution: (*i*) To find the surface temperature: Heat generated = heat convected

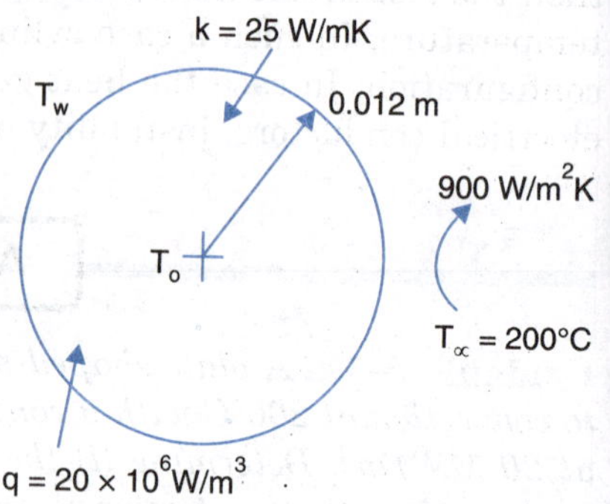

Fig. AE-3.2. Solid cylinder

Therefore $\pi r^2 \times 1.\dot{q} = 2\pi r \times 1\,.\,h(T_w - T_\infty)$.

$$\mathbf{T_w} = T_\infty + r\dot{q}/2h$$
$$= 200 + 0.012 \times 20 \times 10^6/2 \times 900$$
$$= \mathbf{333.3C.}$$

(*ii*) $$T_o = T_w + (\dot{q}R^2/4k)$$
$$= 333.3 + \frac{20 \times 10^6 \times 0.012^2}{4 \times 25}$$
$$= 333.3 + 28.8 = \mathbf{362.1°C}$$

(*iii*) Temperature gradient: $\pi r^2 \times \dot{q} \times 1 = -k.2\pi r(dT/dr)$

Therefore $$\textbf{(dT/dr)} = -\frac{0.012 \times 20 \times 10^6}{2 \times 25} = \mathbf{-4800°C/m.}$$

Example A-3.3: (*a*) *Using other data as in example A-3.1, determine in the case of a sphere of 0.012 m radius (i) the surface temperature (ii) the centre temperature and (iii) the temperature gradient at the surface.* $\dot{q} = 20 \times 10^6\ W/m^3,\ R = 0.012\ m,\ k = 25\ W/mK,\ T_\infty = 200°C,\ h = 900\ W/m^2K.$

(*b*) *Summarise the results of the three problems in a tabular form and comment on the results.*

Solution: The given data are shown in Fig. AE-3.3.

To find the surface temperature, heat generated = heat convected.

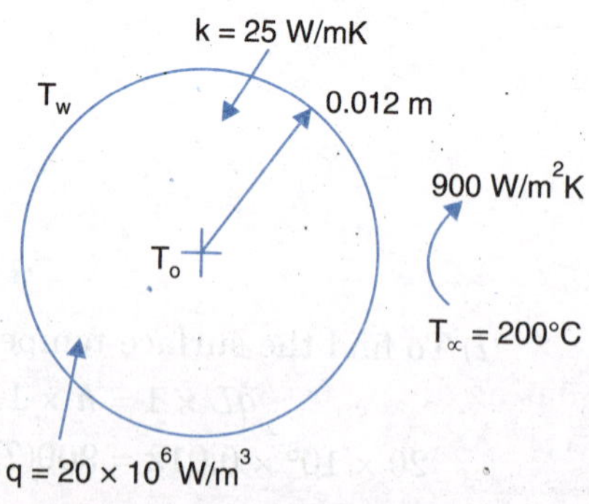

Fig. AE-3.3. Solid sphere

$$(4/3)\pi R^3 \dot{q} = 4\pi R^2\,.\,h(T_w - T_\infty)$$

Therefore $$T_w = T_\infty + (R\dot{q}/3h)$$
$$= 200 + \frac{0.012 \times 20 \times 10^6}{3 \times 900}$$
$$= \mathbf{288.9°C}$$

(*ii*) $T_o = T_w + (\dot{q}R^2/6k)$

$$= 288.9 + \frac{20 \times 10^6 \times 0.012^2}{6 \times 25}$$

$$= 288.9 + 19.2$$

Maximum temperature = 308.1°C

(*iii*) Temperature gradient at the surface

$$Q = -kA\, dT/dr. \text{ Therefore } (4/3)\,\pi R^3 \times \dot{q} = -k4\pi R^2\,(dT/dr)$$

Therefore $\quad \mathbf{(dT/dr)} = (-R/3)\,(\dot{q}/k) = -\frac{0.012 \times 20 \times 10^6}{3 \times 25} = \mathbf{-3200°C/m}$

(*b*) The values are tabulated below for comparison.

Shape	***Surface temp. °C***	***Maximum °C***	***temp. diff. ΔT°C***	***Temp. gradient at surface °C/m***
Plate	466.7	524.3	57.6	– 9600
Cylinder	333.3	362.1	28.8	– 4800
Sphere	288.9	308.1	19.2	– 3200

COMMENTS

1. The temperature increase is highest in the plate type of geometry and lowest for spherical shape. Discuss why ?
2. If the total temperature drop is the criterion then either lower values of heat generation or higher values of convection coefficient or smaller linear dimension has to be used.
3. There is another aspect to the problem. In case the same amount of heat generation (not rate) is required the number of cylinders required will be 53. If the spherical shape is adopted 331 spheres will be needed.
4. Volume/surface area in these cases will also be a guidance in deciding about the shape. Smaller this ratio, lower will be the maximum temperature.

Example A-3.4: *Assuming that the maximum temperature above the fluid temperature to be the same, compare for the same radius, the heat generation rates of cylinder and spherical configurations.*

Solution: $(T_o - T_\infty)$ should be the same for both cylinder and sphere. Using equation 3.15 and 3.24 (*c*).

$$(T_o - T_\infty)_{cyl} = \frac{\dot{q}_{cyl}}{4k}R^2 + \frac{\dot{q}_{cyl}\,.R}{2h}$$

$$(T_o - T_\infty)_{sp} = \frac{\dot{q}_{sp}}{6k}R^2 + \frac{\dot{q}_{sp}\,.R}{3h}$$

equating and dividing by $\dot{q}_{cyl}$ and cancelling R

$$\frac{R}{4k} + \frac{1}{2h} = \frac{\dot{q}_{sp}}{\dot{q}_{cyl}}\left[\frac{R}{6k} + \frac{1}{3h}\right]$$

$$\therefore \quad \frac{\dot{q}_{sp}}{\dot{q}_{cyl}} = \frac{\dfrac{R}{4k} + \dfrac{1}{2h}}{\dfrac{R}{6k} + \dfrac{1}{3h}} = \frac{9}{4} \times \frac{2Rh + 4k}{3Rh + 6k}$$

Let $R = 0.012$ m, $k = 30$ W/m²k and $h = 900$ W/m²

and $\dot{q}_{cyl}$ = **20 × 10⁶ W/m³.**

$$\frac{\dot{q}_{sp}}{20 \times 10^6} = \frac{9}{4} \frac{(2 \times 0.012 \times 900) + (4 \times 30)}{(3 \times 0.012 \times 900) + (6 \times 30)} = 1.5$$

∴ heat generation in the **sphere** for the given condition is **30 × 10⁶ W/m³.**

For the same radius and same temperature drop spherical shape will dissipate more heat due to the (volume/area) ratio of (r/3) and r.

Example A-3.5: *The heat generation rate in a plane wall of 0.24 m thickness is 0.4 MW/m³. The wall is exposed on both sides to convection at 30°C.*

(a) Determine and compare the maximum temperatures for k = 25, k = 50, k = 200 and k = 410 W/mK assuming h = 250 W/m²K.

(b) Determine and compare the maximum temperature for h = 50, 250, 500 and 1000 W/m²K with k = 25 W/mK.

Solution: (a) In all cases in this part $h = 250$ W/m²K. Therefore T_w can be found using: $1 \times 0.12 \times 4 \times 10^5 = 250 \times 1\ (T_w - 30)$

Therefore **T_w = 222°C**

$(T_o - T_w) = qL^2/2k$. Therefore for $k = 25$, $T_o - T_w = 115.2$°C

Therefore **T_o = 337.2°C.**

For other values, similar calculations yield the results in the tabulation.

k, W/mK	T_w °C	$(T_o - T_w)$ °C	T_o °C
25	222	115.2	337.2
50	222	57.6	279.6
200	222	14.4	236.4
410	222	7.0	229.0

(b) Sample calculation: In this case k is constant at 25 W/mK.

heat generated = heat convected

$$\dot{q} \times L \times 1 = h \times 1\ (T_w - T_\infty)$$

Therefore $T_w = T_\infty + \dot{q}\ L/h$, in the case **h = 50 W/m²K,**

$$T_w = 30 + \left[\frac{4 \times 10^5 \times 0.12}{50}\right] = \mathbf{990°C}$$

$$T_o - T_w = (\dot{q}/2k)\ (L^2) = \frac{4 \times 10^5 \times 0.12^2}{2 \times 25} = 115.2°\text{C},$$

and is constant for all h values.

For $\quad$ **h = 50 W/m²K, T_o = 1105.2°C**

For other values of *h*, values are calculated as per the sample and tabulated below.

h, W/mK	T_w °C	$T_o - T_w$ °C	T_o °C
50	990	115.2	1105.2
250	222	115.2	337.2
500	126	115.2	241.2
1000	78	115.2	193.2

COMMENTS

1. The value of convection coefficient influences the maximum temperature and whenever heat generation is encounterd, it is desirable to design for high value of convection coefficient.
2. Heat generation rate and thickness also play an important role on the value of maximum temperature.
3. Convection coefficient has a more pronounced effect in reducing maximum temperature, as compared to conductivity.

Example A-3.6: *A nuclear fuel rod is in the form of solid cylinder. The heat generation rate is 50 MW/m³. The conductivity of the material is 24.2 W/mK. The heat generated is absorbed by a fluid at 200°C. Determine the convective heat transfer coefficient required for diameter of (i) 25 mm and (ii) 16 mm.*

Solution: Using the equation 3.15(*b*)

Case 1:

$$T_o - T_\infty = \frac{\dot{q}R^2}{4k} + \frac{\dot{q}R}{2h}$$

$$360 - 200 = \frac{50 \times 10^6 \times 0.0125^2}{4 \times 24.2} + \frac{50 \times 10^6 \times 0.0125}{2h}$$

Solving $\quad$ **h = 3941 W/m²K.**

Case 2:

$$360 - 200 = \frac{50 \times 10^6 \times 0.008^2}{4 \times 24.1} + \frac{50 \times 10^6 \times 0.008}{2h}$$

Solving $\quad$ **h = 1575 W/m²K.**

COMMENT

Smaller diameter means lower convection coefficient, but the heat generated is also lower. An optimum has to be worked out balancing the size and number of fuel rods.

Example A-3.7: *A cylindrical nuclear fuel rod of 25 mm dia solid cylinder is converted to hollow cylinder of 5 mm thickness. If the maximum temperature in the rod should not exceed 360°C when the heat is absorbed on the outside by a fluid at 200°C, determine the convection coefficient required. Conductivity of the material is 24.2 W/mK.*

Solution: Equating the volumes with inside radius r_i

$$\pi \times 0.0125^2 \times 1 = \pi \times 1\ [(r_i + 0.005)^2 - r_i^2]$$

Solving $\quad$ **r_i = 0.013125 m** and **r_o = 0.018125 m.**

Inside is adiabatic. All the heat flows at the outer surface.

Using equation 3.20(*a*)

$$T_{max} - T_o = (\dot{q}/4k)\,(r_o^2 - r_i^2) + (\dot{q}/2k)\,r_i^2\,.\,\ln(r_i/r_o),\ T_o - \text{Surface temp.}$$

$$= \frac{50 \times 10^6}{4 \times 24.2}\,(0.0181252 - 0.0131252) + \frac{50 \times 10^6}{2 \times 24.2}\,.\,0.0131252 \ln \frac{0.013}{0.018}$$

$$= 23.26,$$

∴ $\mathbf{T_o = 336.24}$°C, as T_{max} = 360°C

$\mathbf{Q = hA(336.24 - 200)}$

∴ $$\mathbf{h} = \frac{50 \times 10^6 \times (0.018125^2 - 0.013125^2) \times 1}{0.018125 \times 2 \times 136.24} = \mathbf{1581.8\ W/m^2K.}$$

COMMENTS

1. This is better compared to 3940.7 for solid cylinder, Example 3.6.
2. By increasing the surface area for heat collection, the hollow cylindrical shape helps in adopting lower value of convection coefficient.
3. Space required and cost of production will be higher.

Example A-3.8: *A nuclear reactor fuel rod of solid cylinder of 25 mm is converted into + shaped form of thickness 6 mm. The heat generation rate is 50 MW/m³. Thermal conductivity is 24.2 W/mK. The heat generated is absorbed by a fluid at 200°C. Determine the convective heat transfer coefficient on the surface for maximum temperature of 360°C.*

Solution: Assuming B as side length (See Fig. AE-3.8)

$$\pi \times 0.0125^2 = B \times 0.006 + (B - 0.006)\,0.006$$
$$= B \times 2.0 \times 0.006 - 0.006^2$$

Therefore $B = 43.91$ mm

The maximum temperature can be found using a slab of 6 mm thickness, $L = 3$ mm

$$T_o - T_w = (\dot{q}/2k)L^2 = \frac{50 \times 10^6 \times 0.003^2}{2 \times 24.2}$$

$$= 9.3°\text{C}$$

Fig. AE-3.8. Problem model

Therefore $\mathbf{T_w}$ = 360 – 9.3 = **350.7°C**

$Q = hA$ (350.7 – 200) considering one arm, Assuming unit length

$Q = 1 \times 0.003 \times 0.04391 \times 50 \times 10^6 = h \times 0.04391 \times (350.7 - 200)$

Therefore **h = 995.4 W/m²K.**

COMMENTS

1. This is also a good configuration for heat removal.
2. The basic requirement is that for a given volume, generating heat, the surface area should be increased to reduce the convection coefficient and maximum temperature.

Example A-3.9: *A nuclear fuel element is in the form of a solid sphere of 12.5 mm radius. Its thermal conductivity is 24.2 W/mK. The heat generation rate is 50 × 10⁶ W/m³. The heat generated is absorbed by a fluid at 200°C. If the maximum temperature is limited to 360°C, determine the convection coefficient required.*

Solution: $T_o - T_w = (q/6k)\,(R^2)$

$$360 - T_w = \frac{50 \times 10^6 \times 0.0125^2}{6 \times 24.2} = 53.8°\text{C}$$

(compare 80.7°C for solid cylinder)

Therefore **surface temperature**

$$= 360 - 53.8 = \mathbf{306.2°C}$$

Heat generated = Heat convected

$$(4/3)R^3\,\dot{q}\pi = h.4\pi R^2\,(T_w - T_\infty)$$

$$h = (R/3)\,q/(T_w - T_\infty)$$

$$= \frac{0.0125}{3} \times \frac{50 \times 10^6}{(306.2 - 200)}$$

$$= \mathbf{1961.7\ W/m^2K}$$

Also on the higher side.

Hollow cylinder with inside adiabatic conditions appears to be the one requiring least value of convection coefficient.

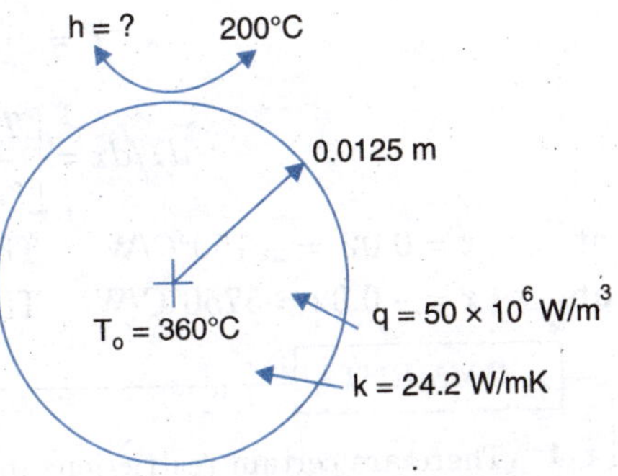

Fig. AE-3.9. Solid sphere

Example A-3.10: *A wall 40 mm thick has its surfaces maintained at 0°C and 100°C. The heat generation rate is 3.25 × 10⁵ W/m³. If the thermal conductivity of the material is 2 W/mK, determine the temperature at the mid plane, the location and value of the maximum temperature and the heat flow at either end.*

Solution: The data are shown in Fig. AE-3.10.

Equation (3.8), (3.9), (3.10) and are applicable for this problem.

$$T_x = \frac{T_{w2} + T_{w1}}{2} + \frac{T_{w2} + T_{w1}}{2}\,.\,(x/L) + (\dot{q}/2k)\,(L^2 - x^2)$$

$$x_{max} = (k/2\dot{q}L)\,(T_{w2} - T_{w1})$$

The temperature at $x = 0$, (mid plane)

$$T_o = \frac{100 + 0}{2} + \frac{100 - 0}{2}\,.\,\frac{0}{L} + \frac{3.25 \times 10^5}{2 \times 2} \times 0.02^2$$

$$= 82.5°\text{C}$$

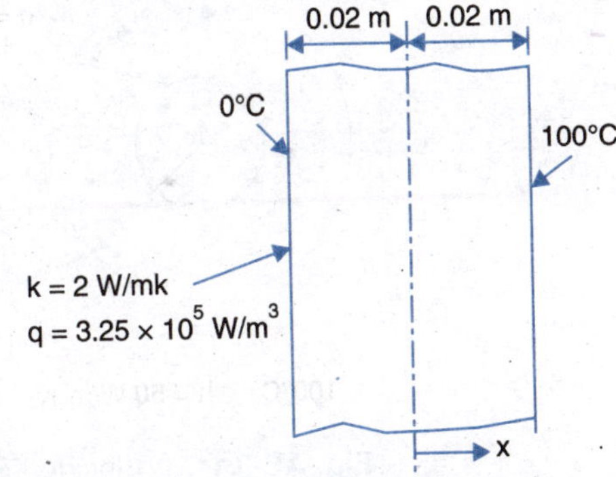

Fig. AE-3.10. Slab with different boundary condition

$$\mathbf{x_{max}} = \frac{2}{2 \times 3.25 \times 10^5 \times 0.02}\,(100 - 0) = \mathbf{0.01538\ m}$$

(positive and so on the right of centre)

$$\text{Therefore } \mathbf{T_{max}} = \frac{100 + 0}{2} + \frac{100 - 0}{2}\,.\,\frac{0.01538}{0.02} + \frac{3.25 \times 10^5}{2 \times 2} \times (0.022 - 0.015382)$$

$$= 50 + 38.46 + 13.27 = \mathbf{101.74°C}$$

heat flow is found either by using $- kA\,dT/dx$ or by calculation of the heat generated from the maximum temperature plane.

On the 100°C side, $(0.2 - 0.01538) \times q = 1500\ \text{W/m}^2$

On the 0°C side $(0.02 + 0.01538)$ $\quad q = 11500\ \text{W/m}^2$

Total = 13000 W/m²

By differentiating equation (3.8)

$$T = \frac{T_{w_1} + T_{w_2}}{2} + \frac{\dot{q}}{2k}(L^2 - x^2) + \frac{T_{w_2} - T_{w_1}}{2L}x$$

$$\therefore \qquad dT/dx = \left[\frac{T_{w_2} - T_{w_1}}{2L}\right] - \left[\frac{\dot{q}}{k}(x)\right]$$

at $x = 0.02$, = – 750°C/W Therefore $Q = -2 \times 1 \times -750 = 1500$ W/m²

at $x = -0.02$ = 5750°C/W Therefore $Q = 2 \times 5750 = 11500$ W/m² checks.

COMMENTS

1. There are certain restrictions in this boundary. For example if $(T_{w2} - T_{w1}) > (2\dot{q}/k)(L^2)$ then the solution will indicate that the maximum temperature occurs at a distance greater than L.
2. The data in this problem is academic. But in real arrangement of fuel elements, it can happen due to fluid flow differences that the two sides may be at different temperatures. This may cause distorsion of the element and possible damage. So fluid flow around elements may have to be analysed very carefully to avoid this problem.

Example A-3.11: *Using numeric methods solve for the temperature for the models shown in Fig. AE-3.11(a) and AE-3.11(b).*

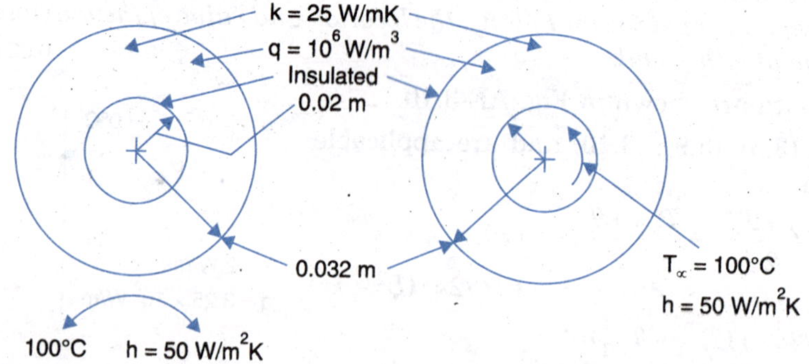

Fig. AE-3.11. (a) Inside insulated **Fig. AE-3.11.** (b) Outside insulated

Solution: (*a*) Consider nodes as indicated [Fig. AE-3.11(*c*)]

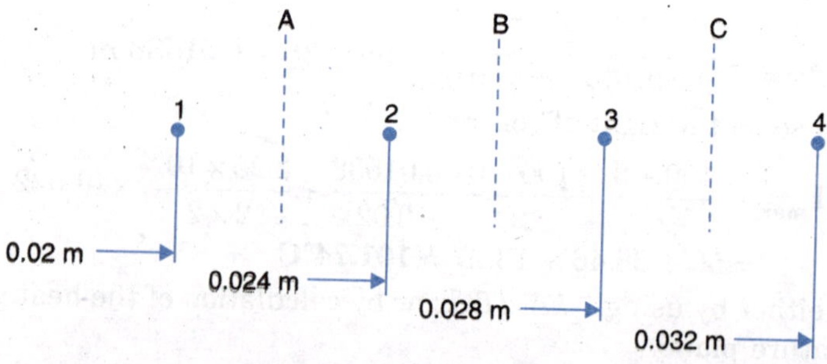

Fig. AE-3.11. (c) Nodes

T_4 can be determined using heat generated = heat conducted assuming 1 m length,

$$\dot{q} \times \pi (0.032^2 - 0.02^2) = 2\pi rh(T_4 - T_\infty)$$

$$10^6 \times (0.0932^2 - 0.02^2) = 2 \times 0.032 \times 50\,(T_4 - 100)$$

Therefore $\mathbf{T_4 = 295°C}$

Considering the heat flow between node 3 and 4, the heat generated up to plane C is conducted with temperature drop between 3 and 4.

Therefore $10^6.\pi \times (0.03^2 - 0.02^2) = \dfrac{25 \times 2 \times 0.03\,(T_3 - 295)\,\pi}{0.004}$

Therefore $\mathbf{T_3 = 296.33}$

Similarly $10^6 \times (0.026^2 - 0.02^2) = \dfrac{25 \times 2 \times 0.026\,(T_2 - 296.33)}{0.004}$

(Cancelling π on both sides)

Therefore $\mathbf{T_2 = 297.18°C}$

Similarly $10^6 \times (0.022^2 - 0.02^2)$

$$= \frac{25 \times 2 \times 0.022\,(T_1 - 297.18)}{0.004} \quad \therefore\ \mathbf{T_1 = 297.49°C}$$

Therefore $\Delta T = T_1 - T_4 = \mathbf{2.49°C}$

Check: use equation (3.20)

$$T_{max} - T_o = \frac{10^6}{4 \times 25}(0.032^2 - 0.02^2) + \frac{10^6}{2 \times 25}(0.02^2)\ln(0.02/0.032)$$

$$= 6.24 - 3.76 = \mathbf{2.48°C}\text{ checks.}$$

(*b*) By using similar procedure: **(for outside adiabatic)**

$10^6 \times (0.032^2 - 0.02^2) = 2 \times 0.02 \times 50\,(T_1 - 100)$ $\quad \mathbf{T_1 = 412°C}$

$10^6 \times (0.032^2 - 0.22^2) = 2 \times 0.022 \times 25\,(T_2 - 412)/0.004,$ $\quad \mathbf{T_2 = 413.96°C}$

$10^6 \times (0.032^2 - 0.026^2) = \dfrac{2 \times 0.026 \times 25\,(T_2 - 413.96)}{0.004},$ $\quad \mathbf{T_3 = 415.03°C}$

$10^6 \times (0.032^2 - 0.03^2) = \dfrac{2 \times 0.03 \times 25\,(T_4 - 415.03)}{0.004},$ $\quad \mathbf{T_4 = 415.36°C}$

Therefore $\Delta T = T_4 - T_1 = \mathbf{3.36°C}$

Check: use equation (3.22)

$$T = (\dot{q}/2k)(R_o^2)\ln(R_o/R_i) - (\dot{q}/4k)(R_o^2 - R_i^2)$$

$$= \frac{10^6}{2 \times 25} \times 0.032^2 \ln(0.032/0.02) - \left[\frac{10^6}{4 \times 25}(0.032^2 - 0.02^2)\right]$$

$$= 9.63 - 6.24 = \mathbf{3.39°C}\text{ checks.}$$

COMMENT

The numerical method is very powerful in solving even odd shaped elements.

Example A-3.12: *Nuclear fuel rods are to be clad with aluminium or stainless steel or zirconium with melting temperatures in the range of 650°C, 1400°C, and 1800°C. The diameter is 30 mm and the heat generation rate is 7 × 10^8 W/m^3. The convection on the surface is to a fluid at 100°C with h = 7000 W/m^{2}K. The conductivity of the material is 52 W/mK. At shut down times, coolant may not circulate and the surface temperature may reach the mean between the steady state surface and centre temperature. Check for the material to be used.*

Solution: The data given are shown in Fig. AE-3.12.

The surface temperature is found using heat generated = heat conducted

$$\pi r^2 \times \dot{q} = h 2\pi r\,(T_w - T_\infty)$$

$$0.015 \times 7 \times 10^8 = 2 \times 7000\,(T_w - 100)$$

Therefore $\mathbf{T_w = 850°C.}$

$$T_o - T_w = (q/4k)\,R^2$$

$$= \frac{7 \times 10^8 \times 0.015^2}{4 \times 52}$$

$$= 757.2°C$$

Therefore $\mathbf{T_o = 1607.2°C}$

Mean temperature $= \dfrac{1607.2 + 850}{2} = 1228.6°C.$

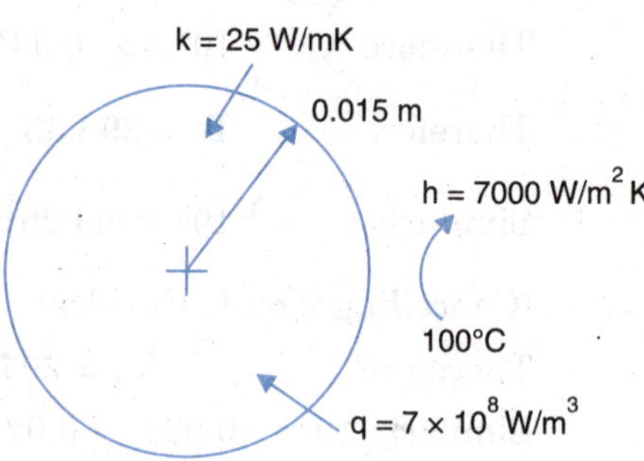

Fig. AE-3.12. Problem model

1. Aluminium cannot be used.
2. To check for stainless steel: mean temp. = 1228°C

Therefore use of stainless steel is risky as it is near 1400°C. **Hence Zirconium cladding can be safely used**. (look for its exact melting point).

The assumption is that after shut down, heat generation will cease which in actual practise is not the case.

Example A-3.13: *A copper cable of 30 mm diameter carries 300A when exposed to air at 30°C with a convection coefficient of 20 W/m^{2}K. The resistance is 5 × 10^{-3} ohm/m. If k = 395 W/mK, determine the surface and center temperatures.*

Solution: Heat generated in 1 m length = I^2R = $300^2 \times 5 \times 10^{-3}$ W = 450 W

$$\dot{q} = 450/\pi r^2 \times 1 = 450/\pi \times 0.015^2 = \mathbf{636620\ W/m^3}$$

To find surface temperature:

$$450 = 2\pi \times 0.015 \times 20\,(T_w - 30) \qquad \text{Therefore } \mathbf{T_w = 268.73°C}$$

$$T_o - T_w = \left[\frac{\dot{q}R^2}{4k}\right] = \frac{636620 \times 0.015^2}{4 \times 395} = 0.09°C$$

Therefore **center temperature = 268.82°C.**

The cable is more or less at uniform temperature. This is generally the case with materials of high conductivity and small diameters.

Example A-3.14: *In a slab of thickness as shown in Fig. AE-3.14 insulated at x = 0 plane, heat is generated at any plane as per $\dot{q} = q_o[x/L]$. Determine the temperature distribution, heat flow at any section and the heat flow at the surface.*

Solution: The model is shown in Fig. AE-3.14.

The differential equation reduces to (Ref. equation 3.1)

$$d^2T/dx^2 = (-\,q_o/k)\,(x/L)$$

Integrating: $dT/dx = (-\,q_o/k)\,.\,(x^2/2L) + C_1$

as $dT/dx = 0$ at $x = 0$, (insulated) $\therefore\ C_1 = 0.$

Integrating $T = -\dfrac{q_o}{6kL}\,(x^3) + C_2$

at $x = L,\ \ T = T_w.$ Therefore $T_w + \left[\dfrac{q_o}{6kL}.L^3\right] = C_2$

Therefore $T = \left[\dfrac{q_o}{6kL}.(L^3 - x^3)\right] + T_w$

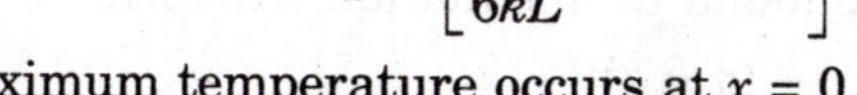

or $T - T_w = \left[\dfrac{q_o}{6kL}.(L^3 - x^3)\right]$ ...(AE-14.1)

Fig. AE-3.14. Problem model

The maximum temperature occurs at $x = 0$

Therefore $T_o = T_w + \left[\dfrac{q_o L^2}{6k}\right]$ ...(AE-14.2)

The heat flow at any location x is given as

$$Q = -\,kA\,dT/dx = (-\,k)\,.\left[\frac{-\,q_o}{k}.x^2/2L\right] = \frac{\mathbf{q_o x^2}}{\mathbf{2L}}\ \mathbf{W/m^2}$$

At the surfae $x = L$, $\therefore\ \ Q = \dfrac{q_{oL}}{2}.\ \mathrm{W/m^2}.$

COMMENT

In actual situation, the heat generation in most cases (including nuclear fuel rod) will not be uniform. It may be higher near the surface compared to the centre.

Example A-3.15: *In a slab the heat generation rate at x is given by $\dot{q} = 10^6\,e^{-50x}\ W/m^3$ where x is in m. Consider a slab insulated at x = 0, and the thickness at which $\dot{q}$ = 1000 W. Determine the temperature difference between the planes and the heat flow at the surface.*

Solution: This variation can be considered as $\dot{q} = q_o e^{-ax}$

(**Note:** a has a dimension of 1/m)

Substituting this in the differential equation.

$$(d^2T/dx^2) + (q/k) = 0 \text{ is modified as } \frac{d^2T}{dx^2} + \frac{q_o e^{-ax}}{k} = 0.$$

Integrating it $dT/dx = \left[\dfrac{q_o}{ka}e^{-ax} + C_1\right]$

As at $x = 0$, dT/dx

$$\therefore\quad C_1 = \left[\frac{-\,q_o}{ka}e^{-ao}\right] \text{ or } C_1 = \left[\frac{-\,q_o}{ka}\right]$$

Therefore $\frac{dT}{dx} = \left[\frac{q_o}{ka} e^{-ax} - \frac{q_o}{ka}\right]$

Integrating $T = \frac{-q_o}{ka^2} . e^{-ax} - \frac{q_o x}{ka} + C_2$

at $x = L$ $T = T_w$

Therefore ∴ $T_w = \frac{-q_o}{ka^2} . e^{-aL} - \frac{q_o L}{ka} + C_2$

Therefore $C_2 = (q_o/ka^2)\, e^{-aL} + (q_o/ka)(L) + T_w$

Therefore $T = [q_o/ka^2]\,(e^{-aL} - e^{-ax}) + (q_o/ka)\,(L - x) + T_w$

or $T - T_w = (q_o/ka^2)\,(e^{-aL} - e^{-ax}) + (q_o/ka)\,(L - x)$ ...(AE-15.1)

Maximum temperature occurs at $x = 0$.

$$T_o - T_w = (q_o/ka^2)\,(e^{-aL} - 1) + (q_o/ka)\,(L)$$

Heat flow = $-\, kA\, dT/dx$, dT/dx can be found from equation AE-15.1.

as $dT/dx = (q_o/ka)\,(e^{-ax}) - (q_o/ka)$ ...[AE-15.2(*a*)]

Therefore $Q\,|\,$Surface $= -\frac{q_o}{a}\,[1 - e^{-aL}]$ ...(AE-15.2)

For $\mathbf{q_o = 10^6,\ a = 50,\ k = 15\ W/mK.}$

The thickness is found using the conditions that at $x = L$,

$$\dot{q} = 1000 = q_o e^{-ax} = 10^6 e^{-50x}.\ \text{Solving } \mathbf{x = 0.1382\ m.}$$

$$T_o - T_w = (q_o L/ka^2)\,(e^{-aL} - 1) + (q_o L/ka)$$

$$= \frac{10^6}{15 \times 50^2} . [e^{(-50 \times 0.1382)} - 1] + \left[\frac{10^6}{15 \times 50} \times 0.1382\right]$$

$$\Delta T = -\,26.64 + 184.2 = 157.6°\text{C}$$

(dt/dx) surface $= \frac{10^6}{15 \times 50} . e^{-50 \times 0.1382} - \frac{10^6}{15 \times 50}$ (equation [15.2(*a*)] is used)

$= 1.33 - 1333.33 =$ **– 1332°C/m**

Therefore $Q = (-\,1332) \times (-\,15) =$ **19980 W/m²**

Also from equation AE-15.2, $Q = (q_o/a)\,(1 - e^{-aL}) = \frac{10^6}{50}\,(1 - e^{50 \times 0.1382}) =$ **19980 W.**

Example A-3.16: *In a plane wall heat flux $\beta\, q_o$ enters the wall at $x = 0$. In the wall there is a uniform heat generation $\dot{q} = \dot{q}_o\,(1 - \beta)\,\alpha\, e^{-\alpha x}$. Under steady state conditions all the heat is convected at $x = L$. Determine the temperature distribution, the temperature drop $(T_o - T_L)$ and the heat flow at any section as well as at $x = L$.*

Solution: The model is shown in Fig. AE-3.16.

The differential equation reduces to $(d^2T/dx^2) + (q_o/k)\,(1 - \beta)\,\alpha\, .\, e^{-\alpha x} = 0$.

Integrating $dT/dx = (q_o/k)\,(1 - \beta)\, e^{-ax} + C_1$ ...(1)

At the boundary at $x = 0$ heat is to be conducted in at $\beta\, q_o$, $A = 1$

$$-kA\, dT/dx\,\big|_{x=0} = \beta\, q_o$$

Therefore $\quad dT/dx$ at $x = 0, = -(\beta\, q_o/k)$

Therefore $\quad (q_o/k)\,(1-\beta)\, e^{-\alpha\times 0} + C_1 = -\beta\, q_o/k$

Therefore $\quad C_1 = [-(\beta q_o/k) - (q_o/k)]/(1-\beta) = -q_o/k.$

Therefore $\quad (dT/dx) = (q_o/k)\,(1-\beta)\, e^{-\alpha x} - q_o/k \qquad \ldots(2)$

Heat flux at any section can be obtained by multiplying this equation by $-k$. Since $(Q/A = -k\, dT/dx)$

Therefore $\quad Q/A = -q_o(1-\beta)\, e^{-\alpha x} - q_o$

$$= q_o\,[1-(1-\beta)\, e^{-\alpha x}] \qquad \ldots(3)$$

Fig. AE-3.16

Integrating equation (2),

$$T = (-q_o/k)\,\frac{(1-\beta)}{\alpha}\, e^{-\alpha x} - (q_o x/k) + C_2$$

The boundary condition is at $x = L$, $T = T_w$

$$C_2 = T_w + (q_o/k)\,\frac{(1-\beta)}{\alpha}\,.\, e^{-\alpha L} + (q_o L/k)$$

Therefore $$\mathbf{T - T_w = (q_o/k)\,\frac{(1-\beta)}{\alpha}\,[e^{-\alpha L} - e^{-\alpha x}] + [q_o/k]\,[L - x]} \qquad \ldots(4)$$

and $$T_o - T_w = [q_o/k]\,\frac{(1-\beta)}{\alpha}\,[e^{-\alpha L} - 1] + [q_o/k]\,[L] \qquad \ldots(5)$$

Note: q_o – W/m^2, α = 1/m, β dimensionless.

The heat flow can also be obtained by integrating the heat generated and adding the flux at $x = 0$

Heat generated $\quad = \int_0^L q_o\,(1-\beta)\,\alpha\, e^{-\alpha x}\, dx = -q_o\,(1-\beta)\, e^{-\alpha L} + q_o\,(1-\beta)$

Heat entering $\quad = \beta\, q_o \qquad \ldots[5(a)]$

Adding, total flow $\quad = q_o\,[(1-(1-\beta)\, e^{-\alpha L})]$

From equation (3) at $\quad x = L$

$$Q/A = q_o\,[1-(1-\beta)\, e^{-\alpha L}] \qquad \ldots(6) \quad \text{checks.}$$

Example A-3.17: *At an instant the temperature distribution in a nuclear fuel rod of 25 mm radius at steady state is given by the equation $T = A - Br^2$ when it was generating heat at a steady state. The properties are k = 30 W/mK, density = 1800 kg/m^3, c = 0.8 kJ/kgK. If A 800°C and B = 4.167 × 10^5 °C/m^2 at that condition, determine the heat generation rate/m and also/m^3 basis. If the heat generation rate is doubled, determine the initial rate of temperature rise at r = 0 and at surface.*

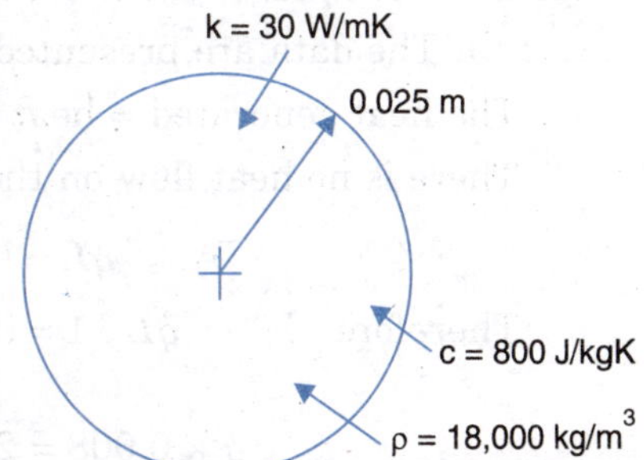

Fig. AE-3.17. Problem model

Solution: The known data are shown in Fig. AE-3.17. The heat generated can be found by calculating the heat flow at the surface.

$$-kA\,(dT/dr)\ |\ \text{at}\ (r = R) = Q$$

$$T = 800 - 4.167 \times 10^5\, r^2$$

Therefore $dT/dr = -2\beta r = -4.167 \times 10^5 \times 2r$, $\dfrac{\partial^2 T}{\partial r^2} = -2\beta = -4.167 \times 10^5 \times 2$

Therefore $Q = -30 \times 2\pi \times 0.025 \times -4.167 \times 10^5 \times 2 \times 0.025$

$\mathbf{Q = 98182.624}$ **W/m length**

The volume is $\pi r^2 \times 1$

Therefore $\dot{q} = 98182.624/\pi \times 0.025 \times 0.025 = \mathbf{5 \times 10^7}$ **W/m³**

The rate of temperature rise is given by $\partial T/\partial \tau$

The general differential equation under unsteady conditions in cylindrical co-ordinates can be simplified using the given relation as

$$\partial^2 T/\partial r^2 + (1/r)\, \partial T/\partial r + q/k = (\rho c/k)\, (\partial T/\partial \tau) = \frac{1}{\alpha} . \frac{\partial T}{\partial r}$$

$\partial^2 T/\partial r^2 = -2B$, $\partial T/\partial r = -2Br$ substituting in the equation

$$-2\text{B} + -2Br/r + q/k = \frac{1}{\alpha}\, (\partial T/\partial \tau)$$

$$\alpha\, [-4B + q/k] = \partial T/\partial \tau$$

$$\dot{q} = 5 \times 10^7 \times 2 = 10^8 \text{ W/m}^3$$

$$\alpha = (k/\rho c) = 30/(1800 \times 800)$$

Substituting the values $(\partial T/\partial \tau) = \dfrac{30}{1800 \times 800}\, [-4 \times 4.167 \times 10^5 + (10^8/30)] =$ **34.72°C/s**

This rate of temperature rise is irrespective of the radius. But as time proceeds, the rate will change.

COMMENT

When the heat generated is not collected completely by a fluid, the temperature of the material will rise continuously. Steady state will not be achieved.

Example A-3.18: *It is proposed to heat the window glass panes in a living space to 26°C. A company offers resistance embedded glasses with uniform heat generation. The outside is at – 15°C, and the convection coefficient on the outside is 20 W/m²K. The pane is 8 mm thick and has a conductivity of 1.4 W/mK. What should be heat generation rate if the inside surface temperature is equal to the room temperature.*

Solution: The data are presented in Fig. AE-3.18.

The heat generated = heat convected

There is no heat flow on the inner surface as temperatures are equal

$$\dot{q}L = (T - T_\infty)\, hA, \text{ but } T = 26 - \dot{q}L^2/2k$$

Therefore $\dot{q}L \,.\, 1 = (26 - \dot{q}L^2/2k - T_\infty)\, h \,.\, 1.$

$$\dot{q} \times 0.008 = 20 \times 1 \times \left[26 - \frac{\dot{q} \times 0.008^2}{2 \times 1.4} - (-15)\right]$$

$$q \times 0.008 = 20\, [41 - 2.28 \times 10^{-5}\, \dot{q}]$$

$8.457 \times 10^{-3} \times \dot{q} = 20 \times 41$

Therefore $\dot{q} = 97 \text{ kW/m}^3$

For the pane: 1m^2 basis: 775.7 W/m^2

Outside surface temperature, T is found using heat generated $Q = hA(\Delta T)$

$775.7 = 20(T - (-15))$

Therefore **T = 23.785°C**

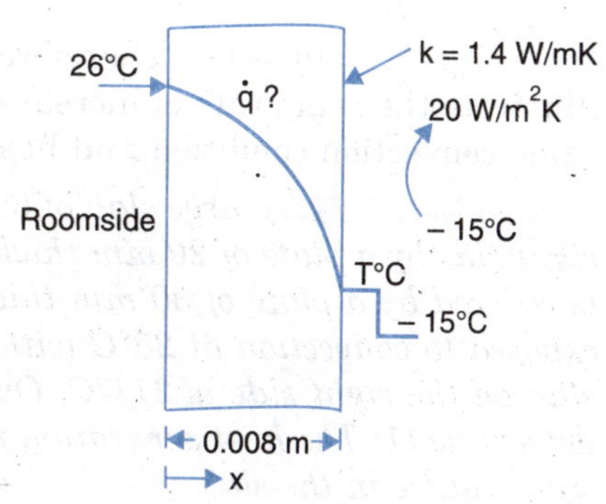

Fig. AE-3.18

COMMENTS

1. If the inside wall temperature is reduced, the power requirement will be lower. But, comfort condition is that wall temperature and room temperature should be equal.
2. This arrangement reduces surface condensation and better comfort as the radiation from body to the glass is reduced, but it is expensive.

Example A-3.19: *In the example A-3.18 if the outside surface temperature can be adjusted to be 4°C and if the heat generation rate is reduced to 1/4th the value, determine the heat loss from the room air. Also find the convection coefficient that has to exist inside the room.*

Solution: The total heat loss is now known by using

$$Q = hA(T - T_\infty) = 20 \times 1 \times [(4 - (-15)] = 380 \text{ W}$$

The heat generation rate now is 775.7/4 W/m^2 = 193.925 W

Therefore heat from room air = 186.075 W

Present heat generation rate, $q = 97 \times 1000/4 = 24250 \text{ W/m}^3$

Therefore inside surface temperature: $4 + \left[\frac{24250}{2 \times 1.4} \times 0.008^2\right] = \mathbf{4.6°C}$

Therefore $h_i\,(26 - 4.6) = 186.075$

$\mathbf{h_i = 8.7\ W/m^2K.}$

The heat loss is reduced but occupants will feel chilled by radiation to 4.6°C.

Example A-3.20: *Compare the temperatures at the centre line for circular rods with heat generation rate of 50 MW/m^3, when exposed to a fluid at 150°C with a convection coefficient of 1000 W/m^{2}K. The conductivity is 25 W/mK. Take r = 0.025, 0.05, 0.1 and 0.2.*

Solution: First the surface temperatures are to be found using

$$Q = hA\,(T_w - T_\infty) \qquad Q = \pi r^2 1 \times \dot{q} \qquad A = 2\pi r \times 1$$

$$\pi r^2 q/h 2\pi r \times 1 = (T_w - T_\infty)$$

Therefore $T_w = (\dot{q}r/2h) + T_\infty$

r, m	T_w, °C	T_o, °C
0.025 m	775°C	1087.5 °C
0.05 m	1400°C	2650°C
0.1 m	2650°C	7650 °C
0.2 m	5150°C	25150°C

$T_o - T_w = (q/4k)\, r^2$. Therefore, $T_o = T_w + (q/4k)\, r^2$ is used in the above. This problem is to illustrate the rapid rate of increase in the value of maximum temperature with radius for the same convection condition and heat generation rate.

Example A-3.21: *A large slab of 60 mm thickness, generating heat uniformly is covered on the right side by a plate of 20 mm thickness of thermal conductivity 50 W/mK. On the other side it is covered by a plate of 30 mm thickness with thermal conductivity 25 W/mK. Both sides are exposed to convection at 25°C with h = 1000 W/m²K. The temperature of the heat generating slab on the right side is 211°C. On the left the temperature is 261°C. Assuming steady state, determine (1) The heat generation rate and (2) Thermal conductivity of the slab (3) Maximum temperature in the slab.*

Solution: The known data are shown in Fig. AE-3.21, Assume unit area.

On the right side heat conducted = heat convected

$$\frac{(211 - T_R)\, 50}{0.02} = (T - 25)\, 1000$$

Solving: $T_R = 157.57°\text{C}$

$Q_R = (157.57 - 25)\, 1000 = 132571.4$ W

Similarly on the left side:

$$\frac{(261 - T_L) \times 25}{0.03} = (T_L - 25)\, 1000, \text{ Solving } T_L = 132.27°\text{C}$$

$Q_L = (132.27 - 25)\, 1000 = 101772.72$ W

Total heat flow = 239844 W

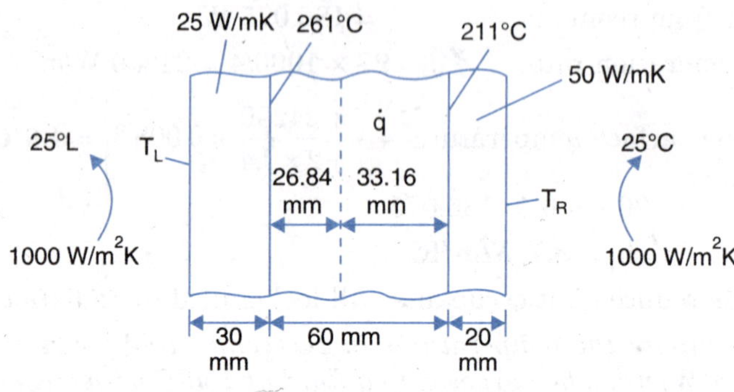

Fig. AE-3.21

This is generated by $1 \times 1 \times 0.06$ m³,

$$\therefore \qquad \dot{q} = \frac{239844}{0.06} = \mathbf{3.9974 \times 10^6\ W/m^3}$$

Considering the right side: The heat 132571.4 W comes out of x thickness of slab.

$x = 132571.4 / 3.9974 \times 10^6 = 0.03316$ m.

The heat comes out on the left side from 0.06 – 0.03366 = 0.02684 m

No heat flows across this plane.

$$\text{Considering the sides:} \quad T_o - 211 = \frac{\dot{q}}{2k} L_R^2 = \frac{3.9974 \times 10^6 \times 0.03316^2}{2k} \qquad ...(A)$$

$$T_o - 261 = \frac{\dot{q}}{2k} L_L^2 = \frac{3.3974 \times 10^6}{2k} \times 0.02684^2 \quad \text{...(B)}$$

(A) – (B), gives

$$50 = \frac{\dot{q}}{2k} (0.03316^2 - 0.02684^2)$$

Solving **k = 15.174 W/mK**

$$T_o = T_{wR} + \frac{\dot{q}}{2k} L_R^2, \text{ solving } T_{max} = T_o = \mathbf{355.83°C.}$$

Example A-3.22: *A fuse wire of diameter 2 mm with a resistivity of 5 micro ohm cm is to be used in a circuit. The convection coefficient to surroundings at 30°C is 6 W/m²K. If the wire melts at 800°C, determine the rating. Thermal conductivity = 357 W/mK.*

Solution: The resistance for 1 m length is found using . $R = \rho \frac{L}{A}$ where ρ is the resistivity, L is the length in cm and A is the area in cm²

$$R = \frac{5 \times 10^{-6} \times 100}{\pi \times 0.1^2} = \mathbf{0.016 \text{ ohm}}$$

Heat convected $= hA\,\Delta T = 6 \times \pi \times 0.002 \times 1\,(800 - 30) = 29.03$ W

This is equal to I^2R. Therefore **I = 42.7 amps**

Heat generation rate **q** $= 29.03/V = 29.03/\pi \times r^2 \times 1$ (V-Volume)

$= \mathbf{9.24 \times 10^6\ W/m^3}$

$T_o - T_w = qR^2/4k = 0.0065°C$ (R – radius here)

When radius is small, the whole wire is at near uniform temperature.

REVIEW QUESTIONS

3.1 Distinguish between the variation of heat flux and heat rate along the radius of a solid cylinder with uniform heat generation.

3.2 Distinguish between the variation of heat flux and heat rate along the radius of a solid sphere with uniform heat generation.

3.3 Distinguish between the variation of heat flux and heat rate along the thickness of a solid slab with uniform heat generation with one surface insulated.

3.4 Can steady state condition prevail when the surface of a heat generating body is insulated?

3.5 Distinguish between the way food material is cooked on a stove and the same is cooked in micro-wave oven.

OBJECTIVE QUESTIONS

3.1 Choose the correct statement from each question.

(*i*) In a slab generating heat uniformly and at steady state convecting equally on both sides, the temperature gradient will

(*a*) flatten out as the distance from the centre increases

Chapter 3

(*b*) become steeper as the distance from the centre increases

(*c*) will remain constant

(*d*) can be any of (*a*, *b*) or (*c*) depending the heat generation rate.

(*ii*) For the same linear size (*i.e.* L, r) and heat generation rate, the temperature drop from centre to surface is highest in

(*a*) spherical shape (*b*) plate shape

(*c*) cylindrical shape (*d*) rod of square section.

(*iii*) The temperature drop in a heat generating solid under steady state conduction depends to a greater extent on

(*a*) linear dimension (*b*) thermal conductivity

(*c*) heat generation rate (*d*) convection coefficient at the surface.

(*iv*) The most effective way to reduce the temperature drop in a heat generating solid is to

(*a*) reduce the linear dimension

(*b*) reduce the thermal conductivity

(*c*) reduce the convection coefficient on the surface

(*d*) reduce the heat generation rate.

(*v*) The thermal gradient in a heat generating cylinder under steady conduction, at half the radius location will be

(*a*) One half of that at surface (*b*) One fourth of that at surface

(*c*) Twice that at surface (*d*) Four times that at surface.

(*vi*) In a sphere under steady state conduction with uniform heat generation, the temperature gradient at half the radius location will be

(*a*) one half of that at the surface (*b*) one fourth of that at the surface

(*c*) one eight of that at the surface (*d*) 2 times of that at the surface.

Answers

3.1 (*i*) (*b*), (*ii*) (*b*), (*iii*) (*a*), (*iv*) (*a*), (*v*) (*a*), (*vi*) (*a*).

EXERCISE PROBLEMS

3.1 An exothermic reaction in a slab of material generates heat uniformly at a rate of 2×10^6 W/m^3. The material has a thermal conductivity of 6.5 W/mK and the thickness is 80 mm. The slab is exposed on both sides. If the surface is maintained at 150°C by convection, determine the centre temperature. Also determine the value of convection coefficient if heat is extracted by fluid at 100°C.
[**Ans.** 396°C, 1600 W/m^{2}K]

3.2 An exothermic reaction takes place in a long cylinder of radius 40 mm with a thermal conductivity of 6.5 W/mK which leads to the generation of heat uniformly in the volume. If the allowed temperature difference between the surface and the centre is 246°C, determine the heat generation rate possible. Also find the required value of convection coefficient to dissipate the heat to a fluid at 100°C, with the surface temperature to be at 150°C. [**Ans.** 4×10^6 W/m^3, 1600 W/m^{2}K]

3.3 Heat is generated uniformly by chemical reaction in a sphere of radius 40 mm. The conductivity of the material is 6.5 W/mK. If the allowable temperature rise at the centre over the surface is 246°C, determine the heat generation rate. If the surface temperature is 150°C and if a fluid at 100°C is used to collect the heat, determine the required value of convection coefficient.
[**Ans.** 6×10^6 W/m^3, 1600 W/m^{2}K]

3.4 In a slab of thickness 0.1 m, insulated on one face heat is generated as per the rate $\dot{q} = 10^6 \times \cos \dfrac{\pi x}{0.2}$ where x is the distance in m from the insulated face. If the maximum temperature is not to exceed 150°C and the thermal conductivity of the material is 52 W/mK, determine the surface temperature. Also find the heat transfer rate at the surface and the convection coefficient required to dissipate the heat to a fluid at 40°C. [**Ans.** 72°C, 63622 W, 989 W/m^{2}K]

3.5 Heat generation in a cylinder of radius 0.04 m is non-uniform and varies with the radius r by the relation $q = q_o (1 - cr)$ where c is a constant having a unit of $1/m$. The conductivity of the material is 12 W/mK. If c = 20/m. $q_o = 10^6$ W/m^3 and if the centre temperature is to be 170°C, determine the temperature at which the surface should be maintained. Also find the heat flow rate at the surface and the convection coefficient required to maintain the temperature if the heat is extracted by fluid at 60°C.

[**Ans.** $T - T_w = (q_o/4k)(R^2 - r^2) - (q_o/9kR)(R^3 - r^3)$, 148.5°C, 2346 W/m^2, 105.5 W/m^{2}K]

3.6 In a sphere of radius R, the heat generation rate varies with the radius r as $q = q_o [1 - (r/R)^2]$. If the thermal conductivity is k, derive the expression for the temperature variation with radius.

[**Ans.** $T - T_w = (q_o/6k)(R^2 - r^2) - (q_o/20kR^2)(R^4 - r^4)$]

3.7 In problem 3.6, if $q_o = 1 \times 10^6$, $R = 0.04$, $k = 12$ and if the centre temperature is 170°C, determine the surface temperature. Also find the heat flow rate at the surface. Also find the value of h if the fluid extracting heat is at 60°C. [**Ans.** 154.4°C, 107.2 W, 56 W/Km^2]

3.8 It is proposed to generate power in a nuclear installation at the rate of 10 MW/m^3 of fuel. The fuel has a thermal conductivity of 25 W/mK. Each fuel element has a sectional area of 20 cm^2. The convection coefficient has a value of 2000 W/m^{2}K. The heat extraction is at 200°C. Compare the maximum temperatures of solid cylinder and hollow cylinder of 10 mm thickness with heat extraction from the outside surface. [**Ans.** 518 °C, 504 °C]

3.9 A plate of 0.1 m thickness insulated on one side and generating heat uniformly has to have the maximum temperature gradient limited to 1000°C/m to avoid thermal distortion. If the thermal conductivity of the material has a value of 200 W/mK, determine the maximum heat generation rate. Also find the centre temperature if the surface is at 100°C. [**Ans.** 2×10^6 W/m^3, 150°C]

3.10 A cylinder of radius 0.2 m generates heat uniformly at 2×10^6 W/m^3. If the thermal conductivity of the material has a value of 200 W/mK, determine the maximum temperature gradient. Also find the centre temperature if the surface is at 100°C. What is the value of heat flux at the surface and heat flux per m length. [**Ans.** 1000°C/m, 200°C, 2×10^5 W/m^2, 2.5×10^5 W/m length]

3.11 A slab generating heat at the rate of 3.8×10^6 W/m^3 is 30 mm thick and has its surfaces maintained at 100°C and 300°C. The thermal conductivity of the material is 1.24 W/mK. Determine the location and the value of maximum temperature. Also find the temperature at the mid point.

[**Ans.** 2.2 mm from centre, 566°C, 544.8°C]

3.12 The heat generation rate in a solid cylinder of radius R varies as $q_o e^{-ar}$. Derive an expression for the temperature distribution.

3.13 A slab of 0.10 m thickness generating heat uniformly at 2×10^6 W/m^3 has a thermal conductivity of 25 W/m°C. On both sides a layer of stainless steel of 5 mm thickness with k = 12.5 W/mK is laid to protect the surface. The outside wall of the stainless steel is exposed to a fluid at 210°C with a convection coefficient of 500 W/m^{2}K. Determine the centre temperature and also the temperature on both sides of the stainless steel plate. Compare the temperature gradients at the surface of the heat generating wall and also the stainless steel material.

3.14 A cylindrical rod generating heat at a rate of 2×10^6 W/m^3 and having a radius of 0.05 m has a thermal conductivity of 25 W/mK. It is clad with stainless steel layer of 5 mm thickness with a conductivity of 12.5 W/mK and the surface is exposed to fluids at 210°C with a convection coefficient of 500 W/m^{2}K. Determine the centre temperature, heat transfer rate at the surface and all surface temperatures. Calculate also the temperature gradients at the interface for both materials.

3.15 A heat generating wall of 0.1 m thickness with a heat generation rate of 5 MW/m^3 is of materi with a conductivity of 25 W/mK is exposed on both sides to convection as well as radiation surroundings at 100°C. The convection coefficient is 200 W/m^{2}K. Determine the centre and su face temperatures.

3.16 A concrete dam of thickness 15 m, having a thermal conductivity of 5 W/mK has heat generatic due to the setting of cement at a rate of 20 W/m^3. The surfaces are at 20°C. Determine the centr temperature. [**Ans.** 132.5°C

(**Note:** Concrete is laid after chilling the aggregate and cooling is also done for some period time when laying a large volume of concrete).

3.17 Using numerical method solve for the centre temperature in the case of a slab of 0.1 m thicknes exposed on both sides to convection to fluid at 180°C with h = 200 W/m^{2}K. The conductivity the material is 25 W/mK. The heat generation rate is 2 MW/m^3.

3.18 Solve problem 3.18 for a cylinder of radius 0.05 m.

3.19 Solve problem 3.18 for a sphere of radius 0.05 m.

3.20 A computer chip made of silicon and of thickness 4 mm has a heat generation rate of 1.7 × 10^6 W m^3. (This is equal to 1W for a chip of 4 mm × 10 mm × 15 mm). The chip is exposed to cold air o one side at 20°C with a convection coefficient of 25 W/m^{2}K. Determine the centre temperatur (The other side may be assumed to be insulated). k = 95 W/mK.

3.21 The heat generation in a wall varies as $q_x = q_o \cos ax$, where q_o is the heat generation at $x = 0$. is a constant with unit $\frac{1}{m}$. The wall thickness is L and the wall is insulated at $x = 0$. The surfac temperature is T_w. Derive an expression for the heat flux at the wall surface.

3.22 An aluminium cable of diameter 30 mm carries an electrical current of 250 A. The thermal con ductivity is 200 W/mK. The electrical resistivity is 3 μΩ cm. Calculate the temperature drop fo the cable (**Note:** Resistance in ohm = Resistivity × length/area).

3.23 Derive an expression for the temperature distribution in a hollow cylinder with heat sourc which varies as $\dot{q} = a + br$ when $\dot{q}$ is the generation rate per unit volume at radius r.

The boundary conditions are $T = T_i$ at $r = r_i$. and $T = T_o$ at $r = r_o$.

3.24 A 2 kW resistance heater wire of thermal conductivity 15 W/m^2 °C is 0.5 m long and 4 mm in diameter. The outer surface temperature is 105 °C. Assuming uniform heat generation determin the temperature at the center of the wire. [**Ans.** 126 °C

Heat Transfer with Extended Surfaces (FINS)

4

CHAPTER

4.0 INTRODUCTION

There are numerous situations where heat is to be transferred between a fluid and a surface. In such cases the heat flow depends on three factors namely (*i*) area of the surface (*ii*) temperature difference and (*iii*) the convective heat transfer coefficient.

The base surface area is limited by design of the system. The temperature difference depends on the process and cannot be altered. The only choice appears to be the convection heat transfer coefficient and this also cannot be increased beyond a certain value particularly if the fluid is a gas. Any such increase will be at the expense of power for fans or pumps. Thus the possible option is to increase the base area by the use of extended surfaces or fins. The situation is depicted in Fig. 4.1. The fins extend, form the base surface and provide additional convection area for the heat conducted into the fin at the base. Fins are thus used whenever the available surface area is found insufficient to transfer the required quantity of heat with the available temperature drop. In the case of fins, the direction of heat transfer by convection is perpendicular to the direction of conduction flow. The conduction in fins is considered to be one dimensional though it is essentially two dimensional. This is acceptable, as the length along the fin is much larger compared to the transverse length. The process of heat transfer with fins is often termed as combined conduction convection systems.

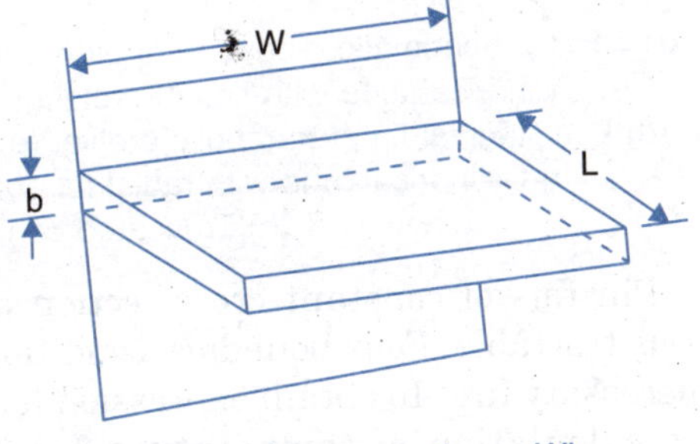

Bare area for convection = Wb
Fin area for convection = W(2L + b)

Fig. 4.1

Common examples of the use of extended surfaces are in cylinder heads of air cooled engines and compressors and on electric motor bodies. In air conditioners and radiators, tubes with circumferential fins are used to increase the heat flow. Electronic chips cannot function without use of fins to dissipate the heat generated. Several shapes of fins are in use. These are (*i*) Plate fins of constant sectional area (*ii*) Plate fins of variable sectional area (*iii*) Annular or Circumferential fins of constant thickness (*iv*) Annular fins of variable thickness (*v*) Pin fins of constant sectional area and (*vi*) Pin fins of variable sectional area. Some of these are shown in Fig. 4.1(A).

The main aim of the study is to design fins to optimise the use of a given amount of material to maximise heat transfer. For this purpose it will be desirable that the fin surface temperature is closer to the base surface temperature. This can be achieved by the use of materials of high thermal conductivity like copper or aluminium. In terms of weight and eas of fabrication, aluminium will score over copper though its thermal conductivity is lower. will be shown later that there are limitations about the length of the fin in terms of effectivenes of the material used. In order to increase the area for a given volume, thinner fins should b chosen. Fins are found to be more valuable when the convective heat transfer coefficient i low. This is the case in gas flow and natural convection and fins are more commonly used i these cases.

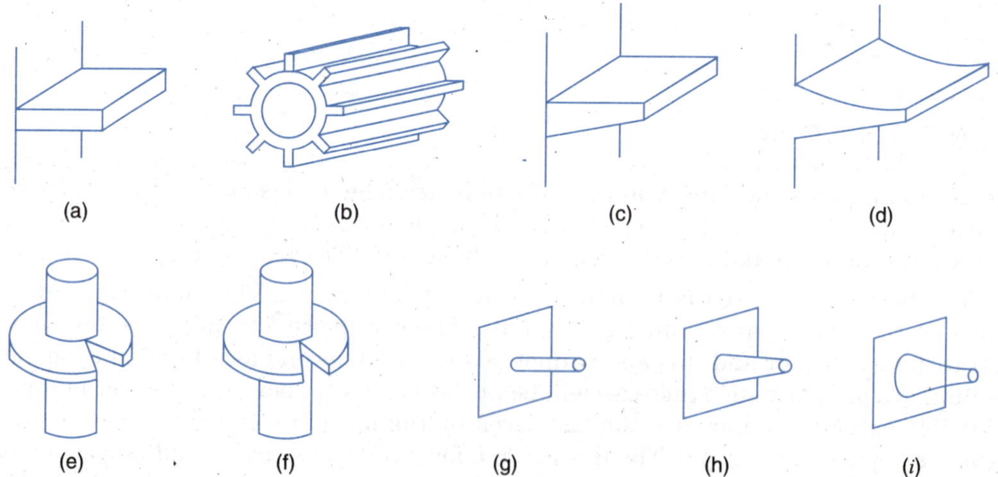

Fig. 4.1 (A). Schematic diagrams of different types of fins: (a) Longitudinal fin of rectangular profile; (b) Cylindrical tube with fins of rectangular profile; (c) Longitudinal fin of trapezoidal profile; (d) Longitudinal fin of parabolic profile; (e) Cylindrical tube with radial fin of rectangular profile; (f) Cylindrical tube with radial fin of truncated conical profile; (g) Cylindrical pin fin; (h) Truncated conical spine; (i) Parabolic spine

Pin fins of constant cross-section are chosen first for analysis in order to make the analysis tractable. Four boundary conditions are analysed. In order to calculate the heat flow, it is necessary first to obtain expression for temperature variation along the fin. This will lead to the calculation of temperature gradient at the base and heat flow. In the case of circumferential and variable area pin fins, the formulation is more complex. These are designed using charts plotting efficiency against physical parameters. As in more and more applications, circumferential fins are used, learning the use of the charts is more valuable. As convective heat transfer coefficient is rather low in gas flows, fins are invariably used when heat transfer from or to gases is involved. In liquid flows, as convective heat transfer coefficient is higher, fins are rarely used.

4.1 FIN MODEL

Most widely used fins are of the radial type with heat exchangers of tubular construction.

In these fins both the temperature and heat flow area vary along the heat flow direction. The governing differential equation is of the form given below:

$$\frac{d^2}{dr^2} + \frac{1}{r}\frac{d}{dr} - \frac{2h}{kt} = 0.$$

where $\theta = T - T_\infty$, and t is fin thickness.

The solution is rather complex, expressed in terms of Bessel functions. Hence charts with fin efficiency against a combination of physical parameters are generally used in the design of such variable area fins. The same is discussed later in article 4.5.

To start with simplified fin with constant area along heat flow direction is analysed in this article.

A simplified model for the analysis of heat transfer with extended surfaces is shown in Fig. 4.2. The area of cross section is assumed to be constant along the heat flow direction (variable area will be discussed later). The surface area of the slab from which heat is to be dissipated to the fluid surrounding it is extended by a fin on it. Heat is transferred from the surface to the fin at its base by conduction. This heat is convected to the surrounding fluid over the fin surface (it may be radiated also). The energy balance under steady conditions for the fin will yield

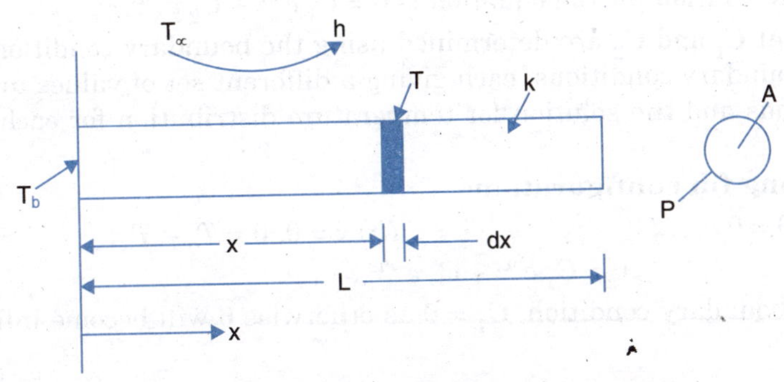

Fig. 4.2. Pin fin

heat conducted in at the base – heat convected over the fin surface upto the section x

= heat conducted out at the section x into the remaining fin material.

The process indicates that the temperature continuously varies along the length and the heat flow also varies along the length. The main quantities required to be calculated are (*i*) the temperature along the length and (*ii*) the total heat flow.

4.2 TEMPERATURE CALCULATION

The parameters involved in the analysis are:

T_b — the base temperature, T_∞—the fluid temperature,

k — thermal conductivity of the material, (considered as constant)

h — convective heat transfer coefficient,

A — the sectional area perpendicular to the base surface

P — perimeter of the fin and

L — fin length.

The analysis can be done by considering the energy balance for a small elemental volume at a distance x as shown in Fig. 4.3.

The heat balance under steady conditions gives

heat conducted in, at section x (1, 2) — heat conducted out at section $x + dx$ (3, 4)

— heat convected over the surface of the element = 0

$$-kA\,(dT/dx) - (-kA\,dT/dx + (d/dx)(-kA\,dT/dx)\,dx) - hP\,dx\,(T - T_\infty) = 0 \quad ...(4.1)$$

assuming k and A are constants and not influenced by temperature or location and $P\,dx$ is the surface area for convection

$$d^2T/dx^2 - (hP/kA)(T - T_\infty) = 0 \quad ...(4.2)$$

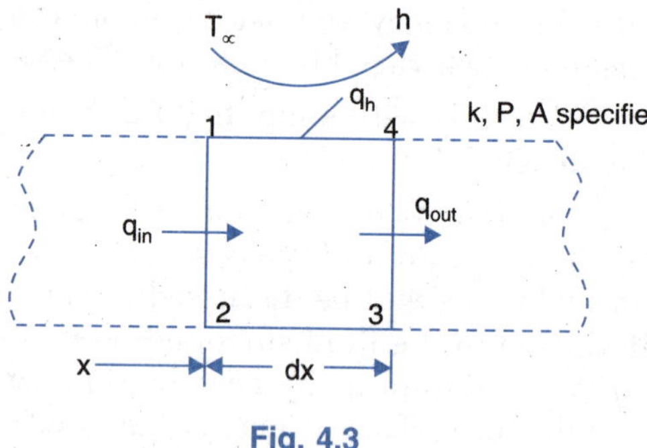

Fig. 4.3

In order to solve the equation, a new variable θ is introduced.

$$\theta = T - T_\infty,\ d^2T/dx^2 = d^2\theta/dx^2$$

Also a variable m is defined as $\sqrt{(hP/kA)}$

The equation reduces to $\mathbf{d^2\theta/dx^2 - m^2\theta = 0}$ ...(4.3)

The general solution for the equation is $\theta = C_1 e^{mx} + C_2 e^{-mx}$ ...(4.4)

The constant C_1 and C_2 are determined using the boundary conditions. There are four possible sets of boundary conditions, each giving a different set of values to C_1 and C_2. These boundary conditions and the solution for temperature distribution for each case is shown in Table 4.1.

Case 1: Long fin configuration:

(*i*) $x \to \infty$, $\theta = 0$ (*ii*) $x = 0$, $\theta = T_b - T_\infty$

$$\theta = C_1 e^{mx} + C_2 e^{-mx}.$$

From first boundary condition, $C_1 = 0$ as otherwise θ will become infinite which is not possible.

$$\therefore \quad \theta = C_2 e^{-mx},\ \text{at } x = 0,\ \theta = T_b - T_\infty$$

$$T_b - T_\infty = C_2 e^{-mo} = C_2.$$

$$\therefore \quad \frac{\theta}{\theta_o} = \frac{T - T_\infty}{T_b - T_\infty} = e^{-mx} \quad ...(4.5)$$

In this case the variation of temperature is exponential.

A question arises as to what should be the limiting length for this equation to apply. Assuming that the temperature ratio $(T - T_\infty)/(T_o - T_\infty) = 0.001$, the corresponding **value of mL = 4.6.** Hence as long as $L = 4.6/m = 4.6\left(\frac{kA}{hP}\right)^{1/2}$ this equation can be applied without significant error. For example if $A = 1$ cm², $k = 80$ W/mK, $h = 100$ W/m²K and $p = 2$ cm, then

$$L_{max} = 4.6 \times \sqrt{(8 \times 1 \times 10^{-4})/(100 \times 0.02)} = 0.29 \text{ m}.$$

So upto the length of about 29 cm, long fin model can be opted for this data.

If $(T - T_\infty)/(T_o = T_\infty) = 0.01$ then $mL = 2.65$

then $L_{max} = 0.167$ m

Case 2. Short fin end insulated:

At $x = 0$, $\theta = \theta_o = T_b - T_\infty$, At $x = L$, $\dfrac{d\theta}{dx} = 0$ as the surface is insulated. From the first condition, $\theta = C_1 e^{mx} + C_2 e^{-mx}$ leads to

$$\therefore \quad \theta_o = C_1 + C_2 \quad ...(1)$$

$$\left.\frac{d\theta}{dx}\right|_L = m(-C_1 e^{mL} + C_2 e^{-mL}) = 0 \quad ...(2)$$

$$\therefore \quad C_1 e^{mL} = C_2 e^{-mL} \quad \text{or} \quad C_2 = C_1 e^{2mL} \quad ...(3)$$

Using equation (1) $\theta_o = C_1 + C_1 e^{2mL}$ or $C_1 = \dfrac{\theta_o}{1+e^{2mL}}$

Using equation (3)

$$\therefore \quad C_2 = \frac{\theta_o}{1+e^{2mL}} . e^{2mL} = \frac{\theta_o}{1+e^{-2mL}}$$

$$\therefore \quad \frac{\theta}{\theta_o} = \frac{e^{mx}}{1+e^{2mL}} + \frac{e^{-mx}}{1+e^{-2mL}}$$

$$= \frac{e^{mx}\, e^{-mL}}{e^{-mL} + e^{mL}} + \frac{e^{-mx} . e^{mL}}{e^{-mL} + e^{mL}} = \frac{e^{-m(L-x)} . e^{m(L-x)}}{e^{-mL} + e^{mL}}$$

$$\frac{T - T_\infty}{T_b - T_\infty} = \frac{\cosh m(L-x)}{\cosh mL} \quad ...(4.6)$$

In this case the heat convected at the tip is neglected. The error due to this can be reduced by increasing the length by ΔL equal to $t/2$ where the thickness of the fin is t. In the case of circular fins $\Delta L = D/4$.

The temperature ratio at the tip is

$$\frac{T_L - T_\infty}{T_o - T_\infty} = \frac{1}{\cosh mL} \quad ...(4.7)$$

This model is applicable in most of the situations for initial estimation.

Case 3. Short fin with convection, h_L at the tip.

The boundary conditions are

$$\text{at } x = 0, \quad \theta = \theta_o, \quad \text{at} \quad x = L, -kA \left.\frac{dT}{dx}\right|_L = h_L A (T_L - T_\infty)$$

The algebra is more involved. The resulting equation is

$$\frac{T - T_\infty}{T_b - T_\infty} = \frac{\cosh m(L-x) + \dfrac{h_L}{mk} \sinh m(L-x)}{\cosh mL + \dfrac{h_L}{mk} . \sinh mL} \quad ...(4.8)$$

At the tip, the temperature ratio is

$$\frac{T_L - T_\infty}{T_b - T_\infty} = \frac{1}{\cosh mL + \dfrac{h_L}{mk} . \sinh mL} \quad ...(4.9)$$

Equation (4.6) can be obtained from equation (4.8) with $h_L = 0$.

Case 4. Specified end temperatures.

At $x = 0$, $\theta = T_{b1} - T_\infty$ where T_{b1} is the temperature at end 1

At $x = L$, $\theta = T_{b2} - T_\infty$ where T_{b2} is the temperature at end 2.

In this case also the algebra is involved. The resulting solution is

$$\frac{T - T_\infty}{T_{b1} - T_\infty} = \frac{[(T_{b2} - T_\infty)/(T_{b1} - T_\infty)] \sinh mx + \sinh m(L - x)}{\sinh mL} \quad ...(4.10)$$

This type of fins are used in liquid flows and where there is space limitation.

The boundary conditions and solutions are given in Table 4.1.

Table 4.1. Temperature distribution in constant area fins for different boundary conditions, $m = \sqrt{hP/kA}$

Boundary condition and general nomenclature	*Temperature distribution*	
1. **Long fin** $x = 0, \theta = T_o - T_\infty$	$(T - T_\infty)/(T_o - T_\infty) = e^{-mx}$	...(4.5)
2. **Short fin end insulated** (length L) $x = 0, \theta = T_o - T_\infty$ $x = L, dT/dx = 0$	$\frac{T - T_\infty}{T_o - T_\infty} = \frac{\cosh m(L - x)}{\cosh (mL)}$	...(4.6)
3. **Short fin (convection at the tip h_L, considered),** $x = 0, \theta = T_o - T_\infty$ $x = L, (-kA\, dT/dx)\,\lvert L = h_L A(T_L - T_\infty)$ $\frac{T - T_\infty}{T_o - T_\infty} = \frac{\cosh m(L - x) + \left(\frac{h_L}{mk}\right) \sinh m(L - x)}{\cosh (mL) + \left(\frac{h_L}{mk}\right) \sinh mL}$		...(4.8)
4. **Fixed end temperature** $x = 0, \theta = T_{b1} - T_\infty$ $x = L, \theta = T_{b2} - T_\infty$ $\frac{T - T_\infty}{T_{b1} - T_\infty} = \frac{[(T_{b2} - T_\infty)/(T_{b1} - T_\infty)] \sinh (mx) + \sinh m(L - x)}{\sinh (mL)}$		...(4.10)

The first three examples illustrate the difference between the first three end conditions for similar fins at similar surroundings.

Example 4.1: *A long rod 12 mm square section made of low carbon steel protrudes into air at 35°C from a furnace wall at 200°C. The convective heat transfer coefficient is estimated at 22W/m²K. The conductivity of the material is 51.9 W/mK. Determine the location from the wall at which the temperature will be 60°C. Also calculate the temperature at 80 mm from base.*

The data are shown on Fig. Ex. 4.1.

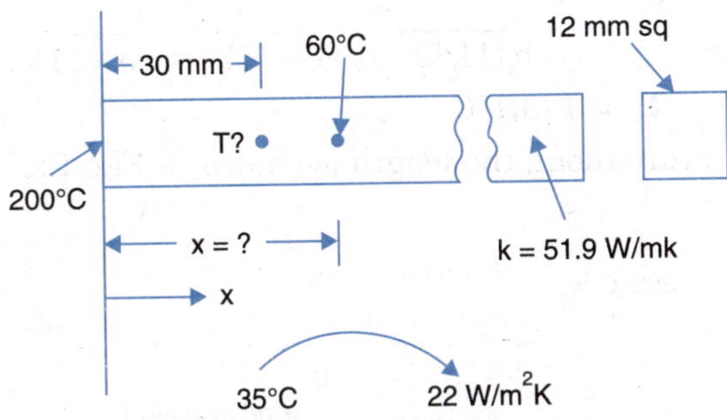

Fig. Ex. 4.1. Problem model

Solution: The long fin model will suit the problem situation.

$$\frac{T - T_\infty}{T_b - T_\infty} = \frac{60 - 35}{200 - 35} = e^{-mx}$$

The distance x is to be calculated:

$$\mathbf{m} = \sqrt{hP/kA} = \sqrt{(22 \times 4 \times 0.012/51.9 \times 0.012 \times 0.012)}$$

$$= \mathbf{11.89},\ \mathrm{m}^{-1}$$

Taking logarithm and solving x = **0.159 m or 159 mm.**

Temperature at 80 mm: $(T - 35)/(200 - 35) = e^{-11.89 \times 0.08}$ $\therefore$ **T = 98.74°C.**

Example 4.2: *In the example 4.1 if the length of the rod is 159 mm determine the end temperature. Also find the temperature at 80 mm distance. All the data in example 4.1 are used as such, except that the length is 159 mm and end insulated (or heat convected at the tip is negligible).*

Solution: The temperature distribution is given by

$$\frac{T - T_\infty}{T_b - T_\infty} = \frac{\cosh m(L - x)}{\cosh mL} \qquad m \text{ (from example 1)} = 11.89$$

To find end temperature $x = L$.

$$\frac{T_L - 35}{200 - 35} = \frac{1}{\cosh(11.89 \times 0.159)} = 0.295, \quad \therefore \quad T_L = \mathbf{83.72°C}$$

(compare with that of long fin at this location, 60°C)

Temperature at 80 mm (or 0.08 m)

$$\frac{T - 35}{200 - 35} = \frac{\cosh 11.89\,(0.159 - 0.08)}{\cosh(11.89 \times 0.159)} = 0.4354 \quad \therefore \quad \mathbf{T = 106.84°C}$$

(compare at the same location in long fin, 98.74°C).

Example 4.3. *In the example 4.1, consider the fin to be 80 mm long and end face convection also exists. Determine the end temperature.*

Solution: This problem model is the short fin situation.

$$\frac{T - T_\infty}{T_b - T_\infty} = \frac{\cosh m(L - x) + (h_L/km)\sinh m(L - x)}{\cosh(mL) + (h_L/km)\sinh mL}$$

Chapter 4

as $x = L$ this reduces to

$$\frac{T_L - 35}{200 - 35} = \frac{1}{\cosh(11.89 \times 0.08) + (22/(51.9 \times 11.89))\sinh(11.89 \times 0.08)}$$

$$\mathbf{T_L = 143.1°C}$$

The plot of temperature along the length is shown in Fig. Ex. 4.3.

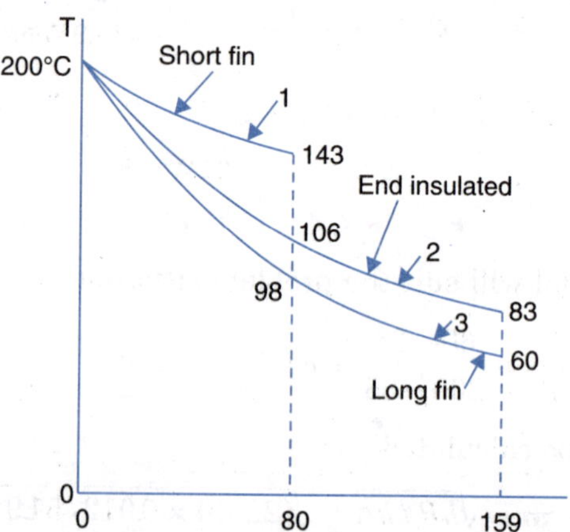

Fig. Ex. 4.3. Temperature variation along fins of different boundary

The heat flow in these cases will be discussed in examples 4.4, 4.5 and 4.6.

The examples bring out the point that it is important to decide about the model to be adopted from the data provided.

4.3 HEAT FLOW CALCULATION

Referring to Fig. 4.2, the following two methods can be used to determine the heat flow:

(*i*) All the heat has to come into the fin at the base by conduction or (*ii*) the total heat flow into the fin is the sum of heat convected over the surface of the fin.

Both methods lead to the same expression.

(*i*) Using the first method **for long fin boundary**

$$Q = -kA\,(dT/dx)x = 0$$

For long fin $\quad (T - T_\infty) = (T_o - T_\infty)\,e^{-mx}$

$$dT/dx = -m\,(T_o - T_\infty)\,e^{-mx},\; Q = (-kA)\,\{-m(T_o - T_\infty)\}\,e^{-m0}$$

$$Q = kA\,\sqrt{hP/kA}\,.\,(T_o - T_\infty)\,.\,1 = \sqrt{hP\,kA}\,.\,(T_b - T_\infty) \qquad ...(4.11)$$

Using the second method: for long fin . L ..... > ∞

$$Q = \int_0^\infty hP\,dx\,(T - T_\infty),\text{ as } (T - T_\infty) = (T_b - T_\infty)\,e^{-mx}$$

$$Q = \int_0^\infty hP\,(T_b - T_\infty)\,.\,e^{-mx}\,dx = \left[-(1/m)\,.\,hP\,(T_b - T_\infty)\,e^{-mx}\right]_0^\infty$$

$$= [0 - (-(1/\sqrt{kA/hP}))\,.\,hP\,(T_b - T_\infty)\,.\,e^{-m0}]$$

$$= \sqrt{hP\,kA}\,(T_o - T_\infty) \qquad ...(4.11)$$

(*ii*) **Short fin end insulated boundary:**

$$Q = -kA \left.\frac{dT}{dx}\right|_{x=0}$$

$$T - T_\infty = (T_b - T_\infty) \cdot \frac{\cosh m(L-x)}{\cosh mL}$$

$$\left.\frac{dT}{dx}\right|_{x=0} = \frac{(T_b - T_\infty)}{\cosh mL} \times -m \sinh mL$$

$$\therefore \quad Q = -kA \times -m \times (T_b - T_\infty) \cdot \frac{\sinh mL}{\cosh mL}$$

or

$$Q = \sqrt{hPkA}\ (T_b - T_\infty) \tanh(mL) \quad ...(4.12)$$

The method of integration will lead to the same expression.

The results for the other two involve more algebraic manipulation. The expressions for heat flow is shown in Table 4.2.

Table 4.2: Expressions for heat flow for fins with various boundaries

Boundary condition	Heat flow	
1. **Long fin,** $x = 0, T = T_b$ $x = \infty, T = T_\infty$	$Q = \sqrt{(hPkA)} \cdot (T_b - T_\infty)$	...(4.11)
2. **Short fin end insulated** $x = 0, T = T_b$	$Q = \sqrt{(hPkA)} \cdot (T_b - T_\infty) \cdot \tanh(mL)$	...(4.12)
3. **Short fin** $x = L, -kA\,(dT/dx) = hA\,(T_b - T_\infty)$ $Q = \sqrt{(hPkA)} \cdot (T_b - T_\infty) \left[\frac{\sinh mL + (h_L/mk) \cdot \cosh mL}{\cosh mL + (h_L/mk) \cdot \sinh mL}\right]$		...(4.13)
4. **Fixed end temperature** $x = 0, T = T_{b1}.$ $x = L, T = T_{b2},$ $Q = \sqrt{(hPkA)}\ [(T_{b1} - T_\infty) + (T_{b2} - T_\infty)] \frac{\cosh(mL) - 1}{\sinh(mL)}$		...(4.14)

Note that equation (4.12) can be obtained from (4.13) by assuming $h_L = 0$, or end insulated condition. So also equation (4.11) from (4.12) by taking $L = \infty$, then $\tanh(mL) = 1$.

Examples 4.4—4.7 compare the heat flow characteristics of fins under the four different boundary conditions.

Example 4.4: *Calculate in the case of problem 4.1, the total heat flow and heat convected upto 159 mm and 80 mm lengths. (Fig. 4.3)*

Use equation (4.11) and using the data of example 4.1.

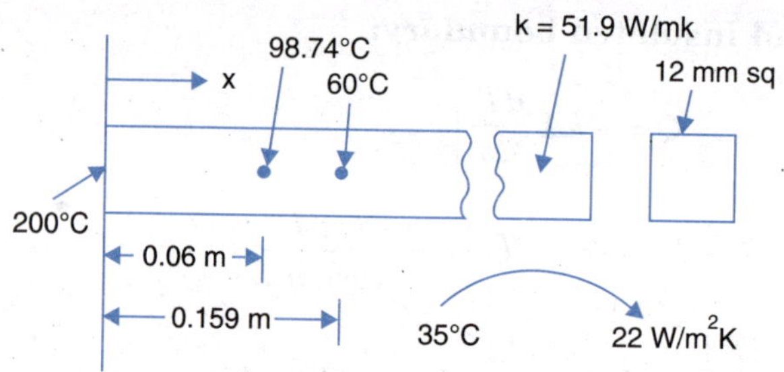

Fig. Ex. 4.4. Problem model

Solution: Equation 4.11 is applicable:

$$Q = \sqrt{(hPkA)}\ .\ (T_b - T_\infty)$$

$$Q = A(22 \times 4 \times 0.012 \times 51.9 \times 0.012^2)\ (200 - 35)$$

$$= \mathbf{14.66\ W}$$

The heat convected upto 0.159 m length can be found by taking the difference of total heat flow and heat conducted at 0.159 m length: From 4, 1, $T_{0.159}$ = 60°C, $T_{0.08}$ = 98.74°C.

Heat convected upto 0.159 m

$$\mathbf{Q_{0.159}} = 14.66 - \sqrt{(hPkA)}\ (T_{0.159} - T_\infty) = 14.66 - 2.22 = \mathbf{12.44\ W}$$

or 84.86% of long fin

Similarly, heat convected upto 0.08 m is given by

$$\mathbf{Q_{0.08}} = (14.66) - \sqrt{(h\text{P}k\text{A})}\ (T_{0.08} - T_\infty) = 14.66 - 5.66 = \mathbf{9\ W}$$

or 61.4% of long fin.

The results indicate that **most of the heat is dissipated in a short length of** the fin and extending the fin length beyond a certain value is uneconomical. Hence beyond a certain length the fin will not contribute significantly to heat transfer.

Example 4.5: *Using the data in examples 4.1 and 4.2, determine the heat dissipated by a fin of 0.159 m length with end insulated conditions. Also find the heat dissipated upto 0.08 m length in this case.*

Solution: The equation suitable for this situation is given by 4.12.

$$Q = \sqrt{(hPkA)}\ (T_b - T_\infty)\ .\ \tanh\ (mL)$$

Using the data and calculated value of m from problem 1, m = 11.89

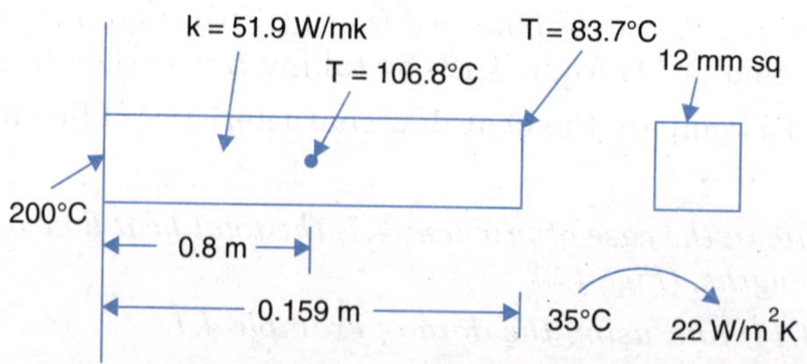

Fig. Ex. 4.5. Problem model

$$Q = \sqrt{(22 \times 4 \times 0.012 \times 51.9 \times 0.012 \times 0.012)}\ (200 - 35) \times \tanh \times (11.89 \times 0.159)$$

$$= \mathbf{14\ W}$$

Only 0.66 W less than that of long fin. (example 4.4)

Heat dissipated upto 0.08 m can be found by difference of original fin and fin of L = (0.159 – 0.08) m end insulated with the base temperature as calculated.

$$Q_{0.08} = 14 - \sqrt{(hPkA)} \times (106.84 - 35) . \tanh (11.89 \times 0.079) = 14 - 4.69 = 9.31 \text{ W}$$

or 66.5%.

Example 4.6: *In example 4.3, determine the heat flow through the fin using expression 4–13, for a length 80 mm.*

Solution:

$$Q = \sqrt{(hPkA)}\ (T_o - T_\infty) . \frac{\sinh (mL) + (h_L / mk) . \cosh (mL)}{\cosh (mL) + (h_L / mk) . \sinh (mL)}$$

$$= \sqrt{(22 \times 4 \times 0.012 \times 51.9 \times 0.012 \times 0.012)\ (200 - 35)}$$

$$\times \frac{\sinh (11.89 \times 0.08) + (22 / 11.89 \times 51.9) \cosh (11.89 \times 0.08)}{\cosh (11.89 \times 0.08) + (22 / 11.89 \times 51.9) \sinh (11.89 \times 0.08)}$$

= **11.08 W** or 75.6% of long fin or 79% of fin of double its length.

Hence use of long fin generally is wasteful of material.

Example 4.7: *Consider the data in example 4.1. The ends of the rod 159 mm long is held at 200°C and 100°C. Determine the temperature at mid location. Also find the minimum temperature and its location.*

The data are presented in Fig. Ex. 4.7.

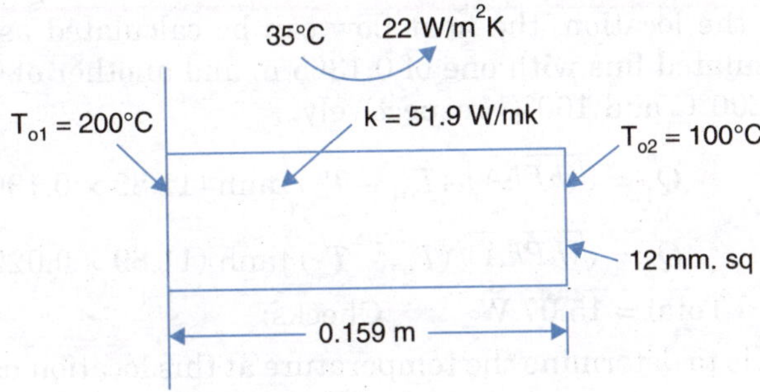

Fig. Ex. 4.7. Problem model

Solution: This situation corresponds to the end conditions being held at specified temperatures.

Using equations (4.10) and (4.14).

$$\frac{T - T_\infty}{T_{b1} - T_\infty} = \frac{[(T_{b2} - T_\infty) / (T_{b1} - T_\infty)] \sinh (mx) + \sinh m(L - x)}{\sinh (mL)}$$

Temperature at the mid location

$$\frac{T - 35}{200 - 35} = \frac{[(100 - 35) / (200 - 35)] \sinh (11.89 \times 0.159/2) + \sinh 11.89\ (0.159 - 0.159/2)}{\sinh (11.89 \times 0.159)}$$

$$= 0.47 \quad \therefore \quad \mathbf{T = 112.65°C}$$

$$Q = \sqrt{(hPkA)}\ .\ ((T_{b1} - T_\infty) + (T_{b2} - T_\infty))\ ((\cosh (mL) - 1)/\sinh (mL))$$

$$= \sqrt{(22 \times 4 \times 0.012 \times 51.9 \times 0.012 \times 0.012)}\ \{(200 - 35)$$

$$+ (100 - 35)\}\ \{(\cosh (11.89 \times 0.159) - 1)/\sinh (11.89 \times 0.159)\} = \mathbf{15.07\ W}$$

To find the location of the minimum temperature, the condition that $(dT/dx)_x = 0$ at the point is used.

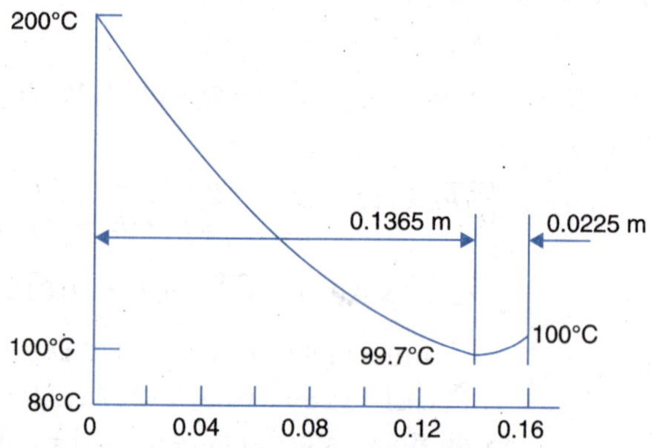

Fig. 4.9. Temperature variation along fin

Differentiating equation (4.10) and equating to zero.

$\{(T_{b2} - T_\infty)/(T_{b1} - T_\infty)\} \times \cosh (mx) = \cosh m(L - x)$. Solving by trial

$x_{min} = 0.1365$ m and substituting in equation (4.8), $\mathbf{T_{min} = 99.7°C}$

The temperature plot is shown in Fig. 4.9.

As a check for the location, the heat flow can be calculated as the sum of heat flow from two short end insulated fins with one of 0.1365 m and another of 0.0225 m lengths with base temperatures of 200°C and 100°C respectively.

$$Q_1 = \sqrt{(hPkA)}\ (T_{01} - T_\infty) \tanh (11.89 \times 0.1365) = 13.56 \text{ W}$$

$$Q_2 = \sqrt{(hPkA)}\ (T_{02} - T_\infty) \tanh (11.89 \times 0.0225) = 1.51 \text{ W}$$

Total = 15.07 W Checks:

Another check is to determine the temperature at this location using the above concept of two fins.

$$(T - 35)/(200 - 35) = 1/(\cosh (11.89 \times 0.1365)) \quad \therefore \quad T = 97.67°C$$

$$(T - 35)/(100 - 35) = 1/(\cosh (11.89 \times 0.0225)) \quad \therefore \quad T = 97.74°C$$

The value obtained using the single unit concept is 97.7°C. This also checks the location of minimum temperature. Note that such checks are very useful.

4.4 FIN PERFORMANCE

It is necessary to evaluate the performance of fins to achieve minimum weight or maximum heat flow etc. Fin effectiveness, fin efficiency and total efficiency are some methods used for performance evaluation of fins.

4.4.1. Fin Effectiveness, ε_f

Fins are used to increase the heat transfer from a surface by increasing the effective surface area. When fins are not present, the heat **convected by the base** area is given by $Ah(T_b - T_\infty)$, where A is the base area. When fins are used the heat transferred by the fins, q_f, is calculated using equations (4.11) to (4.14). The ratio of q_f to the heat convected at the base area is defined as fin effectiveness.

$$\varepsilon_f = \frac{q_f}{hA(T_b - T_\infty)} \quad ...(4.15)$$

Fin effectiveness should be as large as possible for effective use of material. Use of fins with effectiveness less than 2 can be rarely justified considering the long fin boundary,

$$\varepsilon_f = \frac{\sqrt{hPkA}\,(T_b - T_\infty)}{hA\,(T_b - T_\infty)} = \left(\frac{kP}{hA}\right)^{0.5} \quad ...(4.16)$$

Some conclusions from equation (4.16) are

1. Thermal conductivity of the fin material should be high to give large fin effectiveness. This leads to the choice of aluminium and its alloys.

2. The ratio $\frac{P}{A}$ should be large. This requirement can be achieved by the **use of thinner fins. Use more thin fins at closer pitch than fewer thicker fins at longer pitch.**

3. Effectiveness will be higher if h is lower. Generally convection in gas flow and heat flow under free convection lead to lower values of heat transfer coefficient, h. **Hence fins are** invariably **used on the gas side of heat exchangers. The condition that $\varepsilon_f > 2$ leads to the relation $\frac{kP}{hA} > 4$.**

4. About the choice of length of fins, it is seen that 98% of the maximum possible fin heat tranfer takes place in end insulated fins when $mL = 2.3$ or $L = 2.3/m$. It will make little sense in extending the length beyond this.

Concept of fin thermal resistance can also be used to express effectiveness of fins. The resistance to heat flow by the fin can be expressed as

$$R_f = \frac{T_b - T_\infty}{q_f} \quad ...(4.17)$$

The convection resistance of the base area is (A is fin base area)

$$R_b = \frac{1}{hA} \quad ...(4.18)$$

$$\therefore \quad \varepsilon_f = \frac{R_b}{R_f} = \frac{1}{hA} \cdot \frac{q_f}{T_b - T_\infty} \quad ...(4.19)$$

same as equation (4.15).

Higher values of effectiveness can result from higher values of convection resistance compared to the fin resistance. In effect it can be said that **fins reduce to resistance to heat flow at the surface or when convection coefficient is lower.**

Chapter 4

4.4.2. Fin Efficiency, η_f

This quantity is more often used to determine the heat flow when variable area fins are used. **Fin efficiency is defined as the ratio of heat transfer by the fin to the heat transfer that will take place if the whole surface area of the fin is at the base temperature. This situation is not possible as the temperature has to be continuously lower in fins.**

$$\eta_f = \frac{q_f}{hA_s(T_b - T_\infty)} \qquad \text{...(4.20)}$$

where A_s is the surface area of the fin equals $P \times L$.

For constant area fin with adiabatic tip,

$$\eta_f = \frac{\sqrt{hPkA}\,(T_b - T_\infty)\,.\,\tanh(mL)}{hPL\,(T_b - T_\infty)} = \frac{\tanh(mL)}{mL} \qquad \text{...(4.21)}$$

In order to compensate for convection at the end of fin, the length of the fin, is increased as indicated below. In the case of plate fins new length $L_c = L + \frac{t}{2}$ where t is the thickness. In the case of circular fins $L_c = L + D/4$, where D is the diameter of the fin. Error associated with this assumption is negligible if $\frac{ht}{k}$ or $\frac{hD}{2k} \le 0.0625$.

From equation (4.21), it is seen that fin efficiency is a function of the term "mL_c".

$$mL_c = \left(\frac{hP}{kA}\right)^{0.5} L_c \qquad \text{...(4.22)}$$

Assuming plate fins $P \simeq 2W$, where W is the width of the fin and $A = Wt$. Substituting

$$mL_c = \left(\frac{2h}{kt}\right)^{0.5} L_c \qquad \text{...(4.23)}$$

Multiplying and dividing the RHS of (4.23), by L_c

$$mL_c = \left(\frac{2h}{ktL_c}\right)^{0.5} .\, L_c^{3/2} = \left(\frac{2h}{kA_p}\right)^{0.5} L_c^{1.5} \qquad \text{...(4.24)}$$

where A_p is called the profile area = $L_c \times t$.

Hence $$\eta_f = f\left[L_c^{1.5} . \left(\frac{2h}{kA_p}\right)^{0.5}\right]. \qquad \text{...(4.25)}$$

This relationship is used to plot the fin efficiency of variable area fins and annular fins. Such a chart for plate fins is given below. Similar charts are available far annular fins and pin fins.

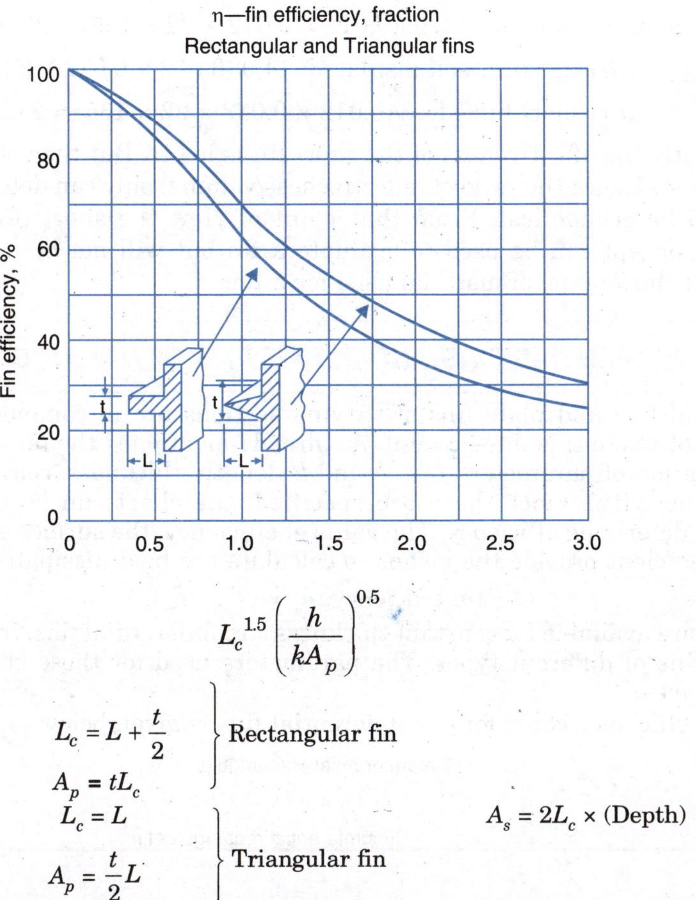

$$L_c^{1.5}\left(\frac{h}{kA_p}\right)^{0.5}$$

$$\left.\begin{array}{l} L_c = L + \dfrac{t}{2} \\ A_p = tL_c \end{array}\right\} \text{Rectangular fin}$$

$$\left.\begin{array}{l} L_c = L \\ A_p = \dfrac{t}{2}L \end{array}\right\} \text{Triangular fin} \qquad A_s = 2L_c \times (\text{Depth})$$

.4.3. Overall Surface Efficiency or Total Efficiency, η_T

'in efficiency gives the performance of a single fin. Overall efficiency gives the performance of n array of fins and the surface on which these fins are provided. Let there be N fins each of rea A_f. Let the total area be A_t. The free surface over which convection takes place is $_b = A_t - NA_f$.

The total heat flow = heat flow by fins + heat flow by base area

$$q_t = [N\eta_f\, hA_f + h(A_t - N_A A_f)]\,(T_b - T_\infty)$$

$$= hA_t\left[1 - \frac{NA_f}{A_t}(1-\eta_f)\right](T_b - T_\infty)$$

The ideal or maximum value is

$$= hA_t\,(T_b - T_\infty)$$

$$\therefore \qquad \eta_t = 1 - \frac{NA_f}{A_t}(1-\eta_f) \qquad \text{...(4.26)}$$

This equation can be used to calculate the heat flow in a surface array provided η_t is vailable.

Ɛxample 4.8: *Determine the value of fin effectiveness for the fins in examples 4.4, 4.5 and 4.6.*

Solution: In all these cases the base area = 0.012 × 0.012 m², h = 22 W/m²K, $(T_o - T_\infty)$ = (200 - 35) = 165. The heat dissipated are 14.66, 14 and 11.08, using equation (4.15).

Chapter 4

Effectiveness: (*i*) long fin: 14.66/(0.012 × 0.012 × 22 × 165) = 28.0

(*ii*) short fin end insulated: 14.0/(0.012 × 0.012 × 22 × 165) = 26.7

(*iii*) short fin: 11.08/(0.012 × 0.012 × 22 × 165) = 21.2

Apparently the effectiveness of the short fin is lower. But the volume of material used also is the lowest. Hence the ratio of (effectiveness/volume) only can determine the form of the fin which will be economical. From that point of view the short fin has the advantage. Effectiveness concept can be used to evaluate a fin but will not be useful by itself for the optimisation of the volume of material used for a fin.

4.5 CIRCUMFERENTIAL FINS AND PLATE FINS OF VARYING SECTIONS

Circumferential fins and plate fins of varying sections are in common use. The preceding analysis has not taken this into account. As already mentioned the fin efficiency is correlated to the combination of parameters L, t, h and k (length, thickness, convection coefficient and thermal conductivity). Once these are specified, the chart can be entered by using the parameters to determine efficiency. The value of efficiency, the surface area, temperature and convection coefficient provide the means to calculate the heat dissipated.

$$Q = \text{fin efficiency}.\, A_s\, h\, (T_b - T_\infty) \quad ...(4.18)$$

Charts are available for constant thickness circumferential fins, triangular section plate fins and pin fins of different types. The parameters used for these charts are given in the charts themselves.

The fin efficiency chart for circumferential fins is given below:

Circumferential Rectangular Fins

$Q = \eta\, A_s h\, (T_b - T_\infty)$

Circumferential rectangular fin

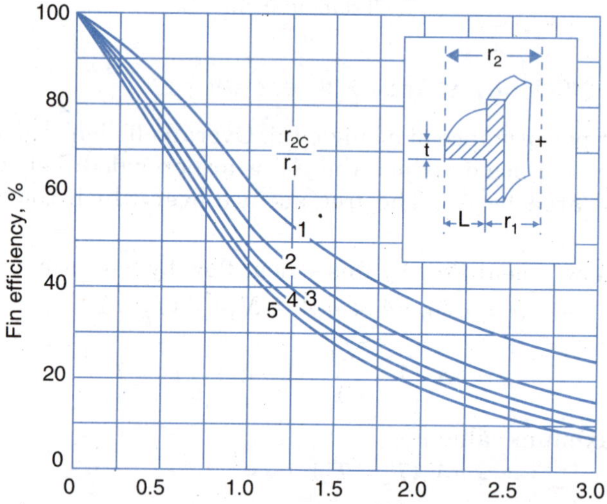

$$L_c^{1.5}\left(\frac{h}{kA_p}\right)^{0.5}$$

$$L_c = L + \frac{t}{2} \qquad r_{2c} = r_1 + L_c$$

$$A_p = t(r_{2c} - r_1) \qquad A_s = 2\pi(r_{2c}^2 - r_1^2)$$

Example 4.9: *Determine the heat flow for (i) flat rectangular fins and (ii) triangular fins of 20 mm length and 3 mm base thickness. Thermal conductivity = 45 W / mK. Convection coefficient = 100 W / m²K, base temperature = 120°C surrounding fluid temperature = 35°C Determine also the fin effectiveness. Use the charts.*

The data are shown in Fig. Ex. 4.9(a). Using the nomenclature in the chart (Skeleton chart shown in Fig. Ex. 4.9(b). Refer chart under article 4.4.2

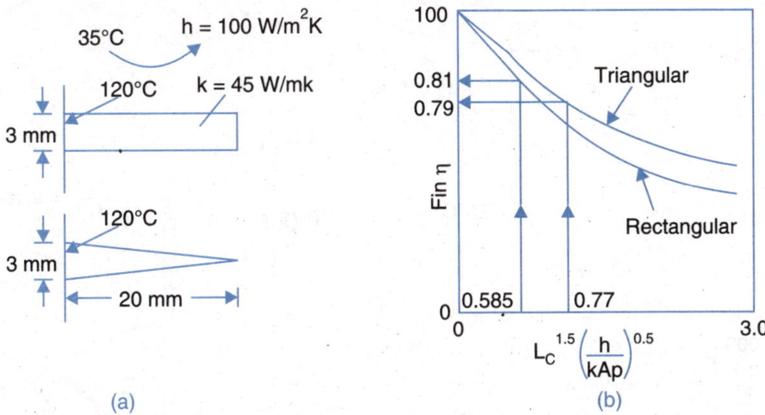

Fig. Ex. 4.9

Solution: (*i*) Heat transfer at the root area

$$= 0.003 \times 1 \times 100\,(120 - 35) = 25.5 \text{ W}$$

$$L_c = 0.02 + 0.003/2 = 0.0215$$

$$A_p = 0.003 \times 0.0215,\ A_s = 2 \times 0.0215 \times 1$$

$$\mathbf{L_c^{1.5}\,(h/k\,A_p)^{0.5}} = 0.0215^{1.5}\,(100/(45 \times 0.0215 \times 0.003))^{0.5} = \mathbf{0.585}$$

Entering the chart at 0.585, **fin efficiency** is read as **0.81** (As in Fig. Ex. 4.9(*b*))

$$\mathbf{Q} = 0.81 \times 2 \times 0.0215 \times 1 \times 100\,(120 - 35)$$

$$= \mathbf{296.06 \text{ W/m depth.}}$$

Check: $Q = \sqrt{(hPkA)}\ .\ (T_o - T_\infty)\tanh\,(mL),$

$$\sqrt{(hP / kA)} = \sqrt{[(100 \times 2) / (45 \times 0.003)]}$$

$$m = 38.49,\ mL = 0.77,\quad \therefore\quad Q = 285.73 \text{ W},\quad \text{checks.}$$

Effectiveness = 296/25.5 = **11.61**

(*ii*) For triangular fin:

$$L_c = L = 0.02,\ A_s = 2 \times 0.02 \times 1,\ A_p = (0.003/2) \times 0.02$$

$$\mathbf{L_c^{1.5}\,(h/kA)^{0.5}} = 0.02^{1.5}\,[(100)/(45 \times 0.0015 \times 0.02)]^{0.5} = \mathbf{0.77}$$

Entering the chart at 0.77, efficiency is read as 0.79. (as in Fig. Ex. 4.9 (*b*))

$$Q = 0.79 \times 2 \times 0.02 \times 1 \times 100 \times (120 - 35) = \mathbf{268.6 \text{ W}}$$

Effectiveness = 268.6/ 25.5 = **10.53**

COMMENTS

1. Though the effectiveness is less, in case (*ii*) note that the volume of material used is only half of that used for the rectangular fin.
2. If heat flow/volume is considered, the ratio for the triangular fin will be twice that of rectangular fin.

Chapter 4

Example 4.10: *A circumferential fin on a pipe of 50 mm OD is 3 mm thick and 20 mm long. Using the property values and other parameters in example 4.9, determine the (i) heat flow and effectiveness (ii). If the pitch is 10 mm, determine the increase in heat flow for 1 m length of pipe. Also determine the total efficiency using the nomenclature in the efficiency chart [Skeleton chart shown in Fig. Ex. 4.10(b)].*

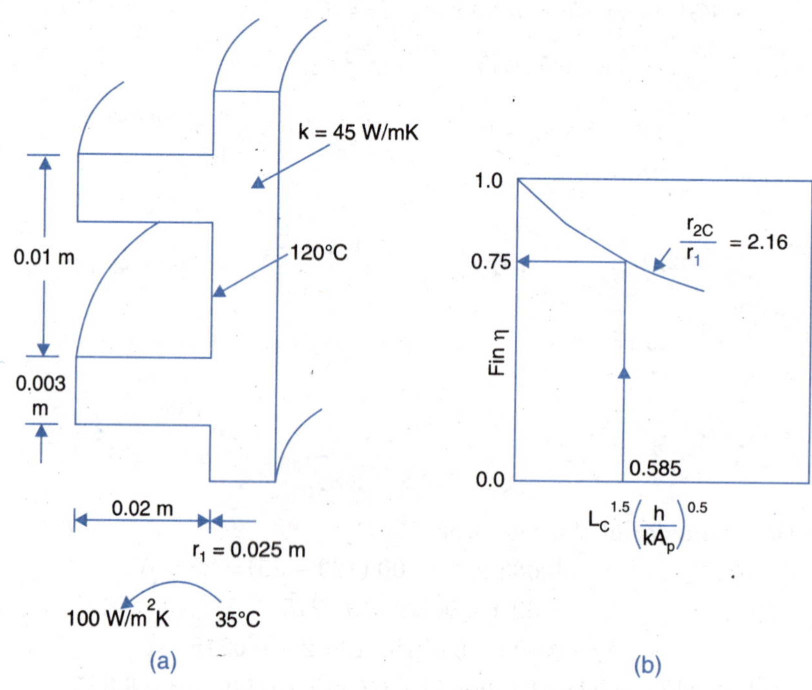

Fig. Ex. 4.10

Solution:

$$L_c = L + t/2 = 0.02 + 0.003/2 = 0.0215 \text{ m}$$

$$r_{2c} = r_1 + L_c = 0.025 + 0.0215 = 0.0465 \text{ m}$$

$$A_p = 0.003\,(0.0465 - 0.0215) = 0.0255 \times 0.003$$

$$A_s = 2\pi\,(0.0465^2 - 0.0215^2)$$

(*i*) $$\mathbf{L_c^{1.5}\,(h/kA_p)^{0.5}} = 0.0215^{1.5}\,(100/(45 \times 0.003 \times 0.0255))^{0.5} = \mathbf{0.585}$$

$$\mathbf{r_{2c}/r_1} = 0.0465/0.0215 = \mathbf{2.16}$$

Entering the chart at 0.585 and using the curve corresponding to $r_{2c}/r_1 = 2.16$ (interpolation) the **fin efficiency** is read as **0.75** [As in Fig. Ex. 4.10(*b*)]

$$\mathbf{Q} = 0.75 \times 2\pi\,(0.0465^2 - 0.0215^2)\,100 \times (120 - 35) = \mathbf{68.09\ W/fin}$$

Effectiveness = 68.09/2(π × 0.025 × 0.003 × 100 × (120 – 35)) = **17**

(*ii*) Considering 1 m length the heat flow **without the fins**

Heat flow = 2π × 0.025 × 1 × 100 × (120 – 35) = **1335.2 W/m length.**

As the pitch is 10 mm, the number of fins per m length will be 100 fins remaining base length will be 0.7 m.

Q with fins = 100 × 68.09 + 1335.2 × 0.7 = **7743.6 W**

This is about 5.8 times that of bare pipe.

Total area considering fin tip

$$= (\pi \times 0.025 \times 1) + 100 \times \pi\,(0.045^2 - 0.025^2) = 0.94247 \text{ m}^2$$

Maximum heat flow = $hA\Delta T = 100 \times 0.94247 \times (120 - 35) = 8011$ W.

Total efficiency = $\frac{7743.6}{8011}$ = **0.9665 or 96.65%.**

4.6 OPTIMISATION

If fins of constant cross sectional area are used, then there is a wastage of material because, the section required for heat conduction for the same temperature gradient will be continuously decreasing with length. Several possibilities are there for the choice of section. Some of these are trapesoidal, triangular, convex parabolic and concave parabolic as shown in Fig. 4.4.

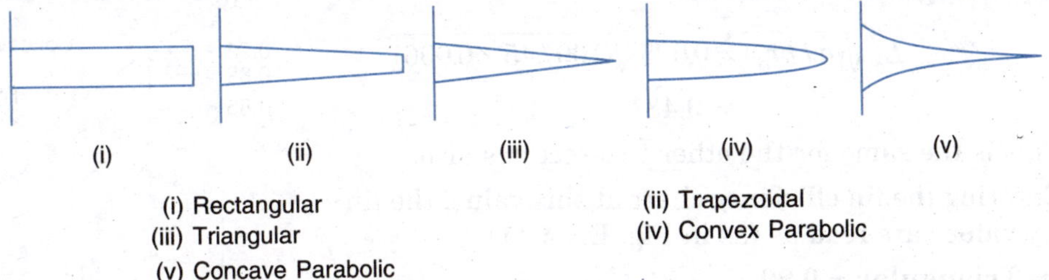

Fig. 4.4. Fin sections

Out of these sections, the concave parabolic has the highest fin efficiency for a given set of parameters, and this will also require the least weight for a given amount of heat dissipation.

Fin efficiency chart for pin fins is given below

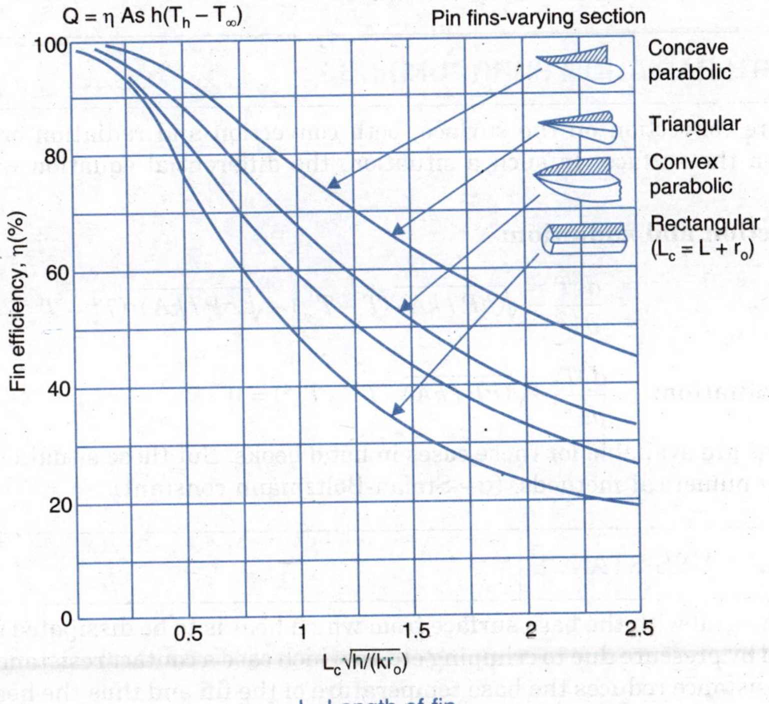

Example 4.11: *Considering pin fins of constant section, triangular, concave and convex parabolic sections of base dia 12 mm and length 25 mm, determine the fin efficiency, k = 45 W/mK, h = 100 W/m²K referring to efficiency chart.*

Solution: $L_c \sqrt{(h/kr_o)}$ is to be determined for these cases.

(*i*) **Constant section:** $L_c = L + r_o = 0.025 + 0.006 = 0.031$

$$L_c \sqrt{(h/kr_o)} = 0.031 \sqrt{(100/45 \times 0.006)} = 0.597$$

Entering the **fin efficiency** chart at 0.597, the **fin efficiency** is read as **0.74** (as in Fig. Ex. (4.11)

(*ii*) **Triangular:** $L_c = L$

$$L_c \sqrt{(h/kr_o)} = 0.025 \sqrt{(100/45 \times 0.006)}$$

$$= 0.481$$

This is the same for the other two sections also.

Entering the fin efficiency chart at this value, the fin efficiency values are read as (as in Fig. Ex. 4.11)

(*i*) **Triangular = 0.89**

(*ii*) **Convex parabolic = 0.85**

(*iii*) **Concave parabolic = 0.93**

The material used is the least in the case of concave parabolic section. The efficiency is the highest. The surface areas are vary nearly the same for the last three.

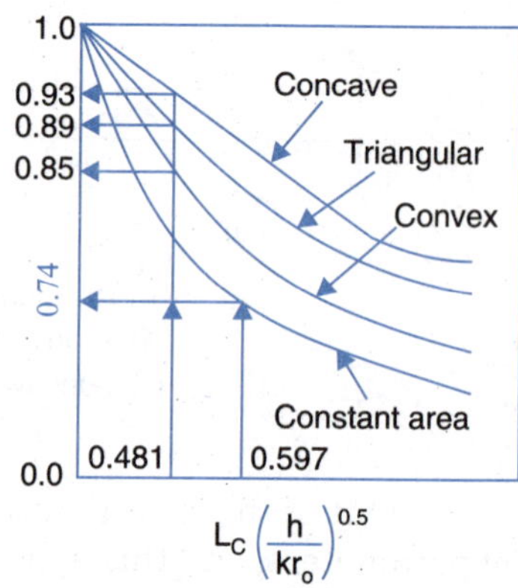

Fig. Ex. 4.11. Skeleton

4.7 FIN WITH RADIATION SURROUNDINGS

Instead of pure convection on the surface, both convection and radiation or pure radiation may prevail on the surface. In such a situation, the differential equation will be as shown below:

Convection and radiation:

$$\frac{d^2T}{dx^2} - \sqrt{(hP/kA)}\,(T - T_\infty) - \sqrt{(\sigma P/kA)}\,(T^4 - T_\infty^{\ 4}) = 0 \qquad ...(4.21)$$

Pure radiation:

$$\frac{d^2T}{dx^2} - \sqrt{(\sigma P/kA)}\,(T^4 - T_\infty^{\ 4}) = 0 \qquad ...(4.22)$$

Solutions are available for these cases in hand books. But these situations can be solved more easily by numerical methods. (σ—Stefan-Boltzmann constant).

4.8 CONTACT RESISTANCE

Fins may be integral with the base surface from which heat is to be dissipated or these may be welded or fixed by pressure due to crimping etc. in which case a contact resistance is introduced. The contact resistance reduces the base temperature of the fin and thus the heat dissipated by the fin. If loosely fitted this drop may be high. Manufacture of integral fins is not possible in many cases, but care should be taken to reduce the contact resistance. A trial solution will be necessary in this case as the base temperature at the contact depends on the heat flow which again is a function of the base temperature. An example is worked out later in the chapter for this situation. (Problems 4.25, 4.28)

4.9 NUMERICAL METHODS

Numerical methods can be conveniently adapted to solve for the temperature distribution and the heat flow in fins. Considering the fin shown in Fig. 4.5, the energy equation for i^{th} node can be written as

heat conducted into the volume – heat conducted out of the volume = heat dissipated at the surface.

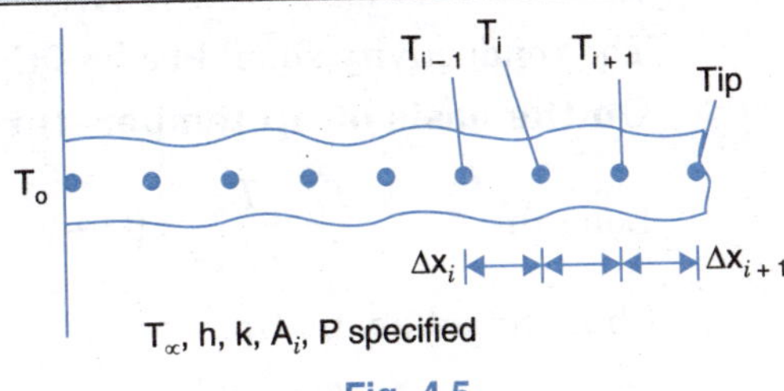

Fig. 4.5

$$\frac{kA_i\,(T_{i-1}-T_i)}{\Delta x_i}-\frac{kA_{i+1}\,(T_i-T_{i+1})}{\Delta x_{i+1}} = hP_i\,(T_i - T_\infty)\,(\Delta x_i + \Delta x_{i+1})/2$$

or

$$(kA_i/\Delta x_i) \times T_{i-1} + \{(kA_{i+1}/(\Delta x_{i+1})\} \times T_{i+1} - \left[\frac{kA_i}{\Delta x_i}+\frac{kA_{i+1}}{\Delta x_{i+1}}+\frac{hP_i\,(\Delta x_i+\Delta x_{i+1})}{2}\right] T_i + \frac{hP_i\,(\Delta x_i+\Delta x_{i+1})}{2}\,T_\infty = 0 \quad ...(4.23)$$

If Δx and A and P are taken to be constants, then this reduces to

$$T_{i-1}+T_{i+1}-\left[2+\frac{hP}{kA}.\Delta x^2\right]T_i+\frac{hP}{kA}.\Delta x^2.T_\infty = 0 \quad ...(4.24)$$

For the last node alone area is $P\Delta x/2$ and the equation is

$$T_{i-1}-\left[1+\frac{hP}{kA}.\frac{\Delta x^2}{2}\right]T_i+\frac{hP}{kA}.\frac{\Delta x^2}{2}.T_\infty = 0 \quad ...(4.25)$$

The resulting simultaneous equations can be solved by solving the coefficient matrix equation. All variations including that in conductivity and convection coefficient in addition to the sectional area and perimeter can be taken into account. Of course a computer is needed to solve the matrix. Now softwares are available using which the problem can be solved by inputting the values of various parameters like the dimensions, conductivity, area, convection coefficient etc. In case radiation is involved, the term convection is replaced by the term radiation.

Example 4.12: *A rod of 5 mm dia and 100 mm length is used as a fin. The convection coefficient is 30 W/m²K. Determine for what value of thermal conductivity above which the fin can be considered as an infinite or long fin.*

Solution: On the basis of heat flow: heat flow in long fin, $Q = \sqrt{hPkA}\,(T_o - T_\infty)$.

Adiabatic tip, $Q = \sqrt{hPkA}\,(T_o - T_\infty)\tanh(mL)$.

Assuming 1%, difference is allowable,

Chapter 4

It tanh $(mL) = 0.99$, then $mL = 2.65$

If $L \geq 2.65/m$, then it can be considered as a long fin.

Using the equality $mL = 2.65$, As $L = 0.1$ m, $m = 26.5$

$$\therefore \quad m = 26.5 = \sqrt{[(30 \times \pi \times 0.005 \times 4)/(k \times \pi \times 0.005^2)]}$$

$$\therefore \quad k = 34.18 \text{ W/mK.}$$

The conductivity should be less than or equal to 34.18 W/mK.

On the basis of tip temperature,

Long fin, $$\frac{T_L - T_\infty}{T_o - T_\infty} = 0,$$

Short fin adiabatic tip,

$$\frac{T_L - T_\infty}{T_o - T_\infty} = \frac{1}{\cosh(mL)}$$

Assuming 2% difference, between these two

$$\frac{1}{\cosh(mL)} = 0.02 \quad \therefore \quad mL = 4.6.$$

In this case, as $L = 0.1$, $m = 46$,

$$\therefore \quad 46 = \sqrt{[(30 \times \pi \times 0.005 \times 4)/(k \times \pi \times 0.005^2)]}$$

$$\therefore \quad \mathbf{k = 11.34 \text{ W/mK.}}$$

Example 4.13: *In the example 4.12, if $T_b = 200°C$, and $T_\infty = 30°$, determine the end temperature assuming (i) long fin model, (ii) short fin end insulated model m = 26.5 (from example 4.12).*

(i) long fin model: $(T - 30)/(200 - 30) = e^{-mx} = e^{-26.5 \times 0.1} = 0.07065$, $T = 42°C$

(ii) assuming short fin end insulated model.

Solution: $(T - 30)/(200 - 30) = 1/\cosh(mL) = 1/\cosh(26.5 \times 0.1)$ $\therefore$ $T = 53.9°C$

It can be checked that the heat flow will be less only by 1%

i.e. $1 - \tanh(mL) = (1 - \tanh(26.5 \times 0.1) = 0.01$

ADVANCED EXAMPLES

Example A-4.1: *One end of long rod 1 cm dia. having a thermal conductivity of 45 W/mK is placed in a furnace. The rod is exposed to air at 30°C over its surface and the convection coefficient is estimated at 35 W/m²K. If the temperature is read as 265°C at a distance of 39.3 mm from the furnace end, determine the base temperature of the rod.*

Solution: Equation 4.5 for long fin model can be used.

$$(T - T_\infty)/(T_b - T_\infty) = e^{-mx}$$

$$m = \sqrt{(hP/kA)} = \sqrt{[(35 \times \pi \times 0.01 \times 4)/(45 \times \pi \times 0.01^2)]} = 17.64$$

$$x = 0.0393, \; mx = 0.693$$

$$(265 - 30)/(T_b - 30) = e^{-0.693} = 0.50 \quad \therefore \quad \mathbf{T_b = 500°C.}$$

Example A-4.2: *One end of a long rod of 1 cm dia. is maintained at 500°C by placing it in a furnace. The rod is exposed to air at 30°C with a convection coefficient of 35 W/m²K. The temperature measured at a distance of 78.6 mm was 147.5°C. Determine the thermal conductivity of the material.*

Solution: Equation 4.5 can be used as this is a long fin configuration.

$$(T - T_\infty)/(T_b - T_\infty) = e^{-mx},\ (147.5 - 30)/(500 - 30) = 0.25 = e^{-m \times 0.0786}$$

Solving $m = 17.64 = \sqrt{(hP/kA)} = \sqrt{[(35 \times \pi \times 0.01 \times 4)/(k \times \pi \times 0.1^2)]}$

Solving **k = 45 W/mK.**

COMMENT

This idea is used in determining the thermal conductivity of materials.

Example A-4.3: *One end of a rod of 1 cm dia. and 80 mm length is maintained at a temperature T when the rod is exposed to air at 30°C. The conductivity of the material is 12.5 W/mK and the convection coefficient over the surface is 25 W/m²K. If the temperature at the tip was measured as 80°C, determine the value of temperature T.*

The data are presented in Fig. AE-4.3. The short fin end insulated model is suitable for the problem. So equation (4.6) is used.

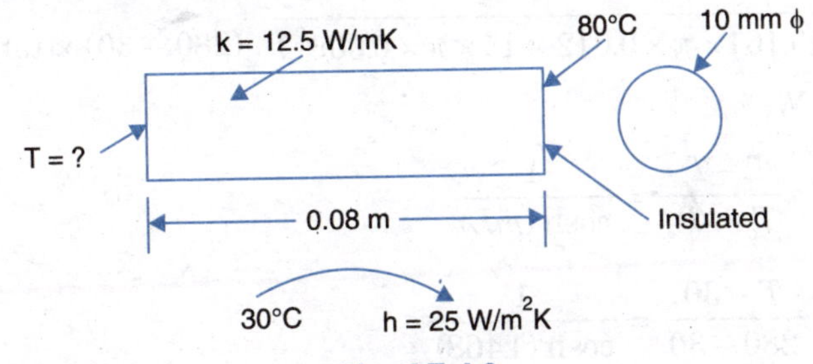

Fig. AE-4.3

Solution: $(T_L - T_\infty)/(T_b - T_\infty) = 1/\cosh(mL)$

$$m = \sqrt{(hP/kA)} = \sqrt{[(25 \times \pi \times 0.01 \times 4)/(12.5 \times \pi \times 0.1^2)]}$$

$$= 28.28,\ mL = 2.26$$

$(80 - 30)/(T - 30) = 1/\cosh(2.26) = 0.206$. Solving **T = 272.84°C.**

Example A-4.4: *A short fin of 0.08 m length and diameter 12 mm is exposed to air at 30°C. Thermal conductivity is 15 W/mK. The base temperature is 280°C. The heat dissipated by the fin is 7W. Determine the value of convection coefficient and also the tip temperature.*

The data are shown in Fig. AE-4.4.

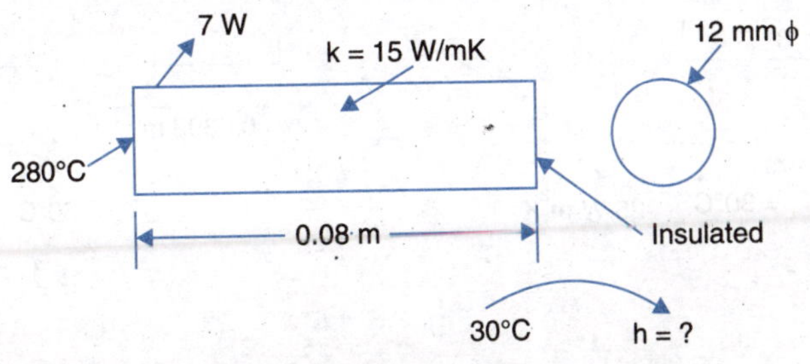

Fig. AE-4.4. Problem model

Chapter 4

Short fin, end insulated condition is suitable for the problem and heat flow is given. So equation (4.11) is used.

Solution: $Q = \sqrt{(hPkA)}\ (T_b - T_\infty) \tanh (mL) = 7$ W

Convection coefficient is involved in tanh (mL) as well as in $\sqrt{(hPkA)}$. Hence a trial solution is attempted. Assume: $h = 20$ W/m²K,

$$m = \sqrt{(hP/kA)} = \sqrt{[(20 \times \pi \times 0.012 \times 4)/(15 \times \pi \times 0.012^2)]}$$

$$= 21.08,\ mL = 1.6864$$

$$Q = \sqrt{(20 \times \pi \times 0.012 \times 15 \times \pi \times 0.012^2/4)}\ (280 - 30) \tanh (1.6864) = 8.35 \text{ W}$$

Assume: $h = 15$, $m = 18.26$, $mL = 1.46$.

$$Q = \sqrt{(15 \times \pi \times 0.012 \times 15 \times \pi \times 0.012^2/4)}\ (280 - 30) \tanh (1.46) = 6.952 \text{ W}$$

Assume: $h = 16$, $m = 18.86$, $mL = 1.508$, $Q = 7.25$ W

Interpolating, **h** = 15 + (7 – 6.952)/(7.25 – 6.952) = **15.161 W/m²K**

$$m = 18.355,\ mL = 1.468,$$

$$Q = \sqrt{hPkA}\ (T_b - T_\infty) \tanh (mL)$$

$$= \sqrt{(15.161 \times \pi \times 0.012 \times 15 \times \pi \times 0.006^2)}\ .\ (280 - 30) \times \tanh (1.468)$$

$$= 7.0 \text{ W}.$$

$$\frac{T - T_\infty}{T_b - T_\infty} = \frac{1}{\cosh (mL)}$$

$$\frac{T - 30}{280 - 30} = \frac{1}{\cosh (1.468)}$$

Solving, tip temperature = 139.3°C

Example A-4.5: *One end of long rod of diameter 10 mm is inserted into a furnace. The temperatures measured at two points A and B, 39.3 mm apart gave 265°C and 147.5°C respectively. If the convection coefficients is 35 W/m²K, when exposed to air at 30°C, determine the conductivity of the material.*

The data are shown in Fig AE-4.5(a). ***The location A itself can be taken as the base*** *and then the fin can be treated as a long fin. This is the easier method. Fig. AE-4.5(b).*

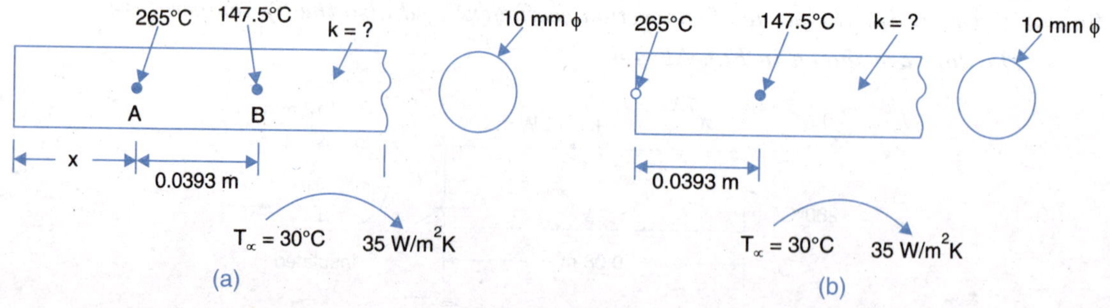

Fig. AE-4.5

Solution: $\dfrac{T - T_\infty}{T_b - T_\infty} = e^{-mx}$

$(147.5 - 30)/(265 - 30) = e^{-m \times 0.0393}$ $\therefore$ **m = 17.64**

$$m = 17.64 = \sqrt{[(hP)/(kA)]} = \sqrt{[(35 \times \pi \times 0.01 \times 4)/(k \times \pi \times 0.01^2)]}$$

k = 45 W/mK

[Or $(T_A - T_\infty)/(T_b - T_\infty) = e^{-mx}$, $(T_B - T_\infty)/(T_b - T_\infty) = e^{-m(x+0.393)}$

dividing, $(T_B - T_\infty)/(T_A - T_\infty) = e^{-m \times 0.0393}$, same as above].

COMMENT

This method can also be used to experimentally estimate the value of thermal conductivity.

Example A-4.6: *One end of a long rod of dia. 12 mm is inserted into a furnace and the temperatures measured at points A and B 91.2 mm apart are found to be 147.5°C and 77°C respectively. If the conductivity of the material is 45 W/mK and the surrounding is at 30°C determine the convection coefficient.*

Solution: The known data are shown in Fig. AE-4.6. Considering the section A itself as the base and using long fin model.

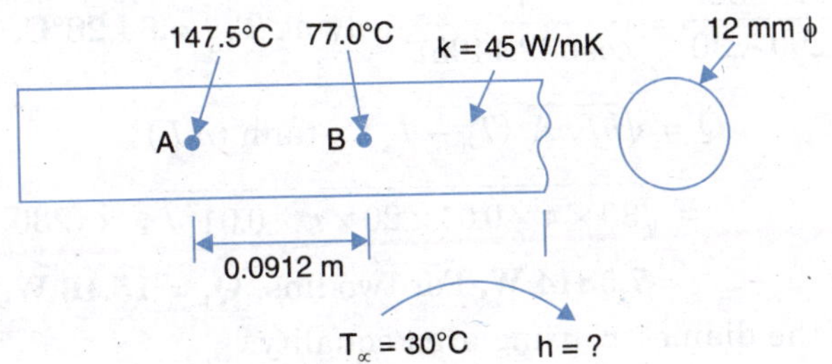

Fig. AE-4.6

$(77 - 30)/(147.5 - 30) = e^{-m \times 0.0912}$, solving $m = 10.05$

$$\sqrt{[(hP)/(kA)]} = \sqrt{[(h \times \pi \times 0.012 \times 4)/(45 \times \pi \times 0.012^2)]} = 10.05,$$

$\therefore$ **h = 13.62 W/m²K.**

Example A-4.7: *Two rods of dia. D mm and length L mm have one of the ends at 120°C and are exposed to air at 30°C. The conductivity of the material of one rod is 45 W/mK and the temperature of the rod at the end is measured as 80°C, while the end temperature of the other rod is 60°C. Determine the conductivity of the other material.*

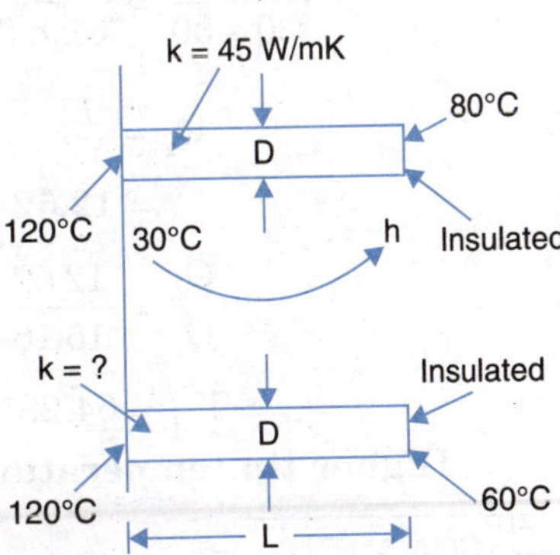

Fig. AE-4.7

Solution: The known data are shown in Fig. AE-4.7. Short fin end insulated condition suits the problem

$(80 - 30)/(120 - 30) = 1/\cosh(m_1 L)$,

$\therefore$ $m_1 L = 1.193$

$(60 - 30)/(120 - 30) = 1/\cosh(m_2L)$,

$\therefore \quad m_2L = 1.763$, dividing L cancels

$m_1/m_2 = 0.677$

$= \sqrt{(hP/k_1A)(k_2A/hP)}$,

$k_2/k_1 = 0.677^2, \quad \therefore \quad \mathbf{k_2 = 20.62\ W/mK.}$

Example A-4.8: *Two fins of 0.01 m diameter and 0.1 m length are used to dissipate heat from a surface at 230°C. The surrounding is at 30°C. The thermal conductivity of the material is 20 W/mK. The convection coefficient is 30 W/m²K. To check alternate arrangement it is proposed to replace the fins by a single fin of the total area of the two fins with length and other conditions being the same as before. Compare the heat transferred and the tip temperatures.*

Solution: **Case 1.** $D = 0.01$ m, $k = 20$ W/mK, $h = 30$ W/m²K

$L = 0.1$ m, $T_b = 230°C$, $T_\infty = 30°C$

To calculate the tip temperature, assuming short fin end insulated condition,

$$m = \sqrt{hP/kA} = \sqrt{\frac{30 \times \pi \times 0.01 \times 4}{20 \times \pi \times 0.01^2}}$$

$$= 24.495, \quad \therefore \quad \mathbf{mL = 2.4495}$$

$$\frac{T - 30}{200 - 30} = \frac{1}{\cosh(2.4495)}, \text{ solving } \mathbf{T_{L1} = 64.28°C.}$$

$$Q = \sqrt{hPkA}\,(T_b - T_\infty) \times \tanh(mL)$$

$$= \sqrt{30 \times \pi \times 0.01 \times 20 \times \pi \times 0.01^2/4} \times (230 - 30) \tanh(2.4495)$$

$$= \mathbf{7.5814\ W}, \text{ For two fins, } \mathbf{Q_1 = 15.16\ W.}$$

Case 2. To find the diameter, using area equality,

$$(\pi D^2/4) = 2\pi \times 0.01^2/4, \quad \therefore \quad D = 0.014142 \text{ m}$$

$$m = \sqrt{\frac{30 \times \pi \times 0.014142 \times 4}{20 \times \pi \times 0.014142^2}} = 20.6, \quad \therefore \quad \mathbf{mL = 2.06.}$$

$$\frac{T - 30}{230 - 30} = \frac{1}{\cosh(2.06)}, \text{ Solving } \mathbf{T_{L2} = 80.18°C}$$

$$\mathbf{Q_2} = \sqrt{30 \times \pi \times 0.014142 \times 20 \times \pi \times 0.014142^2/4} \times (230 - 30) \times \tanh(2.06)$$

$$= \mathbf{12.52\ W}$$

$$\frac{Q_2}{Q_1} = \frac{12.52}{15.16} = 0.826, \text{ reduced to } 82.6\%$$

$\mathbf{T_{L1}} = 64.28°C$, $T_{L2} = 80.18°C$.

Higher tip temperature.

COMMENT

Thinner fins are more effective interms of material used.

Example A-4.9: *A plate fin of 8 mm thickness of 60 mm length is used on a wall at 200°C. The convection coefficient is 25 W/m²K. The conductivity is 210 W/mK. (i) If the surrounding is at 35°C, determine the heat flow. (ii) if the same fin is split into 4 mm thick fins determine the total heat flow.*

Short fin end insulated condition can be used or chart can also be used. Consider 1 m width of fin.

Solution: (*i*) **8 mm fin**

$$Q = \sqrt{(hPkA)}(T_b - T_\infty)\tanh(mL),$$

$$m = \sqrt{((hP)/(kA))} = \sqrt{[(25 \times 2.016)/(210 \times 0.008)]} = 5.48 \quad mL = 0.3286$$

$$\mathbf{Q} = \sqrt{(25 \times 2.016 \times 210 \times 0.008)}\ (200 - 35)\tanh(0.3286) = \mathbf{481.74}\ \text{W/m width}$$

Fin effectiveness

$$= \frac{q}{hA\Delta T} = \frac{481.74}{25 \times 1 \times 0.008 \times 165} = \mathbf{14.6}$$

(*ii*) **4 mm fin**

$$m = \sqrt{[(25 \times 2.008)/(210 \times 0.004)]} = 7.73,\ mL = 0.464$$

$$\mathbf{Q} = \sqrt{(25 \times 2.008 \times 210 \times 0.004)}\ (200 - 35)\tanh(0.464)$$

$= 465.2$ W/fin. For two fins $\mathbf{Q} =$ **930.40 W/m width.**

Fin effectiveness

$$= \frac{465.2}{25 \times 1 \times 0.004 \times 165} = \mathbf{28.2}$$

about 97% increase over single fin.

COMMENT

As already indicated in the section 4.4.1, the use of thinner fins with closer pitch will provide higher effectiveness, as seen in this case.

Example A-4.10: *A plate fin of 8 mm thickness and 60 mm length is used on a wall at 200°C. The convection coefficient is 25 W/m²K and the surrounding is at 35°C. The conductivity of the material is 210 W/mK. Determine the change in heat flow if the fin volume is used as a fin of 4 mm thickness and 120 mm length.*

(1) The heat flow in 8 mm fin is obtained as 481.74 W/m width from example A-4.9.

(2) For 4 mm thick fin, 0.12 m long, using the short fin end insulated model.

Solution:

$$m = \sqrt{[(25 \times 2.008)/(210 \times 0.004)]} = 7.73,\ mL\ 0.928$$

$$\mathbf{Q} = \sqrt{(hPkA)}\ (T_b - T_\infty)\tanh(mL)$$

$$= \sqrt{[(25 \times 2.008 \times 210 \times 0.004)]}\ (200 - 35)\tanh(0.928) = \mathbf{781.8\ W}$$

From example AE-4.9, Q = 481.74 W

There is an **improvement of only 62% as against 97% in the case of two fins of the same length. Increasing the length is less effective.**

Example A-4.11: *A motor body's outside dia. is 30 cm and its surface temperature should not exceed 50°C when dissipating 0.15 kW. The length of the body is 20 cm. Longitudinal fins of 12 mm thickness and 30 mm height are proposed. The conductivity of the material is 40 W/mK and the convection coefficient is 40W/m²K. Determine the number of fins required. Atmospheric temperature is 35°C.*

This is a short fin situation. Equation (4.13) is to be used.

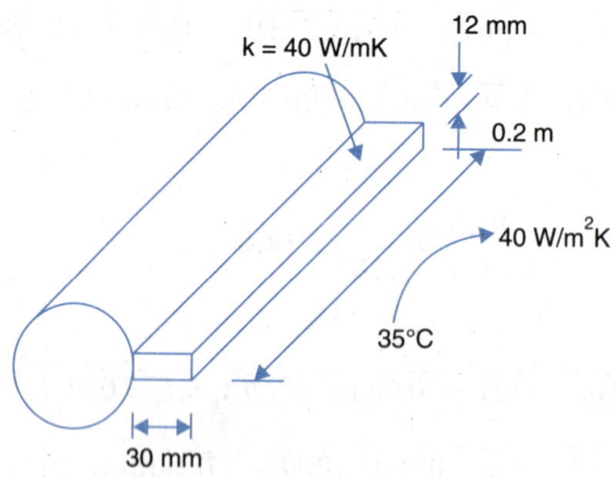

Fig. AE-4.11. Problem model

Solution: $Q = \sqrt{(hPkA)}\,(T_b - T_\infty)\ .\ \{[(\sinh (mL) + (h/mk) \cosh (mL)]/[(\cosh (mL) + (h/mk) \sinh (mL))]\}$

$$m = \sqrt{[(hP)/(kA)]} = \sqrt{(40/40)\,[(0.2 + 0.012) \times 2/(0.2 \times 0.012)]} = 13.292,$$

$$mL = 0.39875$$

$$\mathbf{Q} = \sqrt{(40 \times 0.424 \times 40 \times 0.2 \times 0.012)}\ (50 - 35)\ (\sinh (0.39875) + (40/13.292 \times 40) \cosh (0.39875))/(\cosh (0.39875) + (40/13.292 \times 40) \sinh (0.39875)) = \mathbf{8.45\ W}$$

Number of fins required = 150/8.45 = 18 fins

The pitch will be, assuming that fins are arranged over 270°

$$(\pi D \times 270)/(360) \times (1)(18) = 40 \text{ mm}.$$

COMMENT

In case fin thickness is reduced to 6 mm, it can be shown that 21 fins only will be required. Try to workout the volume reduction.

Example A-4.12: *Frying requires oil to be heated to about 350°C. A laddle is used in the frying. The section of the handle is 4 mm × 15 mm. The surroundings are at 35°C. The conductivity of the material is 210 W/mK. If the temperature at a distance of 40 cm from the oil should not reach 45°C, determine the convection heat transfer coefficient required.*

The situation can be modelled as shown in Fig. AE-4.12.

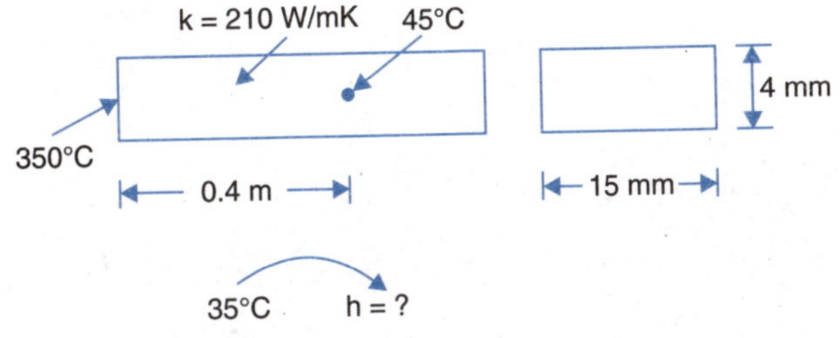

Fig. AE-4.12

Solution: Long fin assumption can be made:

$$P = (15 + 4)\ 2\ \text{mm} = 0.038\ \text{m}.$$

$$A = 0.015 \times 0.004\ \text{m}^2,\ x = 0.4\ \text{m}.$$

$$(T - T_\infty)/(T_b - T_\infty) = e^{-mx},$$

$$(45 - 35)/(350 - 35) = e^{-m\times 0.4} \quad \therefore \quad m = 8.625 = \sqrt{[(h \times 0.038)/(210 \times 0.015 \times 0.004)]}$$

$\therefore$ **h = 24.67 W/m²K,** Some air circulation is required.

It instead of aluminium, stainless steel with k = 22 W/mK is used, then,

$$8.625 = \sqrt{(h \times 0.038)/(22 \times 0.015 \times 0.004)},$$

$$\mathbf{h = 2.58\ W/m^2K}$$

This is a better proposition as this low value of h is easier to obtain.

Example A-4.13: *A rectangular fin has to be designed for maximum heat flow per given volume. The sectional area available is 5 cm². The convective heat transfer coefficient has a value of 100 W/m²K. The conductivity of the material is 200 W/mK. Using short fin end insulated conditions determine the thickness for maximum heat flow.*

Assuming a thickness t m, the length will be (5 × 10⁻⁴/t) m, Considering depth of 1 m, P = 2 m, and A = t × 1 m²

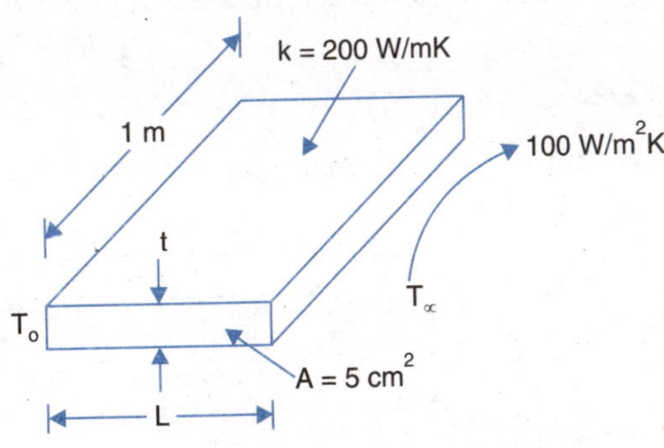

Fig. AE-4.13. Problem model

Solution:

$$Q = \sqrt{(hPkA)}\,(T_b - T_\infty)\tanh(mL)$$

$$= \sqrt{(hPk)}\,.\,(T_b - T_\infty)\sqrt{t}\,.\tanh\left\{\sqrt{[(hP)/(kt)]}\,.\,5 \times 10^{-4}/t\right\}$$

letting $\sqrt{hpk}\ (T_b - T_\infty) = C$, and substituting the values for the remaining

$$Q = C\ \sqrt{t}\ .\ \tanh\left(\sqrt{(100 \times 2/200)\,5 \times 10^{-4}(1/t^{3/2})}\right)$$

$$= C\ \sqrt{t}\ \tanh\ \sqrt{5 \times 10^{-4}/t^{3/2}}$$

Differentiating Q with respect to t

$$(dQ)/(dt) = \text{constant}\ [(t^{1/2}) \times 5 \times 10^{-4} \times (-3/2)\ t^{(-3/2)-1}/(\cosh^2(5 \times 10^{-4}/t^{3/2})]$$
$$+ \text{constant}\ t^{-1/2}\ .\ \tanh\ (5 \times 10^{-4}/t^{3/2})^{1/2}$$

Equating to zero,

$$3 \times 5 \times 10^{-4}\ t^{(1/2)-(5/2)+(1/2)} = (\cosh^2(5 \times 10^{-4}/t^{3/2})/(\tanh\ (5 \times 10^{-4}/t^{3/2}))$$

$$15 \times 10^{-4}/t^{3/2} = \cosh\ (5 \times 10^{-4}/t^{3/2})\ \sinh\ (5 \times 10^{-4}/t^{3/2})$$

This is solved by trial starting from 0.01 m and going down.

Assumed:

Thickness mm:	10	8	6	**5**	4	3
RHS value:	0.59	0.95	2.21	**4.215**	13.01	3.88
LHS value:	1.5	2.1	3.23	**4.243**	5.93	4.118

The optimum thickness is 0.005 m or **5 mm.**

check : using $\quad Q = \sqrt{(hPkA)}\ (T_b - T_\infty)\ \tanh\ (mL)$

Assumed

thickness mm:	10	8	6	**5**	4	3
Length mm:	50	62.5	83.33	**100**	125	166.6
Heat flow/ΔT*:*	9.24	10.79	12.26	**12.56**	12.17	10.90

The value of heat flow is seen as maximum at t = 5 mm (about).

The length of the fin will be 10 cm.

Example A-4.14: *A volume of 5 cm³ is available for a circular pin fin. Determine the optimum diameter. Conductivity = 200 W/mK, convection coefficient = 200 W/m²K. Assume end insulated fin.*

Solution:

$$Q = \sqrt{(hPkA)}\ .\ (T_b - T_\infty)\ .\ \tanh\ (mL)$$

Assuming a diameter D and noting $(\pi D^2/4)\ .\ L = 5 \times 10^{-6}$

$$Q = \sqrt{(hk)}\ (T_b - T_\infty)\ \sqrt{(\pi D \pi D^2/4}$$
$$\times \tanh 0.5\ [(h/k)\ .\ (\pi D\ .\ 4/\pi D^2)\ .\ (5 \times 10^{-6} \times 4)/(\pi D^2)]^{0.5}$$
$$= (\pi/2)\ \sqrt{(hk)}\ (T_b - T_\infty)\ .\ D^{3/2}\ \tanh\ [c/D^{5/2}]$$

where
$$c = \sqrt{4h/k} \times 4 \times 5 \times 10^{-6}/\pi$$

Let $(\pi/2)\quad \sqrt{(hk)}\ (T_b - T_\infty) = C_1$

$$Q = C_1\ D^{3/2}\ .\ \tanh\ [c\ D^{-5/2}]$$

$$(dQ/dD) = C_1\ [D^{3/2}\ c(-5/2)\ D^{(-5/2)-1}/\cosh^2(cD^{-5/2})]$$
$$+ (3/2)\ D^{(3/2-1)}\ \tanh\ (cD^{-5/2})\}$$

Equating to zero, $(5/3)\ cD^{-5/2} = \cosh (cD^{-5/2}) \sinh (cD^{-5/2})$

$$c = \sqrt{[(100 \times 4)/(200)]} \times 4 \times 5 \times 10^{-6}/\pi = 9 \times 10^{-6}$$

Substituting and solving by trial **D = 0.01** or **10 mm.**

As a check Q is calculated using

$Q = \sqrt{(hPkA)}\ (\Delta T) \tanh (mL)$, Assuming D, L can be calculated as $(\pi D^2/4)$

$L = 5 \times 10^{-6}$

Assumed Value of D mm	5	6	7	8	9	**10**	11
Heat flow / ΔT	0.079	0.109	0.127	0.146	0.156	**0.159**	0.156

Maximum heat flow at D = 10 mm, in this case the length will be 6.37 cm.

Example A-4.15: *A copper fin 12 mm dia. and 200 mm long spans two walls one at 200°C and the other at 120°C. The thermal conductivity of the material is 330 W/mK. Air at 30°C flows between the walls and the convection coefficient over the rod was 56.8 W/m²K. Determine the heat flow through the rod. Also find the temperatures at the mid section and the location and value of the minimum temperature.*

The data are presented in Fig. AE-4.15.

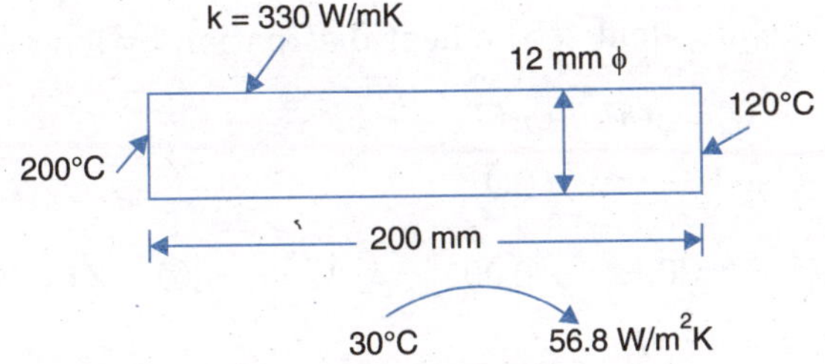

Fig. AE-4.15

Solution: The model suitable is a fin with specified end temperatures. Using equations (4.10) and (4.14).

$$(T - T_\infty)/(T_b - T_\infty) = [\{(T_L - T_\infty)/(T_b - T_\infty)\} \sinh (mx) + \sinh m (L - x)] / [\sinh (mL)]$$

at the centre: $x = 0.1$ m, $L = 0.2$ m.

$$m = \sqrt{[(hP)/(kA)]} = \sqrt{[(56.8 \times \pi \times 0.012 \times 4)/(330 \times \pi \times 0.012^2)]} = 7.575$$

$$(T - 30)/(200 - 30) = ([(120 - 30)/(200 - 30)] \sinh (7.575 \times 0.1) + \sinh (7.575)(0.2 - 0.1))/(\sinh (7.575 \times 0.2)$$

Centre temperature, **T = 129.9°C**

To find the minimum temp., location, find $(dT)/(dx)$ and equate to zero

$$\{(T_L - T_\infty)/(T_b - T_\infty)\} \cosh (mx)\, m + (-m) \cosh m (L - x) = 0$$

or $\{(T_L - T_\infty)/(T_b - T_\infty)\} \cosh (mx) = \cosh m (L - x)$, solving by trial

The minimum temperature location is x = 0.169 m.

$(T_{min} - 30)/(200 - 30) = [((120 - 30)/(200 - 30)) \sinh (7.575 \times 0.169))$
$+ \sinh (7.575 \times (0.02 - 0.169))]/\sin h\ (7.575 \times 0.2$

$= 0.5153$

$\therefore \quad \mathbf{T_{min} = 117.61°C}$

Heat flow: $Q = \sqrt{(hPkA)}\ \{(T_b - T_\infty) + (T_L - T_\infty)\}\ (\cosh mL - 1)/(\sinh mL)]$

$= \sqrt{(56.8 \times \pi \times 0.012 \times 330 \times \pi \times 0.012^2 / 4)}\ \{(200 - 30)$
$+ (120 - 30)\}\ [(\cosh (m \times 0.2) - 1)/\sinh (m \times 0.2)] = \mathbf{47\ W.}$

Check: Considering as two fins of $L = 0.169$ and 0.031

$Q = Q_1 + Q_2 = \sqrt{(hPkA)}\ (T_b - T_\infty) \tanh (mx) + \sqrt{(hPkA)}\ (TL - T_\infty) \tanh m(L - x)$
$= 41.16 + 5.87 = \mathbf{47.03\ W.}$

Check for temp: $(L = 0.169)$

$(T - T_\infty)/(T_b - T_\infty) = 1/\cosh (mx) = 0.156 \qquad \mathbf{T = 117.74°C.}$

Example A-4.16: *For the purpose of thermal process requirements the junction of two long rods of 5 mm dia. are to be held at 700°C. The rods are exposed to air to 30°C with a convection coefficient of 12 W/m²K. Determine the power input needed in the case of materials with following conductivity values (i) 360 W/mK, (ii) 218 W/mK and (iii) 149 W/mK.*

Solution: The situation is equivalent to the heat dissipation by long fins

$$Q = \sqrt{(hPkA)}\ (T_b - T_\infty)$$

As two rods are involved, power = 2*Q*

(*i*) $\mathbf{Q} = \sqrt{(12 \times \pi \times 0.005 \times 360 \times \pi \times 0.005^2 / 4)}\ (700 - 30) = \mathbf{24.5\ W}$

Power = 2 × 24.5 = 49 W

(*ii*) $\mathbf{Q} = \sqrt{(12 \times \pi \times 0.005 \times 218 \times \pi \times 0.005^2 / 4)}\ (700 - 30) = \mathbf{19.03\ W}$

Power = 38.06 W

(*iii*) $Q = \sqrt{(12 \times \pi \times 0.005 \times 149 \times \pi \times 0.005^2 / 4)}\ (700 - 30) = \mathbf{15.73\ W}$

Power = 31.46 W

COMMENTS

1. This condition is for steady state requirements. But the practical situation requires quick heating and the power requirements in the process will be several times more.
2. Such a situation prevails in welding of two rods.

Example A-4.17: *A fin in the form of a ring of 0.25 mm thickness and 15 mm OD and 15 mm long is used on an electric device to dissipate heat. Consider the outer surface alone to be effective and exposed to air at 25°C with a convection coefficient of 40 W/m²K. The conductivity of the material is 340 W/mK. If the heat output is 0.25 W and if the device is also of the same OD, determine the device temperature with and without the fin.*

Solution: The heat is lost from the surface of the device by convection **without fin:**

$$0.25 = 40 \times (\pi \times 0.015^2/4)\,(T_s - 25)$$

$\therefore$ $\mathbf{T_s = 60.36°C}$

With the addition of the fin: Fig. AE-4.17.

$$Q = \sqrt{(hPkA)}\,(T_o - T_\infty)\tanh(mL)$$

$P = \pi \times 0.015,$

$A = (\pi/4)\,(0.015^2 - 0.0145^2),$

$L = 0.015$ m

Fig. AE-4.17

$$m = \sqrt{(40 \times \pi \times 0.015)/(340 \times (0.015^2 - 0.0145^2)/4)} = \mathbf{8.876.}$$

$\therefore$ $\mathbf{mL} = 28.876 \times 0.015 = \mathbf{0.3281}$

Substituting

$$\therefore \quad \sqrt{hPkA} = \sqrt{(40 \times \pi \times 0.015 \times 340 \times \pi \times (0.015^2 - 0.0145^2)/4}$$

$$= \mathbf{0.08616}$$

$$Q = 0.25 = 0.08616 \times \tan(0.3281)\,(T_o - 25)$$

$$\therefore \quad \mathbf{T_o = 9.16 + 25 = 34.16°C.}$$

There is a drop of 26.2°C due to the fin.

Example A-4.18: *A handle fixed to a chemical process equipment is shown in Fig. AE-4.18. The handle is exposed to air at 37°C with a convection coefficient of 15 W/m²K. The conductivity of the material is 20 W/mK. Determine the heat lost and also the temperature at the mid location.*

The data given are shown in Fig. AE-4.18(a).

Solution: The handle can be considered as equal to a fin of length 190 mm with specified end conditions. [Fig. AE-4.18(*b*)].

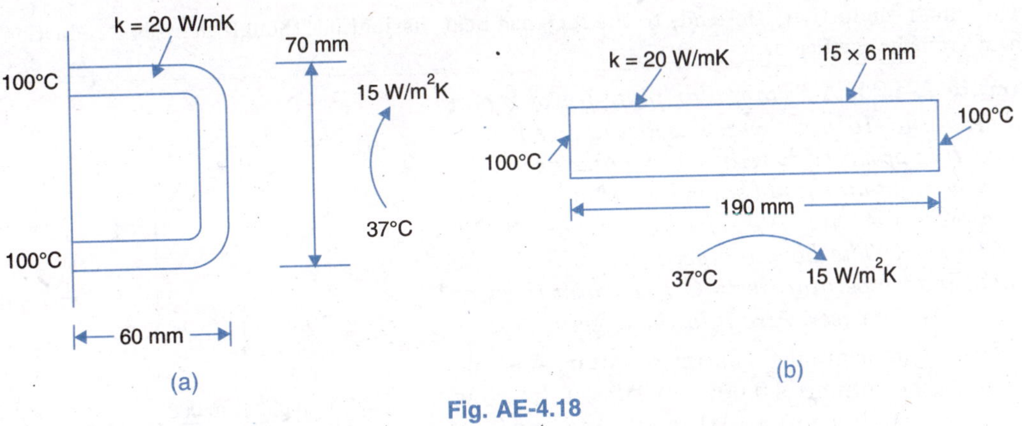

Fig. AE-4.18

Using equation (4.10)

$$\mathbf{m} = \sqrt{(hP/kA)} = \sqrt{(15 \times 0.042/20 \times 0.015 \times 0.006)} = \mathbf{18.7}$$

$$\mathbf{mL = 3.555}$$

$$(T - T_\infty)/(T_b - T_\infty) = [((T_L - T_\infty)/(T_o - T_\infty)) \sinh (mx) + \sinh m\,(L - x)]/(\sinh (mL)$$

$x = 0.095$ m, substituting and calculating

$(T - 37)/(100 - 37) = 0.329$; temperature at mid location, $T =$ **57.7°C**

$$Q = \sqrt{(hPkA)}\ \{(T_o - T_\infty) + (TL - T_\infty)\}\ \{(\cosh (mL) - 1)/(\sinh (mL))\}$$

$$= \sqrt{(15 \times 0.042 \times 20 \times 0.015 \times 0.006)}\ \{(100 - 37) + (100 - 37)\}$$

$\times$ (cosh (3.555) – 1)/sinh (3.555) = **4.01W** (Check considering two fins)

Example A-4.19: *A plane wall exposed to a fluid with a convection coefficient is found insufficient to dissipate the heat. One mm thick plate fins of 40 mm length are added with a pitch of 10 mm. The addition of the fins causes a reduction in the convection coefficient to 30 W/m²K from the original value of 40 W/m²K. If the conductivity of the material is 210 W/mK determine the percentage increase in heat dissipation.*

The data are shown in Fig. AE-4.19.

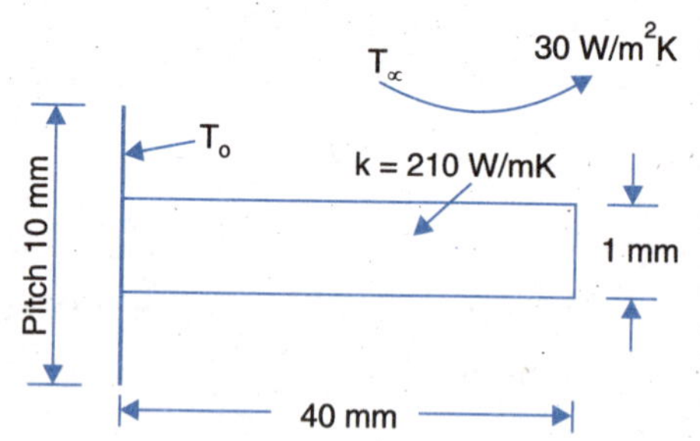

Fig. AE-4.19

Solution: Heat dissipated by base surface of one pitch width, (considering 1 m depth)

$$\mathbf{Q} = 40 \times 0.01 \times 1 \times (T_b - T_\infty) = \mathbf{0.4\ \Delta T}$$

Heat dissipation with fin:

$$m = \sqrt{(hP/kA)} = \sqrt{(30 \times 2/210 \times 0.001)} = 16.9,\ mL = 6.76$$

$$\mathbf{Q} = 0.009 \times 30 \times 1 \times (T_b - T_\infty) + (\sqrt{(hPkA)}\ (T_b - T_\infty) \tanh (mL)$$

$$= 0.27\Delta T + \sqrt{30 \times 2 \times 210 \times 0.001}\ \tanh (6.76) = \mathbf{3.82\ \Delta T.}$$

The **increase** is about **855%.**

COMMENT

Thin, high conductivity fin leads to the increase heat dissipation, though there is a reduction in heat transfer coefficient.

Example A-4.20: *A rectangular plate fin of 2 mm thickness and 15 mm length is being used on a surface. It is proposed to reduce the material to 70% and use a triangular fin of same base thickness. The convection coefficient is 100 W/m²K, the conductivity is 200 W/mK. The base temperature is 5°C. The surrounding temperature is 35°C. Determine the heat gain in both the cases. Also find the effectiveness.*

Solution: The original fin is rectangular. Assume 1 m depth, the volume is $0.002 \times 0.015 \times 1$ m³. Now using 70% of the volume this is made into a triangular fin: length, $0.7 \times 0.002 \times 0.015 \times 1 = 0.001 \times L \times 1 \quad \therefore \quad L = 0.021$ m.

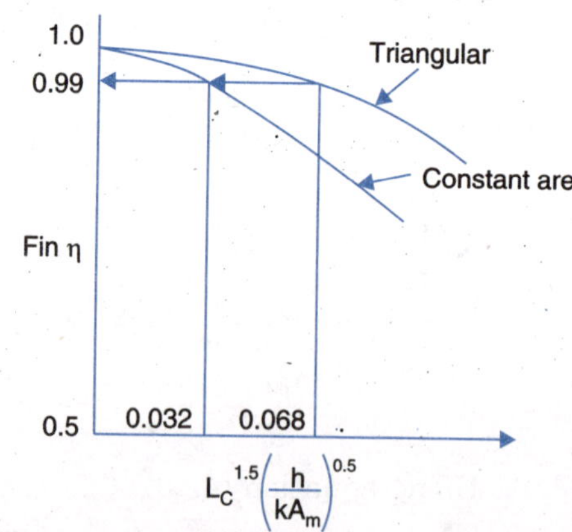

Fig. AE-4.20. Skeleton chart

The heat flow is found using the chart

Rectangular fin	**Triangular fin**
$L_c = L + (t/2) = 0.016$ m	$L_c = L = 0.021$ m
$A_p = 0.002 \times 1$ m^2	$A_p = 0.001 \times 1$ m^2
$A_s = 2 \times 0.016 \times 1$ m^2	$A_s = 2 \times 0.021 \times 1$ m^2
$L_c^{1.5}\,[h/kA_m]^{0.5} = 0.032$	$L_c^{1.5}\,[h/kA_m]^{0.5} = 0.068$

Entering the chart with the values of efficiencies are read as 0.99 and 0.99 for both triangular and rectangular fins. (Method is indicated in Fig. AE-4.20)

Heat flow: **rectangular fin:**

$$Q = \text{fin eff. } h\, A_s\, (T_b - T_\infty)$$
$$= 0.99 \times 100 \times 2 \times 0.016\, (5 - 35)$$
$$= \mathbf{95.04\ W\ (inwards)}$$

Triangular fin: $Q = 0.99 \times 100 \times 2 \times 0.021\,(5 - 35) =$ **124.74 W (inwards)**

An increase of 23.8% after reduction in volume of material of 30%

Effectiveness: $\dfrac{\text{heat flow by fin}}{\text{heat flow over area replaced by fin}}$

(*i*) **Rectangular:** 95.04/100 × 0.002 (5 – 35) = **15.84**

(*ii*) **Triangular:** 124.74/100 × 0.002 (5 – 35) = **20.79.**

Example A-4.21: *A thermometer well is made of 1 mm thick material of thermal conductivity 55 W/mK and the inner diameter is 8 mm. The convection coefficient on the surfaces is 50 W/m²K. The wall temperature is 120°C. The thermometer placed in contact with well bottom reads 380°C. Determine the fluid temperature and the error as a percentage of true temperature in °C. Discuss the possible methods to reduce the error. The length of the well is 6 cm.*

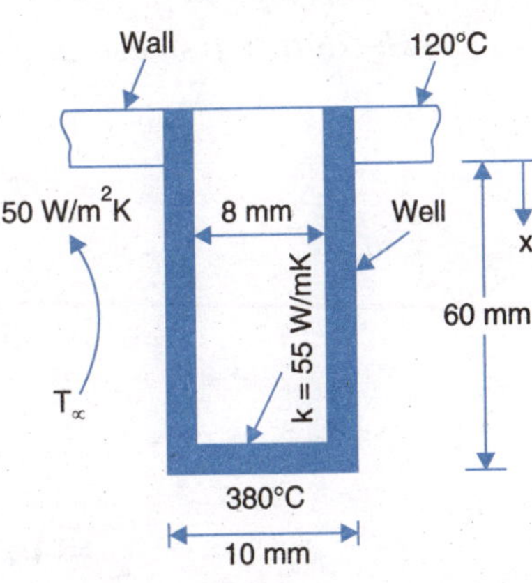

Fig. AE-4.21. Thermometer well

Solution: The data are shown in Fig. AE-4.21. Let the fluid temperature be T_∞. The thermometer well can be considered as a short fin (end insulated). Only the outer surface is effective for convection. Measuring x from wall surface, $L = 0.006$ m, $P = \pi \times 0.01$ m, $A = (\pi/4)\,(0.010^2 - 0.008^2)$

$$m = \sqrt{(hP/kA)} = \sqrt{[(50 \times 0.01 \times \pi \times 4)/(55 \times \pi \times (0.01^2 - 0.008^2)]}$$
$$= 31.782,\ mL = 1.907$$
$$(T - T_\infty)/(T_b - T_\infty) = (380 - T_\infty)/(120 - T_\infty) = 1/(\cosh mL) = 0.291$$
$$380 - T_\infty = 0.291\,(120 - T_\infty) = 34.88 - 0.291\,T_\infty$$
$$380 - 34.88 = (1 - 0.291)\,T_\infty\ ;\ \mathbf{T_\infty = 487°C}$$

The **error involved** is 487 – 380 = **107°C or 22%.** The reasons can be

1. The wall temperature is too different from the fluid temperature which may not be the case most of the time
2. Low value of thermal conductivity
3. Short length and
4. Low convection coefficient.

This arrangement is often used to measure the temperature of the fluid flowing in a pipe or duct.

Chapter 4

Example A-4.22: *Consider the data in example A-4.21. If the wall temperature is 300°C and if h = 300 W/m²K and L = 0.08 m determine the error.*

Solution: $\mathbf{m} = \sqrt{(hP/kA)} = \sqrt{(300 \times \pi \times 0.01 \times 4 / 55 \times \pi \times (0.01^2 - 0.008^2))} = \mathbf{77.88},$

$$\mathbf{mL = 6.228}$$

$$(T - T_\infty)/(T_b - T_\infty) = (380 - T_\infty)/(300 - T_\infty) = 1/\cosh (6.228) = 0.0039$$

$$380 - 1.184 = 0.9061\, T_\infty$$

$$\mathbf{T_\infty = 380.32°C \quad Error\ is\ 0.32°C}$$

COMMENTS

1. This shows that the well should be longer and the convection coefficient higher to reduce the error.
2. It may not be generally possible to increase convection coefficient and so using longer well is the practice. The well is often placed at an angle to wall so that longer well can be accomodated in the pipe.

Example A-4.23: *A thermometer well of 10 mm OD and 8 mm ID is of material with thermal conductivity of 25 W/mK. It is exposed to gases at 2°C, and the temperature of the wall in which the well is fixed is 12°C. If an error of not more than 0.6°C is to be involved, determine the length of the well. h = 50 W/m²K.*

The data are shown in Fig. AE-4.23.

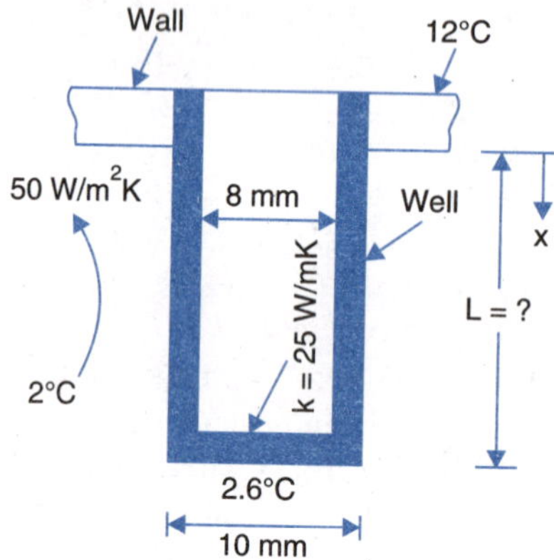

Fig. AE-4.23. Thermometer wall

Solution: Assuming end insulated boundary,

$$P = \pi \times 0.01 \text{ m}$$

$$A = (\pi/4)\,(0.01^2 - 0.008^2) \text{ m}^2$$

$$(T - T_\infty)/(T_b - T_\infty) = (2.6 - 2)/(12 - 2) = 1/\cosh (mL) = 0.6/10 = 0.06$$

$$\therefore \quad mL = 3.506\,;\ m = \sqrt{(hP/kA)} = \sqrt{(50 \times \pi \times 0.01 \times 4 / 25 \times \pi \times (0.01^2 - 0.008^2))}$$

$$= 47.14$$

$$\therefore \quad \mathbf{L = 0.0744\ m\ or\ 7.44\ cm.}$$

Example A-4.24: *Circumferential fins of constant thickness of 1 mm are fixed on a 50 mm pipe at a pitch of 4 mm. The fin length is 20 mm. The wall temperature is 130°C. The thermal conductivity of the material is 210 W/mK. The convection coefficient is 50W/m²K. Determine the heat flow and effectiveness.*

Solution: The data are shown in Fig. AE-4.24(*a*)

$$r_1 = 0.025 \text{ m}, r_2 = 0.045 \text{ m}$$

$$r_{2c} = 0.045 + 0.0005 = 0.0455 \text{ m}$$

$$r_{2c}/r_1 = 0.0455/0.025 = 1.82$$

$$L_c = L + t/2 = 0.0205 \text{ m}$$

$$A_p = L_c \times t = 0.0205 \times 0.001$$

$$A_s = 2\pi\,(r_{2c}^2 - r_1^2)$$

$$L_{c1.5}\,(h/kA_m)^{0.5} = 0.0205^{1.5}\,[50/(210 \times 0.0205 \times 0.001)]^{0.5} = 0.316$$

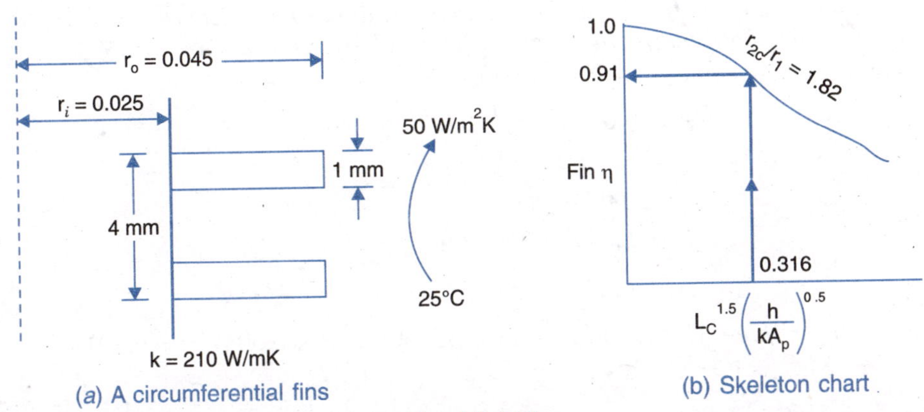

(*a*) A circumferential fins (b) Skeleton chart

Fig. AE-4.24

Entering the chart at this value [Fig. AE-4.24(*b*)] the efficiency is read as 0.91

$$\mathbf{Q_f} = \text{fin eff. } A_s\, h(\Delta T)$$

$$= 0.91 \times 2\pi\,(0.0455^2 - 0.025^2) \times 50 \times (130 - 25) = \mathbf{43.38\ W.}$$

Heat flow on the **bare area is**

$$\mathbf{Q_c} = 2 \times \pi \times 0.025 \times 0.003 \times 50\,(130 - 25) = \mathbf{2.47\ W}$$

Total heat flow per pitch distance = **45.85 W**

Without fin: 2 × π × 0.025 × 0.004 × 50 (130 – 25) = **3.3 W,** an increase of **14 fold.**

Effectiveness = 55.59 (try this out).

Example A-4.25: *In the example A-4.24, if there is a contact resistance of 0.88×10^{-4} m²C/W, find the heat flow through the fin.*

The contact resistance reduces the base temperature and so the heat flow is reduced.

Solution: The heat flow is 43.38 W,

Resistance for the area of the fin contact surface is given by

$$(0.88 \times 10^{-4} \text{ m}^2\text{C/W})(\pi \times 0.025 \times 0.001 \text{ m}^2) = 0.56° \text{ C/W}$$

Therefore temperature drop = 43.38 × 056 = 24.3°C or T_o = (130 – 24.3) = 105.7°C

Corrected heat flow = 33.3 W, temp. drop at contact = 33.3 × 0.56 = 18.65°C

This process is repeated till the difference is small

$$T_o = 111.4°C$$

Corrected $Q = 35.7$ W

$\Delta T = 20°C$ corrected $Q = 35.1$ W, and so on.

The **final answer is Q = 35.2 W** as against 43.38 W **and ΔT at base = 19.7°C** or the actual base temperature is 110.3°C.

COMMENT

It can be seen that contact resistance drastically reduces the heat flow in fins and hence every attempt should be made to obtain good contact between surfaces.

Example A-4.26: *The outer diameter of a small engine cylinder is 56 mm. Determine the heat dissipation by a circumferential fin of 4 mm thickness and 40 mm length if k = 210 W/mK and h = 115 W/m²K. The base temperature is 200°C and surroundings are at 35°C.*

Solution: The data are shown in Fig. AE-4.26(*a*).

$$r_1 = 0.028 \text{ m}, r_2 = 0.068 \text{ m}$$

$$r_{2c} = r_2 + t/2 = 0.07 \text{ m}$$

$$r_{2c}/r_1 = 0.07/0.028 = 2.5$$

$$L_c = L + t/2 = 0.042 \text{ m}$$

$$A_p = 0.042 \times 0.004, As = 2\pi\,(r_{2c}^2 - r_1^2)$$

$$L_c^{3/2}\,[h/kA_p]^{0.5} = 0.042^{1.5}\,[(115)(210 \times 0.042 \times 0.004)]^{0.5} = 0.491$$

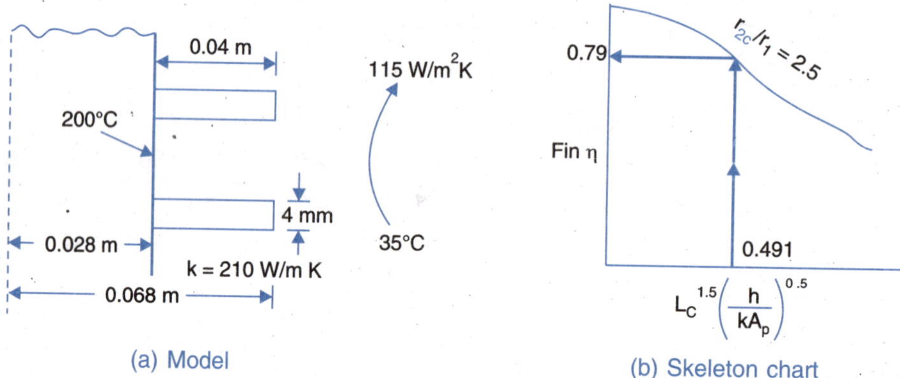

(a) Model (b) Skeleton chart

Fig. AE-4.26

Entering the chart at 0.491 reading on curve for 2.5 as in [Fig. AE-4.26(*b*)]. **Fin eff. = 0.79. Q** = 0.79 × 2π (0.07² – 0.028²) (100 – 35) × 115 = **152.7 W per fin.**

Example A-4.27: Annular aluminium fins of 1 mm thickness and 15 mm length are fixed on copper tubes of *OD* 16 mm at 0°C. Air at 30°C passes over the fins with a convection coefficient of 100 W/m²K. If k = 225 W/mK, determine the heat gain per fin, for a capacity of 2 kW, determine the tube requirements if pitch is 5 mm.

Solution: The data are shown in Fig AE-4.27. Using Table 4.4:

$$r_1 = 0.008 \text{ m}, r_2 = 0.023 \text{ m}$$

$$r_{2c} = r_2 + t/2 = 0.0235 \text{ m}.$$

$$L = 0.015 \text{ m}, L_c = 0.0155 \text{ m}$$

$$A_p = 0.0155 \times 0.001$$
$$A_s = 2\pi\,(0.0235^2 - 0.008^2)$$
$$Lc^{1.5}\,(h/kA_m)^{0.5} = 0.0155^{1.5}\,[100/(225 \times 0.0155 \times 0.001)]^{0.5} = 0.327$$
$$r_{2c}/r_1 = 0.0235/0.008 = 2.94$$

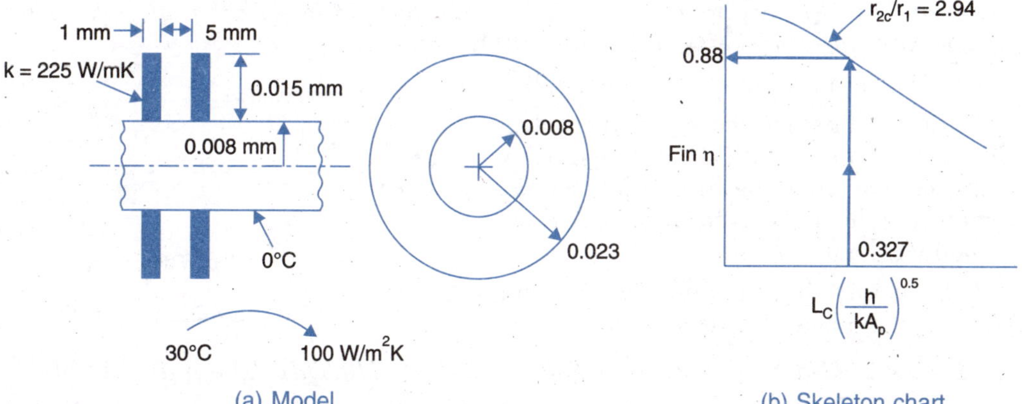

(a) Model (b) Skeleton chart

Fig. AE-4.27

Entering the chart at this value the fin efficiency is read as 0.88 [as in Fig. AE-4.27(*b*)]

$$\mathbf{Q} = 0.88 \times 100 \times 2\pi\,(0.0235^2 - 0.008^2)\,(0 - 30) = \mathbf{8.1\ W}$$

Per pitch length $Q = 8.1 + \pi Dh\,\Delta Tl$

$$= 8.1 + \pi \times 0.016 \times 100 \times 30 \times 0.004 = 8.7\ \text{W}$$

No. of pitch distances 2000/8.7 = 229, length = **1.145 m.**

Example A-4.28: *Annular fins, 2 mm thick and 15 mm length are fixed on a 30 mm tube at 100°C. Air at 30°C flows over the fins with a convection coefficient of 75 W/m²K. k = 235 W/mK. A contact resistance of 2 × 10⁻⁴ m² K/W is encountered. Determine the reduction in heat flow due to the contact resistance.*

Solution: *Using the chart.*

$$r_1 = 0.015\ \text{m},\ r_2 = 0.03\ \text{m},\ r_{2c} = 0.031\ \text{m},\ t = 0.002\ \text{m}$$
$$L = 0.015\ \text{m}\ ;\ L_c = 0.016\text{m},\ r_{2c}/r_1 = 0.031/0.015 = 2.07$$
$$A_p = Lc.t = 0.016 \times 0.002,\ As = 2\pi\,(0.031^2 - 0.015^2)$$
$$Lc^{1.5}\,(h/kA_p)^{0.5} = 0.016^{1.5}\,[75/(235 \times 0.016 \times 0.002)^{0.5} = 0.2021$$

Entering the chart at 0.2021 and reading on curve for $r_{2c}/r_1 = 2.07$ (interpolation) fin **efficiency is 0.94** (as in figures of previous example AE-4.27)

$$\mathbf{Q} = 2\pi\,(0.031^2 - 0.015^2) \times 75 \times 0.94\,(100 - 30) = \mathbf{22.82\ W.}$$

The temperature drop at contact depends on heat flow and vice versa. So a trial solution is necessary.

The contact resistance for the area of contact.

$$= (2 \times 10^{-4})/(\pi \times 2 \times 0.015 \times 0.002) = 1.061°\text{C/W}$$

Contact temperature drop = $Q.R$ = 22.82 × 1.061 = 24.2°C

Base temp. = 75.8°C. corrected Q = 14.93 W

Chapter 4

Final values are ΔT = 18°C

Base temp. = 82.01°C. Q = 16.95 W

A reduction of 26%.

Example A-4.29: *A flat aluminium plate 0.2 m wide and 3 mm thick and fairly long has its top exposed to solar radiation of 800 W/m², all of which is absorbed. Heat is collected at the side edges at 60°C. The bottom is well insulated and top end losses are negligible. Determine the maximum temperature in the plate. k = 240 W/mK.*

Solution: The data are show in Fig. AE-4.29.

The x coordinate is chosen at the centre. Considering a strip of width dx at a distance x from the centre, heat balance yields heat conducted in + heat received at this surface –heat conducted out = 0.

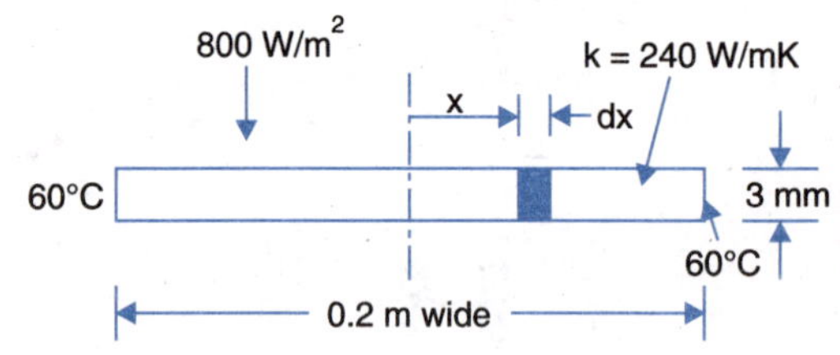

Fig. AE-4.29. Schematic

Taking unit length and considering thickness as t and heat flux as q W/m²

$$-k \times t \times 1\,(dT/dx) - (-k \times t \times 1(dT/dx) - (d/dx)\,[-kt(dT/dx)dx] + \dot{q}\,dx \times 1 = 0$$

$$(d^2T/dx^2) + \dot{q}/kt = 0. \quad \text{...(4.29.1)}$$

Integrating, the solution is

$$T = \frac{-\dot{q}x^2}{2kt} + C_1x + C_2 \quad \text{...(A)}$$

at $x = L$, $T = T_w$, at $x = -L$, $T = T_w$.

$$T_w = \frac{-\dot{q}L^2}{2kt} + C_1L + C_2 \quad \text{...(B)}$$

$$T_w = \frac{-\dot{q}L^2}{2kt} + C_1L + C_2 \quad \text{...(C)}$$

Subtracting (B) from (C)

$$0 = -2C_1L, \quad \therefore \quad C_1 = 0.$$

Using equation (B),

$$T_w = \frac{-\dot{q}L^2}{2kt} + C_2$$

$$\therefore \qquad C_2 = T_w + \frac{-\dot{q}L^2}{2kt}$$

Using this value in equation (A)

$$T = \frac{-\dot{q}x^2}{2kt} + \frac{\dot{q}L^2}{2kt} + T_w$$

$$(T - T_w) = (\dot{q}/2kt)\,(L^2 - x^2) \quad \text{...(4.29.2)}$$

$\therefore$ at $x = 0_1$ $T = T_0$ $\therefore$ $(T_0 - T_w) = (\dot{q}/2kt)\, L^2$

$$T_0 - 60 = (800/(2 \times 240 \times 0.003))\, [0.1^2] \quad ...(4.29.3)$$

$\therefore$ $\mathbf{T_0 = 65.56°C.}$

Check: $Q = -\, kA\, (dT)/(dx)\ |x = L$ from equation 4.29.2, $dT/dx = (-\, q/kt)\, x$.

Substituting $Q = [-\,240 \times 0.003 \times (-\,800)/(0.003 \times 240)] \times 0.1 = 80$ W checked. This is the heat received by the strip of 0.1 m width and 1 m length. (0.1×800) and so the values found are checked.

Example A-4.30: *Using the data in example AE-4.29, if heat is lost by combined convection and radiation with a total coefficient of h, determine the temperature distribution and the maximum temperature.*

Solution: Considering a strip of dx width at a distance x and writing the energy equation (Fig. AE-4.29).

heat conducted in + heat received by heat flux – heat conducted out – heat convected = 0.

$$-\, kA\, (dT/dx) + dx \times 1 \times q - \{-\, kA(dT/dx) + (d/dx)\, (-\, kA(dT/dx))dx\} - hdx\, (T - T_\infty) = 0$$

which reduces to (taking $A = t \times 1$)

$$d^2T/dx^2 + q/kt - (h/kt)\, (T - T_\infty) = 0.$$

Now introducing a new variable $\theta = T - T_\infty$, this reduces to

$$(d^2\theta/dx^2) - (h\, /\, kt)\theta + q/kt = 0$$

Setting $m^2 = h/kt$, $\dfrac{d^2\theta}{dx^2} - m^2\,\theta + \dfrac{q}{kt} = 0$ ...(4.30.1)

This equation is of the standard form

$(d^2y/dx^2) - m^2y + S = 0$ for which the general solution is

$$y = c_1e^{mx} + c_2e^{-mx} + S/m^2$$

The solution for equation 4.30.1 is

$$\theta = c_1e^{mx} + c_2e^{-mx} + (q/kt)\, (kt/h) = c_1e^{mx} + c_2e^{-mx} + (q/h)$$

$$\theta = T - T_\infty = c_1e^{mx} + c_2e^{-mx} + q/h$$

at $x = 0$, $dT/dx = 0$, $mc/e^{mx} - mc_2e^{-mx} = 0$, and $x = 0$, so $c_1 = c_2$

$$T - T_\infty = c_1(e^{mx} + e^{-mx}) + q/h$$

At $x = L$, $T = T_w$

$$(T_w - T_\infty) = c_1(e^{mL} + e^{-mL}) + q/h$$

$$c_1 = ((T_w - T_\infty)/(e^{mL} + e^{-mL})) - (q/h)/(e^{mL} + e^{-mL})$$

$$T - T_\infty = (T_w - T_\infty)((e^{mx} + e^{-mx})/(e^{mL} + e^{-mL})) - (q/h)\, ((e^{mx} + e^{-mx})/\, (e^{mL} + e^{-mL})) + q/h$$

or $T - T_\infty = (T_w - T_\infty)\, (\cosh\, (mx)/\cosh\, (mL)) + (q/h)\, (1 - (\cosh\, mx)/\, (\cosh\, mL))$

maximum temperature is at $x = 0$

$$\mathbf{T_0 - T_\infty = (T_w - T_\infty)\, (1/\cosh\, mL) + (q/h)\, [1 - (1/\cosh\, mL)].}$$

Example A-4.31: *In the example A-4.30, consider the following data and find the maximum temperature : $q = 800\ W/m^2$, Total width 15 cm, $k = 240\ W/mK$, $h = 10\ W/m^2K$, $t = 0.003$ m, $T_w = 60°C$, $T_\infty = 30°C$. Also find the heat flow by each mode.*

Chapter 4

Solution: $m = \sqrt{h/kt} = \sqrt{(10/240 \times 0.003)} = 3.73$, As $L = 0.075$ m $mL = 0.28$

(*i*) $T_0 - T_\infty = (60 - 30)/\cosh(0.28) + (800/10)(1 - 1/\cosh(0.28))$

$= 28.86 + 3.04 = 31.90$

$\mathbf{T_0 = 61.90°C}$

(*ii*) **Heat flow by conduction** $= -kA[(dt/dx)]$ at $x = L$

$T - T_\infty = (T_w - T_\infty)/(\cosh(mx)/[(\cosh(mL))] + (q/h)(1 - (\cosh mx)/\cosh mL))$

$[dt/dx]\ L = ((T_w - T_\infty)/(\cosh mL))\ m \sinh mL - (q/h)\ m\ (\sin mL)/(\cosh mL)$

$= m\ (T_w - T_\infty) \tanh(mL) - (q/h)\ m \tanh(mL)$

$= 3.73\ (60 - 30) \tanh(0.28) - (800/10)\ 3.73 \tanh(0.28) = -50.89$

$-kA[dt/dx]\,|\,L = -240 \times 0.003 \times -50.89 =$ **36.64 W.**

(*iii*) **Heat flow by convection:**

$$\int_0^L h(T - T_\infty)\,dx = \int_0^L h((T_w - T_\infty)/(\cosh mL)) \cosh(mx)\,dx + \int_0^L (q/h)\,(1 - (\cosh mx)/(\cosh mL))\,dx$$

$= h\,((T_w - T_\infty)/(m) \tanh(mL) + q[L - (1/m) \tanh(mL)]$

$= 10 \times ((60 - 30)/3.73) \tanh(0.28) + 800[0.075 - (1/3.73) \tanh(0.28)]$

$= 21.95 + 1.47 =$ **23.42 W**

Total $= 36.64 + 23.42 =$ **60.08 W**

Total input per half the width $= 0.075 \times 800 = 60$ W. So checks.

Example A-4.32: *A circular strip of aluminium 3 mm thick and 0.2 m diameter is exposed to solar radiation of 800 W/m² at the top. The bottom is insulated and the top losses are negligible. If heat is collected around the circumference at 60°C, determine the centre temperature. k = 240 W/mK.*

Solution: Refer Fig. AE-4.29 Considering the strip of width dr at r, the energy balance is

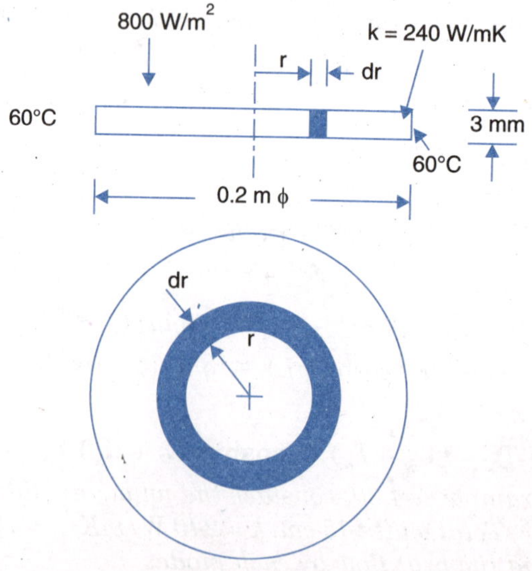

Fig. AE-4.32. Schematic

$- k2\pi rt\ (dT)/(dr) - [- k2\pi rt\ (dT)/(dr) + (d/dr)\ (- k2\ \pi rt\ (dT/dr)\ dr] + q\ 2\pi r dr = 0$

On simplification $(d/dr)\ (rdT/dr) = (- q/kt)r$ ...(1)

Integrating and using boundary conditions,

$dT/dr = 0$ at $r = 0$ and $T = T_w$ at $r = R$

$$\therefore \quad r\frac{dT}{dr} = \frac{\dot{q}r^2}{2kt} + C_1$$

at $r = 0$, $$\frac{dT}{dr} = 0, \quad \therefore \quad C_1 = 0,$$

$$\frac{dT}{dr} = \frac{-\dot{q}r}{2kt}$$

Integrating $$T = \frac{-\dot{q}r^2}{4kt} + C_2 \qquad ...(A)$$

At $r = R$, $T = T_w$

$$\therefore \quad T_w = \frac{-\dot{q}R^2}{4kt} + C_2$$

$$\therefore \quad C_2 = T_w + \frac{\dot{q}R^2}{4kt}$$

Substituting in (A) $$T = \frac{-\dot{q}r^2}{4kt} + \frac{\dot{q}R^2}{4kt} + T_w$$

$\therefore \quad T - T_w = (q/4\ kt)\ (R^2 - r^2)$ ...(2)

at $r = 0$, $T = T_o$ (*i.e.* the centre)

Substituting in (2),

$$T_o - T_w = \frac{\dot{q}R^2}{4kt}$$

$\therefore \quad \mathbf{T_o} = T_w + (q/4\ kt)R^2 = T_w + (800 \times 0.1^2)/(4 \times 240 \times 0.003) = 60 + 2.78$

$= \mathbf{62.78°C}$

Check: heat flow $= - kA\ (dT)/(dr)\,|\, r = R$; $(dt/dr) = (- q/2kt)\ r$

$Q = - k2\pi\ Rt \times (- q/2kt)R = \pi R^2 q$ which is the heat received.

Example A-4.33: *Heat is generated at q (W/m³) in a thin rod of diameter D and length 2 L with a thermal conductivity of k and is exposed at the surface to convection to a fluid at T_∞ with a convection coefficient of h. The ends are also maintained at T_∞ using a heat sink. With x coordinate origin at mid point, derive an expression for the temperature distribution.*

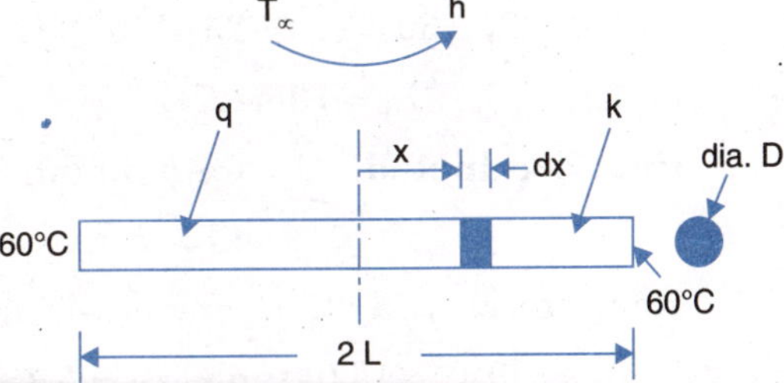

Fig. AE-4.33. Schematic

Solution: Considering the slice of thickness dx at a distance of x from origin, and taking the heat balance, (Fig. AE-4.33).

Net heat conducted + heat generated – heat convected = 0

$$-kA\,dT/dx - [-kA\,dT/dx + (d/dx)(-kA\,dT/dx)\,dx] + q\,dx\,A - hP\,dx\,(T - T_\infty) = 0$$

$$d^2T/dx^2 - (hP/kA)(T - T_\infty) + q/k = 0$$

Introducing a new variable, $\theta = T - T_\infty$, and also defining $m^2 = hP/kA$, this equation reduces to

$$d^2\theta/dx^2 - m^2\theta + q/k = 0$$

This is a standard form for which the solution is

$$\theta = c_1e^{mx} + c_2e^{-mx} + (q/km^2),$$

at $x = 0$, the boundary condition is $(dT/dx) = 0$ so $c_1 = c_2$

$$T - T_\infty = c_1(e^{mx} + e^{-mx}) + q/km^2 \text{ ; at } x = L,\ T = T_\infty$$

$$0 = c_1(e^{mL} + e^{-mL}) + q/km^2 \text{ ; } c_1 = -(q/km^2)(1/(e^{mL} + e^{-mL}))$$

$$T - T_\infty = (q/km^2)[1 - \cosh mx/\cosh mL]$$

$$\therefore \quad \mathbf{T_o - T_\infty = (q/km^2)[1 - (1/\cosh mL)]} \qquad \text{...(4.33.1)}$$

$$(T - T_\infty)/(T_o - T_\infty) = (\cosh(mL) - \cosh(mx))/(\cosh(mL) - 1)$$

this equation gives the temperature distribution.

Heat conducted $= -k\,A(dT/dx)\,|L = kA\,(q/km^2)\,m\tanh mL$

$$= \mathbf{(qA/m) \times \tanh(mL)\ at\ x = L} \qquad \text{...(4.33.2)}$$

Heat convected $= \int_0^L h(T - T_\infty)\,P\,dx = \int_0^L (hqP/km^2), [1 - \cosh mx/\cosh mL)]\,dx$

$$= \mathbf{(hq\,P/km^2)(L - (1/m)\tanh(mL))} \qquad \text{...(4.33.3)}$$

Example A-4.34: *A square rod of side 10 mm and 0.2 m length has a uniform heat generation rate of 10^6 W/m^3. The thermal conductivity of the material is 25 W/mK and it is exposed to air at 30°C with a convective heat transfer coefficient of 25 W/m^{2}K. The ends are maintained at 30°C. Determine the temperature at the centre and also the heat conducted at the ends and the heat convected.*

Reading from example A-4.33, using equation (4.33.1).

Solution: $T_0 - T_\infty = \{q/km^2\}[1 - 1/\cosh(mL)],\ m = \sqrt{(hP/kA)}$

$$h = 25 \text{ W/m}^2\text{ K},\ P = 0.04 \text{ m},\ k = 25 \text{ W/mK},\ A = 1 \text{ cm}^2 = 1 \times 10^{-4} \text{ m}^2$$

$$m = \sqrt{(25 \times 0.04/25 \times 1 \times 10^{-4})} = 20,\ \text{m}^2 = 400,\ \text{mL} = 20 \times 0.1 = 2$$

$$T_0 - 30 = (10^6/(25 \times 400))[1 - 1/\cosh(2)] = 73.42°\text{C}$$

$$\therefore \quad \mathbf{T_0 = 103.42°C}$$

Heat conducted : $Q = (qA/m)\tanh(mL) = (10^6 \times 1 \times 10^{-4}/20 \tanh(2))$

= **4.82** W (one half length)

Heat convected : $(hqP/\text{km}^2)[L - (1/\text{m})\tanh(mL)] = (25 \times 10^6 \times 0.04/(25 \times 400))$

$[0.1 - (1/20)\tanh(2)]$ = **5.18 W** (One half length)

Total = 10 W. Heat generated = $10^6 \times 1 \times 10^{-4} \times 0.1$ = 10 W. Checks.

Example A-4.35: *A copper rod heated at one end is placed in an air stream. The diameter of the rod is 10 mm and the length is 80 mm. The base temperature is 80°C and the surrounding temperature is 30°C. It is found that 12 W power is needed to maintain steady conditions in the rod. Taking k = 385 W/mK, determine the value of convective heat transfer coefficient over the surface.*

Solution:

$$Q = \sqrt{(hPkA)}\ (T_0 - T_\infty) \tanh (mL) \text{ (end insulated)}$$

$$12 = \sqrt{h(\pi \times 0.01 \times 385 \times \pi \times 0.01^2/4)}\ (80 - 30) \tanh (m \times 0.08)$$

$$m = \sqrt{h}\ \sqrt{(P/kA)},$$

h is involved in the hyperbolic function also. Hence trial solution is made.

Q is found using various values of h. The trial values are listed.

Hence **h = 120 W/m² K** corresponding to 12 W.

h	Q
60	6.65
100	10.36
120	**12.00**
140	13.61

Chapter 4

Example A-4.36: *In a gas turbine stage the gas temperature at the blade surface is found to be 400°C. The convection coefficient over the surface is 260 W/m²K. The conductivity of the material is 25 W/mK. The root of the blade is at 300°C. The blade section has an area of 2.5 cm² and the average circumference is 10 cm. The height of the blade is 6 cm.*

Determine the heat in flow at the root.

Solution: The blade acts as a fin. The section is not of simple geometic shape. However P and A are known

$$m = \sqrt{(hP/kA)} = \sqrt{((260 \times 0.1)/(25 \times 2.5 \times 10^{-4}))} = 64.5,\ mL = 3.87$$

Assuming end insulated condition.

$$Q = \sqrt{(hPkA)}\ (T_0 - T_\infty) \tanh (mL)$$

$$\sqrt{(260 \times 0.1 \times 25 \times 2.5 \times 10^{-4})}\ (400 - 300) \tanh (3.87)$$

= **40.3 W/blade.** (Heat flow at tip neglected).

This situation helps in blade cooling.

Example A-4.37: *Formulate the set of simultaneous equations to be solved for a plate fin shown in Fig. AE-4.37. The depth is 1m.*

Solution: Considering **element 1**,

heat conducted in $= (115 - T_1)/\{(0.01)/(50 \times 0.019)\}$

heat conducted out $= (T_1 - T_2)/(0.01)/(50 \times 0.017)$

heat convected $= 10(2 \times 0.01005) \times (T_1 - 20)$

using the energy balance this reduces to

$$85\ T_2 - 180.201\ T_1 + 10929 = 0$$

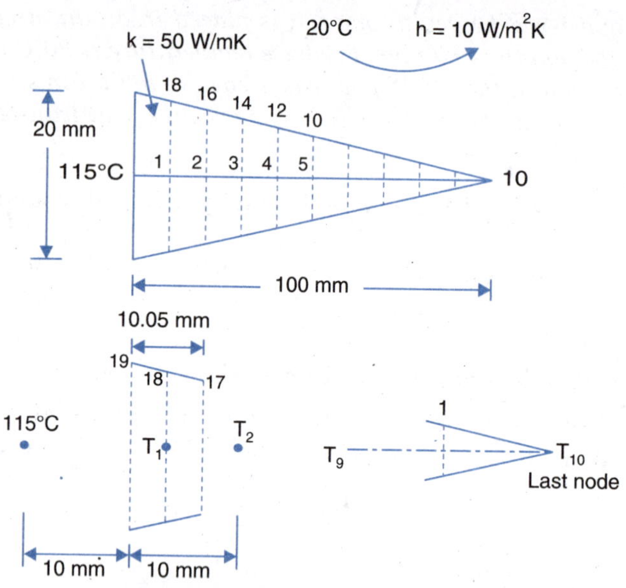

Fig. AE-4.37. Model of nodes

Similarly for other elements

$$85\,T_1 - 160.201\,T_2 + 75\,T_3 + 4.02 = 0$$
$$75\,T_2 - 140.201\,T_3 + 65\,T_4 + 4.02 = 0$$
$$65\,T_3 - 120.201\,T_4 + 55\,T_5 + 4.02 = 0$$
$$55\,T_4 - 100.201\,T_5 + 45\,T_6 + 4.02 = 0$$
$$45\,T_5 - 80.201\,T_6 + 35\,T_7 + 4.02 = 0$$
$$35\,T_6 - 60.201\,T_7 + 25\,T_8 + 4.02 = 0$$
$$25\,T_7 - 40.201\,T_8 + 15\,T_9 + 4.02 = 0$$
$$15\,T_8 - 20.201\,T_9 + 5\,T_{10} + 4.02 = 0$$
$$5\,T_9 - 5.1005\,T_{10} + 2.01 = 0$$

For the last node

$$(T_9 - T_{10})/[(0.01)/(50 \times 0.001)] + 10 \times 0.01005\,(20 - T_{10}) = 0$$

This reduces to the last equation above.

These equations can be solved for temperature distribution using matrix inversion or other such methods.

REVIEW QUESTIONS

4.1 Explain the function of fins attached to surfaces.

4.2 Discuss the conditions that will necessitate the use of fins.

4.3 Define fin effectiveness. Explain the parameters that may lead to higher value of effectiveness.

4.4 Define fin efficiency. What are the conditions that will lead to higher fin efficiency?

4.5 Define thermal resistance of a fin.

4.6 Discuss how the conductivity of the fin material affects fin effectiveness and fin efficiency.

4.7 Discuss how convection coefficient affects fin effectiveness and fin efficiency.

4.8 For effective use of a given volume of material, explain what arrangement will lead to higher effectiveness.

4.9 Define "Total fin efficiency".

4.10 In a pipe carrying liquid chemical to be cooled should fins be arranged on the inside or outside?

4.11 Discuss how contact resistance affects fin efficiency.

OBJECTIVE QUESTIONS

4.1 Choose the correct statement from each question.

(*i*) In a long fin if the thermal conductivity is increased with other parameters maintained constant

(*a*) the temperature will drop at a faster rate along the length

(*b*) the temperature will drop at a lower rate along the length

(*c*) the temperature gradient is not strongly influenced by the conductivity

(*d*) the temperature gradient is dependent on the heat flow only.

(*ii*) In a long fin if the convection coefficient is increased with other parameters maintained constant

(*a*) the temperature drop along the length is not strongly influenced by the convection coefficient

(*b*) the temperature gradient depends only on heat flow rate

(*c*) the temperature drop will be faster along the length

(*d*) the temperature drop along the length will be at a lower rate.

(*iii*) In a long fin if the parameter $m = \sqrt{(hP/kA)}$ increases, other parameters being maintained constant then

(*a*) the temperature drop along the length will be at a lower rate

(*b*) the temperature drop along the length will be steeper

(*c*) the parameter m influences the heat flow only

(*d*) the temperature profile will remain the same.

(*iv*) For a given sectional area of fin if the circumference is increased by adopting different geometric shape, then

(*a*) the temperature variation along the fin length will be steeper

(*b*) the temperature variation along the fin length will be flatter

(*c*) the circumference length does not affect the temperature change

(*d*) the circumference will only influence the heat convected.

(*v*) In a given fin configuration increase in conductivity will

(*a*) decrease the total heat flow

(*b*) will affect only the temperature gradient

(*c*) increase the total heat flow

(*d*) heat flow is influenced only by the base temperature and sectional area.

(*vi*) An increase in convection coefficient over a fin will

(*a*) increase effectiveness (*b*) decrease effectiveness

(*c*) does not influence effectiveness (*d*) influences only the fin efficiency

Chapter 4

(*vii*) In the case of fins, it is desirable to have
  (*a*) area of section maintained constant along the length
  (*b*) area of section reduced along the length
  (*c*) area of section increased along the length
  (*d*) better to vary the convection coefficient than the area.

(*viii*) Fin effectiveness will be increased more by
  (*a*) having a higher value of convection coefficient
  (*b*) higher sectional area
  (*c*) higher thermal conductivity
  (*d*) longer circumference.

(*ix*) If a square section fin is split longitudinally and used as two fins
  (*a*) the total heat flow will decrease
  (*b*) the total heat flow will increase
  (*c*) the toal heat flow will remain constant
  (*d*) heat flow may increase or decrease depending on the material used.

(*x*) For a given volume of material for use in a pin fin
  (*a*) longer the fin, better the total heat flow
  (*b*) shorter the fin, better the heat flow
  (*c*) as the volume is constant, the heat flow will not change
  (*d*) as length is increased heat flow will increase and after some length will decrease.

(*xi*) (*a*) If the convection coefficients is low, it is not desirable to use a fin.
  (*b*) If the conductivity is large, a longer fin will be more effective.
  (*c*) Plate fins of smaller thickness is better from the point of view of heat dissipation.
  (*d*) Finned surface is desirable under conditions of boiling.

(*xii*) (*a*) Aluminium fins are better because the material is light.
  (*b*) A constant area fin provides the best (heat flow/weight) ratio.
  (*c*) On rare occasions the heat flow may be reduced by the addition of fins.
  (*d*) If conductivity is high, a short fin will be a good proportion.

(*xiii*) (*a*) A constant temperature gradient along the length of a constant area, fin is not possible.
  (*b*) The temperature gradient, in circular section fin dissipating heat will increase along the length.
  (*c*) Longitudinal fins are less effective compared to annular fins.
  (*d*) Fins at close pitch will give a high heat dissipation rate.

(*xiv*) (*a*) An aluminium fin will be cooler at a given distance compared to a copper fin of identical section and other parameters.
  (*b*) An aluminium fin will be hotter at a given distance compared to a copper fin of identical section and other parameters.
  (*c*) An aluminium fin of same configuration will dessipate more heat compared to copper fin.
  (*d*) A aluminium fin of same configuration will dessipate less heat compared to copper fin.

Answers

(*i*) (*b*)	(*ii*) (*c*)	(*iii*) (*b*)	(*iv*) (*a*)	(*v*) (*c*)
(*vi*) (*a*)	(*vii*) (*b*)	(*viii*) (*d*)	(*ix*) (*b*)	(*x*) (*d*)
(*xi*) (*c*)	(*xii*) (*a, d*)	(*xiii*) (*a*), (*b*) (*d*)	(*xiv*) (*b*), (*d*).	

EXERCISE PROBLEMS

4.1 A copper rod extends from a surface at 300°C. The diameter is 12 mm and length is 9 cm. The rod is exposed to air at 30°C with convection coefficient of 35 W/m^2 K. Assuming end is insulated determine the tip temperature. Assume k = 340 W/mK. If an aluminium fin of the same diameter is used, what should be the length of the fin, if the tip temperature is to be the same as the copper fin. Assume k = 210 W/mK.

4.2 A surface is at 200°C. Pin fins of diameter 6 mm and length 12 mm are used at 12 mm spacing between centres. The surroundings are at 20°C. The convection coefficient is 30 W/m^{2}K. The conductivity of the material is 131 W/m K. Determine the increase in heat loss from the surface. Another proposal is to use fins of half the length at the same spacing, but the diameter increased to 8 mm. Compare the heat dissipation for this proposal, workout the heat dissipation/unit volume of fin.

4.3 A poker used for stirring hot coal in a furnace is in the from of a pipe of 20 mm ID and 24 mm OD. One end is at 350°. The outside of the poker is exposed to air at 35°C with a convection coefficient of 20 W/m^{2}K. The conductivity of the material is 45 W/mK. Determine the temperature at a distance of 0.35 m from the end.

4.4 A stainless steel vessel of circular shape of inside diameter 20 cm has a plate thickness of 2 mm. It is filled with hot gravy at 90°C up to a depth of 8 cm from the top. Determine the temperature at the top of the vessel. The surroundings are at 30°C and the convection coefficient is 30 W/m^{2}K. The conductivity of the material is 20 W/mK. Assume only the outside surface is effective in convection.

4.5 The size of a transformer tank is 0.6 m × 0.4 m × 0.6 m. Plate fins are used to dissipate heat at the four sides. Fins are 5 mm thick and 6 cm long and run along the vertical direction, 0.6 m deep. The conductivity of the material is 45 W/mK. The convection coefficient is 45 W/m^{2}K. Totally 100 fins are used. Determine whether it can meet the requirements of the unit if maximum wall temperature is 60°C the surrounding temperature is 38°C. Heat to be dissipated is 10 kW.

4.6 In an experiment to determine conductivity of materials, rods of different materials extend from a vessel containing boiling water, and the rods are coated with a thin layer of wax and the length upto which the wax melts is used to determine the conductivity of materials. The melting point of wax is 40°C, the base temperature is 99°C, the diameter of rods is 8 mm. The length of rods are 60 mm, determine the length upto which wax will melt if the rod is of copper with k = 330 W/mK. The surroundings are at 25°C and the convection coefficient is 20 W/m^{2}K. State whether the length of melt in the case of Aluminium rod will be longer or shorter and justify your answer.

4.7 A hollow hexagonal pipe of sides 10 mm (inside) and 1.5 mm wall thickness protrudes from a surface at 200°C. The length of protrusion is 9 cm. The material has a thermal conductivity of 45 W/mK and the surface is exposed to a fluid at 35°C with a convection coefficient of 15 W/m^{2}K. Determine the heat loss and the temperatures at the tip and at mid point along the length. Convection is absent inside of pipe.

4.8 A square section pipe of 18 mm side and 1.55 mm wall thickness is used as a fin from a surface at 200°C. The outside is exposed to air at 35°C, with a convection coefficient of 15 W/m^{2}K. Determine the heat loss and the temperatures at the tip and the mid point. The length of the pipe is 9 cm. Conductivity = 55 W/mK.

4.9 A rectangular section of thickness 3 mm and depth 37.5 mm and length 9 cm is used as a fin on a wall at 200°C . The surroundings are at 35°C and the convection coefficient is 15 W/m^{2}K. The conductivity of the material is 45 W/mK. Determine the heat loss from the fin and the tip and mid point temperatures.

4.10 Circular pipe of 23.9 mm OD and 1.5 mm wall thickness and length 90 mm is used as a fin on a wall at 200°C. The outside is exposed to air at 35°C with a convection coefficient of 15 W/m^{2}K. The conductivity of the material is 45 W/mK. Determine the heat loss from the fin and also the mid section and tip temperatures.

4.11 A plate fin having section of an angle of 5 mm thickness and 12 mm side is proposed to be fixed at 90° to the surface. Determine the heat loss. T_0 = 110°C, h = 30 W/m^{2}K. k = 200 W/mK. The surrounding is at 35°C.

4.12 A thermometer well used to measure temperature of a fluid flowing in a pipe is of 6 mm ID, 1 mm wall thickness and 65 mm long. The fluid is at 600°C. The convection coefficient over the well surface is 50 W/m^{2}K. The conductivity of the material is 50 W/mK. The uninsulated wall was at 80°C lower than the fluid temperature. Determine the temperature that will be indicated by the thermometer. As the error in measurement was found unacceptable, the pipe wall was insulated on the outside so that the wall temperature was now 20°C lower than the fluid temperature. Estimate the reduction in the error.

4.13 A turbine blade is of hollow section of 2 mm thickness with the outside periphery of 100 mm. The blade is exposed to gasess at 400°C with a convection coefficient of 80 W/m^2 K. The conductivity of the material is 32 W/mK. If the blade root is at 300°C, determine the heat flow into the root section.

4.14 A rod of 1 cm square section and 6 cm length with a thermal conductivity of 50 W/mK protrudes from a surface at 180°C and is exposed to air at 35°C. The tip temperature is measured as 90°C. Determine the value of convection coefficient and the heat dissipated by the rod.

4.15 The area of section available for a rectangular fin is 2 cm^2. The conductivity of the material is 130 W/mK and the convection coefficient is 65 W/m^{2}K. The wall temperature is 200°C and the surrounding is at 35°C. Determine the thickness for maximum heat flow and the heat dissipated for 1 m depth of such a fin.

4.16 The handle of a stainless coffee cup is in the form of a circular ring of 3 cm mean dia. and wire diameter of 4 mm. It is welded to the cup surface at a circumferential point. The coffee is at 80°C. The surroundings are at 30°C and the convection coefficient is 8 W/m^{2}K. The conductivity of the material is 30 W/mK. Determine the temperature of the ring opposite to the point of weld. Also determine the heat dissipated by the handle.

4.17 Two hot surfaces one at 300°C and the other at 200°C are connected by a plate 3 cm wide and 0.5 cm thick, the distance between the surfaces being 10 cm. The plate is of material with conductivity of 130 W/mK and is exposed to gases at 80°C with a convective heat transfer coefficient of 65 W/m^{2}K. Determine the location and value of the minimum temperature in the plate and also the heat dissipated by the plate.

4.18 The volume of material available for a pin fin of square section is 4 cm^3. The conductivity of the material is 140 W/MK and the convection coefficient is 60 W/m^{2}K. Determine the size of the fin for maximum heat flow.

4.19 A plate fin of 2 mm thickness and 20 mm length is dissipating heat from a surface at 200°C. The fin is exposed to air at 30°C with a convection coefficient of 85 W/m^{2}K. If the conductivity of the material is 235 W/mK, determine the heat dissipated in 1 m depth of fin. If a contact resistance of 2.5×10^{-4} m^2 °C/W is involved determine the reduction in heat dissipation.

[**Ans.** 53,392.4, 28.8%]

4.20 A circumferential fin of thickness 1 mm and length 16 mm is used on a 15mm dia. pipe used to cool air. The pipe surface temperature is 3°C and the air temperature is 28°C. If the convection coefficient is 80 W/m^{2}K and the conductivity is 200 W/mK determine the heat dissipated by a fin. If the cooling load is 2 kW and if the pitch is 4 mm, determine the length of finned pipe necessary to meet the load.

4.21 The cylinder head of an engine is fitted with plate fins of 4 mm thickness and 25 mm length the material having a thermal conductivity of 50 W/mK. The convection coefficient is 85 W/m^{2}K. If the total fin depth is 0.8 m, determine the heat dissipation for a wall to surrounding temperature difference of 290°C.

4.22 A cast iron cylinder of an air cooled engine is provided with circumferential fins of 4 mm thickness and 25 mm length. The *OD* of the cylinder is 90 mm. The conductivity of the metal is 45 W/mK. The convection coefficient is 80 W/m^{2}K. The surface is at 280°C and the surroundings are at 38°C. Determine the heat loss per fin.

4.23 Circumferential fin of a water preheater is 2 mm thick and 40 mm long and are fixed on pipes of 60 mm outside diameter. The conductivity of the material is 55 W/mK and the fins are exposed to hot gases at 200°C. The wall temperature is 120°C. The convection coefficient is 45 W/m^{2}K. If fins are fixed at 12 mm pitch and if total length of pipe is 40 m, determine the total heat collection rate.

4.24 A room heater uses steam in pipes of 60 mm *OD* and the steam temperature is 130°C. The pipe is provided with fins of 1 mm thickness and 30 mm length at a pitch of 8 mm. The room air is at 25°C and the convection coefficient is 22 W/m^{2}K. The conductivity of the material is 210 W/mK. If the heat loss to the surroundings from the room is 4 kW, determine the length of pipe required.

4.25 An electronic device is of 15 mm dia and space is available for three circumferential fins of 0.8 mm thickness and 8 mm length. If the heat to be dissipated by the fins is 12 W and if the base temperature is not to exceed 60°C when the surrounding is at 20°C, check the suitability of silver, copper, gold and aluminium with conductivity values of 407, 386, 317 and 204 W/mK. The convection coefficient has value of 25 W/m^{2}K.

4.26 Solar radiation is incident on a plate of 2 mm thickness and conductivity of 204 W/mK at 800 W/m^2. Energy is collected by water flowing in pipes fixed at a pitch distance of 12 cm at a temperature of 48°C. Determine the maximum temperature in the plate. Assume that losses are negligible.

4.27 Radiation flux of 5000 W/m^2 is incident on an annular plate of 150 mm inner diameter and 500 mm outer diameter. The plate is 4 mm thick and has a conductivity of 210 W/mK. The inner and outer edges are maintained at 10°C by collection of heat. There is a loss to outside air at 40°C by convection with h = 40 W/m^{2}K. Determine the maximum temperature in the plate and also the heat flow at either of the edges.

4.28 Heat is generated in a bar of rectangular section of 6 mm × 24 mm at a uniform rate of 10^6 W/m^3. The conductivity of the material is 30 W/mK. The bar is surrounded by a fluid at 180°C with a convection coefficient of 200 W/m^{2}K. The bar is 1 m long and the ends are fixed to walls maintained at 180°C. Determine the temperature at mid lenght and the heat flow at the ends and over the surface.

4.29 A heating surface has resistance rods 20 mm dia. and 1 m length generating 4 KW of heat. The ends are fixed to supports at 300°C. The combined convection and radiation heat transfer coefficient on the rod is 500 W/m^{2}K and the surroundings are at 700°C. If the conductivity of the material is 15 W/mK, determine the maximum temperature in the rod and the heat flow at the ends.

4.30 A solar collector plate is exposed to a flux of 900 W/m^2. Heat is collected by water pipes fixed at 12 cm pitch with a water temperature of 48°C. The plate is 2 mm thick and has a conductivity of 204 W/mK. If the losses over the plate is accounted by a convection coefficient of 15 W/m^{2}K to air at 30°C, determine the maximum temperature in the plate and also the rate of heat collection by the water per pitch width and 1 m length.

4.31 Circumferential fins of 2 mm thickness are fixed with a pitch of 5 mm on a 3 cm pipe. The outside diameter of fins is 6 cm. Thermal conductivity of the fin material is 180 W/m°C. The pipe surface is at 120°C. Air at 25°C flows across producing a convection coefficient of 60 W/m^2°C. If 200 fins are fixed per m length of pipe, determine the increase in heat transfer rate due to the fins.

[**Ans.** 4783 W per m length]

4.32 Pin fins or 2.5 mm dia. and 3 cm length are fixed at 6 mm pitch on a surface at 100°C. Thermal conductivity of fin material is 237 W/m°C. Air at 30°C flows over the fins producing h = 35 W/m^2 °C. Determine the heat transfer rate per 1m square area of the surface. Also calculate the overall effectiveness. [**Ans.** 17.4 kW, 7.1]

4.33 Circumferential fins are fixed on a pipe of 5 cm diameter at a pitch of 4 mm. The outer diameter of the fin is 6 cm. The fin thickness is 1 mm. The thermal conductivity of the fin material is 186 W/m°C. The pipe surface is at 180°C. Air at 25°C flows over the fins producing a convection coefficient of 40 W/m^2°C. Determine the increase in heat transfer rate per m length of pipe due to the fins. [**Ans.** 2639 W per m length]

4.34 Flanges each of 10 mm thickness are fixed on pipes of OD 100 mm. The outer diameter of the flanges is 200 mm. The pipe surface is at 178.4°C. The flanges are exposed to air at 12°C with a convection coefficient of 25 W/m^2°C. Determine the heat flow rate through the flanges.

[**Ans.** 214 W]

5 CHAPTER

Two Dimensional Steady Heat Conduction

5.0 INTRODUCTION

In several situations one dimensional conduction approximation provides reasonably acceptable answers. However there are situations where the heat conduction in two dimensions has to be considered. One example is corners in a rectangular furnace. The solution required is the temperature at various locations and the heat flow. If the temperature field is established, heat flow can be determined by Fourier's conduction equation. Mainly four methods have been in use for solving two dimensional conduction problems. These are (1) Analytical method solving the differential equation using the boundary conditions (2) Graphical method of sketching equal temperature lines and then the equal heat flow paths (3) Use of electrical analogy and use of conduction paper to plot equal temperature lines and (4) Finite difference method. Rigorous analytical solutions are available only for very simple boundary conditions and these are not amenable for complex boundaries. With computer software becoming popular, the fourth method is replacing the graphical and analogy methods.

5.1 METHOD OF SOLUTION OF DIFFERENTIAL EQUATION

From chapter 2, the differential equation in cartesian coordinates for steady two dimensional conduction, without heat generation is given as [refer equation (2.3)]

$$\frac{\partial^2 T}{\partial x^2} + \frac{\partial^2 T}{\partial y^2} = 0 \quad ...(5.1)$$

This equation is solved by the method of separation of variables. Solutions are available only for a limited sets of boundary conditions. However limited, these solutions provide an insight into the problem.

5.1.1. One of the simpler problem is illustrated here. Consider a rectangular plate as shown in Fig. 5.1. The plate is W m wide and H m high. The origin of the cartesian coordinates is considered to be at the left bottom corner of the plate. The two sides and the bottom side are at temperature T_1. The top side temperature varies as in equation (5.2).

$$T = T_m \sin \frac{\pi x}{W} + T_1 \quad ...(5.2)$$

This leads to the condition that at the top corners the temperature is T_1 in whichever way the corner is reached. Thus singularity is avoided in this boundary set as compared to the next boundary set considered. The essential feature of this method is that the solution for this equation is assumed to take the product from.

Let $$T = XY \qquad ...(5.3)$$

where $X = f(x)$, $Y = f(y)$

Substituting (5.3) in (5.1), the equation 1 reduce to ordinary differential equation

$$\frac{1}{X}\frac{d^2X}{dx^2} + \frac{1}{Y}\cdot\frac{d^2Y}{dy^2} = 0 \qquad ...(5.4)$$

As these two terms are functions of independent variables, this equation can be reduced to two ordinary differential equations. These are

$$\frac{d^2X}{dx^2} + \lambda^2 X = 0 \qquad ...(5.5)$$

$$\frac{d^2Y}{dy^2} - \lambda^2 Y = 0 \qquad ...(5.6)$$

where λ^2 **is called separation constant.** Since the sine function boundary condition is to be satisfied, λ^2 cannot be zero. The solutions are

$$X = C_1\, e^{-\lambda x} + C_2\, e^{\lambda x} \qquad ...(5.7)$$

$$Y = C_3 \cos \lambda x + C_4 \sin \lambda x \qquad ...(5.8)$$

Hence, $$T = XY = (C_3 \cos \lambda x + C_4 \sin \lambda x)\,(C_1 e^{-\lambda x} + C_2 e^{\lambda x}) \qquad ...(5.9)$$

The boundary conditions are: (Fig. 5.1)

$$T = T_1 \text{ at } y = 0,\ T = T_1 \text{ at } x = 0,\ T = T_1 \text{ at } x = W$$

at $y = H,\ T = T_1 + T_m \sin\left(\frac{\pi x}{W}\right)$

It is easier to use a new variable $\theta = T - T_1$

The boundary conditions reduce to

$\theta = 0$ at $y = 0$, $\theta = 0$, at $x = 0$, $\theta = 0$ at $x = W$,

$$\theta = T_m \sin \frac{\pi x}{W} \text{ at } y = H. \qquad ...(5.10)$$

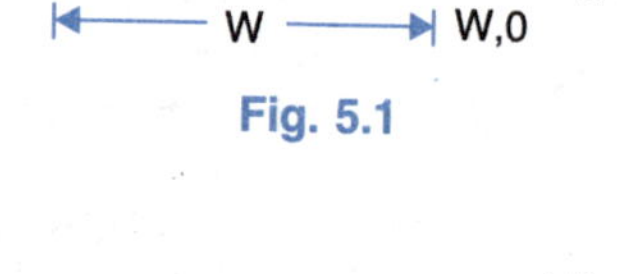

Fig. 5.1

Applying these conditions, we get

$$(C_3 \cos \lambda x + C_4 \sin \lambda x)\,(C_1 - C_2) = 0 \qquad ...(a)$$

$$C_3\,(C_1 e^{-\lambda x} + C_2\, e^{\lambda x}) = 0 \qquad ...(b)$$

$$(C_3 \cos \lambda W + C_4 \sin \lambda W)\,(C_1 e^{-\lambda x} + C_2 e^{\lambda x}) = 0 \qquad ...(c)$$

$$(C_3 \cos \lambda\, x + C_4 \sin \lambda y)\,(C_1 e^{-\lambda H} + C_2 e^{\lambda H}) = T_m \sin\left(\frac{\pi x}{W}\right) \qquad ...(d)$$

From a, $c_1 = c_2$. From b, $c_3 = 0$

From c, $c_3 c_1 \sin \lambda W\,(e^{-\lambda x} - e^{\lambda x}) = 0 \qquad ...(5.11)$

$$c \sin \lambda W\,(e^{-\lambda x} - e^{\lambda x}) = 0.$$

$$\therefore \qquad \sin \lambda W = 0$$

This can be satisfied for all values of

$$\lambda = \frac{\pi x}{W}.$$

This solution reduces to as $\dfrac{(e^{\lambda y} - e^{-\lambda y})}{2} \to \sinh(\lambda\, y)$

$$\theta = T - T_1 = \sum_{n=1}^{\infty} c_n \sin\frac{n\pi x}{W} \sinh\frac{n\pi y}{W} \qquad ...(5.12)$$

This should satisfy the fourth boundary condition also *i.e.* at $y = H$.

$$\therefore \qquad T_m \sin\frac{\pi x}{W} = \sum_{n=1}^{\infty} c_n \sin\frac{n\pi x}{W} \sinh\frac{n\pi H}{W} \qquad ...(5.13)$$

As $c_n = 0$ for $\sin n > 1$, its reduces only to c_1. c_1 is obtained from (5.13) and substituted in (5.12). obtaining.

$$T - T_1 = \mathbf{T_m} \frac{\mathbf{\sinh(\pi y / W)}}{\mathbf{\sinh(\pi H / W)}} \cdot \mathbf{\sin\frac{\pi x}{W}} \qquad ...(5.14)$$

5.1.2. The other boundary condition is slightly different from the previous one. Instead of a sine function in the temperature at $y = H$, a constant temperature is assumed. This leads to singularity at the top corners.

$$T = T_1 \text{ at } y = 0,\ T = T_1 \text{ at } x = 0,\ T = T_1 \text{ at } x = W.$$

$$T = T_2 \text{ at } y = H \qquad ...(5.15)$$

As the first three boundary conditions are the same, the result of these boundaries give (equation 5.11)

$$T - T_1 = \sum_{n=1}^{\infty} c_n \sin\frac{n\pi x}{W} . \sinh\frac{n\pi y}{W}$$

The last boundary condition gives

$$T_2 - T_1 = \sum_{n=1}^{n=\infty} c_n \sin\frac{n\pi x}{W} . \sinh\frac{n\pi H}{W} \qquad ...(5.16)$$

Expanding this in a Fourier series in the interval $0 < x < W$

$$T_2 - T_1 = (T_2 - T_1)\frac{2}{\pi}\sum_{n=1}^{\infty}\frac{(-1)^{n+1}+1}{n}\sin\frac{n\pi x}{W} \qquad ...(5.17)$$

Upon comparison between (5.16), (5.17)

$$c_n = \frac{2}{\pi}(T_2 - T_1)\frac{1}{\sin(n\pi H / W)}\frac{(-1)^{n+1}+1}{n}$$

The final solution thus is

$$\frac{T - T_1}{T_2 - T_1} = \frac{2}{\pi}\sum_{n}^{\infty}\frac{(-1)^{n+1}+1}{n} . \sin\frac{n\pi x}{W} . \frac{\sinh(n\pi y / W)}{\sinh(n\pi H / W)} \qquad ...(5.18)$$

Some more boundary conditions are solved in textbooks exclusively on conduction. For complex boundaries the use of computers softwares will provide the temperature distribution and heat flow at surfaces at a reasonable cost and time.

Example 5.1: *A plate 1 m × 2 m side has both its 2 m sides and one 1 m side at 100°C. The temperature along the fourth side is given by* $T = 400 \sin\left(\frac{\pi x}{W}\right) + 100$*, where x is in m from the corner and T is in °C. Determine the temperatures taking 1 m on the x direction and 2 m on the y direction at the following locations. W = 1 m, H = 2 m*

(i) (0.25, 0.5), (ii) (0.5, 0.5), (iii) (0.25, 1), (iv) (0.5, 1), (v) (0.25, 1.5), (vi) (0.5, 1.5), (vii) (0.25, 2) and (viii) (0.5, 2).

Solution: These points are shown in Fig. Ex. 5.1 (Use radian mode in calculations).

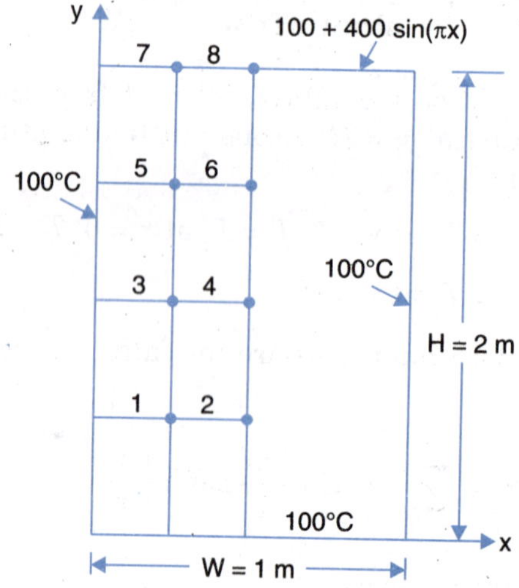

Fig. Ex. 5.1

$$T_{(x,\,y)} = T_1 + T_m \frac{\sinh(\pi y/W)}{\sinh(\pi H/W)} \sin\left(\frac{\pi x}{W}\right)$$

(i) $$\mathbf{T}_{(0.25,\,0.5)} = 100 + 400 \frac{\sinh(\pi \times 0.5/1)}{\sinh(\pi \times 2/1)} \sin\left(\frac{\pi \times 0.25}{1}\right) = \mathbf{102.43°C}$$

(ii) $$\mathbf{T}_{(0.5,\,0.5)} = 100 + 400 \frac{\sinh(\pi \times 0.5/1)}{\sinh(\pi + 2/1)} \sin\left(\frac{\pi \times 0.5}{1}\right) = \mathbf{103.44°C}$$

(iii) $$\mathbf{T}_{(0.25,\,1)} = 100 + 400 \frac{\sinh(\pi \times 1).\sin(\pi \times 0.25/1)}{\sinh(\pi \times 2/1)} = \mathbf{112.2°C}$$

(iv) $$\mathbf{T}_{(0.5,\,1)} = 100 + 400 \frac{\sinh(\pi \times 1/1).\sin(\pi \times 0.5/1)}{\sinh(\pi \times 2/1)} = \mathbf{117.25°C}$$

(v) $$\mathbf{T}_{(0.25,\,1.5)} = 100 + 400 \frac{\sinh(\pi \times 1.5/1).\sin(\pi \times 0.25)}{\sinh(\pi \times 2/1)} = \mathbf{158.79°C}$$

(*vi*) $\mathbf{T}_{(0.5,\,1.5)} = 100 + 400\,\dfrac{\sinh(\pi \times 1.5/1).\sin(\pi \times 0.5)}{\sinh(\pi \times 2/1)} = \mathbf{183.15°C}$

(*vii*) $\mathbf{T}_{(0.25,\,2)} = 100 + 400\,\dfrac{\sinh(\pi \times 2/1).\sin(\pi \times 0.25)}{\sinh(\pi \times 2/1)} = \mathbf{382.15°C}$

(*viii*) $\mathbf{T}_{(0.5,\,2)} = 100 + 400\,\dfrac{\sinh(\pi \times 2/1).\sin(\pi \times 0.5)}{\sinh(\pi \times 2/1)} = \mathbf{500°C}$

The calculated values are shown in Fig. Ex. 5.1 (*a*). These values can be used to plot contours of temperature also. Heat flow at any section can be also calculated using these values.

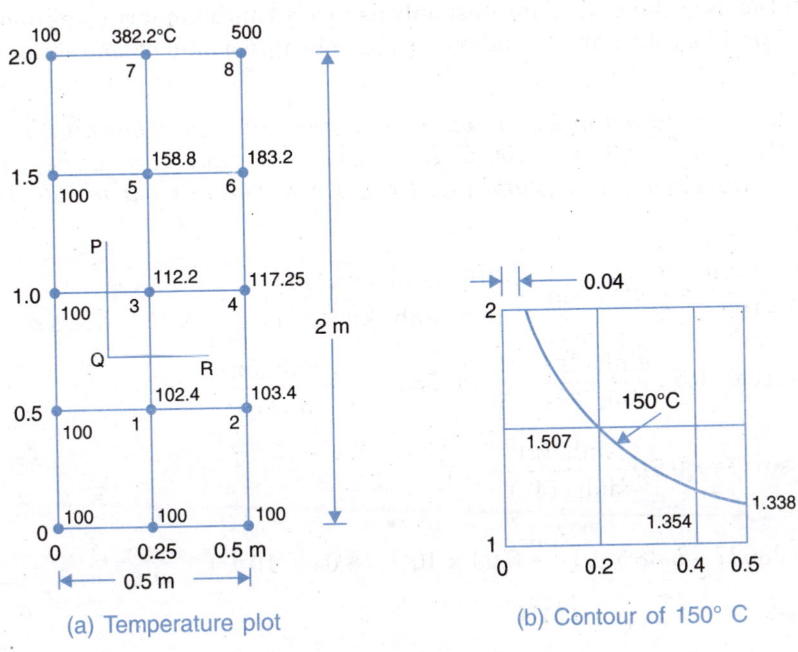

(a) Temperature plot

(b) Contour of 150° C

Fig. Ex. 5.1

For example at the 1 m level, the heat flow in the x direction across PQ (Fig. Ex. 5.1(*a*)) can be roughly found by using dT/dx as $-(112.2 - 100)/0.25 = -48.8$°C/m. Q_1 = kA 48.8 W. If area and conductivity are known, then Q_x can be found. Q_y can be found by using similar method. Across QR, $Q_y = -$ kA. $(102.4 - 112.2)/0.25$ = kA 39.2 W.

Example 5.2: *In example 5.1, determine the locations of 150°C in the plate.*

Solution: This is solved using

$$T = T_1 + T_m \frac{\sinh(\pi y/W)}{\sinh(\pi H/W)} \sin(\pi x/W)$$

In this case $W = 1,\ H = 2,\ T_m = 400,\ T_1 = 100,\ T = 150$

$$\therefore \quad \frac{(150 - 100)}{400} \times \sinh(\pi \times 2/1) = \sinh(\pi y/W).\ \sin(\pi x/W)$$

First at $y = 2$ the x value is found. After that at intervals of 0.1 for x and y values are found and tabulated. It is shown plotted in Fig. Ex. 5.1 (*b*).

Chapter 5

x	y
0.04	**2.000**
0.2	1.507
0.3	1.406
0.4	1.354
0.5	1.338

Similar contours can be worked out for other values of temperature and a complete picture can be obtained.

This solution is of theoretical interest only as rarely this boundary conditions may prevail in any practical problem. But any boundary can be split into a sum of sine series and then this method may be used.

Example 5.3: *A rectangle of 1 m × 2 m (with 1 m on the x direction) has all edges except y = H edge at 100°C. The y = H edge is at 300°C. Determine the temperature at the points (0.5, 1.0), and (0.5, 1.5). Using equation (5.5) and Fig. Ex. 5.3 (use radian mode in calculations).*

Solution:

$$\frac{T_{(0.5,1)}-100}{300-100}=\frac{2}{\pi}(2/1)\sin\frac{\pi\times 0.5}{1}.\frac{\sinh(1.\pi.1/1)}{\sinh(1.\pi\times 2/1)}$$

$$+\frac{2}{3}\sin(3\pi\times 0.5).\frac{\sinh(3\pi)}{\sinh(6\pi)}+\frac{2}{5}\sin(5\pi\times 0.5)\frac{\sinh(5\pi)}{\sinh(10\pi)}$$

$$+\frac{2}{7}\sin(7\pi\times 0.5).\frac{\sinh(7\pi)}{\sinh(14\pi)}+\ldots]$$

$$=\frac{2}{\pi}(0.08627-5.38\times 10^{-5}+6.03\times 10^{-8}-8.04\times 10^{-11})$$

$= 0.0549 \quad \therefore$ $\mathbf{T_{(0.5,1)} = 111°C}$

$$\frac{T_{(0.5,1.5)}-100}{300-100}=\frac{2}{\pi}[2.\sin\left(\frac{\pi\times 0.5}{1}\right).\frac{\sinh(1.5\times\pi/1)}{\sinh(2\pi)}$$

$$+\frac{2}{3}\sin(3\pi\times 0.5).\frac{\sinh(3\times 1.5\times\pi)}{\sinh(6\pi)}$$

$$+\frac{2}{5}\sin(5\pi\times 0.5).\frac{\sinh(5\times 1.5\pi)}{\sinh(10\pi)}+\ldots]$$

$$=\frac{2}{\pi}[0.4157-6\times 10^{-3}+1.55\times 10^{-4}-4.8\times 10^{-6}]=0.2609$$

$\therefore$ $\mathbf{T_{(0.5,1.5)} = 152°C}$

Fig. Ex. 5.3

In this case it is difficult to locate position for a given temperature. Analytical solutions have limited application. However, these methods are basic and can be used for at least validation purpose.

Also the problem can be split up into 4 problems with all except one edge at 0°C and the solutions can be super imposed.

5.2 GRAPHICAL METHOD

This method was very popular when computing aids were rather primitive. The method consists of plotting by trial, constant temperature contours and constant heat flow paths, using the boundary conditions as the initial guide. The orthogonality of these two sets of lines help in plotting orthogonal squares. An example is shown in Fig. 5.2. In most cases by the use of symmetry it is necessary to plot these only for part of an object. In the example, the temperatures are specified for the inner wall and outer walls of a square duct. Due to symmetry the plot is required for only one eight of the wall.

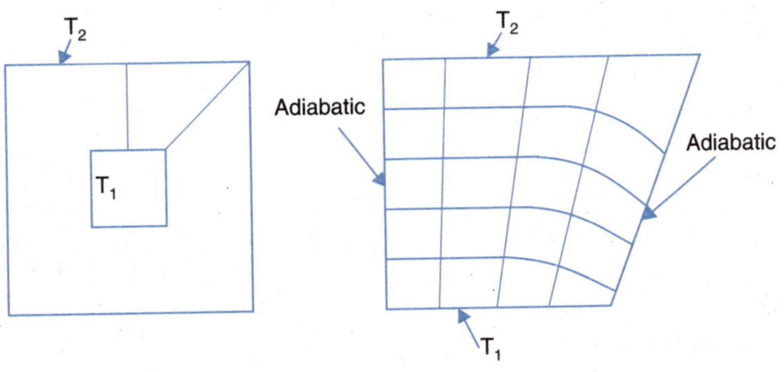

Fig. 5.2

The important requirement is that these intersecting lines should form curvilinear squares. Also the isothermals should be perpendicular to the adiabatic surfaces. The distances have to be adjusted repeatedly to obtain such squares. The number of temperature intervals (N) and the number of flow paths (M) are counted and the heat flow is calculated for unit depth as

$$Q = \frac{M}{N} k (T_1 - T_2) \quad \text{...(5.19)}$$

This is because $\frac{(T_1 - T_2)}{N}$ is the temperature slope in each square and $M\Delta x$ is the total length of the path. If any square is taken in the flow path i, depth perpendicular to paper is l. The heat flow is given by

$$Q_i = k\,\Delta x\,l \cdot \frac{(T_1 - T_2)}{N} \cdot \frac{1}{\Delta x} \quad \text{...(5.20)}$$

By the principle adopted in drawing flow line the flow is the same along all the paths. As there are M such paths $Q = Q_i\,M$ or

$$Q = Mk\Delta x\,l\,\frac{(T_1 - T_2)}{N} \cdot \frac{1}{\Delta x} = k\,l\,\frac{M}{N}(T_1 - T_2)$$

$$= k \times S(T_1 - T_2) \quad \text{...(5.21)}$$

The quantity ($M\,l\,/N$) is called the conduction shape factor, S. Note that the unit for S is meter. For several shapes the shape factor has been determined and these are listed in hand books and textbooks. A few of these are shown in Fig. 5.3.

These shape factors can be used to determine heat flow in such cases by the use of equation (5.21).

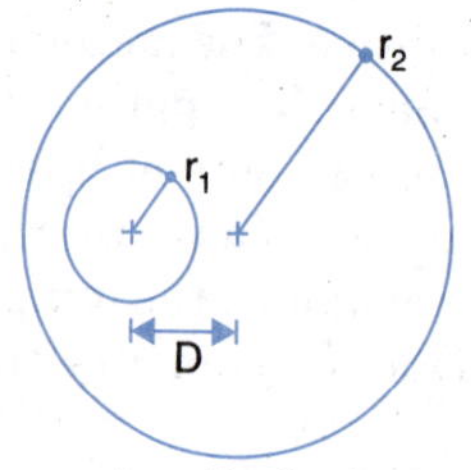

Eccentric cylinder length, L

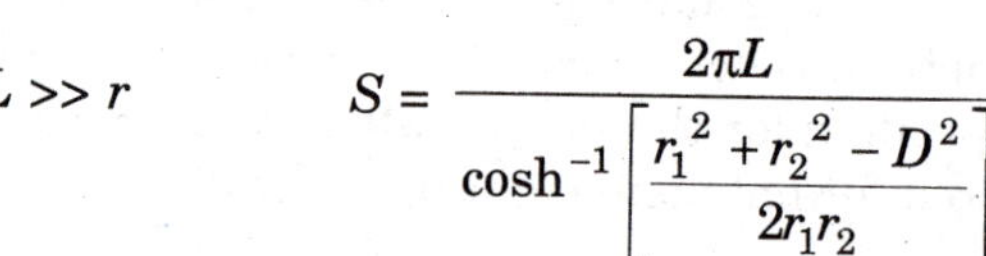

$L >> r \qquad S = \dfrac{2\pi L}{\cosh^{-1}\left[\dfrac{r_1^2 + r_2^2 - D^2}{2r_1 r_2}\right]}$

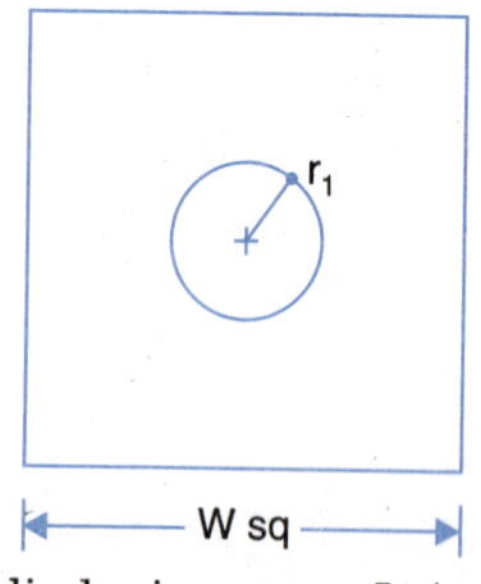

Cylinder in square, Length L

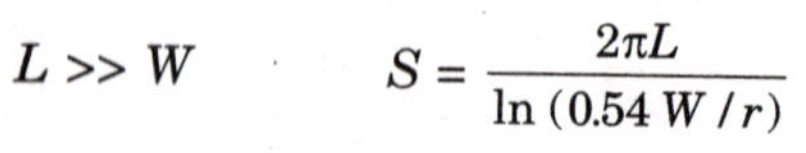

$L >> W \qquad S = \dfrac{2\pi L}{\ln(0.54\,W/r)}$

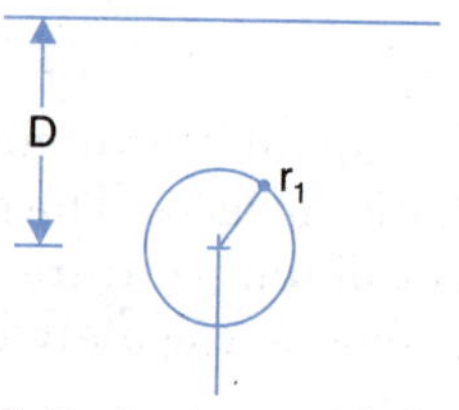

Cylinder in semi infinite medium, Length L

$L >> r \qquad S = \dfrac{2\pi L}{\cosh^{-1}(D/r)}$

$D > 3r \qquad S = \dfrac{2\pi L}{\ln(2D/r)}$

$D >> r \qquad S = \dfrac{2\pi L}{\ln(L/r)\left[\dfrac{\ln(L/D)}{\ln(L/r)}\right]}$

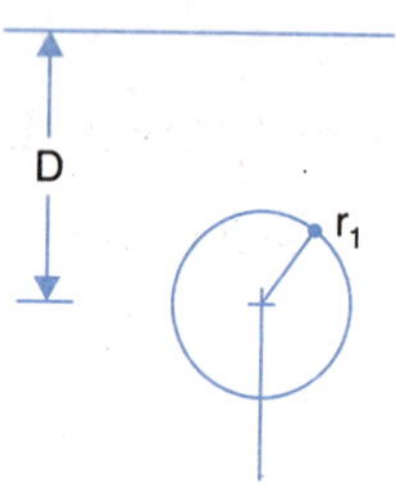

Sphere in semi infinite medium

$D > r \qquad S = \dfrac{4\pi r}{1-(r/2D)}$

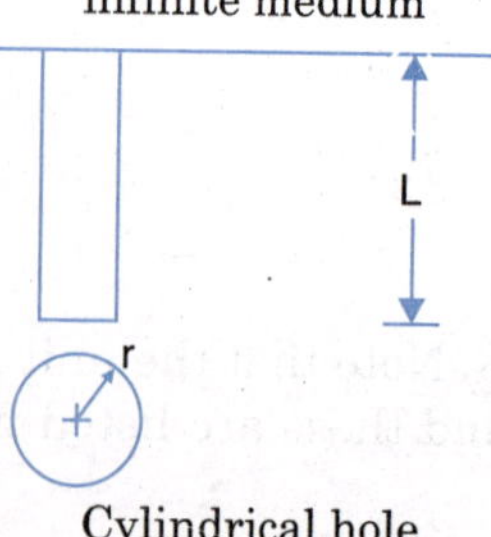

Cylindrical hole

$L >> 2r \qquad S = \dfrac{2\pi L}{\ln(2L/r)}$

Fig. 5.3. Shape factors

Example 5.4: *A cylinder of dia 0.6 m with surface temperature of 200°C is enclosed in a square of 1.2 m side, the material having a thermal conductivity of 2.5 W/mK. The outside surface is at 5°C. Determine the heat loss per 15 m length.*

Solution: The data are shown in Fig. Ex. 5.4. Using the tabulation in Fig. 5.3, the shape factor is read as

$$S = 2\pi L/\ln(0.54\ W/r)$$

Substituting the values,

$$S = 2\pi \times 15/\ln(0.54 \times 1.2/0.3) = \mathbf{122.382}$$

$$Q = k\,S\,(T_1 - T_2) = 2.5 \times 122.382 \times (200 - 5)$$

$$= \mathbf{59661.5\ W}$$

The shape factor can be graphically determined by taking 1/4 of the insulation and plotting equal temperature and flow lines. The shape factor values are available for many other configurations. (Ref. Heat Transfer J.P. Holman).

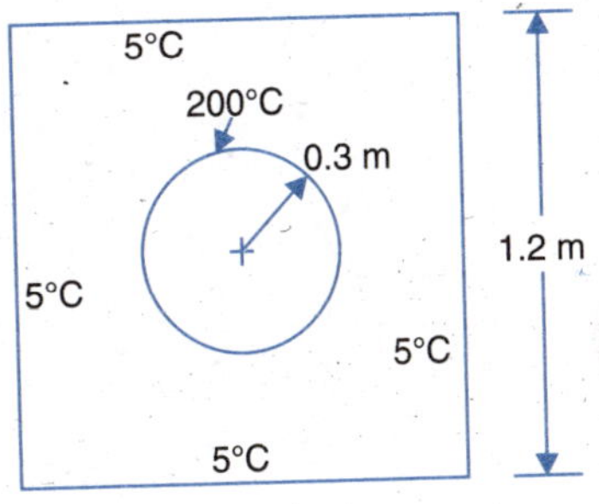

Fig. Ex. 5.4 Problem model

5.3 NUMERICAL METHOD

There are three methods used under this heading namely (*i*) the energy balance method, where the sum of energy flowing into volume is equated to the energy stored or depleted (*ii*) the finite difference method, where the differential equation is converted to difference equation. The resultant equations often time will be the same as obtained using the energy balance method and (*iii*) the finite element method which is more complicated due to the basic nature of approximating the function rather than the derivative. By adopting numerical method, softwares have been prepared to solve complex situations even taking care of variation of properties with temperature.

5.3.1. The Application of Numerical Method

This method for two dimensional flow is attempted in this article. Let us consider the two dimensional flow situation for which analytical solution was obtained in the article, 5.1.1. The solid can now be divided into small elements as shown in Fig. 5.4 (*a*). As the surface temperatures are given, nodes are selected at the boundary. In this case three types of nodes are encountered. These are (*i*) internal nodes (numbered as 1 for example) (*ii*) surface nodes (numbered as 2 for example) and (*iii*) corner node (numbered as 3).

Later we shall see some other typical nodes also.

Consider node 1. Let the node be in the mth column and nth row. The node is shown enlarged in Fig. 5.4 (*b*). All the nodal distances are taken as equal to $\Delta x = \Delta y$ for convenience of simplification. For steady state conduction, the sum of the heat flow into the volume marked $ABCD$ should be zero. The total heat flow is the sum of flow through each of the faces AB, BC, CD and DA. Considering depth to be unity and using the important approximation that the temperature gradient is equal to the temperature difference divided by the distance, the following relationship is obtained.

$$k\Delta y \cdot 1 \frac{(T_{m-1,n} - T_{m,n})}{\Delta x} + \frac{T_{m,n+1} - T_{m,n}}{\Delta y} k\,\Delta \cdot x \cdot 1 + \frac{T_{m+1,n} - T_{m,n}}{\Delta x} k\Delta y \cdot 1$$

$$+ \frac{T_{m,n-1} - T_{m,n}}{\Delta y} \cdot k\Delta x \cdot 1 = 0 \qquad ...(5.22)$$

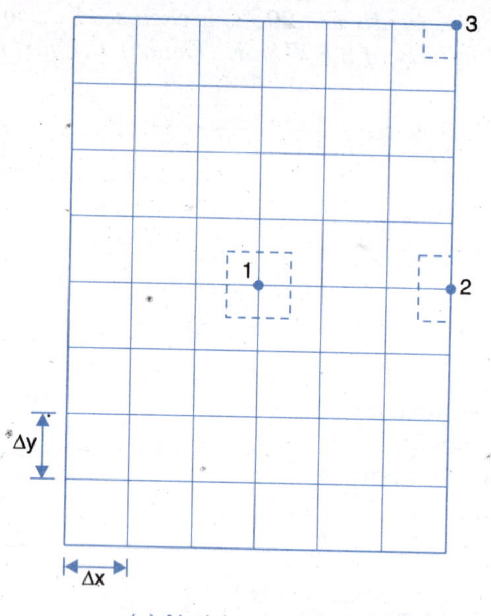

(a) Nodal representation

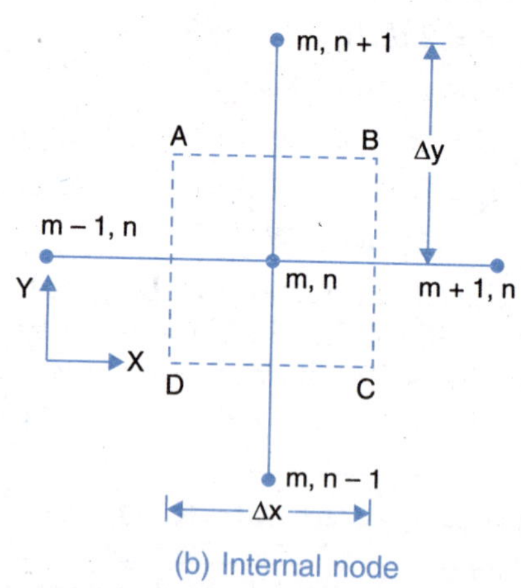

(b) Internal node

Fig. 5.8

Simplification of (5.22) yields

$$T_{m-1,\,n} + T_{m,\,n+1} + T_{m+1,\,n} + T_{m,\,n-1} - 4T_{m,\,n} = 0 \quad \text{...[5.22(a)]}$$

For internal nodes, when equal nodal distances are used the nodal temperature is the average of the adjacent temperatures. It is also assumed that the properties like conductivity are uniform and independent of temperature or location or direction. The equation [5.22(*a*)] is applicable for all the nodes except those on the surface and corner.

Let us now consider the surface node (2). The node is shown in an enlarged view in Fig. 5.4 (*c*). As in the previous case assuming unit depth and considering the volume enclosed by *ABCD*, the total heat flow into the volume should be zero. There is no heat flow to the surface. Note now that the distances *AB* and *CD* are equal to $\Delta x/2$. The sum of heat conducted is given by

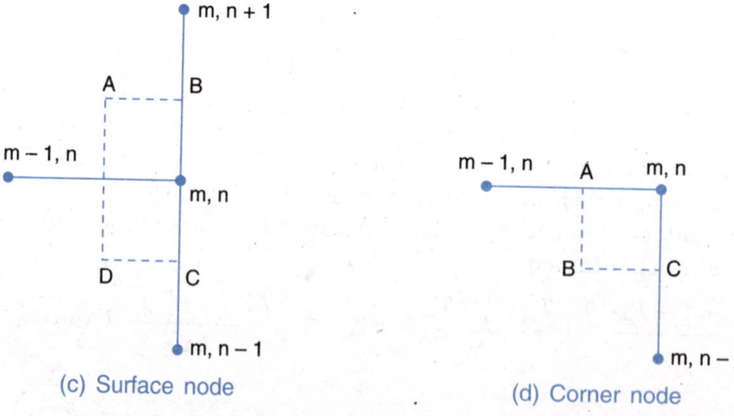

(c) Surface node (d) Corner node

Fig. 5.4

$$k\Delta y.1\,\frac{(T_{m-1,n}-T_{m,n})}{\Delta x}+\frac{T_{m,n+1}-T_{m,n}}{\Delta y}\frac{\Delta x}{2}k.1+\frac{T_{m,n-1}-T_{m,n}}{\Delta y}\frac{\Delta x}{2}k.1=0$$

or
$$2T_{m-1,n}+T_{m,n+1}+T_{m,n-1}-4T_{m,n}=0 \qquad \text{...[5.22 (b)]}$$

The other type of node in this problem is the external corner node, shown enlarged in Fig. 5.4 (*d*). In this case heat flow is through faces *AB* and *BC*. The net heat flow is

$$\frac{T_{m-1,n}-T_{m,n}}{\Delta x}\frac{\Delta y}{2}k.1+\frac{T_{m,n-1}-T_{m,n}}{\Delta y}.\frac{\Delta x}{2}k.1=0$$

Simplification gives
$$T_{m-1,\,n}+T_{m,\,n-1}-2T_{m,\,n}=0 \qquad \text{...[5.22 (c)]}$$

Now as many equations as there are nodes can be obtained in this manner. The solution of these equations can be obtained using standard computer programmes.

The other types of nodes are (*i*) internal corner node with and without convection (*ii*) surface and corner nodes with heat convection and (*iii*) irregularly shaped node. But practically many other nodes are possible and the method to be used to obtain the equation connecting the temperatures is similar to the method adopted in the above case. Some of the important cases are given in data books and hand books. Some of these are also dealt with in the worked examples.

Example 5.5: *Derive the nodal equation for the element shown in Fig. Ex. 5.5.*

Solution: The total heat flow through the surfaces into the volume containing the node is equated to zero. The conduction surfaces are *AB*. *BC*, *DE* and *EA*. The convection surface is *D-m, n*. The convection temperature difference is assumed as $(T_\infty - T_{m,n})$. This yields

$$\frac{(T_{m-1,n}-T_{m,n})}{\Delta x}.\Delta y.k1+\frac{T_{m,n+1}-T_{m,n}}{\Delta y}\Delta xk.1+\frac{T_{m+1,n}-T_{m,n}}{\Delta x}\frac{\Delta y}{2}k1$$
$$+\frac{T_{m,n-1}-T_{m,n}}{\Delta y}.\frac{\Delta x}{2}.k1+h\frac{\Delta y}{2}1.(T_\infty-T_{m,n})=0$$

Simplifying, we get

$$2T_{m-1,n}+2T_{m,n+1}+T_{m+1,n}+T_{m,n-1}-T_{m,n}\left(6+\frac{h\Delta y}{k}\right)+\frac{h}{k}\Delta y.T_\infty=0 \qquad \text{...[5.22 (d)]}$$

Similarly even radiation boundary can be considered.

The numerical method is flexible and can be used to analyse any type of boundary and also three dimensional flow. In three dimensional flow, two more surfaces for heat flow will be there and here also the nodal temperature will be the average of adjacent six nodal temperatures in the case of internal nodes. For other nodes, the areas, and flows have to be considered carefully to arrive at the nodal equations. This tedius process has been to some extent simplified in the finite element softwares and hence time need not be wasted in detailing. Very powerful softwares are replacing the present methods of solution of two and three dimensional heat conduction problems.

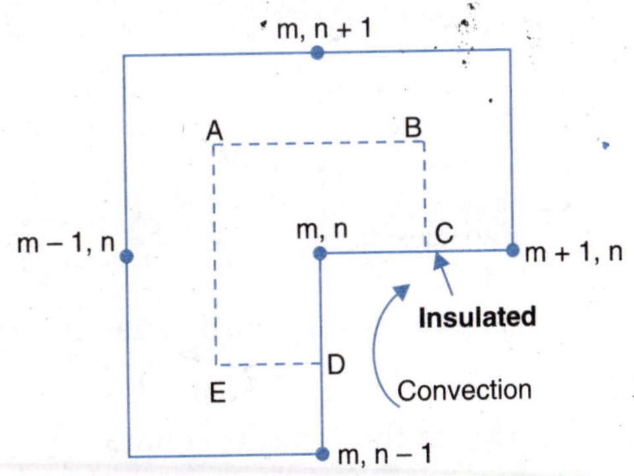

Fig. Ex. 5.5. Internal corner node

Chapter 5

5.4 ELECTRICAL ANALOGY

The flux plot by judgement is a tedious process and is to some extent subjective. As any potential is similar to temperature potential, electrically conducting papers and measuring instruments can be used to plot equivalent voltage and flow lines which can be used to calculate the shape factor more accurately. But this method is also obsolete.

5.5 THE FINITE DIFFERENCE FORMULATION

The temperature gradients are written as follows. Considering node (m,n) surrounded by nodes $(m + 1, n)$, $(m -1, n)$, $(m, n+1)$ and $m\ (n - 1)$ with nodal distances of $\Delta x = \Delta y$.

$$\left.\frac{\partial T}{\partial x}\right|_{m+\frac{1}{2}n} \simeq \frac{T_{m+1,n} - T_{m,n}}{\Delta x}$$

$$\left.\frac{\partial T}{\partial x}\right|_{m-\frac{1}{2},n} \simeq \frac{T_{m,n} - T_{m-1,n}}{\Delta x}$$

$$\left.\frac{\partial T}{\partial y}\right|_{m,n+\frac{1}{2}} \simeq \frac{T_{m,n+1} - T_{m,n}}{\Delta y}$$

$$\left.\frac{\partial T}{\partial y}\right|_{m,n-\frac{1}{2}} \simeq \frac{T_{m,n} - T_{m,n-1}}{\Delta y}$$

$$\left.\frac{\partial^2 T}{\partial x^2}\right|_{m,n} \simeq \frac{\left.\frac{\partial T}{\partial x}\right|_{m+\frac{1}{2},n} - \left.\frac{\partial T}{\partial x}\right|_{m-\frac{1}{2},n}}{\Delta x} = \frac{T_{m+1,n} + T_{m-1,n} - T_{m,n}}{\Delta x^2}$$

$$\left.\frac{\partial^2 T}{\partial y^2}\right|_{m,n} \simeq \frac{\left.\frac{\partial T}{\partial y}\right|_{m,n+\frac{1}{2}} - \left.\frac{\partial T}{\partial y}\right|_{m,n-\frac{1}{2}}}{\Delta y} = \frac{T_{m,n+1} + T_{m,n-1} - 2T_{m,n}}{\Delta y^2}$$

The equation $\quad \dfrac{\partial^2 T}{\partial x^2} + \dfrac{\partial^2 T}{\partial y^2} = 0$

Reduces to $\quad \dfrac{T_{m+1,n} + T_{m-1,n} - 2T_{mn}}{\Delta x^2} + \dfrac{T_{m,n+1} + T_{m,n-1} - 2T_{m,n}}{\Delta y^2} = 0$

As $\Delta x = \Delta y$ (assumed), the final result is

$$T_{m+1,\,n} + T_{m-1,\,n} + T_{m,\,n+1} + T_{m,\,n-1} - 4T_{m,\,n} = 0.$$

This is the same as equation [5.6 (*a*)]

If heat generation is involved or unsteady conditions prevail, the proper differential equation should be converted to difference equation as in the above example.

ADVANCED EXAMPLES

Example A-5.1: *For the boundary conditions for the plate shown in Fig. AE-5.1 determine using analytical method the temperature at the midpoint p, under steady two dimensional conduction. (use upto 5 terms in the series summation).*

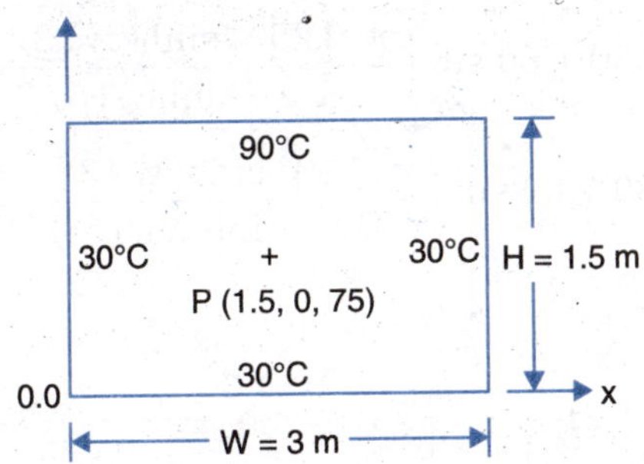

Fig. AE-5.1. Problem model

Solution: Using equation (5.5)

$$\frac{T-T_1}{T_2-T_1}=\frac{2}{\pi}\sum_{n=1}^{\infty}\frac{(-1)^{n+1}+1}{n}\sin\frac{n\pi x}{W}.\frac{\sinh(n\pi y/W)}{\sinh(n\pi H/W)}$$

$$\frac{x}{W}=0.5,\frac{y}{W}=0.25,\frac{H}{W}=0.5$$

Substituting and noting that odd terms are zero

$$\frac{T-30}{90-30}=\frac{2}{\pi}\left[2\sin\frac{\pi}{2}.\frac{\sinh(\pi/4)}{\sinh(\pi/2)}+\frac{2}{3}\sin(1.5\pi).\frac{\sinh(0.75\pi)}{\sinh(1.5\pi)}\right.$$

$$\left.+\frac{2}{5}\sin(2.5\pi).\frac{\sinh(1.25\pi)}{\sinh(2.5\pi)}+\frac{2}{7}\sin(3.5\pi)\frac{\sinh(1.75\pi)}{\sinh(3.5\pi)}+\frac{2}{9}\sin(4.5\pi).\frac{\sinh(2.25\pi)}{\sinh(4.5\pi)}\right]$$

$$=\frac{2}{\pi}[0.7549-0.0626+0.0078-0.0012+0.0002]$$

$$=0.4451$$

$\therefore$ $\mathbf{T_p = 56.71°C}$

More terms should be considered for points near the surface.

Example A-5.2: *Use the data in example A-5.1 except that the top surface temperature is given by T(x) = 30 + 60 sin (πx/3) where x is in m and T(x) is in °C. Draw the contour for T = 60°C.*

Solution: Fig. AE-5.1 and equation (5.14) are used.

First the location for 60°C is found on the edge $y = 1.5$.

$$T_{(x,y)}=T_1+T_m\sin\frac{\pi x}{W}.\frac{\sinh(\pi y/W)}{\sinh(\pi H/W)}$$

$$\mathbf{y = 1.5, H = 1.5,}\ W=3,\ T_1=30°C,\ T_m=60°C$$

$$60=30+60\sin\frac{\pi x}{3}.\frac{\sinh(\pi 1.5/3)}{\sinh(\pi 1.5/3)}\quad\therefore\ \mathbf{x = 0.5\ m}$$

Now taking x values of 0.75. 1, 1.25 and 1.5. y values are determined

$$\mathbf{x = 0.75:}\ 60 = 30 + 60 \sin\left(\frac{\pi.0.75}{3}\right)\frac{\sinh(\pi y/3)}{\sinh(\pi 1.5/3)} \qquad \therefore \quad \mathbf{y = 1.21\ m}$$

$$\mathbf{x = 1.00:}\ 60 = 30 + 60 \sin\left(\frac{\pi.\ 1}{3}\right)\frac{\sinh(\pi y/3)}{\sinh(\pi\ 1.5/3)} \qquad \therefore \quad \mathbf{y = 1.05\ m}$$

$$\mathbf{x = 1.25:}\ 60 = 30 + 60 \sin\left(\frac{\pi\ .\ 1.25}{3}\right)\frac{\sinh(\pi y/3)}{\sinh(\pi\ 1.5/3)} \qquad \therefore \quad \mathbf{y = 0.97\ m}$$

$$\mathbf{x = 1.5:}\ 60 = 30 + 60 \sin\left(\frac{\pi\ .\ 1.5}{3}\right)\frac{\sinh(\pi y/3)}{\sinh(\pi\ 1.5/3)} \qquad \therefore \quad \mathbf{y = 0.94\ m}$$

This is ploted in Fig. AE-5.2.

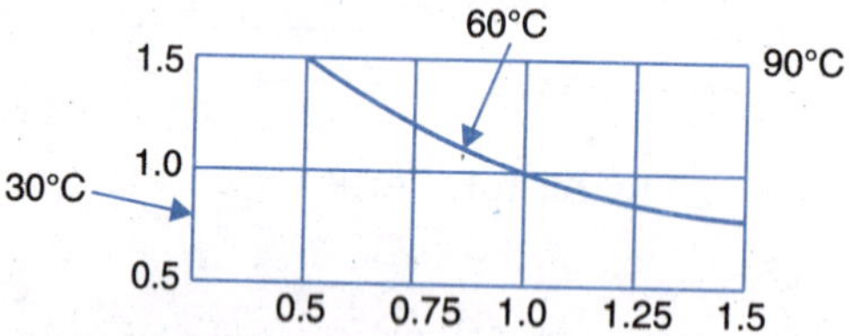

Fig. AE-5.2. Contour for 60°C

Example A-5.3: *A long pipe of outside diameter of 0.5 m is buried in earth with axis at a depth of 1.5 m. The surface of the pipe is at 85°C and the surface temperature of the earth is 20°C. Determine the heat loss from the pipe per m length. The conductivity of the earth is 0.52 W/mK.*

The data are shown in Fig. AE-5.3.

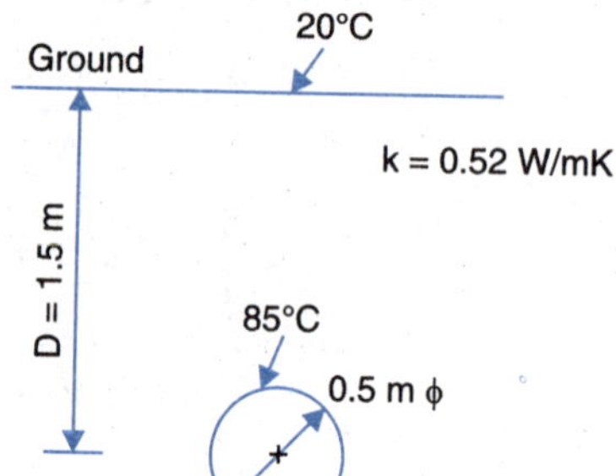

Fig. AE-5.3. Problem model

Solution: This problem is solved by using the shape factor concept. The shape factor is available in handbooks and texts. In this case the shape factor is given by

As $D > 3r$
$$S = \frac{2\pi L}{\ln(2D/r)} = \frac{2\pi \times 1}{\ln(2 \times 1.5/0.25)}$$

$$= 2.529\ \text{m} = \mathbf{2.529\ m}$$

$$\therefore \quad Q = kS\,\Delta T = 0.52 \times 2.529 \times (85 - 20)$$

$$= \mathbf{85.48\ W/m\ length.}$$

Example A-5.4: *If the pipe of example A-5.3 is buried with axis at a depth of 0.5 m, determine the heat loss/m length.*

Solution: In this case $D < 3r$, From handbook

$$\therefore \quad S = 2\pi L/\cosh^{-1}(D/r) = 2\pi \times 1/\cosh^{-1}(0.5/0.25) = \mathbf{4.77}$$

$$\therefore \quad Q = kS\,\Delta T = 0.52 \times 4.77 \times (85 - 20) = \mathbf{161.23\ W/m\ length}$$

If the equation as in example A-5.3 is used, Q = 153.19 W, an under estimate.

xample A-5.5: *If the pipe in Fig. AE-5.3 buried at a much deeper level like 4 m, then determine ıe heat loss.*

olution: In this case $D >> r$ and so also $L >> D$

$$\therefore \quad S = \frac{2\pi L}{\ln\left(\frac{L}{r}\right)\left\{\frac{\ln(L/2D)}{1-\ln(L/r)}\right\}} = \frac{2\pi . 1}{\ln\left(\frac{1}{0.25}\right)\left\{\frac{\ln(1/2\times 4)}{1-\ln(1/0.25)}\right\}} = \mathbf{2.05\ m}$$

$$\mathbf{Q} = kS\,\Delta T = 2.05 \times 0.52 \times 65 = \mathbf{69.19\ W/m.}$$

xample A-5.6: *A pipe of OD.0.5 m carrying steam at 250°C is insulated with glass wool f k = 0.058 W/mK for a thickness of 0.1 m and is buried in soil with centre line at a depth of .5 m. The soil surface is at 0°C. Determine the heat flow per m length under steady onditions. Also determine the temperature on the outside of insulation. k = 0.52 W/mK for the oil.*

olution: The data are shown in Fig. AE-5.6.

In this case two resistances are involved one due to the shape factor for the outer cylinder nd the other resistance due to the insulation.

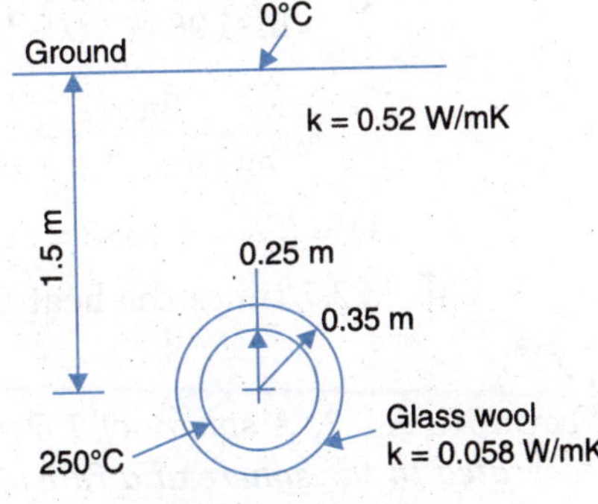

Fig. AE-5.6. Problem model

$$Q = \frac{250-0}{R_1 + R_2}$$

$$R_2 = \frac{1}{ks},\ R_1 = \frac{\ln\frac{r_2}{r_1}}{2\pi k_w L}$$

$$S = 2\pi l/\ln(2D/r_2)$$
$$= 2\pi \times 1/\ln(2 \times 1.5/0.35) = 2.925$$

$$\therefore \quad R_2 = \frac{1}{0.52\times 2.925} = 0.6576°\text{C/W}$$

$$R_1 = \ln\frac{0.35}{0.25} / 2\times\pi\times 0.058\times 1 = 0.9233°\text{C/W}$$

$$\therefore \quad Q = \frac{250-0}{0.6576+0.923} = \mathbf{158.14\ W/m\ length}$$

The temperature on the outside surface of insulation:

$$Q = \frac{250-T}{\ln\frac{r_2}{r_1}/2\pi k_w l} = \frac{250-T}{0.9233}$$

$$\therefore \quad \mathbf{T = 104°C.}$$

Example A-5.7: *A long pipe of OD 0.6 m passes centrally inside in a concrete wall of 1.2 m thickness, the wall height is large compared to thickness. If the pipe is at 120°C and the concrete surfaces are at 5°C, determine the heat loss per m length. The conductivity of the concrete is 0.8 W/mK.*

Chapter 5

Solution: In this problem also, the shape factor is to be used. The data are shown in Fig. AE-5.7

From handbooks

$$\text{for} \quad \frac{D}{2} > r, S = \frac{2\pi L}{\ln(4D/2r)}$$

$$= 2\pi \times 1/\ln(1.2 \times 4/0.6)$$

$$= \mathbf{3.02}$$

$$\therefore \quad \mathbf{Q} = kS\,\Delta T = 0.8 \times 3.02 \times (120 - 5)$$

$$= \mathbf{277.84\ W/m.}$$

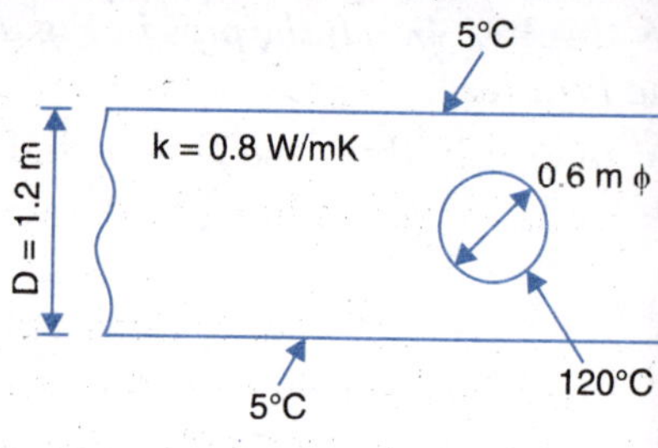

Fig. AE-5.7. Problem model

Example A-5.8: *If in example A-5.8, the pipe is buried in concrete 1.2 × 1.2 m square section determine the heat loss/ m length.*

Solution: The situation is represented in Fig. AE-5.8

$$S = \frac{2\pi L}{\ln(1.08\,W/D)}$$

$$= \frac{2\pi \times 1}{\ln(1.08 \times 1.2/0.6)} = \mathbf{8.1588}$$

$$\therefore \quad \mathbf{Q} = 0.8 \times 8.1588 \times (120 - 5) = \mathbf{750.61\ W/m}$$

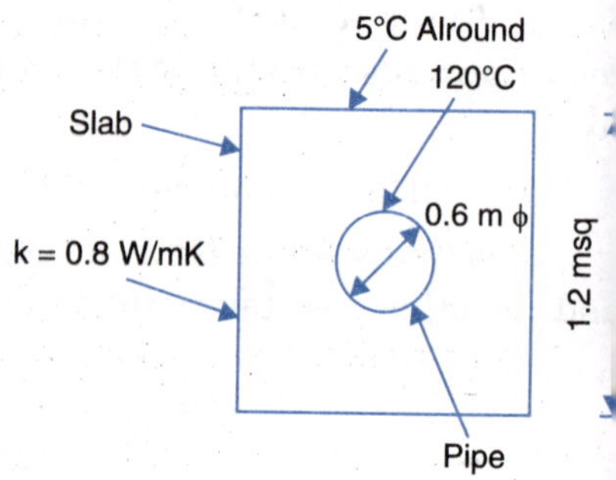

Fig. AE-5.8. Problem model

This is 2.7 times the heat flow of Fig. AE-5.7 due to the reduction of resistance on two sides.

Example A-5.9: *A sphere of 1.5 m dia is buried in soil with centre at a depth of 6 m. Heat is generated in the sphere at a rate of 600 W. The soil surface is at 5°C and the conductivity of the soil is 0.52 W/mK. Determine the surface temperature of the sphere under steady state condition.*

Solution: The situation is represented in Fig. AE-5.9. Under steady state, heat generated = head conducted away = $kS\Delta T$

From handbooks:

$$S = 4\pi r/(1 - (r/2D)) = \frac{4 \times \pi \times 0.75}{(1 - (0.75/2 \times 6))}$$

$$= \mathbf{10.053}$$

$$\therefore \quad \mathbf{Q} = kS\,\Delta T = 0.52 \times 10.053 \times (T - 5)$$

$$= 600\ \text{W}$$

∴ The surface temperature

$$\mathbf{T = 119.8°C.}$$

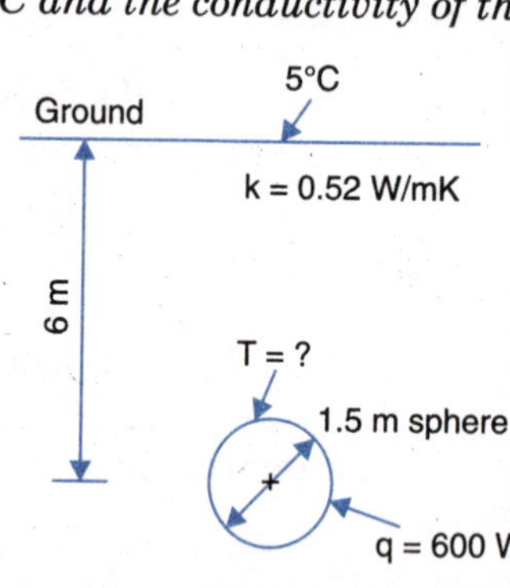

Fig. AE-5.9. Problem model

Example A-5.10: *A rectangular pipe with a surface temperature of 120°C is buried in soil as shown in Fig. AE-5.10. Determine the heat loss per m length. The soil surface is at 0°C and its conductivity is 0.52 W/mK.*

Solution: The shape factor is given by

$$S = 1.675L\left[\log\left(1+\frac{b}{a}\right)\right]^{-0.59}\left[\frac{D}{C}\right]^{-0.078}$$

$$= 1.685 \times 1\left[\log\left(1+\frac{1.0}{0.4}\right)\right]^{-0.59}\left[\frac{1}{0.3}\right]^{-0.078}$$

$$= \mathbf{2.1967}$$

$$\therefore \quad \mathbf{Q} = 0.52 \times 2.1967\,(120 - 0)$$

$$= \mathbf{137.08\ W/m\ length.}$$

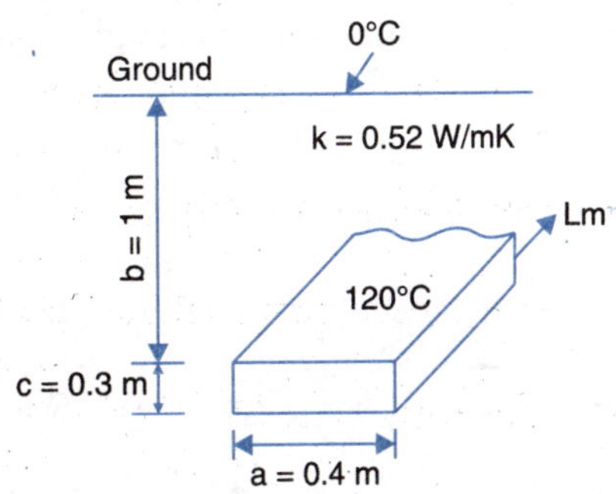

Fig. AE-5.10. Problem model

Example A-5.11: *A casting in the form of a hemisphere of radius 0.4 m is in a sand mould on the ground with the circular face parallel to the ground and on the surface as shown in Fig. AE-5.11. The surface temperature of the casting is 900°C and the soil temperature is 60°C. Determine the heat loss to the soil.*

Fig. AE-5.11. Problem model

Solution: $k = 0.6$ W/mK.

The shape factor $S = 2\pi r$

$$\therefore \quad \mathbf{Q} = 0.6 \times 2 \times \pi \times 0.4\,(900 - 60)$$

$$= \mathbf{1266.7\ W}$$

Example A-5.12: *A spent nuclear material in the form of a thin disk and dia 0.4 m generating heat at 1 kW is buried in soil at a depth of 2 m. Determine the surface temperature of the disk. Conductivity of the soil is 0.6 W/mk. Soil surface is at 30°C.*

Solution: The shape factor when depth is greater than diameter is given by $8R$. Where R is the radius of the disk $S = 8R$

$$\therefore \quad 1000 = 0.6 \times 8 \times 0.4\,(T - 30) \quad \therefore \mathbf{T = 550.83°C}$$

(**Note:** Depth of burial does not figure in the shape factor)

Example A-5.13: *Two long pipes one of OD 0.12 m carrying steam at 200°C and another of OD 0.08 m carrying water at 30°C are buried in soil at a centre distance of 0.6 m. The conductivity of the soil is 0.52 W/mK. Determine the heat flow between the pipes for 1 m length.*

Solution: The situation is shown in Fig. AE-5.13.

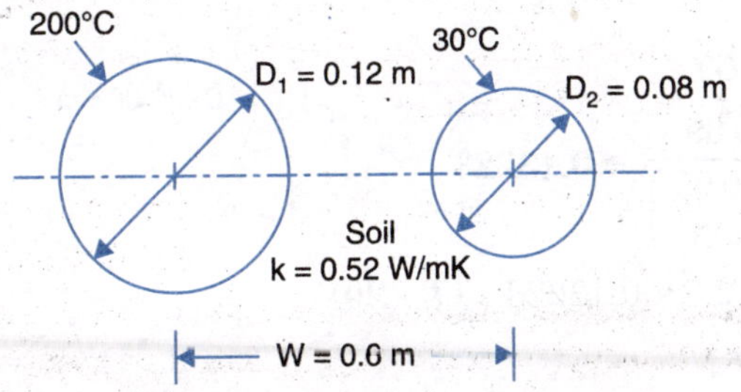

Fig. AE-5.13. Problem model

The shape factor is given by

$$S = \frac{2\pi L}{\cosh^{-1} \dfrac{(4W^2 - D_1^2 - D_2^2)}{2D_1 D_2}}$$

$$L = 1,$$

$$\therefore \quad S = \frac{2\pi \times 1}{\cosh^{-1} \dfrac{4 \times 0.6^2 - 0.12^2 - 0.08^2}{2 \times 0.12 \times 0.08}} = \mathbf{1.25763}$$

$$\therefore \quad \mathbf{Q} = 1.25763 \times 0.52\,(200 - 30) = \mathbf{111.17\ W/m}$$

Example A-5.14: *Two spheres of 0.2 m dia and 0.1 m dia with surface temperatures of 400°C and 40°C are buried in soil at a centre distance of 1 m. The conductivity of the soil is 0.52 W/mK. Determine the heat exchange between the spheres.*

Solution: The distance between the spheres D is greater than $5 \times r_1$, the radius of the larger sphere. Under this condition the shape factor

$$S = \frac{4\pi r_2}{(r_2/r_1)\left[1 - \dfrac{(r_1/D)^4}{1 - (r_2/D)^2}\right] - [2r_2/D]}$$

Where r_2 is the radius of smaller sphere.

$$\therefore \quad S = \frac{4\pi\, 0.05}{(0.1/0.2)\left[1 - \dfrac{(0.1/1)^4}{1 - \left(\dfrac{0.05}{1}\right)^2}\right] - [2 \times 0.05/1]} = \mathbf{1.571\ m}$$

$$\therefore \quad \mathbf{Q} = 0.52 \times 1.571\,(400 - 40) = \mathbf{294.09\ W.}$$

Example A-5.15: *A heater of 5 mm dia and 90 mm length is fitted in a vertical hole in a metal with a conductivity of 12.5 W/mK, whose surface is at 20°C. If the heater dissipates 60 W, determine the temperature of the heater surface.*

Solution: The data is shown in Fig. AE-5.15.

From handbook, the shape factor S is obtained

$$S = \frac{2\pi L}{\ln(4L/D)}$$

$$= \frac{2\pi \times 0.09}{\ln\left(\dfrac{4 \times 0.09}{0.005}\right)} = \mathbf{0.13223}$$

and $$Q = kS\Delta T = 12.5 \times 0.13223 \times (T - 20)$$

$$= 60$$

$$\therefore \quad \mathbf{T = 56.3°C}$$

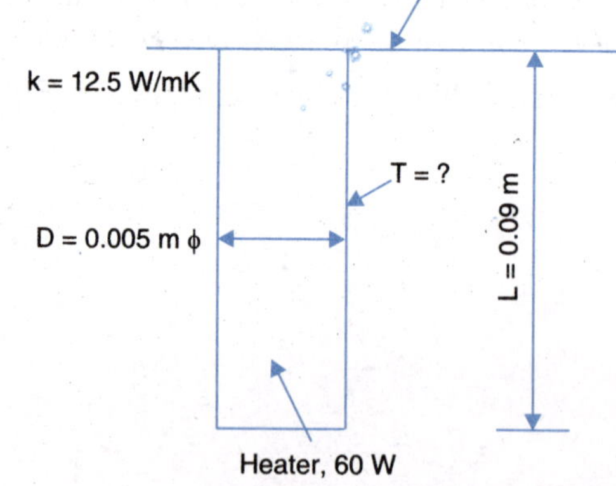

Fig. AE-5.15. Model

Example A-5.16: *A disk 40 mm dia and generating 200W is placed on a semi infinite medium with thermal conductivity of 200W/mK. If the temperature of the medium is 30°C, determine the temperature of the disk surface in contact with the medium.*

Solution: The shape factor for this situation is given as $2D$ where D is the diameter of the disk.

$$\therefore \quad Q = k2D\Delta T = 200 = 200 \times 2 \times 0.04 \times (T - 30)$$

$$\therefore \quad \mathbf{T = 42.5°C.}$$

Example A-5.17: *If in the example A-5.16, there is a contact resistance of 5×10^{-5} m^2 °C/W, determine the surface temperature.*

Solution: There are two resistances in this case, (*i*) contact resistance and (*ii*) conduction resistance.

These are: $\dfrac{1}{kS}$ and $5.0 \times 10^{-5}/(\pi D^2/4)$

As $\qquad \mathbf{S = 2D,}$

$$\therefore \quad 200 = \frac{T-30}{\dfrac{1}{200 \times 2 \times 0.04} + \dfrac{5 \times 10^{-5} \times 4}{\pi \times 0.04^2}} = \frac{T-30}{0.0625 + 0.0398}$$

$$\therefore \quad \mathbf{T = 50.46°C.}$$

Example A-5.18: *A furnace has inside dimensions of 1 m × 1.2 m × 1.5 m. The walls are 0.25 m thick. The inside surface is at 750°C while the outside surface is at 80°C. If the conductivity of the material is 0.45 W/mK determine the heat loss, taking into account the corner and edge effects and also the bottom. (shape factors are taken from handbooks).*

Solution: (*i*) The wall areas are: $2 \times 1 \times 1.2 + 2 \times 1.2 \times 1.5 + 2 \times 1 \times 1.5\ m^2 = 9\ m^2$

Shape factor for walls $= \dfrac{A}{L}$, (L – thickness)

$\therefore$ Heat flow through walls $= 0.45 \times \dfrac{9}{0.25} \times (750 - 80) = \mathbf{10854\ W}$

(*ii*) There are four edges each of 1 m, 1.2 m and 1.5 m length and all of 0.25 × 0.25 section. Total length = 4 + 4.8 + 6 = 14.8 m. The shape factor is $0.54D$ where D is the length of edges.

$$\therefore \quad Q_{edges} = 14.8 \times 0.54 \times 0.25 \times 0.45 \times (750 - 80) = \mathbf{602\ W}$$

(*iii*) There are 8 corners each of cubical shape of 0.25 m side. The shape factor is $0.15L$, where L is the length of side.

$$\therefore \quad Q_{corners} = 8 \times 0.15 \times 0.25 \times 0.45\ (750 - 80) = \mathbf{90.45\ W}$$

Total heat flow $= \mathbf{11546\ W}$

considering one dimensional flow with average area

$$2 \times 1.25 \times 1.45 + 2 \times 1.45 \times 1.75 + 2 \times 1.25 \times 1.75 = 13.075$$

$$Q = 0.45 \times \frac{13.075}{0.25}\ (750 - 80) = \mathbf{15768.45\ W}$$

Over estimates the heat flow by 27%.

Chapter 5

Example A-5.19: *Write down the nodal equation for an inside corner node with inside surface exposed to convection.*

Solution: The node is shown in Fig. AE-5.19.

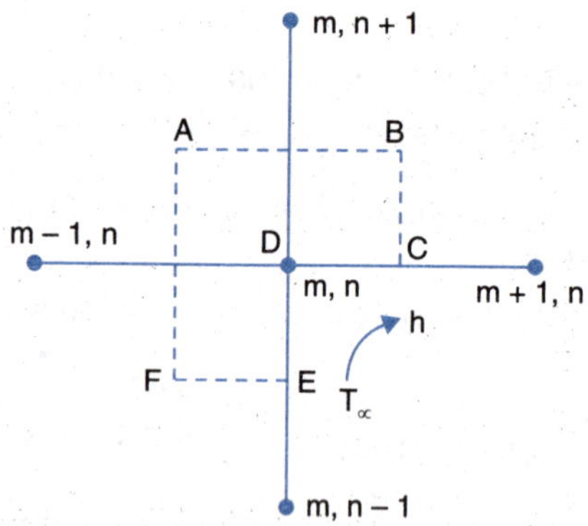

Fig. AE-5.19

Considering the node m,n

Heat flow across AB, BC, CD, DE, EF and FA are to be summed up. There is conduction across AB, BC, EF and FA and convection over CD and DE. Considering unit depth,

$$\therefore \quad \frac{k\Delta x(T_{m,n+1} - T_{m,n})}{\Delta y} + \frac{k\Delta y}{2}\frac{(T_{m+1,n} - T_{m,n})}{\Delta x} + \frac{h\Delta x}{2}(T_\infty - T_{m,n})$$

$$+ \frac{h\Delta y}{2}(T_\infty - T_{m,n}) + k\frac{\Delta x}{2}\frac{(T_{m,n-1} - T_{m,n})}{\Delta y} + k\Delta y\frac{(T_{m-1,n} - T_{m,n})}{\Delta x} = 0$$

Noting $\Delta x = \Delta y$. The equation is simplified as

$$2T_{m,\,n+1} + 2T_{m-1,\,n} + T_{m+1,\,n} + T_{m,\,n-1} - T_{m,\,n}\left[\frac{2h}{k}\Delta x + 6\right] + 2\left[\frac{h\Delta x}{k}\right]T_\infty = 0$$

This can be verified with the equation in texts.

$h\ \Delta x/k$ is considered as Biot number.

Example A-5.20: *Derive the nodal equation for an exterior corner node with convection on one face, the other face being adiabatic. Indicate modification for heat generation.*

Solution: The node is shown in Fig. AE-5.20.

Considering heat flow across AB, BC and CD, (AD is adiabatic)

$$k\frac{\Delta y}{2}\frac{T_{m-1,n} - T_{m,n}}{\Delta x} + k\frac{\Delta x}{2}\frac{T_{m,n-1} - T_{m,n}}{\Delta y} + h\frac{1}{2}\Delta y(T_\infty - T_{m,n})$$

$$= 0$$

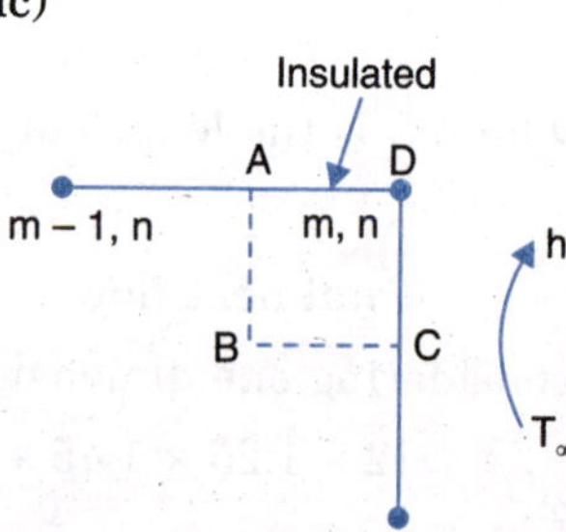

Fig. AE-5.20 Corner node

Noting $\Delta x = \Delta y$, simplification gives

$$2T_{m-1,n} + 2T_{m,n-1} - T_{m,n}\left[4 + \frac{h\Delta x}{k}\right] + T_\infty\frac{h\Delta x}{k} = 0 \quad ...(1)$$

If q is the volume heat generation rate, the heat generated in the element

$$= \frac{\Delta x}{2} \times \frac{\Delta y}{2} \times 1\, q = \frac{\Delta x^2}{4} q$$

This quantity should be added to the LHS of the equation 1 to take care of heat generation.

Example A-5.21: *Derive the nodal equation for the node m, n shown in Fig. AE-5.21.*

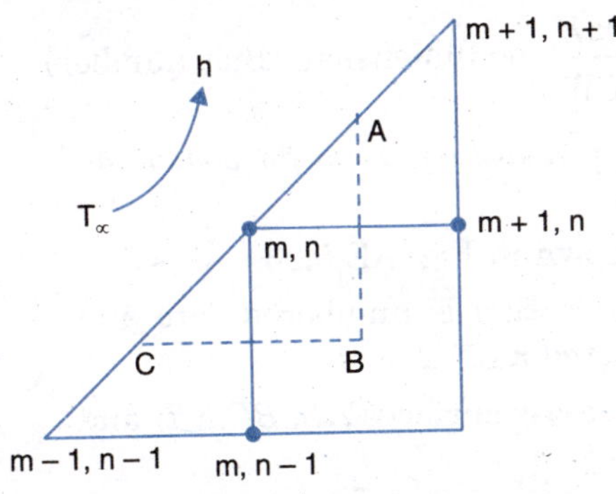

Fig. AE-5.21

Solution: The node m,n is enclosed by surfaces AB, BC, and CA. The surface CA is exposed to convection.

$$\frac{(T_{m+1,n} - T_{m,n})}{\Delta x} k\Delta y + \frac{(T_{m,n-1} - T_{m,n})}{\Delta y} .k\Delta x + h\sqrt{\Delta x^2 + \Delta y^2}\,(T_\infty - T_{m,n}) = 0$$

Noting $\Delta x = \Delta y$

$$T_{m+1,n} + T_{m,n-1} - T_{m,n}\left(2 + \frac{\sqrt{2}\,\Delta x h}{k}\right) + \frac{\sqrt{2}\,\Delta x h}{k} T_\infty = 0$$

If heat generation is present add $\frac{\Delta x . \Delta y}{2} q$ to the LHS.

Example A-5.22: *A tool tip as shown in Fig. AE-5.22 has a uniform heat flux of q W/m^2 on the top surface. Formulate the nodal equation for the tip. Assume a constant thickness.*

Solution: The node is enclosed by surface AB, BC, and CA

AB receives uniform flux q W/m^2

BC-conduction is present across this face

CA-convection over the face.

Heat balance yields: (assuming thickness t)

$$t \times \frac{\Delta x}{2} . q + \frac{T_{m+1,n} - T_{m,n}}{\Delta x} . \frac{\Delta y}{2} t\, k$$

$$+ h.t \frac{1}{2} \sqrt{\Delta x^2 + \Delta y^2}\,(T_\infty - T_{m,n}) = 0$$

Cancelling t and noting $\Delta x = \Delta y$

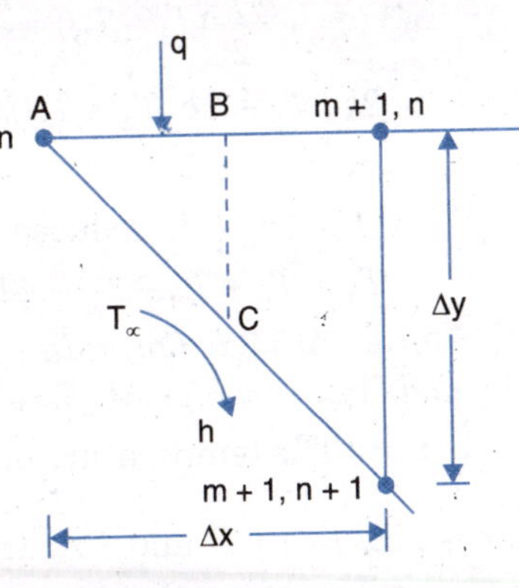

Fig. AE-5.22. Model

Chapter 5

$$T_{m+1,n} + \frac{q\Delta x}{k} - T_{m,n}\left(1 + \frac{\Delta x\sqrt{2}.h.}{k}\right) + \frac{\Delta xh\sqrt{2}}{k}.T_\infty = 0$$

Note: Check dimensions in each term for example

$$\frac{q}{k}\Delta x \rightarrow \frac{W}{m^2}.\frac{mK}{W}.m \text{ leading to } K$$

$$\frac{\Delta xh}{k} \rightarrow \frac{m.W\,mK}{m^2\,K\,W} \text{ no dimension. (Biot number)}$$

Example A-5.23: *Formulate a nodal equation for a node on the interface of two materials.*

Solution: The situation is shown in Fig. AE-5.23

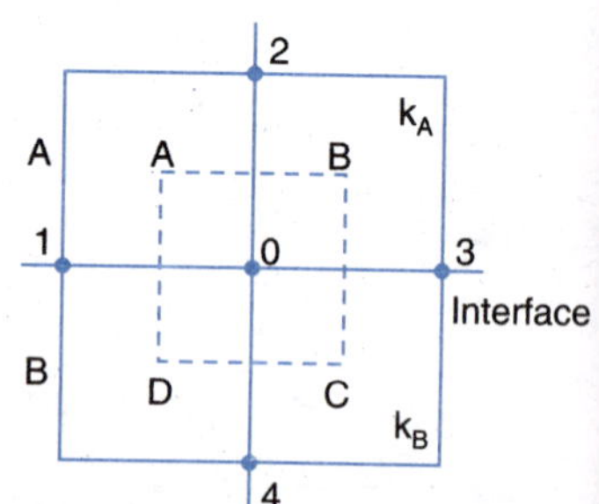

Fig. AE-5.23. Model of node

The node under consideration is numbered zero and the adjacent nodes are numbered as 1, 2, 3, 4.

Considering heat flow across surfaces *AB, BC, CD* and *AD* (Unit depth)

(*i*) Conduction across *AB* is in a single material *A*

(*ii*) Conduction across *BC* is half in material *A* and half in *B*

(*iii*) Conduction across *CD* is in material *b*

(*iv*) Conduction across *DA* is half in material *A* and half in *B*.

$$k_A\frac{\Delta x}{\Delta y}(T_2 - T_0) + k_A\frac{\Delta y}{2\Delta x}(T_3 - T_0) + k_B\frac{\Delta y}{2\Delta x}(T_3 - T_0)$$

$$+ k_B\frac{\Delta x}{\Delta y}(T_4 - T_0) + k_B\frac{\Delta y}{2\Delta x}(T_1 - T_0) + k_A\frac{\Delta y}{2\Delta x}(T_1 - T_0) = 0$$

Cancelling Δx and Δy as equal,

$$k_A(T_2 - T_0) + \frac{k_A}{2}(T_3 - T_0) + \frac{k_B}{2}(T_3 - T_0) + k_B(T_4 - T_0)$$

$$+ \frac{k_B}{2}(T_1 - T_0) + \frac{k_A}{2}(T_1 - T_0) = 0$$

$$2k_A T_2 + 2k_B T_4 + T_3(k_A + k_B) + T_1(k_A + k_B) - T_0(4k_A + 4k_B) = 0$$

If $k_A = k_B$ this reduces to

$T_1 + T_2 + T_3 + T_4 - 4T_0 = 0$ as a check.

Example A-5.24: *Formulate from basics the nodal equation for node m, n in Fig. AE-5.24.*

Solution: The temperature on the slant face is T_s. Considering faces *AB, BC, CD* and *DA*, these are all of length $\frac{3}{4}\Delta x$ and distances from T_s to $T_{m,n}$ is $\frac{1}{2}\Delta x$.

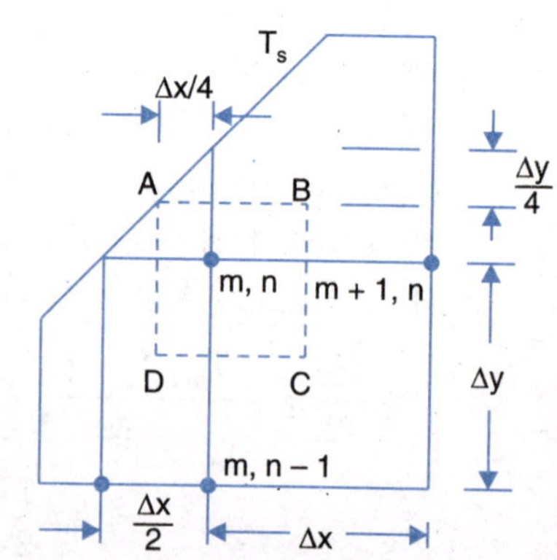

Fig. AE-5.24. Node

Writing the energy balance equation,

$$k\frac{3}{4}\Delta x\frac{(T_s - T_{m,n})}{(\Delta y/2)} + k\frac{3}{4}\Delta y\frac{(T_{m+1,n} - T_{m,n})}{\Delta x}$$

$$+ k\frac{3}{4}\Delta x\frac{(T_{m,n-1} - T_{m,n})}{\Delta y} + k\frac{3}{4}\Delta y\frac{(T_s - T_{m,n})}{(\Delta x/2)} = 0$$

Cancelling k, Δx and $\frac{3}{4}$

$$2(T_s - T_{m,n}) + (T_{m+1,n} - T_{m,n}) + (T_{m,n-1} - T_{m,n}) + 2(T_s - T_{m,n}) = 0$$

$\therefore$ $$T_{m+1,n} + T_{m,n-1} + 4T_s - 6T_{m,n} = 0$$

This can be verified using tabulated values for a general node the slant face cutting the axes at a. Δx and b. Δy.

Example A-5.25: *The temperature distribution and boundary condition in part of a solid is shown in Fig. AE-5.25. Determine the temperatures at nodes marked A, B and C. Determine the heat convected over surface exposed to convection. k = 1.5 W/mK.*

Solution: Considering A

$$k\,0.1\frac{(137 - T_A)}{0.1} + \frac{(172.9 - T_A)}{0.1}k \times 0.1$$

$$+ \frac{(200 - T_A)}{0.1} + k \times 0.1 + \frac{(132.8 - T_A)}{0.1} \times k \times 0.1 = 0$$

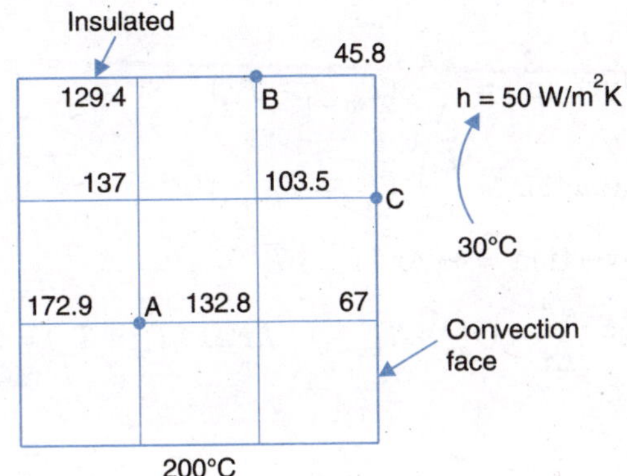

Fig. AE-5.25

$\therefore$ $$137 + 172.9 + 200 + 132.8 = 4T_A$$

$\therefore$ $$\mathbf{T_A = 160.68°C}$$

Considering B

$$k \times 0.05\frac{(129.4 - T_B)}{0.1} + \frac{(103.5 - T_B)}{0.1}k \times 0.1 + k \times 0.05\frac{45.8 - T_B}{0.1} = 0$$

$\therefore$ $129.4 + 2 \times 103.5 + 45.8 = 4T_B$ $\quad\therefore$ $\mathbf{T_B = 95.55°C}$

Chapter 5

Considering C

$$\frac{1.5 \times 0.05\,(45.8 - T_C)}{0.1} + \frac{(103.5 - T_C) \times 0.1 \times 1.5}{0.1} + \frac{(67 - T_C) \times 1.5 \times 0.05}{0.1} + (30 - T_C)\,50 \times 0.1 = 0$$

$\therefore \quad T_C = \mathbf{48.7°C}$

Heat flow by convection over the surface

$$= 50\,(200 - 30) \times 0.05 + 50 \times 0.1\,(67 - 30) + 50 \times 0.1\,(48.7 - 30) + 50 \times 0.05\,(45.8 - 30) = \mathbf{743.15\ W.}$$

Example A-5.26. *Show that for the node m shown in Fig. AE-5.26, the nodal equation can be written as* $T_{m-1} + T_{m+1} - \left(\frac{hP\Delta x^2}{kA} + 2\right) T_m + \frac{hP\Delta x^2}{kA} T_\infty = 0$

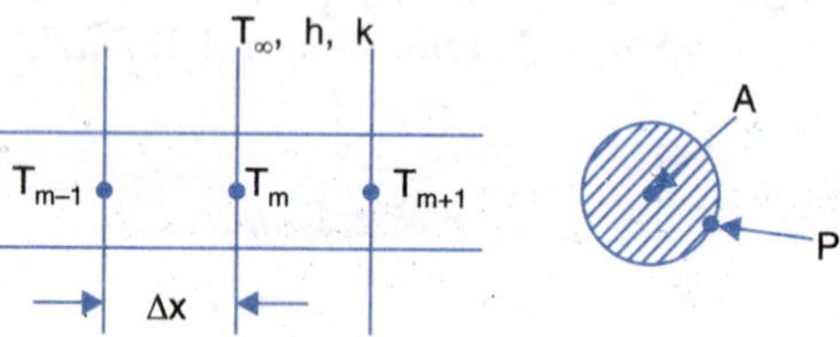

Fig. AE-5.26

Solution:

Heat flow in by conduction $= \frac{kA}{\Delta x}(T_{m-1} - T_m)$

Heat flow out by conduction $= \frac{kA}{\Delta x}(T_m - T_{m+1})$

Heat flow out by convection $= hP\Delta x\,(T_m - T_\infty)$

$$\therefore \quad \frac{kA}{\Delta x}(T_{m-1} - T_m) - \frac{kA}{\Delta x}\,.\,(T_m - T_{m+1}) - hP\Delta x\,(T_m - T_\infty) = 0$$

Dividing by $(kA/\Delta x)$

$$T_{m-1} - T_m - T_m + T_{m+1} - \frac{hP\Delta x^2}{kA} T_m + \frac{hP\Delta x^2}{kA} T_\infty = 0$$

or

$$T_{m-1} + T_{m+1} - T_m\left(\frac{hP\Delta x^2}{kA} + 2\right) + \frac{hP\Delta x^2}{kA}\,.\,T_\infty = 0.$$

REVIEW QUESTIONS

5.1 List and compare the various methods used to analyse two dimensional conduction.

5.2 What are constant temperature lines or isotherms?

5.3 What are adiabatic lines? How do they intersect isotherms?

5.4 What are heat flow lanes?

5.5 What is the advantage numerical method for use in two dimensional conduction?

5.6 What is conduction shape factor? How is it used to determine heat flow in two dimensional heat conduction?

EXERCISE PROBLEMS

5.1 A rectangle 0.5 m × 1 m has both the 1 m sides and one 0.5 m side at 200°C. The other side is having a temperature distribution given by $T = 200 + 400 \sin (\pi x/0.5)$ where x is in m and T in °C. Locate the y values at $x = 0.5$ m at which the temperatures will be 300, 400, 500°C. Also locate the values of x for $y = 1$ m at which these temperatures occur.

5.2 For the plate in problem 5.1 draw the contour for 350°C.

5.3 In problem 5.1 determine the temperatures at point p (0.45, 0.5) and determine the heat conducted out across the plane $x = 0.475$ between the locations $y = 0.45$ and $y = 0.55$. Take $k = 40$W/mK.

5.4 A plate of size 0.5 m × 1 m has three of the sides at 200°C while the fourth side of 0.5 m length is held at 600°C. Determine the temperatures at location P (0.45, 0.5) and calculate the amount of heat conducted along the x direction across the plane $x = 0.475$ between the locations $y = 0.45$ and $y = 0.55$.

5.5 A pipe carrying oil and having a diameter of 0.6 m is buried in soil of conductivity 0.6 W/mK at a depth of 6 m. The surface temperature of the pipe is 80°C. The surface of the soil is at –10°C. Determine the heat loss from the pipe for 1 m length. If the velocity is 2 m/s and the density is 900 kg/m^3 and specific heat 2000 J/kgK determine the temperature drop in flow through a distance of 100 m.

5.6 In problem 5.5, if the pipe is insulated with a 10 cm layer of insulation of thermal conductivity 0.04 W/mK determine the heat loss per m and also the temperature on the out side surface of insulation.

5.7 A sphere containing spent nuclear material and of diameter 0.6 m is buried in soil at a depth of 10 m. The soil thermal conductivity is 0.6 W/mK. If the heat generated in the sphere is 10^6 W/m^3, determine the surface temperature of the sphere. The soil surface temperature is 10°C.

5.8 A pipe carrying steam and of *OD* 0.15 m is embedded centrally in a concrete wall of thickness 0.4 m. The conductivity of the material is 1.4 W/mK. If the surface temperature of the pipe is 200°C and the surface temperature of the wall is 10°C, determine the heat loss from the pipe for a length of 6 m.

5.9 If in the problem 5.8 the pipe is encased in a concrete block of 0.4 × 0.4 m centrally, determine the heat loss per m length.

5.10 A pipe carrying steam at 250°C with and *OD* of 0.2 m is insulated using a material with conductivity of 0.085 W/mK. The insulation is off-set from the axis by 0.05 m and the *OD* of the insulation is 0.4 m. Determine the heat loss from the pipe if the outside surface temperature of the insulation is 60°C.

5.11 A pipe carrying hot water at 90°C and having an outer diameter of 0.4 m is buried parallel with a water pipe of diameter of 0.3 m at a centre distance of 0.8 m. The conductivity of the soil is 0.65 W/mK. If the water is at 15°C determine the heat gain by the water pipe.

5.12 A hole in a large metal piece is to be heat treated. The hole is 5 mm in dia and 65 mm in depth. The hole surface should be maintained at 220°C. Determine the heat input necessary. The metal surface is at 40°C and the conductivity of the metal is 30 W/mK.

5.13 A sphere of 0.35 m dia and another of 0.25 m dia are buried at a centre distance of 0.85 m in soil. The larger sphere is at 250°C and the smaller one is at 30°. If the conductivity of the soil is 0.52 W/mK, determine the heat flow from the larger to the smaller sphere.

5.14 An area of dia 50 mm in a large metal surface is to be maintained at 300°C. The metal is at 30°C. Determine the rating for the heater. The conductivity of the metal is 130 W/mK.

5.15 A furnace has internal dimensions of 2 m × 1.5 m × 1.2 m and the wall thickness is 0.2 m. The inner surfaces are at 650°C and the outer surface is at 80°C. If the conductivity of the wall material is 0.4 W/mK. Determine the heat loss through the walls and the floor.

5.16 Write down the nodal equation for the nodes shown in Fig. 5.16 (*a*), (*b*), and (*c*).

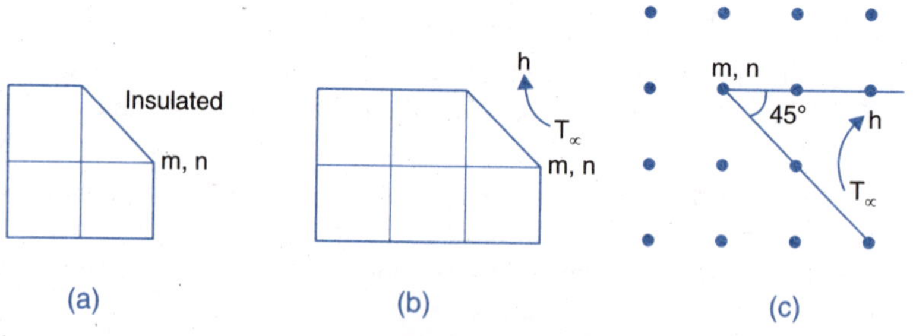

Fig. 5.16

5.17 Write the nodal equation for a corner node when corner is in contact with another material as in Fig. 5.17.

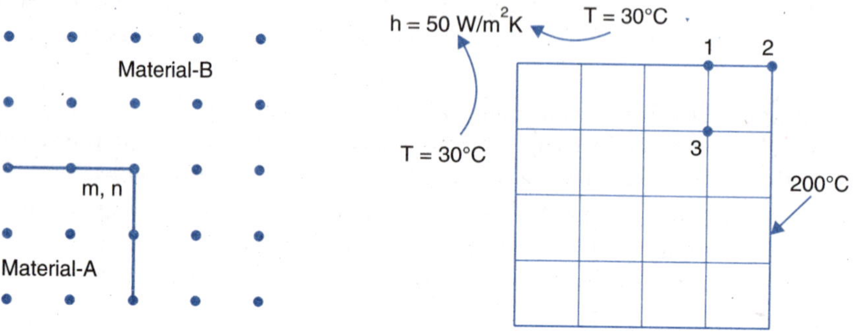

Fig. 5.17 **Fig. 5.18**

5.18 Write down the nodal equations for the nodes shown in Fig. 5.18 for nodes 1, 2 and 3.

5.19 The temperature in a plate at the nodes indicated are shown in Fig. 5.19. Determine the heat flow over the surfaces.

The conductivity is 3 W/m°C, Top surface is exposed to convection at 50°C with h = 50 W/m²K

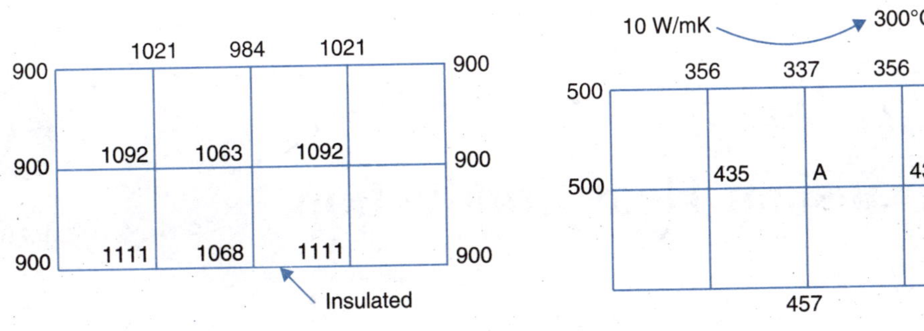

Fig. 5.19 Fig. 5.20

5.20 A part of a solid with temperatures at the nodes and the boundaries are shown in Fig. 5.20. Determine the temperature at node A and also the heat flow over the convecting surface. The top surface is exposed to convection at 300°C with h = 10 W/m^{2}K. [**Ans.** 416°C, 3490 W/m^2]

5.21 A rod 15 mm dia and 200 mm long has a heat generation rate of 10^6 W/m^3. The rod is exposed to convection over the surface to air at 40°C with h = 20 W/m^{2}C. The conductivity of the material is 5 W/mK. Both ends are at 40°C. Write down nodal equations taking 20 mm nodal distances and solve for the temperatures.

6 CHAPTER

Transient Heat Conduction

6.0 INTRODUCTION

Heat transfer equipments operating at steady state is only one phase of their functioning These have to be started and shut down as well as their performance level may have to be altered as per external requirements. A heat exchanger will have to operate at different capacities. This changes the conditions at the boundary of heat transfer surfaces. Before a barrier begins to conduct heat at steady state the barrier has to be heated or cooled to the temperature levels that will exist at steady conditions. In production processes requiring heat treatment, metal parts have to be heated at a required rate and then cooled, also at a specified rate. The time required for heating or cooling is an important information for production purposes. Thus the study of transient conduction situation is an important component of conduction studies. This study is a little more complicated due to the introduction of another variable namely time to the parameters affecting conduction. This means that temperature is not only a function of location but also a function of time, τ, *i.e.* $T = T(x, y, z, \tau)$. In addition heat capacity and heat storage (as internal energy) become important parameters of the problem. The rate of temperature change at a location and the spatial temperature distribution at any time are the important parameters to be determined in this study. This will automatically provide information about the heat conduction rate at any time or position through the application of Fourier's law.

6.1 HEATING OR COOLING

A wall exposed to the sun has to first get heated before it conducts heat into an airconditioned room. A vegetable taken out of a refrigerator and left in air gets heated up. How long will it take for the material to reach room conditions? A vegetable is placed in refrigerator. How long will it take for it to cool down? What is the load on the refrigerator at any instant? Questions similar to these are answered in this chapter.

6.1.1. Types of Heating or Cooling

There are two distinguishable ways in which things may heat up or cool down. (*i*) The temperature within the body remains the same at all locations at any one time but this value will vary continuously with time [Fig. 6.1 (*a*)]. This can be observed in heating up or cooling down of materials with high conductivity, small size or low surface convection. The temperature gradient within the body is nearly zero all over the body or the temperature at any location and at any instant is nearly the same. (*ii*) The other type is when there is an observable and large temperature variation within a body. This is illustrated in Fig. 6.1 (*b*). In Fig. 6.1(*a*), the temperature is shown to vary with time, but within the body the temperature is the same at ny instant. In Fig. 6.1(*b*) temperature is shown to vary not only with time but also with ocation.

In the first type, conduction within the body is not a problem. The body as a whole can e considered to be at a given temperature at any instant. The model for this study is known s 'Lumped Parameter or Lumped Capacitance model. The body in this case need not be studied s made up of different layers with heat being conducted and stored at different rates at ifferent layers. In this case an energy balance equation is sufficient to predict the temperature t any instant. This phenomenon is discussed first before taking up the more complex–the econd type.

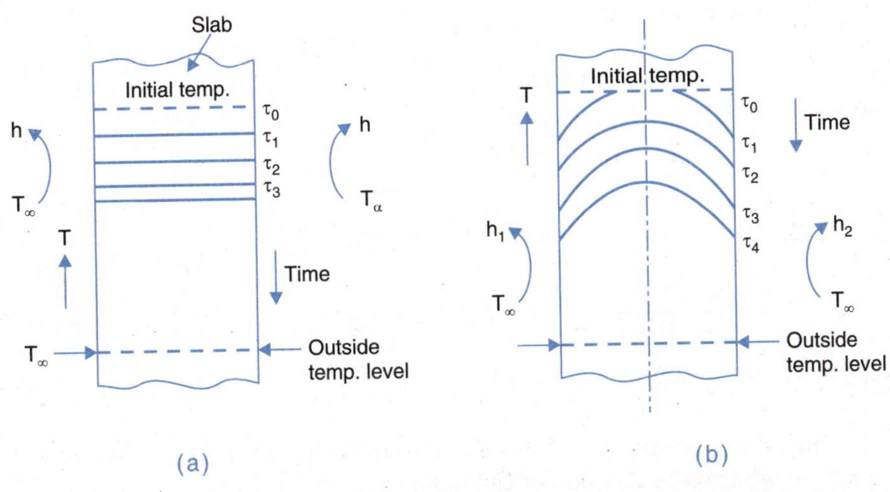

Fig. 6.1. (a) Cooling down without large temperature gradient in the body
(b) Cooling down with a large temperature gradient within the body

6.2 LUMPED CAPACITANCE METHOD

It is also known as lumped heat capacity system. This model is applicable when a body with a known or specified temperature level is exposed suddenly to surroundings at a different temperature level and when the temperature level in the body as a whole increases or decreases without any difference of temperature within the body *i.e.*, $T = T(\tau)$ only. Heat is received from or given to the surroundings at the surface and this causes a temperature change instantly all through the body. The model is shown in Fig. 6.2.

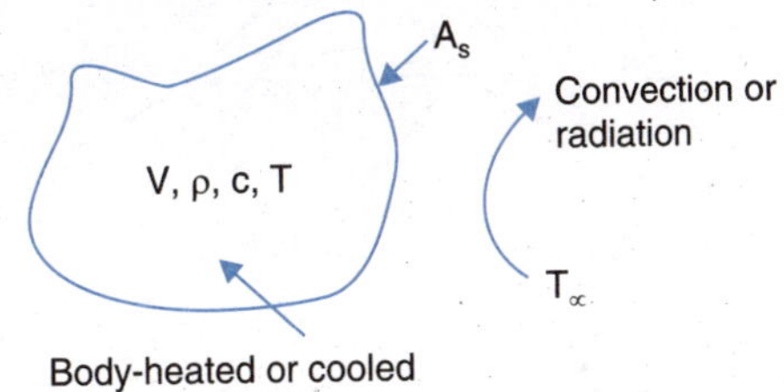

Fig. 6.2. Lumped heat capacity system

The body with surface area A_s, volume V, density ρ, specific heat c and temperature T at the time zero is exposed suddenly to the surroundings at T_∞ with a convection coefficient h (may be radiation coefficient h_r). This causes the body temperature T to change to $T + dT$ in the time interval $d\tau$. The relationship between dT and $d\tau$ can be established by the application of the energy conservation principle.

Heat convected over the boundary over a time period $d\tau$ = Change in internal energy during this time

Chapter 6

If dT is the temperature change during the time period $d\tau$ then the following relationshi results: (A_s-Surface area)

$$hA_s (T - T_\infty) \, d\tau = \rho c V \, dT \quad \text{...(6.1}$$

This equation can be integrated to obtain the value of T at any time. The integration i possible after introducing a new variable,

$$\theta = T - T_\infty \quad \text{...(6.2}$$

The equation (6.1) now becomes

$$hA_s \theta \, d\tau = \rho c V \, d\theta \quad \text{...(6.3}$$

$$\frac{hA_s}{\rho c V} d\tau = d\theta \quad \text{...[6.3(a)}$$

Separating the variables and integrating and using the initial conditions that at $\tau = 0$ $\theta = \theta_i$ and denoting $V/A_s = L$, the equivalent length of the body, we get

$$\ln \frac{\theta}{\theta_i} = \frac{hA_s}{\rho c V} . \tau$$

Substituting for θ and θ_i and taking the antilog

$$\frac{T_\tau - T_\infty}{T_i - T_\infty} = e^{-\frac{hA_s}{\rho c V}.\tau} = e^{-\frac{h}{\rho c L}\tau} \quad \text{...(6.4}$$

Heat flow out up to time τ for cooling,

$$\theta_\tau = \rho c V (T_i - T_\tau) \quad \text{...[6.4(a)]}$$

where T_i is the initial temperature of the body. The value of V/A_s for cylinder, sphere and cube are $r/2$, $r/3$ and $a/6$ where a is the side of the cube.

Equation (6.4) shows that the temperature decays or builds up exponentially and the rate depends on the parameter ($hA_s/\rho c V$) as shown in Fig. 6.3 (a, b).

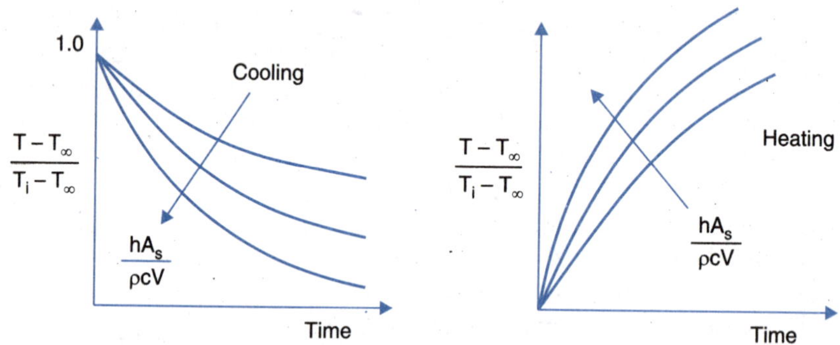

Fig. 6.3. (a) Cooling, (b) Heating

For radiation boundary refer advanced example A-6.9.

6.2.1. Applicability of the Model

In equation (6.4) depends on several factors, but the condition can be checked by using the relative temperature drop within the solid compared to the temperature drop from the outside surface to the fluid. Heat is conducted within the solid and this can be obtained using the equations derived in the case of a slab as

$$Q = \frac{kA(\Delta T)_s}{L}$$

where ΔT_s is the temperature drop in the solid, L is the thickness and k thermal conductivity.

This heat is convected at the surface and is given by

$Q = hA(\Delta T)_c$ where $(\Delta T)_c$ is the convection drop.

Equating these two RHS

$$\frac{(\Delta T)_s}{(\Delta T)_c} = \frac{hL}{k} = \frac{(L/k)}{(1/h)} = \frac{\text{Conduction resistance}}{\text{Convection resistance}}$$

It can be checked that the term **hL/k** is dimensionless. This dimensionless quantity is named as **Biot number.** Biot number provides a measure of temperature drop in the solid compared to the difference between the surface temperature and fluid temperature. If this number is small then $(\Delta T)_s$ will be small and so the temperature difference in the body is small and so it can be taken as the check for the applicability of lumped parameter model. If Biot number is less than 0.1, it has been proved that this model can be used without appreciable error. As this model is simpler to apply, it is worthwhile checking in all cases of transient conduction for the value of Biot number before attempting the solution. **If Bi < 0.1, then the simpler model can be adopted.**

6.2.2. Electrical Analogy

Now going back to equation (6.4), the term $(hA_s/\rho Vc)$ can be used to visualise an **electrical analogy in terms of a capacitance resistance circuit** and the process of heating or cooling as charging and discharging the capacitor $\frac{1}{hA_s}$ is the convection resistance, ρVc is the heat capacity of the system and hence the analogy. In the case of transient conduction, the heat capacity to the material is taken as equivalent to the capacitance in an electrical circuit. The equivalent circuit is shown in Fig. 6.4.

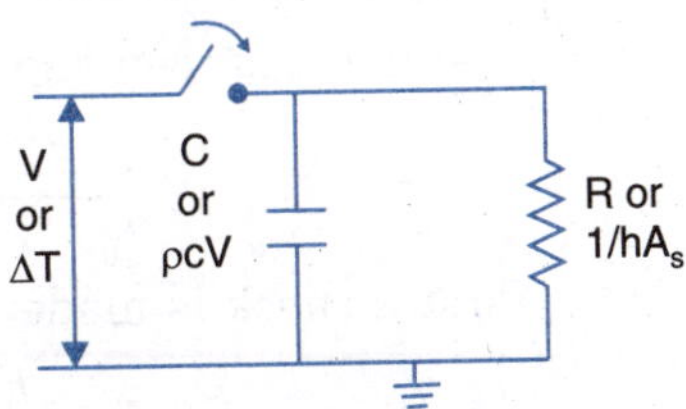

Fig. 6.4. Resistance capacitance circuit

6.2.3. Thermal Time Constant

The quantity $(hA_s/\rho Vc)$ can also be expressed as a product of two dimensionless numbers by rearranging the terms as below, taking $A_s/V = 1/L$ an equivalent length

$$\frac{hA_s\tau}{\rho Vc} = \frac{hL}{k}\left(\frac{k\tau}{\rho cL^2}\right) = \frac{hL}{k}\cdot\frac{\alpha\tau}{L^2} = Bi.Fo. \qquad ...(6.5)$$

The Biot number has already been defined. It should be noted here that $L = V/A_s$ = Volume/Surface area. The value of L can be obtained from the shape of the solid. The quantity $(k/\rho c)$ has already been defined as thermal diffusivity. $\alpha\tau/L^2$ is defined as Fourier number a dimensionless time. This number signifies the heat penetration depth in time τ, with respect to the body dimension. Fourier number should be large for quick heating or cooling. Leaving out the time, higher the value of thermal diffusivity or smaller the body dimension, quicker will be the heating up or cooling down.

The response characteristics of a body with respect to heating or cooling is compared by using the time required for the temperature difference to drop to $1/e$ times the original value

(e-base of natural logarithm). This time is called the **thermal time constant** for the condition specified. This leads to the condition that the time constant

$$\tau_c = \frac{\rho Vc}{hA_s} \quad ...(6.6)$$

Equation 6.4 (a) can be modified as below:

$$Q_\tau = \rho cV\,(T_i - T_\infty)\,[1 - \exp\,(\tau/\tau_c)] \quad ...[6.4(b)]$$

For quick response as in the case of probes for temperature mesurements the time constant should be small.

Example 6.1: *A thermocouple is formed by soldering end-to-end wires of 0.5 mm dia. The thermal diffusivity of the material is 5.3 × 10^{-6} m^2/s. The conductivity of the material is 19.1 W/mK. The probe initially at 30°C is placed in a fluid at 600°C to measure the temperature of the fluid. If the convective heat transfer coefficient between the wire and the fluid is 85 W/m^{2}K, determine the time constant for the probe and also the time taken for it to read 598°C.*

Solution: Using equation (6.4)

Known data: wire dia = 0.5 mm, $\alpha = 5.3 \times 10^{-6}$ m^2/s

k = 19.1 W/mK, T_i = 30°C, T_∞ = 600°C

h = 85 W/m^{2}K.

Find: Time constant τ_c, Time to reach 598°C.

Assume: Lumped capacitance model.

$$\frac{T - T_\infty}{T_0 - T_\infty} = e^{-(hA_s/\rho Vc)\tau} \text{ and noting } A_s/V = \frac{2\pi\, rl}{\pi\, r^2 l} = \frac{2}{r}$$

First a check is made to find the applicability of the Lumped parameter model

$$\frac{hL}{k} = \frac{85 \times 0.00025}{19.1 \times 2} = 0.55 \times 10^{-3} << 0.1.$$

So the model can be used. Noting $\rho c = k/\alpha$

(i) $\boldsymbol{\tau_c}$ = Time constant (eqn. 6.6)

$$(\rho Vc/hA_s) = \frac{k}{\alpha}\cdot\frac{r}{2}\cdot\frac{1}{h} = \frac{19.1}{5.3 \times 10^{-6}}\cdot\frac{0.00025}{2}\cdot\frac{1}{85} = \mathbf{5.355\ s}$$

(ii)
$$\frac{598 - 600}{30 - 600} = e^{-\left(\frac{85 \times 2 \times 5.3 \times 10^{-6}}{0.00025 \times 19.1}\right)\tau} = e^{-0.1887\tau}$$

$\therefore$ τ = **30 seconds.**

COMMENT

This is rather long and attempts should be made to reduce the value, if the instrument is to be useful. This can be achieved by reducing heat capacity and or increasing the value of h.

Example 6.2: *Assuming (hL/k) = 0.05, determine the diameter of spherical balls of (i) Alloy steel of k = 39.8 W.mK (ii) Aluminium, of k = 204.2 W/mK and (iii) Ceramic material of k = 1.5 W/mK. In all cases convection coefficient is 80 W/m^{2}K.*

Solution: Alloy steel: $$L = \frac{V}{A_s} = \frac{4}{3}\pi r^3/4\pi r^2 = \frac{r}{3} = \frac{D}{6}$$

$$0.05 = 80 \times \left(\frac{D}{6}\right) \times \left(\frac{1}{39.8}\right) \quad \text{Solving } \boldsymbol{D = 0.14925 \text{ m.}}$$

Aluminium: $L = D/6$.

$$0.05 = 80 \times \left(\frac{D}{6}\right) \times \left(\frac{1}{204.2}\right) \quad \therefore \quad \boldsymbol{D = 0.76575 \text{ m.}}$$

Ceramic material:

$$0.05 = 80 \times \frac{D}{6} \times \frac{1}{1.5} \quad \therefore \quad \boldsymbol{D = 0.005625 \text{ m.}}$$

COMMENT

Higher the thermal conductivity even larger dimension component can be analysed using lumped capacitance model.

Example 6.3: *A slab of high thermal conductivity at temperature T_i at zero time receivers on one side a heat flux of $q W/m^2$. On the other side there is convection at a temperature T_∞ with convection coefficient h W/m^2K. The thickness of the slab is L m. Drive an expression for the temperature at time τ. Assume that lumped heat capacity model can be used.*

Solution: The energy equation is

$$Aq + Ah\,[T_\infty - T_\tau] = \rho c A L \frac{dT_\tau}{d\tau}$$

or

$$q + h\,(T_\infty - T_\tau) = \rho c L \frac{dT_\tau}{d\tau}$$

Defining $\theta_\tau = T_\tau - T_\infty$, and $Q = q/\rho c L$

$$\frac{dT_\tau}{d\tau} + \frac{h}{\rho c L}\theta_\tau = \frac{q}{\rho c L} = Q$$

The solution in this equation is, defining $\frac{h}{\rho c L}$ as m,

$$\theta_\tau = Ce^{-m\tau} + \theta_p$$

C is the constant of integration and $\theta_p = \frac{Q}{m}$ is the particular solution.

$$\theta_\tau = \theta_0 \text{ at } \tau = 0, \text{ i.e. } \theta_\tau = T_i - T_\infty.$$

$$\therefore \quad \theta_0 = C + \frac{Q}{m} \quad \text{...(1)}$$

$$\therefore \quad \theta_\tau = \theta_0\, e^{-m\tau} + (1 - e^{-m\tau})\,\frac{Q}{m}$$

The final result being

or

$$\theta_\tau = T_\tau - T_\infty = \theta_0 e^{-m\tau} + (1 - e^{-m\tau})\,\frac{Q}{h}$$

when the time is long, or $\tau \longrightarrow \infty$

$$\theta_\infty \times (T_{\tau\infty} - T_\infty) = \frac{q}{h} \quad \text{...(2)}$$

Chapter 6

Example 6.4: *Alloy steel balls of 20 mm diameter initially at 30°C are to be heated to 350°C using hot fluid at 600°C. The convective heat transfer coefficient is 80 W/m²K. The conductivity, specific heat and density are 39.8 W/mK, 4.61 J/kgK, and 7833 kg/m³.*

Determine the time required and time constant.

Solution: Known data: $D = 0.02$ m, $h = 80$ W/m^{2}K, $k = 39.8$ W/mK, $\rho = 7833$ kg/m^3, $c = 4.61$ J/kgK.

Assumption: Lumped capacitance model.

First check for value of hL/k. , $L = D/6 = (0.02/6)$

$$\therefore \quad \frac{hL}{k} = \frac{80 \times 0.02}{6 \times 39.8} = 0.0067$$

$\therefore$ model chosen can be adopted.

Using equation (6.4),

$$\frac{350 - 600}{30 - 600} = e^{\frac{-80 \times 6}{0.02 \times 7833 \times 461} \times \tau} = e^{-0.006646 \times \tau}$$

Solving $\tau = 124$ s.

Time constant: $$\tau_c = \frac{\rho V c}{h A_s} = \frac{7833 \times 0.02 \times 461}{80 \times 6}$$

$$= \mathbf{150.5\ s.}$$

Example 6.5: *A solar* ***concentrator*** *provides a heat flux of 8000 W/m² to heat a 10 mm plate on one side. The other side of the plate is exposed to convection at 40 W/m²K at 20°C. The plate material has a density of 800 kg/m³ and specific heat of 500 J/kg K. The initial temperature of the plate is 20°C. Thermal conductivity of the material is 80 W/m°C.*

(i) Determine the time for the plate to reach 100°C.

(ii) Determine the plate temperature after 6 minutes.

Solution:

$$Bi = \frac{hL}{k} = \frac{40 \times 0.01}{80} = 0.005$$

Hence lumped parameter model can be used.

From equation (1)

$$\theta_\tau = \theta_0 e^{-m\tau} + (1 - e^{-m\tau})\,\frac{q}{h}$$

This can be reordered as

$$\tau = \frac{1}{m} \ln \frac{\left(\theta_0 - \frac{q}{h}\right)}{\left(\theta_\tau - \frac{q}{h}\right)}$$

$$\theta_0 = 20 - 20 = 0, \quad \theta_\tau = 100 - 20 = 80$$

$$\frac{q}{h} = \frac{8000}{40} = 200, \quad m = \frac{k}{\rho c L} = \frac{80}{800 \times 500 \times 0.01} = 0.02$$

$$\therefore \qquad \tau = \frac{1}{0.02} \ln \frac{-\frac{8000}{40}}{80 - \frac{8000}{40}} = \mathbf{25.5\ seconds}$$

The plate will reach 100°C after 25.5 seconds or 0.43 min.

To find the temperature after 6 minutes or 360 sec

$$\theta_\tau = 0 + (1 - e^{-0.002 \times 360}) \frac{8000}{40} = 102.6\ °\text{C}$$

Temperature after 6 min.

$$\therefore \qquad T_\tau = 102.6 + 20 = \mathbf{122.6°C}$$

Example 6.6: *A steel cube of 30 mm side at 600°C is allowed to cool in air at 35°C, with convection coefficient of 65 W/m²K. The conductivity, density and specific heat are 36.3 W/mK, 7753 kg/m³, and 486 J/kgK. Determine the temperature of the cube after 2 minutes.*

Solution: $\frac{hL}{k}$, L = Side/6 for cube.

$\therefore$ **(hL/k) = 65 × 0.03/6 × 36.3 = 0.00895.**

So lumped capacitance model can be adopted.

$$\frac{T - 600}{35 - 600} = e^{\frac{-65 \times 6 \times 120}{7753 \times 486 \times 0.03}} = e^{-0.414}$$

$$= 0.661$$

Solving, $T = 226.5$°C. **After 2 minutes, the temperature of the cube = 226.5°C.**

6.3 SEMI INFINITE SOLID

Another model or system for which analytical solution may be obtained is the semi infinite solid model. Theoretically a solid which extends in both the positive and negative y and z directions to infinity and in the positive x direction to infinity is defined as a semi infinite body. There can be no such body in reality. If one surface of a solid with a particular temperature distribution is suddenly exposed to convection conditions or has its surface temperature changed suddenly, conduction will produce a change in the temperature distribution along the thickness of the body. If this change does not reach the other side or surface of the solid under the time under consideration, then the solid may be modelled as semi infinite solid. A thick slab with a low value of thermal diffusivity exposed to a different environment on its surface can be treated as semi infinite body, provided heat does not penetrate to the full depth in the time under consideration. A road surface exposed to solar heat or chill winds can be cited as an example of a semi infinite body. There are a number of practical applications in engineering for the semi infinite medium conduction. This model provides a useful idealisation for many practical problems. It may be used to determine transient heat transfer near the surface of roads exposed to solar heat or snow.

The differential equation applicable is the simplified general heat conduction equation: in rectangular coordinates, (excluding heat generation) equation 2.6(*c*).

$$\frac{\partial^2 T}{\partial x^2} = \frac{1}{\alpha}\frac{\partial T}{\partial \tau} \quad \text{...(6.7)}$$

The equation is solved by using a similarity variable $\eta = x/(4\alpha\tau)^{0.5}$. The equation (6.7) is transformed to $\frac{d^2T}{d\eta^2} = -2\eta \frac{dT}{d\eta}$ for which the solution is equation (6.8).

There are three types of boundary conditions for which solutions are available in a simple form. These are (*i*) at time $\tau = 0$, the surface temperature is changed and maintained at a specified value, (*ii*) at time $\tau = 0$, the surface exposed to convection at T_∞ and (*iii*) at time $\tau = 0$, the surface is exposed to a constant heat flux q.

6.3.1. Surface Temperature Suddenly Changed and Maintained

The conditions can be stated as follows:

at $\tau = 0$, $T(x, \tau) = T_i$, or $T(x, 0) = T_i$

For $\tau > 0$, $T(0, \tau) = T_s$ *i.e.* at $x = 0$, $T = T_s$ at all times.

The analytical solution for this case is given by derivation available in specialised texts on conduction

$$\frac{T_{x,\tau} - T_s}{T_i - T_s} = \text{erf}\left(\frac{x}{2\sqrt{\alpha\tau}}\right) \quad \text{...(6.8)}$$

where, erf indicates "error function of" and the definition of error function is generally available in mathematical texts. Usually tabulations of error function values are available in handbooks. (Refer appendix).

The heat flow at the surface at any time is obtained using Fourier's equation $-kA\,(dT/dx)$. The surface heat flux at time τ is

$$q_s(\tau) = k(T_s - T_i)/\sqrt{\pi\alpha\tau} \quad \text{...(6.9)}$$

The total heat flow during a given period can be obtained by integrating $q_s(\tau)\,d\tau$ between the limits of 0 and τ

$$Q_\tau = 2k \times A\,(T_s - T_i)\,\sqrt{\tau/\pi\alpha} \quad \text{...(6.10)}$$

The heat flow at any section x at a specified time is given by

$$q_x(\tau) = \frac{k\,(T_s - T_i)}{\sqrt{\pi\alpha\tau}} \exp\,[-x^2/4\alpha\tau] \quad \text{...(6.11)}$$

The temperature distribution with time in this case is shown schematically in Fig. 6.5.

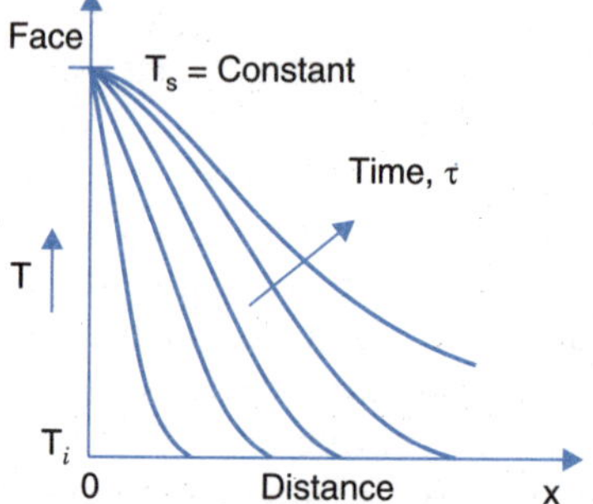

Fig. 6.5. Temperature distribution in semi infinite solid whose surface is suddenly raised and maintained at T_s

Example 6.7: *A thick concrete wall fairly large in size initially at 30°C suddenly has its surface temperature increased to 600°C by an intense fire which lasted for 25 minutes. The material will disintegrate upto a depth where the temp. reaches 400°C. Determine the thickness which may disintegrate. The thermal diffusivity is 4.92 × 10⁻⁷ m²/s; k = 1.28 W/mK. Also determine the total heat flow/m² during the time.*

Solution: As it is mentioned that it is a thick wall, without any thickness dimension, semi infinite solid model is applicable. Also the surface temperature is suddenly increased and maintained at 600°C. Using equation (6.8)

$$\frac{T_{(x,\tau)} - T_s}{T_i - T_s} = \text{erf}\left(\frac{x}{2\sqrt{\alpha\tau}}\right)$$

Substituting the values

$$\frac{400 - 600}{30 - 600} = \text{erf}\left(\frac{x}{2\sqrt{\alpha\tau}}\right), \text{ using tables of error function}$$

$$\frac{x}{2\sqrt{\alpha\tau}} = 0.32$$

$$\therefore \quad x = 0.32 \times 2\sqrt{\alpha\tau} = 0.32 \times 2\sqrt{4.92 \times 10^{-7} \times 25 \times 60} = \mathbf{0.017\ m}$$

The concrete will disintegrate to a depth of 17 mm.

Using equation (6.10)

$$\mathbf{Q(\tau)} = 2k\,(T_s - T_i)\left(\frac{\sqrt{\tau}}{\pi\alpha}\right)$$

For 25 minutes, $\therefore$ $\mathbf{Q} = 2 \times 1.28\,(600 - 30)\sqrt{25 \times 60 / \pi \times 4.92 \times 10^{-7}}$

$= \mathbf{45.5 \times 10^6\ J/m^2}$

6.3.2. The Boundary Condition More Commonly Met with is Convective Boundary

The initial and boundary conditions are:

$$\tau = 0, T(x, 0) = T_i$$

$$-k\left.\frac{dT}{dx}\right|_{x=0} = h(T_\infty - T_{0,\tau}) \qquad ...(6.12)$$

The temperature at any location and time is obtained as

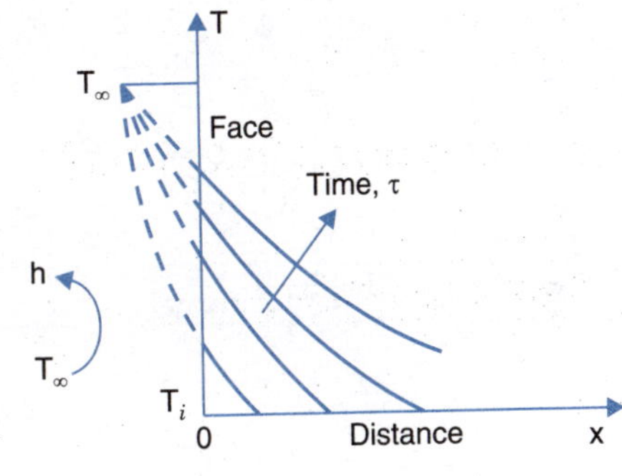

Fig. 6.6. Temperature variation with time–Semi infinite solid-convection boundary

$$\frac{T_{x,\tau} - T_i}{T_\infty - T_i} = \left[1 - \text{erf}\frac{x}{2\sqrt{\alpha\tau}}\right] - \left[1 - \text{erf}\left(\frac{x}{2\sqrt{\alpha\tau}} + \frac{h\sqrt{\alpha\tau}}{k}\right)\right] \times \left[\exp\left(\frac{hx}{k} + \frac{h^2\alpha\tau}{k^2}\right)\right] \qquad ...[6.13(a)]$$

Also $$\frac{T_{x,\tau} - T_\infty}{T_i - T_\infty} = \text{erf}\frac{x}{2\sqrt{\alpha\tau}} + \exp\left[\frac{hx}{k} + \frac{h^2\alpha\tau}{k^2}\right]\left[1 - \text{erf}\left(\frac{x}{2\sqrt{\alpha\tau}} + \frac{h\sqrt{\alpha\tau}}{k}\right)\right] \qquad ...[6.13(b)]$$

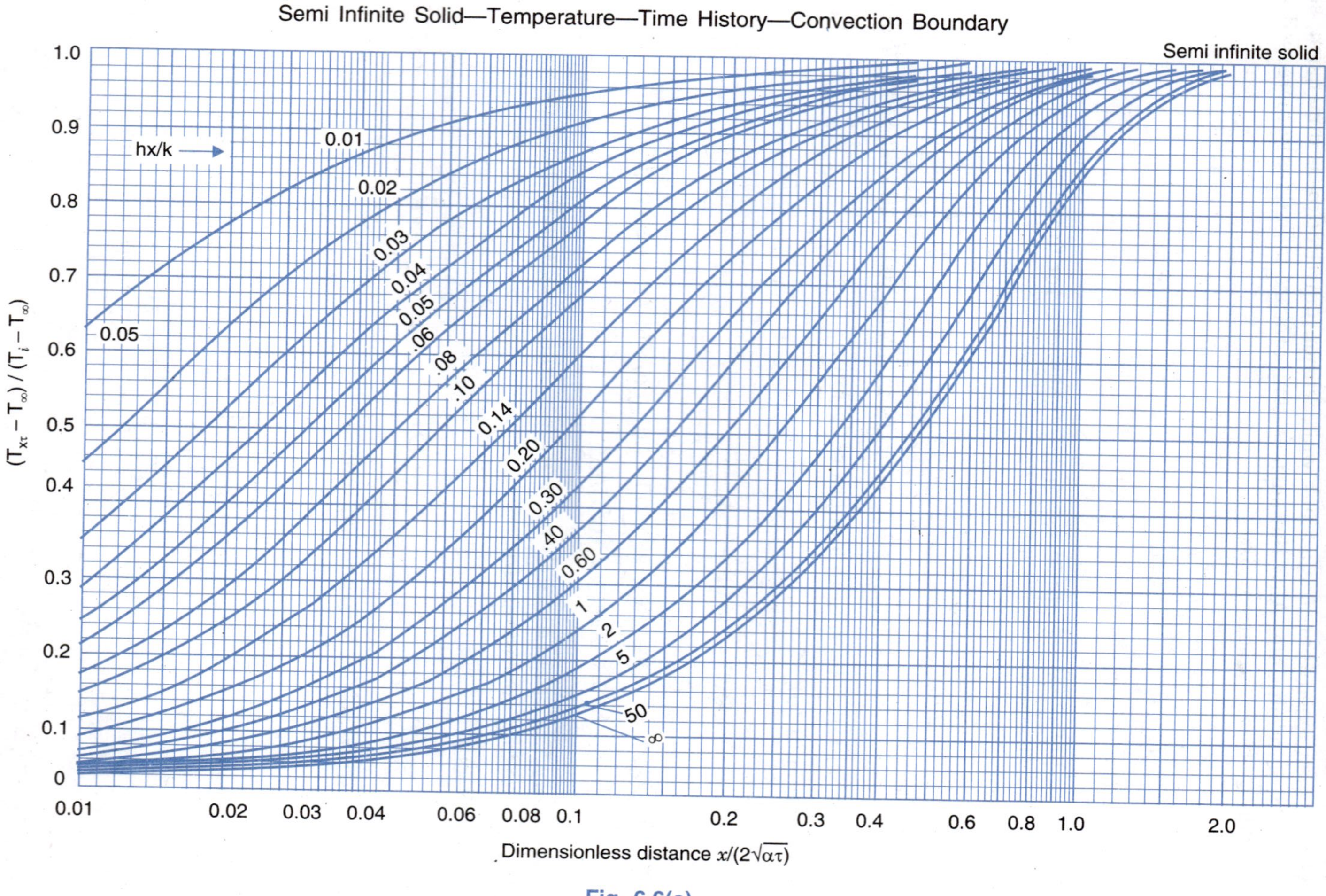
Semi Infinite Solid—Temperature—Time History—Convection Boundary
Semi infinite solid
hx/k
0.05
0.01
0.02
0.03
0.04
0.05
.06
.08
.10
0.14
0.20
0.30
.40
0.60
1
2
5
50
∞
$(T_{x\tau} - T_\infty) / (T_i - T_\infty)$
1.0
0.9
0.8
0.7
0.6
0.5
0.4
0.3
0.2
0.1
0
0.01
0.02
0.03
0.04
0.06
0.08
0.1
0.2
0.3
0.4
0.6
0.8
1.0
2.0
Dimensionless distance $x/(2\sqrt{\alpha\tau})$

Fig. 6.6(a)

As the equation is rather, long charts Fig. 6.6(*a*) are available to calculate the temperature ratio with parameters $x/2\sqrt{\alpha\tau}$, and hx/k. Such a chart is shown in the previous page. The heat flow can be found by obtaining $\partial T/\partial x$ at any section.

For surface temp. $T_{o\tau}$ these equations can be simplified as

$$\frac{T_{o,\tau}-T_i}{T_\infty-T_i}=1-\left[1-\operatorname{erf}\frac{h\sqrt{\alpha\tau}}{k}\right]\left[\exp\frac{h^2\alpha\tau}{k^2}\right] \qquad ...[6.13(c)]$$

$$\frac{T_{o\tau}-T_\infty}{T_i-T_\infty}=\exp\frac{h^2\alpha\tau}{k^2}\left[1-\operatorname{erf}\frac{h\sqrt{\alpha\tau}}{k}\right] \qquad ...[6.13(d)]$$

The temperature variation with time is shown schematically in Fig. 6.6. Note that the surface temperature drops quickly.

Example 6.8: *A concrete wall initially at 30°C is exposed to gases at 900°C with a convective heat transfer coefficient of 85 W/m²K. The thermal diffusivity of the material is 4.92 × 10⁻⁷ m²/s. The thermal conductivity of the material is 1.28 W/mK. Determine the temperature of the surface and temperatures at 1 cm depth and also 5 cm depth after 1 hr. Also estimate the heat flow at the surface at the instant.*

Solution: The temperature is determined using equation (6.13) at the surface $x = 0$.

$$\frac{T-T_i}{T_\infty-T_i}=[1-\operatorname{erf} 0]-\left[1-\operatorname{erf}\left(0+\frac{h\sqrt{\alpha\tau}}{k}\right)\right]\times\left[\exp\left(0+\frac{h^2\alpha\tau}{k^2}\right)\right]$$

$$=1-\left[1-\operatorname{erf}\frac{85\sqrt{4.92\times10^{-7}\times3600}}{1.28}\right]\left[\exp\left(\frac{85^2\times4.92\times10^{-7}\times3600}{1.28^2}\right)\right]$$

$$= 1 - [1 - 0.999924]\,[2466] = 1 - 0.1875 = 0.8125$$

∴ Surface temperature, **T = 737°C**

Use of chart on page 210 will be more convenient than the use of equation (6.13).

The chart is plotted with

$$[(T_{x\tau} - T_\infty)/(T_i - T_\infty)] \text{ against } x/2\sqrt{\alpha\tau} \text{ with } (hx/k) \text{ as parameter.}$$

For 1 cm depth, after one hour or 3600 seconds

$$\frac{x}{2\sqrt{\alpha\tau}}=\frac{0.01}{2\sqrt{4.92\times10^{-7}\times3600}}=0.1188$$

$$\frac{hx}{k}=\frac{85\times0.01}{1.28}=0.664$$

Entering the x axis at 0.1188 and reading against the value of hx/k = 0.664, the temperature ratio is obtained as 0.3.

$$\therefore \qquad \frac{T_{x\tau}-900}{30-900}=0.3, \text{ solving, } T_{x\tau}=\mathbf{639°C}$$

Chapter 6

By equation it is obtained as 638°C.

For **5 cm depth after one hour,**

$$\frac{x}{2\sqrt{\alpha\tau}} = \frac{0.05}{2\sqrt{4.92 \times 10^{-7} \times 3600}} = 0.594$$

$$\frac{hx}{k} = \frac{85 \times 0.05}{1.28} = 3.32$$

Entering at 0.594 and reading at $\frac{hx}{k} = 3.32$, the temperature ratio is obtained as 0.715.

$$\frac{T_{x\tau} - 900}{30 - 900} = 0.715. \text{ Solving } \mathbf{T_{x\tau} = 278°C}$$

By **equation** the value is **267°C.**

Heat flow is obtained assuming linear variation of temp. up to 0.01

$$\therefore \qquad \mathbf{Q} = \frac{737 - 639}{0.01} \times 1.28 \times 1 = \mathbf{12672\ W/m^2}$$

6.3.3. The Third Initial and Boundary Condition Set for Semi Infinite Solid is Exposure to Uniform Heat Flux, q W/m²

$$\tau = 0,\ T_{(x,0)} = T_i$$

$$-k\left.\frac{\partial T}{\partial x}\right|_{x=0} = q \qquad \text{...(6.14)}$$

The solution for temperature distribution is given by

$$(T_{x,\tau} - T_i) = \frac{2q(\alpha\tau/\pi)^{1/2}}{k} \exp\left(\frac{-x^2}{4\alpha\tau}\right) - \frac{qx}{k}\left[1 - \text{erf}\left(\frac{x}{2\sqrt{\alpha\tau}}\right)\right] \qquad \text{...(6.15)}$$

An example for this is solar radiation being incident on a road surface or heating in a radiation furnace.

Example 6.9: *Solar radiation falls on a concrete wall starting from 10 am and the flux is 800 W/m². The initial temperature was 30°C. Determine the surface temperature and the temperature at a depth of 5 cm at 12 noon. Thermal diffusivity = 4.92 × 10⁻⁷ m²/s and conductivity = 1.28 W/mK.*

Solution: At the surface $x = 0$, using equation (6.15).

$$(T_o - T_i) = (T_0 - 30) = \frac{2 \times 800\,(4.92 \times 10^{-7} \times 2 \times 3600/\pi)^{1/2}}{1.28}.\ \exp(0) - 0$$

$$= 42 \quad \therefore \quad \mathbf{T_o = 72°C} \text{ at surface}$$

At $x = 0.05$ m,

$$T - 30 = \frac{2 \times 800\,(4.92 \times 10^{-7} \times 2 \times 3600/\pi)^{1/2}}{1.28} \exp \frac{-0.05^2}{4 \times 4.92 \times 10^{-7} \times 2 \times 3600}$$

$$-\frac{800 \times 0.05}{1.28}\left[1-\text{erf}\,\frac{0.05}{2\sqrt{4.92 \times 10^{-7} \times 2 \times 3600}}\right]$$

$$= 35.19 - \frac{800 \times 0.05}{1.28}[1-0.4475] = 35.19 - 17.26 = 17.92$$

$\therefore$ **T = 47.92°C**

The temperature variation in this case with time is shown schematically in Fig. 6.7.

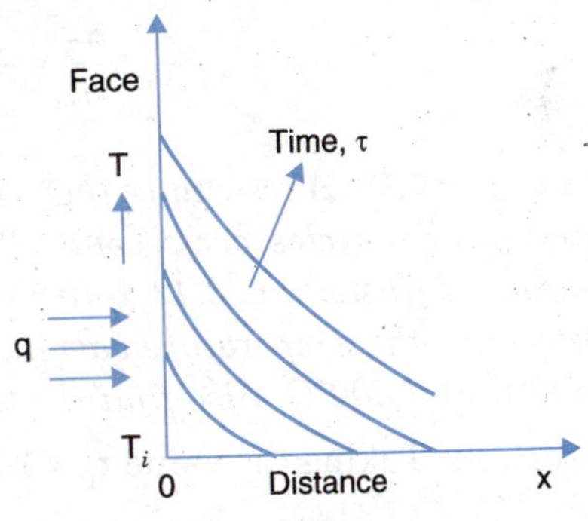

Fig. 6.7. Temperature variation constant surface heat flux

The learner is often confused about when to use the semi infinite solid model. Generally when the thickness is large this can be tried. If the temperature penetrates to the other surface, then the infinite slab model is to be used (discussed later).

As the equations for the convection environment is long and calculations are tedious, recourse may be had to the use of charts which plot the temperature ratio in terms of the parameter $(x/2\sqrt{\alpha\tau})$ and either (hx/k) or $h^2\,\alpha\tau/\kappa^2$.

6.4 PERIODIC HEAT CONDUCTION

There are occasions when the surrounding temperature varies periodically as in the case of an IC engine cylinder. In these cases it is necessary to know (*i*) the variation of temperature of the solid and (*ii*) the time lag between the variation in the surrounding and the variation in the solid. In the case of semi infinite solid it is also necessary to know upto what depth the variation will penetrate at steady state in addition to the study of these variations at different depths.

6.4.1. Lumped Capacitance System

In this case any temperature variation in the surroundings causes a temperature variation throughout the solid. The variation is shown in Fig. 6.8.

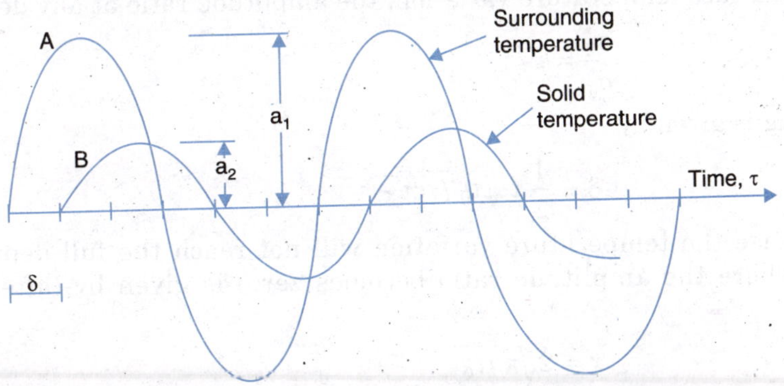

Fig. 6.8. Temperature variation with time-lumped capacitance system

Chapter 6

The surrounding temperature varies with time as shown in curve A. The solid temperature will vary as shown in curve B. The two quantities of interest are (*i*) amplitude ratio a_2/a_1, and (*ii*) the lag δ in terms of degree or radian. For sinusoidal variations, the following relations held. The lag δ is given by, where τ_0 is the period per cycle in seconds

$$\delta = \tan^{-1} [(2\pi/\tau_0).\ (c\rho V/hA_s)] \qquad ...(6.16)$$

The amplitude ratio is given by

$$\frac{a_2}{a_1} = \frac{1}{\sqrt{1+\tan^2\delta}} \qquad ...(6.17)$$

Example 6.10: *A resistance thermometer wire of 0.5 mm dia is used to measure the temperature varying at 2 cycles per second. The thermal conductivity of the material is 69 W/mK. The thermal diffusivity is 2.42 × 10⁻⁵ m²/s. The convective heat transfer coefficient is 280 W/m²K. Determine the temperature variation in the wire if the surrounding temperature varies between 1500°C and 500°C. Also find the time lag between these two.*

Solution: Taking the value $\tau_0 = 1/2 = 0.5$ as there are two cycles/sec and $\rho cV/hA_s$ is rearranged as $\left(\frac{\rho c}{k}\frac{k}{h}\frac{V}{A_s}\right)$ and $V/A_s = \pi r^2\ l/2\pi rl = r/2 = 0.0005/4$ m and using equation (6.16), the lag angle is given by

$$\delta = \tan^{-1}\left[(2\pi/0.5)(1/2.42\times10^{-5}).\left(\frac{69}{200}\right).\left(\frac{0.0005}{4}\right)\right]$$

∴ **δ = 1.526 radians or 87.44° or in terms of time 0.12 sec.**

Amplitude ratio is found using equation (6.17)

$$\frac{\mathbf{a_2}}{\mathbf{a_1}} = \frac{1}{\sqrt{1+\tan^2\delta}} = \frac{1}{\sqrt{1+498}} = \mathbf{0.045}$$

The wire temperature will fluctuate by only 45°C. This shows that this measurement is off by 4.5% only.

6.4.2. For Semi Infinite Solid

For sinusoidal surface temperature variation, the amplitude ratio at any depth x is given by

$$\frac{a_2}{a_1} = e^{-x\sqrt{\pi/\alpha\tau_0}} \qquad ...(6.18)$$

and the time lag is given by

$$\therefore \qquad \delta = \frac{1}{2}x\sqrt{\tau_0/\alpha\pi} \qquad ...(6.19)$$

In this case the temperature variation will not reach the full depth. The depth of penetration (where the amplitude ratio becomes zero) is given by (after equilibrium is established)

$$x = 2\sqrt{\pi\alpha\tau_0} \qquad ...(6.20)$$

The derivations are involved but the results are usable.

Example 6.11: *The surface temperature of the soil varies with 24 hr period, the minimum temperature being –40°C. Determine the minimum temperature at a depth of 0.2 m and also the time of occurrence of the minimum temperature. Surface minimum occurs at 3 am. The thermal diffusivity of the material is 1.26×10^{-6} m^2/s. The mean temperature is 0°C.*

Solution: Using equation (6.18), and taking $x = 0.2$, $\tau_0 = 24 \times 3600$ sec

$$\frac{\mathbf{a_2}}{\mathbf{a_1}} = e^{-x\sqrt{\pi/\alpha\tau_0}} = e^{-0.2\sqrt{\pi/1.26\times10^{-6}\times24\times3600}} = \mathbf{0.3415}$$

∴ The minimum temperature at this depth is –13.66°C.

Using equation (6.19) to find the time lag,

$$\therefore \quad \delta = \frac{1}{2}\sqrt{\tau_0/\alpha\pi}\;x = \frac{1}{2}\sqrt{\left\{\frac{24\times3600}{1.26\times10^{-6}\times\pi}\right\}}\times0.2 = \mathbf{14774\ s}$$

$$= \mathbf{4.1\ hours} \text{ (in terms of angle } \frac{4.1}{24}\times360 = 61.6°\text{)}$$

This occurs at 7.1 am.

Using equation (6.20), the **depth of penetration** is found.

$$x = 2\sqrt{\pi\alpha\tau_0} = 2\sqrt{\pi\times1.26\times10^{-6}\times24\times3600} = \mathbf{1.17\ m}$$

The seasonal variation will be felt upto 1.17 m depth only.

6.5 TRANSIENT HEAT CONDUCTION IN LARGE SLAB OF LIMITED THICKNESS, LONG CYLINDERS AND SPHERES EXPOSED TO CONVECTION

This model is the one which has a large number of applications in heating and cooling processes a special case being heat treatment. The general solution process attempts to estimate the temperature at a specified location in a body (which was at a specified initial temperature) after exposure to a different temperature surroundings for a specified time. The other quantity of interest is the change in the internal energy of the body after such exposure.

6.5.1. The Differential Equation Applicable for a Slab

Extending to ∞ in the y and z directions and thickness $2L$ in the x direction with both surfaces suddenly exposed to the surroundings is equation (2.6)

$$\frac{\partial^2 T}{\partial x^2} = \frac{1}{\alpha}\frac{\partial T}{\partial \tau}$$

The situation for slab is shown in Fig. 6.9.

The initial condition at time zero is

$T = T_i$ all through the solid. *i.e.* $x = -L$ to $x = L$.

The boundary condition is

$$h\,(T_\infty - T_L) = -k\frac{\partial T}{\partial x} \text{ at } x = L \text{ and } x = -L$$

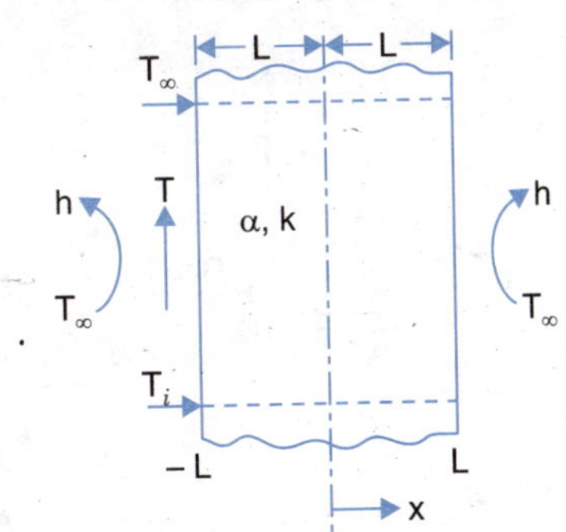

Fig. 6.9. Slab exposed to convective surroundings.

The equation is solved using a set of new variables X and θ defining $T = X.\theta$ (X is a function of x only and θ is a function of τ only). The algebra is long and tedious.

The solution obtained is given below:

$$\frac{T_{x,\tau} - T_\infty}{T_i - T_\infty} = \sum_{n=1}^{\infty} \frac{2Bi \sin(\delta_n) \cos(\delta_n x/L)}{\delta_n (Bi + \sin^2 \delta_n)} . e^{-\delta_n^2 . Fo} \quad ...(6.21$$

The temperature essentially is a function of Bi, Fo and $\frac{x}{L}$ or $T = f\left(Bi, Fo, \frac{x}{L}\right)$

where $T_{x,\tau}$—the temperature at x and time τ

T_i—initial temperature

T_∞—surrounding temperature

Bi—$\frac{hL}{k}$—biot number

Fo—Fourier number $\alpha\tau/L^2$

δ_n—roots of the equation $\delta_n \tan \delta_n = Bi$

The solution using calculating devices is rather tedius and the results in a graphica form, was first published by Heisler in 1947, using the parameters Biot number and Fourier number. As there is an additional parameter (x/L) which cannot be represented in a single chart, the solution was split into two quantities. The first solution is for the centre temperature at $x = 0$. In the centre temperature chart $(T_{0\tau} - T_\infty)/(T_i - T_\infty)$ is plotted against Fourier number with hL/k as parameter. Then the location temperature and centre temperature excess ratio is obtained using the parameters Biot number and (x/L). In the location temperature chart $(T_{x\tau} - T_\infty)/(T_{o,\tau} - T_\infty)$ is plotted against Biot number with x/L as parameter. The charts Fig. 6.11 (a–f) are shown in the next pages.

The product of these two temperature ratios gives the required location temperature ratio.

$$\frac{T_{x,\tau} - T_\infty}{T_i - T_\infty} = \frac{T_{o,\tau} - T_\infty}{T_i - T_\infty} . \frac{T_{x,\tau} - T_\infty}{T_{o,\tau} - T_\infty} \quad ...(6.22)$$

The suffix, $x\tau$ indicates at location x and time τ. The suffix, 0τ indicates at centre at time τ.

The centre temperature ratio $\frac{T_{o,\tau} - T_\infty}{T_i - T_\infty}$ is obtained from a chart the skeleton form of which is shown in Fig. 6.10.

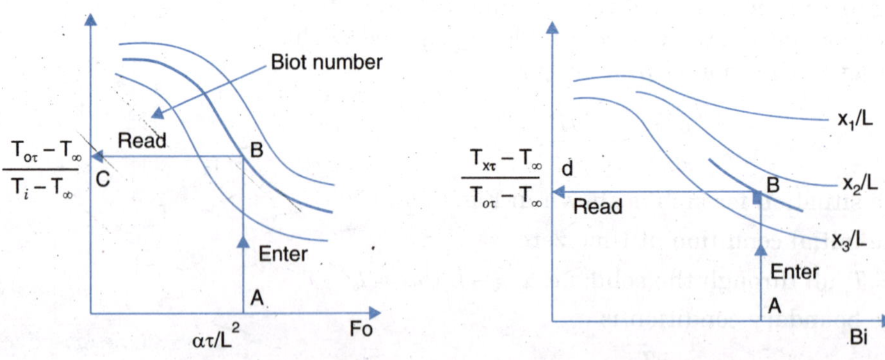

Fig. 6.10

Fig. 6.11

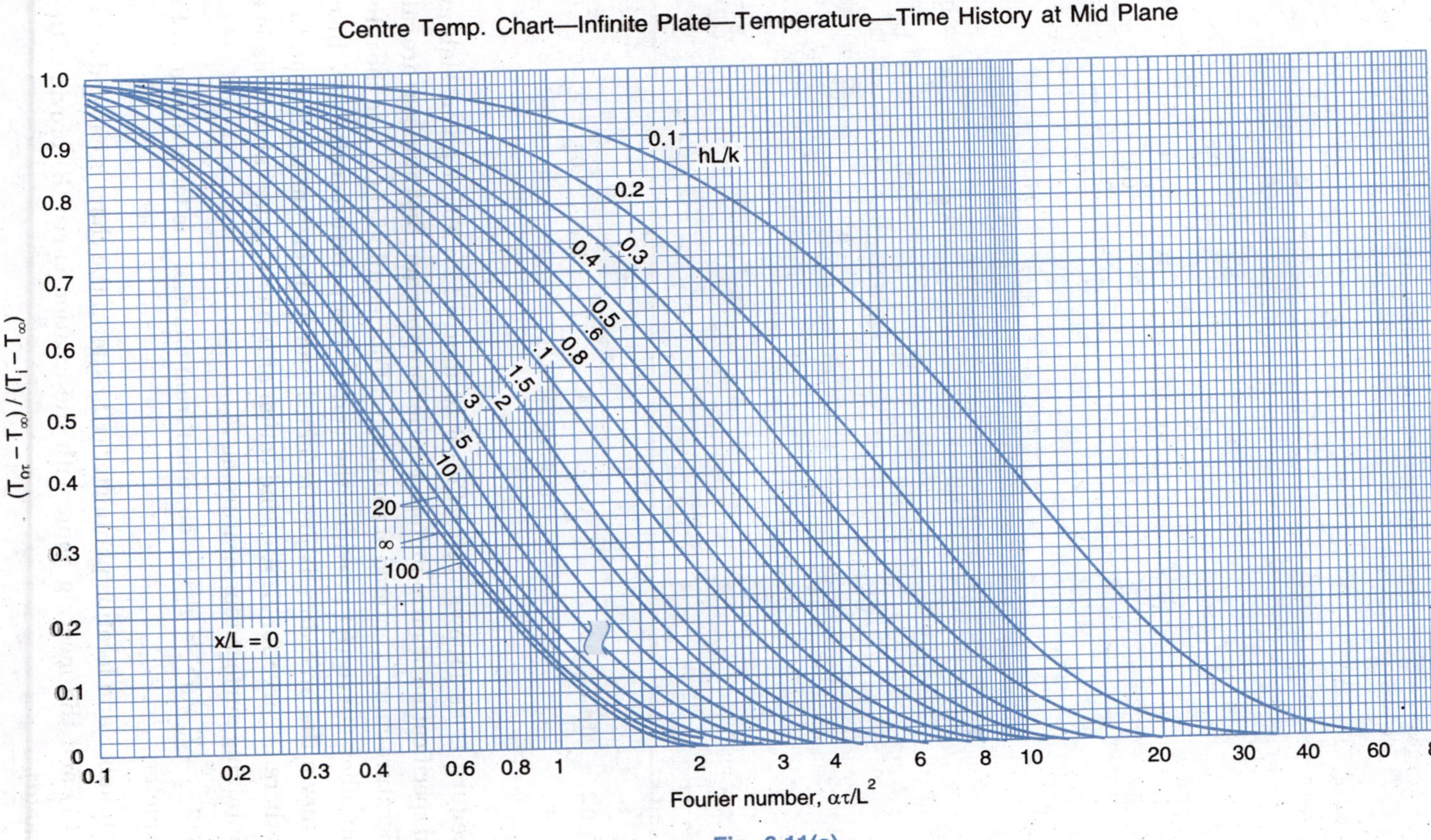
Centre Temp. Chart—Infinite Plate—Temperature—Time History at Mid Plane
$(T_{0\tau} - T_\infty) / (T_i - T_\infty)$
1.0
0.9
0.8
0.7
0.6
0.5
0.4
0.3
0.2
0.1
0
hL/k
0.1
0.2
0.3
0.4
0.5
.6
0.8
.1
1.5
2
3
5
10
20
∞
100
x/L = 0
0.1
0.2
0.3
0.4
0.6
0.8
1
2
3
4
6
8
10
20
30
40
60
80
Fourier number, $\alpha\tau/L^2$

Fig. 6.11(a)

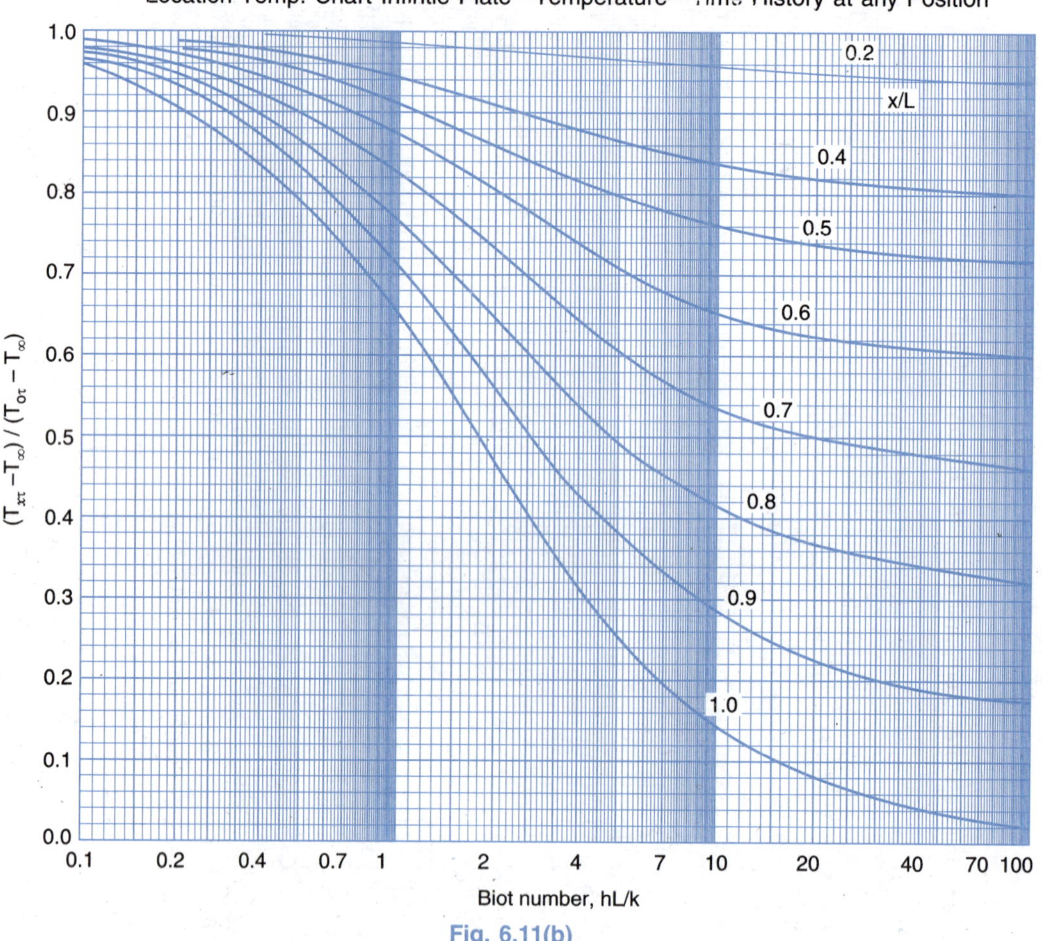

Fig. 6.11(b)

The procedure is as follows (refer Fig. 6.10 and 6.11): For a given slab and time specification and specification of surroundings Fourier number and Biot numbers are calculated. The centre temperature chart is entered at the value of Fourier number (*A*) and the intersection with the Biot number curve (*B*) provides the centre temperature ratio (*C*). In case the temperature is known the procedure is just reversed and Fourier number is read to find the time. Note that there is curve for $Bi = \infty$, indication of h has a large value. This is equal to raising and maintaining surface temperature at T_∞.

The ratio $(T_{x,\tau} - T_\infty)/(T_{o,\tau} - T_\infty)$ is obtained from chart of the form shown in Fig. 6.11, with x/L as parameter.

With Biot number and location parameter (x/L) known the chart is entered at the value of Biot number (*A*) and the meeting point with the location curve (*B*) provides the excess temperature ratio (*d*).

In case the temperature is known then the meeting point of line from *d* and line from *A* gives the location parameter. Now the product of these two ratios as in equation 6.22 will give

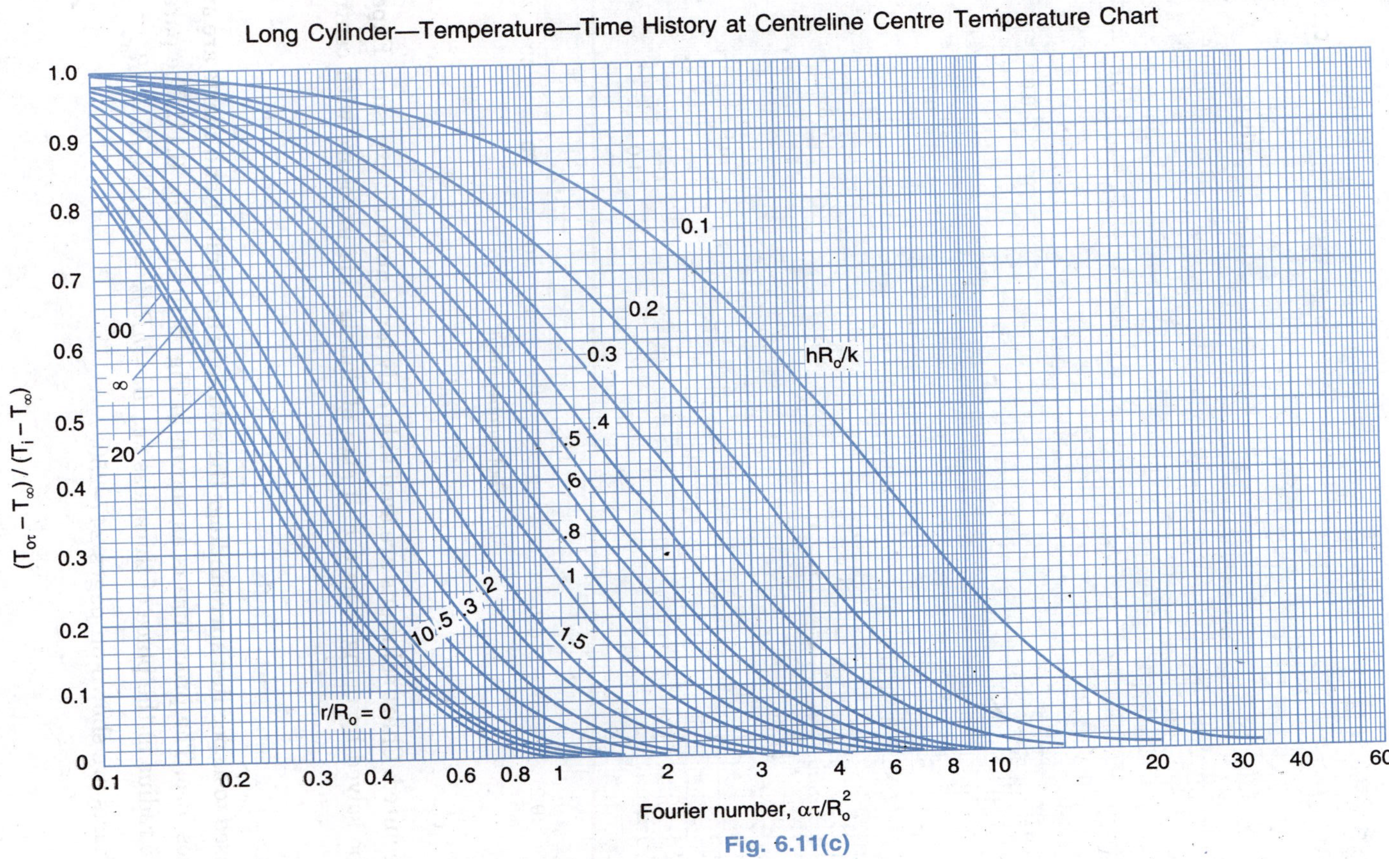
Long Cylinder—Temperature—Time History at Centreline Centre Temperature Chart
$(T_{o\tau} - T_\infty) / (T_i - T_\infty)$
1.0
0.9
0.8
0.7
0.6
0.5
0.4
0.3
0.2
0.1
0
0.1
0.2
0.3
0.4
0.6
0.8
1
2
3
4
6
8
10
20
30
40
60
Fourier number, $\alpha\tau/R_o^2$
hR_o/k
0.1
0.2
0.3
.4
.5
.6
.8
1
1.5
2
3
5
10
$r/R_o = 0$
00
∞
20

Fig. 6.11(c)

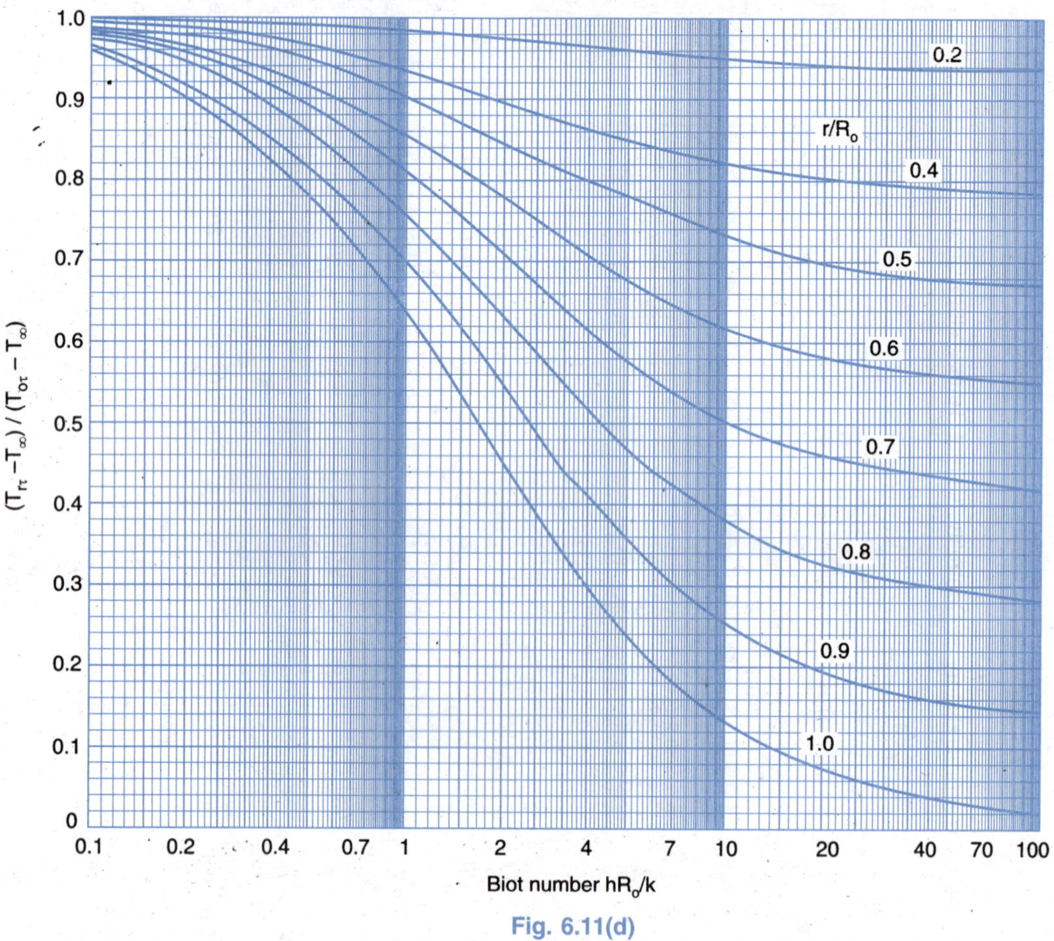

Fig. 6.11(d)

the solution required. Chart solutions are available only for symmetric exposure. Exposure on one side can be solved by using it as a half slab of double side exposure with centre plane insulated.

6.5.2. For Long Cylinders and Spheres

Suddenly exposed to a specified surroundings the solutions are similar and charts are available for these shapes also. The procedure is as shown in Fig. 6.10 and Fig. 6.11. The parameters will involve the radius in the place of thickness *i.e.*, $Bi = hR_o/k = \alpha\, \tau/R_o^2$, $x/L = r/R_o$.

These charts are shown on pages 247–250.

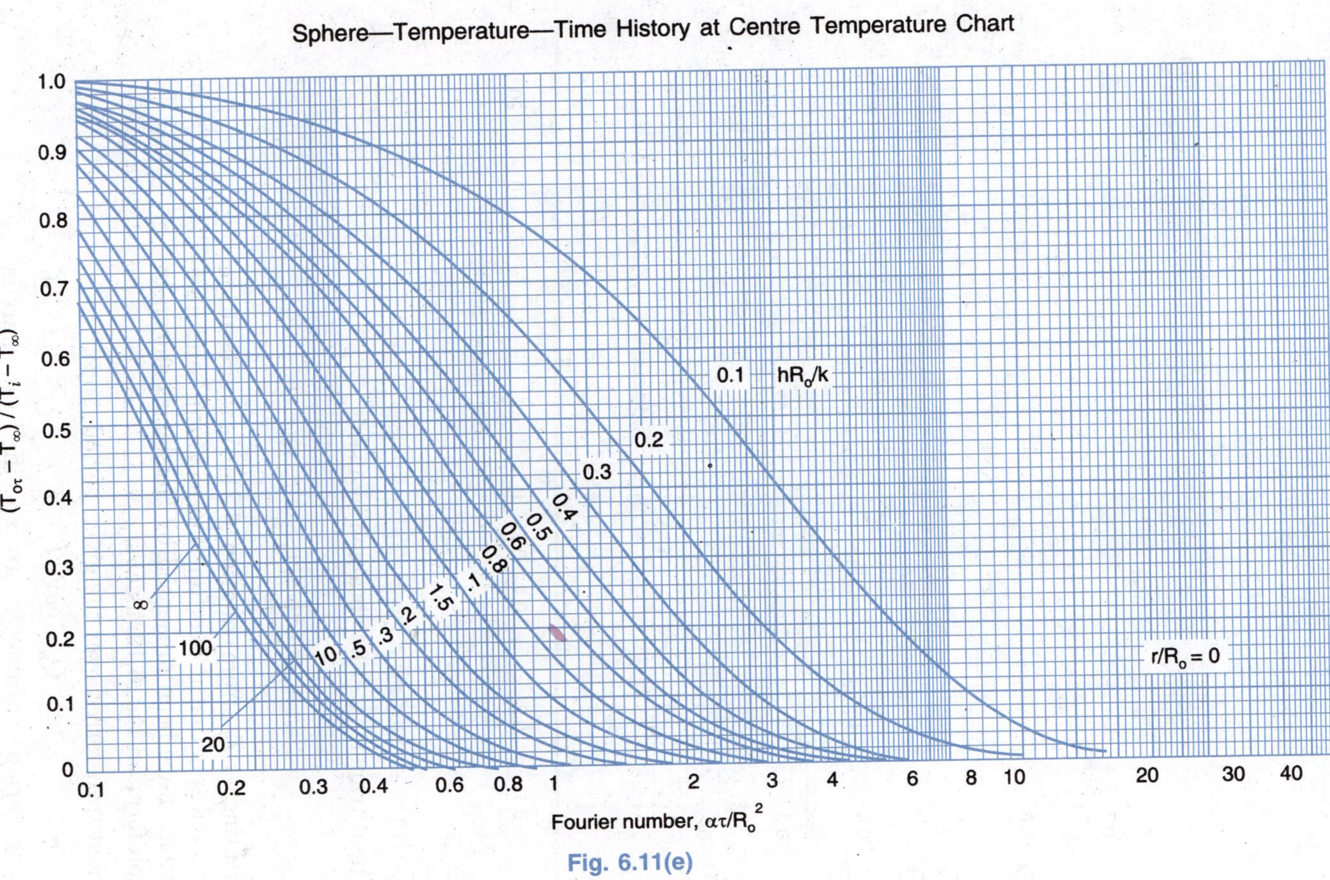
Sphere—Temperature—Time History at Centre Temperature Chart
(Toτ − T∞) / (Ti − T∞)
1.0
0.9
0.8
0.7
0.6
0.5
0.4
0.3
0.2
0.1
0
0.1 hRo/k
0.2
0.3
0.4
0.5
0.6
0.8
1
1.5
2
3
5
10
20
100
∞
r/Ro = 0
0.1
0.2
0.3
0.4
0.6
0.8
1
2
3
4
6
8
10
20
30
40
Fourier number, ατ/Ro²

Fig. 6.11(e)

Sphere—Temperature—Time History at any Radius Location Chart

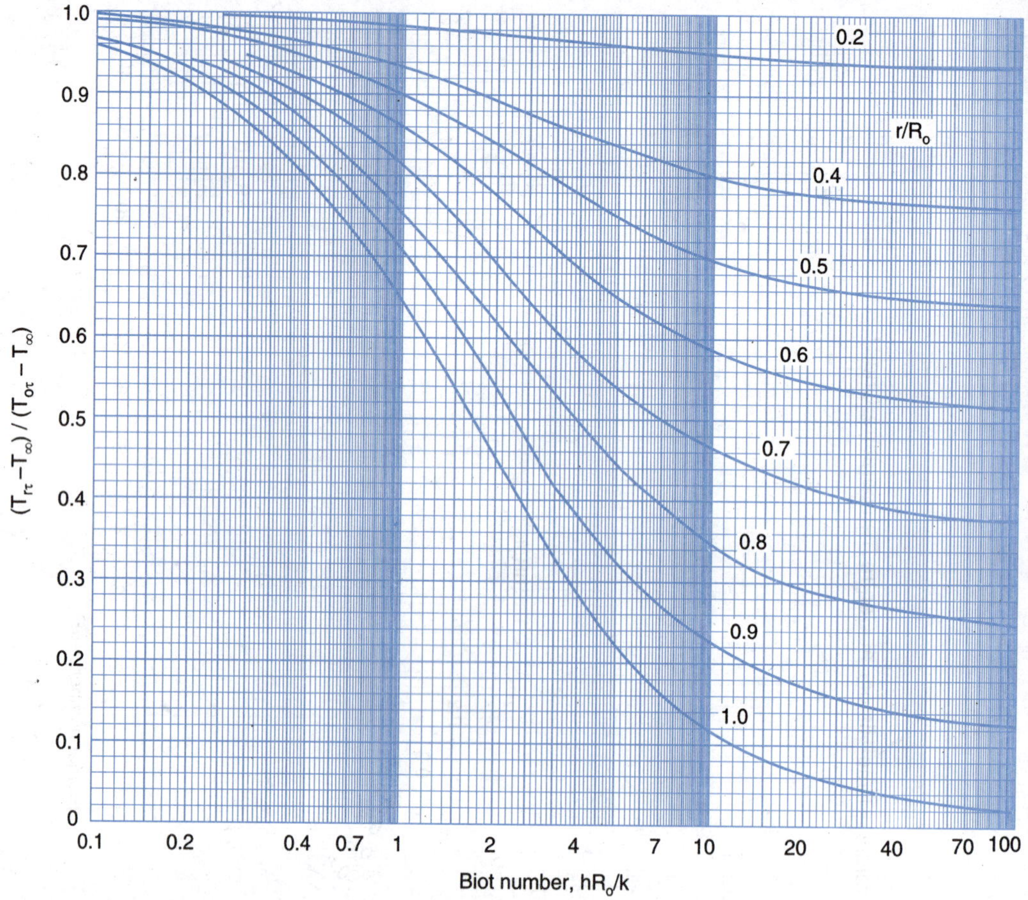

Fig. 6.11(f)

6.5.3. Heat Transfer During a given Time Period

The total heat transfer can be obtained by using

$$Q = \int_0^{\tau} h(T_{L,\tau} - T_{\infty})\, d\tau$$

and substituting for $T_{L,\tau}$ from equation (6.21). As the resulting expression indicates that it is a function of $h^2\alpha\tau/k^2$ and hL/k these solutions have been presented by Heisler as shown in the skeleton form in Fig. 6.12. as Q/Q_o.

where Q-heat transferred over the given period, and

$$Q_o = \rho c V\,(T_i - T_{\infty}) \qquad \text{...(6.23)}$$

As the various parameters are specified the chart is entered at A after calculating $h^2\ \alpha\tau/k^2$. The meeting point with the Biot number curve at B provides the value of Q/Q_o (C). The heat flow upto the specified time is the product of the value read and $\rho c V(T_i - T_{\infty})$. Separate charts are available for the shapes slab, long cylinder and sphere. These are given in page 251.

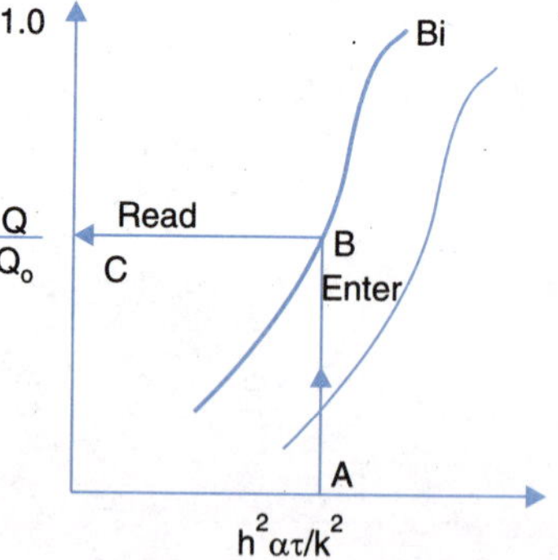

Fig. 6.12. Internal energy ratio

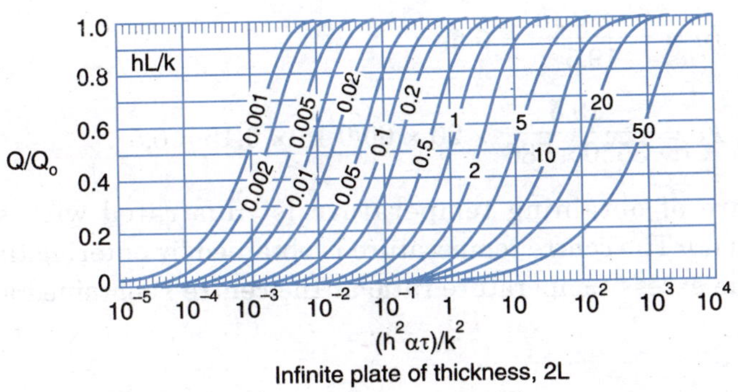

Infinite plate of thickness, 2L

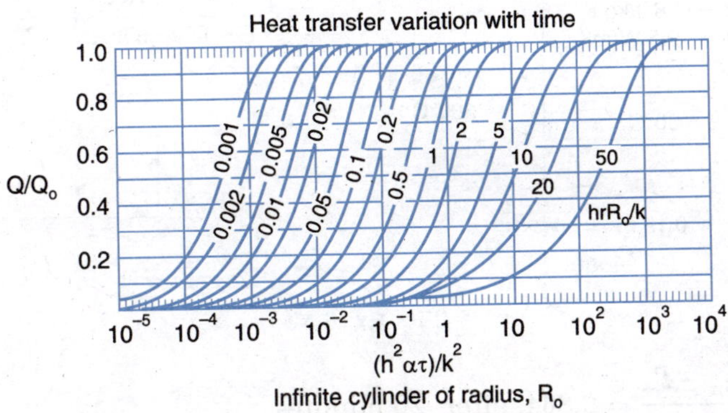

Infinite cylinder of radius, R_o

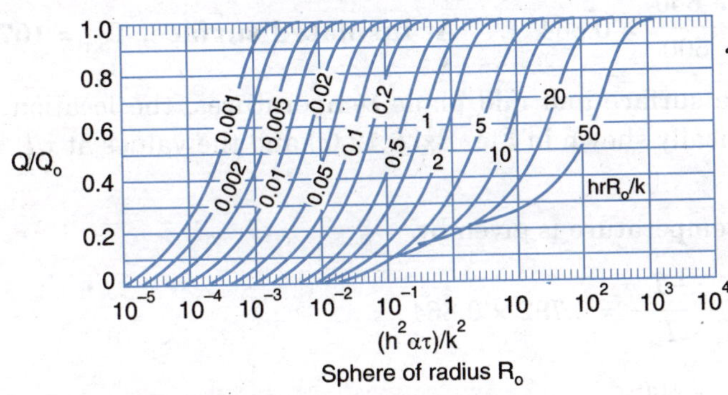

Sphere of radius R_o

Fig. 6.12. Internal energy ratio

Example 6.12: *A slab of thickness 15 cm initially at 30°C is exposed on one side to gases at 600°C with a convective heat transfer coefficient of 65 W/m²K. The other side is insulated. Using the following property values determine the temperatures at both surfaces and the centre plane (x = 0.075 m) after 20 minutes, density: 3550 kg/m³, sp. heat = 586 J/kgK, conductivity = 19.5 W/mK. Also calculate the heat flow upto the time into the solid.*

Solution: The data is presented in Fig. Ex. 6.12(*a*). The slab model with the centre plane at zero and thickness 0.15 m is used. As inside is insulated this can be considered as half slab with $x = 0$ at insulated face.

The quantities *Bi* and *Fo* are calculated using

$$Bi = \frac{65 \times 0.15}{19.5} = 0.5,$$

$$Fo = \frac{19.5}{3550 \times 586} \times 20 \times 60/0.15 \times 0.15 = 0.5$$

The procedure of obtaining temperature is illustrated with skeleton charts in Fig. Ex. 6.12 (*b*) and (*c*). The centre temperature is obtained by entering the chart as shown in Fig. Ex. 6.12 (*b*). The excess temperature ratio at the centre is obtained as 0.864.

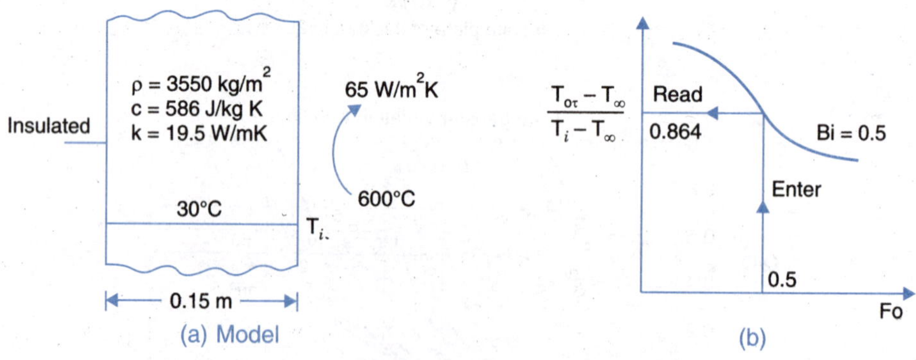

Fig. Ex. 6.12

$$\frac{T_{o,\tau} - T_\infty}{T_i - T_\infty} = 0.864, \text{ after } 20 \text{ minutes}$$

$$\frac{T_{o,\tau} - 600}{30 - 600} = 0.864 \quad \therefore \quad \textbf{At insulated surface, } \mathbf{T_{o,\tau} = 107.52°C}$$

To obtain the surface and mid plane temperatures, the location chart is entered at *Bi* = 0.5 as schematically shown in Fig. Ex. 6.12 (*c*) and the values at *x/L* = 1 and 0.5 are read as 0.792 and 0.948.

The surface temperature is given by

$$\frac{T_{L,\tau} - T_\infty}{T_i - T_\infty} = 0.792 \times 0.864$$

$$\frac{T_{L,\tau} - 600}{30 - 600} = 0.6843$$

∴ **Surface temperature T_L = 210°C**

The mid plane temperature:

$$\frac{T_{x,\tau} - 600}{30 - 600} = 0.864 \times 0.948$$

∴ **T = 133.13°C**

The heat flow is determined using the heat flow chart as shown schematically in Fig. Ex. 6.12(d). First the parameter is calculated:

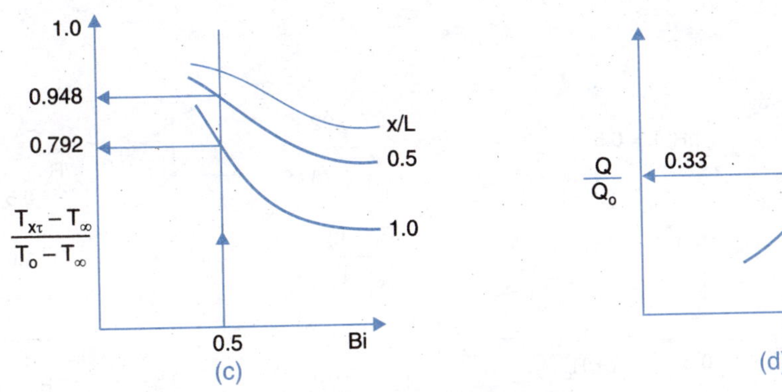

Fig. Ex. 6.12

$$\frac{h^2\alpha\tau}{k^2} = \frac{65 \times 65 \times 19.5 \times 20 \times 60}{3550 \times 586 \times 19.5^2} = 0.125$$

Entering the chart at this point and finding the meeting of point with $Bi = 0.5$, the ratio Q/Q_o is read as 0.33.

$\therefore$ $\quad$ **Q** = 0.33 × 3550 × 586 × 0.15 × 1(600 – 30)

$\quad\quad$ **= 58.7 × 10⁶ J/m²**

A rough check can be made by using an average temperature increase and finding the change in internal energy. The average temperature rise is (107.52 + 210 + 133.13)/3 – 30 = 120.22°C.

Q = 3550 × 0.15 × 586 × 120.22 = **37.51 × 10⁶**

This is of the same order of magnitude and hence checks.

Example 6.13: *A long cylinder of radius 15 cm initially at 30°C is exposed over the surface to gases at 600°C with a convective heat transfer coefficient of 65 W/m²K. Using the following property values determine the temperatures at the centre, mid radius and outside surface after 20 minutes. Density = 3550 kg/m³, sp. heat = 586 J/kg K, conductivity = 19.5 W/mK. Also calculate the heat flow.*

Solution: The procedure is described in articles 6.1.2 and 6.1.3.

$$Bi = \frac{hR}{k} = \frac{65 \times 0.15}{19.5} = 0.5,\ Fo = \frac{\alpha\tau}{R^2} = \frac{19.5}{3550 \times 586} \times \frac{20 \times 60}{0.15 \times 0.15} = 0.5$$

Entering the chart for centre temperature as schematically shown in Fig. Ex. 6.13 (a) the temperature ratio is read as 0.72.

$\therefore$ Centre temperature is found using

$$\frac{T_{o,\tau} - 600}{30 - 600} = 0.72$$

$\therefore$ $\quad$ $\mathbf{T_{o,\tau} = 189.6°C}$

Chapter 6

To calculate the temperatures at the surface and mid radius, the location chart as schematically shown in Fig. Ex. 6.13 (*b*) is entered at $Bi = 0.5$ and values are read at $r/R = 1$ and 0.5, as 0.78 and 0.92.

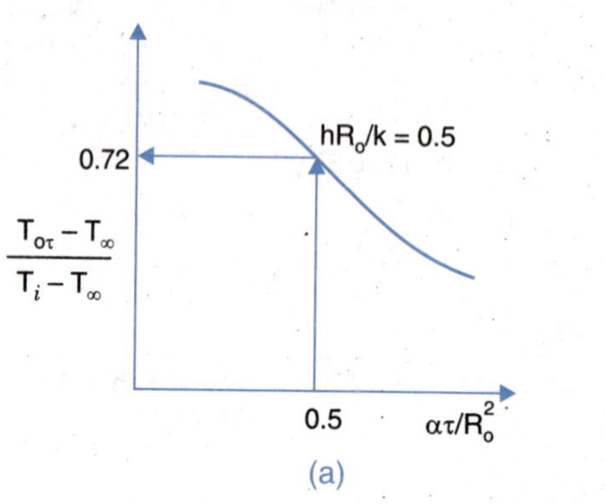

(a)

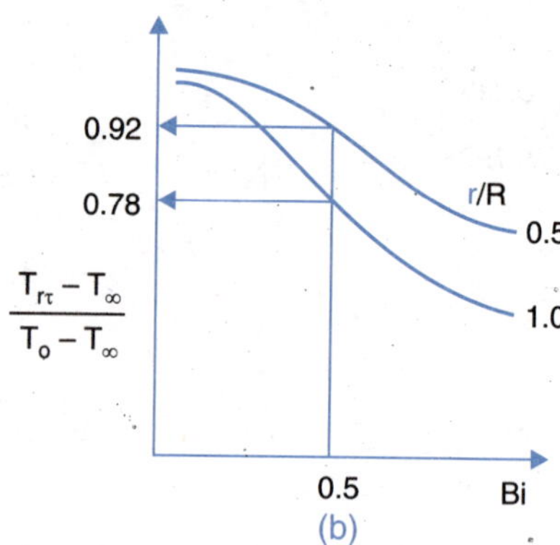

(b)

Fig. Ex. 6.13

∴ **Surface temperature** is obtained using

$$\frac{T_{R,\tau} - T_\infty}{T_i - T_\infty} = \frac{T_{R,\tau} - 600}{30 - 600}$$

$$= 0.72 \times 0.78$$

∴ $$\mathbf{T_{R,\tau} = 279.9°C}$$

Mid radius temperature is obtained using

$$\frac{T_{r,\tau} - T_\infty}{T_i - T_\infty} = 0.72 \times 0.92$$

∴ $$\mathbf{T_{r,\tau} = 222.4°C}$$

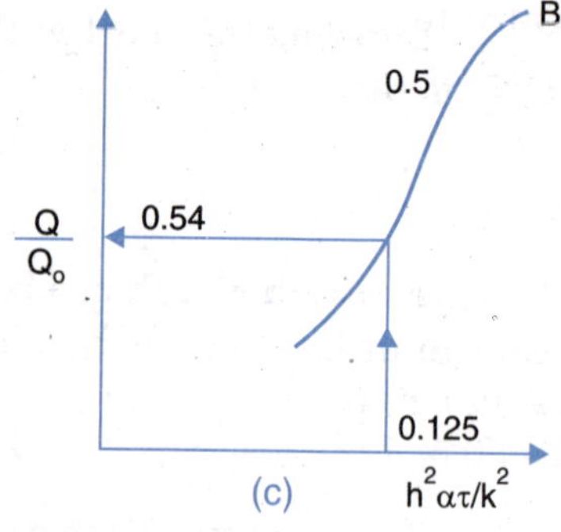

(c)

Fig. Ex. 6.13

Heat flow: Calculating the value of $h^2\alpha\tau/k^2$, as schematically shown the heat flow chart is entered at this value 0.125 and the meeting point with $Bi = 0.5$ is read of as Q/Q_o is equal to 0.54 [Fig. Ex. 6.13 (*c*)]

$$h^2\ \alpha\tau/k^2 = \frac{65 \times 65 \times 19.5}{3550 \times 586} \times \frac{30 \times 60}{19.5 \times 19.5} = 0.125$$

∴ $$\mathbf{Q} = \pi \times 0.15^2 \times 1 \times 3550 \times 586 \times (600 - 30) \times 0.54$$

$$= \mathbf{45[illegible]6 \times 10^6\ J/m\ length}$$

Note that the cylinder of equal dimension gets heated up quicker due to larger surface area for a given volume.

Example 6.14: *Use the data in example 6.13 for a sphere of radius 0.15 m. Determine the centre, surface and mid radius temperatures and also the heat flow after 20 minutes.*

Solution: The values of Bi, Fo, $h_2\alpha\tau/k^2$ are read from the example 6.8 as 0.5, 0.5 and 0.125.

The centre excess temperature chart is entered at $Fo = 0.5$ and read at $Bi = 0.5$ (ref. figures in example 6.14) and the value is 0.58. Also the location charts are entered at $Bi = 0.5$ for $r/R = 1$ and 0.5 and the temperature ratios are read as 0.782 and 0.942.

Centre temperature:

$$\frac{T-600}{30-600} = 0.58 \qquad \therefore \mathbf{T_{mid} = 269.4°C}$$

Surface temperature:

$$\frac{T-600}{30-600} = 0.58 \times 0.782 \qquad \therefore \mathbf{T_{sur} = 341.5°C}$$

Mid radius temperature:

$$\frac{T-600}{30-600} = 0.58 \times 0.942 \qquad \therefore \mathbf{T = 288.6°C}$$

Q/Q_o is obtained from the chart as 0.73

$$\therefore \qquad \mathbf{Q} = 0.73 \times \frac{4}{3} \times \pi \times 0.15^3 \times 3550 \times 586\ (600-30) = \mathbf{12.24 \times 10^6 J}$$

Note: It can be seen that the sphere heats up quicker compared to the other two geometries.

Example 6.15: *In the case of the sphere in example 6.14, determine the time when the centre temperature will reach 372°C.*

Solution: Direct solution is possible only for specified centre temperature. For temperature specifications at other locations, a trial solution is needed.

Bi = 0.5, Temperature ratio is

$$\frac{372-600}{30-600} = \frac{228}{570} = 0.4$$

Entering the centre temperature chart for sphere as shown schematically at 0.4 and noting the cut off for *Bi* = 0.5, *Fo* is read as 0.77 (Fig. Ex. 6.15)

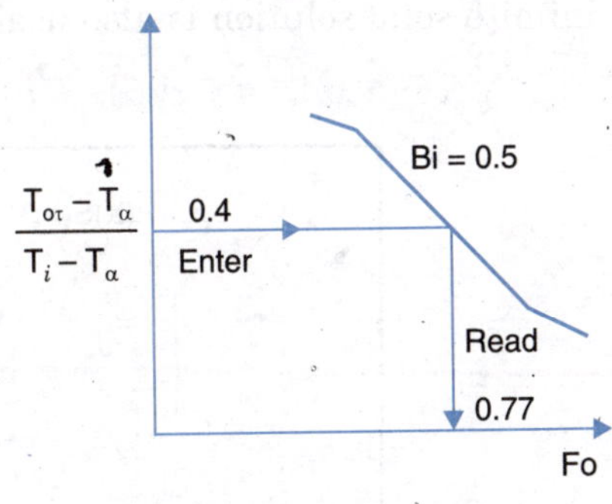

Fig. Ex. 6.15

$$\therefore \qquad \frac{19.5}{3550 \times 586} \times \frac{\tau}{0.15^2} = 0.77$$

$$\therefore \qquad \mathbf{\tau = 1848.27\ sec\ or\ 30.8\ min.}$$

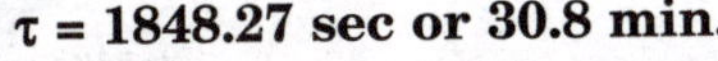

COMMENT

Trial and error method may be necessary to find the time to reach a given temperature at radii other than the centre.

6.6 PRODUCT SOLUTION

Most of the practical cases involve shapes of specified sizes and infinite slab or long cylinder solution cannot be used directly. In the case of a short cylinder, in addition to the curved surface the flat surfaces are also exposed to the surroundings and hence the cooling or heating will be faster. Such cases can be treated as a combination of parts of two or more solids and temperature ratio is obtained as the product of temperature ratios obtained considering it as part of each of these solids. In the case of a short cylinder the situation is shown in Fig. 6.13.

For exposures at base and top the short cylinder is taken as part of slab of $2L$ = height of cylinder and the temperature ratio at any axial distance is determined as described earlier.

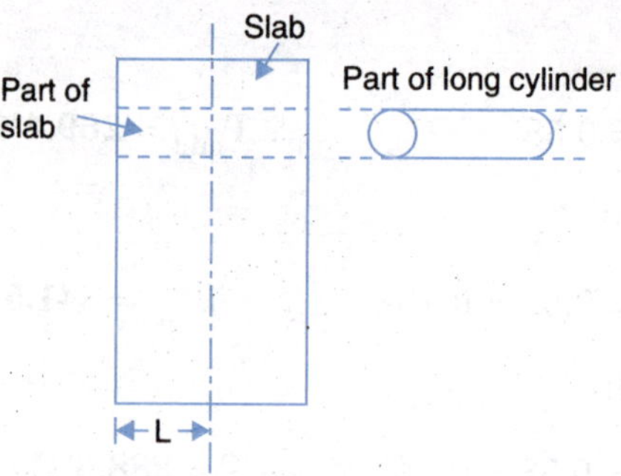

Fig. 6.13

For the exposure of the curved surface the short cylinder is treated as part of a long cylinder and the temperature ratio at any radial position is calculated as described earlier.

The temperature ratio at the given point is the product of these two ratios determined. The heat flow also can be calculated as a combination of individual heat flows. Some of the possible combinations and the equations are shown in Table 6.1. A brick like object is taken as a combination of three slabs. A long rectangular prism is taken as a combination of two slabs and a semi infinite solid. A hemisphere is taken as a combination of sphere and slab. Semi-infinite solid solution is also usable in such combinations.

Table. 6.1. Product solutions for temperatures in multidimensional systems

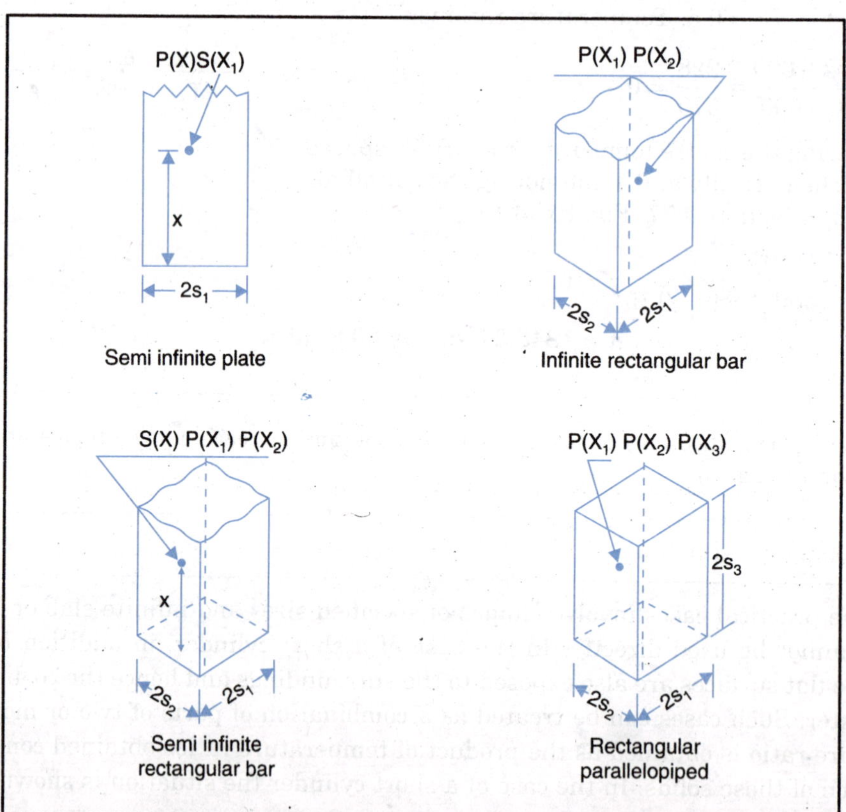

(Contd...)

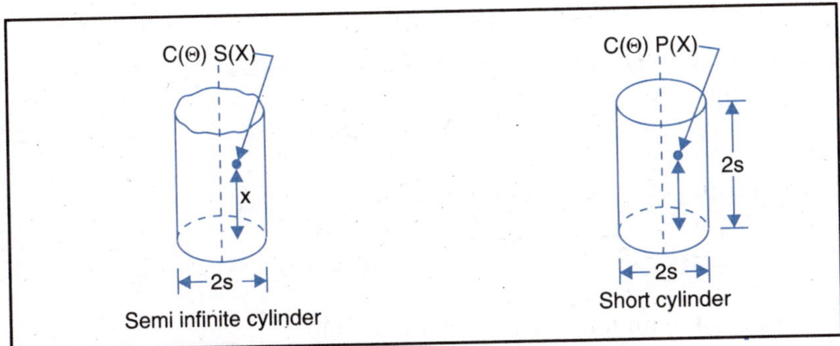

P(X) Solution for infinite plate

S(X) Solution for semi-infinite bodies

C(θ) Solution for infinite cylinder

Intersection of two bodies:

$$\left(\frac{Q}{Q_o}\right)_{\text{Total}} = \left(\frac{Q}{Q_o}\right)_1 + \left(\frac{Q}{Q_o}\right)_2 \left[1 - \left(\frac{Q}{Q_0}\right)_1\right]$$

Intersection of three one dimensional systems:

$$\left(\frac{Q}{Q_o}\right)_{\text{Total}} = \left(\frac{Q}{Q_o}\right)_1 + \left(\frac{Q}{Q_o}\right)_2 \left[1 - \left(\frac{Q}{Q_0}\right)_1\right] + \left(\frac{Q}{Q_o}\right)_3 \left[1 - \left(\frac{Q}{Q_o}\right)_1\right]\left[1 - \left(\frac{Q}{Q_o}\right)_2\right]$$

Example 6.16: *A cylinder of radius 0.15 m and height 0.30 m having properties as in examples 6.12 and 6.13 initially at 30°C is exposed to gases at 600°C with a convective heat transfer coefficient of 65 W/m²K. Determine the temperatures after 20 minutes at locations 1 to 9 shown in Fig. Ex. 6.16. This problem can be solved using the solutions of examples 6.12 and 6.13. Using these solutions, the temperature ratios at various locations are tabulated below:*

	Mid plane axis	*Half thickness mid radius*	*Surface*
Slab	*0.864*	*0.864 × 0.948*	*0.864 × 0.792*
Cylinder	*0.72*	*0.72 × 0.92*	*0.72 × 0.78*

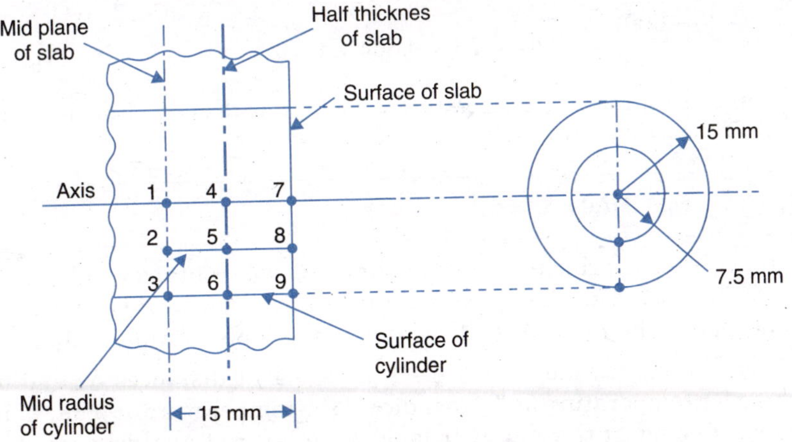

Fig. 6.16

Solution: The temperatures at the various locations from 1 to 9 are found as below:

Point 1: Axis of cylinder and mid plane of slab

$$\therefore \quad \frac{T_1 - T_\infty}{T_i - T_\infty} = \frac{T_1 - 600}{30 - 300} = 0.72 \times 0.864 \qquad \therefore \quad \mathbf{T_1 = 245.4°C}$$

Point 2: Mid radius of cylinder and mid plane of slab

$$\frac{T_2 - T_\infty}{T_i - T_\infty} = \frac{T_2 - 600}{30 - 600} = 0.72 \times 0.92 \times 0.864 \qquad \therefore \quad \mathbf{T_2 = 273.8°C}$$

Point 3: Surface of cylinder and mid plane of slab

$$\frac{T_3 - 600}{30 - 600} = 0.72 \times 0.78 \times 0.864 \qquad \therefore \quad \mathbf{T_3 = 323.4°C}$$

Point 4: Axis of cylinder and half thickness of slab

$$\therefore \quad \frac{T_4 - 600}{30 - 600} = 0.72 \times 0.864 \times 0.948 \qquad \therefore \quad \mathbf{T_4 = 263.9°C}$$

Point 5: Mid radius of cylinder and half thickness of slab

$$\frac{T_5 - 600}{30 - 600} = 0.72 \times 0.92 \times 0.864 \times 0.948 \qquad \therefore \quad \mathbf{T_5 = 290.7°C}$$

Point 6: Surface of cylinder and half thickness of slab

$$\frac{T_6 - 600}{30 - 600} = 0.72 \times 0.78 \times 0.864 \times 0.948 \qquad \therefore \quad \mathbf{T_6 = 337.8°C}$$

Point 7: Axis of cylinder and surface of slab

$$\therefore \quad \frac{T_7 - 600}{30 - 600} = 0.72 \times 0.864 \times 0.792 \qquad \therefore \quad \mathbf{T_7 = 319.2°C}$$

Point 8: Mid radius of cylinder and surface of slab

$$\therefore \quad \frac{T_8 - 600}{30 - 600} = 0.72 \times 0.92 \times 0.864 \times 0.792 \qquad \therefore \quad \mathbf{T_8 = 341.6°C}$$

Point 9: Surface of cylinder and surface of slab

$$\frac{T_9 - 600}{30 - 600} = 0.72 \times 0.78 \times 0.864 \times 0.792 \qquad \therefore \quad \mathbf{T_9 = 381°C}$$

Note: It can be seen that the temperature is highest at the edge. This is because the point is nearest to the heat transfer surfaces.

6.7 NUMERICAL METHOD

Using energy conservation principle, and dividing the solid into several layers, expressions can be obtained for the increase in temperature during any given period at any location subjected to transient conditions. The method is illustrated using a plane wall as shown in Fig. 6.14.

Considering the planes numbered $n - 1$, n, $n + 1$ at distances Δx and using a double subscript notation for temperature as T_n^p to indicate the temperature at plane n at time interval p each time interval being $\Delta\tau$ the energy balance for a layer of Δx thickness at plane n can be written as net **heat energy conducted into the layer during a time period = energy stored in the layer = heat capacity × temperature increase during the time period**

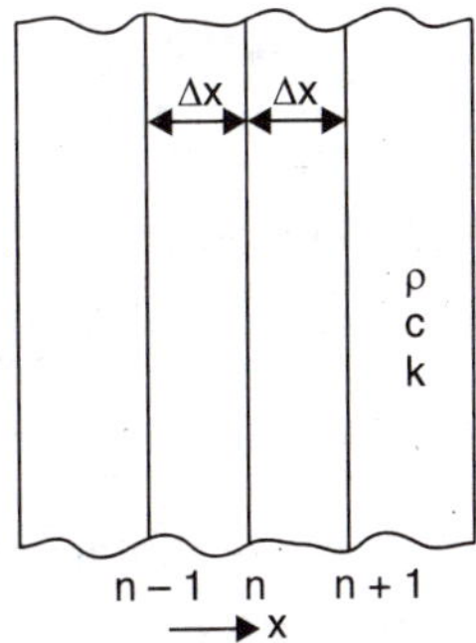

Fig. 6.14. Mode representation

or
$$\Delta\tau \times \frac{kA(T_{n-1}^p - T_n^p)}{\Delta x} + \frac{kA(T_{n+1}^p - T_n^p)}{\Delta x} \times \Delta\tau = \rho c A\, \Delta x\, (T_n^{p+1} - T_n^p)$$

or
$$T_{n-1}^p + T_{n+1}^p - 2T_n^p = \frac{\Delta x^2}{\alpha \Delta\tau}(T_n^{p+1} - T_n^p)$$

$$\text{denoting } \frac{\Delta x^2}{\alpha\Delta\tau} = M \qquad ...(6.24)$$

$$\frac{1}{M}[T_{n-1}^p + T_{n+1}^p - (2-M)T_n^p] = T_n^{p+1} \qquad ...(6.25)$$

Note that $(\alpha\Delta\tau/\Delta x^2)$ is a form of Fourier number. The equation (6.25) can be expressed as

$$T_n^{p+1} = Fo\ [T_{n-1}^p + T_{n+1}^p] + (1 - 2Fo)\ T_n^p \qquad ...[6.25(a)]$$

This method is called explicit method, because the unknown quantity T_n^{p+1} is expressed in terms of all known quantities. For convection boundary (surface node) the equation is

$$T_o^{p+1} = 2Fo\ (T_1^p + Bi\ T_\infty) + (1 - 2Fo - 2BiFo)\ T_o^p \qquad ...[6.25(b)]$$

Where o refers to the surface node. Bi = Biot number $(h\Delta x/k)$.

For equation 6.25(b) to be applicable $Fo\ (1 + Bi) \le 0.5$

This equation is the basis for numerical method of temperature estimation under transient conditions in a slab. In the use of the method it is found that Δx and $\Delta\tau$ should be chosen in such a way that $M > 2$. Otherwise the solution does not converge.

A special case is when the values of Δx, $\Delta\tau$ are chosen for a given material with property values specified, such that $M = 2$. In this case

$$T_n^{p+1} = \frac{1}{2}(T_{n-1}^p + T_{n+1}^p) \qquad ...(6.26)$$

The present nodal temperature is equal to the average of the temperatures at the adjacent nodes during the previous time interval. Later this will be used in graphical method. In this case if time interval $\Delta\tau$ is chosen for a given material the distance interval Δx is automatically specified. Though this may appear as a restriction, it is convenient for both graphical method and for calculator solutions. An advantage of the numerical method is that any initial temperature distribution can be dealt with. Shapes like cylinder, sphere or even odd shapes can be analysed, except that in these case the area, volumes etc. will vary with the nodal plane and so each coefficient has to be determined specifically.

Example 6.17: *Given $T_{n-1}^p = T_{n+1}^p = 120°C$ and $T_n^p = 60°C$ show that the application of equation 6.25 (a) will contradict the second law of thermodynamics if $Fo > 0.5$.*

Solution:
$$T_n^{p+1} = Fo\ (120 + 120) + (1 - 2Fo)\ 60$$
$$= 240\ Fo + 60 - 120\ Fo = 120Fo + 60.$$

If $Fo = 0.5$, $T_n^{p+1} = 120°C$, same as adjacent nodes.

If $Fo = 0.6$, $T_n^{p+1} = 132°C$.

Heat flow has to be from the adjacent nodes. As $T_n^p < T_{n-1}^p$ or T_{n+1}^p heat cannot flow from 120°C to 132°C, it contradicts the second law.

Example 6.18: *A semi infinite slab with a thermal diffusivity of 1.67 × 10^–6^ m²/s initially at 100°C has its surface suddenly raised to 200°C and maintained at the level. Taking a time interval of 120 seconds and nodal thickness of 2 cm, write down the nodal equation and solve for the temperatures at the nodes upto 6 time intervals.*

Solution: M is determined in the equation (6.25) using equation (6.24).

$$M = \frac{\Delta x^2}{\alpha \Delta \tau} = \frac{0.02^2}{1.67 \times 10^{-6} \times 120} = 2$$

This is convenient for the solution of the problem. Equation (6.26) results if $M = 2$

$$\therefore \quad T_n^{p+1} = (T_{n-1}^p + T_{n+1}^p)/2$$

The calculation is done in the tabulation below.

The present nodal temperature is the average of the previous nodal temperatures at the adjacent nodes.

At node zero the temperature is maintained at 200°C and at any location, it can be verified that the temperature is the mean of the adjacent nodal temperatures at the previous period. For example, take node 2 time interval 2. The temperature is 125 = (150 + 100)/2.

Node No. Time interval	***0***	***1***	***2***	***3***	***4***	***5***	***6***	***7***	***8***
0	200	100	100	100	100	100	100	100	100
1	200	150	100	100	100	100	100	100	100
2	200	150	125	100	100	100	100	100	100
3	200	162.5	125	112.5	100	100	100	100	100
4	200	162.5	137.5	112.5	106.25	100	100	100	100
5	200	168.75	137.5	121.875	106.25	103.175	100	100	100
6	**200**	**168.75**	**145.19**	**121.875**	**112.5**	**103.175**	**101.56**	**100**	**100**

This can be plotted for better visualisation.

6.7.1. Heat Generation

If heat generation is present, equation (2.25) becomes (q – heat generation rate W/m³)

$$\frac{1}{M}\left[T_{n-1}^p + T_{n+1}^p - (2 - M)T_n^p + \frac{q\Delta x^2}{k}\right] = T_n^{p+1} \quad ...(6.27)$$

and if $\Delta\tau$ and Δx are chosen such that $M = 2$

$$\frac{1}{2}\left[T_{n-1}^p + T_{n+1}^p + \frac{q\Delta x^2}{k}\right] = T_n^{p+1} \quad ...(6.28)$$

Example 6.19: *In the example 6.18 other things remaining the same if the heat generation is 1.25 × 10^6^ W/m³ and conductivity is 50 W/mK, determine the nodal temperatures upto 6 or 7 times intervals.*

Solution: The equation to be used is 6.28

$$\frac{q\Delta x^2}{k} = \frac{1.25 \times 10^6 \times 0.02^2}{50} = 10$$

So the present nodal temperature is the average of the adjacent nodes at the end of previous time interval plus 10/2 °C. The values are tabulated using

$$T_n^{p+1} = \frac{1}{2}\,[T_{n-1}^p + T_{n+1}^p + 10]$$

For example node 3 and time interval 3, $T_{3,3} = \dfrac{135 + 110 + 10}{2} = 127.5.$

Nodal temperatures

Node No. Time	0 Surface	1	2	3	4	5	6	7	8	9
0	200	100	100	100	100	100	100	100	100	100
1	200	155	105	105	105	105	105	105	105	105
2	200	157.5	135	110	110	110	110	110	110	110
3	200	172.5	138.75	**127.5**	115	115	115	115	115	115
4	200	174.4	155	131.9	126.3	120	120	120	120	120
5	200	182.5	158.1	145.6	130.9	128.1	125	125	125	125
6	200	184	169	149.5	141.9	133	131.6	130	130	130

A plot will be very graphic in bringing out the result-try it.

6.7.2. In The Case of Convection Boundary

The nodal equation for the zeroth node or the surface will be different. For all other nodes equations (6.25), (6.26) and (6.28) will apply as per the conditions specified. For the zeroth or surface node with

$$M > (2 + h\Delta x/2)$$

$$T_0^{p+1} = \frac{2}{M}\left[\frac{h\Delta x}{k} T_\infty + T_1^p + T_o^p\,(M-1)\right] \qquad \text{...(6.29)}$$

This equation can be used to solve for the nodal temperatures, but this does not converge and so $\Delta x^2/\alpha\Delta x$ should be taken to have values higher than $2 + 2\,[(h\Delta x)/k]$.

6.7.3. The Implicit Method

In this method, the present temperature at a node is expressed in terms of the future temperatures of the adjacent nodes. In the explicit method the nodal temperatures can be calculated after every time interval. In the implicit method all the equations for the entire number of time intervals have to be solved simultaneously. This is not difficult with the use of computers. The advantage of the method is that this is entirely stable as different from the other method where $Fo < 0.5$ for stability.

For interior nodes, (one dimensional)

$$(1 + 2Fo)\,T_n^{p+1} - Fo\,(T_{n-1}^{p+1} + T_{n+1}^{p+1}) = T_n^p. \qquad \text{...(6.30)}$$

For surface nodes

$$(1 + 2Fo + 2FoBi)\,T_o^{p+1} - 2Fo\,T_1^{p+1} = 2FoBi\,T_\infty + T_o^p \qquad \text{...(6.31)}$$

Zero is the surface node 1 is the next node.

Bi and T_∞ are due to the convection resistance. It can be seen the present temperature at the given node is expressed in terms of future temperatures of all the three nodes.

Chapter 6

6.8 GRAPHICAL METHOD

Graphical method can be adopted for both semi-infinite and infinite slabs (without heat generation) using the equation 6.26. If the nodal temperatures at any time is known, then choose Δx and so that $\Delta x^2/\alpha\Delta\tau = 2$. Referring to Fig. 6.15, if the nodal temperatures at $n-1$ and $n+1$ nodes at the previous time are (to scale) marked as T_{n-1}^{p} and T_{n+1}^{p} at the present time the temperature at the nth node is obtained by joining the two temperatures and taking the point on the nth nodal plane.

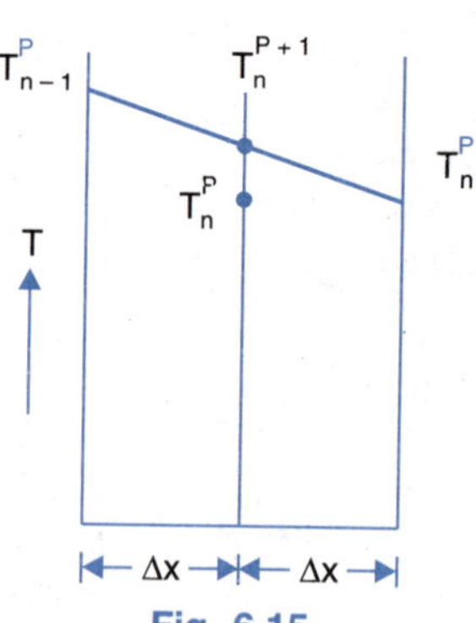

Fig. 6.15

The full picture can be obtained by continuing the process. Convection boundary can also be taken care of by adding a layer of thickness

$$\Delta x = \frac{k}{h} \text{ m.}$$

This layer should be used as such with the single node at its surface with $T = T_\infty$.

Example 6.20: *Work out the example 6.18 by the graphical method. The nodes are laid as in Fig. Ex. 6.20. The first line is from A to x_2 which fixes*

$$T_1^1 = (T_0^0 + T_2^0)/2.$$

Next T_1^1 is joined with x_3 giving $T_2^2 = (T_1^1 + T_3^1)/2$

Now joining T_2^2 to A we get $T_1^3 = (T_0^2 + T_2^2)/2$. This is continued as in figure.

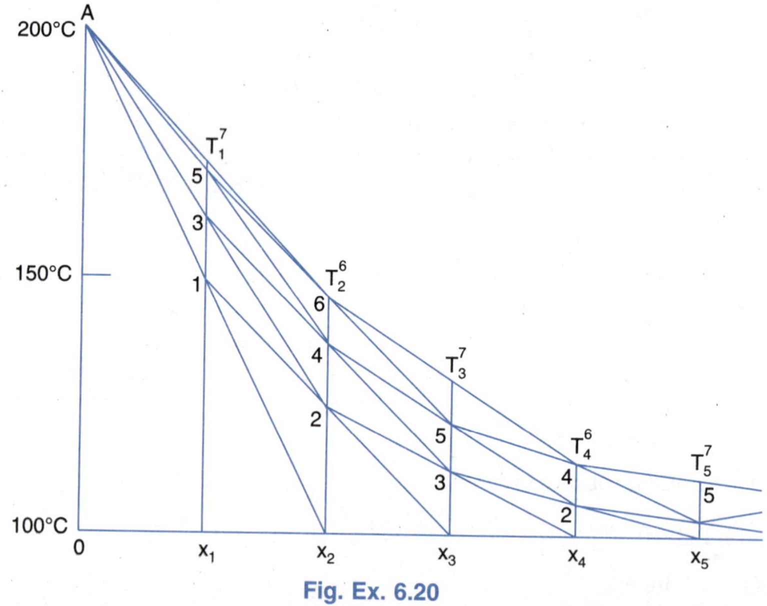

Fig. Ex. 6.20

Solution: It may be noted that at node 1 or odd numbered nodes temperatures at 1, 3, 5, 7 *i.e.*, odd time interval ends will be obtained. At the even numbered nodes, temperatures at even number time intervals will be obtained. Interpolation will provide the temperature at the end of any interval at all nodes.

This ingenious idea of Schmidt can be very easily executed by simple computer programme.

ADVANCED EXAMPLES

Lumped Parameter System (A-6.1 to A-6.14)

Example A-6.1: *A diecast component has a mass of 1.2 kg and density 7150 kg / m^3 with surface area of 0.075 m^2. The thermal conductivity of the material is 95 W / mK and the specific heat is 385 J / kg K. It comes out of the machine at 345°C and is exposed to air at 20°C with a convective heat transfer coefficient of 56.8 W / m^{2}K. Determine (i) The temperature of the part after 5 minutes (ii) The time required to reach 50°C (iii) The time constant (iv) The value of convective heat transfer coefficient upto which the lumped parameter model can be used (v) The volume/area ratio upto which the lumped parameter model can be used.*

Solution: First a check has to be made to ascertain whether lumped parameter model can be used. For that purpose the Biot number is to be calculated

$$Bi = hL/k,\ L = \text{Volume/Surface area, Volume} = \frac{\text{mass}}{\text{density}}$$

$$L = 1.2/(7150 \times 0.075) = 2.24 \times 10^{-3}\ \text{m}$$

$$\therefore \qquad Bi = 56.8 \times \frac{1.2}{7150} \times \frac{1}{0.075} \times \frac{1}{95} = 1.34 \times 10^{-3}$$

This is much less than 0.1. So the model can be adopted.

The calculations are made using equation (6.4).

$$\frac{T - T_\infty}{T_o - T_\infty} = e^{-(hA/\rho cV)\,\tau} = e^{(-h/\rho c)\,(1/L)\,\tau}$$

$$(i) \qquad \frac{T - 20}{345 - 20} = \exp\left[\frac{-56.8}{7150 \times 385} \times \frac{1}{2.24 \times 10^{-3}} \times 5 \times 60\right] = 0.063$$

$\therefore$ **Temperature after 5 minutes is T = 40.5°C**

$$(ii) \qquad \frac{50 - 20}{345 - 20} = \exp\left[\frac{-56.8}{7150 \times 385} \times \frac{1}{2.24 \times 10^{-3}} \times \tau\right]$$

Taking ln,
$$2.383 = \frac{56.8}{7150 \times 385 \times 2.24 \times 10^{-3}} \times \tau$$

$\therefore$ τ = **258.7 sec** or 4.31 minutes

(*iii*) The time constant is the time at which the temperature ratio (equation 6.6)

$$\frac{T - T_\infty}{T_o - T_\infty} = \frac{1}{e} \text{ or } \frac{h\tau}{\rho c L} = 1$$

$$\therefore \qquad \tau = \rho c L/h = \frac{7150 \times 385 \times 2.24 \times 10^{-3}}{56.8} = \mathbf{108.6\ seconds}$$

(*iv*) For the lumped parameter model to be used

$$\frac{hL}{k} < 0.1 \text{ converting to equality}$$

Chapter 6

$$\mathbf{h} = \frac{0.1\,k}{L} = \frac{0.1 \times 95}{2.24 \times 10^{-3}} = \mathbf{4241\ W/m^2K}$$

For any value less than this the model can be used

(*v*) $$\frac{hL}{k} = 0.1, \frac{V}{A} = \mathbf{L} = \frac{95 \times 0.1}{56.8} = \mathbf{0.1672\ m}$$

the present value is 0.00224 m.

Note: This problem brings out all the various nuances of the model.

Example A-6.2: *Cylindrical pieces of size 60 mm dia and 60 mm height with density = 7800 kg/m³, specific heat = 486 J/kgK and conductivity 43 W/mK are to be heat treated. The pieces initially at 35°C are placed in a furnace at 800°C with convection coefficient at the surface of 85 W/m²K. Determine the time required to heat the pieces to 650°C. If by mistake the pieces were taken out of the furnace after 300 seconds, determine the shortfall in the requirements.*

Solution: First it is necessary to check for the use of lumped parameter model by calculating Biot number.

$$\text{Biot number} = \frac{hL}{k},\ L = \text{Volume/Surface area}$$

$$\text{Volume} = \pi r^2 h,\ \text{surface area} = 2\pi r^2 + 2\pi r h$$

$$\therefore \quad L = \frac{\pi r^2 h}{2\pi r^2 + 2\pi r h} = \frac{rh}{2r + 2h} = \frac{0.03 \times 0.06}{2(0.03 + 0.06)} = 0.01\ \text{m}$$

$$\therefore \quad Bi = \frac{85 \times 0.01}{43} = 0.02$$

This is less than 0.1 and hence the lumped parameter model is applicable. To determine the time needed for the heating, equation (6.4) is used.

$$\frac{650 - 800}{35 - 800} = \exp\left[\frac{-85}{7800 \times 486} \times \frac{1}{0.01} . \tau\right]$$

Solving τ = **726.6 seconds or 12 min 7 seconds**

If the piece is taken out after 300 seconds, then

$$\frac{T - 800}{35 - 800} = \exp\left[\frac{-85 \times 300}{7800 \times 486 \times 0.01}\right]$$

$\therefore$ **T = 409.6°C. This falls short by about 240°C.**

Example A-6.3: *A bearing piece in the form of half of a hollow cylinder of 60 mm ID, 90 mm OD and 100 mm long is to be cooled to –100°C from 30°C using a cryogenic gas at –150°C with a convection coefficient of 70W/m²K. density = 8900 kg/m³, specific heat = 444 J/kgK, conductivity = 17.2 W/mK. Determine the time required.*

Solution: The piece to be cooled is shown in Fig. AE-6.3.

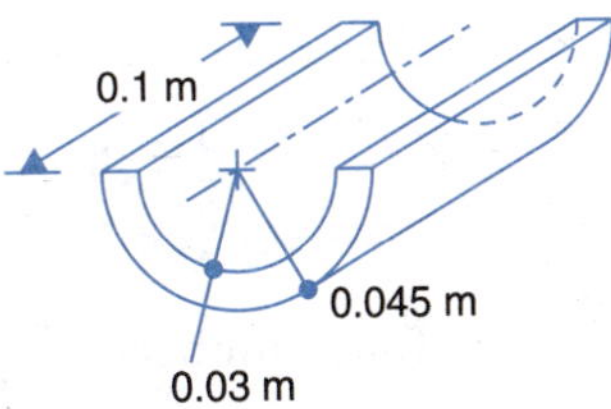

Fig. AE-6.3

The volume of the piece

$$V = \pi \frac{(0.045^2 - 0.03^2)}{2} \times 0.1 \text{ m}^3$$

$$= 1.76715 \times 10^{-4} \text{ m}^3$$

Surface area = $(\pi \times 0.03 \times 0.1)$
$+ (\pi \times 0.045 \times 0.1) + (2 \times 0.1 \times 0.015)$

$$+ \frac{2\pi(0.045^2 - 0.03^2)}{2} = 0.030096 \text{ m}^2$$

$$\therefore \quad L = \frac{V}{A} = \frac{1.76715 \times 10^{-4}}{0.030096} = 5.872 \times 10^{-3} \text{ m}$$

$$Bi = \frac{hL}{k} = \frac{70 \times 5.872 \times 10^{-3}}{17.2} = 0.02389.$$

This is smaller than 0.1 hence, lumped parameter model can be used. To determine the time, equation (6.4) is used.

$$\frac{-100-(-150)}{30-(-150)} = \exp\left[\frac{-70 \times \tau}{8900 \times 444 \times 5.872 \times 10^{-3}}\right]$$

Solving for **τ,** time required for the cooling is obtained as **424.6 sec** or 7.08 minutes.

Example A-6.4: *A thermocouple in the form of a long cylinder of 2 mm dia initially at 30°C is used to measure the temperature of a cold gas at –160°C. The convection coefficient over the surface is 60 W/m²K. The material properties are; density = 8922 kg/m³, specific heat = 410 J/kgK, conductivity = 22.7 W/mK. Determine the time it will take to indicate –150°C. Also calculate the time constant.*

Solution: Considering 1 m length, the characteristic length for the solid is

$$V/A_s = \pi R^2/2\pi R = R/2 = 0.0005 \text{ m.}$$

$$Bi = \frac{60 \times 0.0005}{22.7} = 1.322 \times 10^{-3}.$$

Hence lumped parameter model can be used.

$$\frac{-150-(-160)}{30-(-160)} = \exp\left[\frac{-60 \times \tau}{8922 \times 410 \times 0.0005}\right]$$

∴ **τ = 89.76** seconds, rather high. This can be reduced by using smaller wire diameter or higher value of h.

The time constant $\tau_c = \frac{\rho c L}{h} = \frac{8922 \times 410 \times 0.0005}{60}$ = **30.5 seconds.**

Example A-6.5: *It is desired to estimate the batch time for a heat treatment process involved in cooling alloy steel balls of 15 mm dia from 820°C to 100°C in oil at 40°C with a convection coefficient of 18 W/m²K. The material properties are: density = 7780 kg/m³, specific heat = 526 J/kgK, conductivity = 45 W/mK. Determine the time required. If this is to be achieved in 10 minutes, determine the value of convection coefficient.*

Solution: The characteristic length in the case of a sphere is

$$\frac{4}{3}\pi R^3/4\pi R^2 = \frac{R}{3}\text{ m} = \frac{0.015}{6}$$

$\therefore$ Biot number $= \dfrac{18 \times 0.015}{6 \times 45} = 1 \times 10^{-3}$ m. So lumped parameter model can be used.

The time is determined using equation (6.4).

$$\frac{100-40}{820-40} = \exp\left[\frac{-18 \times \tau \times 6}{7780 \times 526 \times 0.15}\right] \qquad \therefore \quad \boldsymbol{\tau = 1457.8 \text{ seconds}}$$

If the cooling is to be achieved in 10 min. or 600 seconds

$$\frac{100-40}{820-40} = \exp\left[\frac{-h \times 600 \times 6}{7780 \times 526 \times 0.015}\right] \qquad \therefore \quad \mathbf{h = 43.74\ W/m^2K}$$

Check whether $\quad hL/k < 0.1,\ \dfrac{43.74 \times 0.015}{6 \times 45} = 2.43 \times 10^{-3}$.

Example A-6.6: *In example A-6.5, if the convection coefficient is increased to 5000 W/m²K using evaporating liquid at 40°C, determine the time required to cool to 100°C.*

Solution: Using data from example, 6.5, ($L = R/3$)

$$\frac{hL}{k} = \frac{5000 \times 0.015}{6 \times 45} = 0.28$$

So the lumped parameter model cannot be used. Recourse should be taken to the use of Heisler chart. Assuming that the centre to reach 100°C. Refer procedure in section 6.1.2.

$$\text{excess temperature ratio} = \frac{T_o - T_\infty}{T_i - T_\infty} = \frac{100-40}{820-40} = 0.077$$

Entering the centre temperature chart at 0.077 and reading on $Bi = 0.28$, $Fo = 3.4$

$$\therefore \qquad \alpha\tau/R^2 = 3.4 \quad \therefore \tau = 3.4 \times R^2/\alpha$$

$$\alpha = \frac{k}{\rho c} = 45/(7780 \times 526),\ R = 0.015/2$$

$$\therefore \qquad \tau = \frac{3.4 \times 0.015^2 \times 7780 \times 526}{4 \times 45} = \mathbf{17.5\ s}$$

By using location chart, the surface temperature can be obtained. The correction factor is 0.87.

$$\therefore \qquad \frac{T - T_\infty}{T_o - T_\infty} = 0.077 \times 0.87 \quad \therefore \quad \mathbf{T = 92.3°C}$$

If lumped parameter method is used (equation 6.4) the time is estimated as 5.25 seconds.

Example A-6.7: *A copper sphere of 10 mm dia at 80°C is placed in an air stream at 30°C. The temperature dropped to 65°C after 61 seconds. Calculate the value of convection coefficient. Assume property values as follows: density = 8925 kg/m³, specific heat = 397 J / kg K, conductivity = 393 W/mK. The characteristic length dimension = R/3 = 0.005/3 m.*

In order to decide about the model to be used, the maximum value of convection coefficient for the use of lumped conductance model is calculated

for $$\frac{hL}{k} = 0.1 \quad \therefore \quad h = \frac{0.1 \times 393 \times 3}{0.005} = 23580 \text{ W/m}^2\text{K}.$$

Solution: The actual value of h will be within this range and so lumped parameter model can be used. The equation (6.4) is used.

$$\frac{65-30}{80-30} = \exp\left[\frac{-h \times 61 \times 3}{8925 \times 397 \times 0.005}\right] \quad \therefore \mathbf{h = 34.53 \text{ W/m}^2\text{K}}$$

This is well within 23580 and so lumped parameter model can be used.

Note: This method can be used for estimating convection coefficient in a process.

Example A-6.8: *Glass spheres of radius of 2 mm at 600°C are to be cooled in an air stream at 30°C to a temperature of 80°C without any surface crack. Estimate the maximum value of convection coefficient that may be used. Also determine the minimum time required for the cooling. The property values are density = 2225 kg/m³, specific heat = 835 J/kgK, conductivity = 1.4 W/mK.*

Solution: Surface crack occurs due to temperature difference within the solid. The lumped conductance model assumes that the temperature all over the body at any point of time is the same. So if $Bi < 0.1$, then the cooling should not cause cracking. The condition is

$$\therefore \quad \frac{hL}{k} = 0.1, \quad \frac{h \times 0.002}{3} \times \frac{1}{1.4} = 0.1$$

$\therefore$ **h = 210 W/m²K.** The maximum value.

There may be an error of about 5% $\therefore$ safe value of h is 200 W/m² K.

Time for cooling is obtained using equation 6.4

$$\frac{80-30}{600-30} = \exp\left[\frac{-200 \times 3\tau}{2225 \times 835 \times 0.002}\right], \quad \mathbf{\tau = 15.07 \text{ s}}$$

COMMENT

It will be desirable to adopt still lower value of convection coefficient, so that sufficient safety margin can be provided.

Chapter 6

Example A-6.9: *A coal pellet of 1 mm dia sphere is to be heated by radiation with a source at 1200 K from 300 K to 900 K. Determine the time required. Take density = 1350 kg/m³. Specific heat 1260 J/kgK, conductivity = 0.26 W/mK.*

Solution: The equations available cannot be used as such. The heat balance can be represented by (assuming black body conditions)

$$\sigma A(T_\infty^4 - T^4)\, d\tau = \rho V c\, dT$$

Separating variables

$$\frac{\sigma A}{\rho V c} d\tau = \frac{dT}{T_\infty^4 - T^4}$$

A standard form of integral available is used:

$$\int \frac{dx}{a^4 - x^4} = \frac{1}{4a^3} \ln\left|\frac{a+x}{a-x}\right| + \frac{1}{2a^3}\tan^{-1}\frac{x}{a}$$

noting that $x \longrightarrow T$ and $a \longrightarrow T_\infty$

$$\frac{\sigma A}{\rho V_c} \cdot \tau = \left[\frac{1}{4T_\infty^3} \ln \left| \frac{T_\infty + T}{T_\infty - T} \right| + \frac{1}{2T_\infty^3} \tan^{-1} \frac{T}{T_\infty} \right]_{T_1}^{T_2} \qquad ...(1)$$

(if calculators are used in taking $\tan^{-1}$ (T/T_∞) radian mode should be used)

Noting **A/V** = (3/*R*), $\mathbf{T_1}$ = 300, $\mathbf{T_2}$ = 900 **K**

$$\frac{5.67 \times 10^{-8} \times 3}{1350 \times 1260 \times 0.0005} \cdot \tau = \frac{1}{4 \times 1200^3} \left[\ln \left(\frac{1200 + 900}{1200 - 900} \right) - \ln \left(\frac{1200 + 300}{1200 - 300} \right) \right]$$

$$+ \frac{1}{2 \times 1200^3} \left[\tan^{-1} \frac{900}{1200} - \tan^{-1} \frac{300}{1200} \right]$$

Solving, $\boldsymbol{\tau}$ **= 1.6147 sec**

(hint taking 10^{-8} to the RHS will help in the working).

A check: Applicability of pumped conductance model is checked by calculating the Biot number.

$$h_r = (T_\infty + T)\,(T_\infty^2 + T^2) \text{ Assuming an average of 600°K for } T$$

$$h_r = 5.67 \times 10^{-8}\,(1200 + 600)\,(1200^2 + 600^2) = 183.7 \text{ W/m}^2\text{K}$$

$$\therefore \qquad \frac{hL}{k} = \frac{182.7 \times 0.005}{3 \times 0.26} = 0.118$$

Note: Just on the border.

A check for heat transfer: Q = heat capacity × temp. change

$$\therefore \qquad Q = \frac{4}{3}\pi\, 0.0005^3 \times 1350 \times 1250\,(900 - 300) = 0.534 \text{ J} = 0.559 \text{ J}$$

Using the convection coefficient calculate using 600 K as average

$$\mathbf{Q} = hA\,(\Delta T)\,\tau = 183.7 \times 4\pi \times 0.0005^2 \times (1200 - 600) \times 1.6147$$

$$= \mathbf{0.559 \text{ J. Checks}}$$

Example A-6.10: *A metal sphere 20 mm radius at 900 K is allowed to cool in a room at 310 K by (i) convection only (ii) radiation only.*

Determine in each case the time required for the sphere to reach 450 K, Density = 2700 kg/m³, specific heat = 1110 J/kgK, conductivity = 218 W/mK. Convection coefficient = 18 W/m²K.

Solution:

(*i*) Check for applicability of lumped conductance model.

$$Bi = \frac{hL}{k} = \frac{18}{218} \frac{0.02}{3} = 0.00055. \text{ So applicable.}$$

Using equation (6.4)

$$\frac{450 - 310}{900 - 310} = \exp\left[\frac{-18 \times 3\tau}{2700 \times 1110 \times 0.02} \right]$$

Solving $\boldsymbol{\tau}$ **= 1597 seconds**

(*ii*) Cooling by radiation: refer, equation (1) – ve sign due to cooling

$$\tau = -\frac{\rho c V}{\sigma A}\left[\frac{1}{4T_\infty^3}\ln\left|\frac{T_\infty + T}{T_\infty - T}\right| + \frac{1}{2T_\infty^3}\tan^{-1}\frac{T}{T_\infty}\right]_{T_2}^{T_1}$$

$$\tau = \frac{-2700 \times 1110 \times 0.02}{5.67 \times 3} \times 10^8 \left[\frac{1}{4 \times 310^3}\left\{\ln\left|\frac{310+900}{310-900}\right| - \ln\left|\frac{310+450}{310-450}\right|\right\} + \frac{1}{2 \times 310^3}\left\{\left(\tan^{-1}\frac{900}{310}\right) - \left(\tan^{-1}\frac{450}{310}\right)\right\}\right]$$

= **1273 seconds**

Example A-6.11: *A metal plate 10 mm thick at 30°C is suddenly exposed on one face to a heat flux of 3000 W/m² and the other side is exposed to convection to a fluid at 30°C with a convective heat transfer coefficient of 50 W/m²K.*

Determine the temperature variation with time and also the steady state temperatures and the time to reach 1°C less than the steady state temperature on the hot face. Density = 8933 kg/m³, specific heat = 385 J/kgK, conductivity = 380 W/mK.

Solution: Steady state condition is solved for first.

The heat flow through the wall and the convection over the face is given

$$3000 = \frac{T_1 - 30}{\frac{1}{50} + \frac{0.01}{380}} \qquad \therefore \; \mathbf{T_1 = 90.08°C}$$

and

$$3000 = (T_2 - 30)50 \qquad \therefore \; \mathbf{T_2 = 90°C}$$

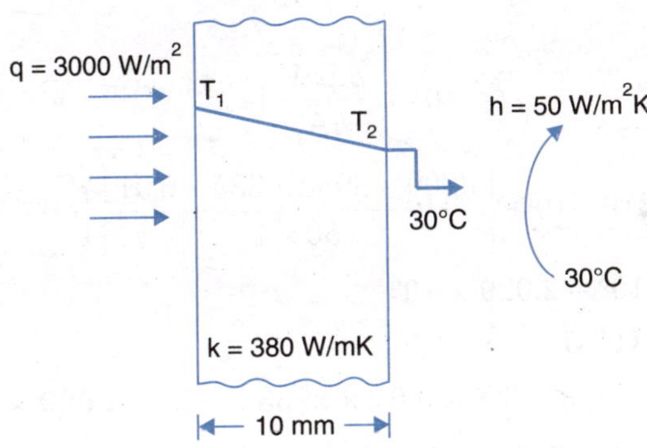

Fig. AE-6.11. Problem model

The slab is almost at constant temperature over the thickness. During the transient conditions, the energy balance at any instant when the plate temperature is T is

$$A_s\,[q - h(T - T_\infty)]\,d\tau = \rho c V\,dT$$

Now introducing a new variable, $\theta = T - T_\infty$

Separating the variables

$$\frac{A_s}{\rho c V}\,d\tau = \frac{dT}{q - h\,(T - T_\infty)} = \frac{d\theta}{q - h\theta}$$

Integrating from 0 time to time τ

$$\frac{A_s\tau}{\rho cV} = -\frac{1}{h}\ln\left[\frac{q-h\theta_2}{q-h\theta_1}\right] \quad ...(1)$$

or
$$\frac{q-h\theta_2}{q-h\theta_1} = \exp\,[-hA_s\tau/\rho cV]$$

or
$$\frac{q-h\,(T-T_\infty)}{q-h\,(T_1-T_\infty)} = \exp\,[-hA_s\tau/\rho cV] \quad ...(2)$$

This provides the variation of temperature T with time. The temperature to be reached is 89°C. (note: $(T_1 = T_\infty)$, $A_s/V = \frac{1}{L}$

$$\frac{3000-50\,(89-30)}{3000\times 1} = \exp\,[-50\times\tau/8933\times 385\times 0.01]$$

Solving: τ = 2816 seconds

Check for $\frac{hL}{k}$, $\frac{50\times 0.01}{380}$ = 0.0013 so use of lumped conductance model is justified.

A check can also be made for energy flow:

As $T_1 - T_\infty = 30 - 30 = 0$, rearranging energy balance equation,

$$h\,(T-T_\infty)\,d\tau = q\,[1-\exp\,(-hA_s/\rho cV)]\,d\tau$$

Integrating the LHS, we get the energy convected over the surface from T_∞ to T and this is given by integrating the RHS from 0 time to time τ.

∴ **Energy convected** $= \int_o^\tau q(1-e^{-(hA_s/\rho cV)\,\tau})\,d\tau$

$$= q\,(\tau-0) + \frac{q.\rho cV}{hA_s}\,[e^{-(hA_s/\rho cV)\tau} - e^0)$$

$$= 3000\,(2816-0) + \left[\frac{3000\times 8933\times 385\times 0.01}{50\times 1}\right]\left[e^{-(50\times 1/8933\times 385\times 0.01)\,2816} - 1\right]$$

$= 8.448\times 10^6 - 2.029\times 10^6$

$= \mathbf{6.419\times 10^6\ J}$

Energy stored = (59 – 30) × 0.01 × 8933 × 385 = $\mathbf{2.029\times 10^6\ J}$

Energy in = energy convected + energy stored

$= 6.419\times 10^6 + 2.029\times 10^6$ J = $\mathbf{8.448\times 10^6\ J}$

From heat flux ⟶ 3000 × 2816 = $\mathbf{8.448\times 10^6\ J}$

Checks to the required accuracy.

COMMENT

This type of counter checking is necessary particularly in nonstandard situations. Any incorrect assumptions etc. can be highlighted by such checking.

Example A-6.12: *A device has a mass of 0.5 kg and specific heat of 750 J/kgK and a surface area of 0.04 m^2. The device is initially at 30°C and is exposed to convection to a fluid at 30°C with a convective heat transfer coefficient of 12 W/m^{2}K. If the temperature of the device should increase to 120°C in 120 seconds, using an embedded heater, determine its capacity.*

Solution: The example can be modelled on the same basis as Example A-6.11. The energy equation is

$$[q - hA\,(T - T_\infty)\,d\tau] = \rho c V\,dT, \text{ letting } \theta = (T - T_\infty)$$

$$\therefore \quad \frac{1}{\rho c V}.d\tau = \frac{dT}{q - hA(T - T_\infty)} = \frac{d\theta}{q - hA\theta}$$

Integrating from 0 to time τ

$$\frac{1}{\rho c V}\tau = -\frac{1}{hA}\ln\left(\frac{q - hA\theta_2}{q}\right) \text{ as } hA\theta_1 = 0$$

$$\therefore \quad \frac{q - hA\theta_2}{q} = \exp\left[\frac{-hA\tau}{\rho c V}\right], \text{ as } \rho V = \text{mass} = 0.5$$

$$\frac{q - 12 \times 0.04\,(120 - 30)}{q} = \exp\left[\frac{-12 \times 0.04 \times 120}{0.5 \times 750}\right] = 0.8576$$

Solving **$q = 303.4$ W**

Note: As in the Example A-6.11, energy flow check can be made.

Example A-6.13: *A piece of material is to be dried in an oven. The mass is 0.5 kg, sp. heat = 2999 J/kg K. The surface area is 0.085 m^2. In the process the material gets heated and additionally 45 J/s of heat is used up in evaporation of moisture. The oven is at 180°C and the convection coefficient is 16 W/m^{2}K. Determine the time required to heat up the material from 30°C to 90°C.*

Solution: The energy equation at the instant when the temperature is T is as below:

$$hA(T_\infty - T)\,d\tau - 45\,d\tau = mc\,dT, \text{ as } (m = \rho V)$$

Separating variables

$$\frac{1}{mc}.d\tau = \frac{dT}{hA(T_\infty - T) - 45} = \frac{d\theta}{-hA\theta - 45}$$

(Letting $\theta = T - T_\infty$). Integrating between limits,

$$\therefore \quad \frac{1}{mc}\tau = -\frac{1}{hA}\ln\frac{-hA(T_\infty - T_2) - 45}{-hA(T_\infty - T_i) - 45}$$

$$\therefore \quad -\frac{hA}{mc}\tau = \ln\frac{-h(T_\infty - T_2) - 45}{-hA(T_\infty - T_i) - 45}$$

$$\frac{-hA(T_\infty - T_2) - 45}{-hA(T_\infty - T_i) - 45} = \exp\left[\frac{-hA}{mc}\tau\right] \quad ...(1)$$

Chapter 6

For the given data

$$\frac{-16 \times 0.085\,(180-90)-45}{-16 \times 0.085\,(180-30)-45} = \exp\left[\frac{-16 \times 0.085}{0.5 \times 2999}\tau\right]$$

$$\frac{-167.4}{-249} = \exp\,[-9.07 \times 10^{-4}\,\tau]$$

$\therefore$ $\quad$ **τ = 438 seconds.**

Example A-6.14: *An electronic component generating heat at 50 W is mounted on a sink of mass 0.4 kg and at steady state the temperature of the sink is 115°C. When the surrounding is at 30°C determine the temperature of the component 6 minutes after power supply is switched on. The initial and surrounding temperature are 30°C. Specific heat = 949 J/kgK.*

Solution: Let the steady state temperature be T_s and surrounding at T_∞. Then $hA_s\,(T_s - T_\infty) = q$ as energy generated = energy convected. Now the energy balance any time after power supply is switched on is given by

$$q.d\tau - hA_s\,(T - T_\infty)\,d\tau = mc\,dT$$

Substituting for q from steady state condition,

$$[hA_s\,(T_s - T_\infty) - hA_s\,(T - T_\infty)]\,d\tau = mc\,dT$$

$\therefore$

$$hA_s\,(T_s - T)\,d\tau = mc\,dT$$

Separating Variables, defining $\qquad \theta = T - T_s$

$$\frac{hA_s}{mc}\,d\tau = \frac{dT}{(T_s - T)} = \frac{d\theta}{\theta}$$

Integrating and rearranging (As at start $T_i = T_\infty$)

$$\frac{-hA_s}{mc}\,\tau = \ln\frac{T_s - T_2}{T_s - T_i} \text{ or } \ln\frac{T_s - T_2}{T_s - T_\infty}$$

$\therefore$

$$\frac{T_s - T_2}{T_s - T_i} = \exp\left[\frac{-hA_s}{mc}\,\tau\right]$$

From steady state condition:

$$hA_s = q/(T_s - T_\infty) = \frac{50}{115-30} = \frac{50}{85}, \text{ substituting this value}$$

$$\frac{115 - T_2}{115 - 30} = \exp\left[\frac{-50 \times 6 \times 60}{85 \times 0.4 \times 949}\right]$$

Solving $\qquad$ **T_2 = 66.34°C.**

COMMENT

How long will it take to reach steady state? Will it ever?

Semi infinite Solids (A-6.15 to A-6.23)

Example A-6.15: *On a hot day the wood surface gets heated to 50°C to a considerable depth. Sudden sharp showers cool the surface to 20°C and maintain the surface at this temperature level. Determine the temperature at 2 cm depth after 40 minutes. The material properties are:*

density = 2115 kg/m³. Specific heat = 920 J/kgK. Conductivity = 0.062 W/mK. Also calculate the heat flow from the surface upto the time, instantaneous heat flow at the surface and 2 cm depth.

Solution: The semi infinite solid model is suitable for this example. The equation (6.8) is used to find the temperature. $T_s = 20°C$, $T_i = 50°C$,

$$\frac{T_{x,\tau} - T_s}{T_i - T_s} = erf\left(\frac{x}{2\sqrt{\alpha\tau}}\right), \alpha = \frac{k}{\rho c} = \frac{0.062}{2115 \times 920}$$

$$\therefore \quad \frac{x}{2\sqrt{\alpha\tau}} = \frac{0.02}{2 \times \sqrt{(0.062/2115 \times 920) \times 40 \times 60}} = 1.1435$$

From error function Tables, erf (1.1435) = 0.89414 $\therefore$ $\mathbf{T_{x,\tau} = 46.82°C}$

Total heat flow upto the time: (equation 6.10)

$$\mathbf{Q} = 2k\,(T_s - T_i)\,\sqrt{\tau/\pi\alpha}$$

$$= 2 \times 0.062\,(20 - 50)\,\sqrt{40 \times 60 \times 2115 \times 920/\pi \times 0.062}$$

$$= \mathbf{-\,576006\ J/m^2}$$

Instantaneous surface flow: (equation 6.9)

$$\mathbf{q_0} = k(T_s - T_i)/\sqrt{\pi\alpha\tau}$$

$$= 0.062 \times (20 - 50)/\sqrt{\pi \times 0.062 \times 40 \times 60/(2115 \times 920)}$$

$$= \mathbf{-\,120\ W/m^2} \text{ (flows out of the wood)}$$

Instantaneous heat flow at 0.02 m depth: equation (6.11)

$$\mathbf{q_x} = \frac{k\,(T_x - T_i)}{\sqrt{\pi\alpha\tau}}\ \exp\left[-x^2/4\alpha\tau\right]$$

$$= \frac{0.062(46.82 - 50)}{\sqrt{\pi \times 0.062 \times 40 \times 60/(2115 \times 920)}}$$

$$\times \exp\left[-0.02^2/(4 \times 0.0062 \times 40 \times 60/2115 \times 920)\right]$$

$$= \mathbf{-\,3.44\ W/m^2} \text{ (flows out of the wood)}$$

Example A-6.16: *The soil temperature was 5°C. Suddenly the surface temperature drops to – 10°C. Determine the depth at which the temperature will reach 0°C after 12 hours. Also find the temperature at 0.1 m and 0.05 m. Calculate also the total heat removed per m² area during the period. Thermal diffusivity = 3.097 × 10⁻⁷ m²/s, conductivity = 0.657 W/mK.*

Solution: Semi infinite slab model can be used for the example. Using equation (6.8).

$$\frac{T - T_s}{T_o - T_s} = \text{erf}\,(x/2\sqrt{\alpha\tau}), \frac{0-(-10)}{5-(-10)} = \frac{10}{15} = 0.667$$

$$\text{erf}\,(x/2\sqrt{3.097 \times 10^{-7} \times 12 \times 60 \times 60}) = 0.667$$

Using error function tables, this value corresponds to

$$\therefore \quad x/2\sqrt{3.097 \times 10^{-7} \times 12 \times 60 \times 60} = 0.68$$

$$\therefore \quad \mathbf{x = 0.1573\ m}$$

at **x = 0.1 m** $x/2\sqrt{\alpha\tau} = 0.43225$

$\therefore$ $$\frac{T-(-10)}{5-(-10)} = \text{erf}\,(0.43257) = 0.45912 \quad \therefore \mathbf{T = -3.11°C}$$

at **x = 0.05 m** $x/2\sqrt{\alpha\tau} = 0.2161$ erf (0.2161) = 0.24010

$\therefore$ **T = – 6.4°C**

Total heat flow $= 2k(T_s - T_i)\sqrt{\tau/\pi\alpha} = 2 \times 0.657\,(-10-5)\sqrt{12 \times 3600/\pi \times 3.097 \times 10^{-7}}$

$= \mathbf{4.15 \times 10^6\ J/m^2}$

Example A-6.17: *Still water of some depth initially at 37°C has its surface raised to 100°C and maintained at this level. Determine the depth at which the temperature will reach 48°C after 10 seconds. Also find the heat flow during this time into the water. If the surface temperature was raised to 50°C determine these quantities.* $\alpha = 1.595 \times 10^{-7}\ m^2/s$; $k = 0.656\ W/mK$.

Solution: Semi infinite slab model only suits the example, with usual notations.

$$\frac{T_{x,\tau} - T_o}{T_i - T_o} = \text{erf}\left(\frac{x}{2\sqrt{\alpha\tau}}\right) \quad \therefore \quad \frac{48-100}{37-100} = \text{erf}\left(\frac{x}{2\sqrt{10 \times 1.595 \times 10^{-7}}}\right)$$

$$0.8254 = \text{erf}\,(395.86\,x)$$

Using error function tables, and solving,

x = 2.43 mm

Total heat flow $\mathbf{Q} = 2k(T_o - T_i)\sqrt{\tau/\pi\alpha} = 2 \times 0.656\,(100-37)\sqrt{10/\pi \times 1.595 \times 10^{-7}}$

$= \mathbf{369249\ J/m^2}$

For $\mathbf{T_o}$ **= 50°C, x is obtained as 0.354 mm**

and **Q = 76194 J/m².**

Note: This situation is similar to some one gripping a hot body with hand and 48°C may cause death of cells. So burn depth can be estimated by this process.

Example A-6.18: *A large slab at 350°C suddenly has its surface exposed to convection at 30°C with h = 80 W/m²K. Determine the temperatures at a depth of 40 mm and at the surface after 300 seconds, given thermal diffusivity = 5.6 × 10⁻⁶ m²/s and conductivity = 25 W/mK.*

Solution: This is a semi infinite slab with convection boundary [equation 6.13(*b*)]

$$\frac{T_{x,\tau} - T_i}{T_\infty - T_i} = [1 - \text{erf}(x/2\sqrt{\alpha\tau}] - \exp[(hx/k) + (h^2\,\alpha\tau/k^2)] \times [1 - \text{erf}\,\{(x/2\sqrt{\alpha\tau}) + (h\sqrt{\alpha\tau}/k)\}]$$

(*i*) $x = 0.04$ m $\tau = 300$s, $\alpha = 5.6 \times 10^{-6}$ m²/s, $h = 80$ W/m²K, $k = 25$ W/mK. Substituting and simplifying

$$\frac{T_{x,\tau} - 350}{30 - 350} = 1 - 0.5117\,(1.1563)\,(1 - 0.61941) = 0.0482$$

$\therefore$ $\mathbf{T_{x,\tau} = 334.6°C}$

(*ii*) For the surface temperature $x = 0$.

$$\therefore \quad \frac{T_{o,\tau} - T_i}{T_\infty - T_i} = 1 - [\exp (h^2\alpha\tau / k^2)] \times [1 - \text{erf} (h\sqrt{\alpha\tau} / k)]$$

$$= 1 - 1.01735 (1 - 0.15) = 0.1353.$$

$$\therefore \quad \mathbf{T_{surface} = 306.7°C}$$

This example can also be solved using charts. (page 238)

The skeleton charts are shown in Fig. AE-6.18(*a*) and AE-6.18(*b*)

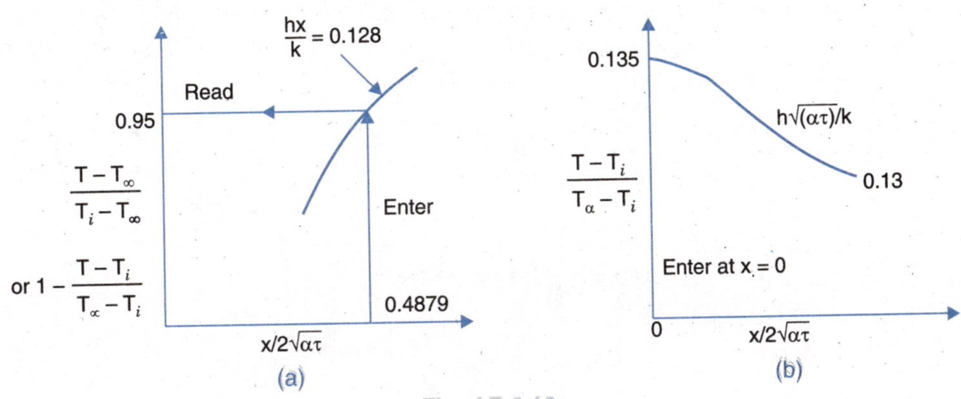

Fig. AE-6.18

In the case 1 Fig. AE-6.18(*a*) is used.

$$\frac{T - T_i}{T_\infty - T_i} = 0.05 \text{ compared to } 0.0482 \text{ by equation}$$

In the case 2, Fig. AE-6.18(*b*) is used as $x = 0$, the intersection with y axis is taken as the value. The value read is 0.135 as compared to 0.1352. (This chart is available in handbooks)

Note: As the equation is long, the charts (refer p. 238) may be used for quick calculation and also for checking. It is not easy to determine heat flow in this case. But $\int_o^\tau h(T - T_\infty)\, d\tau$ will provide the value. T is the surface temperature at $x = 0$, at any time. Instantaneous heat flow can be obtained by hA $(T - T_\infty)$ where T is the temperature at the surface at the instant.

Example A-6.19: *A thick wooden slab at 30°C is suddenly exposed to combustion gases at 850°C with convection coefficient of 40 W/m²K. Determine the time elapsed before the surface reaches 400°C. Thermal diffusivity = 1.472 × 10⁻⁷ m²/s. Conductivity = 0.207 W/mK.*

Solution: This is also a semi infinite slab model with surface convection. In this case $x = 0$ plane is considered.

$$\frac{T - T_i}{T_\infty - T_i} = \frac{400 - 30}{850 - 30} = \frac{370}{820} = 0.4512$$

$$= 1 - \exp (h^2\, \alpha\tau/k^2) \times [1 - \text{erf}(h\sqrt{\alpha\tau} / k)]$$

as both terms involve τ, a trial solution is attempted. The value of RHS for τ = 55, 60, 66 and 75 yields the temperature ratio as 0.4091, 0.4228, 0.4322 and 0.4517. So the **time elapsed is near 75 seconds**, corresponding to the last trial.

The chart similar to one on Fig. AE-6.18(*b*) available in data books gives by intersectior on the *x* axis, $h\sqrt{\alpha\tau} / k = 0.65$

$\therefore$ $\tau = 76.9$ seconds. The values agree within reasonable limits.

COMMENT

The time taken for the wood to get ignited in a fire can be determined by this process.

Example A-6.20: *The soil temperature to a good depth was 5°C. Suddenly chill winds at −20°C start and blow over the surface producing a convection coefficient of 65 W/m²K. Determine the depth at which the temperature will be zero after 12 hours. Thermal diffusivity = 3.097 × 10⁻⁷ m²/s. Thermal conductivity = 0.657 W/mK.*

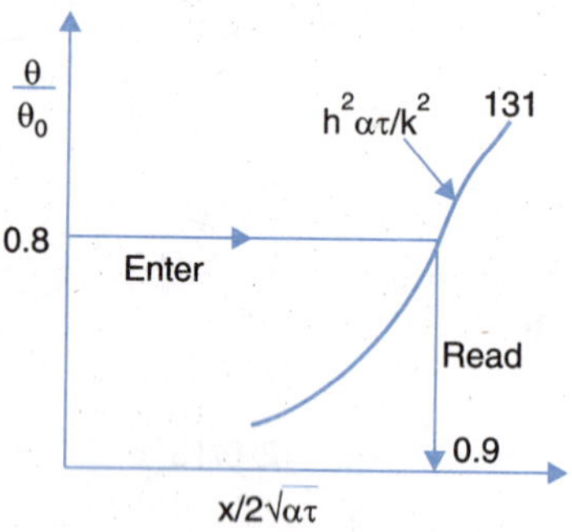

Fig. AE-6.20

Solution: This is a semi infinite slab model with surface convection. As *x* is involved is exponential as well as error function in a complex equation, it is easier to use chart solution, rather than trial solution. A chart of the type as shown in Fig. AE-6.20 is available in handbooks. Reading from such chart at

$$\frac{T_x - T_\infty}{T_i - T_\infty} = \frac{0-(-20)}{5-(-20)} = 0.8$$

$$(h^2\alpha\tau/k^2) = 65 \times 65 \times 3.097 \times 10^{-7} \times 12 \times 3600/0.657^2 = 131$$

Entering the chart as shown at $\frac{\theta}{\theta_0} = 0.8$, $x / 2\sqrt{\alpha\tau}$ is read as 0.9

$$\therefore \quad \mathbf{x} = 0.9 \times 2\sqrt{\alpha\tau} = 0.9 \times 2\sqrt{3.097 \times 10^{-7} \times 12 \times 3600}$$

$$= \mathbf{0.2082\ m}$$

This can also be checked by substitution in the equation (6.13) as an exercise.

COMMENT

In places with cold weather (Alaska for example) the safe depth at which water mains can be located to avoid freezing can be determined by this process.

Example A-6.21: *A cylindrical refractory combustion chamber wall internal diameter 25 cm and wall thickness of 25 mm initially at 35°C is suddenly exposed to combustion products at 1450°C with a convection coefficient at 225 W/m²K. Considering the wall as a semi infinite slab, determine the temperature at various locations 60 seconds after gas exposure. Thermal diffusivity = 5.16 × 10⁻⁷ m²/s, Conductivity = 1.04 W/mK.*

Solution: The assumption of semi infinite solid model is valid due to a large radius, short exposure time and low diffusivity.

The value of $x/2\sqrt{\alpha\tau}$ at x = 5, 10, 15, 20 and 25 mm are calculated. Using the chart similar to Fig. AE-6.18(a) (page 238) the values of temperature ratio are read off corresponding to the (hx/k) value. These are tabulated below:

$x(m)$	0.005	0.01	0.015	0.02	0.025
$x/2\sqrt{\alpha\tau}$	0.4488	0.898	1.347	1.795	2.24
hx/k	1.082	2.163	3.245	4.327	5.41
$(T_x - T_\infty)/(T_i - T_\infty)$	0.73	0.9	0.97	0.99	1.0
T_x(°C)	417	176.5	77.5	49	35

The inside surface temperature (x = 0) has to be calculated by using the equation

$$\frac{T_s - T_i}{T_\infty - T_i} = 1 - \exp\left(\frac{h^2\alpha\tau}{k^2}\right) \times \left[1 - \text{erf}\left(h\frac{\sqrt{\alpha\tau}}{k}\right)\right] = 0.618$$

$$\frac{T_s - 35}{1450 - 35} = 1 - \exp\left[\frac{225^2 \times 5.16 \times 10^{-7} \times 60}{1.04^2}\right] \times \left[1 - \text{erf}\left(\frac{225}{1.04}\sqrt{5.16 \times 10^{-7} \times 60}\right)\right]$$

$$\frac{T_s - 35}{1450 - 35} = 0.618$$

$\therefore$ $\quad$ $\mathbf{T_s = 909.5°C}$

COMMENT

From the tabulation it can be noted that at the end of this time the temperature on the outside surface at 0.025 m is still at the initial temperature. *i.e.* the heat has not reached the surface. Hence semi infinite solid model is applicable.

Example A-6.22: *The temperature of road surface at 9 am was 5°C to a good depth. From 9 am to 12 noon the surface is exposed to solar radiation of 650 W/m² of which 0.4 times is absorbed. Determine the surface temperature and the temperature at a depth of 4 cm at 12 noon. α = 8.26 × 10⁻⁷m²/s, k = 0.552 W/mK.*

Solution: The example can be modelled as semi infinite solid with uniform heat flux at the surface. (equation 6.15)

$$T_x - T_i = \frac{2q\,(\alpha\tau/\pi)^{0.5}}{k}\exp\left[\frac{-x^2}{4\alpha\tau}\right] - \frac{qx}{k}\left[1 - \text{erf}\,(x/2\sqrt{\alpha\tau})\right]$$

$$x = 0,\ \tau = 3 \times 3600 \text{ seconds},\ q = 0.4 \times 650 \text{ W/m}^2$$

Chapter 6

$$\therefore \qquad T_s - 5 = \frac{2 \times 0.4 \times 650\,(8.26 \times 10^{-7} \times 3 \times 3600/\pi)^{0.5}}{0.552}.1 \text{ as } x = 0,$$

$$= 50.2$$

$$\therefore \qquad \mathbf{T_s = 55.2°C}$$

at $x = 0.04$ m,

$$T_x - 5 = \frac{2 \times 0.4 \times 650}{0.552}(8.26 \times 10^{-7} \times 3 \times 3600/\pi)^{0.5}.\exp\left(\frac{-0.04^2}{4 \times 8.26 \times 10^{-7} \times 3 \times 3600}\right)$$

$$- \frac{0.4 \times 650 \times 0.04}{0.552}\,[1 - \text{erf}\,(0.04/2\sqrt{(8.26 \times 10^{-7} \times 3 \times 3600)}]$$

$$= 48.0 - 14.4 = 33.6$$

$$\therefore \qquad \mathbf{T_x = 38.6°C.}$$

COMMENT

At the same time, the surface will be exposed to air flow at a lower temperature. So the temperature calculated is the worst case value.

Example A-6.23: *Compare the response of a thermocouple and mercury in glass thermometers when exposed to surroundings with a sinusoidal temperature variation and convection coefficient of 95 W/m²K.*

The dimensions and property values are:

(i) Thermocouple: density 7600 kg/m³, specific heat: 502 J/kg K dia: 0.0016 m

(ii) Mercury: density 13600 kg/m³, specific heat: 1360 J/kg K, dia = 12.8 mm.

Solution: Thermometers are to be used to measure the temperature in a furnace with time period of 20 minutes. This is a lumped parameter system with periodic temperature change of surroundings.

The angle of lag is given by (equations 6.16, 6.17)

$$\delta = \tan^{-1}[(2\pi/\tau_0)(\rho cV/hA)], \text{ amplitude ratio} = \frac{1}{\sqrt{1 + \tan^2 \delta}}$$

$$\frac{V}{A} = \frac{\pi r^2 . l}{2\pi r l} = \frac{r}{2}$$

(*i*) Thermocouple $\quad \delta = \tan^{-1}[(2\pi/20 \times 60)(7600 \times 502 \times 0.0004/95)]$

$= 0.084$ radians or 4.8° or 16 seconds

Amplitude ratio: $1/\sqrt{1 + \tan^2 0.084}$ = **0.9965**

(*ii*) Glass thermometer: $\delta = \tan^{-1}[(2\pi/1200)(13600 \times 1360 \times 0.0032/95)]$

$= 127$ radians

$\therefore$ **Amplitude ratio = 0.23** (check by substitution)

COMMENTS

1. The thermocouple responds much better, following the variation closely and with very small error.
2. This is because of its small diameter and lower heat capacity.

Cyclic Temperature Variation—Semi Infinite Solid

Example A-6.24: *A thick furnace wall is exposed to hot gases in a furnace and the surface temperature varies sinusoidally every 20 minutes from a maximum to minimum value. The property of the materials are: thermal diffusivity = 5.16 × 10⁻⁷ m/s, conductivity = 1.04 W/mK. Determine the angle of lag and amplitude ratio at a depth of 0.05 m. Also find the depth of penetration under equilibrium.*

Solution: This is a semi infinite slab with periodic surface temperature change.

The time of lag is given by $\frac{x}{2}\sqrt{\tau_0 / \pi\alpha}$ ...(6.18)

The amplitude ratio is given by $e^{-x\sqrt{\pi/\alpha\tau_0}}$ ...(6.19)

The depth of penetration is given by $x_0 = 2\sqrt{\pi\alpha\tau_0}$ ...(6.20)

$$\tau_0 = 20 \times 60 = 1200 \text{ seconds}$$

$$\therefore \quad \textbf{Time lag} = \frac{0.05}{2}\sqrt{1200/\pi \times 5.16 \times 10^{-7}}$$

$$= \textbf{680.2 seconds} \text{ or } 1.78 \text{ radians or } 102°$$

$$\textbf{Amplitude ratio} = \exp\left[-0.05\sqrt{\pi/5.16 \times 10^{-7} \times 1200}\right]$$

$$\therefore \quad \frac{T}{T_{max}} = \mathbf{0.0284}$$

Depth of penetration: $x_0 = 2\sqrt{\pi\alpha\tau} = 2\sqrt{\pi \times 5.16 \times 10^{-7} \times 1200}$

$$= 0.088 \text{ m.}$$

COMMENT

Such a variation will not affect the initial condition beyond 0.088 m thickness of wall. Also the amplitude ratio is low. See Fig. AE-6.24. As the amplitude ratio is small, the wall material will not suffer thermal fatique and spalling.

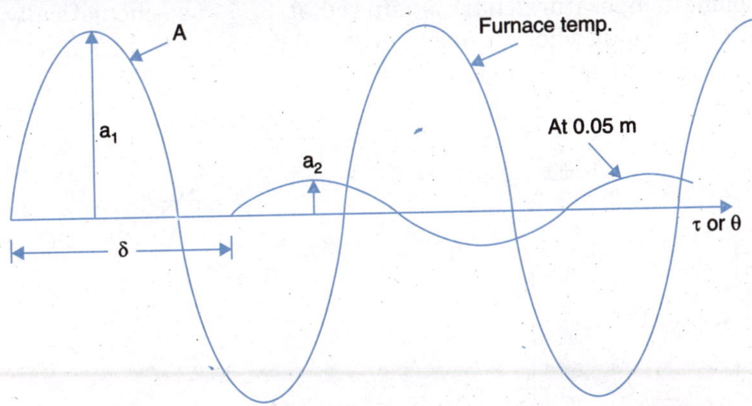

Fig. AE-6.24. Temperature variation

Examples on Finite Thickness Solids

Example A-6.25: *A plane wall made of material of density 7800 kg/m^3, thermal conductivity 45 W/mK and specific heat 465 J/kgK is 0.12 m thick. After it was heated to a uniform temperature of 310°C, the wall is exposed to convection on both sides to surroundings at 30°C with a convective heat transfer coefficient of 450 W/m^{2}K. Determine the temperature after 8 minutes at (i) mid plane (ii) 0.03 m from centre plane and (iii) 0.054 m from centre plane and (iv) at the surface. also calculate (i) the heat flow out during this period for 1 m^2 area on one side and (ii) The instantaneous flow rate at the surface.*

Solution: The data are presented in Fig. AE-6.25(*a*). This corresponds to infinite slab model and use of Heisler charts for solution is necessary. Also the mid plane excess temperature ratio is to be found using the mid plane chart and then location chart has to be used.

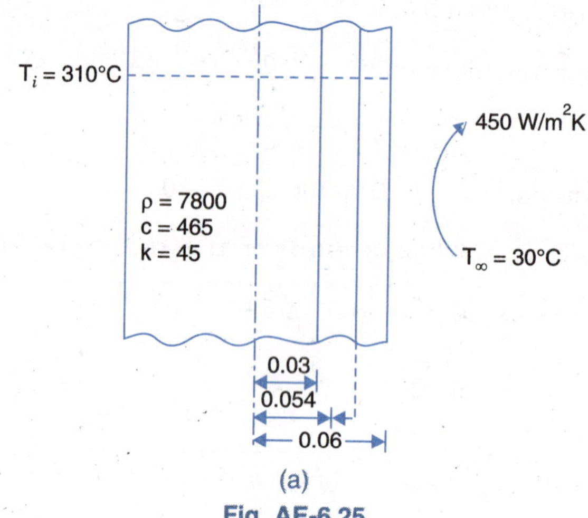

(a)

Fig. AE-6.25

The parameters required are $\dfrac{hl}{k} = \dfrac{450 \times 0.06}{45} = 0.6$

(This indicates lumped parameter model cannot be used)

$$\alpha\tau/L^2 = \frac{45 \times 8 \times 60}{7800 \times 465 \times 0.06^2} = 1.654$$

The mid plane temperature chart is entered at 1.654 as schematically shown and the reading is taken at the point where it meets $hL/k = 0.6$.

[Fig. AE-6.25(*b*)] the value read is 0.478.

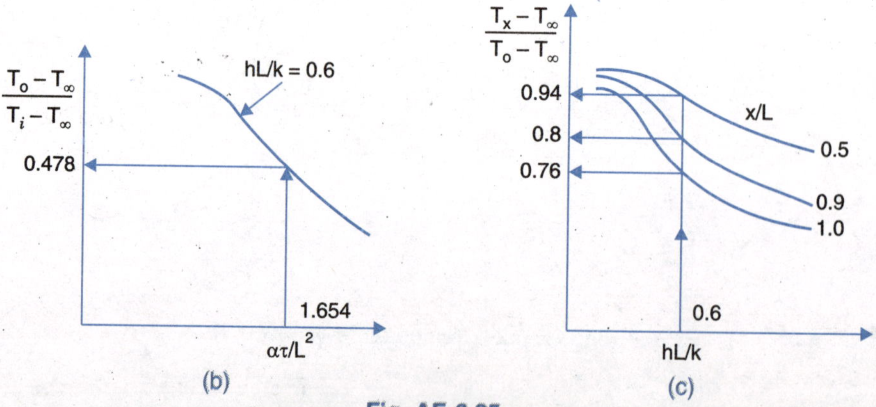

(b) (c)

Fig. AE-6.25

$$\frac{T_o - 30}{310 - 30} = 0.478 \quad \therefore \mathbf{T_o = 163.84°C}$$

To find temperatures at 0.03, 0.054 and 0.06 m planes $\frac{x}{L}$ values are found as 0.5, 0.9 and 1.

Entering the location chart at $hL/k = 0.6$, values are read at $x/L = 0.5$, 0.9 and 1 as 0.94, 0.8 and 0.76 respectively [Fig. AE-6.25(*c*)] using the product

$$\therefore \quad \frac{T - T_\infty}{T_i - T_\infty} = \frac{T - T_\infty}{T_o - T_\infty} \cdot \frac{T_o - T_\infty}{T_i - T_\infty}$$ the temperatures are calculated as

155.81, 137.07 and 131.72°C respectively.

For 0.03 m (sample calculation)

$$\frac{T - 30}{310 - 30} = 0.478 \times 0.94 = 0.44932$$

$$\therefore \quad \mathbf{T = 155.81°C}$$

To calculate the total heat flow, the heat flow chart is to be used. This requires the parameter

$$\frac{h^2 \alpha \tau}{k^2} = \frac{450^2 \times 45 \times 8 \times 60}{7800 \times 465 \times 45 \times 45} = 0.596$$

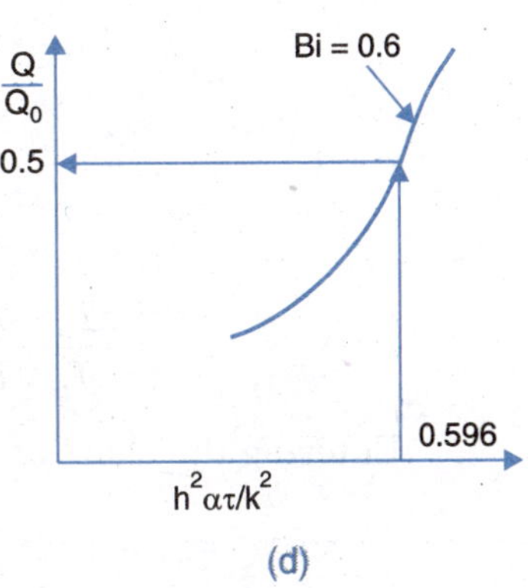

(d)

Fig. AE-6.25

Entering the chart as schematically shown in Fig. AE-6.25(*d*) at this value and reading on $hL/k = 0.6$, Q/Q_o is 0.5.

$$\therefore \quad Q = 0.5 \times Q_o = 0.5 \times 7800 \times 465 \times 0.06\,(310 - 30)$$

$$\mathbf{= 30.46 \times 10^6\ J/m^2}$$

Instantaneous flow at the surface $= hA(T_s - T_\infty)$

$= 450 \times 1 \times (131.71 - 30) = \mathbf{4.58 \times 10^4\ W/m^2}$

COMMENT

If the temperature level is plotted to scale, the type of variation can be visualised.

Example A-6.26: *A large plate of thickness 0.2 m initially at 40°C is exposed to hot furnace gases at 530°C with a convection coefficient of 420 W/m²K. The density, specific heat and thermal conductivity of the material are 2700 kg/m³, 1110 J/kgK and 218 W/mK. Determine the time required for the centre plane to reach 410°C. At this instant calculate the surface temperature and the temperature at 0.02 m depth from surface. Also calculate the heat flow upto the time. If the surface is maintained at 530°C with h ⟶ ∞, then determine the centre temperature after 150 seconds.*

Solution: The infinite slab model is applicable and Heisler Charts are used as in example A-6.25. The entry points and parameter to be read are reversed. In this case the centre temperature ratio is known. Enter the chart corresponding to this value.

$$\frac{hL}{k} = \frac{420 \times 0.1}{218} = 0.193$$

(This also indicates that lumped parameter model cannot be used)

The procedure is shown in Fig. AE-6.26.

$$\alpha = \frac{k}{\rho c} = \frac{218}{2700 \times 1110} = 72.74 \times 10^{-6}$$

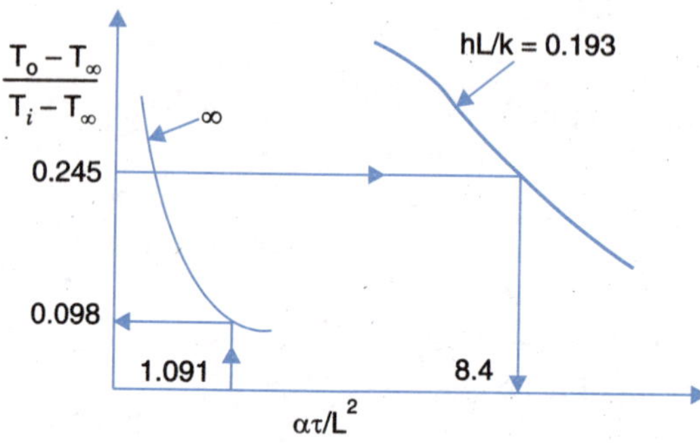

Fig. AE-6.26

$$\frac{T_o - T_\infty}{T_i - T_\infty} = \frac{410 - 530}{40 - 530} = 0.245$$

Entering the chart at 0.245 on the y axis $\alpha\tau/L^2$ is read as 8.4 (Fig. AE-6.26)

$\therefore$ $$\tau = 8.4 \times L^2/\alpha = \frac{8.4 \times 0.1^2}{72.74 \times 10^{-6}} = \mathbf{1155\ sec.}$$

(ii) To determine temperature at surface and 0.02 from surface $\frac{x}{L} = 1$ and 0.8.

Reading from location chart $(T - T_\infty)/(T_o - T_\infty)$ are obtained as 0.91 and 0.95. $\therefore$ These **temperatures are 420.8 and 416°C** (method as in example A-6.25)

(iii) The total heat flow is calculated as the method in example A-6.25, using the parameter $h^2\alpha\tau/k^2$ calculated as 0.30 and hL/k = 0.193 reading from the chart,

$$\frac{Q}{Q_o} = 0.72 \quad \therefore \mathbf{Q} = 0.72 \times 0.1 \times 2700 \times 1110\ (530 - 40)$$

$$= \mathbf{105.7 \times 10^6\ J/m^2}$$

(iv) Reading the centre temperature chart at $hL/k = \infty$, at the time of 150 s the Fourier number is $72.74 \times 10^{-6} \times 150/0.1^2 = 1.091$

$$Fo = 1.091, \quad \frac{T_0 - 530}{40 - 530} = 0.098, \quad \therefore \mathbf{T_o = 481.98°C.}\ \text{(rapid heating)}$$

COMMENT

When the surface is exposed to convection, the surface temperature rises rather slowly and may not reach the fluid temperature, whereas the surface temperature equals the fluid temperature in the second case. This is the reason for the rapid heating. The convection resistance is zero in the second case and only unduction resistance exists.

Example A-6.27: *A vegetable piece in the form of a slab of thickness of 6 cm at 8°C is allowed to warm up in air at 35°C with a convection coefficient of 11 W/m^2K. Determine the time required for the centre to reach 22°C. The material properties are Thermal diffusivity = 1.42×10^{-7} m^2/s. Thermal conductivity = 0.5978 W/mK, density = 1000 kg/m^3, specific heat = 4218 J/kgK. Assume one dimensional flow and exposure on both sides.*

Solution: Using infinite slab model and Heisler Chart for slab and entering the chart at (22 – 35)/(8 – 35) = 0.4815 and reading on $hL/k = 11 \times 0.03/0.5978) = 0.552$, Fourier number is read as 1.75 [Refer Fig. AE-6.25 (*a*), (*b*), (*c*), (*d*)]

$$\therefore \qquad \tau = 1.75 \times 0.03^2/1.42 \times 10^{-7} = \mathbf{11091} \text{ seconds or 3.08 hours.}$$

The heat transfer can be determined using the parameter $h^2\alpha\tau/k^2 = 0.532$ (as calculated). Entering the chart at this value and reading against $hL/k = 0.552$, $Q/Q_o = 0.6$.

Assuming a slab 20 × 20 cm area, thickness being 0.03 m

$$\mathbf{Q} = 0.6 \times 0.2 \times 0.2 \times 1000 \times 4218 \times 0.03\ (8 - 35) \times 2 = \mathbf{164000\ J} \text{ or } \mathbf{164\ kJ}$$

COMMENT

For heating up under free convection it takes a long time. This is the reason for taking out frozen meat from the freezer long before cooking.

Example A-6.28: *A plane wall with material properties as follows is 0.12 m thick. Density = 7800 kg/m^3. Specific heat 0.465 kJ/kgK. Thermal conductivity = 45 W/mK. It is at 420°C when the surroundings are at 20°C. It is proposed to allow it to cool by convection to the surroundings with convective heat transfer coefficient of 450 W/m^2K. If 50% of energy above the surroundings is to be dissipated, how long should the wall be exposed.*

Solution:

In this example $Q/Q_o = 0.5$, [see Fig. AE-6.25(*d*)]

The parameters are $\dfrac{hL}{k} = \dfrac{450 \times 0.06}{45} = 0.6$

and $h^2\alpha\tau/k^2$ is to be determined.

So entering the heat flow chart at $Q/Q_o = 0.5$ and cutting the $hL/k = 0.6$ line $h^2\alpha\tau/k^2$ is read as 0.596.

$$\therefore \qquad 450^2 \times \frac{45}{7800 \times 465} \times \tau \times \frac{1}{45^2} = 0.596$$

$$\therefore \qquad \tau = \mathbf{480\ seconds}$$

Long Cylinder

Example A-6.29: *To heat treat alloy steel rods, these are placed in a gas stream at 800°C with a convective heat transfer coefficient of 120 W/m^2K. The rod is of 19 mm radius and is at 30°C to start with. Determine the time required for the rod centre line to reach 610°C. Density = 7978 kg/m^3. Specific heat = 0.585 kJ/kgK, thermal conductivity = 21.9 W/mK. Also determine at this time surface and mid radius temperatures and the heat flow.* (Refer charts on pp. 247 and 248).

Chapter 6

Solution: This is a long rod with surface convection. The Heisler Charts for cylinders are to be used.

$$\frac{T_o - T_\infty}{T_i - T_\infty} = \frac{610 - 800}{30 - 800} = 0.247,\ hR/k = 120 \times 0.019/21.9 = 0.104$$

As illustrated in Fig. AE-6.29 entering the chart on the temperature ratio side at 0.247 and cutting the Biot number curve 0.104 (interpolated) $\alpha\tau/R^2$ or Fourier number is read as 7.4.

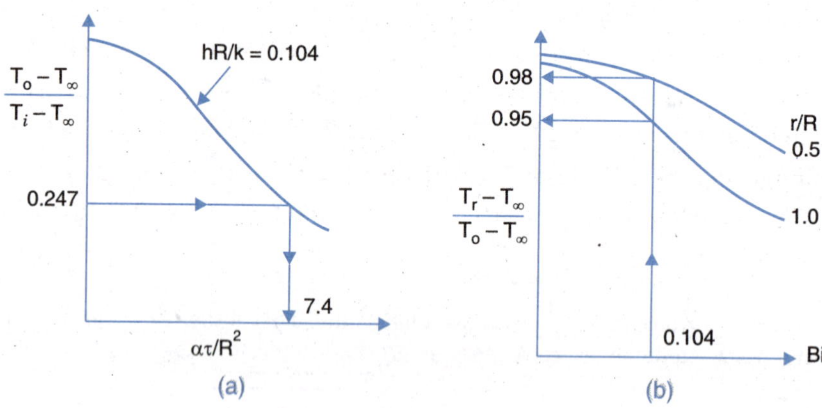

Fig. AE-6.29

$$\tau = \frac{7.4 \times R^2}{\alpha} = \frac{7.4 \times 0.019^2}{(21.9/7978 \times 585)}$$

$\therefore$ $= \mathbf{569.3\ seconds}$

The surface and mid radius temperatures are read using location chart as illustrated and entering at $Bi = 0.104$

$\frac{T - T_\infty}{T_o - T_\infty}$ is read as 0.95 and 0.98 for $r/R = 1$ and 0.5 [Fig. AE-6.30(*b*)]

$\therefore$ $\mathbf{T_s = 619.5°C \quad T_{0.5} = 613.8°C}$

The heat flow is determined using the parameter $h^2\ \alpha\tau/k^2 = 0.0802$ and entering the chart at this value and reading at $Bi = 0.104$. $Q/Q_o = 0.75$ (chart p. 251)

$\therefore$ $$\mathbf{Q} = 0.75\ \rho Vc\ (T_i - T_\infty) = 0.75 \times 7978 \times \pi \times \frac{0.019^2}{4} \times 585$$

$= \mathbf{3.06 \times 10^6\ J/m\ length.}$

COMMENT

Quality control in processes require such estimation of time, temp., heat required etc.

Example A-6.30: *A metallic rod of 25 mm radius initially at 800°C (density = 3970 kg/m³, conductivity = 10.5 W/mK, specific heat = 1225 J/kgK) is exposed to cooling fluid at 300°C with h = 1500 W/m²K. Determine the bulk average temperature, the center, mid radius and surface temperatures, after 40 seconds.*

Solution: This is a long rod with convection boundary problem and Heisler Charts for cylinders should be used (for use of charts ref Examples A-6.25 and A-6.29). The Parameters are:

$$\frac{hR}{k} = \frac{1500 \times 0.025}{10.5} = 3.57$$

$$\frac{\alpha\tau}{R^2} = \frac{10.5 \times 40}{3970 \times 1225 \times 0.025^2} = 0.1382$$

$$\frac{h^2\alpha\tau}{k^2} = \frac{1500^2 \times 10.5 \times 40}{3970 \times 1225 \times 10.5^2} = 1.762$$

Entering the center temperature chart at 0.1382 and reading against $hR/k = 3.57$, the center temperature ratio is 0.85

i.e., $$\frac{T_o - T_\infty}{T_i - T_\infty} = 0.85 \quad \therefore \quad \mathbf{T_o = 725°C}$$

For the mid radius and surface temperatures, the location charts are used. Entering at $hR/k = 3.57$, and reading on $r/R = 1$ and $r/R = 0.5$, the temperature ratios are 0.30 and 0.80

$\therefore$ $$\mathbf{T_s = 428°C \text{ and } T_{0.5} = 640°C.}$$

To determine the bulk average temperature, the heat loss has to be calculated. Entering the heat flow chart at $h^2\alpha\tau/k^2 = 1.762$, Q/Q_o is read as 0.48.

The heat loss/heat capacity provides the average temperature drop which is directly equal to the product of Q/Q_o and the temperature difference at the beginning.

$\therefore$ **Average temperature after exposure** $= T_i - 0.48\,(T_i - T_\infty) = 800 - 0.48\,(800 - 300)$

$= \mathbf{560°C.}$

COMMENT

Though the convection coefficient is high, the low thermal conductivity and short time of exposure cause slow cooling of the material.

Chapter 6

Example A-6.31: *A cylindrical piece of pudding is to be cooked in 802 seconds when exposed to convection at 120°C with a convective heat transfer coefficient of 68.6 W/m²K. It is specified as cooked if the centre line temperature reaches 98°C. The initial temperature is 25°C. Using the following properties determine the maximum radius of the piece, density = 840 kg/m³, specific heat 3400 J/kgK, thermal conductivity = 1.2 W/mK.*

Solution: As length is not specified long cylinder configuration is assumed. The centre temperature ratio is known but, both parameters involve (hR/k and $\alpha\tau/R^2$), the radius value. Hence a trial solution has to be attempted. The centre temperature ratio $= \dfrac{98 - 120}{25 - 120} = 0.232.$

The chart is read as illustrated in Fig. AE-6.31. *Y*-axis value is known 0.232. As radius is not specified, the Fourier number is read for several values of Biot number. The values are tabulated below.

hR/k	=	20	10	5	1	0.5
$\alpha\tau/R^2$	=	0.38	0.42	0.49	1.1	2.2
R, m	=	0.35	0.175	0.0875	0.0175	0.00875

For these Biot numbers, as h and k are known the R value is calculated. As Fourier number is available the time can be calculated. The times are tabulated below.

Time, $s = 0.11 \times 10^6$ 30.6×10^3 8.9×10^3 802 401

$$\alpha = \frac{1.2}{840 \times 3400} = 0.42 \times 10^{-6}$$

Sample calculation: For value of Biot number 20. (refer Fig. AE-6.31)

$$Fo = 0.38 = \frac{0.42 \times 10^{-6} \times \tau}{0.35^2} \quad \therefore \tau = 0.11 \times 10^6 \text{ sec}$$

As can be seen, the time **requirement is satisfied for radius of 0.0175 m or diameter of 35 mm.**

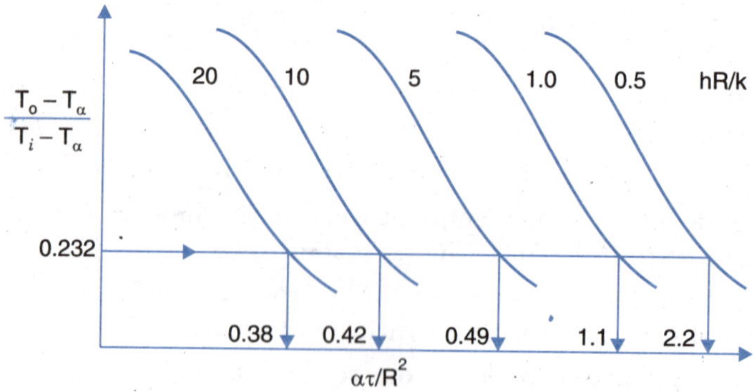

Fig. AE-6.31

COMMENT

Engineering design is always a trial process because of complexity of the requirement. Though tedius one should not be shy of trial solutions if he desires to be a professional engineer.

Example A-6.32: *A long cylinder of 0.2 m dia initially at 600°C is exposed to a fluid at 100°C with a convection coefficient of 76.0 W/m²K. The material properties are density = 7835 kg/m³, specific heat = 485 J/kgK and thermal conductivity = 19.0 W/mK. Plot for time periods of 10, 20, 30 and 40 minutes, the temperature profile. The example will help visualise the temperature variation with time.*

Solution:
$$\frac{hR}{k} = \frac{76 \times 0.1}{19} = 0.4$$

Fourier Number for 10, 20, 30 and 40 minutes are calculated as

$$\frac{\alpha\tau}{R^2} = \frac{19 \times \tau}{0.1^2 \times 7835 \times 485} = 0.30, 0.6, 0.9, 1.2$$

For plotting, the temperature at various radius should be calculated.

$\frac{r}{R}$ values used are 0.4, 0.6, 0.8 and 1.0

Using the center temperature chart, as illustrated in Fig. AE-6.32(*a*) the center temperature ratios are read at $\frac{hR}{k} = 0.4$.

Time min	10	20	30	40
Fo	0.3	0.6	0.9	1.2
Temp. ratio	0.88	0.712	0.57	0.47

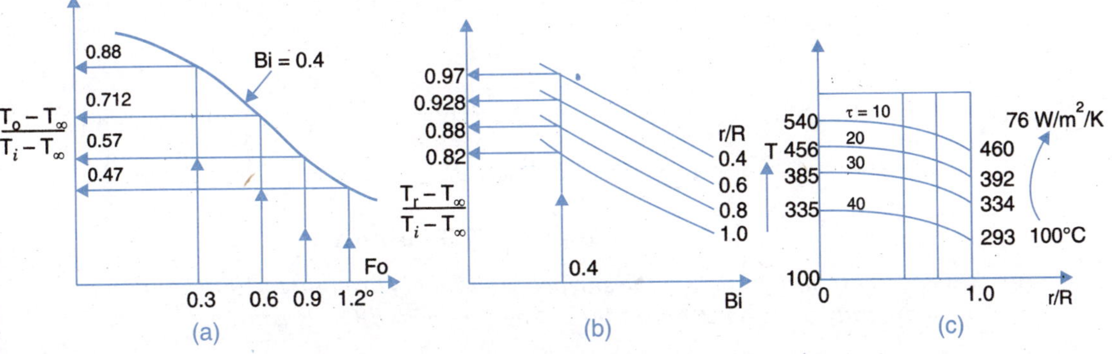

Fig. AE-6.32

The location temperature chart is read at $hR/k = 0.4$ and for $r/R = 0.4, 0.6, 0.8$, and 1 are 0.97, 0.928, 0.88 and 0.82 as schematically shown in Fig. AE-6.32(*b*).

The temperatures are tabulated as below: sample calculation of centre temperature after 10 minutes and shown below:

$$\frac{T_o - T_\infty}{T_i - T_\infty} = 0.88 = \frac{T_o - 100}{600 - 100} \qquad \therefore \ \mathbf{T_o = 540°C}$$

Temperature at $r/R = 0.4$, after 10 minutes

$$\frac{T - T_\infty}{T_i - T_\infty} = 0.88 \times 0.97 \qquad \therefore \ \mathbf{T = 526.8°C}$$

The results from table is shown plotted in Fig. AE-6.32(*c*)

Time Min	*Centre*	$\frac{r}{R}$ *= 0.4*	*0.6*	*0.8*	*1.0*
10	540	526.8	508.3	487.2	460.8
20	456	445.3	430.4	413.2	391.9
30	385	376.5	364.5	350.8	333.7
40	335	328	318	306.8	292.7

COMMENT

The reason for the slope of the temperature lines being flatter near the centre and steeper near the surface is the area being larger at the surface and smaller at the inside. (refer Fourier's equation)

Example A-6.33: *In example A-6.32 determine the time when the temperature at the surface reaches 350°C.*

Solution: This requires the use of the charts in the reverse order. The location chart is used first. The reading is 0.82 (at Bi = 0.4, at surface r/R = 1). See Fig. AE-6.32(*b*)

$$\frac{350-100}{600-100} = \frac{T_o - T_\infty}{T_i - T_\infty} \cdot \frac{T - T_\infty}{T_o - T_\infty}$$

$$0.5 = 0.82.\ \frac{T_o - T_\infty}{T_i - T_\infty}$$

$$\therefore \quad \frac{T_o - T_\infty}{T_i - T_\infty} = 0.6098$$

Now entering the centre temperature chart at this value, Fourier number read as 0.82

$$\frac{\alpha\tau}{R^2} = 0.82 \quad \therefore \quad \tau = \frac{0.82 \times R}{\alpha} = \frac{0.82 \times 0.1 \times 0.1 \times 7835 \times 485}{19}$$

= 1640 seconds or 27.3 minutes.

Example A-6.34: *In example A-6.32 if the surface is suddenly brought to 100°C and maintained determine the temperatures after 20 minutes at the center and 0.6 R radius.*

Solution: This is a special case where $\frac{hR}{k} = \infty$

∴ Using center temperature chart, and reading on line $hR/k = \infty$, $\alpha\tau/R^2 = 0.6$

Temperature ratio is = 0.06 (as read from chart page 247)

$$\therefore \quad T_o = 0.06 \times 500 + 100 = \mathbf{130°C}$$

The location temperature ratio as read from chart for r/R = 0.6 is 0.54.

$$\therefore \quad \mathbf{T = 0.06 \times 0.54 \times 500 + 100 = 116.2°C.}$$

COMMENT

From the table in Example A-6.32, the surface temperature is much higher (even after 40 minutes) that the fluid temperature of 100°C. In this case the surface temperature is 100°C. Hence the rapid cooling. Only conduction resistance exists in this case.

Sphere

Example A-6.35: *A sphere of 32 mm diameter made of material with density = 400 kg/m³, specific heat = 1600 J/kgK and thermal conductivity = 1.7 W/mK initially at 800°C is suddenly exposed to convection over the surface to surroundings at 300°C with a convective heat transfer coefficient of 80 W/m²K. Determine the time required for the centre to reach 470°C and also determine the surface and mid radius temperatures and the total heat transfer and the instantaneous rate of heat transfer at the surface. If the sphere is removed from the surrounding and kept insulated from the instant estimate its equilibrium temperature.*

Solution: This example is under transient heat conduction from a sphere exposed suddenly to convection. Use of Heisler Chart is necessary for the solution. (pages 249, 250 and 251)

The parameters are:

$$\text{Biot number} = \frac{hR}{k} = \frac{80 \times 0.016}{1.7} = 0.753$$

$$\text{Centre temperature ratio} = \frac{470 - 300}{800 - 300} = 0.34$$

Entering the centre temperature chart at this value as illustrated in Fig. AE-6.35(a) he Fourier number is read as 0.68.

$$\therefore \quad \frac{\alpha\tau}{R^2} = 0.68 \quad \therefore \quad \tau = \frac{0.68 \times R^2}{\alpha}$$

$$= 0.68 \times 0.016^2/(1.7/400 \times 1600)$$

$$\mathbf{= 65.54\ seconds}$$

To determine the surface and mid radius temperatures, location temperature chart is o be used. Entering the chart as illustrated in Fig. AE-6.35(b) at Biot number = 0.753, and reading at r/R = 0.5 and 1.0.

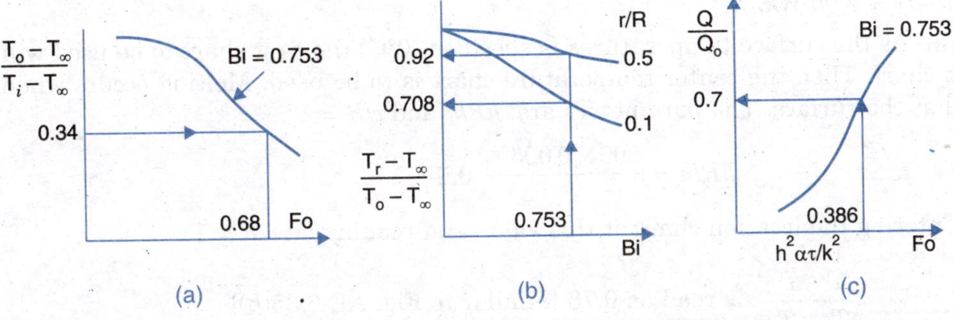

Fig. AE-6.35

The values obtained are 0.92 and 0.708.

$\therefore$ mid radius temperature:

$$\frac{T - T_\infty}{T_i - T_\infty} = \frac{T_o - T_\infty}{T_i - T_\infty} \cdot \frac{T - T_\infty}{T_o - T_\infty}$$

$$\frac{T - 300}{800 - 300} = 0.34 \times 0.92 \quad \therefore \quad \mathbf{T = 456.4°C}$$

Surface temperature:

$$\frac{T_s - 300}{800 - 300} = 0.34 \times 0.708 \quad \therefore \quad \mathbf{T_s = 420.4°C}$$

In order to determine the heat flow, heat flow chart is to be used. The parameter required is

$$\frac{h^2\alpha\tau}{k^2} = \frac{80 \times 80 \times 1.7 \times 65.54}{400 \times 1600 \times 1.7^2} = 0.386$$

Entering the chart at this value and reading against biot number = 0.753 Q/Q_o is read as 0.7 as illustrated in Fig. AE-6.35(c).

Chapter 6

$$\therefore \quad \mathbf{Q} = 0.7 \times \frac{4}{3} \times \pi \times 0.016^3 \times 400 \times 1600 \times (800 - 300) = \mathbf{3843J}$$

Instantaneous heat flow rate $= hA(T_s - T_\infty)$

$= 80 \times 4 \times \pi \times 0.016^2 (420.4 - 300) = \mathbf{31\ W}$

The heat removed **= 3843 J.**

The average temperature drop due to this heat removal = $\rho Vc \times \Delta T$

$$\therefore \quad \Delta T \times 400 \times \frac{4}{3} \pi r^3 \times 1600 = 3843$$

$$\therefore \quad \Delta T = 350°C.$$

$\therefore$ **Average temperature** = 800 – 350 = **450°C.**

Example A-6.36: *Ice balls of 10 mm dia at – 32°C are exposed to an air current at 15°C with a convection heat transfer coefficient of 200 W/m²K. Determine the time when the surface layer will begin to melt. Also determine the center, midsurface temperatures and heat removed. Use the following property values, density = 920 kg/m³, specific heat 2040 J/kgK. Thermal conductivity = 2.00 W/mK.*

Solution: As the surface temperatures is specified (0°C) the first chart to be used will be the location chart. Then the center temperature chart is to be used. Melting occurs when 0°C is reached at the surface. The parameters are: hR/k and $r/R = 1$

$$hR/k = \frac{200 \times 0.005}{2} = 0.5$$

Entering the location chart at this value and reading on $r/R = 1$

$\dfrac{T - T_\infty}{T_o - T_\infty}$ is read as 0.78 [Similar to Fig. AE-6.35(*b*)]

Using $$\frac{T - T_\infty}{T_i - T_\infty} = \frac{T_o - T_\infty}{T_i - T_\infty} \frac{T - T_\infty}{T_o - T_\infty}$$

$$\frac{0 - 15}{-32 - 15} = 0.319 = \frac{T_o - T_\infty}{T_i - T_\infty} \times 0.78$$

$$\therefore \quad \frac{T_o - T_\infty}{T_i - T_\infty} = 0.4092$$

Entering the center temperature chart (as illustrated in Fig. AE-6.35) at this value and reading against $hR/k = 0.5$, Fourier number is read as 0.76.

$$\therefore \quad \frac{\alpha\tau}{R^2} = 0.76 \quad \therefore \tau = \frac{0.76 \times 0.005^2 \times 2040 \times 920}{2.0} = 17.83 \text{ s}$$

The time to reach 0°C at surface = **17.83 sec.**

In order to determine the mid radius plane temperature, the location chart is entered at $hR/k = 0.5$, and reading against $r/R = 0.5$ the ratio is read as 0.942, using it center temperature ratio is calculated,

$$\therefore \quad \frac{T - T_\infty}{T_i - T_\infty} = \frac{T - 15}{-32 - 15} = 0.4092 \times 0.942 \quad \therefore \quad \mathbf{T = -3.12°C}$$

For center temperature $\frac{T_o - T_\infty}{T_i - T_\infty}$ = 0.4092 (already read)

$\therefore$ $T_o = -4.23°C$

To determine the heat flow:

$\frac{h^2 \alpha \tau}{k^2}$ is calculated $= \frac{200^2 \times 2 \times 17.83}{920 \times 2040 \times 2^2} = 0.19$

Entering the heat flow chart at this value and reading against $Bi = 0.5$, Q/Q_o **= 0.65**

$\therefore$ $Q = 0.65 \times \frac{4}{3} \times \pi \times 0.005^3 \times 920 \times 2040\ (-32 - 15) =$ **– 30.02 J.**

COMMENT

If h = 20 W/m²K, then, hR/k = 0.05. The problem can be solved with lumped capacitance method. The time will be 178 s.

Example A-6.37: *Steel balls with density of 7900 kg/m³, specific heat of 582 J/kgK and thermal conductivity of 24.0 W/mK initially at 800°C are cooled in an oil bath at 100°C, with a convection heat transfer coefficient of 400 W/m²K. If the centre temperature after 103.5 seconds was 450°C, determine the radius. Also determine the surface temperature and the percentage of internal energy removed (above the bath temperature).*

Solution: The radius is not specified. The parameter Biot number and Fourier number both involve the unknown quantity. So trial solution is necessary.

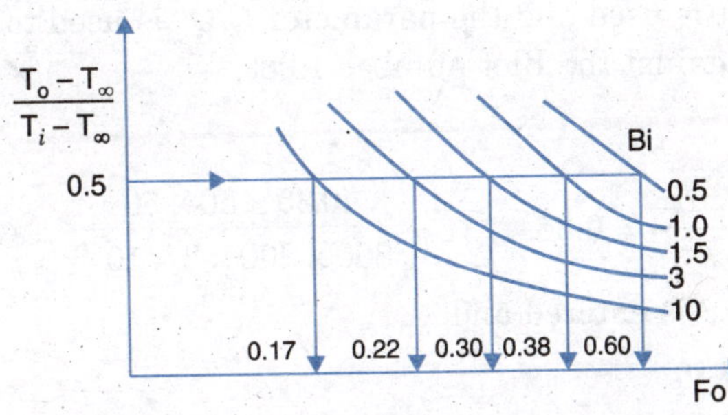

Fig. AE-6.37

Given, $\frac{T_o - T_\infty}{T_i - T_\infty} = \frac{450 - 100}{800 - 100} = \frac{350}{700} = 0.5$

Entering the centre temperature chart as illustrated in Fig. AE-6.37 at this value, Fourier numbers are read for various values of Biot numbers and are tabulated below.

Calculating radius from Biot number and using the radius and the Fourier number the time is calculated and tabulated.

Bi	10	3	1.5	1	0.5
Fo	0.17	0.22	0.30	0.38	0.6
R	0.6	0.18	0.09	0.06	0.03
τ, s	11726	1365	465	262	103.5

Chapter 6

The Bi = 0.5 and Fo = 0.60 the time is 103.5 seconds as required ∴ Radius of the steel balls = **0.03 m.**

The surface temperature is determined using the location chart. $r/R = 1$ and temperature ratio is read entering at Bi = 0.5 against r/R = 1, as 0.78

∴ Surface temperature is given by

$$\frac{T - 100}{800 - 100} = 0.5 \times 0.78 \quad \therefore \mathbf{T = 373°C}$$

To determine the heat flow:

$h^2 \alpha\tau/k^2$ is determined as $\dfrac{400 \times 400 \times 24 \times 103.5}{7900 \times 582 \times 24 \times 24} = 0.1475$

Entering the heat flow chart at this value $\mathbf{Q/Q_0}$ is read against Bi = 0.5 as **0.7.**

∴ Percentage internal energy removed (above the surroundings) is **70.0%.**

Example A-6.38: *Determine the time required to remove 50% of internal energy above the surroundings at 30°C from a sphere of 16 cm dia at 650°C suddenly exposed to surroundings at 30°C with a convection coefficient of 800 W/m²K. The property values are: Thermal conductivity = 50 W/mK, specific heat = 450 J/kgK, thermal diffusivity = 2 × 10⁻⁵ m²/s. Also determine the centre and surface temperatures at this instant. How much more time is required to remove 30% more energy.*

Solution: Biot number = $hR/k = \dfrac{800 \times 0.08}{50} = 1.28$

The heat flow chart is used and the parameter Q/Q_o is used to enter the chart and the value of $h^2 \alpha\tau/k^2$ is read against the Biot number 1.28.

The value is 0.35.

$$\therefore \quad \frac{h^2\alpha\tau}{k^2} = 0.35 \quad \therefore \tau = \frac{0.35 \times 50 \times 50}{800 \times 800 \times 2 \times 10^{-5}} = \mathbf{68.35\ seconds}$$

The value Q/Q_o = 0.8 is entered and

$h^2 \alpha\tau/k^2$ is read as 1.0

$$\therefore \text{ The time is } \frac{68.35 \times 1.0}{0.35} = \mathbf{195.35\ seconds}$$

Additional time required is 195.29 – 68.35 = **124.94 s**

Almost twice more. This is due to the continuous reduction in temperature difference.

To determine the temperature fourier number is calculated and the center temperature chart is entered at this value and the temperature ratio is read against Bi = 1.28.

$$Fo = \frac{\alpha\tau}{R^2} = 2 \times 10^{-5} \times 68.35/0.08^2 = 0.214$$

The value read is 0.73

$$\frac{T_o - T_\infty}{T_i - T_\infty} = 0.73 \quad \therefore \quad \mathbf{T_o = 482.6°C}$$

For the surface temperature, the location chart is read against $r/R = 1$, entering at $Bi = 1.28$. The ratio read is 0.57

$$\therefore \quad \frac{T_s - T_\infty}{T_i - T_\infty} = \frac{T_s - 30}{650 - 30} = 0.73 \times 0.57 \quad \therefore \mathbf{T_s = 280°C.}$$

COMMENT

Because of the higher temperature difference at the beginning, the heat removal rate is quicker at the start. As time proceeds the temperature difference becomes lower and lower leading to slower heat removal rate.

Example A-6.39: *A steel sphere of radius 0.1 m with density = 7900 kg / m³, specific heat = 586 J/kgK and thermal conductivity of 40.0 W/mK initially at 100°C is exposed to convection at 1000°C with convective heat transfer coefficient of 400 W/m²K. Determine the temperatures at r/R = 0, 0.4, 0.6, 0.8 locations at 8, 16, 24 and 32 minutes times. Plot the temperature to scale.*

Solution: The Fourier number are calculated for the various times specified. The center temperature ratios are found using the center temperature chart entering at the Fourier numbers corresponding to the specified Times [Fig. AE-6.39(*a*)]. Then location chart is used at the Biot number for the given radius ratios [Fig. AE-6.39(*b*)]. These are tabulated.

$$\text{Biot Number} = \frac{hR}{k} = \frac{400 \times 0.1}{40} = 1.0, \ \alpha = \frac{k}{\rho c} = \frac{40}{7900 \times 586} = 8.64 \times 10^{-6}$$

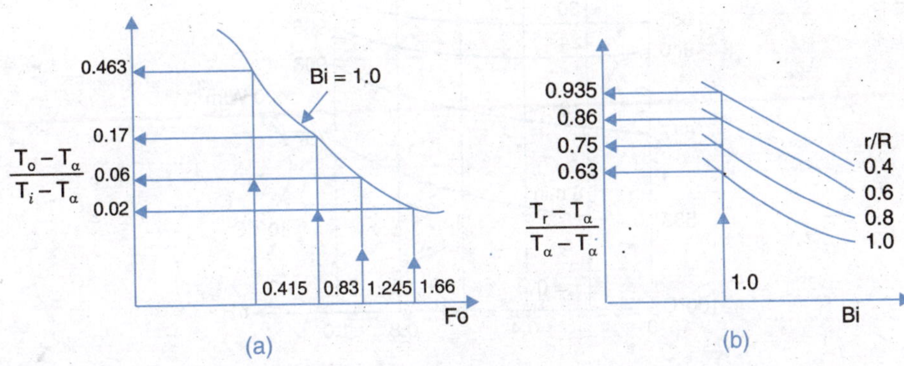

Fig. AE-6.39

Time min	8	16	24	32
Fo, ($\alpha\tau/R^2$)	0.415	0.83	1.245	1.66
Centre temp. ratio	0.463	0.17	0.06	0.02

Location temperature ratios read are:

r/R	0	0.4	0.6	0.8	1.0
Modifier	1	0.935	0.86	0.75	0.63

Chapter 6

Center temp. ratio $\dfrac{T_o - T_\infty}{T_i - T_\infty} = 0.463$

$$\frac{T_o - 1000}{100 - 1000} = 0.463 \quad \therefore \mathbf{T_o = 583.3°C}$$

To find surface: temperature;

$$\frac{T - T_\infty}{T_i - T_\infty} = \frac{T - 1000}{100 - 1000} = 0.463 \times 0.63 \quad \mathbf{T_s = 737.5°C}$$

The calculated temperatures are tabulated below and shown plotted in Fig. AE-6.39(*c*)

Time	*Location, r/R*				
	0	*0.4*	*0.6*	*0.8*	*1*
8	583.3	610.4	641.6	687.5	737.5
16	847	857	868.4	885.3	903.6
24	946	949.5	953.6	959.5	966
32	982	983.2	984.5	986.5	988.7

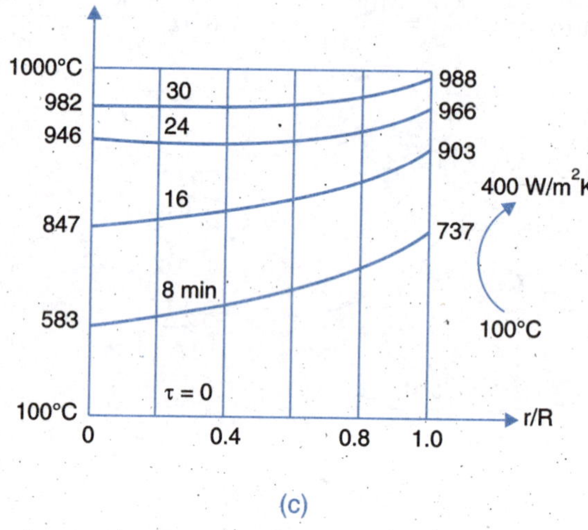

(c)

Fig. AE-6.39

COMMENT

The slopes of temperature curves are steeper near the surface and flatter near the centre. This is due to the variation of heat flow area available.

Example A-6.40: *A brinjal in the shape of a sphere of radius 5 cm initially at 25°C is cooked in water at 100°C with convection coefficient of 240 W/m²K. Taking the properties as below calculate the time for the temperature at the centre to reach 40, 50, 60, 70 and 80°C. Density = 1100 kg/m³, specific heat = 4100 J/kgK, Thermal conductivity = 0.6 W/mK.*

Solution: The centre temperature ratios and also the Biot number are specified. So the corresponding Fourier numbers are read from the centre temperature chart (Fig. AE-6.40).

$$Bi = \frac{hR}{k} = \frac{240 \times 0.05}{0.6} = 20, \ \alpha = \frac{0.6}{1100 \times 4100} = 0.133 \times 10^{-6} \text{ m}^2/\text{s}$$

Center Temp. °C	*40*	*50*	*60*	*70*	*80*
Temperature ratio $(T_o - T_\infty)/(T_i - T_\infty)$	0.8	0.67	0.53	0.4	0.267
Fo, $\alpha\tau/R^2$ *(from chart)*	0.09 (extrapolated)	0.118	0.145	0.18	0.228
Time, s	**1691**	**2217**	**2725**	**3383**	**4245**

Sample calculation for 40°C, centre temperature ratio is

$$\frac{T_o - T_\infty}{T_i - T_\infty} = \frac{40 - 100}{25 - 100} = 0.8$$

Fourier number read is 0.09

$$\therefore \quad 0.09 = \alpha\tau/R^2 = 0.133 \times 10^{-6} \times \tau \times \frac{1}{0.05^2} \quad \therefore \quad \tau = \mathbf{1691.25\ sec}$$

COMMENT

1. Obviously cooking as a whole piece is not desirable as it may take more than an hour.
2. This is the reason why vegetables cut to pieces for cooking.

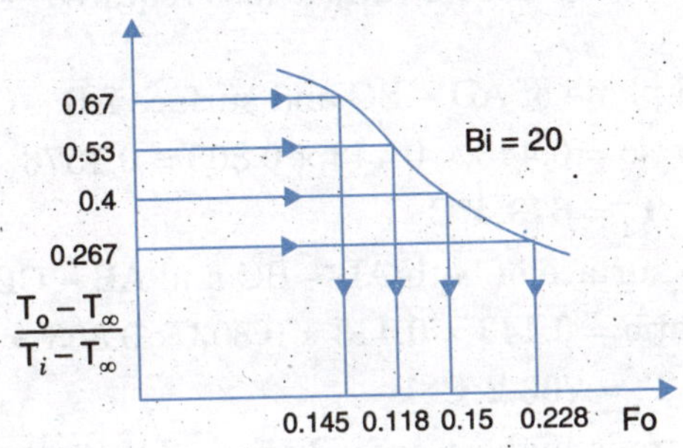

Fig. AE-6.40

Example A-6.41: *A long steel billet of 35 × 35 cm section initially at 35° is exposed to 800°C gases with h = 110 W/m²K in a furnace. Determine the time needed for the centre to reach 650°C. Also determine the temperatures at the points 1 and 3, 4 shown in Fig. AE-6.41 at this time. The properties are: density = 7833 kg/m³, specific heat: 465 J/kgK. Thermal conductivity: 41.5 W/mK. Also calculate the heat input per 1 m length.*

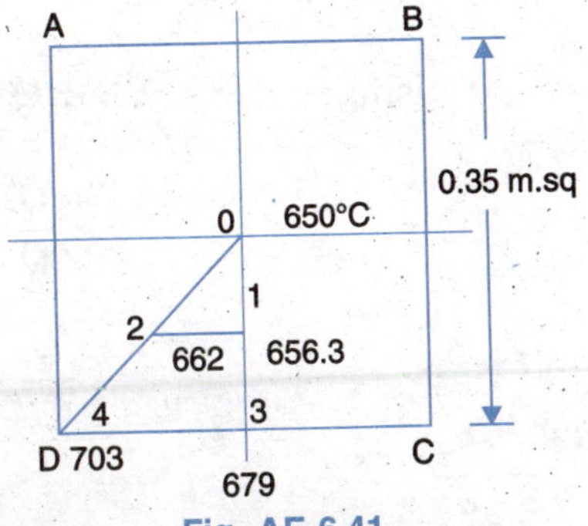

Fig. AE-6.41

Solution: The billet is exposed on four faces to convection. Hence a product solution with two slabs of $2L = 0.35$ m or $L = 0.175$ m is to be used, using Heisler charts.

$$\frac{hL}{k} = \frac{110 \times 0.35}{2 \times 41.5} = 0.464$$

The centre temperature is obtained as the product of solutions for slab $AD - BC$ and slab $AB - CD$.

$$\left(\frac{T - T_\infty}{T_i - T_\infty}\right)_{AD-BC} \times \left(\frac{T - T_\infty}{T_i - T_\infty}\right)_{AB-CD} = \frac{650 - 800}{35 - 800} = 0.196.$$

∴ Solution for one slab = $\sqrt{0.196}$ = 0.443

Because both slabs are identical.

Entering the chart at this value and reading on $Bi = 0.464$ the Fourier number $\alpha\tau/L^2 = 2.2$.

$$\tau = \frac{2.2 \times (0.35/2)^2}{41.5} \times 7833 \times 465 = \mathbf{5913\ sec = 1\ hour\ 38.6\ min}$$

(Note for rectangular shapes, time for a given center temperature cannot be directly determined. Trial solution is necessary. The trial is to be in terms of assuming several time intervals and then finding the center temperature to obtain a match.)

To obtain the temperature **at point 1.**

This point is on the mid plane of $AD - BC$ and $L/2$ plane of $AB - CD$. Mid plane temperature ratio is 0.443 for $x/L = 0.5$, the ratio is 0.957. Reading from chart

Using these values, temperature ratio is 0.443 × (0.443 × 0.957).

∴ $\mathbf{T_1 = 656.3°C.}$

For **point 3, 4** surface temperature ratio is also required: This ratio from location chart for $x/L = 1$ is 0.803.

∴ For **point 3,** mid plane of AD – BC and surface AB – CD

∴ Temperature ratio = 0.443 × 0.443 × 0.803 = 0.1576

∴ $\mathbf{T_3 = 679.4°C}$

For **point 4**, it is the surface of both AB – BC and AB – CD:

∴ Temperature ratio = 0.443 × 0.443 × 0.803 × 0.803 = 0.12654

∴ $\mathbf{T_4 = 703.2°C}$

The corner temperature is highest as it is exposed on both sides to heating.

The heat flow is determined

Using $$h^2 \alpha\tau/k^2 = \frac{110^2 \times 41.5}{7833 \times 465} \times \frac{5913}{41.5^2} = 0.473 \text{ and } hL/k = 0.464$$

$$\frac{Q}{Q} = 0.78. \quad \text{From Table 6.1}$$

$$\frac{Q}{Q_o} = \left[\frac{Q}{Q_o}\right]_i + \left[\frac{Q}{Q_o}\right]_2 \left[1 - \left(\frac{Q}{Q_o}\right)_1\right]$$

$$= 0.78 + 0.78 \times 0.22 = 0.9516$$

$\therefore$ $$Q = 0.9516 \times 0.35 \times 0.35 \times 1 \times 7833 \times 465\ (800 - 35)$$
$$= \mathbf{324.8 \times 10^6\ J\ per\ m\ length.}$$

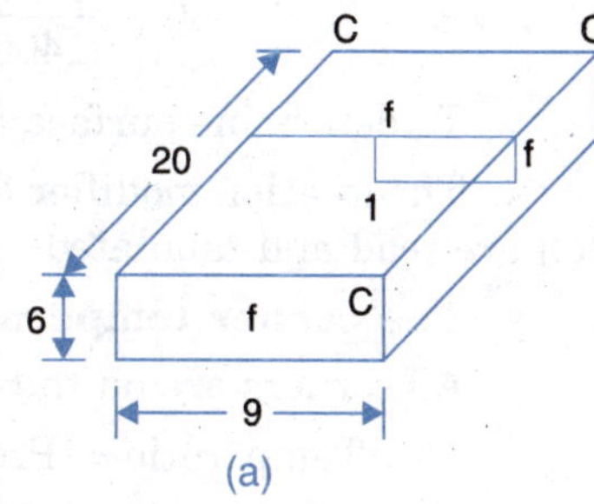

(a)

Fig. AE-6.42

Example A-6.42: *A refractory brick of size 6 cm × 9 cm × 20 cm at 1400°C is exposed suddenly to convection at 30°C with h = 60 W/m²K. Determine (i) the center temperature, (ii) corner temperatures and (iii) temperature at the centre point of all faces after 45 minutes.*

Solution: Property values are k = 1 W/mK, density = 2645 kg/m³, Specific heat = 960 J/kgK.

The solid can be considered as a combination of three infinite slabs of

(*i*) 2*L* = 6 cm, (*ii*) 2*L* = 9 cm & (*iii*) 2*L* = 20 cm

L, m	0.03	0.045	0.1
hL/k	1.8	2.7	6
$\alpha\tau/L^2$	1.182	0.525	1063
Centre temperature ratio, from chart	0.33	0.6	0.96
Location Temp. ratio	0.5	0.4	0.22

For 0.03 value $hL/k = 60 \times 0.03/1 = 1.8$

Using the centre temperature chart the temperature ratios are read as illustrated in [Fig. AE-6.42(*b*)] and tabulated

$\therefore$ The centre temperature ratio is

$$\frac{\theta}{\theta_i} = \frac{\theta}{\theta_i}\bigg|_1 \times \frac{\theta}{\theta_i}\bigg|_2 \times \frac{\theta}{\theta_i}\bigg|_3$$

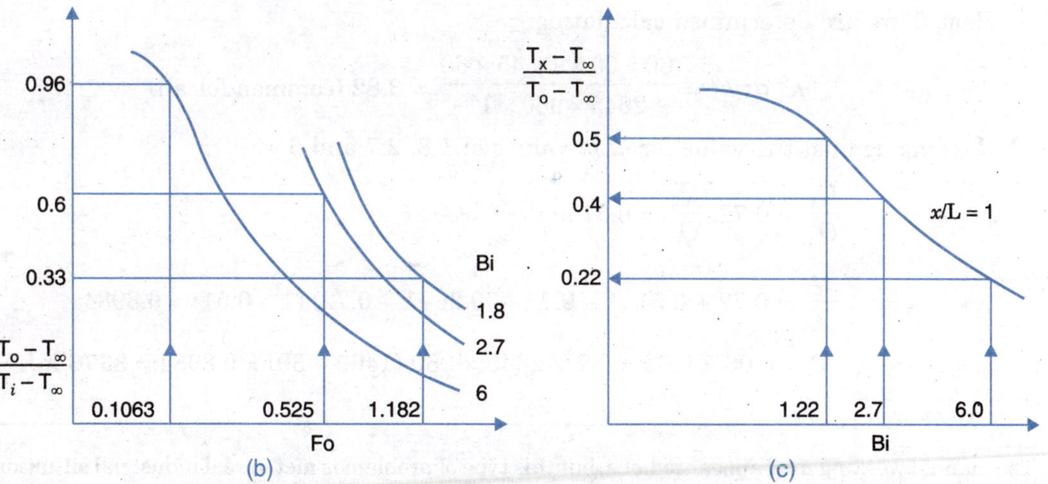

Fig. AE-6.42

For the centre: Temperature ratio = 0.33 × 0.6 × 0.96 = 0.19

$$\therefore \quad \frac{T - 30}{1400 - 30} = 0.19 \quad \therefore \ \mathbf{T = 290.3°C}$$

To determine surface and corner temperatures:

The location modifier for the three slabs at Bi = 1.8, 2.7 and 6 and x/L = 1 [Fig. AE-6.42 (*c*)] are read and tabulated.

The corner temperatures:

All corners are on the face of the three slabs considered:

∴ Temp. ratio = (Products or center temp. ratios) × (Products or surface temp. ratios)

= 0.33 × 0.6 × 0.96 × 0.22 × 0.4 × 0.5

∴ $\mathbf{T_{corner} = 41.45°C}$

For the centre point of faces:

(*i*) **6 × 9 face:** This point is on center plane of 6 cm slab × center plane of 6 cm slab × surface of 20 cm slab:

∴ Temp. ratio = 0.33 × 0.6 × 0.96 × 0.22

T = 87.29°C

(*ii*) **6 × 20 face:** Center plane of 20 cm × center plane of 6 cm × surface of 9 cm

Temp. ratio = 0.96 × 0.33 × 0.6 × 0.4

∴ **T = 134.2°C**

(*iii*) **9 × 20 face:** Center plane of 9 cm × center plane of 20 cm × surface of 6 cm = 0.6 × 0.96 × 0.33 × 0.5

∴ **T = 160.2°C**

The temperatures at other locations can be calculated in a similar manner. The total heat flow is given by

$$\frac{Q}{Q_0} = \frac{Q_1}{Q_0} + \frac{Q_2}{Q_0}\left[1 - \frac{Q_1}{Q_0}\right] + \frac{Q_3}{Q_0}\left[1 - \frac{Q_1}{Q_0}\right]\left[1 - \frac{Q_2}{Q_0}\right]$$

Heat flows are determined calculating

$$h^2\,\alpha\tau/k^2 = \frac{60 \times 60 \times 1 \times 45 \times 60}{2645 \times 960 \times 1^2} = 3.82 \text{ (common for all)}$$

Reading against the value for hL/k values of 1.8, 2.7 and 6

$$\therefore \quad \frac{Q_1}{Q_0} = 0.72,\ \frac{Q_2}{Q_0} = 0.51 \text{ and } \frac{Q_3}{Q_0} = 0.26$$

$$\therefore \quad \frac{Q}{Q_0} = 0.72 + 0.51\,(1 - 0.72) + 0.26\,(1 - 0.72)\,(1 - 0.51) = 0.8984$$

∴ **Q** = 0.06 × 0.09 × 0.2 × 2645 × 960 (1400 – 30) × 0.8984 = **3376 kJ.**

COMMENT

The manual working may appear tedious, but this type of problem is met in real industrial situation. Presently with softwares, the solution is fairly easy even for more complex shapes.

Example A-6.43: *A vegetable piece at 20°C and of size 30 mm × 60 mm × 60 mm rests on 60 × 60 mm face while being heated by exposure to 190°C with h = 18.4 W/m²K. Determine the time needed for the centre point of the face resting on support to reach 88°C. Also determine the temperatures at points 1, 2, 3, 4, 5, 6, 7, 8, 9, 10 as shown in figure. Assume that the base is adiabatic. The property values are density = 1000 kg/m³, specific heat 4180 J/kgK, conductivity = 0.552 W/mK. Also calculate the heat added.*

Solution: The example can be solved as the product solution of three identical slabs of $L = 0.03$ m. The face on which the piece rests, is not exposed to convection. This face is considered as insulated. On the shorter side the slab is assumed to be half of 0.06 m thickness.

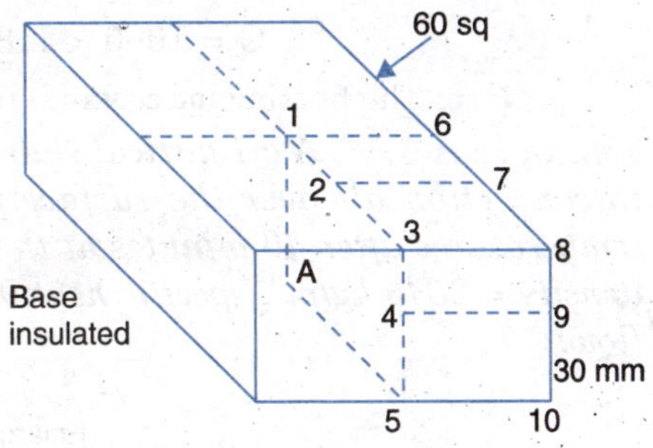

Fig. AE-6.43

$$\frac{hL}{k} = 18.4 \times 0.03/0.552 = 1.0$$

The temperature at 'A' is given and time is required:

$$\frac{T_o - T_\infty}{T_i - T_\infty} = \frac{88 - 190}{20 - 190} = 0.6$$

This is the product of three equal temperature ratios of three identical solutions

∴ The centre temperature ratio considering one slab is = $(0.6)^{1/3}$ or = 0.8434

The chart is entered at this value and Fourier number is read against $Bi = 1$ and the value read is 0.37.

$$\therefore \qquad \tau = 0.37 \times L^2/\alpha = \frac{0.37 \times 0.03 \times 0.03 \times 1000 \times 4180}{0.552} = \mathbf{2521.6\ sec.}$$

To determine the temperatures at points noted, the location temperature modifier for $x/L = 0.5$ and 1 are to be determined. Using the location chart and entering at $Bi = 1$. These values are read as 0.908 and 0.64.

To calculate temperatures ratio. In all cases the centre temperature ratio is involved

Point 1, 5: center of two, surface of one plane

$$\therefore \qquad \text{Temp. ratio} = 0.8434^2 \times (0.8434 \times 0.64)$$

$$= 0.6 \times 0.64 \qquad \therefore \mathbf{T_{1,5} = 124.7°C}$$

Point 2, 4: centre of one, surface of one, quarter plane ($x/L = 0.5$) of one

$$\text{Temp. ratio} = 0.6 \times 0.64 \times 0.908 \qquad \therefore \mathbf{T_{2,4} = 130.7°C}$$

Point **3, 6, 10:** centre of one, surface of two

$$\therefore \qquad \text{Temp. ratio} = 0.6 \times 0.64 \times 0.64 \qquad \therefore \mathbf{T_{3,6,10} = 148.2°C}$$

Point **7, 9:** surface of two and one quarter plane ($x/L = 0.5$)

$$\therefore \qquad \text{Temp.. ratio} = 0.6 \times 0.64 \times 0.64 \times 0.908 \qquad \mathbf{T_{7,9} = 152.1°C}$$

Point 8: Surface of three

$$\therefore \qquad \text{Temp. ratio} = 0.6 \times 0.64 \times 0.64 \times 0.64 \qquad \therefore \mathbf{T_8 = 163.3°C}$$

To determine the heat flow Q/Q_o, the parameter $h^2\ \alpha\tau/k^2$ is calculated

$$= \frac{18.4^2 \times 0.552 \times 2521.6}{4180 \times 1000 \times 0.552^2} = 0.369$$

Chapter 6

Entering the heat flow chart at this value and reading on biot number = 1, $\frac{Q}{Q_o} = 0.23$

This is the same for all three slabs.

$$\therefore \qquad \frac{Q}{Q_o} = 0.23 + 0.23\ (0.77) + 0.23\ (0.77) \times (0.77) = 0.5435$$

$$\therefore \qquad Q = 1000 \times 4180 \times 0.06 \times 0.06 \times 0.03 \times 0.5435 \times (190 - 20) = \mathbf{41.71\ kJ}$$

Note: The heating (or cooking) time is long. The corner temperature (8) is highest

Example A-6.44: *A cylindrical slab 20 cm dia and 20 cm thick at 1400°C is suddenly exposed to convection all over the surface to a fluid at 30°C with h = 60 W/m²K. Determine the temperatures after 40 minutes at the points marked in Fig. AE-6.44. The property values are. density = 2645 kg/m³, specific heat 960 J/kgK, conductivity = 2 W/mK. Also calculate the heat flow:*

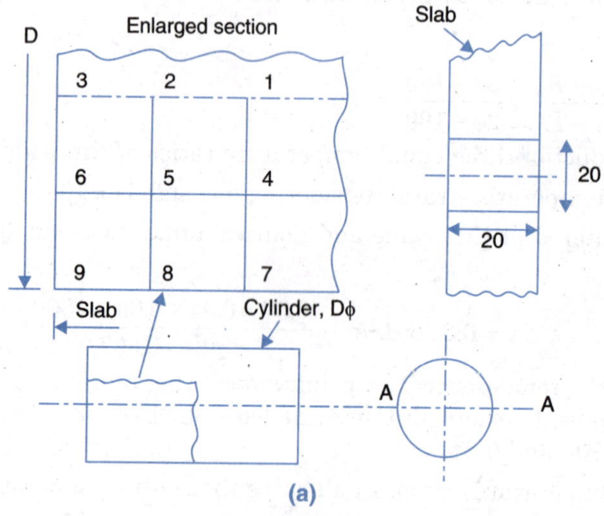

(a)

Fig. AE-6.44

Solution:

A product solution used.

(*i*) a slab of $2L$ = 20 cm or L = 0.1 m

(*ii*) Cylinder of radius 0.1 m.

(*i*) $\frac{hL}{k} = \frac{60 \times 0.1}{2} = 3,$ (*ii*) $\frac{hR}{k} = \frac{60 \times 0.1}{2} = 3$

$$\text{Fourier number} = \alpha\tau/L^2 = \frac{2}{2645 \times 960} \times \frac{40 \times 60}{0.1^2} = 0.189 \text{ in both cases.}$$

The temperature ratios at center and the location modification factors are determined (*i*) by entering the charts at Fo = 0.189 and reading against hL/k = 3 and (*ii*) by entering the charts at hR/k = 3 and reading against x/L or r/R = 0.5 as illustrated in Fig. AE-6.44. The values are indicated in the figures.

Slab: Center temp. ratio: 0.902, location factor 0.5 ⟶ 0.825, 1 ⟶ 0.37

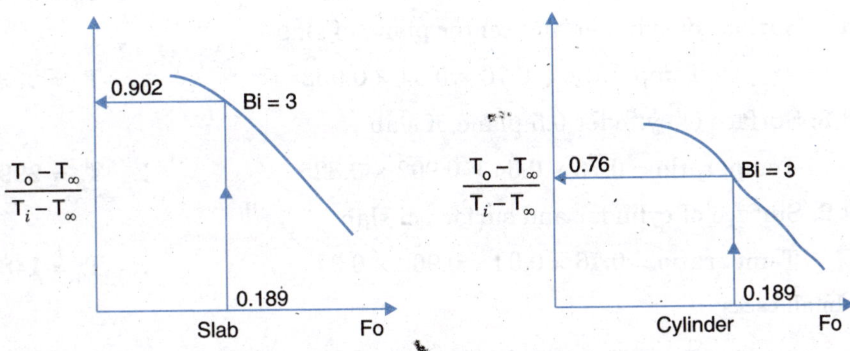

Cylinder center temp. ratio: 0.76, location factor 0.5 ⟶ 0.81, 1 ⟶ 0.34
Slab center temp. ratio: 0.902, location factor ⟶ 0.825, 0.37

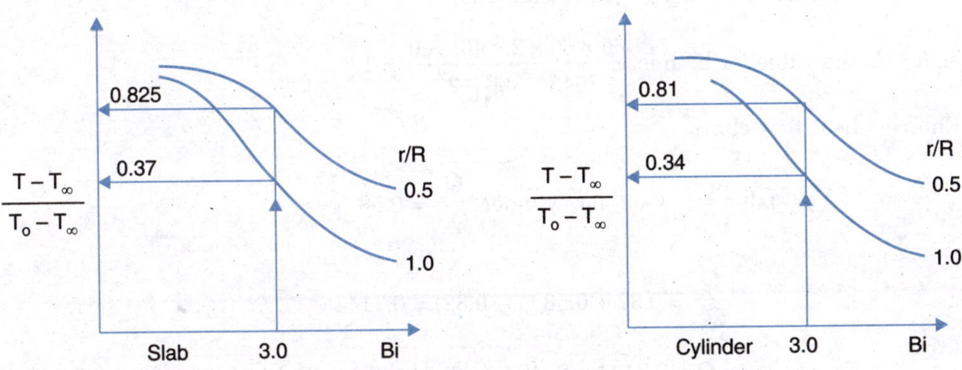

Fig. AE-6.44

Temperatures:

Point 1: Centre line of cylinder and center plane of slab

∴ Temp. ratio = 0.76 × 0.902 ∴ **T_1 = 969.2°C**

Point 2: Centre line of cylinder, 0.5 surface of slab

∴ Temp. ratio = 0.76 × 0.902 × 0.825 ∴ **T_2 = 804.8°C**

Point 3: Centre line of cylinder, and surface of slab

∴ Temp. ratio = 0.76 × 0.902 × 0.37 ∴ **T_3 = 377.5°C**

Point 4: Mid radius of cylinder and center plane of slab

∴ Temp. ratio = 0.76 × 0.81 × 0.902 ∴ **T_4 = 790.7°C**

Point 5: Mid radius of cylinder and 0.5 plane of slab

Temp. ratio = 0.76 × 0.81 × 0.902 × 0.825 ∴ **T_5 = 657.6°C**

Chapter 6

Point 6: Mid radius of cylinder and surface of slab

Temp. ratio = 0.76 × 0.81 × 0.902 × 0.37 ∴ **T_6 = 311.5°C**

Point 7: Surface of cylinder and center plane of slab

∴ Temp. ratio = 0.76 × 0.34 × 0.902 ∴ **T_7 = 293.4°C**

Point 8: Surface of cylinder 0.5 plane of slab

Temp. ratio = 0.76 × 0.34 × 0.902 × 0.825 ∴ **T_8 = 299.8°C**

Point 9: Surface of cylinder and surface of slab

Temp. ratio = 0.76 × 0.34 × 0.902 × 0.37 ∴ **T_9 = 148.9°C**

Note: In all cases:

$$\frac{T - T_\infty}{T_i - T_\infty} = \frac{T - 30}{1400 - 30} = \text{Temperature ratio}$$

As it is to be expected, the edge cooled quicker.

To determine heat flow $h^2\ \alpha\tau/k^2$ is calculated.

$$\text{as } R = L\text{, the value for both is } \frac{60 \times 60 \times 2 \times 40 \times 60}{2645 \times 960 \times 2^2} = 1.7$$

Entering heat flow chart

$$\text{For slab } \frac{Q}{Q_o} = 0.32\text{, for cylinder } \frac{Q}{Q_o} = 0.58$$

$$\therefore \quad \frac{Q}{Q_o} = 0.32 + 0.58\,(1 - 0.32) = 0.7144$$

$$\therefore \quad Q = 0.7144 \times 2645 \times 960 \times \pi \times 0.1^2 \times 0.2 \times (1400 - 30) = \mathbf{15.62 \times 10^6\ J.}$$

COMMENT

There are a large number of industrial situations where short cylindrical components are heat treated. Hence this is a real life problem.

Example A-6.45: *A hemispherical piece of radius 0.1 m initially at 1400°C is exposed to convection all over its surface to a fluid at 30°C with convective heat transfer coefficient of 60 W/m²K. Determine the temperatures at points 1, 2, 3, 4, 5, 6 as shown in Fig. AE-6.45(a). The elapsed is 40 minutes.*

Solution: The property values are: density = 2645 kg/m³

Specific heat: 960 J/kgK, conductivity = 2 W/mK.

The example can be solved taking the solid as a composite of a sphere and slab, the dimensions being 0.1 m

$$\frac{hL}{k} = \frac{60 \times 0.1}{2} = 3, \frac{hR}{k} = 3$$

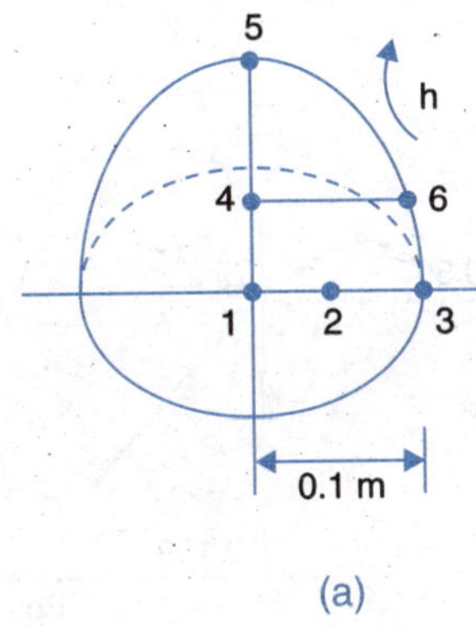

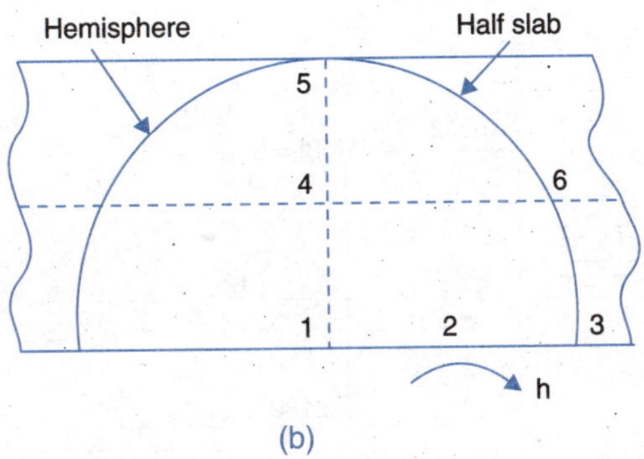

Fig. AE-6.45

$$\frac{\alpha\tau}{L^2} = \frac{\alpha\tau}{R^2} = \frac{2 \times 40 \times 60}{2645 \times 960 \times 0.1^2} = 0.189$$

as $R = L$

$h^2\ \alpha\tau/k^2 = 1.7$ (see example A-6.44)

Reading the temperature ratio charts and heat flow charts as diagrammatically shown in Fig. AE-6.45

Plane:	Center: 0.902	Mid plane: 0.825	Surface = 0.37
Sphere:	Center: 0.59	Mid radius: 0.798	Surface = 0.345

Temperature ratio: $\frac{T - T_\infty}{T_i - T_\infty} = \frac{T - 30}{1400 - 30}$

Point 1: Center of sphere and surface of slab

∴ Temperature ratio = 0.59 × 0.902 × 0.37 ∴ $\mathbf{T_1 = 299.8°C}$

Point 2: Mid radius of sphere and surface of slab

Temp. ratio = 0.59 × 0.798 × 0.902 × 0.37 ∴ $\mathbf{T_2 = 245.27°C}$

Point 3: Surface of sphere, surface of slab

Temp. ratio = 0.59 × 0.345 × 0.902 × 0.37 ∴ $\mathbf{T_3 = 123°C}$

Point 4: Mid radius of sphere and 0.5 plane of slab

Temp. ratio = 0.59 × 0.798 × 0.902 × 0.825 ∴ $\mathbf{T_4 = 510°C}$

Point 5: Surface of sphere and surface of slab

Temp. ratio = 0.59 × 0.345 × 0.902 × 0.37 ∴ $\mathbf{T_5 = 123°C}$

Point 6: Surface of sphere and 0.5 plane of slab

Temp. ratio = 0.59 × 0.345 × 0.902 × 0.825 ∴ $\mathbf{T_6 = 237.5°C}$

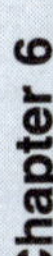

The lowest temperature is at the points 3 and 5 as it should be.

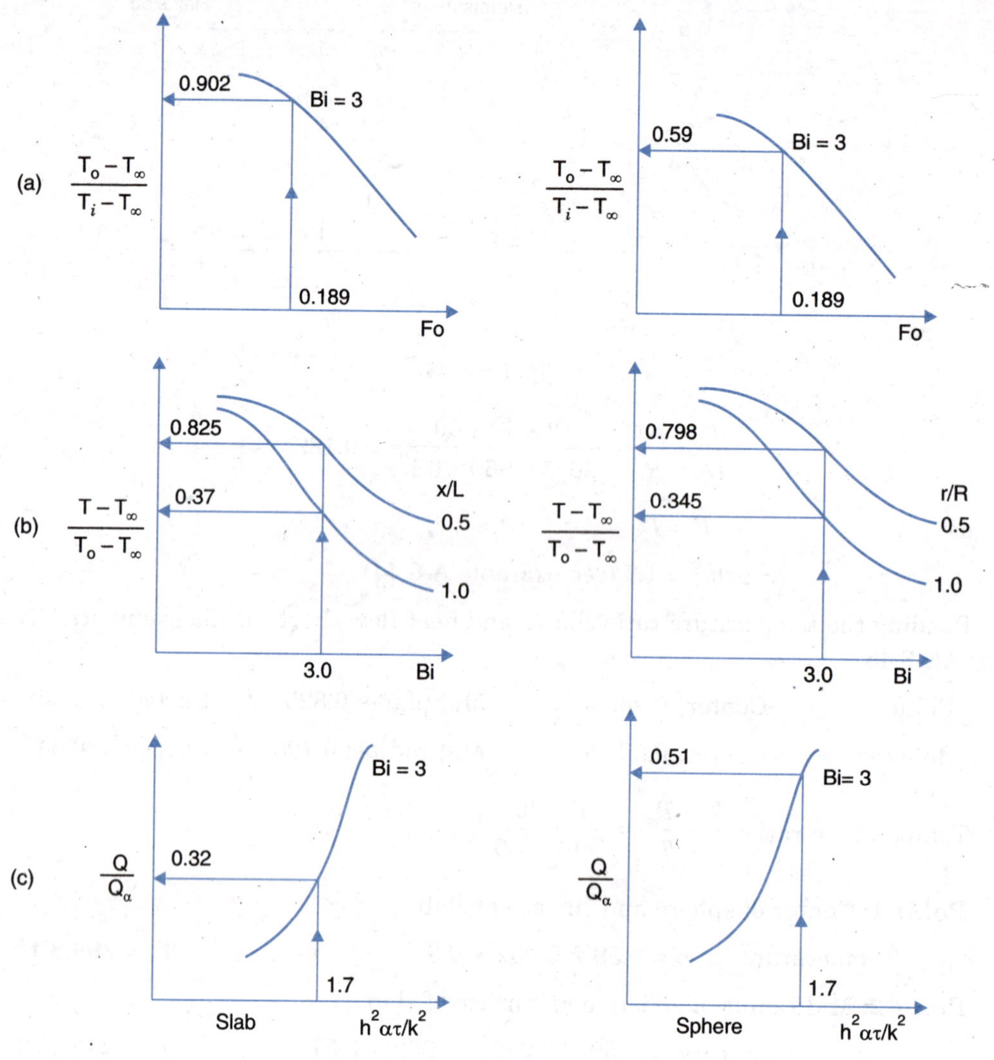

Fig. AE-6.45

The heat flow is calculated by using the parameter $h^2 \alpha\tau/k^2 = 1.7$ and entering the chart at this value and reading against the Biot number 3 [Fig. AE-6.45(*c*)]

For slab $\dfrac{Q}{Q_o} = 0.32$, for sphere $\dfrac{Q}{Q_o} = 0.51$

$$\frac{Q}{Q_o} = 0.32 + 0.51\,(1 - 0.32) = 0.6668$$

$$\therefore \quad \mathbf{Q} = 0.6668 \times \frac{1}{2} \times \frac{4}{3} \times \pi \times 0.1^3 \times 2645 \times 960 \times (1400 - 30)$$

$$= \mathbf{4.86 \times 10^6\ J.}$$

Example A-6.46: *One half of a cylinder of 0.1 m radius and 0.2 m length initially at 1400°C is exposed all over to convection at 30°C with h = 60 W/m²K. The property values are as in example A-6.45. Determine the temperatures after 40 minutes at points 1, 2, 3, 4, 5, 6, 7, 8, 9 shown in Fig. AE-6.46.*

Solution: This solid can be taken as combination of cylinder and two slabs. All are of $L = 0.1$ m dimension.

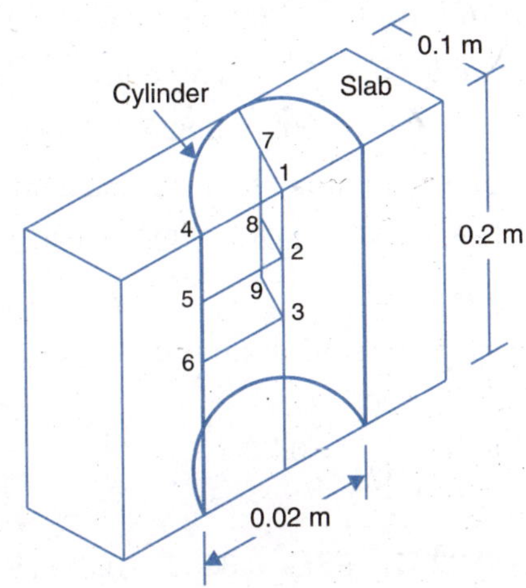

Fig. AE-6.46

Taking the temperature ratio values for plane and cylinder from example A-6.44. (as the values of R, L and properties and time elapsed are the same)

Temperature ratios at locations $\dfrac{r}{R}$ and $\dfrac{x}{L}$ are

Location	0	0.5	1.0
Slabs	0.902	0.825	0.37
Cylinder	0.76	0.81	0.34

Temperature Ratio, in all cases,

$$\frac{T - T_\infty}{T_i - T_\infty} = \frac{T - 30}{1400 - 30}$$

Point 1: axis of cylinder and surface of both slabs

$\therefore$ Temp. ratio = 0.76 × 0.902 × 0.37 × 0.902 × 0.37 $\quad \therefore$ **T_1 = 146°C**

Point 2: axis of cylinder, surface of one slab and 0.5 of another slab

$\therefore$ Temp. ratio = 0.76 × 0.902 × 0.37× 0.902 × 0.825 $\quad \therefore$ **T_2 = 289°C**

Point 3: axis of cylinder, surface of one slab and center plane of another slab

Temp. ratio = 0.76 × 0.902 × 0.37 × 0.902 $\quad \therefore$ **T_3 = 343.4°C**

Point 4: surface of cylinder, surface of both slab

Temp. ratio = 0.76 × 0.34 × 0.902 × 0.37 × 0.902 × 0.37 $\quad \therefore$ **T_4 = 69.4°C**

Point 5: surface of cylinder, surface of one slab and 0.5 of other slab

Temp. ratio = 0.76 × 0.34 × 0.902 × 0.37 × 0.902 × 0.825 ∴ $\mathbf{T_5 = 118°C}$

Point 6: surface of cylinder, surface of one slab and center plane of slab

Temp. ratio = 0.76 × 0.34 × 0.902 × 0.37 × 0.902 ∴ $\mathbf{T_6 = 136.6°C}$

Point 7: mid radius of cylinder, mid lengths of one slab and surface of 1 slab

Temp. ratio = 0.76 × 0.81 × 0.902 × 0.825 × 0.902 × 0.37 ∴ $\mathbf{T_7 = 239.5°C}$

Point 8: mid radius of cylinder 0.5 of one slab and 0.5 of another

Temp. ratio 0.81 × 0.34 × 0.902 × 0.825 × 0.902 × 0.825 ∴ $\mathbf{T_8 = 497°C}$

Point 9: mid radius cylinder 0.5 plane of a slab and center plane of another

Temp. ratio = 0.76 × 0.81× 0.902 × 0.825 × 0.902 ∴ $\mathbf{T_9 = 596.1°C}$

Note that point 4 has cooled fastest.

The heat flow can be calculated using (Example A-6.48)

$$\left.\frac{Q}{Q_o}\right|_{cyl} = 0.58 \text{ for slab } \frac{Q}{Q_o} = 0.32$$

$$\therefore \quad \frac{Q}{Q_o} = 0.58 + 0.32\,(1 - 58) + 0.32\,(1 - 0.58)\,(1 - 0.32) = \mathbf{0.806}$$

$$\therefore \quad \mathbf{Q} = 0.806 \times 0.5 \times \pi \times 0.1^2 \times 0.2 \times 2645 \times 960 \times (1400 - 30)$$

$$= \mathbf{8.81 \times 10^6\ J.}$$

Example A-6.47: *A circular slab 20 cm dia and 20 cm thickness at 1400°C is exposed over the curved surface to 30°C with h = 600 W/m²K and has the flat surfaces maintained by intense cooling at 30°C. Determine the temperature at mid point. Properties are: density = 2645 kg/m³, specific heat = 960 J/kgK: k = 2W/mK.*

Solution: This is a combination of an infinite slab and a long cylinder. In the case of the slab intense coling means, h is very large and so $hL/k = \infty$ is to be noted.

For slab and cylinder $\alpha\tau/L^2 = \dfrac{2 \times 40 \times 60}{2645 \times 960 \times 0.1^2} = 0.189$ as $L = R = 0.1$ m

Cylinder $hR/k = \dfrac{60 \times 0.1}{2} = 3$

For slab entering the center temperature chart at 0.189 and reading against $hL/k = \infty$, temperature ratio = 0.79. For cylinder reading against $hR/k = 3$, temp. ratio = 0.76

∴ Total temperature ratio = 0.79 × 0.76 = 0.6004

$$\frac{T - 30}{1400 - 30} = 0.6004 \quad \therefore \mathbf{T = 852.5°C}$$

In case temperature at other locations are needed, the location modification chart has to be used.

Example A-6.48: *A long cylinder of 0.2 m dia initially at 1400°C has its surface and the base exposed to convection at 30°C with h = 60 W/m²K. Using the property values as in example A-6.47 determine the temperature at the centre of the base and at 0.1 m from the base on the axis of the cylinder after 40 minutes.*

Solution: This is a combination of long cylinder and semiinfinite slab exposed to convection. The solution for the centre line of cylinder is used from example A-6.46. Temperature ratio = 0.76.

For the simi infinite slab

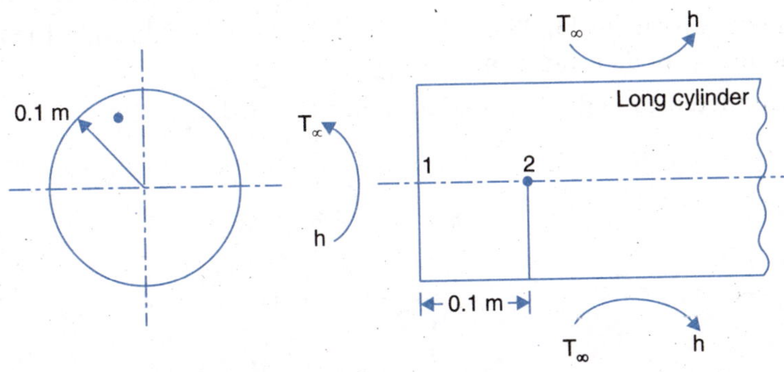

Fig. AE-6.48

$$\frac{T_x - T_i}{T_\infty - T_i} = (1 - \text{erf}\,\frac{x}{2\sqrt{\alpha\tau})} - \exp\left[\frac{hx}{k} + \frac{h^2\alpha\tau}{k^2}\right]\left[1 - \text{erf}\left(\frac{x}{2\sqrt{\alpha\tau}} + \frac{h\sqrt{\alpha\tau}}{k}\right)\right]$$

(*i*) **x = 0,**

$$\frac{T - 1400}{30 - 1400} = 1 - \exp\,\frac{h^2\alpha\tau}{k^2}\left[1 - \text{erf}\left(\frac{h\sqrt{\alpha\tau}}{k}\right)\right]$$

$$= 1 - \exp\left[\frac{60 \times 60 \times 2 \times 40 \times 60}{2^2 \times 2645 \times 960}\right]\left[1 - \text{erf}\,\frac{60\sqrt{(2/(2645 \times 960)) \times 40 \times 60}}{2}\right]$$

$$= 1 - 5.481 \times [1 - 0.93401]$$

$$= 1 - 0.6383 \quad \therefore \quad \frac{\mathbf{T - T_\infty}}{\mathbf{T_i - T_\infty}} = 1 - 0.6383 = \mathbf{0.3617}$$

(*ii*) **x = 0.1**

$$\frac{T_x - T_i}{T_\infty - T_i} = \left(1 - \text{erf}\,\frac{0.1}{2\sqrt{\alpha\tau}}\right) - \exp\left[\frac{0.1h}{k} + \frac{h^2\alpha\tau}{k^2}\right]\left[1 - \text{erf}\left(\frac{0.1}{2\sqrt{\alpha\tau}} + \frac{h\sqrt{\alpha\tau}}{k}\right)\right]$$

$$= (1 - 0.89612) - 110.09\,[1 - \text{erf}\,(1.15 + 1.304)] = 0.0467$$

$$\therefore \quad \frac{\mathbf{T - T_\infty}}{\mathbf{T_i - T_\infty}} = 1 - 0.0467 = \mathbf{0.9533}$$

Temperature at the **centre of base:**

$$\text{Temp. ratio} = \frac{T_1 - 30}{1400 - 30} = 0.76 \times 0.3617 = 0.2749 \quad \therefore \quad \mathbf{T_1 = 406.6°C}$$

At 0.1 m from base: on centre line:

$$\text{Temp. ratio} = \frac{T_2 - 30}{1400 - 30} = 0.76 \times 0.9533 \quad \therefore \ \mathbf{T_2 = 1022.6°C}$$

Example A-6.49: *A slab 0.24 m thick initially at 80°C has both surfaces suddenly lowered to 20°C and surfaces maintained at the level. Thermal diffusivity = 1.5 × 10⁻⁶ m²/s. Using numerical method estimate the temperature at various locations of the slab at the end 30 minutes. The slab is shown in Fig. AE-6.49.*

Solution: One half of the slab can be considered for solution as both sides are identical.

Choosing nodal distance of 0.03 m. The nodal temperature at node 1 is 20°C at all times after, 0 sec. Considering node other than, 5

net heat conducted to node 2 = heat stored in the node.

For node 2:

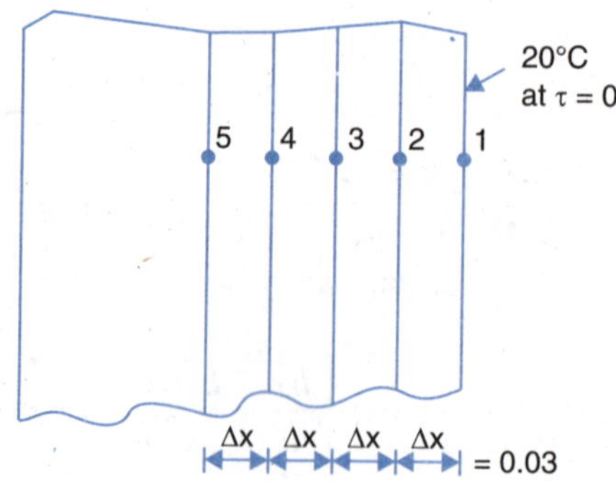

Fig. AE-6.49. Nodes

$$\Delta\tau \times kA\left(\frac{T_1^p - T_2^p}{\Delta x}\right) + \Delta\tau \times kA\left(\frac{T_3^p - T_2^p}{\Delta x}\right) = \rho cA\ \Delta x \times (T_2^{p+1} - T_2^p)$$

where p denotes time interval number.

Simplifying and assuming $\Delta x^2/\alpha\Delta\tau = 2$, for the chosen value of $\Delta x = 0.03$

$$\Delta\tau = 300 \text{ sec.}$$

$$T_2^{p+1} = (T_1^p + T_3^p)/2$$

Similarly $\quad T_3^{p+1} = (T_4^p + T_2^p)/2, \quad T_4^{p+1} = (T_5^p + T_3^p)/2$

For node, 5, there is heat flow only from node 4

$$\Delta\tau \times kA\ (T_4^p - T_5^p) = \rho cA\ \frac{\Delta x}{2}\ (T_5^{p+1} - T_5^p)$$

$$\therefore \qquad T_4^p = \frac{\rho c}{k} \cdot \frac{\Delta x^2}{2\Delta\tau}\ (T_5^{p+1} - T_5^p) + T_5^p$$

as $$\frac{\rho c}{k} \cdot \frac{\Delta x^2}{2\Delta\tau} = 1; \quad T_5^{p+1} = T_4^p$$

The present (p) last nodal temperature equals the previous (p + 1) temperature of the last but one node.

The solution is given as a tabulation. The temperature at any time intervals are the mean of the adjacent temperature at the previous level for nodes 2, 3, and 4, and for T_5 it is equal to the temperature at the previous interval at T_4.

Time, end of minute	T_1	T_2	T_3	T_4	T_5
0	20	80	80	80	80
5	20	50	80	80	80
10	20	50	65	80	80
15	20	42.5	65	72.5	80
20	20	42.5	57.5	72.5	72.5
25	20	38.75	57.5	65	72.5
30	20	38.75	51.875	65	65
35	20	35.94	51.875	58.43	65

At 30 min the nodal temperatures are 20, 37.35, 51.9, 61.72 and 65°C (at nodes 2 and 4, mean value is taken).

REVIEW QUESTIONS

6.1 State the conditions under which lumped capacitance model can be adopted to determine the temperature response of a solid suddenly exposed to convection at a different temperature.

6.2 Explain the physical significance of Biot number.

6.3 Explain why lumped capacitance model may not suit when a solid is cooled by water under forced convection conditions.

6.4 Explain why in cooling by forced convection in air or natural convection in air the temperature response of the solid can be analysed by adopting lumped capacitance model.

6.5 Same size of spheres of copper, aluminium, and ceramic are cooled by convection. In which case lumped capacitance model can be used to investigate the temperature response.

6.6 Explain how the increase in value of the following properties accelerate or decelerate the response when lumped capacitance model is used for analysis. (1) density (2) specific heat (3) size (4) conductivity (5) convection coefficient.

6.7 Explain the physical significance of Fourier number.

6.8 State under water condition can semi infinite solid model can be adopted.

6.9 Discuss the difference between explicit and implicit method of finite difference analysis and state the advantage of each.

6.10 What is stability in finite difference method of solution? In explicit method state the condition for stability.

OBJECTIVE QUESTIONS

6.1 Choose the correct statement from each question.

(*i*) (*a*) Transient conduction means very little heat transfer

(*b*) Transient conduction means conduction when the temperature at a point varies with time

(*c*) Transient conduction means heat transfer for a short time

(*d*) Transient conduction means heat transfer with very small temperature difference.

(*ii*) Lumped parameter model can be used when

(*a*) the thickness is small

(*b*) when the conductivity is high

(*c*) when the convective heat transfer coefficient is low

(*d*) when conditions (*a*), (*b*) and (*c*) are true

(*iii*) In the lumped parameter model, the temperature variation is

(*a*) linear with time (*b*) sinusoidal with time

(*c*) exponential with time (*d*) cubic with time.

(*iv*) The response time of a thermocouple is the time taken for the temperature change to be

(*a*) 0.5 of original temperature difference

(*b*) 1/1.414 of original temperature difference

(*c*) 1/*e* of original temperature difference

(*d*) 99% of the original temperature difference.

(*v*) To make a thermocouple to respond quickly

(*a*) the wire diameter should be large

(*b*) convective heat transfer coefficient should be high

(*c*) density should be very small

(*d*) specific heat should be high.

(*vi*) Semi infinite model can be adopted when

(*a*) thickness of the solid is very large (*b*) heat diffusion is very slow

(*c*) short time period (*d*) all of these.

(*vii*) The temperature distributions in a semi infinite solid with time are shown in Fig. Q.(*vii*) (*a*), (*b*), (*c*) and (*d*) match these with the statements and figures

(1) sudden surface temperature change (2) cooling

(3) convection boundary (4) constant heat flux.

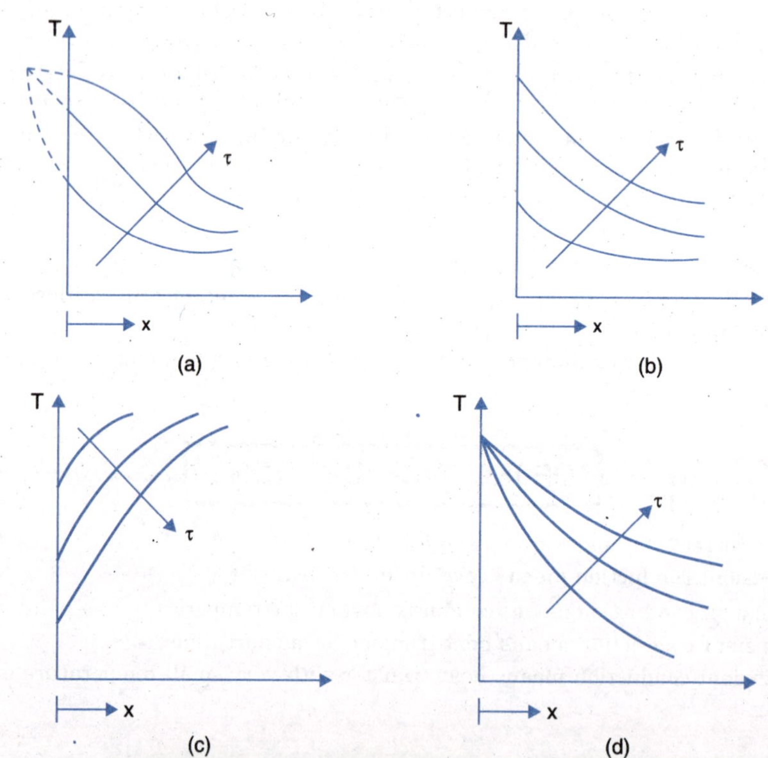

(*viii*) Heating or Cooling of a road surface can be analysed using

(*a*) lumped parameter model (*b*) Infinite slab model

(*c*) Semi infinite slab model (*d*) none of these.

(*ix*) Match the figures and the statements under transient conditions.

(1) cooling (2) heating

(3) steady state (4) heat generation.

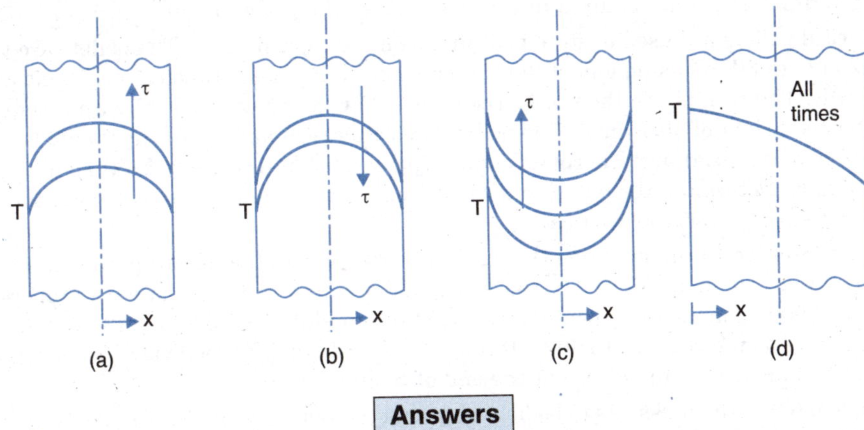

Answers

6.1 (*i*) (*b*), (*ii*) (*d*), (*iii*) (*c*), (*iv*) (*c*), (*v*) (*b*), (*vi*) (*d*), (*vii*) 1(*d*), 2(*c*), 3(*a*), 4(*b*), (*viii*) (*c*), (*ix*) 1(*b*), 2(*c*), 3(*d*), 4(*a*).

EXERCISE PROBLEMS

6.1 For the materials with the following values of conductivity and a convective heat transfer coefficient of 60 W/m^{2}K, determine the minimum diameter for cylindrical shape for which lumped parameter model can be applied

(*a*) Aluminium k = 204.2 W/mK (*b*) Silumin with k = 136.8 W/mK

(*c*) Steel k = 43.3 W/mK (*d*) Nickel steel with k = 19.1 W/mK

(*e*) Inwar with k = 10.7 W/mK.

6.2 In the problem 6.1 determine the diameter if spherical shape is used.

6.3 In the problem 6.2 determine the thickness if flat slab shape is used.

6.4 Determine the maximum value of convective heat transfer coefficient for the following materials of spherical shape of diameter 0.05 m for the application of lumped parameter model

(*a*) k = 204.2 W/mK (*b*) k = 136.8 W/mK

(*c*) k = 43.3 W/mK, and (*d*) k = 107 W/mK.

6.5 Determine the time constant for the following materials if the wire is of 1 mm dia and exposed to convection at 50 W/m^{2}K. The property values are:

Material	*density, kg/m^3*	*Specific heat, J/kgK*	*Thermal conductivity W/mK*
a	2700	896	204.2
b	2660	867	136.8
c	8196	461	10.4
d	8954	381	386
e	8922	394	24.9
f	19350	134	162.7
g	10525	235	407

Chapter 6

6.6 Pellets of 3 mm × 10 mm × 8 mm size are to be cooled from 25°C to –90°C in atmosphere at – 176°C with a convection coefficient of 45 W/m^2K. The material properties are density = 10525 kg/m^3, specific heat = 235 J/kg/K thermal conductivity = 407 W/mK. Determine the time required.

6.7 Plate glass 5 mm thick at 200°C is to be cooled from both sides using air at 40°C. If surface cracks are to be avoided, determine the maximum value of convective heat transfer coefficient. Material properties are density = 2500 kg/m^3, specific heat 670 J/kgK. Thermal conductivity 0.744 W/mk. Also determine the time required in this case to cool the plate to 80°C.

6.8 Cylindrical pellets are used to store heat in a heat recovery process. The pellets are to be heated from 200°C to 300°C using gas at 400°C. The convective heat transfer coefficient attainable is 40 W/m^2K. The pellets are then to be cooled to 200°C using air at 60°C with a convective heat transfer coefficient of 50 W/m^2K. The pellets are cylindrical and 15 mm dia and 20 mm long. The properties of material are: Thermal conductivity = 12 W/mK, density 2400 kg/m^3, specific heat = 456 J/kgK. Determine the batch time (1 heating and 1 cooling). Also calculate the heat recovered for 1 kg of material per batch.

6.9 In an air preheater of rotary storage type the mass moves over the cold air portion during 6 seconds and is exposed to air at 35°C. The mass is made of wires of 3 mm dia with property values of density = 7900 kg/m^3, specific heat = 394 J/kgK and conductivity of 25 W/mK. The material when entering the cooling zone is at 260°C. If convective heat transfer coefficient has a value of 245 W/m^2K. Determine the temperature at the end of cooling cycle.

6.10 In problem 6.9, if the mass has to be heated from a temperature of 155°C to 260°C by using gases at 320°C, determine the time required. Also determine the heat collected per kg of material.

If these two processes are to be accommodated in 300° of the rotation, calculate angles of each sector and the rpm.

6.11 Ice flakes at 0°C are to be cooled to –20°C for long time storage by passing these through a cold chamber at –40°C with a convection coefficient of 25 W/m^2K. The flakes are 2 mm thick and 5 mm square and are exposed on one side to convection. Determine the time required; density = 920 kg/m^3, k = 2.25 W/mK, specific heat = 2261 J/kg.

6.12 A constantine thermocouple wire of 0.4 mm dia is to be used to measure a temperature of 600°C. The wire is initially at 35°C. If the thermocouple should read 595°C in 2 seconds, determine the value of convection coefficient required. Density = 8922 kg/m^3, specific heat = 410 J/kgK; Thermal conductivity = 22.7 W/mK.

6.13 Ice in the form of long cylinders of dia 15 mm at –20°C are exposed to air at 30°C with a convection coefficient of 15 W/m^2K. How long can it be kept exposed before the ice begins to melt. The property values are: density = 920 kg/m^3, specific heat = 2261 J/kgK, conductivity = 2.25 W/mK.

6.14 A tool bit in the shape of 15 mm square and 5 mm thickness is to be heated from 40°C to 1200°C using gases at 1500°C, without much of distortion with Biot number to be restricted to 0.05. Calculate the maximum value of convective heat transfer coefficient that can be used. How long will the heating take? Density = 9350 kg/m^3, specific heat = 134 J/kgK, k = 162.7 W/mK.

6.15 A thick billet initially at 40°C has its surface temperature increased and maintained at 900°C. Determine the heat input needed for maintaining the surface temperature for 20 minutes, over an area of 0.5 m^2. Also determine the temperature at a depth of 0.05 m at this time. Thermal diffusivity = 1.17×10^{-5} m^2/s, conductivity = 28 W/mK, specific heat = 460 J/kgK.

6.16 Sudden flow of lava raises and maintains the surface temperature of soil at 1400°C. Determine the temperature after 12 hours at the following depths

(*i*) 0.2 m (*ii*) 0.4 m and 1 m. Thermal diffusivity = 2.6×10^{-6} m^2/s; k = 0.325 W/mK. Also determine the heat flow into the ground. Specific heat = 796 J/kgK.

6.17 Sandy soil heated by lava flow to 400°C to a considerable depth has its surface suddenly exposed to snow which reduced the surface temperature to 0°C. Determine taking the property values from exercise 6.16, the depth at which the temperature will be 200°C after 12 hours from the time it started snowing.

6.18 Hot water from a process at 60°C is continuously let into the top layer of a pond. There is little chance of mixing of layers. Determine the temperature at a depth of 40 mm from the hot water layer after exposure to the hot water for 6 hours. Thermal diffusivity = 1.44 × 10^{-10} m²/s. Also calculate the heat going into the cold water over this period k = 0.5978 W/mK. Specific heat = 4180 J/kgK.

6.19 Warm water at a constant temperature of 60°C is kept in a concrete container of wall thickness 0.2 m, initially at 25°C. Estimate the time it will take for the outside surface temperature of the wall to begin to increase. Thermal diffusivity = 4.9 × 10^{-7} m²/s. Also determine the temperature at 5 cm from inside at this time.

6.20 A furnace starts operation at 6 am. The inside surface temperature increases immediately to 900°C. The wall was at 100°C to start with. If the wall is 25 cm thick determine the time elapsed before the outside surface temperature will begin to increase. Thermal diffusivity = 1.67 × 10^{-6} m²/s. At this point of time determine the temperature at 12.5 cm from inside.

6.21 Chill wind at –20°C flows over a thick wall initially at 18°C and the convective heat transfer coefficient over the surface is 65 W/m²K. Determine the surface temperature and the temperature at a depth of 0.05 m from surface after 1 hour. Also calculate the heat loss from the wall during the period. Thermal diffusivity = 2.2 × 10^{-6} m²/s. Thermal conductivity = 1.35 W/mK.

6.22 Process gases at 800°C are let into a chimney lined with 20 cm layer of chrome brick which was initially at 60°C. The convective heat transfer coefficient has a value of 65 W/m²K. Determine assuming the radius to be large, the surface temperature and the temperature at 5 cm depth after 45 minutes. Thermal diffusivity = 1.67 × 10^{-6} m²/s.

6.23 An ice block at –15°C has air at 20°C flowing over the surface with a convective heat transfer coefficient of 25 W/m²K. Determine the time at which the surface will begin to melt. Thermal diffusivity = 1.08 × 10^{-6} m²/s, k = 2.25 W/mK. At this instant determine the temperature at a depth of 3 cm.

6.24 An ice block at –15°C kept open has a heat flux of 600 W/m² incident over the ice. Determine when the surface will begin to melt. Thermal diffusivity = 1.08 × 10^{-6} m²/s, k = 2.15 W/mK. Also calculate the temperature at a depth of 20 cm at that instant.

6.25 Radiant heating is used to heat a thick plate at 30°C with a surface flux of 5 kW/m². Determine the surface temperature after 30 minutes. Thermal diffusivity = 6.7 × 10^{-5} m²/s. Also determine the temperature at 15 mm depth at this time. k = 22.5 W/mK.

6.26 A thermocouple of 0.2 mm dia wire is used to measure the temperature of a gas which varies sinusoidally with a period of 5 minutes. Determine the percentage of error expected. h = 60 W/m²K; density = 8200 kg/m³. Specific heat = 560 J/kgK. Conductivity = 12 W/mK.

6.27 A thick metal plate is exposed to a cyclic variation of temperature at the surface with a period of 10 minutes. Determine the lag and amplitude of variation at a depth of 2 cm. Thermal diffusivity = 1.67 × 10^{-6} m²/s. Also determine the depth upto which this variation will be felt.

6.28 A metal surface is exposed to periodic temperature variation of the surface which is sinusoidal. If the depth of peneration is more than 1 cm distortion will occur. If the period is 5 minutes check whether distortion will occur. Thermal diffusivity = 6.6 × 10^{-6} m²/s.

6.29 A large plate of 0.16 m thickness initially at 600°C is allowed to cool in air at 30°C with surface convection coefficient of 35 W/m²K. Thermal diffusivity = 9.6 × 10^{-6} m²/s: conductivity = 43.5 W/mK. Determine the temperature at the centre, surface and at (x/L) = 0.5 plane after 30 minutes. Also estimate the heat flow during the time.

6.30 A refractory wall 0.2 m thick, initially at 60°C is exposed on one side to gases at 600°C with a convection coefficient of 60 Wm²K. Determine the time elapsed for the temperature at 0.1 m from surface to reach 300°C. Assume one side is adiabatic. Thermal diffusivity = 1.16 × 10^{-6} m²/s. Conductivity = 2.2 W/mK. Also determine the surface temperatures at this instant. Calculate also the heat conducted into the wall during this time.

6.31 A large slab of meat is to be cooked by heating on both sides in an oven at 180°C. The meat is 4 cm thick and is initially at 30°C. The convective heat transfer coefficient over the surface is 80 W/m^{2}K. The meat is well cooked when the centre temperature is 150°C. Determine the time required for the cooking. Thermal diffusivity = 2.69 × 10^{-7} m^2/s. Also estimate the surface temperature at this time. k = 0.7 W/mK.

6.32 A large slab of thickness 0.18 m at 800°C is allowed to cool in air at 40°C with a convective heat transfer coefficient of 45 W/m^{2}K. Determine the time for the surface to reach 600°C. Estimate the centre temperature at this time. Thermal diffusivity = 8.6 × 10^{-6} m^{2}s, k = 12 W/mK. What will be the average temperature of the slab if no further cooling is allowed.

6.33 Estimate the time required to remove 60% energy over the datum of 60°C in a slab of 0.18 m thickness exposed to convection at 60°C with h = 85 W/m^{2}K. The thermal diffusivity = 8.6 × 10^{-6} m^2/s. Thermal conductivity = 12 W/mK. Calculate the centre and surface temperatures at this time.

6.34 The surfaces of a slab of 0.16 m thickness initially at 60°C is suddenly brought to 300°C by intense heating and maintained at this value. Determine the time for the centre to reach 280°C. Thermal diffusivity = 6.6 × 10^{-6} m^2/s. Also estimate the temperature at 0.04 m depth from the surface.

6.35 Estimate and plot the temperatures at 10, 20, 30 and 40 minutes, along the thickness of a large wall of 0.26 m thickness exposed on both sides to convection to 600°C with h = 75W/m^{2}K. The initial temperature was 100°C. Thermal diffusivity = 6.6 × 10^{-6} m^2/s. k = 15 W/mK.

6.36 A long cylinder of radius 0.09 m initially at 700°C is allowed cool by convection over the surface to surroundings at 30°C with h = 42 W/m^{2}K. The material has a conductivity of 19.2 W/mK and a thermal diffusivity of 5.3 × 10^{-6} m^2/s. Determine the surface and centre temperatures and also the temperature at mid radius after 20 minutes.

6.37 A concrete column of 0.4 m dia initially at 15°C has its surface exposed to chill winds at –30°C with a convective heat transfer coefficient of 45 W/m^{2}K. Estimate the time for the centre to reach 0°C. Thermal diffusivity = 4.9 × 10^{-7} m^2/s. Conductivity = 1.3 W/mK. Estimate the surface temperature also. Calculate the heat loss during the period. Sp. heat = 1130 J/kgK.

6.38 A pudding in the shape of a cylinder of 5 cm dia is to be cooked for the centre to reach 80°C using steam at 120°C with a convection coefficient of 165 W/m^{2}K. The properties are: density 1100 kg/m^3. Sp. heat 3800 J/kgK, conductivity 0.8 W/mK.

Determine the time required.

6.39 A long cylinder of 50 mm dia at 30°C is to be heated in hot air at 180°C till the surface temperature increases to 100°C. The convection coefficient over the surface is 25 W/m^{2}K. Determine the time required. Thermal diffusivity = 4.8 × 10^{-6} m^2/s. Conductivity = 45 W/mK.

6.40 In a batch process long cylinders are to be heated in 10 minutes from 30°C to a centre temperature of 600°C in hot surroundings at 800°C with a convective heat transfer coefficient of 80 W m^{2}K. If the conductivity is 18 W/mK and thermal diffusivity is 3.6 × 10^{-6}m^2/s, determine the maximum possible diameter.

6.41 A sphere of 0.6 m dia initially at 600°C is allowed to cool in air at 30°C with a convective heat transfer coefficient of 35 W/m^{2}K. Determine the temperatures at the centre, mid radius and surface after 30 minutes. Thermal diffusivity = 3.3 × 10^{-6} m^2/s. Thermal conductivity = 7.2 W/mK. Specific heat = 837 J/kgK. Determine also the heat removed.

6.42 A vegetable piece of spherical shape and of dia 6 cm is to be cooked from 30°C till the centre reaches 75°C using steam at 110°C with a convection coefficient of 40 W/m^{2}K. Determine the time required. Thermal diffusivity = 1.7 × 10^{-7} m^2/s. Conductivity = 0.67 W/mK. Also calculate the surface temperature at this instant.

6.43 Steel balls of dia 0.2 m for a large bearing are to be cooled in a salt bath after heating to 720°C. The bath is at 160°C. The convection coefficient is 35 W/m^{2}K. Determine the time required for the ball surface to reach 230°C. Density = 7865 kg/m^3, specific heat = 461 J/kgK. Thermal conductivity 12.8 W/mK. Also calculate the centre temperature and average temperature for this period of cooling.

6.44 Meat balls of dia 10 cm are cooked from initial temperature of 30°C to 90°C (at surface) in a steam cooker at 130°C with a convection coefficient of 85 W/m^{2}K. Determine the time required. Thermal diffusivity = 1.7×10^{-7} m^2/s. Conductivity = 0.7 W/mK. Also calculate the centre temperature and heat conducted. Specific heat = 3800 J/kgK.

6.45 A cube of 10 cm side initially at 30°C is heated in a furnace at 1200°C with a convective heat transfer coefficient of 125 W/m^{2}K. Determine the time for the centre to reach 900°C. Thermal properties are: density = 2400 kg/m^3; conductivity = 1.965 W/mK ; specific heat = 1130 J/kgK.

6.46 A brick like object of sizes 6 × 9 × 15 cm at 30°C is heated in a furnace at 1200°C with a convection coefficient of 85 W/m^2 K. Determine the centre, corner and mid surface temperatures after 30 minutes. Density = 2400 kg/m^3; conductivity = 1.965 W/mK. specific heat = 1130 J/kgK.

6.47 A long billet of 9 cm × 15 cm section initially at 30°C is heated all around and at the base by gases at 1200°C with a convection coefficient of 125 W/m^{2}K. Determine the corner and base centre temperatures after 20 minutes. Use property values of example 6.46.

6.48 A short cylinder 10 cm radius and 20 cm height initially at 630°C is suddenly exposed to air at 30°C with a convective heat transfer coefficient of 25 W/m^{2}K. Determine the centre and edge temperatures after 20 minutes. Use property value of example 6.46.

6.49 A quarter sector of a cylinder of radius 0.1 m and height 0.2 m at 630°C is suddenly exposed to air at 30°C with a convection coefficient of 25 W/m^{2}K. Determine the centre temperature. Use property values of example 6.46.

6.50 A quarter of a sphere of radius 0.1 m initially at 600°C is exposed on all surfaces to convection at 30°C with a convective heat transfer coefficient of 25 W/m^{2}K. Determine the temperature at corners, middle of edges and middle point of solid after 20 minutes. Use property values of example 6.46.

6.51 A 1/8 part of a sphere of radius 0.1 m initially at 600°C is exposed on all surfaces to convection at 30°C with a convective heat transfer coefficient of 25 W/m^{2}K. Determine the temperatures at corners and mid point of edges and the centre of the solid after 20 minutes. Use property values of example 6.46.

6.52 A slab at 200°C has a heat generation of 2×10^{-6} W/m^3 started at time zero. The slab is 0.18 m thick and is exposed on one side to convection to 200°C a convection coefficient of 150 W/m^{2}K. Using numerical method determine the temperature at various thicknesses after 30 minutes. Thermal diffusivity = 8.58×10^{-6} m^2/s. k = 19.9 W/mK.

6.53 Using the data in example 6.52 for a long cylinder to diameter 0.18 m. Determine using numerical method the temperature after 30 minutes from the start of heat generation at various radial positions.

7 CHAPTER

Introduction to Convection

7.0 INTRODUCTION

In the previous chapters during the discussion of conduction problems convection was applied as a boundary condition only. The basic mechanism of convection was not discussed so far. In this chapter it is attempted to explain the basic mechanism of convection. In industrial processes involving heat transfer, convection plays a very important role. When heat is to be transferred from one fluid to another through a barrier convection is involved on both sides of the barrier. In most cases the main resistance to heat flow is by convection.

The process of heat transfer between a surface and a fluid flowing in contact with it is called convection. If the flow is caused by an **external device** like a pump or blower, it is termed as **forced convection.** If the flow is caused by the **buoyant forces** generated by heating or cooling of the fluid by the surface, the process is called as **natural or free convection.**

In the previous chapters the heat flux by convection was determined using equation.

$$q = h\,(T_s - T_\infty) \qquad ...(7.1)$$

q is the heat flux in W/m^2, T_s is the surface temperature and T_∞ is the fluid temperature of the free stream, the unit being °C or K. Hence the unit of convective heat transfer coefficient h is W/m^2 K or W/m^2 °C both being identically the same.

In this chapter the basic mechanism of convection and the method of analysis that leads to the correlations for convection coefficient are discussed. In this process the law of conservation of mass, First law of thermodynamics and Newtons laws of motion are applied to the system.

Convective heat transfer coefficient is influenced by (*i*) fluid properties like density, viscosity and other thermal properties like specific heat, conductivity (*ii*) the flow velocity and (*iii*) the surface geometry. It may be noted that the physical or thermal properties of the surface material play no part in the process of convective heat transfer. As the fluid properties vary with temperature and location the value of convective heat transfer coefficient will vary from point to point. This leads to the situation that analytically derived equations are applicable only to a limited extent. Practical correlations obtained through experiments are discussed in the next chapters. Property values for various fluids available in handbooks is necessary for obtaining numerical value for convective heat transfer coefficient.

7.1 MECHANISM OF CONVECTION

In conduction, energy is transferred as heat either due to free electron flux or lattice vibration. There is no movement of mass in the direction of energy flow. In convection, energy flow occurs at the surface purely by conduction. But in the next layers both conduction and diffusion-mass movement in the molecular level or macroscopic level occurs. Due to the mass movement the rate of energy transfer is higher. Higher the rate of mass movement, higher will be the heat flow rate.

The velocity variation and the temperature variation at a point in the flow of a fluid over a surface is shown in Fig. 7.1.

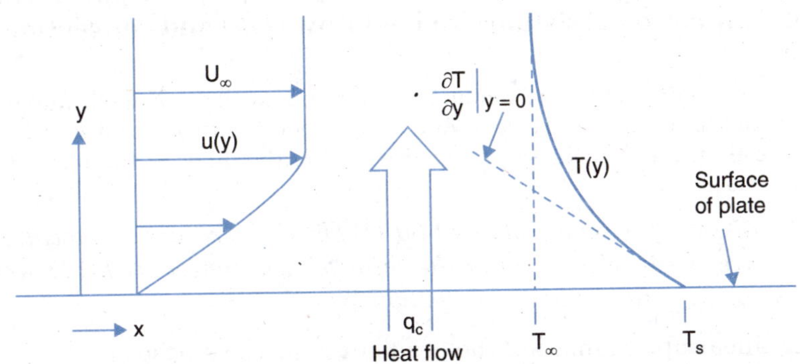

Fig. 7.1. Velocity and temperature variation at a point in a fluid flowing over a surface, laminar flow, plate hotter

The fluid coming in contact with the surface receives heat by conduction. This is passed on by diffusion and mass movement to the fluid in the free stream.

Considering conduction at the surface layer the heat flux is given by the equation

$$q = -k \left.\frac{\partial T}{\partial y}\right|_{y=0} \quad \text{...(7.2)}$$

where k is the thermal conductivity of the fluid. Combining equations (7.1) and (7.2) we obtain equation. (7.3) for Convection coefficient.

$$h = -k \left.\frac{\partial T}{\partial y}\right|_{y=0} \Big/ (T_s - T_\infty) \quad \text{...(7.3)}$$

As the values of thermal conductivity of the fluid and the temperatures of the surface and the fluid, T_s and T_∞ are known, the value of convection coefficient can be determined if the temperature gradient at the surface is known. The model postulated is used to solve for temperature as a function of y so that the temperature gradient can be calculated. The gradient is influenced by the fluid stream and the surface. Boundary layer theory is applied for the determination of the gradient.

Example 7.1: *Convert equation (7.3) into dimensionless form.This is done first by replacing T by a dimensionless temperature $T^* = (T - T_s)/(T_\infty - T_s)$ and y by a dimensionless distance $y^* = y/L$.*

Solution: ∴ $$\left.\frac{\partial T}{\partial y}\right|_{y=0} = \frac{(T_\infty - T_s)}{L} \left.\frac{\partial T^*}{\partial y^*}\right| y = 0$$

Chapter 7

Substituting and rearranging

$$\frac{hL}{k} = \frac{\partial\,[(T - T_s)/(T_\infty - T_s)]}{\partial\,(y/L)}\bigg|\, y = 0 \text{ or } \frac{\partial T^*}{\partial y^*}\bigg|\, y = 0 \qquad ...(7.4)$$

It can be seen that the quantity (hL/k) is directly related to the dimensionless temperature gradient at the surface and thus plays an important role in convection heat transfer.

It may be noted that hL/k is dimensionless and this group is known as **Nusselt number (Nu), a ratio of conduction resistance to heat flow (L/k) and convection resistance to heat flow ($1/h$).**

(**Note:** In the previous chapter on conduction similar quantity to Nusselt number was used. The number was called Biot number. The basic difference between the two is that k in Biot number is the conductivity of the solid while k in Nusselt number is that of the fluid. So there is a distinct difference between the two numbers.)

Example 7.2: *Air at 20°C flows over plate at 60°C. The temperature at a location 0.5 mm from the surface is measured as 40°C. Determine the value of local convective heat transfer coefficient. The thermal conductivity may be taken as 0.2656 W/mK.*

Solution: Using linear approximation the gradient near the surface is

$$\frac{\partial T}{\partial y} = \frac{-20}{0.0005}\ \text{°C/m.}$$

Using equation (7.3)

$$\mathbf{h} = -\,0.02656 \times (-\,20/0.0005)/(60 - 20) = \mathbf{26.56\ W/m^2K.}$$

As the distance is small, linear variation of temperatures is assumed. Hence this value becomes an estimate rather than an exact value.

Example 7.3: *Air at 20°C flows over a surface at 80°C. The local heat flow was measured at a point as 1000 W/m². Esimate the value of local convective heat transfer coefficient, temperature gradient at the surface and the temperature at a distance of 0.5 mm from the surface. Thermal conductivity of air can be taken as 0.02826 W/mK.*

Solution: The value of local convection coefficient is determined using equation (7.1).

$$1000 = h(80 - 20), \quad \therefore \quad \mathbf{h = 16.67\ W/m^2K}$$

The temperature gradient at the surface is determined using equation (7.2).

$$1000 = -\,0.02826\, \frac{\partial T}{\partial y}\bigg|_{y=0} \quad \therefore \quad \frac{\partial T}{\partial y}\bigg|_{y=0} = \mathbf{-\,35385.7°C/m.}$$

The **temperature at 0.0005 m** is $80 + \dfrac{\partial T}{\partial y}\bigg|_{y=0} \times \Delta y$ (approximately)

$$= 80 - 0.0005 \times 35385.7 = \mathbf{62.3°C}$$

7.2 THE CONCEPT OF VELOCITY BOUNDARY LAYER

We have seen that in the determination of the convective heat transfer coefficient the key is the determination of the temperature gradient in the fluid at the solid-fluid interface. The velocity gradient at the surface is also involved in the determinations. This is obtained using the boundary layer concept to solve for $u = f(y)$, $T = f\,'(y)$. The simplest situation is the flow over a flat plate. The fluid enters with a uniform velocity of u_∞ as shown in Fig. 7.2. When fluid particles touch the surface of the plate the velocity of these particles is reduced to zero due to viscous forces. These particles in turn retard the velocity in the next layer, but as these two are fluid layers, the velocity is not reduced to zero in the next layer. This retardation process continues along the layers until at some distance y the scale of retardation becomes negligible and the velocity of the fluid is very nearly the same as free stream velocity u_∞ at this level. The retardation is due to shear stresses between the layers along planes parallel to the flow.

The value of y where velocity $u = 0.99\ u_\infty$ is called hydrodynamic boundary layer thickness denoted by δ. The velocity profile in the boundary layer depicts the variation of u with y, through the boundary layer. This is shown in Fig. 7.2.

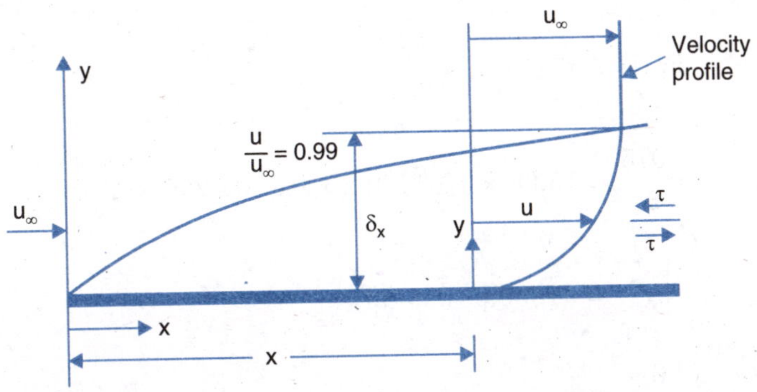

Fig. 7.2. Velocity boundary layer on a flat plate

The model characterises the flow as consisting of two distinct regions (*i*) a thin boundary layer in which the velocity gradients and shear stresses are large and (*ii*) the remaining region outside of the boundary layer where the velocity gradients and shear stresses are negligibly small. This is also called potential flow. The boundary layer thickness increases along the direction of flow over a flat plate as effects of viscous drag is felt farther into the free stream.

This is called the velocity boundary layer model as this describes the variation of velocity in the boundary layer.

The direct application of velocity boundary layer is in fluid mechanics for the determination of the wall shear stress and then the dimensionless drag coefficient.

The net shear over the plate in flow is the wall shear and shear stress beyond the boundary layer is zero.

The wall shear is given by the equation

$$\tau_s = \mu \left.\frac{\partial u}{\partial y}\right|_{y=0} \qquad ...(7.5)$$

It may be seen that the velocity gradient can be determined if a functional relationship such as $u = f(y)$ is available. Such a relationship is obtained using the boundary layer model and applying the continuity and Newtons laws of motion to the flow. The friction coefficient C_f is defined as below.

$$C_f = \tau_s/(\rho u_\infty^{\ 2}/2) \qquad ...(7.6)$$

There are local and average values for both τ_s and C_f denoted as $\tau_{s.x}$. $\overline{\tau}_s C_{fx}$ and $\overline{C}_f$. The average value is obtained by integrating the local value between the limits 0 and L where L is the distance upto which the average value is required. In heat transfer the friction coefficient by analogy is found to provide a value for Nusselt number and hence its importance. Measured values of C_f are also available for various values of an important parameter, namely Reynolds number. Curve fitted equations are also available for C_f.

Example 7.4: *Air at 20°C flows over both sides of a surface measuring 0.2 m × 0.2 m. The drag force was 0.075 N. Determine the velocity gradient at the surface. Kinematic viscosity has a value of 15.06 × 10⁻⁶ m²/s. Density = 1.205 kg/m³. Also determine the friction coefficient if the free stream velocity has a value of 40 m/s.*

Solution: Using equation (7.5)

$$\text{Force} = \tau \,.\, A = \mu A \left.\frac{\partial u}{\partial y}\right|_{y=0} \quad ; \mu = \nu \,.\, \rho \qquad \text{...}(a)$$

$$\frac{0.075}{2} = 15.06 \times 10^{-6} \times 0.2 \times 0.2 \times 1.205 \times \left.\frac{\partial u}{\partial y}\right|_{y=0}$$

Solving $\left.\dfrac{\partial u}{\partial y}\right|_{y=0} = 51660.6$ (m/s)/m

Using equation (7.6)

$$C_f = \frac{0.075}{2 \times 0.2 \times 0.2} \times \frac{2}{1.205 \times 40 \times 40} = 0.973 \times 10^{-3} \qquad \text{...}(b)$$

Note: Check for dimensional consistency of equation (a)

$$\text{Force} = N,\ \mu = \nu \cdot \rho = \frac{m^2}{s} \cdot \frac{\text{kg}}{m^3} = \frac{\text{kg}}{ms},\ A = m^2$$

$$\frac{\partial u}{\partial y} = \frac{m}{s} \cdot \frac{1}{m}$$

$$\therefore \quad N = \frac{m^2}{s}\frac{\text{kg}}{m^3}\, m^2\, \frac{m}{s} \cdot \frac{1}{m} = \text{kg} \,.\, \frac{m}{s^2} \text{ checks}$$

Considering equation (b) C_f is dimensionless

$$\frac{N}{m^2} \cdot \frac{m^3 s^2}{\text{kg.m}^2} = N \,.\, \frac{s^2}{\text{kg.m}} = 1. \text{ Also checks.}$$

7.3 THERMAL BOUNDARY LAYER

Velocity boundary layer automatically forms when a real fluid flows over a surface, but thermal boundary layer will develop only when the fluid temperature is different from the surface temperature. Considering the flow over a flat plate with fluid temperature of T_∞ and surface temperature T_s ($T_s > T_\infty$) the temperature of the fluid is T_∞ all over the flow till the fluid reaches the leading edge of the surface. The fluid particles coming in contact with the surface are slowed down to zero velocity and the fluid layer reaches equilibrium with the surface and reaches temperature T_s. These particles in turn heat up the next layer and a temperature gradient develops. At a distance y, the temperature gradient becomes negligibly small or the temperature equals the free stream temperature. **The distance *y* at which the ratio $[(T_s - T)/(T_s - T_\infty)] = 0.99$ is defined as thermal boundary layer thickness δ_t.** The flow can now be considered to consist of two regions. A thin layer of thickness δ_t in which the temperature gradient is large and the remaining flow where the temperature gradient is negligible. As the distance from the leading edge increases the effect of heat penetration increases and the thermal boundary layer thickness increases. The heat flow from the surface to the fluid can be calculated using the temperature gradient at the surface (equation 7.2). The temperature gradient is influenced by the nature of free stream flow. The convective heat transfer coefficient can be calculated using equation (7.3). Examples 7.2. and 7.3 illustrate the principles involved. The development of the thermal boundary layer is shown in Fig. 7.3.

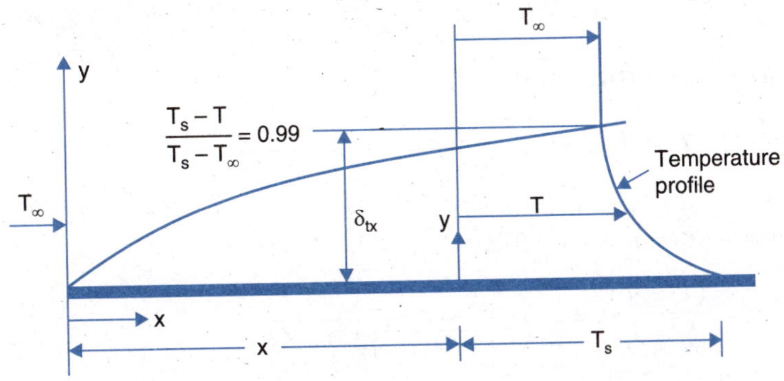

Fig. 7.3. Thermal boundary layer on a flat plate

The thermal and velocity boundary layers will not be identical except in a case where Pr $(C\mu/k) = 1$. Additional influencing factors change the thickness of the thermal boundary layer as compared to the thickness of the velocity boundary layer at any location. Note that both boundary layers exist together. Similar development of boundary layer is encountered in convective mass transfer also.

Example 7.5: *The temperature ratio $[(T_s - T)/(T_s - T_\infty)] = y/0.0075$ in the boundary layer of a flow over a flat plate. If $k = 0.03$ W/mK, determine the value of convective heat transfer coefficient.*

Solution: The equation (7.2) is applicable for this case. The equation is

$$h = -k \left.\frac{\partial T}{\partial y}\right|_{y=0} \Big/ (T_s - T_\infty), \text{ from the given relationship}$$

$$\frac{\partial\, [(T_s - T)/(T_s - T_\infty)]}{\partial\, y} = -\frac{1}{(T_s - T_\infty)} \cdot \frac{\partial T}{\partial y} = \frac{1}{0.0075}$$

$$\therefore \quad \left.\frac{\partial T}{\partial y}\right|_{y=0} = -\frac{T_s - T_\infty}{0.0075}$$

Substituting $\quad \mathbf{h} = -0.03 \times -\left(\dfrac{T_s - T_\infty}{0.0075}\right) / (T_s - T_\infty) = \mathbf{4\ W/m^2\ K.}$

Example 7.6: *The temperature ratio $[(T_s - T)/(T_s - T_\infty)] = \sin(\pi y/0.015)$ in the boundary layer of a flow over a flat plate. If $k = 0.03$ W/mK, determine the value of convective heat transfer coefficient.*

Chapter 7

Solution: $[(T_s - T_\infty)/(T_s - T_\infty)] = \sin(\pi y/0.015)$

$$\frac{\partial [(T_s - T)/(T_s - T_\infty)]}{\partial y} = \frac{\partial}{\partial y} [\sin(\pi y/0.015)]\, (T_s - T_\infty)$$

$$= \frac{\pi}{0.015} \,.\, \cos(\pi y/0.015)\;(T_s - T_\infty)$$

or
$$\left.\frac{\partial T}{\partial y}\right| y = 0 = -\frac{\pi(T_s - T_\infty)}{0.015} \,.\, \cos(0) = \frac{-\pi(T_s - T_\infty)}{0.015}$$

$$\therefore \qquad \mathbf{h} = -0.03 \times \left[\frac{-\pi(T_s - T_\infty)}{0.015}\right] / (T_s - T_\infty) = \mathbf{6.28\ W/m^2K.}$$

Example 7.7: *The temperature ratio*

$$[(T_s - T)/(T_s - T_\infty)] = \frac{1}{2}\left(\frac{y}{0.0075}\right)^3 + \frac{3}{2}\frac{y}{0.0075}$$

in the boundary layer of a flow over a flat plate is as above. If k = 0.03 W/mK, determine the value of convective heat transfer coefficient, given.

Solution:
$$\frac{(T_s - T)}{T_s - T_\infty} = \frac{1}{2}\left(\frac{y}{0.0075}\right)^3 + \frac{3}{2}\frac{y}{0.0075}$$

$$\frac{\partial [(T_s - T)/(T_s - T_\infty)]}{\partial y} = -\frac{1}{T_s - T_\infty}\frac{\partial T}{\partial y}$$

$$= \frac{3}{2}\frac{y^2}{0.0075} + \frac{3}{2} \cdot \frac{1}{0.0075} \text{ at } y = 0, \frac{\partial T}{\partial y} = -\frac{(T_s - T_\infty) \times 3}{2 \times 0.0075}$$

$$\therefore \qquad \mathbf{h} = -0.03 \times \left[\frac{-(T_s - T_\infty) \times 3}{2 \times 0.0075}\right] / (T_s - T_\infty) = \mathbf{6\ W/m^2K}$$

Note that in this case we may have to assume that beyond $y = 0.0075$, the temperature equals T_∞.

These are some of the simple temperature profiles that may be used in the analysis to estimate the value of convection coefficient.

7.4 LAMINAR AND TURBULENT FLOW

The formation of the boundary layer starts at the leading edge. In the starting region the flow is well ordered. The streamlines along which particles move is regular. The velocity at any point remains steady. This type of flow is defined as laminar flow. There is no macroscopic mixing between layers. The momentum or heat transfer is mainly at the molecular diffusion level. After some distance in the flow, macroscopic mixing is found to occur. Large particles of fluid is found to move from one layer to another. The motion of particles become irregular. The velocity at any location varies with respect to a mean value. The flow is said to be turbulent. Due to the mixing the boundary layer thickness is larger. The energy flow rate is also higher. The velocity and temperature profiles are flatter, but the gradient at the surface is steeper due to the same reason. This variation is shown in Fig. 7.4.

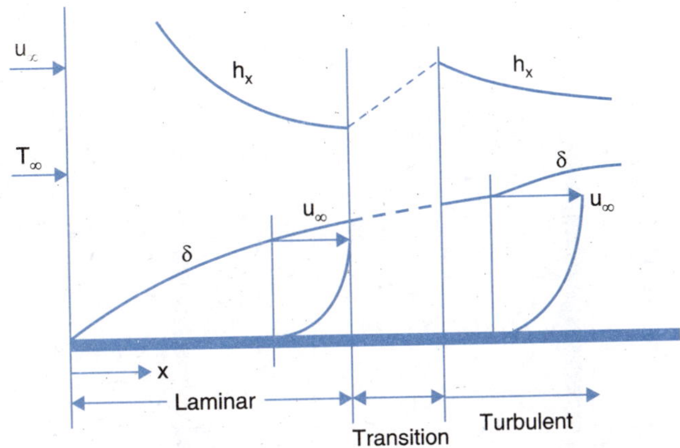

Fig. 7.4. Building up of boundary layer over a flat plate—Laminar—Turbulent

The changeover does not occur at a sharp location. However for calculations some location has to be taken as the change over point. In the velocity boundary layer, this transition is determined by a dimensionless group, Reynolds number-defined for flow over a plate by the equation

$$Re_x = \rho\, u_\infty\, x/\mu \quad \text{or} \quad u_\infty\, x/\nu \qquad \text{...[7.7(a)]}$$

For flow in a tube or across a tube or sphere it is given by the equation.

$$Re = \rho\, u_\infty\, D/\mu \quad \text{or} \quad u_\infty\, D/\nu \qquad \text{...[7.7(b)]}$$

The grouping represents the ratio of inertia and viscous forces. Upto a point the inertia forces keep the flow in order and laminar flow exists. When the viscous forces begin to predominate, movement of particles begin to be more random and turbulence prevails.

The **transition Reynolds number** for flow over a flat plate depends on many factors and may be anywhere from 10^5 to 3×10^6. Generally the value **is taken as** $\mathbf{5 \times 10^5}$ unless otherwise specified. For **flow through tubes the transition value is 2300, unless otherwise specified.**

In the quantitative estimation of heat flow, the correlation equations for the two regions are distinctly different and hence it becomes necessary first to establish whether the flow is laminar or turbulent.

Turbulent flow is more complex and exact analytical solutions are difficult to obtain. Analogical model is used to obtain solutions.

7.5 FORCED AND FREE CONVECTION

When heat transfer occurs between a fluid and a surface, if the flow is caused by a fan, blower or pump or a forcing jet, the process is called **forced convection.** The boundary layer development is similar to the descriptions in the previous section. When the temperature of a surface immersed in a stagnant fluid is higher than that of the fluid, the layers near the surface get heated and the density decreases in these layers.

The surrounding denser fluid exerts buoyant forces causing fluid to flow upwards near the surface. This process is called **free convection flow** and heating is limited to a layer, as shown in Fig. 7.5. The heat transfer rate will be lower as the velocities and temperature gradients are lower. If the surface temperature is lower, the flow will be in the downward direction.

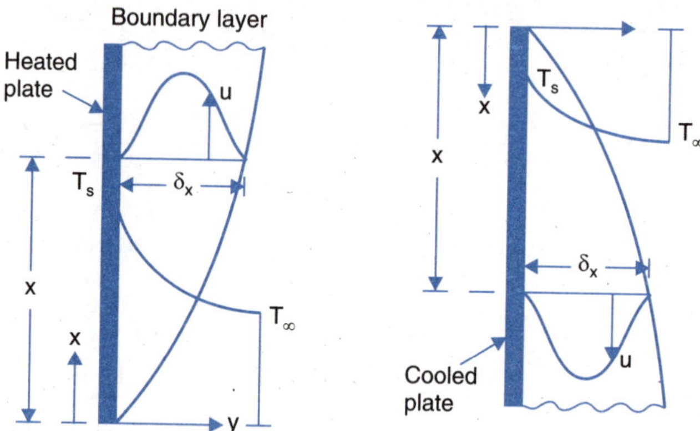

Fig. 7.5. Free convection boundary layer near a vertical surface

7.6 METHODS USED IN CONVECTION STUDIES

The important methods of analysis used in convection studies are

1. Analytical method which can again be subdivided as
   (*i*) Formulating and solving the differential equation, also known as exact method
   (*ii*) Formulating and solving integral equations using assumed profiles for velocity and temperature–(approximate analysis)
   (*iii*) Numerical finite element method
2. Dimensional analysis
3. Analogical methods which apply solutions from one transport phenomenon to another
4. Empirical correlations developed from experimental results guided by the results of analytical methods and dimensional analysis.

The analytical methods provide a basis as well as validation for any empirical correlation postulated using experimental results. Dimensional analysis also provides an insight into the possible ways in which empirical relations can be attempted. Analogical method is very helpful in extending the solutions from one transport phenomenon to another. This is particularly found useful in treating turbulent flow.

When it comes to quantitative estimation one has to largely depend today on correlations obtained from many experimental results. However such correlations are based on a strong foundation laid by analytical methods and as such are reliable for design. At first glance the number of correlations may overwhelm a learner but one has to understand that the complexity of the phenomenon and the variety of materials encountered make it necessary to use different correlations for different range of values of variables.

The future may prove to be completely different because the availability of softwares (of course based on the basic principles) and powerful computers may lead to the use of the finite element method extensively, and the correlations may be used only for a check of values so obtained. It is necessary for a learner with the future in mind to expose himself to the use such softwares for covection studies also.

7.6.1. Analytical Method

Formulation of differential equations: We have seen that to solve for the shear stress and friction coefficient the velocity gradient at the surface has to be determined. Similarly to solve for the convection coefficient the temperature gradient at the surface has to be determined. This can be done by considering control volume element in the boundary layer and applying the law of conservation of mass, Newtons second law and first law of thermodynamics. The resulting differential equations are solved for the given boundary to obtain $u = f(y)$ and $T = f_1(y)$. The gradients can be obtained by differentiation of these functions. As the temperature and velocity gradients are negligible beyond the respective boundary layers, it is found sufficient to restrict the solution to the boundary layer. This also provides a boundary condition required for the solution.

The control volume in the boundary layer considered and the various quantities flowing in and out of the element are shown in Fig. 7.6.

First the law of conservation of mass is applied to the volume of section dx, dy and depth unity. The application of the law leads to the statement below:

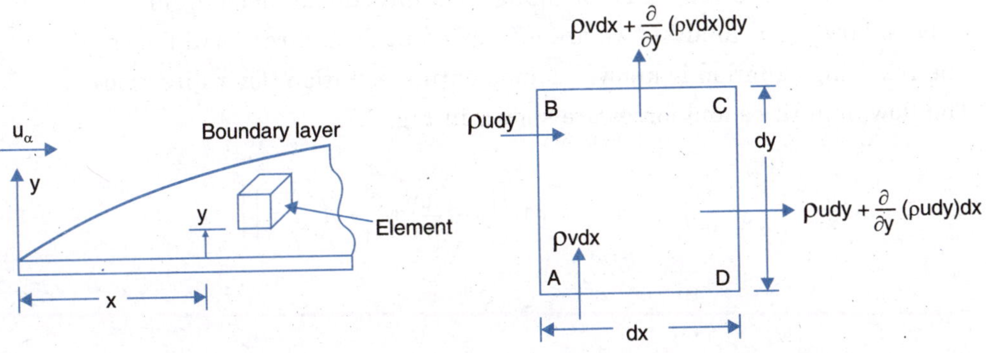

Fig. 7.6. Control volume considered for derivation of continuity equation

The net mass flow through the bounding surfaces of the volume = mass stored in the volume. Considering a time interval of $d\tau$,

Flow through face $AB = \rho u\, dy\, .\, 1\, .\, d\tau$

Flow through face $CD = \rho u.dy.1.d\tau + \dfrac{\partial}{\partial x}(\rho u\, dy.1)\, dx\, d\tau$

Flow through face $AD = \rho v\, dx.1.\, d\tau$

Flow through face $BC = \rho v\, dx.1.\, d\tau + \dfrac{\partial}{\partial y}(\rho v\, dy.1)\, dy\, d\tau$

Taking the difference, Net mass flow

$$= \frac{\partial}{\partial x}(\rho u)\, dx\, dy\, d\tau + \frac{\partial}{\partial y}(\rho v)\, dx\, dy\, d\tau$$

Mass stored in the volume = $dx.dy.1.d\rho$

simplifying, we get

$$\frac{\partial(\rho u)}{\partial x} + \frac{\partial(\rho v)}{\partial y} = \frac{\partial \rho}{\partial \tau} \quad ...[7.8(a)]$$

Chapter 7

For steady flow, no storage is possible,

$$\therefore \quad \frac{\partial(\rho u)}{\partial x} + \frac{\partial(\rho v)}{\partial y} = 0 \qquad ...[7.8(b)]$$

For incompressible flow or constant density conditions ρ = constant.

$$\frac{\partial u}{\partial x} + \frac{\partial v}{\partial y} = 0 \qquad ...[7.8(c)]$$

This equation is known as continuity equation. The physical meaning of the equation is that in the boundary layer the mass flow change in x direction is balanced by the mass flow change in the y direction or vice versa.

7.6.2. x-Directional Momentum Equation

Newtons second law of motion is applied to the control volume. The statement resulting from the application is

x directional forces = rate of change of x directional momentum.

In boundary layer analysis we are interested in the x directional forces.

The resulting equation is known as momentum equation (for x direction)

The flow quantities and forces are shown in Fig. 7.7.

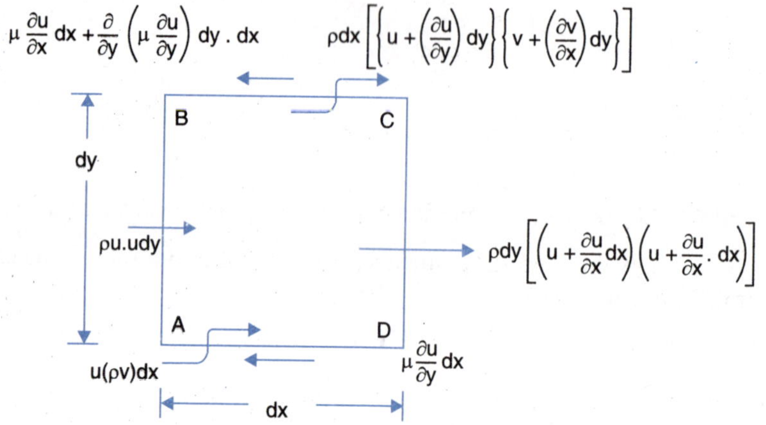

Fig. 7.7. x-directional momentum flow through the control volume

Again an element of unit depth is considered. **The pressure and body forces are neglected** as these do not contribute significantly for the boundary layer flow discussed.

Considering unit depth, the x-directional momentum flow through various faces is listed below. (constant density conditions)

face AB: $u\ (\rho u\ dy.1)$

face CD: $\left(u + \frac{\partial u}{\partial x} dx\right) \rho \left(u + \frac{\partial u}{\partial x} dx\right) dy.1$

face AD: $u\ (\rho v\ dx.1)$

face BC: $\left(u + \frac{\partial u}{\partial y} dy\right)\left(v + \frac{\partial v}{\partial y} dy\right) \rho\, dx.1$

Neglecting higher order terms, the net momentum flow is given by

$$\rho u \frac{\partial u}{\partial x} dx\, dy + \rho u \frac{\partial u}{\partial x} dx\, dy + \rho v \frac{\partial u}{\partial y} dx\, dy + \rho u \frac{\partial v}{\partial y} dx\, dy$$

Rearranging this

$$\rho u \frac{\partial u}{\partial x} dx\, dy + \rho v \frac{\partial u}{\partial y} dx\, dy + \rho u \left(\frac{\partial u}{\partial x} + \frac{\partial v}{\partial y}\right) dx\, dy$$

As $\frac{\partial u}{\partial x} + \frac{\partial v}{\partial y} = 0$ as per the continuity equation, the net momentum flow is equal to

$$\rho u \frac{\partial u}{\partial x} dx\, dy + \rho \nu \frac{\partial u}{\partial y} dx\, dy$$

The net forces on the element are only the shearforces as other forces are negligible. This is equal to

$$\frac{\partial}{\partial y}\left(\mu \frac{\partial u}{\partial y}\right) dy\, dx$$

Rearranging,

$$\rho u \frac{\partial u}{\partial x} + \rho v \,.\, \frac{\partial u}{\partial y} = \mu \,.\, \frac{\partial^2 u}{\partial y^2} \qquad \text{...[7.9 (a)]}$$

Dividing by ρ results in equation (7.9).

$$u \frac{\partial u}{\partial x} + \nu \frac{\partial u}{\partial y} = \nu \frac{\partial^2 u}{\partial y^2} \qquad \text{...(7.9)}$$

Where ν is the kinematic viscosity equal to μ/ρ. This is also called momentum diffusivity.

In the equation [7.9(a)] ρu equals the x-directional mass flow. $\frac{\partial u}{\partial x}$ equals the rate of change of x-directional velocity for the mass flow. The product is the rate of change of momentum for the mass flow in x-direction.

The term ρv equals the mass moving from one layer to another (in y-direction). $\frac{\partial u}{\partial y}$ equals the rate of change of x-directional velocity for the mass moving from one layer to another. The product equals the rate of change of x directional momentum for the mass moving from one layer to another. The sum equals the rate of change of x-direction momentum for the mass moving from on layer to another. The sum equals the rate of change of x-direction momentum for the flow through the element. The RHS term $\frac{\partial}{\partial y}\left(\mu \frac{\partial u}{\partial y}\right)$ equals the net shear force in the element.

These two equations can be written in the dimensionless form by defining the following terms. Non-dimensionalysing leads to the identification of a dimensionless parameter affecting the flow.

$$x^* = \frac{x}{L}, y^* = \frac{y}{L}, u^* = \frac{u}{u_\infty}, v^* = \frac{v}{u_\infty}$$

Then the dimensionless forms are

$$\frac{\partial u^*}{\partial x^*} + \frac{\partial v^*}{\partial y^*} = 0 \qquad ...(7.10)$$

$$u^* \frac{\partial u^*}{\partial x^*} + v^* \frac{\partial v^*}{\partial y^*} = \frac{1}{Re_L} \frac{\partial^2 v^*}{\partial y^{*2}} \qquad ...(7.11)$$

This also indicates that Reynolds number Re_L (u^*L/v) plays an important role in boundary layer flow. Equation (7.6) for coefficient of friction, C_f similarly can be expressed as

$$C_{fx} = \frac{2}{Re_L} \left.\frac{\partial u^*}{\partial y^*}\right|_{y=0} \qquad ...(7.12)$$

The continuity equation and momentum equation have to be solved simultaneously to obtain the velocity profile in the boundary layer. This leads to the determination of boundary layer thickness and coefficient of friction, through velocity gradient.

These partial differential equations are solved by first defining a stream function ψ (x, y) that automatically satisfies the continuity equation which leads to the conversion to ordinary differential equation

$$u = \frac{\partial \psi}{\partial y} \quad \text{and} \quad v = -\frac{\partial \psi}{\partial x}.$$

Introducing a new variable

$$\eta = y\sqrt{\frac{u_\infty}{vx}}, \text{ we can let } \psi = \sqrt{vx\, u_\infty}\; f(\eta)$$

where $f(\eta)$ denotes a dimensionless stream function

$$u = \frac{\partial \psi}{\partial y} = \frac{\partial \psi}{\partial \eta} \cdot \frac{\partial \eta}{\partial y} = u_\infty \frac{d[f(\eta)]}{d\eta}$$

$$v = -\frac{\partial \psi}{\partial x} = \frac{1}{2}\sqrt{\frac{vu_\infty}{x}} \left\{\frac{d[f(\eta)]}{d\eta} . \eta - f(\eta)\right\}$$

Expressing the partial differentials in terms of η and substituting in the equation a non linear ordinary differential equation is obtained

$$f(\eta) . \frac{d^2[f(\eta)]}{d\eta^2} + 2\frac{d^3[f(\eta)]}{d\eta^3} = 0$$

The boundary conditions to be applied are

$$f(\eta) = 0 \text{ and } \frac{d[f(\eta)]}{d\eta} = 0 \text{ when } \eta = 0. \quad \frac{d[f(\eta)]}{d\eta} = 1 \text{ at } \eta = \infty.$$

The equation was numerically solved by Blasius in 1908. The results show that when $\frac{u}{u_\infty}$ is plotted against $\frac{y}{x}\sqrt{Re_x}$ a single curve results for the velocity profile. This is shown in Fig. 7.8. The results are validated by experimental results by Hansen. Recent computerised solutions also validate the results obtained.

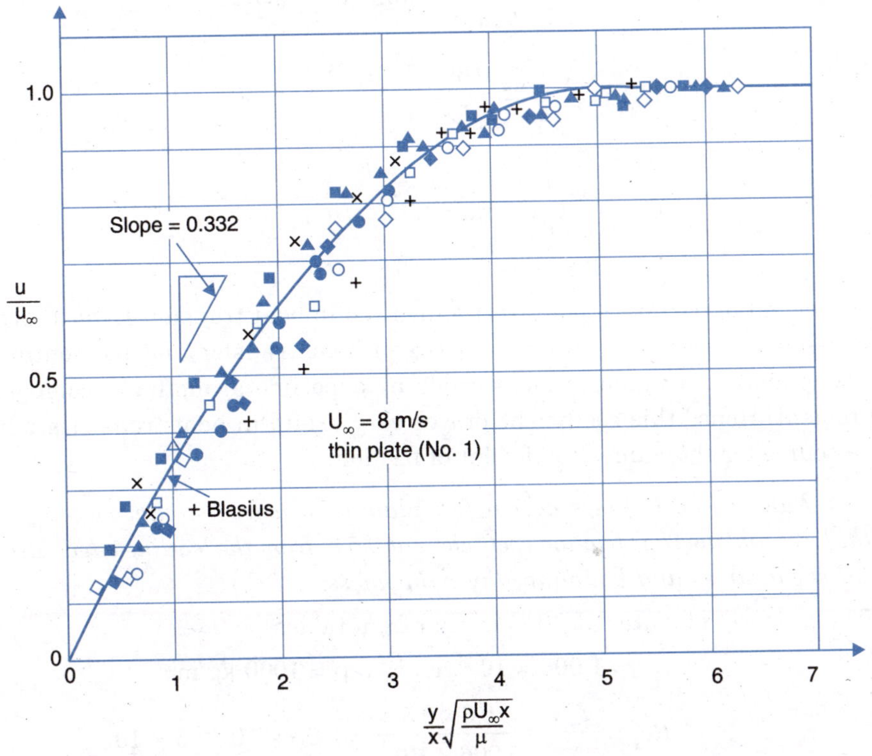

Fig. 7.8. Velocity profile in a laminar boundary layer according to Blasius, with experimental data of Hansen

Two **significant conclusions** are arrived at from the Blasius solution.

The ratio $\frac{\mathbf{u}}{\mathbf{u}_\infty}$ **reaches a value of 0.99 at**

$$\frac{\mathbf{y}}{\mathbf{x}}\sqrt{\mathbf{Re_x}} = \mathbf{5}$$

As per the definition of boundary layer thickness $y = \delta$ at this condition. This leads to the very important relation

$$\delta_x = \frac{5x}{\sqrt{Re_x}} \qquad \text{...(7.13)}$$

The other result is that the slope at $y = 0$ for the velocity profile was calculated as 0.332. When applied to the graph.

$$\left.\frac{\partial (u/u_\infty)}{\partial\left(\frac{y}{x}\right)\sqrt{Re_x}}\right|_{y=0} = 0.332 \quad \therefore \quad \left.\frac{\partial u}{\partial y}\right|_{y=0} = 0.332 \,.\, \frac{u_\infty}{x}\sqrt{Re_x}$$

$$\therefore \qquad \tau_s = \mu \left.\frac{\partial u}{\partial y}\right|_{y=0} = 0.332 \,.\, \mu\, \frac{u_\infty}{x}\sqrt{Re_x} \qquad \text{...(7.14)}$$

$$C_{fx} = \frac{\tau_s}{\frac{1}{2}u_\infty^2} = \frac{0.664}{\sqrt{Re_x}} \qquad \text{...(7.15)}$$

$$\overline{C}_f = \frac{1}{L}\int_0^L C_{fx}\, dx = 1.33 / \sqrt{Re_L} \qquad \text{...(7.16)}$$

or $\qquad = [2\, C_f]_{x=L}$.

The coefficient of friction is not directly involved in heat transfer. Later this quantity is related to Nusselt number using analogy between heat transfer and momentum transfer. This quantity is more conveniently measurable by experimetal methods, compared to heat flow measurement. Hence this coefficient proves very useful in heat transfer studies. Moody diagram is a source for the values coefficient of friction.

Example 7.8: *Water at 30°C flows over a flat plate with a free stream velocity of 0.6 m/s. Determine the local and average friction coefficient at 0.5 m from the leading edge. Also determine the local wall shear stress and boundary layer thickness.*

Solution: The property of water at 30°C is read from tables.

$$\nu = 1.006 \times 10^{-6}\ \text{m}^2/\text{s} \qquad \rho = 1000\ \text{kg/m}^3$$

$$Re_x = \frac{u_\infty x}{\nu} = \frac{0.6 \times 0.5}{1.006 \times 10^{-6}} = 2.98 \times 10^5 < 5 \times 10^5$$

So the flow is laminar

$$\therefore \qquad C_{fL} = 0.664/(Re_L)^{0.5} = 1.22 \times 10^{-3},\ \overline{\mathbf{C}}_\mathbf{f} = 2 \times C_{fL}. = \mathbf{2.44 \times 10^{-3}}$$

From equation (7.14) $\boldsymbol{\tau_{sx}} = \mu \left.\dfrac{du}{dy}\right| y = 0 = \mu\, 0.332 \,.\, \dfrac{u_\infty}{x}\sqrt{Re_x}$

$$= 1.006 \times 10^{-6} \times 1000 \times 0.332 \times \frac{0.6}{0.5}\sqrt{2.98 \times 10^5} = \mathbf{0.219\ N/m^2}$$

$$\delta_{0.5} = \frac{5 \times 0.5}{\sqrt{2.98 \times 10^5}} = 0.00458\ \text{m or } 4.58\ \text{mm.}$$

7.7 ENERGY EQUATION

This equation is obtained by applying the first law of thermodynamics to the element in the boundary layer as shown in Fig. 7.8(*a*). The specific heat is denoted by *c*.

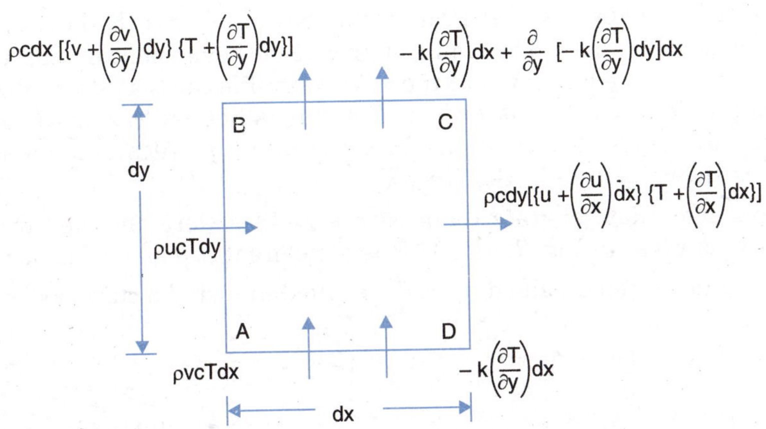

Fig. 7.8 (a) Energy flow through the elemental control volume

Considering unit depth and neglecting viscous work, the application of the first law leads to the statement-**under steady conditions, the heat flow in = heat flow out.**

The energy flow with the mass flow through various faces are:

face *AB*: $\rho\, u\, c\, T\, dy.1$

face *CD*: $\rho c \left(u + \frac{\partial u}{\partial x} dx\right)\left(T + \frac{\partial T}{\partial x} dx\right) dy\ .\ 1$

face *DA*: $\rho\, c\, v\, T\ .\ dx\ .\ 1$

face *BC*: $\rho\, c \left(v + \frac{\partial v}{\partial y} dy\right)\left(T + \frac{\partial T}{\partial y} dy\right) .\ dx\ .\ 1$

The net energy flow with the mass is given by (neglecting higher order terms)

$$\rho\, c\, u \frac{\partial T}{\partial x} dx\, dy\ .\ 1 + \rho\, c\, v \frac{\partial T}{\partial y} dx\, dy\ .\ 1 + \rho\, c\, T \left(\frac{\partial u}{\partial x} + \frac{\partial v}{\partial y}\right) dx\, dy\ .\ 1$$

The third term is zero as the quantity within the bracket is zero.

The energy conducted into the volume along the x direction is negligible. Summing the heat conducted through face *AD* and *BC*, the net heat flow by conduction is

$$k\ .\ \frac{\partial^2 T}{\partial y^2} dx\, dy\ .\ 1$$

Summing up we get equation (7.17) known as energy equation

$$u\ .\ \frac{\partial T}{\partial x} + v\ .\ \frac{\partial T}{\partial y} = \alpha \frac{\partial^2 T}{\partial y^2} \qquad ...(7.17)$$

where $\alpha = k/\rho c$ known as thermal diffusivity, as already defined in chpater 1.

The non-dimensional form of the equation which identifies dimentionless parameters relevent to the flow is obtained by defining

$$T^* = \frac{T - T_s}{T_\infty - T_s},\ \frac{u^*}{u_\infty}, v^* = \frac{v}{u_\infty}$$

$$u^* \frac{\partial T^*}{\partial x^*} + v^* \frac{\partial T^*}{\partial y^*} = \frac{1}{Re\ Pr} \frac{\partial^2 T^*}{\partial^2 y^*} \qquad ...(7.18)$$

Chapter 7

This equation indicates that Prandtl number influences the heat transfer situation in addition to Reynolds number. The similarity between the momentum and energy equations is obvious. If $\alpha = \upsilon$, then we can say that the velocity profile calculated will directly apply for the temperature profile $\alpha = \upsilon$ means $Pr = 1$. But this is not the case as Prandtl number have higher values for liquids. The solutions for temperature profiles in laminar flow obtained by Pohlhausen is shown plotted in Fig. 7.8(*b*).

The result is that separate temperature profiles were obtained for different Prandtl numbers, as illustrated in Fig. 7.8(*b*). This is of no great use as unification is not possible.

These values when modified by $Pr^{1/3}$ resulted in a single curve as shown in Fig. 7.8(*c*).

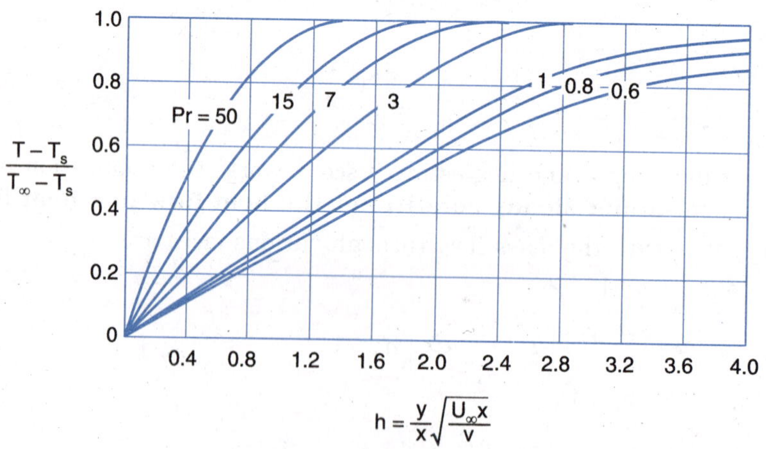

Fig. 7.8 (b) Dimensionless temperature distributions in a fluid flowing over a heated plate for various Prandtl numbers

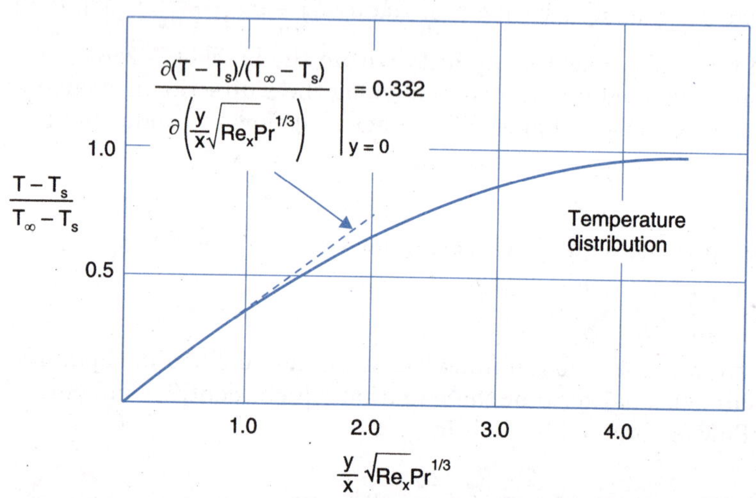

Fig. 7.8 (c) Dimensionless temperature distribution for laminar flow over a heated plate at uniform temperature

In this case the temperature ratio $(T - T_s)/(T_\infty - T_s)$ reached a value of 0.99 at $\frac{y}{x} Re_x^{1/2} Pr^{1/3} = 5$.

Hence the thermal boundary layer thickness was obtained as

$$\delta_{th} = \frac{\delta_h}{Pr^{1/3}} = \frac{5x}{Re_x^{1/2}\, Pr^{1/3}} \qquad ...(7.19)$$

Also the slope at $y = 0$ was obtained as 0.332

i.e.,
$$\left.\frac{\partial\,[(T - T_s)/(T_\infty - T_s)]}{\partial\left[\left(\frac{y}{x}\right).Re_x^{1/2}\, Pr^{1/3}\right]}\right|_{y=0} = 0.332.$$

$$\therefore \quad \left.\frac{\partial T}{\partial y}\right|_{y=0} = 0.332\,\frac{Re_x^{1/2}\, Pr^{1/3}\,(T_\infty - T_s)}{x}$$

Heat flux
$$q_x = -k\left.\frac{\partial T}{\partial y}\right|_{y=0} = -\,0.332\,k\,\frac{Re_x^{1/2}\, Pr^{1/3}}{x}\,(T_\infty - T_s)$$

$$h_x = \frac{q_x}{T_s - T_\infty} = 0.332\,\frac{k}{x}\,.\;Re_x^{1/2}\, Pr^{1/3} \qquad ...(7.20)$$

$$\therefore \quad \frac{h_x x}{k} = Nu_x = 0.332\, Re_x^{1/2}\, Pr^{1/3} \qquad ...(7.21)$$

The value of Nusselt number at length L is

$$Nu_L = 0.332\, Re_L^{1/2}\, Pr^{1/3} \qquad ...(7.22)$$

$$\overline{Nu}_L = 0.664\, Re_L^{1/2}\, Pr^{1/2} \qquad ...[7.22(b)]$$

The average value of h upto L is

$$\bar{h} = 2\,h\,(x = L) \qquad ...(7.23)$$

From experimental results it is found that these equations are applicable if the property values are used at film temperature $(T_s + T_\infty)/2$.

Example 7.9: *Show that* $\overline{Nu}_L = 2\, Nu_L$

Solution. Nu_L is the local Nusselt number at $x = L$

$$\overline{Nu}_L = \text{Average Nusselt number from } x = 0 \text{ to } x = L$$

$$\therefore \quad \overline{Nu}_L = \frac{1}{L}\int_0^L Nu_L\, dx = \frac{1}{L}\int_0^L 0.332\, Re_x^{1/2}\, Pr^{1/3}\, dx$$

As
$$Re_x = \frac{ux}{\upsilon},\; \overline{Nu}_L = \frac{1}{L} \times 0.332\,.\,Pr^{1/3}\,\frac{u^{1/2}}{\upsilon^{1/2}}\cdot\int_0^L x^{1/2}\, dx$$

Chapter 7

$$= \frac{1}{L} \times 0.332 \,.\, Pr^{1/3} \frac{u^{1/2}}{v^{1/2}} \,.\, L^{1.5} / 0.5$$

$$= 0.664 \;.\; Pr^{1/3} \,.\left(\frac{uL}{v}\right)^{0.5}$$

$$= 0.664 \,.\, Re_L^{\;0.5} \,.\, Pr^{1/3}$$

$\therefore$ $$\overline{\mathbf{Nu}}_{\mathbf{L}} = 2\,\mathbf{Nu}_{\mathbf{L}}.$$

Example 7.10: *Air at 20°C and one atmosphere flows over a surface at 100°C with a free stream velocity of 6 m / s. Determine the values of Reynolds number, thermal and hydrodynamic boundary layer thicknesses and the local value and average values of convective heat transfer coefficients at distances of 0.1, 0.25, 0.5, 0.75, 1, 1.25 m from the leading edge. Also determine the length at which the flow turns to turbulent taking critical Reynolds number as 5 × 10⁵.*

Solution: The temperature at which properties are to be read is (20 + 100)/2 = 60°C for air at one atmospheric pressure. The following values are read from tables:

$\nu = 18.97 \times 10^{-6}$ m²/s, $Pr = 0.696$,

$k = 0.02896$ W/mK (these being the required ones for the problem)

$$Re_x = \frac{u_\infty\, x}{\nu}, \; \delta_x = 5x/Re_x^{\,1/2},\; \delta_{th\,x} = \delta_x/Pr^{1/3}$$

$$Nu_x = 0.332\, Re_x^{\,1/2}\, Pr^{1/3}\; h_x = Nu_x\, k/x,\; \overline{h}_L = 2h_L$$

The calculated values are given below

x, m	0.1	0.25	0.5	0.75	1.0	1.25	1.5
Re_x	31629	79072	1.58×10^5	2.37×10^5	3.16×10^5	3.95×10^5	4.74×10^5
δ_x, mm	2.81	4.45	6.29	7.7	8.89	9.94	10.89
$\delta_{th\,x}$, mm	3.17	5.02	7.10	8.69	10.03	11.22	12.29
Nu_x	52.32	82.73	117	143.3	165.5	185	202.7
h_L, W/m²K	15.15	9.58	6.78	5.53	4.79	4.29	3.91
$\overline{h}_L$, W/m²K	30.30	19.16	13.56	11.06	9.58	8.58	7.82

The distance at which $Re_x = 5 \times 10^5$ is determined using

$$\frac{6 \times x}{18.97 \times 10^{-6}} = 5 \times 10^5, \text{ solving } x = 1.58 \text{ m}$$

plotting these tabulated result to scale, (particularly h) will be useful for understanding the phenomenon. The reason for the reduction in the value of convective heat transfer coefficient can be roughly explained as due to thicker layer through which diffusion has to take place along the direction of flow. As $Pr < 1$, $\delta_{th\,x} > \delta_x$.

COMMENTS

1. Re_x, δ, δ_{th} and Nu_x increase continuously as these are functions of x and so as x increases, these values increase.
2. Average value of convection coefficient is larger than the local value. This is because the values at low values of x are higher.

Example 7.11: *Air at 8 bar flows over an isothermal plane surface at a velocity of 6 m/s. The temperature of air is 120°C. The plane is at 80°C. Determine the boundary layer thickness at a distance of 0.2 m. Also determine the average value of convection coefficient upto this length.*

Solution: The property values have to be at the mean temperature = (80 + 120)/2 = 100°C.

Property values from handbook at 1 bar are

$$\nu = 23.13 \times 10^{-6} \text{ m}^2/\text{s},\ Pr = 0.688,\ k = 0.0321 \text{ W/mk.}$$

While Pr and k do not vary with pressure, ν does vary with pressure, corrected to 8 bar,

$$\therefore \quad \nu = 23.13 \times 10^{-6} \times (1/8) = 2.819 \times 10^{-6}$$

At 0.2 m, $Re = 6 \times 0.2/2.819 \times 10^{-6} = 4.25 \times 10^5$

$$\delta = 6 \times 0.2/(4.25 \times 10^5)^{1/2} = \mathbf{1.84\ mm.}$$

$$\delta_{th} = 1.84/0.688^{1/3} = \mathbf{2.08\ mm.}$$

$$\overline{Nu} = 0.664 \times Re^{1/2} Pr^{1/3} = 0.664 \times (4.25 \times 10^5)^{1/2} \times 0.688^{1/3} = \mathbf{382.45}$$

$$\therefore \quad \overline{h} = 382.45 \times 0.0321/0.2 = \mathbf{61.38\ W/m^2K}$$

7.8 INTEGRAL METHOD

The integral method uses the idea of control volume large enough to contain the boundary layer and the net flows are summed by integration rather than by using the rates. The details are shown in Fig. 7.9.

The shear force along surface BC and the conduction across it is zero as it is beyond the boundary layer.

The mass or energy or momentum flow is obtained by considering a small layer by at y from the wall and integrating it over the whole face. For example, the energy flowing over face AB is given by

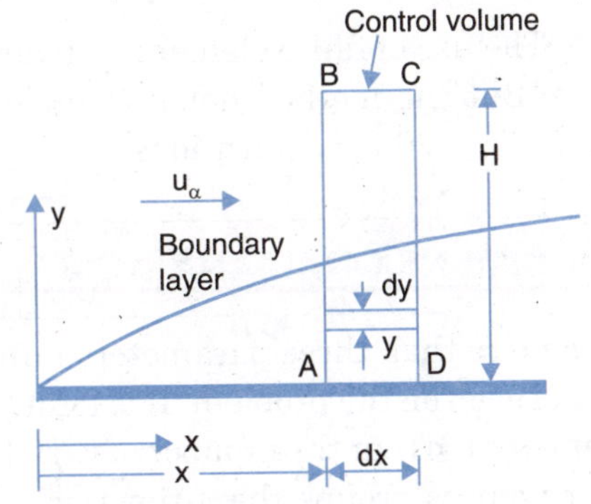

Fig. 7.9. Control volume used in the integral method

$$\text{Energy flow along} \quad AB = \int_0^H \rho\, u\, c\, T\, dy$$

and the flow out across CD is given by

$$\int_0^H \rho\, u\, c\, T\, dy + \frac{\partial}{\partial x}\left[\int_0^H \rho\, u\, c\, T\, dy\right] . dx$$

The heat flow is by conduction at the surface AD, and this is calculated using Fourier's law. (Refer solved examples A-7.20 and A-7.22 for detailed working).

The integration is possible if the velocity and temperature profiles satisfying the boundary condition is assumed. A suitable velocity profile satisfying boundary conditions at $y = 0$ and $y = \delta$ is

$$\frac{u}{u_\infty} = \frac{3}{2}\frac{y}{\delta} - \frac{1}{2}\left(\frac{y}{\delta}\right)^3 \qquad \text{...(7.24)}$$

Chapter 7

This yields boundary layer thickness as

$$\delta_x = 4.64\, x/\sqrt{Re_x} \qquad ...(7.25)$$

The friction coefficient is given by

$$C_{fx} = 0.647\, (Re_x)^{-0.5} \qquad ...(7.26)$$

The values differ from the values obtained from exact analysis only by a small percentage. An important strength of the method is that it is possible to use it for more exacting boundary conditions. Also it has been proved that it is possible to obtain a universal velocity profile using dimensionless quantities. This profile can be used to obtain good approximations when other methods fail.

The thermal boundary layer thickness can be determined using a temperature profile of the form

$$\frac{T_s - T}{T_s - T_\infty} = \frac{3}{2}\frac{y}{\delta_{th}} - \frac{1}{2}\left(\frac{y}{\delta_{th}}\right)^3 \qquad ...(7.27)$$

This yields a local Nusselt number given by equation (7.28), which differs very little from equation (7.20).

$$Nu_x = 0.33\, Re_x^{\,0.5}\, Pr^{0.33} \qquad ...(7.28)$$

The use of finite element softwares for analysis will replace this method also in the near future. But the method helps in understanding the basic principles involved. See advanced example A-7.20 for derivations.

7.9 DIMENSIONAL ANALYSIS

When more than three parameters influence a problem it becomes very difficult to analyse the effect of each on the problem. It is found desirable to group these parameters into dimensionless parameters so that the number of variables can be reduced to three. In this attempt to formulate dimensionless groups the π-theorem proposed by Buckingham is useful. The theorem states that if there are m physical quantities involved in a problem requiring n primary dimensions to express them, then the number of independent dimensionless groups that can be formed is equal to $(m-n)$. In case there are seven quantities and four dimensions, it is possible to group these as

$$\pi_1 = f(\pi_2, \pi_3)$$

where π_1, π_2 and π_3 are dimensionless groups. The dimensional analysis aids in the identification of the groups. This approach provides a method to represent or correlate experimental results. This method does not lead to exact functional relationship and it also does not ensure that all the variables are correctly chosen. Fluid mechanics and Heat transfer studies involve a large number of dimensionless parameters like Reynolds number, Nusselt number etc which have been first identified using this method. Almost all empirical correlations are based on these parameters so identified. The application of the method is illustrated by a problem under the section advanced examples (A-7.16 and A-7.17).

Example 7.12: *Using the method of dimensional analysis obtain the dimensionless numbers in the case of forced convection.*

After listing out the influencing parameter with dimensions repeating variables are chosen. The number of repeating variables equals the number of dimensions involved in representing the influencing parameters.

Parameter	Symbol	Unit	Dimensions
Flow length	L	m	L
Flow velocity	u_∞	m/s	L/t
Fluid density	ρ	kg/m^3	M/L^3
Fluid viscosity	μ	kg/ms	M/Lt
Conductivity	k	W/mK	ML/t^3T
Specific heat	C	$J/kg\ K$	$L^2/t^2\ T$
Convection coefficient	h	$W/m^2\ K$	$M/t^3\ T$

L–length, M—mass, t–time, T—temperature.

Solution: There are seven variables and four dimensions. So there can be three π terms.

The length L, density ρ, conductivity k and viscosity μ are chosen as repeating variables. These should contain all the four dimensions and one should not be a simple product of the other. With the repeating variables each of the other parameters form the π terms. For example,. first

$$\pi_1 = L^a \rho^b k^c \mu^d h = L^a \frac{M^b}{L^{3b}} \frac{M^c L^c}{t^{3c}\ T^c} \cdot \frac{M^d}{L^d\ t^d} \cdot \frac{M}{t^3\ T} \quad ...(a)$$

As the π terms are dimensionless, the dimensions on the R.H.S. should also be zero. This leads to four equations, solving which the dimensionless parameter is obtained. In this case, taking the dimensions on the R.H.S. one by one

$L \longrightarrow a - 3b + c - d = 0$

$M \longrightarrow b + c + d + 1 = 0$

$t \longrightarrow -3c - d - 3 = 0$

$T \longrightarrow -c - 1 = 0$

Solving $a = 1, b = 0, c = -1$ and $d = 0$

Substituting in equation (a),

$$\pi_1 = L^1 \rho^0 k^{-1} \mu^0 h = \frac{\mathbf{hL}}{\mathbf{k}} \quad \text{(Nusselt Number)}$$

$$\pi_2 = L^a \rho^b k^c \mu^d c = L^a \frac{M^b}{L^{3b}} \cdot \frac{M^c L^c}{t^{3c}\ T^c} \cdot \frac{M^d}{L^d t^d} \cdot \frac{L^2}{t^2 T} \quad ...(b)$$

As before considering each dimension on the R.H.S.

$L \longrightarrow a - 3b + c - d + 2 = 0$

M $\longrightarrow b + c + d = 0$

Chapter 7

$$t \longrightarrow -3c - d - 2 = 0$$

$$T \longrightarrow -c - 1 = 0.$$

Solving $a = 0,\quad b = 0,\quad c = -1,\quad d = 1.$

Substituting in equation (*b*)

$$\pi_2 = L^0\rho^0k^{-1}\mu^1\, c = \frac{\mathbf{c}\mu}{\mathbf{k}} \qquad \text{(Prandtl number)}$$

$$\pi_3 = L^a\rho^b k^c\mu^d u_\infty = L^a\,\frac{M^b}{L^{3b}}\cdot\frac{M^cL^c}{t^{3e}\,T^e}\cdot\frac{M^d}{L^d\,t^d}\cdot\frac{L}{t} \qquad \text{...(c)}$$

Summing up the indices for each dimension.

$$L \longrightarrow a - 3b + c - d + 1 = 0$$

$$M \longrightarrow b + c + d = 0$$

$$t \longrightarrow -3c - d - 1 = 0$$

$$T \longrightarrow -c = 0.$$

Solving, $a = 1,\quad b = 1,\quad c = 0,\quad d = -1$

$$\therefore \qquad \pi_3 = L^1\rho^1k^0\mu^{-1}u_\infty = \frac{u_\infty\,\rho L}{\mu}. \qquad \text{(Reynolds number)}$$

We can now write $\quad Nu = f(Re, Pr)$;

But the exact function has to be determined by experiments.

Two more problems are worked out by Rayleigh method under advanced examples in 7.16 and 7.17.

Example 7.13: *Convert the energy equation (7.15) into dimensionless form:*

The equation is given by

$$u\,\frac{\partial T}{\partial x} + v\,\frac{\partial T}{\partial y} = \alpha\,\frac{\partial^2 T}{\partial y^2}.$$

Solution: u, v, T, x, y are the dimensional quantities. These can be replaced by using dimensionless quantities defined below:

$$u^* = u/u_\infty,\ v^* = v/u_\infty,\ x^* = x/L,\ y^* = y/L \text{ and}$$

$$T^* = (T - T_s)/(T_\infty - T_s)$$

$$\therefore \qquad u\,\frac{\partial T}{\partial x} = \frac{u^*u_\infty}{L(T_\infty - T_s)}\,\frac{\partial T^*}{\partial x^*},\quad v\,\frac{\partial T}{\partial y} = \frac{v^*u_\infty}{L(T_\infty - T_s)}\,\frac{\partial T^*}{\partial y^*}$$

$$\alpha\,\frac{\partial^2 T}{\partial y^2} = \frac{k}{\rho c}\cdot\frac{1}{L^2}\cdot\frac{1}{(T_\infty - T_s)}\cdot\frac{\partial^2 T^*}{\partial y^{*2}}$$

Substituting

$$\frac{u^*u_\infty}{L(T_\infty - T_s)}\cdot\frac{\partial T^*}{\partial x^*} + \frac{v^*u_\infty}{L(T_\infty - T_s)}\cdot\frac{\partial T^*}{\partial y^*} = \frac{k}{\rho c}\cdot\frac{1}{L^2}\,\frac{1}{(T_\infty - T_s)}\cdot\frac{\partial^2 T^*}{\partial y^{*2}}$$

Cancelling common terms and multiplying and dividing R.H.S. by μ, we get

$$u^*\,\frac{\partial T^*}{\partial x^*} + v^*\,\frac{\partial T^*}{\partial y^*} = \frac{k}{\mu c}\cdot\frac{\mu}{\rho u_\infty L}\cdot\frac{\partial^2 T^*}{\partial y^{*2}}$$

This reduces to

$$u^* \frac{\partial T^*}{\partial x^*} + v^* \frac{\partial T^*}{\partial y^*} = \frac{1}{Re_L \, Pr} \cdot \frac{\partial^2 T^*}{\partial y^{*2}} \quad ...(7.29)$$

This shows that as all the quantities are dimensionless. The solutions will be similar if $Re_L \, Pr$ product is equal irrespective other geometric or property values. This is easily realised by looking at equation (7.20). Thus nondimensionalising helps in generalising experimental results obtained with one set of parameters.

7.10 ANALOGICAL METHODS

So far we have discussed certain relationships applicable to laminar flow. Because of the irregularity of flow it is not easy to use the exact method for the analysis of turbulent flow. In laminar flow momentum transfer between layers is due to molecular level diffusion. In turbulent flow large particles move from one layer to another speeding up the momentum transfer. Similar explanation can be given for heat transfer also. Molecular movement is random. If such randomness is assumed for particle movement also, then except for a difference in scale both laminar and turbulent flows can be said to be similar. The values of ν and α are increased in turbulent flow by additive quantities defined as eddy diffusivity of momentum and eddy diffusivity of heat to enhance the rate of transfer. This leads to the conclusion that similar correlations will result except that the constants c, m and n have to be determined separately.

$$\delta_x = c_1 \, Re_x^{\,n1} \quad ...(7.30)$$

$$Nu = c_2 \, Re^{m2} \, Pr^n \quad ...(7.31)$$

This is also established by using 1/7 power law for velocity variation and also using the universal velocity profile correlation equations for turbulent flow which will be discussed in the next chapters.

7.10.1. Reynolds-Colburn Analogy

Using what is known as mixing length theory, it is shown that

$$\tau = \rho \, (v + \varepsilon_m) \frac{d\bar{u}}{dy}$$

and

$$q = - c_p \rho \, (\alpha + \varepsilon_H) \frac{d\bar{T}}{dy}$$

where ε_m is called eddy momentum diffusivity and ε_H is called eddy diffusivity of heat.

From equation (7.15)

$$\frac{C_{fx}}{2} = \frac{0.332}{Re_x^{\,1/2}} \quad ...(a)$$

From equation (7.21),

$$Nu_x = 0.332 \, Re^{1/2} \, Pr^{1/3} \quad ...(b)$$

This equations (a), (b) can be manipulated such that

$$\frac{Nu_x \, Pr^{2/3}}{Re_x \, Pr} = \frac{0.332}{Re_x^{\,1/2}} = \frac{C_{fx}}{2}$$

Chapter 7

The group $\frac{Nu}{Re\,Pr}$ reduces to $\frac{hx}{\rho C\,u_\infty}$ which is dimensionless and is defined as Stanton number. Hence

$$\frac{hx}{\rho C\,u_\infty}\,.\,Pr^{2/3} = St_x\,Pr^{2/3} = \frac{C_{fx}}{2} \quad ...(7.32)$$

This equation connects the friction factor and convective heat transfer coefficient. So if one quantity is known, the other quantity can be determined. The equation (7.32) is known as Reynolds–Colburn analogy. This has been found by experiments to be applicable in turbulent flow conditions also. In general

$$St\,Pr^{2/3} = \frac{C_f}{2} \quad ...(7.33)$$

7.11 CORRELATION OF EXPERIMENTAL RESULTS

The validation of the equations derived is essential before these could be confidently applied in design situations. An experimental set up like the one shown in Fig. 7.10 can be used to measure average values of convection coefficients.

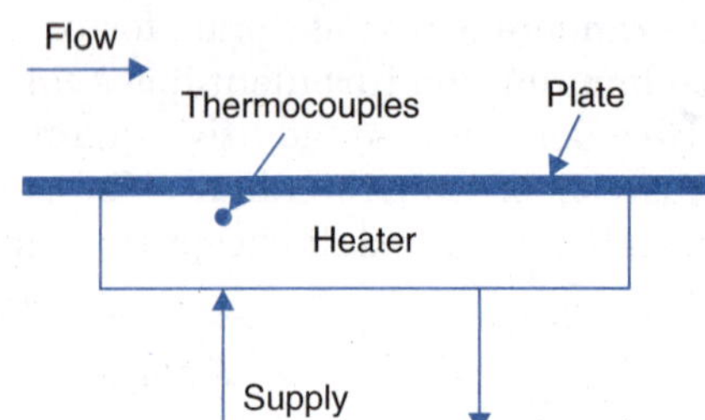

Fig. 7.10. Experimental set up for convection measurements

The energy supplied and temperatures can be measured for various conditions of flow. The average value of convection coefficient can be readily calculated from measurements. The dimensionless parameters can be calculated and the experimental results can be plotted as in Fig. 7.11. Using the three identified parameters.

After obtaining these curves, the equations for these lines will yield the constant c, m and n of equation of the form (7.32). This can be checked by plotting all the results as in Fig. 7.12.

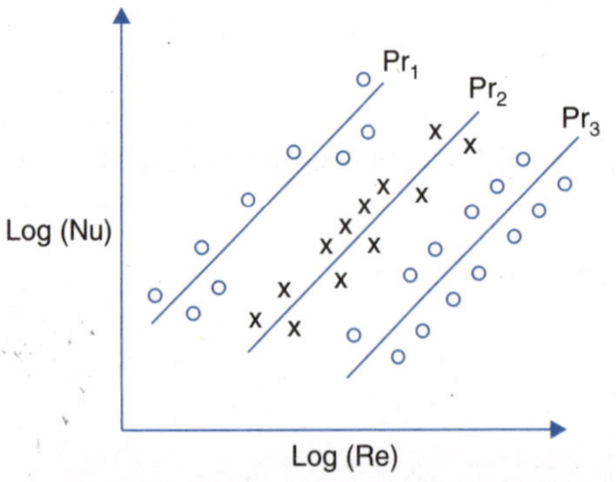

Fig. 7.11. Plot of experimental results using dimensionless parameters

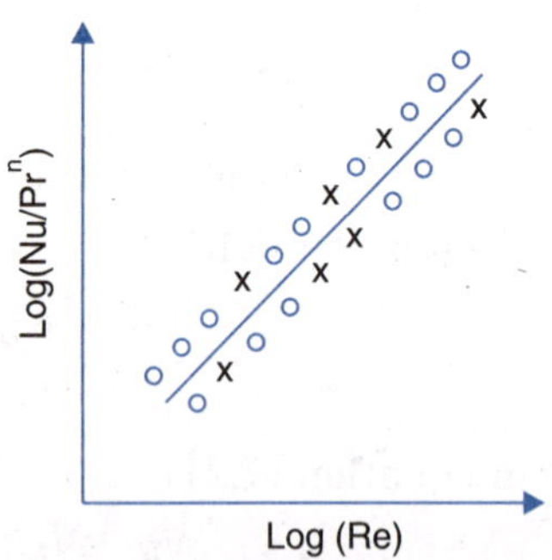

Fig. 7.12. Unified plot of experimental results

The resulting single curve will establish the relationship required.

This procedure is described easily. But the actual work requires a lot of trial and also the experimental points for extreme values of parameters like *Pr* may not fall on lines similar to the ones in Fig. 7.11. A curve approximated by multiple straight lines may result in place of single line as in Fig. 7.12.

Most of the usable correlations are the ones established by this method. Hence in convection one has to depend to a very great extent on such experimental correlations. It is to be mentioned that such correlations generally give results, only within about ± 25% of true value. For more accurate determination, one has to use handbooks rather than textbooks to look up more suitable correlations.

There are many other boundary conditions in external flow. These are discussed in the next Chapter 8 and the available correlations are listed. Internal flow is discussd in Chapter 9.

ADVANCED EXAMPLES

In the first three examples a comparative study of various materials is attempted. This will help in better understanding the basics.

Example A-7.1: *Consider the flow of the following fluids at a film temperature of 40°C and a free stream velocity of 3 m/s over a flat plate. (a) Determine the lengths at which the flow turns turbulent (b) Determine the thickness of boundary layer (hydrodynamic and thermal) at the location where Re = 10^5. Fluids : (i) air (ii) Carbon dioxide (iii) Hydrogen (iv) water (v) mercury and (vi) glycerine.*

Solution: The property values required for these calculations are (*i*) kinematic viscosity and (*ii*) Prandtl number. These are looked up from property tables at 40°C–in some cases interpolation is necessary. These are tabulated below:

Fluid	*Kinematic viscosity m^2/s*	*Prandtl number*
Air	16.960×10^{-6}	0.699
Carbon dioxide	9.294×10^{-6}	0.76
Hydrogen	118.600×10^{-6}	0.684
Water	0.657×10^{-6}	4.34
Mercury	0.109×10^{-6}	0.0252
Glycerine	223.000×10^{-6}	2450

The distance is calculated using the equation [7.8 (*a*)] Turbulence starts of $Re = 5 \times 10^5$.

$$Re_x = \frac{u_\infty x}{\nu} \quad \therefore \quad x = \frac{5 \times 10^5}{3.0} \cdot \nu$$

The hydrodynamic boundary layer thickness is found by first finding x as above at $Re = 10^5$ and then using $\delta_h = 5x/Re_x^{0.5}$ equation (7.11) and the thermal boundary layer thickness is determined using equation (7.17).

$\delta_{th} = \delta_h / Pr^{0.33}$. The results are tabulated below.

$T_f = 40°C$, $u_\infty = 3$ m/s, critical Reynolds no. $= 5 \times 10^5$

Fluid	*Critical distance, m Re = 5 × 10⁵*	*Distances at Re = 10⁵ m*	*Hydrodynamic boundary layer thickness mm*	*Thermal boundary layer thickness mm*
Air	2.83	0.565	8.94	10.00
Carbon dioxide	1.55	0.310	4.90	5.37
Hydrogen	19.77	3.950	62.50	71.00
Water	0.11	0.0219	0.346	0.212
Mercury	0.018	0.00363	0.0057	0.0196
Glycerine	37.17	7.43	117.5	8.72

COMMENTS

(*i*) Higher the value of viscosity longer it takes for the flow to become turbulent as the viscous forces dampen out any disturbance.

(*ii*) If the Prandtl number is less than unity, thermal boundary layer is thicker than velocity boundary layer and vice versa.

(*iii*) Because of the large value of *Pr* for Glycerine, the thermal boundary layer thickness is much lower compared to hydrodynamic boundary layer thickness. Thermal boundary layer is well inside of hydro dynamic boundary layer.

Example A-7.2: *For the flow situations considered in example A-7.1 determine the local and average friction coefficients up to the location where Re = 10⁵. Also determine the local and average convection coefficients.*

Solution: Equation (7.13) is to be used for the determination of friction coefficient

$$C_{fx} = 0.664/Re_x^{0.5}$$

As Reynolds number is specified the value will be the same for all cases.

Local friction coefficient $= 0.664/\sqrt{(10^5)} = 2.1 \times 10^{-3}$

Average friction coefficient $= 2 \times C_{fx} = 4.2 \times 10^{-3}$

But this friction coefficient is for different lengths from 0.00363 m for mercury to 7.43 m for glycerine. Equation (7.20) is to be used to determine the value of convection coefficient.

$$Nu_x = 0.332 \,.\, Re_x^{0.5}\, Pr^{0.33}$$

$$\therefore \qquad h_x = \frac{Nu_x \,.\, k}{x}$$

The thermal conductivity of the materials are read from tables (in some cases interpolation is necessary).

These are tabulated below:

Fluid	*Thermal conductivity at 40°C, W/mK*
Air	26.56×10^{-3}
Carbon dioxide	26.004×10^{-3}
Hydrogen	191.256×10^{-3}
Water	0.628
Mercury	8.1725
Glycerine	0.2861

Using the above equation and the value of x from previous problem, local and average convection coefficient are calculated and tabulated.

$$u_{\infty} = 3 \text{ m/s}, T = 40°\text{C}$$

Fluid	*x at $Re = 10^5$*	*Nu_x*	*h_x W/m²K*	*$\overline{h_1}$, W/m²K*
Air	0.565	93.17	4.38	8.76
Carbon dioxide	0.310	95.80	8.04	16.08
Hydrogen	3.950	92.5	4.48	8.96
Water	0.0219	171.26	4911	9822
Mercury	0.00363	30.78	69298	138596
Glycerine	7.43	1415.3	54.5	109

COMMENT

The value of convection coefficient is high for Mercury, glycerine and water. Also note from example A-7.1 the very thin boundary layers in these cases.

Example A-7.3: *Consider the flow of the following fluids over a plate 1 m × 1 m. Determine the free stream velocity so that flow turns turbulent just at leaving the plate. The plate is at 60°C and the fluid is at 20°C and gases at 1 atm pressure: (i) air (ii) Helium (iii) Hydrogen (iv) Water (v) Engine oil (vi) Freon 12 (liquid). Also determine the heat loss in this process over the plate area.*

Solution: Turbulence may be assumed to set in at $Re = 5 \times 10^5$. From the definition of Reynolds number

$$Re = \frac{u_{\infty} x}{\nu}, u_{\infty} = \frac{Re \,.\, \nu}{x} = \frac{5 \times 10^5}{1} \times \nu$$

The temperature at which properties are to be read are (60 + 20)/2 = 40°C. (film temperature). These are read from property tables. Thermal conductivity and Prandtl number values are necessary to determine convection coefficient. These values are tabulated below (with some interpolation).

Chapter 7

Fluid	*Kinematic viscocity* m^2/s	*Thermal conductivity* W/mK	*Prandtl number*
Air	16.96×10^{-6}	26.56×10^{-3}	0.699
Helium	133.4×10^{-6}	157.464×10^{-3}	0.6772
Hydrogen	118.6×10^{-6}	191.256×10^{-3}	0.684
Water	0.657×10^{-6}	628×10^{-3}	4.34
Engine oil	241×10^{-6}	144.2×10^{-3}	2870
Freon 12	200×10^{-6}	69.2×10^{-3}	3.5

The calculated values as described above is tabulated. $\bar{h} = 0.664 \times \frac{k}{x}$.

$Re^{0.5}\, Pr^{0.33}$ is used to calculate $\bar{h}$.

$$Q = h\, A\Delta T = h \times 1 \times 40$$

$$L = 1 \text{ m},\ Re = 5 \times 10^5,\ T = 40°\text{C},\ A = 1 \times 1 \text{ m}^2$$

Fluid	u_∞ **m/s**	$\bar{h}$**, W/m²K**	**Q, W**
Air	8.48	11.07	443
Helium	66.70	64.92	2597
Hydrogen	59.3	79.12	3164.8
Water	0.3285	480.96	19238.4
Engine oil	120.5	962.2	38486
Freon 12	100.0	49.33	1973.2

COMMENTS

1. Higher the value of kinematic viscosity, higher the initial velocity for a given Reynolds number at a given location.
2. $\bar{h}$ depends both on Prandtl number and conductivity.
3. Engine oil appears to dissipate more heat. But look at the velocity required.
4. For high/low values of *Pr* this equation is not suitable.
5. Velocities of 100 m/s or 120.5 m/s are not practicable. The values are only for comparison.

Example A-7.4: *Air flows over a plate at a free stream velocity of 5 m/s. The plate temperature is 100°C and the air temperature is 20°C. If the average value of convective heat transfer coefficient is 15 W/m²K, determine the length of the plate in the direction of flow. Using equations (7.20) and (7.21), we get*

Solution:

$$\bar{h} = 0.664 \times \frac{k}{L} Re_L^{0.5}\, Pr^{0.333}$$

$$= 0.664 \times \frac{k}{L} \sqrt{u_\infty L/\nu}\, Pr^{0.333}$$

or

$$L = 0.664^2 \frac{k^2}{h^2} \cdot \frac{u_\infty}{\nu} \cdot Pr^{0.667}$$

k, ν and *Pr* are to be found at film temperature of (100 + 20)/2 = 60°C. The values are read from data book: $k = 28.96 \times 10^{-3}$, $\nu = 18.97 \times 10^{-6}$ m²/s, $Pr = 0.696$

Substituting $$\mathbf{L} = 0.664^2 \frac{0.02896^2}{15^2} \cdot \frac{5}{18.97 \times 10^{-6}} \times 0.696^{0.667} = \mathbf{0.34\ m}$$

Check: The validity $Re_L = \dfrac{0.34 \times 5}{18.97 \times 10^{-6}} = 89615$

This is less than 5×10^5. So the flow is laminar and the equation used is valid.

$$\overline{Nu} = 0.664 \,.\, Re_L{}^{0.2} \times Pr^{0.033} = 0.664 \times 89615^{0.5} \times 0.696^{0.333} = \mathbf{176.155}$$

$$\therefore \quad \bar{\mathbf{h}} = \frac{176.155 \times 28.96 \times 10^{-3}}{0.34} = \mathbf{15\ W/m^2K.}$$

Example A-7.5: *Air at 1 atm pressure flows over a plate 0.5 long with a free stream velocity of 5 m/s. The film temperature is 60°C. If the friction coefficient on the plate is found as 3.68×10^{-3} for one side, determine the average convective heat transfer coefficient.*

Solution: The example can be solved by using equation (7.14) and then solving for Re_L and using the same in equations (7.20) and (7.21) or (7.22) directly to determine the Stanton number and then the Nusselt number.

$$\overline{C_f} = 1.334/(Re_L)^{0.5} \quad \text{...(7.14)}$$

$$\therefore \quad Re_1 = (1.334/\overline{C_f})^2 = (1.334/3.68 \times 10^{-3})^2 = 131406.25, \text{ so laminar}$$

$$Nu = 0.664\, Re_L{}^{0.5}\, Pr^{0.333} = 0.664 \times 131406.25^{0.5} \times 0.696^{0.333} = 213.31$$

$$\mathbf{h} = \frac{213.31 \times 28.96 \times 10^{-3}}{0.5} = \mathbf{12.36\ W/m^2K}$$

(property values are taken from example 7.4)

$$St.\, Pr^{0.667} = C_f/2 = 3.68 \times 10^{-3}/2$$

$$\therefore \quad St = 2.343 \times 10^{-3}/2$$

$$St = h/c\,\rho\, u_\infty$$

$$\therefore \quad \mathbf{h} = 2.343 \times 10^{-3} \times 1005 \times 1.060 \times 5 = \mathbf{12.48\ W/m^2K}$$

The difference is due to truncation in calculation.

Example A-7.6: *Nitrogen at a pressure of 0.1 atm flows over a flat plate with a free stream velocity of 8 m/s. The temperature of the gas is – 20°C. The plate temperature is 20°C. Determine the length for the flow to turn turbulent. Assume 5×10^5 as critical Reynolds number. Also determine the thickness of thermal and velocity boundary layers and the average convection coefficient for a plate length of 0.3 m.*

Solution: Properties are to be found at film temperature.

Film temperature = (– 20 + 20)/2 = 0°C

As density and kinematic viscosities will vary with pressure, dynamic viscosity is read from tables.

$$\mu = 16.67 \times 10^{-6}\ \text{Ns/m}^2,\ k = 24.31 \times 10^{-3}\ \text{W/mK}$$

$$Pr = 0.705,\ \rho = 1.250 \times 0.1 = 0.125\ \text{kg/m}^3$$

$$Re = \frac{u_\infty \rho x}{\mu} = 5 \times 10^5$$

$$\therefore \quad x = \frac{5 \times 10^5 \times 16.67 \times 10^{-6}}{8 \times 0.125} = 8.335 \text{ m}$$

Check for dimensions:

$$u_\infty = \frac{\text{m}}{\text{s}}, \rho = \frac{\text{kg}}{\text{m}^3}, x = m, \mu = \text{Ns/m}^2$$

$$\therefore \quad \frac{\text{m}}{\text{s}} \frac{\text{kg}}{\text{m}^3} \cdot \text{m} \frac{\text{m}^2}{\text{Ns}} = \frac{\text{kg} \cdot \text{m}}{\text{s}^2} \frac{1}{\text{N}} = \text{dimensionless}$$

As the density is low, the kinematic viscosity is higher and hence turbulence is suppressed for a longer distance.

At $x = 0.3$ m, the Reynolds number is less than 5×10^5 and so the flow is laminar.

$$\delta_h = 5x/Re^{0.5} = 5 \times 0.3/(8 \times 0.125 \times 0.3/16.67 \times 10^{-6})^{0.5}$$

$$= 5 \times 0.3/\sqrt{17996.4} = 0.01118 \text{ m or } \mathbf{11.18\ mm}$$

Thermal boundary

layer thickness $= \delta_h/Pr^{0.33} = 11.18/0.705^{0.33} = $ **12.56 mm**

Average convection coefficient:

$$\bar{h} = \frac{0.644 \times 24.31 \times 10^{-3}}{0.3} (8 \times 0.125 \times 0.3/16.67 \times 10^{-6})^{0.5} (0.705)^{0.33}$$

$$= \mathbf{6.23\ W/m^2K,}$$

If the **pressure was atmospheric**, then boundary layers thickness is

$$\delta_h = 5 \times 0.3/(8 \times 1.25 \times 0.3/16.67 \times 10^{-6})^{0.5} = \mathbf{3.54\ mm}$$

$$\delta_t = 3.54/(0.705)^{0.3} = \mathbf{3.98\ mm}$$

$$\bar{\mathbf{h}} = \frac{0.664 \times 24.31 \times 10^{-3}}{0.3} [(8 \times 1.25 \times 0.3/16.67 \times 10^{-6})^{0.5} (0.705)^{0.33}]$$

$$= \mathbf{20.32\ W/m^2K.}$$

COMMENT

Instead of using μ and ρ, the kinematic viscosity can be corrected by $v_{P_2} = v_{P_1} \times \frac{P_1}{P_2}$.

Example A-7.7: *An aircraft flies at an altitude where the temperature is – 60°C and the pressure is 0.08 atm. The air speed is 900 km/hr. The wing tank contains fuel at 20°C, keeping the surface at this temperature. Assuming the effect of curvature to be small, determine the Reynolds number at the wing tip if the wing is 3 m wide. Also determine the location from the leading edge where Re = 5 × 10⁵. If $C_{fx} = 0.059 \times Re_x^{-0.2}$, determine the value of local heat transfer coefficient.*

Solution: $Re = \frac{\rho u_\infty x}{\mu}$, The values of ρ and μ are to be found at (– 60 + 20)/2 = – 20°C and 0.08 atm,

$$\rho = 1.395 \times 0.08 \text{ kg/m}^3,\ \mu = 16.18 \times 10^{-6} \text{ Ns/m}^2,\ x = 3 \text{ m}$$

$$k = 22.79 \times 10^{-3} \text{ W/mK},\ Pr = 0.716,\ u = 900 \times 1000/60 \times 60 = 250 \text{ m/s}$$

$$Re = 1.395 \times 0.08 \times 250 \times 3/16.18 \times 10^{-6} = \mathbf{5.173 \times 10^6}$$

Flow is turbulent

The location at which $Re = 5 \times 10^5$ is found using

$$x = \frac{5 \times 10^5 \times 16.18 \times 10^{-6}}{1.395 \times 0.08 \times 250} = \mathbf{0.29\ m}$$

$$C_{fx} = 0.059 \times Re_x^{-0.2} = 0.059/(5.173 \times 10^6)^{0.2} = 2.68 \times 10^{-3}$$

Using equation (7.35)

$$St \times Pr^{2/3} = C_{fx}/2$$

$$\frac{Nu_x}{Re_x\ Pr} . Pr^{2/3} = 2.68 \times 10^{-3}/2$$

$$\therefore \quad Nu_x = \frac{2.68 \times 10^{-3}}{2} \times Re_x\ Pr^{0.33}$$

$$\therefore \quad \mathbf{h_x} = \frac{22.79 \times 10^{-3}}{3} . \frac{2.68 \times 10^{-3}}{2} \times (5.173 \times 10^6) \times (0.716)^{0.333}$$

$$= \mathbf{47.11\ W/m^2K}$$

There are other correlations available which are discussed in the next chapter.

Example A-7.8: *A low speed wind tunnel is to be designed for tests upto Re = 5 × 10⁵ with air at 0.8 atm and 20°C. The model length is restricted to 0.3 m. If the test section is 0.45 m square, determine the mass flow required. Check whether the boundary layer on the wall will affect the test. The tunnel length is 1.5 m.*

Solution: The properties are to be obtained at 20°C

$$\rho = 1.205,\ \mu = 18.14 \times 10^{-6} \text{ Ns/m}^2$$

$$Re_x = \frac{\rho u_\infty x}{\mu},\ 5 \times 10^5 = \frac{1.205 \times 0.8 \times u_\infty \times 0.3}{18.14 \times 10^{-6}}$$

$$\therefore \quad u_\infty = 31.36 \text{ m/s}$$

$$\therefore \quad \textbf{flow rate} = (31.36 \times 0.45 \times 0.45) \times (1.205 \times 0.8) = \mathbf{6.12\ kg/s}$$

The boundary layer thickness at $x = 1.5$ m is to be found.

The flow will be turbulent (see next chapter for correlations)

The Reynolds number $= 2.5 \times 10^6$

$\therefore$ The boundary layer thickness is given by

$$\delta = 0.381 \times Re_x^{-0.2} = \mathbf{0.02\ m}$$

In a width of 0.45 m, 0.04 m should not cause any problem.

Note: Examples 7.9 and 7.10 investigates the effect of individual parameters on convection coefficient in laminar and turbulent flow.

Example A-7.9: *Investigate the effect of various parameters on the average value of convection coefficient in laminar flow over a flat plate.*

$$\overline{Nu}_L = 2Nu_x = 0.664.\ Re_L^{0.5}\ Pr^{0.33}.$$

Solution: Substituting for the dimensionless number,

$$\frac{\bar{h}L}{k} = 0.664\left[\frac{u_\infty \rho L}{\mu}\right]^{0.5}\left[\frac{c\,\mu}{k}\right]^{0.33}$$

using 1/2 and 1/3 in place of 0.5 and 0.33

$$\bar{h} = 0.664\,\frac{k}{L}\cdot\frac{u_\infty^{1/2}\,\rho^{1/2}\,L^{1/2}}{\mu^{1/2}}\,\frac{c^{1/3}\,\mu^{1/3}}{k^{1/3}}$$

$$= 0.664\,k^{2/3}\,u_\infty^{1/2}\rho^{1/2}\,c^{1/3}/L^{1/2}\,\mu^{1/6}$$

$$\frac{h_2}{h_1} = \left(\frac{k_2}{k_1}\right)^{2/3}\left(\frac{\rho_2}{\rho_1}\right)^{1/2}\left(\frac{u_{\infty 2}}{u_{\infty 1}}\right)^{1/2}\left(\frac{c_2}{c_1}\right)^{1/3}\left(\frac{L_1}{L_2}\right)^{1/2}\left(\frac{\mu_1}{\mu_2}\right)^{1/6}$$

The relationships can be summarised as below: In the laminar flow over a flat plate:

1. Convection coefficient varies directly as (2/3) power of thermal conductivity. Higher the value of thermal conductivity higher the value of convection coefficient, but at a reduced ratio of increase.
2. Convection coefficient varies directly as the square root of free stream velocity. If the free stream velocity is increased four times, the convection coefficient will double.
3. The effect of density variation is similar to that of free stream velocity. Increased density will increase the convection coefficient. But to double the coefficient density has to increase four times.
4. The convection coefficient is directly proportional to 1/3 power of specific heat. *i.e.* to double the value of convection coefficient, specific heat should increase 8 times.
5. The convection coefficient reduces with length, as the square root of the ratio. At four times the length the average coefficient will be reduced to half the value.
6. Dynamic viscosity has also an inverse relation to 1/6th power. Increased viscosity will reduce the convection coefficient.

Though the relations are obvious, it is better to organise these influencing parameters.

Example A-7.10: *Investigate the effect of various physical parameters on the average value of convection coefficient in fully turbulent flow from leading edge in flow over a flat plate. Using the following equation.*

$$\overline{Nu}_L = 0.037\,Re_L^{4/5}\,Pr^{1/3}$$

Solution: Expanding, *i.e.*, substituting the values of physical parameters for the dimensionless numbers.

$$\frac{\bar{h}_L L}{k} = 0.037\,\frac{u_\infty^{4/5}\,\rho^{4/5}\,L^{4/5}}{\mu^{4/5}}\,\frac{c^{1/3}\,\mu^{1/3}}{k^{1/3}}$$

$$\bar{h}_L = 0.037\,k^{2/3}\,L^{-1/5}\,\rho^{4/5}\,c^{1/3}\,u_\infty^{4/5}\,\mu^{-7/15}$$

$$\frac{h_2}{h_1} = \left(\frac{k_2}{k_1}\right)^{2/3}\left(\frac{\rho_2}{\rho_1}\right)^{4/5}\left(\frac{c_2}{c_1}\right)^{1/3}\left(\frac{u_{\infty 2}}{u_{\infty 1}}\right)^{4/5}\left(\frac{L_1}{L_2}\right)^{1/5}\left(\frac{\mu_1}{\mu_2}\right)^{7/15}$$

The influences are similar to that in laminar flow except for the degree.

The effect of conductivity and specific heat are the same as in laminar flow.

1. Convection coefficient proportional to the 2/3 power of thermal conductivity *i.e.* h increases as $k^{2/3}$.
2. Convection coefficient is proportional to 1/3 power of specific heat *i.e.* h increases as $c^{1/3}$.
3. Convection coefficient increases as the 4/5th power of density (as compared 1/2 power) *i.e.* density increase has more pronounced effect in turbulent flow.
4. Convection coefficient increases as the 4/5th power of free stream velocity. (as compared 1/2 power) *i.e.* velocity increase has a more pronounced effect on convection coefficient in turbulent flow.
5. As the distance increases, convection coefficient decreases (inverse effect), but to a very much reduced effect *i.e.* 1/5th power only. With distance, the convection coefficient does not decrease rapidly.
6. Dynamic viscosity has a more pronounced effect in reducing convection coefficient (7/15 power instead of 1/6).

Note. Density and free stream velocity have a pronounced influence in increasing the value of h and the dynamic viscosity reduces the value of h more drastically.

With distance, the value of h does not drop rapidly in turbulent flow.

Example A-7.11: *Air at 1 atm at a temperature of 50°C flows over a plate 0.2 m long and 0.1 m wide. The Reynolds number is 40,000. (flow is along the 0.2 m side). Determine the rate of heat transfer from the plate at 100°C to air 50°C. If the velocity of flow is doubled and the pressure is increased to 5 atm, determine the percentage change.*

The properties of air are read from tables and interpolated for film temperature of 75°C.

Solution: Density: 1.0145, kg/m^3 Sp. heat: 1.009 kJ/kg K

Prandtl No: 0.693 $k = 30.065 \times 10^{-3}$ W/mK

$\mu = 20.835 \times 10^{-5}$ kg/ms or Ns/m^2

In the first case, the flow is laminar

$$\therefore \quad \bar{\mathbf{h}} = 0.664 \frac{k}{L} (Re_L)^{0.5} Pr^{0.33}$$

$$= \frac{0.664 \times 30.065 \times 10^{-3}}{0.20} (40000)^{0.5} (0.693)^{0.33} = \mathbf{17.67\ W/m^2K}$$

$$\mathbf{Q_1} = 0.2 \times 0.1 \times 17.67 \times (100 - 50) = \mathbf{17.66\ W.}$$

If necessary we can determine the velocity also

$$40000 = \frac{u_\infty \times 1.0145 \times 0.2}{20.835 \times 10^{-6}} \quad \therefore \quad u_\infty = 4.1 \text{ m/s}$$

From example 7.9.

$$\frac{h_2}{h_1} = \left(\frac{k_2}{k_1}\right)^{2/3} \left(\frac{\rho_2}{\rho_1}\right)^{1/2} \left(\frac{u_{\infty 2}}{u_{\infty 1}}\right)^{1/2} \left(\frac{c_2}{c_1}\right)^{1/2} \left(\frac{L_1}{L_2}\right)^{1/2} \left(\frac{\mu_1}{\mu_2}\right)^{1/6}$$

as c, L, μ and k remain the same and

Chapter 7

$$u_{\infty 2} = 2\, u_{\infty 1} \text{ and } \frac{\rho_2}{\rho_1} = 5 \text{ (as temp. is constant)}$$

$$\mathbf{h_2} = 17.66 \times 5^{0.5}\, 2^{0.5} = \mathbf{55.85\ W/m^2\ K}$$

$$\mathbf{Q_2} = 0.2 \times 0.1 \times 55.85 \times (100 - 50) = \mathbf{55.85\ W}$$

Percentage change $= 100\ (55.85/17.67) = \mathbf{316.2\%}$

Check for laminar flow

$$Re = \frac{0.2 \times 1.0145 \times 5 \times 4.1 \times 2}{20.835 \times 10^{-6}} = 4 \times 10^5. \text{ Laminar.}$$

COMMENT

It is better to solve the problem from basics, calculating *Re* etc., as it is rather difficult to remember the long equation.

Example A-7.12: *A rectangular plate measuring L × 2L has air flowing over the surface at a film temperatures of T_f and free stream velocity of u_∞. The heat transfer from the plate to air is the same irrespective of the flow being parallel to L or parallel to 2L. Assuming critical Reynolds number to be 5×10^5, and assuming that the flow is laminar when the flow direction is along L direction, determine the outlet Reyonlds number for flow in the L direction.*

Solution: As the heat flow and temperature difference and area are the same the value of average convective heat transfer coefficient has to be the same in both cases.

If both flow are laminar, this is not possible as the average convection coefficient in the laminar region goes on decreasing with distance. So the correlation for average convection coefficient for both laminar and turbulent flows have to be used. Let us assume that the Reynolds number in the *L* direction is *Re*. Then the Reynolds number in the 2*L* direction will be 2 *Re*. The correlations are:

L direction:

Laminar $$Nu_L = \frac{hL}{k} = 0.664\ Re^{0.5}\ Pr^{0.33}$$

2*L* direction combined laminar (from chapter 8)

and turbulent $$Nu_{2L} = \frac{h\,2L}{k} = [0.037\ (2\ Re)^{0.8} - 871)]\ Pr^{0.33} \text{ (data book)}$$

Equating

$$2 \times 0.664\ Re^{0.5}\ Pr^{0.33} = [0.037\ (2\ Re)^{0.8} - 871)]Pr^{0.33}$$

$$1.328\ Re^{0.5} = 0.037\ (2\ Re)^{0.8} - 871$$

Solving by Trial: by assuming values for *Re*,

(2×10^5 the residue is – 343

4×10^5 the residue is + 241).

By trial between these values, $\mathbf{Re = 3.161 \times 10^5}$

Check: $\mathbf{Nu_L} = 0.664 \times (3.161 \times 10^5)^{0.5}\ Pr^{0.33} = \mathbf{373.33\ Pr^{0.33}}$

$\mathbf{Nu_{2L}} = [0.037\ (2 \times 3.161 \times 10^5)^{0.8} - 871]\ Pr^{0.33} = \mathbf{746.65\ Pr^{0.33}}$

As $Nu_{2L} = 2Nu_L$, the value checks.

Example A-7.13: *Water at 20°C flows over a plate 0.4 m long at 60°C with a free stream velocity of 0.6 m/s. Check the change in the value of h if instead of the film temperature, the free stream temperature property values are used.*

Solution: Film temperature = (60 + 20)/2 = 40°C.

at 20°C: density ρ = 1000 kg/m^2 Pr = 7.020, $\mu = 1.006 \times 10^{-3}$ kg/ms

c = 4178 J/kg K, k = 0.5978 W/mK

at 40°C: ρ = 995 kg/m^3 Pr = 4.340, k = 0.628 W/mK

c = 4178 J/kg K, $\mu = 0.654 \times 10^{-3}$ kg/ms

Using the expression derived in example 7.9 in the laminar region

$$\frac{h_2}{h_1} = \left(\frac{k_2}{k_1}\right)^{2/3} \left(\frac{\rho_2}{\rho_1}\right)^{1/2} \left(\frac{u_{\infty 2}}{u_{\infty 1}}\right)^{1/2} \left(\frac{c_2}{c_1}\right)^{1/2} \left(\frac{L_1}{L_2}\right)^{1/2} \left(\frac{\mu_1}{\mu_2}\right)^{1/6}$$

or preferably we can also use the direct calculation of h.

Substituting the values (as u_∞ c and L remains constant)

$$\frac{\mathbf{h_2}}{\mathbf{h_1}} = \left[\frac{0.628}{0.5978}\right]^{2/3} \left[\frac{995}{1000}\right]^{1/2} \left[\frac{1.006 \times 10^{-3}}{0.654 \times 10^{-3}}\right]^{1/6} = \mathbf{1.1075}$$

or under estimation by about 10% (using property values at 20 °C)

First check for laminar flow: (By direct method)

$$Re = \frac{\rho u_\infty x}{\mu} = 1000 \times 0.6 \times 0.4/(1.006 \times 10^{-3}) = 2.39 \times 10^5$$

So laminar:

$$\overline{\mathbf{h}}_1 = \frac{2 \times 0.664 \times 0.5978}{0.4} (2.39 \times 10^5)^{0.5} (7.02)^{0.333} = \mathbf{1856.15\ W/m^2K}$$

$$\overline{\mathbf{h}}_2 = \frac{2 \times 0.664 \times 0.628}{0.4} \left(\frac{995 \times 0.6 \times 0.4}{0.654 \times 10^{-3}}\right)^{0.5} . (4.34)^{0.333} = \mathbf{2055.05\ W/m^2K}$$

$$\textbf{\% under estimate} = 100 \left(\frac{2055.05 - 1856.15}{2055.05}\right) = \mathbf{9.68\%.}$$

Example A-7.14: *A thin conducting plate separates two parallel air streams. The hot stream is at 200°C and 1 atm pressure. The free stream velocity is 15 m/s. The cold stream is at 20°C and 2 atm pressure and the free stream velocity is 5 m/s. Determine the heat flux at the mid point of the plate of 1 m length.*

Solution: The plate temperature has to be assumed to determine the film temperature. As the temperature drop on each side will be inversely proportional to the value of convection coefficient, the ratio of convection coefficients can be a guidance in assuming the plate temperature.

Using example 7.9, as the pressure is 2 atm, $p_2/p_1 = 2$

$$\frac{h_2}{h_1} = \left(\frac{u_{\infty 2}}{u_{\infty 1}} \frac{\rho_2}{\rho_1}\right)^{1/2} = \left(\frac{5}{15} \frac{2}{1}\right)^{1/2} = 0.8 \text{ (approx.)}$$

Hence an increase of 100°C on the cold side and a drop of 80°C on the hot side are assumed, fixing the plate temperature as 120°C. The film temperatures are:

(200 + 120)2 = 160°C, (120 + 20)/2 = 70°C. The property values are read and are given below. On the cold side, the pressure is 2 atm. So dynamic viscosity μ value should be used with density value at 2 atm.

at 160°C: $\nu = 30.09 \times 10^{-6}$ m²/s, $Pr = 0.682$, $k = 36.4 \times 10^{-3}$ W/mK,

at 70°C: $\rho = 1.029 \times 2$, $Pr = 0.694$, $k = 29.66 \times 10^{-3}$ W/mK,

$\mu = 20.59 \times 10^{-6}$ kg/ms

Check for laminar flow at 0.5 m : $Re_h = \dfrac{15 \times 0.5}{30.09 \times 10^{-6}} = 2.49 \times 10^5$, laminar

$$Re_c = \frac{5 \times 0.5 \times 2 \times 1.029}{20.59 \times 10^{-6}} = 2.5 \times 10^5, \text{ laminar}$$

$$h_{hot} = \frac{0.664 \times 36.4 \times 10^{-3}}{0.5} (2.49 \times 10^5)^{0.5} (0.682)^{0.333} = 21.24 \text{ W/m}^2\text{K}$$

$$h_{cold} = \frac{0.664 \times 29.66 \times 10^{-3}}{0.5} (2.5 \times 10^5)^{0.5} (0.694)^{0.333} = 17.43 \text{ W/m}^2\text{K}$$

Equating the heat flow

$$21.24 (200 - T_1) = 17.43 (T_1 - 20)$$

Plate temperature ∴ $\mathbf{T_1 = 118.9°C}$

This is close to the value assumed and hence, no need for repeating the calculation with this assumed value.

Heat flux on the hot side = 21.24 (200 – 118.9) = **1723 W**

Check: On the cold side 17.43 (118.9 – 20) = 1723.8 W.

Example A-7.15: *Air at 200°C flows over a plane at 120°C. The air pressure is 1.8 atm. The free stream velocity is 15 m/s. The plate measures 0.2 m × 0.4 m. The air flow was designed to be in the direction of 0.4 m side. Due to some misunderstanding the equipment was fabricated with the flow in the 0.2 m direction. Determine the change in performance.*

Solution: Originally as per design $L = 0.4$ m. Now in the fabrication $L = 0.2$ m.

The film temperature is (200 + 120)/2 = 160°C

As pressure is different from 1 atm, density and dynamic viscosity should be used in the determination of Reynolds number.

The property values are read from tables.

Density $\rho = 0.815 \times 1.8$ kg/m³, $Pr = 0.682$

$k = 36.4 \times 10^{-3}$, $\mu = 24.53 \times 10^{-6}$ kg/ms

$L = 0.4$

$$\overline{\mathbf{h}}_{0.4} = \left[\frac{2 \times 0.644 \times 36.4 \times 10^{-3}}{0.4}\right]\left[\frac{15 \times 0.4 \times 0.815 \times 1.8}{24.52 \times 10^{-6}}\right]^{0.5} 0.682^{0.33}$$

$= \mathbf{61.81\ W/m^2K}$ $\quad Re = 3.59 \times 10^5$ ∴ laminar

$$\overline{\mathbf{h}}_{0.2} = \left[\frac{2 \times 0.644 \times 36.4 \times 10^{-3}}{0.2}\right]\left[\frac{15 \times 0.2 \times 0.815 \times 1.8}{24.52 \times 10^{-6}}\right]^{0.5} 0.682^{0.33} = \mathbf{87.41\ W/m^2K}$$

The heat transfer rate will therefore increase.

The % increase is $\dfrac{100\,(87.41 - 61.81)}{61.81} = 41.43\%$

Now the flow rate has to be doubled because the flow width is doubled if the velocity is to be maintained. If the flow rate remains the same, the $u_\infty = 15/2$. In that case,

$$\overline{\mathbf{h}}_{0.2} = \frac{2 \times 0.644 \times 36.4 \times 10^{-3}}{0.2}\left[\frac{15 \times 0.2 \times 0.815 \times 1.8}{24.52 \times 10^{-6}}\right]^{0.5} 0.682^{0.33} = \mathbf{61.81\ W/m^2K}$$

There is no change in the value of heat transfer. (Refer example 7.12)

This can be **checked** using equation in example 7.9.

$$\frac{h_2}{h_1} = \left(\frac{u_{\infty 1}}{u_{\infty 2}}\right)^{1/2}\left(\frac{L_1}{L_2}\right)^{1/2} = \left(\frac{2}{1}\right)^{1/2}\left(\frac{1}{2}\right)^{1/2} = 1.$$

Example A-7.16: *Using the method of dimensional analysis, obtain the relevant dimensionless groups in forced convection, to obtain convection coefficient.*

The physical quantities pertinent to forced convection are listed below, together with dimensions.

Physical quantity	*Primary dimension*
(1) Flow length, L, m	L
(2) Free stream velocity u_∞, m/s	L/t
(3) Fluid density ρ, kg/m³	m/L^3
(4) Fluid viscosity μ, kg/ms	m/Lt
(5) Thermal conductivity k, W/mK $\left(N = kg\,\dfrac{m}{s^2}\right)$	mL/t^3T ($W = Nm/s\ mL^2/t^3$)
(6) Specific heat c, J/kg K	$L^2/t^2\,T$ ($J = Nm \rightarrow mL^2/t^2$)
(7) Convection coefficient h, W/m²K	m/t^3T

(1) t–time (2) m–mass, (3) T–Temperature (4) L–Length.

Solution: There are seven physical quantities and four dimensions. Hence three π quantities can be identified.

This can be done by two methods, the simpler of which is used here. All the physical parameters together has to also form a dimensionless quantity. Calling this as π, we can write

$$\pi = L^a\, u_\infty^{\ b}\, \rho^c\, \mu^d\, k^e\, c^f\, h^g \qquad ...(a)$$

$$= L^a \left[\frac{L}{t}\right]^b \left[\frac{m}{L^3}\right]^c \left[\frac{m}{Lt}\right]^d \left[\frac{mL}{t^3T}\right]^e \left[\frac{L^2}{t^2T}\right]^f \left[\frac{m}{t^3T}\right]^g$$

Chapter 7

Collecting the indices of each primary dimension.

$$L \rightarrow a + b - 3c - d + e + 2f = 0$$
$$m \rightarrow c + d + e + g = 0$$
$$t \rightarrow -b - d - 3e - 2f - 3g = 0$$
$$T \rightarrow -e - g - f = 0$$

There are seven unknowns and only four equations. So three values should be arbitrarily chosen. This should be done in such a way that by this assumption, two, of the equations are not reduced to be the same. Only then the four unknowns can be solved for.

Some general rules are:

1. The values arbitrarily fixed should be such that the primary quantity occurs only once. In this case h is the primary quantity. So the value of g should have a non-zero value only once.
2. The remaining parameters (after the choice of arbitrary values) should result in four equation *i.e.* they should involve all the primary dimensions used. At least one of the chosen value should be non-zero.
3. The choice should not lead to the same equation or multiples of it, *i.e.* the coefficient matrix should be non singular.

h is the quantity to be solved for. For the first dimensionless number let the index for h be assumed as 1 *i.e.* $g = 1$. Let us assume $b = 0$ and $c = 0$. The equations reduce to

$$a - d + e + 2f = 0$$
$$d + e + 1 = 0 \qquad \text{eqns. not repeated,}$$
$$-d - 3e - 2t - 3 = 0$$
$$-e - f - 1 = 0$$

solving $a = 1, \quad e = -1, \quad f = d = 0$

$$\therefore \qquad \pi_1 = L\, k^{-1}\, h = \frac{\mathbf{hL}}{\mathbf{k}} \quad \textbf{(Nusselt number)}$$

Not to repeat h, the next choice is made with $g = 0$, and $a = 1$ and $f = 0$ to obtain

$$1 + b - 3c - d + e = 0$$
$$c + d + e = 0$$
$$-b - d - 3c = 0 \qquad \text{no equation repeated}$$
$$-e = 0$$

This gives $\quad e = 0, b = 1, c = 1$ and $d = -1, (a = 1, f = 0, g = 0)$

$$\therefore \qquad \pi_2 = L.\, u_\infty\, \rho\, \mu^{-1} = \rho\, \mathbf{u}_\infty\, \mathbf{L}/\mu \qquad \textbf{(Reynolds numbers)}$$

Finally we let $d = 1, b = 0, g = 0$

The equations reduce to

$$a - 3c - 1 + e + 2f = 0$$
$$c + 1 + e = 0 \qquad \text{no equation repeated}$$
$$-1 - 3e - 2f = 0$$
$$-e - f = 0$$

This gives $f = 1, e = -1$, and $d = 1$ and other terms $a = 0$ and $c = 0, b = 0, g = 0$

This gives $\pi_3 = \frac{c\mu}{k}$ **(Prandtl number)**

In the case of flow through pipe another physical parameter namely diameter is involved. This will give another dimensionless number D/L, which will be later seen in the correlations. In equation (a) add the parameter D to obtain the solution.

Obviously skill is required in the choice of the first three terms and their values.

Example A-7.17: *Using dimensional analysis obtain a correlation for free convection.*

The physical parameters involved are:

(1) A length parameter L,	L
(2) density r, kg/m^3	m/L^3
(3) Specific heat c, J/kg K	L^2/t^2T
(4) Thermal conductivity k, W/mK $\left(N = kg\,\frac{m}{s^2}\right)$	mL/t^3T
(5) Viscosity m, kg/ms	m/Lt
(6) Coefficient of cubical expansion b, 1/°C	$1/T$
(7) Temperature difference causing density change ΔT, °C	T
(8) gravitational acceleration, g, m/s^2	L/t^2
(9) Convection coefficient h, W/m^{2}K	m/t^3T

Solution: The physical parameters are nine in number. The primary dimensions involved are four. Hence five π quantities can be found. The primary quantity is h. The dimensions involved are tabulated below.

	L	ρ	c	k	μ	β	ΔT	g	h
Mass	–		–	√	√	–	–	–	À
Length	√	À	À	√	√	–	–	√	–
time	–	–	√	√	√	–	–	√	√
temp.	–	–	√	√	–	À	√	–	

We can now write $\pi = L^a \rho^b c^c k^d \mu^e \beta^f \Delta T^g g^h h^i$

$$= L^a \left[\frac{m}{L^3}\right]^b \left[\frac{L^2}{t^2T}\right]^c \left[\frac{mL}{t^3T}\right]^d \left[\frac{m}{Lt}\right]^e \left[\frac{1}{T}\right]^f T^g \left[\frac{L}{t^2}\right]^h \left[\frac{m}{t^3T}\right]^i$$

Combining for each dimension,

for L $\quad a - 3b + 2c + d - e + h = 0$ ...(i)

for m $\quad b + d + e + i = 0$ ...(ii)

for t $\quad - 2c - 3d - e - 2h - 3 = 0$ ...(iii)

for T $\quad - e - d - f + g - i = 0$ ...(iv)

As h is the primary quantity $i = 1$ first and $a = 1$, $d = -1$, $g = 0$, $h = 0$ are arbitrarily chosen.

Chapter 7

The equations (*i*)–(*iv*) reduce to

$$1 - 3b + 2c - 1 - e = 0$$
$$b - 1 + e + 1 = 0$$
$$-2c + 3 - e - 3 = 0$$
$$-c + 1 - f - 1 = 0$$

Solving $b = 0$, $e = 0$ and $f = 0$

This gives $\pi_1 = L^1\, k^{-1}\, h^1 = \frac{\mathbf{hL}}{\mathbf{k}}$, Nusselt number

$i = 0$ in all other cases. The following arbitrary values are chosen.

$i = 0, f = 0, e = 1, d = -1, c = 1$

Then the set of equations reduce to

$$a - 3b + 2 - 1 - 1 = 0$$
$$b - 1 + 1 = 0$$
$$-2 + 3 - 1 - 2h = 0$$
$$-1 + 1 + g = 0$$

Solving $a = 0, b = 0, g = 0, h = 0$

This gives $\pi_2 = c^1\, k^{-1}\, \mu^1 = \frac{\mathbf{c}}{\mathbf{k}}\, \mu$, **Prandtl number**

other choices like $a = 0, i = 0, b = 0, f = 0, c = 1$ will also yield the same result. Try this.

Next the following choice is made

$$i = 0, f = 1, g = 1, a = 0, b = 0$$

This gives

$$2c + d - e + h = 0$$
$$d + e = 0$$
$$-2c - 3d - e - 2h = 0$$
$$-c - d - 1 + 1 = 0$$

Solving $c = 0, d = 0, e = 0, h = 0$

This gives $\pi_3 = \beta\, \Delta T$

Next we choose $i = 0, f = 0, g = 0, c = 0, h = 1$

The equations for L, m, t and T reduce to

$$a - 3b + d - e + 1 = 0$$
$$b + d + e = 0$$
$$-3d - e - 2 = 0$$
$$-d = 0$$

Solving $a = 3, b = 2, e = -2, d = 0$

This gives $\pi_4 = \frac{L^3 \rho^2}{\mu^2}\, g$. This can be checked to be dimensionless

The product $\pi_3 \times \pi_4 = \pi_5 = g\, \beta\, \Delta T L^3\, \rho^2/\mu^2$ **known as Grashof number**

The product $\pi_2 \times \pi_3$ is known as **Rayleigh number** and is given by

$$\text{Ra} = g\, \beta\, \Delta T L^3 \rho^2\, c/\mu k$$

The work may appear tedious, but such work only has led to the correlation of xperimental results into usable relationships for design.

Example A-7.18: *Using the following relationships 1, 2 and 3 for the velocity boundary layer, derive a general expression for the y directional velocity v and obtain therefrom the value at y = δ.*

Solution: Assume continuity equation: $\frac{\partial u}{\partial x} + \frac{\partial v}{\partial y} = 0$ ...(1)

Velocity distribution: $$\frac{u}{u_\infty} = \frac{3}{2}\frac{y}{\delta} - \frac{1}{2}\left(\frac{y}{\delta}\right)^3$$ ...(2)

Boundary layer thickness $$= 4.64\, x/\sqrt{Re_x} = \frac{4.64 x^{1/2}\, \nu^{1/2}}{u_\infty^{1/2}}$$ ...(3)

Also obtain a relationship for the flow out of the boundary layer.

First equation (3) is substituted in equation (2) to obtain a relationship between u_∞ and x.

$$\frac{u}{u_\infty} = \frac{3}{2} \cdot \frac{y u_\infty^{1/2}\, x^{-1/2}}{4.64\, \nu^{1/2}} - \frac{1}{2} \cdot \frac{y^3 . u_\infty^{3/2}\, x^{-3/2}}{4.64^3\, \nu^{3/2}}$$

$$\frac{\partial u}{\partial x} = u_\infty \left[\frac{3}{2} \cdot \left(-\frac{1}{2}\right) \cdot \frac{y\, u_\infty^{1/2}}{4.64 \nu^{1/2}\, x^{3/2}} - \frac{1}{2}\left(-\frac{3}{2}\right) \cdot \frac{y^3 . u_\infty^{3/2}}{4.64^3\, \nu^{3/2}} . x^{-5/2}\right]$$

Substituting in equation (1)

$$\frac{\partial v}{\partial y} = -\frac{\partial u}{\partial x} = u_\infty \left[\frac{3}{4} \cdot \frac{u_\infty^{1/2}\, x^{-3/2}}{4.64\, \nu^{1/2}} . y - \frac{3}{4} \cdot \frac{u_\infty^{3/2}\, x^{-5/2}}{4.64^3\, \nu^{3/2}} . y^3\right]$$

Integrating $$\frac{v}{u_\infty} = \left[\frac{3}{4} \cdot \frac{1}{2} \frac{u_\infty^{1/2}\, x^{-3/2}}{4.64\, \nu^{1/2}} . y^2 - \frac{3}{4} \cdot \frac{1}{4} \frac{u_\infty^{3/2}\, x^{-5/2}}{4.64\, \nu^{3/2}} . y^4\right]$$

Using the value of $$\delta = 4.64 \frac{x^{1/2}\, \nu^{1/2}}{u_\infty^{1/2}}$$

$$\frac{v}{u_\infty} = \frac{3}{8} \cdot \frac{y^2}{x\delta} - \frac{3}{16} \cdot \frac{y^4}{\delta^3 x}$$ ...(A)

This is the general expression for y directional velocity v.

at $y = \delta$, $$\frac{v}{u_\infty} = \frac{3}{16}\frac{\delta}{x} = \frac{3}{16} \frac{4.64\, v^{1/2}\, x^{-1/2}}{u_\infty^{1/2}}$$

$$\therefore \quad v_\delta = \frac{3}{16} . 4.64 . u_\infty^{1/2}\, \nu^{1/2}\, x^{-1/2}$$

This is dependent only on x.

The flow out of the boundary layer *i.e.* at $y = \delta$, up to L can be determined by integrating the expression:

$$m_y = \int_0^L \rho\, v_\delta\, dx = \frac{3}{16} \times 4.64 \times u_\infty^{1/2}\, \nu^{1/2} \int_0^L x^{-1/2}\, dx$$

$$= \frac{3}{16} 4.64\, L\, \rho\, u_\infty^{1/2}\, \nu^{1/2} \left[\frac{x^{1/2}}{(1/2)}\right]_0^L$$

$$= \frac{3}{8} 4.64\, L\, \rho\, u_\infty \left[\frac{\nu^{1/2}}{L^{1/2} u_\infty^{1/2}}\right]$$

$$= \frac{3}{8} \rho\, u_\infty\, \delta_L, \text{ (as } \delta_L = 4.64\, L/Re_L^{0.5}) \qquad \text{...(B)}$$

This can also be verified using the velocity distribution to determine the flow through the boundary layer. This expression (B) gives the amount of fluid moving out of the boundary layer into the free stream.

At $x = L$, along x

$$m_\delta = \int_0^\delta \rho\, u\, dy = \rho\, u_\infty \int_0^\delta \left[\frac{3}{2}\frac{y}{\delta} - \frac{1}{2}\left(\frac{y}{\delta}\right)^3\right] dy$$

$$= \rho\, u_\infty \left[\frac{3}{4}\frac{y^2}{\delta} - \frac{1}{8}\frac{y^4}{\delta^3}\right]$$

at $y = \delta$, $m_\delta = \dfrac{5}{8} \rho\, u_\infty\, \delta$

Flow in the free stream of depth $\delta = \rho\, u_\infty\, \delta$

$\therefore$ difference $= \dfrac{3}{8} \rho\, u_\infty\, \delta$. As already seen this is the flow in the y direction into the free stream.

Example A-7.19: *Show that in the case of incompressible boundary layer flow with negligible pressure gradient, $\partial^3 u/\partial y^3 = 0$ at $y = 0$.*

The momentum equation for the boundary layer is

$$u\frac{\partial u}{\partial x} + v\frac{\partial u}{\partial y} = \nu\frac{\partial^2 u}{\partial y^2}.$$

Solution: Differentiating with respect to y

$$\frac{\partial u}{\partial y}\cdot\frac{\partial u}{\partial x} + u\,.\frac{\partial^2 u}{\partial x \partial y} + \frac{\partial v}{\partial y}\cdot\frac{\partial u}{\partial y} + v\frac{\partial^2 u}{\partial y^2} = \nu\frac{\partial^3 u}{\partial y^3}$$

$$\frac{\partial u}{\partial y}\cdot\left(\frac{\partial u}{\partial x} + \frac{\partial v}{\partial y}\right) + u\frac{\partial^2 u}{\partial x \partial y} + v\frac{\partial^2 u}{\partial y^2} = \nu\frac{\partial^3 u}{\partial y^3}$$

As $\dfrac{\partial u}{\partial x} + \dfrac{\partial v}{\partial y} = 0$ and as both $u = 0$ and $v = 0$ at $y = 0$ the L.H.S. is zero.

So $$\frac{\partial^3 u}{\partial y^3} = 0 \text{ at } y = 0$$

That $v = 0$ at $y = 0$ can also be seen from equation A in example A-7.18. The assumed velocity profile leads to

$$\frac{u}{u_\infty} = \frac{3}{2}\frac{y}{\delta} - \frac{1}{2}\left(\frac{y}{\delta}\right)^3$$

$$\frac{\partial u}{\partial y} = u_\infty\left[\frac{3}{2}\cdot\frac{1}{\delta} - \frac{1}{2}3y^2\frac{1}{\delta^3}\right]$$

$$\frac{\partial^2 u}{\partial y^2} = u_\infty\left[-\frac{6}{2}y\frac{1}{\delta^3}\right]$$

$$\frac{\partial^3 u}{\partial y^3} = -3\,u_\infty/\delta^3$$

This is not zero. This shows that the profile assumed is approximate.

Example A-7.20: *Derive the integral momentum equation for the boundary layer over a flat plate.*

Solution: Considering the volume *ABCD*, with unit depth, and with *BC* well removed from boundary layer

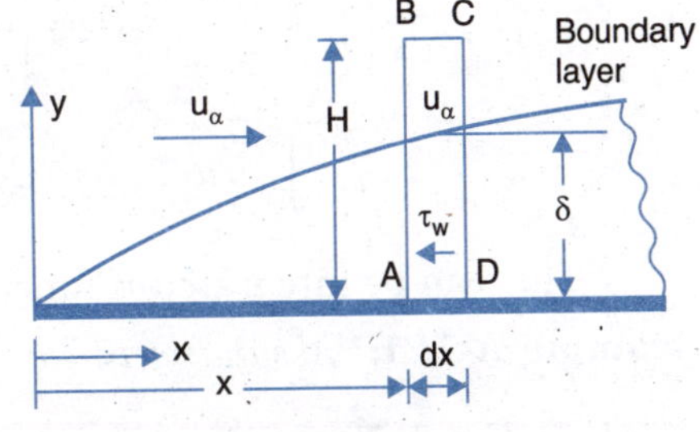

Fig. AE-7.20

Mass flow in through $AB = \int_0^H \rho\, u\, dy$

Mass flow out through

$$CD = \int_0^H \rho\, u\, dy + \frac{d}{dx}\left[\int_0^H \rho\, u\, dy\right] dx$$

The difference has to come through only *BC*, as *AD* is the wall surface

$$\therefore \quad \text{Mass flow in through } BC = \frac{d}{dx}\left[\int_0^H \rho\, u\, dy\right] dx$$

x directional momentum flow in through

$$AB = \int_0^H \rho\, u\,.\,u\, dy$$

x directional momentum flow in through *BC*, as x directional velocity well removed from boundary layer is u_∞.

$$= \frac{d}{dx}\left[\int_0^H \rho\, u\,.\,u_\infty\, dy\right] dx$$

x directional momentum flow out through *CD*

$$= \int_0^H \rho\, u^2\, dy + \frac{d}{dx}\left[\int \rho\, u^2\, dy\right] dx$$

Chapter 7

The net momentum flow

$$= \frac{d}{dx}\left[\int_0^H \rho u^2\, dy\right] dx - \frac{d}{dx}\left[\int_0^H \rho u u_\infty\, dy\right] dx$$

$$= -\frac{d}{dx}\left[\int_0^H \rho u(u_\infty - u)\, dy\right] dx$$

As $(u_\infty - u) = 0$ or negligible beyond, δ, the limit for integration can now be δ instead of H.

$\therefore$ Net momentum flow $= -\frac{d}{dx}\left[\int_0^\delta \rho u(u_\infty - u)\, dy\right] dx$

As the net force (neglecting pressure gradient) is $-\tau_\omega$

$$-\mu \left.\frac{du}{dy}\right|_{y=0} dx$$

$\therefore$ The momentum equation is given by

$$\frac{d}{dx}\left[\int_0^\delta \rho u(u_\infty - u)\, dy\right] dx = \mu \left.\frac{du}{dy}\right|_{y=0} dx$$

or

$$\frac{d}{dx}\left[\int_0^\delta \rho u(u_\infty - u)\, dy\right] = \mu \left.\frac{du}{dy}\right|_{y=0}$$

This can be integrated if function $u = f(y)$ is known or specified.

Example A-7.21: *Assuming linear variation of velocity in the boundary layer i.e.* $\frac{u}{u_\infty} = \frac{y}{\delta}$, *upto* $y = \delta$, *and then* $u = u_\infty$, *determine using the integral momentum equation, the boundary layer thickness.*

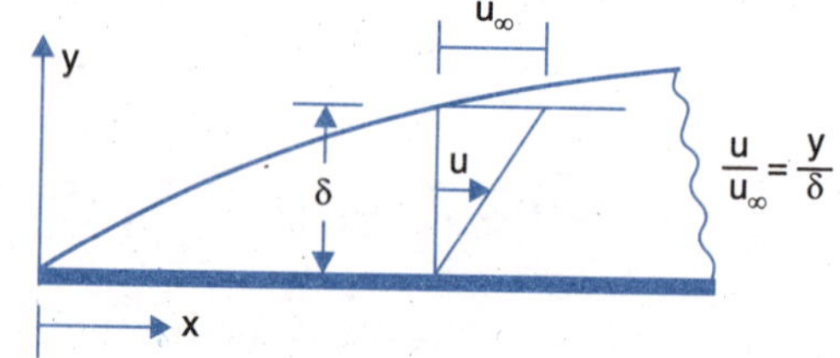

Fig. AE-7.21

Solution: $\frac{u}{u_\infty} = \frac{y}{\delta}$, $\therefore$ $u = \frac{u_\infty y}{\delta}$.

Substituting in the integral momentum equation

$$\frac{d}{dx}\left[\int_0^\delta \rho u(u_\infty - u)\, dy\right] = \mu \left.\frac{du}{dy}\right|_{y=0} \quad \text{...(A)}$$

Evaluating the integral

$$\int_0^\delta \rho \frac{u_\infty y}{\delta}\left(u_\infty - \frac{u_\infty y}{\delta}\right) dy = \int_0^\delta \frac{\rho u_\infty^2}{\delta} y\, dy - \int_0^\delta \frac{\rho u_\infty^2}{\delta^2} y^2\, dy$$

$$= \left[\frac{\rho u_\infty y^2}{2\delta} - \frac{\rho u_\infty^2 y^3}{3\delta^2}\right]_0^\delta = \frac{1}{6}\rho u_\infty^2 \delta$$

The R.H.S. $\left.\frac{du}{dy}\right|_{y=0} = \frac{u_\infty}{\delta}$

Substituting in A

$$\frac{1}{6}\,\rho\,u_\infty^2\,\delta\,.\,d\delta = \mu\,u_\infty\,dx$$

Integrating $$\frac{1}{12}\,\rho\,u_\infty^2\delta^2 = \mu\,u_\infty\,x + C$$

As $\delta = 0$ at $x = 0$, $C = 0$

$$\delta^2 = \frac{12\,\mu x^2}{\rho u_\infty x} = 12x^2\,(\nu/u_\infty\,x)$$

$\therefore$ $$\delta = \sqrt{12}\;x\,/\,\sqrt{Re_x} = \mathbf{3.46}\;\boldsymbol{x/\sqrt{Re_x}}$$

The constant is 3.46 in the place of 5 for exact solution. A cubic fit will increse this to 4.64

$$\mathbf{C_{fx}} = \tau_s/(\rho u^2/2) = \mu\left.\frac{du}{dy}\right|_{y=0}\cdot\frac{2}{\rho u_\infty^2}$$

$$= \mu\,\frac{u_\infty}{\delta}\cdot\frac{2}{\rho u_\infty^2} = \frac{\mu\,.\,2\sqrt{Re_x}}{\rho u_\infty\times 3.46\,x} = \mathbf{0.55/Re_x^{0.5}}$$

compare with 0.664 $Re_x^{-0.5}$ by the exact analysis.

Example A-7.22: *Derive the integral energy equation for the thermal boundary layer over a flat plate.*

Solution: Considering the volume *ABCD* with *BC* well removed from the thermal boundary layer, and neglecting viscous work.

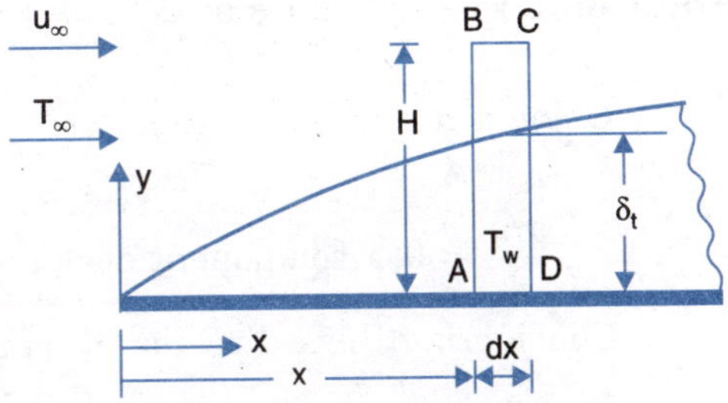

Mass flow through

$$BC = \frac{d}{dx}\left[\int_0^H \rho u\,dy\right]dx$$

(Ref. previous example)

Fig. AE-7.22

Energy flow in through $AB = \int_0^H c\rho\,u\,T\,dy$

Energy flow through *BC* (T_∞ is used as the entering fluid is at T_∞)

$$= \frac{d}{dx}\left[\int_0^H c\rho u\,T_\infty\,dy\right]dx$$

Energy flow through *CD*

$$= \int_0^H c\rho\,u\,T\,dy + \frac{d}{dx}\left[\int_0^H c\rho u\,T\,dy\right]dx$$

Net energy flow out of the element

$$\frac{d}{dx}\left[\int_0^H c\rho u\,T\,dy\right]dx - \frac{d}{dx}\left[\int_0^H c\rho u\,T_\infty\,dy\right]dx$$

$$= -\frac{d}{dx}\left[\int_0^H c\,\rho u(T_\infty - T)\,dy\right]dx$$

As $T_\infty - T = 0$ beyond δ_t upper limit for inegration can be changed to δ_t.

This energy flow is equal to the heat conducted at the wall surface AD, as there is no temperature gradient along the other faces.

The heat conducted $= -k \left.\dfrac{dT}{dy}\right|_{y=0} dx$

$\therefore$ The energy equation reduces to

$$\frac{d}{dx}\left[\int_0^{\delta_t} c\,\rho u(T_\infty - T)\,dy\right] = k\left.\frac{dT}{dy}\right|_{y=0}$$

This equation can be integrated if $u = f_1(y)$ and $T = f_2(y)$ are known or specified. (See standard texts for the same, a simple example is given in the next example).

Example A-7.23: *The flow of oil in a journal bearing can be approximated as a parallel flow between two large plates with one plate stationary and one plate moving. Such a flow is known as Couelte flow. Two large parallel plates are separated by 2 mm thick oil film. The upper plate moves with a constant velocity of 12 m/s. While the lower is stationary. Both plates are maintained at 20°C. Derive an expression for the velocity and temperature distribution in the oil. Also determine the location and value of maximum temperature in the oil and the heat flow to the plates.*

Solution: Property values at 20°C: k = 0.145 W/m°C, μ = 0.80 kg/ms,

Using continuity equation $\dfrac{\partial u}{\partial x} + \dfrac{\partial v}{\partial y} = 0$, with coordinates as shown, and as $v = 0$,

$\therefore \quad \dfrac{du}{dx} = 0$, x component does not change in the x direction.

Considering the x directional momentum equation,

$$\rho\left(u\frac{\partial u}{\partial x} + v\frac{\partial u}{\partial y}\right) = \mu\frac{\partial^2 u}{\partial y^2} - \frac{\partial P}{\partial x}.$$

As flow is due to the motion of plate and not due to pressure, $\dfrac{\partial P}{\partial x} = 0$,

as $\dfrac{\partial u}{\partial x} = 0$, $v = 0$, the eqn. reduces to $\mu\dfrac{\partial^2 u}{\partial y^2} = 0$, Integrating twice, $u(y) = C_1 y + C_2$.

at $y = 0$, $u = 0$, at $y = L$, $u = V$, $\therefore$ $\mathbf{u = \dfrac{y}{L} \times V}$ ...(A)

Heat is generated by friction heating due to viscosity.

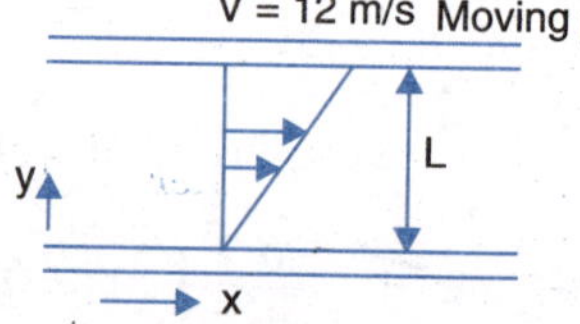

Fig. AE-7.23

Energy equation reduces to $k\dfrac{\partial^2 T}{\partial y^2} = -\mu\left(\dfrac{V}{L}\right)^2$

Integrating after dividing both sides by $\dfrac{V}{L}$,

$$T(y) = -\frac{\mu}{2k}\left(\frac{y}{L}V\right)^2 + C_3 y + C_4.$$ As $T = T_0$ at both $y = 0$, $y = L$,

$$C_4 = T_0 \text{ and } C_3 = \mu\frac{V^2}{2k}\cdot\frac{y}{L}$$

$$\mathbf{T(y) = T_0 + \mu\frac{V^2}{2k}\left(\frac{y}{L} - \frac{y^2}{L^2}\right)}, \textbf{ to get the heat flow,}$$

$$\frac{dT}{dy} = \frac{\mu V^2}{2kL}\left(1 - \frac{2y}{L}\right), \text{ This is zero at } y = \frac{L}{2} \quad \therefore \quad \text{Maximum is at } \left(\frac{L}{2}\right).$$

$$\mathbf{T_{max} = T_0 + \frac{\mu V^2}{2k}\left[\frac{(L/2)}{L} - \left(\frac{L/2}{L}\right)^2\right]} = T_0 + \frac{\mu V^2}{8k} = 20 + \frac{0.8\times 12^2}{8\times 0.145}$$

$$= \mathbf{119°C.}$$

at $\quad y = 0 \; q = -k\left.\frac{dT}{dy}\right|_{y=0} = -k\frac{\mu V^2}{2kL}(1-0) = -\frac{\mu V^2}{2L} = \frac{0.8\times 12^2}{2\times 0.002}$

$= \mathbf{-\,28800\ W/m^2}$

at $\quad y = L$, This works as **28,800 W/m²**.

The heat flux at the two plates are equal in magnitude but flow direction is opposite.

Example A-7.24: *Assuming the following velocity and temperature profiles and the integral energy equation derive an expression for the local heat transfer coefficient*

(i) $u = u_\infty$ *for all values*

(ii) $\dfrac{T - T_\infty}{T_w - T_\infty} = \dfrac{y}{\delta}$, δ – *thermal boundary layer thickness.*

Solution: (*i*) The energy equation is

$$\frac{d}{dx}\left[\int_0^\delta c\,\rho u(T - T_\infty)\,dy\right] = k\left.\frac{dT}{dy}\right|_{y=0}$$

Substituting the values of u and T in the L.H.S. integral

(*ii*) $$\int_0^\delta \frac{c\,\rho u_\infty\,(T_w - T_\infty)}{\delta}\,.\,y\,.\,dy = \left[\frac{c\,\rho u_\infty\,(T_w - T_\infty)}{\delta}\frac{y^2}{2}\right]_0^\delta$$

$$= \frac{c\rho u_\infty\,(T_w - T_\infty)\,\delta}{2}$$

$$\text{R.H.S., } k\left.\frac{dT}{dy}\right|_{y=0} = k\,\frac{(T_w - T_\infty)}{\delta}$$

$$\frac{d}{dx}\left[\frac{c\rho u_\infty\,(T_w - T_\infty)\,\delta}{2}\right] = k\,\frac{T_w - T_\infty}{\delta}$$

$$\therefore \qquad \delta\,d\delta = \frac{2k}{c\rho\,u_\infty}\,dx, \text{ integrating}$$

$$\delta^2 = \frac{4k}{c\rho\, u_\infty} x + C \quad \text{or} \quad \text{as } d = 0 \text{ at } x = 0, C = 0$$

$$\therefore \quad \delta^2 = 4x^2 \frac{k}{\rho c} \cdot \frac{\mu}{\mu\, u_\infty x} = 4 \times x^2/(Re_x Pr)$$

$$\therefore \quad \delta = 2x/(Re_x\, Pr)^{0.5}$$

$$h_x\,(T_w - T_\infty) = \frac{k(T_w - T_\infty)}{2\delta} \,.\, (Re_x\, Pr)^{0.5}$$

$$\therefore \quad \mathbf{\frac{hx}{k} = Nu_x = 0.5\ (Re_x\ Pr)^{0.5}}$$

Compare with relations for liquid metal given by

$Nu_x = 0.565\ (Re_x\ Pr)^{0.5}$ for $Pr < 0.05$ and $Re < 5 \times 10^5$ (for liquid metals, as the velocity boundary layer is very thin $u = u_\infty$ is valid).

Example A-7.25: *Assuming the following velocity and temperature profiles and the integral form of the energy equation, determine the value of the local convection coefficient*

(*i*) $u = u_\infty$ *for all y alues*

(*ii*) $\dfrac{T - T_\infty}{T_w - T_\infty} = \dfrac{3}{2}\dfrac{y}{\delta} - \dfrac{1}{2}\left(\dfrac{y}{\delta}\right)^3$, *δ– Thermal boundary layer thickness.*

Solution: (*i*) The integral form of the energy equation is

$$\frac{d}{dx}\left[\int_0^\delta c\,\rho u(T - T_\infty)\,dy\right] = k\left.\frac{dT}{dy}\right|_{y=0}$$

The L.H.S. after substitution of the profiles is (the integral only)

$$\left[\int_0^\delta c\,\rho u_\infty\ (T_w - T_\infty)\right]\left[\frac{3}{2}\frac{y}{\delta} - \frac{1}{2}\left(\frac{y}{\delta}\right)^3\right] dy$$

$$c\,\rho\, u_\infty\,(T_w - T_\infty)\left[\frac{3}{4\delta}y^2 - \frac{1}{8}\frac{y^4}{\delta^3}\right]_0^\delta = c\,\rho\, u_\infty\,(T_w - T_\infty)\,\frac{5}{8}\,\delta$$

$$k\left.\frac{dT}{dy}\right|_{y=0} = k\,(T_w - T_\infty)\,.\,\frac{3}{2}\frac{1}{\delta} \quad \text{only the first term contributes}$$

$$\therefore \quad \delta\frac{d\delta}{dx} = \frac{8}{5}\frac{k}{c\rho\, u_\infty}\frac{3}{2}, \text{ integrating}$$

$$\delta^2 = \frac{48}{10}\frac{k}{c\mu}\frac{\mu}{\rho\, u_\infty\, x}x^2 \quad \therefore \quad \delta = \frac{2.191\,x}{(Re_x\ Pr)^{0.5}}$$

$$\frac{3}{2}k\,(T_w - T_\infty)\,\frac{1}{\delta} = h_x\,(T_w - T_\infty)$$

$$\therefore \qquad \frac{\mathbf{h_x x}}{\mathbf{k}} = \frac{3}{2}\frac{(Re_x\ Pr)^{0.5}}{2.191} = \mathbf{0.685\ (Re_x\ Pr)^{0.5}}$$

Compare with 0.565 $(Re_x\ Pr)^{0.5}$ correlation for liquid metals for $Pr < 0.05$ and $Re_x < 5 \times 10^5$.

REVIEW QUESTIONS

7.1 Explain the physical significance of continuity equation.

7.2 Explain, how Fourier's law is involved in the calculation of convection coefficient?

7.3 What is the important quantity that is obtained from the analysis of (*i*) hydrodynamic boundary layer (*ii*) thermal boundary layer?

7.4 Distinguish between local and average convection coefficient.

7.5 Define friction coefficient. Explain, how this is involved in heat transfer calculations?

7.6 Show that Prandtl number is equal to (υ/α).

7.7 Explain the physical significance of Reynold's number.

7.8 What is critical Reynold's number?

7.9 Explain the mechanism of momentum transfer in (*i*) Laminar flow and (*ii*) Turbulent flow.

7.10 What is eddy diffusivity of momentum and eddy diffusivity of heat? Explain, how these account for the increased rate of transport in turbulent flow.

7.11 Describe the three zones in turbulent boundary layer.

7.12 Define Nusselt number. Explain its significance.

7.13 Define Stanton number.

7.14 What is the difference between Reynolds analogy and Colburn analogy?

7.15 What is the condition for (*i*) $\delta_h > \delta_t$, (*ii*) $\delta_h = \delta_t$ and (*iii*) $\delta_h < \delta_t$?

OBJECTIVE QUESTIONS

7.1 Choose the correct statement from each question.

(*i*) The convective heat transfer coefficient in laminar flow over a flat plate

(*a*) increases if a lighter fluid is used

(*b*) increases if a higher viscosity fluid is used

(*c*) increases if higher velocities are used

(*d*) increases with distance.

(*ii*) In the boundary layer over a flat plate in laminar flow the velocity is

(*a*) zero at the boundary layer thickness

(*b*) slowly decreases from the free stream to the solid surface

(*c*) slowly increases from the free stream to the wall

(*d*) only temperature in the boundary layer will be different from that of free stream.

(*iii*) As viscosity of fluid increases the boundary layer thickness

(*a*) will increase (*b*) will decrease

(*c*) will not change

(*d*) will increase at medium values and then will decrease.

(*iv*) The temperature gradient in the fluid flowing over a flat plate

(*a*) will be zero at the surface (*b*) will be positive at the surface

(*c*) will be very steep at the surface (*d*) will be zero at the top of the boundary layer.

(*v*) The ratio of thermal to hydrodynamic boundary layer thickness varies as

(*a*) root of Reynolds number (*b*) root Nusselt number

(*c*) root of Prandtl number (*d*) one third power of Prandtl number.

(*vi*) In liquid metal flow over a flat plate ($T_s > T_\infty$)

(*a*) The hydrodynamic boundary layer will be very thick and thermal boundary layer will be very thick

(*b*) The hydrodynamic boundary layer will be very thin and thermal boundary layer will also be thin

(*c*) The two will be more or less equal

(*d*) The thermal boundary layer will be thick and hydrodynamic boundary layer will be thin.

(*vii*) In forced convection molecular diffusion causes

(*a*) momentum flow in turbulent region

(*b*) momentum flow in the laminar region

(*c*) heat flow in the turbulent region

(*d*) diffusion has no part in energy transfer.

(*viii*) Heat transfer rate

(*a*) will be higher in turbulent flow (*b*) will be lower in turbulent flow

(*c*) will depend only on the fluid (*d*) will depend only on viscosity.

(*ix*) Nusselt number is

(*a*) ratio of viscous to inertia forces

(*b*) dimensionless heat transfer coefficient

(*c*) ratio of conduction to convection resistance

(*d*) signifies the velocity gradient at the surface.

(*x*) Reynolds number is

(*a*) ratio of conduction to convection resistance

(*b*) ratio of buoyant to inertia forces

(*c*) ratio of viscous to inertia forces

(*d*) ratio of heat conducted to the heat capacity.

(*xi*) Prandtl number is

(*a*) ratio of buoyant force to inertia force

(*b*) ratio of conduction to convection resistance

(*c*) signifies the temperature gradient at the surface

(*d*) ratio of Molecular momentum diffusivity to thermal diffusivity.

(*xii*) The Stanton number is

(*a*) the dimensionless temperature gradient at the surface in convection

(*b*) mass diffused to heat diffused

(*c*) dimensionless convection coefficient

(*d*) wall heat transfer/heat transfer by convection.

Answers

7.1 (*i*) (*c*), (*ii*) (*b*), (*iii*) (*a*), (*iv*) (*d*), (*v*) (*d*), (*vi*) (*d*), (*vii*) (*b*), (*viii*) (*a*), (*ix*) (*c*), (*x*) (*c*), (*xi*) (*d*), (*xii*) (*c*)

EXERCISE PROBLEMS

7.1 Nitrogen at 1 atm and – 20°C flows over a flat plate at 20° at a free stream velocity of 15 m/s. Determine the boundary layer thickness, (both hydrodynamic and thermal) coefficient of drag and local convection heat transfer coefficient at a distance of 0.4 m from the leading edge.

7.2 Air at 10 atm and 25°C flows with free stream velocity of 6 m/s over a flat plate at 75°C. Determine the heat flux at 0.05 m intervals up to a distance where the Reynolds number is 5×10^5.

7.3 A van with a roof length of 3.6 m travels in air at 30°C. Determine the speed of travel at which the flow over the roof will turn just turbulent.

7.4 Air flows parallel to the sides of a house at – 20°C with a speed of 12 km/hr. The size of the surface is 4 m × 3 m, the flow being along the 4 m side. The wall surface is at 20°C. Determine the heat loss to air from the wall.

7.5 Compare the boundary layer thicknesses at 0.4 m for the flow of

(*i*) air and (*ii*) water, with a free stream velocity of 1 m/s. The film temperature is 40°C.

7.6 Compare the value of local heat transfer coefficient at 0.2 m for flow of (*i*) water (*ii*) engine oil at a film temperature of 40°C, with a free stream velocity of 1.5 m/s.

7.7 Compare the value of local convective heat transfer coefficient for water flow over a flat plate at a film temperature of 40°C at $Re = 5 \times 10^5$ assuming the flow to be (*i*) laminar (*ii*) turbulent.

7.8 Assuming $u/u_\infty = y/\delta$, derive an expression for the y component of velocity in the boundary layer. Use the continuity equation and $\delta = 3.64\, x/\sqrt{Re_x}$.

7.9 Assuming $u/u_\infty = \sin(\pi\; y/2\delta)$ derive an expression for the y variation of velocity in the boundary layer. Use the continuity equation and $\delta = 4.5\, x/\sqrt{Re_x}$.

7.10 Derive an expression for the y direction momentum flow at the boundary layer. Use the continuity and momentum equations.

7.11 Experimental results for convection coefficient for flow of air at a film temperature 30°C predicted a value of 15 W/m²K at a location x. If the results are to be used for predictions for flow of water at a film temperature of 30°C, for velocities of 0.1 and 0.2 times that of air, determine the value of x at which the convection coefficient will have the same value as of that of air.

7.12 The local convection coefficient for flow of air at a film temperature of 20°C and pressure of 0.6 atm at a distance of 0.5 m was determined as 12 W/m²K. Using the result predict the value of local convection coefficient for Hydrogen at a film temperature of 100°C and at 2 atm at a distance of 0.4 m. Assume equal velocity.

7.13 Helium at 0.2 atm and a film temperature of 50°C was used to cool a plate specimen at 100°C by flow over the surface evolving heat at 1 kW. It is proposed to change the medium to Hydrogen keeping the film temperature to be the same. Determine the pressure at which Hydrogen is to be used to achieve the purpose.

7.14 It is proposed to use air flow to predict the laminar flow heat transfer characteristics of Hydrogen in the pressure range of 0.1 to 8 atm with flow velocities ranging from 10 m/s to 60 m/s at a film temperature of 30°C. If the air pressure variation to be limited to 0.8 to 1.2 atm and if the film temperature is to be 20°C and the plate length is to remain the same, determine the air velocity range suitable for the test.

7.15 Hot gases at 120°C and 1 atm with property values approximating to that of air flows over a 1 m × 1 m plate parallel to a side with a free stream velocity of 30 m/s. One side of the plate is maintained at 60°C. The heat flow through the plate is no correction to be taken up by air at

20°C, flowing parallel to the hot stream with velocity of 10 m/s. Determine the pressure of the air to be used.

7.16 A mild breeze at 7 km/hr flows over the flat roof of a building. The surface temperature is 10°C while the air is at – 30°C. Determine the heat loss by convection over the roof of 3 m × 3 m size.

7.17 Water at 20°C flows over a plate at 80°C with a velocity of 3 m/s. The length of the plate is 1.2 m. Determine the value of local and average convection coefficients at the trailing edge. Also determine the location at which turbulence sets in.

7.18 In the example 7.17, determine the location in the laminar region at which the local heat transfer coefficient has the same value as that at the trailing edge.

7.19 At a location the ground water temperature varies between 15°C to 25°C. This water is to be chilled by flow over a flat plate below which brine at – 5°C flows. The rate of flow of brine is large so that the plate temperature can be taken to be constant at – 5°. The water flow is at a free stream velocity of 1 m/s. Determine the change in the value of local convection coefficient during the seasons if the plate length is 0.8 m.

7.20 The average friction coefficient in the laminar region for flow over a flat plate 0.6 m long is 3.32×10^{-6}. Determine the value of average convection coefficient for flow of air at a film temperature of 40°C. Also determine the boundary layer thickness.

7.21 A 10 cm diameter potato at 20°C is cooled using air at 5°C at a velocity of 1 m/s. The thermal conductivity of the potato is 0.49 W/m°C. At the instant the convection coefficient is 19.1 W/m^{2}K. Determine the heat flow rate and the temperature gradientat the surface at the instant.

[**Ans.** 9 W, – 585°C/m]

7.22 An average man has a surface area of 1.8 m^2 and skin temperature of 33°C. When clothed, the temperature outside is 30°C. When exposed to 10°C air. The heat transfer coefficient by convection when walking is given as $h = 8.6\ V^{0.53}$, where V is the walking velocity. Calculate the values of h and heat flow for walking velocities of 0.5, 1, 1.5 and 2 m/s.

Walking velocity m/s	***h, W/m^{2}K***	***Q, W***
0.5	5.956	214.4
1.0	8.6	309.6
1.5	10.66	383.8
2.0	12.42	447

7.23 A 4 m × 4 m plate maintained at constant temperature of 80°C is subjected to parallel flow of air at 20°C with a velocity of 10 m/s. The total force acting on the upper surface of the plate is 2.4 N. Using momentum and heat transfer analogy, determine the average convection coefficient to air. Assume ρ = 1.204 kg/m^3 and c_p = 1007 J/K°C and Pr = 0.7309. [**Ans.** h = 46.54 W/m^2 °C]

7.24 The value of friction coefficient for a surface over which air flows at 800 km/h was measured as 0.0016. The plane area is 75 m^2. Using momentum-heat transfer analogy, determine the value of convective heat transfer coefficient. Assume c_p = 999 J/kg°C and ρ = 0.4141 kg/m^3.

[**Ans.** 89.6 W/m^2°C]

Convective Heat Transfer–Practical Correlations–Flow Over Surfaces

8

CHAPTER

8.0 INTRODUCTION

In chapter 7 the basics of convection was discussed and the methods of analysis were enumerated, correlations were obtained for laminar flow over flat plate at uniform temperature, starting from basic principles and using the concept of boundary layer. The application of these correlations are limited. However these equations provide a method of correlation of experimental results and extension of these equations to practical situations of more complex nature. Though the basic dimensionless numbers used remain the same, the constants and power indices are found to vary with ranges of these parameters and geometries. In this chapter it is proposed to list the various types of boundaries, ranges of parameters and the experimental correlations found suitable to deal with these situations, as far as flow over surfaces like flat plates, cylinders, spheres and tube banks are concerned.

8.1 FLOW OVER FLAT PLATES

Equations for heat transfer in laminar flow over flat plate were derived from basics in Chapter 7. In this chapter additional practical correlations are introduced. Though several types of boundary conditions may exist, these can be approximated to three basic types. These are (*i*) constant wall temperature, (as may be obtained in evaporation, condensation etc., phase change at a specified pressure) (*ii*) constant heat flux, as may be obtained by electrical strip type of heating and (*iii*) flow with neither of these quantities remaining constant, as when two fluids may be flowing on either side of the plate.

Distinct correlations are available for constant wall temperature and constant heat flux. But for the third case it may be necessary to approximate to one of the above two cases. As it happens most of the practical situations fall in this category and the constant wall temperature correlations are adopted most of the time.

8.1.1. Laminar Flow

The condition is that the Reynolds number should be less than 5×10^5 or as may be stated otherwise. For the condition that the plate temperature is constant the following equations are valid with fluid property values taken at the film temperature.

Hydrodynamic boundary layer thickness (from Chapter 7)

$$\delta_x = 5x/Re_x^{0.5} \qquad \text{...(8.1)}$$

Thermal boundary layer thickness

$$\delta_{tx} = \delta_x\, Pr^{-0.333} \quad ...(8.2$$

Displacement thickness and Momentum thickness are not directly used in heat transfer calculations. However, it is desirable to be aware of these concepts.

Displacement thickness is the difference between the boundary layer thickness and the thickness with uniform velocity equal to free stream velocity in which the flow will be the same as in the boundary layer. For laminar flow displacement thickness is defined as,

$$\delta_d = \int_0^{\delta}\left(1 - \frac{u}{u_\infty}\right) dy$$

Assuming the velocity variation as $\left(\frac{u}{u_\infty}\right) = \left(\frac{2y}{\delta}\right) - \left(\frac{y}{\delta}\right)^2$ displacement thickness is obtained as.

$$\delta_d = \delta_x/3 \quad ...(8.3)$$

Momentum thickness is the difference between the boundary layer thickness and the layer thickness which at the free stream velocity will have the same momentum as in the boundary layer.

Momentum thickness δ_m in the laminar region is defined by

$$\delta_m = \int_0^{\delta}\left[\frac{u}{u_\infty} - \left(\frac{u}{u_\infty}\right)^2\right] dy$$

Assuming the velocity profile as in the previous case, momentum thickness is obtained as

$$\delta_m = \delta_x/7 \quad ...(8.4)$$

Friction coefficient defined as $\tau_s/(\rho u_\infty^2/2)$ is given by

$$C_{fx} = 0.664\, Re_x^{-0.5} \quad ...(8.5)$$

The average value of C_f in the laminar region for a length L from leading edge is given by (Chapter 7)

$$C_{fL} = 1.328\, Re_L^{-0.5} \quad ...(8.6)$$

The value of **local Nusselt number** is given by (Chapter 7)

$$Nu_x = 0.332\, Re_x^{0.5}\, Pr^{0.33} \quad ...(8.7)$$

$$\overline{Nu}_L = 2Nu_L = 0.664\, Re_L^{0.5}\, Pr^{1/3} \quad ...[8.7(a)]$$

This is valid for Prandtl number range of 0.6 to 50.

For low values of Prandtl numbers as in the case of **liquid metals, the local Nusselt number is**

$$Nu_x = 0.565\, (Re_x\, Pr)^{0.5} \quad ...(8.8)$$

The average value up to length L will be twice the local value at L.

This is valid for Prandtl number less than 0.05 (liquid metals). A more general expression applicable for both low and high values of Prandtl number is given by

$$Nu_x = \frac{0.3387\, Re_x^{0.5}\, Pr^{0.333}}{[1+(0.0468/Pr)^{0.67}]^{0.25}} \quad ...(8.9)$$

In this case also the average value up to length L will be twice the local value at L. This is valid for $Pr < 0.05$ and $Pr > 50$ and $Re_x\, Pr > 100$. (liquid metals and silicones).

Note: The modification for very high values of Prandtl number is very little as can be seen in the worked out problems.

It may be seen that there is gap in the range of Prandtl number 0.6 to 0.1. If one goes through property values of various fluids in practical application, it will be seen that no fluid is having Prandtl numbers in this range.

8.1.2. Constant Heat Flux

The local Nusselt number is given by

$$Nu_x = 0.453\, Re_x^{0.5}\, Pr^{0.333} \quad ...(8.10)$$

This is also valid in the range of Prandtl numbers 0.6 to 50. In constant heat flux boundary the plate temperature varies along the lengths. Hence the temperature difference between the plate and the free stream varies continuously. The average **difference in temperature between the fluid and surface at *x*** is given by

$$T_{wx} - T_\infty = (qx/k)/[0.6795\, Re_x^{0.5}.\, Pr^{0.33}] \quad ...(8.11)$$

For low as well as high values of Prandtl numbers the relationship is (For $Pr < 0.05$ and $Pr > 50$)

$$Nu_x = \frac{0.453\, Re_x^{0.5}\, Pr^{0.333}}{[1+(0.0207/Pr)^{0.67}]^{0.25}} \quad ...(8.12)$$

The property values are at film temperature.

In all cases, the average Nusselt number is given by

$$\overline{\mathbf{Nu}}_{\mathbf{L}} = \mathbf{2Nu_L} \quad ...(8.13)$$

This is applicable in all cases when $\mathbf{Nu \propto Re^{0.5}}$

Using the analogy between heat and momentum transfer the **Stanton number** is given by

$$St\, Pr^{0.67} = C_f/2 \quad ...(8.14)$$

The equations (8.1) to (8.14) are applicable for laminar flow over flat plates. The choice of the equation depends upon the values of Prandtl number and Reynolds numbers (laminar flow).

Property values should be at the film temperature, $(T_s + T_\infty)/2$.

Eight examples follow, using different fluids at different conditions.

Example 8.1: *In a process water at 30°C flows over a plate (1 m × 1 m) maintained at 10°C with a free stream velocity of 0.3 m/s. Determine the hydrodynamic boundary layer thickness, thermal boundary layer thickness, local and average values of friction coefficient, heat transfer coefficient and refrigeration necessary to maintain the plate temperature. Also find the values of displacement and momentum thicknesses.*

Solution: The film temperature = (30 + 10)/2 = 20°C

Chapter 8

The property values are:

Kinematic viscosity = 1.006×10^{-6} m²/s

Thermal conductivity = 0.5978 W/mK

Prandtl number = 7.02, at 1 m

$$Re_x = \frac{u_\infty x}{\nu} = \frac{0.3 \times 1}{1.006 \times 10^{-6}} = 2.982 \times 10^5 \quad \therefore \text{ laminar}$$

$$\therefore \quad \delta_x = 5x/Re_x^{0.5} = 9.156 \times 10^{-3} \text{ m} = \mathbf{9.156 \text{ mm}}$$

$$\delta_{tx} = \delta_x \cdot Pr^{-0.33} = 9.156(7.02)^{-0.33} = \mathbf{4.782 \text{ mm}}$$

Thermal boundary layer will be thinner if Pr > 1

Displacement thickness

$$\delta_d = \delta_x/3 = 9.156/3 = \mathbf{3.052 \text{ mm}}$$

Momentum thickness

$$\delta_m = \delta_x/7 = 9.156/7 = \mathbf{1.308 \text{ mm}}$$

$$C_{fx} = 0.664/Re^{0.5} = 0.664/(2.982 \times 10^5)^{0.5} = \mathbf{1.216 \times 10^{-3}}$$

$$C_{fL} = 2 \times C_{fL} = 2 \times 1.216 \times 10^{-3} = \mathbf{2.432 \times 10^{-3}}$$

$$Nu_x = 0.332 \times Re_x^{0.5}\, Pr^{0.33} = 0.332 \times (2.982 \times 10^5)^{0.5} \times 7.02^{0.33}$$

$$= \mathbf{347.15}$$

$$h_x = Nu_x \frac{k}{L} = 347.15 \times 0.5978/1 = \mathbf{207.52 \text{ W/m}^2\text{K}}$$

$$\bar{h} = 2\, h_L = \mathbf{415.04 \text{ W/m}^2\text{K}}$$

cooling required = $hA\,\Delta T$ = 415.04 × 1 × 1 × (30 – 10) = **8301 W or 8.3 kW.**

Example 8.2: *Sodium potassium alloy (25% + 75%) at 300°C flows over a 20 cm long plate element at 500°C with a free stream velocity of 0.6 m/s. The width of plate element is 0.1 m. Determine the hydrodynamic and thermal boundary layer thicknesses and also the displacement and momentum thicknesses. Determine also the local and average value of coefficient of friction and convection coefficient. Also find the heat transfer rate.*

Solution: The film temperature is (300 + 500)/2 = 400°C

The property values are:

Kinematic viscosity = 0.308×10^{-6} m²/s, $Pr = 0.0108$,

Thermal conductivity = 22.1 W/mK, at 0.2 m,

$$Re_x = 0.6 \times 0.2/0.308 \times 10^{-6} = 3.9 \times 10^5 \quad \therefore \text{ laminar}$$

$$\therefore \quad \delta_x = 5x/Re_x^{0.5} = \mathbf{1.6 \text{ mm}}$$

$$\delta_{tx} = \delta_x \cdot Pr^{-0.33} = \mathbf{7.25 \text{ mm}}$$

This is larger by several times. So most of the thermal layer is outside the velocity boundary layer.

Displacement thickness:

$$\delta_d = 1.6/3 = \mathbf{0.53 \text{ mm}}$$

Momentum thickness

$$\delta_m = 1.6/7 = \mathbf{0.229 \text{ mm}}$$

It can be seen that thermal effect is predominant

$$\mathbf{C_{fx}} = 0.664/Re^{0.5} = 0.664/(3.9 \times 10^5)^{0.5} = \mathbf{1.064 \times 10^{-3}}$$

$$\overline{\mathbf{C}}_{\mathbf{fL}} = \mathbf{2.128 \times 10^{-3}}$$

Using equation (8.8) as the Prandtl number is very low (less than 0.05)

$$Nu_x = 0.565 \times (Re_x\, Pr)^{0.5} = 36.65$$

$$\mathbf{h_x} = Nu_x \frac{k}{L} = 36.65 \times 22.1/0.2 = \mathbf{4050\ W/m^2K}$$

$$\overline{\mathbf{h}} = 2 \times h_L = \mathbf{8100\ W/m^2K}$$

$$\textbf{Heat flow} = 8100 \times 0.2 \times 0.1 \times (500 - 300) = \mathbf{32{,}399\ W} \quad \text{or} \quad \mathbf{32.4\ kW}$$

Alternately **using equation** (8.9)

$$Nu_x = \frac{0.3387\, Re_x^{\,0.5}\, Pr^{0.333}}{[1 + (0.0468 / Pr)^{0.67}]^{0.25}}$$

$$= \frac{0.3387 \times (3.9 \times 10^5)^{0.5} (0.0108)^{0.33}}{[1 + (0.0468 / 0.0108)^{0.67}]^{0.25}} = 33.79$$

$$\therefore \qquad \mathbf{h_x} = \frac{33.79 \times 22.1}{0.2} = \mathbf{3734\ W/m^2K}$$

$$\overline{\mathbf{h}} = \mathbf{7468\ W/m^2K} \quad Q = 29.87\ \text{kW}$$

If equation (8.7) had been used $Q = 40.5$ kW, an over estimate.

Equation 8.9 is a more recent correlation and so the value obtained by the same should be more reliable.

Example 8.3: *Engine oil at 80°C flows over a flat surface at 40°C for cooling purpose, the flow velocity being 2 m/s. Determine at a distance of 0.4 m from the leading edge the hydrodynamic and thermal boundary layer thickness. Also determine the local and average values of friction and convection coefficients.*

Solution: The film temperature is (80 + 40)/2 = 60°C

The property values are read from tables at 60°C as kinetic viscosity = 83×10^{-6} m²/s, **Pr = 1050.** Thermal conductivity = 0.1407 W/mK

$$Re_x = \frac{u_\infty x}{\nu} = \frac{2 \times 0.4}{83 \times 10^{-6}} = 9639, \text{ laminar}$$

$$\therefore \qquad \boldsymbol{\delta_x} = 5x/Re_x^{\,0.5} = 0.02037\ \text{m} = \mathbf{20.37\ mm}$$

$$\boldsymbol{\delta_{tx}} = \delta_x Pr^{-1/3} = 20.37 \times 1050^{-0.333} = \mathbf{2\ mm}$$

Thermal boundary layer is very thin as different from liquid metal—in this case viscous effect is predominant.

$$C_{fx} = 0.664/Re_x^{\,0.5} = 6.76 \times 10^{-3}$$

$$\overline{\mathbf{C}}_{\mathbf{fL}} = \mathbf{0.0135} \text{ (rather large)}$$

As the values of Prandtl number is very high equation (8.9) can be used

$$Nu_x = \frac{0.3387\, Re_x^{\,0.5}\, Pr^{0.33}}{[1 + (0.0468 / Pr)^{0.67}]^{0.25}}$$

$$= \frac{0.3387 \times 9639^{0.5} \times 1050^{0.33}}{[1+(0.0468/1050)^{0.67}]^{0.25}} = 337.97/1.0003 = 337.87$$

$$\mathbf{h_x} = \frac{Nu_x k}{x} = \frac{337.87 \times 0.1407}{0.4} = \mathbf{118.85\ W/m^2K}$$

$$\overline{\mathbf{h}} = 2h_x = 118.85 \times 2 = \mathbf{237.69\ W/m^2K}$$

For 1 m width the heat flow is given by

$$Q = 237.69 \times 0.4 \times 1\ (80 - 40) = 3803 \text{ W or } 3.803 \text{ kW}$$

If **equation (8.7)** is used $Nu = 331.3$ and $\overline{\mathbf{h}}$ **= 233.01 W/m²K.** The difference is very little.

Example 8.4: *Air at 20°C flows over a flat plate having* ***a uniform heat flux*** *of 800 W/m². The flow velocity is 4 m/s and the length of the plate is 1.2 m. Determine the value of heat transfer coefficient and also the temperature of the plate as the air leaves the plate.*

Solution: As the plate temperature varies, the value of film temperature cannot be determined. For the first trial, the properties of air at 20°C are used.

$$\nu = 15.06 \times 10^{-6},\ k = 0.02593 \text{ W/mK},\ Pr = 0.703$$

First, a check for laminar flow:

$$Re = \frac{u_\infty L}{\nu} = \frac{4 \times 1.2}{15.06 \times 10^{-6}} = 3.187 \times 10^5 \quad \therefore \text{ laminar.}$$

For constant heat flux, the **average temperature difference** can be found by using equation (8.11)

$$\overline{\mathbf{T_x - T_\infty}} = (qL/k)/[0.6795\ Re_x^{\ 0.5}\ Pr^{0.33}] = \mathbf{108.54°C}$$

Now properties may be found at (108.54 + 20)/2 = 64.27°C

T°C	*ν, m²/s*	*k, W/mK*	*Pr*
60	18.97×10^{-6}	0.02896	0.696
70	20.02×10^{-6}	0.02966	0.694
64.27	19.42×10^{-6}	0.02926	0.695

Using the equation again

$$T_x - T_\infty = \frac{800 \times 1.2}{0.02926} \cdot \frac{1}{0.6795\,(4 \times 1.2/19.42 \times 10^{-6})^{0.5}(0.695)^{0.333}}$$

$$= 109.644°C$$

∴ Film temperature = 64.82°C

It does not make much of a difference.

To determine the value of convection coefficient, equation (8.11) is used.

$$\mathbf{Nu_x} = 0.453\ [Re_x\ Pr]^{0.5} = 0.453 \left[\frac{4 \times 1.2 \times 0.695}{19.42 \times 10^{-6}}\right]^{0.5} = \mathbf{187.75}$$

$$\therefore \quad \mathbf{h_x} = \frac{187.75}{1.2} \times 0.02926 = \mathbf{4.58\ W/m^2K}$$

$$\overline{\mathbf{h}} = \mathbf{9.16\ W/m^2K}$$

To find the temperature at the trailing edge the basic heat flow equation is used:

$$(T_w - T_\infty) = \frac{qx}{kNu_x} \text{ as } (h = Nuk/x)$$

$$= \frac{800 \times 1.2}{0.02926 \times 187.75} = 174.75°\text{C}$$

$$\therefore \quad \mathbf{T_w = 194.75°C.}$$

Example 8.5: *Water at 10°C flows over a flat plate with a uniform heat flux of 8.3 kW/m². The velocity of flow is 0.3 m/s. Determine the value of convective heat transfer coefficient and also the temperature at a distance of 1 m from the leading edge.*

Solution: As the film temperature cannot be specified the properties will be taken at 10°C for the first trial

$$\nu = (1.788 + 1.006) \times 10^{-6}/2 = 1.393 \times 10^{-6} \text{ m}^2/\text{s}$$

$$\mathbf{Pr} = (13.6 + 7.03)/2 = \mathbf{10.31}$$

$$k = (0.5524 + 0.5978)/2 = 0.5751 \text{ W/mK}$$

at 1 m, $Re_x = 0.3 \times 1/1.393 \times 10^{-6} = 2.154 \times 10^5 \quad \therefore$ laminar.

The **average temperature difference**

$$= \frac{q.L}{k} \cdot \frac{1}{0.6795\, Re^{0.5} Pr^{0.333}}$$

$$= \frac{8300 \times 1}{0.5751} \cdot \frac{1}{0.6795 \times (2.154 \times 10^5)^{1/2}} \cdot \frac{1}{10.31^{0.333}} = \mathbf{21.03°C}$$

The property values can now be taken at 15.1°C and results refined.

The heat transfer coefficient can be determined using equation (8.10).

$$Nu_x = 0.453\, Re_x^{0.5} Pr^{0.333}$$

taking property values at 15.51°C

$$Nu_x = 465.9$$

$$\therefore \quad \mathbf{h_x} = 465.9 \times 0.58762/1 = \mathbf{273.8\ W/m^2K}$$

Average value = 547.5 W/m²K (compare with example 8.1)

Temp. difference at 1 m: (assuming 1 m width)

$$h\Delta T = q \qquad \therefore \quad \Delta T = \frac{q}{h}$$

$$\therefore \quad h = Nu.\, k/x \qquad \therefore \quad \Delta T = \frac{qx}{Nu\, k}$$

$$\Delta T = \frac{qx}{k} \cdot \frac{1}{Nu_x} = \frac{8300 \times 1}{0.58762 \times 465.9} = \mathbf{30.32°C}$$

Example 8.6: ***Sodium potassium alloy (25% + 75%)*** *at 300°C flows over a plate element with free stream velocity of 0.6 m/s. The plate has* ***a uniform heat generation*** *rate of 1600 kW/m². Determine the value of average convection coefficient for a length of 0.2 m. Also determine the plate temperature at this point.*

Solution: The Prandtl number has a value less than 0.05 and there is no equation to determine the temperature difference. Equation (8.12) is used, starting with property values at 300°C

$$\nu = 0.336 \times 10^{-6} \text{ m}^2/\text{s}, \mathbf{Pr = 0.0134}, k = 22.68$$

$$\mathbf{Re_x} = 0.6 \times 0.2/0.366 \times 10^{-6} = 3.279 \times 10^5$$

∴ Laminar. For low value of Pr using equation (8.12)

$$Nu_x = \frac{0.453 Re_x^{\ 0.5} Pr^{0.33}}{[1 + (0.0207/Pr)^{0.67}]^{0.25}}$$

$$= \frac{0.453 \times 3.279 \times 10^5 (0.0134)^{0.333}}{[1 + (0.0207/0.0134)^{0.67}]^{0.25}} = 49.83$$

$$\mathbf{h_x} = \frac{49.83 \times 22.68}{0.2} = \mathbf{5651\ W/m^2K}$$

$$\bar{\mathbf{h}} = \mathbf{11302.1\ W/m^2K}$$

The average temperature difference:

$$\Delta T = \frac{q}{h} = \frac{1600000}{11302} = 141.6°\text{C}$$

∴ Plate temperature = 300 + 141.6 = 441.6°C

Compare with example 8.2. The results can be refined now taking property values at 300 + (141.6)/2 = 370.8°C (film temperature). Interpolating

$$\nu = -(0.366 - 0.308) \times 0.708 + 0.366 = 0.325 \times 10^{-6} \text{ m}^2/\text{s}$$

$$Pr = -(0.0134 - 0.0108) \times 0.708 + 0.0134 = 0.0116$$

$$k = -(22.68 - 22.10) \times 0.708 + 22.68 = 22.27 \text{ W/mK}$$

$$\therefore \quad Nu_x = \frac{0.453 \times (0.6 \times 0.2/0.325 \times 10^{-6})^{0.5} (0.0116)^{0.33}}{[1 + (0.0207/0.0116)^{0.67}]^{0.25}}$$

= **49.7** as compared to 49.83. Values are not very different.

Using **equation (8.8), $\mathbf{Nu_x}$** = 0.565 $(Re\ Pr)^{0.5}$ = **36.98**, compared with 49.7.

Example 8.7: *Engine oil at 60°C flows over a flat surface with a velocity of 2 m/s, the length of the surface being 0.4 m. If the plate has* ***a uniform heat flux*** *of 10 kW/m², determine the value of average convective heat transfer coefficient. Also find the temperature of the plate at the trailing edge.*

Solution: As the film temperature cannot be determined, the property values are taken at free stream temperature of 60°C

Kinematic viscosity $= 83 \times 10^{-6}$ m²/s, $Pr = 1050$, $k = 0.1407$ W/mK

$$Re_x = \frac{u_\infty x}{\nu} = \frac{2 \times 0.4}{83 \times 10^{-6}} = 9639 \quad \therefore \text{ laminar}$$

Using equation (8.12)

$$Nu_x = \frac{0.453\, Re_x^{\ 0.5} Pr^{0.33}}{[1 + (0.0207/Pr)^{0.67}]^{0.25}} = \frac{0.453 \,.\, 9639^{0.5} \,.\, 1050^{0.333}}{[1 + (0.0207/1050)^{0.67}]^{0.25}} = 451.95$$

$$\mathbf{h_x} = \frac{451.95 \times k}{x} = \frac{451.95 \times 0.1407}{0.4} = \mathbf{158.97\ W/m^2K}$$

$$\overline{\mathbf{h}} = h_L \times 2 = \mathbf{317.94\ W/m^2K}$$

The average temperature difference:

$$\Delta T = \frac{q}{h} = \frac{100000}{317.94} = 31.45°\text{C}$$

Now the film temperature can be taken as

$$\frac{31.45}{2} + 60 = 75.73°\text{C}$$

Using property tables

	ν	*Pr*	*k*
80°C	37×10^{-6}	490	0.1384
60°C	83×10^{-6}	1050	0.1407

at 75.73°C, $\nu = 46.82 \times 10^{-6}$ m²/s, $Pr = 609.6$, $k = 0.1389$ W/mK

$$Nu_x = \frac{0.453(2 \times 0.4 / 46.82 \times 10^{-6})^{0.5} (609.6)^{0.33}}{[1 + (0.0207 / 609.6)^{0.67}]^{0.25}} = 501.95$$

$$h_x = \frac{Nu_x k}{x} = \frac{501.95 \times 0.1389}{0.4} = 174.3\ \text{W/m}^2\text{K}$$

$$\overline{h} = \mathbf{348.6\ W/m^2K}$$

$$\Delta T = \frac{10000}{348.6} = 28.7°\text{C}$$

The value can be refined further using new value of film temperature.

To determine the **plate temperature at the edge:**

$$\Delta T = \frac{10000}{174.3} = 57.4°\text{C}$$

$$\therefore \quad T = 60 + 57.4 = \mathbf{117.4°C}$$

Compare with example 8.3.

8.1.3. Other Special Cases

Laminar constant wall temperature, with heating starting at a distance x_0 from the leading edge.

The correlation is obtained as below

$$Nu_x = 0.332.\ Re_x^{0.5}\ Pr^{0.33} \left[1 - \left(\frac{x_o}{x}\right)^{0.75}\right]^{-0.333} \quad ...(8.15)$$

At $x_o = 0$, this will reduce to the regular expression given by equation (8.7). The average value in this case will not be $2Nu_x$ and the above expression has to be integrated over the length to obtain the value.

The average value up to length L is given by

$$\overline{Nu}_L = Nu_L \times \frac{L}{L-x}\left[1-\left(\frac{x}{L}\right)\frac{2P+1}{2P+2}\right]^{2P/(2P+1)} \quad ...[8.15(a)]$$

where $P = L$ for laminar flow and $P = 4$ for turbulent flow.

Example 8.8: *Considering water at 30°C flowing over a flat plate 1 m × 1 m at 10°C with a free stream velocity of 0.3 m/s, plot the variation of h_x along the length if heating starts from 0.3 m from the leading edge. Also find the average value up to 1 m.*

Solution: The film temperature = (30 + 10)/2 = 20°C

The property values are: $\nu = 1.006 \times 10^{-6}$ m²/s, $Pr = 7.02$, $k = 0.5978$ W/mK

The maximum value of $Re_x = \frac{0.3 \times 1}{1.006 \times 10^{-6}} = 2.98 \times 10^5$

∴ Laminar flow exists all along.

$$Nu_x = 0.332\, Re_x^{0.5}\, Pr^{0.33}\left[1-\left(\frac{x_o}{x}\right)^{0.75}\right]^{-0.333}$$

$$h_x = k.Nu_x/x$$

at x = 0.35:

$$\mathbf{h_x} = \frac{0.5978}{0.35} \times 0.332\left(\frac{0.3 \times 0.35}{1.006 \times 10^{-6}}\right)^{0.5}(7.02)^{0.333}$$

$$\left[1-\left(\frac{0.3}{0.35}\right)^{0.75}\right]^{-0.333} = \mathbf{733.93\ W/m^2K}$$

Similarly for other values at 0.4, 0.5, 0.6, 0.7, 0.8, 0.9, and 1.0.

Distance x m	***h_x with heating from x = 0 W/m²K***	***h_x with heating from x = 0.3 m W/m²K***
0.3	367.47	0
0.35	340.21	733.93
0.4	318.24	549.67
0.5	284.64	416.90
0.6	259.84	315.09
0.7	240.57	309.34
0.8	225.03	279.70
0.9	212.16	257.20
1.0	**201.27**	**239.35**

The average value over the heated length can be found only by equation [8.15(a)]

At 1 m, with heating from 0.3 m. Average value up to 1 m

$$Re_L = 2.98 \times 10^5.$$

$$Nu_L = 0.332 \times (2.98 \times 10^5)^{0.5} \,.\, (7.02)^{0.333}$$
$$= 347$$

$$\overline{Nu}_L = 347 \times \frac{1}{1-0.35}\left[1-\left(\frac{0.3}{1}\right)^{0.75}\right]^{0.667}$$
$$= 377.53$$

$\therefore$ $\overline{h}_L = (377.53 \times 0.5978)/1 =$ **225.68 W/m²K.**

8.2 TURBULENT FLOW

$Re_x > 5 \times 10^5$ or as specified. In flow over flat plate, the flow is initially laminar and after some distance turns turbulent, the value of Reynolds number at this point being near 5×10^5. However, there are circumstances under which the flow turns turbulent at a very short distance, due to higher velocities or due to disturbances, roughness etc. The critical reynolds number in these cases is low and has to be specified. In the turbulent region the velocity **boundary layer thickness** is given by $(10^5 > Re < 10^7)$

$$\delta_x = 0.37x \times Re_x^{-0.2} \quad \text{...(8.16)}$$

$$\delta_t \approx \delta_x \quad \text{...(8.17)}$$

The **displacement and momentum thickness** are much thinner. The **displacement thickness is**

$$\delta_d = \delta_x/8 \quad \text{...(8.18)}$$

Momentum thickness is

$$\delta_m = (7/72)\ \delta_x \quad \text{...(8.19)}$$

The local friction coefficient defined as $\tau_w/(\rho\ u_\infty{}^2/2)$ is given for the range Re_x from 5×10^5 to 10^7 by

$$C_{fx} = 0.0592\ Re_x^{-0.2} \quad \text{...(8.20)}$$

For higher values of Re in the range 10^7 to 10^9

$$C_{fx} = 0.37\ [\log_{10} Re_x]^{-2.584} \quad \text{...(8.21)}$$

The **local Nusselt number** is given by

$$Nu_x = 0.0296\ Re_x{}^{0.8} Pr^{0.33} \quad \text{...(8.22)}$$

The **average Nusselt number** is given by

$$\overline{Nu} = 0.037\ Re^{0.8}\ Pr^{0.33} \quad \text{...(8.23)}$$

$\overline{\textbf{Nu}}$ **= Nu$_x$/0.8** in this case as **Nu is dependent on Re$_x^{0.8}$.** Using analogy between momentum and heat transfer

$$Nu_x = \frac{(C_{fx}/2) Re_x Pr}{1 + 12.8\,(C_{fx}/2)^{0.5}(Pr^{0.68} - 1)} \quad \text{...(8.24)}$$

To obtain the average value, this expression has to be integrated from $x = 0$ to $x = L$. But this is more complex.

For constant heat flux, the Nusselt number is found to increase by 4% over the value for constant wall temperature.

$\therefore$ **Nu$_x$ (constant heat flux) = 1.04 Nu$_x$(Constant wall temperature)** ...(8.25)

Chapter 8

Example 8.9: *Air at –10°C flows over a flat surface at 10°C with a free stream velocity of 80 m/s. The length of the plate is 3.1 m. Determine the* ***location at which the flow turns turbulent.*** *Also determine the local and average value of convection coefficient assuming that the flow is turbulent although. Compare the value of local heat transfer coefficient calculated using the equation obtained by analogy. (8.24).*

Solution: The film temperature is (–10 + 10)/2 = 0°C

The property values are

$$\nu = 13.28 \times 10^{-6} \text{ m}^2/\text{s}, Pr = 0.707, k = 0.02442 \text{ W/mK}$$

Re_x at = 3.1 m

$$Re = (80 \times 3.10/13.28 \times 10^{-6}) = 1.8675 \times 10^7$$

∴ turbulent flow exists

location at which $Re = 5 \times 10^5$ is

$$\mathbf{x} = \frac{5 \times 10^5 \times 13.28 \times 10^{-6}}{80} = \mathbf{0.083\ m}$$

This length is much shorter (2.7% of the total length) and so the assumption that flow is turbulent all through is valid. Hydrodynamic boundary layer thickness

$$\delta_x = 0.37 \times 3.1 \times (1.8675 \times 10^7)^{-0.2} = \mathbf{40.34\ mm}$$

Thermal boundary layer thickness is also

$$\delta_t = \mathbf{40.34\ mm}$$

Displacement thickness

$$\delta_d = 41.54/8 = \mathbf{5.04\ mm}$$

Momentum thickness

$$\delta_m = \frac{7}{72} \times 41.54 = \mathbf{3.92\ mm}$$

As Re is in the border (< 10^8), we can calculate C_{fx} using equations (8.20) or (8.21).

$$C_{fx} = 0.0592 \times Re_x^{-0.2} = \mathbf{2.08 \times 10^{-3}}$$

Using equation (8.21) $C_{fx} = 0.37\ [\log_{10} Re_x\ ]^{-2.584} = 2.197 \times 10^{-3}$

Local Nusselt number is given by equation (8.22)

$$Nu_x = 0.0296\ Re_x^{0.8}\ Pr^{0.33}$$

$$= 0.0296 \times (1.8675 \times 10^7)^{0.8}\ (0.707)^{0.33} = \mathbf{17302.2}$$

∴ $$h_x = \frac{Nu_x k}{L} = \mathbf{136.3\ W/m^2K}$$

Average value $$\overline{Nu} = 0.037\ Rex^{0.8} \times Pr^{0.33}$$

∴ $$\bar{h} = \mathbf{170.4\ W/m^2K}$$

If constant heat flux prevails, this value is increased by 4%. Using Analogy: using C_{fx} by equation (8.20)

$$Nu_x = \frac{(C_{fx}/2)\,Re_x . Pr}{1 + 12.8\,(C_{fx}/2)^{0.5}(Pr^{0.68} - 1)} = \mathbf{15035}$$

By using C_{fx} for higher range:

$$Nu_x = 15922$$

These values are not very much different from the one using equation (8.22) (8%).

The values of convection coefficients calculated may be out by as much as 25% in certain cases and as such these estimates are acceptable. It is always desirable to evaluate the required quantity by all available correlation so that a safe value can be obtained.

Example 8.10: *Water at 30°C flows over a flat plate 1 m × 1 m at 10°C with a free stream velocity of 4 m/s. Determine the thickness of boundary layers, displacement thickness, momentum thickness, local and average value of drag coefficient and convection coefficient.*

Solution: The film temperature = (30 + 10)/2 = 20°C. Property values at this temperature are $\nu = 1.006 \times 10^{-6}$ m²/s, $Pr = 7.02$, $k = 0.5978$ W/mK.

The maximum value of Reynolds number at 1 m is

$$= 4 \times 1/1.006 \times 10^{-6} = 3.976 \times 10^{6} \quad \therefore \text{ Turbulent}$$

The length at which flow turns turbulent:

$$(4 \times x/1.006 \times 10^{-6}) = 5 \times 10^{5} \quad \therefore \mathbf{x = 0.12575\ m.}$$

This is 12.5% of total length. As such the assumption that the flow is turbulent althrough is not unacceptable.

Boundary layer thickness:

$$\boldsymbol{\delta_x} = 0.37 \times Re_x^{-0.2} \times 1 = 0.37 \times (3.976 \times 10^{6})^{-0.2}$$

$$= 0.1771 \text{ m} = \mathbf{17.71\ mm}$$

Thermal boundary layer thickness is also the same = **17.71 mm**

Displacement thickness:

$$\boldsymbol{\delta_d} = \delta_x/8 = \mathbf{2.21\ mm}$$

Momentum thickness

$$\boldsymbol{\delta_m} = \frac{7}{72} \times \delta_x = \mathbf{1.72\ mm}$$

$$C_{fx} = 0.0592 \times Re_x^{-0.2} = 2.83 \times 10^{-3}$$

$$\overline{\mathbf{C_f}} = C_{fx}/0.8 = \mathbf{3.54 \times 10^{-3}}$$

$$Nu_x = 0.0298\, Re_x^{0.8} Pr^{0.33} = 0.0296 \times (3.976 \times 10^{6})^{0.8} \times (7.02)^{0.3}$$

$$= 10788.8$$

$$\mathbf{h_x} = \frac{Nu_x k}{x} = \frac{10788.8 \times 0.5978}{1} = \mathbf{6449.6\ W/m^2K}$$

$$\overline{Nu} = 0.037 \times Re^{0.8} \times Pr^{0.33} = 13486$$

$$\overline{\mathbf{h}} = \mathbf{8062\ W/m^2K}$$

For constant heat flux, the average value is increased by 4%.

By using analogy

$$Nu_x = \frac{(C_{fx}/2) Re_x . Pr}{1 + 12.8 (C_{fx}/2)^{0.5} (Pr^{0.68} - 1)} = 16967$$

∴ $\mathbf{h_x = 10143\ W/m^2K}$

This is on the higher side.

8.2.1. Combined Laminar and Turbulent Flow

The assumption that the flow is turbulent althrough (from start) may not be acceptable in many situations.

The average values are now found by integrating the local values up to the location where Re = 5 × 10⁵ using laminar flow relationship and then integrating the local value beyond this point using the turbulent flow relationship and then taking the average. This leads to the following relationship for constant wall temperature.

$$\delta_x = 0.37x \times Re_x^{-0.2} - 10256x \times Re_x^{-1.0} \quad ...(8.26)$$

$$C_{fL} = 0.074\ Re_L^{0.2} - 1742\ Re_L^{-1.0} \quad ...(8.27)$$

This is for critical Reynolds number of 5×10^5. A more general relationship can be used for other values of critical Reynolds number.

$$C_{fL} = \frac{0.455}{(\log_{10} Re_L)^{2.584}} - \frac{A}{Re_L} \quad ...(8.28)$$

Where A is given in the tabulation below

Re_{cr}	A
3×10^5	1050
5×10^5	1700
1×10^6	3300
3×10^6	8700

$$Nu_x = Pr^{0.333}\ [0.037Re_L^{0.8} - 871] \quad ...(8.29)$$

for $Re_{cr} = 5 \times 10^5$, or more generally

$$Nu_L = Pr^{0.333}\ [0.037Re_L^{0.8} - A] \quad ...(8.30)$$

where
$$A = 0.037\ Re_{cr}^{\ 0.8} - 0.664\ Re_{cr}^{\ 0.5} \quad ...(8.31)$$

By analogy $St_x . Pr^{0.666} = C_{fx}/2$ ...(8.32)

For large temperature differences, the estimates may be off the mark by as much as 25%. For low or high Prandtl numbers, these expressions are to be used with some reservations.

For constant heat flux, the value of h is to be increased by 4 per cent.

Example 8.11: *Considering the data of Example 8.10, determine the average value of convection coefficient and C_f values taking into consideration the laminar region. Compare with example 8.10.*

Plate length 1 m, velocity = 4 m/s, plate temperature = 10°C, Water temperature = 30°C, Film temperature = 20°C. The property values are ν = 1.006 × 10⁻⁶ m²/s, Pr = 7.02, k = 0.5978 W/mK.

Solution: The maximum value of Reynolds number

$$= 4 \times 1/1.006 \times 10^{-6} = 3.976 \times 10^6 \quad \therefore \text{ turbulent}$$

Assuming $Re_{cr} = 5 \times 10^5$

$$\delta_x = 0.37 \times Re_x^{-0.2} - 10256 \times Re_x^{-1.0}$$

$$= 0.37 \times 1\ [4/1.006 \times 10^{-6}]^{-0.2} - 10256 \times 1/[4/1.006 \times 10^{-6}]$$

$= 0.01513$ m or **15.13 mm (compared to 18.24 mm)**

$$C_{fL} = 0.074\, Re_L^{-0.2} - 1742\, Re_L^{-1.0}$$

$$= \mathbf{3.10 \times 10^{-3}} \text{ (compared to } \mathbf{3.54 \times 10^{-3}} \text{ in example 8.10)}$$

$$\mathbf{Nu} = Pr^{0.33}\, [0.037\, Re_L^{0.8} - 871]$$

$$= \mathbf{11818} \text{ (compared to 13486).}$$

Example 8.12: *Air at –10°C flows over a flat plate at 10°C with a free stream velocity of 10 m/s, the length of the plate being 3.1 m. Determine the average value of friction coefficient and convection coefficient taking into account the laminar length and compare the values with those obtained assuming turbulent flow throughout. (example 8.9)*

Solution: The film temperature = (– 10 + 10)/2 = 0°C

The property values are: $\nu = 13.28 \times 10^{-6}$ m²/s, $Pr = 0.707$

$k = 0.02442$ W/mK

The maximum value of Reynolds number

$= 3.1 \times 10/(13.38 \times 10^{-6}) = 2.33 \times 10^6$ ∴ turbulent

Critical length: 0.664 m ∴ necessary to consider laminar region.

Assuming turbulent flow throughout:

$$\overline{C_f} = \frac{0.0592}{0.8}\, Re_L^{-0.2} = \mathbf{3.94 \times 10^{-3}}$$

Taking laminar region into account

$$\overline{C_f} = 0.074\, Re_L^{-0.2} - 1742/Re_L = \mathbf{3.195 \times 10^{-3}}$$

Heat transfer coefficient **(turbulent all through)**

$$\overline{Nu_L} = 0.037 \times Re_L^{0.8}\, Pr^{0.33} = 4098$$

$$\therefore \quad \mathbf{h} = \frac{Nu_L \times k}{L} = \mathbf{32.28\ W/m^2K}$$

Considering laminar region

$$\overline{Nu_L} = Pr^{0.333}\, [0.037 \times Re_L^{0.8} - 871] = 3321$$

$$\overline{\mathbf{h}} = \mathbf{26.17\ W/m^2K}$$

Note that at low velocities it will be better to consider the laminar region in taking averages.

8.3 FLOW ACROSS CYLINDERS

The other type of flow over surfaces is **flow across cylinders** often met within heat exchangers and hot or cold pipe lines in the open. An important difference is the velocity distribution along the flow. The obstruction by the cylinder causes a closing up of the streamlines and an increase in pressure at the stagnation point. The velocity distribution at various locations in the flow differs from the flow over a flat plate as shown in Fig. 8.1.

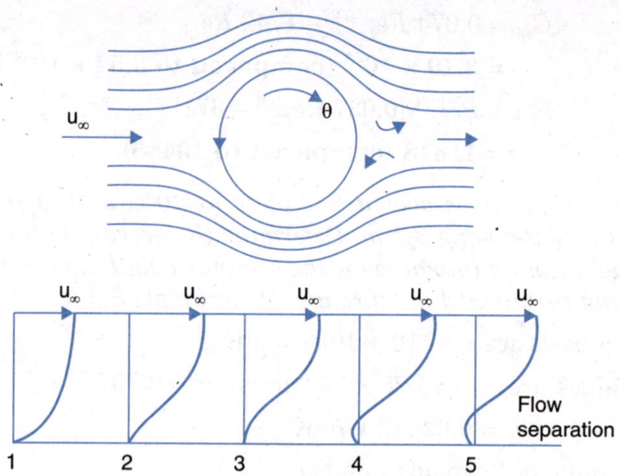

Fig. 8.1. Velocity distribution at various angular locations in flow across cylinders

As the flow pattern affects the heat transfer, it is found to be difficult to provide a generalised analytical solution for the problem. The drag coefficient C_D is defined by

Drag force = $C_D\, A_f \dfrac{\rho u_\infty^{\ 2}}{2}$. Where A_f is the frontal or projected area. (for a cylinder of length of L it is equal to $L.D$). It is not based on the wetted area. The nature of variation of drag coefficient for cylinder and sphere with Reynolds number is shown in Fig. 8.2. Reynolds number should be calculated with diameter D as the length parameter and is some times referred as Re_D.

Thus a simple and single correlation for C_D is difficult. The variation of local heat transfer coefficient with angular location for two values of Reynolds number is shown in Fig. 8.3.

For angles up to 80°, the variation of Nusselt number can be represented by

$$h_\theta = 1.14\, Re_D^{\ 0.5}\, Pr^{0.4} \left[1 - \left(\frac{\theta}{90}\right)^3\right] \qquad \text{...[8.32(a)]}$$

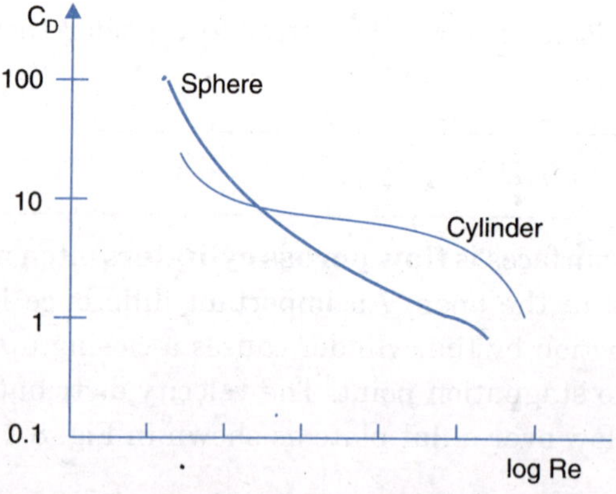

Fig. 8.2. Variation of C_D with Reynolds number for flow over cylinders and spheres

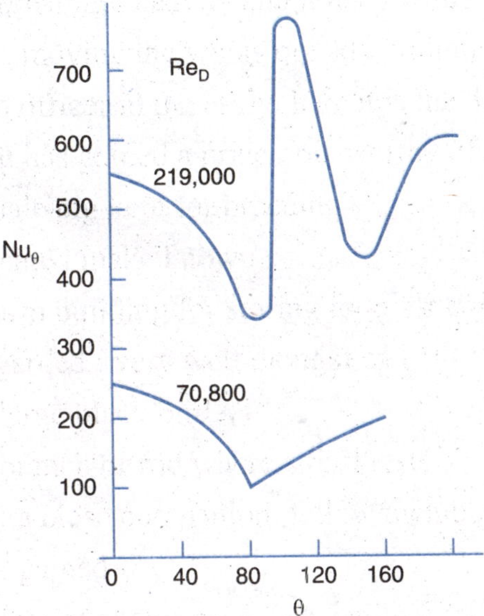

Fig. 8.3. Variation of Nusselt number with angular location

This also shows that averaging out the convection coefficient is difficult. The experimental values measured by various researchers plotted using common parameters Re_D and Nu_D (log log plot) is shown in Fig. 8.4. It can be seen that scatter is high at certain regions and several separate straight line correlations are possible for various ranges. Some researchers have limited their correlations for specific ranges and specific fluids. Thus a number of correlations are available and are listed below.

A very widely used correlation is of the form (1958)

$$Nu_D = CRc_D{}^m\ Pr^{0.333} \qquad ...(8.33)$$

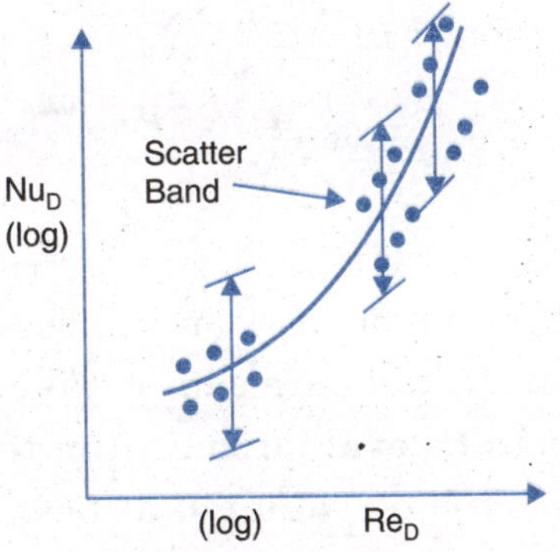

Fig. 8.4. Variation of Nu_D with Re_D for flow across cylinders

Where C and *m* are tabulated below. The applicability of this correlation for very low values of Prandtl number is doubtful. The length parameter in Nusselt number is diameter *D* and Nusselt number is referred as Nu_D.

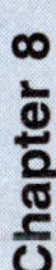

The properties are to be evaluated at the film temperature.

Re_D	C	m
0.4–4.0	0.989	0.330
4–40.0	0.91	0.385
40–4000	0.683	0.466
4000–40000	0.193	0.618
40000–400000	0.0266	0.805

A more recent (1972) generalised form is

$$Nu_D = C\, Re_D{}^m Pr^n \left(\frac{Pr_\infty}{Pr_w}\right)^{0.25} \quad ...(8.34)$$

The validity for this correlation is for

$0.7 < Pr < 500;\ 1 < Re_D < 10^6$ and

with $n = 0.36$ for $Pr < 10$ and $n = 0.37$ for $Pr > 10$

The values of C and m are tabulated below:

Re_D	C	m
1–4.0	0.75	0.4
40–10^3	0.51	0.5
$10^3 - 2 \times 10^5$	0.26	0.6
$2 \times 10^2 - 10^6$	0.076	0.7

The properties for Re and Pr should be at free stream temperature.

A two range (1972) correlation is given below: (f-film temp.)

$$Nu_D = [0.43 + 0.50\, Re_D{}^{0.5}]\, Pr^{0.38} \left(\frac{Pr_f}{Pr_w}\right)^{0.25} \quad ...[8.35(a)]$$

$1 < Re_D < 10^3$

$$Nu_D = 0.25\, Re_D{}^{0.6}\, Pr^{0.38} \left(\frac{Pr_f}{Pr_w}\right)^{0.25} \quad ...[8.35(b)]$$

$10^3 < Re_D < 2 \times 10^5$

This equation is applicable both for gases and liquids with the following conditions:

1. For gases the ratio of Pr numbers can be taken as unity.

2. For gases properties to be evaluated at film temperature.

3. For liquids properties to be evaluated at free stream temperature.

This is not suitable for very low and very high values of Prandtl numbers.

A correlation for liquids is given by (1965)

$$Nu_D = [0.35 + 0.56\, Re_D{}^{0.52}]\, Pr^{0.33} \quad ...(8.36)$$

$10^{-1} < Re_D < 10^5$ and properties at film temperature.

No indication is available for the applicability at low values of *Pr*. Another correlation 1972) applicable over wider range is

$$Nu_D = [0.4\, Re_D^{0.5} + 0.06\, Re_D^{0.67}]\, Pr^{0.4} \left[\frac{\mu_\infty}{\mu_w}\right]^{0.25} \quad ...(8.37)$$

$$10 < Re_D < 10^5;\ 0.67 < Pr < 300$$

$$0.25 < \frac{\mu_\infty}{\mu_w} < 5.2.$$

The properties are to be evaluated at free stream temperature T_∞. Another set of equations (1977) suitable for a wider range of parameters both Reynolds and Prandtl is

$$Nu_D = 0.3 + \frac{0.62\, Re_D^{0.5}\, Pr^{0.333}}{\left[1+\left(\frac{0.4}{Pr}\right)^{0.67}\right]^{0.75}}\left[1+\left(\frac{Re_D}{282000}\right)^{0.625}\right]^{0.8} \quad ...[8.38(a)]$$

for $100 < Re_D < 10^7,\ Pe = Re_D\ Pr > 0.2$

The properties are to be evaluated at film temperature.

A modification of this equation for limited range of Reynolds number is

$$Nu_D = 0.3 + \frac{0.62 Re_D^{0.5}\, Pr^{0.333}}{\left[1+\left(\frac{0.4}{Pr}\right)^{0.67}\right]^{0.25}}\left[1+\left(\frac{Re_D}{282000}\right)^{0.5}\right] \quad ...[8.38(b)]$$

$$2 \times 10^4 < Re_D < 4 \times 10^5,\ Pr > 0.2$$

This equation use properties at film temperature and is applicable for all fluids. Finally **for liquid metals another correlation.** (1975) is obtained as (for $Pe < 0.2$)

$$Nu_D = [0.8237 - \ln (Pe_D^{0.5})]^{-1} \quad ...[8.39(a)]$$

where $Pe = RePr$

Other correlations for liquid metals over cylinder are (1979)

$$Nu_D = 1.125\, (Re_D\, Pr)^{0.413} \quad ...[8.39(b)]$$

$$1 \le Re_D\, Pr \le 100.$$

Analytical results are also available for **constant wall temperature.**

$$Nu_D = 1.015\, (Re_D Pr)^{0.5}.$$

For constant **heat flux** $Nu_D = 1.145\, (Re_D Pr)^{0.5}$

This is applicable only for very low values of *Pr* and *Pe*. *Nu* will become negative for higher values of *Pe* in equation [8.39(*a*)]. Equations (8.33) to (8.39) are obtained from various experimental results, the difference being that each one of these is dividing the spectrum into different ranges of parameters. However a common warning is that most of these may give results varying as much as 25% from experimental results. A single correlation applicable for various ranges will be easier to use in computer application (say 8.38).

In actual application one has to weight carefully the parameter ranges before choosing the equation to be used.

8.3.1. Flow Across non Circular Shapes

The general correlation used for gases, including the $Pr^{0.333}$ in the constant is (1949)

$$Nu = C\, Re_D^{\,n} \quad ...[8.40(a)]$$

$$Nu = C_1 Re_D^{\,n} Pr^{0.33} \quad ...[8.40(b)]$$

The values of C, C_1 and n various shapes are tabulated below. The properties are at film temperature.

Shape	*Re_D range*	*n*	*C*	*C_1*
Square along diagonal, length diagonal distance	5000–100000	0.588	0.222	0.246
Square along diagonal, length diagonals distance	2500–7500	0.624	0.267	0.292
Square along face, side	2500–8000	0.699	0.160	0.178
Square along face, side	5000–100000	0.675	0.092	0.102
Plane, perpendicular, width	4000–15000	0.731	0.205	0.228
hexagon perpendicular to flats, corner distance	5000–19500	0.638	0.144	0.160
hexagon perpendicular to flats, corner distance	19500–100000	0.782	0.035	0.039
hexagon along flats, length, between flats	5000–100000	0.638	0.138	0.153
ellipse along minor axis, length, major axis	3000–15000	0.804	0.085	0.095
Ellipse along major axis, length minor axis	2500–15000	0.612	0.224	0.250

Example 8.13: *Air at 30°C flows across a steam pipe of 0.2 m dia. at a surface temperature of 130°C, with a velocity of 6 m/s. Determine the value of convective heat transfer coefficient using equations (8.33), (8.34), (8.35), (8.37), and (8.38).*

Solution: Property values are required both at T_∞, T_f and T_w

i.e. 30°C, (30 + 130)/2 = 80°C and 130°C

	ν	*Pr*	*k*	μ
30°C:	16×10^{-6} m²/s	0.701	0.02675	18.63×10^{-6} Ns/m²
80°C:	21.09×10^{-6} m²/s	0.692	0.03047	21.08×10^{-6} kg/ms
130°C:	26.625×10^{-6} m²/s	0.687	0.034135	23.29×10^{-6} kg/ms

Equation (8.33), properties at film temp.:

$$Re = uD/\nu = 6 \times 0.2/(21.09 \times 10^{-6}) = 56{,}899$$

$$\therefore \quad Nu = C Re^m Pr^{0.33}$$

From tables **C = 0.0266, m = 0.805**

$\therefore \quad Nu = 0.0266\ (56899)^{0.805}\ (0.692)^{0.333} = 158.29$

$\therefore \quad \mathbf{h} = Nu \times \dfrac{k}{D} = \mathbf{21.11\ W/m^2K}$

Equation (8.34) $\quad Nu = CRe^m\, Pr^n \left(\dfrac{Pr_\infty}{Pr_w}\right)^{0.25}$

Properties at free stream temp.: at 30°C

$$Re = (6 \times 0.2/16 \times 10^{-6}) = 75000 = 0.75 \times 10^5$$

From tables, $\quad$ **C = 0.26, m = 0.6** as $Pr < 10$, $n = 0.36$,

$$\therefore \quad Nu = 0.26\ (75000)^{0.6}\ (0.701)^{0.36} \left(\frac{0.701}{0.687}\right)^{0.25}$$

$= 192.8 \quad \therefore \quad \mathbf{h = 25.79\ W/m^2K}$ (k at 30°C, 0.02675)

Equation [8.35(*b*)], is applicable

$$Nu = 0.25\ Re^{0.6}\ Pr^{0.38} \left(\frac{Pr_f}{Pr_w}\right)^{0.25}$$

or gases: $(Pr_f/Pr_w)^{0.25} = 1$. for gases properties at film temp. at 80°C

$$Nu = 0.25\ (56899)^{0.6}\ (0.692)^{0.38}$$

$$= 154.97 \quad \therefore \quad \mathbf{h} = \frac{154.9 \times 0.03047}{0.2} = \mathbf{23.61\ W/m^2K}$$

Equation (8.37) Properties at T_∞, 30°C

$$Nu = [0.4\ Re^{0.5} + 0.06\ Re^{0.67}]\ Pr^{0.4} \left[\frac{\mu_\infty}{\mu_w}\right]^{0.25}$$

$$= \left[0.4 \times \left(\frac{6 \times 0.2}{16 \times 10^{-6}}\right)^{0.5} + 0.06 \left(\frac{6 \times 0.2}{16 \times 10^{-6}}\right)^{0.67}\right] [0.701]^{0.4} \left[\frac{18.63 \times 10^{-6}}{23.29 \times 10^{-6}}\right]^{0.25}$$

$= 180.76 \quad \therefore \quad \mathbf{h = 24.18\ W/m^2K}$, ($k = 0.02675$)

Equation [8.38(*b*)] properties at film temperature

$$Nu = 0.3 + \frac{0.62 Re^{0.5} Pr^{0.333}}{\left[1 + \left(\dfrac{0.4}{Pr}\right)^{0.67}\right]^{0.25}} \left[1 + \left(\frac{Re}{282000}\right)^{0.5}\right]$$

$$= 0.3 + \frac{0.62\ (56899)^{0.5} (0.701)^{0.333}}{\left[1 + \left(\dfrac{0.4}{0.701}\right)^{0.67}\right]^{0.25}} \left[1 + \left(\frac{56899}{282000}\right)^{0.5}\right]$$

$= 167.36 \quad \therefore \quad \mathbf{h = 25.5\ W/m^2K}$

In this example all the various equation provide answers within a small band. This i only fortitious and not necessarily so in all cases. The parameters are not in the extrem range.

Example 8.14: ***Liquid sodium at 300°C*** *flows across a tube 0.05 m outside dia. at 500°C with a velocity of 8 m / s. Determine the value of convective heat transfer coefficient using suitabl correlations.*

Solution: Property values may be required at all the three temperature T_∞, T_f and T_w.

T°C	ν, *m²/s*	*Pr*	*k, W/mK*	μ = νρ, *kg/ms*
300	0.394×10^{-6}	0.0063	70.94	878×10^{-6}
400	0.330×10^{-6}	0.0056	63.97	854×10^{-6}
500	0.289×10^{-6}	0.0053	56.99	829×10^{-6}

The possible correlations are **only 8.33** and 8.38(*a*). Equation (8.33) properties at film temp., 400°C, $Nu = CRe^m\, Pr^{0.333}$

$Re = 8 \times 0.05/0.330 \times 10^{-6} = 1212121$, (1.212×10^6) values of *C* and *m* are only up to 400,000, $C = 0.0266$, $m = 0.805$

$$\therefore \quad Nu = 0.0266\,(1.212 \times 10^6)^{0.805} \times 0.0056^{0.333} = 372.85,$$

$$\therefore \quad \mathbf{h} = \frac{372.85 \times 63.97}{0.05} = \mathbf{477022} \quad \text{or} \quad 0.477 \times 10^6 \text{ W/m}^2\text{K}$$

equation [8.38(*a*)]
$$Nu = 0.3 + \frac{0.62\, Re^{0.5}\, Pr^{0.333}}{\left[1+\left(\dfrac{0.4}{Pr}\right)^{0.67}\right]^{0.75}} \left[1+\left(\frac{Re}{282000}\right)^{0.625}\right]^{0.8}$$

$$= 0.3 + \frac{0.62\,(1.212 \times 10^6)^{0.5}(0.0056)^{0.333}}{\left[1+\left(\dfrac{0.4}{0.0056}\right)^{0.67}\right]^{0.25}} \left[1+\left(\frac{1.212 \times 10^6}{282000}\right)^{0.625}\right]^{0.8}$$

$$= 159.16 \quad \therefore \quad \mathbf{h = 203236\ W/m^2K}$$

The correlation (8.33) is an older one and hence the values obtained from the more recent correlation [8.38(*a*)] has to be taken as more reliable.

Example 8.15: *Water at 30°C flows across a pipe 10 cm OD at 50°C with a velocity of 0.6 m / s. Determine the value of convection coefficient using applicable correlations.*

	v, m²/s	*Pr*	*k, W/mK*	ρ, *kg/m³*
T_∞, 30°C	0.8315×10^{-6}	5.68	0.6129	997.5
T_f, 40°C	0.657×10^{-6}	4.34	0.628	995
T_w, 50°C	0.5675×10^{-6}	3.68	0.63965	990

Solution: Equation (8.33): Properties at T_f

$$Re = 0.1 \times 0.6/0.657 \times 10^{-6} = 91{,}324,\ (9.13 \times 10^4)$$

$$Nu = C.\, Re^m\, Pr^{0.333}$$

From tables $C = 0.0266,\ m = 0.805$

$\therefore$ $Nu = 427.22 \quad \therefore \mathbf{h} = \dfrac{Nu\, k}{D} = \mathbf{2683\ W/m^2K}$

Equation (8.34): Properties at free stream temperature, T_∞

$$Re = 0.1 \times 0.6/0.8315 \times 10^{-6} = 72159$$

$$Nu = CRe^m\, Pr^n\, (Pr_f/Pr_w)^{0.25}$$

$$= 0.26\ (72159)^{0.6}\ (5.68)^{0.37}\ (5.68/3.68)^{0.25}$$

$$= 453.1,\ \mathbf{h} = \mathbf{2777\ W/m^2K}$$

Equation (8.36): Properties at film temperature

$$Nu = [0.35 + 0.56\ Re^{0.52}]\ Pr^{0.333}$$

$Re < 10^5$, so applicable

$\therefore$
$$Nu = [0.35 + 0.56\ (91324)^{0.52}]\ [4.34]^{0.333}$$

$$= 347.5,\ \mathbf{h} = \mathbf{2182\ W/m^2K}$$

Equation [8.35(*b*)] Properties at free stream temperature

$Re = 72159$ applicable

$$Nu = 0.25\ Re^{0.6}\ Pr^{0.38}\ (Pr_f/Pr_w)^{0.25}$$

$$= 0.25\ (72159)^{0.6}\ (5.68)^{0.38} \left(\frac{5.68}{3.68}\right)^{0.25}$$

$$= 443.3 \quad \therefore \quad \mathbf{h} = \mathbf{2717\ W/m^2K}$$

(note the similarity between equation (8.34) and this)

Equation (8.37): Properties at T_∞

$$Nu = (0.4\ Re^{0.5} + 0.06\ Re^{0.67})\ Pr^{0.4} \left[\frac{\mu_\infty}{\mu_w}\right]^{0.25}$$

$Re = 72159$

$Nu = 475.6 \quad \therefore \quad \mathbf{h} = \mathbf{2915\ W/m^2K}$

Equation [8.38(*b*)]: Properties at film temperature

$$Nu = 0.3 + \frac{0.62\ Re^{0.5}\, Pr^{0.333}}{\left[1+\left(\dfrac{0.4}{Pr}\right)^{0.67}\right]^{0.25}} \left[1+\left(\frac{Re}{282000}\right)^{0.5}\right]$$

$Re = 91324 \quad Pr = 4.34,\ k = 0.628$

$Nu = 458.4 \quad \therefore\ \mathbf{h} = \mathbf{2877\ W/m^2K}$

COMMENT

In this case also, the values by various correlations fall in a small band varying from 2181-2915 with fewer lower than 2700. For water and air, the correlations give less scatter. But it is essential that in real life design, to investigate all possible methods so that a safe and optimum design can be produced.

Example 8.16: *Air flows across an elliptical tube 0.1 m by 0.15 m perpendicular to the minor axis with a velocity of 2.4 m/s. Air is at 20°C and the tube surface is at 40°C. Determine the value of convection coefficient.*

Solution: The properties are required at the film temperature *i.e.* 30°C

$$\nu = 16 \times 10^{-6}\ \text{m}^2/\text{s},\ Pr = 0.701,\ k = 0.02675$$

$$Re = 0.1 \times 2.4/16 \times 10^{-6} = 15000$$

$Nu = C.\ Re^n$, From tables for noncircular shapes

C = 0.224, n = 0.612

$$\therefore \quad Nu = 0.224 \times 15000^{0.612} = 80.54$$

$$\mathbf{h} = \frac{80.54 \times 0.02675}{0.1} = \mathbf{21.54\ W/m^2K.}$$

8.4 FLOW ACROSS SPHERES

There are a number of applications for flow over spheres in industrial processes. As in the case of flow across cylinders, the flow development has a great influence on heat transfer. Various correlations have been obtained from experimental measurements and these are listed in the following paras.

The following three relations are useful for air with $Pr = 0.71$ (1954)

$$Nu = 0.37\ Re^{0.6} \quad 17 < Re < 7000 \qquad ...(8.41)$$

With Properties evaluated at film temperature.

The next correlation can be used for higher values of Re (1978)

$$Nu = 2 + [0.25\ Re + 3 \times 10^{-4}\ Re^{1.6}]^{0.5} \qquad ...(8.42)$$

$$100 < Re < 3 \times 10^5$$

For still higher values (1978)

$$Nu = 430 + 5 \times 10^{-3}\ Re + 0.025 \times 10^{-9}\ Re^2 - 3.1 \times 10^{-17}\ Re^3 \qquad ...(8.43)$$

$$3 \times 10^5 < Re < 5 \times 10^6$$

The next correlation is **for liquids** (1946)

$$Nu Pr^{-0.3} = 0.97 + 0.68\ Re^{0.5} \qquad ...(8.44)$$

$1 < Re < 2000$ with properties at T_f

For **oils and water** and for higher values of Re (1961)

$$Nu Pr^{-0.3} \left(\frac{\mu_w}{\mu_\infty}\right)^{0.25} = 1.2 + 0.53\ Re^{0.54} \qquad ...(8.45)$$

$1 < Re < 200000$ with properties at T_∞

A more **recent and a general equation** is (1972)

$$Nu = 2 + (0.4\ Re^{0.5} + 0.06\ Re^{0.67})\ Pr^{0.4} \left(\frac{\mu_w}{\mu_\infty}\right)^{0.25} \qquad ...(8.46)$$

$3.5 < Re < 8 \times 10^4$, $0.7 < Pr < 380$ and properties at T_∞

$$1 < \left(\frac{\mu_w}{\mu_\infty}\right) < 3.2$$

For a sphere falling in a fluid like quenching in hot bath,

$$Nu = 2 + 0.6 \; Re^{0.5} \, Pr^{0.333} \, [25 \, (x/D)]^{-0.7} \quad \text{...(8.47)}$$

For low values of **Pr (liquid metals)**

$$Nu = 2 + 0.386 \, (Re \, Pr)^{0.5} \quad \text{...(8.48)}$$

$3.56 \times 10^4 < Re < 1.525 \times 10^5$ with properties at film temperature.

These relations also provide values in the range of ± 25%.

Example 8.17: *Air at 30°C flows over a sphere of 0.1 m dia. with a velocity of 8 m / s. The solid surface is at 50°C. Determine the value of convection coefficient.*

Solution: The property values are

	°C	$\nu \times 10^6$ m^2/s	*Pr*	*k, W/mK*	$\mu \times 10^6$, *kg/ms*
T_∞	30	16	0.701	0.02675	18.63
T_f	40	16.96	0.699	0.02756	19.12
T_w	50	17.95	0.698	0.02820	19.61

At film temp.: $Re = 0.1 \times 8/16.96 \times 10^{-6} = 47170$

∴ Equation (8.41) can be used

$$Nu = 0.37. \, Re^{0.6} = 235.72, \quad \therefore \; \mathbf{h = 64.96 \; W/m^2K}$$

Using eqn. (8.42)

$$Nu = 2 + (0.25 \, Re + 3 \times 10^{-4} \, Re^{1.6}]^{0.5} = 146.25$$

$$\therefore \quad \mathbf{h = 40.3 \; W/m^2K}$$

Another possible equation is (8.46) (properties at T_∞)

$$Nu = 2 + (0.4 \, Re^{0.5} + 0.06 \, Re^{0.67}) \, Pr^{0.4} \left(\frac{\mu_\infty}{\mu_w}\right)^{0.25}$$

$$= 2 + \left[0.4\left[\frac{0.1 \times 8}{16 \times 10^{-6}}\right]^{0.5} + 0.06\left[\frac{0.1 \times 8}{16 \times 10^{-6}}\right]^{0.67}\right](0.701)^{0.4}\left[\frac{18.63}{19.61}\right]^{0.25}$$

$$= 151.88$$

$$\therefore \quad \mathbf{h = 40.62 \; W/m^2K}$$

COMMENT

The equation (8.46) being the latest correlation, it is safer to consider the value of 40.62 W/m²K for convection coefficient.

Example 8.18: ***Engine oil*** *flows over a sphere of 4 cm dia. with a velocity of 0.31 m/s. The oil is at 40°C and the ball is at 80°C. Determine the value of convection coefficient.*

Solution: Two possible correlations are (8.45) and (8.46).

Equation (8.45),

$$Nu \, Pr^{-0.3} \left(\frac{\mu_w}{\mu_\infty}\right)^{0.25} = 1.2 + 0.53 \, Re^{0.54}$$

with properties at T_∞

Chapter 8

The property values are:

	°C	ν, m^2/s	Pr	k, W/mK,	ρ kg/m^3
T_∞	40	241×10^{-6}	2870	0.1442	876
T_f	60	83×10^{-6}	1050	0.1407	864
T_w	80	37×10^{-6}	490	0.1384	852

$$Re = \frac{0.31 \times 0.04}{241 \times 10^{-6}} = 51.45 \quad \therefore \text{ The equation (8.45) is valid}$$

$$Nu\,(2870)^{-0.3} \left[\frac{37 \times 10^{-6} \times 852}{241 \times 10^{-6} \times 876}\right]^{0.25} = 1.2 + 0.53 \left[\frac{0.31 \times 0.04}{241 \times 10^{-6}}\right]^{0.54}$$

$$Nu \times 0.057 = 1.2 + 4.45 = 5.65$$

$$\therefore \quad Nu = 99.07$$

$$\therefore \quad \mathbf{h = 357.15\ W/m^2K}$$

The other correlation is given by equation (8.46)

$$Nu = 2 + (0.4\, Re^{0.5} + 0.06\, Re^{0.67})\, Pr^{0.4} \left[\frac{\mu_\infty}{\mu_w}\right]^{0.25}$$

with properties at T_∞

$$\therefore \quad Nu = 2 + 144.2 = 146.2$$

$$\therefore \quad \mathbf{h = 527.12\ W/m^2K}$$

Here one of the conditions $\mu_\infty/\mu_w < 3.2$ is not satisfied and the ratio is about 6.5.

The other possible correlation is 8.44 with properties at T_f

$$Nu\, Pr^{-0.3} = 0.97 + 0.68\, Re^{0.5}$$

$$Nu \times 0.124 = 0.97 + 0.68\, (0.31 \times 0.04/83 \times 10^{-6})^{0.5}$$

$$= 0.97 + 0.68\, (149.4)^{0.5}$$

$$Nu = 74.81 \quad \therefore \mathbf{h = 263.15\ W/m^2K}$$

Note that the scatter is a little more than 25% between 263.15 and 357.15.This is generally the case with fluids of higher Prandtl numbers.

Example 8.19: ***Liquid sodium at 200°C** flows over a sphere at 400°C, the diameter of the sphere being 4 cm. The velocity of flow is 0.8 m/s. Determine the value of convection coefficient.*

	°C	ν, m^2/s	Pr	k, W/mK	ρ, kg/m^3
T_∞	200	0.506×10^{-6}	0.0075	81.41	903
T_f	300	0.394×10^{-6}	0.0063	70.94	878
T_w	400	0.330×10^{-6}	0.0056	63.97	854

Considering properties at T_f

$$Re = 0.04 \times 0.8/0.394 \times 10^{-6} = 0.8122 \times 10^5$$

Solution: $\therefore$ Equation (8.48) is applicable in this range of Reynolds number

$$Nu = 2 + 0.386\,(Re\,Pr)^{0.5} = 10.73$$

$\therefore$ $$\mathbf{h = 19032\ W/m^2K}$$

8.5 FLOW OVER BLUFF BODIES

Perpendicular to flat plate:

$$Nu = 0.2\,\text{Re}^{2/3} \qquad ...[8.49(a)]$$

$$St.\,Pr^{2/3} = 0.930\,Re_L^{-0.5} \text{ for all angles of attack} \qquad ...[8.49(b)]$$

$$1 < Re < 10^5 \text{ with properties at } T_f$$

For half cylinder with flat surface at the rear

$$Nu = 0.16\,Re^{2/3} \qquad ...(8.50)$$

$$1 < Re < 10^5 \text{ with properties at } T_f.$$

8.5.1. Packed Beds

Definition of two terms are necessary in dealing with packed beds. These are

(*i*) Packing diameter $D_P = \dfrac{6V}{A}$ where V is the volume and A is the surface area.

(*ii*) Void fraction ε = the empty volume/total volume of bed. The equation to find the rate **of heat flow from gas to the packing is**

$$\frac{h_c D_p}{k} = \frac{1-\varepsilon}{\varepsilon}\,[0.5\,Re_{DP}^{0.5} + 0.2\,Re_{DP}^{0.67}]\,Pr^{1/3} \qquad ...(8.51)$$

$$10 < Re_{DP} < 10^4 \quad 0.34 < \varepsilon < 0.78$$

$$Re = \frac{D_P.V_s}{\nu(1-\varepsilon)},$$ where V_s is the fluid velocity if the bed is empty.

For heat flow from wall to gas (for cylinders)

$$Nu_{DP} = 2.58\,Re_{DP}^{0.33}\,Pr^{0.33} + 0.094\,Re_{DP}^{0.8}\,Pr^{0.4} \qquad ...[8.52(a)]$$

for spheres

$$Nu_{DP} = 0.208\,Re_{DP}^{0.33}\,Pr^{0.33} + 0.220\,Re_{DP}^{0.8}Pr^{0.4} \qquad ...[8.52(b)]$$

$$40 < Re_{DP} < 2000 \quad \text{and} \quad Re_{DP} = \frac{u_\infty D_P}{\nu}$$

See under advanced examples.

8.6 FLOW ACROSS BANK OF TUBES

In most heat exchangers in use, tube bundles are used with one fluid flowing across tube bundles. First it is necessary to define certain terms before discussing heat transfer calculations. Two types of tube arrangement are possible:

(*i*) inline and (*ii*) staggered. The distance between tube centres is known as pitch. The pitch along the flow is known as (S_n) and the pitch in the perpendicular direction is called (S_p). These are shown in Fig. 8.5. Other notations like S_T and S_L in place of S_P and S_n, T denoting transverse and L denoting longitudinal are also used.

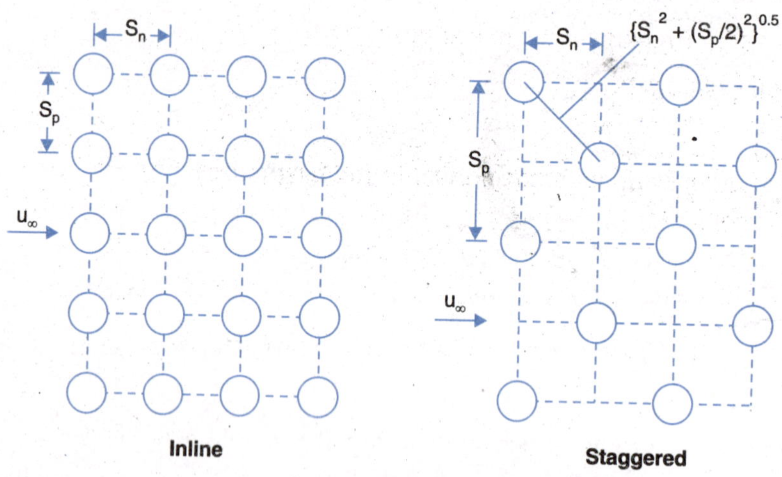

Fig. 8.5. Tube arrangements in tube banks

Due to the obstruction caused by the tubes, the velocity near the tube increases and this increased value has to be used in the calculation of Reynolds number. In the case of in line the actual velocity near the tubes

$$V_{max} = [S_p/(S_p - D)]u_\infty \qquad ...(8.53)$$

In the case of staggered arrangement the larger of the value given by (8.53) and (8.54) is to be used

$$V_{max} = [S_p/2\ (S_D - D)]u_\infty \qquad ...(8.54)$$

where

$$S_D = \left[S_n^{\ 2} + \left(\frac{S_p}{2}\right)^2\right]^{0.5} \qquad ...(8.55)$$

This is because of the larger obstruction possible in the staggered arrangement.

For number of rows of tubes of 10 or more

$$Nu = 1.13\ C\ Re^n.\ Pr^{0.33} \qquad ...(8.56)$$

$$N \geq 10,\ 2000 < Re < 40000$$

Reynolds number to be calculated based on V_{max}: The property values should be at T_f. The value of C and n are tabulated below in Table 8.1. For larger values of S_p/D, tubes can be considered as individual tubes rather than tube bank.

Table 8.1. Values of constant and index for equation (8.56).

S_p/D	1.25		1.5		2.0		3.0	
	C	*n*	*C*	*n*	*C*	*n*	*C*	*n*
Inline								
1.25	0.348	0.592	0.275	0.608	0.100	0.704	0.0633	0.752
1.5	0.367	0.586	0.250	0.620	0.101	0.702	0.0678	0.744
2.0	0.418	0.570	0.299	0.602	0.229	0.632	0.1980	0.648
3.0	0.290	0.601	0.357	0.584	0.374	0.581	0.2860	0.608
Staggered								
0.6	—	—	—	—	—	—	0.213	0.636
0.9	—	—	—	—	0.446	0.571	0.401	0.581
1.0	—	—	0.497	0.558	—	—	—	—
1.125	—	—	—	—	0.478	0.565	0.518	0.560
1.25	0.518	0.556	0.505	0.554	0.519	0.556	0.552	0.562
1.5	0.451	0.568	0.460	0.562	0.452	0.568	0.488	0.568
2.0	0.404	0.572	0.416	0.568	0.482	0.556	0.449	0.570
0.3	0.310	0.592	0.356	0.580	0.440	0.562	0.421	0.574

In case the number of tubes is less than 10 in depth, the value of convection coefficient should be multiplied by the fraction given in table below (Table 8.2).

Table 8.2. Correction factors for tube banks with rows less than 10 values of (h_N/h_{10})

N	1	2	3	4	5	6	7	8	9	10
Inline	0.68	0.75	0.83	0.89	0.92	0.95	0.97	0.98	0.99	1.0
Staggered	0.64	0.30	0.87	0.90	0.92	0.94	0.96	0.98	0.99	1.0

The calculation may appear tedius, but such flows are very common in practical equipments and hence the importance.

If the number of rows in the flow direction is greater than 20, then the following correlation is applicable. This equation is found applicable for less than 20 rows deep also (up to 4)

$$Nu = C\,Re^m Pr^{0.36}\left[\frac{Pr_\infty}{Pr_w}\right]^{0.25} \qquad ...(8.57)$$

$$N_L \geq 20,\ 0.7 < Pr < 500,\ 1000 < Re < 2 \times 10^6$$

Property values at T_∞ *Re* based on V_{max}

The values of *C* and *m* are tabulated in Table 8.3.

Chapter 8

The convection factors for number of rows in the bank being less than 20 is tabulated below.

Correction factor for equation (8.57) for number of rows less than 20.

N_L	1	2	3	4	5	7	10	13	16
Aligned	0.70	0.80	0.86	0.90	0.92	0.95	0.97	0.98	0.99
Staggered	0.64	0.76	0.84	0.89	0.92	0.95	0.97	0.98	0.99

Table 8.3. Constants for equation (8.57)

Conditions	*C*	*m*
Inline, $10^3 < Re < 2 \times 10^5$	0.27	0.63
Inline, $2 \times 10^5 < Re < 2 \times 10^6$	0.021	0.84
Staggered: $10^3 < Re < 2 \times 10^5$		
$S_p/S_n < 2$	$0.35 \left(\frac{S_p}{S_n}\right)^{0.2}$	0.60
$S_p/S_n > 2$	0.4	0.60
$2 \times 10^5 < Re < 10^6$	0.022	0.82

Liquid metals are now in use in heat exchangers.

The available correlation for tube bank is

$$Nu = 403 + 0.228\,(Re)^{0.67} \qquad ...(8.58)$$

$$2000 < Re < 80000$$

For finned tubes correlations are more complicated but are available in handbooks.

As the temperature varies along the flow, ΔT varies along the flow. Instead of arithmetic average of $(\Delta T)_{in}$ and $(\Delta T)_{exit}$, LMTD should be used, $(\Delta T)_{in} = T_s - T_i$, where T_s is tube surface temperature and T_i is the fluid inlet temperature. $(\Delta T)_{exit} = T_s - T_0$, where T_o is the fluid outlet temperature.

$$(\Delta T)_{mean} = \frac{(\Delta T)_{in} - (\Delta T)_{exit}}{\ln\,[(\Delta T_{in})/(\Delta T_{exit})]}$$

To determine T_o, the following equation (8.59) is used.

$$\frac{T_s - T_o}{T_s - T_i} = \exp\left[-\frac{\pi D N \bar{h}}{\rho V N_T S_P c}\right] \qquad ...(8.59)$$

Where N is the total number of tubes in the bank and N_T is the number of tubes in the transverse plane. T_o is determined by equation (8.59).

$$q = N\,(\bar{h}\pi D\,\Delta T_{mean}) \qquad ...(8.60)$$

q is the heat transfer for unit height of bank.

Example 8.20: *20 mm OD copper tubes are arranged in line at 30 mm pitch perpendicular to flow and 25 mm pitch along the flow. The entry velocity of air is 1 m/s, and the air temperature is 20°C. The tube wall is at 40°C. Determine the value of convection coefficient if the number of tubes along the flow is 6 (or Bank is 6 rows deep).*

Solution:

$S_p = 30$ mm, $D = 20$ mm, $S_p/D = 1.5$

$S_n = 25$ mm $S_n/D = 1.25$

From tables, $C = 0.367$, $n = 0.586$ (look carefully for S_p/D and S_n/D)

Property values at $T_f = 30°C$ are

$$\nu = 16 \times 10^{-6} \text{ m}^2/\text{s},\ Pr = 0.701,\ k = 0.02675 \text{ W/mK}$$

$$V_{max} = [S_p/(S_p - D)]\, u_\infty = \frac{30}{10}.1 = 3 \text{ m/s}$$

$$\therefore \quad Nu = 0.367\, Re^{0.586}$$

$$Re = 3 \times 0.02/16 \times 10^{-6} = 3750 > 2000$$

Hence equation is applicable:

$$Nu = 45.61 \quad \therefore \quad \mathbf{h = 61.0\ W/m^2K}$$

But the bank is only 6 rows deep

$$\therefore \quad h_6 = h_{10} \times 0.95 = \mathbf{57.95\ W/m^2K}$$

(Value 0.95 is read from Table 8.2).

Example 8.21: *Work out the value of h for* ***staggered arrangement*** *using data of example 8.20.*

Solution: From tables for $S_p/D = 1.5$ and $S_n/D = 1.25$

$$C = 0.451,\ m = 0.568$$

$$V_{max} = [S_p/2(S_D - D)]\, u_\infty = 3 \text{ m/s}$$

$$S_D = \left[Sn^2 + \left(\frac{S_p}{2}\right)^2\right]^{0.5} = [25^2 + 15^2]^{0.5} = 29.15$$

$$\therefore \quad V_{max} = [20/2\,(29.15 - 20)] = 1.6385, \text{ lower}$$

$\therefore$ 3 m/s is taken as from $[S_p/(S_p - D)]\, u_\infty$

$$\therefore \quad Nu = 0.451\,(3750)^{0.568} = 48.33$$

$$\therefore \quad \mathbf{h = 64.64\ W/m^2K}$$

For 6 rows: correction factor is 0.94

$$\mathbf{h_6} = 64.64 \times 0.94 = \mathbf{60.76\ W/m^2K}$$

For increasing the value of h, S_p should be reduced.

Example 8.22: *Using the data of example 8.20 and assuming 10 rows perpendicular to the flow, determine the heat transfer rate for 1 m height of tube bank.*

Solution: Using equation (8.59),

$$\frac{T_s - T_o}{T_s - T_i} = \exp\left[-\frac{\pi D N h}{\rho V N_T S_P C}\right]$$

$$\frac{40 - T_o}{40 - 20} = \exp\left[-\frac{\pi \times 0.02 \times 60 \times 57.95}{1.25 \times 3 \times 10 \times 0.03 \times 1005}\right]$$

Solving $T_o = 23.51°C$.

$$\text{LMTD} = \frac{(40-20)-(40-23.51)}{\ln\left[\frac{40-20}{40-23.51}\right]} = 18.19.$$

$$\mathbf{q} = 60 \times 57.95 \times \pi \times 0.02 \times 18.19 = \mathbf{3974\ W/m\ height.}$$

COMMENT

In case arithmetic average of T_i to T_o was used, $\Delta T = 21.755°C$, which will give a higher value of q.

ADVANCED EXAMPLES

Example A-8.1: *The local Nusselt number in the case of rough plate was correlated to give*

$$Nu_x = 0.04\ Re_x^{0.9} Pr^{1/3}.$$

Determine the average value up to a length L.

Solution:

$$\frac{h_x x}{k} = 0.04\left(\frac{u_\infty}{\nu}\right)^{0.9} x^{0.9} Pr^{1/3}$$

$$h_x = 0.04k.\ Pr^{1/3}\left(\frac{u_\infty}{\nu}\right)^{0.9} x^{-0.1}$$

to determine the average length up to L,

$$\bar{h} = \frac{1}{L}\int_0^L h_x dx = \frac{1}{L} \times 0.04\ k\ Pr^{1/3}\left(\frac{u_\infty}{\nu}\right)^{0.9}\int_0^L x^{-0.1} dx$$

$$= \frac{0.04\ k\ Pr^{1/3}(u_\infty / v)^{0.9}}{L}.\left[\frac{x^{-0.1+1}}{-0.1+1}\right]_0^L$$

or

$$\bar{h} = \frac{0.04\ k\ Pr^{1/3}}{L}\left(\frac{u_\infty}{\nu}\right)^{0.9}[L^{0.9}/0.9]$$

or

$$\frac{\bar{h}L}{k} = \frac{0.04}{0.9} Pr^{1/3} Re_L^{0.9}$$

$$\overline{Nu} = \frac{0.04}{0.9} Re_L{}^{0.9}\, Pr^{1/3} \text{ or } \overline{Nu} = \frac{1}{0.9} Nu_L$$

generally average will be $(1/n)$ times Nu_L where n is the index of Reynolds number.

Example A-8.2: *In testing a model, the following measurements were made*

Average Nusselt No.	*Reynolds number*
355	*80,000*
424	*100,000*
587	*150,000*
739	*200,000*

The test is done using air at film temperature of 30°C. If the results can be correlated by an equation of the form $\overline{Nu} = C\, Re^m\, Pr^{1/3}$, *determine C and m. For flow of air over a similar surface of length 6 cm at 70°C with a velocity of 40 m / s, determine the average value of convection coefficient. Air is at 110°C and 1 atm.*

Solution: Two readings are sufficient to determine the two unknowns, namely C and m. However, a check can be made to determine the average values of C and m

At 30°C, $Pr = 0.701$ for air

$$355 = C\,(80{,}000)^m\,(0.701)^{0.333} \quad \text{...(1)}$$

$$424 = C\,(100{,}000)^m\,(0.701)^{0.333} \quad \text{...(2)}$$

$$587 = C\,(150{,}000)^m\,(0.701)^{0.333} \quad \text{...(3)}$$

$$739 = C\,(200{,}000)^m\,(0.701)^{0.333} \quad \text{...(4)}$$

Dividing these expressions one by the other and taking **logarithms** and solving, the **average value of m = 0.8.**

Substituting this value and finding C, the average is

$$\mathbf{C = 0.04777}$$

As an example dividing equation (1) by equation (2)

$$\frac{355}{424} = (0.8)^m$$

taking **logarithms** and solving $m = 0.796$, and

$$355 = C\,(80{,}000)^{0.796} \times (0.70)^{0.333}$$

$$= C\,(7995.76 \times 0.8883)$$

$$\therefore \quad C = 0.04997$$

$$\therefore \quad \overline{\mathbf{Nu}} = \mathbf{0.04777\ Re^{0.8}\ Pr^{1/3}}$$

For the given data, $T_f = (70 + 100)/2 = 80°C$, $u = 40$ m/s, $L = 0.06$ m

$\nu = 21.09 \times 10^{-6}$ m²/s, $Pr = 0.692$, $k = 0.03047$ W/mK

$$\therefore \quad Re = 0.06 \times 40/21.06 \times 10^{-6} = 113{,}798$$

$$\therefore \quad \overline{Nu} = 0.04777 \times (113798)^{0.8}\,(0.692)^{0.333} = 468.56$$

$\therefore$ $\quad$ **h** = 468.56 × 0.03047/0.06 = **237.95 W/m²K**

Both Nu and Re lie between data 2 and 3 as a check.

Example A-8.3: *In flow over a wedge, the local Nusselt number is given by*

$$Nu_x = C.\ Re_x^{1/2} \text{ where}$$

$$Re_x = \left(\frac{Vx^m}{\nu}.x\right) \text{ and}$$

$m = \beta/(2 - \beta)$. Where V is the approach velocity and $\pi\beta$ is the wedge angle. Determine the relation between average and local Nusselt number for a length L for $\beta = 0.2$, $\beta = 0.5$ and $\beta = 1.0$.

Solution:

$$h_x = \frac{k}{x}.C\left(\frac{V}{\nu}\right)^{0.5} x^{(m/2)+0.5} = k.C\left(\frac{V}{\nu}\right)^{0.5} x^{(0.5m-0.5)}$$

$$\bar{h} = \frac{1}{L}\int h_x dx = k.C\left(\frac{V}{\nu}\right)^{0.5}.\frac{1}{L}.\int_0^L x^{0.5(m-1)}dx$$

$$= \frac{k.C}{L}.\left(\frac{V}{\nu}\right)^{0.5}\frac{1}{0.5\,(m+1)}.L^{0.5(m-1)+1}$$

the general expression is:

$$\therefore \quad \frac{\bar{h}L}{k} = \frac{C}{0.5(m+1)}\left(\frac{VL^m.L}{\nu}\right)^{0.5}$$

$$\text{As, } C\,Re_L^{0.5} = Nu_L,\ \overline{Nu}_L = \frac{C}{0.5(m+1)}Re_L^{0.5} = \frac{1}{0.5(m+1)}.Nu_L = \overline{Nu}_L$$

$$\text{For } \beta = 0.2,\quad m = \frac{0.2}{1.8} \quad \therefore \frac{1}{0.5(m+1)} = \frac{1}{0.5\left(\frac{0.2}{1.8}+1\right)} = 1.8$$

$$\therefore \quad \overline{Nu} = 1.8\ Nu_L$$

$$\text{For } \beta = 0.5,\quad m = \frac{0.5}{1.5} = \frac{1}{3},\ \frac{1}{0.5(m+1)} = \frac{1}{0.5(4/3)} = \frac{3}{2}$$

$$\overline{Nu} = 1.5\ Nu_L$$

$$\beta = 1.0,\quad m = 1,\ \frac{1}{0.5(1+1)} = 1$$

$$\therefore \quad \overline{Nu} = Nu_L$$

Note: If $\beta = 0$, $\overline{Nu} = Nu_L$ same as flat plate.

Example A-8.4: *A plate 5 m long at 470°C has air flowing over it with a velocity of 2.5 m/s. The air is at 30°C. Determine the heat to be supplied for every m length. Assume unit width.*

Solution: The values to be determined are the heat transfer in the first, second and subsequent *m* lengths. This can be done by calculating the average heat transfer coefficient for 1 m, 2 m, 3 m, 4 m and 5 m lengths and then finding the heat transfer in each of the case and then taking the difference.

The film temperature = (470 + 30)/2 = 250°C

The property values are: $\nu = 40.61 \times 10^{-6}$ m²/s, $Pr = 0.677$, $k = 0.04268$ W/mK

Distance, m	*1*	*2*	*3*	*4*	*5*
$Re \times 10^5$	0.615	1.231	1.847	2.462	3.078
Nu	144.06	204.58	250.56	289.32	323.47
$\bar{h}$	6.174	4.366	3.565	3.087	2.761

The flow is completely in the laminar region:

$\therefore$ $\overline{Nu} = 0.664\, Re^{0.5} Pr^{0.33}$. The values and $\bar{h}$ are tabulated above.

The heat flow: (width 1 m) $Q = hA\,(T_w - T_\infty)$, $T_w = 470°C$, $T_\infty = 30°C$

L, m		*Q, W*	*ΔQ, W*
1	1 × 6.174 × (470 – 30)	= 2716.6	2716.6
2	2 × 4.366 × (470 – 30)	= 3842.1	1225.5
3	3 × 3.565 × (470 – 30)	= 4705.8	863.7
4	4 × 3.087 × (470 – 30)	= 5433.1	727.3
5	5 × 2.761 × (470 – 30)	= 6074.2	641.1

COMMENTS

1. The heat transfer in the extended length decreases as it should.
2. In case the data is such that the flow is turbulent before the end of the plate, the heat flow will increase from the location where the flow turns turbulent.

Example A-8.5: *A motor cycle travels at 100 kmph. On the engine head a fin of 0.16 m length and 0.04 m width is exposed to convection on both sides. The fin surface is at 300°C and air is at 20°C. Determine the rate of heat removal from the fin assuming turbulent flow prevails all through.*

Solution: In this case it is assumed that due to disturbances turbulent flow starts even at low Reynolds numbers.

The film temperature = (300 + 20)/2 = 160°C

The property values are $\nu = 30.09 \times 10^{-6}$ m²/s, $Pr = 0.682$, $k = 0.0364$ W/mK

$$\overline{Nu} = 0.037\, Re_L^{\,0.8}\, Pr^{0.33}$$

$$\therefore \quad \overline{Nu} = 0.037\, ((100000/3600) \times 0.16/30.09 \times 10^{-6})^{0.8}\, (0.682)^{0.333} = 444.95$$

$$\therefore \quad \mathbf{\bar{h} = 101.23\ W/m^2K}$$

$$\therefore \quad \mathbf{Q} = hA\,\Delta T = 101.23 \times 0.16 \times 0.04 \times 2 \times (300 - 20) = \mathbf{362.8\ W/fin}$$

The Reynolds number is 1.477×10^5 $\therefore$ **laminar flow region.** Under such **assumption**

$$\overline{Nu} = 0.664\, Re^{0.5} Pr^{0.33} = 224.6, \quad \therefore \quad \mathbf{h = 51.1\ W/m^2}$$

The heat transfer is only about 50% of that assuming turbulent conditions.

Example A-8.6: *An aircraft travelling at 300 kmph has a wing span of 2 m and is at an altitude where the pressure is 0.7 bar and temperature is –10°C. The wing absorbs solar radiation at 800 W/m². Determine the wing surface temperature under this condition.*

Solution: This example may be modelled as uniform heat flux model. The film temperature is not known. Assuming 0°C, the property values are

$$\nu = 13.28 \times 10^{-6}\ \text{m}^2/\text{s},\ Pr = 0.707,\ k = 0.02442\ \text{W/mK}$$

$$u_\infty = 300 \times 1000/3600\ \text{m/s}$$

As pressure is reduced to 0.7 bar, ν will increase by 1/0.7, due to the change in density

$$\therefore \qquad Re_L = \left(\frac{300 \times 1000}{3600} \times \frac{2 \times 0.7}{13.26 \times 10^{-6}}\right) = 8.8 \times 10^6$$

This is in the turbulent region. (Fully turbulent condition is assumed as L_{cr} = 0.11 m) The local Nusselt number is calculated using

$$Nu_x = 0.0296\ Re_x^{0.8}\ Pr^{0.33} = 9476$$

$$h_x = 115.7\ \text{W/m}^2\text{K}$$

For **constant heat flux = $\mathbf{h_x = h_x \times 1.04 = 120.3}$ W/m²K**

ΔT at this location:

$$800 = 120.3 \times 1 \times \Delta T = \mathbf{6.65°C}$$

∴ Wing temp. = **– 3.35°C** at the trailing edge.

The assumption of 0°C as T_f is in error. The new value can be now used and the results refined.

At a distance of 1 m from leading edge.

$Re = (8.8/2) \times 10^6$, fully turbulent condition is assumed.

$$Nu_x = 0.0296\ Re_x^{0.8}\ Pr^{0.33} = 5442$$

$$h_x = 132.9\ \text{W/m}^2\text{K}$$

Constant heat flux

$$h_x = 1.04 \times 132.9 = 138.2\ \text{W/m}^2\text{K}$$

$$\therefore \qquad \Delta T = 5.79°\text{C} \quad \text{or} \quad \mathbf{T = -4.21°C}$$

The minimum wing surface temperature is **–10°C** and the maximum **about – 3.35°C.**

Example A-8.7: *A surface 1 m × 1 m size has one half very rough and the other half smooth. The surface is at 100°C. Air at 72 kmph and 20°C flows over the surface. If the flow direction is reversed, is there a possibility of change in the average value of convection coefficient? If initially the rough area is at the leading edge, determine the change in the value.*

Solution: If the rough surface is at the leading edge, then the flow is turbulent all through. However, if the smooth surface is at the leading edge, turbulence will begin only at $Re = 5 \times 10^5$ or at the beginning of the rough surface. So, the convection coefficient will depend on the direction of flow.

The film temperature is (100 + 20)/2 = 60°C.

The property values are: $\nu = 18.97 \times 10^{-6}$ m²/s, Pr = 0.696, k = 0.02896 W/mK. Initially the rough surface is at the leading edge. So the flow is turbulent all through. $Re_x = 1.054 \times 10^6$ (calculated).

$$\therefore \quad \overline{\mathbf{h}} = \frac{k}{L} 0.037\,(Re_L)^{0.8}\,Pr^{0.333}$$

$$= \frac{0.02896}{1} \times 0.037 \times \left(\frac{72000}{3600} \times 1 \times \frac{1}{18.97 \times 10^{-6}} \right)^{0.8} 0.696^{0.333}$$

$$= \mathbf{62.5\ W/m^2K}$$

When the smooth surface is at leading edge,

The Reynolds number at mid location

$$= \left(\frac{72000}{3600} \times \frac{0.5}{18.97 \times 10^{-6}} \right) = 5.27 \times 10^5$$

$\therefore$ Turbulence starts even before the rough surface is met

$$\therefore \quad \overline{h} = \frac{k}{L}\,(0.037\,Re_L{}^{0.8} - 871)Pr^{0.33} = \mathbf{40.15\ W/m^2K}$$

A change of about 36%.

It may so happen that rough surface may start even before the value of $Re = 5 \times 10^5$. In such a case the critical Reynolds number should be taken as the Reynolds number at the location where the rough area begins. Then the number 871 will be different. Equation (8.30) and (8.31) should be used in such a case

i.e. $\quad \overline{Nu} = Pr^{0.333}\,(0.37\,Re_L{}^{0.8} - A)$

where $\quad A = 0.037\,Re_{cr}{}^{0.8} - 0.664\,Re_{cr}{}^{0.5}$.

Example A-8.8: *Wind blows at 20 kmph parallel to the wall of adjacent rooms. The first room extends to 10 m and the next one to 5 m. The wall is 3.2 m high. The room inside is at 20°C and the ambient air is at 40°C. The walls are 25 cm thick and the conductivity or the material is 1.2 W/mK. On the inside convection coefficient has a value of 6 W/m²K. Determine the heat gain through the walls of each room.*

Solution: The film temperature is not known. But it has to be between 40°C and 20°C. A choice is made as 35°C.

The property values are: $\upsilon = 16.48 \times 10^{-6}$ m²/s, $Pr = 0.70$, $k = 0.02716$ W/mK

Reynolds numbers at 10 m and 15 m locations are:

10 m: $$Re = \left(10 \times \frac{20000}{3600} \times \frac{1}{16.48 \times 10^{-6}} \right) = 3.371 \times 10^6$$

15 m: $$Re = \left(15 \times \frac{20000}{3600} \times \frac{1}{16.48 \times 10^{-6}} \right) = 5.0566 \times 10^6$$

Average values of h are:

10 m: $$\overline{\mathbf{h}} = \frac{k}{L}.Pr^{0.33}\,(0.037\,Re^{0.8} - 871) = \mathbf{12.783\ W/m^2K}$$

Similarly, **for 15 m:** $\overline{\mathbf{h}}$ **= 12.324 W/m²K.**

∴ Average for **last 5 m** = (12.324 × 15 – 12.783 × 10)/5 = **11.406 W/m²K**

Heat gain in the first room

$$= \frac{A\,\Delta T}{\frac{1}{h_1}+\frac{L_1}{k_1}+\frac{1}{h_2}} = \frac{20\times 3.2\times 10}{\frac{1}{12.783}+\frac{0.25}{1.2}+\frac{1}{6}} = \mathbf{1412.1\ W}$$

Heat gain in the second room

$$= \frac{20\times 3.2\times 5}{\frac{1}{11.406}+\frac{0.25}{1.2}+\frac{1}{6}} = \mathbf{691.6\ W}$$

The surface temperature can now be found using the heat flow quantity at the surface

$$10 \times 3.2 \times 12.783\,(40 - T_s) = 1412.1 \quad \therefore\ \mathbf{T_s = 36.55°C}$$

The value of T_f can now be refined and calculations repeated to obtain more accurate values.

Example A-8.9: *Icebergs 1 km long by 0.8 km wide and 0.3 km thick at 0°C are proposed to be towed to arid regions for obtaining supply of fresh water. If the average water temperature is 10°C and if the iceberg is to travel at 1.2 km/hour, determine the thickness of ice melted per hour. The latent heat of ice is 334 kJ/kg. Assume that the iceberg is towed along the 1 km direction.*

Solution: The temperature of the ice is taken as 0°C. So the film temperature = 5°C.

Properties of water are: (using 0°C and 20°C values)

$$\nu = (1.788 - 1.006) \times 0.75 + 1.006 \times 10^{-6}\ \text{m}^2/\text{s}$$

$$Pr = (13.600 - 7.020) \times 0.75 + 7.020$$

$$k = (0.552 - 0.5978) \times 0.75 + 0.5978$$

$$\nu = 1.5925 \times 10^{-6}\ \text{m}^2/\text{s},\ Pr = 11.955,\ k = 0.56345\ \text{W/mK}$$

$$L = 1000\ \text{m},\ u_\infty = 1.2 \times 1000/3600 = 0.333\ \text{m/s, latent heat} = 3.34 \times 10^5\ \text{J/kg.}$$

∴ $\mathbf{Re_L = 0.2093 \times 10^9}$

∴ A suitable correlation is

∴ $St\ Pr^{2/3} = C_f/2$ can be used

Using equation (8.28) $C_{fL} = \dfrac{0.495}{(\log_{10} Re_L)^{2.584}} - \dfrac{1700}{Re_L} = 2.066 \times 10^{-3}$

$$\frac{Nu}{Re\ Pr}\ .\ Pr^{2/3} = 2.066 \times 10^{-3}/2$$

$$\therefore\quad Nu = \frac{C_f}{2} \times Re.\ Pr^{1/3} = 494478.6$$

$$\therefore\quad \bar{\mathbf{h}} = \frac{Nu \times k}{L} = \frac{494478.6}{1000} \times 0.56345 = \mathbf{278.61\ W/m^2K}$$

∴ **heat gain for each 1 m²** = $h\ (\Delta T)$. Time

$$= 278.61 \times 10 \times 3600 = \mathbf{10.03 \times 10^6\ J/h}$$

∴ **mass of ice melted** = $10.03 \times 10^6/3.34 \times 10^5$ = **30.03 kg/h**

The thickness melted = 30.03 mm/hr or 3 cm/hr.

Example A-8.10: *Water flows over a flat plate having a uniform heat generation rate. The plate is 15 cm × 15 cm side. Water is at 20°C and the flow velocity is 3 m/s. Determine the heat that may be carried away by the water if the maximum temperature of the plate is not to exceed 80°C.*

Solution: The property values can be evaluated at a mean temperature, *i.e.*, 50°C in this case to start with

Property values are:

$$\nu = (0.657 + 0.478) \times 10^{-6}/2 = 0.5675 \times 10^{-6} \text{ m}^2/\text{s}$$

$$Pr = (4.34 + 3.02)/2 = 3.68$$

$$k = (0.628 + 0.6513)/2 = 0.63956 \text{ W/mK}$$

$$Re = 3 \times 0.15/0.5675 \times 10^{-6} = 7.929 \times 10^5$$

∴ The flow is turbulent as $Re > 5 \times 10^5$

$$\therefore \qquad \overline{Nu} = (0.037 \, . \, Re^{0.8} - 871) \, Pr^{0.33} = 1649$$

$$\therefore \qquad \mathbf{h} = \frac{Nu \, . \, k}{L} = \frac{1649 \times 0.63956}{0.15} = 7032 \text{ W/m}^2\text{K}$$

For constant heat flux: h = $1.04 \times h$ (constant temp.)

$$= \mathbf{7313.25 \text{ W/m}^2\text{K}}$$

ΔT is taken as 60, *i.e.,* (80 – 20) as the temperature at the edge should not exceed 80°C. This is conservative prediction as the temperature difference varies over the length.

$$\therefore \qquad \mathbf{Q} = h\,A\,\Delta T = 7313.25 \times 0.15 \times 0.15 \times 60 = \mathbf{9873 \text{ W} = 9.873 \text{ kW}}$$

This assumes an average value of h. But the maximum temperature is to be at edge. So the local Nusselt number should be used for better estimate.

$$Nu_x = 0.0296 \, Re_x^{0.8} \, Pr^{0.33} = 2395 \quad \therefore \quad h_x = 10213$$

∴ **h** (constant heat flux) = **10621.6 W/m²K.**

$$q_x = h\,\Delta T = 10601.6 \times 60 = 637295 \text{ W/m}^2$$

∴ **Heat carried from plate** = $A.q$ = **14.34 kW**

There is a difference of 31%.

As the heat generation at every location is the same this method may be used.

These calculations can serve only as a first estimate, and refinements are necessary to get at more accurate values.

Example A-8.11: ***Glycerine at 30°C** flows past a 30 cm square flat plate at a velocity of 1.5 m/s. The drag force measured was 8.9 N. Determine the value of convection coefficient for such a system.*

Solution: This example has to use the analogy method. C_f can be calculated using the drag force. The film temperature is taken as 30°C.

density = 1258 kg/m^3, ν = 501 × 10^{-6} m^2/s, **Pr = 5380**, k = 0.2861 W/mK

$$\text{Force on } 1\text{ m}^2 = \frac{8.9}{0.3 \times 0.3}\text{ N}$$

$$C_f = \tau / \frac{\rho u_\infty^2}{2} = \frac{8.9}{0.3 \times 0.3} / \frac{1258 \times 1.5^2}{2} = 0.06987$$

$$St\, Pr^{2/3} = C_f/2 = 0.034935$$

$$Nu = (C_f/2) \times Re\, Pr^{1/3},\ Re = 1.5 \times 0.3/501 \times 10^{-6}$$

$$\therefore \quad Nu = 549.86 \quad \therefore \quad \mathbf{h} = \frac{Nu\,.\,k}{L} = \mathbf{524.4\ W/m^2/K}$$

Measurement of drag force is much easier compared to the measurement of heat flow. Hence the analogy method is very useful in estimating convection coefficients.

Example A-8.12: *Helium at a pressure of 0.15 atm and 30°C flows over a flat plate at 70°C at a velocity of 50 m/s. The plate is 1 m long. Calculate the value of convection coefficient.*

Solution: The film temperature is (70 + 30)/2 = 50°C

The property values are: density = (0.178 + 0.130)/2 kg/m^3

$$\nu = ((105 + 176)/2) \times 10^{-6}\text{ m}^2/\text{s},\ Pr = (0.684 + 0.667)/2$$

$$k = (0.14304 + 0.1791)/2$$

As the pressure is not 1 atm., the value of

$$\mu = \nu\rho \text{ only remains constant} \quad \therefore \quad \nu_1\rho_1 = \nu_2\rho_2$$

$$\therefore \quad \nu_2 = \nu_1 \frac{\rho_1}{\rho_2} = \nu_1 \times \frac{P_1}{P_2} \text{ as temperature is constant}$$

$$= 936.7 \times 10^{-6}\text{ m}^2/\text{s}$$

$$\therefore \quad Re = 50 \times 1/936.7 \times 10^{-6} = 0.53338 \times 10^5 \quad \therefore \quad \text{laminar}$$

$$\therefore \quad Nu = 0.664\, Re^{0.5}\, Pr^{0.333} = 134.6$$

$$\therefore \quad \mathbf{h = 21.68\ W/m^2K}$$

Example A-8.13: ***Liquid ammonia** at – 20°C flows with a **velocity of 5 m/s** over a plate 0.45 m length at 20°C. Determine the value of average convection coefficient.*

Solution: The film temperature is (– 20 + 20)/2 = 0°C

The property values are

$$\rho = 640\text{ kg/m}^3,\ \nu = 0.373 \times 10^{-6}\text{ m}^2/\text{s}$$

$$Pr = 2.050,\ k = 0.5396\text{ W/mK}$$

$$Re = 5 \times 0.45/0.373 \times 10^{-6} = 6.03 \times 10^6$$

Turbulent flow prevails.

$$\overline{Nu} = (0.037\, Re^{0.8} - 871)\, Pr^{0.333} = 11381.7$$

$$\therefore \quad \mathbf{h = 13648\ W/m^2K}$$

The value is high as it is liquid flow at a high velocity.

Example A-8.14: *Calculate the value of convection coefficient for flow of the following fluids at 10°C across a pipe 20 mm dia at 30°C, the flow velocity being 5 m/s*

(a) air (b) water (c) engine oil (d) liquid ammonia.

Solution: (*a*) The property values **for air are**

T°C	ν × 10⁶ m²/s	Pr	k, W/mK
10	14.16	0.705	0.02512
20	15.06	0.703	0.02593
30	16.0	0.701	0.02675

$$Re_D = 5 \times 0.02/15.06 \times 10^{-6} = 6640 \text{ (film temp.)}$$

Using $Nu_D = C\, Re^m\, Pr^{0.333}$ from table under equation (8.33)

$$C = 0.193,\ m = 0.618$$

∴ $Nu_D = 0.193 \times 6640^{0.618}\, (0.0703)^{0.333} = 39.5$

∴ $\mathbf{h = 51.22\ W/m^2K}$

(*b*) **Water**

$$20°C,\ \nu = 1.006 \times 10^{-6}\ m^2/s,\ Pr = 7.020,\ k = 0.5978\ W/mK$$

$$Re_D = 5 \times 0.02/1.006 \times 10^{-6} = 99403$$

Using $Nu_D = C.\, Re^m\, Pr^{0.333}$, $C = 0.0266$, m = 0.805

∴ $Nu = 536.9$ ∴ $\mathbf{h = 16048\ W/m^2K}$

(*c*) Engine oil at 20°C

$$\nu = 901 \times 10^{-6}\ m^2/s,\ Pr = 10400,\ k = 0.1454\ W/mK$$

The Prandtl number value is very high. One suitable correlation can be

$$Nu = [0.35 + 0.56\, Re^{0.52}]\, Pr^{0.333}$$

$$Re = 5 \times 0.02/901 \times 10^{-6} = 111$$

∴ $Nu = 149.14$ ∴ $\mathbf{h = 1084\ W/m^2K}$

If the general equation is used

$$C = 0.683,\quad m = 0.466$$

∴ $Nu = C\, Re^m\, .\, Pr^{0.333} = 133.82$

∴ $\mathbf{h = 972.9\ W/m^2\ K}$

(*d*) Liquid ammonia at 20°C

$$\nu = 0.358 \times 10^{-6}\ m^2/s,\ Pr = 2.02,\ k = 0.521\ W/mK$$

$$Re = 5 \times 0.02/0.358 \times 10^{-6} = 2.79 \times 10^5$$

Using $Nu = C\, Re^m\, Pr^{0.333}$, $C = 0.0266$, $m = 0.805$

∴ $Nu = 814.3$, ∴ $\mathbf{h = 21213\ W/m^2K}$

Note that for liquids higher value (of one or two orders of magnitude) of convection coefficient is obtained for the same velocity of flow.

Example A-8.15: *A wire 0.5 mm dia is at 40°C in a cross flow of air at 20°C while dissipating 35 W/m. Determine the velocity of the air stream.*

The convective heat transfer coefficient can be determined from the data.

Solution: $35 = h \times \pi \times 0.0005 \times 1 \times (40 - 20)$

∴ $h = 1114.08\ W/m^2K$

$$Nu = C.\, Re^m\, Pr^{0.333}$$

Chapter 8

$$\frac{h.D}{k} = C\, Re^m\, Pr^{0.333}$$

m and C are to be found. Properties are: at 30°C

$$\nu = 16 \times 10^{-6}\ \text{m}^2/\text{s},\ Pr = 0.701,\ k = 0.02675\ \text{W/mK}$$

$$Re = \frac{0.0005 \times u_\infty}{16 \times 10^{-6}}$$

$$\frac{h.D}{k} = \frac{1114.08 \times 0.0005}{0.02675} = 20.82$$

For range of values of Re around 1000 – 2000, $C = 0.683$, $m = 0.466$ By trial

R.H.S.	Assumed Velocity	L.H.S.
20.82 (constant)	20 m/s	12.18
	40 m/s	16.8
	100 m/s	25.79
	80 m/s	23.25
	70 m/s	21.84
	65 m/s	21.10
	60 m/s	20.33
	63 m/s	20.80

∴ **The velocity is 63 m/s (nearest) R_e = 1968**

Check by another equation

$$Nu = C\, Re^m\, Pr^n \left(\frac{Pr_\infty}{Pr_w}\right)^{0.25}$$

Properties at $T_\infty = 20°\text{C}$, $K = 0.02593$

$u_\infty = 63$ m/s, $\nu = 15.06 \times 10^{-6}$ m²/s

$Pr_\infty = 0.703$, $Pr_w = 0.699$

∴ $$Re = \frac{63 \times 0.0005}{15.06 \times 10^{-6}} = \mathbf{2091.63}$$

From table under equation (8.34) for this value of Re

C = 0.26, m = 0.6, n = 0.37

$$Nu = 0.26 \times (2091.63)^{0.6}\, (0.703)^{0.37} \left(\frac{0.699}{0.703}\right)^{0.25} = 22.4$$

∴ **h = 1161 W/m²K**

$Q = 36.5$ W as against 35 W

∴ The estimate is acceptable.

Example A-8.16: *Air at 20°C flows over a cylinder 10 cm dia at 60°C with a velocity of 10 m/s. Compare the value of convective heat transfer coefficient with a plate of length $\pi D/2$ with other parameters remaining the same.*

Solution: The properties at $T_f = (20 + 60)/2 = 40°C$ are

$$\nu = 16.96 \times 10^{-6} \text{ m}^2\text{/s } Pr = 0.699, k = 0.02756 \text{ W/mK}$$

Cylinder: $Re_D = 0.1 \times 10/16.96 \times 10^{-6} = 58962$

Using $Nu = C \,.\, Re^m \,.\, Pr^{0.333}$, from table under equation (8.33)

$$C = 0.0266, \text{m} = 0.805$$

$\therefore$ $Nu = 163.43 \quad \therefore \quad$ **h = 45.04 W/m²K**

Taking it as a plate: $L = \pi \times 0.1/2$

$\therefore$ $Re = (\pi \times 0.05) \times 10/16.96 \times 10^{-6} = 92617.7 \quad \therefore$ Laminar

$\overline{Nu} = 0.664\, Re^{0.5}\, Pr^{0.333} = 172.86 \quad \therefore \quad$ **h = 30.32 W/m²K**

The curved surface provides greater turbulence and higher convection coefficient.

Example A-8.17: *Approximating a human body as a cylinder of 0.3 m dia and 1.75 m long, at surface temperature of 30°C exposed to winds at 15 kmph at 10°C, determine the rate of heat loss.*

Solution: This problem may give an idea about the chilling in cold winds.

The film temperature $= (30 + 10)/2 = 20°C$

Property values are $\nu = 15.06 \times 10^{-6}$ m²/s

$$Pr = 0.703, \quad k = 0.02593 \text{ W/mK}$$

$$Re = \frac{15000}{3600} \times 0.3 \times \frac{1}{15.06 \times 10^{-6}} = 83{,}001$$

Using the common equation

$$Nu = C\, Re^m\, Pr^{0.333}, C = 0.0266, \text{m} = 0.805$$

$\therefore$ $Nu = 215.64$, **h = 18.64 W/m²K**

$\therefore$ **Q** $= 18.64 \times \pi \times 0.3 \times 1.75\,(30 - 10) =$ **614.8 W,**

rather high. Of course clothing reduces this loss to a very small value over larger fraction of the area.

Also we can use $Nu = 0.25\, Re^{0.5}\, Pr^{0.38}$ (for gases properties at T_1)

$= 195.5, \quad \therefore \quad$ **h = 16.9 W/m²K**

A difference of 8%.

Example A-8.18: ***Carbon dioxide*** *at 1 atm and 100°C flows across a pipe 0.2 m dia at 300°C with a velocity of 50 m/s. Determine the value of convection coefficient.*

Solution: The property values at the film temperature of 200°C are

$$\nu = 19.2 \times 10^{-6} \text{ m}^2\text{/s}, Pr = 0.715, k = 0.03094 \text{ W/mK}$$

$$Re = 60 \times 0.2/19.2 \times 10^{-6} = 6.25 \times 10^5$$

The equation that can be used is

$$Nu = 0.25\, Re^{0.6}\, Pr^{0.38}\ (10^3 < Re < 10^5)$$

$= 660.85 \quad \therefore \quad$ **h = 102.2 W/m²K**

The **other suitable equation** is

$$Nu = C \,.\, Re^m\, Pr^{0.37} \left(\frac{Pr_\infty}{Pr_w}\right)^{0.25}$$

with data at T_∞

Property values at T_∞, 100°C, $\nu = 12.6 \times 10^{-6}$ m²/s

$$Pr = 0.733,\ k = 0.02279 \text{ W/mK}$$

$$Re = 50 \times 0.2/12.6 \times 10^{-6} = 7.93 \times 10^5,\ Pr_w = 0.712$$

so values of $C = 0.076$, m = 0.7 ∴ $Nu = 920.0$ ∴ **h = 104.8 W/m²K** about 1.2% change.

Example A-8.19: *A tube with 4 cm square cross section has air blown across it at 20 m/s. The air is at 30°C and the pipe surface is at 50°C. Determine the convective heat transfer coefficient when (i) the flow is along diagonal and (ii) the flow is perpendicular to a face.*

Solution: The film temperature is (50 + 30)/2 = 40°C

The property values are:

$$\nu = 16.96 \times 10^{-6} \text{ m}^2/\text{s},\ Pr = 0.699,\ k = 0.02756 \text{ W/mK}$$

(*i*) flow along diagonal

$$D = \sqrt{(0.04^2 \times 2)} = 0.0566 \text{ m}$$

∴ $Re = 20 \times 0.0566 / 16.96 \times 10^{-6} = 66708$

Using $Nu = C.\ Re^n$, from tables under Section 8.3.1 $C = 0.222$, $n = 0.588$

∴ $Nu = 152.40$ ∴ **h = 74.27 W/m²K**

(*ii*) Perpendicular to face

$$Re = 20 \times 0.04/16.96 \times 10^{-6} = 47170$$

Using $Nu = C\ Re^n$, $C = 0.092$, $n = 0.675$

∴ $Nu = 131.4$ ∴ **h = 90.5 W/m² K**

Example A-8.20: *A spherical chemical container of 4 m dia has a surface temperature of 15°C while the outside air is at 25°C. Determine the heat gain when a 20 kmph wind blows across it.*

Solution: The film temperature is (15 + 25)/2 = 20°C

The property values are:

$$\nu = 15.06 \times 10^{-6} \text{ m}^2/\text{s},\ Pr = 0.703$$

$$k = 0.02593 \text{ W/mK}$$

$$Re = \frac{2000}{3600} \times 4 \times \frac{1}{15.06 \times 10^{-6}} = 1.4755 \times 10^6$$

suitable equation for this Reynolds number is (equation 8.43)

$$Nu = 430 + 5 \times 10^{-3} Re + 0.025 \times 10^{-9} Re^2 - 3.1 \times 10^{-17} Re^3$$

$$= 7762.7 \quad \therefore \quad \mathbf{h = 50.94\ W/m^2K}$$

$$\therefore \quad \textbf{heat gain} = 4\pi \times \left(\frac{4}{2}\right)^2 \times 50.94 \times (25 - 15)\text{W} = \mathbf{26.04\ kW.}$$

Example A-8.21: *In order to preheat air for a process from 50°C to 350°C packed bed of 10 cm dia pipe filled with spherical particles of 1 cm dia is used. The flow rate of air is 20 kg/hr. The pipe surface is at 400°C. Determine the height of bed required. (Refer Section 8.5.1)*

Solution: The film temperature at entry = (50 + 400)/2 = 225°C

The film temperature at exit = (350 + 400)/2 = 375°C

∴ Average film temperature = (225 + 375)/2 = 300°C

Taking the properties of air at this condition

$$\nu = 48.2 \times 10^{-6} \quad \rho = 0.615 \text{ kg/m}^3, Pr = 0.674,$$
$$k = 0.04605 \text{ W/mK}, c_p = 1047 \text{ J/kg K}$$

Eq. particle dia. $= 6 \times$ volume/surface area

$$= 6 \times \frac{4}{3} \pi r^3/4\pi r^2 = 2r = D = 0.01 \text{ m}$$

$\therefore$ Superficial velocity $= \dfrac{20}{0.615} \times \dfrac{1}{\pi \times 0.05^2} \times \dfrac{1}{3600}$ m/s = 1.15 m/s

$\therefore$ $Re_{DP} = uD/v = 238.6$ Using equation [8.52(*b*)].

From equation [8.52(*b*)] $Nu = 0.208\, Re_{DP}^{1/3}\, Pr^{1/3} + 0.220\, Re_{Dp}^{0.8}\, Pr^{0.4}$

$= 16.1 \quad \therefore$ **h = 74.14 W/m²K**

Q = 20 × 1047 × (350 – 50)/3600 W = **1745 W**

Heat transfer is also equal to the pipe area × h × LMTD

LMTD = 154.2°C

$\pi DL \times h \times \Delta T = \pi \times 0.1 \times L \times 74.14 \times 154.2 \quad \therefore$ **L = 0.486 m.**

Example A-8.22: *Sodium potassium alloy (25% + 75%) flows across a nuclear fuel rod at 500°C. The temperature of the alloy is 300°C. The rod dia is 5 cm and the flow velocity is 6 m/s. Determine the value of convection coefficient.*

Solution: At film temperature 400°C

$$\nu = 0.308 \times 10^{-6} \text{ m}^2\text{/s}, Pr = 0.0108, k = 22.10 \text{ W/mK}$$

Reynolds number $= 6 \times 0.05/0.308 \times 10^{-6} = 9.74 \times 10^5$

$Re\,Pr = 10519$ (properties at T_f)

$\therefore$ possible correlation is, 8.38(*a*)

$$Nu = 0.3 + \frac{0.62\, Re^{0.5}\, Pr^{0.333}}{\left[1+\left(\frac{0.4}{Pr}\right)^{0.666}\right]^{0.25}}\left[1+\left(\frac{Re}{282000}\right)^{0.625}\right]^{0.8} = 182.77$$

$\therefore$ **h = 80785.6 W/m²K** (Liquid Metal)

Example A-8.23: *Dry air at 80°C blows over a copper sphere at 320°C at a velocity of 14 m/s. The diameter of the sphere is 25 mm. Determine the heat transfer rate at that instant.*

Solution: The film temperature is 200°C.

The property values are:

$$\nu = 34.85 \times 10^{-6} \text{ m}^2\text{/s}, Pr = 0.68, k = 0.03931 \text{ W/mK}$$
$$Re = 0.025 \times 14/34.85 \times 10^{-6} = 1 \times 10^4$$

The suitable equation is, (8.42)

$$Nu = 2 + (0.25\, Re + 3 \times 10^{-4}\, Re^{1.6})^{0.5} = 59.18$$

$\therefore$ **h = 93.05 W/m²K**

$\therefore$ heat transfer rate $= 4\,\pi\, r^2\, h\, \Delta T = 43.85$ W

We can also use: $Nu = 0.37 \times Re^{0.6} = 93.17$, **h = 146.5 W/m²K**

This is an older correlation (1953) as compared to the previous (1978).

Example A-8.24: *An incandescent bulb can be considered as a sphere of 0.05 m dia. The bulb surface is at 140°C. Air at 20°C flows over it with a velocity of 0.6 m/s. Determine the heat loss from the bulb surface.*

Solution: The film temperature = (140 + 20)/2 = 80°C

The property values are:

$$\nu = 21.09 \times 10^{-6} \text{ m}^2\text{/s}, Pr = 0.692,$$

$$k = 0.03047 \text{ W/mK}$$

$$\therefore \quad Re = 0.6 \times 0.05/21.09 \times 10^{-6} = 1422.5$$

Using $\mathbf{Nu = 0.37 \times Re^{0.6} = 28.84,\ h = 17.58\ W/m^2K}$

Using $Nu = \mathbf{2 + (0.25\ Re + 3 \times 10^4\ Re^{1.6})^{0.5} = 21.72}$

Not very much different, $\mathbf{h = 13.24\ W/m^2K}$

$$\text{Heat loss} = h\, 4\, \pi\, r^2\, \Delta T = 16.56 \text{ W, for } h = 17.58 \text{ W/m}^2\text{K}$$

$$= 12.48 \text{ W for } h = 13.24 \text{ W/m}^2\text{K.}$$

Example A-8.25: *Liquid sodium at 200°C flows over a sphere of 5 cm dia at a velocity of 1.2 m/s. The surface is at 400°C. Determine the rate of heat loss from the sphere.*

Solution: The film temperature is (200 + 400)/2 = 300°C

The property values are

$$\nu = 0.394 \times 10^{-6} \text{ m}^2\text{/s}, Pr = 0.0063, k = 70.94 \text{ W/mK}$$

$$Re = 0.05 \times 1.2/0.394 \times 10^{-6} = 1.52 \times 10^5$$

The applicable equation is

$$Nu = 2 + 0.386\, (Re\ Pr)^{0.5} = 13.96$$

$$\therefore \quad \mathbf{h = 19800\ W/m^2K}$$

$$\therefore \quad Q = 4\pi\, r^2 h\, \Delta T = \mathbf{31102\ W}$$

This shows that liquid sodium can extract heat at a high rate and so is used in breeder reactors.

Example A-8.26: *A bank of tubes consists of 10 mm tubes 100 numbers arranged in a square array of 15 mm pitch. The tube wall is at 40°C and air enters at 20°C with a velocity of 4 m/s. Determine the value of convection coefficient and the heat transferred to the air for 1 m length of tubes in the array.*

Solution: The property values at T_∞, T_f and T_w are

T°C	ρ, kg/m³	ν × 10⁶, m²/s	Pr	k, W/mK	C_p, J/kg K
20	1.205	15.06	0.703	0.02593	1005
30	1.165	16.00	0.707	0.02675	
40	1.128	16.96	0.699	0.02756	

$$V_{max} = 4 \times \frac{15}{15 - 10} = 12 \text{ m/s}$$

Using properties at film temperature

$$Re = 12 \times 0.01/16 \times 10^{-6} = 7500$$

Using the tables for the constants in the equation:

$$Nu = 1.13\ Pr^{0.33}\ C\ Re^n$$

$$S_n/D = 1.5 \text{ and } S_p/D = 1.5$$

$$C = 0.25,\ n = 0.62$$

$$\therefore \quad Nu = 63.41$$

$$\therefore \quad \mathbf{h = 169.61\ W/m^2K}$$

S_n 15 mm

S_p 15 mm

Square Array

No correction necessary as it is 10 rows deep.

Area for heat flow: $\pi \times 0.01 \times 100 \times 1\ m^2$

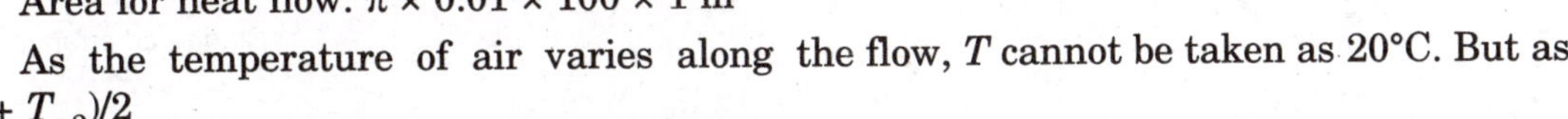

As the temperature of air varies along the flow, T cannot be taken as 20°C. But as $(T_{\infty 1} + T_{\infty 2})/2$

The heat transfer due to the temperature rise for the air is equated to the heat transfer by convection: (10 × 10 array)

$$\text{mass flow} = 10 \times 0.005 \times 1 \times 12 \times 1.205 \text{ kg/s}$$

$$\therefore \quad = 169.61 \times \pi \times 0.01 \times 100 \left[40 - \frac{20 + T_{\infty 2}}{2}\right]$$

$$= 10 \times 0.005 \times 1 \times 12 \times 1.205 \times 1005\ (T_{\infty 2} - 20)$$

$$15985.4 - T_{\infty 2} \times 266.42 = 726.615\ T_{\infty 2} - 14532.3$$

$$\therefore \quad \mathbf{T_{\infty 2} = 30.73°C}$$

$$Q = 169.61 \times \pi \times 0.01 \times 100 \left[40 - \frac{20 + 30.73}{2}\right]$$

$$= \mathbf{7798\ W\ or\ 7.8\ kW/m\ height}$$

A more rigorous method is by the use of equation (8.59) to (8.60).

$$\frac{T_s - T_o}{T_s - T_i} = e^{-\left[\frac{\pi D N \bar{h}}{\rho V N S_P C}\right]} = e^{-\left[\frac{\pi \times 0.01 \times 100 \times 169.61}{1.165 \times 12 \times 10 \times 0.015 \times 1005}\right]}$$

$$\frac{40 - T_o}{40 - 20} = 0.7766$$

Solving $\quad T_o = 24.47°C$

$$\Delta T_i = 20, \quad \Delta T_o = 15.53 \quad \therefore \quad \text{LMTD} = \frac{20 - 15.53}{\ln\left(\frac{20}{15.53}\right)} = 17.67°C$$

$$Q = 100 \times 169.61 \times \pi \times 0.1 \times 17.67$$

$$= \mathbf{9415\ W/m.}$$

Example A-8.27: *For the data in example 8.26, assume that the pitch parallel to flow direction is reduced to 10 mm, determine the value of convection coefficient for staggered arrangement.*

Solution: To find V_{max}

Check: $S_p - D < 2\sqrt{((S_p/2)^2 + S_n^{\ 2})} - 2D$

$$S_p - D = 15 - 10 = 5 \text{ mm},$$

$$2\sqrt{(S_p/2)^2 + S_n^{\,2}} - 2D = 2\sqrt{7.5^2 + 10^2} - 20 = 25 - 20 = \mathbf{5}$$

These two are equal. So eitherway V_{max} has the same value.

By equation (8.54) for staggered arrangement,

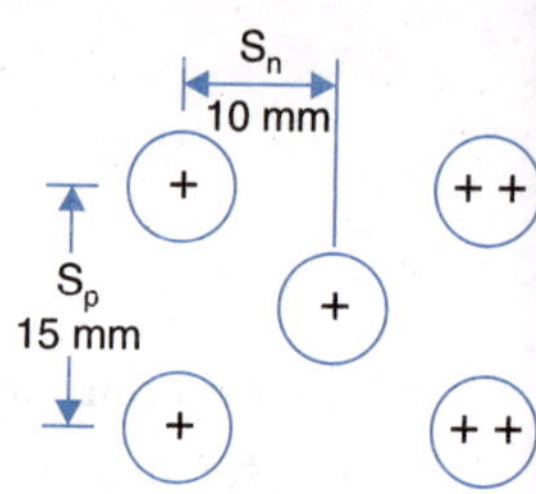

Staggered arrangement

$$V_{max} = \left(\frac{S_p}{2(S_D - D)}\right)u_\infty, \text{ where } S_D = \left[S_n^2 + \left(\frac{S_p}{2}\right)^2\right]^{0.5}$$

$$S_D = (10^2 + 7.5^2)^{0.5} = 12.5$$

$$V_{max} = \frac{15}{2(12.5 - 10)} = 3 \times 4 = 12 \text{ m/s.}$$

In the example given, the value of V_{max} is the same as in 8.26. But the values of C and n are different for staggered arrangement. These are: 0.497 and 0.558. ($S_p/D = 1.5$, $S_n/D = 1$)

$$\therefore \quad Nu = 0.497\,(7500)^{0.558}\ 1.13 \times Pr^{0.333} \qquad \text{(eqn. 8.56)}$$

$$= 72.49 \quad \therefore \quad \mathbf{h = 193.91\ W/m^2K}$$

No correction as the bank is 10 rows deep

Heat flow can be calculated using equations (8.59) and (8.60).

$$\frac{40 - T_o}{40 - 20} = e^{-\left(\frac{\pi \times 0.01 \times 100 \times 193.91}{1.165 \times 12 \times 10 \times 0.015 \times 1005}\right)} = 0.749$$

$$40 - T_o = 14.98, \quad T_o = 25.02$$

$$\text{LMTD} = \frac{(40 - 20.00) - (40 - 25.02)}{\ln\left(\frac{40 - 20}{40 - 25.02}\right)} = 17.37°\text{C.}$$

$$\therefore \quad Q = 100 \times \pi \times 0.01 \times 193.91 \times 17.37$$

$$= \mathbf{10581.6\ W/m.}$$

Example A-8.28: *Water flows over 20 mm tubes 6 rows deep inline arrangement with 40 mm square pitch. Water enters at 20°C and the tube wall is at 60°C. The velocity of water is 1.2 m/s at enetry. Determine the value of convection coefficient.*

Solution: The property values are:

T	$\nu \times 10^6$	Pr	k	C_p
20	1.006	7.020	0.5978	41.78
40	0.657	4.340	0.628	
60	0.478	3.020	0.6513	

$$V_{max} = u_\infty \cdot \frac{40 - 20}{20} = 2.4 \text{ m/s}$$

Using property values at $T_f = 40°C$

for $\quad S_n/D = 2$ and $S_p/2 = 2$,

the values of C and n in the equation

$$Nu = 1.13\, Pr^{0.333} \times C \,.\, Re^n \text{ are } C = 0.229,\ n = 0.632$$

$\therefore$ $Nu = 500.32$ $\therefore$ $\mathbf{h_{10} = 15710\ W/m^2K}$

As the bank is 6 rows deep, this is corrected by 0.94 (see tables)

$\therefore$ $\mathbf{h_6 = 14768\ W/m^2K}$

Assuming 1 m × 1 m, the number of rows perpendicular to flow direction is 1/0.04 = 25 rows. Total tubes = 25 × 6 = 150.

Using the relationship that the increase in the internal energy of water = heat convected, and also assuming water outlet temperature as $T_{\infty 2}$

$$1 \times 1 \times 1.2 \times 1000 \times 4178\,(T_{\infty 2} - 20) = \pi \times 0.02 \times 150 \times 1 \times 14768 \times \left(60 - \frac{20 + T_{\infty 2}}{2}\right)$$

$$\mathbf{T_{\infty 2} = 21.1°C}$$

$\therefore$ **Heat transfer = 5491 kW**

Heat transfer can be calculated more rigorously by equations (8.59) and (8.60).

$$\frac{T_s - T_o}{T_s - T_i} = e^{-\left[\frac{\pi D N h}{\rho V N_t \,.\, S_P C}\right]}$$

$$\frac{60 - T_o}{60 - 20} = e^{-\left[\frac{\pi \times 0.02 \times 150 \times 14768}{1000 \times 2.4 \times 25 \times 0.04 \times 4178}\right]} = 0.9862$$

Solving $T_o = 20.55°C.$

$$\text{LMTD} = \frac{(60 - 20) - (60 - 20.55)}{\ln\left(\dfrac{60 - 20}{60 - 20.55}\right)} = 39.72°C$$

$$Q = 150 \times 14768 \times \pi \times 0.02 \times 39.72 \text{ W/m}$$

$$= \mathbf{5528433\ W/m}$$

$$= \mathbf{5528.4\ kW}$$

Example A-8.29: *Using the data or example 8.28 except that the bank is of staggered arrangement with the longitudinal pitch reduced to 18 mm, calculate the value of convection coefficient.*

Solution: In this case, distance

$$P = (18^2 + 20^2)^2 - 20 = 6.9 \text{ mm}$$

$$2P = 13.81 \text{ mm}$$

This is less than (40 – 20) = 20

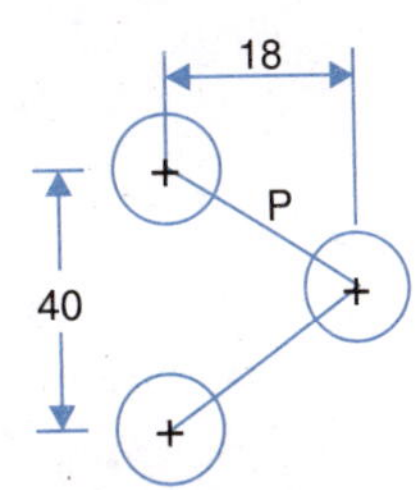

Tube arrangement

$$V_{max} = 1.2 \times \frac{40}{13.81} = 3.475 \text{ m/s}$$

$$Re = 3.475 \times 0.02/0.657 \times 10^{-6}$$

$$= 105806,\ (1.05 \times 10^5)$$

Using the relation: $Nu = 1.13 \times (Pr)^{0.333}\, C\, Re^n$

In this arrangement, (2, 0.9) **C = 0.446, *n* = 0.571**

$$\therefore \qquad Nu = 608.0 \quad \therefore \quad \mathbf{h_{10} = 19091\ W/m^2K}$$

For 6 rows, $\mathbf{h_6}$ **= 17946 W/m²K** as compared to 14768 in the inline arrangement (21.5% increase)

If the other correlation is used, with properties at T_∞

$$Re = 3.475 \times 0.02/1.006 \times 10^{-6} = 69085$$

$$C = 0.27,\ m = 0.63$$

$$Nu = C\ .\ Re^m\ Pr^{0.36} \left(\frac{Pr_\infty}{Pr_w}\right)^{0.25}$$

$$= \mathbf{752.38} \quad \therefore \quad \mathbf{h_{10} = 22489\ W/m^2K}$$

h_6 = 22140 W/m²K (about 18% more than the other correlation). For water flow this may be better

Example A-8.30: *Hot air is used to heat up rocks in a container 1 m dia and 2 m long. The rocks are approximately spherical with a diameter of 30 mm. The bed has void space of 42%. The rock is at 20°C. The air is at 80°C and flows in the axial direction at a flow rate of 1 kg/s. Calculate the value of convection coefficient.*

Solution: The case is heat transfer from gas to fillings as compared to wall to the gas. (Refer Section 8.5.1)

$$\text{Reynolds} = \frac{D\,V_S}{\nu(1-\varepsilon)}$$

$$D = 6\ \text{V/surface area} = 6 \times \frac{4}{3}\,\pi\,r^3/4\,\pi\,r^2 = 2r = D = 0.03\ \text{m}$$

V_S = superficial velocity based on the area of empty container

$$= 1 \times \text{sp vol}/(\pi \times 1^2/4) = 4/\pi = 1.273\ \text{m/s}$$

At 80°C, 1 atm, sp. vol. = 1 m³/kg

The property values at 100/2 = 50°C, $\nu = 17.95 \times 10^{-6}$ m²/s, $Pr = 0.698$

$$k = 0.02826\ \text{W/mK}$$

$$Re = \frac{0.03 \times 1.273}{(1-0.42) \times 17.95 \times 10^{-6}} = 3668.2$$

By equation (8.51) $$\frac{h.D}{k} = \frac{1-\varepsilon}{\varepsilon}\,[0.5\,Re^{0.5} + 0.2\,Re^{0.67}]\,Pr^{1/3}$$

$$= \frac{0.58}{0.42}\,[0.5 \times (3668.2)^{0.5} + 0.2\,(3668.2)^{0.67}]\,0.698^{0.333}$$

$$Nu = 96.98,$$

$$\mathbf{h} = 96.98 \times 0.02826/0.03 = \mathbf{91.36\ W/m^2K.}$$

OBJECTIVE QUESTIONS

8.1 Choose the correct statement from each question.

(*i*) In flow over a flat plate the convection coefficient

(*a*) Always increases along the flow

(*b*) Decreases, increases and decreases

(*c*) Increases upto critical Reynolds number and then decreases

(*d*) Decreases upto critical Reynolds number and then increases.

(*ii*) In laminar flow over a flat plate,

(*a*) The thermal boundary layer and hydrodynamic boundary layers are of equal thickness

(*b*) The thermal boundary layer is thicker if the Prandtl number is greater than one

(*c*) The thermal boundary layer is thicker if the Prandtl number is less than one

(*d*) The thermal boundary layer is always thinner in the laminar region.

(*iii*) The friction factor in the laminar region is proportional to

(*a*) $Re^{0.5}$ (*b*) $Re^{0.2}$ (*c*) $Re^{-0.2}$ (*d*) $Re^{-0.5}$.

(*iv*) The boundary layer thickness in laminar flow is proportional to

(*a*) $Re^{0.2}$ (*b*) $Re^{0.5}$ (*c*) $Re^{-0.5}$ (*d*) $Re^{-0.2}$.

(*v*) In laminar flow over a flat plate the convection coefficient is proportional to

(*a*) $x^{-0.5}$ (*b*) $x^{0.5}$ (*c*) $x^{-0.2}$ (*d*) x.

(*vi*) The boundary layer thickness in laminar flow over a flat plate, is proportional to

(*a*) x (*b*) $x^{0.5}$ (*c*) $x^{-0.5}$ (*d*) x^{-1}.

(*vii*) (*a*) Higher the value of kinematic viscosity thinner will be the boundary layer

(*b*) Higher the Prandtl number lower will be the thickness of thermal boundary layer.

(*c*) The convective heat transfer coefficient will be lower in turbulent flow as compared to laminar flow.

(*d*) The boundary layer will thicken as the free stream velocity increases.

(*viii*) In flow across a cylinder, the local Nusselt number will be highest at

(*a*) 90° from the stagnation point

(*b*) At the stagnation point

(*c*) At 80° from stagnation point

(*d*) At 135° from stagnation point.

(*ix*) In banks of tubes heat transfer will be highest

(*a*) In linear arrangement with square pitch

(*b*) Linear arrangement with lower longitudinal pitch

(*c*) Staggered arrangement with equal pitch

(*d*) Staggered arrangement with lower longitudinal pitch.

Answers

8.1 (*i*) (*b*) (*ii*) (*c*) (*iii*) (*d*) (*iv*) (*c*)
(*v*) (*a*) (*vi*) (*b*) (*vii*) (*b*) (*viii*) (*d*)
(*ix*) (*d*)

EXERCISE PROBLEMS

8.1 Nitrogen at 30°C flows over a plate maintained at 70°C with a free stream velocity of 10 m/s.

(*a*) Determine the local and average values of convective heat transfer coefficient for $L = 0.7$ m.

(*b*) Also calculate the values of the boundary layer thickness (velocity, thermal) and momentum and displacement thicknesses at this location.

(*c*) Determine also the location at which the flow turns turbulent, considering $Re_{cr} = 5 \times 10^5$.

8.2 Air at 20°C was heated by flow over a flat plate at 100°C, the flow velocity being 16 m/s. The plate is 1 m wide and 1.2 m long. Due to deterioration of performance of the fan and the heating medium, the plate temperature is reduced to 80°C while the flow velocity is reduced to 10 m/s. Determine the percentage reduction in heat flow.

8.3 It is desired to predict the performance of heat transfer in an equipment using liquid ammonia at a film temperature of – 30°C upto a Reynolds number of 10^5, using heat transfer studies on similar flat surface, with air as the test fluid. The maximum velocity of air is 10 m/s and the film temperature is 40°C. Determine the length of the plate to be used for the experiment.

8.4 A plate air heater 1 m wide and 2 m long at 120°C is designed to heat air at 20°C with a flow velocity of 8 m/s, with flow along 1 m direction. During fabrication, misunderstanding the instruction, the flow direction was fixed along the 2 m direction. Determine whether the heating rate will be lower or higher than the design value, and by what percentage.

8.5 Water at 20°C flows over a flat plate at 80°C with a velocity of 1.2 m/s. Determine the incremental heat transfer for every 0.2 m length after the leading edge up to a length of 1.2 m.

8.6 Air at 10 atm and 60°C flows over a plate at a velocity of 4 m/s, the plate temperatrue being 100°C. Determine at a location 0.5 m from the leading edge, the local and average values of convection coefficient. Also calculate the thermal boundary layer thickness at this location.

8.7 In a wind tunnel air flow is at 0.1 atm and 20°C. For heat transfer studies a plate 1 m length is kept at 80°C in the air stream. The flow velocity has a maximum value of 30 m/s. Determine the maximum Reynolds number at which observations can be made. Also calculate the heat input for the plate if it is 0.4 m wide. It is desired to modify the tunnel to operate at 0.8 atm and 40°C with the plate being at 80°C. Calculate the heat input if the maximum velocity is now reduced to 18 m/s.

8.8 A thin metal plate has hot air at 180°C flowing on one side at 20 m/s and cold air at 30°C flowing on the other side, both streams being parallel. Determine the heat flux at 0.4 m from the leading edge.

8.9 Air at 20°C flows over a flat plate at 60°C at a velocity of 60 m/s. Determine the value of average convection coefficient up to a Reynolds number of 10^7 (*i*) considering the laminar region (*ii*) considering that roughness causes turbulence even at the leading edge.

8.10 Air at 20°C flows over a flat plate with a velocity 10 m/s. The plate temperature is maintained at 60°C from a point 0.5 m from the leading edge. Determine the value of convection coefficient at a location 1 m from the leading edge.

8.11 A plate 40 cm square has a uniform heat generation rate of 8 kW/m^2. Water at 20°C flows over it with a velocity of 1.2 m/s. Determine the value of convection coefficient at the trailing edge and also the plate temperature at this location.

8.12 If air at 2 atm and 20°C flows over a plate 0.4 m square with a velocity of 8 m/s and if the plate has uniform heat generation of 8 kW/m^2, determine the temperature at the trailing edge. Also determine the value of convection coefficient at this location.

8.13 Air at 10°C flows **across** a road surface at 10°C with a speed of 25 kmph. The road surface is 10 m wide and rough enough to cause turbulence even at the leading edge. Determine the value of convection coefficient at the trailing edge and also the average convection coefficient. Calculate also the thickness of hydrodynamic boundary layer at the trailing edge.

8.14 Liquid sodium at 200°C flows over a plate at 400°C with velocity of 6 m/s. Determine the average value of convection coefficient if the plate is (*i*) 0.6 m long (*ii*) 2 m long.

8.15 Engine oil at 80°C is cooled by flow over a plate maintained at 20°C with water flow below it. The flow velocity is 0.66 m/s. Determine the heat absorbed by a plate 0.25 m × 0.25 m size.

8.16 The value of local Nusselt number is given by the following equation

$$Nu_x = \frac{(C_{fx}/2)\, Re\, Pr}{1 + 12.8\sqrt{C_{fx}/2}\,(Pr^{0.68} - 1)}$$

where $C_{fx} = 0.059\, Re_x^{-02}$ and $Re = \dfrac{u_\infty\, x}{\nu}$ and

$$Nu_x = \frac{u_\infty\, x}{k}$$

Derive an expression for the average value of h.

8.17 A surface well insulated at the bottom absorbs heat at 650 W/m² on the top face. The top is exposed to air flow at 0.1 atm and – 10°C, with a velocity of 300 kmph. If the plate is 3 m long in the direction of flow, determine the average plate temperature.

8.18 A truck travels at 130 kmph in air at 50°C and its surface is at 10°C. The truck may be approximated to a rectangular box of 3 m × 2 m × 6 m. Assuming that there is no flow separation and the flow is turbulent althrough, determine the heat loss from four surfaces. (neglect front and back). Also calculate the frictional drag.

8.19 Water at 20°C flows over a slab of ice at 0°C of size 1 m × 1 m with a velocity of 0.8 m/s. Determine the rate of thickness reduction of the ice (mm per min) if the latent heat of melting has a value of 3.34×10^5 J/kg. Density of ice = 900 kg/m³

8.20 A steam pipe of 0.4 m *OD* at 95° passes in air at 25° C over a distance of 15 m. Air flows across the tube at 5 kmph. Determine the heat loss through the exposed length of pipe.

8.21 A thermometer at 30°C having a cylindrical shape 8 mm dia is inserted into an air stream at 40°C flowing with a velocity of 25 m/s. Estimate the value of convective heat transfer coefficient over the bulb.

8.22 Investigate the variation of current carrying capacity of a fuse wire exposed to air flow, with the variation of velocity of flow across the wire. Assume dimensions and property values.

8.23 A fin of cylindrical shape 1 cm dia extending from a base at 100°C is exposed to air flow across it at 20°C with a velocity of 5 m/s. Determine the value of average heat transfer coefficient over the fin and the heat dissipated by the fin. The fin is 8 cm long.

8.24 An electrical heater rod 25 mm diameter is exposed to air at 20°C, the velocity of flow across it being 25 m/s. If the rod surface temperature is not to exceed 180°C, determine the capacity of heater rod, that can be used.

8.25 A nuclear fuel rod is to be cooled by using sodium potassium alloy (25% + 75%) at a temperature of 200°C, with a maximum velocity of 16 m/s. If the rod of 25 mm dia should have its surface maintained at temperatures not exceeding 400°C, determine the heat generation rate for *m* length of fuel rod.

8.26 Pipe carrying oil in the arctic region, has outside diameter of 0.8 m, a temperature of 60°C. If winds blow across it at 40 kmph and – 20°C, determine the heat loss per 1 m length of pipe.

8.27 Helium at 1 atm and 50°C flows across a 5 mm dia cylinder at 150°C, with a velocity of 9 m/s. Determine the heat transfer rate per 1 m length of pipe.

8.28 Air flows across a 4 cm square cylinder with a velocity of 10 m/s. The air is at 25°C while the surface is at 75°C. Determine the heat transfer rate if (*i*) flow is along the diagonal (*ii*) flow is along face (or perpendicular to face).

8.29 Air at 25°C flows across a 4 × 6 cm elliptical pipe at 75°C with a velocity of 10 m/s. Estimate the heat flow/m length if the flow is (*i*) across minor axis (*ii*) across major axis.

8.30 A chemical storage tank of spherical shape of 2 m dia has its surface maintained at 80°C by steam heating. Air at 20°C flows at 20 kmph across the tank. Determine the required of heating capacity.

8.31 Water flows across a sphere of 25 mm dia at 40°C with a velocity of 1.4 m/s. The sphere surface is at 80°C. Determine the heat loss rate from the sphere at the instant.

8.32 A nuclear element is in the form of sphere of 25 mm dia. Liquid sodium at 200°C is used to extract heat. A flow velocity of 12 m/s is used. If the surface temperature is not to exceed 400°C determine the heat generation rate possible (per sphere and also per m^3).

8.33 Heated steel balls of 15 mm dia at 260°C are quenched in water at 40°C by free fall. If the balls travel in the water at 8 m/s, determine the value of convection coefficient.

8.34 A bank of tubes consists of 20 mm dia tubes at 30 mm pitch in the linear and transverse directions and arranged in line. The surface temperature of the tubes is 60°C. Air at 20°C flows across the tubes at an entry velocity of 8 m/s. If the tube bank is 8 rows deep and 20 rows across and if the length of tubes is 1 m, determine the heat transferred to the air.

8.35 If the tube arrangement in the bank of problem 8.34 is changed to staggered arrangement with the longitudinal pitch changed to 18 mm, determine the heat transferred to the air.

8.36 Water is preheated in a tube bank with flue gases at 220°C flowing across the tubes at a velocity of 10 m/s. The tubes are 50 mm dia and are arranged staggered with a transverse pitch of 100 mm and lognitudinal pitch of 45 mm. If the bank is 6 rows deep and the tubes are 2 m long and the there are 20 rows across, determine the heat picked up by the water. Tube surface is at 30°C.

8.37 An airconditioning coil has 15 mm dia tubes arranged staggered with a transverse pitch of 30 mm and longitudinal pitch of 13.55 mm and is 4 rows deep. Determine the value of convective heat transfer coefficient. Air at 30°C enters the bank with a velocity of 10 m/s. The tube surface is at 10°C.

8.38 A packed bed is used to transfer heat from the vessel wall to the gas flowing through the bed. The bed is a pipe of 0.25 m dia and 0.8 m length filled with spherical filling of 8 mm dia. The tube wall is at 230°C. The gas enters at 30°C.

Determine the heat transfer from the surface to the gas. The rate of flow of air is 2 kg/s at 1 atm pressure.

8.39 For solar heat storage in pebbles solar heated air at 70°C is passed through a bed of pebbles of average diameter of 12 mm. The pebbles are at 30°C. The flow velocity is (superficial) 8 m/s. The void fraction is 42%. Determine the value of convective heat transfer coefficient.

8.40 Engine oil at 60°C flows over the upper surface of a 5 m long flat plate whose temperature is 20°C. The flow velocity is 2 m/s. Determine the total drag force and rate of heat transfer per unit width. Property values at 40°C for the oil are

$$\rho = 876 \text{ kg/m}^3, \; Pr = 2870, \; k = 0.144 \text{ W/m°C}$$

$$\upsilon = 242 \times 10^{-4} \text{ m}^2\text{/s.}$$

[**Ans.** 181 N, 11,040 W]

8.41 Air at 20°C flows with a velocity of 8 m/s over a flat plate of size 1.5 m × 6 m at a temperature of 140°C. Determine the rate of heat transfer from the plate if the air flow along (*a*) 6 m side (*b*) 1.5 m side. Assume the following property values.

$$k = 0.02953 \text{ W/m°C}, \; Pr = 0.7154$$

$$\upsilon = 2.548 \times 10^{-5} \text{ m}^2\text{/s}$$

[**Ans.** (*a*) 1.43 × 10^4 W, (*b*) 8670 W]

8.42 Water at 15°C flows across a pipe of 22 mm OD with a velocity of 4 m/s. Determine the drag force over 1 m length. Assume for water at 15°C, $\rho = 999.1$ kg/m^3, $\mu = 1.138 \times 10^{-3}$ kg/ms, $C_D = 1.0$.

[**Ans.** 175.8 N]

8.43 A 25 cm diameter stainless steel ball (ρ = 8055 kg/m^3, C = 480 J/kg°C) is cooled from 300°C to 200°C by flow of air at 25°C with a velocity of 3 m/s. Determine the average value of convection coefficient and the time taken for cooling. Assume that the ball is at uniform temperatures both at 300°C and 200°C. Consider variation viscosity of air during the process. Assume for air : At 250°C, μ = 2.76 × 10^{-5} kg/ms. At 25°C, μ = 1.849 × 10^{-5} kg/ms, υ = 1.562 × 10^{-5} m^2/s, k = 0.02551 W/m°C, Pr = 0.7296. [**Ans.** h = 13.8 W/m^2°C, time taken = 1 hr and 26 min]

8.44 During a cold winter day wind at 54 km/h flows parallel to a 4 m high wall of 10 m length. The air temperature is 5°C. The wall temperature is 12°C. Determine the rate of heat loss from the wall. Also calculate the loss in case the wind velocity is doubled. [**Ans.** 9081 W, 16200 W]

8.45 An average person generates heat at the rate of 82 W while resting. Assuming that the quarter of the heat is lost from the head and also assuming the head can be approximated to a sphere of 30 cm dia, determine the average surface temperature of the head if winds at 10°C flows at 35 km/h over the head. [**Ans.** 12.7°C]

8.46 Air flows across a square duct of side 0.2 m and 1.5 m length at a velocity of 200 m/min. The temperature of air is 30°C and that of the duct surface is 65°C. Determine the heat flow rate. Use $Nu = 0.102\ Re^{0.675}\ Pr^{0.333}$, k = 0.02717 W/m°C, υ = 1.774 × 10^{-5} m^2/s, Pr = 0.7235. [**Ans.** 640 W]

9 CHAPTER
Convective Heat Transfer Practical Correlations–Internal Flow

9.0 FLOW THROUGH TUBES

The internal flow configuration is the most convenient and popularly used geometry for heating or cooling of fluids in various thermal and chemical processes. There are basic differences in the development of boundary layer between the external flow geometry discussed in Chapter 8 and internal flow geometry. In the case of internal flow, the fluid is confined by a surface, and the boundary layer after some distance cannot develop further. **This region is called entrance region** and the length up to the section is called **entry length.** The region beyond this point is known as **fully developed region.** Another important difference is that the flow does not change over at a location froam laminar to turbulent conditions, but is **laminar or turbulent from the start**, depending upon the value of Reynolds number (based on diameter) being greater or less than about 2300.

A third difference is that in many cases there is no well defined free stream velocity as in the case of flow over a flat plate. So the mean velocity using the mass flow is to be adopted in place of u_∞ in all equations. The mean velocity u_m is defined by the following equation.

$$u_m = \frac{1}{A}\int_A u\,dA$$

The flow rate is defined by

$$G = \rho A u_m = \frac{\rho \pi D^2}{4} \,.\, u_m$$

In the case of circular section,

$$A = \frac{\pi D^2}{4}.$$

Using these in the definition of Reynolds number leads, in the case of circular pipes, to the following additional expression for Reynolds number.

$$\mathbf{Re = 4G/\pi D\mu} \qquad ...(9.1)$$

Where G is the flow through pipe in kg/s.

This expression (9.1) may be often used in place of $u_m \rho D/\mu$ in the analysis.

A fourth difference is that as the fluid flows through the pipe, its mean temperature increases and there is no free stream temperature as such. The concept of free stream temperature has to give way to a new term called bulk mean temperature or **mixing cup temperature.** The **bulk mean temperature** at a location x along the pipe is the average temperature of the fluid if the fluid at the location is well mixed to attain a common temperature. The mean temperature is defined by:

$$T_m = \frac{1}{u_m A}\int_A uT\,dA.$$

Hence the convection correlations will be different in the case of internal flow. However the dimensionless parameters used are the same because the basic phenomenon does not change.

Detailed derivations are given under advanced examples after the end of general discussions.

9.1 HYDRODYNAMIC BOUNDARY LAYER DEVELOPMENT

The development of hydrodynamic boundary layer in a pipe, together with velocity distributions at various sections for laminar and turbulent flows are shown in Figs. 9.1(*a*) and 9.1(*b*) (for isothermal flow). Refer advanced example 9.1 for the shape of the profile in laminar flow given by

$$\frac{u_r}{u_{max}} = 1 - \left(\frac{r}{R}\right)^2$$

where u_{max} is the velocity at the centre line. Also $u_{max} = 2u_m$.

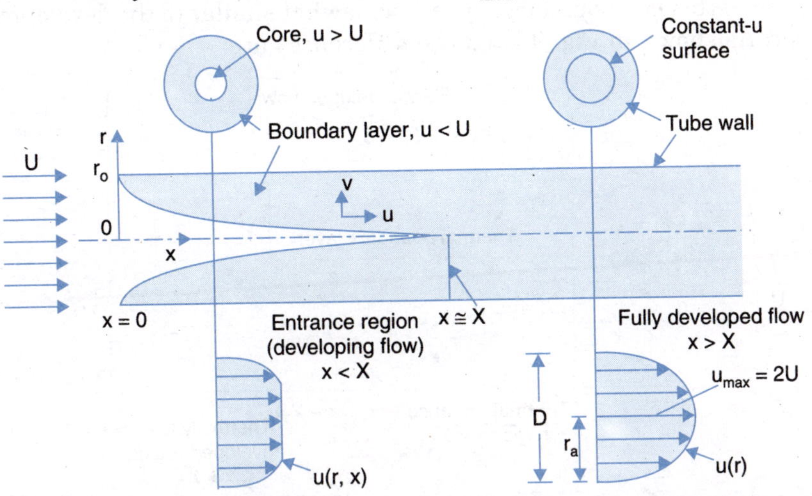

(a) Entrance region and fully developed flow region of laminar flow through a tube

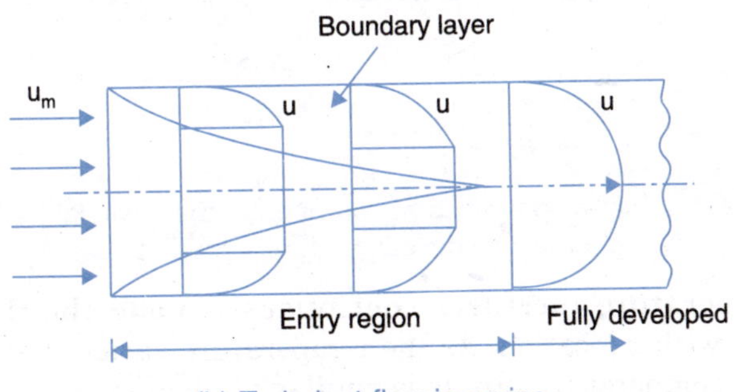

(b) Turbulent flow in a pipe

Fig. 9.1

Chapter 9

The velocity distribution beyond the entry region will remain invariant because there is no further momentum transfer between layers. But the actual distribution will be affected by the fluid property variation during heating or cooling. If heating or cooling causes reduction in the viscosity near the wall, the velocity profile flattens out as compared to isothermal flow. If viscosity increases, then the velocity near the wall will be reduced further and the velocity distribution will be more peaked. This is shown in Fig. 9.2(a). Such distortion will affect the heat transfer correlations to some extent.

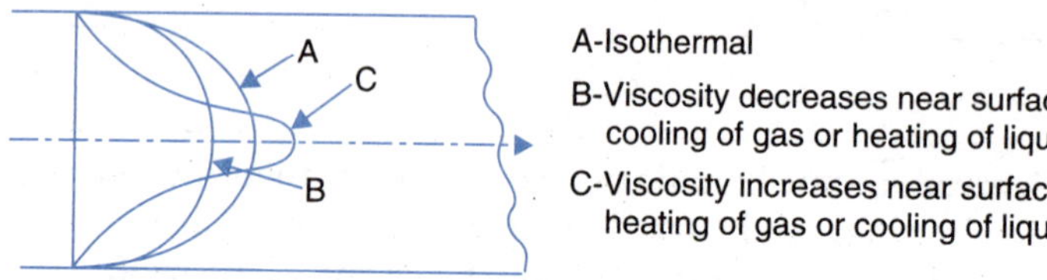

Fig. 9.2. (a) Distortion in velocity profile due to heating/cooling

9.2 THERMAL BOUNDARY LAYER

The development of thermal boundary layer is somewhat similar to the development of velocity boundary layer. As shown in Fig. 9.2(b). The differences are:

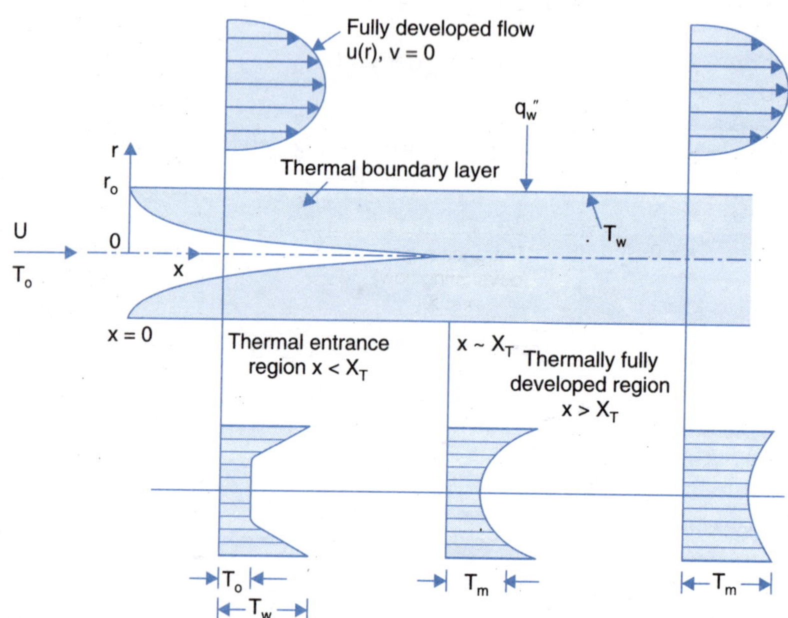

Fig. 9.2. (b) Thermal entrance region and the thermally fully developed region of fully developed flow through a tube

(i) As the temperature increases continuously along the flow, the direct plot of temperature will vary with x location. As the temperature varies in the same ratio at all the radii, the shape of the temperature profile is similar at all along the length. Thus, the plot of dimensionless temperature ratio will provide a constant profile in the fully developed region. The bulk mean temperature T_m varies along the length as heat is added/removed along the length. The ratio $(T_w - T_r)/(T_w - T_m)$ remains constant along the x direction in the fully developed flow. T_r is the temperature at radius r and T_m is the bulk mean temperature.

$$\frac{\partial}{\partial r}\left(\frac{T_w - T}{T_w - T_m}\right)_{r = r_0} = \frac{-(\partial T/\partial r)_{r = r_0}}{T_w - T_m} \neq f(x)$$

As $\left(\frac{\partial T}{\partial r}\right)$ is not a function of x, the local convection coefficient is constant and independent of x. Refer Fig. 9.3. (Page 429)

(*ii*) The entry length will be different as compared to the velocity boundary layer.

(*iii*) Boundary conditions are also different–constant wall temperature and constant heat flux.

(*iv*) The development of both boundary layers may be from entry or heating may start after the hydrodynamic boundary layer is fully developed.

These are in addition to the laminar and turbulent flow conditions.

Thus it is not possible to arrive at a limited number of correlations for convection coefficient.

In the case of internal flow, there are four different regions of flow namely (*i*) Laminar entry region (*ii*) Laminar fully developed flow (*iii*) Turbulent entry region and (*iv*) Turbulent fully developed region.

There are two possible boundary conditions in all the four regions listed, namely constant wall temperature or constant heat flux. **As such 16 separate situations arise in this type of flow.**

Another complicating factor is whether velocity and thermal boundary layers develop simultaneously or thermal boundary starts (heating) after velocity boundary layer has developed. As correlations are different for such different conditions, a large number of correlation have to be used.

Another factor is the fluid property values and their variation with temperature. Due to this liquid metals having very low value of Prandtl number require quite different correlations.

A third situation is non circular section of the flow duct, which again requires different correlations.

In the following articles the well accepted correlations are discussed.

9.3 LAMINAR FLOW

Constant Wall Temperature: (Re_d < 2300) Reynolds number is defined as below:

$$Re = Du_m/\nu = 4G/\pi D\mu \quad ...(9.2)$$

Friction factor f is defined as below

$$f = \frac{\Delta P}{\rho(L/D)(u_m^2/2)} \quad ...(9.3)$$

where ΔP is the pressure drop in length L, the other terms defined as usual. This factor f (Darcy friction factor) is also equal to $4C_f$. Refer advanced example 9.2 for proof. Where C_f is Fanning friction coefficient defined in Chapters 6 and 7 as $C_f = \tau_w / \frac{1}{2}\rho u^2$.

For laminar flow $f = 64/Re$ ...(9.4)

Refer advanced example 9.3 for proof.

The entry length depends on several factors. Varied correlations are available in literature. A fairly well accepted correlation is given by

Hydrodynamic $\mathbf{x/D = 0.05\ Re}$ ...[9.4(*a*)]

Thermal $\mathbf{(x_t/D) = 0.05\ Re.Pr}$

9.3.1. Entry Region

For short lengths (x/D small), with hydrodynamic layer already developed: the average Nusselt number is (1943)

$$Nu = 1.67\ [Re\ Pr/(x/D)]^{0.333} \qquad ...(9.5)$$

For $\left(\frac{x}{D}\right)/Re\ Pr < 0.01$, $Re\ Pr\ \frac{D}{x}$ is also termed as **Graetz number Gz.**

The property values are to be taken at $(\mathbf{T_{mi} + T_{mo}})/2$ where T_{mi} and T_{mo} are bulk mean temperatures at entry and at exit. Iterative working may become necessary if the outlet temperature is not specified.

For longer tubes for length x, with **hydrodynamic boundary layer already developed,** the average value of convection coefficient is given by Hassen as (1943) (applicable for $Pr > 07$)

$$Nu = 3.66 + \frac{0.0668\ (D/x)\ Re\ Pr}{1 + 0.04\ [D/x]\ Re\ Pr]^{0.67}} \qquad ...(9.6)$$

For $\frac{x}{D}/Re\ Pr < 0.01$

It is to be noted that **after the entry length Nusselt number** does not vary with length and **is constant** as given by equation (9.7) ($Nu \neq f(x)$)

$$\mathbf{Nu = 3.66} \qquad ...(9.7)$$

Equations (9.5) and (9.6) are applicable assuming velocity boundary layer is fully developed before heating begins. This assumption has practical use, as in the case of highly viscous fluids.

9.3.2. Both Boundary Layers Simultaneously Developing—Constant Wall Temperature

For short lengths for liquid in tubes (1936)

$$Nu = 1.86\ (Re\ Pr\ D/x)^{0.333} \left(\frac{\mu}{\mu_w}\right)^{0.14} \qquad ...(9.8)$$

for $(x/D)\ (Re\ .\ PrD/x)^{0.333}\ (\mu/\mu_w)^{0.14} \geq 2$, $0.5 < Pr < 1600$, $(\mu/\mu_w) < 9.75$

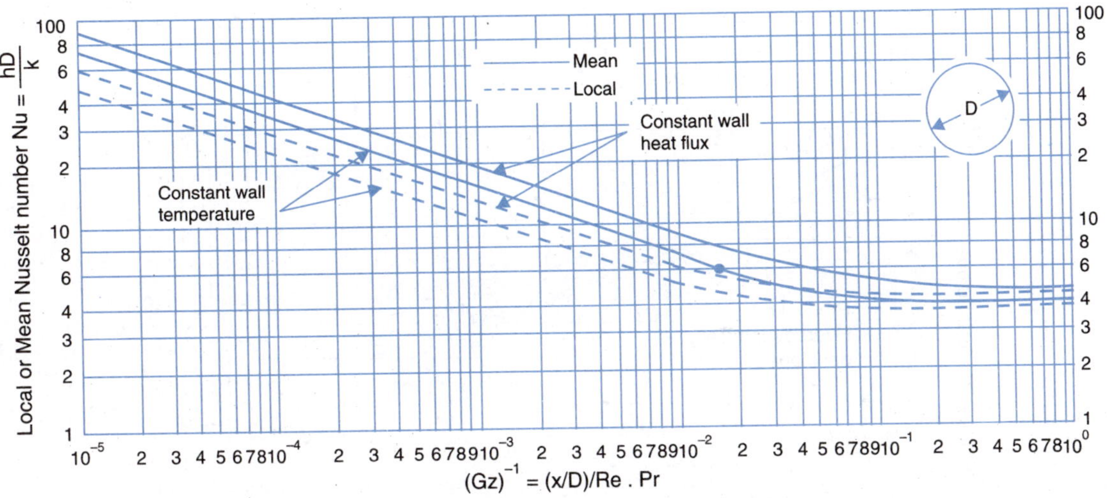

Fig. 9.3. Mean and local Nusselt numbers for thermally developing, hydrodynamically developed laminar flow inside a circular tube

More generally given by Kays (1950)

$$Nu = 3.66 + \frac{0.104\,(Re\,Pr\,D/x)}{1 + 0.16\,(Re\,Pr\,D/x)^{0.8}} \quad ...(9.9)$$

For $Pr > 0.6$

This also reduces after the entry length

$$Nu = 3.66 \qquad L >> D \quad ...(9.10)$$

The Nusselt number variation in the entry region for constant wall temperature and constant heat flux (with Hydrodynamic boundary layer already developed) is shown in Fig. 9.3.

9.3.3. For Laminar, Constant Heat Flux

The equation obtained for fully developed hydrodynamic flow is given below: (1956)

$$Nu = 1.30\left[\frac{Re\,Pr}{(x/D)}\right]^{0.33} \quad ...(9.11)$$

applicable for $(x/D)/Re\,Pr < 0.01$

Beyond entry region as in the other case **Nusselt number is independent of length and is a constant** given by equation (9.12).

$$\mathbf{Nu = 4.364} \quad ...(9.12)$$

$$Pr > 0.6$$

The analogy between heat and momentum transfer also provides

$$St\,Pr^{0.67} = f/8 \quad ...(9.13)$$

The values of f can be obtained from equation (9.4) for laminar region.

The Nusselt number variation in the entry region at constant wall temperature is shown in Fig. 9.4. (Both boundary layers developing simultaneously).

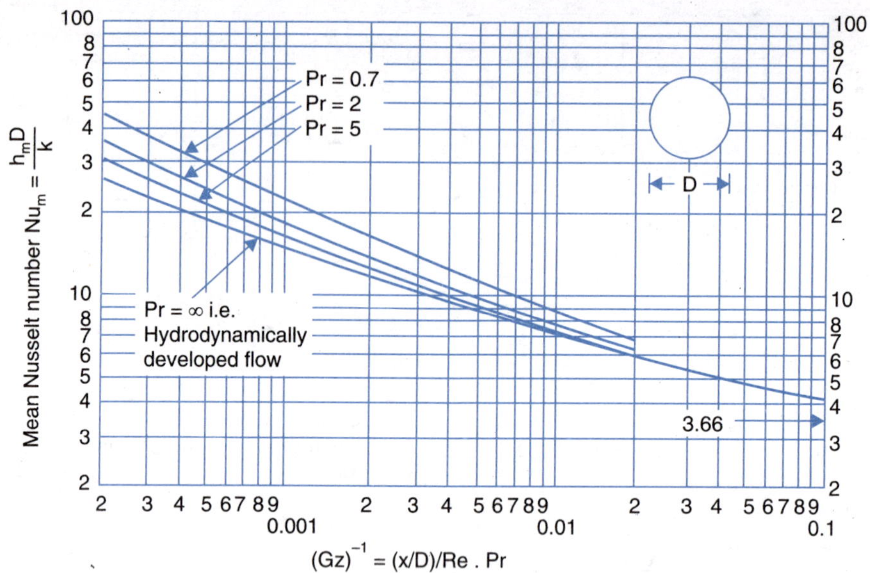

Fig. 9.4. Mean Nusselt numbers for simultaneously developing laminar flow inside a circular tube subjected to constant wall temperature

Example 9.1: *Water at a bulk mean temperature of 40°C flows through a tube of 0.05 m inner diameter with a velocity of 0.025 m/s. The tube is 2 m long and its surface is at a constant temperature of 60°C. Determine the heat transfer and the mean temperature rise during the flow. Assume hydrodynamic boundary layer already developed.*

Solution: The properties of water at 40°C are

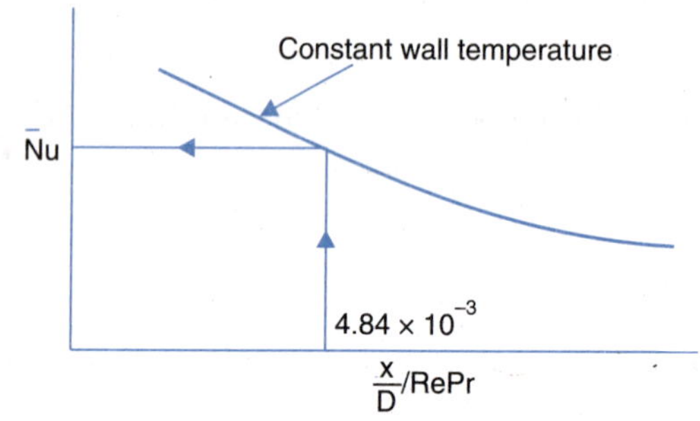

Fig. Ex. 9.1

$$\rho = 995 \text{ kg/m}^3, \quad \nu = 0.657 \times 10^{-6} \text{ m}^2\text{/s},$$

$$Pr = 4.34, \ C = 4178 \text{ J/kgK}, \ k = 0.628 \text{ W/mK}.$$

$$\textbf{Re} = \frac{0.05 \times 0.025}{0.657 \times 10^{-6}} = \textbf{1902.6.} \text{ laminar flow}$$

$$\frac{x}{D} / Re\ Pr = \frac{2}{0.05} / (1902.6 \times 4.34) = 4.844 \times 10^{-3} < 0.01$$

Using the chart in Fig. 9.3 (schematically shown in Fig. Ex. 9.1) the value of $\overline{Nu}$ is read as 9.

$$\bar{h} = \frac{9 \times 0.628}{0.05} = \mathbf{113.04\ W/m^2\ K}$$

Heat transferred, $\mathbf{Q} = hA(T_S - T_M)$

$$= 113.04 \times \pi \times 0.05 \times 2(60 - 40) = \mathbf{710\ W.}$$

Temperature rise $= \dfrac{Q}{mc}$

$$m = \rho AV = 995 \times \pi \times \frac{0.05^2}{4} \times 0.025 = 0.04884 \text{ kg/s}$$

Temperature rise $= 710/(0.04884 \times 4178) = 3.48°C$

By equation (9.5), Nu $= 1.67 \left(Re\ Pr \dfrac{D}{x} \right)^{1/3} = 1.67 \left(1902.6 \times 4.34 \times \dfrac{0.05}{2} \right)^{1/3} = 9.86$

By equation (9.6), Nu $= 3.66 + \dfrac{0.668 \left(Re\ Pr \dfrac{D}{x} \right)}{1 + 0.04 \left(Re\ Pr \dfrac{D}{x} \right)^{0.67}} = \mathbf{9.35}$

The values are close.

Example 9.2: *Water at 30°C enters a pipe of 25 mm ID with a mean velocity of 0.06 m/s. The pipe surface temperature is 50°C. Determine the outlet temperature for lengths of (i) 1 m (ii) 4 m and (iii) 10 m. Assume hydrodynamic boundary layer fully developed.*

Solution: As the bulk mean temperature is not known, property values at the entry temperature is first considered. At 30°C

$$\rho = 997.5 \text{ kg/m}^3,\ \nu = 0.8315 \times 10^{-6} \text{ m}^2/\text{s},\ Pr = 5.68$$

$$c = 4178 \text{ J/kgK},\ k = 0.6129 \text{ W/mK}.$$

Case (i) 1 m length

$$\mathbf{Re} = \frac{u_m D}{\nu} = 0.06 \times 0.025/0.8315 \times 10^{-6} = \mathbf{1804}$$

∴ **Flow is laminar.**

$$\frac{x}{D} / Re\ Pr = \frac{1}{0.025} / (1804 \times 5.68) = 0.004 < 0.01$$

Using the chart Fig. 9.3, the average value of *Nu* is read as 10.

Using equation (9.5), $Nu = 1.67 \left(1804 \times 5.68 \times \dfrac{0.025}{1} \right)^{0.333} = 10.6.$

∴ $$\mathbf{h} = 10.6 \times 0.6129/0.025 = \mathbf{260\ W/m^2\ K.}$$

As the temperature difference varies continuously, logarithmic mean temperature should be used in heat flow calculation.

Energy balance gives

$$Q = hA(LMTD) = mc(T_{mo} - T_{mi}) \quad ...(a)$$

where $A = \pi dL$

Chapter 9

$$LMTD = \frac{(T_w - T_{mi}) - (T_w - T_{mo})}{\ln \dfrac{T_w - T_{mi}}{T_w - T_{mo}}} = \frac{T_{mo} - T_{mi}}{\ln \dfrac{T_w - T_{mi}}{T_w - T_{mo}}}$$

Substituting in (*a*), and rearranging

$$\frac{T_w - T_{mi}}{T_w - T_{mo}} = e^{-\frac{hA}{mc}}, \quad m = \rho AV \qquad ...(b)$$

$$\frac{hA}{mc} = \frac{260 \times \pi \times 0.025 \times 1}{997.5 \times \dfrac{\pi \times 0.0025^2}{4} \times 0.06 \times 4178} = 0.1664$$

$$\therefore \qquad \frac{50 - 30}{50 - T_{mo}} = e^{-0.1664}$$

Solving $\mathbf{T_{mo} = 33.07°C}$

Another trial with properties at (30 + 33.07)/2 will improve the result.

Case (ii) For 4 m length, with the properties as in case (*i*)

$$Re = 1804, \ \frac{x}{D} / Re\,Pr = \frac{4}{0.025} / (1804 \times 5.68) = 0.016.$$

From chart Fig. 9.3, at this value $\overline{Nu}$ is read as 6. (Try by equation (9.6), $Nu = 6.27$)

$$\therefore \qquad \mathbf{h} = \frac{6 \times 0.6129}{0.025} = \mathbf{147.1\ W/m^2\ K}$$

From equation (*b*)

$$\frac{T_w - T_{mi}}{T_w - T_{mo}} = e^{-\frac{hA}{mc}}$$

For 4 m length $\dfrac{hA}{mc} = 0.3765$

$$\frac{50 - 30}{50 - T_{mo}} = e^{-0.3765}$$

Solving $\mathbf{T_m = 36.3°C}$

Another trial will improve results.

Case (iii) 10 m length,

$$\frac{x}{D} / Re\,Pr = 0.04$$

Reading from chart Fig. 9.3, at this value,

$$Nu = 4.9 \qquad \text{(By equation (9.6), } Nu = 4.91)$$

$$\therefore \qquad \mathbf{h = 120.13\ W/m^2\,K.}$$

Using equation (*b*)

$$\frac{50 - 30}{50 - T_{mo}} = e^{-\frac{hA}{mc}}, \ \frac{hA}{mc} = 0.7687.$$

∴ $\mathbf{T_{mo} = 40.72°C}$

Another trial with properties at 35°C will improve the results.

Example 9.3: *Water at 60°C flows through a smooth pipe of 2.54 cm ID at a velocity of 0.02 m/s. The pipe wall is at 80°C. Assuming both boundary layers to develop simultaneously, determine the exit temperature at 3 m from entry.*

Solution: As bulk mean temperature is not known, assume property values at 65°C.

$$\rho = 982.25 \text{ kg/m}^3, \quad C = 4186 \text{ J/kg K}, Pr = 3.0,$$

$$k = 0.65565 \text{ W/m K}, \quad \mu = 446.43 \times 10^{-6}$$

$$Re_d = \frac{\rho u_m d}{\mu} = \frac{982.25 \times 0.02 \times 0.0254}{446.43 \times 10^{-6}} = 1117.7.$$

$(Re_d \cdot Pr\ D/x)^{1/3} \times (\mu/\mu_w)^{0.14} > 2$, μ_w at 80°C = 3.55×10^{-4} kg/ms

$$(1117.7 \times 3 \times 0.0254/3)^{1/3} \times \left(\frac{446.43 \times 10^{-4}}{3.55 \times 10^{-4}}\right)^{0.14} = 3.15 > 2.$$

∴ **Equation (9.8) can be used.**

$$Nu = 1.86(Re \cdot Pr\ D/x)^{1/3} \left(\frac{\mu}{\mu_w}\right)^{0.14} = 1.86 \times 3.15 = 5.859$$

$$h = \frac{5.859 \times 0.65565}{0.0254} = 151.24 \text{ W/m}^2\text{K}$$

Using heat balance:

$$h \cdot \pi d \cdot L\left(T_w - \frac{T_{m1} + T_{m2}}{2}\right) = mc\,(T_{m2} - T_{m1})$$

$$T_w = 80°\text{C}, T_m = 60°\text{C}, m = \rho A u_m = 982.25 \times \pi \times \frac{0.0254^2}{4} \times 0.02$$

$$m = 0.009954 \text{ kg/s}.$$

$$\pi \times 151.24 \times 0.0254 \times 3\left(80 - 30 - \frac{T_{m2}}{2}\right) = 0.009954 \times 4186\,(T_{m2} - 60)$$

Solving: $$50 - \frac{T_{m2}}{2} = 1.1509\,(T_{m2} - 60),$$

∴ $\mathbf{T_{m2} = 72.11°C.}$

∴ Bulk mean temperature is 66.06°C, as compared to 65°C

Another trial will give **71.9°C** as exit temperature.

If equation (a) of example 9.2 is used,

$$\frac{80 - T_o}{80 - 60} = e^{-\frac{151.24 \times \pi \times 0.0254 \times 3}{0.009954 \times 4186}} = e^{-0.41868.}$$

$$T_o = 71.62°\text{C}.$$

Again another iteration may be necessary.

Chapter 9

Example 9.4: *Engine oil enters at 35°C into a tube of 10 mm ID at the rate of 0.05 kg/s and is to be heated to 45°C. The tube wall is maintained at 100°C by condensing steam. Determine the length of the tube required.*

Solution: Bulk mean temperature = $(T_{mi} + T_{mo})/2 = \dfrac{35+45}{2} = 40°C$.

The property value from data book:

$$\rho = 876 \text{ kg/m}^3,\ c = 1964 \text{ J/kg K},\ \mu = 0.210 \text{ Ns/m}^2,\ k = 0.144 \text{ W/mK},$$

$$Pr = 2870.$$

Using equation (9.1)

$$\mathbf{Re} = \frac{4G}{\pi D\mu} = \frac{4\times 0.05}{\pi\times 0.01\times 0.210} = \mathbf{30.32}$$

∴ Laminar.

Assuming long tube, *Nu* = 3.66. (Beyond entry length)

$$\therefore \qquad \mathbf{h} = \frac{3.66\times 0.144}{0.01} = \mathbf{52.7\ W/m^2\ K}$$

Heat transferred: $mc\,\Delta T$

$$= 1966 \times 0.05 \times (45 - 35) = 983 \text{ W}.$$

The temperature varies as shown in figure.

100°C
ΔT$_1$ = 65
ΔT$_2$ = 55
45
35

Fig. Ex. 9.4

$$\therefore \qquad LMTD = \frac{65-55}{\ln\dfrac{65}{55}} = \mathbf{59.9°C}$$

$$\pi DL\, h(LMTD) = Q$$

$$\therefore \qquad L = Q/\pi Dh(LMTD) = \frac{983}{\pi\times 0.01\times 52.7\times 59.9} = \mathbf{9.92\ m}$$

Tube length required = 9.92 m say 10 m.

Example 9.5: *Water at a bulk mean temperature of 40°C flows in a tube of 0.05 m diameter with a velocity of 0.025 m/s. The tube length is 2 m. In the surface there is a uniform heat flux of 1500 W/m².*

Determine the heat transfer coefficient.

Solution: At the bulk mean temperature of 40°C, the property values are

$$\rho = 995 \text{ kg/m}^3, \qquad \nu = 0.657 \times 10^{-6} \text{ m}^2\text{/s},$$

$$c = 4178 \text{ J/kg K}, \quad Pr = 4.34,\ k = 0.628 \text{ W/mK}.$$

Reynolds number $= \dfrac{0.025\times 0.05}{0.657\times 10^{-6}} = \mathbf{1902.6}$

∴ Flow is laminar.

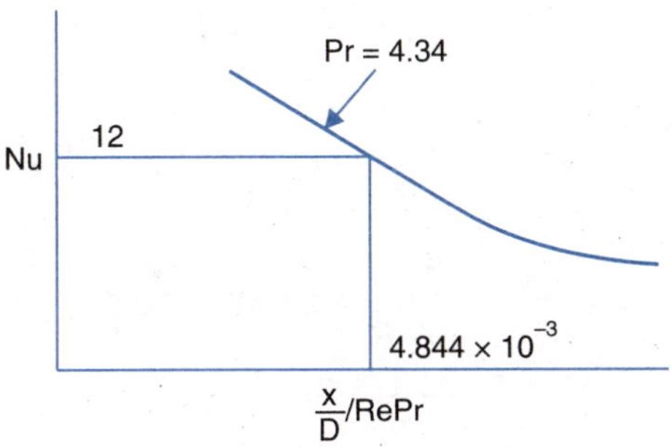

Fig. Ex. 9.5

Check for $\frac{x}{D} / Re\ Pr = \frac{10}{0.025} / 1902.6 \times 4.34 = 4.844 \times 10^{-3} < 0.01$

The values can be read from chart, Fig. 9.3. (shown schematically)

$$Nu = 12 \quad \therefore \quad \mathbf{h} = \frac{12 \times 0.628}{0.05} = \mathbf{150.72\ W/m^2\ K}$$

Heat transfer rate $= qA = q\pi\ DL = 1500 \times \pi \times 0.05 \times 2 = \mathbf{471.24\ W}$

Average temperature difference $= \frac{Q}{hA} = \frac{471.24}{150.72 \times \pi \times 0.05 \times 2} = \mathbf{9.95°C}$

Example 9.6: *Water at 10°C enters at the rate of 0.01 kg/s into a tube of 20 mm inner diameter. It is to be heated to 40°C. There is a uniform heat flux of 15 kW/m² over the surface. Determine the (i) Reynolds number, (ii) length of pipe needed, (iii) heat transfer coefficient, (iv) Tube temperature at outlet, (v) Friction factor, and (vi) Pressure drop over the length.*

(Assume fully developed flow.)

Solution: Bulk mean temperature = (40 + 10)/2 = 25°C.

Property values at 25°C are

$$\rho = 997\ \text{kg/m}^3,\ c_p = 4180\ \text{J/kg K},\ k = 0.608\ \text{W/m K},$$

$$\mu = 910 \times 10^{-6}\ \text{Ns/m}^2,\ \text{using equation (9.1)}$$

(*i*) **Reynolds number** $= \frac{4G}{\pi D \mu} = \frac{4 \times 0.01}{\pi \times 0.02 \times 910 \times 10^{-6}} = \mathbf{699.6}$

$\therefore$ **Laminar flow.**

(*ii*) Length of pipe needed. Under uniform heat flux, the heat balance yields

$$q \times \pi DL = m\ C_p(T_{mo} - T_{mi})$$

$$\mathbf{L} = \frac{m\ C_p}{\pi q D}(T_{mo} - T_{mi}) = \frac{0.01 \times 4180}{\pi \times 15000 \times 0.02} \times 30 = \mathbf{1.33\ m.}$$

(*iii*) Heat transfer coefficient: under fully developed condition, **Nu = 4.364.** (eqn. 9.12)

$$\mathbf{h} = \frac{k}{D} Nu = \frac{0.608}{0.02} \times 4.364 = \mathbf{132.67\ W/m^2\ K.}$$

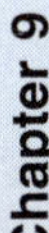

(*iv*) Tube surface temperature at exit:

$$\mathbf{T_{so}} = \frac{q}{h} + T_{mo} = \frac{1500}{132.67} + 40 = \mathbf{153.1°C}$$

(*v*) Friction factor:

$$\mathbf{f} = \frac{64}{Re} = \frac{64}{699.6} = \mathbf{91.48 \times 10^{-3}}$$

(*vi*) Pressure drop:

$$\mathbf{\Delta P} = \frac{\rho . f\, Lu^2}{2g_e D}, \; u = \frac{4m}{\rho\pi\, D^2}$$

$$\Delta P = \frac{997 \times 91.48 \times 10^{-3} \times 1.33}{2 \times 1 \times 0.02} \times \left(\frac{4 \times 0.01}{997 \times \pi \times 0.02^2}\right)^2 = \mathbf{3.09\ N/m^2}$$

Example 9.7: *Water is to be heated at the rate of 0.01 kg/s from bulk temperature of 20°C to 60°C as it flows through a tube of 20 mm ID by means of electrical resistance heating at the rate of 15 × 10³ W/m². Calculate the length required. Also find the heat transfer coefficient.*

Solution: Bulk mean temperature = (60 + 20)/2 = 40°C

The property values are:

$\rho = 995$ kg/m³, $\nu = 0.657 \times 10^{-6}$ m²/s, $Pr = 4.34$,

$c = 4178$ J/kg K, $k = 0.628$ W/m K, $\mu = \rho\nu$

Using equation (9.1),

$$\mathbf{Re} = 4G/\pi D\mu = \frac{4 \times 0.01}{\pi \times 0.02 \times 995 \times 0.657 \times 10^{-6}} = \mathbf{974}$$

$\therefore$ **Laminar.**

Constant heat flux condition: assuming fully developed flow

$$Nu = 4.364, \quad \therefore \quad \mathbf{h} = \frac{4.364 \times 0.628}{0.02} = \mathbf{136.0\ W/m^2\ K}$$

Using energy balance,

$$\pi DLq = mc\,\Delta T_m \quad \therefore \quad L = \frac{mc\,\Delta T_m}{\pi Dq}$$

$$L = \frac{0.01 \times 4178 \times (60 - 20)}{\pi \times 0.02 \times 15 \times 10^3} = \mathbf{1.773\ m.}$$

Example 9.8: *A solar concentrator causes a heat flux of 2000 W/m² on tube of 60 mm ID. Pressurised water flows through the tube at a rate of 0.01 kg/s. If the bulk temperature at inlet is 20°C, what will be the length required to heat the water to a bulk temperature of 80°C? Also find the wall temperature at exit.*

Solution: Bulk mean temperature = (80 + 20)/2 = 50°C.

The property values are

$\rho = 990$, $\nu = 0.5675 \times 10^{-6}$, $Pr = 3.68$, $k = 0.64$ W/m K, $c = 4181$ J/kg K.

$$\textbf{Re} = \frac{4G}{\pi D\mu} = \frac{4 \times 0.01}{\pi \times 0.06 \times 990 \times 0.5675 \times 10^{-6}} = \textbf{377.7}$$

∴ **Flow is laminar.**

$$\pi DLq = mc\,\Delta T_m, \quad \therefore\ L = \frac{mc\,\Delta T_m}{\pi Dq}$$

$$\textbf{L} = \frac{0.01 \times 4181 \times (80 - 20)}{\pi \times 0.06 \times 2000} = \textbf{6.65 m.}$$

Assuming fully developed condition,

$$Nu = 4.364, \quad \therefore\ h = \frac{4.364 \times 0.64}{0.06} = \textbf{46.55 W/m}^2\ \textbf{K}$$

To find the surface temperature at exit:

$$\textbf{T}_{\textbf{so}} = \frac{q}{h} + T_{mo} = \frac{2000}{46.55} + 80 = \textbf{122.97°C.}$$

Example 9.9: *Air at 20°C flows inside a pipe 18 mm ID having a uniform heat flux of 150 W/m² on the surface, the average flow velocity at entry being 1.0 m/s. The air pressure is 2 bar. Determine the value of convection coefficient. If the pipe is 2.5 m long, determine the air exit temperature and the wall temperature at the exit. Assume fully developed hydrodynamic boundary layer.*

Solution: The exit temperature of air can be obtained using energy balance relationship

Area × heat flux = mass × sp. heat × temp. rise,

(mass flow = area × velocity × density)

$$\pi \times 0.018 \times 2.5 \times 150 = \frac{\pi \times 0.018^2}{4} \times 1.0 \times \frac{2 \times 10^5}{287 \times 293} \times 1005 \times \Delta T$$

$$\therefore \quad \Delta T = 34.86°\text{C} \quad \therefore \quad \textbf{T}_{\textbf{mo}} = \textbf{54.86°C}$$

Note that the mass flow is calculated with the density at entrance.

Average bulk temperature = (54.86 + 20)/2 = **37.43°C**

The properties at this temperature and pressure of 2 bar are density = 2.275 kg/m³, Pr = 0.7, c = 1005 J/kg K, μ = 18.994 × 10^{-6} kg/ms, k = 27.35 × 10^{-3} W/m K,

$$\therefore \quad \textbf{Re} = \frac{1.0 \times 2.275 \times 0.018}{18.994 \times 10^{-6}} = \textbf{2156} < 2300 \quad \therefore \quad \text{flow is laminar}$$

Note that Re is determined using velocity at entrance

$\frac{x}{D}/Re\,Pr = 0.092 > 0.01$. So long pipe value is used.

$$\therefore \quad Nu = 4.364$$

From chart Fig. 9.3, $Nu = 5$.

$$\therefore \quad \textbf{h} = \frac{5 \times 0.02735}{0.018} = \textbf{7.597 W/m}^2\ \textbf{K}$$

As $q \cdot A = hA\,\Delta T$, cancelling A,

$$150 = 7.597 \times \Delta T, \quad \therefore \quad \Delta T = 19.74°\text{C}.$$

Wall temperature at exit, $\mathbf{T_{so}}$ = 54.86 + 19.74 = **74.6°C.**

Chapter 9

9.4 TURBULENT FLOW

The development of boundary layer is similar except that the **entry region length in practice is found to lie between 10 to 60 times the diameter.** The convective heat transfer coefficient has a higher value as compared to laminar flow.

The friction factor for smooth pipes is given by equation (9.14) for $Re > 10^4$

$$f = 0.184\ Re^{-0.2} \qquad ...[9.14(a)]$$

$$f = [0.7 \ln Re - 1.64]^{-2} \qquad ...[9.14(b)]$$

$$f = 4[1.58 \ln Re - 3.28]^{-2} \qquad ...[9.14(c)]$$

For transition range

$$f = 0.316\ Re^{-0.25} \qquad ...(9.15)$$

For rough pipes, the Moody chart can be used. Also for relative roughness of ε/D

$$f = \frac{1.325}{\left[\ln \dfrac{\varepsilon}{3.7\,D} + 5.74/Re^{0.9}\right]^2} \qquad ...(9.16)$$

The importance of the calculation of the friction factor in heat transfer is in the use of its value for the determination of convection coefficient by analogy.

The following two relations are useful for the entrance region.

$$Nu = 0.036\ Re^{0.8}\ Pr^{0.33} \left(\frac{D}{x}\right)^{0.055} \qquad ...(9.17)$$

$$\overline{Nu} = Nu\left[1 + \frac{C}{(x/D)}\right] \qquad ...(9.18)$$

$C = 1.4$ hydrodynamic layer developed

$C = 6$ hydrodynamic layer not developed

$x/D > 10$

Also $$\overline{Nu} = Nu\ [1 + (D/x)^{0.7}] \qquad ...(9.19)$$

$2 < x/D < 20$, In (9.18) and (9.19) Nu is calculated using (9.20).

In the equations (9.18) and (9.19), Nu is calculated for fully developed flow. The property values are to be at average bulk temperature.

The more popular correlation for fully developed flow in smooth tubes is due to **Dittus and Boelter (1930) (modified Colburn)**

$$Nu = 0.023\ Re^{0.8}\ Pr^{n} \qquad ...(9.20)$$

$n = 0.3$ for cooling and 0.4 for heating of fluids.

A modified form when the property variation due to temperature change was large proposed by **Sieder and Tate (1936) is**

$$Nu = 0.027\ Re^{0.8}\ Pr^{0.33} \left[\frac{\mu_m}{\mu_w}\right]^{0.14} \qquad ...(9.21)$$

$$0.7 < Pr < 16700,\ Re > 10^4,\ \frac{L}{D} > 60$$

A more recent correlation (1970) which fits experimental values more closely is given by (properties at film temp., except μ)

$$Nu = \frac{(f/8)\, Re\, Pr}{1.07 + 12.7\,(f/8)^{0.5}\,[Pr^{0.67} - 1]} \left[\frac{\mu_m}{\mu_w}\right]^n \qquad ...(9.22)$$

$n = 0.11$ for heating of fluids

$n = 0.25$ for cooling of fluids

$n = 0$ for constant heat flux

Also $\frac{\mu_m}{\mu_w}$ to be replaced by $\frac{T_w}{T_m}$ for gases, (temp. in K)

For Rough tubes

$$St.\, Pr^{0.67} = f/8 \qquad ...(9.23)$$

In the place of the constants 1.07 and 12.7 in the equation (9.22) more refined values are also available in handbooks, based on f and Pr.

In the choice of the equation to be used, there will be some ambiguity. But equation (9.22) is more recent and found to fit the experimental values more closely.

The correction factors $\left[\frac{\mu_m}{\mu_w}\right]^n$ for liquids or $\left[\frac{T_w}{T_m}\right]^n$ for gases is used to correct for large variation in properties.

Example 9.10: *Air at 20°C flows through a tube 8 cm dia with a velocity of 9 m/s. The tube wall is at 80°C. Determine for a tube length of 5 m, the exit temperature of air.*

Solution: For the first trial property value taken at 20°C. $\rho = 1.205$, $\nu = 15.06 \times 10^{-6}$ m²/s, $Pr = 0.703$. $k = 0.02593$ W/mK, $C_p = 1005$ J/kg K, $\mu = 18.14 \times 10^{-6}$ kg/ms

$$\mathbf{Re} = 9 \times 0.08/15.06 \times 10^{-6} = \mathbf{47809}$$

∴ flow is turbulent

Using equation (9.20) (heating, $n = 0.4$)

$$Nu = 0.023\, Re^{0.8}\, Pr^{0.4} = 110.69, \quad \therefore \quad \mathbf{h = 35.88\ W/m^2\ K}$$

The bulk temperature at exit can be found by energy balance.

As the temperature difference varies along the length *LMTD* should be used in place of ΔT.

$$mc(T_{mo} - T_{mi}) = hA\,.\,LMTD = hA\, \frac{(T_w - T_{mo}) - (T_w - T_{mi})}{\ln \dfrac{T_w - T_{mo}}{T_w - T_{mi}}}$$

This leads to the equation. (Ref. Example 9.1)

$$\frac{T_w - T_{mi}}{T_w - T_{mo}} = e^{+\frac{hA}{mc}} \qquad ...(A)$$

$$A = \pi DL,\ m = \rho\, VA_c = 1.205 \times 9 \times \pi \times \frac{0.08^2}{4} = 0.05451 \text{ kg/s}$$

Chapter 9

$$\frac{hA}{mc} = \frac{35.88 \times \pi \times 0.08 \times 5}{0.05451 \times 1005} = 0.823$$

$$\therefore \qquad \frac{80-20}{80-T_{mo}} = e^{0.823}, \text{ Solving } \mathbf{T_{mo} = 53.65°C}$$

The mean temperature = (53.65 + 20)/2 = 36.825°C

Taking properties of 40°C and reworking.

$$T_{mo} = 35.47.$$

Using equation (9.22), with $f = [0.79 \ln \text{Re} - 1.64]^{-2} = 0.021736$

$$Nu = 87.91, \mathbf{h = 30.28\ W/m^2\ °C}$$

Using analogy, $St.\, Pr^{2/3} = \frac{f}{8}$, $Nu = 102.37$, $\mathbf{h = 35.27\ W/m^2\ K}$

The correlations yield in the case of air, value not for removed from each other.

Example 9.11: *Air at 20°C flows through an 8 cm dia pipe with a velocity of 9 m/s, and the pipe wall is at 80°C. The length of the pipe is 1 m. Determine the exit temperature of air.*

Solution: $\frac{L}{D} = \frac{1}{0.08} = 12.5$. So it is better to use entry region correlations.

Considering properties at 20°C

$\rho = 1.205$ kg/m³, $\nu = 15.06 \times 10^{-6}$ m²/s, $Pr = 0.703$,

$k = 0.02593$ W/m K, $C_p = 1005$ J/kg K, **Re** $= 9 \times 0.08/15.06 \times 10^{-6} =$ **47809.**

Using equation (9.17)

$$Nu = 0.036\, Re^{0.8}\, Pr^{0.33}\, (D/x)^{0.055}$$

$$= 154.37, \quad \therefore \quad \mathbf{h = 50.04\ W/m^2\ K}$$

Using equation (9.19)

$$\overline{Nu} = Nu\left[1+\left(\frac{D}{x}\right)^{0.7}\right], \text{ where } Nu = 0.023\, Re^{0.8}\, Pr^{0.4} = 113.32$$

$$\therefore \qquad \overline{Nu} = 113.32\left[1+\left(\frac{0.08}{1}\right)^{0.7}\right] = 132.64$$

$$\therefore \qquad \mathbf{h = 42.99\ W/m^2\ K}$$

Using equation (9.18)

$$\overline{Nu} = Nu\,[1 + 6/(x/D)] = 113.32\,[1 + 6/12.5] = 167.71$$

$$\therefore \qquad \mathbf{h = 54.36\ W/m^2\ K}$$

Adopting the value of 50.04 W/m² K using the heat balance

The temperature variation is shown in Fig. Ex. 9.11. So *LMTD* should be used in place of ΔT.

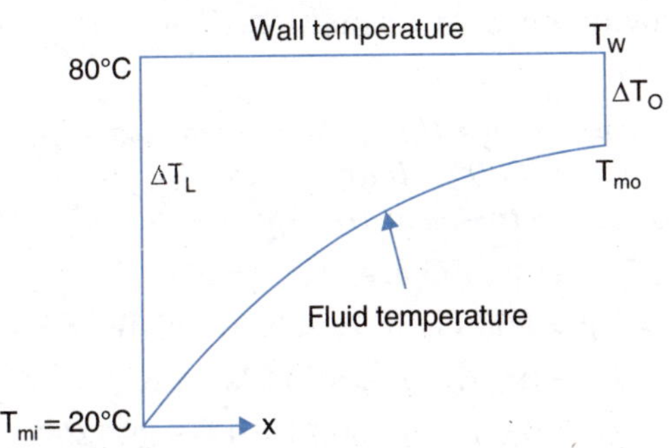

Fig. Ex. 9.11. Fluid temperature along the flow

From advanced example 9.10,

$$\frac{T_w - T_{mi}}{T_w - T_{mo}} = e^{-\frac{hA}{mc}} = \frac{80-20}{80-T_{mo}} = e^{-0.1653}$$

as $$\frac{hA}{mc} = \frac{50.04 \times \pi \times 0.08 \times 1}{1.205 \times \pi \times \frac{0.08^2}{4} \times 9 \times 1005} = 0.1653$$

Solving $\mathbf{T_{mo} = 29.1°C.}$

9.5 LIQUID METAL FLOW

Liquid metals are characterised by low values of Prandtl number and higher values of conductivity. Liquid metals provide large values of convection coefficient and are used to extract heat in breeder type of nuclear reactors where the volume heat generation rate is very high. Constant heat flux situation is more common in these cases.

9.5.1. Liquid Metals—Constant Heat Flux

The following three relations are applicable for **constant heat flux.** Properties are to be evaluated at average of bulk temperatures.

Entry region:

$$Nu = 3.0\, Re^{0.0833} \qquad Re\, Pr < 100 \qquad ...(9.24)$$

A simple relation, but with large deviations is given by

$$Nu = 0.625\, (Re\, Pr)^{0.4} \qquad ...(9.25)$$

$$10^2 < Re\, Pr < 10^4, \ \frac{L}{D} > 60$$

More recent correlation now in use is

$$Nu = 4.82 + 0.0185\, (Re\, Pr)^{0.827} \qquad ...(9.26)$$

$$10^2 < Re\, Pr < 10^4 \text{ and } 3.6 \times 10^3 < Re < 9.05 \times 10^8.$$

9.5.2. Liquid Metals—Constant Wall Temperature

For constant wall temperature the correlation is

$$Nu = 5 + 0.025\, (Re\, Pr)^{0.8} \qquad ...(9.27)$$

$$Re\, Pr > 100, \ \frac{L}{D} > 60$$

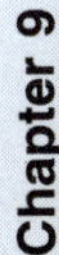

These equations provide only an estimation. For more dependable and closer values, one has to go into literature.

Example 9.12: *Liquid sodium flows through a 5 cm dia pipe at a velocity of 4 m/s. The temperature of the fluid at entry is 300°C. If at the pipe wall the fluid is exposed to uniform heat flux, determine the value of convective heat transfer coefficient.*

Solution: The property values at 300°C are

$$\rho = 903 \text{ kg/m}^3, \ \nu = 0.506 \times 10^{-6} \text{ m}^2\text{/s},$$

$$Pr = 0.0075, \ k = 81.41 \text{ W/m K}, \ c = 1327.2 \text{ J/kg K}$$

$$\mathbf{Re} = 0.05 \times 4/0.506 \times 10^{-6} = \mathbf{395257}$$

Using equation (9.26)

$$Nu = 4.82 + 0.0185\ (Re\ Pr)^{0.827} = 18.58, \quad \therefore \quad \mathbf{h = 30244\ W/m^2\ K}$$

Using equation (9.25)

$$Re = 0.625\ (Re\ Pr)^{0.4} = 15.3 \quad \therefore \quad \mathbf{h = 24904\ W/m^2\ K}$$

The values are quite different.

The heat flux value influences the value of convection coefficient only marginally. The wall temperature is decided by the heat flux value. Higher the heat flux, larger will be the temperature difference between the fluid and the surface.

Example 9.13: *Liquid sodium at 200°C is heated in a tube of 5 cm dia to 400°C with a uniform heat flux of 2500 kW/m². The flow rate is 9 kg/s. Determine the length of pipe required and also the mean temperature difference between the wall and fluid.*

Solution: The property vlaue can be evaluated at (200 + 400)/2 = 300°C. The values are

$$\rho = 878 \text{ kg/m}^3, \ \nu = 0.394 \times 10^{-6} \text{ m}^2\text{/s}, \quad Pr = 0.0063,$$

$$k = 70.94 \text{ W/mK}, \qquad c = 1281.2 \text{ J/kg K}$$

The length required can be calculated using the energy balance.

mass flow × sp. heat × temp. rise = Area × heat flux

$$9 \times 1281.2 \times 200 = \pi \times 0.05 \times L \times 2500{,}000$$

$$\therefore \qquad \mathbf{L = 5.873\ m}$$

To estimate the wall temperature, the value of convection coefficient is to be determined. (G-mass flux) (as $\mu - \nu\rho$, $u = m/\rho A$)

$$\mathbf{Re} = 4G/\pi D\mu = 4 \times 9/\pi \times 0.05 \times 0.394 \times 10^{-6} \times 878$$

$$= \mathbf{662509} \text{ (check using calculated velocity)}$$

Using equation (9.26),

$$Nu = 4.82 + 0.0185\ (Re\ Pr)^{0.827} = 23.07$$

$$\therefore \qquad \mathbf{h = 32737\ W/m^2\ K}$$

The wall temperature can be determined by using energy balance.

$$h \times \pi DL \left(T_s - \frac{T_{mi} + T_{mo}}{2} \right) = q\ \pi DL,$$

Substituting $32737 \times \Delta T = 2500{,}000 \quad \therefore \quad$ **ΔT(average) = 76.37°C.**

Example 9.14: *Liquid sodium potassium alloy (25 + 75) at 200°C is to be heated in a pipe of 0.020 m dia to 400°C at a rate of 0.5 kg/s the pipe wall being at 500°C. Determine the length of pipe required.*

Solution: The property values are evaluated at (200 + 400)/2 = 300°C

$$\rho = 799 \text{ kg/m}^3,\ \nu = 0.366 \times 10^{-6} \text{ m}^2\text{/s},\ Pr = 0.0134,\ k = 22.68 \text{ W/m K}$$

$$c = 1038.3 \text{ J/kg K}$$

$$\textbf{Re} = 4m/\pi D\mu = 4m/\pi D\nu\rho = 4 \times 0.5/\pi \times 0.02 \times 0.336 \times 10^{-6} \times 799$$

$$= \mathbf{118567}, \quad \therefore \quad \text{turbulent flow.}$$

Using equation (9.27),

$$Nu = 5 + 0.025\ (Re\ Pr)^{0.8} = 14.095$$

$$\therefore \qquad \mathbf{h = 15983\ W/m^2\ K}$$

$$15983 \times \pi \times 0.02 \times L\left(500 - \frac{200+400}{2}\right) = 0.5 \times 1038.3 \times (400 - 200)$$

$$\therefore \qquad L = 0.52 \text{ m}, \frac{L}{D} = 26,$$

This is shorter than the entry region $\frac{L}{D} = 60$. Hence this correlation may deviate from actual value. However, this is safe, as the value of h will be higher in the entry region and the length required may be even shorter.

9.6 FLOW THROUGH NON-CIRCULAR SECTIONS

For non-circular sections, the value of Reynolds number should be calculated using hydraulic diameter D_h in place of D. The hydraulic diameter is defined as the ratio of four times the sectional area A to the wetted perimeter P.

$$D_h = 4\,A/P \qquad \text{...(9.28)}$$

For circular section $D_h = D$

Reynolds number should be calculated using the velocity and not the mass flow as flow does not depend on D_h.

In turbulent flow the correlations already mentioned in section 9.4 (equations 9.14 to 9.23) can be used with this value of D_h given by equation (9.28).

In the case of laminar flow, the values of Nusselt number for fully developed flow is shown in Table 9.1. For short tubes of non circular sections correlations are more involved and are available in literature.

Table 9.1 *Nus* self number for fully developed laminar flow-noncircular sections

Shape	*Constant wall temp.*	*Constant heat flux all over*	*f. Re*
Square	2.98	3.09	56.41
Rectangles and Long Parallel Plates			
2:1	3.39	3.02	62.19
4:1	4.44	2.93	72.93
8:1	5.59	2.90	82.34
Triangle (60°)	2.47	1.89	53.13
Ellipse 0.9	3.66	4.35	74.80
Hexagon	3.34	3.86	60.22

A third type of boundary is also possible, and the same is uniform heat flux along length and uniform peripheral temperature at any location, but this is not considered here.

9.6.1. A Special Case of Non-circular Flow—Flow through an Annulus

A number of possible situations are there but one of important application is with outside surface adiabatic and inside surface with either constant heat flux or constant temperature.

$$D_h = D_o - D_i \qquad ...(9.29)$$

For laminar flow, fully developed, Nusselt number varies with D_i, D_o as shown below for constant wall temperature and constant heat flux (both inner tubes)

$D_o - D_i$	0.05	0.1	0.25	0.5	
Nu_T	17.46	11.56	7.37	5.74	Const. wall temp.
Nu_H	17.81	11.91	8.5	6.58	Const. heat flux

For turbulent flow the correlations of Section 9.4 can be used with D replaced by D_h.

Example 9.15: *n-butyl alcohol flows through a square duct of 0.1 m side with a velocity of 30 mm/s. The duct is 4 m long. The walls are at constant temperature of 27°C. The bulk mean temperature is 20°C. Determine the heat transfer coefficient and friction factor.*

Solution: For a square the hydraulic diameter is the side length itself.

The property values at the bulk mean temperature are

$\rho = 810$ kg/m³, $\mu = 29.5 \times 10^{-4}$ Ns/m², $Pr = 50.8$, $k = 0.167$ W/m K.

Reynolds number

$$\textbf{Re} = \frac{v D_h \rho}{\mu} = \frac{0.03 \times 0.1 \times 810}{29.5 \times 10^{-4}} = \mathbf{823.7}$$

From Table 9.1

$\overline{\textbf{Nu}}$ for square duct $= \mathbf{2.98}$

$$\therefore \qquad \textbf{h} = \frac{2.98 \times 0.167}{0.1} = \mathbf{4.97\ W/m^2\ K}$$

From Table 9.1, $f.\ Re = 56.91$, $\therefore$ $\mathbf{f} = \dfrac{56.91}{823.7} = \mathbf{69.09 \times 10^{-3}}$.

Example 9.16: *Water is cooled in the annular section of double pipe heat exchanger with inner tube containing evaporating refrigerant vapour at 5°C. The inner tube is 2.5 cm outer dia and the outer pipe is 5 cm inner dia. The average bulk temperature of water is 10°C. For the flow rates of water of (i) 0.055 kg/s (ii) 0.400 kg/s, determine the convection coefficient. The outside is insulated. Also calculate the same for flow of water in the inside of tube with refrigerant outside for circular section. Assume thin tube.*

Solution: The property values are taken at 10°C

$\rho = 1001$ kg/m³, $\nu = 1.397 \times 10^{-6}$ m²/s, $Pr = 10.31$, $k = 0.5751$ W/m K,

$c = 4197$ J/kg K, $D_h = D_o - D_i = 0.025$ m, $D_i/D_o = 0.5$

Case (i) Flow = 0.055 kg/s (const. wall temp.-evaporation)

Velocity of flow in the annulus

$$= \frac{0.055}{1001} \times \frac{4}{\pi(D_o^2 - D_i^2)} = 0.0374 \text{ m/s}$$

$$\therefore \quad \text{Re} = \frac{0.025 \times 0.0374}{1.397 \times 10^{-6}} = \mathbf{668.36} \quad \therefore \quad \textbf{laminar}$$

Fully developed flow: (From tabulation)

$$\therefore \quad Nu = 3.66 \quad \therefore \quad \mathbf{h = 84.19\ W/m^2\ K}$$

Case (ii) Flow = 0.4 kg/s

$$\therefore \quad \text{Velocity} = \frac{0.4}{1001} \times \frac{4}{\pi \times (0.05^2 - 0.025^2)} = 0.2714 \text{ m/s}$$

$$\text{Re} = \frac{0.2714 \times 0.025}{1.397 \times 10^{-6}} = \mathbf{4856, \ turbulent}$$

(n = 0.3, cooling)

$$\therefore \quad Nu = 0.023 \times Re^{0.8}\, Pr^{0.3} = 41.12, \quad \therefore \quad \mathbf{h = 945.8\ W/m^2\ K}$$

for circular section:

$$u = \frac{0.4}{1001} \times \frac{4}{\pi \times 0.025^2} = 0.814 \text{ m/s}$$

$$\therefore \quad \text{Re} = \frac{0.025 \times 0.814}{1.397 \times 10^{-6}} = 14568, \text{ Nu} = 99.18$$

$$\therefore \quad \mathbf{h = 2281\ W/m^2\ K.}$$

9.7 THE VARIATION OF TEMPERATURE ALONG THE FLOW DIRECTION

Constant Wall Temperature: The change in temperature (bulk) of the fluid for length dx is equal to the heat convected in the section divided by the heat capacity. This can be written as

$$mc\,(dT_b) = P\,dx\,h(T_s - T_b) \qquad \text{...(9.30)}$$

P is the perimeter,

for constant surface temperature, (T_s = constant), this leads to

$$\frac{(T_s - T_{bo})}{(T_s - T_{bi})} = \exp\left[\frac{-\bar{h}\,A_s}{m\,C_p}\right] \qquad \text{...(9.31)}$$

where A_s = surface area = πdL, T_{bo}, T_{bi} = bulk temperatures.

This is shown plotted in Fig. 9.5

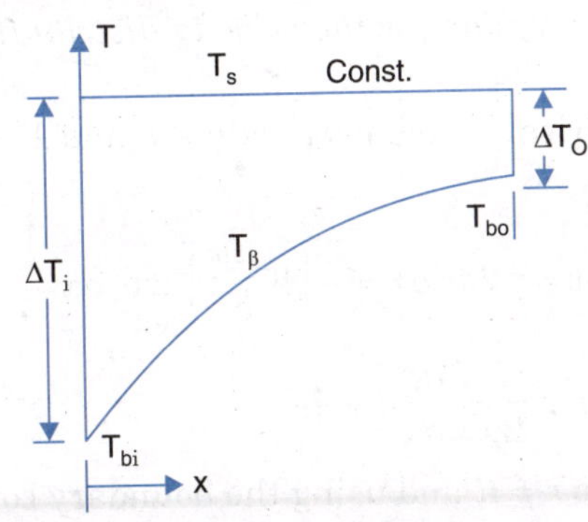

Fig. 9.5. Temperature variation of the fluid along flow-constant wall temperature

The heat flow can be more accurately calculated by

$$q = \bar{h}\, A_s \left[\frac{\Delta T_i - \Delta T_o}{\ln (\Delta T_i / \Delta T_o)} \right] \quad ...(9.32)$$

$A_s = \pi DL$. Length L can be more accurately calculated using equation (9.32) rather than using the arithmetic average as below:

$$q = h\, A_s \left[T_s - \frac{T_{bi} + T_{bo}}{2} \right]$$

Constant heat flux, q W/m²: The bulk temperature of the fluid varies linearly as equal amounts of heat is added along the length.

$$T_{mx} = T_{mi} + \frac{q\, A_s}{m\, c_p} \quad ...(9.33)$$

The surface temperature is calculated using the heat flow by convection at the location. It is given by (9.34)

$$T_s = T_{mi} + \frac{4qD}{k} \left[\frac{1}{4Nu} + \frac{x/D}{Re\, Pr} \right] \quad ...(9.34)$$

The variation of surface and fluid temperatures along the flow direction is shown in Fig. 9.6.

The variation of surface temperature T_s becomes linear in the fully developed region as in the relation (9.35), h and q are constant

$$(T_s - T_b)\, hA_s = qA_s \quad ...(9.35)$$

The determination of convection coefficient in internal flow is of practical importance. Several other configurations (internally finned tubes, coiled tubes etc.) may be met in practice.

For solutions to these situations one has recourse to a rather extensive literature available in this area.

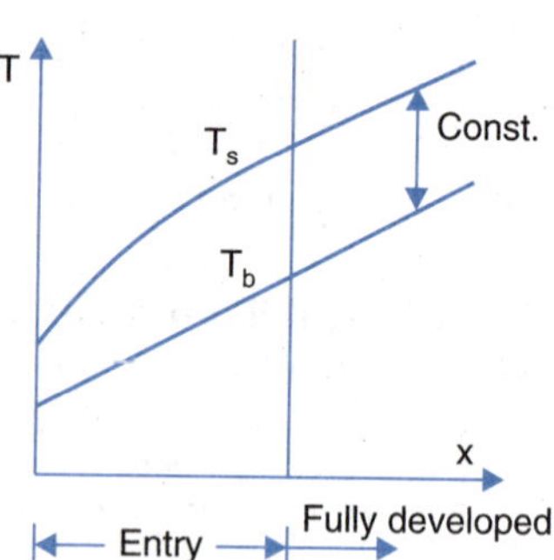

Fig. 9.6. Variation of surface and fluid temperatures-uniform heat flux

ADVANCED EXAMPLES

Example A-9.1: *Derive the expression for the velocity distribution in fully developed laminar flow in a pipe. Show that U_{max} = 24 m.*

Solution: Considering cylindrical element of radius r and length dx in the flow, the force balance gives (Fig. AE-9.1)

$$\pi r^2\, [P - (P + dp)] = \tau\, 2\pi r\, dx = -\left(\mu \frac{du}{dr} \right) 2\pi r\, dx$$

$$\therefore \qquad du = \frac{1}{2\mu} \left(\frac{dp}{dx} \right) r\, dr$$

Integrating from $r = 0$ to $r = R$ and using the boundary conditions at $r = R$, $u = 0$ and at $r = 0$, $u = u_{\text{max}}$.

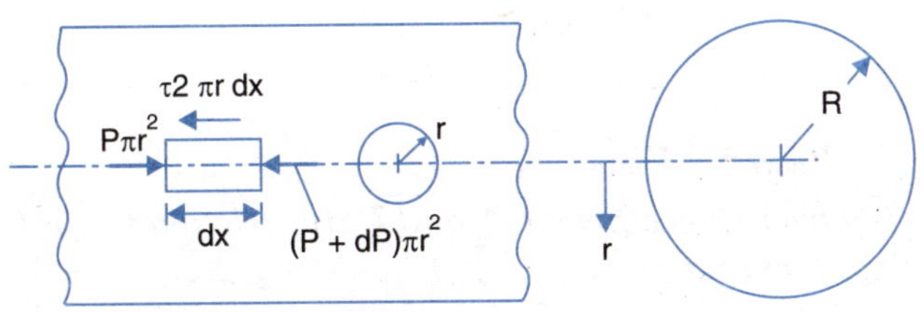

Fig. AE-9.1

$$u = \frac{1}{4\mu} \cdot \frac{dp}{dx} r^2 + c$$

From the first boundary condition

$$c = -\frac{1}{4\mu} \frac{dp}{dx} . R^2$$

$$\therefore \qquad u = \frac{r^2 - R^2}{4\mu} \cdot \frac{dp}{dx} \qquad ...(1)$$

at $r = 0$

$$u_{max} = -\frac{R^2}{4\mu} \cdot \frac{dp}{dx}, \text{ dividing and simplifying}$$

(as $\frac{dp}{dx}$ is –ve in the x direction, u_{max} is positive in the x direction)

$$\frac{\mathbf{u}}{\mathbf{u_{max}}} = \mathbf{1} - \left(\frac{\mathbf{r}}{\mathbf{R}}\right)^{\mathbf{2}} \qquad ...(2)$$

The velocity distribution is parabolic in fully developed laminar flow.

Consider an annular area at r with thickness dr as shown in the Fig. AE-9.1(a).

Volume flow through the area

$$dV = u \times 2\pi r \, dr$$

From equation (1), $u = \frac{r^2 - R^2}{4\mu} \cdot \frac{dp}{dx}$

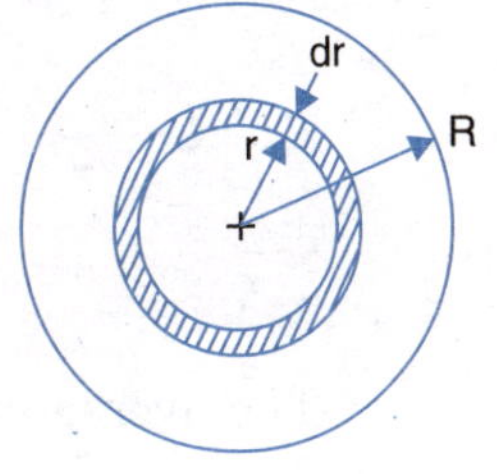

Fig. AE-9.1(a)

$$\therefore \qquad dV = \frac{2\pi}{4\mu} \cdot \frac{dp}{dx} [r^3 - R^2 r] \, dr .$$

Integrating between 0 and R

$$V = \frac{2\pi}{4\mu} \cdot \frac{dp}{dx} \left[\frac{R^4}{4} - \frac{R^4}{2}\right] = -\frac{2\pi}{4\mu} \cdot \frac{dp}{dx} \cdot \frac{R^4}{4} = -\frac{\pi R^4}{8\mu} \cdot \frac{dp}{dx}$$

Average velocity $\qquad u_m = \frac{V}{\pi R^2}$, or $V = u_m . \pi R^2$.

$$\therefore \qquad u_m = -\frac{\pi R^4}{8\mu} \frac{dp}{dx} / \pi R^2 = -\frac{R^2}{8\mu} \cdot \frac{dp}{dx}$$

Chapter 9

$$u_{max} = -\frac{R^2}{4\mu} \cdot \frac{dp}{dx}$$

$\therefore$ $$\mathbf{u_{max} = 2\, u_m.}$$

Example A-9.2: *Fanning friction coefficient C_f and Darcy friction factor f are defined as below. Show that $C_f = f/4$ using flow in a pipe. ($g_c = g_o$)*

$$C_f = \tau_s/(\rho u^2/2g_c), \quad f = \Delta P \cdot \frac{D}{L}\frac{1}{(\rho u^2/2g_c)}$$

Solution: Considering the fluid element of length L shown in figure

$$\Delta P \pi R^2 = 2\pi R \tau_s L$$

or $$\Delta P = 2\tau_s L/R$$

Substituting for ΔP in the definition of f

$$f = \frac{2\tau_s \cdot L}{R} \cdot \frac{D}{L} \cdot \frac{1}{(\rho u^2/2g_c)}$$

noting $$R = \frac{D}{2}$$

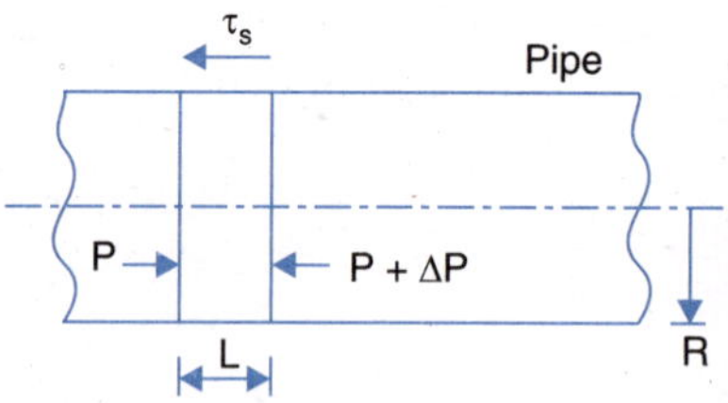

$$f = 4\tau_s/(\rho u^2/2g_c) = 4C_f \text{ or } \mathbf{C_f = f/4}$$

Fig. AE-9.2

Example A-9.3: *Show that for laminar fully developed flow in a pipe $f = 64/Re$.*

Solution: The mass flow through a pipe under this condition is calculated using

$$m = \rho \int_0^R u\, 2\pi r\, dr$$

substituting for u from equation (1), $u = \frac{r^2 - R^2}{4\mu} \cdot \frac{dp}{dx}$.

and also using $-\Delta p/L$ in place of dp/dx (fully developed)

$$\dot{m} = -\frac{\Delta P \cdot \pi}{2L\mu} \rho \int_0^R (r^2 - R^2) r\, dr$$

$$\dot{m} = \frac{\Delta P\, \pi R^4 \cdot \rho}{8L\mu} = \frac{\Delta P\, \pi D^4 \rho}{128\, L\mu}$$

The average velocity = volume flow/area

$\therefore$ $$\overline{U} = \frac{m}{\rho \pi R^2} = \frac{\Delta P D^2}{32\, L\mu}$$

$\therefore$ $$\Delta P = \frac{32\, \mu L}{D^2}\overline{U}$$ Dividing and multiplying by $\overline{U}$, ρ and 2

$$\Delta P = 32 \times \frac{2\, \rho\, \overline{U}}{2\, \rho\, \overline{U}} \cdot \frac{\overline{U}\, \mu L}{D^2}$$

$$= \frac{64}{(\rho \overline{U} D/\mu)} \cdot \frac{L}{D} \cdot \frac{\rho \overline{U}^2}{2}$$

$$= \left[\frac{64}{Re}\right] \cdot \left[\frac{L}{D} \cdot \frac{\rho \bar{U}^2}{2}\right] \quad ...(3)$$

Compare with definition of f

$$f = \Delta P \, \frac{D}{L} \cdot \frac{1}{(\rho U^2/2g_c)}$$

$$\Delta P = f \times \left(\frac{L}{D} \cdot \frac{\rho U^2}{2g_c}\right)$$

Compare with equation (3) and note g_c is a force conversion factor with value 1 in S.I. system and in others g/g_e will be used in place of $1/g_c$.

$$\therefore \qquad \mathbf{f = \frac{64}{Re}}.$$

Example A-9.4: *In the flow of liquid metal in a pipe of radius R the velocity is constant (u) all through the section (slug flow). The temperature variation at a section is parabolic given by* $T - T_s = C\left[1 - \left(\frac{r}{R}\right)^2\right]$ *where* T_s *is the wall temperature and C is a constant and r is the radius at which temperature is T. Derive the value of Nusselt number (hD/k) at this location.*

Solution: From basic convection equation at any location the heat flow at the wall is given by, where T_b is the fluid temperature

$$q = h(T_s - T_b) \quad ...(1)$$

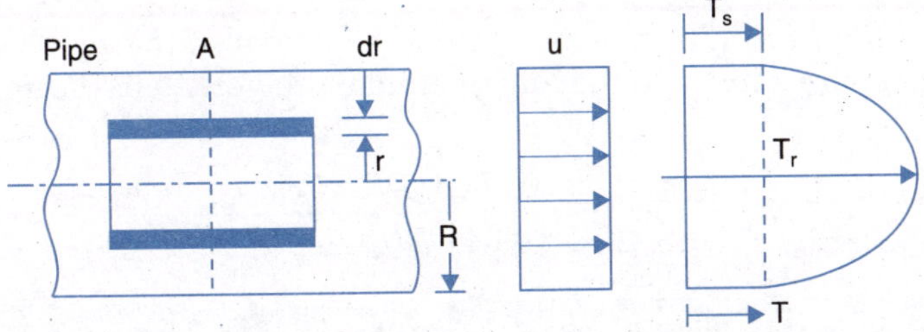

Fig. AE-9.4

Considering a small element at radius r, the average temperature is found by

$$T_b = \frac{1}{c\rho\pi R^2 u_m} \int_0^R 2\pi r \, dr \, u\rho \, cT = \frac{2}{u_m R^2} \int_0^R uT \, r \, dr$$

In this case $\qquad u = u_m$ and $T - T_s = C\left[1 - \left(\frac{r}{R}\right)^2\right]$

Substituting

$$T_b = \frac{2}{R^2} \int_0^R \left(T_s + C\left[1 - \left(\frac{r}{R}\right)^2\right]\right) r \, dr$$

Chapter 9

$$= \frac{2}{R^2}\left[T_s \frac{r^2}{2} + C\frac{r^2}{2} - \frac{C}{4R^2} \cdot r^4\right]_0^R$$

$$\therefore \qquad T_b = T_s + (C/2) \qquad \text{...(2)}$$

The heat flux at the surface is found from the distribution

$$q = k \left.\frac{\partial T}{\partial r}\right|_{r=R} = -kC\,2 \cdot \left.\frac{r}{R^2}\right|_{r=R} = -2C\frac{k}{R}$$

Substituting
$$h = \frac{q}{T_s - T_b} = \frac{-2C\dfrac{k}{R}}{(-C/2)} = \frac{4k}{R}$$

$$\therefore \qquad Nu = \frac{hD}{k} = \frac{(4k/R) \times 2R}{k} = 8$$

For the given distributions **Nu = 8 for slug flow.**

Example A-9.5: *Engine oil at 20°C flows through a 2 cm dia tube at a rate of 0.01 kg/s. The tube wall is at 100°C. Determine the temperature at 0.5 m, 1 m, 2.5 m, 5 m lengths.*

Solution: As the mean of bulk temperature is not specified, the properties are first evaluated at 20°C

$$\rho = 888 \text{ kg/m}^3, \quad \nu = 901 \times 10^{-6} \text{ m}^2/\text{s}, \quad Pr = 10400, \quad k = 0.1454 \text{ W/m K}$$

The velocity of flow
$$= \frac{\text{mass}}{\text{density}} \times \frac{1}{\text{Area}} = \frac{0.01}{888} \times \frac{4}{\pi \times 0.02^2}$$

$$= 0.0358 \text{ m/s}$$

$$\therefore \qquad \textbf{Re} = 0.358 \times 0.02/901 \times 10^{-6} = \mathbf{0.8}$$

In the cae of high Prandtl numbers, the development of thermal boundary layer takes longer distance compared to hydrodynamic boundary layer. (0.04 *Re Pr* = 332.8 m) using equation (9.6),

$$Nu = 3.66 + \frac{0.0668\,(D/L)\,Re\,Pr}{1 + 0.04\,[(D/L)\,Re\,Pr]^{0.67}}$$

L, m	0.5	1	2.5	5
Nu	11.25	8.67	6.33	5.23
$\bar{h}$ W/m² K	81.79	63.03	46.02	38

The exit temperature can be determined (using arithmetic mean)

$$mc(T_{mo} - T_{mi}) = h\,\pi\,DL\left(T_s - \frac{T_{mo} + T_{mi}}{2}\right)$$

at 0.5 m

$$0.01 \times 1880\,(T_{mo} - 20) = \frac{81.79 \times \pi \times 0.02 \times 0.5}{2}\,[200 - (20 + T_{mo})]$$

$T_{mo} = 30.23$ ∴ New trial becomes necessary

1 m $T_{mo} = 35.24$ -do-

2.5 m $T_{mo} = 45.8°C$ -do-

5 m $T_{mo} = 58.56°C$ -do-

The property values are now evaluated as the new estimated mean temperatures

0.5 m	(20 + 30.23)/2	= 25°C
1 m	(20 + 35.24)/2	= 27.5°C
2.5 m	(20 + 45.8)/2	= 32.5°C
5 m	(20 + 58.56)/2	= 39°C

Temp.	ρ	ν	*Pr*	*k*	*c*
25	885	736×10^{-6}	8517.5	0.1451	1901
27.5	883.5	653.5×10^{-6}	7576.25	0.1450	1911.5
32.5	880.5	488.5×10^{-6}	5693.75	0.1447	1932.5
39.0	876.6	274×10^{-6}	3246.5	0.1443	1959.8

Length	*Reynolds*	*Nu*	*h*	T_{mo} °C	*Previous estimate* °C
0.5	0.974	11.26	81.68	30.12	30.23
1.0	1.09	8.686	62.97	35.0	35.24
2.5	1.467	6.352	45.96	45.18	45.8
5.0	2.616	5.259	37.94	57.3	58.56

The outlet temperatures are very near the assumed values and hence can be taken as good.

Example A-9.6: *An air precheater used hot gases (property values equal to that of air) at 350°C to heat air at 30°C. The gas pressure is 1 bar and the flow velocity is 1 m/s. Compare 5 cm tubes and 10 cm tubes for the unit. Determine the length of pipe required in each case and also the friction pressure drop. The gas has to be cooled to 200°C.*

Solution: The bulk mean temperature = (350 + 200)/2 = 275°C

The property values at this temperature are density = 0.6445 kg/m^3,

$\nu = 44.4 \times 10^{-6}$ m^2 s, $Pr = 0.6755$, $k = 0.044365$ W/m K, $c = 1053$ J/kg K

0.5 m dia

$$Re = 1 \times 0.05/44.4 \times 10^{-6} = 1126$$

0.1 m dia

$$Re = 1 \times 0.1/44.4 \times 10^{-6} = 2252$$

The flow is in the laminar region. The assumption that the boundary layers develop simultaneously is better. (equation 9.9)

$$Nu = 3.66 + \frac{0.104\,(Re\,Pr\,D/x)}{1 + 0.16\,(Re\,Pr\,D/x)^{0.8}}$$

Chapter 9

For 0.05 m dia:

The length is not known. **Assuming 5 m**

$$Nu = 4.10, \quad h = 3.64 \text{ W/m}^2 \text{ K}$$

Using heat balance

$$\frac{\pi \times 0.05 \times 0.05}{4} \times 1 \times 0.6445 \times 1053\,(350 - 200)$$

$$= h \times \pi \times 0.05 \times L \left(\frac{350 + 200}{2} - 30\right)$$

$\therefore$ **L = 1.43 m**

So the value is to be recalculated using **L = 1.43 m**

$$Nu = 4.523 \quad \therefore \quad h = 4.01 \text{ W/m}^2 \text{ K}$$

$$\frac{\pi \times 0.05 \times 0.05}{4} \times 1 \times 0.6445 \times 1053 \times (350 - 200)$$

$$= h \times \pi \times 0.05\, L \left(\frac{350 + 200}{2} - 30\right)$$

$\therefore$ **L = 1.29 m or 1.3 m.**

Another calculation based on 1.3 m yields $Nu = 4.56$, $h = 4.05$

$\therefore$ **L = 1.284 m.** $\therefore$ 1.30 m can be used

For 0.1 m dia: Assuming 5 m length,

$$Nu = 4.57 \quad \therefore \quad h = 2.03$$

Using energy balance:

$$\frac{\pi \times 0.1 \times 0.1}{4} \times 1 \times 0.6445 \times 1053\,(350 - 200)$$

$$= h \times \pi \times 0.1 \times L \left(\frac{350 + 200}{2} - 30\right)$$

$\therefore$ **L = 5.12 m**

This is close enough to the assumed value and repeated calculations yields

$\therefore$ $Nu = 4.57$. Adopt 5.12 m

$$f = \frac{64}{Re}$$

$$\therefore \quad \Delta P = \frac{f \times \rho (L/D)\, u_m^2}{2}$$

$$\textbf{0.05 m} \text{ dia, } \mathbf{\Delta P} = \frac{64}{1126} \cdot \frac{0.6445 \times 1.3 \times 1^2}{2 \times 0.05} = \mathbf{0.476 \text{ N/m}^2}$$

$$\textbf{0.1 m} \text{ dia, } \mathbf{\Delta P} = \frac{64}{2252} \cdot \frac{0.6445 \times 5.12 \times 1^2}{2 \times 0.1} = \mathbf{0.469 \text{ N/m}^2}$$

A comparison of various quantities is given in the table below:

D, m	*L, m*	*Volume flow*	*Surface area*	*h W/m² K*	Δp
0.05	1.3	0.00196 m³	0.2042 m²	4.01	0.476
0.10	5.12	0.00785	1.6085 m²	2.03	0.469

4 times volume flow is heated in 8 times the area in the case of 0.1 m dia tubes. As such 0.1 m diameter appears to be disadvantageous, for this flow velocity. Pressure drop is less for the 0.1 m dia pipe even though its length is more. Several factors should be considered before a choice could be made.

Example A-9.7: *In the example A-9.6, assume the velocity to be 8 m/s and compare 0.1 m dia and 0.05 m dia in terms of frictional drop and length requirements.*

Solution: At 275°C, (T_m), $\rho = 0.6445$ kg/m³, $\nu = 44.4 \times 10^{-6}$ m²/s, $Pr = 0.6755$, $k = 0.044365$ W/mK, $c = 1053$ J/kg K.

0.05 m dia

$$Re = 8 \times 0.05/44.4 \times 10^{-6} = 9009, \text{ flow is turbulent}$$

∴ The length required is not known

Assuming 1 m length (equation 9.17)

$$Nu = 0.036\, Re^{0.8}\, Pr^{0.33}\, (D/L)^{0.055} = 39$$

$$\therefore \quad \mathbf{h = 34.66\ W/m^2\ K}$$

Using energy balance equation

$$\frac{\pi \times 0.05 \times 0.05}{4} \times 8 \times 0.6445 \times 1053 \times 150$$

$$= 34.66 \times \pi \times 0.05 \times L \times \left(\frac{350 + 200}{2} - 30\right)$$

$$\therefore \quad \mathbf{L = 1.2\ m}$$

Reworking with this value: $Nu = 38.66$, ∴ 34.3 W/m² K

$L = 1.21$ m So 1.2 m is used

$$\Delta P = f\rho\,\frac{L}{D}\cdot\frac{u^2}{2},\quad f = 0.316 \times Re^{-0.2}$$

$$\therefore \quad \mathbf{\Delta P} = 0.316 \times \frac{1}{9009^{0.2}} \times \frac{0.6445 \times 1.21}{0.05 \times 2} \times 8^2 = \mathbf{25.52\ N/m^2}$$

0.1 m dia: Assuming a length of 4 m, $Re = 18018$, turbulent

Using equation (9.17),

$$Nu = 65.46, \qquad h = 29.04 \text{ W/m}^2 \text{ K}$$

using energy balance

$$\frac{\pi \times 0.1 \times 0.1}{4} \times 8 \times 0.6445 \times 1053(350 - 200)$$

$$= 29.04 \times \pi \times 0.1 \times L \left(\frac{350 + 200}{2} - 30\right)$$

$\therefore$ **L = 2.86 m**

Reworking using 2.9 m

$$Nu = 66.63 \quad h = 29.56 \text{ W/m}^2 \text{ K}$$

$\therefore$ **L = 2.6 m**

The value is used for reworking,

$$Nu = 67.03, h = 29.74 \text{ W/m}^2 \text{ K}$$

$\therefore$ **L = 2.8 m**

This value can be adopted.

$$f = 0.184 \times Re^{-0.2} = 0.0259$$

$$\Delta P = \frac{0.0259 \times 0.6445 \times 2.8 \times 8^2}{2 \times 0.1} = \mathbf{14.97 \text{ N/m}^2}$$

Various quantities are shown in the table for comparison:

D, m	*L, m*	*Volume flow, m³/s*	*Surface area*	*h*	*Δp*
0.05	1.2	0.0157	0.1885 m²	34.64	25.52
0.1	2.8	0.0628	0.8796 m²	29.04	14.97

4 times the volume is heated by 6 times the area by 0.1 m pipe. The only advantage is lower pressure drop. Hence 0.05 m pipe appears to be advantageous even at the higher velocity.

However several factors have to be considered before a final choice.

Example A-9.8: *Water is heated in an economiser (under pressure) from 40°C to 160°C. The tube wall is at 360°C. Determine the length of 0.05 m dia tube, if the flow velocity is 1 m/s.*

Solution: $T_m = 100°\text{C}$, $\rho = 961 \text{ kg/m}^2$, $\nu = 0.293 \times 10^{-6} \text{ m}^2\text{/s}$

$Pr = 1.74$, $k = 0.6804 \text{ W/m K}$, $c = 4216 \text{ J/kg K}$

$\mu_w = 909 \times 0.189 \times 10^{-6} \text{ kg/ms}$

$Re = 0.05 \times 1/0.293 \times 10^{-6} = 0.1706 \times 10^6$ $\therefore$ Turbulent

Using the more recent expression (equation 9.22)

$$Nu = \frac{(f/8)\, Re\, Pr}{1.07 + 12.7\,(f/8)^{0.5}\,[Pr^{0.67} - 1]} \left(\frac{\mu_m}{\mu_w}\right)^{0.11}$$

$$f = (1.82 \log_{10} Re - 1.64)^{-2} = 0.016$$

$$\therefore \quad Nu = 476.2 \quad \therefore \quad h = 6480.5 \text{ W/m}^2 \text{ K}$$

Heat balance is used to determine the length:

$$\frac{\pi \times 0.05^2}{4} \times 1 \times 961 \times 4216(160 - 40)$$

$$= \pi \times 0.05 \times L \times 6480.5 \left(360 - \frac{160 + 40}{2}\right)$$

$$\therefore \quad \mathbf{L = 3.6 \text{ m}}, \frac{L}{D} = 72 \quad \therefore \quad \text{Use of equation is justified.}$$

$$\Delta P = f \,.\, \rho\, u^2\, L/2D = \mathbf{553.5 \text{ N/m}^2}.$$

If we use equation (9.20)

$$Nu = 0.023\ Re^{0.8}\ Pr^{0.33}, \quad Nu = 424.1, \qquad h = 5773\ \text{W/m}^2\ \text{K}$$

L = 4.05 m

LMTD can be used to assess length more accurately

$$Q = m\,AC\,(160 - 40) = h\,A.\,LMTD = 5773 \times \pi \times 0.05 \times L\left[\frac{320 - 200}{\ln\left(\frac{320}{200}\right)}\right]$$

∴ **L = 4.12 m, 13% longer.**

Example A-9.9: *For example A-9.8, if 10 cm dia pipes are used, determine the length required.*

Solution: Using the property values as in 9.8,

$$Re = 0.3413 \times 10^6$$

$$Nu = \frac{(f/8)\ Re\ Pr}{1.07 + 12.7\,(f/8)^{0.5}\ [Pr^{0.67} - 1]}\left[\frac{\mu_m}{\mu_w}\right]^{0.11}$$

$$f = (1.82 \log_{10} Re - 1.64)^{-2} = 0.014$$

∴ $Nu = 843.2 \quad \therefore \quad h = 5737\ \text{W/m}^2\ \text{K}$

∴ **L = 8.15 m**

$$\mathbf{\Delta P} = \frac{f\rho\,.\,u^2\,.\,L}{2D} = \frac{0.014 \times 961 \times 1^2 \times 8.15}{2 \times 0.1} = \mathbf{548\ N/m^2}$$

For the same temperature rise, longer length is required for higher diameter. This is due to the larger mass flow for the same velocity (4 times). The surface area has also increased by 4 times. The pressure drop also appear to be equal. From these factors above definite, choice is not possible.

Example A-9.10: *Dry compressed air at 7 bar and 125°C is to be cooled to 75°C in a tubular cooler, the diameter being 5 cm and the wall temperature at 20°C. Determine the length of tube required. The velocity used is 1.2 m/s.*

Solution: The property values are to be evaluated at (125 + 75)/2 = 100°C,

$\rho = 0.946 \times 7, \qquad Pr = 0.688, \qquad k = 0.0321\ \text{W/m K},$

$c = 1009\ \text{J/kg K}, \qquad \mu_m = 21.87 \times 10^{-6}\ \text{kg/ms}$

∴ $\nu = 21.87/0.946 \times 7\ \text{m}^2/\text{s}, \qquad \mu_w = 18.14 \times 10^{-6}\ \text{kg/ms}$

$$Re = 1.2 \times 0.05 \times 0.946 \times 7/21.87 \times 10^{-6} = 18167$$

∴ Flow is turbulent (cooling index = 0.25)

∴ $$Nu = \frac{(f/8)\ Re\ Pr}{1.07 \times 12.7\,(f/8)^{0.5}\ [Pr^{0.67} - 1]}\left[\frac{\mu_m}{\mu_w}\right]^{0.25}$$

$$f = (1.82 \log_{10} Re - 1.64)^{-2} = 0.0268$$

∴ $Nu = 48.2 \quad \therefore \quad h = 30.99\ \text{W/m}^2\ \text{K}$

As the temperatures are known, *LMTD* can be used

$$\frac{\pi \times 0.05^2}{4} \times 1.2 \times 7 \times 0.964 \times 1009\,(125 - 75)$$

$$= \pi \times 0.05 \times L \times h \left[\frac{105 - 55}{\ln\left(\frac{105}{55}\right)}\right]$$

$\therefore$ $\mathbf{L = 2.13\ m,\ \frac{L}{D} = 42.6 < 60}$

So another trial using *Nu* in the entry region can be used

$$Nu = 0.036\, Re^{0.8}\, Pr^{0.33}\, (D/L)^{0.55}$$

$$Nu = 66.07, \quad \therefore \quad h = 42.48\ \text{W/m}^2\ \text{K}$$

L = 1.55 m

A later correlation is

$$\overline{Nu} = Nu\left(1 + \frac{C}{x/D}\right) \quad C = 6.0$$

$$= 48.2\left(1 + \frac{6}{(1.55/0.05)}\right) = 57.53,$$

$\therefore$ $h = 37$ W/m² K, $\therefore$ **L = 1.79 m**

The length ranges from about 1.5 m to 2.1 m

Less than ± 20% about 1.8 m.

Alternately using

$$\frac{T_s - T_o}{T_s - T_i} = e^{-\frac{hA_s}{me}},$$

$$h = 37\ \text{W/m}^2\ \text{K},\ A_s = \pi \times 0.05 \times L,$$

$$m = \frac{\pi \times 0.05^2}{4} \times 1.2 \times 7 \times 0.946 = 0.015603$$

$$\frac{20 - 75}{20 - 125} = e^{-\frac{37 \times \pi \times 0.05\, L}{0.015603 \times 1009}} = e^{-0.3692\, L}$$

Solving $L = 1.75$ m

with constant wall temperature, this route of solving for L is easier.

Example A-9.11: *Steam at 1 bar and 200°C flows inside a 5 cm dia tube at a velocity of 15 m/s. The tube wall temperature is 500°C. Determine the temperature of steam at the outlet if the pipe is 6 m long.*

Solution: The property values at 1 bar and 200°C are (steam which is superheated can be considered as gas).

$\rho = 0.464$ kg/m³, $\quad \nu = 30.6 \times 10^{-6}$ m²/s, $\quad Pr = 0.94, \quad C_p = 1976$ J/kg K

$k = 0.0334$ W/m² K, $\quad \mu = 15.89 \times 10^{-6}$ kg/ms

$Re = 0.05 \times 15/30.6 \times 10^{-6} = \mathbf{24510}$

So the flow is turbulent

Using **equation (9.22),** ($T_m = 200°C$)

$$Nu = \frac{(f/8)\, Re\, Pr}{1.07 + 12.7\,(f/8)^{0.5}\,[Pr^{0.67} - 1]} \left[\frac{T_w}{T_m}\right]^{0.11}$$

Note: Temperature in Kelvin to be used

$$f = (0.79 \ln Re - 1.64)^{-2} = 0.0248$$

$$Nu = 72.42 \quad \therefore \quad \mathbf{h = 48.5\ W/m^2\ K}$$

If the **equation (9.20)** is used

$$Nu = 0.023\, Re^{0.8}\, Pr^{0.4} = 72.85 \quad \therefore \quad \mathbf{h = 48.67\ W/m^2\ K}$$

These values are close.

Using equation (9.21)

$$Nu = 0.027\, Re^{0.8}\, Pr^{0.33} \left[\frac{\mu_m}{\mu_w}\right]^{0.14}$$

μ_m at 200°C = 15.89×10^{-6}

μ_w at 500°C = 28.64×10^{-6}

$$\therefore \quad Nu = 79.08 \quad \therefore \quad \mathbf{h = 52.83\ W/m^2\ K}$$

Not far removed from other values.

The exit temperature is determined using energy balance. (Assuming arithmetic mean)

mass flow × sp. heat × temp. rise

= Area × convect coefficient × average temp. difference

Taking the value of h as 48.5 W/m² K

$$0.464 \times 15 \times \frac{\pi \times 0.05^2}{4} \times 1976\,(T_{mo} - 200)$$

$$= 48.5 \times \pi \times 0.05 \times 6 \left(500 - \frac{200 + T_{mo}}{2}\right)$$

$$\therefore \quad \mathbf{T_{mo} = 475°C}$$

This is large enough. So another trial is necessary taking property values at (475 + 200)/2 = 337.5°C.

The property values are:

	ρ	ν	*Pr*	*c*	*k*	μ
300	0.384	44.3×10^{-6}	0.91	2014	0.44419	24.32
400	0.326	60.5×10^{-4}	0.90	2073	0.05594	28.64

Chapter 9

Taking at 337.5°C

$\rho = 0.362$ kg/m³, $\nu = 50.375 \times 10^{-6}$ m²/s, $Pr = 0.906$, $k = 0.0486$ W/m K,

$\mu = 25.94 \times 10^{-6}$ kg/ms

$Re = 0.05 \times 15/50.375 \times 10^{-6} = 14888$ ∴ Turbulent

Using equations (9.22) and (9.14)

$$f = (1.82 \log_{10} Re - 1.64)^{-2} = 0.0282$$

$$Nu = \frac{(f/8)\, Re\, Pr}{1.07 \times 12.7\, (f/8)^{0.5}\, [Pr^{0.67} - 1]} \left[\frac{T_w}{T_m}\right]^{0.11} = 47.76$$

Temperature in Kelvin

∴ $h = 46.41$ W/m² K

Using equation (9.20)

$$Nu = 0.023\, Re^{0.8}\, Pr^{0.333} = 48.5, \quad \therefore \quad h = 47.14 \text{ W/m}^2 \text{ K}$$

Using the value and substituting in the energy balance (check please) taking mass flow rate as before

T_{mo} = 470.8°C, less than 475°C previously calculated

Another trial is not warranted in the academic level. But for design purposes, another trial will be useful.

Example A-9.12: *Air flows through a 5 cm ID rough pipe with a relative roughness (ε/D) of 0.006, the mean velocity of air being 12 m/s. The air enters at 150°C and the tube wall is at 30°C. Determine the exit temperature of air. The pipe is 3 m long.*

Solution: Considering that the air may be cooled to about 90°C and taking a mean temperature of 120°C, the property values are $\rho = 0.898$, $\nu = 25.45 \times 10^{-6}$ m²/s, $Pr = 0.686$, $c = 1009$, $k = 0.03338$ W/m K, $\rho = 0.8345$ (150°C).

$Re = 0.05 \times 12/25.45 \times 10^{-6} =$ **23576**, ∴ Turbulent.

using equation (9.16)

$$\mathbf{f} = \frac{1.325}{\left[\ln\left(\frac{\varepsilon}{D} \cdot \frac{1}{3.7}\right) + 5.74/Re^{0.9}\right]^2} = \mathbf{0.0321}$$

(for smooth tubes this works out to 0.025, equation 9.14(*b*))

(from Moody diagram $f = 0.035$)

Using analogy $St.\, Pr^{2/3} = f/8$

$$Nu = \frac{0.0321}{8} \times Re\, .\, Pr^{0.333} = 83.45$$

∴ **h = 55.72 W/m² K**

To determine the outlet temperature, energy balance is used

$$\frac{\pi \times 0.05^2}{4} \times 12 \times 0.8345 \times 1009\, (150 - T_{mo})$$

$$= \pi \times 0.05 \times 3 \times 55.72 \left(\frac{150 + T_{mo}}{2} - 30\right)$$

$\therefore \qquad \mathbf{T_{mo} = 54.43°C}$

In case *LMTD* method is used, the outlet temperature comes as **61.8°C**

So another trial is necessary. Considering T_{mo} = 60°C

The mean temperature = (150 + 60)/2 = 105°C

The property values at 105°C are (in the usual units)

$$\rho = 0.934, \quad \nu = 23.71 \times 10^{-6}\ m^2/s, \quad Pr = 0.689, \quad k = 0.03242, \quad c = 1009$$

$$\mathbf{Re} = 12 \times 0.05/23.71 \times 10^{-6} = \mathbf{25306}$$

$$\mathbf{f} = \frac{1.325}{\left[\ln \frac{0.006}{3.7} + \frac{5.74}{(25306)^{0.9}}\right]^2} = \mathbf{0.03211}$$

h value is calculated as before comes as **58.17 W/m² K.**

Example A-9.13: *Water at 20°C flows through smooth tubes of 25 mm dia at a mean velocity of 2.2 m/s. The surface of the tube is at 5°C and the water is chilled in the process. The tube is 4 m long. Determine the outlet temperature of water.*

Solution: Considering property values at 20°C

$$\rho = 1000\ kg/m^3, \quad \nu = 1.006 \times 10^{-6}\ m^2/s, \quad Pr = 7.02, \quad k = 0.5978\ W/m\ K, \quad c = 4216\ J/kg\ K$$

$Re = 2.2 \times 0.025/1.006 \times 10^{-6} = 54672$, Turbulent flow

Using equation (9.22)

$$Nu = \frac{(f/8)\ Re\ Pr}{1.07 + 12.7\,(f/8)^{0.5}\ [Pr^{0.67} - 1]} \left[\frac{\mu_m}{\mu_w}\right]^{0.25}$$

$$f = [1.82 \log_{10} Re - 1.64]^{-2} = 0.0205, \text{ [equation 9.14(b)]}$$

$$\mu_m = \nu_m \rho_m, \quad \mu_m = \nu_w \times \rho_w = 1.5925 \times 10^{-6} \times 1000.5$$

$$\mu_m = 1.006 \times 10^{-6} \times 1000$$

substituting $Nu = 314.98 \quad \therefore \quad \mathbf{h = 7532\ W/m^2\ K}$

Using the simpler expression (cooling)

$$Nu = 0.023\ Re^{0.8}\ Pr^{0.3} = \mathbf{254.6}, \ h = 6088\ W/m^2\ K$$

There is a larger difference in this case.

To determine the outlet temperature, energy balance is made, using average temperature, instead of *LMTD*.

$$\frac{\pi \times 0.025^2}{4} \times 2.2 \times 1000 \times 4216(20 - T_{mo})$$

$$= 7532 \times \pi \times 0.025 \times 4 \times \left(\frac{20 + T_{mo}}{2} - 5\right)$$

$\therefore \qquad \mathbf{T_{mo} = 13.81°C}$

In case *LMTD* method is used T_{mo} = **14.9°C**

Alternately, using $h = 7532\ W/m^2\ K$.

Chapter 9

$$\frac{T_s - T_o}{T_s - T_i} = e^{-\frac{h \times \pi \times D \times L}{m \times c}}$$

$$\frac{5 - T_o}{5 - 20} = e^{-\frac{7532 \times \pi \times 0.025 \times 4}{1.08 \times 4216}}$$

Solving, $T_o = 13.92°\text{C}$.

Example A-9.14: *Air is heated using an electrically heated pipe of 15 mm dia from 20°C to 180°C, with constant heat flux maintained over the wall. The flow rate is 2 kg/hr. The average temperature difference is 20°C, so that the wall temperature does not exceed 200°C. Determine the length required and the heat flux.*

Solution: The property values are evaluated at (20 + 180)/2 = 100°C

$\rho = 0.946 \text{ kg/m}^3$, $\upsilon = 23.13 \times 10^{-6} \text{ m}^2/\text{s}$

$Pr = 0.688$, $k = 0.0321$ W/m K, $c = 1009$ J/kg K

$\mu = 21.87 \times 10^{-6}$ kg/ms, velocity = 3.32 m/s. (Calculated)

Using mass flow,

$$\textbf{Re} = 4\,G/\pi D\mu = \frac{4 \times 2}{3600} \times \frac{1}{\pi \times 0.015} \times \frac{1}{21.87 \times 10^{-6}}$$

$$= \mathbf{2156.2}$$

So laminar flow (assume fully developed flow)

$$Nu = 4.36, \left(\text{From chart for } \frac{x}{D} \cdot \frac{1}{Re\,Pr} = 0.34, Nu = 4.36 \text{ so checks}\right)$$

$\therefore$ $h = 9.33 \text{ W/m}^2 \text{ K}$

$$Q = mc\,\Delta T$$

$$Q = \frac{2}{3600} \times 1009 \times (180 - 60) = 67.267 \text{ W}$$

Heat convected = $9.33 \times \pi \times 0.015 \times L \times 20 = 67.267$ W

$\therefore$ **L = 7.65 m**

This is also equal to the product of heat flux and area

$\therefore$ Heat flux = $67.267/\pi DL$ = **186.6 W/m² or 8.8 W/m length.**

Example A-9.15: *Nitrogen at – 20°C gets heated as it flows through a pipe of 25 mm dia at a flow rate of 13.725 kg/hr at 2 atm pressure. The pipe is heated with a uniform heat flux of 500 W/m². The pipe is 4 m long. Determine the value of the pipe temperature at the exit.*

Solution: As the pipe length and heat flux are specified the exit temperature of nitrogen can be found

$$c \approx 1030 \text{ J/kg K}$$

$$\pi \times 0.025 \times 4 \times 500 = \frac{13.725}{3600} \times 1030 \times (T_{mo} - (-20))$$

$\therefore$ $T_{mo} = 20°\text{C}$

Property values are taken at 0°C

$\rho = 1.25 \times 2$ kg/m³ $\quad Pr = 0.705, \quad c = 1030$ J/kg K

$k = 0.02431$ W/m K $\quad \mu = 16.67 \times 10^{-6}$ kg/ms

$$\textbf{Re} = \frac{4\,G}{\pi\,D\mu} = \frac{4 \times 13.725}{3600 \times \pi \times 0.025 \times 16.67 \times 10^{-6}} = \mathbf{11648}$$

Turbulent flow

The equations applicable are

$$Nu = 0.023\, Re^{0.8}\, Pr^{0.4} \qquad \text{...(A)}$$

$$Nu = \frac{(f/8)\,Re\,Pr}{1.07 + 12.7\,(f/8)^{0.5}\,[Pr^{0.67} - 1]} \cdot \quad \text{(const. heat flux)} \qquad \text{...(B)}$$

$$f = (1.82 \log_{10} Re - 1.62)^{-2} = 0.03$$

Using A: $Nu = 35.8$, **h = 34.81 W/m² K**

Using B: $Nu = 34.07$ **h = 33.13 W/m² K**

To determine the temperature at exit, using $h = 34.81$ W/m² K

$qA = hA\,\Delta T \qquad \therefore \quad \Delta T = 14.4$°C

$\therefore$ **Pipe temperature** = – 20 + 14.4 = **5.6°C**

Example A-9.16: *Exhaust from a truck engine enters the exhaust pipe at 400°C and the pipe is 5 cm in dia with a constant wall temperature of 120°C. The pipe is 3 m long. The flow rate is 500 kg/hr. Determine the exit temperature.*

Assume properties as that of air.

Solution: Making a first assumption of 200°C exit, the property values are evaluated at 300°C.

$\rho = 0.615$ kg/m³ $\quad \upsilon = 48.2 \times 10^{-6}$ m²/s $\quad Pr = 0.6/4.$

$k = 0.04605$ W/m K $\quad \mu = 29.71 \times 10^{-6}$ kg/ms $\quad c = 1047$ J/kg K

$$Re = 4\,G/\pi D\mu = \frac{4 \times 500}{3600} \times \frac{1}{\pi \times 0.05 \times 29.71 \times 10^{-6}}$$

$$= 119043 \quad \therefore \quad \text{Turbulent}$$

for a first estimate:

$$Nu = 0.023\, Re^{0.8}\, Pr^{0.3} = 234.4, \quad \therefore \quad h = 215.34 \text{ W/m}^2\text{ K}$$

Using heat balance,

$$\frac{500}{3600} \times 1047 \times (400 - T_{mo}) = 216.34 \times \pi \times 0.05 \times 3 \left(\frac{400 + T_{mo}}{2} - 120\right)$$

$$\mathbf{T_{mo} = 254.6°C}$$

The mean temperature can now be taken as 325°C

The properties are:

$\rho = 0.5905$ kg/m³ $\quad \upsilon = 51.83 \times 10^{-6}$ m²/s, $\quad Pr = 0.675$

$c = 1053$ J/kg K $\quad k = 0.047565$ W/m K $\quad \mu_m = 30.545 \times 10^{-6}$ kg/ms

$\mu_w = (120°\text{C}) = 22.85 \times 10^{-6}$ kg/ms

Chapter 9

$Re = 4\ m/\pi D\mu = 115789$

Using $$Nu = \frac{(f/8)\,Re\,Pr}{107 + 12.7\,(f/8)^{0.5}\,[Pr^{0.67} - 1]}\left(\frac{\mu_m}{\mu_w}\right)^{0.25}$$

$$f = (1.82 \log_{10} Re - 1.64)^{-2} = 0.01742$$

∴ $Nu = 196.1$, and $h = 186.55$ W/m² K

$$\frac{500}{3600} \times 1047\,(400 - T_{mo}) = 186.55 \times \pi \times 0.05 \times 3\left(\frac{400 + T_{mo}}{2} - 120\right)$$

$$\mathbf{T_{mo} = 270°C}$$

In case *LMTD* method is used T_{mo} = 273°C

Another iteration with properties at 335°C yields

$\rho = 0.5805$ kg/m³, $\upsilon = 53.282 \times 10^{-6}$ m²/s, $Pr = 0.675$, $k = 0.0482$ W/m K

$\mu = 30.879 \times 10^{-6}$ kg/ms

$Re = 114537$, $f = 0.0175$, $Nu = 194.98$, $h = 187.96$ W/m² K

$\mathbf{T_{mo}}$ **= 269.3°C**. The outlet can be taken to be at 270°C.

Example A-9.17: *A water pipe of dia 7.5 cm absorbs solar radiation at 800 W/m² over the projected area as it passes along an open space for a distance of 10 m. The flow rate of water entering at 25°C is 100 kg/hr. Determine the pipe temperature at the outlet.*

Solution: The temperature rise can be found using the heat flux and flow rate (projected area = length × diameter)

$$800 \times 10 \times 0.0785 = \frac{100}{3600} \times 4178\,(T_{mo} - 25)$$

∴ $T_{mo} = 30.17$°C

Taking property values at (25 + 30.17)/2 = 27.585°C

$\rho = 998.1$ m³/kg $\upsilon = 0.87364$ m²/s, $Pr = 6.004$

$k = 0.6093$ W/m K $c = 4178$ J/kg K

$$\mathbf{Re} = 4\,G/\pi\,D\,\mu = \frac{4 \times 100}{3600} \times \frac{1}{\pi \times 0.075 \times 0.87364 \times 10^{-6} \times 98.1} = \mathbf{564.88}$$

Laminar, constant heat flux, fully developed

∴ $Nu = 4.36$, ∴ **h = 35.42 W/m² K**

∴ $35.42 \times \pi \times D \times 1(T_w - T_\infty) = D \times 1 \times 800$

∴ $T_w - T_\infty = 7.19$°C, **At outlet** $\mathbf{T_w}$ = 30.17 + 7.19 = **37.36°C.**

Example A-9.18: *Saturated steam at 300°C enters a radient super heater tube of 7.5 cm dia of 4 m length exposed to uniform heat flux of 60 kW/m². Determine the tube wall temperature at exit. The steam flow rate is 0.5 kg/s.*

Solution: The average property values may be taken as below:

$\rho = 46.2$ kg/m³ $\mu = 21.28 \times 10^{-6}$ kg/ms $c = 4424$ J/kg K

$k = 0.0442$ W/mK, $Pr = 2.13$

Re $= 4\,G/\pi\,D\,\mu = 4 \times 0.5/\pi \times 0.075 \times 21.28 \times 10^{-6} = 398885$ or $\mathbf{3.99 \times 10^5}$

Flow is turbulent, constant heat flux,

$$Nu = \frac{(f/8)\, Re\, Pr}{1.07 + 12.7\,(f/8)^{0.5}\,[Pr^{0.67} - 1]}$$

$$f = (1.82 \log_{10} Re - 1.64)^{-2} = 0.0137$$

$\therefore$ $Nu = 1026.5$ $\therefore$ **h = 605 W/m² K**

$$q \times A = hA(\Delta T)$$

$\therefore$ $60000 = 605 \times \Delta T$, $\therefore$ $\Delta T = 99.17°C$

(wall temperature above steam temperature)

Exit steam temperature $= 0.5 \times 4424 \times \Delta T = 60000 \times \pi \times 0.075 \times 4$

$\therefore$ $\Delta T = 25.56°C$ (Temp. rise of steam)

$\therefore$ **Surface temp. at exit** = 300 + 25.56 + 99.17 = **424.73°C.**

Example A-9.19: *In example A-9.18 if the flow rate is increased to 2 kg/s, maintaining other parameters, determine the tube wall temperature at the outlet. Use the same average value of properties.*

Solution: $\mathbf{Re} = 4\, G/\pi\, D\, \mu = \dfrac{4 \times 2}{\pi \times 0.075 \times 21.28 \times 10^{-6}}$

$= \mathbf{1.6 \times 10^6}$ Turbulent flow

$$f = (1.82 \log Re - 1.64)^{-2} = 0.01074$$

$$Nu = \frac{(f/8)\, Re\, Pr}{1.07 + 12.7\,(f/8)^{0.5}\,[Pr^{0.67} - 1]} = 3318.2$$

$\therefore$ $h = \mathbf{1955.5}$ W/m² K

$$q = h\, \Delta T$$

ΔT is wall temp. above fluid temp.

$\therefore$ $60000 = 1955.5 \times \Delta T$, $\therefore$ **ΔT = 30.68°C**

Fluid temp. rise:

$$q \times \pi\, D\, L = m\, c_p\, (T_{mo} - T_{mi})$$

$$60000 \times \pi \times 0.075 \times 4 = 2 \times 4424\, (T_{mo} - 300)$$

$\therefore$ $\mathbf{T_{mo} = 306.39°C}$

$\therefore$ **Wall temp. at exit** = 306.39 + 30.68 = **337.07°C** (compare with 424 in 9.18)

Larger flow rate keeps lower wall temperature.

At low loads, the pipes may over heat due to reduced flow rates and proper precautions become necessary to prevent tube failure due to overheating.

Example A-9.20: *Liquid ammonia flows in a duct of section of equilateral triangle of 1 cm side. The average bulk temperature is 20°C and the duct wall is at uniform temperature of 50°C. Fully developed laminar flow with Reynolds number of 1200 is maintained. Calculate the average value of heat transfer for 1 m length.*

Solution: For fully developed flow, the triangular section, **laminar flow, Nu = 2.47** (from Table 9.1)

$$Nu = \frac{h \,.\, D_h}{k}, D_h = \frac{4A}{P}$$

$$A = 0.01 \times 0.01 \sin 60/2 \text{ m}^2 \quad P = 3 \times 0.01$$

$$\therefore \qquad D_h = 5.77 \times 10^{-3} \text{ m}$$

$$k = 0.521 \text{ W/m}^2 \text{ K} \quad \therefore \quad h = \frac{2.47 \times 0.521}{5.77 \times 10^{-3}} = \mathbf{222.89 \text{ W/m}^2 \text{ K}}$$

$$\mathbf{Q} = 0.03 \times 1 \times 222.89 \times (50 - 20) = \mathbf{200.6 \text{ W/m.}}$$

Example A-9.21: *Water at mean bulk temperature of 20°C flows in a duct having cross-section of 10 mm × 20 mm with uniform wall temperature of 60°C. For fully developed laminar flow, calculate the heat transfer for 1 m length.*

Solution: From Table 9.1 for rectangle of 1 : 2 sides, the value of *Nu* for fully developed laminar flow is read as 3.39.

k at 20°C for water is 0.5978 W/m K

Hydraulic radius,

$$D_h = \frac{4 \times 0.01 \times 0.02}{2\,(0.01 + 0.02)} = 0.01333 \text{ m}$$

$$\therefore \qquad h = \frac{3.39 \times 0.5978}{0.01333} = \mathbf{152.04 \text{ W/m}^2 \text{ K}}$$

$$\mathbf{Q} = 1 \times 0.06 \times 152.04 \times (60 - 20) = \mathbf{364.9 \text{ W/m.}}$$

Example A-9.22: *Water flows in a pipe of 50 mm ID at a rate of 3 kg/s and is heated from 15°C to 25°C over a length of 2 m. Determine the temperature of the tube wall if uniform wall temperature conditions prevail.*

Solution: The property values are at 20°C, (15 + 25)/2

$$\rho = 1000 \text{ kg/m}^3, \qquad \upsilon = 1.006 \times 10^{-6} \text{ m}^2\text{/s} \quad Pr = 7.02$$

$$k = 0.5978 \text{ W/m K}, \qquad c = 4178 \text{ J/kg K}$$

$$\mathbf{Re} = 4 \text{ G}/\pi D \mu = 4 \text{ m}/\pi D \upsilon \rho = 4 \times 3/\pi \times 0.5 \times 1000 \times 1.006 \times 10^{-6} = \mathbf{75938}$$

$\therefore$ Flow is turbulent, heating of fluid

$$Nu = \frac{(f/8)\, Re\, Pr}{1.07 + 12.7\,(f/8)^{0.5}\,[Pr^{0.67} - 1]} \left(\frac{\mu_m}{\mu_w}\right)^{0.11}$$

As the wall temperature is not known, it is **assumed as 60°C**

$$\mu_w = 985 \times 0.478 \times 10^{-6} \text{ kg/ms}$$

$$f = (1.82 \log_{10} Re - 1.64)^{-2} = 0.01906$$

$$\therefore \qquad Nu = 507.18 \therefore \quad \mathbf{h = 6064 \text{ W/m}^2 \text{ K}}$$

$$3 \times 4178 \times 10 = 6064 \times \pi \times 0.05 \times 2\,(T_w - 20)$$

$$\mathbf{T_w = 85.8°C}$$

This is near enough to our assumption of 60°C to warrant one more iteration. Another older correlation is

$$Nu = 0.023\ Re^{0.8}\ Pr^{0.4} = 402.4,\ \mathbf{h = 4810\ W/m^2\ K}$$

This leads to a wall temperature of: **102.94°C.**

Example A-9.23: *The pressure drop in flow of water at the rate of 1 kg/s in a pipe of 25 mm dia over a length of 1.5 m was measured as 7000 N/m². The water inlet temperature is 20°C and the wall temperature is maintained at 50°C. Determine the water temperature at the exit.*

Solution: The property values are taken at 20°C

$$\rho = 1000\ \text{kg/m}^3, \quad \upsilon = 1.006 \times 10^{-6}\ \text{m}^2/\text{s},\ Pr = 7.02,$$

$$k = 0.5978\ \text{W/m K}, \quad c = 4178\ \text{J/kg K} \quad \rho_{50} = 990\ \text{kg/m}^3,$$

$$\nu_{50} = 5675 \times 10^{-6}\text{m}^2/\text{s}$$

Friction factor is found from pressure drop and analogy is used to solve from in

$$\Delta P = f \cdot \rho \frac{L}{D}\frac{u^2}{2},\ u = 0.001 \times 4/\pi \times 0.025^2\ \text{m/s} = 2.0372\ \text{m/s}$$

Substituting, $f = 7000 \times 0.025 \times 2/1000 \times 1.5 \times 2.0372^2 = 0.05622$

$$Re = u\ D/\nu = 2.0372 \times 0.025/1.006 \times 10^{-6} = 50626$$

If we assume smooth pipe,

$$f = (1.82 \log Re - 1.64)^{-2} = 0.02087$$

so the pipe considered should be a rough pipe,

Using Colburn analogy

$$\therefore \quad St.\ Pr^{2/3} = f/8$$

$$Nu = \left(\frac{f}{8}\right) Re\ .\ Pr^{1/3} = \frac{0.05622}{8} \times 50626 \times 7.02^{0.333}$$

$$= 681.22$$

$$\therefore \quad h = 16289\ \text{W/m}^2\ \text{K}$$

$$1 \times 4178\ (T_{mo} - 20) = \pi \times 0.025 \times 1.5 \times 16289\left(50 - \frac{T_{mo} + 20}{2}\right)$$

$$\therefore \quad \mathbf{T_{mo} = 31.2°C}$$

Another trial can be made with $T_m = 25.6$°C

This may be tried.

Example A-9.24: *Water is heated at the rate of 1 kg/s from 30°C to 50°C in a smooth pipe of 25 mm ID. The tube wall temperature is 9.1°C above the fluid temperature all along the length. Estimate the length of tube required and also the heat flux and the temperature of wall at exit.*

Solution: The mean bulk temperature = (30 + 50)/2 = 40°C

$$\rho = 995\ \text{kg/m}^3,\ \upsilon = 0.657 \times 10^{-6}\ \text{m}^2/\text{s},\ Pr = 4.34,$$

$$k = 0.628\ \text{W/m K} \quad c = 4178\ \text{J/kg K}$$

As the temperature difference between the wall and fluid is constant, this should be a case of constant heat flux.

$$Re = 4\ G/\pi\ D\ \mu = 4 \times 1/\pi \times 0.025 \times 995 \times 0.657 \times 10^{-6}$$

$$= 77908 \quad \text{So the flow is turbulent}$$

Using $$Nu = \frac{(f/8)\,Re\,Pr}{1.07 + 12.7\,(f/8)^{0.5}\,[Pr^{0.67} - 1]}$$

$$f = (1.82 \log Re - 1.64)^{-2} = 0.01896$$

$\therefore$ $Nu = 382.16 \quad \therefore \quad h = 9600 \text{ W/m}^2 \text{ K}$

Using $mc\,(T_{mo} - T_{mi}) = h \times \pi\, DL\, \Delta T$

$$1 \times 4178 \times (50 - 30) = 9600 \times \pi \times 0.025 \times L \times 9.1$$

Length of tube, **L = 12.18 m**

$$Q \times A = hA\,\Delta T \quad \therefore \quad Q = 9600 \times 9.1 = 87.36 \text{ kW/m}^2, (A = 1)$$

heat flux = 87.36 kW/m² or 6.861 kW/m length

Wall temperature at exit = 59.1°C. (50 + 9.1)

Example A-9.25: *Water flows at the rate of 6 kg/s in a pipe 5 cm dia and 9 m length. The relative roughness is 0.002. The pipe wall is at a uniform temperature of 70°C. Water enters at 40°C. Determine the exit temperature.*

Solution: As a first trial, consider properties at 40°C

$\rho = 995$ m³/kg $\quad v = 0.657 \times 10^{-6}$ m²/s, $Pr = 4.34$,

$k = 0.628$ W/m K $\quad c = 4178$ J/kg K

As the pipe is rough, it is desirable to use Colburn analogy

$$\varepsilon/D = 0.002$$

$$St.\,Pr^{2/3} = f/8$$

$$Re = 4 \times 6/\pi \times 0.05 \times 995 \times 0.657 \times 10^{-6} = 233724$$

$$f = \frac{1.325}{\left[\ln \frac{\varepsilon}{3.7D} + 5.74/Re^{0.9}\right]^2} = 0.0234$$

$$Nu = (f/8)\,Re\,Pr^{0.333}$$
$$= 1115.7$$

$\therefore$ $h = 14014 \text{ W/m}^2 \text{ K}$

Using energy balance, T_{mo} = 57°C

Another trial will be desirable taking property values say at 50°C.

Example A-9.26: *A constant heat flux is provided from the inside of a cylinder of diameter 2.5 cm and water at a high pressure flows over it in an annulus between the cylinder and 5 cm dia pipe. Water enters the annulus at 120°C and is to be heated to 240°C. The flow rate is 6 kg/s. The heat flux is 8100 kW/m². Determine the length required and also the temperature of the inner cylinder wall at the exit. The outer surface can be taken as insulated.*

Solution: The mean temperature is (120 + 240)/2 = 180°C

Considering property values at this temperature.

$\rho = 889$ kg/m³ $\quad v = 0.173 \times 10^{-6}$ m²/s, $\quad Pr = 1.044$,

$k = 0.6757$ W/m K $\quad c = 4417$ J/kg K $\quad D_h = D_o - D_i = 0.025$ m

$$\text{Velocity of flow} = \frac{6}{1000} \times \frac{4}{\pi (0.05^2 - 0.025^2)} = 4.074 \text{ m/s.}$$

The velocity and heat flow rate should be used to calculate the length.

Reynolds number

$$\textbf{Re} = 4.074 \times 0.025/0.173 \times 10^{-6}$$

$$= \textbf{588728} \quad \therefore \quad \text{Flow is turbulent}$$

Using the general equation:

$$\textbf{Nu} = 0.023\, Re^{0.8}\, Pr^{0.4} = \textbf{964.9, h = 26079} \text{ W/m}^2\text{ K}$$

Using $$Nu = \frac{(f/8)\, Re\, Pr}{1.07 + 12.7\,(f/8)^{0.5}\,[Pr^{0.67} - 1]}$$

and $$f = (1.82 \log Re - 1.64)^{-2} = 0.01274$$

$$Nu = 902$$

Taking the later value,

$$\textbf{h = 24374 W/m}^2\textbf{ K}$$

The length required can be determined using the heat flux

$$6 \times (240 - 120)\ 4417 = \pi \times 0.025 \times L \times 8100{,}000$$

Length required $\therefore$ **L = 5 m**

The temperature difference determined using $h = 24379$

Using $h \times \Delta T = q, \quad \therefore \quad \Delta T = 332.3°\text{C}$

wall temp. at exit $= 240 + 332.3 =$ **572.3°C.**

Example A-9.27: *Dry air at a pressure of 8 bar and 20°C is chilled in an annulus between a 5 cm tube and 2.5 cm tube whose walls are kept at 0°C by evaporating refrigerant. The length is 6 m. Determine the exit temperatures for flow rates of*

(i) 7.6 kg/hr *(ii) 75 kg/hr.*

Solution: Considering property values at 10°C (8 bar)

$\rho = 1.247 \times 8$ kg/m^3, $\mu = 17.65 \times 10^{-6}$ kg/ms, $k = 0.02512$ W/m K

$c = 1005$ J/kg K $Pr = 0.705$, $\therefore \nu = 1.80 \times 10^{-6}$ m^2/s

μ at 0°C $= 17.16 \times 10^{-6}$ kg/ms

Reynolds should be found using the velocity and D_h

$$\mathbf{D_h = 0.05 - 0.025 = 0.025 \text{ m}}$$

(*i*) **Flow rate of 7.6 kg/hr**

$$u = \frac{7.6}{3600 \times 1.247 \times 8} \times \frac{4}{\pi (0.05^2 - 0.025^2)} = 0.1437 \text{ m/s}$$

$$\textbf{Re} = \frac{0.1437 \times 0.025}{1.80 \times 10^{-6}}$$

$$= \textbf{1996}$$

$\therefore$ Flow is laminar.

Chapter 9

Assuming fully developed flow $D_i/D_o = 0.5$.

$$Nu = 5.74 \text{ (from tables section 9.6.1)} \quad \therefore \quad \mathbf{h = 5.77\ W/m^2\ K}$$

$$\frac{7.6}{3600} \times 1005 \times (20 - T_{mo}) = \pi \times 0.025 \times 6 \times 5.77 \left(\frac{T_{mo} + 20}{2} - 0\right)$$

$$\mathbf{T_{mo} = 4.38°C}$$

Another trial with $T_m = (20 + 4.38)/2$ will improve the result. But this will be marginal.

(*ii*) **Flow rate of 75 kg/hr**

Velocity $$u = \frac{75}{3600} \times \frac{1}{1.247 \times 8} \times \frac{4}{\pi\,(0.05^2 - 0.025^2)} = 1.4181 \text{ m/s}$$

$$\therefore \quad Re = \frac{1.4181 \times 0.025}{1.8 \times 10^{-6}} = 19696, \text{ flow is turbulent}$$

General correlations can be used with $D = D_h$

$$Nu = \frac{(f/8)\, Re\, Pr}{1.07 + 12.7\,(f/8)^{0.5}\,[Pr^{0.67} - 1]} \left(\frac{\mu_m}{\mu_w}\right)^{0.25}$$

$$f = (1.82 \log Re - 1.64)^{-2} = 0.026$$

Substituting

$$\therefore \quad Nu = 50.84 \quad \therefore \quad \mathbf{h = 51.09\ W/m^2\ K}$$

Using energy balance,

$$\therefore \quad \mathbf{T_{mo} = 5.39°C}$$

Another trial can be done but the average value is very near the value chosen and hence only marginal improvement can be expected.

Example A-9.28: *Ethylene glycol is cooled from 60°C to 40°C in a 30 mm dia tube, the tube wall temperature being maintained constant at 20°C. The average velocity at entry is 10 m/s. Determine the length required.*

Solution: The average of bulk temperature is 50°C. The property values are

$\rho = 1094$ kg/m³, $\mu_w = 19.174 \times 10^{-6} \times 1116$, $k = 0.25765$ W/m K

$c = 2518$ J/kg K $Pr = 72$, $\nu = 6.7165 \times 10^{-6}$ m²/s, $\mu_m = \rho\nu$

$Re = 10 \times 0.03/6.7165 \times 10^{-6} = 44666$

Flow is turbulent

Using the more recent equation

$$Nu = \frac{(f/8)\, Re\, Pr}{1.07 + 12.7\,(f/8)^{0.5}\,[Pr^{0.67} - 1]} \left(\frac{\mu_m}{\mu_w}\right)^{0.25}$$

$$f = [1.82 \log Re - 1.64]^{-2} = 0.0215$$

Substituting

$$Nu = 552.5, \quad \therefore \quad h = 4810 \text{ W/m}^2\text{ K}$$

Using heat balance: ρ at 60°C = 1087 kg/m^3

$$\frac{\pi \times 0.03^2}{4} \times 10 \times 1087 \times 2518\,(60 - 40) = 4810 \times \pi \times 0.03 \times L(50 - 20)$$

$\therefore$ **L = 28.45 m.**

Example A-9.29: *Oil at 20°C flows at an average velocity of 2 m/s through a 30 cm dia pipe. The pipe is passing through water at 0°C for a length of 200 m. Neglecting the pipe conduction resistance and assuming the pipe surface is at 0°C throughout, determine (a) oil exit temperature; (b) heat transfer rate from oil; and (c) pumping power required for this length.*

Solution: Assumptions: Steady state condition, pipes surface at constant temperature of 0°C and flow is hydrodynamically developed.

As exit temperature is not known, properties at 20°C is used for the first trial.

ρ = 888 kg/m^3, υ = 901 × 10^{-6} m^2/s,

k = 0.145 W/m K, Pr = 10400, c = 1880 J/kg°C.

$$\text{Reynold's number} = \frac{VD}{\upsilon} = \frac{2 \times 0.3}{901 \times 10^{-6}} = 666, \text{ laminar flow.}$$

To check for entry length,

$L = 0.05.\ Re.\ Pr.\ D = 0.05 \times 666 \times 10400 \times 0.3 = 104000$ m.

For entry length:

$$\mathbf{Nu} = 3.66 + \frac{0.065\,(D/L).\,Re\,Pr}{1 + 0.04\,[(D/L)\,Re\,Pr]^{0.667}}$$

$$= 3.66 + \frac{0.065\,(0.3/200) \times 666 \times 10400}{1 + 0.04\,[(0.3/200) \times 666 \times 10400]^{0.67}} = \mathbf{37.3}$$

$$\mathbf{h} = \frac{k}{D}.\,Nu = \frac{0.145}{0.3} \times 37.3 = \mathbf{18.0\ W/m^2\,K.}$$

Surface Area $= \pi DL = \pi \times 0.3 \times 200 = \mathbf{188.5\ m^2}$, $\dot{m} = 888 \times \pi \times \frac{0.3^2}{4} \times 2$

$= \mathbf{125.5\ kg/s}$

Using equation (9.31), $T_e = T_s - (T_s - T_i)\exp(-hA/mc_p)$

$$\mathbf{T_e} = 0 - (0 - 20)\exp\left(\frac{18 \times 188.5}{125.5 \times 1880}\right) = \mathbf{19.71°C.}$$

This is near 20°C. No second trial necessary.

$$\mathbf{Q = hA\ LMTD},\ LMTD = \frac{20 - 19.71}{\ln \dfrac{0 - 19.71}{0 - 20}} = \mathbf{-\,19.85°C}$$

$\therefore$ $\mathbf{Q} = 18 \times 188.5 \times (-\,19.85) = \mathbf{-\,6.74 \times 10^4\ W}$

Heat flow out at 67.4 kW.

Chapter 9

$$\mathbf{f} = 64/Re = 64/666 = \mathbf{0.0961}, \Delta P = f\frac{L}{D}\cdot\frac{\rho V^2}{2}$$

$$\mathbf{\Delta P} = \frac{1}{2}\left(0.0961 \times \frac{200}{0.3} \times 888 \times 2^2\right) = \mathbf{1.14 \times 10^5\ N/m^2}$$

$$\mathbf{Power} = \frac{\dot{m}\,\Delta P}{\rho} = \frac{125.5 \times 1.4 \times 10^5}{888} = \mathbf{16100\ W\ or\ 16.1\ kW.}$$

Example A-9.30: *Liquid sodium potassium alloy (25 : 75) at 300°C is heated to 500°C in a tube of 6m length and of diameter 0.05 m, using uniform heat flux. Determine the average wall temperature and heat flux. Flow rate is 3.0 kg/s.*

Solution: The property values are taken at 400°C, (T_m = (300 + 500)/2)

$$\rho = 775\ \text{kg/m}^3,\ \nu = 0.308 \times 10^{-6}\ \text{m}^2/\text{s},\ Pr = 0.0108,\ k = 22.1\ \text{W/m K}$$

$$c = 1000.6\ \text{J/kg K}$$

$$Re = 4\ G/\pi\ D\mu = 4 \times 3.0/\pi \times 0.05 \times 0.308 \times 10^{-6} \times 775.73$$

$$= 320043$$

∴ Flow is turbulent

For constant heat flux for liquid metals the suitable correlation is (equation 9.26)

$$Nu = 4.82 + 0.0185\ (Re\ Pr)^{0.827}$$

$$= 4.82 + 0.0185\ (320043 \times 0.0108)^{0.827} = 20.44$$

∴ $\mathbf{h = 9034\ W/m^2\ K}$

$$q \times \pi \times D \times L = mc\ \Delta T = 3 \times 1000.6 \times (500 - 300)$$

∴ $q = 637\ \text{kW/m}^2$

$$mc\ (T_{mo} - T_{mi}) = h\ \pi\ DL\ \Delta T$$

$$3 \times 1000.6 \times (500 - 300) = 9034 \times \pi \times 0.05 \times 6.\Delta T$$

∴ $\Delta T = 70.51°\text{C}$

∴ **Average wall temperature** = (300 + 70.51 + 500 + 70.51)/2 = **470.51°C.**

Another estimate is

$$Nu = 0.625\ (Re\ Pr)^{0.4} = 16.26$$

$$\mathbf{h = 7190\ W/m^2\ K}$$

This is on the lower side.

Example A-9.31: *A square duct of 0.2 m side and 8 m long passes through an altic with air flowing at 80°C at a rate of 0.15 m³/s in it. The duct surface temperature under steady conditions is 60°C. Determine the air exit temperature and the rate of heat flow out.*

Solution: Assuming steady condition and air as ideal gas. As exit temperature is not known, properties are taken at inlet temperature of 80°C. If necessary another trial will be made.

$$\text{Density} = 0.9994\ \text{kg/m}^3,\ C_p = 1008\ \text{J/kg°C},\ Pr = 0.7154$$

$$k = 0.02953\ \text{W/m K},\ \upsilon = 2.097 \times 10^{-5}\ \text{m}^2/\text{s}.$$

Reynolds number (VD/v),

$$\mathbf{V} = \frac{v}{A} = \frac{0.15}{0.2 \times 0.2} = \mathbf{3.75\ m/s,}\ \mathbf{D_h} = \frac{4A_C}{P} = \frac{4a^2}{4a} = a = \mathbf{0.2\ m,}$$

$$\mathbf{Re} = \frac{3.75 \times 0.2}{2.097 \times 10^{-5}} = \mathbf{35765,}\ \text{Turbulent flow}$$

Entry length $= 10D = 10 \times 0.2 = 2$ m, < 8 m

$$\therefore \quad \mathbf{Nu} = \frac{hD_h}{k} = 0.023 \times (35765)^{0.8}\,(0.7154)^{0.3} = \mathbf{91.4.}$$

$$\mathbf{h} = \frac{Nu\,.\,k}{D_h} = \frac{91.4 \times 0.02953}{0.2} = \mathbf{13.5\ W/m^2\,°C.}$$

Surface area for heat flow, $A_s = 8 \times 4 \times 0.2 = 6.4\ m^2$

Mass flow $= \rho v = 1.009 \times 0.15 = 0.151$ kg/s

Using equation (9.31), $\mathbf{T_e} = T_s - (T_s - T_c) \exp\left(\frac{-hA_s}{mC_p}\right)$

$$= 60 - (60 - 80)\,.\,\exp\,(-13.5 \times 6.4/0.151 \times 1.008) = \mathbf{71.3°C}$$

$$LMTD = (80 - 71.3)/\ln\left(\frac{60 - 71.3}{60 - 80}\right) = -15.2°C$$

$$\therefore \quad \mathbf{Q} = -13.5 \times 6.4 \times 15.2 = \mathbf{-1313\ W.}$$

As average temperature $(71.3 + 80)/2 = 75.7$°C, nearer to 80°C

Another trial may not improve results further.

Example A-9.32: *Liquid sodium potassium alloy (25:75) at 200°C is heated to 400°C in a pipe of 50 mm ID with a constant wall temperature of 600°C. Determine the length required if the flow rate is 3 kg/s.*

Solution: The property values are to be evaluated at $(200 + 400)/2 = 300$°C

$\rho = 799\ kg/m^3$ $\nu = 0.366 \times 10^{-3}\ m^2/s$, $k = 22.68$ W/m K,

$c = 1038.3$ J/kg K $Pr = 0.0134$

$Re = 4\ m/\pi\,D\mu = 4 \times 3/\pi \times 0.05 \times 799 \times 0.366 \times 10^{-6}$

$= 261236$, (2.61×10^5), Turbulent

For constant wall temperature (equation 9.27)

$$Nu = 5 + 0.025\,[Re\,Pr]^{0.8} = 22.11$$

$$\therefore \quad \mathbf{h = 10029\ W/m^2\,K}$$

Using energy balance

$$mc\,(T_{mo} - T_{mi}) = h\,\pi\,DL\left(T_w - \frac{T_{mo} + T_{mi}}{2}\right)$$

Chapter 9

$$3 \times 1038.3 \times (400 - 200) = 10029 \times \pi \times 0.05 \times L\left(600 - \frac{400+200}{2}\right)$$

$\therefore$ **L = 1.32 m.**

Example A-9.33: *Compare the value of convective heat transfer coefficient for a circular pipe of diameter 5 cm with uniform wall temperature, with a square section of the same area. The fluid is air at an average bulk temperature of 40°C.*

The flow rate is 0.06 kg/s. The air heated.

Solution: The property values at 40°C are:

$\rho = 1.128$ kg/m^3, $\nu = 16.96 \times 10^{-6}$ m^2/s, $Pr = 0.699$

$k = 0.02756$ W/m K, $c = 1005$ J/kg K

The value of Re for cylinder

$$= 4\,G/\pi\,D\mu = 4 \times 0.06/\pi \times 0.05 \times 1.128 \times 16.96 \times 10^{-6}$$
$$= 79865$$

$\therefore$ Turbulent

$$Nu = 0.023\,Re^{0.8}\,Pr^{0.4} \qquad (Pr^{0.4}\text{–heating})$$
$$= 166.5,$$

$\therefore$ $\mathbf{h = 91.77\ W/m^2\,K}$

For square section: side a is calculated using

$$\frac{\pi D^2}{4} = a^2$$

$$\therefore \quad a = \frac{\sqrt{\pi \times 0.05^2}}{2} = 0.0443 \text{ m}$$

$$D_h = \frac{4.a^2}{4a} = a,\ u = 0.06 \times \frac{1}{1.128} \times \frac{1}{a^2} = 27.28 \text{ m/s}$$

$$Re = 27.28 \times 0.0443/16.96 \times 10^{-6} = 71284, \text{ Turbulent}$$

$\therefore$ $Nu = 0.023\,Re^{0.8}\,Pr^{0.4} = 155.7$, $\mathbf{h = 96.8\ W/m^2\,K}$

This comparison is for the same area and mass flow and so the velocity in the square section is the same. Use of non-circular section is not desirable in heat transfer or for construction purposes, but may be dictated by other considerations.

REVIEW QUESTIONS

9.1 Explain the differences, in boundary layer development, between internal flow and external flow.

9.2 Explain why internal flow is treated in two different regions namely entry region and fully developed region?

9.3 State in which region local convection coefficient is higher? Explain the reason for the same.

9.4 What factors govern the value of entry length in hydrodynamic flow and thermal flow? State the values of each.

9.5 Discuss the differences between thermal boundary layer and hydrodynamic boundary layer.

9.6 In laminar flow, what is the shape of velocity profile in fully developed flow?

9.7 Is the temperature profile in fully developed flow invariant along the length? Explain by what process the same can be made invariant?

9.8 What is the velocity which replaces u_∞ in internal flow? Define the same.

9.9 Define bulk mean temperature. Explain the need for the use of this temperature.

9.10 Write the expression in terms of mass flow for Reynolds number.

9.11 Discuss the effect of variation of viscosity with temperature on the velocity profile in fully developed flow.

9.12 List the two types of boundary condition used in internal flow.

9.13 Explain the relevance of friction factor in heat transfer study.

9.14 Discuss the reason for treating heat transfer in liquid metal separate from other fluids.

9.15 Define hydraulic diameter and indicate its application in internal flow heat transfer.

OBJECTIVE QUESTIONS

9.1 Choose the correct statement in flow through pipes:

(*i*) (*a*) In laminar flow, the exit Nusselt number in the fully developed condition is lower than at entry.

(*b*) In laminar flow, the exit Nusselt number in the fully developed condition is higher than at entry.

(*c*) In laminar flow, the Nusselt number remains constant.

(*d*) In laminar flow the Nusselt at constant wall temperature is higher as compared to Nusselt at constant heat flux.

(*ii*) In flow through pipes for the same Reynolds number,

(*a*) the thermal entry length is longer for low Prandtl number fluids

(*b*) the thermal entry length is longer for high Prandtl number fluids

(*c*) Prandtl number does not influence the thermal entry length

(*d*) the thermal entry length effect is more pronounced only in turbulent flow.

(*iii*) (*a*) Flow of air can be considered as hydrodynamic layer fully developed and thermal layer developing

(*b*) Flow of oils can be considered as *HFD* and thermal layer developing

(*c*) Flow of liquid metals can be considered as hydrodynamic layer fully developed and thermal layer developing

(*d*) Flow of water can be considered as hydrodynamic layer fully developed and thermal layer developing.

(*iv*) In pipe flow, the average convection coefficient

(*a*) will be higher in rough pipes

(*b*) will be higher in smooth pipes

(*c*) roughness affects only pressure drop and not the convection coefficient

(*d*) only Reynolds and Prandtl numbers influence the convection coefficient and not the roughness.

(*v*) In pipe flow

(*a*) for constant heat flux, the initial length is more effective compared to the end length (*b*) for constant wall temperature the initial length are less effective compared to the end lengths

(*c*) in fully developed flow and constant wall temperature, the effectiveness increases with length

(*d*) in fully developed flow and constant heat flux, the effectiveness decreases with length.

(*vi*) (*a*) In smooth pipes a laminar flow remains laminar all through the length

(*b*) In smooth pipes a laminar flow turns turbulent after a certain length

(*c*) The temperature profile in fully developed layer remains the same

(*d*) The velocity profile in a pipe flow is established at the entry.

Answers

(*i*) (*a*) (*ii*) (*b*) (*iii*) (*c*) (*iv*) (*a*) (*v*) (*a*) (*vi*) (*a*).

EXERCISE PROBLEMS

9.1 Show that $(1.2 \log Re - 1.64)^{-2} = 4\,(1.58 \ln Re - 3.28)^{-2}$

9.2 Show that

$$\sqrt{2/f} = 2.46 \ln (Re\,.\,\sqrt{f/2} + 0.292) \text{ and } \sqrt{1/4\,f} = 2 \log\,[Re\,\sqrt{4\,f}\,] - 0.8.$$

9.3 For air flow at constant wall temperature of 100°C and bulk mean temperature of 40°C through a 4 cm ID pipe, determine the value of average convection coefficients for a length of 1 m if the entrance velocity is (*i*) 0.6 m/s, (*ii*) 0.8 m/s, (*iii*) 10 m/s.

9.4 For air flow at a constant wall temperature of 100°C and average bulk temperature of 40°C through a 4 cm ID pipe, determine the value of average convection coefficient for an inlet velocity of 0.8 m/s if the pipe length is (*i*) 1 m, (*ii*) 3 m, (*iii*) 10 m.

9.5 Air is to be heated at a rate of 5 kg/hr through a 40 mm ID pipe from 20°C to 60°C, the pipe wall being maintained at 100°C by condensing steam. Determine the length required.

9.6 5 m length of 50 mm dia pipe has its wall maintained at 100°C. Air at 20°C enters the pipe. If the exit temperature of air is to be 60°C, determine the air flow rate.

9.7 Air at 20°C is to be heated to 40°C using constant heat flux over a pipe of 40 mm ID at a flow rate of 5 kg/hr. Determine the pipe length required and also the average pipe wall temperature. The heat flux is 500 W/m^2.

9.8 100 kg of water per hour is to be heated by flow through a 2.5 cm ID pipe from 20°C to 60°C. The pipe wall is at 100°C. Determine the length of pipe required. Also determine the mean temperature at half this length.

9.9 Water is to be heated from 20°C at a rate of 500 kg/hr by flow through a pipe of 25 mm ID with a wall temperature of 100°C. Determine the length required. Also determine the temperature at half this length.

9.10 Water flows at a rate of 500 kg/hr through a pipe uniformly heated. It is heated from 20°C to 60°C through a length of 5.6 m. The pipe diameter is 25 mm. Determine the average pipe temperature.

9.11 Water flows at a rate of 500 kg/hr through a 40 mm dia pipe with uniform wall heat flux and is heated from 20°C to 40°C over a length of 8 m. Determine the wall temperature at the exit.

9.12 Engine oil is cooled from 120°C by flow through a tube of 5 mm ID, with uniform wall temperature of 40°C at a rate of 0.08 kg/s. Estimate the temperature of the oil at the outlet for lengths of 1 m, 5 m and 10 m.

9.13 Furnace oil is to be heated from 30°C to 70°C for proper atomisation. Steam is used at 110°C for the heating. If the flow rate through a 15 mm ID pipe is 40 kg/hr, determine the length of pipe required. (use the property values of engine oil for the estimate).

9.14 Liquid ammonia flows in a duct of diameter 2 cm. The average bulk temperature is 20°C. The duct wall is at 50°C. If fully developed laminar flow prevails, determine the heat transfer per m length.

9.15 Liquid refrigerant $R12$ ($CCl_2\,F_2$) flows through a 4 mm dia tube at 30°C, the Reynolds number being 1000. Calculate the length necessary to cool the fluid to 20°C. The wall temperature is 10°C.

9.16 Air at 14 atm pressure flows through a 75 mm ID pipe at a rate of 0.55 kg/s, the duct wall being at 20°C. The average air temperature at inlet is 60°C. The duct is 6 m long. Estimate the temperature of air as it leaves the duct.

9.17 Determine the pressure drop and pumping power required to overcome the pressure drop when water flows at 5 l/s through a horizontal pipe 30 m long and 4 cm diameter. The water temperature is 15°C. The density and viscosity values are 999.91 kg/m^3 and 1.38×10^{-3} kg/ms.

[**Ans.** 101.5 kPa, 0.508 kW]

9.18 Water is to be heated from 10°C to 80°C by passing through a 2 cm ID tube of 7 m length. The rate of flow is 8 l/min. Heat is supplied uniformly by an electrical heat all through length. Assuming that all the electrical energy is absorbed by the water determine (a) the inner surface temperature of the pipe and the heater power. [**Ans.** 113.3°C, 38627 W]

9.19 Hot air at atmospheric pressure and 85°C flows at a rate of 0.1 m^3/s through a 10 m long square duct of 0.15 m side. The pipe is situated in the attic of a house. The pipe wall temperature is constant at 70°C. Determine the exit temperature and the rate of heat loss from pipe.

[**Ans.** 75.7°C, 941 W]

9.20 Water at 10°C flows through a 2 mm ID pipe of 15 m length. The average flow velocity is 1.2 m/s. Determine the pressure drop and the power required to overcome the pressure. Assume density as 999.7 kg/m^3. Viscosity as 1.307×10^{-3} kg/ms. [**Ans.** 188 kPa, 0.71 W]

9.21 Air at 0.9 atm and 30°C flows through a 10 mm dia pipe at a rate of 4.2×10^{-4} kg/s. Constant heat flux is maintained over the wall surface. The tube wall temperature is 50°C above the average fluid temperature. For a tube length 0.5 m, determine the outlet temperature of the air.

9.22 Water at 10°C flows with an average velocity of 1 m/s through a pipe of 20 mm dia and 1 m length, with uniform wall heat flux of 100 kW/m^2. Determine the average wall temperature.

9.23 Water is heated in a triangular tube of 2 cm side from 10°C to 50°C. The tube wall is at 80°C. The flow velocity is 1.1 m/s. Determine the length required.

9.24 A square duct of 25 cm side carries cool air at 10°C over length 20 m. The average velocity at entrance is 1.5 m/s. The duct wall is at 25°C. Determine the outlet temperature of the air.

9.25 A rectangular duct of 15 cm × 30 cm section carries cool air at 10°C over a length of 20 m. The average velocity at entrance is 1.5 m/s. The duct wall is at 25°C. Determine the outlet temperature of the air.

9.26 Water flows through the annular area formed between a 2 cm and a 4 cm circular pipes (coaxial) at a velocity of 2 m/s. The outside surface is insulated. The inside surface is at 80°C. Water enters at 20°C. Determine the outlet temperature of the water if the length is 4 m.

9.27 Oil at 160°C is to be cooled to 80°C by flow in an annulus formed between 15 mm and 30 mm tubes. The flow velocity is 0.8 m/s. The inside tube wall is maintained at 30°C by larger quantity of water flow. Determine the length required.

9.28 In a heat exchanger 4 cm dia pipes of 3 m length are used to heat air with a flow rate of 0.02 kg/s. The air inlet is at 30°C. The tube wall temperature is 200°C. Determine the outlet temperature of the air. A proposal is made to use 2 cm dia pipes with mass flow per unit sectional area remaining the same. Determine the exit temperature of air in this case.

9.29 Water flows through a pipe with a Reynolds number of 2000. Compare the values of convection coefficients for (*i*) both boundary layers developing together and (*ii*) hydrodynamic layer fully developed and thermal layer developing.

9.30 Water flows in a rough pipe with a relative roughness 0.004. The pipe diameter is 5 cm and the length is 3 m. Water at 20°C enters at a flow velocity of 1.2 m/s. Determine the outlet temperature. Also determine the pressure drop over this length.

9.31 Water flows in a pipe of 5 cm dia and 3 m length with an inlet velocity of 1 m/s. The pressure drop was observed as 8000 N/m^2. The water enters at 60°C. The tube wall is at 15°C. Determine the water outlet temperature.

9.32 An air conditioning duct has a section of 45 cm × 90 cm and the average velocity of air is 7 m/s and the air entry is at 10°C. The wall is at 20°C. Determine the heat gain over a length of 1 m.

9.33 A tubular heat exchanger has a total mass flow of 10,000 kg/hr. The maximum velocity is to be limited to 1 m/s. The water is heated from 30°C to 35°C with a tube wall temperature of 45°C. Keeping the length of the tube as the influencing parameter compare 37.5 mm and 50 mm ID tubes.

9.34 Air at 50°C enters a rectangular duct or sides 0.15 m × 0.20 m and 7 m length with a velocity of 7 m/s. The walls of the duct are at a uniform temperature of 10°C.

Determine the outlet temperature of the air and the rate of heat flow out.

[**Ans.** 32.8°C, 3674 W]

10 CHAPTER

Natural Convection

10.0 INTRODUCTION

In the previous chapters, we discussed heat transfer under forced flow conditions. Though it is desirable to use forced convection as higher rates of heat transfer coefficient may be obtained by this method, leading to compact equipments, it is not possible and sometimes not desirable to use forced convection in the design of several equipments. In these situations, free or natural convection has to be adopted in designs. Medium size electrical transformers erected in open space cannot use forced convection. So also steam radiators used for comfort heating.

When a surface is maintained in still fluid at a temperature higher or lower than that of the fluid, a layer of fluid adjacent to the surface gets heated up or cooled. A density difference is created between this layer and the still fluid surrounding it. The density difference introduces a buoyant force causing flow of the fluid near the surface. Heat transfer under such conditions is known as natural or free convection. Usually a thin layer of flowing fluid forms over the surface. The fluid beyond this layer is essentially still, and is at a constant temperature of T_∞.

The flow velocities encountered in free convection is lower compared to flow velocities in forced convection. Consequently the value of convection coefficient is lower, generally by one order of magnitude. Hence for a given rate of heat transfer larger area will be required. As there is no need for additional devices to force the fluid, this mode is used for heat transfer in simple devices as well as for devices which have to be left unattended for long periods.

The basic nature of flow and various correlations available for the calculation of convection coefficient for different flow situations and geometries are discussed in this chapter.

Boundary layer theory is used for the analysis of natural convection also. The example 10.1 brings out the difference in area requirements between natural and forced convection conditions.

The heat transfer rate is calculated using the general convection equation given below:

$$Q = h\,A(T_w - T_\infty) \qquad \text{...(10.1)}$$

Q—heat transfer in W, h—convection coefficient – W/m^{2}K.

A—area in m^2, T_w—surface temperature

T_∞—fluid temperature at distances well removed from the surface (here the stagnant fluid temperature).

Example 10.1: *A transformer of 50 kW capacity dissipates by natural convection 1.20 kW of heat generated inside it. If the value of convective heat transfer coefficient is 5 W/m^{2}K, and if the temperature of the unit should not exceed 30°C above the ambient level, determine the area required. In case of forced convection h = 50 W/m^{2}K. Determine the area under this condition.*

Solution: Equation (10.1) is suitable for the calculation

$$Q = h\,A\,(T_w - T_\infty)$$

or $$A = Q/h\,(T_w - T_\infty)$$

Substituting $Q = 1200$ W, $h = 5$ W/m² K, $(T_w - T_\infty) = 30°C$

the area is obtained as **8 m²**. This is a fairly large area and hence the equipment has to be bulky. Under **forced convection**, the area required will be **0.8 m²**.

The methods of analysis of the boundary layer and the basic physical laws applied in the analysis are the same as in forced convection. The exception is that buoyant forces are to be taken into account in the formulation of momentum equation. The boundary conditions are also different. The velocity is zero both at the surface and also at the edge of boundary layer, the bulk fluid remaining still. There is no free stream velocity as in forced convection. So the velocity term in Reynolds number has to be eliminated as it can be seen in Section 10.2 (equation 10.10).

Analytical methods using exact differential equations as well as approximate integral methods can be used to study simple cases. However, these methods are useful mainly in the validation of experimental results. The dimensional analysis method is useful in identifying dimensionless groups that can be used in interpreting the experimental results. Some of the derivations are given under the section advanced examples.

For design situations, correlations obtained from experimental results using the identified dimensionless numbers are used more often. These correlations are constantly reviewed and improved by researchers.

10.1 BASIC NATURE OF FLOW UNDER NATURAL CONVECTION CONDITIONS

The layer of fluid near the surface gets heated or cooled depending on the temperature of the solid surface. A density difference is created between the fluid near the surface and the stagnant fluid. This causes as in a chimney a flow over the surface.

Similar to forced convection a thin boundary layer is thus formed over the surface. Inertial, viscous and buoyant forces are predominant in this layer. Temperature and velocity gradients exist only in this layer. The velocity and temperature distributions in the boundary layer near a hot as well as cold vertical surfaces are shown in Fig. 10.1.

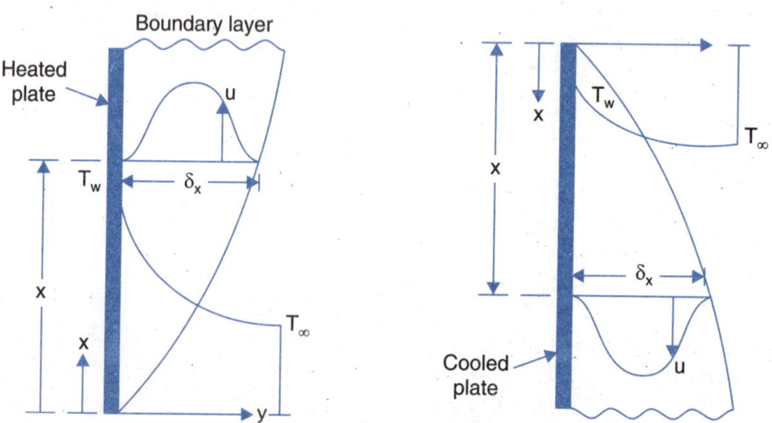

Fig. 10.1. Velocity and temperature distributions in the boundary layer

The velocity is zero at the surface and also at the edge of the boundary layer. As in the case of forced convection the temperature gradient at the surface is used in the determination of heat flow (heat is transferred at the surface by conduction mode). If $T_w > T_\infty$ the flow will be upward and flow will be downward if $T_w < T_\infty$. Also the measurement of x will be upwards in the first case and downwards in the second. This is because the flow is gravity driven.

$$h = -k \left. \frac{\partial\left(\dfrac{T - T_\infty}{T_w - T_\infty}\right)}{\partial\left(\dfrac{y}{\delta}\right)} \right|_{y=0}$$

The temperature gradient at the surface can be evaluated using either the solution of differential equations or by assumed velocity and temperature profiles in the case of integral method of analysis. This leads to the identification of Nusselt number and Prandtl number as in the case of forced convection. These numbers have the same physical significance as in forced convection.

The buoyant forces play an important role in this case, in addition to the viscous and inertia forces encountered in forced convection. This leads to the identification of an additional dimensionless group called Grashof number.

Grashof number is defined by equation (10.2)

$$Gr = \frac{g\beta\,(T_w - T_\infty)\,.\,L^3}{\nu^2} \quad ...(10.2)$$

In the case of horizontal cylinders, the height L will be replaced by diameter D.

where β is the coefficient of cubical expansion having a dimension of 1/Temperature. For gases $\beta = 1/T$ where T is in K. For liquids β can be calculated if variation of density with temperature at constant pressure is known. The other symbols carry the usual meaning.

The physical significance of this number is given by,

$$\mathbf{Gr} = \frac{\textbf{Inertia force}}{\textbf{Viscous force}} \cdot \frac{\textbf{Buoyant force}}{\textbf{Viscous force}} \quad ...(10.3)$$

The flow turns turbulent for value of $Gr\,Pr > 10^9$. As in forced convection the microscopic nature of flow and convection correlations are distinctly different in the laminar and turbulent regions. The example 10.2 illustrates the statement of equation (10.3).

Example 10.2: *Check the dimensions for Grashof number and rearrange it to express it as ratio of forces as noted in equation (10.3).*

$$Gr = \frac{g\beta\,(T_w - T_\infty)\,.\,L^3}{\nu^2} \to \frac{m}{s^2} \cdot \frac{1}{K} \cdot \frac{Ks^2}{m^4} \cdot m^3 = 1$$

Solution: So the group is dimensionless (y and d are length dimensions and can be upresented by L)

Viscous force $\qquad \tau = \mu \dfrac{du}{dy} \to \mu \,.\, \dfrac{u}{y}$

Inertia force is proportional to ρv^2 or $\rho\, u^2$

Buoyant forces are proportional to $d\rho \,.\, dH \,.\, g$

the group $\dfrac{g\beta\,(T_w - T_\infty)\,.\,L^3}{\nu^2}$ is rearranged as, $\nu = \mu/\rho$

$$= g\beta \frac{\rho u^2}{\mu^2} . (T_w - T_\infty) \frac{\rho}{u^2} . L^3$$

$$= \left(\frac{\rho u^2}{\mu u/L}\right)\left(\frac{1}{(\mu u/L)} \cdot g\beta\rho (T_w - T_\infty) . L\right) = \left[\frac{\text{Inertia force}}{\text{Viscous force}}\right]\left[\frac{\text{Buoyant force}}{\text{Viscous force}}\right]$$

From the definition $\beta = -\frac{1}{\rho} \cdot \frac{\rho_\infty - \rho_w}{T_\infty - T_w}$ and as $\rho_\infty - \rho_w = d\rho$.

$\rho\, \beta\, (T_w - T_\infty) = d\rho, \quad L = dH$

so the expression $g\, \beta\rho\, (T_w - T_\infty)\, L$ represents buoyant force.

Example 10.3: *A vertical plate is maintained at 40°C in 20°C still air. Determine the height at which the boundary layer will turn turbulent if turbulence sets in at Gr Pr = 10^9. Repeat the problem for water flow at film temperature of 30°C. This illustrates the difference between gases and liquids.*

Solution: The property **values of air should** be evaluated at T_f, (40 + 20)/2 = 30°C

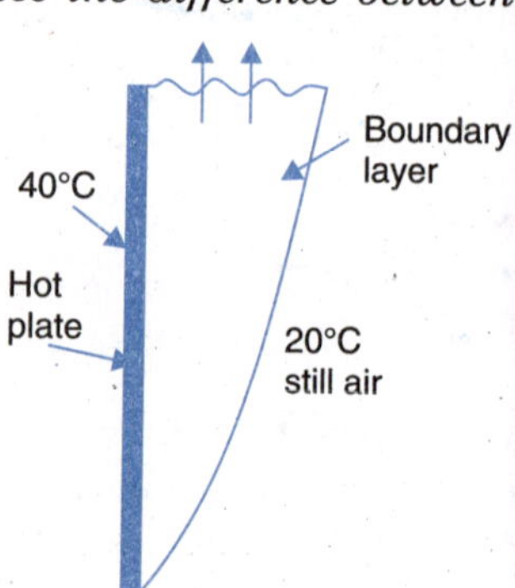

Fig. Ex. 10.3

The property values are read from tables,

$\nu = 16 \times 10^{-6}$ m²/s, $\rho = 1.165$ kg/m³, $Pr = 0.701$

$$\beta = \frac{1}{273 + 30} = \frac{1}{303}$$

The flow becomes turbulent when $Gr.\, Pr = 10^9$

$$\therefore \quad 10^9 = 9.81 \times \frac{1}{303} \times \frac{(40 - 20)x^3}{(16 \times 10^{-6})^2} \times 0.701$$

solving x = 0.8262 m. The flow turns turbulent at this height

For water the property values at 30°C are

$\nu = 0.8315 \times 10^{-6}$ m²/s, $Pr = 5.68$, $\beta = 0.31 \times 10^{-3}$/K

Turbulence sets in when $Gr\, Pr = 10^9$

$$10^9 = 9.81 \times 0.31 \times 10^{-3} \times \frac{(40 - 20)x^3}{(0.8315 \times 10^{-6})^2} \times 5.68$$

solving x = 0.126 m, as against 0.8262 m.

In water turbulence sets in at a much shorter distance due to higher inertia which overcomes viscous forces quickly.

10.2 METHODS OF ANALYSIS

Derivation of continuity and momentum equations. Consider the element in the boundary layer of size $dx\, dy$ 1 as shown in Fig. 10.2.

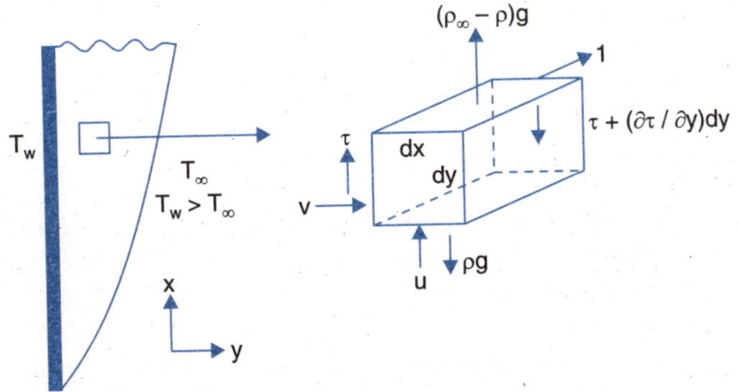

Fig. 10.2. Forces on an element in the boundary layer

The **continuity equation** is found identical to that in forced convection

$$\frac{\partial u}{\partial x} + \frac{\partial v}{\partial y} = 0 \quad \text{...(10.4)}$$

In the momentum equation, an additional force term is involved namely the buoyant force term on the volume considered, namely $dx \cdot dy \cdot 1$ given by $g\,\beta\,(T - T_\infty)$.

The resulting momentum equation is

$$u\,\frac{\partial u}{\partial x} + v\,\frac{\partial u}{\partial y} = \nu\,\frac{\partial^2 u}{\partial y^2} + g\,\beta\,(T - T_\infty) \quad \text{...(10.5)}$$

where β is the coefficient of cubical expansion (equal to $1/T$ for gases–T in absolute units).

These two equations are to be solved to obtain the velocity distribution and hydrodynamic boundary layer thickness, the boundary conditions being

$u = 0$ at $y = 0$ and $u = 0$ at $y = \delta$ (boundary layer thickness)

The energy equation is the same as that in forced convections, with no additional terms involved.

$$u\,\frac{\partial T}{\partial x} + v\,\frac{\partial T}{\partial y} = \alpha\,\frac{\partial^2 T}{\partial y^2} \quad \text{...(10.6)}$$

The boundary conditions are $T = T_w$ at $y = 0$ and $T = T_\infty$ at $y = \delta$

An idea about the dimensionless numbers correlating the phenomenon can be obtained by non-dimensionalising these equations. In the Natural convection no quantity like free stream velocity u_∞ is encountered. So a reference velocity term u_x is introduced. The following definitions are used for non-dimensionalising these equations (also refer Chapter 7).

$$x^* = x/L,\; y^* = y/L,\; u^* = u/u_x,\; v^* = v/u_x$$

and

$$T^* = \frac{T - T_\infty}{T_w - T_\infty} \quad \text{...(10.7)}$$

The momentum equation reduces to

$$u^*\,\frac{\partial u^*}{\partial x^*} + v^*\,\frac{\partial v^*}{\partial y^*} = \frac{g\,\beta\,(T_w - T_\infty)\,L}{u_x^2}\,T^* + \frac{1}{Re}\cdot\frac{\partial^2 u^*}{\partial {y^*}^2} \quad \text{...(10.8)}$$

Chapter 10

The energy equation reduces to

$$u^* \frac{\partial T^*}{\partial x^*} + v^* \frac{\partial T^*}{\partial y^*} = \frac{1}{Re\,Pr} \cdot \frac{\partial^2 T^*}{\partial y^{*2}} \qquad ...(10.9)$$

where Re = Reynolds number = $u_x L/\nu$

Three dimensionless parameters identified in the process are *Re*, *Pr* and

$$\frac{g\,\beta\,(T_w - T_\infty)\,L}{u_x^2}$$

As u_x cannot be determined it is found convenient to eliminate the quantity by multiplying this term with Re^2

$$\frac{g\,\beta\,(T_w - T_\infty)\,L}{u_x^2} \cdot \frac{u_x^2\,L^2}{\nu^2} = \frac{g\,\beta\,(T_w - T_\infty)\,L^3}{\nu^2} \qquad ...(10.10)$$

This group can be recognised as Grashof number already defined.

The heat conducted at the fluid surface is equal to the heat convected to the fluid. This principle leads to the formulation of Nusselt number.

$$-k \cdot \left.\frac{\partial T}{\partial y}\right|_{y=0} = h(T_w - T_\infty)$$

$\frac{\partial T}{\partial y}$ is solved for by using the energy equation involving u and T. The solution for u is obtained from the momentum and continuity equations. This analysis determines that the heat transfer correlations for natural convection of the form

$$Nu = f\,(Gr.\,Pr) \qquad ...(10.11)$$

The dimensional analysis method also confirms this conclusion (see the section advanced example 7.17).

10.3 INTEGRAL METHOD

Continuity, momentum and energy equations can be also obtained in the integral form considering **volume ABCD** as shown in Fig. 10.3 the volume extending beyond the boundary layer (see also **Chapter 7 on forced convection Section 7.8 and advanced Example 7.17)**

The momentum integral can be written as

$$\frac{d}{dx}\left[\int_0^\delta u^2\,dy\right] = -\nu \left.\frac{\partial u}{\partial y}\right|_{y=0},\ + g\,\beta \int_0^\delta (T - T_\infty)\,dy$$

The energy integral can be written as

$$\frac{d}{dx}\left[\int_0^\delta u(T - T_\infty)\,dy\right] = -\alpha \left.\frac{\partial T}{\partial y}\right|_{y=0}.$$

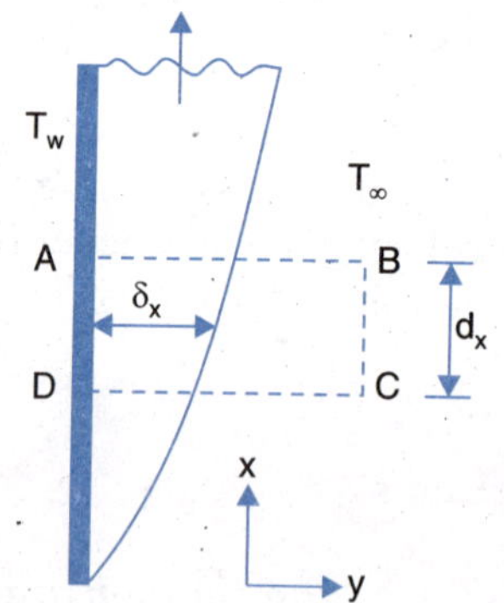

Fig. 10.3. Elemental volume for integral analysis

The integration limits can be restricted to δ as no contribution to energy or momentum flow beyond $y = \delta$ exists. The integral formulations are similar to those in forced convection. The integral equations can be solved only if relations like $u = u(y)$ and $T = T(y)$ are available. A set of suitable **distributions satisfying the boundary conditions are:**

$$\frac{u}{u_x} = \frac{y}{\delta}\left(1 - \frac{y}{\delta}\right)^2 \quad \text{...[10.12(a)]}$$

and

$$\frac{T - T_\infty}{T_w - T_\infty} = \left(1 - \frac{y}{\delta}\right)^2 \quad \text{...[10.12(b)]}$$

The boundary conditions $u = 0$ at $y = 0$ and $y = \delta$, $T = T_w$ at $y = 0$ and $T = T_\infty$ at $y = \delta$ are satisfied by these equations.

From such analysis the local boundary layer thickness in laminar flow is obtained as

$$\delta_x = \frac{3.93\, x\, (0.952 + Pr)^{0.25}}{Gr_x^{\ 0.25}\ Pr^{0.5}} \quad \text{...(10.13)}$$

$$-kA \left.\frac{dT}{dy}\right|_{y=0} = hA\,(T_w - T_\infty)$$

From [10.12(b)], $\left.\dfrac{dT}{dy}\right|_{y=0} = -\dfrac{2}{\delta}\,(T_w - T_\infty)$

$$\therefore \qquad \frac{h}{k} = \frac{2}{\delta} \quad \text{or} \quad Nu_x = \frac{hx}{k} = \frac{2x}{\delta}$$

Substituting in (10.13), we get (10.14).

and

$$Nu_x = \frac{0.508\, Pr^{0.5}\, Gr_x^{\ 0.25}}{(0.952 + Pr)^{0.25}} \quad \text{...(10.14)}$$

The average value up to L works out as

$$\overline{\mathbf{Nu}} = \mathbf{(4/3)\ Nu_L} \quad \text{...(10.15)}$$

($x^{3/4}$ is involved in the integration from $(L^3)^{0.25}$ in Gr_x)

Note that this is different from the correlation in forced convection over flat plate, where $\overline{Nu} = 2\,Nu_L$. ($x^{1/2}$ is involved in the integration)

The same method applied to turbulent flow with the velocity and temperature variations as below leads to the boundary layer thickness for turbulent region as in equation (10.17).

The **velocity distribution and temperature distribution in the turbulent region** is assumed as satisfying the boundary condition, $u = 0$ at $y = 0$ and $y = \delta$.

$$\frac{u}{u_x} = 1.862\left(\frac{y}{\delta}\right)^{1/7}\left(1 - \frac{y}{\delta}\right)^4 \quad \text{...[10.16(a)]}$$

and

$$\frac{T - T_\infty}{T_w - T_\infty} = \left(1 - \frac{y}{\delta}\right)^{1/7} \quad \text{...[10.16(b)]}$$

This leads to

$$\delta_{turbulent} = \frac{0.565\, x\,(1 + 0.494\, Pr^{2/3})}{Gr^{0.1}\ Pr^{8/15}} \quad \text{...(10.17)}$$

Equations (10.14) and (10.17) form the basis for correlations of experimental results in free convection. For design purposes correlations have been formulated using experimental results and these will be discussed in the next sections.

Example 10.4: *Illustrates the values of heat transfer coefficient in the laminar region for a gas and a liquid. Note the large difference. For the flows in example 10.3 determine the value of boundary layer thickness and average convection coefficient at the location where flow turns turbulent.*

Solution: Air: The property values required are: k = 0.02675 W/m K, Pr = 0.701

$$\delta_x = 3.93\, x[0.952 + Pr]^{0.25}/Gr^{0.25}\, Pr^{0.5},\ Gr\,Pr = 10^9 \qquad ...[10.17(a)]$$

$$\therefore \quad \mathbf{Gr} = 10^9/Pr = 10^9/0.701 = \mathbf{1.4265 \times 10^9}$$

$$x = 0.8262 \text{ m (Example 10.3)}$$

Substituting in 10.17(a)

$$\therefore \quad \boldsymbol{\delta_x = 0.0226 \text{ m or } 22.6 \text{ mm}}$$

From equation (10.14),

$$Nu_x = \frac{0.508\, Pr^{0.5}\, Gr^{0.25}}{(Pr + 0.952)^{0.25}} = 72.9$$

$$\mathbf{h_x} = \frac{Nu_x \times k}{x} = \frac{72.9 \times 0.02675}{0.8262} = \mathbf{2.36 \text{ W/m}^2 \text{ K}}$$

$$\bar{\mathbf{h}} = \frac{4}{3} h_L = \mathbf{3.15 \text{ W/m}^2 \text{ K}}$$

Water: $Gr = 10^9/Pr = 10^9/5.68$, k = 0.6129 W/m K, Pr = 5.68

$$\mathbf{Gr = 1.76 \times 10^8}$$

$$x = 0.126 \text{ m (Example 10.3)}$$

$$\boldsymbol{\delta_x} = \frac{3.93\, x\, 0.126\, [0.952 + 5.68]^{0.25}}{(1.76 \times 10^8)^{0.25}\, (5.68)^{0.5}} = \mathbf{2.89 \times 10^{-3} \text{ m} = 2.89 \text{ mm}}$$

$$Nu_x = \frac{0.508 \times 5.68^{0.5}\, [1.76 \times 10^8]^{0.25}}{(0.952 + 5.68)^{0.25}} = 86.9$$

$$\mathbf{h_x} = \frac{86.9 \times 0.6129}{0.126} = \mathbf{422.7 \text{ W/m}^2 \text{ K},}$$

$$\bar{\mathbf{h}} = \frac{4}{3} h_L = \mathbf{563.6 \text{ W/m}^2 \text{ K.}}$$

Example 10.5: *Illustrates the variation of boundary layer thickness as well as the variation of local and average heat transfer coefficients. A vertical plate maintained at 40°C is placed in still air at 20°C. The plate is 0.80 m high and 1 m wide. Tabulate the values of δ_x and h_x at 0.2 m intervals. Also determine the heat input required to maintain the plate at 40°C.*

Solution: The property values are to be evaluated at 30°C.

$$\rho = 1.165,\ \nu = 16 \times 10^{-6} \text{ m}^2/\text{s},\ Pr = 0.701,\ k = 0.02675 \text{ W/m K},\ \beta = 1/303,$$

$$Gr_x = \frac{g\beta\,(T_w - T_\infty)\,x^3}{\nu^2} = \frac{9.81}{303} \times \frac{20}{(16 \times 10^{-6})^2} \,.\, x^3 = 2.53 \times 10^9 \,.\, x^3$$

$$\delta_x = \frac{3.93\, x\, [0.952 + Pr]^{0.25}}{Gr_x^{\ 0.25}\, Pr^{0.5}} = 5.32\, x/Gr_x^{\ 0.25}$$

$$Nu_x = \frac{0.508\,.\, Pr^{0.5}\, Gr_x^{0.25}}{[0.952 + Pr]^{0.25}} = 0.3751\,.\, Gr_x^{\ 0.25}$$

$$h_x = 0.3751 \frac{Gr_x^{0.25}}{x} \times 0.02675 = \frac{1}{x} \times 0.010034 \,.\, Gr_x^{\,0.25}$$

$$\overline{h}_L = \frac{4}{3} h_L$$

The values calculated are given below:

Distance x, m	*Gr*	δ_x *(mm)*	Nu_x	h_x	$\overline{h}$ W/m² K
0.2	2.024×10^7	15.871	25.16	3.365	**4.49**
0.4	1.619×10^8	18.874	42.31	2.83	**3.77**
0.6	5.463×10^8	20.887	57.35	2.56	**3.41**
0.8	1.295×10^9	22.445	71.16	2.38	**3.17**

The heat input required for maintaining the plate at 40°C on one side is

$$Q = hA\,\Delta T = 3.17 \times 0.8 \times 1 \times (40 - 20) \text{ W} = 50.72 \text{ W}$$

on both sides $\mathbf{Q = 101.44\ W.}$

10.4 CORRELATIONS FROM EXPERIMENTAL RESULTS

In Natural convection also most of the correlations used in design have been formulated from experimental results. The parameters used are the three dimensionless numbers, *Nu*, *Pr* and *Gr*. A plot of *Nu* against (*Gr Pr*) on log-log scale results in curves as shown in Fig. 10.4. This curve has been divided into a number of straight lines in some cases and just two straight lines in some cases.

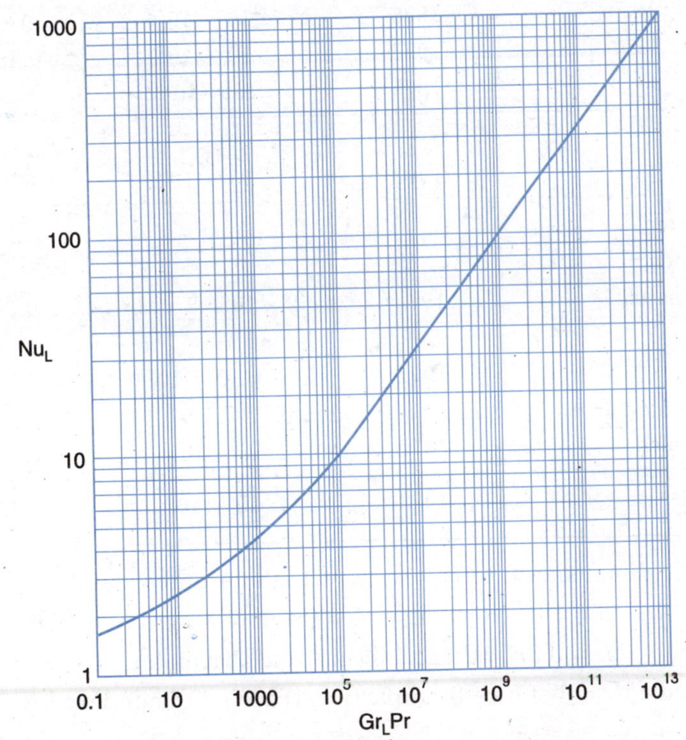

Fig. 10.4. Free convection experimental results

For the **average Nusselt number** a correlation of the form given in (10.18) below is found suitable

$$Nu = C\,(Gr\,Pr)^n \quad ...(10.18)$$

The values of C and n for different configurations and selected ranges of parameters is given in Table 10.1. These are for constant wall temperature conditions. The property values are evaluated at the film temperature of

$$T_f = (T_w + T_\infty)/2$$

The product $Gr\,Pr$ is known as Rayleigh number (Ra)

Table 10.1. Values of C and n in the equation $Nu = C\,(Gr\,Pr)^n$ $(GrPr)^n$ in natural convection

Geometry	*Year of publication*	*Application range of values of Gr Pr*	C	n
Vertical planes and cylinders with $D/L > 35/Gr_L^{0.25}$	1954	10^4—10^9	**0.59**	**0.25**
Characteristic dimension—L	1951	10^9—10^{13}	**0.021**	**0.4**
Preferred eqn.	1968	10^9—10^{13}	**0.1**	**0.333**
Horizontal cylinder	1975	10^{-2}—10^2	**1.02**	**0.148**
	1975	10^2—10^4	**0.85**	**0.188**
Characteristic	1975	10^4—10^7	**0.480**	**0.25**
dimension–D–	1975	10^7—10^{12}	**0.125**	**0.33**
Vertical cylinder $D = H$	1983	10^4—10^6	**0.775**	**0.21**
Horizontal plate -different shapes Upper surface heated	1972	2×10^4—8×10^6	0.54	0.25
or lower surface cooled	1972	8×10^6—10^{11}	0.15	0.33
Lower surface heated or upper surface cooled	1972	10^5—10^{11}	0.27	0.25
Irregular solids	1973	10^4—10^9	0.52	0.25

Characteristic dimension L to be used in Grashof number for different shapes:

Square: L = side, Rectangle: L = (Sum of length and width/2)

Cylinder: $L = 0.9D$, other shapes: Area/ Perimeter.

Irregular solids: L = distance fluid particle travels in boundary layer.

Majority of practical cases can be solved using the listed correlations.

However, it may be noted that it does not cover all situations.

More recent correlations are discussed in Section 10.5.

Examples 10.6 to 10.12 illustrate the application of the above correlations. Note the difference in values of heat transfer coefficient for the same situation, using different correlations applicable in the same range of *Gr Pr*.

Example 10.6: *A vertical plate 0.80 m high and 1 m wide is maintained at 35°C in still air at 25°C. Determine the value of average convection coefficient using correlation given in Table 10.1 and compare with the answer of Example 10.5.*

Solution: The property values are evaluated at (25 + 35)/2 = 30°C with usual units

$\rho = 1.165 \quad \upsilon = 16 \times 10^{-6} \quad Pr = 0.701 \quad k = 0.02675, \quad \beta = 1/303$

$$\mathbf{Gr} = 9.81 \times \frac{1}{303} \cdot \frac{(35-25)0.8^3}{(16 \times 10^{-6})^2} = \mathbf{0.6475 \times 10^9}$$

From Table 10.1, $C = 0.59, \quad n = 0.25$

$\therefore \quad Nu = 0.59\,(Gr\,Pr)^{0.25} = 86.12$

$\therefore \quad$ **h = 2.88 W/m² K** (about 3.2, in Example 10.5).

Example 10.7: *A vertical plate 4 m high and 1 m wide is maintained at 60°C in still air at 0°C. Determine the value of convection coefficient.*

Solution: The average value of temperature = (0 + 60)/2 = 30°C. Property values from 10.6 can be used.

$$\mathbf{Gr} = 9.81 \times \frac{1}{303} \cdot \frac{(60-0) \times 4^3}{(16 \times 10^{-6})^2} = \mathbf{4.86 \times 10^{11}}$$

From Table 10.1 $\quad C = 0.021, n = 0.4$

$Nu = 0.021\,(Gr\,Pr)^{0.4} = 861$, **h = 5.76 W/m² K**

The **alternate set of values are**

$C = 0.1, n = 0.333$

using these values, $\quad Nu = 698.25$ **h = 4.67 W/m² K**

There is a variation of 20–25%. The literature indicates the later correlation is more desirable. So the value of $h = 4.67$ W/m² K is more suitable.

Later we shall see some more correlations also for this range.

Example 10.8: *Check above what diameters these correlations used in Examples 10.6 and 10.7 can be used for vertical cylinders.*

Solution: The condition is $(D/L) \geq \dfrac{35}{Gr^{0.25}}$

For Example (10.6), $\quad Gr = 0.6475 \times 10^9$

$\therefore \quad (D/L) \geq 0.22$ m

For Example (10.7) $\quad Gr = 4.86 \times 10^{11}$

$\therefore \quad (D/L) \geq 0.042$ m

For large values of Gr, even small dia cylinders can be considered as vertical plate of width πD.

Example 10.9: *A plate heater with graded heat flux so that the plate temperature remains constant is used to heat water in a large tank. The plate is 0.6 m × 0.30 m with the longer side vertical. The plate temperature is 60°C while the water temperature is 20°C. Determine the value of convection coefficient.*

Solution: The film temperature is (60 + 20)/2 = 40°C

Property values are with usual units

$$\rho = 995, \quad \nu = 0.657 \times 10^{-6}\ \text{m}^2/\text{s}, \quad Pr = 4.34, \quad k = 0.628, \quad \beta = 0.41 \times 10^{-3}$$

$$\textbf{Gr} = \frac{9.81 \times 0.41 \times 10^{-3}\,(60 - 20) \times 0.6^3}{(0.657 \times 10^{-6})^2} = \mathbf{8.05 \times 10^{10}}$$

so the flow is turbulent.

Two correlations are possible and values of h are (Table 10.1)

$$Nu = 0.021\,(Gr\,Pr)^{0.4} = 870.06, \quad \mathbf{h = 910.66\ W/m^2\,K}$$

or

$$Nu = 0.1\,(Gr\,Pr)^{0.333} = 704.33, \quad \mathbf{h = 737.2\ W/m^2\,K}$$

Between the two there is a difference of 20–25% in this case also. Some later correlations are available for this situation, and the same will be discussed subsequently.

Example 10.10: *Water is heated in a tank using horizontal pipes of diameter 50 mm with a wall temperature of 60°C maintained by steam condensing on the inside of the tubes. The water in the tank is at 20°C. Calculate the value of convection coefficient if the water is stagnant.*

Solution: This is a case of free convection over a horizontal pipe.

The film temperature is (60 + 20)/2 = 40°C

The property values are as in Example 10.9

$$\textbf{Gr} = 9.81 \times 0.41 \times 10^{-3}\,(60 - 20) \times 0.05^3/(0.657 \times 10^{-6})^2 = \mathbf{4.66 \times 10^7}$$

Possible correlation is (Table 10.1)

$$Nu = 0.125\,(Gr\,Pr)^{0.33} = 0.125\,(4.66 \times 10^7 \times 4.34)^{0.33}$$

$$\therefore \quad Nu = 68.83, \quad \mathbf{h = 864.6\ W/m^2\,K}$$

Example 10.11: *A horizontal plate 1 m × 0.8 m is kept in a water tank with the top surface at 60°C providing heat to warm stagnant water at 20°C. Determine the value of convection coefficient. Repeat the problem for heating on bottom surface.*

Solution: This is a case of natural convection.

The film temperature is (60 + 20)/2 = 40°C

The property values are taken as in Example 10.9

In this case $L = (a + b)/2 = (1 + 0.8)/2 = 0.9$ m

Upper surface heated: (Table 10.1 is used to find n and C)

$$\textbf{Gr} = 9.81 \times 0.41 \times 10^{-3}\,(60 - 20) \times 0.9^{0.3}/(0.657 \times 10^{-6})^2$$

$$= \mathbf{2.717 \times 10^{11}}, \quad n = 0.33,\ C = 0.15$$

$$Nu = 0.15\ (Gr\ Pr)^{0.33} = 1444.5 \qquad \therefore\ \mathbf{h = 1007.9\ W/m^2\ K}$$

Lower surface heated: $n = 0.25$, $C = 0.27$

$$Nu = 0.27\ (Gr\ Pr)^{0.25} = 281.4, \qquad \therefore\ \mathbf{h = 196.32\ W/m^2\ K}$$

As the flow is obstructed when the bottom surface is heated, the value of convection coefficient is lower. If (A_s/P) is taken for L, higher values of h is obtained.

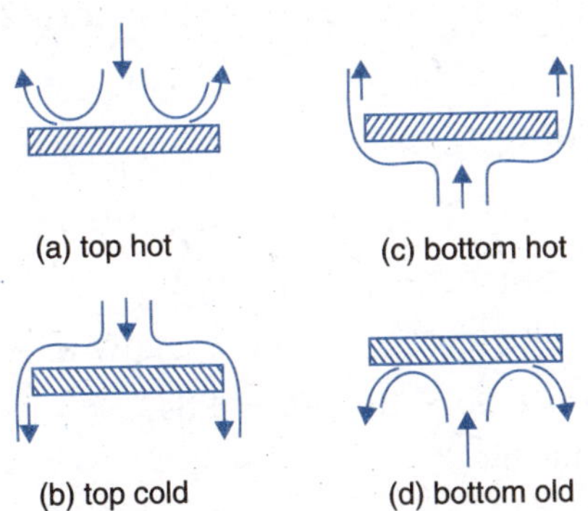

Fig. Ex. 10.11. Flow pattern for horizontal plates

10.5 A MORE RECENT SET OF CORRELATIONS

Churchill and Chu (1975) for vertical plate and vertical cylinders in natural convection have given the equations (10.19) and (10.20) below.

$$Nu = 0.68 + \frac{0.67\ (Gr\ Pr)^{0.25}}{[1 + (0.492/Pr)^{9/16}]^{4/9}} \qquad ...(10.19)$$

This is valid for $Gr\ Pr < 10^9$

A correlation (which gives somewhat better accuracy) for larger range of $Gr\ Pr$ (from 10^{-1} to 10^{12}) is

$$Nu = \left[0.825 + \frac{0.387\ (Gr\ Pr)^{1/6}}{[1 + (0.492/Pr)^{9/16}]^{8/27}}\right]^2 \qquad ...(10.20)$$

These are for constant wall temperature: Equations (10.19) and (10.20) are valid for constant flux also. In this case, the value 0.492 is to be replaced by 0.437.

As the temperature difference varies with length, a better accuracy is possible if $(T_w - T_\infty)$ at mid-length is used in the calculation of Grashof number. This can be done only by trial.

These correlations can be used for vertical cylinders if $(D/L) \geq 35/Gr^{0.25}$.

Example 10.12: *Repeat Examples 10.6, 10.7 and 10.9 using the recent correlations given by (10.19) and (10.20).*

Chapter 10

Solution: For Example 10.6:

$$Gr = 0.6475 \times 10^9,\ Pr = 0.701,\quad k = 0.02675 \text{ W/m K},\ L = 0.8 \text{ m}$$

Equation (10.19) is applicable:

$$Nu = 0.68 + \frac{0.67[0.6475 \times 10^9 \times 0.701]^{0.25}}{[1 + (0.492/0.701)^{9/16}]^{4/9}} = 75.63$$

$\therefore$ **h = 2.53 W/m² K (compared to 2.88)**

For Example 10.7: $Gr = 4.86 \times 10^{11}$, $Pr = 0.701$, $k = 0.02675$ W/mk, $L = 4$ m

Equation (10.20) is applicable:

$$Nu^{0.5} = \left[0.825 + \frac{0.387\,(4.86 \times 10^{11} \times 0.701)^{1/6}}{[1 + (0.492/0.701)^{9/16}]^{8/27}}\right] = 27.91$$

$\therefore$ $Nu = 779.05$, $\therefore$ **h = 5.21 W/m² K (compared to 5.76)**

for Example 10.9 $Gr = 8.05 \times 10^{10}$, $Pr = 4.34$, $k = 0.628$ W/mk, $L = 0.6$ m

Equation (10.20) is applicable

$$Nu^{1/2} = \left[0.825 + \frac{0.387\,(Gr\ Pr)^{1/6}}{[1 + (0.492/Pr)^{9/16}]^{8/27}}\right] = 30.92$$

$Nu = 955.81$ $\therefore$ **h = 1000.4 W/m² K (compared to 910.66)**

10.6 CONSTANT HEAT FLUX CONDITION—VERTICAL SURFACES

Here the value of wall temperature is not known. So ΔT is unspecified for the calculation of Grashof number. Though a trial solution can be attempted, it is found easier to eliminate ΔT by q which is known in most cases. This is done by multiplying Grashof number by Nusselt number and using $q = h\Delta T$.

This product is known as **modified Grashof number, Gr***

$$Gr_x^* = Gr_x\, Nu_x = \frac{g\,\beta\,\Delta t\, x^3}{\nu^2} \cdot \frac{hx}{k} = \frac{g\,\beta\, q\, x^4}{k\,\nu^2} \qquad \text{...(10.21)}$$

The correlation for laminar range is given by

$$Nu_x = 0.60\ [Gr_x^*\ Pr]^{0.2} \qquad \text{...(10.22)}$$

$$10^5 < Gr^* < 10^{11}$$

The correlation for the turbulent region above $Gr^* > 10^{11}$ is

$$Nu_x = 0.17\ (Gr_x^*\ Pr)^{0.25} \qquad \text{...(10.23)}$$

The **average value for** the laminar region is obtained by integration as

$$h = \frac{5}{4}\, h_L = 1.25\, h_L \qquad \text{...(10.24)}$$

$x^{4/5}$ is involved in the integration

For the turbulent region, it can be proved that h_x is

$$\therefore \quad \mathbf{h = 1.136 \times h_L} \qquad ...(10.25)$$

Example 10.13: *A flat electrical heater of 0.4 m × 0.4 m size is placed vertically in still air at 20°C. The heat generated is 1200 W/m². Determine the value of convective heat transfer coefficient and the average plate temperature.*

Solution: The film temperature is not known. Assuming an average value of h = 5 W/m² K,

ΔT = 1200/5 = 240°C

∴ T_f = (240 + 20)/2 = 130°C. From tables, the property values are read as

ρ = 0.876, ν = 26.625 × 10^{-6} m²/s, Pr = 0.685, k = 34.135 × 10^{-3} W/m K

c = 1011 J/kg K

$$\mathbf{Gr^*} = \frac{g\beta q x^4}{k\nu^2} = 9.81 \times \frac{1}{273+130} \times \frac{1200 \times 0.4^4}{34.135 \times 10^{-3}\,(26.625 \times 10^{-6})^2}$$

$$= \mathbf{3.09 \times 10^{10}}$$

This is in the laminar range. (for G^*)

∴ $Nu_x = 0.60\,[Gr_x^*\,Pr]^{0.2} = 69.71$, ∴ $\mathbf{h_x = 5.95\ W/m^2\,K}$

$\bar{\mathbf{h}} = \frac{5}{4}h = \mathbf{7.44\ W/m^2\,K}$, ∴ T = 20 + 1200/7.44 = 181.3°C.

∴ T_f = (181.3 + 20)/2 = 100°C. **Another trial** with properties at 100°C should be made.

Taking average temp. 100°C and plate temp. of 180°C.

$$\mathbf{Gr_x^*} = \frac{g\beta q x^4}{k\nu^2} = 9.81 \times \frac{1}{273+100} \times 1200 \times \frac{0.4^4}{(23.13 \times 10^{-6})} \times \frac{1}{32.1 \times 10^{-3}}$$

$= \mathbf{4.7 \times 10^{10}}$, using eqn. (10.22), Nu = 75.9, $\bar{\mathbf{h}}$ **= 7.62 W/m² K**

Plate temp. = 177.6°C. Hence acceptable

If equation (10.19) was used with 0.492 replaced by 0.437

$Gr = 0.503 \times 10^9$

$$Nu = 0.68 + \frac{0.67\,(Gr\,Pr)^{0.25}}{[1+(0.437/Pr)^{9/16}]^{4/9}} = 71.6$$

$$\therefore \quad \mathbf{h} = \frac{71.6 \times 32.1 \times 10^{-3}}{0.4} = \mathbf{5.74\ W/m^2\,K}$$

$\mathbf{\Delta T}$ = 208, T_w = 228, T_f = **128°C.**

Example 10.14: *If in the Example 10.13, the heat flux has a value of 10,000 W/m², determine the value of h and the average plate temperature.*

Solution: The film temperature is not specified.

Assuming h = 20 W/m² K, ΔT = 500°C. So taking property values at 250°C

ν = 40.61 × 10^{-6} m²/s, Pr = 0.677, k = 42.68 × 10^{-3} W/m K, L = 0.4 m.

$$\therefore \quad \mathbf{Gr^*} = 9.81 \times \frac{1}{273+250} \times 10^4 \times 0.4^4 \times \frac{1}{42.68 \times 10^{-3}} \times \frac{1}{(40.61 \times 10^{-6})^2}$$

$$= \mathbf{6.82 \times 10^{10}}$$

∴ **laminar (reference to Gr*)**

$$\therefore \quad Nu_x = 0.60\ (Gr^*\ Pr)^{0.2} = 81.48$$

$$\therefore \quad \mathbf{h = 8.70\ W/m^2\ K}$$

∴ Reworking is necessary as

$$\Delta T = 1150°C \quad \therefore \quad T_f \approx 600°C$$

$$\nu = 96.89 \times 10^{-6}\ m^2/s \quad k = 62.22 \times 10^{-3}\ W/m\ K, \quad Pr = 0.699$$

$$\mathbf{Gr^*} = 981 \times \frac{1}{273+600} \times 10^4 \times 0.4^4 \times \frac{1}{62.22 \times 10^{-3}} \times \frac{1}{(96.89 \times 10^{-6})^2}$$

$$= \mathbf{4.92 \times 10^9}$$

∴ laminar

$$\therefore \quad Nu_x = 0.60\ (Gr_x^*\ Pr)^{0.2} = 48.48$$

$$\therefore \quad \mathbf{h = 7.54\ W/m^2\ K, \Delta T = 1326°C.}$$

The value of h will be lower than 7.54. The plate temperature will be around 1400°C. The value of heat flux is not suitable for free convection as it leads to an unusually high plate temperature. In case of water this flux may lead to boiling.

10.6.1. Constant Heat Flux, Horizontal Surfaces

For horizontal surfaces, the correlations are given in Table 10.1 for constant wall temperature conditions. Example 10.11 illustrates the situation. For constant heat flux conditions the following correlations are available. The property values except β in these cases is to be evaluated at

$$T_e = T_w - 0.25\ (T_w - T_\infty) \qquad ...(10.26)$$

β is evaluated at T_∞.

The characteristic length L = Area/perimetre generally. For circle 0.9 D and for rectangle $(L + W)/2$

T_w is to be estimated using the basic relation.

$$\bar{h}\ (T_w - T_\infty) = q \qquad ...(10.27)$$

It may be noted that iteration becomes necessary in solving these problems.

For heated face facing upwards or cooled face facing downwards:

laminar conditions,

$$Nu = 0.54\ (Gr\ Pr)^{1/4},\ Gr\ Pr \to 10^5 \text{ to } 2 \times 10^7 \qquad ...[10.28(a)]$$

$$\overline{Nu} = 0.14\ (Gr\ Pr)^{1/3} \qquad ...(10.28)$$

$$Gr\ Pr \to 2 \times 10^7 \text{ to } 3 \times 10^{10}.$$

For the turbulent range

$$\overline{Nu} = 0.16\ (Gr\ Pr)^{1/3} \qquad ...(10.29)$$

$$2 \times 10^8 < Gr\ Pr < 10^{11}$$

For heated surface facing downward

$$Nu = 0.27\ (Gr\ Pr)^{1/4}$$

$$Gr\ Pr \to 3 \times 10^5 \text{ to } 3 \times 10^{10}$$

$$\therefore \quad Nu = 0.58\ (Gr\ Pr)^{0.2}$$

$$10^6 < Gr\ Pr < 10^{11}. \qquad ...(10.30)$$

Example 10.15: *A plate heater 0.4 × 0.4 m using electrical elements, has a constant heat flux of 1.2 kW/m². It is placed in room air at 20°C with the hot side facing up. Determine the value of h and average plate temperature.*

Solution: The property values have to be evaluated at

$$T_e = T_w - 0.25\ (T_w - T_\infty),\ T_\infty = 20°C,\ T_w \text{ is not specified.}$$

Assuming a value of $h = 10$ W/m² K,

$$T_w - T_\infty = 1200/10 = 120°C, \quad \therefore \quad T_w = 140°C$$

$$\therefore \quad T_e = 140 - 0.25\ (140 - 20) = 110°C$$

Properties of air at 110°C are: $\rho = 0.922$, $\nu = 24.29 \times 10^{-6}$, $Pr = 0.687$, $k = 32.74 \times 10^{-3}$ (usual units).

L for square = length of side = 0.4 m

$$\mathbf{Gr} = \frac{g\beta \, .\, \Delta T\ L^3}{\nu^2} = 9.81 \times \frac{1}{293} \times \frac{120 \times 0.4^3}{(24.29 \times 10^{-6})^2} = \mathbf{4.36 \times 10^8}$$

$\therefore$ The equation applicable is **(10.29)**

$$\overline{Nu} = 0.16\ (Gr\ Pr)^{1/3} = 107.38, \qquad \therefore \quad \mathbf{h = 8.76\ W/m^2\ K}$$

Now using this value of h, $\Delta T = 137°C$ $\qquad \therefore \quad T_w = 157°C$

$$T_e = 157 - 0.25\ (157 - 20) = 122.75°C$$

Properties at this condition are:

$$\rho = 0.8925,\ \nu = 25.74 \times 10^{-6} \text{ m}^2\text{/s},\ Pr = 0.685,\ k = 33.57 \times 10^{-3} \text{ W/m K}$$

$$\mathbf{Gr} = 9.81 \times \frac{1}{273} \times \frac{137 \times 0.4^3}{(25.74 \times 10^{-6})^2} = \mathbf{0.476 \times 10^9}$$

$\therefore$ Equation (10.29) is applicable

$$Nu = 0.16\ (0.476 \times 10^9 \times 0.685)^{1/3} = 110.09$$

$$\therefore \quad \mathbf{h = 9.24\ W/m^2\ K}$$

using this value, $\quad \mathbf{\Delta T = 130°C} \quad \therefore \quad \mathbf{T_w = 150°C}$

$$T_e = 150 - 0.25\ (150 - 20) = 117.5°C$$

Taking property values at this temperature (usual units)

$$\rho = 0.904,\ \upsilon = 25.16 \times 10^{-6}, \quad Pr = 0.687,\ k = 33.22 \times 10^{-3}$$

$$Gr = 9.81 \times \frac{1}{273} \times \frac{130 \times 0.4^3}{(25.16 \times 10^{-6})^2} = 4.72 \times 10^8$$

$$\therefore \quad Nu = 0.16\ (Gr\ Pr)^{1/3} = 109.94 \quad \therefore \quad \mathbf{h = 9.13\ W/m^2\ K}$$

Further iteration may not improve the accuracy further.

Average plate temperature difference = 1200/9.13 = **131.4°C**

$\therefore$ **Average plate temperature** = **151.4°C.**

Example 10.16: *A circular disk of 0.2 m diameter with a constant heat generation rate of 1 kW/m² is kept with its heated surface facing down in air at 20°C. Determine the value of convection coefficient and the average plate temperature.*

Solution: This is a problem of free convection. The property values are to be taken at

$$T_e = T_w - 0.25\,(T_w - T_\infty).$$

As T_w is not specified, a suitable value is assumed and then iterative calculations is made. As the value of convection coefficient will be lower, assuming h = 4 W/m² K,

$$\Delta T = 250°\text{C } i.e.\ (1000/4), \quad \therefore \quad T_w = 270°\text{C}$$

$$T_e = 270 - 0.25\,(270 - 20) = 207.5°\text{C}$$

Taking property values at 200°C (nearer)

$$\rho = 0.746,\ \nu = 34.85 \times 10^{-6}\ \text{m}^2/\text{s}, \quad Pr = 0.68$$

$$k = 39.31 \times 10^{-3}\ \text{W/m K}, \quad L = 0.9 \times 0.2 = 0.18\ \text{m},\ \beta = \frac{1}{293}$$

$$\therefore \qquad \text{Gr} = \frac{9.81 \times 250 \times 0.18^3}{293 \times (34.85 \times 10^{-6})^2} = \mathbf{4.019 \times 10^7}$$

Using equation (10.30), (suitable for this range)

$$Nu = 0.58\,(Gr\,Pr)^{0.2} = 17.81 \quad \therefore \quad \mathbf{h = 3.89\ W/m^2\,K}$$

This is very near the value. However **another trial** can be made:

$$\Delta T = 257°\text{C} \qquad \therefore \quad T_w = 277°\text{C}$$

$$T_e = 277 - 0.25\,(257) = 212.75°\text{C}$$

property values at 212.75°C

$$\rho = 0.728,\ \nu = 36.32 \times 10^{-6},\ Pr = 0.679,\ k = 40.17 \times 10^{-3}\ \text{W/m K}$$

$$\text{Gr} = \frac{9.81}{293} \times \frac{257 \times 0.18^3}{(36.32 \times 10^{-6})^2} = \mathbf{3.804 \times 10^7}$$

$$\therefore \qquad Nu = 0.58\,(Gr\,Pr)^{0.2} = 17.61 \qquad \therefore \quad \mathbf{h = 3.93\ W/m^2\,K}$$

$$\therefore \qquad \mathbf{\Delta T} = 1000/3.93 = \mathbf{254.4°C} \qquad \therefore \quad T_w = 274.4°\text{C}$$

When the heated surface faces downward, the surface temperature is higher.

10.7 FREE CONVECTION FROM INCLINED SURFACES

The previous correlations for vertical plates at constant wall temperature conditions can be applied simply by **multiplying Gr by cos θ** where θ is the angle the surface makes with the vertical (or by sin θ if the angle is from horizontal). In the *Gr* calculation, *g* is to be multiplied by cos θ). This is valid up to an angle of 60° from vertical. In the case of **constant heat flux Gr* is multiplied by cos θ** in the correlations.

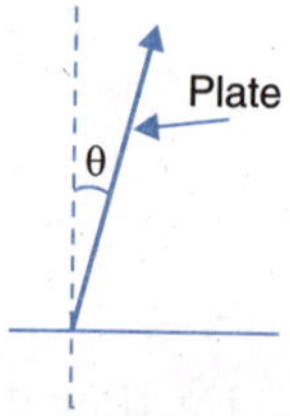

Plate

θ

Fig. 10.5

For horizontal surfaces with constant heat flux with heated surface facing down, the correlation is

$$Nu = 0.56\,(Gr\,Pr\cos\theta)^{0.25}$$

$$\theta < 88° \text{ and } 10^5 < Gr\,Pr\cos\theta < 10^{11} \qquad \ldots(10.31)$$

Properties to be evaluated at T_e (equation 10.26) and β alone at T_∞.

For heated surface facing upwards, the correlations are more involved. A simple one is

$$Nu = 0.56\ [Gr\ Pr \cos\theta]^{1/4} + 0.14\ [(Gr\ Pr)^{1/3} - (Gr_c\ Pr)^{1/3}] \quad ...(10.32)$$

where Gr_c is critical value of Grashof number **causing flow separation**. This value depends on the angle and is tabulated below:

Angle	***15***	***30***	***60***	***75***
Gr_c	5×10^9	2×10^9	10^8	10^6

Simplified formulations are available for air. Equation (10.22) can be used with $Gr^* \cos\theta$ in place of Gr^*.

In the turbulent region, heated surface facing upwards.

$$Nu_x = 0.17\ (Gr_x^*\ Pr)^{1/4} \quad ...(10.33)$$

$$10^{10} < Gr^*\ Pr < 10^{15}$$

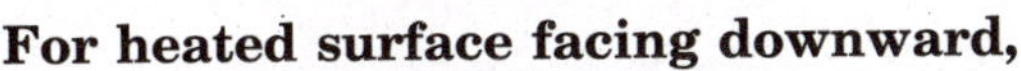

Fig. 10.5 (a) Fluid flow over inclined surface–showing flow separation

For heated surface facing downward,

$Gr^* \cos^2\theta$ replaces Gr_x^*. The relation will be nearly the same as in Table 10.1

Equation (10.33) approximates to the equation given in Table 10.1.

Example 10.17: *Consider a surface 0.8 m high, kept at an angle of 55° from the horizontal at a constant wall temperature of 40°C in air at 20°C. Determine the value of convection coefficient and compare the same with that of similar vertical plate.*

Solution: The film temperature is (40 + 20)/2 = 30°C, θ = 35° (vertical)

The property values are:

$$\rho = 1.165,\ \nu = 16 \times 10^{-6}\ \text{m}^2/\text{s},\ Pr = 0.701,\ k = 26.75 \times 10^{-3}\ \text{W/m K}$$

For vertical plate:

$$\mathbf{Gr} = \frac{9.81 \times 20 \times 0.8^3}{303 \times (16 \times 10^{-6})^2} = \mathbf{1.295 \times 10^9}$$

Equation applicable is (equation 10.20)

$$Nu = \left[0.825 + \frac{0.387\ (Gr\ Pr)^{1/6}}{[1 + (0.492/Pr)^{9/16}]^{8/27}}\right]^2$$

$$Nu = 119.03 \quad \therefore \quad \mathbf{h = 3.98\ W/m^2\ K}$$

Inclined surface: (equation 10.20 modified)

$$Nu = \left[0.825 + \frac{0.387\ (Gr \cos\theta\ Pr)^{1/6}}{[1 + (0.492/Pr)^{9/16}]^{8/27}}\right]^2$$

$$Nu = 111.91 \quad \therefore \quad \mathbf{h = 3.74\ W/m^2\ K.}$$

Using another possible correlation:

Vertical plate: $Nu = 0.1\ (Gr\ Pr)^{1/3} = 96.83, \quad \therefore \quad h = 3.24\ \text{W/m}^2\ \text{K}$

Inclined plate: $Nu = 0.1\ (Gr\ Pr \cos\theta)^{1/3} = 90.5 \quad \therefore \quad h = 3.03\ \text{W/m}^2\ \text{K}$

In terms of numbers the difference is about 23% lower by this correlation.

Equation 10.20 being the more recent and well experimented, the value as per the eqn. should be more reliable.

Example 10.18: *Consider a vertical plate of height 0.8 m with a constant heat flux of 2 kW/m². The still air surrounding it is at 20°C. Determine the average surface temperature. If during the installation, the plate is inclined to the vertical at 30°, determine the change in temperature.*

Solution: The still air is at 20°C. As the wall temperature is not specified, an estimate is made **assuming**

$$\mathbf{h = 10\ W/m^2\ K} \quad \therefore \quad \Delta T = 2000/10 = 200°C$$

$\therefore$ The film temperature can be taken as (200 + 20)/2 = 110°C

$$\rho = 0.898,\ \nu = 25.45 \times 10^{-6}\ m^2/s, \quad Pr = 0.686,\ k = 33.38 \times 10^{-3}\ W/m\ K$$

using equation (10.21), vertical location:

$$\mathbf{Gr^*} = \frac{g\,\beta\,q\,x^4}{k\,\nu^2} = \frac{9.81}{273+110} \times \frac{2000 \times 0.8^4}{33.38 \times 10^{-3}\,(25.45 \times 10^{-6})^2}$$

$$= \mathbf{9.71 \times 10^{11},\ turbulent}$$

The suitable correlation is equation (10.23)

$$\therefore \quad Nu = 0.17[Gr^*\ Pr]^{0.25} = 153.56 \quad \therefore \quad \mathbf{h = 6.40\ W/m^2\ K}$$

This is lower than the assumed value: using this value ΔT = 314°C, $\therefore$ T_w = 334°C

$$\therefore \quad T_f = (334 + 20)/2 = 177°C.$$

Taking property values at 180°C

$$\rho = 0.779,\ \nu = 32.49 \times 10^{-6}\ m^2/s,\ Pr = 0.681,\ k = 37.8 \times 10^{-3}\ W/m\ K$$

$$\mathbf{Gr^*} = \frac{9.81}{273+180} \times \frac{2000 \times 0.8^4}{37.8 \times 10^{-3}\,(32.49 \times 10^{-6})^2} = \mathbf{4.45 \times 10^{11}}$$

$$\therefore \quad Nu = 0.17\ [4.45 \times 10^{11} \times 0.681]^{0.25} = 126.1$$

$$\therefore \quad \mathbf{h = 5.96\ W/m^2\ K}$$

Another iteration is now made

$$\Delta T = 336°C \quad \therefore \quad T_w = 356°C$$

$$\therefore \quad T_f = (356 + 20)/2 = 187.83$$

Taking property values at 190°C

$$\rho = 0.763,\ \nu = 33.67 \times 10^{-6}\ m^2/s,\ Pr = 0.68,\ k = 38.56 \times 10^{-3}\ W/m\ K$$

$$Gr^* = \frac{9.81}{273+190} \times \frac{2000 \times 0.8^4}{38.56 \times 10^{-3}\,(33.67 \times 10^{-6})^2} = 3.97 \times 10^{11}$$

$$\therefore \quad Nu = 0.17\ (Gr^*\ Pr)^{0.25} = 122.54$$

$$\therefore \quad \mathbf{h = 5.91\ W/m^2\ K}$$

$$\therefore \quad T_w = (2000/5.91) + 20 = 358.6°C$$

In case of inclined surface,

$$Nu = 0.17\ (Gr\ Pr \times \cos\theta)^{0.25} = 118.21 \quad \therefore \quad \mathbf{h = 5.7\ W/m^2\ K}$$

$$\therefore \quad T_w = 371°C.$$

Example 10.19: *A flat heater of circular shape of 0.2 m dia with a heat generation rate of 1.2 kW/m² is kept in still air at 20°C with the heated surface facing downward and the plate inclined at 15° to the horizontal. Determine the value of convection coefficient.*

Solution: The plate surface temperature is not specified.

Assuming $h = 5$ W/m² K, $\Delta T = 240°C$, $T_w = 260°C$

Taking properties at $T_e = T_w - 0.25\,(T_w - T_\infty) = 200°C$

$$\rho = 0.746,\ \nu = 34.85 \times 10^{-6}\ \text{m}^2/\text{s},\ Pr = 0.677,\ k = 39.31 \times 10^{-3}\ \text{W/m K}$$

$$L = 0.9 \times 0.2 = 0.18\ \text{m},\ \theta = 75°$$

$$\mathbf{Gr} = \frac{9.81}{273+20} \times \frac{240 \times 0.18^3}{(34.85 \times 10^{-6})^2} = \mathbf{3.86 \times 10^7}$$

$$Gr\,Pr \cos\theta = 6.76 \times 10^6$$

Equation (10.31) is used

$$Nu = 0.56\,(Gr\,Pr\cos\theta)^{0.25} = 28.56 \quad \therefore \quad \mathbf{h = 6.24\ W/m^2\,K}$$

Another trial may be made using this value

$$\Delta T = 192°C \quad \therefore \quad T_w = 212°C,\ T_e = 212 - 0.25\,(212 - 20) = 164°C$$

Property values are now taken at 160°C

$$\rho = 0.815,\ \nu = 30.09 \times 10^{-6}\ \text{m}^2/\text{s},\quad Pr = 0.682,\quad k = 36.4 \times 10^{-3}\ \text{W/m K}$$

$$Gr\,Pr\cos\theta = \frac{9.81}{273+20} \times \frac{192 \times 0.18^3}{(30.09 \times 10^{-6})^2} \times 0.682 \times \cos 75 = 7.30 \times 10^6$$

$$\therefore \quad Nu = 29.12,\ h = \mathbf{5.89\ W/m^2\,K}$$

The temperature difference based on this value is 204°C

$$\therefore \quad T_w = 224°C,\ T_e = 224 - 0.25\,(224 - 20) = 173°C$$

Taking property values at 175°C,

$$\rho = 0.788,\quad \nu = 31.89 \times 10^{-6}\ \text{m}^2/\text{s},\ Pr = 0.681,\quad k = 37.45 \times 10^{-3}\ \text{W/m K}$$

Using these values, $Nu = 28.78$, $h = 5.99$ W/m² K

$$\Delta T = 200°C \quad \therefore \quad \mathbf{T_w = 220°C}$$

(compare with Example 10.16 and discuss the results).

Chapter 10

10.8 HORIZONTAL CYLINDERS

The general form of the equation $\overline{Nu} = cRe^n$ can be used with the values of c and n given in Table 10.1 can be used for smooth horizontal cylinders.

Another equation for the complete range of $Gr\,Pr$ is given in equation (10.34).

$$Nu^{0.5} = 0.60 + 0.387 \left[\frac{(Gr\,Pr)}{[1 + (0.559/Pr)^{9/16}]^{16/9}}\right]^{1/6} \qquad ...(10.34)$$

valid in the range $\quad 10^{-5} < Gr\,Pr < 10^{12}$

An alternate form of the equation (10.34), which gives the same value for h is given below.

$$\overline{Nu}_D = \left[0.60 + \frac{0.387 \times (Gr\,Pr)^{1/6}}{[1 + (0.559/Pr)^{9/16}]^{8/27}}\right]^2 \qquad \text{...[10.34(a)]}$$

This equation gives $\overline{Nu}_D$ directly.

for the laminar range, $Gr\,Pr < 10^9$ the correlation is

$$Nu = 0.36 + \frac{0.518\,(Gr\,Pr)^{0.25}}{[1 + (0.559/Pr)^{9/16}]^{4/9}} \qquad \text{...(10.35)}$$

The correlation for **horizontal cylinders** in **liquid metals** is given by

$$Nu = 0.53\,(Gr\,Pr^2)^{1/4} \qquad \text{...(10.36)}$$

Inclined cylinders: The expression in the case of inclined cylinders is rather long and may be looked up in data books. An example on this topic is given under advanced examples.

Example 10.20: *Water in a tank is heated by a horizontal steam pipe of 0.25 m dia, maintained at 60°C. The water is at 20°C. Calculate the value of convective heat transfer coefficient by different correlations and compare the results.*

Solution: The film temperature is (60 + 20)/2 = 40°C

The property values are:

$\rho = 995$, $\nu = 0.657 \times 10^{-6}$ m²/s, $Pr = 4.34$, $k = 0.628$ W/m² K

$\beta = 0.41 \times 10^{-3}$/K

$$\text{Gr} = \frac{9.81 \times 0.41 \times 10^{-3} \times 0.25^3 \times 40}{(0.657 \times 10^{-6})^2} = \mathbf{5.82 \times 10^9}$$

$$Gr\,Pr = 2.53 \times 10^{10}$$

The appropriate correlation is

$$Nu^{0.5} = 0.60 + 0.387\left[\frac{(Gr\,Pr)}{[1 + (0.559/Pr)^{9/16}]^{16/9}}\right]^{1/6}$$

$\therefore \quad Nu = 397.12 \quad \therefore \quad \mathbf{h = 997.6\ W/m^2\,K}$

Using [10.34(a)]

$$\overline{Nu}_D = \left[0.60 + \frac{0.387 \times (5.84 \times 10^9 \times 4.34)^{1/6}}{[1 + (0.559/4.34)^{9/16}]^{8/27}}\right]^2 = 397.5$$

The difference is negligible.

The other correlation is (Table 10.1)

$Nu = 0.125\,(Gr\,Pr)^{1/3} = 366.83 \quad \therefore \quad \mathbf{h = 921.5\ W/m^2\,K}$

Another possibility is

$Nu = 0.13\,(Gr\,Pr)^{1/3} = 381.51 \quad \therefore \quad \mathbf{h = 958.4\ W/m^2\,K}$

All estimates are close.

10.9 OTHER GEOMETRIES

For spheres: The general correlation is

$$Nu = 2 + 0.43\,(Gr\,Pr)^{0.25} \qquad \text{...(10.37)}$$

Properties to be evaluated at film temperature

For higher values ($3 \times 10^5 < Gr\,Pr < 8 \times 10^{10}$) 0.43 is replaced by 0.50

A more recent (1983) correlation for spheres is

$$\overline{Nu}_D = 2 + \frac{0.589\,(Gr\,Pr)^{0.25}}{[1 + (0.464/Pr)^{9/16}]^{4/9}} \quad \text{...[10.37(a)]}$$

$$Pr \ge 0.7 \text{ and } Gr\,Pr \le 10^{11}$$

Irregular Solids:

Short cylinder $D = H$ $\quad Nu = 0.775\,(Gr\,Pr)^{0.208}$ ...(10.38)

For other solids $\quad Nu = 0.52\,(Gr\,Pr)^{0.25}$ ...(10.39)

where the characteristic length is the distance travelled by a particle in the boundary layer.

For rectangular solids, the characteristic length L is calculated using

$$\frac{1}{L} = \frac{1}{L_H} + \frac{1}{L_V} \quad \text{...(10.40)}$$

where L_H = Average horizontal length, and L_V is the vertical length. Based on this length

$$Nu = 0.55\,(Gr\,Pr)^{0.25} \quad \text{...(10.41)}$$

Equations (10.37) and (10.41) should give similar estimates.

Example 10.21: *A spherical heater of dia 0.2 m dia with surface at 60°C is used to heat water at 20°C in a tank. Determine the value of convective heat transfer coefficient.*

This is a case of free convection.

Solution: The film temperature is 40°C. The property values are

$\rho = 995$, $\nu = 0.657 \times 10^{-6}$ m²/s, $Pr = 4.34$, $k = 0.628$ W/m K

$\beta = 0.41 \times 10^{-3}$/K

The correlation is $Nu = 2 + 0.43\,(Gr\,Pr)^{0.25}$

$$\text{Gr} = \frac{9.81 \times 0.41 \times 10^{-3} \times 0.2^3 \times 40}{(0.657 \times 10^{-6})^2} = \mathbf{2.98 \times 10^9}$$

$$Nu = 2 + 0.43(2.98 \times 10^9 \times 4.34)^{0.25} = 147.03$$

$\therefore$ **h = 461.7 W/m² K.**

Example 10.22: *A small cylindrical steam heater of diameter 0.2 m and height 0.2 m maintained at 60°C is immersed in a tank of water at 20°C. Determine the value of convection coefficient.*

This is an example in free convection. The film temperature is 40°C. The property values are as in Example 10.21.

Solution: An available correlation is given by equation (10.38).

$$Nu = 0.775\,(Gr\,Pr)^{0.208}$$

$$\text{Gr} = \frac{9.81 \times 0.41 \times 10^{-3} \times 0.2^3 \times 40}{(0.657 \times 10^{-6})^2} = \mathbf{2.98 \times 10^9}$$

$Nu = 0.775\,(2.98 \times 10^9 \times 4.34)^{0.208} = 98.31$ $\quad\therefore$ **h = 308.7 W/m² K.**

Another possible method is to use the length as the length of travel in boundary layer = $D + h = 2D$ in this case = 0.4 m

$Nu = 0.52\,(Gr\,Pr)^{0.25}$, here Gr is based on length $2D$

$= [0.52]\ [\{9.81 \times 0.41 \times 10^{-3} \times 40 \times (0.4)^3/(0.657 \times 10^{-6})^2\}\ 4.34]^{0.25}$

$= 294.96$, $\mathbf{h = 463.1\ W/m^2\ K}$

Another way in which the length can be found is

$$\frac{1}{L} = \frac{1}{L_H} + \frac{1}{L_V} = \frac{1}{D} + \frac{1}{D} = \frac{1}{0.2} + \frac{1}{0.2} = 10$$

∴ $L = 0.1$ m. Using this value of L

$$\mathbf{Gr} = \frac{9.81 \times 0.41 \times 10^{-3} \times 0.1^3 \times 40}{(0.657 \times 10^{-6})^2} = \mathbf{0.373 \times 10^9}$$

Using equation (10.41),

$$Nu = 0.55\ (Gr\ Pr)^{0.25} = 110.32 \quad \therefore \quad \mathbf{h = 692.8\ W/m^2\ K}$$

There is a wider spread. However, there is no other way by which such estimates can be made.

10.10 SIMPLIFIED EXPRESSIONS FOR AIR

Using average property values, simplified (but less accurate) correlations for air is given in Table 10.2. These are applicable for 1 atm pressure and moderate temperature differences.

Table 10.2. Simplified expressions for convective heat transfer coefficient for free convection in air

Constant wall temperature $\Delta T = T_w - T_\infty$

Geometry	*Laminar* $10^4 < Gr\ Pr < 10^9$	*Turbulent* $Gr\ Pr > 10^9$	*Eqn. No*
1. Vertical plane or cylinder	$h = 1.42\ (\Delta T/L)^{1/4}$	$h = 1.31\ (\Delta T)^{1/3}$	(10.42) (*a, b*)
2. Horizontal cylinder	$h = 1.32\ (\Delta T/D)^{1/4}$	$h = 1.24\ (\Delta T)^{1/3}$	(10.43) (*a, b*)
3. Horizontal plate heated face up or cooled face down	$h = 1.32\ (\Delta T/L)^{1/4}$	$h = 1.52\ (\Delta T)^{1/3}$	(10.44) (*a, b*)
4. Heated face down or cooled face up	$h = 0.59\ (\Delta T/L)^{1/4}$		(10.45)

For pressures other than 1 atm, multiply the RHS of these expressions as below, where *p* is in bar.

Laminar $(P/1.0132)^{0.25}$ ...(10.46)

Turbulent $(P/1.0132)^{2/3}$ (*a, b*)

Example 10.23: *(Read Example 10.3) A vertical plate is maintained at 40°C in stagnant air at 20°C. The flow turns turbulent at a height of 0.8262 m. It is found that h = 3.15 W/m² K (Example 10.5). In Example 10.6, the average value of h at 0.2, 0.4, 0.6, 0.8 are found as 4.49, 3.77, 3.41 and 3.17 W/m² K.*

Solution: Compare the values that may be obtained using simplified expressions equation (10.42)

The equation (10.42) is

$$h = 1.42 \left(\frac{\Delta T}{L}\right)^{0.25}, \quad \Delta T = 20°C$$

Distance: m	*0.2*	*0.4*	*0.6*	*0.8*	*0.8262*
$\bar{h}$ by classical method W/m² K	**4.49**	**3.77**	**3.41**	**3.17**	**3.15**
Simplified expression	**4.49**	**3.78**	**3.41**	**3.18**	**3.15**

It is very interesting that there is very close agreement.

Example 10.24: *The following results using accurate correlations for free convection in air on a vertical surface are available.*

(*a*) *L = 0.8 m* $\quad \Delta T = 10°C \quad$ **h = 2.88 W/m² K** $\quad Gr = 0.65 \times 10^9$

(*b*) *L = 4 m* $\quad \Delta T = 60°C \quad$ **h = 5.76 W/m² K** $\quad Gr = 4.86 \times 10^{11}$

Check using approximate relations.

Solution: (*a*) $h = 1.42 (\Delta T/L)^{1/4} = 1.42 (10/0.8)^{1/4} =$ **2.67 W/m² K** (2.88)

(*b*) $h = 1.31 (\Delta T)^{1/3} = 1.31 (60)^{1/3} =$ **5.13 W/m² K** (5.76)

It can be seen that the **approximate relations provide very good estimates in the laminar region** and good estimates in the turbulent region.

Example 10.25: *A steam pipe with an outer diameter of 0.4 m at 110°C passes through still air at 30°C. Determine the value of convection coefficient by the accurate and approximate methods. film temperature = (110 + 30)/2 = 70°C.*

Solution: Property values are:

$$\rho = 1.029,\ \nu = 20.02 \times 10^{-6}\ m^2/s,\ Pr = 0.694,\ k = 29.66 \times 10^{-3}\ W/m\ K$$

Using horizontal cylinder correlation

$$\mathbf{Gr} = \frac{9.81 \times 80 \times 0.4^3}{(273 + 70)(20.02 \times 10^{-6})^2} = \mathbf{3.65 \times 10^8}$$

$$\mathbf{Gr\ Pr} = 2.54 \times 10^8$$

$$Nu = 0.36 + \frac{0.518\,(Gr\ Pr)^{0.25}}{[1 + (0.559/Pr)^{9/16}]^{4/9}} = 49.67$$

$\therefore$ **h = 3.68 W/m² K**

Approximate method:

$$h = 1.32\,(\Delta T/D)^{0.25} = \mathbf{4.97\ W/m^2\ K}$$

Using equation of Table 10.1

$$Nu = 0.125\,(Gr\ Pr)^{0.33} \quad \text{or} \quad 0.53\,(Gr\ Pr)^{0.25}$$

$$= 79.12 \qquad Nu = 66.88$$

h = 5.87 W/m² K $\qquad$ **h = 4.96 W/m² K**

Another correlation is

$$Nu^{0.5} = 0.60 + 0.387 \left[\frac{(Gr\ Pr)}{[1 + (0.559/Pr)^{9/16}]^{16/9}}\right]^{1/6}$$

Nu = 75.14 h = **5.57 W/m² K.**

Chapter 10

This shows that the approximate method is not too far off from other methods and actually coincides with one of the correlations and may be used with confidence for first estimates.

10.11 FREE CONVECTION IN ENCLOSED SPACES

Double glazed windows used in buildings to reduce heat loss is an example for this situation. This is shown in Fig. 10.6. Inclined spaces are encountered in solar collectors. In these cases, the fluid receives and transfers the heat between the surface by free convection with the average temperature of the fluid remaining the same. The heat transfer will be higher than in pure conduction.

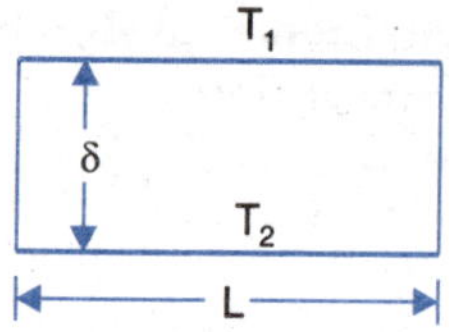

The space (rather short) between surface 1 and 2 is stagnant.

$$Q = h\,A\,(T_1 - T_2)$$

is to be used to calculate the heat flow, h is evaluated as described below. In this case, the Grashof number is calculated as

Fig. 10.6. Model of enclosed space

$$Gr = \frac{g\beta(T_1 - T_2)\delta^3}{v^2} \quad \text{...(10.47)}$$

$$Nu = \frac{h\delta}{k}, \ \delta = \text{Distance between surfaces} \quad \text{...(10.48)}$$

one of the correlations available is:

$$Nu = 0.42\,(Gr\,Pr)^{0.25}\,Pr^{0.012}\left(\frac{L}{\delta}\right)^{-0.30} \quad \text{...(10.49)}$$

valid for $10^4 < Gr\,Pr < 10^7$, q_w = Constant

$1 < Pr < 20,000 \quad 10 < L/\delta < 40$

$$Nu = 0.046\,(Gr\,Pr)^{1/3} \quad \text{...(10.50)}$$

$10^6 < Gr\,Pr < 10^9 \quad 1 < Pr < 20, \quad 1 < L/\delta < 40.$

Example 10.26: *A double glazed window 1 m × 1 m used in an airconditioned space has the two plates at a distance of 1 cm. One plate is at 10°C while the other is at 30° C. Determine the convection coefficient between the plates.*

Solution: This can be considered as enclosed space:

$T_f = 20°C$, $\nu = 15.06 \times 10^{-6}$ m²/s,

$Pr = 0.703$, $k = 25.93 \times 10^{-3}$ W/m K

Using equations (10.47), (10.48), (10.49)

$$Gr = \frac{9.81 \times 20 \times 0.01^3}{293 \times (15.06 \times 10^{-6})^2} = 2952.4$$

$$Gr\,Pr = 2075.57$$

Using equation (10.49), a possible fit

$$Nu = 0.42\,(Gr\,Pr)^{0.25}\,Pr^{0.012}\,(L/\delta)^{-0.30} = 0.709$$

$$\mathbf{h = 1.84\ W/m^2\,K}$$

Comparing conduction and convection resistances: (1m²)

Convection resistance $= 1/h = 0.544$ K/W

Conduction: $l/k = \dfrac{0.01}{25.93 \times 10^{-3}} = 0.386$ K/W

Conduction heat flow is higher.

For closed vertical or horizontal cylinders (length as linear dimension)

$$Nu = 0.55\,(Gr\,Pr)^{0.25} \quad \text{...(10.51)}$$

The free convection in the **annular space of cylinders and spheres** are discussed under advanced examples.

For a **horizontal cavity heated from below:**

$$Nu = 1 + 1.44\left[1 - \frac{1708}{Gr\,Pr}\right] + \left[\left(\frac{Gr\,Pr}{5830}\right)^{1/3} - 1\right] \quad \text{...(10.52)}$$

In case the quantities in square brackets are – ve these should be taken as zero.

10.12 ROTATING CYLINDERS, DISKS AND SPHERES

A flow is caused due to rotation of axisymmetric objects like motor rotors etc. The flow is neither due to density difference nor due to devices like pumps, but due to the centripetal action.

A new Reynolds number called **peripheral speed Reynolds number (Re_ω)** is defined and used in the analysis as this quantity is found to influence the setting in of turbulence (ω–rotational speed. radians/s)

$$\mathbf{Re_\omega = \omega\,\pi\,D^2/\nu} \quad \text{...(10.53)}$$

For cylinders $$Nu = \frac{h_c D}{k} = 0.11\,(0.5\,Re_\omega^2 + Gr_D\,Pr)^{0.35} \quad \text{...(10.54)}$$

For rotating disk $$Nu = 0.35\left(\frac{\omega D^2}{4\nu}\right)^{0.5} \quad \text{...(10.55)}$$

for $\omega D^2/\nu < 5 \times 10^5$

For turbulent condition, the local value of Nusselt number at radius is (d is the diameter)

$$Nu_r = \frac{hr}{k} = 0.0195\left(\frac{\omega d^2}{4\nu}\right)^{0.8} \quad \text{...(10.56)}$$

and the average value is (D_0–outside dia)

$$Nu_r = 0.015\left(\frac{\omega D_0^2}{4\nu}\right)^{0.8} - 100\left(\frac{D_c}{D_0}\right)^2 = \frac{hr}{k} \quad \text{...(10.57)}$$

when D_c corresponds to $\omega D^2/\nu = 5 \times 10^5$

For spheres, in the laminar region

$$Nu_D = 0.43\,Re_\omega^{0.5}\,Pr^{0.4} \quad \text{...(10.58)}$$

for $$Re_\omega = \frac{\omega D^2}{\nu} < 5 \times 10^4$$

$$Nu = 0.066\ Re_{\omega}^{0.67}\ Pr^{0.4} \qquad ...(10.59)$$

for $\quad 5 \times 10^4 < Re_{\omega} = \omega\ D^2/\nu < 7 \times 10^5$

These relations are useful in calculating the heat loss in rotating machine elements.

Example 10.27: *A hot metal cylinder of dia 0.4 m is machined at a speed of 250 rpm. The cylinder surface is at 60°C and air is at 20°C. Determine the value of convection coefficient over the surface.*

Solution: The film temperature = 40°C

The property values are:

$$\rho = 1.128,\ \nu = 16.96 \times 10^{-6}\ \text{m}^2/\text{s},\ Pr = 0.699,$$
$$k = 27.56 \times 10^{-3}\ \text{W/m K}$$

Using equation (10.53),

$$\mathbf{Re}_{\omega} = \omega\ \pi\ D^2/\nu = \frac{2\pi \times 250}{60} \times \frac{\pi \times 0.4^2}{16.96 \times 10^{-6}} = \mathbf{775912}$$

$$Gr\ Pr = \frac{9.81 \times 40}{(273 + 40)} \times \frac{0.4^3}{(16.96 \times 10^{-6})^2}\ .\ 0.699 = \mathbf{1.95 \times 10^8}$$

Using equation (10.54)

$$Nu = 0.11\ [0.5\ Re^2 + Gr\ Pr]^{0.35} = 1145.5,$$
$$\mathbf{h = 78.93\ W/m^2\ K.}$$

Example 10.28: *A thin disk of 0.3 m dia is being ground on the face and rotates at 3000 rpm. The disk surface is at 60°C while the air is at 20°C. Determine the value of convection coefficient. Taking the property values from Example 10.27.*

Solution:
$$\frac{\omega D^2}{\nu} = \frac{2\pi \times 3000}{60} \times \frac{0.3 \times 0.3}{16.96 \times 10^{-6}} = 1.668 \times 10^6 > 5 \times 10^5$$

Using equation (10.57)

$$Nu_r = 0.015 \left(\frac{\omega D_0^2}{4\nu}\right)^{0.8} - 100 \left(\frac{D_c}{D_0}\right)^2$$

To find D_c $\quad \dfrac{2\pi \times 3000}{60} \times \dfrac{D_c^2}{16.96 \times 10^{-6}} = 5 \times 10^5$

$\therefore \qquad D_c = 0.1643$ m

$$\therefore \qquad Nu_r = 0.015\ (1.668 \times 10^6)^{0.8} - 100 \left(\frac{0.1643}{0.3}\right)^2 = 1394.5$$

$$\mathbf{h = 256.2\ W/m^2\ K.}$$

10.13 COMBINED FORCED AND FREE CONVECTION

When flow velocities are low, natural convection will also contribute in addition to forced convection. This can be checked using the following:

If $Gr/Re^2 >> 1$ free convection prevails

If $Gr/Re^2 << 1$ forced convection prevails

If $Gr/Re^2 \approx 1$ both should be considered

In such cases, for flow in pipes in the **laminar region**

$$Nu = 1.75 \left[\frac{\mu}{\mu_m}\right]^{0.14} [G_z + 0.012\,(G_z\, Gr^{0.333})^{1.333}]^{0.333} \qquad ...(10.60)$$

where $$\mathbf{G_z = Re\ Pr\ .\ \frac{D}{L}}$$

For turbulent conditions

$$Nu = 4.69\, Re^{0.27}\, Pr^{0.21}\, Gr^{0.07}\, (D/L)^{0.36} \qquad ...(10.61)$$

$$Re > 2000,\ Gr\ Pr\ \frac{D}{L} < 5000$$

or $$Re > 800,\ Gr\ Pr\ \frac{D}{L} > 2 \times 10^4$$

The combined $\overline{Nu}_C$ is calculated by

$$\overline{Nu}_c = \overline{Nu}_n \pm \overline{Nu}_f \qquad ...(10.62)$$

where $\overline{Nu}_c$ is the combined value, $\overline{Nu}_n$ is the value calculated as per applicable equation for natural convection and $\overline{Nu}_f$ is the value as per forced convection. Plus sign is used if forced convection assists natural convection and minus sign when forced convection flow is opposed to the flow in natural convection.

Natural or free convection has extensive applications as equipments based on this phenomenon are simple and will operate without breakdown over long periods of time.

It is to be kept in mind that estimates may be ± 25% removed from actual values and the design dimensions should be arrived at based on the performance assurance required.

Example 10.29: *A vertical plate 2 m high is maintained at 60°C in air at 20°C. What is the limit of flow velocity for the heat flow to be considered as (i) free convection (ii) both free and forced convection and (iii) forced convection?*

Solution: T_f = 40°C. The property values are: $\nu = 16.96 \times 10^{-6}$ m²/s

$$\mathbf{Gr} = \frac{9.81 \times 40}{313} \times \frac{2^3}{(16.96 \times 10^{-6})^2} = \mathbf{3.486 \times 10^{10}}$$

$$\mathbf{Re^2} = \left(\frac{u_\infty . L}{\nu}\right)^2 = \frac{u_\infty^2\, .\, 2^2}{(16.96 \times 10^{-6})^2} = \mathbf{1.39 \times 10^{10}\ u_\infty^{\ 2}}$$

The conditions are: (*i*) $Gr/Re^2 >> 1$ (*ii*) $Gr/Re^2 \approx 1$ (*iii*) $Gr/Re^2 << 1$

(*i*) Say $$Gr/Re^2 = 10 = \frac{3.486 \times 10^{10}}{1.39 \times 10^{10} \times u_\infty^2} \qquad \therefore\ \mathbf{u_\infty = 0.5\ m/s}$$

(*ii*) $$Gr/Re^2 = 1 = \frac{3.486 \times 10^{10}}{1.39 \times 10^{10} \times u_\infty^2} \qquad \therefore\ \mathbf{u_\infty = 1.58\ m/s}$$

(*iii*) $$Gr/Re^2 = 0.05 = \frac{3.486 \times 10^{10}}{1.39 \times 10^{10} \times u_\infty^2} \qquad \therefore\ \mathbf{u_\infty = 7.08\ m/s}$$

(0.05 is chosen as a small value limit and 10 as high limit.)

ADVANCED EXAMPLES

Example A-10.1: *Derive an expression for the velocity distribution in the boundary layer in free convection that will satisfy the boundary conditions. Determine from the same the y-location at which the velocity is maximum and the value of the maximum velocity.*

Solution: The Boundary conditions are

(*i*) $u = 0$ at $y = 0$, (*ii*) $u = 0$ at $y = \delta$, 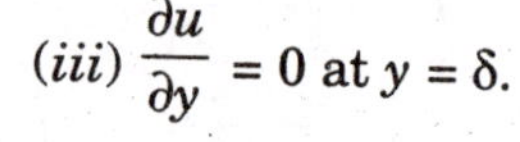 (*iii*) $\frac{\partial u}{\partial y} = 0$ at $y = \delta$.

From the momentum equation, at a given location,

$$u\frac{\partial u}{\partial x} + v\frac{\partial u}{\partial y} = g\,\beta(T - T_\infty) + \nu\frac{\partial^2 u}{\partial y^2}$$

The LHS is zero as $u = 0$ at $y = 0$ and $v = 0$ at $y = 0$

(*iv*) $\therefore$ $$\frac{\partial^2 u}{\partial y^2} = -\frac{g\beta\,(T_w - T_\infty)}{\nu} \text{ at } y = 0$$

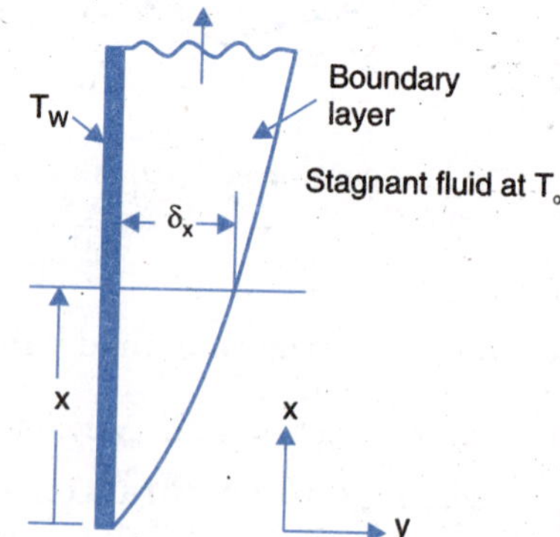

Fig. AE-10.1

A cubic distribution can be used as four conditions are available. Also taking a reference velocity u_x which can be eliminated later, it may be assumed that

$$\frac{u}{u_x} = a + b\,.\,y + c\,.\,y^2 + d\,.\,y^3$$

where *a*, *b*, *c* and *d* have to be determined from the known conditions.

condition (*i*) leads to $a = 0$, using the second condition, $y = \delta$

$$\frac{u}{u_x} = b\,\delta + c\,\delta^2 + d\,\delta^3 = 0$$

$$\therefore \quad b = -(c\,\delta + d\,\delta^2) \quad \text{...(1)}$$

Substituting for *b* and cancelling u_x

$$-c\,\delta - d\,\delta^2 + 2c\,\delta + 3\,\delta^2 = 0$$

$$c\,\delta + 2\,d\,\delta^2 = 0$$

$$\therefore \quad d = -c/2\delta \quad \text{...(2)}$$

Using the other boundary condition, substituting the result in (*iv*)

$$\frac{\partial^2 u}{\partial y^2} = u_x\,(2c + 6d\,y),\text{ at } y = 0,\ 2c\,u_x = -\frac{g\beta\,(T_w - T_\infty)}{\nu} \quad \text{...(3)}$$

$$\therefore \quad c = -g\,\beta\,(T_w - T_\infty)/2\nu\,u_x$$

substituting this in equation (2)

$$d = -c/2\delta = g\,\beta(T_w - T_\infty)/4\nu\,u_x\delta \quad \text{...(4)}$$

substituting for *c* and *d* in equation (1)

$$b = -(c\delta + d\delta^2) = \frac{g\beta(T_w - T_\infty)\delta}{2\nu u_x} - \frac{g\beta(T_w - T_\infty)\delta^2}{4\nu u_x \cdot \delta} = \frac{g\beta(T_w - T_\infty)\delta}{4\upsilon u_x}$$

Substituting for a, b, c and d in the assumed profile

$$\frac{u}{u_x} = 0 + \frac{g\beta(T_w - T_\infty)\delta}{4\nu u_x} \cdot y - \frac{g\beta(T_w - T_\infty)}{2\nu u_x} \cdot y^2 + \frac{g\beta(T_w - T_\infty)}{4\nu u_x \delta} \cdot y^3$$

$$= \frac{g\beta(T_w - T_\infty)}{4\nu u_x} [\delta \cdot y - 2y^2 + y^3/\delta] \quad \text{...(5)}$$

$$= \frac{g\beta(T_w - T_\infty)\delta y}{4\nu u_x}\left[1 - 2\frac{y}{\delta} + \frac{y^2}{\delta^2}\right] = \frac{g\beta(T_w - T_\infty)y}{4\nu u_x} \cdot \delta\left[1 - \frac{y}{\delta}\right]^2$$

or

$$u = \frac{g\beta(T_w - T_\infty)\delta^2}{4\nu} \cdot \frac{y}{\delta}\left[1 - \frac{y}{\delta}\right]^2 \quad \text{...(6)}$$

The dimensions can be checked on the RHS to be m/s.

This is the result of an assumed cubic profile satisfying the boundary conditions.

As u varies along y at any given x location from zero at $y = 0$ to zero again at $y = \delta$, a maximum value has to occur at some y location. (Ref. Fig. AE-10.1)

To determine the y location at which u is u_{max} [refer equation (5)]

$$\frac{\partial u}{\partial y} = 0$$

$\therefore$ Constant $[\delta - 4y + 3y^2/\delta] = 0$

$$\therefore \quad \delta^2 - 4\delta y + 3y^2 = 0,\ y = \frac{4\delta \pm \sqrt{(16\delta^2 - 12\delta^2)}}{6}$$

solving for y, $y = (1/3)\,\delta$ or δ, the second solution is trivial.

$\therefore$ **u_{max} occurs at y = (1/3) δ**

Substituting in [equation (5)]

$$u_{max} = \frac{g\beta(T_w - T_\infty)}{4\nu} \cdot \delta^2 \frac{\delta}{3\delta}\left[1 - \frac{1}{3}\right]^2,\ u_{max} = \frac{g\beta(T_w - T_\infty)\delta^2}{27\,\nu} \quad \text{...(7)}$$

and this expression together with equations (6) and (7) can be used to determine u and u_{max} in the laminar region.

Example A-10.2: *A plate maintained at 60°C is kept vertical in still air at 20°C. Determine the velocity profile at x = 0.4 m. Also determine the value of u_{max} at 0.2, 0.4 and 0.6 m heights.*

Solution: The film temperature is 40°C. The property values are:

$$\rho = 1.128,\ \nu = 16.96 \times 10^{-6}\ \text{m}^2/\text{s},\ Pr = 0.699,$$

$$k = 27.56 \times 10^{-3}\ \text{W/m K},\ \beta = 1/313$$

Chapter 10

The value of δ_x is involved in all the calculations.

$$\mathbf{Gr} = \frac{9.81 \times 40}{313} \cdot \frac{x^3}{(16.96 \times 10^{-6})^2} = \mathbf{4.36 \times 10^9 \cdot x^3}$$

$$\delta_x = 3.93\, Pr^{-0.5}\, (0.952 + Pr)^{0.25}\, x \,.\, Gr_x^{-0.25}$$

$$= 5.328x\, Gr^{-0.25} = \mathbf{5.328 \times Gr^{-0.25}}$$

$$\mathbf{u_{max}} = \frac{g\beta\,(T_w - T_\infty)\,\delta^2}{27\,\nu} = \mathbf{2737.7\, \delta^2}$$

All cases are laminar

x, m	*Gr*	δ_x *m*	u_{max}, *m/s*
0.2	3.487×10^7	0.013868	0.527
0.4	2.789×10^8	0.016492	0.745
0.6	$\mathbf{9.414 \times 10^8}$	**0.018252**	**0.912**

At the location 0.4 m $\delta = 0.016492$ m or 16.5 mm

$$u = \frac{g\beta\,(T_w - T_\infty)\,\delta^2}{4\,\nu}\, \frac{y}{\delta}\left(1 - \frac{y}{\delta}\right)^2 = 5.026\, \frac{y}{\delta}\left(1 - \frac{y}{\delta}\right)^2$$

Taking

y/δ	0.1	0.2	0.333	0.4	0.5	0.6	0.7	0.8	0.9
u, m/s	0.41	0.64	0.745	0.724	0.628	0.483	0.317	0.161	0.045

Note that the maximum value of $u = 0.745$ *i.e.* u_{max} at $y/\delta = 1/3$

A plot of these values will be similar to the profile in Fig. 10.1

Example A-10.3: *Using the situation in example 10.2 and values at x = 0.6 m.*

(*i*) *Calculate the value of average free convection heat transfer coefficient for x = 0.6 m and*

(*ii*) *Compare this value with that of forced convection with* $u_\infty = 0.912$ *m/s.*

(*iii*) *For free convection*

Solution:

$$Nu_x = \frac{0.508\, Pr^{0.5}\, Gr^{0.25}}{(0.952 + Pr)^{0.25}}$$

$\mathbf{Gr = 9.414 \times 10^3}$, $Pr = 0.699$ $k = 27.56 \times 10^{-3}$ W/m K,

$\nu = 16.96 \times 10^6$ m²/s

$Nu_x = 65.631$, $h_x = 3.02$ W/m² K,

$$\mathbf{\bar{h} = 4.03\ W/m^2\,K}, \left(i.e. \frac{4}{3} h_L\right)$$

Under forced convection $Re = \dfrac{u_\infty L}{\nu} = \dfrac{0.912 \times 0.6}{16.96 \times 10^{-6}} = 32264$

$$Nu = 0.644\, Re^{0.5}\, Pr^{1/3} = 102.66,$$

$\therefore$ $\mathbf{h = 4.716\ W/m^2\ K.}$

An increase of only 17% at these conditions of low velocity.

Example A-10.4: *Banks of plates are used to dissipate the heat from a transformer. The plates are 0.6 m high and 0.18 m wide. The plate surface is at 80°C and the air is at 40°C. Determine the distance between the plates so that the boundary layers do not interfere. Also calculate the number of plates required to dissipate 2 kW.*

Solution: This is a case of free convection. For the boundary layers not to interfere the distance between the plates should be greater than twice the boundary layer thickness.

The film temperature is (40 + 80)/2 = 60°C

The property values are:

$$\rho = 1.06,\ \nu = 18.97 \times 10^{-6}\ \text{m}^2/\text{s},$$

$$Pr = 0.696,\ k = 28.96 \times 10^{-3}\ \text{W/m K}$$

$$\text{Gr} = \frac{9.81 \times 1 \times 40 \times 0.6^3}{333 \times (18.97 \times 10^{-6})^2} = \mathbf{0.707 \times 10^9}, \quad \therefore \quad \textbf{laminar}$$

$$\delta_x = 3.93 \times Pr^{-0.5}\,(0.952 + Pr)^{0.25}\, Gr_x^{-0.25} = \mathbf{0.01794\ m}$$

$\therefore$ **Distance between plates** = $2 \times \delta_x$ = **39 mm** or say 5 cm for safe operation.

$$Nu_x = \frac{0.508\, Pr^{0.5}\, Gr_x^{0.25}}{(0.952 + Pr)^{0.25}} = 61.00 \quad \therefore \quad h_x = 2.94\ \text{W/m}^2\,\text{K}$$

$\therefore$ $\mathbf{\bar{h} = 3.93\ W/m^2\ K}$

$$A = 2 \times N \times 0.18 \times 0.6,\ Q = 2000$$

$\therefore$ $2000 = 2 \times N \times 0.18 \times 0.6 \times 3.93\,(80 - 40)$

$\therefore$ **N = 59 plates or say 60 plates.**

Example A-10.5: *A horizontal cylinder of 0.4 m dia at a surface temperature of 40°C is placed in air at 80°C. Compare the heat gain by the cylinderical surface with that of a vertical plate of height* $\dfrac{\pi \times 0.4\ m}{2}$ *under the same conditions.*

Solution: This is a situation of free convection. The film temperature is 60°C. The property values of air are

$$\rho = 1.06,\ \nu = 18.97 \times 10^{-6}\ \text{m}^2/\text{s},\ Pr = 0.696,\ k = 28.96 \times 10^{-3}\ \text{W/m K}$$

Cylinder: $Gr = \dfrac{9.81 \times 1 \times 40 \times 0.4^3}{333 \times (18.97 \times 10^{-6})^2} = \mathbf{2.096 \times 10^8}$

$$Gr\, Pr = 1.459 \times 10^8$$

The correlations applicable are: (From Table 10.1)

$$Nu = 0.125\ (Gr\ Pr)^{1/3} = 65.8$$

$$\mathbf{h = 4.76\ W/m\ K}$$

Approximate sol. for air $\bar{h}$:

$$h = 1.32\ (\Delta T/d)^{0.25} = \mathbf{4.17\ W/m^2\ K}$$

Other correlations are also available. However, these provide a fair estimate.

Plane:

$$\mathbf{Gr} = \frac{9.81 \times 1 \times 40 \times (\pi \times 0.4/2)^3}{333 \times (18.97 \times 10^{-6})^2} = \mathbf{8.12 \times 10^8}$$

$$Gr\ Pr = 5.653 \times 10^8$$

The correlations available are: (Table 10.1)

$$Nu = 0.59\ (Gr\ Pr)^{1/4} = 90.98,\ h = \mathbf{4.193\ W/m^2\ K}$$

$$h = 1.42\ (\Delta T/L)^{0.25}$$

$$h = 1.42\ (40/\pi \times 0.2)^{0.25} = \mathbf{4.01\ W/m^2\ K.}$$

The heat transfer by the vertical plate is marginally lower and the cylindrical surface appears to be better.

Example A-10.6: ***A hot plate kept vertical*** *in air develops a boundary layer by free convection. The thickness of the boundary layer at a distance of 0.4 m was 19 mm. (a) Estimate the location where it will be 22.5 mm. (b) If the gas was carbon dioxide with $\nu = 12.6 \times 10^{-6}\ m^2/s$ as compared to air with $\nu = 23.13 \times 10^{-6}\ m^2/s$, determine the boundary layer thickness at 0.4 m. Assume that the Prandtl number is nearly the same and all other conditions remain unchanged.*

Solution: For a given situation, δ_x is proportional to $xGr^{-0.25}$ or $x^{0.25}$ (equation 10.13)

$$\frac{22.5}{19} = \frac{x^{0.25}}{0.4^{0.25}} \qquad \therefore\ \mathbf{x = 0.787\ m}$$

δ_x is proportional to $x.\ Gr^{-0.25}$ as x is the same and other conditions remain unchanged,

δ_x is proportional to $\dfrac{1}{(\nu^2)^{0.25}}$ or $\dfrac{1}{\nu^{0.5}}$

$$\frac{\delta}{19} = \left(\frac{23.13 \times 10^{-6}}{12.6 \times 10^{-6}}\right)^{0.5} \qquad \therefore\ \boldsymbol{\delta = 25.74\ \mathbf{mm}}$$

It is assumed that laminar conditions prevail in both cases.

Example A-10.7: *An athlete lies still on the ground in cool air at 24°C. His body temperature is 36°C. Approximating his body to be a cylinder of 0.3 m dia and 2 m long, determine the heat loss from his body.*

Compare this with the heat loss when he runs a 400 m distance in 55 sec, the other surrounding conditions remaining the same. Neglect end losses.

Solution: Considering free convection over horizontal cylinder. The film temperature is 60/2 = 30°C. The property values are $\rho = 1.165\ kg/m^3$, $\nu = 16 \times 10^{-6}\ m^2/s$, $Pr = 0.701$, $k = 26.75 \times 10^{-3}$ W/m K.

$$\mathbf{Gr} = \frac{9.81}{303} \times \frac{12 \times 0.3^3}{(16 \times 10^{-6})^2} = \mathbf{4.098 \times 10^7}$$

$$Gr\ Pr = 2.872 \times 10^7$$

A suitable correlation is (equation 10.35)

$$\therefore \quad Nu = 0.36 + \frac{0.518\,(Gr\ Pr)^{0.25}}{[1 + (0.559/Pr)^{9/16}]^{4/9}} = 29$$

$$\therefore \quad \mathbf{h = 2.586\ W/m^2\,K}$$

$\therefore$ **Heat loss rate** = $\pi \times 0.3 \times 2 \times 2.586\,(12) =$ **58.5 W**

If the athlete runs, then it is forced convection

$$u_\infty = 400/55 \text{ m/s}$$

$$\therefore \quad Re = \frac{0.3 \times 400}{55 \times 16 \times 10^{-6}} = 1.36 \times 10^5$$

The correlation is (over cylinder) (Refer Chapter 8, Section 8.3)

$$Nu = 0.0266\ Re^{0.805}\ Pr^{1/3} = 321.28$$

$$\therefore \quad \mathbf{h = 28.65\ W/m^2\,K}$$

This is more than 10 times the value for free convection and the athlete will loose more than 0.5 kW while running.

Example A-10.8: *Cans of tomato sauce are to be cooled in a refrigerator. The cans are cylindrical and 20 cm in length and 15 cm in diameter. The can surface is at 30°C and the air in the refrigerated space is at 10°C. Two alternate stacking are possible—namely vertical stacking and horizontal stacking. Determine the method that will be better. Neglect cooling on the end surfaces.*

Solution: The film temperature = 20°C. The property values are $\rho = 1.205$ kg/m^3,

$$\nu = 15.06 \times 10^{-6} \text{ m}^2\text{/s},\ Pr = 0.703,\ k = 25.93 \times 10^{-3} \text{ W/m K}.$$

For vertical stacking:

$$\mathbf{Gr} = \frac{9.81 \times 20 \times 0.2^3}{293 \times (15.06 \times 10^{-6})^2} = \mathbf{2.362 \times 10^7}$$

$$Gr\ Pr = 1.6605 \times 10^7 \quad \therefore \text{ laminar}$$

$$\frac{D}{L} \geq 35/Gr^{0.25} = 0.50$$

The ratio is less than this value as $(D/L) = 0.75$.

So it is necessary to consider this as short cylinder with $D = H$ (Table 10.1)

$$Nu = 0.775\,(Gr\ Pr)^{0.208}$$

$$\frac{1}{L} = \frac{1}{L_G} + \frac{1}{L_V} = \frac{1}{0.15} + \frac{1}{0.2}$$

$$\mathbf{L = 0.086}$$

$$\therefore \quad \mathbf{Gr} = \frac{9.81 \times 20}{293} \times \frac{0.086^3}{(15.06 \times 10^{-6})^2} = \mathbf{1.88 \times 10^6}$$

$$\therefore \quad Nu = 14.534 \quad \therefore \quad \mathbf{h = 4.381\ W/m^2\,K}$$

Horizontal stacking:

$$Gr = \frac{9.81 \times 20}{293} \times \frac{0.15^3}{(15.06 \times 10^{-6})^2} = 9.964 \times 10^6$$

$$Gr\ Pr = 7.005 \times 10^6$$

The correlation is (Table 10.1)

$$Nu = 0.48\ (Gr\ Pr)^{0.25} = 24.694$$

$$\therefore \quad \mathbf{h = 4.27\ W/m^2\ K}$$

There is no distinct advantage due to the method of stacking for this size and conditions, as the linear dimensions are close.

Example A-10.9: *Plate type of coolers are used to cool the oil in a transformer. The design specified 0.6 m high 0.2 m wide plates. The surface temperature was 80°C and air temperature was 20°C. During installation, 0.2 m side was placed vertical by mistake. Determine the change in the convection coefficient and the heat loss.*

Solution: The film temperature is 50°C. The property values are:

$$\rho = 1.093,\ \nu = 17.95 \times 10^{-6},\ Pr = 0.698,\ k = 28.26 \times 10^{-3}\ \text{W/m K}$$

Case (i): 0.6 m vertical:

$$\mathbf{Gr} = \frac{9.81 \times 60 \times 0.6^3}{323 \times (17.95 \times 10^{-6})^2} = \mathbf{1.22 \times 10^9}$$

$$Gr\ Pr = 0.8527 \times 10^9, \text{ laminar}$$

one of the applicable correlation is equation (10.19)

$$Nu = 0.68 + \frac{0.670\ (Gr\ Pr)^{0.25}}{[1 + (0.492/Pr)^{9/16}]^{4/9}} = 88.39$$

$$\mathbf{h = 4.163\ W/m^2\ K}$$

Case (ii): 0.2 m vertical $Gr = \dfrac{9.81 \times 60 \times 0.2^3}{323 \times (17.95 \times 10^{-6})^2} = 4.52 \times 10^7$

$$Gr\ Pr = 3.158 \times 10^7$$

using the same correlations, $Nu = 39.16$, **h = 5.533 W/m² K**

Heat flow increases by 33%

Short lengths provide higher values of heat transfer coefficient as boundary layer is thinner.

Example A-10.10: *A vertical surface 5 m high is at 80°C and is surrounded by still air at 20°C. Determine the average value of convective heat transfer coefficient at 1 m intervals and also at the location where $Gr = 10^9$.*

Solution: The film temperature is 50°C. The property values are:

$$\rho = 1.093,\ \nu = 17.95 \times 10^{-6}\ \text{m}^2/\text{s},\ Pr = 0.698,\ k = 28.26 \times 10^{-3}\ \text{W/m K}$$

At Gr = 10⁹

$$10^9 = \frac{9.81 \times 60}{323} \times \frac{L^3}{(17.95 \times 10^{-6})^2} \quad \therefore \quad L = 0.561\ \text{m}.$$

$$Nu = 0.59\ (Gr\ Pr)^{1/4} = 95.9,\ \mathbf{h = 4.83\ W/m^2\ K}$$

For other lengths:

Using the simpler correlation for the turbulent region, beyond 0.561 m (Table 10.1)

$$Nu = 0.1\ (Gr\ Pr)^{1/3}$$

The calculated values are shown in the table below.

Length, m	*1*	*2*	*3*	*4*	*5*
Gr Pr	3.948×10^9	3.158×10^{10}	1.066×10^{11}	2.526×10^{11}	4.93×10^{11}
Nu	158.05	316.09	474.14	632.18	790.23
h	4.466	4.466	4.466	4.466	4.466

This shows that in the turbulent region, the value of convection coefficient remains the same.

COMMENT

Other values remaining constant $Gr \propto L^3$ or $Gr^{1/3} \propto L$.

$\therefore \quad (Gr_{2m}) = 2^3\ Gr_{1m},\ Gr_{3m} = 3^3\ Gr_{1m}$ etc.

This can be verified from the table values. This leads to $Nu_{2m} = 2Nu_{1m}$, $Nu_{3m} = 3Nu_{1m}$. In the expression for Nu, $= hL/k$, as L increases in the same ratio as Nu, h remains constant.

Example A-10.11: *A vertical water heater element is in the form of a cylinder of 6 cm dia and 45 cm length and its surface is maintained at 80°C. Water at 20°C is to be heated. Investigate the vertical and horizontal configurations.*

Solution: The film temperature is (80 + 20)/2 = 50°C, (usual units)

$$\rho = 990,\ \nu = 0.5675 \times 10^{-6},\ k = 0.63965,\ \beta = 0.48 \times 10^{-3},\ Pr = 3.68$$

Vertical:
$$\mathbf{Gr} = \frac{9.81 \times 0.48 \times 10^{-3} \times 60 \times 0.45^3}{(0.5675 \times 10^{-6})^2} = \mathbf{7.994 \times 10^{10}}$$

$$\frac{D}{L} = \frac{6}{45} = 0.133$$

$$0.133 > \frac{35}{Gr^{0.25}} = 0.065$$

$\therefore$ Vertical plate configuration can be used (Table 10.1)

$$Nu = 0.1\ (Gr\ Pr)^{1/3} = 665.07,\ \mathbf{h = 945.37\ W/m^2\ K}$$

Horizontal:
$$Gr = \frac{9.81 \times 0.48 \times 10^{-3} \times 60 \times 0.06^3}{(0.5675 \times 10^{-6})^2} = 0.1895 \times 10^9$$

$$Nu = 0.125\ (Gr\ Pr)^{0.333} \text{ (Table 10.1)}$$

$$Nu = 110.85,\ \mathbf{h = 1181.7\ W/m^2\ K}$$

Horizontal configuration is found better by about 25%.

Example A-10.12: ***A long horizontal cylindrical heater*** *rod of diameter 2.5 cm is used to heat liquid sodium. The heater surface is at 400°C and the liquid is at 200°C. Determine the value of convection coefficient.*

Solution: The film temperature is 300°C

Property values are $\rho = 878,\ \nu = 0.394 \times 10^{-6},\ Pr = 0.0063,\ k = 70.94$ (usual units)

Chapter 10

$$\beta = [(903 - 854)/200]\ \frac{1}{878} = 2.79 \times 10^{-4}$$

The correlation available is (eqn. 10.36)

$$Nu = 0.53\ (Gr\ Pr^2)^{1/4}$$

$$\textbf{Gr} = \frac{9.81 \times 2.79 \times 10^{-4} \times 0.025^3 \times 200}{(0.394 \times 10^{-6})^2} = \mathbf{55.1 \times 10^6}$$

$\therefore$ $$Nu = 3.624 \quad \therefore \quad \mathbf{h = 10284\ W/m^2\ K.}$$

High value due to liquid metal.

Example A-10.13: *A long horizontal cold water tube* *of 2.5 cm OD is immersed in hot oil bath at 100°C. The tube surface is at 20°C. Determine the value of convection coefficient.*

Solution: The film temperature is 60°C, oil is on the outside.

$$\rho = 864,\ \nu = 83 \times 10^{-6}\ \text{m}^2/\text{s},\ Pr = 1050,\ k = 0.1407\ \text{W/m K}$$

$$\beta = \frac{876 - 852}{40} \times \frac{1}{864} = 6.944 \times 10^{-4}/\text{K}$$

$$\textbf{Gr} = \frac{9.81 \times 6.944 \times 10^{-4} \times 80 \times 0.025^3}{(83 \times 10^{-6})^2} = \mathbf{1236.12}$$

Using equation (10.35), $Nu = 0.36 + \dfrac{0.518\,(Gr\ Pr)^{1/4}}{[1 + (0.559/Pr)^{9/16}]^{4/9}} = 17.733$

$\therefore$ $$\mathbf{h = 99.8\ W/m^2\ K.}$$

Example A-10.14: *A sphere of dia 2 m* *storing chilled brine at – 10°C is insulated by having it enclosed in another sphere of 2.1 m dia. The gap is filled with air at 0.05 atm. The outside sphere is at 30°C. Determine the heat convection across the space.*

Solution: The film temperature is 10°C

$$\rho = 1.247 \times 0.05,\ \mu = 17.65 \times 10^{-6}\ \text{kg/m s},$$

$Pr = 0.705$, $k = 25.12 \times 10^{-3}$ W/m K. Available correlation is

From reference handbook

$$\frac{k_{eff}}{k} = 0.74 \left[\frac{b^{1/4}}{D_o D_i\ (D_i^{-7/5} + D_o^{-7/5})^{5/4}}\right] (Gr\ Pr)^{1/4} \left[\frac{Pr}{0.861 + Pr}\right]^{1/4}$$

$$2b = D_o - D_i,\ b = 0.1/2 = 0.05$$

K_{eff} = The thermal conductivity that a motionless fluid (with k) in the gap should have to transfer the same amount of heat as the moving fluid.

$$\frac{\mathbf{k_{eff}}}{\mathbf{k}} = 0.74 \left[\frac{0.05^{0.25}}{2.1 \times 2(2^{-7/5} + 2.1^{-7/5})^{5/4}}\right]$$

$$\times \left[\frac{9.81 \times 1 \times 40 \times 0.05^3}{283 \times [(17.65 \times 10^{-6}/(1.247 \times 0.05)]^2} \times 0.705\right]^{0.25} \times \left[\frac{0.705}{0.861 \times 0.705}\right]^{0.25} = \mathbf{0.629}$$

considering the heat flow as though by conduction,

$$\mathbf{Q} = k_{eff}\,\pi\,(D_iD_o/b)\,(T_i - T_o)$$
$$= 0.629 \times 25.12 \times 10^{-3} \times \pi \times (2.1 \times 2/0.05) \times 40 = \mathbf{166.8\ W.}$$

Example A-10.15: ***A weather baloon*** *of dia 4 m is at an altitude where the pressure is 0.1 atm and the temperature is – 40°C. The baloon surface is at 0°C. (a) Determine rate of heat loss from the surface at the instant. (b) If air flows over the balloon at 0.3 m/s, determine the value of convective heat transfer coefficient due to forced convection. (c) What is the velocity of flow which will give the same value of convection coefficient as in free convection?*

Solution: The properties are to be evaluated at the film temperature of – 20°C

$$\rho = 1.395,\ Pr = 0.716,\ k = 22.79 \times 10^{-3},\ \mu = 16.18 \times 10^{-6}\ \text{kg/m s}$$

As the pressure is 0.1 atm

$$\rho = 0.1 \times 1.395 = 0.1395\ \text{kg/m}^3,\ \nu = (16.18 \times 10^{-6}/0.1395)\ \text{m}^2\text{/s}$$

$$Gr = 9.81 \times \frac{1}{253} \times \frac{40 \times 4^3}{(16.18 \times 10^{-6}/0.1395)^2} = 7.379 \times 10^9$$

$$Gr\,Pr = 5.283 \times 10^9$$

(*a*) For sphere, (equation 10.37)

$$Nu = 2 + 0.50\,(Gr\,Pr)^{0.25} = 136.8$$

$$\therefore \quad \mathbf{h = 0.779\ W/m^2\ K}$$

$$\therefore \quad \mathbf{Q} = 4\pi r^2 h(T_w - T_\infty) = 4\pi \times 2^2 \times 0.779 \times 40 = \mathbf{1567\ W}$$

By using [10.37(*a*)]

$$\overline{Nu}_D = 2 + \frac{0.589\,(5.283 \times 10^9)^{0.25}}{[1 + (0.469/0.716)^{9/16}]^{4/9}} = 124.64$$

$$h = 124.64 \times 22.79 \times 10^{-3}/4$$
$$= 0.71\ \text{W/m}^2\ \text{K. Compared to } 0.779\ \text{W/m}^2\ \text{K.}$$

(*b*) Forced convection with u = 0.3 m/s

$$Re = \frac{u\,D}{\nu} = \frac{0.3 \times 4}{(16.18 \times 10^{-6}/0.1395)} = 10346.1$$

$$Nu = 2 + [0.25\,Re + 3 \times 10^{-4}\,Re^{1.6}]^{0.5} = 60.16 \text{ equation (8.42)}$$

$$\therefore \quad \mathbf{h = 0.343\ W/m^2\ K}$$

(*c*) h = 0.779, $Nu = \dfrac{h\,D}{k} = \dfrac{0.779 \times 4}{22.79 \times 10^{-3}} = 139.4$

$$139.4 = 2 + (0.25\,Re + 3 \times 10^{-4}\,Re^{1.6})^{0.5}$$

$$\therefore \quad 0.25\,Re + 3 \times 10^{-4}\,Re^{1.6} = 18868$$

solving by trial

$$Re = 43600 \quad \therefore \quad \mathbf{u = 1.27\ m/s.}$$

Example A-10.16: ***In a solar flat plate collector*** *the plate is of 1 m square and its temperature is 160°C and the glass cover plate at a distance of 8 cm from the collector surface is at 40°C. The space between is evacuated and is at 0.1 atm. Determine the heat transfer coefficient if the collector is inclined at 20°C to the horizontal.*

Solution: The film temperature is 100°C (usual units)

$$\rho = 0.946 \times 0.1,\ Pr = 0.688,\ k = 32.10 \times 10^{-3},\ \mu = 21.87 \times 10^{-6}$$

$$\textbf{Gr Pr} = \frac{9.81 \times 120 \times 0.08^3 \times 0.688}{373 \times (21.87 \times 10^{-6}/0.946 \times 0.1)^2} = 20801$$

Using equation (10.49),

$$Nu_\delta = 0.42\ (Gr\ Pr\ \cos\theta)^{0.25}\ Pr^{0.012}\ (L/\delta)^{-0.3} = 2.317$$

$$\mathbf{h = 0.998\ W/m^2\ K}$$

If pressure is 1 atm: $GrPr = 2081 \times 10^6$

$$\therefore \quad Nu_\delta = 0.42\ (Gr\ Pr\ \cos\theta)^{0.25}\ Pr^{0.012}\ (L/\delta)^{-0.30} = 7.33$$

$$\therefore \quad \mathbf{h = 2.94\ W/m^2\ K.}$$

Example A-10.17: *Determine* ***the spacing between two plates of 1 m*** *square for heat transfer by free convection to be the same as heat transfer by conduction. One plate is at 40°C while the other is at 20°C. Assume that all four edges are closed.*

Solution: **This is a case of enclosed space.** It can be shown that

$$\mathbf{Nu_\delta} = \frac{\mathbf{k_{eff}}}{\mathbf{k}}$$

If $k_{eff} = k$, then the heat transfer by convection will equal the heat transfer by conduction. Considering the laminar region.

$$Nu_\delta = 0.42\ (Gr_\delta\ Pr)^{0.25}\ Pr^{0.012}\ (L/\delta)^{-0.30}$$

$$1 = 0.42 \left[\frac{g\,\beta\,\delta^3}{\upsilon^2}\,Pr\right]^{0.25} Pr^{0.012}\ (L)^{-0.3}\ \delta^{0.30}$$

$$1 = \delta^{1.05}\ 0.42 \left[\frac{g\,\beta\,Pr}{\upsilon^2}\right]^{1/4} Pr^{0.012}\ L^{-0.30}$$

$$L^{0.3}/0.42 \left[\frac{g\,\beta\,Pr}{\nu^2}\right]^{1/4} Pr^{0.012} = \delta^{1.05}$$

In this case $\quad \nu = 16 \times 10^{-6}\ m^2/s \quad Pr = 0.701,\ k = 26.75 \times 10^{-3}\ W/m\ K$

$$\therefore \quad \delta^{1.05} = 1^{0.3}/0.42 \left[\frac{9.81 \times 0.701}{303\ (16 \times 10^{-6})^2}\right]^{0.25} (0.701)^{0.012} = 0.02464$$

$$\therefore \quad \boldsymbol{\delta = 0.02939\ \mathbf{m\ or\ 29.4\ mm}}$$

Conduction through an air gap of 29.4 mm will be the same as due to free convection in the enclosed space.

Example A-10.18: ***A cylinder of dia*** *0.04 m and length 1.2 m is* ***inclined to vertical*** *at 30° its surface is maintained at 80°C in air at 20°C. Determine the value of convection coefficient.*

Solution: The film temperature is 50°C

$$\nu = 17.95 \times 10^{-6}\ m^2/s,\ Pr = 0.698,\ k = 28.26 \times 10^{-3}\ W/m\ K$$

$$Gr_L\, Pr = \frac{9.81 \times 60 \times 1.2^3}{323 \times (17.95 \times 10^{-6})^2} \times 0.698 = 6.82 \times 10^9$$

$$\mathbf{Gr_D = 3.62 \times 10^5} \text{ (within applicable limits, } 6.9 \times 10^5)$$

Using available **data book** correlations:

$$(Gr_L\, Pr)_{cr} = 2.6 \times 10^9 + 1.1 \times 10^9 \times \tan\theta = 3.23 \times 10^9$$

∴ Flow is turbulent

$$Nu_L = [0.47 + 0.11\,(\sin\theta)^{0.8}]\; Gr_D^{-1/12}\,(Gr_L Pr)^{1/3}$$

$$= [0.47 + 0.11\,(\sin 30)^{0.8}]\,(Gr_L\, Pr)^{1/3} \,.\, Gr_D^{-1/12} = 347.94$$

∴ $\mathbf{h = 8.19\ W/m^2\,K.}$

Example A-10.19: *A cylinder of 0.04 m dia and 0.6 m length is inclined at 30° to the vertical and its surface is maintained at 80°C in air at 20°C. Determine the value of convection coefficient.*

Solution: The film temperature is 50°C, $\nu = 17.95 \times 10^{-6}$ m²/s

$$Pr = 0.698\ k = 28.26 \times \text{W/m K}$$

$$\mathbf{Gr_L\, Pr} = \frac{9.81 \times 60 \times 0.6^3}{323 \times (17.95 \times 10^{-6})^2} \times 0.698 = \mathbf{0.8527 \times 10^9}$$

This is less than the critical value of

$$2.6 \times 10^9 + 1.1 \times 10^9 \tan\theta = 3.23 \times 10^9 \text{ for this case.}$$

From handbook,

∴ $$Nu_L = [2.9 - 2.32\,(\sin\theta)^{0.8}]\,(Gr_D)^{-1/12}\,(Gr_L\, Pr)^{(1/4 + (1/12)(\sin\theta)^{1.21})}$$

$Gr_D = 3.62 \times 10^5$ ∴ within applicable limits

∴ $Nu_L = 194.37$ ∴ $\mathbf{h = 9.15\ W/m^2\,K}$

This is reasonable as h in the laminar region in free convection is generally higher than that in the turbulent region.

Example A-10.20: ***A square duct of 0.3 m*** *side carrying conditioned air at 10°C passes through a room at 30°C. Determine the heat gain for 1 m length.*

Solution: The film temperature is 20°C

$$\nu = 15.06 \times 10^{-6} \text{ m}^2/\text{s},\ Pr = 0.703,\ k = 25.93 \times 10^{-3} \text{ W/m K}$$

In this case, the length of travel of a particle along the boundary layer is to be taken as the characteristic length.

$$L = 0.3 + 0.3 = 0.6 \text{ m}$$

$$Gr\, Pr = \frac{9.81 \times 20 \times 0.6^3}{293 \times (15.06 \times 10^{-6})^2}\ 0.703 = 4.48 \times 10^8$$

Using equation (10.39),

$$Nu = 0.52\,(Gr\, Pr)^{0.25} = 75.666$$

∴ $\mathbf{h = 3.27\ W/m^2\,K}$

$$\mathbf{Q = 2 \times 0.6 \times 3.27\,(30 - 10) = 78.48\ W/m.}$$

COMMENT

Another way is to consider convection on the sides as vertical wall and convection from bottom as horizontal plate with cooling below and horizontal plate with cooling at top.

Example A-10.21: ***A ceramic block*** *at 480°C is 0.2 m × 0.2 m × 0.1 with 0.1 m vertical. It is exposed to air at 20°C. Determine the rate of heat loss to the air.*

Solution: The film temperature is (480 + 20)/2 = 250°C

The property values are

$$\nu = 40.61 \times 10^{-6}\ \text{m}^2/\text{s},\ Pr = 0.677,\ k = 42.68 \times 10^{-3}\ \text{W/m K}$$

The characteristic length is given by

$$\frac{1}{L} = \frac{1}{L_H} + \frac{1}{L_V} = \frac{1}{0.2} + \frac{1}{0.1} \quad \therefore \quad L = 0.067\ \text{m}$$

$$\therefore \quad Gr\,Pr = \frac{9.81 \times 460 \times 0.067^3}{(273 + 250)\,(40.61 \times 10^{-6})^2} \,.\, 0.677 = 1.05 \times 10^6$$

$$Nu = 0.52\,(Gr\,Pr)^{0.25} = 16.74 \quad \therefore \quad \mathbf{h = 10.60\ W/m^2\,K}$$

$\therefore$ **heat loss** $= h\,A\,\Delta T = 10.6 \times \{4 \times 0.2 \times 0.1 + 2 \times 0.2 \times 0.2) \times 460 =$ **780.3 W.**

Example A-10.22: ***A small copper block*** *having a square bottom of 3 cm side and vertical height of 6 cm at 100°C cools in air at 20°C.*

Calculate the convection coefficient.

Solution: The film temperature is (100 + 20)/2 = 60°C

The property values are:

$$\nu = 18.97 \times 10^{-6}\ \text{m}^2/\text{s},\ Pr = 0.696,\ k = 28.96 \times 10^{-3}\ \text{W/m K}$$

The length parameter is

$$\frac{1}{L} = \frac{1}{L_V} + \frac{1}{L_H} = \frac{1}{0.03} + \frac{1}{0.06} \quad \therefore \quad L = 0.02\ \text{m}$$

$$\therefore \quad \mathbf{Gr} = \frac{9.81 \times 80 \times 0.02^3}{(273 + 60)(18.97 \times 10^{-6})^2} = \mathbf{52393}$$

$$Nu = 0.52 \times (Gr\,Pr)^{0.25} = 7.186$$

$$\therefore \quad \mathbf{h = 10.40\ W/m^2\,K}$$

if the length of travel is considered:

$$L = 0.03 + 0.06 = 0.09\ \text{m}$$

$$\therefore \quad Gr = 4774312$$

$$\therefore \quad Nu = 22.2, \quad \therefore \quad \mathbf{h = 7.14\ W/m^2\,K.}$$

Example A-10.23: ***A triangular plate of equal sides*** *of 0.6 m is maintained horizontally in air at 20°C. The plate temperature is 80°C. The hotter side of the plate faces down. Determine the value of convection coefficient.*

Solution: The film temperature = 50°C

Property values are: $v = 17.95 \times 10^{-6}$ m²/s, $Pr = 0.698$

$k = 28.26 \times 10^{-3}$ W/m K

$$L = A/P = \left(\frac{0.6 \times 0.6 \sin 60°}{2}\right) / 3 \times 0.6 = 0.0866 \text{ m}$$

$$Gr = \frac{9.81}{323} \times \frac{60 \times 0.0866^3}{(17.95 \times 10^{-6})^2} = 3.67 \times 10^6$$

Using values from Table 10.1

$$Nu = 0.27 \times (Gr\ Pr)^{0.25} = 10.8$$

$\therefore$ **h = 3.5 W/m² K.**

Example A-10.24: ***A cylindrical vessel*** *of dia 1 m contains chilled brine at – 20°C. It is surrounded by a another vessel of 1.1 m dia. The space between contains air at 0.05 atm pressure. The outside vessel wall is at 20°C. Determine the heat flow for 1 m length of vessel.*

Solution: The average temperature = 0°C

The property values are $\rho = 1.293 \times 0.05$

$Pr = 0.707$, $\mu = 17.16 \times 10^{-6}$ kg/ms, $k = 24.42 \times 10^{-3}$ W/m K, $b = 0.05$ m

From data book,

$$\frac{k_{eff}}{k} = 0.386 \left[\frac{Pr}{0.861 + Pr}\right]^{0.25} [R_{ac}]^{0.25} \left[\frac{\ln (D_0/D_i)}{b^{0.75} (D_i^{-0.6} + D_0^{-0.6})^{1.25}}\right]$$

$\mathbf{k_{eff}/k = 0.809}$

$$\mathbf{Q} = \frac{2\pi\, k_{eff}}{\ln D_0/D_1} [T_1 - T_0] = \frac{2\pi \times 0.809 \times 24.42 \times 10^{-3}}{\ln 1.1/1} [40] = \mathbf{52\ W.}$$

Example A-10.25: ***Two vertical plates*** *1.2 m × 1.2 m enclose an air space of 4 cm thickness. One plate is at 40°C, while the other is at 20°C. Determine the value of k_{eff}/k for various pressures from 1 atm to 0.05 atm.*

Solution: The film temperature = 30°C,

The property values are,

$\rho = 1.165$, $Pr = 0.701$, $k = 26.75 \times 10^{-3}$ W/m K,

$\mu = 18.63 \times 10^{-6}$ kg/m s

at 1 atm, $\nu = 18.63 \times 10^{-6}/1.165 = 15.99 \times 10^{-6}$

$$\mathbf{Gr_\delta} = \frac{9.81}{303} \times 20 \times \frac{0.04^3 \times (1.165)^2}{(18.63 \times 10^{-6})^2} = \mathbf{1.621 \times 10^5}$$

Chapter 10

$$\frac{\mathbf{k_{eff}}}{\mathbf{k}} = Nu_\delta = 0.42\,(Gr_\delta\, Pr)^{1/4}\, Pr^{0.012}\,(L/\delta)^{-0.30} = \mathbf{2.77}$$

0.9 atm, $\nu = 18.63/(1.165 \times 0.9) = 17.768 \times 10^{-6}$ m²/s

$$\frac{k_{eff}}{k} = 0.42\,(Gr\,Pr)^{1/4}\, Pr^{0.012}\,(L/\delta)^{-0.30} = 0.13794\,(Gr)^{1/4} = 2.7676.\ P^{0.5}$$

$$Gr = \frac{9.81 \times 20 \times 0.04^3 \times (1.165)^2\,(P^2)}{303 \times (18.63 \times 10^{-6})^2} = 162055\ P^2$$

The results are shown tabulated:

Pressure atm	k_{eff}/k	*Pressure atm*	k_{eff}/k
1	2.77	0.3	1.516
0.9	2.63	0.2	1.238
0.8	2.475	0.15	1.072
0.7	2.316	0.14	1.036
0.6	2.144	0.13	0.998
0.5	1.957	0.10	0.875

When the density is reduced to 0.13 atm, the heat transfer by convection is equal to heat transfer by pure conduction.

Example A-10.26: ***Two glass plates 1 m × 1 m enclose air at a pressure of 0.1 atm.*** *The plate temperatures are 30°C and 10°C. Determine the thickness of air space so that heat transferred by convection will be the same as heat conducted by the air film.*

Solution: For this condition $k_{eff}/k = 1$

Film temperature = 20°C, $\rho = 1.025 \times 0.1$, $Pr = 0.703$

$\mu = 18.14 \times 10^{-6}$ kg/m s $\quad k = 25.93 \times 10^{-3}$ W/m K

$\nu = (18.14 \times 10^{-6}/1.205 \times 0.1) = 150.53 \times 10^{-6}$ m²/s, $L = 1$

$$k_{eff}/k = 0.42\,(Gr\,Pr)^{0.25}\, Pr^{0.012}\,(L/\delta)^{-0.3} = 1$$

$$0.42 \times \left[\frac{9.81 \times 20}{293 \times (150.53 \times 10^{-6})^2}\right]^{0.25} 0.703^{0.262}\ \delta^{1.05}\ 1^{-0.3} = 1$$

∴ $\quad$ **δ = 0.0415 m or 4.15 cm.**

Example A-10.27: *A sphere of 30 mm dia has its surface maintained at 80°C while being immersed in still (i) air (ii) water (iii) oil at 20°C. Determine the power required to maintain the temperature.*

Solution: The property values evaluated at 50°C are

	ν	*Pr*	*k*	β
air	17.95×10^{-6}	0.698	28.26×10^{-3}	1/323
oil	162×10^{-6}	1960	0.14245	6.9×10^{-3}
water	0.5675×10	3.68	0.63965	0.48×10^{-3}

Air: $$\textbf{Gr Pr} = \frac{9.81 \times 60 \times 0.03^3}{323 \times (17.95 \times 10^{-6})^2} \times 0.698 = \mathbf{1.066 \times 10^5}$$

$$Nu = 2 + 0.43\,(Gr\,Pr)^{0.25} = 9.77 \text{ (Equation 10.37)}$$

$$\therefore \quad \mathbf{h = 9.2\ W/m^2\,K}$$

Water: $$Gr\,Pr = \frac{9.81 \times 0.48 \times 10^{-3} \times 60 \times 0.03^3}{(0.5675 \times 10^{-6})^2} \times 3.68 = 8.716 \times 10^7$$

$$Nu = 2 + 0.50\,(Gr\,Pr)^{0.25} = 50.31$$

$$\therefore \quad \mathbf{h = 1072.7\ W/m^2\,K}$$

Oil: $$Gr\,Pr = \frac{9.81 \times 6.9 \times 10^{-3} \times 60 \times 0.03^3 \times 1960}{(162 \times 10^{-6})^2} = 8.19 \times 10^6$$

$$Nu = 28.75 \quad h = \mathbf{136.5\ W/m^2\,K}$$

Power required $= h\,A\,\Delta T$, $4\pi\,(0.03/2)^2 \times 60 \times h = 0.16965\,h$

Air: 1.56 W, Water: 181.98 W, Oil: 23.16 W.

Example A-10.28: *Water in a pan of 20 cm diameter to depth of 10 cm and at 20°C is heated by an electrical heater on its bottom which maintains a constant temperature of 80°C. Determine the initial rate of heating. Neglect all losses.*

Solution: This is a case where a limited amount of fluid (enclosed) is heated from bottom. The correlation available is (data book).

$$Nu_\delta = 1 + 1.44\left[1 - \frac{1708}{Ra_\delta}\right] + \left[\left(\frac{Ra_\delta}{5830}\right)^{1.3} - 1\right]$$

$$Ra_\delta = Gr_\delta\,Pr \text{ based on thickness } \delta$$

The property values are: at 50°C $v = 0.5675 \times 10^{-6}$ m²/s

$$Pr = 3.68,\ k = 0.63965,\quad \beta = 0.48 \times 10^{-3}$$

$$\mathbf{Ra_\delta} = Gr\,Pr = \frac{9.81 \times 0.48 \times 10^{-3} \times 60 \times 0.1^3}{(0.5675 \times 10^{-6})^2} \times 3.68 = \mathbf{3.228 \times 10^9}$$

$$Nu = 1 + 1.44 + 81.12 = 83.56$$

$$\therefore \quad \mathbf{h = 535\ W/m^2\,K}$$

$$\textbf{Heating rate} = \frac{\pi D^2}{4}\,.\,h\,\Delta T = \mathbf{1008.5\ W} \text{ or about 1 kW.}$$

Chapter 10

Table 10.3. Summary of free convection correlations for different geometrics

Nos.	*Geometry*	*Recommended Correlation and Restrictions*
1.	Vertical plate. Also for vertical cylinder if $(D/L) \geq 35/Gr_L^{0.25}$	$\overline{Nu} = \left[0.825 + \frac{0.387\,(Gr\,Pr)^{1/6}}{[1+(0.492/Pr)^{9/16}]^{8/27}}\right]^2$ No restriction.
2.	Inclined plate.	Same as above except, $Gr\,Pr$ to be multiplied by cos θ, θ – angle with vertical.
3.	Horizontal surface – upper surface heated/lower surface cooled L = Area/Perimeter.	$\overline{Nu}_L = 0.27\,(Gr\,Pr)^{0.25}$ $10^4 \leq Gr\,Pr \leq 10^7$ $Nu_L = 0.15\,(Gr\,Pr)^{0.333}$ $10^7 \leq Gr\,Pr \leq 10^{10}$.
4.	Horizontal surface upper surface cooled/ lower surface heated.	$\overline{Nu}_D = \left[0.60 + \frac{0.387\,(Gr\,Pr)^{1/6}}{[1+(0.559/Pr)^{9/16}]^{8/27}}\right]^2$ or $\overline{Nu}_D = C\,(Gr\,Pr)^n$ C and n from Table 10.1 for various ranges.
5.	Sphere	$\overline{Nu}_D = 2 + \frac{0.589\,(Gr\,Pr)^{0.25}}{[1+(0.469/Pr)^{9/16}]^{4/9}}$ $Gr\,Pr \leq 10^{11}$, $Pr \geq 0.7$
6.	Horizontal cylinder in liquid metal.	$\overline{Nu}_D = 0.53\,(Gr\,Pr^2)^{0.25}$

REVIEW QUESTIONS

10.1 Explain what is meant by 'buoyancy driven' or 'gravity driven' flow?

10.2 State the differences between the velocity profiles of forced convection flow over a flat plate and natural convection over a flat plate.

10.3 Explain the physical significance of Grashof number.

10.4 Write down the buoyancy force term to be used in the momentum equation for natural convection.

10.5 Explain how by non-dimensionalising the differential equations, the various dimensionless numbers affecting the flow can be identified?

10.6 Sketch the flow pattern in natural convection in the following cases:
(*a*) hot vertical plate (*b*) cold vertical plate

(*c*) horizontal cylinder with hot surface (*d*) horizontal cylinder with cold surface
(*e*) horizontal plate with heated top side (*f*) horizontal plate with bottom side heated
(*g*) inclined hot plate.

10.7 State the criterion for spacing hot parallel planes.

OBJECTIVE QUESTIONS

10.1 Choose the correct statement in the following cases:

(*i*) (*a*) Buoyant forces and inertia forces only influence free convection heat transfer.
(*b*) Viscous and buoyant forces only influence free convection heat transfer.
(*c*) Viscous and inertia forces only influence free convection heat transfer.
(*d*) Viscous, inertia and buoyant forces influence the heat transfer in free convection.

(*ii*) The heat transfer rate in free convection in the laminar region depends on
(*a*) ΔT (*b*) $\Delta T^{1.25}$ (*c*) $\Delta T^{1.33}$ (*d*) $\Delta T^{0.25}$.

(*iii*) The heat transfer rate in free convection in the turbulant region depends on
(*a*) $\Delta T^{1.33}$ (*b*) ΔT (*c*) $\Delta T^{0.33}$ (*d*) $\Delta T^{1.25}$.

(*iv*) The convection coefficient in free convection over a vertical plate in the laminar region depends on
(*a*) $L^{-0.25}$ (*b*) $L^{-1.25}$ (*c*) $L^{-0.33}$ (*d*) $L^{+0.25}$.

(*v*) The convection coefficient in the case of a vertical plane in free convection in turbulent region depends on
(*a*) $L^{-0.25}$ (*b*) $L^{0.25}$ (*c*) $L^{0.0}$ (*d*) $L^{0.33}$.

(*vi*) When some flow velocity is superimposed on free convection, the predominance of either is determined by
(*a*) $Gr >> Re^2$ means forced convection
(*b*) $Gr \approx Re^2$ means none of the two
(*c*) $Gr << Re^2$ means free convection
(*d*) $Gr = Re$ means purely free convection.

(*vii*) In free convection, the slope of the curve *Nu vs Gr Pr*
(*a*) Increases with increasing *Gr Pr*
(*b*) Decreases with increasing *Gr Pr*
(*c*) Increases and then decreases with increasing *Gr Pr*
(*d*) Decreases and then increases with increasing *Gr Pr*.

Answers

(*i*) (*d*) (*ii*) (*b*) (*iii*) (*a*) (*iv*) (*a*) (*v*) (*c*) (*vi*) (*b*)
(*vii*) (*a*).

EXERCISE PROBLEMS

10.1 A large chemical process tank is in the form of a cylinder of 2 m dia and 4.5 m high. The surface temperature is 80°C and the surrounding air is at 40°C. Determine the rate of heat loss over the curved surface under these conditions.

10.2 100 m³ of chemical at 80°C is to be stored in a process plant. Examine the heat loss by free convection from the following shapes (*i*) cubical (*ii*) vertical cylinder of $D : L = 1 : 3$ (*iii*) horizontal cylinder of the same length to diameter ratio (*iv*) sphere. The surrounding is at 20°C. In the case of cylinder neglect the end losses and for the cubical shape from the top and bottom.

10.3 A cylindrical tank of 1 m dia and 2 m height has its surface maintained at 120°C. It contains water at 20°C. Determine the free convection heat transfer coefficient at the surface.

10.4 A hot water tank contains water at a bulk temperature of 60°C. It is of 1 m × 1 m section and 2 m high. The sides are exposed to air at 20°C. Determine the values of convection coefficients on the air side.

10.5 A room heater radiator (really convector) has steam panels of 0.6 m × 1 m size. The surface temperature is 50°C and the room air is at 10°C. Investigate which way of positioning the panel will be better

(*i*) 0.6 m side vertical (*ii*) 1 m side vertical (*iii*) horizontal with heating from both sides.

10.6 The following two options are available for heating room air at 0°C:

(*i*) Horizontal steam pipes of 20 cm dia runing along the wall of the room (5 m length)

(*ii*) Vertical pipes of the same dia running along the wall of the room (two 2.5 m height)

Determine in which case the heat transfer will be higher.

10.7 In winter the walls of a room are at 5°C. The room air is maintained at 25°C. Determine the convection coefficient over a wall of 3 m height.

10.8 In an air conditioned room the air is at 20°C. The wall is heated by direct solar radiation to 35°C. Determine the value of convection coefficient over a wall 3 m high.

10.9 A horizontal black surface 1 m × 1 m receives radiant flux of 1.5 kW/m^2. The bottom is insulated. The air is at 20°C. Determine the value of average convective heat transfer coefficient and the temperature of the surface.

10.10 A horizontal surface receives heat at 1.5 kW/m^2 on the bottom side and has its top insulated. The bottom is exposed to air at 20°C. Determine the surface temperature at equilibrium.

10.11 Vertical plate heaters are used to heat milk in a tank. The plates are 0.2 m wide and 0.6 m long and the surface is at 90°C. The milk is at 30°C. Determine the value of average convective heat transfer coefficient for the following positions:

(*i*) 0.2 m side vertical (*ii*) 0.6 m side vertical

(*iii*) Horizontal, both sides heated

(*iv*) Plates with 0.6 m side inclined at 30° with vertical.

Assume properties of water for the milk.

10.12 A horizontal steam pipe with a surface temperature of 90°C is used to heat water in a large tank, the water being at 30°C. Determine the heat flow per 1 m length of the pipe.

10.13 A surface 1 m × 1 m size is inclined at 45°C to the vertical. The plate is insulated at the top side. The bottom surface is at 100°C and the air is at 40°C. Determine the value of convective heat transfer coefficient over the surface.

10.14 A large tank of oil is kept warm by a circular heater of 0.3 m dia placed horizontally inside. The heater surfaces on both sides is at 120°C. The oil is at 20°C. Determine the heat input required to maintain this surface temperature both at the top and bottom faces.

10.15 A horizontal electrical cable 5 cm dia generates 2.5 W of heat/m length. The surrounding air is at 20°C. Estimate the surface temperature of the cable.

10.16 Estimate the average surface temperature of a 2.5 kW heating element 2.5 cm dia and 0.5 m length used in a water heater, with the water temperature set at a maximum of 50°C.

10.17 Determine the number of plate type of steam heaters of 0.25 × 0.6 m size placed vertically with the 0.6 m side to provide 20 kW of heating to a classroom at 10°C. The plate temperature is 110°C.

10.18 If hot water at 60°C is used instead of steam in the problem 10.17, determine the number of plates required.

10.19 Determine the length of steam pipe of 7.5 cm *OD* at 80°C is required to provide 20 kW of heat to a space at 20°C.

10.20 A block of ice at 0°C 40 × 40 cm × 40 cm is exposed to air at 20°C. Determine the rate of heat loss from the block by free convection. Assume that there is no heat transfer from the bottom.

10.21 A large vertical plate 10 m wide and 4 m high at a constant temperature of 60°C is exposed to still air at 10°C. Determine the heat transfer on one side, using equation (10.20). In case equation from Table 10.1 is used what will be heat flow. [**Ans.** 10781 W, 9676 W]

10.22 A horizontal heater rod of 20 mm diameter with surface temperature of 38°C is submerged in water at 27°C. Determine the heat loss per m length by natural convection. Use the general equation $\overline{Nu} = C(Gr\ Pr)^n$ with $C = 0.53$ and $n = 0.25$. [**Ans.** 836.3 W/m]

10.23 The surface temperature of a wire of 0.02 mm diameter carrying current is 54°C and it is exposed to air at 0°C. Calculate the power lost per m length of the wire, by free convection. [**Ans.** 1.672 W]

10.24 A horizontal pipe carrying steam is at a surface temperature of 250°C. The diameter of the pipe is 30.48 cm. The pipe passes through still air at 15°C. Calculate the heat loss per meter length by free convection. [**Ans.** By older correlation, 1.49 kW/m, By recent correlation 1.62 kW/m]

10.25 A cube of 0.2 m side at a surface temperature of 60°C is exposed to still air at 20°C. Determine the rate of heat flow by free convection. [**Ans.** 108.8 W]

10.26 A vertical surface 3.5 m high and 2 m wide is exposed to a heat flux of 800 W/m^2. It is in still air at 30°C. Determine the average temperature of the surface. [**Ans.** 185°C]

10.27 A 2 cm dia 60 cm long heater is placed horizontally in a glycerine tank. The surface of the heater is at 60°C and glycerine is at 20°C. Estimate the heat transfer rate.

10.28 A cylindrical heater 3.0 cm dia and 0.8 m length with a constant heat flux of 1500 W/m^2 is inclined at 35° to the horizontal in room air at 20°C. Estimate the average surface temperature.

10.29 A sphere of 30 cm dia with its surface maintained at 90°C is exposed to still air at 30°C. Calculate the heat loss.

10.30 A circular heater of 0.2 m dia is used to heat oil at 20°C in a large tank. The heater has a uniform heat generation rate of 10 kW/m^2. Determine the average surface temperature if (*i*) hot surface is facing up and (*ii*) hot surface facing down.

10.31 Two square plates 30 cm side enclose water layer 1 cm thick. The plates are at 40°C and 60°C. Estimate the heat transfer rate. All edges are closed.

10.32 An electrical bulb is of spherical shape of 5 cm dia. The surface is at 120°C and is exposed to air at 40°C. Determine the heat loss by natural convection from the surface.

10.33 A hot metal casting 40 cm × 40 cm × 20 cm at 600°C is exposed to air at 20°C on all the surfaces as it is carried in a conveyor. Determine the rate of cooling.

10.34 A solar water heater plate is at 160°C and the cover plate is at 4 cm with the space evacuated to 0.1 atm. The cover plate is at 40°C. The assemly is inclined at 25° to the horizontal. Determine the rate of heat loss from the plate.

10.35 A spherical vessel of 1 m dia containing liquid refrigerant is at –40°C. An evacuated spherical cover of 1.08 m dia is provided outside with the pressure inside being 0.05 atm. The outside surface is at 20°C. Determine the heat loss. Also determine the heat loss if only pure conduction prevailed.

10.36 If a cylindrical vessel of 1 m dia is used instead of spherical vessel in problem 10.29 estimate the heat loss and the compare the same with pure conduction loss.

10.37 A vertical plate is kept at 40°C in still air at 30°C. The plate is 1 m high. Determine the velocity that may produce the same heat loss by forced convection over the plate as that occurring by free convection.

10.38 A hot plane of height 0.71 m and width 1.02 m is at 232°C. It is exposed to still air at 23°C. Determine the heat transfer by natural convection. [**Ans.** h = 7W/m^2 K, Q = 1060 W]

10.39 Determine the boundary layer thickness at the top when a 0.25 m plane at 70°C is hung vertically in still air at 25°C. [**Ans.** 0.024 m]

10.40 A long hot air duct of rectangular section of 0.75 m × 0.3 m with surface temperature of 45°C, passes through still air at 15°C. The 0.3 m side is vertical. Determine the heat loss at (*a*) side walls (*b*) top surface and (*c*) bottom surface.

[**Ans.** Sides: 78.1 W/m, top 123.1 W/m, bottom 46.6 W/m]

10.41 A horizontal pipe of 10 cm diameter has its outside wall maintained at 165°C by passing steam. It is exposed to still air at 23°C for a length of 10 m. Determine the heat loss by natural convection.

[**Ans.** 3250 W]

10.42 A spherical shaped electrical heater of 6 cm diameter is immersed in water at 30°C. Calculate the convective heat transfer coefficient and heat output when the surface temperature of the heater is at 85°C.

[**Ans.** h = 967.9 W/m^2 K, Q = 602.1 W]

10.43 A horizontal pipe of 6 cm dia carrying hot air passes in a room where the air temperature is 22°C. If the pipe wall temperature is at 65°C, determine the heat loss for a length of 10 m.

[**Ans.** h = 5.893 W/m^2 K, Q = 477.6 W]

10.44 A vertical cylindrical tank of 0.25 m diameter and 0.12 m high contains boiling water. Its surface is at 98°C. The surrounding air is at 25°C. Calculate the rate of heat loss from the sides.

[**Ans.** h = 6.72 W/m^2 K, Q = 46.2 W]

10.45 A horizontal pipe of 30 cm dia carrying brine has its surface at 25°C. The surrounding still air is at 0°C. Calculate the convection coefficient and rate of heat flow for a length of 100 m.

[**Ans.** 4.366 W/m^2 K, Q = 10.287 kW]

10.46 A vertical tank 0.4 m dia and 1.1 m high contains hot fluid such that its surface is at 44°C. Surrounding air is at 20°C. Determine the heat loss rate at this condition.

[**Ans.** h = 4.03 W/m^2 K, Q = 133.6 W]

10.47 A furnace of 2 m × 2 m × 2 m size has its four side walls at 110°C. The surrounding air is at 30°C. Determine the rate of heat flow by free convection from the four vertical walls.

[**Ans.** h = 5.32 W/m^2 K, Q = 6809 W]

Phase Change Processes—Boiling, Condensation Freezing and Melting

11

CHAPTER

11.1 EVAPORATION AND CONDENSATION

The focus in this chapter is on convection process associated with change of phase, namely evaporation (boiling) and condensation. The change of liquid to vapour is due to transfer of heat from the solid surface and condensation is due to heat transfer to the solid surface. In both cases fluid motion is involved and so these are considered as convection mode of heat transfer processes. A unique feature of these processes is that due to the involvement of latent heat, the transfer of heat to or from the fluid can occur without influencing the fluid temperature. Also in both processes large amount of heat transfer is achieved with small temperature difference. This is due to the large value of latent heat as compared to sensible heat involved. Because of this the heat transfer coefficient in these cases are generally much larger compared to forced convection.

In addition, the other phase change processes namely melting and freezing are also considered in this chapter. In these processes also latent heat plays an important role.

Quite a few process equipments and systems involve boiling and condensation. The two well-known and popular systems are the steam power generation and refrigeration. In both cases liquid is evaporated in one component and condensed in another component. In petroleum industry crude oil is separated into useful components by fractional distillation which involves both evaporation and condensation. In the vital industries of metal melting and casting and ice making involve melting and solidification.

In this chapter the objectives are to develop appreciation for the physical conditions associated with these processes and provide a basis for the determination of heat transfer rates for application in the design of the above mentioned equipments. As convection coefficient is determined based on the temperature difference between the surface and the fluid, the surface temperature becomes an important parameter.

$$Q/A = q = h(T_{surface} - T_{fluid}) \quad \text{...(11.1)}$$

T_{fluid} is the fluid temperature and for any fluid at a given pressure this value at boiling is the saturation temperature. The surface temperature has to be higher than the saturation temperature for boiling to take place.

The difference between the surface and liquid temperatures is called the excess temperature. The term $\Delta T = T_w - T_{sat}$ = excess temperature is used in the study of boiling heat transfer as a primary variable (where T_w is the surface temperature and T_{sat} is the saturation temperature).

11.1.1. Nature of Boiling Process

When the temperature of the liquid is lower than the saturation temperature, the liquid is called undercooled. If the temperature is above saturation level it is called superheated liquid. Vapour can form only from superheated or saturated liquid.

Boiling can be classified as pool boiling and flow boiling. Boiling of nonflowing or stationary fluid is known as pool boiling. Boiling in the case of a fire tube of shell boiler will come under this category. Boiling as during the flow of water and steam (two phase fluid) through a tube in water tube boilers with wall heat flux is called flow boiling. Pool boiling is easier to observe by studying what happens when water is heated in a vessel. Compared to free and forced convections some more basic parameters like surface tension, enthalpy of evaporation are involved in the study of boiling process. Because of the large number of variables influencing the process it becomes difficult to arrive at direct correlations for this process, using momentum and energy equations etc. The correlations arrived at from experimental results are based on groups like Nusselt number identified by dimensional analysis explained in Section 11.2.

11.1.2. Regimes of Boiling

The physical phenomenon of pool boiling can be divided basically into four different regimes based on the excess temperature. Some authors subdivide these regimes into more finer ones also. The regimes are:

(values are for water boiling at 100°C)

1. Purely convective region $\Delta T < 5°C$
2. Nucleate Boiling $5 < \Delta T < 30°C$
3. Unstable (nucleate ⇔ film) boiling $30°C < \Delta T < 120°C$
4. Stable film boiling $\Delta T > 120°C$.

Note that the temperature values are indicative only.

Nucleate boiling is the process that is useful in equipment design. Unstable and film boiling regions are avoided except possibly for special applications. Convective boiling is too slow to be useful. Such division of boiling regimes was first identified by Nukiyama using the apparatus shown in Fig. 11.1(*a*)

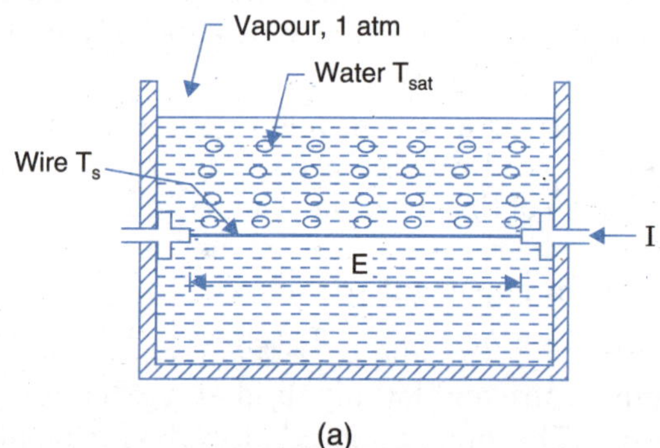

Fig. 11.1 (a) Nukiyama's power controlled heating apparatus for demonstrating boiling curve

The heat flux from a horizontal nichrome wire to saturated water was determined by measuring the current flux *I* and potential *E*. The wire temperature was determined from the variation of electrical resistance of the wire with temperature. This method is known as power ontrolled experiment because the power input is the independent variable and excess emperature is the dependent variable. In the experiment, it was found that when the input xceeded q_{max}, the regime shifted suddenly to the unstable film boiling region and the wire vas burnt out. The condition is known as burn out condition. The unstable film boiling regime ould not be established by this method. Later Drew and Mueller conducted temperature ontrolled experiments using steam at various pressure as heat source and established the unstable film boiling region. Refer Fig. 11.1 for the boiling curve.

Purely Convective Boiling: Boiling of water at low pressures is considered for the study. The excess temperature is less than 5°C. The liquid in contact with the hot surface gets superheated and slowly rises to the surface and vapour formation generally occurs at the surface. This can be observed by allowing water to be heated to near saturation temperature and then heating it further using a slow flame. Vapour will be observed over the water surface. No bubbles will be observed. The heat flux will be of the order of 0.15 MW/m^2.

Nucleate Boiling: As the heating rate is increased bubbles of vapour will be seen to form at favourable locations called nucleation sites on the surface. A favourable location provides a larger surface area by some roughness, notch, etc. The bubbles will grow and will detach themselves from the surface and will beign to rise to the surface. For lower range of values of excess temperature, these bubbles will slowly condense again and will get reduced in size as they rise. This will be the case when the excess temperature is of the order of 10°C.

The heat rate would increase to about three time the pure convection level. Further increase in heat rate and consequently the surface temperature (also the excess temperature) will cause vapour bubbles to rise as a stream to the surface. Bubbles will also form at more number of locations. There will be a steady stream of vapour bubbles rising from a large number of locations. The excess temperature will be about 20°C and the heat rate will be about 0.5 MW/m^2.

Further increase in the heat addition rate increases the surface temperature and individual bubbles combine to form a column of vapour also called slugs of vapour rise directly out of the surface. The columns will also be very close to each other and the liquid movement to fill up the locations where evaporation is taking place is very brisk. The excess temperature will be around 30°C and the heat flux will reach values of 1 MW/m^2 and convection coefficient will be around 20,000 W/m^2 K. Note that the heat rate is the more important quantity from the design point of view. This governs the excess temperature and thus the value of convection coefficient. The heat rate reaches a maximum value beyond which it is found that any further increase quickly increases the excess temperature very rapidly and the process becomes unstable. It may also cause damage to the surface. Hence design is generally limited below this condition.

Film Boiling: A further increase in the heat flux causes a film of vapour to cover the surface. This is due to the merging of favourable locations into an area. This reduces the convection coefficient and excess temperature shoots up to a very high value. In this situation the heat transfer is both by radiation and by conduction to the vapour. If the material is not strong enough for withstanding this temperature, the equipment will fail by damage to the material. This is known as burn out.

Experiment in this regime are conducted by maintaining the surface temperature rather than the heat flux. Experiments are conducted by immersing a tube in the water and passing steam at various saturated pressure conditions. Thus the temperature rather than heat flux becomes the independent variable.

Chapter 11

A typical boiling curve for water boiling at atmospheric pressure is shown in Fig. 11.1. The values indicated are approximate. The q_{min}'' point is called Leidenfrost point. At this condition the surface is completely covered by vapour blanked and heat transfer is by conduction through the vapour film.

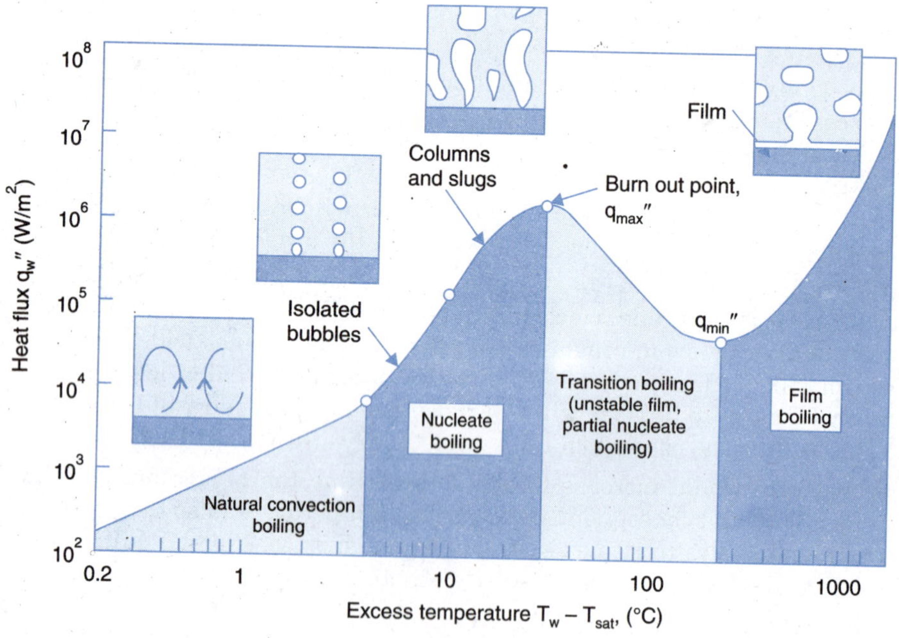

Fig. 11.1. The four regimes of pool boiling in water at atmospheric pressure

11.2 POOL BOILING CORRELATIONS

Dimensional analysis to obtain correlations for boiling process.

Since it is difficult to develop continuity, momentum and energy equations for the boiling process, as in the case of forced and free convection, the method of dimensional analysis comes handy. The influencing parameters are identified as below: $T_s - T_{sat}$ or ΔT, $\rho g\,(\rho_l - \rho_v)$, the buoyant force for flow upwards, latent heat h_{fg}, length L, surface tension σ, (bubble formation), and the fluid properties ρ, c_p, k and μ. Ten variables are identified. It is found that five dimensions are required to express these variables. So five dimensionless numbers can be obtained.

$$h = h[\Delta T, \rho g\,(\rho_l - \rho_v), L, h_{fg}, \sigma, \rho, c_p, \mu]$$

The dimensionless numbers obtained are given below:

$$\frac{h_L}{k} = f\left[\frac{\rho g\,(\rho_l - \rho_v)\,L^3}{\mu^2}, \frac{c_p\,\Delta T}{h_{fg}}, \frac{\mu c_p}{k}, \frac{g\,(\rho_l - \rho_v)\,L^2}{\sigma}\right]$$

If these (h_L/k), $(\mu c_p/k)$ are Nusselt and Prandtl numbers, $(c_p\,\Delta T/h_{fg})$ is called Jacob number and is the ratio of sensible heat flow to latent heat flow. $[g\,(\rho_l - \rho_v)\,L^2]/\sigma$ is called Bond number and is the ratio of gravitational to surface tension force, controlling bubble movement. The remaining one $[\rho g\,(\rho_l - \rho_v)\,L^3]/\mu^2$ is similar to Grashof number. Thus for boiling

$$Nu_L = f\left[\frac{\rho g\,(\rho_l - \rho_v)\,L^2}{\mu^2}, Ja, Pr, Bo\right]$$

It may be noted that these quantities find a place in a modified way in the correlation by Rohsenow, which follows this section.

The correlations obtained by Rohsenow (1952) for nucleate pool boiling is given in equation (11.2).

$$\frac{Q}{A} = q = \left[\frac{c_l\,\Delta T}{h_{fg}\,Pr^n\,C_{sf}}\right]^3 \;{}_l\,h_{fg}\left[\frac{g(\;_l - \;_v)}{g_0}\right]^{0.5} \quad ...(11.2)$$

suffix l denotes liquid properties and v denotes vapour properties.

where c_l — specific heat of **liquid** J/kg K

ΔT — excess temperature °C or K, (difference)

h_{fg} — specific enthalpy of evaporation J/kg

Pr — Prandtl number **of liquid**

n — constant equal to **1 for water and 1.7 for other fluids**

C_{sf} — surface factor shown in tabulation 11.1 and taken as 0.013 for other cases

μ_l — dynamic viscosity of the **liquid** kg/ms or $\frac{\text{Ns}}{\text{m}^2}$

ρ_l — density of the **liquid** kg/m^3

ρ_v — density of **vapour** kg/m^3

σ — surface tension-**liquid-vapour interface** N/m

g — gravitational acceleration m/s^2

g_0 — force conversion factor kg m/Ns^2 = 1 in SI units.

This correlation is the result of a log plot of experimental results with parameters.

$$\frac{q}{{}_l h_{fg}}\left[\frac{g_0}{g(\;_l - \;_v)}\right]^{0.5} \text{ and } \frac{c_l}{h_{fg}}\,\Delta T\,\frac{1}{Pr^{1.7}}.$$

This resulted in a straight line with value of C_{sf} = 0.013 generally leading to the correlation (11.2).

Example 11.1: *Show the equation (11.2) is dimensionally consistent.*

$$\textbf{RHS} \rightarrow q \rightarrow \frac{\textbf{W}}{\textbf{m}^2}$$

Solution: The LHS is taken as three parts:

$$(i)\ \frac{c_l\,\Delta T}{C_{sf}\,Pr^n\,h_{fg}} \rightarrow \frac{\text{J}}{\text{kg K}}.\text{K}\,\frac{\text{kg}}{\text{J}} \text{ nondimensional}$$

$$(ii)\ \mu_l h_{fg} = \frac{\text{kg}}{\text{ms}}\cdot\frac{\text{J}}{\text{kg}} = \frac{\text{J}}{\text{ms}} = \frac{\text{W}}{\text{m}}$$

$$(iii) \left[\frac{g}{g_0}\left(\frac{\rho_l - \rho_v}{\sigma}\right)\right]^{0.5} \rightarrow \left[\frac{\text{m}}{\text{s}^2}\frac{\text{Ns}^2}{\text{kg/m}}\cdot\frac{\text{kg}}{\text{m}^3}\cdot\frac{\text{m}}{\text{N}}\right]^{0.5} = \frac{1}{\text{m}}$$

$$\therefore \quad i \times ii \times iii \; \frac{\text{J}}{\text{ms}}\cdot\frac{1}{\text{m}} \rightarrow \frac{\text{J}}{\text{s}}\cdot\frac{1}{\text{m}^2} \rightarrow \mathbf{W/m^2}$$

So dimensionally consistent

Such equations can be used with any system of units.

Table 11.1. Values of C_{sf} for various surface fluid combinations

Fluid	*Surface combination*	C_{sf}
Water	copper	0.0130
Water	brass	0.0060
Water	emery polished copper	0.0128
Water	scored copper	0.0068
Water	ground polished stainless steel	0.0080
Water	mechanically polished stainless steel	0.0132
Water	chemically etched stainless steel	0.0133
Water	platinum	0.0130
CC14	copper	0.0130
CC14	emery polished copper	0.0070
Ethyl alcohol	chromium	0.0027

Values for more combinations are available in handbooks. For estimates use 0.013 for other combinations. This results in ± 20% accuracy.

The maximum value of heat flux after which boiling becomes unstable is called critical heat flux. The correlation for the same was obtained by Kutateladze by applying dimensional analysis and by Zuber through a hydrodynamic stability analysis as given by equation (11.3).

$$q_{cr} = 0.149\, h_{fg} \left[g g_0 (\ _l - \ _v)\ ^2_v \right]^{1/4} \qquad \text{...(11.3)}$$

The theoretical value of the constant is 0.131 and it is replaced by 0.149 for better agreement with experimental results. The constant in equation (11.3) is applicable for horizontal surface for large characteristic length compared to the mean bubble diameter. Lienhard and Dhir have developed correction factors for other configurations. As h_{fg}, ρ_l, ρ_v and σ are dependent also on pressure, the critical heat flux is found to be influenced by the evaporating pressure. The value of maximum heat flux is found to increase with pressure, the peak value occurring at one-third of critical pressure for the liquid. The value then decreases and reaches a value of zero at P_{cr}. For water, the flux is about 3.8 MW/m^2 at atmospheric pressure. The maximum heat flux dictates the size and the surface temperature of the equipment and hence its importance in the design of evaporating equipments.

Simplified expression for boiling water (as in the case of air in free convection) is given by

$$h = C(\Delta T)^n\ (P/P_a)^{0.4} \qquad \text{...(11.4)}$$

where P is the system pressure and P_a is the standard atmospheric pressure.

The values of *C* and *n* are tabulated below:

Surface	*Applicable value of q, kW/m²*	*C*	*n*
Horizontal	$q < 15.8$	1040	1/3
	$15.8 < q < 236$	5.56	3
Vertical	$q < 3.15$	539	1/7
	$3.15 < q < 63.1$	7.95	3

When the excess temperature is more than that at $q_{\min}$ point a continuous vapour film blankets the surface. There is no contact between the liquid and the surface. This condition is found to be similar to laminar film condensation. The correlation for boiling heat transfer at this condition is based on the results obtained by condensation theory. The result is equation (11.5). The equation is applicable for spheres with the value of the constant as 0.67.

11.2.1. Unstable Film Boiling

Heat transfer occurs both by convection and radiation. The convection component for boiling on horizontal tube of diameter D is given by (with vapour properties at film temperature and ρ_l at saturation temperature).

$$h_c = 0.62\left[\frac{k_v^3\ {}_v(\ {}_l - \ {}_v)g(h_{fg} + 0.68\, c_{pv}\, \Delta T)}{{}_v D\, \Delta T}\right]^{0.25} \qquad \text{...(11.5)}$$

For flat surfaces the correlation is

$$h_c = 0.425\left[\frac{g(\ {}_l - \ {}_v)^{1.5}\ \ {}_v k_v^3 (h_{fg} + 0.68\, c_{pv}\, \Delta T)}{(g_0\ \ )^{0.5}\ .\ \ {}_v .\, \Delta T}\right]^{0.25} \qquad \text{...(11.6)}$$

where suffix v refers to vapour at film temperature $(T_w + T_s)/2$.

The radiation component is given by

$$h_r = \sigma_r\, \varepsilon\left[\frac{T_w^4 - T_{\text{sat}}^4}{T_w - T_{\text{sat}}}\right] \qquad \text{...(11.7)}$$

where σ_r is Stefan Boltzmann constant and ε is the emissivity of the surface.

The combined convection coefficient is given by $(h_r > h_c)$

$$h = h_r + h_c\, (h_c/h)^{1/3} \qquad \text{...(11.8)}$$

This equation has to be solved by trial.

Another suggestion when $h_r < h_c$ is $h = h_c + 0.75\, h_r$ ...[11.9(*a*)]

A correlation is available for the minimum heat flux in film boiling (1958).

$$\left(\frac{q}{A}\right)_{\min} = 0.09\, h_{fg}\, \rho_{vf} \left[\frac{gg_0 \;(\;_l - \;_v)}{(\;_l + \;_v)^2}\right]^{1/4} \qquad \text{...[11.9(b)]}$$

and ρ_{vf} is to be obtained at film temperature and ρ_l at saturation temperature.

11.3 FLOW BOILING

This is nothing but forced convection boiling where the fluid is moved by an external device. The regimes of boiling and the heat flux curve are similar to the ones in pool boiling. The process occurs in modern high pressure forced circulation boilers. As subcooled liquid enters a heated tube with wall heat-flux, initially heat transfer occurs by forced convection and the liquid is heated to saturation condition. After this condition, the flow is considered as two phase flow. The next regime is where bubbles form at the surface and then flow into the core where these may condense. At a later section the flow becomes slug type of flow with alternately liquid and vapour filling the tube. This is followed by what is termed as annular flow with liquid flowing near the surface and vapour forming the core. The heat flux reaches the maximum or critical value in this regime. The flow then becomes unstable with the liquid film breaking out. The heat flux cannot be sustained without the material becoming overheated. Then the flow becomes what is called mist flow with small droplets of liquid floating in the vapour. The heat flux at this regime may be even lower than that in the forced convection regime. At this condition the radiation heat transfer coefficient h_r will start to increase. In steam generators operation is limited to the annular flow regime as otherwise damage to steam pipe may occur. The phenomenon is shown in Fig. 11.2. The heat flux in forced convection boiling is calculated using.

$$q_{\text{Total}} = q_{\text{con}} + q_{\text{boiling}} \qquad \text{...(11.10)}$$

q_{con} is calculated using

$$Nu = 0.019\ \text{Re}^{0.8}\ \text{Pr}^{0.4} \qquad \text{...(11.11)}$$

where $$Re = \frac{G(1-x)D}{\mu_l}$$

where x is dryness G is mass flow rate and

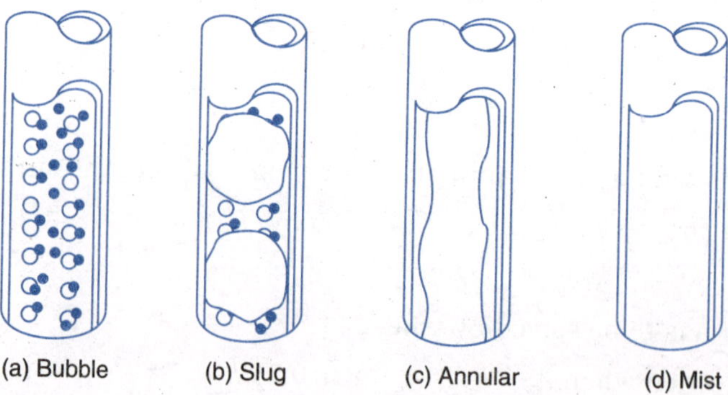

Flow regimes

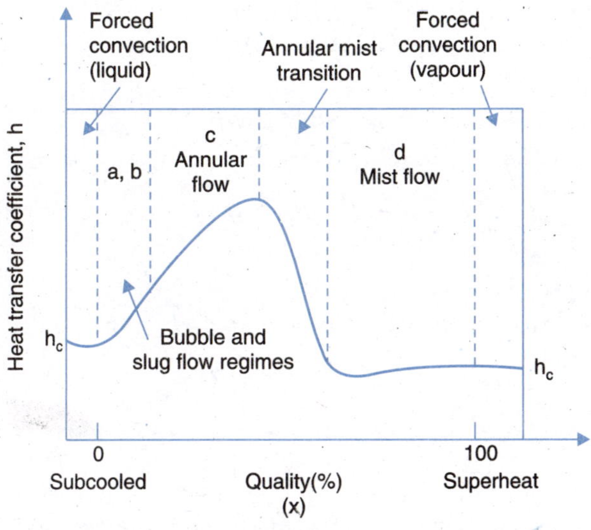

Fig. 11.2. Characteristics of forced-convection vapourization in a vertical tube: Heat transfer coefficient versus quality and type of flow regime

$$q_{boiling} = 2.253\ (\Delta T)^{3.96}\ \text{W/m}^2 \quad ...(11.12)$$

for pressures $\quad 0.2 < P < 0.7$ MPa

$$q_{boiling} = 283.2\ P^{4/3}\ (\Delta T)^3\ \text{W/m}^2 \quad ...(11.13)$$

for $\quad 0.7 < P < 4$ MPa.

Example 11.2: ***Water at*** *atmospheric pressure (saturation temperature = 100°C) is boiling on a brass surface heated from below. If the surface is at 108°C, determine the heat flux and compare the same with critical heat flux.*

Solution: The property values are taken at the liquid temperature

$$\rho_c = 961\ \text{kg/m}^3,\ h_{fg} = 2257\ \text{kJ/kg},\ \rho_v = 0.598\ \text{kg/m}^3,\ c = 4216\ \text{J/kg K}$$

$$\mu_l = 2.816 \times 10^{-4}\ \text{kg/m s},\ \sigma = 58.8 \times 10^{-3}\ \text{N/m},\ Pr = 1.74$$

From Table 11.1 $C_{sf} = 0.0060$, for water $n = 1$

using equation (11.2), (h_{fg} in J/kg)

$$\mathbf{q} = \left[\frac{c_l\ \Delta T}{c_{sf}\ h_{fg}\ Pr_l^n}\right]^3 . \mu_l\ h_{fg} \left[\frac{g(\rho_l - \rho_v)}{g_0\ \sigma}\right]^{0.5}$$

$$= \left[\frac{4216 \times 8}{0.0060 \times 2257 \times 10^3 \times 1.74}\right]^3 . 2.816 \times 10^{-4} \times 2257 \times 10^3 \left[\frac{9.81(961 - 0.598)}{1 \times 58.8 \times 10^{-3}}\right]^{0.5}$$

$$= \mathbf{0.746 \times 10^6\ W/m^2}$$

Critical heat flux is given by equation (11.3)

$$\mathbf{q_{cr}} = 0.149\ h_{fg}\ [\sigma\ g\ g_0\ (\rho_1 - \rho_v)\ \rho_v^{\ 2}]^{1/4}$$

$$= 0.149 \times 2257 \times 10^3\ [58.8 \times 10^{-3} \times 9.81 \times 1\ (961 - 0.598) \times 0.598^2]^{0.25}$$

$$= \mathbf{1.262 \times 10^6\ W/m^2}$$

The actual flux is less than the critical flux at ΔT = 8°C and hence pool boiling exists. The critical flux is found to occur at ΔT = 10.5°C when substituted in equation (11.2). **Using simplified expression,** equation (11.4)

$$\mathbf{h} = 5.56\ (8)^3\ (1)^{0.4} = \mathbf{2846.72\ W/m^2\ K}$$

$$q = h\ \Delta T = 0.0228 \times 10^6\ \text{W/m}^2\text{, A lower prediction.}$$

Example 11.3: *For example 11.2 calculate the flux if the surface was stainless steel mechanically polished.*

Solution: The value of C_{st} = 0.0132

It is noted from equation (11.2), that the **heat flux is inversely proportional to the cubic power of C_{sf}.**

$$\therefore \qquad q = 0.746 \times 10^6 \times \left(\frac{0.0060}{0.0132}\right)^3 = \mathbf{70060\ or\ 0.07 \times 10^6}$$

The heat flux is reduced to about 1/10th value determine with brass. This is more due the roughness as polished surface will not provide large number of nucleation sites.

Example 11.4: *In example 11.2, if the excess temperature is increased to 12°C determine the heat flux and compare with the critical value.*

Solution: It is seen from equation (11.2) that **the heat flux is proportional to $(\Delta T)^3$**

$$\therefore \qquad \mathbf{q} = 0.746 \times 10^6 \left(\frac{12}{8}\right)^3 = \mathbf{2.52 \times 10^6\ W/m^2}$$

This is higher than the critical heat flux and this is not possible. So unstable film boiling may start at this condition.

Hence, it is desirable to calculate the value of q using equation (11.6)

$$h_c = 0.425 \left[\frac{g(\rho_l - \rho_v)^{1.5}\ \rho_v k_v^3\ [h_{fg} + 0.68\, c_{pv}\ \Delta T]}{(g_0\ \sigma)^{0.5}\ \mu_v \,.\, \Delta T}\right]^{0.25}$$

From tables,

k_v = 0.02373 W/mK, $\mu_v = 11.96 \times 10^{-6}$ kg/m s, c_{pv} = 2135 J/kg K

$$h_c = 0.425 \left[\frac{9.81(961 - 0.598)^{1.5} \times 0.02373^3 \times [2257 \times 10^3 + 0.68 \times 2135 \times 12]}{(1 \times 58.8 \times 10^{-3})^{0.5} \times 11.96 \times 10^{-6} \times 12/0.598}\right]^{0.25}$$

$$h = 266\ \text{W/m}^2\ \text{K}$$

$$\mathbf{q = h\ \Delta T = 3186\ W/m^2}$$

Heat flux is considerably reduced.

Note: $\mathbf{h_r} = \sigma\ \varepsilon_s \left[\frac{T_s^4 - T_{sat}^4}{T_s - T_{sat}}\right]$ assuming $\varepsilon_s = 0.6 = 5.67 \times 0.6\ \frac{(3.85)^4 - (3.73)^4}{12}$ = **7.41 W/m² K.**

The contribution due to radiation is negligible at this level.

11.4 CONDENSATION

When saturated vapour comes in contact with a cooler surface, the vapour condenses into liquid. The surface temperature should be lower in this case as compared to the temperature of the vapour. The condensate generally moves down by gravity. **If the liquid wets the surface a thin layer of liquids forms over the surface and the film thickness increases along the downward direction. This type of condensation is known as filmwise condensation and this is the type encountered in most practical situations.** The film introduces a resistance to heat flow between the surface and the vapour. The heat transfer rate is reduced because of this resistance. **If the surface is nonwetting, then droplets form on the surface and these roll down individually.** The vapour is in direct contact with the surface over most of the area and heat transfer rates are much higher as there is very little resistance for heat flow between the vapour and the surface. **This type is known as dropwise condensation.** In some situations special surface coating is used to promote dropwise condensation and obtain higher rates of heat transfer. In practice no surface is found to continue as nonwetting over any length of time. So using the value of heat transfer coefficients assuming dropwise condensation for design purposes is not advisable.

11.4.1. Formation of Condensate Film

As condensate forms on the surface, the surface is fully wetted and the liquid begins to flow down. Under steady conditions additional condensate is added along the length and the film thickness increases in the direction of flow. The velocity and temperature distribution in the film are shown in Fig. 11.3.

An element considered for force analysis is shown in Fig. 11.4. The forces encountered are the viscous shear, gravity force and the buoyant force. The force balance is gravity force acting downwards

= Shear force acting upwards + buoyancy force acting upwards.

$$\rho_l g(\delta - y)\, dx = \mu_l \frac{du}{dy}\, dx + \rho_v g(\delta - y)\, dx \quad \text{...(11.14)}$$

Integrating and using the boundary that $u = 0$ at $y = 0$

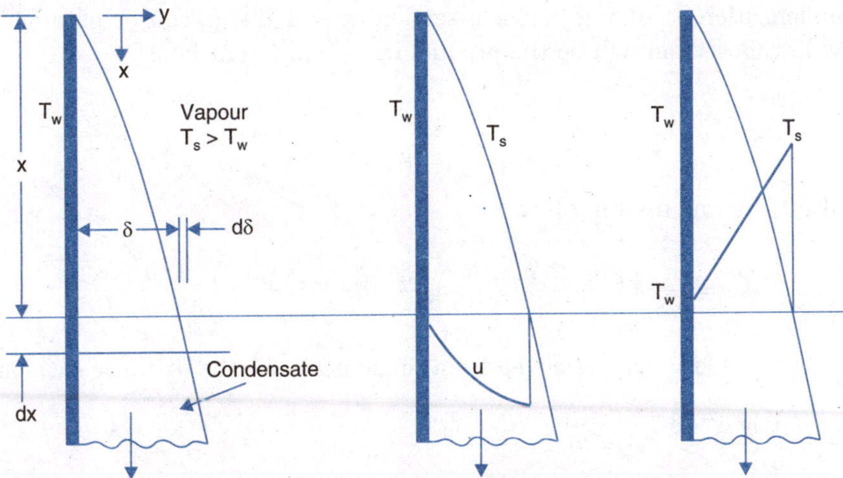

Fig. 11.3. Formation of condensate film

Chapter 11

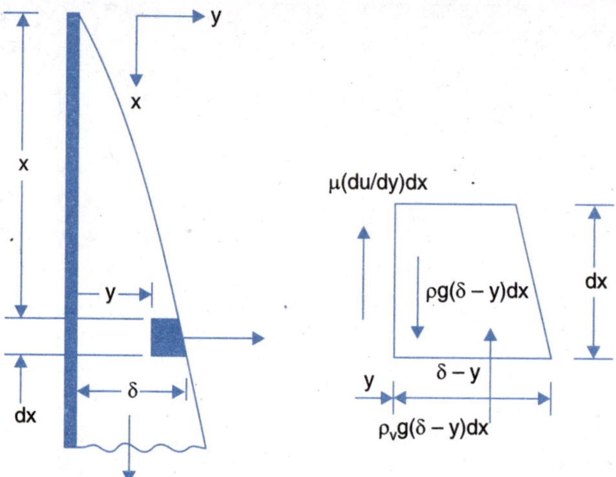

Fig. 11.4. Element considered for force analysis and the forces on the element

$$u = \frac{(\rho_l - \rho_v)g}{\mu_l}\left(\delta\, y - \frac{1}{2}y^2\right) \quad ...(11.15)$$

The total mass flow at any section can be determined by integrating the flow using the velocity distribution.

$$m = \int_0^\delta \rho_l\, u\, dy = \int_0^\delta \rho_l \left[\frac{(\rho_l - \rho_v)}{\mu_l}.g\left(\delta y - \frac{1}{2}y^2\right)\right] dy$$

$$= \frac{\rho_l(\rho_l - \rho_v)g\delta^3}{3\mu_l} \quad ...(11.16)$$

Assuming a linear variation of temperature in the film,

$$q = -k_l\, dx \left.\frac{dT}{dy}\right|_{y=0} = k_l\, dx\, \frac{T_g - T_w}{\delta} \quad ...(11.17)$$

The condensation occurring between sections x and $x + dx$ can be obtained using Δm. The heat flow in this section will be the product mass and latent heat.

$$\Delta m = \frac{dm}{dx}\,.\, dx = \frac{d}{dx}\left[\frac{\rho_l(\rho_l - \rho_v)g\,\delta^3}{3\mu_l}\right] dx \quad ...(11.18)$$

By multiplying and dividing by $d\delta$

$$\Delta m = \frac{d}{d\delta}\left[\frac{\rho_l(\rho_l - \rho_v)g\,\delta^3}{3\mu_l}\right]\frac{d\delta}{dx}\,.\, dx = \frac{\rho_l(\rho_l - \rho_v)g\,\delta^2\, d\delta}{\mu_l} \quad ...(11.19)$$

As heat conducted = heat removed from condensate, for the distance dx along the flow,

$$\frac{\rho_l(\rho_l - \rho_v)g\,\delta^2\, d\delta}{\mu_l}\, h_{fg} = k_l\, dx\, \frac{T_g - T_w}{\delta} \quad ...(11.20)$$

Integrating and using $\delta = 0$ at $x = 0$, yields,

$$\delta_x = \left[\frac{4\mu_l k_l x\,(T_g - T_w)}{g\,h_{fg}\,\rho_l(\rho_l - \rho_v)}\right]^{0.25} \qquad \text{...(11.21)}$$

From basics, heat transfer at the surface by conduction is equal to the heat convected. Assuming linear temperature variation,

$$h\,dx\,(T_w - T_g) = -\,k_l\,dx\,\frac{(T_g - T_w)}{\delta} \qquad \text{...(11.22)}$$

$$\therefore \qquad h = \frac{k_l}{\delta} \qquad \text{...(11.23)}$$

$$\therefore \qquad h_x = \left[\frac{\rho_l(\rho_l - \rho_v)g h_{fg}\,k_l^3}{4\mu_l\,x(T_g - T_w)}\right]^{1/4} \qquad \text{...(11.24)}$$

$$Nu_x = \frac{hx}{k_l} = \left[\frac{\rho_l(\rho_l - \rho_v)g h_{fg}\,x^3}{4\mu_l\,k_l\,(T_g - T_w)}\right]^{1/4} \qquad \text{...[11.25(a)]}$$

The average value of convection coefficient up to lengths L is obtained by

$$\overline{\mathbf{h}} = \frac{1}{L}\int h_x\,dx = \frac{\mathbf{4}}{\mathbf{3}}\,\mathbf{h}\bigg|_{\mathbf{x = L}} \qquad \text{...[11.25(b)]}$$

or

$$\bar{h} = 0.943\left[\frac{\rho_l(\rho_l - \rho_v)g h_{fg}\,k_l^3}{L\mu_l\,(T_g - T_w)}\right]^{1/4} \qquad \text{...(11.26)}$$

For higher values of lengths, 0.943 is replaced by 1.13.

This expression is obtained by direct analysis with the assumption of linear temperature profile. The property values are at film temperature. In order to take care of the actual nonlinear temperature variation and also undercooling the **enthalpy of evaporation is corrected in equation (11.26) by**

$$\mathbf{h_{fg} = h_{fg}\left[1 + 0.68\,\frac{c_l(T_s - T_w)}{h_{fg}}\right]}$$

where c_l is the specific heat of liquid.

It is found that experimental measurements correlate better when properties are evaluated at film temperature. For condensation on horizontal tubes of diameter d the correlation is

$$h = 0.725\left[\frac{\rho_l(\rho_l - \rho_v)g h_{fg}\,k_f^3}{\mu_l\,d(T_g - T_w)}\right]^{1/4} \qquad \text{...(11.27)}$$

and for *n* tube rows in vertical direction, *nd* replaces *d*.

Chapter 11

11.4.2. Turbulent Film

As the condensate film thickness increases, the flow becomes turbulent. In order to define this condition a **film Reynolds number Re_δ** concept is used.

$$\mathbf{Re_\delta} = \frac{\rho_l \mathbf{u_m} \mathbf{D_h}}{\mu_l} \text{ and } \mathbf{D_h = 4A_c/P}$$

where A_c is the flow area $1 \times \delta$ and P is the wetted perimeter (1 m). This can be simplified to

$$\mathbf{Re_\delta} = \frac{\mathbf{4G}}{\mu_l} \; r = \frac{4\rho_l\, u_m \delta}{\mu_l} \qquad \text{...(11.28)}$$

where G is the mass flow per unit depth of plate given by

$$\mathbf{G = \rho_l\, u_m\, \delta}$$

G can be determined using equations (11.16) and (11.21)

For laminar flow, $Re < 30$

$$\frac{\bar{h}_L\,(\nu_l^2/g)^{1/3}}{k_l} = 1.47\, Re^{-1/3} \qquad \text{...[11.28(a)]}$$

For $30 < Re < 1800$

$$\frac{\bar{h}_L\,(\nu_l^2/g)^{1/3}}{k_l} = \frac{Re}{1.08\, Re^{1.22} - 5.2} \qquad \text{...[11.28(b)]}$$

For $Re_\delta > 1800$ the flow is turbulent. The correlation for convection coefficient is

$$\bar{h}_c = 0.0077 \left[\frac{g\rho_l\,(\rho_l - \rho_v)\, k_l^3}{\mu_l^2}\right]^{1/3} Re_\delta^{0.4} \qquad \text{...(11.29)}$$

Another correlation in the turbulent range is

$$\frac{\bar{h}_L\,(\nu_l^2/g)^{1/3}}{k_l} = \frac{Re}{8750 + 58\, Pr^{-0.5}\,(Re^{0.75} - 253)} \qquad \text{...[11.29(a)]}$$

A concept called **condensation number** is also used in the above equation. The condensation number is defined as

$$CO = \bar{h}\left[\frac{\mu_l^2}{k_l^3\,\rho_l\,(\rho_l - \rho_v)\, g}\right]^{1/3} \qquad \text{...(11.30)}$$

This leads to the rewriting of equation (11.29) as

$$CO = 0.0077\, Re_\delta^{0.4} \qquad \text{...(11.31)}$$

After the determination of CO from (11.31), h is calculated using (11.30)

For lower film Reynolds numbers **for vertical plate** ($Re_\delta < 1800$)

$$CO = 1.47\, Re_\delta^{(-1/3)} \qquad \text{...(11.32)}$$

For horizontal cylinders for ($Re_\delta < 1800$)

$$CO = 1.514\, Re_\delta^{-1/3} \qquad \text{...(11.33)}$$

11.4.3. Condensation Inside Tubes

In refrigeration where the condensed refrigent has to be carefully collected condensation occurs inside horizontal tubes. For low vapour velocities the condensate will collect at the bottom and the flow will be along the horizontal direction. Vapour will be occupying the upper half of the pipe. If the vapour velocity is high the liquid will occupy an annular shape.

The correlation is available for the former case of low vapour velocity and is given by equation (11.34). The correlation for higher velocities of flow are available in Hand Book of Heat Transfer by Rohsenow.

$$\bar{h} = 0.555 \left[\frac{\rho_l (\rho_l - \rho_v) g\, k_l^3\, h'_{fg}}{\mu_l\, d (T_g - T_w)}\right]^{1/4} \quad ...(11.34)$$

where $h'_{fg} = h_{fg} + 0.68\, c_l\, (T_g - T_w)$.

Example 11.5: *Check the dimensional consistency of equation (11.24)*

$$h_x = \left[\frac{\rho_l (\rho_l - \rho_v) g\,.\, h_{fg}\, k_l^3}{4\mu x\, (T_g - T_w)}\right]^{1/4}.$$

Solution: LHS W/m² K

$$\textbf{RHS} \left[\frac{\text{kg}}{\text{m}^3} \cdot \frac{\text{kg}}{\text{m}^3} \cdot \frac{\text{m}}{\text{s}^2} \cdot \frac{\text{J}}{\text{kg}} \cdot \frac{\text{W}^3}{\text{m}^3\,\text{K}^3} \cdot \frac{\text{ms}}{\text{kg}} \cdot \frac{1}{\text{m}} \cdot \frac{1}{\text{K}}\right]^{1/4},$$

Cancelling kg, and using $\frac{J}{s}$ as W, RHS $= \left[\frac{\text{W}^4}{\text{m}^8\text{K}^4}\right]^{1/4}$ = **W/m² K.**

The equation is dimensionally consistent. Such equations are independent of the system of units used. Equation (11.34) is also dimensionless may be noted.

Example 11.6: *Saturated steam at a temperature of 65°C condenses on a vertical surface at 55°C. Determine the thickness of the condensate film at locations 0.2, 0.4, 0.6, 0.8, 1 m from the top. Also determine the condensate flow, the film Reynolds number, the local and average values of convective heat transfer coefficients at these locations. Also calculate the condensation numbers.*

Solution: The property values for liquid should be taken at the film temperature = (55 + 65)/2 = 60°C. The liquid property values at 60°C are

$\rho_l = 985$ kg/m³, $k_l = 0.6513$ W/m K, $c = 4183$ J/kg K

$\mu_l = 4.7083 \times 10^{-4}$ kg/ms,

h_{fg} at 65°C = 2346.2 kJ/kg, ρ_v 1/6.197 kg/m³

Considering unit width: using equation (11.21).

$$\delta = \left[\frac{4\mu kx\, (T_g - T_w)}{g h_{fg}\, \rho_l\, (\rho_l - \rho_v)}\right]^{0.25} = \left[\frac{4 \times 4.7083 \times 10^{-4} \times 0.6513 \times 10x}{9.81 \times 2346.2 \times 10^3 \times 985(985 - 1/6.197)}\right]^{0.25}$$

$$= \mathbf{1.53 \times 10^{-4}\,.\, x^{0.25}}$$

$$\text{G} = m = \frac{\rho_l (\rho_l - \rho_v) g\,.\, \delta^3}{3\mu} = \left[\frac{985\,(985 - 1/6.197)9.81}{3 \times 4.7083 \times 10^{-4}}\right] \delta^3$$

$$= 6.73 \times 10^9\, \delta^3 \text{ kg/m width} = \mathbf{0.024176 \times x^{0.75}}$$

Chapter 11

Using equation (11.28)

$$\mathbf{Re}_\delta = \frac{4G}{\mu_l} = \frac{4}{4.7083 \times 10^{-4}} G = 5.718 \times 10^{13} \cdot \delta^3 = \mathbf{205.39 \times x^{0.75}}$$

All through the flow is laminar. Using equation (11.25)

$$h_x = \left[\frac{\rho_l(\rho_l - \rho_v) g\, h_{fg}\, k^3}{4\mu_l x (T_g - T_w)}\right]^{1/4} = 4254.33 \times \left(\frac{1}{x}\right)^{0.25}, \quad \bar{h} = \frac{4}{3} h_L$$

Using equation (11.32) **CO** = 1.47 $Re_\delta^{-1/3}$ = **3.816 × $10^{-5}/\delta$**

These values at various locations are tabulated below:

for flow rate m, use equation (11.16).

Distance m	***0.2***	***0.4***	***0.6***	***0.8***	***1.0***
δ, mm	0.10238	0.12175	0.13474	0.14479	0.1531
m, kg/s/m	0.00723	0.01216	0.0165	0.02045	0.02418
Re_δ	61.42	103.31	140.18	173.74	205.4
h_x	6361.7	5349.5	4833.9	4498.4	4254.3
$\bar{\mathbf{h}}$	**8482.3**	**7132.7**	**6445.2**	**5997.9**	**5672.4**
CO	0.3730	0.3133	0.2830	0.2634	0.2491

The value of h_{fg} can be corrected to take care of under cooling.

11.4.4. Inclined Surfaces

For inclined surfaces, g is replaced by $g \cos \theta$ where θ is the angle with the vertical.

In refrigeration, plate type of condensers are used with horizontal tubes brazed on them. The correlation for the same in terms of condensation number is

$$CO = 1.51\, Re_\delta^{-1/3} \qquad \text{...(11.35)}$$

Some simplified expressions are available using condensation coefficient which depends on the fluid and the temperature. These are applicable for vertical and horizontal tubes. These expressions are to be used with a table of values of condensation coefficients. Some examples are given under advanced examples.

Example 11.7: *Saturated steam at 65°C condenses on horizontal cylinder of 0.2 m dia at 55°C. Determine the value of convection coefficient for (i) single tube and (ii) for a bank of tubes of 10 rows arranged vertically one below the other.*

Solution: The property values are as in example 11.6 as the film temperature is 60°C and the fluid is steam.

Using equation (11.27)

(*i*) For single tube:

$$\mathbf{h} = 0.725 \left[\frac{\rho_l(\rho_l - \rho_v) g\, h_{fg}\, k_f^3}{\mu_f\, d (T_g - T_w)}\right]^{1/4}$$

$$= 0.725 \left[\frac{985\,(985 - 1/6.197)\,9.81 \times 2346.2 \times 10^3 \times 0.6513^3}{985 \times 0.478 \times 10^{-6} \times 0.2 \times 10} \right]^{1/4}$$

$$= \mathbf{6522.4\ W/m^2\ K.}$$

(*ii*) **For a bank of tubes** *nd* is to be used in place of *d* in the above expression

$$\therefore \qquad \mathbf{h = 3667.8\ W/m^2\ K} \qquad \left(\text{Check } 6522.4 \times \frac{1}{10^{0.25}}\right)$$

This reduction is because of the thicker film caused by the condensate from one tube falling and covering the lower tube.

Example 11.8: *Steam at 65°C condenses on vertical tubes of diameter of 0.3 m at 55°C. Determine the location at which the film will become turbulent.*

Solution: Turbulence sets in when

$$\frac{4G}{\mu} = 1800$$

Taking property values from example 11.6,

$$G = \frac{1800}{4} \times \mu = \frac{1800}{4} \times 985 \times 0.478 \times 10^{-6} = 0.2118735 \text{ kg/sm}$$

The length at which turbulence occurs L_T is obtained by considering mass flow for unit depth

$$G = m = \frac{\rho_l(\rho_l - \rho_v)g\,\delta^3}{3\mu}$$

and

$$\delta = \left[\frac{4\,\mu\,kx\,(T_g - T_w)}{g\,h_{fg}\,\rho_l\,(\rho_l - \rho_v)}\right]^{0.25}$$

Solving for *x*,

$$0.2118735 = \frac{\rho_l(\rho_l - \rho_v)g}{3\mu}\left[\frac{4\,\mu\,k\,(T_g - T_w)}{g\,h_{fg}\,\rho_l\,(\rho_l - \rho_v)}\right]^{0.75} x^{0.75}$$

$$\therefore \qquad \mathbf{L_T = x = 18.07\ m.}$$

In this case, it takes considerable length for flow to become turbulent.

11.5 FREEZING AND MELTING

When a liquid in a vessel is exposed on the top to a temperature lower than the freezing temperature, the liquid on the top layer will solidify first and slowly the thickness of the solid layer will increase. The thickness of solid will increase with time. Heat is conducted through the solid and then convected at the interface. The solid and the liquid have completely different thermal properties. If the temperature of the liquid is higher than the freezing temperature it has to be cooled by removal of sensible heat. Similarly, if a solid at freezing temperature is exposed to higher temperature, the solid will begin to melt and if the liquid is not removed, heat has to be conducted through the liquid layer and then convected at the interface. Other configuration like long cylinders may be also encountered. Solidification in a cylinder may be from inside-out or outside-in as required. The model is rather complex and solutions are available only for simpler situations.

11.5.1. Melting of Solid at Freezing Temperature

Melting of solid at freezing temperature when exposed to a higher temperature by convection at the surface.

The physical situation is shown in Fig. 11.5. The assumptions are that the soild is at freezing temperature and the heat required to increase the temperature of the liquid is small compared to the heat of fusion of ice to be melted.

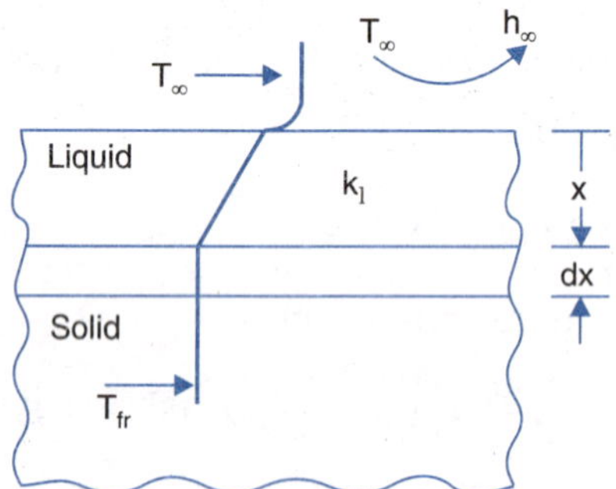

Fig. 11.5. Melting of solid at melting temperature

Then, for unit area, the heat flow at any given instant when the liquid layer is x m thick with conductivity k_l is

$$q = \left[\frac{T_\infty - T_{fr}}{\frac{1}{h_\infty} + \frac{x}{k_l}}\right] \quad \text{...(11.36)}$$

If a layer of thickness dx melts during time $d\tau$, the rate of energy required for melting is given by (for unit area)

$$q = \rho_s\, h_{sf} \frac{dx}{d\tau} \quad \text{...(11.37)}$$

where ρ_s is the density of solid and h_{sf} is the heat of fusion.

Equating and rearranging

$$\left[\frac{h_\infty (T_\infty - T_{fr})}{\rho_s\, h_{sf}}\right] d\tau = \left[1 + \frac{xh_\infty}{k_l}\right] dx \quad \text{...[11.37(a)]}$$

Now introducing a set of new variables

$$\mathbf{x^* = \frac{xh_\infty}{k_l} \text{ and } \tau^* = \left[\frac{h_\infty^{\,2}\,(T_\infty - T_{fr})}{k_l\, \rho_s\, h_{sf}}\right] \tau} \quad \text{...(11.38)}$$

The equation [11.37(*a*)] will become

$$\mathbf{d\tau^* = (1 + x^*)\, dx^*} \quad \text{...(11.39)}$$

As RHS of equation (11.37) becomes

$$x^* = \frac{xh_\infty}{k_l} \quad \therefore \quad \frac{k_l}{h_\infty} \,.\, dx^* = dx,$$

$$\therefore \qquad \left(1 + \frac{xh_\infty}{k_l}\right) dx = \frac{k_l}{h_\infty} (1 + x^*)\, dx^*$$

LHS of equation [11.37(*a*)] becomes, using (11.38)

$$d\tau^* \, (k_l/h_\infty)$$

Equation (11.39) can be integrated from time 0 to time τ or time τ_1 to τ_2

This gives $\tau^* = x^* + 0.5x^{*2}$ ...(11.40)

Equation (11.40) can be used to determine the time required for a given depth of layer to melt or the layer thickness melted during a given time. Solid at a temperature lower than freezing temperature is not considered.

11.5.2. Liquid Solidifying at Freezing Temperature

Equation (11.40) is directly applicable except that k_l should now replace k_s and x is thickness of solid layer and not liquid layer. The model is shown in Fig. 11.6. Here also the time for freezing of certain thickness or the thickness frozen during a given time interval are calculated using equation (11.40).

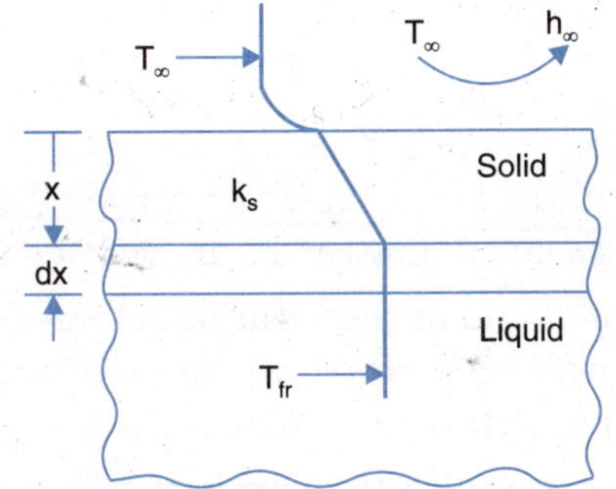

Fig. 11.6. Freezing of liquid at freezing temperature

11.5.3. A More Practical Case of Freezing is when the Liquid is Above Freezing Temperature

The model is shown in Fig. 11.7.

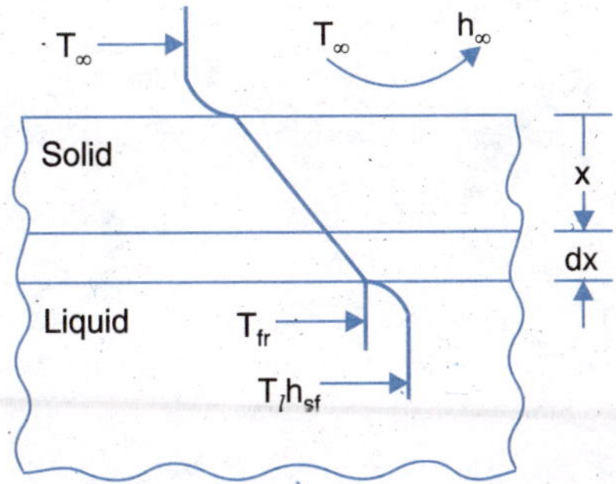

Fig. 11.7. Freezing of liquid at temperature higher than the freezing point

Chapter 11

In this case convection is encountered at the liquid solid interface also.

In this case heat is removed not only for freezing but also for cooling the liquid.

Heat removed for freezing $= \rho_s \, h_{sf} \, dx/d\tau$

Heat removed for cooling the liquid $= h_c \, (T_l - T_{fr})$

The total heat should pass through the solid layer and to the outside.

$$\frac{T_{fr} - T_\infty}{\left(\dfrac{x}{k_s}\right) + \left(\dfrac{1}{h_\infty}\right)} = \frac{T_l - T_{fr}}{(1/h_l)} + \frac{\rho_s \, h_{sf} \, . \, dx}{d\tau} \quad ...(11.41)$$

Defining four new variables

$$T^* = \frac{T_l - T_{fr}}{T_{fr} - T_\infty}, \quad x^* = \frac{x h_\infty}{k_s}$$

$$\tau^* = \left[\frac{h_\infty^2 \, (T_{fr} - T_\infty)}{k_s \, \rho_s \, h_{sf}}\right] \tau \, , \; H^* = \frac{h_l}{h_\infty} \quad ...(11.42)$$

Equation (11.41) reduces to

$$1/(x^* + 1) = H^*T^* + dx^*/d\tau^* \quad ...(11.43)$$

Rearranging and integrating from time zero to time τ,

$$\tau^* = \left(\frac{1}{H^*T^*}\right)^2 \ln\left[\frac{1 - H^*T^*}{1 - H^*T^*(1 + x^*)}\right] - \frac{x^*}{H^*T^*} \quad ...(11.44)$$

In this case, the time required for freezing to a certain depth is directly solvable but the inverse *i.e.*, depth frozen during a given time can be arrived at only by trial or by chart solution.

For the case of freezing of liquid at the saturation temperature, $H^*T^* = 0$ and equation (11.43) should be integrated neglecting H^*T^*, which gives the same equation as equations (11.39) and (11.40) except for properties.

Equation (11.44) should not be directly used in this case, as this will lead to division by zero.

11.5.4. Freezing Inside of Cylinder-Saturated Liquid-Inward and Outward

The model is shown in Fig. 11.8(*a*) and (*b*)

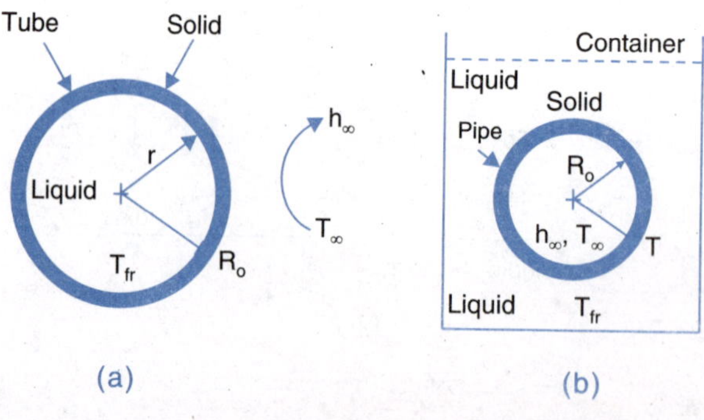

Fig. 11.8. (*a*) Cylinder-Freezing inwards. (*b*) Cylinder-Freezing outwards

Defining $r^* = r/R_0 \quad R^* = h_\infty R_0/k_s$

$$\tau^* = \frac{(T_{fr} - T_\infty)\, k_s \tau}{\rho_s\, h_{sf}\, R_0^2} \quad ...(11.45)$$

For inward freezing the equation is

$$\tau^* = 0.5\, r^{*2} \ln r^* + \left[\frac{1}{2R^*} + 0.25\right](1 - r^{*2}) \quad ...(11.46)$$

For outward freezing the equation is

$$\tau^* = 0.5\, r^{*2} \ln r^* + \left[\frac{1}{2R^*} - 0.25\right](r^{*2} - 1) \quad ...(11.47)$$

Here also it is easier to calculate directly the time required for a given depth of freezing. The inverse is to be solved by trial, *i.e.*, the thickness frozen in a specified time. Charts are available in data books for ease of calculation in this case.

Example 11.9: *Water at 0°C is exposed over the surface to convection at – 11°C with h_∞ = 570 W/m² K. Determine the time required for ice of 3 mm thickness to form.*

Solution: $k_s = 2.22$ W/m K, $h_{sf} = 333700$ J/kg, $\rho_s = 918$ kg/m³

$$\tau^* = x^* + 0.5\, x^{*2}$$

$$\tau^* = \left[\frac{h_\infty^2 (T_{fr} - T_\infty)}{k_s\, \rho_s\, h_{sf}}\right]\tau,\ x^* = \frac{x h_\infty}{k_s},\ x^* = \frac{0.003 \times 570}{2.22} = 0.77027,$$

$\therefore$
$$\tau^* = 1.066928$$

$$\tau = \tau^*\, k_s\, \rho_s\, h_{sf} - h_\infty^{\,2}\, (T_{fr} - T_\infty)$$

$$= 1.066928 \times 2.22 \times 918 \times 333700/(570^2 \times 11) = \mathbf{203\ seconds}$$

It requires about 3.5 minutes to form 3 mm layer of *ice.*

Example 11.10: *In example 11.9, if the water is at 8°C with a convection coefficient of 57 W/m² K at the surface, and if the convection temperature at the surface is – 20°C, determine the time for ice to form to a thickness of 4 mm.*

Solution: This is freezing of water from a temperature above freezing point.

$$T^* = \frac{T_l - T_{fr}}{T_{fr} - T_\infty} = \frac{8}{20} = 0.4$$

$$x^* = 0.004 \times 570/2.22 = 1.027$$

$$H^* = h_l/h_\infty = 57/570 = 0.1$$

$$\tau^* = \left(\frac{1}{H^*T^*}\right)^2 \ln\left[\frac{1 - H^*T^*}{1 - H^*T^*(1 + x^*)}\right] - \frac{x^*}{H^*T^*}$$

$$= \left(\frac{1}{0.4 \times 0.1}\right)^2 \ln\left[\frac{1 - 0.4 \times 0.1}{1 - 0.4 \times 0.1(1 + 1.027)}\right] - \frac{1.027}{0.4 \times 0.1} = 1.66$$

$$\tau = \tau^*\, k_s \,.\, \rho_s\, h_{sf}/h_\infty^{\,2}\, (T_{fr} - T_\infty)$$

$$= 1.66 \times 2.22 \times 918 \times 333700/(570^2 \times 20) = \mathbf{173.6\ s}$$

Note that a lower surface temperature is used in this case, and hence the shorter time requirement.

Example 11.11: *A long cylindrical drum of 0.4 m, dia containing water at 0°C is exposed on the outside to a convection of 600 W/m² K, at – 20°C. Determine the time required for ice to form to a depth of 0.1 m.*

Solution: Using the data from Example 11.9, (Freezing inside cylinder)

$$r^* = r/R_0 = \frac{0.1}{0.2} = 0.5$$

$$R^* = 600 \times 0.2/2.22 = 54.054$$

$$\tau = \tau^* \rho_s h_{sf} R_0^2/k_s (T_{fr} - T_\infty)$$

$$\tau^* = 0.5\, r^{*2} \ln r^* + \left(\frac{1}{2R^*} + 0.25\right)(1 - r^{*2})$$

$$= -0.00866 + 0.194437 = 0.107794$$

∴ τ = **29749 s** (from definition of τ^*, 11.45) or 8.26 hours

Note that the thickness is rather large (10 cm) and hence the longer time period.

ADVANCED EXAMPLES

Example A-11.1: *Water is evaporated under pool boiling conditions at a pressure of 40 bar over a mechanically polished stainless steel surfaces. Calculate the critical heat flux and the excess temperature at which it occurs assuming nucleate boiling to continue up to this point.*

Solution: The temperature corresponding to the pressure is 250.4°C. The property values: at 250.4°C are:

$$\rho_v = 20.09 \text{ kg/m}^3,\ \rho_l = 799.9 \text{ kg/m}^3,\ h_{fg} = 1714.1 \text{ kJ/kg}$$

$$\sigma = 26.1 \times 10^{-3} \text{ N/m. From Table 11.1 } C_{sf} = 0.0132$$

Using equation (11.3) for calculating critical heat flux and substituting the property values.

$$\mathbf{q_{cr}} = 0.149\, h_{fg} (\sigma g g_0 (\rho_l - \rho_v) \rho_v^2)^{1/4}$$

$$= 0.149 \times 1714.1 \times 10^3 (26.1 \times 10^{-3} \times 9.81 \times 1\,(799.9 - 20.09)\, 20.09^2)^{1/4}$$

$$= \mathbf{4.3031 \times 10^6\ W/m^2}.$$

Note: In the text, for water the maximum heat flux is mentioned as 3.8 MW/m².

Assuming that this occurs in the limiting nucleate boiling condition, equation (11.2) is used. Other property values required are c_1 = 4856.36 J/kg K, μ_l = 1.119 × 10⁻⁴ kg/ms

(The viscosity value is determined as $\nu\rho$), Pr = 0.8725

$$q = \left[\frac{C_l\, \Delta T}{h_{fg}\, Pr\, C_{sf}}\right]^3 \mu_l\, h_{fg} \left[\frac{g(\rho_l - \rho_v)}{g_0\, \sigma}\right]^{0.5}$$

$$= \left[\frac{4856.36 \times \Delta T}{1714.1 \times 10^3 \times 0.8725 \times 0.0132}\right]^3 .\ 1.119 \times 10^{-4} \times 1714.1 \times 10^3 \quad ...(1)$$

$$\times \left[\frac{9.81(799.9 - 20.09)}{1 \times 26.1 \times 10^{-3}}\right]^{0.5} = 1545.9 \times \Delta T^3$$

Substituting q_{cr} for q, ΔT = 14.98°C

This value can be verified from plots of q vs ΔT for various pressures available in lterature.

In this case Nucleate pool boiling will occur up to an excess temperature of 14.98°C.

Example A-11.2: *Determine for the situation of example A-11.1, the heat flux for excess temperatures of 2.5, 5, 7.5, 10 and 12.5°C.*

Solution: Using equation (1), $q = 1545.9 \times \Delta T^3$, the calculated values are given below:

ΔT°C	2.5	5	7.5	10	12.5
q, 10^6 W/m²	0.0242	0.1932	0.6522	1.5459	3.0193
h, W/m² K	9680	38640	86960	154590	251544

Example A-11.3: *Determine the heat flux if in example A-11.1 the excess temperature is 400°C.*

Solution: Equation (11.6) is applicable

$$h_c = 0.425 \left[\frac{g(\rho_l - \rho_v)^{1.5}\, \rho_v\, k_v^3\, (h_{fg} + 0.68\, c_{pv}\, \Delta T)}{(g_0\, \sigma)^{0.5}\, \mu_v\, \Delta T}\right]^{1/4}$$

The property values at film temperature (650.4 + 250.4)/2 = 450.4°C

ρ_l = 799.9 kg/m³, h_{fg} = 1714.1 × 10³ J/kg

ρ_v = 15.94, σ = 9 × 10⁻³ N/m, k_v = 0.0627 W/m K, c_{pv} = 5694 J/kg K

μ_v = 22.85 × 10⁻⁶ kg/m s (density and h_{fg} – steam tables)

$$\therefore\ h_c = 0.425 \left[\frac{9.81(799.9 - 15.94)^{1.5}\, 15.94 \times (0.0627)^3\, (1714.1 \times 10^3 + 0.68 \times 400 \times 5694)}{(1 \times 9 \times 10^{-3})^{0.5} \times 22.85 \times 10^{-6} \times 400}\right]^{0.25}$$

$$= 567.7 \text{ W/m}^2\text{ K}$$

$$h_r = \varepsilon\, \sigma\, (T_w^{\,4} - T_s^{\,4})/(T_w - T_s) = 0.65 \times 5.67\, (9.234^4 - 5.234^4)/400 = 59.18 \text{ W/m}^2\text{ K}$$

$$\therefore\ h = 567.1 + 0.75\,(59.18) = \mathbf{612.1\ W/m^2\ K}$$

$$q = h\,\Delta T = 0.2448 \times 10^6 \text{ W/m}^2.$$

This is several times lower than the critical heat flow.

Example A-10.4: *Refrigerant 12 evaporates at – 20°C to provide the cooling of food. Determine the critical heat flux and heat flux when the surface temperature is at 10°C. Also determine the excess temperature at which the critical flux occurs.*

Solution: The property values are:

ρ_l = 1460 kg/m³, ρ_v = 9.15 kg/m³

h_{fg} = 161.78 × 10³, σ = 23 × 10⁻³ N/m, C_{sf} = 0.013 (assumed)

Pr = 4.40, c_l = 909 J/kg K

μ_l = 3.431 × 10⁻⁴ kg/m s

Chapter 11

Using equation (11.3)

$$q_{cr} = 0.149 \times 161.78 \times 10^3\ [23 \times 10^{-3} \times 9.81 \times 1\ (1460 - 9.15)\ 9.15^2]^{1/4}$$

$$= \mathbf{0.310 \times 10^6\ W/m^2}$$

Using equation (11.2)

$$q = \left[\frac{909 \times \Delta T}{161.78 \times 10^3 \times 4.4^{1.7} \times 0.013}\right]^3 3.431 \times 10^{-4} \times 161.78 \times 10^3 \times \left[\frac{9.81(1460 - 9.15)}{1 \times 23 \times 10^{-3}}\right]^{0.5}$$

$\mathbf{q} = 1.84334 \times \Delta T^3$

Substituting q_{cr} for q, $\Delta \mathbf{T} = \mathbf{58.8\ °C}$

When surface temp. is 10°C, ΔT = 30°C *i.e.*, 10 – (– 20)

$$\therefore \qquad \mathbf{q} = 1.84334 \times \Delta T^3 = \mathbf{0.05 \times 10^6\ W/m^2}$$

$$\mathbf{h = 1659\ W/m^2\ K.}$$

Note: Sources for property values of liquids other than water are scarce.

Example A-11.5: *n-Butyl alcohol boils on a copper surface at 1 atm. The saturation temperature is 117.5°C. Determine the critical heat flux and heat flux for excess temperatures of 10, 20, 200°C. Also determine the excess temperature at the critical flux condition.*

Solution: The property values are:

$$\sigma = 0.0183\ \text{N/m},\ h_{fg} = 591500\ \text{J/kg}$$

$$\rho_v = 2.3\ \text{kg/m}^3,\ \mu_v = 0.143 \times 10^{-3}\ \text{kg/m s}$$

$$\rho_l = 737\ \text{kg/m}^3,\ \mu_l = 0.39 \times 10^{-3}\ \text{kg/m s}$$

$$c_l = 2876\ \text{J/kg K},\ Pr = 6.9,\ k_l = 0.163\ \text{W/m K}$$

$$C_{sf} = 0.00305,\ c_{\rho v} = 1706\ \text{J/kg K},\ k_v = 0.02\ \text{W/m K}$$

Using equation (11.3)

$$\mathbf{q_{cr}} = 0.149\ h_{fg}\ [\sigma g g_o (\rho_l - \rho_v)\ \rho_v^{\ 2}]^{1/4}$$

$$= 0.149 \times 591500\ [0.0183 \times 9.81 \times 1\ (737 - 2.3) 2.3^2]^{1/4} = \mathbf{0.453 \times 10^6\ W/m^2}$$

Assuming nucleate boiling and excess temperature ΔT_{cr}

$$0.547 \times 10^6 = \mu_l\, h_{fg} \left[\frac{c_l}{h_{fg}\, pr^{1.7} C_{sf}}\right]^3 \left[\frac{g}{g_o}\frac{(\rho_l - \rho_v)}{\sigma}\right]^{0.5} \Delta T_{cr}^3$$

$$= 0.39 \times 10^{-3} \times 591500 \times \left[\frac{2876}{591500 \times 6.9^{1.7} \times 0.00305}\right]^3 \left[\frac{9.81}{1}\frac{(737 - 2.3)}{0.0183}\right]^{0.5} \Delta T_{cr}^3$$

$$= 30.910 \times \Delta T^3_{cr} \quad \therefore \quad \mathbf{\Delta T_{cr} = 26.06°C}$$

For other excess temperatures

$$\mathbf{q_{10}} = 30.914 \times 10^3 = \mathbf{0.0309 \times 10^6\ W/m^2}$$

$$\mathbf{q_{20}} = 30.914 \times 20^3 = \mathbf{0.2473 \times 10^6\ W/m^2}$$

At 200°C excess temperature, the condition will be film boiling

Using equation (11.6)

$$h_c = 0.425\left[\frac{g(\rho_l - \rho_v)^{1.5}\rho_v k_v^3\left[h_{fg} + 0.68c_{pv}\Delta T\right]}{(g_o\,\sigma)^{0.5}\mu_v\Delta T}\right]^{0.25} = 0.425$$

$$\times\left[\frac{9.81(737 - 2.3)^{1.5}\times 2.3\times 0.02^3\,(591500 + 0.68\times 1706\times 200)}{(1\times 0.0183)^{0.5}\times 0.143\times 10^{-3}\times 200}\right]^{0.25}$$

$= \mathbf{72.9\ W/m^2\ K}$

Let $\varepsilon = 0.6,\ T_w = 317.5 + 273$

$T_{sat} = 117.5 + 273$

$\mathbf{h_r} = \varepsilon\,\sigma\,(T_w^{\ 4} - T_{sat}^{\ 4})/(T_w - T_{sat}) = \mathbf{16.72\ W/m^2\ K}$

$\mathbf{q} = (72.9 + 0.75 \times 16.72)\ 200 = \mathbf{0.017 \times 10^6\ W/m^2}$

lower than at 10°C excess temperature.

Note: The same value of heat flux can occur at more than one excess temperature lying in the nucleate and film boiling regimes.

Example A-11.6: *For water boiling at 1 atm at 100°C determine the value of excess temperature in the film boiling region for a heat flux equal to the critical heat flux for this condition.*

Solution: The property values at 100°C are:

$\rho_l = 961,\ \rho_v = 0.598\ kg/m^3,\ h_{fg} = 2257 \times 10^3\ J/kg.$

$\sigma = 58.8 \times 10^{-3}\ N/m.$

Using equation (11.3), $\mathbf{q_{cr}} = 0.149 \times 2257 \times 10^3$

$\times\ [58.8 \times 10^{-3} \times 9.81(961 - 0.598) \times 0.598^2]^{1/4} = \mathbf{1.262 \times 10^6\ W/m^2}$

In the film boiling region equations (11.6) and (11.7) are to be used.

The convection coefficient h_c is given by

$$h_c = 0.425\left[\frac{g(\rho_l - \rho_v)^{1.5}\rho_v k_v^3\left[h_{fg} + 0.68\times cp_v\Delta T\right]}{(g\sigma)^{0.5}\mu_v\Delta T}\right]^{0.25} \qquad h_r = \varepsilon\,\sigma\left[\frac{T_w^4 - T_{sat}^4}{T_w - T_{sat}}\right]$$

The property values k_v and μ_v should be taken at $(T_w + T_{sat})/2$.

Only a trial solution is possible.

Assuming 1100°C excess temperature, film temp. = 600°C

$\rho_v = 0.252,\ k_v = 0.08176,\ \mu_v = 33.15 \times 10^{-6},\ c_{pv} = 2206\ J/kg\ K$

Assuming an emisivity of 0.8 (rough surface)

$T_w = 1100 + 100 + 273 = 1473\ K,\ T_{sat} = 373\ K$

$$h_r = \frac{0.8\times 5.67}{1100}\ [14.73^4 - 3.73^4] = 193.33\ W/m^2\ K$$

$$h_c = 0.425\times\left[\frac{9.81\,(961 - 0.252)^{1.5}\times 0.252\times(0.08176)^3\,[2257\times 10^3 + 0.68\times 1100\times 2206]}{(1\times 58.8\times 10^{-3})^{0.5}\,3\times 3.15\times 10^{-6}\times 1100}\right]^{1/4}$$

$= 155.18\ W/m^2\ K$

$\therefore\ h = 300.2\ W/m^2\ K \quad \therefore\ q = 0.33 \times 10^6$

Chapter 11

This falls short of the critical flux.

Assuming 1900°C excess temp., T_f = 1000°C

Property values are:

$$\rho_v = 0.172,\ k_v = 0.14072,\ \mu_v = 52.37 \times 10^{-6},$$
$$C_{pv} = 2483$$
$$T_w = 2273 \text{ K},\ T_{sat} = 373 \text{ K}$$
$$h_r = 637.8,\ h_c = 273.5,\ h = 751.85,\ q = 1.43 \times 10^6 \text{ W/m}^2.$$

This is nearer to 1.262 × 10^6.

Note: The material may not be able to withstand this high temperature. The value is near burn out condition.

Example A-11.7: *A copper tube of 2.5 cm dia with its surface temperature maintained at 250°C is immersed in water at 100°C at atmospheric pressure. Determine the convective heat transfer coefficient on the outside.*

Solution: The excess temperature is 150°C. So the regime is film boiling. The property values are at 100°C for liquid and at 175°C for vapour.

$$\rho_l = 961,\ \rho_v = 0.4975,\ k_v = 0.03105$$
$$h_{fg} = 2257 \times 10^3 \text{ J/kg},\ c_{pv} = 2015.8,\ \mu_v = 14.91 \times 10^{-6} \text{ kg/ms}$$

$$\mathbf{h_c} = 0.62 \left[\frac{g(\rho_l - \rho_v)\rho_v k_v^3 (h_{fg} + 0.68\, c_{pv}\, \Delta T)}{D\mu_v\, \Delta T}\right]^{0.25}$$

$$= 0.62 \left[\frac{9.81(961 - 0.4975)0.4975(0.03105)^3\ (2257 \times 10^3 + 0.68 \times 2016 \times 150)}{0.025 \times 14.91 \times 10^{-6} \times 150}\right]^{0.25}$$

$$= \mathbf{173.84\ W/m^2\ K}$$

Let ε = 0.76 (copper oxidised)

$$\mathbf{h_r} = 0.76 \times 5.67 \left[\frac{5.23^4 - 3.73^4}{150}\right] = \mathbf{15.9\ W/m^2\ K}$$

$$\therefore\ \mathbf{h} = 173.8 + 0.75 \times 15.9 = \mathbf{185.8\ W/m^2\ K}$$

$$\therefore\ \mathbf{q} = h\,\Delta T = \mathbf{0.0279 \times 10^6\ W/m^2}$$

(Compare with maximum heat flux of about 4.3031 × 10^6 W/m², example A-11.1)

Example A-11.8: *It is desired to generate 500 kg/hr of steam at 180°C saturated condition using 20 m² area. Determine the surface temperature. Also estimate the value of convection coefficient.*

Solution: The property values required are : h_{fg} = 2015 × 10^3 J/kg (steam tables)

$$\mu_l = 1.538 \times 10^{-4},\ \rho_l = 889,\ \rho_v = 5.1533,\ \sigma = 42.16 \times 10^{-3} \text{ N/m}$$
$$Pr = 1.044,\ c_l = 4417 \text{ J/kg K},$$
$$C_{sf} = 0.013 \text{ (for general application)}$$

Heat flux $= 500 \times 2015 \times 10^3/3600 \times 20 = \mathbf{0.014 \times 10^6\ W/m^2}$

Using equation (11.2)

$$0.014 \times 10^6 = \left[\frac{4417\,\Delta T'}{0.013 \times 2015 \times 10^3 \times 1.044}\right]^3 .\ 1.538 \times 10^{-4} \times 2015 \times 10^3 \times \left[\frac{9.81(889 - 5.15)}{1 \times 42.16 \times 10^{-3}}\right]^{0.5}$$

$$\Delta T = 2.9°C \quad \therefore \quad h = 4876\ W/m^2\ K.$$

(Compare with values of example A-11.3.)

Example A-11.9: *Calculate the flow velocity through a 62.5 mm ID tube for obtaining the value of convective heat transfer coefficient off 4876 W/m² K for water at a bulk mean temperature of 180°C (same as boiling with ΔT = 2.9°C).*

Solution: $Nu = 0.023\ Re^{0.8} Pr^{0.4}$ ...(1)

$$h = \frac{k}{D} \times 0.023 \left(\frac{uD}{v}\right)^{0.8} Pr^{0.4}$$

At 180°C, (bulk mean temperature)

$$v = 0.173 \times 10^{-6}\ m^2/s,\ Pr = 1.044,\ k = 0.6757\ W/mK$$

$$\therefore \quad 4876 = \frac{0.6757}{0.0625} \times 0.023 \left[\frac{0.0625}{0.173 \times 10^{-6}}\right]^{0.8} (1.044)^{0.4} \times u^{0.8}$$

$$\therefore \quad \mathbf{u = 0.63\ m/s},\ Re = 227 \times 10^3\ \text{(Turbulent)}$$

This is for low value of excess temperature.

Example A-11.10: *What should be the velocity in exmaple A-11.9 for heat flux to the equal to the critical heat flux?*

Solution: Critical heat flux (equation 11.3)

$$q_{cr} = 0.149\, h_{fg}\, \rho_v \left[\frac{\sigma g . g_o (\rho_l - \rho_v)}{\rho_v^2}\right]^{0.25}$$

Using the data from example A-11.8

$$q_{cr} = 0.149 \times 2015 \times 10^3 \times 5.1533 \times \left[\frac{42.16 \times 10^{-3} \times 9.81\,(889 - 5.1533)}{(5.1533)^2}\right]^{0.25} = 2.98 \times 10^6\ W/m^2$$

Using equation (1).

$$\therefore \quad 2.98 \times 10^6 = \frac{0.6757}{0.0625} \times 0.023 \left(\frac{0.0625}{0.176 \times 10^{-6}}\right)^{0.8} (1.044)^{0.4}\ u^{0.8}$$

$$\therefore \quad \mathbf{u = 1910\ m/s.}$$

A very high velocity is required to achieve this heat flux in forced convection.

Chapter 11

Example A-11.11: *Determine the value of minimum heat flux in film boiling of water at 100° over a horizontal stainless steel surface.*

Solution: The property values taken at 100°C are $\rho_l = 961$ kg/m³

$$\rho_v = 0.598,\ h_{fg} = 2257 \times 10^3,\ \sigma = 58.8 \times 10^{-3} \text{ N/m},$$

ρ_{vf} is assumed as 0.598 to start with

Using equation [11.9(*b*)], for unit area

$$q = 0.09\, h_{fg}\, \rho_{vf} \left[\frac{g g_o \sigma (\rho_l - \rho_v)}{(\rho_l - \rho_v)^2} \right]^{1/4}$$

Substituting the values **q = 0.019 × 10⁶ W/m²**

Compare with example A-11.6, $\mathbf{q_{cr} = 1.524 \times 10^6}$ **W/m²**

This is about 1.25% of q_{cr}.

This is due to the film resistance.

The value will be still lower as ρ_{vf} corresponding to the film temperature will be lower still.

The value has to be solved for by trial.

Example A-11.12: *In condensation show that the film Reynolds number at location x can be expressed as*

$$Re_\delta = 3.771 \left[\frac{x^3 (T_g - T_w)^3 \rho_f (\rho_f - \rho_v)\, g k_f^3}{\mu_f^5\, h_{fg}^3} \right]^{1/4}$$

Solution: Film Reynolds number is defined as

$Re_\delta = 4G/\mu$ where G is the mass flow per unit depth in the film (equation 11.28), δ is the film thickness.

The mass flow unit depth is given by equation (11.16)

$$G = \frac{\rho_f (\rho_f - \rho_v)\, g\, \delta^3}{3\mu} \qquad \therefore \quad Re_\delta = \frac{4\rho_f (\rho_f - \rho_v)\, g\, \delta^3}{3\mu^2}$$

From equation (11.21)

$$\delta = \left[\frac{4\, \mu k \times (T_g - T_w)}{g\, h_{fg}\, (\rho_f - \rho_v)\, \rho_f} \right]^{0.25}$$

Substituting

$$Re_\delta = \frac{4\rho_f (\rho_f - \rho_v)\, g}{3\mu^2} \cdot \left[\frac{4\, \mu k \times (T_g - T_w)}{g\, h_{fg}\, (\rho_f - \rho_v)\, \rho_f} \right]^{3/4}$$

$$= \left[\frac{4^4\, \rho_f^4\, (\rho_f - \rho_v)^4\, g^4}{3^4 \mu^8} \right]^{1/4} \left[\frac{4^3\, \mu^3\, k^3\, x^3\, (T_g - T_w)^3}{g^3\, {h_{fg}}^3\, (\rho_f - \rho_v)^3\, {\rho_f}^3} \right]^{1/4}$$

$$= \left[\frac{4^7}{3^4} \cdot \frac{\rho_f^4 (\rho_f - \rho_v)^4 \, g^4 \mu^3 \, k^3 \, x^3 \, (T_g - T_w)^3}{\mu^8 g^3 \, h_{fg}{}^3 (\rho_f - \rho_v)^3 \, \rho_f{}^3} \right]^{1/4}$$

$$\therefore \qquad Re_\delta = 3.711 \left[\frac{x^3 \, (T_g - T_w)^3 \, \rho_f \, (\rho_f - \rho_v) \, g k_f^3}{\mu_f^5 \, h_{fg}^3} \right]^{1/4}.$$

Example A-11.13: *Mass flow rate in laminar film condensation is available as seen in the example A-11.12. Derive an expression for the mass flow in a vertical wall at distance x from the top for turbulent film. Use the heat transfer correlation for turbulent region.*

Solution: Two basic relations are used: Assuming unit width and depth distance x, heat transfer as measured by condensate = heat convected

$$m \, h_{fg} = \bar{h} x \Delta T \quad \therefore \quad \bar{h} = m \, h_{fg} / x \Delta T$$

where m is the flow rate kg/s and $\Delta T = (T_g - T_w)$

The heat transfer correlation is given by equation (11.31)

$$CO = 0.0077 \, Re_\delta^{0.4}$$

Substituting for condensation number CO from equation (11.30) and Re

$$\bar{h} \left[\frac{\mu^2}{k^3 \rho_l \, (\rho_l - \rho_v) g} \right]^{1/3} = 0.0077 \left[\frac{4.m}{\mu} \right]^{0.4}$$

Substituting for $\bar{h}$ and rearranging,

$$m \left(\frac{h_{fg}}{x \, \Delta T} \right) \left[\frac{\mu^2}{k^3 \rho_l (\rho_l - \rho_v) g} \right]^{1/3} = \frac{0.0077 \times 4^{0.4}}{\mu^{0.4}} \, m^{0.4}$$

$$m^{3/5} = \frac{0.01341}{\mu^{0.4}} \left[\frac{x^3 \Delta T^3 k^3 \, \rho_l (\rho_l - \rho_v) g}{h_{fg}^3 \, \mu^2} \right]^{1/3}$$

Neglecting ρ_v as small

$$\therefore \qquad m = 7.56585 \times 10^{-4} \left[\frac{x^3 \, \Delta T^3 k^3 \rho_1^2 g}{h_{fg}^3 \, \mu^{3.2}} \right]^{5/9}.$$

This provides a means of direct calculation of mass condensed for a distance of x metre for unit width and unit time in the turbulent region. This value is for unit width and x m from the top where condensation starts for x m².

Example A-11.14: **Derive an expression** *for the average value of convection coefficient in the turbulent region in terms of property values and distance. Use the correlations in terms of* Re_δ *as starting point*

$$CO = 0.0077 \, Re_\delta^{0.4}.$$

Solution: Substituting for CO and for m in Re_δ

$$CO = \bar{h} \left[\frac{\mu^2}{k^3 \rho_l \, (\rho_l - \rho_v) g} \right]^{1/3} \quad \text{and } m \, h_{fg} = \bar{h} \times \Delta T \text{ (unit width)}$$

Chapter 11

$$\bar{h}\left[\frac{\mu^2}{k^3\rho_l(\rho_l-\rho_v)g}\right]^{1/3} = 0.0077 \times 4^{0.4}\,\frac{\bar{h}^{0.4}\,x^{0.4}\,\Delta T^{0.4}}{\mu^{0.4}h_{fg}^{0.4}}$$

$$\therefore \qquad \bar{h}^{0.6} = 0.0077 \times 4^{0.4}\left[\frac{x^{1.2}\,\Delta T^{1.2}k^3\,\rho_l(\rho_l-\rho_v)g}{h_{fg}^{1.2}\,\mu^{3.2}}\right]^{1/3}$$

$$\therefore \qquad \bar{h} = 7.56585 \times 10^{-4}\left[\frac{x^{1.2}\,.\,\Delta T^{1.2}k^3\,\rho_l(\rho_l-\rho_v)g}{h_{fg}^{1.2}\,\mu^{3.2}}\right]^{5/9} \qquad ...(1)$$

This expression is usable directly from property values and temperature difference and plate dimension.

Example A-11.15: *Steam at 100°C condenses on the outside of a horizontal tube of 0.15 m dia maintained at 95°C. Compare the value of convection coefficient of heat transfer with the coefficient obtainable in boiling of water at 100°C with wall temperature at 105°C.*

Solution: Considering condensation and using equation (11.27)

$$h = 0.725\left[\frac{\rho_l(\rho_l-\rho_v)g\,h_{fg}\,k_f^3}{\mu_f d\,(T_g-T_w)}\right]^{1/4}$$

The property values are $\rho_l = 961$, $\rho_v = 0.598$

$k_f = 0.67894$, $\mu_f = 2.907 \times 10^{-4}$, $h_{fg} = 2257 \times 10^3$ J/kg

(f-means at film temperature 97.5°C)

$$h = 0.725\left[\frac{961\,(961-0.598)\,9.81 \times 2257 \times 10^3 \times 0.67894^3}{2.907 \times 10^{-4} \times 0.15 \times 5}\right]^{1/4}$$

$= \mathbf{9488.1\ W/m^2\,K}$

Considering boiling: $\mu_1 = 2.816 \times 10^{-4}$, $\sigma = 58.8 \times 10^{-3}$ N/m

$c_1 = 4216$ J/kg K, $Pr = 1.74$, $C_{sf} = 0.013$

Using equation (11.2)

$$\mathbf{q} = \left(\frac{4216 \times 5}{0.013 \times 2257 \times 10^3 \times 1.74}\right)^3 \times 2.816 \times 10^{-4} \times 2257 \times 10^3 \times \left(\frac{9.81\,(961-0.598)}{58.8 \times 10^{-3}}\right)^{0.5} = \mathbf{0.018 \times 10^6}$$

$\mathbf{h} = q/\Delta T = \mathbf{3582\ W/m^2\,K}$.

COMMENT

For excess temperature of 5°C, in boiling the condition is still in convection zone. Hence, the lower value of convection coefficient.

Example A-11.16: *Compare the value of average convection coefficient in condensation for a horizontal pipe and a vertical plate of height π D/2 for the same conditions.*

Solution: Using equations (11.26) and (11.27)

For a plate: $$\bar{h} = 0.943 \left[\frac{\rho_l(\rho_l - \rho_v) g\, h_{fg}\, k_f^3}{(\pi D/2).\mu_f(T_g - T_w)} \right]^{1/4}$$

For a pipe: $$\bar{h} = 0.725 \left[\frac{\rho_l(\rho_l - \rho_v) g\, h_{fg}\, k_f^3}{D\,.\mu_f(T_g - T_w)} \right]^{1/4}$$

Dividing $$\frac{h_{\text{plate}}}{h_{\text{cyl}}} = \frac{0.943}{(\pi/2)^{0.25}} / 0.725 = \mathbf{1.162}$$

For the same area, plate provides a higher value of convection coefficient in condensation compared to horizontal tube.

Example A-11.17: *A plate type of condenser is of side L × 2L. It was designed to be kept with side of length 2L in the vertical position. But by oversight, the side of length L was kept vertical. If other conditions are the same, determine the change in heat transfer if any. Assume laminar conditions in both cases.*

Solution: Using equation (11.26)

$$h = 0.943 \left[\frac{\rho(\rho - \rho_v) g\, h_{fg}\, .\, k_f^3}{L\, \mu_f(T_g - T_w)} \right]^{1/4}$$

$$\therefore \quad \frac{\bar{h}_{2L}}{h_L} = \left(\frac{1}{2}\right)^{1/4} = 0.8409 \text{ or } h_{2L} = 0.8409\, h_L$$

∴ The heat flow **will increase when the shorter side is kept vertical.** The increase is about 18.92%. It is desirable to have the shorter side vertical. This is due to the thicker liquid film when the length increases. Similar situation is seen in convection over a flat plate.

Example A-11.18: *A plate condenser is designed to be kept vertical. Due to site conditions, it is to be kept at 30° to the vertical. Determine the change in heat transfer, assuming that other conditions remain unchanged.*

Solution: For inclined plate in equation (11.27) g is replaced by $g \cos \theta$ where θ is the angle with the vertical.

$$\therefore \quad \frac{h_{\text{inclined}}}{h_{\text{vertical}}} = (\cos 30)^{1/4} = 0.9647$$

There is a reduction of 3.53 %.

Example A-11.19: *A plate type vertical condenser was designed for a particular fluid with a given temperature difference ΔT. This is used with a new fluid whose density is 0.7 of the original fluid and h_{fg} is 0.5 of the original fluid and the conductivity is double that of the original fluid and the viscosity is 1.2 times that of the original fluid. The temperature difference is reduced to 70% of the original value. Determine the % change in the convection coefficient.*

Solution: $$h_1 = 0.943 \left[\frac{\rho_1^2\, g\, h_{fg1}\, k_{f1}^3}{L\mu_1(T_g - T_w)_1} \right]^{1/4} \qquad h_2 = 0.943 \left[\frac{\rho_2^2\, g\, h_{fg2}\, k_{f2}^3}{L\mu_2(T_g - T_w)_2} \right]^{1/4}$$

$$\rho_2 = 0.7\ \rho_1,\quad h_{fg2} = 0.5\ h_{fg1},\ k_{f2} = 2k_{f1},\quad \mu_2 = 1.2\ \mu_1$$
$$\Delta T_2 = 0.7\ \Delta T_1$$

Substituting

$$h_2 = 0.943 \left[\frac{0.7^2 \rho_1^2\, g\, 0.5 h_{fg1} 2^3 . k_{f1}^3}{L\, 1.2\, \mu_1 \times 0.7\, \Delta T_1}\right]^{1/4}$$

$$= 0.943 \left[\frac{\rho_1^2\, g\, h_{fg1}\, k_{f1}^3}{L\, \mu_1 \times \Delta T_1}\right]^{1/4} \left[\frac{0.7^2 \times 0.5 \times 2^3}{1.2 \times 0.7}\right]^{1/4}$$

$$\therefore \quad \mathbf{\frac{h_1}{h_2}} = 1/\left[\frac{0.7^2 \times 0.5 \times 2^3}{1.2 \times 0.7}\right]^{1/4} = \mathbf{0.8091}$$

$h_2 = h_1 / 0.8091$ or an increase of 23.59%.

COMMENT

This is not a purely academic exercise. There is practical use for this type of calculation. Often, available units with known characteristics for use with a fluid may have to be adopted for use with another fluid. In such cases, the performance can be estimated quickly and accurately by this process. Also, the temperature difference may vary due to seasonal temperature variation. The performance at the changed condition will be required, and this type of calculation will come handy in such cases.

Example A-11.20: *For condensing conditions, compare condensation over a pipe of diameter D with that of 2 pipes having the same total circumference, if (i) Both pipes are horizontal and parallel (ii) If these are one over the other. Assume that other conditions remain the same.*

Solution: $D = 2d \quad \therefore \quad d = D/2$, Using equation (11.28)

$$h = 0.725 \left[\frac{\rho(\rho - \rho_v)\, g\, h_{fg}\, k_f^3}{D\, \mu_f\, (T_g - T_w)}\right]^{1/4}$$

Case (*i*): If D is reduced to half the value, h is increased to $(2)^{0.25}$ times the original value

or $$\mathbf{h_d = 1.1892\ h_D}$$

This is due to a thinner film over the surface when diameter is lowered.

Area being the same, the heat flow will increase by 18.92% if two tubes are used in parallel instead of one tube of larger diameter.

Case (*ii*): If pipes are in vertical row, D is replaced by nD here $D = 2\ (D/2) = D$ $\therefore$ the heat flow remains unchanged. Here in the second pipe, the film becomes thicker and so the change in the first is compensated.

Example A-11.21: *Using the expressions derived in example A-11.13 and A-11.14 determine the flow rate and convection coefficient for a vertical plate at 60°C condensing steam at 100°C. The plate is 5 m high.*

Solution: First a check has to be made to determine whether turbulent conditions prevail.

$$Re_\delta = \frac{4G}{\mu},\quad G\, h_{fg} = \bar{h}\, .\, L\, (T_g - T_w)$$

$$\therefore \qquad Re_\delta = \frac{4\bar{h}L(T_g - T_w)}{h_{fg}\,\mu}$$

assuming laminar conditions to start with)

$$\bar{h} = 0.943\left[\frac{\rho(\rho-\rho_v)g\,h_{fg}\,k_f^3}{L\,\mu_f\,(T_g - T_w)}\right]^{1/4}$$

The data available are

at 100°C, ρ = 961 kg/m^3, ρ_v = 0.598 kg/m^3, h_{fg} = 2257 × 10^3 J/kg at 80°C

k_f = 0.6687 W/m K, μ_f = 3.545 × 10^{-4} kg/m s

Substituting, $\bar{h}$ = 2874 W/m^2 K

nd Re_δ = 2923 ∴ flow is turbulent.

Using expression from example A-11.13

$$\dot{m} = 7.56585 \times 10^{-4}\left[\frac{x^3\,\Delta T^3 k^3\,\rho^2\,g}{h_{fg}^3\,\mu^{3.2}}\right]^{5/9}$$

Substituting **ṁ = 0.68 kg/s m,** $q = \dot{m}\,h_{fg}$ = 1.535 × 10^6 W/m

h̄ = 7673.4 W/m² K

Using expression from example A-11.14, equation (1)

$$\bar{h} = 7.56585 \times 10^{-4}\left[\frac{x^{1.2}\,\Delta T^{1.2}k^3\,\rho(\rho-\rho_v)g}{h_{fg}^{1.2}\,\mu^{3.2}}\right]^{5/9}$$

Substituting the values

h̄ = 7673 W/m² K **A very good agreement.**

$$Re_\delta = \frac{4\times 0.68}{3.545\times 10^{-4}} = 7673 \text{ flow is turbulent.}$$

note that the previous value of Re is approximate)

If other conditions remain the same, determine the distance at which the film turns turbulent. (Re_δ = 1800)

$$\dot{m} = \frac{1800\times 5.545\times 10^{-4}}{4} = 0.25 \text{ kg/ms,}$$

Substitute for m using example A-11.13 to solve for **x, x = 2.74 m.**

Example A-11.22: ***Refrigerant 12 vapour condenses*** *on a vertical plate. The vapour is at 40°C and the plate is at 20°C. Determine the value of convective heat transfer coefficient and the condensate rate for 1 m × 1 m plate.*

Solution: T_f = 30°C, k_f = 0.0709, μ_f = 2.5123 × 10^{-4} kg/m s

at 40°C, ρ = 1257 kg/m^3, h_{fg} = 129.98 × 10^3 J/kg.

ρ_v = 54.65 kg/m^3

Assuming laminar conditions

$$\mathbf{h} = 0.943\left[\frac{\rho(\rho-\rho_v)g\,h_{fg}k_f^3}{L\,\mu_f\,(T_g - T_w)}\right]^{1/4}$$

Chapter 11

$$= 0.943 \left[\frac{1257(1257-54.65)\, 9.81 \times 129.98 \times 10^3 \times (0.0709)^3}{1 \times 2.5123 \times 10^{-4} \times 20}\right]^{1/4}$$

$= \mathbf{573.4\ W/m^2\ K}$

$q = 573.4 \times 1 \times 20 = 11467.8$ W/m²

m $= q/h_{fg} =$ **0.0882 kg/s** or 317.62 kg/hr or 5.29 kg/min.

$$Re_\delta = \frac{4\,m}{\mu} = \frac{4 \times 0.0882}{2.5123 \times 10^{-4}} = 1404 \therefore \text{ laminar flow prevails.}$$

Example A-11.23: *Chilled water pipe of 6.25 cm OD with outside surface at 5°C passes throug an area where air is saturated at 35°C. Determine the condensate drip from the pipe per m length.*

Solution : Here water vapour exists at the partial pressure at saturation at 35°C. $P = 0.0562$ bar.

$$h_{fg} = 2418.6 \text{ kJ/kg}, \rho_v = 0.04 \text{ kg/m}^3$$

The liquid properties at

$T_f = 20$°C are $\rho = 1000$ kg/m³,

$k_f = 0.5978$, $\mu_f = 10.06 \times 10^{-4}$ kg/m s

Using equation (11.27)

$$h = 0.725 \left[\frac{\rho(\rho - \rho_v)\, g\, h_{fg}\, k_f^3}{\mu_f\, d\, (T_g - T_w)}\right]^{1/4}$$

Substituting the values,

$\mathbf{h = 5220\ W/m^2\ K}$

$\mathbf{Q} = \pi D \times 1 \times 5220\ (35 - 5) =$ **30748 W/m**

$\therefore$ **Mass condensed** $= Q/h_{fg} =$ **0.012713 kg/s**

about 45.77 kg/hr of condensate will flow per m length. The high value is because of the air being saturated. This is a practical situation. The drip is undesirable. It may be noted that the drip from the evaporaters of household air conditioners have to be led by a hose pipe to the drain.

Example A-11.24: ***Ammonia vapour*** *at 35°C is to be condensed on the outside of horizontal tubes of OD 2.5 cm with a surface temperature of 25°C. A square array of 10 × 10 tubes of 1.2 m length are used. Determine the rate of condensation of ammonia.*

Solution : The liquid properties at $T_f = 30$°C are

$$\rho = 596, \mu_f = 596 \times 0.35 \times 10^{-6} \text{ kg/m s}, k_f = 0.5071 \text{ W/m K}$$

The vapour properties at 35°C are

$$h_{fg} = 1123.46 \times 10^3 \text{ J/kg}, \rho_v = 1.042 \text{ kg/m}^3$$

Using equation (11.27) with $n = 10$,

$$\mathbf{h} = 0.725 \left[\frac{\rho(\rho - \rho_v)\, g\, h_{fg}\, k_f^3}{n\, D\, \mu_f\, \Delta T}\right]^{1/4}$$

$$= 0.725\left[\frac{596(596-1.042)\,9.81\times 1123.46\times 10^3\times 0.5071^3}{10\times 0.025\times 596\times 0.35\times 10^{-6}\times 10}\right]^{1/4}$$

$= \mathbf{4053.6\ W/m^2\ K}$

$\mathbf{Q} = 4053.6 \times \pi \times 100 \times 0.025 \times 1.2 \times 10$ W.

$= 0.382 \times 10^6$ W or **382 kW.**

Ammonia condensed $= Q/h_{fg} =$ **0.34 kg/s.** or 1224.22 kg/hr.

Example A-11.25: ***Refrigerant 12*** *is to be condensed at the rate of 10,000 kg/hr at 35°C using water at 25°C. The condenser uses a square array of 25 × 25 tubes of 12 mm OD, calculate the length of the tube bundle.*

Solution: The tube surface is assumed to be at 25°C. Property values at

$T_f = 30$°C are $\rho = 1295$, $k_f = 0.0709$ W/m K

$\mu_f = 1295 \times 0.194 \times 10^{-6}$ kg/m s

The vapour properties at 35°C are

$h_{fg} = 133.22 \times 10^3$ J/kg

$\rho_v = 48.08$ kg/m³,

Using equation (11.27) with $n = 25$

$$\mathbf{h} = 0.725\left[\frac{\rho(\rho-\rho_v)\,g\,h_{fg}\,k_f^3}{n\,D\,\mu_f\,\Delta T}\right]^{1/4}$$

$$= 0.725\left[\frac{1295\,(1295-48.08)\,9.81\times 133.22\times 10^3\times 0.0709^3}{25\times 0.012\times 1295\times 0.194\times 10^{-6}\times 10}\right]^{1/4}$$

$= \mathbf{724.62\ W/m^2\ K}$

$$\mathbf{Q} = \frac{10000}{3600} \times 133.22 \times 10^3 = 724.62 \times 10 \times 25 \times 25 \times \pi \times 0.012\,L$$

$\therefore$ **L = 2.17 m.**

Example A-11.26: *Refrigerant 12 at 35°C is condensed inside a 12 mm ID tube with its surface at 25°C. Determine the convective heat transfer coefficient. Compare this value with that of condensation over the 12 mm OD pipe surface with other conditions remaining the same.*

Solution: The property value are: $h_{fg} = 133.22 \times 10^3$ J/kg

$\rho_v = 48.08$ kg/m³

$\rho = 1295$ kg/m³, $k_f = 0.0709$ W/m K,

$\mu_f = 1295 \times 0.194 \times 10^{-6}$ kg/m s

Using equation (11.34) (for condensation inside)

$$h_1 = 0.555\left[\frac{\rho(\rho-\rho_v)\,g\,k^3\,h_{fg}}{\mu_1\,d\,(T_g - T_\infty)}\right]^{1/4}$$

Substituting the values

$\mathbf{h_1 = 1240.4\ W/m^2\ K}$

Using equation (11.27) for outside condensation

$\mathbf{h_o = 1620.3\ W/m^2\ K}$

Chapter 11

The coefficient on the outside is larger by a factor of

$$\frac{0.725}{0.555} = \mathbf{1.306.}$$

COMMENT

In refrigerant condensers, the vapour is under pressure. So, outside condensation is not possible. In steam condensers, the stream is at vacuum pressure (0.1 bar about). Hence outside condensation is used in steam condensers, cold water flowing inside tubes.

Example A-11.27: *In condensation situation compare, in the laminar region, the values of convective heat transfer coefficients when a pipe of diameter 6.25 cm and 1 m length kept (i) vertical and (ii) horizontal.*

Solution: For vertical position

$$\bar{h}_v = 0.943 \left[\frac{\rho(\rho - \rho_v)\, g\, k^3\, h_{fg}}{L\, \mu\, (T_g - T_\infty)}\right]^{1/4} \quad \text{...(A)}$$

For horizontal position

$$\bar{h}_h = 0.725 \left[\frac{\rho(\rho - \rho_v)\, g\, k^3\, h_{fg}}{D\, \mu\, (T_g - T_\infty)}\right]^{1/4} \quad \text{...(B)}$$

Dividing equation (B) by equation (A)

$$\frac{\mathbf{h_h}}{\mathbf{h_v}} = \frac{0.725}{0.943} \times \left(\frac{L}{D}\right)^{0.25} = \mathbf{1.538}$$

horizontal positioning in this case provides 53.8%, more heat transfer. Condensers are thus generally of horizontal type. This is because of larger film thickness as length increases.

Example A-11.28: *A slab of ice is placed in a wooden box.* *The ice is at 0°C. The outside air is at 35° and the convection coefficient over the surface is 25 W/m² K. Determine the thickness melted during 10 min. 60 min. and 120 min.*

Solution: Assume that the water layer is not removed. The property values are h_{sf} = 333700 J/kg, k_l = 0.60, ρ_s = 920 kg/m³.

Using equations (11.40) and (11.38)

$$\tau^* = x^* + 0.5\, x^{*2}$$

$$x^* = \frac{x h_\infty}{k_l}, \quad \tau^* = \left[\frac{h_\infty^2 (T_\infty - T_{fr})}{k_l\, \rho_s\, h_{sf}}\right] \tau$$

For 10 minutes: $\tau^* = \left[\dfrac{25^2 \times 35}{0.60 \times 920 \times 333700}\right] \times 600 = 0.071253$

$$0.5\, x^{*2} + x^* - 0.071253 = 0$$

Using $x = \dfrac{-b \pm \sqrt{b^2 - 4ac}}{2a}$ form of solution for quadratic equation.

$$x^* = \frac{-1 \pm \sqrt{1 + 4 \times 0.5 \times 0.071253}}{1}, \text{ taking the + ve value}$$

∴ $x^* = 0.06888$

∴ $\mathbf{x} = \frac{x^* k_L}{h_\infty} = \frac{0.06888 \times 0.60}{25} = 1.653 \times 10^{-3}$ m or **1.653 mm**

After 60 min. $\tau^* = 0.42752$

∴ $x^* = 0.362$ ∴ $\mathbf{x = 8.69 \times 10^{-3}}$ **m or 8.69 mm**

After 120 min. $\tau^* = 0.855$

∴ $x^* = 0.6462$ ∴ **x = 15.51 mm**

With time the rate of freezing decreases, due to increasing resistance for heat conduction by the water layer above the solid.

Example A-11.29: *Water at 0°C in a lake surface is suddenly exposed to air at – 35°C with a convection coefficient of 25 W/m² K. Determine the depth of freezing after 10 min. 60 min. and 120 min. If this situation continues for 10 days, determine the thickness of ice formed.*

Solution : $h_{sf} = 333700$ J/kg $k_s = 2.22$ W/m K $\rho_s = 920$ kg/m³

Using equation (11.40)

$$\tau^* = x^* + 0.5\, x^{*2} \text{ where } x^* = \frac{xh_\infty}{k_s}$$

and

$$\tau^* = \left[\frac{h_\infty^2 (T_{fr} - T_\infty)}{k_s\, \rho_s\, h_{sf}}\right] \tau$$

For 10 min.

$$\tau^* = \frac{25^2 \times 35}{2.22 \times 920 \times 333700}\, 60 \times 10 = 0.01926$$

$$x^* = -1 \pm \sqrt{1 + 4 \times 0.5 \times 0.01926} = 0.0191$$

∴ $\boldsymbol{x} = \mathbf{1.694 \times 10^{-3}}$ **or 1.694 mm** (melting 1.653 mm)

For 60 min. $\tau^* = 0.1155$

$x^* = 0.10955$

$x = 9.73 \times 10^{-3}$ or **9.73 mm** (melting 8.69 mm)

For 120 min. $\tau^* = 0.2311$

$x^* = 0.2092$

x = 0.0186 m or **18.6 mm** (melting 15.51 mm)

For **10 days,** $\tau^* = 27.73$, $x^* = 6.514$, **x = 0.58 m.**

Compared to melting the increase in freezing thickness is because of the higher conductivity of the ice as compared to that of water.

Example A-11.30: ***Ice candy is frozen in cylindrical moulds*** *of 16 mm dia with a wooden core of 4 mm dia. The water is at its freezing temperature of 0°C. The outside surface of the thin container is exposed to –35°C with a convection coefficient of 25 W/m² K. Determine the time for complete freezing.*

Solution: $\rho_s = 920$ kg/m², $h_{sf} = 333700$ J/kg, $k_s = 2.22$ W/mK

Chapter 11

Using equation (11.46)

$$\tau^* = 0.5r^{*2} \ln r^* + \left(\frac{1}{2R^*} + 0.25\right)(1 - r^{*2}) \quad \text{...(A)}$$

as the radius upto which freezing is to be done is 2 mm (core)

$$r^* = R/R_o = \frac{2}{8} = \frac{1}{4} = 0.25$$

$$R^* = h_\infty R_o/k_s = 25 \times 0.008/2.22 = 0.09009$$

Also $$\tau^* = \frac{(T_{fr} - T_\infty)\,k_s}{\rho_s h_{sf} R_o^2}.\tau \quad \text{(equation 11.45)} \quad \text{...(B)}$$

From (A), $$\tau^* = 0.5\,(0.25)^2 \ln (0.25) + \left(\frac{1}{2 \times 0.09009} + 0.25\right)(1 - 0.25^2) = 5.3942$$

From (B), $$\tau = \frac{5.3942 \times 920 \times 333700 \times 0.008^2}{2.22 \times 35} = \textbf{1364 s or 22.73 min.}$$

Example A-11.31: ***Ice is made in a hollow cylindrical shape*** *by immersing a thin cylinder in a large tank of water. The outer radius is 16 mm. The water is at 0°C. Cold fluid at –35° is circulated inside the thin cylinder with h = 50 W/m² K.*

Determine the time needed for ice to form up to a radius 24 mm (property values taken from example A-11.30).

Solution: The equation applicable is (11.47).

$$\tau^* = 0.5\, r^{*2} \ln r^* + \left(\frac{1}{2R^*} - 0.25\right)(r^{*2} - 1)$$

As freezing is to be done up to R = 24 mm,

$$r^* = R/R_o = \frac{24}{16} = 1.5$$

$$R^* = h_\infty.R_o/k_s = 25 \times 0.016/2.22 = 0.18018$$

$$\tau^* = \frac{(T_{fr} - T_\infty)\tau}{\rho_s\, h_{sf}\, R_o^2} = \frac{35 \times 2.22}{920 \times 333700 \times 0.016^2}\,\tau = 9.886 \times 10^{-4} \times \tau \quad \text{...(A)}$$

Also $$\tau^* = 0.5 \times 1.5^2 \ln (1.5) + \left(\frac{1}{2 \times 0.18018} - 0.25\right)(1.5^2 - 1) \quad \text{...(B)}$$

$$= 3.6124$$

Solving between (A) and (B)

∴ τ = **3654 s = 60.90 min.**

Example A-11.32: *Water at 10°C in a pan is to be frozen by convection over the surface at –20°C. The convection coefficient at the ice water surface is 20 W/m² K. Determine the time needed to freeze 2 cm layer. h = 80 W/m² K at the surface.*

Solution: The equation applicable is (11.44)

$$\tau^* = \left(\frac{1}{H^*T^*}\right)^2 \ln\left[\frac{1-H^*T^*}{1-H^*T^*(1+x^*)}\right] - \frac{x^*}{H^*T^*}$$

$$x^* = \frac{xh_\infty}{k_s} = \frac{0.02\times 80}{2.22} = 0.7207$$

$$H^* = \frac{h_1}{h_\infty} = \frac{20}{80} = 0.25$$

$$t^* = \frac{T_L - T_{fr}}{T_{fr} - T_\infty} = \frac{10}{20} = 0.5$$

$$\tau^* = \frac{h_\infty^2 (T_{fr} - T_\infty)}{k_s \rho_s h_{sf}} \tau = 1.8781 \times 10^4\ \tau \qquad \text{...(A)}$$

Substituting in equation (11.44)

$$\tau^* = \left(\frac{1}{0.25\times 0.5}\right)^2 .\ \ln\left[\frac{1-0.25\times 0.5}{1-0.25\times 0.5\,(1+0.7207)}\right] - \frac{0.7207}{0.25\times 0.5} \qquad \text{...(B)}$$

$= 1.1881$

Using (A) and (B)

$\therefore$ τ **= 6326 = 105.44 min.**

Example A-11.33: *Determine the time needed to freeze (frost bite) a layer of 1 mm thickness of skin exposed to wind at –20°C with a convection coefficient of 50 W/m² K. Assume properties of water and the temp. of the skin as 35°C to start with. The convection coefficient at the interface is 25 W/m² K.*

Solution: Equation (11.44) is applicable.

$$\tau^* = \left(\frac{1}{H^*T^*}\right)^2 \ln\left[\frac{1-H^*T^*}{1-H^*T^*(1+x^*)}\right] - \frac{x^*}{H^*T^*}$$

$$x^* = x \times h_\infty/k_s = 0.001 \times 50/2.22 = 0.02252252$$

$$H^* = \frac{h_1}{h_\infty} = \frac{25}{50} = 0.5$$

$$T^* = \frac{T_L - T_{fr}}{T_{fr} - T_\infty} = \frac{35}{20} = 1.75$$

$$\tau^* = \frac{h_\infty^2 (T_{fr} - T_\infty)}{k_s\, \rho_s\, h_{sf}} \tau = 7.33623 \times 10^{-5}\ \tau \qquad \text{...(A)}$$

Substituting

$$\tau^* = \left(\frac{1}{0.5\times 1.75}\right)^2 \ln\left[\frac{1-0.5\times 1.75}{1-0.5\times 1.75(1+0.02252)}\right] - \frac{0.02252}{0.5\times 1.75} \qquad \text{...(B)}$$

$= 0.19835$

Using (A) and (B), $\therefore$ τ = **2704 s = 45 min.**

Frost bite may start after 45 min.

REVIEW QUESTIONS

11.1 Explain with sketches the conditions at forced convection boiling, nucleate boiling and film boiling.

11.2 Distinguish between pool boiling and flow boiling by giving examples.

11.3 Define excess temperature with respect to boiling. Describe the effect of variation of excess temperature on heat flux.

11.4 With a sketch of boiling curve explain the process of boiling in 'power control' method of experiment.

11.5 Explain the method used for temperature controlled boiling experiment.

11.6 What are the modes of heat transfer in film boiling?

11.7 What is the effect of surface roughening on nucleate boiling?

11.8 Distinguish between drop-wise and film-wise condensation.

11.9 Explain why drop-wise condensation is not adopted in product design, in spite of its advantages.

11.10 What is the ratio of local and average values of convection coefficient in laminar film condensation.

11.11 Define Reynolds number for film condensation. State the limit for laminar flow.

11.12 List the factors affecting freezing/melting of water at saturation temperature.

11.13 Under identical conditions, which process will be faster? Freezing or melting.

11.14 Define τ^* and x^* used in the freezing and melting analysis.

OBJECTIVE QUESTIONS

11.1 Choose the correct statement in each of the following sets:

(*i*) With increase in excess temperature the heat flux in boiling

(*a*) increases continuously

(*b*) decreases and then increases

(*c*) increases then decreases and again increases

(*d*) decreases then increases and again decreases.

(*ii*) The heat flux in nucleate pool boiling is proportional to (where h_{fg} is enthalpy of evaporation)

(*a*) h_{fg} (*b*) $1/h_{fg}$ (*c*) h_{fg}^2 (*d*) $1/h_{fg}^2$.

(*iii*) The heat flux in nucleate pool boiling is proportional to

(where σ is surface tension)

(*a*) σ (*b*) $\sigma^{0.5}$ (*c*) $\sigma^{-0.5}$ (*d*) σ^2.

(*iv*) The critical heat flux in nucleate pool boiling is proportional to (where ρ_v is the density of vapour)

(*a*) ρ_v (*b*) ρ_v^2 (*c*) $\rho_v^{0.5}$ (*d*) $\rho_v^{1/3}$.

(*v*) The critical heat flux in nucleate pool boiling is proportional to (where h_{fg} is the enthalpy of evaporation)

(*a*) h_{fg} (*b*) $h_{fg}^{0.5}$ (*c*) h_{fg}^2 (*d*) $h_{fg}^{1/3}$.

(*vi*) The critical heat flux in nucleate pool boiling is proportional to (where s is the surface tension)
(*a*) σ (*b*) $\sigma^{0.5}$ (*c*) $\sigma^{1/3}$ (*d*) $\sigma^{1/4}$.

(*vii*) The critical heat flux in nucleate pool boiling is proportional to (where ρ_f is the density of liquid)
(*a*) ρ_l^2 (*b*) $\rho_l^{0.25}$ (*c*) $\rho_l^{0.5}$ (*d*) $\rho_l^{1/3}$.

(*viii*) In nucleate pool boiling the heat flux depends on
(*a*) only the material of the surface
(*b*) material and roughness of the surface
(*c*) independent of surface
(*d*) fluid material and surface roughness.

(*ix*) In nucleate pool boiling the heat flux is proportional to (ΔT is the excess temperature)
(*a*) ΔT^3 (*b*) ΔT (*c*) ΔT^2 (*d*) $\Delta T^{0.5}$.

(*x*) In nucleate pool boiling the heat flux for boiling of water is proportional to (μ_l) is the viscosity of liquid)
(*a*) μ_l^{-2} (*b*) μ_l (*c*) $\mu_l^{0.5}$ (*d*) μ_l^2.

(*xi*) In nucleate pool boiling of liquids other than water the heat flux varies as (when μ_l is the viscosity of liquid)
(*a*) $\mu_l^{5.1}$ (*b*) $\mu_l^{-4.1}$ (*c*) μ_l^2 (*d*) μ_l^{-2}.

(*xii*) In nucleate film boiling of water the heat flux varies as (where *k* is the vapour thermal conductivity)
(*a*) $k^{0.33}$ (*b*) $k^{0.75}$ (*c*) k^2 (*d*) $k^{5.1}$.

(*xiii*) In nucleate pool boiling of liquids other than water the heat flux varies as
(*a*) $Pr^{-5.1}$ (*b*) $Pr^{3.1}$ (*c*) Pr^3 (*d*) $Pr^{0.5}$.

(*xiv*) In nucleate pool boiling the convective heat transfer coefficient varies as (where ΔT is excess temperature)
(*a*) ΔT^2 (*b*) ΔT^3 (*c*) $\Delta T^{2.5}$ (*d*) $\Delta T^{0.5}$.

(*xv*) In film boiling the properties that contribute to the value of convection coefficient are
(*a*) h_{fg} μ_l . ρ_l and ΔT (*b*) h_{fg} μ_v, k_v and ρ_l
(*c*) h_{fg}, k_v, μ_l and ΔT (*d*) h_{fg}, k_l, μ_v and ΔT.

(*xvi*) The heat flux in nucleate pool boiling will be higher for
(*a*) horizontal plane (*b*) vertical plane
(*c*) horizontal cylinder (*d*) independent of location.

(*xvii*) In condensing under same conditions, the convection coefficient will be lowest for
(*a*) vertical plate (*b*) vertical pipe
(*c*) horizontal pipe (*d*) row of vertical pipes.

(*xviii*) In condensation over a vertical surface, the value of convection coefficient varies as (*k*-conductivity of liquid)
(*a*) k^3 (*b*) $k^{0.75}$ (*c*) $k^{0.25}$ (*d*) $k^{0.33}$.

(*xix*) The heat transfer rate in laminar film condensation varies as (where $\Delta T = T_g - T_w$)
(*a*) ΔT (*b*) $\Delta T^{0.75}$ (*c*) $\Delta T^{0.25}$ (*d*) $\Delta T^{-0.25}$.

(*xx*) The convective heat transfer coefficient in laminar film condensation varies as (ρ-density of liquid)
(*a*) ρ (*b*) $\rho^{0.5}$ (*c*) $\rho^{-0.5}$ (*d*) ρ^2.

(*xxi*) In laminar film condensation, the average convection coefficient varies as (where μ is the dynamic viscosity of liquid).
(*a*) μ^{-1} (*b*) $\mu^{-0.5}$ (*c*) $\mu^{0.25}$ (*d*) $\mu^{-0.25}$.

Chapter 11

(*xxii*) In laminar film condensation the convection coefficient varies as (where L is the height of plate)

(*a*) L^{-1} (*b*) $L^{-0.5}$ (*c*) L (*d*) $L^{-0.25}$.

(*xxiii*) In laminar film condensation the convection coefficient varies as (h_{fg} is the enthalpy of evaporation).

(*a*) h_{fg} (*b*) $h_{fg}^{0.5}$ (*c*) $h_{fg}^{1/3}$ (*d*) $h_{fg}^{0.25}$.

Answers

11.1 (*i*) (*c*) (*ii*) (*d*) (*iii*) (*c*) (*iv*) (*c*) (*v*) (*a*) (*vi*) (*d*)
(*vii*) (*b*) (*viii*) (*d*) (*ix*) (*a*) (*x*) (*a*) (*xi*) (*b*) (*xii*) (*b*)
(*xiii*) (*a*) (*xiv*) (*a*) (*xv*) (*b*) (*xvi*) (*d*) (*xvii*) (*d*) (*xviii*) (*b*)
(*xix*) (*b*) (*xx*) (*b*) (*xxi*) (*d*) (*xxii*) (*d*) (*xxiii*) (*d*).

EXERCISE PROBLEMS

11.1 Determine the maximum heat flux and the heat flux at an excess temperature of 8°C for water at pressure corresponding to (*i*) 50°C (*ii*) 100°C (*iii*) 180°C (*iv*) 310°C. In all cases assume a surface coefficient of 0.013 and pool boiling in the nucleate regime. Comment on the results.

11.2 Determine for the cases in 11.1 the minimum heat flux in the film boiling regime.

11.3 Determine the heat flux in boiling of water at a pressure corresponding to a saturation temperature of 310°C under film boiling regime if the excess temperature is (*i*) 1000°C (*ii*) 2000°C. Assume boiling to take place over a flat surface.

11.4 Smoke tubes of 6.25 cm *OD* are used in a boiler. Steam is generated at 170°C. The tube surface is at 178°C. There are 100 tubes of length 3 m each. Estimate the rate of possible steam production at saturated conditions. Assume pool boiling conditions and C_{sf} = 0.013.

11.5 Ammonia liquid evaporates in a flooded type of evaporator at – 10°C over tube surfaces immersed in the liquid. Brine at 0°C is circulated through the inside of the tubes and the tube surface is maintained at 0°C. The tube outside dia is 25 mm. 40 tubes of 1 m length are used. Determine the evaporation rate of ammonia. Assume $\sigma = 28 \times 10^{-3}$ N/m, h_{fg} = 1296.5 kJ/kg. C_{sf} = 0.013.

11.6 Boiling heat flux with water over a surface is measured under certain conditions. It is desired to estimate the heat flux for boiling of water at a higher pressure at which $\mu_2 = 0.8\ \mu_l$, $h_{fg2} = 0.8\ h_{fg1}$, $\sigma_2 = 0.6\ \sigma_1$, $P_{r2} = 0.6\ P_{r1}$. There is no significant change in density and specific heat and excess temperature. Determine the % change in the heat flux.

11.7 The maximum heat flux for boiling a liquid was measured under certain conditions. Estimate the critical heat flux for boiling under similar conditions but at a different saturation temperature. The following are the significant changes in property value $\sigma_2 = 0.7\ \sigma_1$, $h_{fg2} = 0.9\ h_{fg1}$, $\rho_{v1} = 2\rho_{v2}$.

11.8 An electrical heater rod of 10 mm dia is immersed in water at atmospheric pressure. The surface temperature reached is 600°C. If the emissivity of the surface is 0.5, estimate the electrical power input.

11.9 Estimate the convection coefficient for steam at 40°C saturation conditions condensing on a vertical surface maintained at 30°C. The plate is 1.5 m high. Also find the film thickness at 1 m and 1.5 m from the top edge.

11.10 Determine the value of convection coefficient for steam at 40°C saturation conditions condensing on a flat surface maintained at 30°C. If the surface is inclined at 40° to the vertical. The plate is 1.5 m high.

11.11 Determine the value of convection coefficient for condensing steam at 40°C saturation conditions over the outside surface of a vertical tube of 6.25 cm *OD* maintained at 30°C. The tube is 1.5 m high. Also determine the film thickness at 1 m and 1.5 m from top.

11.12 Determine the value of convection coefficient for condensing of steam over the outside surface of horizontal tubes of 6.25 cm *OD* with surface temperature maintained at 30°C. Steam temperature is 40°C. Compare the value with that of a vertical surface of 62.5 cm height.

11.13 Determine the value of convection coefficient for steam condensing inside of horizontal pipe of 6 cm *ID* with surface maintained at 30°C. The steam is at 40°C and flows at low velocity.

11.14 Repeat example 11.9 to 11.13 for ammonia condensing at the same temperature and geometric configurations.

11.15 Repeat example 11.9 to 11.13 for refrigerant 12 ($C\ Cl_2F_2$) condensing at the same temperature and geometric configurations.

11.16 Determine for a vertical surface the distance at which the condensate film flow will turn turbulent for steam condensing at 100°C with surface maintained at 40°C. Estimate the value of average convection coefficient for that distance using correlations for turbulent flow. Compare the value with that for laminar flow.

11.17 It is found that the condensate formed over the surface of a tube of 25 cm *OD* is the same whether it is kept vertical or horizontal. Determine the length of the tube [**Ans.** 71 cm]

11.18 A condensation experiment for steam on a vertical surface was conducted and the convection coefficient was determined. If the same test set up was used for condensation at a different conditions where $(\mu_2/\mu_1) = 0.43$ and $\rho_2/\rho_1 = 0.966$ and $(k_2/k_1) = 1.083$ and $h_{fg2}/h_{fg1} = 0.937$, determine the temperature difference to be maintained to obtain the same value of convection coefficient.

11.19 A small steam condenser with horizontal tubes of 6.25 cm *OD* and 6 rows deep with 36 tubes of 1.2 m length condensing steam at 40°C using water at 30°C is proposed to be used to condense ammonia at 50°C with water at 30°C. Assuming material and other compatibilities determine the % change in heat rate.

11.20 Chilled water pipe at 10°C passes through air space where partial pressure of water vapour is 0.04246 bar. The pipe diameter is 0.15 m. Determine the condensate drip per m length of tube per hour.

11.21 Food preparation in the form of horizontal cylinder of 5 cm dia is to be heated by condensation of steam over its surface. Compare the heating rates when the surface is at 40°C for steam temperature of 100°C and 120°C.

11.22 Compare the heat transfer rate for ammonia when condensing at 40°C with tube surface at 30°C on (*i*) inside of tube (*ii*) outside the tube. The tube is thin and the diameter is 2.5 cm and placed in horizontal position. The vapour flow velocity is small.

11.23 A glass tumbler of *OD* 6.5 cm and 12 cm high with ice floating in the liquid inside is kept on a table in a room where the dewpoint temperature is 25°C. Determine the rate of condensate formation on the surface. Assume film-wise condensation. Neglect end effects.

11.24 Compare the values of convective heat transfer coefficients for (*i*) condensation of steam over a horizontal cylinder of 6.25 cm dia at 100°C. The steam is at 110°C and (*ii*) for flow of steam at 110°C over the cylinder at a velocity of 15 m/s.

11.25 At the onset of summer, the air temperature over a large sheet of ice rises to 10°C and the convective heat transfer coefficient due to air flow over the surface is 18 W/m² K. Assuming there is no run off of water. Determine the time taken for melting 4 mm layer of ice at 0°C. Also determine the thickness melted over a period of 6 hours.

11.26 If in the example 11.25, the direct solar radiation has effect equal to an increase in the convection coefficient by 60 W/m² K (the total being 78 W/m² K) determine the increase in the depth melted over a period of 1 hr under this conditions.

11.27 Lake water at 4°C is exposed to chill winds at – 16°C. Determine the depth of formation of ice after 12 hours. Convection coefficient = 60 W/m² K, *h* at interface = 15 W/m² K.

11.28 It is desired to freeze 1 cm dia long cylindrical ice pieces in thin metallic moulds. The water is at 0°C. The outside is exposed to convection coefficient of 40 W/m² K. Determine the time required for (*i*) 2.5 mm thickness of ice to form and (*ii*) for complete freezing. The freezing medium is at – 20°C.

11.29 A long hollow cylindrical piece of ice of *ID* 6 cm and *OD* 10 cm is to be made from water at 0°C. The convection coefficient for the freezing medium at – 20°C is 40 W/m^2 K. Determine the time for forming the ice piece with (*i*) internal mould and (*ii*) external mould.

11.30 A copper pan with polished surface has a diameter of 0.3 m. It is filled with water at 100°C and is open to atmospheric pressure. The bottom is maintained at 118°C. Determine the power required for maintaining this temperature. Also calculate the maximum heat flux.

[**Ans.** 55.8 kW, 1.26 MW/m^2]

11.31 An electrical heater of 6 mm diameter with constant surface temperature of 255°C is immersed in a tank containing water at 100°C, and open to atmosphere. Assuming surface emissivity of unity, calculate the power dissipated per m length.

[**Ans.** h_c = 460 W/m^{2}K, h_r = 21.3 W/m^{2}K, power = 1.39 kW/m length]

11.32 A cylinder of 8 cm outer diameter has cold water flowing inside. When exposed to steam at 100°C its surface is maintained at 50°C. The cylinder is 1 m long. Determine the heat transferred to the cooling water. Also calculate the rate at which steam is condensed at the surface.

[**Ans.** $\bar{h}_L$ = 4094 W/m^2 K, q = 51446 W, $\dot{m}$ = 0.0214 kg/s]

11.33 A square array of 400 tubes of 6 mm *OD* are exposed to steam at 54°C at saturated condition at 0.15 bar. The cooling water flow rate is adjusted that the pipe surface is at 25°C. If the tube bank is 1 m long, calculate the rate of steam condensation. [**Ans.** $\bar{h}$ = 5188 W/m^2 K, m = 0.464 kg/s]

11.34 Water at a pressure of 1 atm and 100°C is boiled in a stainless steel can whose surface is maintained at 108°C. Determine the rate of heat transfer per m^2 of area. Also calculate the rate of evaporation per m^2. Assume c_{sf} = 0.013. [**Ans.** 7.2 × 10^4 W/m^2, 0.032 kg/m^2]

11.35 A 5 mm diameter electrical heater is immersed in a tank of water at 1 atm and 100°C. Determine the maximum heat flux that may be achieved in the nucleate boiling region. Also calculate the surface temperature of the heater at this condition. C_{sf} = 0.006. [**Ans.** 1.02 × 10^6 W/m^2, 119°C]

11.36 An electrical heater of 5 mm diameter is immersed in water at 100°C at 1 atm. Determine the heat transfer per m length if the surface temperature is 350°C, surface emissivity is 0.05.

[**Ans.** q_{con} = 5.93 × 10^4 W/m^2, q_{cr} = 157 W/m^2, Q = 933 W]

11.37 A two meter high vertical plate is kept at 80°C by cooling water on one side. The other side is exposed to saturated steam at 1 atm at 100°C. If the width of the plate is 3 m determine the rate of heat transfer due to condensation. Also calculate the condensate drip rate at the bottom.

[**Ans.** h = 5848 W/m^2 K, Q = 7.02 × 10^5 W, m = 0.303 kg/s]

11.38 In the example 11.37, if the plate is held at 30° to the vertical calculate the values as in the example 11.37. [**Ans.** h = 5641 W/m^2 K, Q = 6.77 × 10^5 W, m = 0.293 kg/s]

(**Hint:** Multiply by (cos 30)$^{o0.25}$)

11.39 A pipe of 3 cm diameter has its surface maintained at 30° C by cooling water. The pipe is exposed to steam at 7.38 kpa and 40°C saturation condition. Calculate the rate of heat transfer due to condensation and the condensate collection per 2 m length.

(**Hint:** Property values are: h_{fg} = 2407 kJ/hg, ρ_r = 0.05 kg/m^3,)

ρ_l = 994 kg/m^3, C_{pl} = 4178 J/kg K, μ_l = 0.720 × 10^{-3} kg/m s, k_l = 0.623 W/m K.

[**Ans.** h = 9292 W/m^2, Q = 17516 W, $\dot{m}$ = 0.0072 kg/s]

11.40 If 12 pipes under similar conditions in example 11.39 are arranged in 4 horizontal rows and three vertical rows, calculate the value of convection coefficient and the heat flow if pipe length is 2 m. (**Hint:** $h = h_{pipe}/N^{0.25}$) [**Ans.** h = 7060 W/m^{2}K, Q = 79850 W]

12 Heat Exchangers

CHAPTER

12.0 INTRODUCTION

In many industrial applications heat has to be transferred from one flowing fluid to another through a solid barrier separating these fluids. The equipments used for this purpose are called Heat Exchangers. The application of convection studies is mainly in the field of design of heat exchangers. The basic design procedures and analysis of performance of heat exchangers are discussed is this chapter. The various correlations discussed in chapters 8, 9, 10 and 11 are used in this chapter to determine convection coefficients.

The analysis starts with single tube type of both parallel flow and counter flow arrangements. The result is the determination of a mean temperature difference for use in equation (12.1). The method is called LMTD method of analysis. Heat exchangers in general are not of single tube type. A correction factor to modify the LMTD for different arrangement is discussed next. LMTD method is directly applicable only if all terminal temperatures are specified. Most cases only flow rates and inlet conditions are available, in addition to the heat exchange area and overall heat transfer coefficient. In these cases LMTD method can be used only by trial and error. A method called NTU method (net transfer unit) comes in handy and this is discussed next.

12.1 OVER ALL HEAT TRANSFER COEFFICIENT

It is desirable and convenient to combine the various thermal resistances that are encountered in the heat transfer from one fluid to another in a heat exchanger. Such a combination is achieved by using the heat transfer equation as in equation (12.1)

$$Q = UA\ (\Delta T) \qquad ...(12.1)$$

where Q is the heat flow rate in W, A is the heat flow area and ΔT is the temperature difference between the fluids.

The quantity U which equals the combined effects of all the resistances is called overall heat transfer coefficient. The unit for overall heat transfer coefficient is the same as the unit for convective heat transfer coefficient (W/m^2 K).

12.1.1. Thermal Resistances

The thermal resistances encountered in heat transfer from one fluid to another are shown in Fig. 12.1.

The resistances are:

(*i*) Convection resistance on the inner surface of the tube for heat flow from fluid to wall. $(1/h_i A_i)$

(*ii*) The resistance due to deposits that will form during operation on the inside of the tube called inside fouling resistance-R_{fi}.

(*iii*) The resistance due to heat conduction through the solid wall separating the fluids.

(*iv*) The resistance due to deposits that will form during operation on the outside of the tube called outside fouling resistance-R_{fo}.

(*v*) Convection resistance on the outer surface for heat flow from wall to the fluid ($1/h_oA_o$).

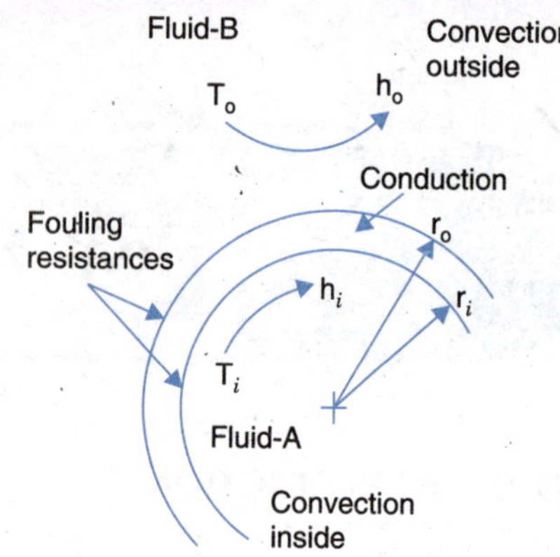

Fig. 12.1 Thermal resistances across heat exchanger surface

The area on the inside will be equal to the area on the outside in the case of flat surfaces. These are not as popularly used in heat exchangers as flow through tubes. **In tube flow the inside area will be different from the outside area. So two values are possible for the area and consequently two values are possible for the overall heat transfer coefficient. These are:**

U_i—overall heat transfer coefficient based on inside area

U_o—overall heat transfer coefficient based on outside area.

The relationship between these two is given by:

$$U_iA_i = U_oA_o$$

This can be proved by application of equation (12.1) to any situation **as $Q = U_iA_i\Delta T = U_oA_o\Delta T$.** In the case of plane wall $A_i = A_o$

It is easier to consider the resistances (*K/W*) rather than conductances (*W/K*) in arriving at the expression for overall heat transfer coefficient.

The total resistances is given by $\dfrac{1}{A_oU_o}$ or $\dfrac{1}{A_iU_i}$, $\dfrac{K}{W}$

depending upon which form is considered. The convection resistances are given by ($1/hA$). Considering flow through pipes, the conduction resistance R_k (*K/W*) is given by $\ln(r_o/r_i)/2\pi kL$, where L is the length of the pipe. The resistances due to deposits are called fouling resistances, R_f. The values of fouling resistances have to be obtained from tabulated values of measurements with actual equipments. The unit used is Km²/W.

Fouling resistance for some cases is given in Table 12.1 in units q Km²/W

For a given area A, the resistance is equal to R_f/A (*K/W*). Hence

$$\frac{1}{U_oA_o} = \frac{1}{h_oA_o} + \frac{R_{fo}}{A_o} + R_k + \frac{R_{fi}}{A_i} + \frac{1}{h_iA_i}$$

$$\therefore \quad \frac{1}{U_o} = \frac{1}{h_o} + R_{fo} + A_o\,R_k + \frac{A_o}{A_i}R_{fi} + \frac{A_o}{A_i}\cdot\frac{1}{h_i} \quad ...(12.2)$$

For pipes $$\frac{1}{U_o} = \frac{1}{h_o} + R_{fo} + \frac{r_o}{k}\ln\left(\frac{r_o}{r_i}\right) + \frac{r_o}{r_i}R_{fi} + \frac{r_o}{r_i}\cdot\frac{1}{h_i} \quad ...[12.2(a)]$$

also $$\frac{1}{U_i} = \frac{A_i}{A_o}\cdot\frac{1}{h_o} + \frac{A_i}{A_o}\cdot R_{fo} + A_iR_k + R_{fi} + \frac{1}{h_i} \quad ...(12.3)$$

For pipes $$\frac{1}{U_i} = \frac{r_i}{r_o} \cdot \frac{1}{h_o} + \frac{r_i}{r_o} R_{fo} + \frac{r_i}{k} \ln\left(\frac{r_o}{r_1}\right) + R_{fi} + \frac{1}{h_i} \quad \text{...[12.3(a)]}$$

The equations (12.2) and (12.3) are basic to heat transfer calculations in heat exchangers. For [12.2(*a*)] and [12.3(*a*)] r_o and r_i are based on the *OD* and *ID* of the pipe. In the example problems in this chapter the value of *U* will be specified.

In practical situations the fouling factor is determined by measurement of heat transfer value for clean surface and surface with deposit.

$$R_f = \frac{1}{U_f} - \frac{1}{U_c}$$

where U_f is overall heat transfer coefficient for surface after the deposits have formed due to usage and U_c is for clean surface.

In actual applications only one of these resistances will be controlling the situation and it will be desirable to improve that value. **Conduction resistances will be often negligibly small** and may be overlooked for first estimates. The estimation of values of '*h*' has been discussed in chapters 8, 9, 10 and 11. It is necessary to use these correlations in this chapter also. In the initial discussions however '*h*' value will be specified.

Table 12.1. Fouling reistance R_f

Type of fluid	*Fouling resistance Km²/W*
Sea water below 52°C	0.0000877
Sea water above 52°C	0.0001754
Treated boiler feed water above 52°C	0.0001754
Fuel oil	0.000877
Quenching oil	0.0007051
Alcohol vapours	0.0000877
Steam, non oil bearing	0.0000877
Industrial oil	0.0003525
Refrigerant vapour	0.0001754
AIR	0.0004

Example 12.1: *In a condenser steam flows outside the tubes of radii 59 mm and 54 mm and sea water flows inside the tubes. The thermal conductivity of the tube material is 60 W/mK. The steam and water temperatures are below 50°C. The convection coefficient on the steam side is 12000 W/m² K and the value on the water side is 650 W/m²K.*

Calculate the values of overall coefficients based on the (i) inside and (ii) outside areas.

Also determine the percentage error involved in neglecting (i) conduction resistance (ii) fouling on the inside and outside and conduction and (iii) considering only the water side resistance.

Solution: From tabulated values the fouling factors are read as

(*i*) sea water below 52°C 0.0000877 Km²/W

(*ii*) steam 0.0000877 Km²/W

Using equation [12.2(*a*)] $\frac{1}{U_o} = \frac{1}{12,000} + 0.0000877 + \frac{0.059 \ln (59/54)}{60}$

$$+ \frac{0.059}{0.054} \cdot 0.0000877 + \frac{0.059}{0.054} \cdot \frac{1}{650}$$

$$\frac{1}{U_o} = 8.333 \times 10^{-5} + 8.77 \times 10^{-5} + 8.708 \times 10^{-5} + 9.582 \times 10^{-5} + 1.681 \times 10^{-3}$$

$$\mathbf{U_o = 491.44\ W/m^2\ K.}$$

In this case it may be noted that all the resistances other than that due to inside convection coefficient are 2 orders of magnitude lower.

Leaving conduction resistance:

$\mathbf{U_o = 513.40\ W/m^2\ K}$ % change: 4.47%

Leaving out fouling and conduction resistances

$\mathbf{U_o = 566.8\ W/m^2\ K}$ % change: 15.4%

Considering only water side resistance,

$\mathbf{U_o = 594.88\ W/m^2\ K}$ % change: 21%

In this case the waterside resistance is controlling, as it forms 80% of the total.

Overall heat transfer coefficient based on inside area:

$$U_oA_o = U_iA_i \,;\; U_o\, 2\pi r_o\, L = U_i\, 2\, \pi\, r_i\, L$$

$$\therefore \qquad \mathbf{U_i} = U_o \frac{r_o}{r_i} = \mathbf{536.94\ W/m^2\ K}$$

The value is higher in proportion to the area ratio.

COMMENT

For preliminary estimates, it may be sufficient to consider the lower convection coefficient only to estimate the area requirement.

12.2 CLASSIFICATION OF HEAT EXCHANGERS

Functionally Heat Transfer Equipments can be classified as

(*i*) **Recuperative type:** This is the commonly used type, the heat being exchanged between fluids separated by a barrier.

(*ii*) **Regenerative or storage type:** In this case some material is heated by a hot fluid. Then the hot fluid flow is stopped. Cold fluid now flows over the hot solid and gets heated. This type is used for air heating in steam plants. This type is also used in solar heating of homes. Compact design is possible by this type when gas to gas heat exchange is encountered.

(*iii*) **Direct mixing type:** In this case the fluids mix and reach a common temperature. This type is rarely used.

12.2.1. Types of Heat Exchangers

The heat transfer surface in heat exchangers can be arranged in several forms. Some of the important arrangements are listed below. These arrangements are schematically shown in Fig. 12.2.

1. Single tube arrangement in which one fluid flows inside the tube and the other fluid flows on the outside. There are three possible directions of fluid flow in this case. **(a) parallel flow,** in which the direction of flow of both fluids is the same. **(b) counter flow** in which the flow directions are opposite to each other **(c) cross flow** in which the fluid on the otuside flows in a perpendicular direction to the pipe axis. These are shown in Figs. 12.2(*a*), (*b*) and (*c*).

The analysis is simpler in the case of single tube heat exchanger and the results obtained for this type is applied with corrective factors to other types.

2. Shell and tube arrangement: Single tube arrangement is suitable only for very small capacities and is not extensively used in industry. Shell and tube type is the most popular arrangement. A number of small bore pipes are fitted between two tube plates and one fluid flows through these tubes. The tube bundle is placed inside a shell and the other fluid flows through the shell and over the surface of the tubes. Compact arrangement is possible with this type.

The sub types of shell and tube arrangement are (*a*) One shell pass and 2, 4 or multiple tube passes. The end chambers adjacent to tube plates are provided with partitions in such a way that the fluid flowing inside the tubes takes several turns within the shell.

(*b*) Two shell passes and 4, 8 or multiple tube passes. Here two shells are used in series and the tubes are arranged for multiple passes. These arrangements are shown in Figs. 12.2 (*d*), (*e*), (*f*). In shell and tube exchangers baffle plates are generally provided to direct the flow across the tubes instead of taking a shorter direct path to the outlet. The baffles also provide support for long tubes. This may be noted in Figs. 12.2(*d*) and (*e*).

3. Cross flow heat exchanger: This arrangement is popular with heat exchangers using air or gas. Plate type as well as tube and plate types are used. The subtype here are (*a*) both fluids unmixed and (*b*) one fluid mixed and the other unmixed. Refer Figs. 12.2(*g*) and 12.2(*h*).

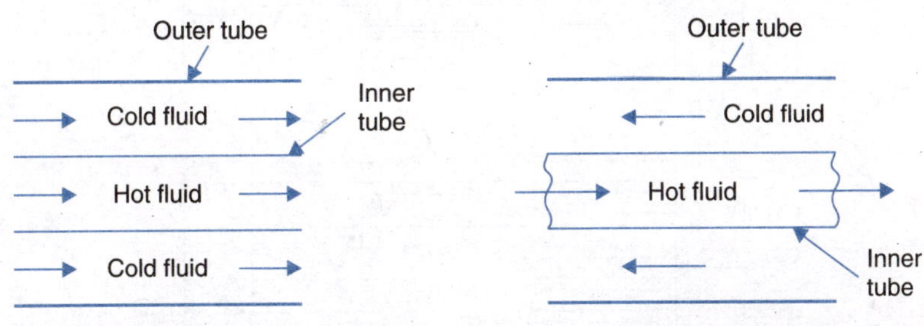

Fig. 12.2. (a) Single tube-parallel flow (b) Single tube-counter flow

Chapter 12

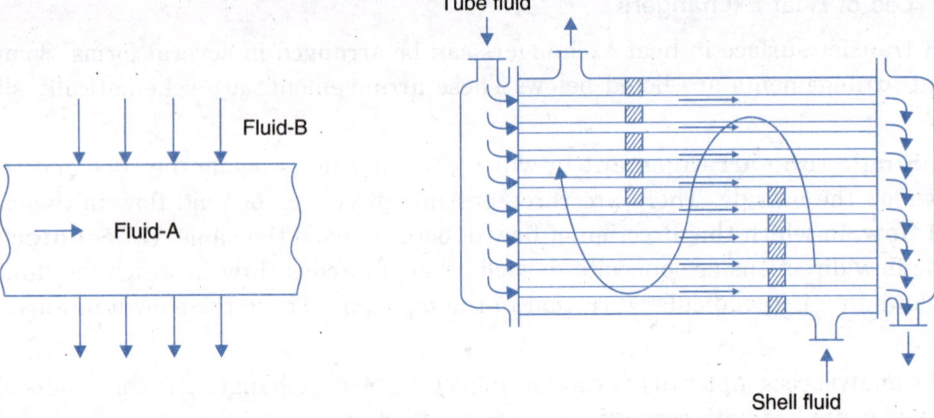

Fig. 12.2. (c) Single tube-cross flow (d) Shell and tube

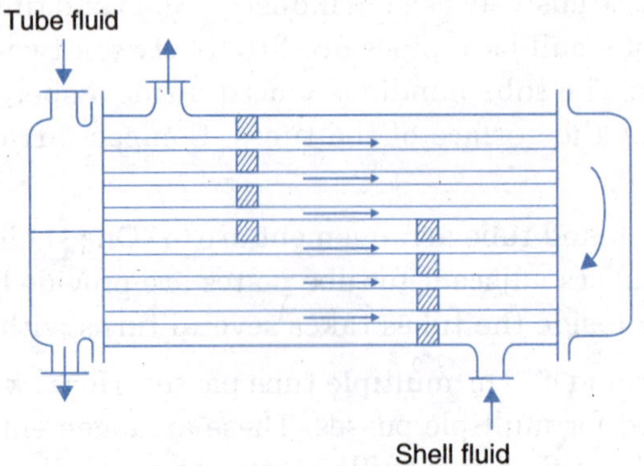

Fig. 12.2. (e) Single shell pass, two tube passes

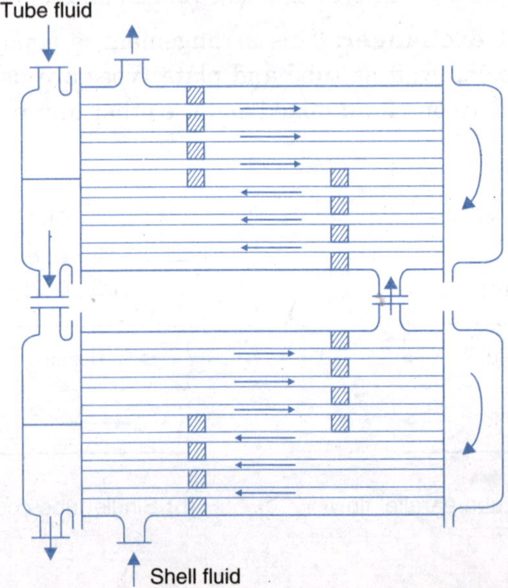

Fig. 12.2. (f) Two shell passes and multiple tube passes

These are shown in Figs. 12.2(g) and (h). When the fluid flows through divided passages without mixing it is called unmixed. For example fluid flowing through pipes in a bank of tubes is said to be unmixed. Along the flow path the fluid in one pipe does not mix with the fluid in the other. Fluid flowing without restriction over a bank of tubes is said to be mixed.

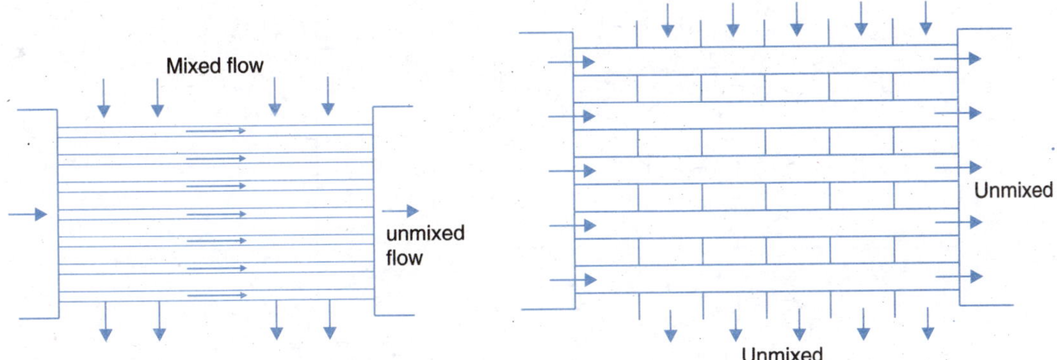

Fig. 12.2. (g) Cross flow—One fluid unmixed

Fig. 12.2. (h) Cross flow—Both fluids unmixed

Compact heat exchangers of the recuperative type is discussed later in the chapter. These are the basic types which are found to be amenable for direct analysis. There exist many other types of peculiar designs, which are not listed here.

12.3 MEAN TEMPERATURE DIFFERENCE-LOG MEAN TEMPERATURE DIFFERENCE

The heat transfer from one fluid to the other at a location is given by equation (12.1). Here the value of overall heat transfer coefficient can be assumed not to vary with location because $\bar{h}$ only is used in the calculation of U. **But the temperature difference continuously varies with location.** In order to determine the total heat flow, either the heat flow should be summed up using elemental areas and the temperature difference at the location or more conveniently an average value of temperature difference should be worked out. A short cut method may be to use the average of the temperature differences at inlet and outlet $((\Delta T_1 + \Delta T_2)/2)$. But this is found to give higher values of heat transfer compared to the actual, which is not desirable.

The temperature variation of the hot and cold fluids along the flow in a parallel flow heat exchanger is shown in Fig. 12.3. It can be seen that the temperature difference between the fluids varies along the flow and the arithmetic average may not be the real average.

One of the important analysis in heat exchanger performance is the determination of the mean temperature difference. The heat transfer is then given by

$$Q = UA\,(\Delta T)_{\text{mean}} \qquad \text{...(12.4)}$$

12.3.1. Determination of Mean Temperature Difference

In the case of single tube parallel flow heat exchanger: Referring to Fig. 12.3, the cold fluid enters at a temperature T_{ci} and leaves at T_{co}. The hot fluid enters at T_{hi} and leaves at T_{ho}. The flow rates and specific heats are m_c, c_c and m_h, c_h respectively. The product **m c** is also known as heat capacity denoted as C_h and C_c.

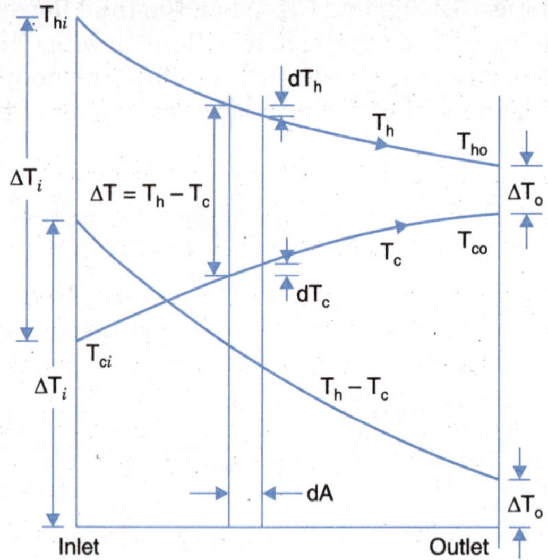

Fig. 12.3. Parallel flow-single tube heat exchanger –Temperature variation along flow

Consider a small area dA at a location. In passing through the area let the hot fluid be cooled by dT_h and let the cold fluid be heated dT_c. Let the heat flow be dQ.

$$dQ = -\, m_h\, C_h\, dT_h = m_c\, c_c\, dT_c \qquad ...(12.5)$$

The –ve sign indicates temperature drop or dT_h is negative.

The total heat flow is calculated using,

$$Q = m_h\, c_h\, (T_{hi} - T_{ho}) = m_c\, c_c\, (T_{co} - T_{ci}) \qquad ...(12.6)$$

This equation is used to determine one of the unknown temperatures if mass flows are specified or the unknown mass flow if all temperatures are specified.

Using equation (12.5)

$$dT_h = -\frac{dQ}{m_h c_h},\; dT_c = \frac{dQ}{m_c c_c}$$

$$dT_h - dT_c = d(T_h - T_c) = -\, dQ\left[\frac{1}{m_h c_h} + \frac{1}{m_c c_c}\right] \qquad ...(12.7)$$

From equation (12.1)

$$dQ = U\, dA\, (T_h - T_c) \qquad ...(12.8)$$

Substituting for dQ in equation (12.7)

$$\frac{d(T_h - T_c)}{(T_h - T_c)} = -\, U\, dA\left[\frac{1}{m_h c_h} + \frac{1}{m_c c_c}\right] \qquad ...(12.9)$$

Integrating from the inlet to the outlet

$$\ln\,(T_h - T_c)\Big|_i^o = -\, U\, A\left[\frac{1}{m_h c_h} + \frac{1}{m_c c_c}\right] \qquad ...(12.10)$$

Substituting the following in equation (12.10),

$$m_h\, c_h = Q/(T_{hi} - T_{ho}) \quad \text{and} \quad m_c c_c = Q/(T_{co} - T_{ci})$$

$$\ln \frac{\Delta T_o}{\Delta T_i} = -\frac{UA}{Q} [(T_{hi} - T_{ho}) + (T_{co} - T_{ci})]$$

$$= \frac{UA}{Q} [(T_{ho} - T_{co}) - (T_{hi} - T_{ci})]$$

$$\therefore \qquad Q = UA \left[\frac{\Delta T_o - \Delta T_i}{\ln \dfrac{\Delta T_o}{\Delta T_i}} \right] \qquad \text{...(12.11)}$$

where ΔT_o and ΔT_i are the difference in temperatures between the hot and cold fluids at outlet and at inlet.

Comparing with equation (12.4), the mean temperature is given by

$$\Delta \mathbf{T}_{\text{mean}} = \frac{\Delta \mathbf{T_o} - \Delta \mathbf{T_i}}{\ln \dfrac{\Delta \mathbf{T_o}}{\Delta \mathbf{T_i}}}$$ and this value is known as Log Mean Temperature Difference or simply LMTD

$$\therefore \qquad Q = UA \text{ (LMTD)} \qquad \text{...(12.12)}$$

Similar derivation is worked out for counter flow under section 12.6.1.

The same expression holds good for counter flow also but the temperature distribution will be different.

The design of a heat exchanger generally involves the calculation of the total heat transfer area required. The flow rate of fluids and the specific heats and also the inlet and outlet temperatures specify the heat transfer rate (equation (12.6)). The overall heat transfer coefficient can be estimated using the fluid properties and flow configuration and dimensions. Hence the remaining unknown namely area A can be estimated. Iterative calculations will be required in actual designs as the distribution of the area can be done by various combinations of tube diameters and lengths.

Example 12.2: *Determine the area required in parallel flow heat exchanger to cool oil from 60°C to 30°C using water available at 20°C. The outlet temperature of the water is 26°C. The rate of flow of oil is 10 kg/s. The specific heat of the oil is 2200 J/kg K. The overall heat transfer coefficient U = 300 W/m² K. Compare the area required for a counter flow exchanger.*

Solution: The temperature variation for parallel flow is shown in Fig. Ex. 12.2(*a*).

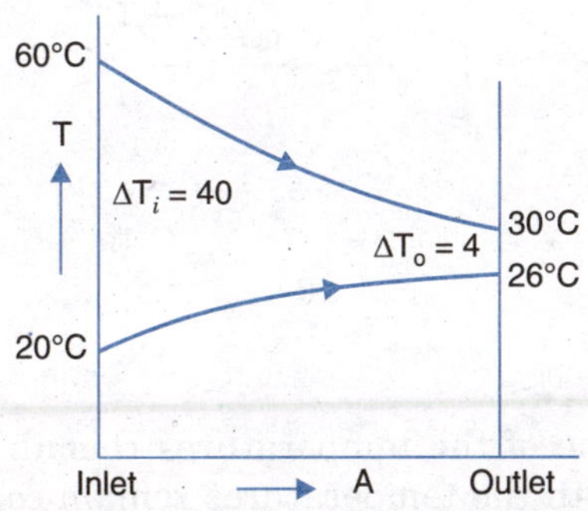

Fig. Ex. 12.2. (a) Parallel flow

Chapter 12

$$Q = m_h\, c_h\,(T_{hi} - T_{ho}) = 10 \times 2200\,(60 - 30)\text{ J/s} = \mathbf{6{,}60{,}000\ W}$$

$$Q = U A\,(\text{LMTD})$$

$$\textbf{LMTD} = \frac{40 - 4}{\ln \dfrac{40}{4}} = \mathbf{15.635°C}$$

$$6{,}60{,}000 = 300 \times A \times 15.635 \quad \therefore \quad \mathbf{A = 140.71\ m^2}$$

As can be seen a single tube arrangement is impractical for this area.

Counter flow:

The temperature variation is shown in Fig. Ex. 12.2(*b*)

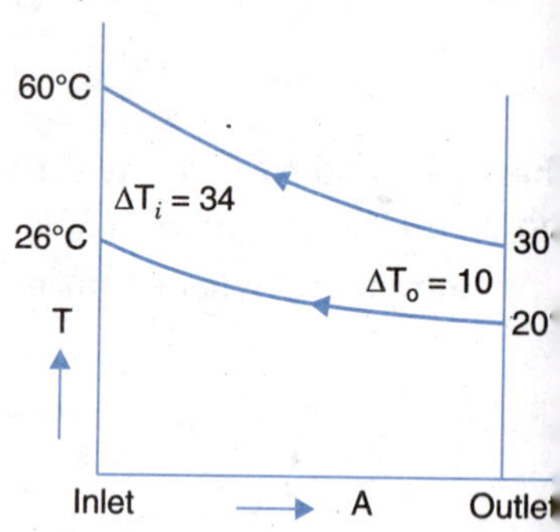

Fig. Ex. 12.2 (b) Counter flow

$$\textbf{LMTD} = \frac{34 - 10}{\ln \dfrac{34}{10}} = \mathbf{19.611°C}$$

$$\therefore \qquad \mathbf{A = 112.18\ m^2}$$

about 20% less.

The flow rate of water can also be determined as it will be a necessary data.

$$Q = m_c\, C_c\,(T_{co} - T_{ci})$$

$$6{,}60{,}000 = m_c \times 4180\,(26 - 20)$$

$$\mathbf{m_c = 26.32\ kg/s}$$

The counter flow arrangement provides more uniform temperature differenc along the flow and hence a better rate of heat flow. The counter flow type can also be used t cool or heat over a wider range of temperatures. In the above case by increasing the area or b reducing flow the hot oil can in the limit be cooled to 20°C. Manipulation in the opposit direction can get the water heated to 60°C. This is not possible in the parallel flow where th exit temperature has to lie somewhere in between the two inlet temperatures. As far as possibl counter flow is always used in heat exchanger designs.

12.3.2. Special Cases of Temperature Variations

(*i*) **Condensing or evaporating of one of the fluids** and one fluid condensing while othe evaporates: The temperature variation is shown in Figs. 12.4(*a*), (*b*) and (*c*). Very large flov rate of one of the fluids will also produce a similar temperature profile. The heat capacity c the fluid is theoretically infinite.

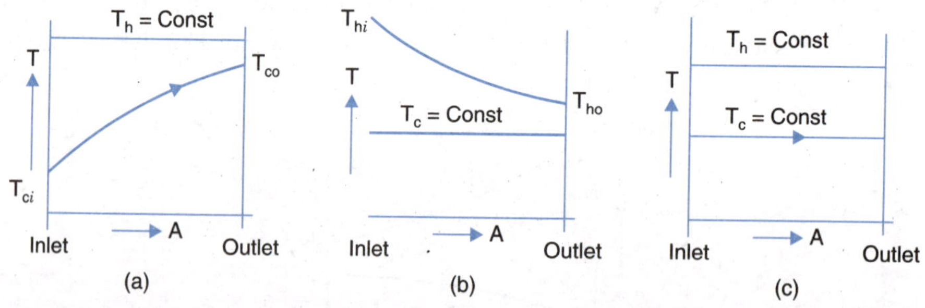

Fig. 12.4. (a) Condensing, (b) Evaporating, (c) Condensing and evaporating

In the first two cases one of the temperatures remain constant throughout the hea exchanger. In the third case both the temperatures remain constant. The direction of flow is material in these cases and the LMTD will be the same for both parallel flow, counter flow and other flow types. In the case of (c) LMTD = $(T_h - T_c)$ itself.

(*ii*) **Cooling, condensing and under cooling a superheated vapour** or heating, evaporating and super heating of under cooled liquid.

The temperature distribution in these cases is shown in Figs. 12.5(*a*) and (*b*).

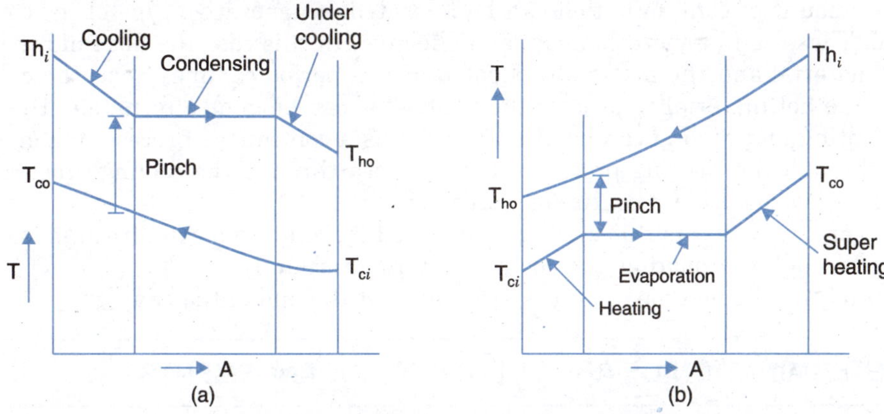

Fig. 12.5 (a) Cooling, condensation and undercooling
(b) Heating, evaporation and superheating

In this case the heat exchanger has to be treated as a combination of three heat exchangers. The area required for evaporation or condensation will be by far the largest. There is another interesting aspect to the design in that the temperature difference marked 'pinch' can be –ve if the areas and flow rates are not properly designed. The application of case (*b*) may be found in boilers of nuclear power plants and case (*a*) in refrigerant condensers.

(*iii*) **Equal heat capacity for both fluids.** $\mathbf{m_h\, c_h = m_c\, c_c}$. The temperature variation is shown in Figs. 12.6(*a*) and (*b*) for parallel and counter flow arrangements.

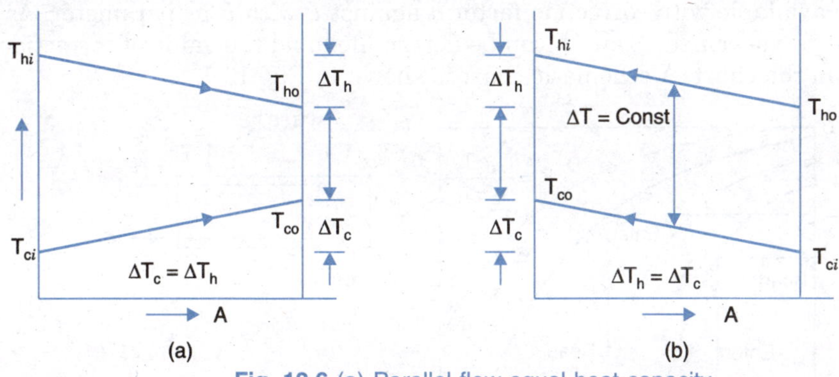

Fig. 12.6 (a) Parallel flow-equal heat capacity
(b) Counter flow-equal heat capacity

In parallel flow the change in temperature will be equal for both fluids ($\Delta T_h = \Delta T_c$). The slope of the lines will be equal but opposite.

In the case of counter flow the temperature difference is the same at all locations and LMTD by equation will be of 0/0 form and so we have to use

$$\text{LMTD} = T_{hi} - T_{co} = \text{constant all along.}$$

The situation is encountered in gas turbine regenerative heaters.

Chapter 12

12.4 REGENERATIVE TYPE

In recuperative heat exchangers the hot fluid and cold fluid exchange heat simultaneous and directly through a barrier. The fluids generally have no chance of mixing and intermediate storage is involved.

In certain cases involving gases this arrangement leads to a bulky exchanger. In ca some small amount of contamination can be tolerated, regenerative type of heat exchang can be used. These can be more compact and effective. In this case the hot fluid is passed some inert material and the material is heated over a period of time. Next, the cold fluid passed over the hot material to pick up the heat. This can be done alternately. By the use two units heating and cooling can be also achieved as a continuous process. When one of th units goes through the heating process the other goes through the cooling process and th operation is cycled to obtain a continuous process.

In larger power plants this cycling is achieved by rotating a cylindrical mass at a slo rate to pass through the heating and cooling sectors alternately.

The analysis of this type of unit is attempted at the end of the chapter.

12.5 DETERMINATION OF AREA IN OTHER ARRANGEMENTS

In the case of two pass, cross flow and other arrangements the LMTD obtained by using singl tube arrangement does not provide the correct average temperature and a correction factor ha to be applied. Two parameters are found to influence the value of correction factor. These are

(*i*) **Capacity ratio (R) given by**

$$R = \frac{\text{hot fluid temperature change}}{\text{cold fluid temperature change}} = \frac{T_{hi} - T_{ho}}{T_{co} - T_{ci}} \quad \text{...[12.13(a)}$$

(*ii*) **The effectiveness (P)**

$$P = \frac{\text{the minimum heat capacity fluid temperature rise}}{\text{the difference between inlet temperatures}} = \frac{t_2 - t_1}{T - t_1} \quad \text{...[12.13(b)}$$

Charts are available with correction factor F against P with R as parameter. As all th four temperatures are specified, P and R can be determined and the value of correction facto F can be read from the chart. A schematic chart is shown in Fig. 12.7.

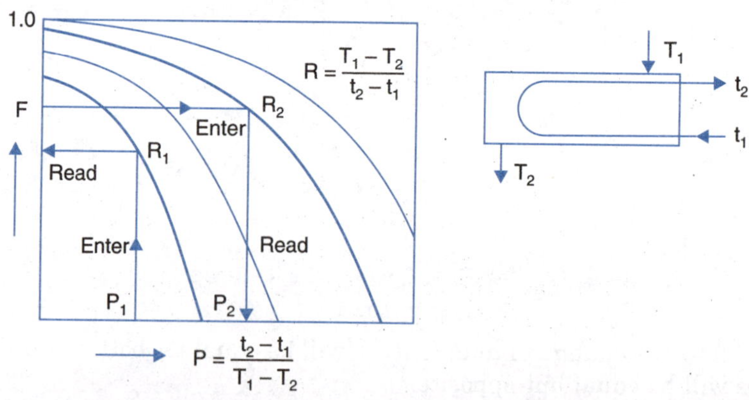

Fig. 12.7. Correction factor chart (schematic)

Corrected average temperature

$$(\overline{\Delta T}) = F\,.\,(\text{LMTD})\ \textbf{counter flow}$$

$$Q = U A (\overline{\Delta T}) \quad \text{...(12.14)}$$

Note: P will be zero for condenser, evaporator etc., and $F = 1$ in these cases. **LMTD should be based on counter flow. Charts for various configurations is shown in the following Figs. 12.8 (a) to (e)**

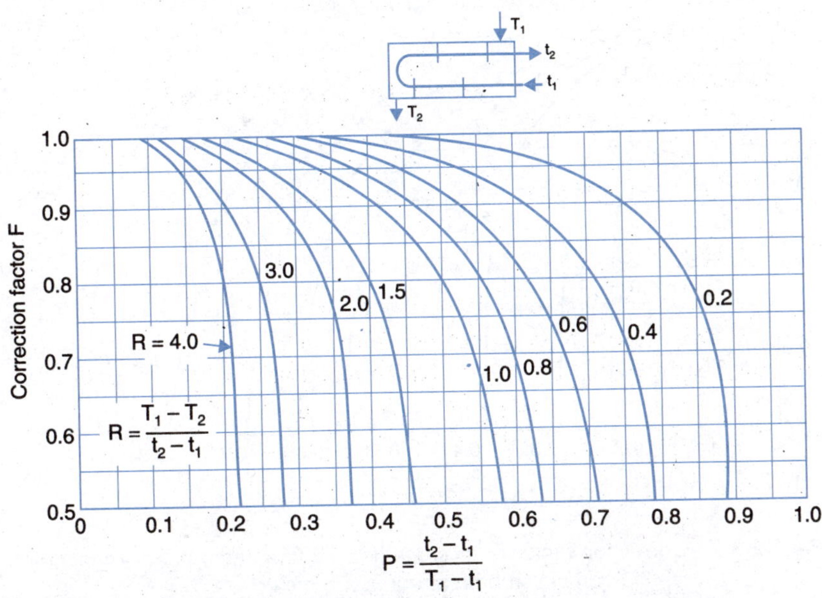

Fig. 12.8 (a) Correction factor F for shell-and-tube heat exchangers with one shell pass and any multiple of two tube passes (2, 4, 6, etc., tube passes)

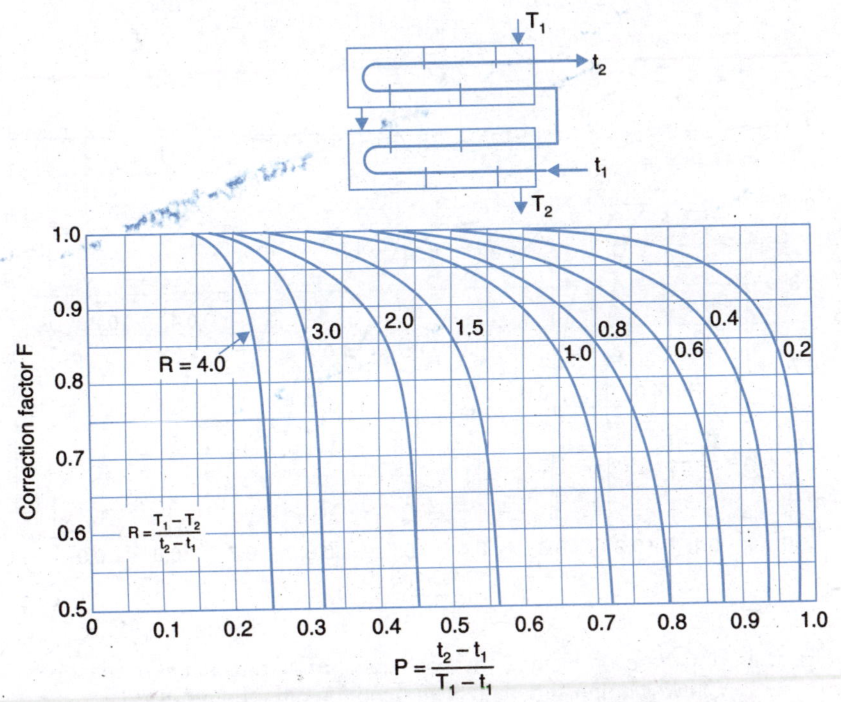

Fig. 12.8 (b) Correction factor F for shell-and-tube heat exchangers with two shell passes and any multiple of four tube passes (4, 8, 12, etc., tube passes)

Chapter 12

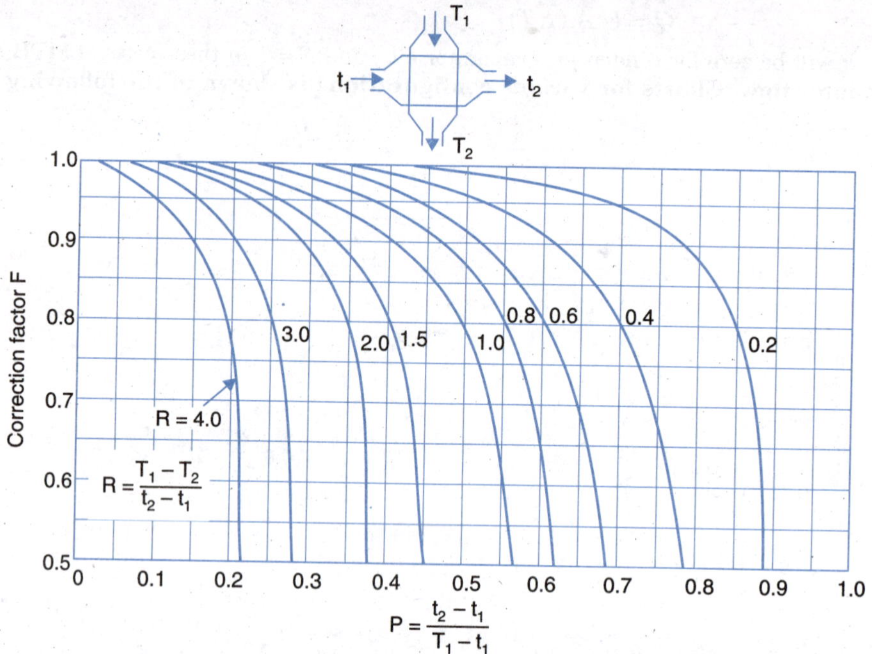

Fig. 12.8 (c) Correction factor F for cross-flow (single-pass) heat exchangers in which both streams are mixed

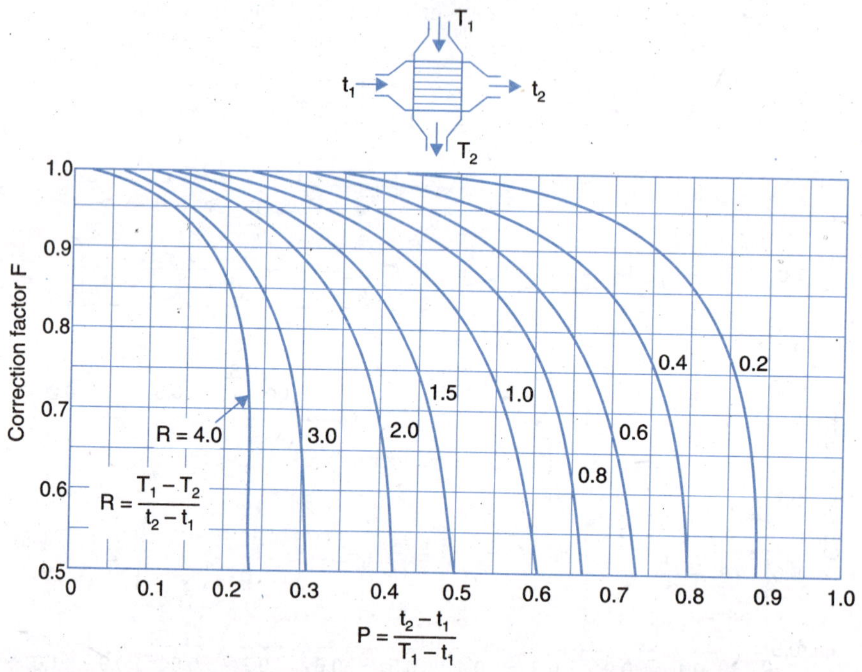

Fig. 12.8 (d) Correction factor F for cross-flow (single-pass) heat exchangers in which one stream is mixed and the other unmixed

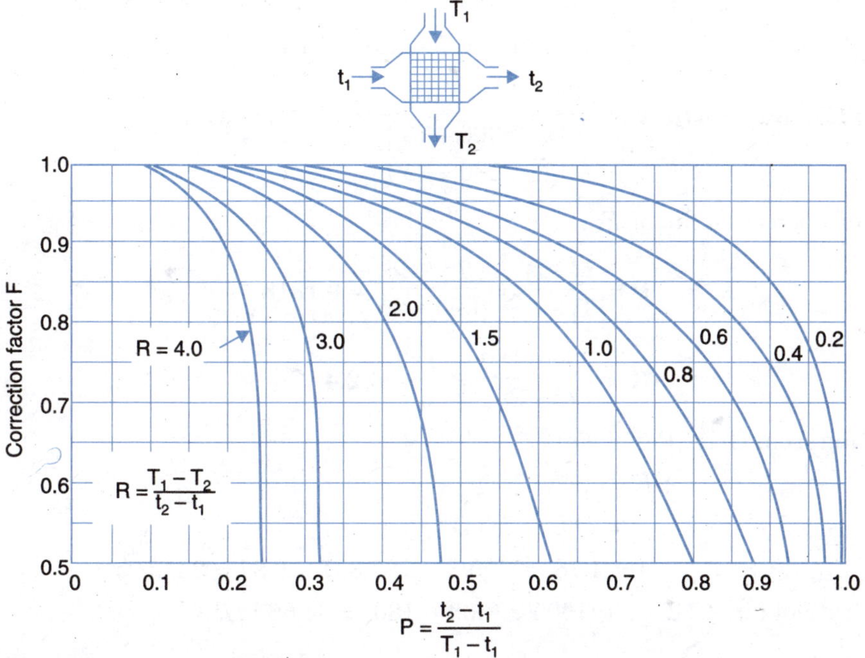

Fig. 12.8 (e) Correction factor F for cross-flow (single-pass) heat exchangers in which both streams remain unmixed

Example 12.3: *Determine the area required for a shell and tube heat exchanger with two tube passes to cool oil at rate of 10 kg/s from 60°C to 30°C flowing in the shell using water at 20°C passing through the tubes and heated up to 26°C. The specific heat of the oil is 2200 J/kg K. The value of overall heat transfer coefficient is 300 W/m² K.*

Solution: The heat flow rate is given by (Refer example 12.2)

$$Q = m_h c_h \Delta T_h = 10 \times 2200 \times (60 - 30) = 6,60,000 \text{ W}$$

$$\text{LMTD counter flow} = \frac{(60-26)-(30-10)}{\ln\left(\dfrac{60-26}{30-10}\right)} = 19.611°\text{C}.$$

$$P = \frac{26-20}{60-20} = \frac{6}{40} = 0.15,\ R = \frac{60-30}{26-20} = 5$$

Reading from chart [see Fig. 12.8(*a*)] $F = 0.875$

$\therefore$ $$6,60,000 = 300 \times 0.875 \times 19.611 \times A$$

$\therefore$ $$\mathbf{A = 128.2\ m^2.}$$

This requirement is between parallel flow and counter flow. See example 12.2.

Example 12.4: *A cross flow heat exchanger with both fluids unmixed is used to heat water flowing at a rate of 20 kg/s from 25°C to 75°C using gases available at 300°C to be cooled to 180°C. The overall heat transfer coefficient has a value of 95 W/m²K. Determine the area required. For gas c_p = 1005 J/kgK.*

Solution. The properties of gas can be taken as equal to that of air. From water flow

Heat transfer rate $Q = 20 \times 4180\,(75 - 25) = \mathbf{4.18 \times 10^6\ W}$

$$\textbf{LMTD counter flow} = \frac{(300 - 75) - (180 - 25)}{\ln\left(\dfrac{300 - 75}{180 - 25}\right)} = \mathbf{187.831°C}$$

To find correction factor F:

$$\mathbf{P} = \frac{t_2 - t_1}{T_1 - t_1} = \frac{75 - 25}{300 - 25} = \mathbf{0.1818}$$

$$\mathbf{R} = \frac{300 - 180}{75 - 25} = \frac{120}{50} = \mathbf{2.4}$$

Reading from chart Fig. 8.9(*c*)

$$F = 0.97$$

$$\therefore \quad A = Q/U.\ \text{LMTD}.$$

$$A = 4.18 \times 10^6/95 \times 187.831 \times 0.97 = \mathbf{241.5\ m^2}$$

Flow rate of air: $4.18 \times 10^6/1005 \times (300 - 180) = 34.66$ kg/s.

12.6 HEAT EXCHANGER PERFORMANCE

It is often necessary to estimate the **performance of a heat exchanger** when the flow rates and other parameters are changed from the design values. For example a heat exchanger may be designed to heat 10 kg/s of water from 30°C to 60°C using gases availabe at 200°C at a flow rate of 40 kg/s. Now it is possible to estimate the value of convective heat transfer coefficients both inside and outside of tubes using the flow rates and once the configurations are finalised, the area can be estimated. During operation, it may become necessary to heat water from other than design temperature and flow rates other than the design values. It will be possible to specify only the new values of flow rates and inlet temperatures. Using these values it will be necessary to determine the exit temperatures.

This can be done by (*i*) **iterative calculations** assuming values of outlet temperature and checking the values of heat flow calculated by equations (12.6) and (12.11) for convergence. This was tedius in the precomputer days. So an alternate method was used. (*ii*) **By using the effectiveness of the heat exchanger and the Net Transfer Units (NTU)** of the heat exchanger. **This is a very powerful method for determination of performance of heat exchangers at off design conditions and when the outlet temperatures are not specified.**

12.6.0. Effectiveness

Effectiveness is defined as the ratio of actual heat transfer to the maximum possible heat transfer in the heat exchanger. The maximum possible heat transfer is determined using the counter flow concept.

The concept is explained using Fig. 12.9.

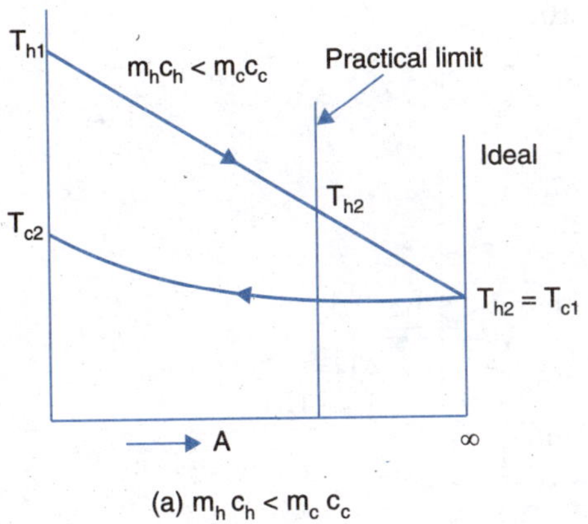

(a) $m_h c_h < m_c c_c$

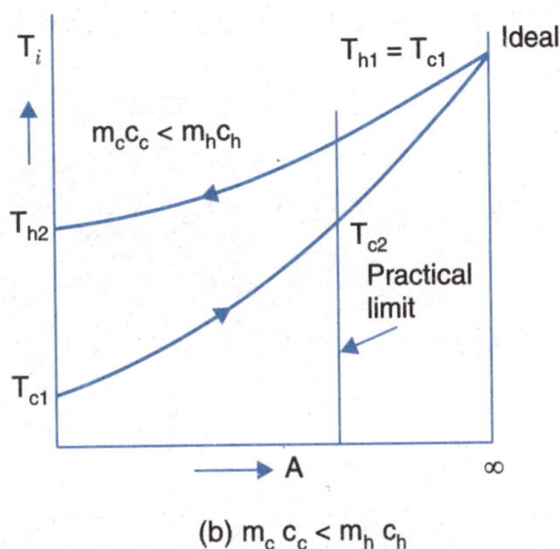

(b) $m_c c_c < m_h c_h$

Fig. 12.9

The maximum possible heat transfer is estimated using the concept that in counter flow arrangement it is possible by proper adjustment of flow rates either to (*i*) **cool the hot fluid to the cold fluid inlet temperature** Fig. 12.9(*a*) or (*ii*) **to heat the cold fluid to the inlet temperature of the hot fluid** Fig. 12.9(*b*). This is achieved by reducing the flow rate of hot fluid in case (*i*) and reducing the flow rate of cold fluid in case (*ii*), keeping the other flow rate constant.

Case (*i*) $m_hc_h < m_cc_c$, Actual heat transfer $= m_hc_h (T_{h1} - T_{h2})$

Maximum possible heat transfer $= m_hc_h (T_{h1} - T_{c1})$

$\therefore$ Effectiveness $= (T_{h1} - T_{h2})/(T_{h1} - T_{c1})$ ...(12.15)

Case (*ii*) $m_cc_c < m_hc_h$

The actual heat transfer $= m_cc_c (T_{c2} - T_{c1})$

The maximum possible heat transfer $= m_cc_c (T_{h1} - T_{c1})$

$$\therefore \quad \text{Effectiveness} = \frac{T_{c2} - T_{c1}}{T_{h1} - T_{c1}} \quad ...(12.16)$$

These definitions can also be written in general as

$$\textbf{effectiveness} = \frac{\textbf{change in temperature of the minimum heat capacity fluid}}{\textbf{difference between the fluid inlet temperatures}}$$

These expressions are simpler because the heat capacities are not involved. Expression like the following also are correct expressions for effectiveness.

$$\varepsilon = \frac{m_h\, c_h\, \Delta T_h}{(mc)_{\min}\,(T_{hi} - T_{ci})} = \frac{m_c\, c_c\, \Delta T_c}{(mc)_{\min}\,(T_{hi} - T_{ci})} \quad ...(12.17)$$

where $(mc)_{min}$ is the lower of the two heat capacities. Simple expressions can be derived for the value of effectiveness in terms of (*i*) C_{min}/C_{max} and

(*ii*) (UA/C_{min}) called NTU or Net Transfer Units. Here $C = mc$.

Chapter 12

12.6.1. Effectiveness: Counter Flow Heat Exchanger

The temperature distribution is shown in Fig. 12.10.

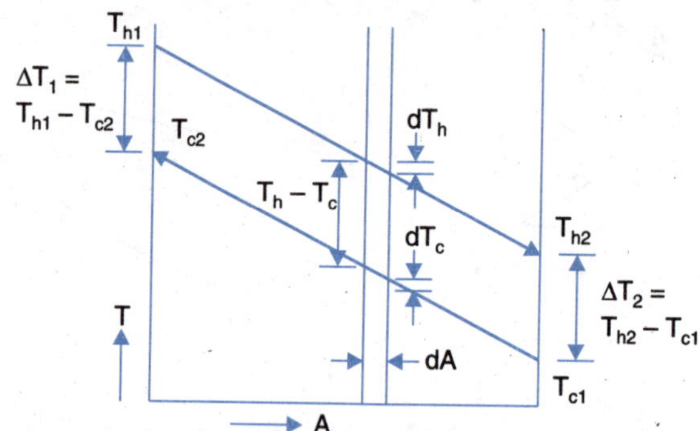

Fig. 12.10. Temperature profile, counter flow

Considering a small area dA at the location indicated, the temperature change dT_c and dT_h are both –ve in the direction considered.

$$dQ = UdA\,(T_h - T_c) \qquad \text{...(A)}$$

$$dQ = -\,C_h\,dT_h = -\,C_c\,dT_c$$

where C = mass × specific heat = heat capacity

$$dT_h = -\,dQ/C_h, \quad dT_c = -\,dQ/C_c, \text{ using equation (A)}$$

$$\therefore \quad dT_h - dT_c = d(T_h - T_c) = -\,UdA\left[\frac{1}{C_h} - \frac{1}{C_c}\right](T_h - T_c)$$

$$\therefore \quad \frac{d(T_h - T_c)}{T_h - T_c} = -\,U\,dA\left[\frac{1}{C_h} - \frac{1}{C_c}\right]$$

Integrating from entry to exist,

$$\ln\left[\frac{T_{h2} - T_{c1}}{T_{h1} - T_{c2}}\right] = -\,UA\left(\frac{1}{C_h} - \frac{1}{C_c}\right) \qquad \text{...(12.18)}$$

Substituting for C_h and C_c as

$$C_h = \frac{Q}{T_{h1} - T_{h2}} \quad \text{and} \quad C_c = \frac{Q}{T_{c2} - T_{c1}}$$

$$\ln\left[\frac{T_{h2} - T_{c1}}{T_{h1} - T_{c2}}\right] = \frac{UA}{Q}\,[(T_{h2} - T_{h1}) + (T_{c2} - T_{c1})]$$

$$= \frac{UA}{Q}\,[(T_{h2} - T_{c1}) - (T_{h1} - T_{c2})]$$

$$\therefore \quad Q = UA\,\{(\Delta T_2 - \Delta T_1)/\ln\,(\Delta T_2/\Delta T_1)\} \qquad \text{...(12.19)}$$

$$= UA\,(\text{LMTD}) \text{ (Refer equation (12.12) also)}$$

Considering equation (12.18) and inverting

$$\frac{T_{h2}-T_{c1}}{T_{h1}-T_{c2}} = \exp\left[-UA\left(\frac{1}{C_h}-\frac{1}{C_c}\right)\right] \quad ...(12.20)$$

Considering C_c as C_{min}

$$\frac{T_{h2}-T_{c1}}{T_{h1}-T_{c2}} = \exp\left[-\frac{UA}{C_{min}}\left(\frac{C_{min}}{C_{max}}-1\right)\right]$$

$$\frac{UA}{C_{min}} = \text{NTU. For simplicity let } \frac{UA}{C_{min}} = N \text{ and } \frac{C_{min}}{C_{max}} = C,$$

$$\frac{T_{h2}-T_{c1}}{T_{h1}-T_{c2}} = \exp\,[-N(C-1)] = 1/\exp\,[-N(1-C)] \quad ...[12.20(a)]$$

Considering RHS, and noting that for obtaining effectiveness we require $T_{h1} - T_{c1}$ in the ratio, add and subtract T_{h1} in the numerator and T_{c1} in the denominator

$$\frac{T_{h2}-T_{c1}}{T_{h1}-T_{c2}} = \frac{(T_{h1}-T_{c1})-(T_{h1}-T_{h2})}{(T_{h1}-T_{c1})-(T_{c2}-T_{c1})}$$

Consider RHS, dividing by $T_{h1} - T_{c1}$, this reduces to

$$(T_{h1}-T_{c1}) = \frac{1-\dfrac{T_{h1}-T_{h2}}{T_{h1}-T_c}}{1-\varepsilon}$$

but $$(T_{h1}-T_{h2}) = \frac{C_c}{C_h}(T_{c2}-T_{c1}) = C(T_{c2}-T_{c1})$$

Substituting in equation [12.20(*a*)] $$\frac{T_{h2}-T_{c1}}{T_{h1}-T_{c2}} = \frac{1-C\,\varepsilon}{1-\varepsilon} = \frac{1}{\exp[-N(1-C)]}$$

Solving for **effectiveness** for counter flow exchanger

$$\varepsilon = \frac{1-\exp[-N(1-C)]}{1-C\exp[-N(1-C)]} \quad ...(12.21)$$

This is a very important equation in using NTU method.

If C_h is considered as C_{min}, we get the same result as below, considering equation (12.20).

$$\frac{T_{h2}-T_{c1}}{T_{h1}-T_{c2}} = \exp\left[-UA\left(\frac{1}{C_h}-\frac{1}{C_c}\right)\right]$$

$$= \exp\left[-\frac{UA}{C_{min}}(1-C)\right] = \exp\,[-N(1-C)]$$

In RHS add and subtract T_{h1} in the numerator and T_{c1} in the denominator.

$$\frac{T_{h2}-T_{c1}}{T_{h1}-T_{c2}} = \frac{(T_{h1}-T_{c1})-(T_{h1}-T_{h2})}{(T_{h1}-T_{c1})-(T_{c2}-T_{c1})}$$

dividing by $T_{h1} - T_{c1}$ and noting $(T_{c2} - T_{c1}) = C\,(T_{h1} - T_{h2})$ this reduces to

$$\frac{1-\varepsilon}{1-C\varepsilon} = \exp\,[-N\,(1-C)]$$

When rearranged, **the same result as equation (12.21) is obtained.**
The derivation for parallel flow is obtained using equation (12.10)

$$\ln \frac{T_{h2}-T_{c1}}{T_{h1}-T_{c1}} = -\,UA\left[\frac{1}{C_h}+\frac{1}{C_c}\right]$$

Inverting this equation

$$\frac{T_{h2}-T_{c1}}{T_{h1}-T_{c1}} = \exp\left[-\frac{UA}{C_{min}}\left(1+\frac{C_{min}}{C_{max}}\right)\right] = \exp\,[-N\,(1+C)] \quad \text{...(A)}$$

This is irrespective of whatever fluid is C_{min}.

Rearranging only the numerator by adding and subtracting T_{h1}, noting C_h is C_{min}

$$\frac{-(T_{h1}-T_{h2})+(T_{h1}-T_{c2})}{T_{h1}-T_{c1}} = \frac{T_{h2}-T_{c2}}{T_{h1}-T_{c1}} - \varepsilon$$

Now adding and subtracting T_{c1} in the numerator of the first term and noting

$$T_{c2} - T_{c1} = C(T_{h1} - T_{h2})$$

$$\frac{(T_{h1}-T_{c1})-(T_{c2}-T_{c1})}{T_{h1}-T_{c1}} - \varepsilon = 1 - C\,\varepsilon - \varepsilon$$

Substituting in equation (A) and rearranging effectiveness for parallel flow is

$$\therefore \quad \varepsilon\,(1 + C) = 1 - \exp\,[-N(1 + C)]$$

$$\therefore \quad \varepsilon = \frac{1-\exp[-N\,(1+C)]}{1+C} \quad \text{...(12.22)}$$

12.6.2. Special Cases

(*i*) When $C_{min} = C_{max}$, $C = 1$

∴ **For parallel flow** substituting in equation (12.22)

$$\varepsilon_p = \frac{1-\exp[-2N]}{2} \quad \text{...(12.23)}$$

For counter flow direct substitution results in zero divided by zero form. So it is obtained by suitable mathematical methods (for limits) as

$$\varepsilon_c = \frac{N}{N+1} \quad \text{...(12.24)}$$

(*ii*) For condensers, evaporators or large value of C, $C_{min}/C_{max} \to 0$ or $C_{min} = 0$, **for both cases**

$$\varepsilon = 1 - e^{-N} \quad \text{...(12.25)}$$

The detailed derivations for these expressions are worked under advanced examples 12.4 to 12.7.

For other configurations also such expressions are available. As these are lengthy, chart solution is generally attempted.

12.6.3. Effectiveness—NTU Chart

A typical effectiveness—NTU chart is shown schematically in Fig. 12.11. Tabulation of equations expressing (*i*) effectiveness in terms of NTU and (*ii*) NTU in terms of effectiveness are available in handbooks and referenced cited.

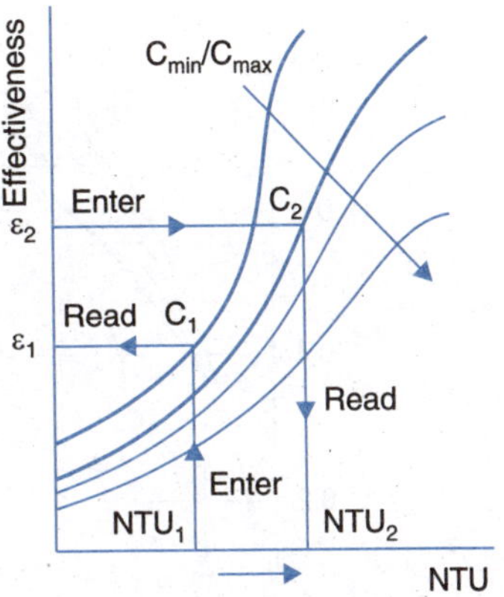

Fig. 12.11. Effectiveness chart (schematic)

The value of NTU can be determined knowing flow rates and specific heats of fluids, area of the exchanger and the overall heat transfer coefficient. C_{min}/C_{max} is used as parameter. This value is also obtainable from the data specified. The chart is entered at the value of NTU and effectiveness is read where this line cuts the parametric line (C_{min}/C_{max}).

Note: $(C_{min}/C_{max}) = 0$ for condensers, evaporators and for very high heat capacity of one fluid.

As effectiveness is defined in terms of the inlet temperatures and one outlet temperature, this outlet temperature is determined in this way. The heat transfer rate and the exit temperature of the other fluid can be worked out using energy balance.

Effectiveness—NTU charts for various configurations is shown in Figs. 12.12(*a*)–(*g*).

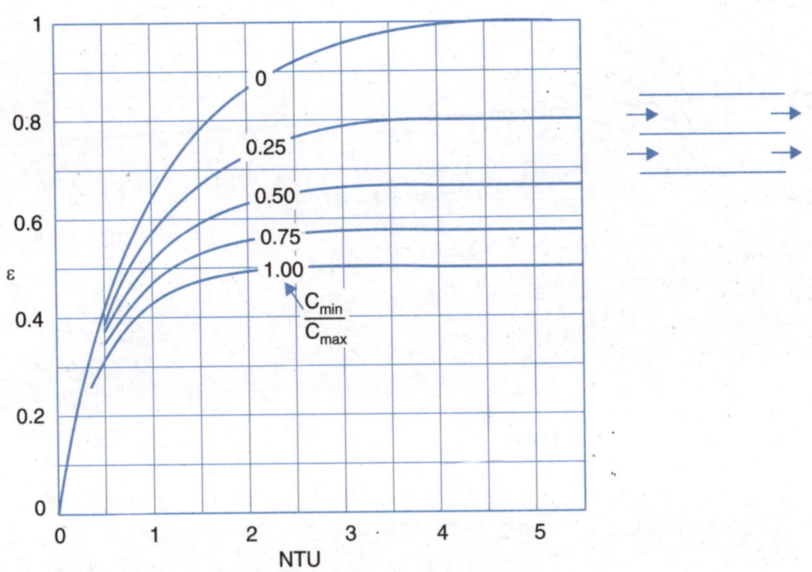

Fig. 12.12 (a) The effectiveness of a parallel flow heat exchanger

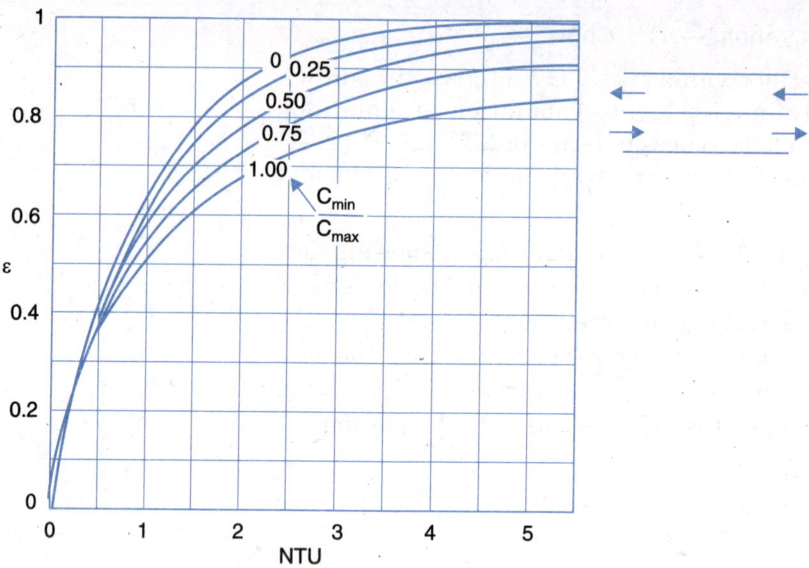

Fig. 12.12 (b) The effectiveness of a counterflow heat exchanger

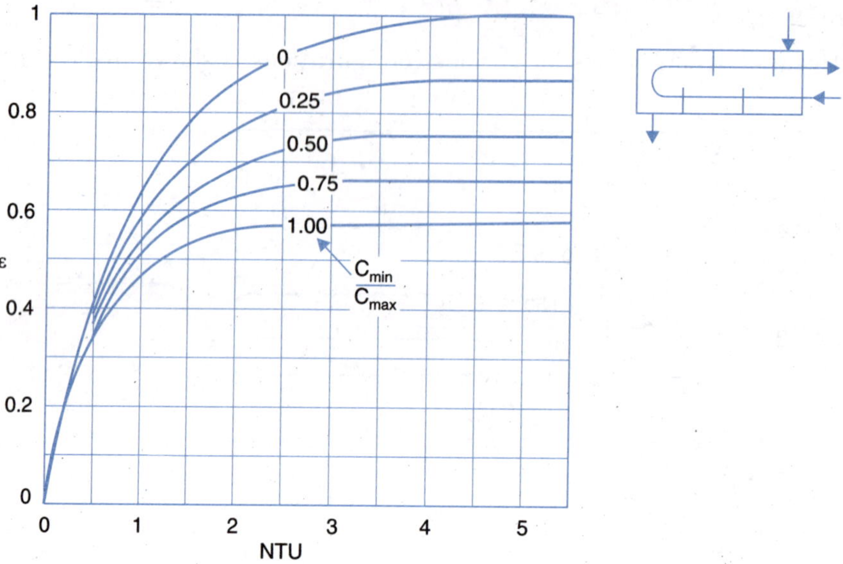

Fig. 12.12 (c) The effectiveness of shell-and-tube heat exchangers with one shell pass and any multiple of two tube passes (2, 4, 6, etc., tube passes)

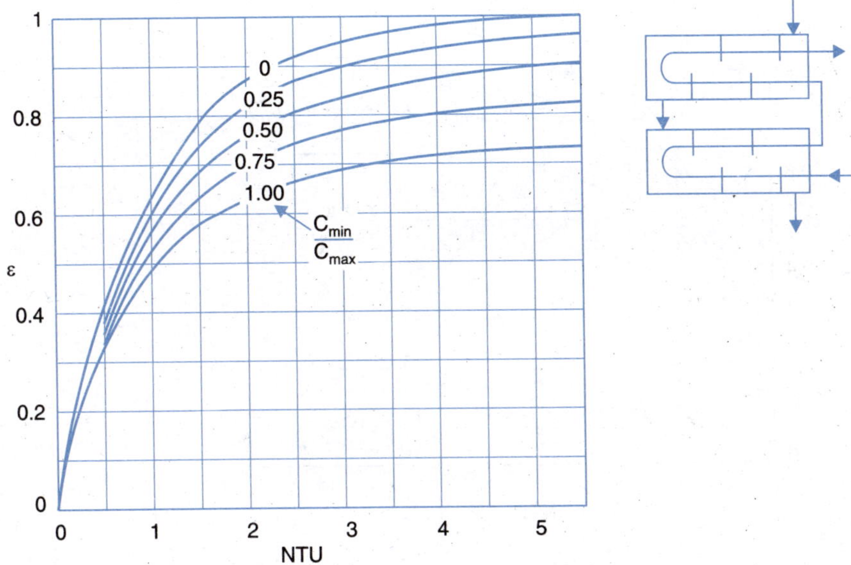

Fig. 12.12 (d) The effectiveness of shell-and-tube heat exchangers with two shell passes and any multiple of four tube passes (4, 8, 12, etc., tube passes)

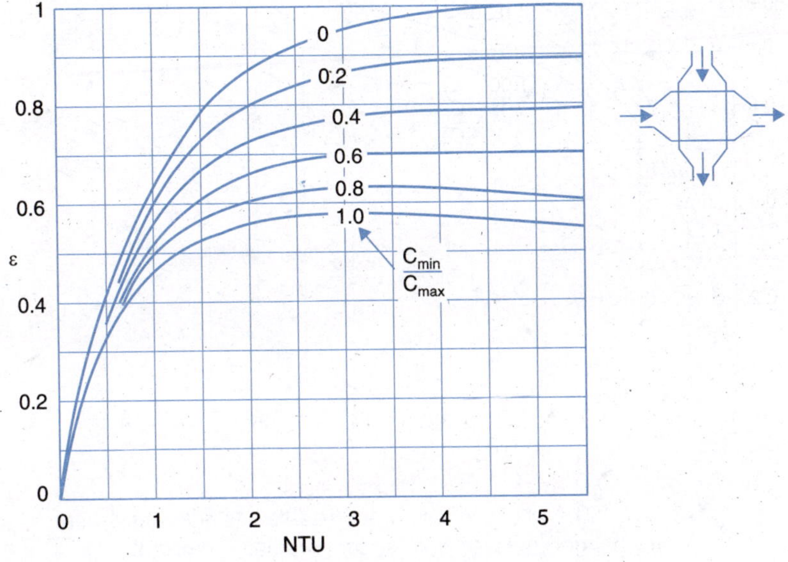

Fig. 12.12 (e) The effectiveness of a cross-flow (single-pass) heat exchanger in which both streams are mixed

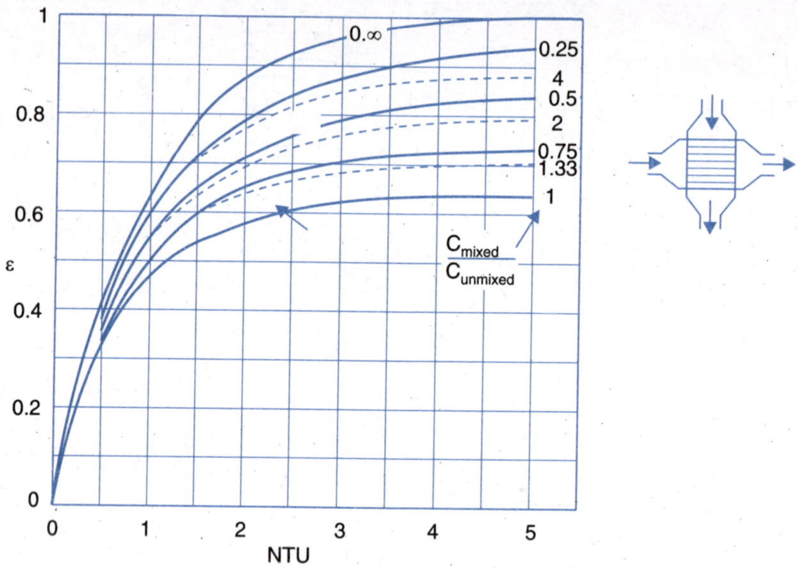

Fig. 12.12 (f) The effectiveness of a cross-flow (single-pass) heat exchanger in which one stream is mixed and the other is unmixed

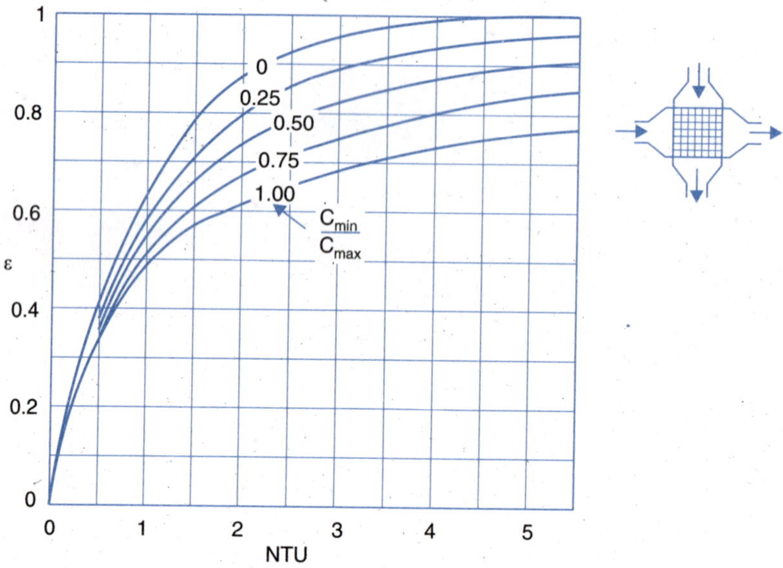

Fig. 12.12 (g) The effectiveness of a cross-flow (single-pass) heat exchanger in which both streams remain unmixed

Example 12.5: *Considering the data of example 12.2, determine the outlet temperatures of the fluids if the oil flow rate is increased to 14 kg/s. The other flow rate remains the same. Workout for parallel flow, counter flow and shell with two tube passes.*

Solution: The unit cools 10 kg/s of oil from 60°C to 30°C using water at 20°C being heated to 26°C. The flow rate was determined as 26.32 kg/s. The areas are (*i*) parallel flow: 140.71 m^2. (*ii*) counter flow: 112.18 m^2 (*iii*) shell with two tube passes 128.2 m^2 (example 12.3).

$$U = 300 \text{ W/m}^2\text{ K.} \quad \text{Specific heat of oil: 2200 J/kg K}$$

Considering the changed flow:

$$m_h C_h = 14 \times 2200 \qquad \therefore C_h \text{ is } C_{min}$$

$$m_c C_c = 26.32 \times 4180$$

$$C = \frac{C_{min}}{C_{max}} = \frac{14 \times 2200}{26.32 \times 4180} = 0.28$$

Parallel flow

$$N = \text{NTU} = UA/C_{min} = \frac{300 \times 140.71}{2200 \times 14} = 1.37$$

Reading from the chart Fig. 12.12(*a*) at this value of NTU and C_{min}/C_{max}

Effectiveness = 0.68

Using equation (12.22)

$$\text{Effectiveness} = \frac{1 - \exp[-N(1+C)]}{1+C} = 0.646$$

It is better to use the equation results

$$\text{Effectiveness} = \frac{60 - T_{h2}}{60 - 20} \quad \therefore \mathbf{T_{h2} = 34.16°C}$$

(as hot fluid is C_{min})

Water outlet temperature: $\dfrac{14(60 - 34.16)\,2200}{26.32 \times 4180} + 20 = \mathbf{27.234°C}$

Check: $\quad \text{LMTD} = \dfrac{(60-20) - (34.16 - 27.234)}{\ln\left(\dfrac{(60-20)}{34.16 - 27.234}\right)} = 18.86°\text{C}$

$$Q = UA \text{ LMTD} = 300 \times 140.71 \times 18.86 = 0.7962 \times 10^6 \text{ W}$$

From energy balance

$$Q = 14 \times 2200 \times (60 - 34.16) = 0.7959 \times 10^6 \text{ W}$$

Checks:

Counter flow: NTU = 300 × 112.18/14 × 2200 = 1.093

$$\text{Effectiveness} = \frac{1 - \exp[-N(1-C)]}{1 - C.\exp[-N(1-C)]} = 0.624$$

From chart Fig. 12.12(*b*): effectiveness = 0.620

Using the equation results

$$0.624 = \frac{60 - T_{h2}}{60 - 20} \qquad \therefore \mathbf{T_{h2} = 35.03°C}$$

Water outlet temperature $= 20 + \dfrac{14\,(60 - 35.03) \times 2200}{26.32 \times 4180} = \mathbf{26.99°C}$

Check $$\text{LMTD} = \frac{(60 - 26.99) - (35.03 - 20)}{\ln\left(\dfrac{60 - 26.99}{35.03 - 20}\right)} = 22.85°\text{C}$$

$$Q = 300 \times 112.18 \times 22.85 = 0.769 \times 10^6$$

From energy balance: $14 \times 2200\,(60 - 35.03) = 0.7691 \times 10^6$

Checks

Shell with two tube passes:

$$\text{NTU} = 300 \times 128.2/14 \times 2200 = 1.249$$

$$C_{min}/C_{max} = 0.28$$

Using chart, Fig. 12.12(*c*) effectiveness $= 0.64 = \dfrac{60 - T_{h2}}{60 - 20}$

$\therefore$ $$\mathbf{T_{h2} = 34.4°C}$$

Water exit temperature using energy balance,

$$= 20 + \frac{14 \times 2200\,(60 - 34.4)}{26.32 \times 4180} = \mathbf{27.17°C}$$

$$Q = 14 \times 2200\,(60 - 34.4) = 0.7885 \times 10^6 \text{ W}$$

Using LMTD and correction factor, (0.875, see example 12.3)

$$\text{LMTD} = \frac{(60 - 27.17) - (34.4 - 20)}{\ln\left(\dfrac{60 - 27.17}{34.4 - 20}\right)} = 22.36°\text{C}$$

$$Q = 128.2 \times 300 \times 22.36 \times 0.875 = 0.7526 \times 10^6 \text{ W}$$

This difference (5% about) is possibly due to the use of chart solution.

COMMENT

The NTU method provides an easy way to determine the performance of a given heat exchanger under different operating conditions. In the above problem the value of U is taken as constant. But it can be adjusted according to the variation of flow rate.

Example 12.6: *An economiser in a boiler has flow of water inside the pipes and hot gases on the outside flowing across the pipes. The flow rate of gases is 2,000 tons/hr and the gases are cooled from 390°C to 200°C. The specific heat of the gas is 1005 J/kg K. Water is heated (under pressure) from 100°C to 220°C. Assuming an overall heat transfer coefficient of 35 W/m² K, determine the area required by NTU method and check the same by LMTD method. Assume that the air flow is mixed.*

Solution: This is a cross flow heat exchanger

The flow rate of gas is, $\dfrac{2000 \times 1000}{3600}$ kg/s = 555.6 kg/s

$$\mathbf{C_h = 5.583 \times 10^5\ W}$$

The flow rate of water $= \dfrac{555.6 \times 1005 \times (390 - 200)}{4180 \times (220 - 100)} = 211.49$ kg/s

$$\mathbf{C_c = 211.49 \times 4180 = 8.84 \times 10^5\ W}$$

$\therefore$ C_h is $C_{\min}$ and $\mathbf{C_{min}/C_{max} = 0.632}$

Effectiveness with C_h as $C_{\min} = \dfrac{T_{h1} - T_{h2}}{T_{h1} - T_{c1}} = \dfrac{390 - 200}{390 - 100} = \mathbf{0.6551}$

From the chart ($C_h = C_{\min}$ mixed), NTU is read as 1.75

$$\therefore \qquad \mathbf{A} = \frac{\text{NTU} \times C_{\min}}{U} = \frac{1.75 \times 5.583 \times 10^5}{35} = \mathbf{27{,}915\ m^2}$$

Check using LMTD:

$$\textbf{LMTD} \text{ counter flow} = \frac{(390 - 220) - (200 - 100)}{\ln\left(\dfrac{390 - 220}{200 - 100}\right)} = \mathbf{131.92°C}$$

To determine correction factor,

$$\mathbf{P} = \frac{220 - 100}{390 - 100} = \frac{120}{290} = \mathbf{0.414},\ \mathbf{R} = \frac{390 - 200}{220 - 100} = \frac{190}{120} = \mathbf{1.583}$$

From chart Fig. 12.8(*d*) $F = 0.8$

$$Q = FUA\ (\text{LMTD})$$

$$\therefore \qquad A = Q/FU\ (\text{LMTD}) = \frac{2000 \times 1000}{3600} \times \frac{1005\,(390 - 200)}{35 \times 131.92 \times 0.8}$$

$\therefore$ $\mathbf{A = 28720\ m^2}$ **checks within reasonable limits.**

The LMTD method is simpler in this case that is when all four terminal temperatures are specified.

Example 12.7: *If in example 12.6, the water flow rate is increased by 20% without altering inlet conditions determine the exit conditions and the percentage change in heat transfer.*

Solution: The value of $C_{\max}$ alone is changed in this case. So the value of NTU remains the same. NTU = 1.75. Taking values from example 12.6 and assuming U is constant

$$\frac{C_{\min}}{C_{\max}} = \frac{0.632}{1.2} = 0.527$$

The value of effectiveness as read from the chart is 0.68. (Note: This can be checked by equations available, 0.67)

$$\therefore \qquad 0.68 = \frac{390 - T_{h2}}{390 - 100} \qquad \therefore\ \mathbf{T_{h2} = 192.8°C}$$

Water outlet temperature:

$$\mathbf{T_{c2}} = \frac{555.6 \times 1005\,(390 - 192.8)}{211.49 - 1.2 \times 4180} + 100 = \mathbf{203.8°C}$$

$$Q = 555.6 \times 1005 \times (390 - 192.8) = 110.112 \times 10^6\ \text{W}$$

Chapter 12

Q at design value = 555.6 × 1005 (390 – 200) = 106.09 × 10^6 W

∴ percentage increase = 3.8%

Note: Though the heat flow has increased, the water outlet temperature is lower because o increased flow rate. The water temperature can be increased if the gas flow rate is increased withou altering the water flow rate.

12.7 STORAGE TYPE HEAT EXCHANGERS

A diagramatic view of a storage type heat exchanger is shown in Fig. 12.13

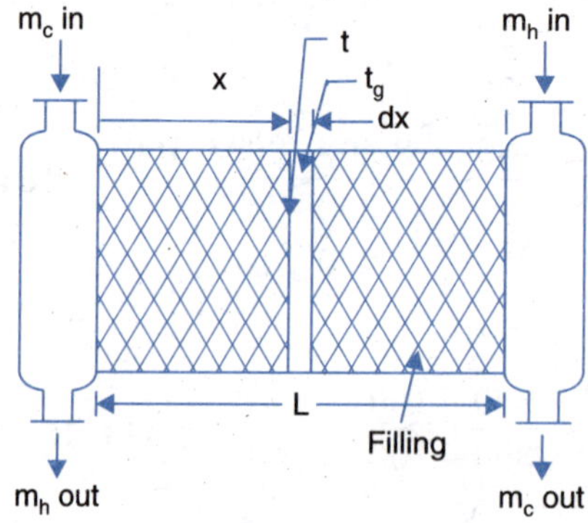

Fig. 12.13. Storage type heat exchanger

Hot gas flows over the matrix filling for a given period, heating the mass. The flow is then stopped and cold fluid flows over the matrix for the next period getting heated in the process.

The energy from the hot fluid is transferred to the cold fluid with intermediate storage in the matrix.

The gas and matrix temperatures at any location can be solved for by writing the energy balance equation for the elemental width dx, using the following notations:

M_c = mass of solid per unit length kg/m

c_s = specific heat of solid

V_1 = free volume per unit length

A_1 = heat transfer surface area per unit length

m = mass flow rate of gas

ρ = density of gas

c = specific heat of gas

L = length of the matrix column

h = convective heat transfer coefficient between gas and matrix

Considering the solid, at the location x for length dx, where the solid temperature is t, and gas temperature t_g, heat transferred by convection at the solid surface = heat stored in the solid.

$$h A_1 (t_g - t)\, dx\, d\tau = M_c\, c_s\, dx \frac{\partial t}{\partial \tau} \,.\, d\tau \qquad ...(12.26)$$

Considering the gas.

Heat transferred by convection = heat in changing the temperature of gas in the free volume + change in the stored energy of the flowing gas.

$$h\,A_1\,(t - t_g)\,dx\,d\tau = \rho\,c\,V_1 dx \frac{\partial t_g}{\partial \tau}\,.\,d\tau + c\,m\,d\tau \frac{\partial t_g}{\partial x}\,.\,dx \quad \text{...(12.27)}$$

These equations after simplification yield

$$\frac{\partial t}{\partial \tau} = \frac{hA_1}{c_s M_c}\,(t_g - t) \quad \text{...(12.28)}$$

$$\frac{\partial t_g}{\partial x} + \frac{\rho V_1}{m}\,.\,\frac{\partial t_g}{\partial \tau} = \frac{hA_1}{cm}\,(t - t_g) \quad \text{...(12.29)}$$

The second term on LHS is very small and is generally neglected. These equations are then solved using the boundary and initial conditions and introducing new variables as below:

$$\text{exi} = \frac{hA_1}{cm}\,.\,x, \quad \text{eta} = \frac{hA_1}{c_s M_c}\,.\,\tau \quad \text{...(12.30)}$$

The results of the solution are available in chart from in data books with (*i*) dimensionless gas temperature ratio and exi with eta as parameter to determine gas temperature at any location and (*ii*) dimensionless solid temperature ratio and exi, with eta as parameter as shown in Fig. 12.14(*a*) and (*b*) (t_0 is initial temperature of solid)

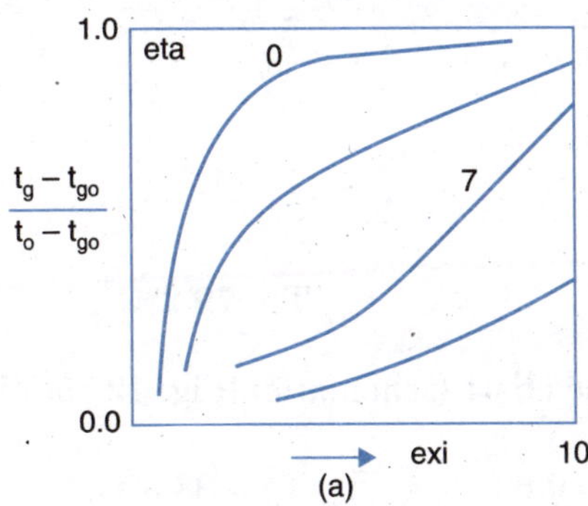

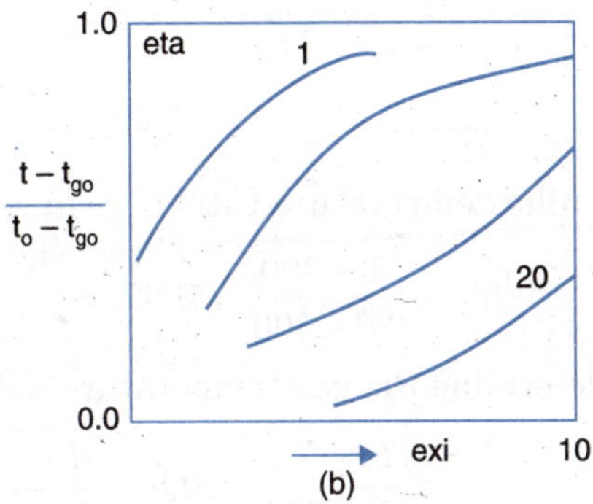

Fig. 12.14. Regenerative exchanger, (a) Gas temperature chart (b) Soild temperature chart (Schematic)

The variables exi and eta are functions of location and time and material properties and flow rate. In this case the solid internal resistance is neglected.

The heat flow over a period of time can be calculated by summing up the heat flow at all the locations.

For continuous rotary type of storage heat exchanger, effectiveness charts are available in handbooks to determine the effectiveness and heat flow.

Example 12.8: *A storage type of heater is 1 m² in section 2.5 m long. It is filled with mesh which provides 30 m²/m length surface area and a void of 60%. The mass of filler per meter length is 2000 kg. The specific heat of the filler material is 475 J/kg K. The convective heat transfer coefficient over the surface is 50 W/m² K.*

Determine the filler and gas temperature at 1.25 m location after 12 minutes. Gas at 300°C flows at a rate of 0.4 kg/s. The initial temperature at the location was 60°C specific heat of gas = 1005 J/kg K.

Solution: This is a storage heat exchanger problem. Use of charts is necessary. The parameters are

$$exi = \frac{hA_1}{cm}.x \quad eta = \frac{hA_1}{c_s M_c}.\tau .$$

$h = 50$ W/m^2 K, $A_1 = 30$ m^2/m, $c = 1005$ J/kg K, $m = 0.4$ kg/s

$x = 1.25$ m, $c_s = 475$ J/kg K, $M_c = 2000$ kg/m, $\tau = 720$ s

$$\therefore \quad \textbf{exi} = \frac{50 \times 30 \times 1.25}{1005 \times 0.4} = 4.66 \quad \textbf{eta} = \frac{50 \times 30 \times 720}{2000 \times 475} = \mathbf{1.14}$$

Using the chart (schematic) [Fig. 12.15(*a*)]

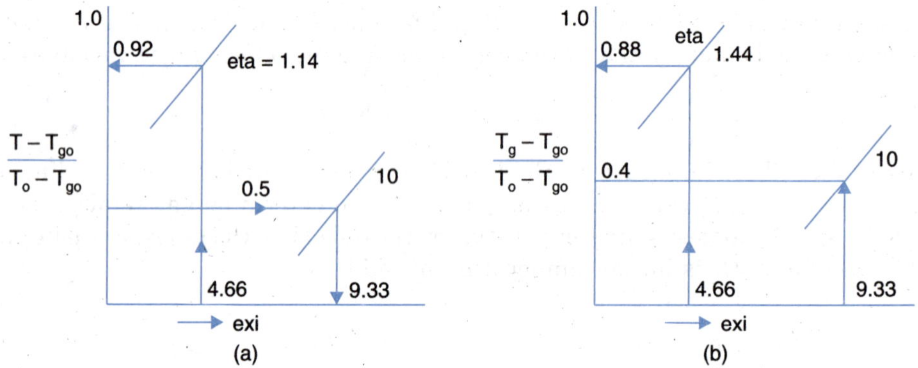

Fig. 12.15

The filler temperature ratio is read as 0.92

$$\frac{T - 300}{60 - 300} = 0.92, \qquad \therefore \quad \mathbf{T = 79.2°C}$$

To determine the gas temperature using the chart (schematic) [Fig. 12.15(*b*)]

$$\frac{T_g - T_{go}}{T_o - T_{go}} = 0.88 \quad \frac{T_g - 300}{60 - 300} = 0.88 \qquad \therefore \quad \mathbf{T_g = 88.8°C.}$$

Example 12.9 : *In example 12.8, determine how long it will take for the filler temperature at the exit position to reach 180°C. Determine the gas temperature at this position at that time.*

Solution:
$$\textbf{exi} = \frac{50 \times 30 \times 2.5}{1005 \times 0.4} = \mathbf{9.33}$$

$$\frac{T - T_{go}}{T_o - T_{go}} = \frac{180 - 300}{60 - 300} = 0.5$$

Reading from the filler temperature chart, eta = 10

$$10 = \frac{50 \times 30 \times \tau}{2000 \times 475}, \quad \therefore \quad \mathbf{\tau = 6333\ s\ or\ 1.76\ hours}$$

reading from gas temperature chart for eta = 10 and exi = 9.33, Fig. 12.14(*b*),

$$\frac{T_g - T_{go}}{T_o - T_{go}} = 0.4 \qquad \therefore \quad t_g = 204°C$$

The mass flow rate of gas and convection coefficient should be increased if the heating time is to be reduced.

12.8 COMPACT HEAT EXCHANGERS

In gas flow, the convective heat transfer coefficients are low and consequently in heat exchangers using gas flow the overall heat transfer coefficient is low. The area required is large. So the heat transfer area has to be increased over and above that of the tubes in order to reduce the heat exchanger size. This is achieved by the use of thin fins. Such heat exchangers are called compact heat exchangers. A heat exchanger with more than 700 m^2 of area per m^3 is considered as compact heat exchanger. Examples are steam to air heaters, automobile radiators and heat exchangers used in small air conditioners. Several types of construction are employed, some of them being circular tubes with circular fins, helical fins, plate fins, flat tubes with plate fin matrix. Test results for several types of matrix arrangement, correlated in terms of Reynolds number against friction factor and Stanton number are available. Reynolds and Stanton numbers are generally calculated using mass velocity.

Kern has proposed a method for heat flow/area calculation for close pitched circular tubes with circumferential fins.

The tube arrangement and fin configuration is shown in Fig. 12.16(*a*). The various terms like *FD*, *OD*, *ID*, *FP*, *t*, *FL* and *TP* are indicated in the figure. Additional terminology are given below:

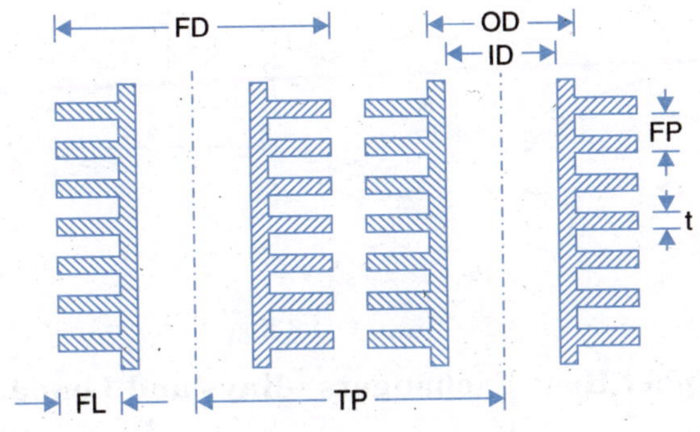

Fig. 12.16 (a) Fin configuration

m = mass flow over bank

NT = Number of tubes in a row

NF = Number of fins in length L of pipe = L/FP

AF = Fin surface area

AB = Bare tube surface area

D_e = equivalent diameter

P = Perimeter in one flow path.

$$AF = \frac{\pi}{4} \times 2 \times NF\,[FD^2 - OD^2]$$

$AB = \pi \times OD \times (L - NF \times t)$, $P = (2 \times 2 \times NF \times FL) + 2L$

$D_e = 2\,(AB + AF)/(\pi \times P)$, $G = m/(NT \times TP \times L)$

$Re = G\,D_e/\mu$. The proposed curve fitted curve for Stanton number is

$$St = \exp\left[\ln 30 + 0.731 \ln\left(\frac{Re}{3000}\right)\right]$$

$$St = \frac{h_o D_e}{k} \times Pr^{-1/3} \quad \therefore \quad h_o = St \times Pr^{1/3} \times k/D_e.$$

Now the fouling conductance h_{fo} is used to find

$$h_{oe} = (h_o \times h_{fo})/(h_o + h_{fo})$$

h_i is found from tube inside flow data.

$$h_{ie} = (h_i \times h_{fi})/(h_i + h_{fi})$$

h based on tube inside area is determined by

$$h = [\{(NF \times AF) + AB\}\, h_{oe}]/(\pi \times ID \times L)$$

$$U_i = (h \times h_{ie})/(h + h_{ie}),\ A = Q/U \times \text{LMTD}.$$

Total number of tubes $N = A/(\pi \times ID \times L)$

From the data specifications area A can be determined.

For other types of compact heat exchangers Kays and Londan have given graphical plots for f and $St.Pr^{2/3}$ as shown in Fig. 12.16(b).

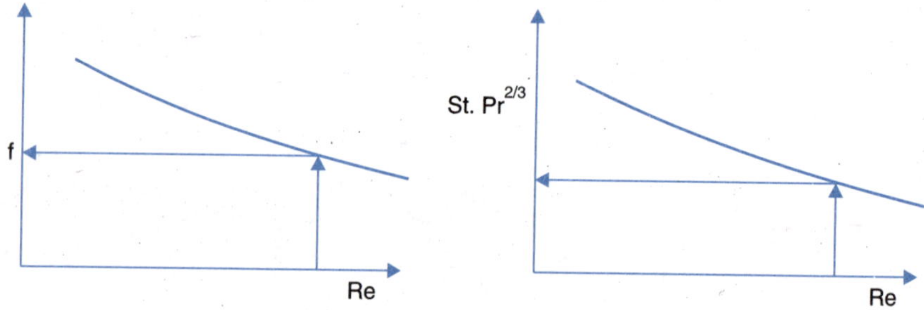

Fig. 12.16. (b) Correlations for compact heat exchanger (Schematic)

(for details refer Compact Heat Exchangers—Kays and Londan, McGraw Hill)

ADVANCED EXAMPLES

Note: Examples A-12.1—A-12.3 are for the determination of overall heat transfer coefficient and are not on heat exchanger performance.

Example A-12.1: *In a refinary fuel oil is to be cooled from 100°C to 40°C by water at 25°C flowing on the outside of the annular tube. The inner diameter is 25 mm and the oil flow rate is 1 kg/s. Water is heated to 45°C. The tube is made of 0.5% carbon steel of thickness 3 mm. The inner diameter of the outer pipe is 62.5 mm. The outside may be considered as insulated. The properties of oil at 70°C are: density = 858 kg/m³, kinematic viscosity υ = 60 × 10⁻⁶ m/s. k = 0.140 W/m K, specific heat = 2100 J/kg K. Determine the overall heat transfer coefficient. Consider good performance even after fairly long usage.*

Solution: In this case, the overall heat transfer coefficient has to be determined. The details required are:

h_o, h_i, R_k, R_{fo} and R_{fi}

(*i*) Determination of h_o: water is flowing in the annular area.

The properties of water at (25 + 45)/2 = 35°C are

$$\text{density} = 992.5 \text{ kg/m}^3, \text{ kinematic viscosity} = 0.61225 \times 10^{-6} \text{ m}^2\text{/s}$$

$$k = 0.634 \text{ W/m K}, \; Pr = 4.01,$$

Hydraulic mean radius = (0.0625 – 0.031) m

The rate of flow is obtained by heat balance.

$$1 \times 2100 \times (100 - 40) = m_{h2o} \times 4180 \times (45 - 25)$$

$$\therefore \quad m_{h2o} = 1.507 \text{ kg/s, density} = 998.5 \text{ kg/m}^3$$

The flow area $= \dfrac{\pi}{4}(0.0625^2 - 0.031^2)$

Average velocity of water $= \left(\dfrac{1.507}{998.5}\right) / [(\pi/4)(0.0625^2 - 0.031^2)] = 0.6574$ m/s

Reynolds number $\quad Re = \dfrac{0.6574 \times (0.0625 - 0.031)}{0.61225 \times 10^{-6}} = 33823$

The correlation applicable for turbulent flow in an annulus is the same as per tube flow using hydraulic mean radius.

$$\therefore \quad Nu = 0.023\, Re^{0.8}\, Pr^{0.4} = 168.4$$

$$\therefore \quad \mathbf{h_o} = \frac{168.5 \times 0.634}{(0.0625 - 0.031)} = \mathbf{3389.5 \text{ W/m}^2 \text{ K}}$$

On the oil side, (within the tube), using mass flow rate,

Reynolds number $\quad Re = 4\, m / \pi\, D\, \mu$

$$= 4 \times 1/\pi \times 0.025 \times 858 \times 60 \times 10^{-6} = 989.31$$

Laminar flow, Constant wall temperature, fully developed (assumed)

$$Nu = 3.66, \quad \therefore \quad \mathbf{h_1} = \frac{3.66 \times 0.14}{0.025} = \mathbf{20.5 \text{ W/m}^2 \text{ K}}$$

As the performance is to be maintained over a long time, fouling resistances should be considered. The values of fouling resistances are read from Table 12.1.

Oil side $\quad \mathbf{R_{fi} = 0.000877 \text{ Km}^2/W}$

Water side $\quad \mathbf{R_{fo} = 0.0001754 \text{ Km}^2/W}$

$$\therefore \quad \frac{1}{U_o} = \frac{1}{h_o} + R_{fo} + \frac{r_o}{k} \cdot \ln\frac{r_o}{r_i} + \frac{r_o}{r_i} \cdot R_{fi} + \frac{r_o}{r_i} \cdot \frac{1}{h_i}$$

For the tube material k = 53.6 W/mK

$$\frac{1}{U_o} = \frac{1}{3389.5} + 0.0001754 + \frac{0.031}{53.6} \cdot \ln\frac{0.031}{0.025} + \frac{0.031}{0.025} \cdot 0.000877 + \frac{0.031}{0.025} \cdot \frac{1}{20.5}$$

$$= 2.95 \times 10^{-4} + 1.754 \times 10^{-4} + 1.244 \times 10^{-4} + 10.87 \times 10^{-4} + 604.9 \times 10^{-4}$$

Chapter 12

∴ $\mathbf{U_o = 16.08\ W/m^2\ K}$

It can be seen that the convection resistance due to oil is controlling and is two orders of magnitude above the others ($2.95 \times 10^{-4} > -600 \times 10^{-4}$)

∴ $\mathbf{U_i = 19.95\ W/m^2\ K.}$

Example A-12.2: *In a shell and tube condenser, the tube bank is 10 rows deep. The tube ID is 20 mm and OD is 25 mm. The tubes are arranged in square array of 50 mm pitch. Water flows across the tubes with a velocity of 0.5 m/s. Sea water flows inside and the flow velocity is 1.0 m/s. The water is cooled form 50°C to 30°C and the sea water temperature in the process increases from 15°C to 25°C. Assuming for sea water, the same property values of water, determine the overall heat transfer coefficient. The tubes are of brass with k = 60.6 W/m K.*

Solution: Assume a tube length 4 m.

The convective heat transfer coefficients should be determined first.

Considering sea water flow, the mean temperature is 20°C. The property values are: density = 1000 kg/m³, kinematic viscosity = 1.006×10^{-6} m²/s, $Pr = 7.02$, $k = 0.5978$ W/m K, $c = 4178$ J/kg K.

Reynolds number $\quad Re = \dfrac{U.D}{\upsilon} = \dfrac{1 \times 0.02}{1.006 \times 10^{-6}} = 19881$

The flow is turbulent

As $L/D = 4/0.02 = 200$ fully developed flow can be assumed.

$$Nu = 0.023\ Re^{0.8}\ Pr^{0.4} = 137.72$$

∴ $\mathbf{h_i = 4117\ W/m^2\ K}$

Considering the flow of water over the tube bank. (refer chapter on external flow)

$$u_{max} = [Sp/(Sp - D)\ u_\infty = 2 \times 0.5 = 1 \text{ m/s}$$

The correlation is $\quad Nu = C\ Re^n$

From tables for $\quad S_p/D = 2,\ S_n/D = 2,\ C = 0.229,\ n = 0.632$

Also for water C should be multiplied by

$1.13 \times Pr^{0.33}$ property values at 40°C are : $k = 0.628$ W/m K

$\rho = 995$ kg/m³, $\quad \upsilon = 0.657 \times 10^{-6}$ m²/s, $\ Pr = 4.34$,

$$Nu = 1.13 \times 4.34^{0.33} \times 0.229 \times [1 \times 0.025/0.657 \times 10^{-6}]^{0.632} = 331.29$$

∴ $h_o = 331.29 \times 0.628 / 0.025 =$ **8322 W/m² K**

The fouling resistances are: Water side $\mathbf{R_{fo} = 0.0001754\ Km^2/W}$

Sea water side $\quad \mathbf{R_{fi} = 0.0000877\ Km^2/W.}$

$$\frac{1}{U_o} = \frac{1}{8322} + 0.0001754 + \frac{0.025}{2 \times 60.6} \cdot \ln \frac{0.025}{0.020} + \frac{0.025}{0.020} \cdot 0.0000877 + \frac{0.025}{0.020} \cdot \frac{1}{4117}$$

∴ $\mathbf{U_o = 1325\ W/m^2\ K.\ U_i = 1656\ W/m^2\ K.}$

Example A-12.3: *In a condenser steam condenses on the outside of a tube bank of pipes of 25 mm OD at 35°C. The bank is 10 rows deep. Sea water flows inside the pipe at a velocity of 1.0 m/s. The water is heated from 15 to 25°C during the flow. The tubes are of brass with a thermal conductivity value of 60.6 W/m K. Determine the value of overall heat transfer coefficient. The tubes are 4 m long. ID = 20 mm.*

Solution: *Refer example A-12.2,* $\mathbf{h_i = 4117\ W/m^2\ K}$

For condensation on the outside: Assuming the tubes to be horizontal, (equation 11.27)

$$h = 0.725\left[\frac{k^3\rho^2 g h_{fg}}{n\mu D(T_f - T_s)}\right]^{0.25} = 0.725\left[\frac{k^3\rho\, g h_{fg}}{n\upsilon D(T_f - T_s)}\right]^{0.25}$$

For a film temperature of 30°C,

the property values are:

$k = 0.6129$ W/m K, $\rho = 997.5$ kg/m^3

$\upsilon = 0.8315 \times 10^{-6}$ m^2/s, $h_{fg} = 2430.5$ kJ/kg, $\mu = \upsilon\,\rho$

$$\therefore \quad \mathbf{h_o} = 0.725\left[\frac{0.6129^3 \times 997.5 \times 9.81 \times 2430.5 \times 10^3}{10 \times 0.8315 \times 10^{-6} \times 0.025 \times 20}\right]^{0.25} = \mathbf{4367.6\ W/m^2\ K}$$

$$\therefore \quad \frac{1}{U_o} = \frac{1}{4367.6} + 0.0001754 + \frac{0.025}{2 \times 60.6}\cdot \ln\frac{0.025}{0.20} + \frac{0.025}{0.020} \times 0.0000877 + \frac{0.025}{0.020}\cdot\frac{1}{4117}$$

$$\therefore \quad \mathbf{U_o = 1158\ W/m^2\ K}$$

$$\mathbf{U_i = 1447\ W/m^2\ K.}$$

Example A-12.4 : ***Derive the expression*** *for the temperature rise of cold fluid in a condenser along the flow area. Also derive the expression for the effectiveness of such a heat exchanger.*

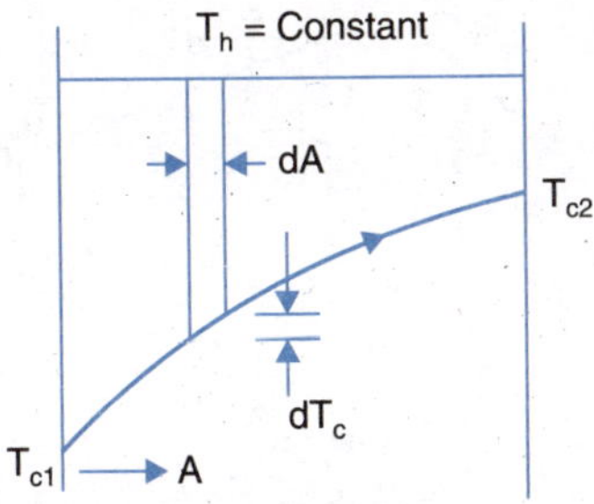

Fig. AE-12.4

Solution: The temperature variation is shown in Fig. AE-12.4. The condensing temperature remains constant.

Consider area dA. Let the heat capacity of the cold fluid be $m_c c_c = C_c$.

The heat transferred in the small area.

$$dq = C_c\, dT_c = UdA\,(T - T_c)$$ (**Note:** dT_c is +ve), T is the hot fluid temperature.

$$\therefore \quad \frac{dT_c}{T - T_c} = \frac{UdA}{C_c}$$

Here C_c is C_{min} and so $\frac{UdA}{C_c}$ is NTU for the small area.

Integrating the expression after separating variables.

$$\int_1^2 \frac{dT_c}{T - T_c} = \int_1^2 UdA/C_c, \quad \therefore \quad \ln\left(\frac{T - T_{c2}}{T - T_{c1}}\right) = \frac{UA}{C_c} = \text{NTU}$$

or

$$\frac{T - T_{c2}}{T - T_{c1}} = e^{-\text{NTU}} \qquad \text{...(1)}$$

(**Note:** NTU = N, $(C_{min}/C_{max}) = C$]

From the definition of effectiveness,

$$\varepsilon = \frac{T_{c2} - T_{c1}}{T - T_{c1}} = \frac{(T - T_{c1}) - (T - T_{c2})}{T - T_{c1}} = 1 - \frac{T - T_{c2}}{T - T_{c1}}$$

Substituting from equation (1) $\varepsilon = 1 - e^{-\text{NTU}}$

This can be also obtained from the general expression for counter flow or parallel flow heat exchanger, substituting $(C_{min}/C_{max}) = 0$. In the case of the condenser C_h tends to ∞.

Example A-12.5: ***Derive the expression for the effectiveness of a parallel flow heat exchanger.***

Solution: The temperature variation is shown in Fig. AE-12.5. Considering a small area dA, The heat transfer through the area.

$$dq = -C_h\, dT_h = C_c\, dT_c = UdA\,(T_h - T_c) \qquad \text{...(1)}$$

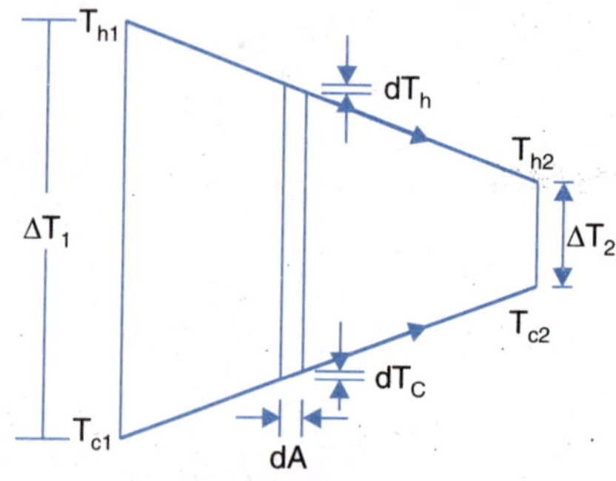

Fig. AE-12.5

$$dT_h = -\frac{dq}{C_h}, \quad dT_c = \frac{dq}{C_c}$$

$$(dT_h - dT_c) = d\,(T_h - T_c) = -dq\left[\frac{1}{C_h} + \frac{1}{C_c}\right]$$

Substituting for $dq = U\,dA\,(T_h - T_c)$

$$d\,(T_h - T_c) = -\,U\,dA\,(T_h - T_c)\left[\frac{1}{C_h} + \frac{1}{C_c}\right]$$

separating variables and integrating, and assuming C_h as C_{min}

$$\ln \frac{T_{h2} - T_{c2}}{T_{h1} - T_{c1}} = -\,UA\left[\frac{1}{C_h} + \frac{1}{C_c}\right] = -\frac{UA}{C_{min}}\left[1 + \frac{C_{min}}{C_{max}}\right]$$

or
$$\frac{T_{h2} - T_{c2}}{T_{h1} - T_{c1}} = \exp\,[-N(1 + C)\,] \quad ...(2)$$

Equation (2) is now modified by adding and subtracting T_{h1} and T_{c1} in the numerator as

$$\frac{T_{h2} - T_{c2}}{T_{h1} - T_{c1}} = \frac{(T_{h1} - T_{c1}) - (T_{h1} - T_{h2}) + (T_{c1} - T_{c2})}{T_{h1} - T_{c1}}$$

as
$$C_c(T_{c2} - T_{c1}) = C_h(T_{h1} - T_{h2})$$

$$T_{c1} - T_{c2} = -\,\frac{C_h}{C_c}\,(T_{h1} - T_{h2}) = -\,C(T_{h1} - T_{h2})$$

Substituting

$$\frac{T_{h2} - T_{c2}}{T_{h1} - T_{c1}} = 1 - \frac{T_{h1} - T_{c2}}{T_{h1} - T_{c1}} - C\,\frac{T_{h1} - T_{h2}}{T_{h1} - T_{c1}} = 1 - \varepsilon - C\varepsilon$$

$\therefore \quad 1 - \exp\,[-N(1 + C)] = \varepsilon\,(1 + C)$

$$\therefore \quad \boldsymbol{\varepsilon = \frac{1 - \exp[-N(1+C)]}{1+C}}$$

The same expression can be obtained assuming $C_c = C_{min}$ (Try this)

If $\quad$ **C** $= C_{min}/C_{max}$ **= 0 then** $\quad$ $\boldsymbol{\varepsilon = 1 - e^{-N}}$

If $\quad$ **C = 1** $\quad$ $\boldsymbol{\varepsilon = (1 - e^{-2N})/2}$.

Example A-12.6: ***Derive from basics the expression for effectiveness for a counter flow heat exchanger with $C_h = C_c$ or $(C_{min}/C_{max}) = 1$.***

Solution: The temperature variation along the exchanger is shown in Fig. AE-12.6.

The temperature difference all along will be constant as $C_h = C_c$

$$\therefore \quad dT_h = dT_c \quad \therefore \quad dT_h - dT_c = 0$$

$$C_h = C_c = C_{min} = C_{max},\; dq = -\,C_h\,dT_h = U\,dA\,(T_h - T_c)$$

As $T_h - T_c$ = constant, this can be integrated as such between the limits giving

$$C_h\,(T_{h1} - T_{h2}) = UA\,(T_h - T_c)$$

Chapter 12

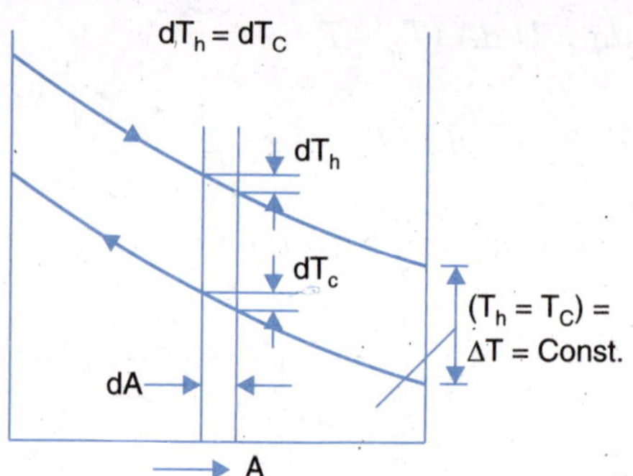

Fig. AE-12.6

$$\therefore \qquad \frac{T_{h1} - T_{h2}}{T_h - T_c} = UA/C_h = NTU = N$$

$$\frac{T_{h1} - T_{h2}}{T_h - T_c} = \frac{T_{h1} - T_{h2}}{T_{h1} - T_{c2}} = \frac{T_{h1} - T_{h2}}{(T_{h1} - T_{c1}) - (T_{c2} - T_{c1})}$$

$$C_c\,(T_{c2} - T_{c1}) = C_h\,(T_{h1} - T_{h2}) \quad \therefore \quad T_{c2} - T_{c1} = T_{h1} - T_{h2}$$

$$\therefore \qquad \frac{T_{h1} - T_{h2}}{T_h - T_c} = \frac{T_{h1} - T_{h2}}{(T_{h1} - T_{c1}) - (T_{h1} - T_{h2})}$$

Dividing by $T_{h1} - T_{c1}$, noting $\varepsilon = (T_{h1} - T_{h2})/(T_{h1} - T_{c1})$

$$\frac{T_{h1} - T_{h2}}{T_h - T_c} = \frac{\varepsilon}{1 - \varepsilon} = \text{N}, \quad \therefore \quad \varepsilon = \frac{N}{N+1}$$

Example A-12.7: ***From basics derive the expression for effectiveness for parallel flow heat exchanger with $C_h = C_c$ or equal heat capacities.***

Solution: The temperature variation is shown in Fig. AE-12.7

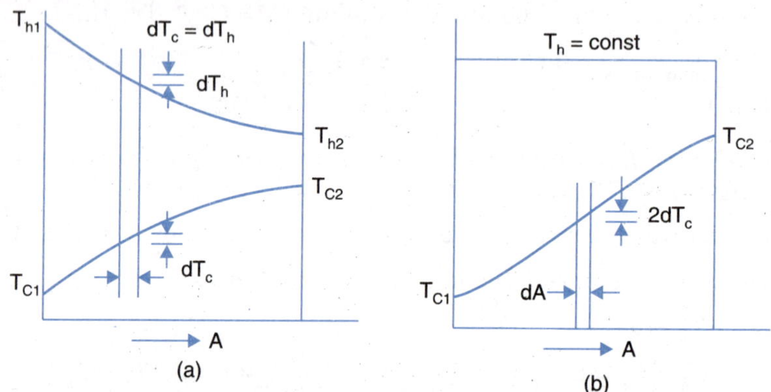

Fig. AE-12.7

As the heat capacities are equal and opposite, $(-\,dT_h = dT_c)$

$$(dT_h - dT_c) = -\,2\,dT_c$$

The temperature variation or ΔT or $(T_h - T_c)$ can also be shown as in Fig. AE-12.7(*b*) keeping one temperature at constant value, with variation at any location at double the variation of any one of the temperatures.

$$dq = C_c\, dT_c = UdA\,(T_h - T_c),\ d(T_h - T_c) = -\,2\, dT_c$$

$$\therefore \quad \frac{d(T_h - T_c)}{T_h - T_c} = -\,\frac{2U\, dA}{C_c}$$

Integrating, $\ln \dfrac{T_{h2} - T_{c2}}{T_{h1} - T_{c1}} = \dfrac{-\,2UA}{C_c} = -\,2\ \text{NTU} = -\,2N$

$$\therefore \quad \frac{T_{h2} - T_{c2}}{T_{h1} - T_{c1}} = \exp\,[-\,2N]$$

Introducing T_{h1} and T_{c1} in the numerator.

$$\frac{T_{h2} - T_{c2}}{T_{h1} - T_{c1}} = \frac{(T_{h1} - T_{c1}) - (T_{h1} - T_{h2}) - (T_{c2} - T_{c1})}{T_{h1} - T_{c1}}$$

As $\quad C_h\,(T_{h1} - T_{h2}) = C_c\,(T_{c2} - T_{c1})$ and as $C_h = C_c$

This reduces to

$$1 - \varepsilon - \varepsilon = \exp\,[-\,2N] \quad \therefore \quad \boldsymbol{\varepsilon} = \frac{\mathbf{1 - \exp[-\,2N]}}{\mathbf{2}}$$

This can also be obtained by substitution in the general expression.

Example A-12.8: *The inlet and outlet temperature of hot and cold fluids in a double pipe heat exchanger are 220°C, 100°C and 80°C and 120°C. Determine whether the exchanger is parallel flow or counter flow. Also determine the LMTD and effectiveness of the exchanger and the capacity ratio. In this case find also the maximum possible heat flow.*

Solution: The temperature are plotted both for parallel flow and counter flow, in Fig. AE-12.8(a) and 12.8(b).

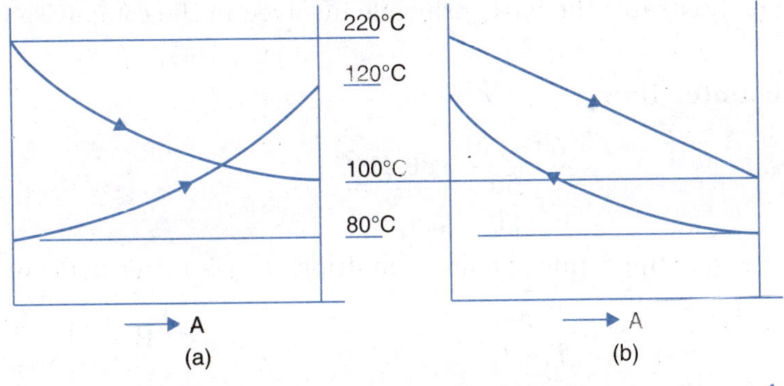

Fig. AE-12.8. (a) Parallel flow, (b) Counter flow

The plot shows that the outlet temperature of the hot fluid is lower than the inlet temperature of the cold fluid. This will lead to heating of end portion of hot fluid by the cold fluid. So the specified condition cannot be achieved in parallel flow.

For the counter flow such a higher temperature for the cold fluid is possible. So the exchanger has to be counter flow type.

$$\text{LMTD} = \frac{(220-120)-(100-80)}{\ln\left(\dfrac{220-120}{100-80}\right)} = 49.7°\text{C}$$

Effectiveness: The fluid having lower heat capacity will have a higher change in it temperature.

here $T_{h1} - T_{h2} = 220 - 100 = 120°\text{C},\ T_{c2} - T_{c1} = 120 - 80 = 40°\text{C}$

So hot fluid is C_{min}. The capacity ratio is given by

$$\frac{C_{min}}{C_{max}} = \frac{\Delta T_{max}\ \text{fluid}}{\Delta T_{min}\ \text{fluid}} = \frac{40}{120} = \frac{1}{3}$$

$$\varepsilon = \frac{T_{h1} - T_{h2}}{T_{h1} - T_{c1}} = \frac{220-100}{220-80} = \frac{120}{140} = 85.71\%.$$

As C_h is C_{min}, the hot fluid can be cooled to the inlet temperature of the cold fluid if large area was available.

∴ Maximum heat flow $= C_h\ (220 - 80) = 140\ C_h$ J/S or W.

Example A-12.9: *In a heat exchanger hot fluid enters at 180°C and leaves at 120°C while the cold fluid enters at 100°C and leaves at 120°C. Determine the LMTD and effectiveness in the following cases. Also find the NTU values.*

Solution: cases: 1. Counter flow

2. One shell pass and multiple tube passes
3. two shell passess and multiple tube passes
4. cross flow, both fluids unmixed
5. cross flow, the cold fluid unmixed

This problem illustrates the basic principle involved in the comparison of various types of heat exchangers.

1. **LMTD counter flow:**

$$= \frac{60-20}{\ln\dfrac{60}{20}} = \mathbf{36.41°C}$$

In the other cases, this value should be multiplied by a factor depending on

$$\mathbf{P} = \frac{t_2 - t_1}{T_1 - t_1} \qquad \mathbf{R} = \frac{T_1 - T_2}{t_2 - t_1}$$

$$\therefore \quad \mathbf{P} = \frac{120-100}{180-100} = \frac{20}{80} = \mathbf{0.25}, \quad \mathbf{R} = \frac{180-120}{120-100} = \frac{60}{20} = \mathbf{3}$$

Entering the chart at $P = 0.25$ the value of F are read at $R = 3$. The values are:

	F	*LMTD, °C*
2. One shell pass and multiple tube passes:	0.8	29.13
3. Two shell passes and 4, 6....8 tube passes:	0.86	31.31

4. Cross flow both fluids unmixed: 0.90 32.8
5. Cross flow cold fluids unmixed: 0.86 31.31

The area will be least where the LMTD is highest.

The hot fluid is C_{min} as it goes through higher temperature drop.

$$\text{Effectiveness : } \frac{T_{h1} - T_{h2}}{T_{h1} - T_{c1}} = \frac{180 - 120}{180 - 100} = \frac{60}{20} = 0.75$$

The effectiveness is the same in all cases as the temperatures are specified. The value of effectiveness is not a direct indicator of the performance of a heat exchanger in terms of area required.

$$\frac{C_{min}}{C_{max}} = \frac{120 - 100}{180 - 120} = \frac{20}{60} = 0.3333$$

The values of NTU can be read from charts

1. Counter flow 1.65
2. One shell and Two tube pass 2.3
3. Two shell pass 1.8
4. Cross flow, both fluids unmixed 1.85
5. Cross flow, Cold fluid unmixed 1.9

Chapter 12

COMMENT

If effectiveness is to be the same for all types of heat exchangers, listed above, the ones having the lowest effective LMTD, should have the highest NTU. The LMTD values and NTU values listed proves the point.

Example A-12.10: *Water flows at a velocity of 1 m/s through a pipe of 25 mm ID and 30 OD and 3 m length. Air at 30°C flows across the tube, with a velocity of 12 m/s. The inlet temperature of the water is 60°C. Determine the exit temperature. The thermal conductivity of the tube material is 47 W/mK.*

Solution: This situation can be modelled as a single tube exchanger with the cold fluid (air) temperature constant.

The value of overall heat transfer coefficient has to be determined. The values of convective heat transfer coefficients have to be calculated.

Inside: Water at 60°C enters the tube. Assuming a bulk mean temperature of 50°C, the property values are read:

$$\rho = 990 \text{ kg/m}^3, \quad \upsilon = 0.5675 \times 10^{-6} \text{ m}^2\text{/s}, \; Pr = 3.68, \; k = 0.63965 \text{ W/mK}$$

$$Re = \frac{1 \times 0.025}{0.5675 \times 10^{-6}} = 44053 \quad \therefore \quad \text{Turbulent}$$

Using $Nu = 0.023\, Re^{0.8}\, Pr^{0.3}$ (cooling)

$Nu = 176.47 \quad \therefore \quad$ **h_i = 4515 W/m^2 K**

Air flows on the outside at 30°C. The property values are taken at 40°C (film temperature, approximate)

$$\rho = 1.128 \text{ kg/m}^3, \quad \upsilon = 16.96 \times 10^{-6} \text{ m}^2\text{/s}, \quad Pr = 0.699, \quad k = 0.02756 \text{ W/mK}$$

$$Re = 12 \times 0.03/16.96 \times 10^{-6} = 21226.4$$

Using the correlation of the form

$Nu = C\ Re^m\ Pr^n$, for this range of Reynolds number

$C = 0.26, \quad m = 0.6, \quad n = 0.37$

$\therefore \quad Nu = 89.85 \qquad \mathbf{h_o = 82.55\ W/m^2K}$

calculating the overall heat transfer coefficient based on outside area,

$$\frac{1}{U_o} = \frac{1}{h_o} + \frac{r_o}{k} . \ln \frac{r_o}{r_i} + \frac{r_o}{r_i} . \frac{1}{h_i}$$

$$= \frac{1}{82.25} + \frac{0.03}{47 \times 2} \ln \frac{30}{25} + \frac{30}{25} . \frac{1}{4515} \qquad \therefore \quad \mathbf{U_0 = 80.03\ W/m^2\ K}$$

The water outlet temperature is not known.

Equating the heat flow calculated using water side heat capacity and temperature change, and also using LMTD approach,

$$\frac{\pi \times 0.025^2}{4} \times 1 \times 1000 \times 4180 \times (60 - T_{ho})$$

$$= 80.03 \times \pi \times 0.03 \times 3 \times \frac{(60 - 30) - (T_{ho} - 30)}{\ln \dfrac{60 - 30}{T_{ho} - 30}}$$

$$90.68\ (60 - T_{co}) = (60 - T_{ho})/\ln \frac{30}{T_{ho} - 30}$$

Solving by trial: $\quad \mathbf{T_{ho} = 59.66°C}$

The temperature drop is small due to the large flow rate and small area.

$$\mathbf{Q} = \text{mass flow} \times C \times \Delta T$$

$$= \frac{\pi \times 0.025^2}{4} \times 1 \times 1000 \times 4180\ (60 - 59.66) = \mathbf{697.6\ W}$$

From LMTD,

$$\mathbf{Q} = UA.\ \text{LMTD}, \quad \text{LMTD} = \frac{30 - 29.66}{\ln \dfrac{30}{29.66}} = 29.83$$

$$= 80.03 \times \pi \times 0.03 \times 3 \times 29.83$$

$$= \mathbf{675\ W} \text{ (correction factor not used)}$$

checks within reasonable limits.

Example A-12.11: *If in Example A-12.10, air flows inside the tube with a velocity of 9 m/s, determine the exit temperature air.*

Solution: Considering a bulk mean temperature of 40°C, and using property values at this temperature (see Example A-12.10)

$$Re = 9 \times 0.025/16.96 \times 10^{-6} = 13266.5$$

Using the correlation

$$Nu = 0.023\ Re^{0.8}\ Pr^{0.3} = 41.05$$

$$\mathbf{h_i} = 41.05 \times 0.02756/0.025 = \mathbf{45.25\ W/m^2\ K}$$

$$\mathbf{h_o} = \mathbf{82.55\ W/m^2\ K} \text{ (Example A-12.10)}$$

$$\therefore \quad \frac{1}{U_o} = \frac{1}{h_o} + \frac{r_o}{k} . \ln \frac{r_o}{r_i} + \frac{r_o}{r_i} . \frac{1}{h_i}$$

$$= \frac{1}{82.55} + \frac{0.03}{2 \times 47} . \ln \frac{30}{25} + \frac{30}{25} . \frac{1}{45.25}$$

$$\therefore \quad \mathbf{U_o = 25.85\ W/m^2\ K.}$$

Equating the heat flow calculated by LMTD and heat capacity methods.

$$\frac{\pi \times 0.025^2 \times 9}{4} \times 1.128 \times 1005\ (6 - T_{ho})$$

$$= 25.85 \times \pi \times 0.03 \times 3 \times \frac{(60 - 30) - (T_{ho} - 30)}{\ln \dfrac{60 - 30}{T_{ho} - 30}}$$

$$0.6854\ (60 - T_{ho}) = \frac{60 - T_{ho}}{\ln \dfrac{60 - 30}{T_{ho} - 30}}$$

solving by trial $\quad \mathbf{T_{ho} = 37°C}$

Check: $\quad Q = mc\, \Delta T = \dfrac{\pi \times 0.025^2 \times 9}{4} \times 1.125 \times 1005\ (60 - 37) = \mathbf{114.88\ W}$

$$Q = UA\ (\text{LMTD}) = 25.85 \times \pi \times 0.03 \times 3 \times \frac{(60 - 30) - (37 - 30)}{\ln \dfrac{60 - 30}{37 - 30}}$$

$= \mathbf{115.51\ W}$ (Correction factor not used)

Agress well. However, another trial with properties of air at (60 + 37)/2 or about 50°C will provide a more closer value.

Example A-12.12: *A heat exchanger is to be designed for a capacity of 100 kW. Water for air conditioning purpose is to be cooled from 15°C to 5°C in a counter flow arrangement using brine available at – 20°C. The value of overall heat transfer coefficient is 650 W/m²K and may be assumed to be constant. Determine the area required if the brine outlet temperature is –10, –5 and 5°C. Comment on the results.*

Solution: The temperature variations are shown in Fig. AE-12.12.

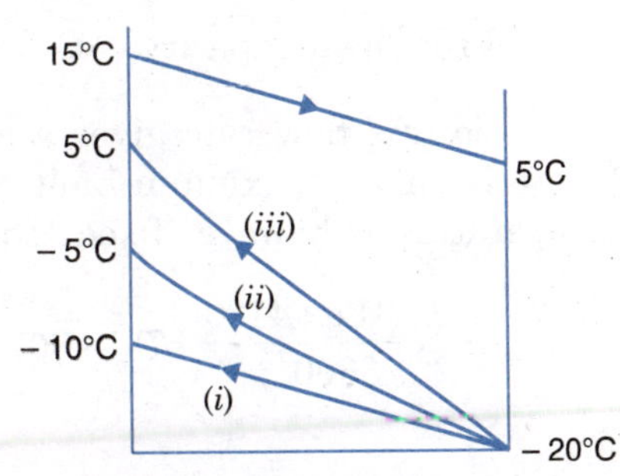

Fig. AE-12.12

The LMTD in the three cases are

(*i*) Equal heat capacity constant temperature difference.

$\therefore$ 25°C

(*ii*) $\dfrac{20 - 25}{\ln \dfrac{20}{25}} = 22.4°C$

(*iii*) $\dfrac{10-25}{\ln\dfrac{10}{25}} = 16.37°C$

Q is the same in all cases and equals 100 kW

∴ Areas required are

(*i*) **6.15 m²** (*ii*) **6.87 m²** (*iii*) **9.4 m²**

Flow rate of brine are:

(*i*) **2.39 kg/s** (*ii*) **1.6 kg/s** (*iii*) **0.96 kg/s.**

COMMENT

As the flow rate of fluid is decreased for the same capacity and for the same heat transfer rate, the area has to be increased as LMTD decreases. In thermodynamic point of view this is better, but in the equipment point of view it becomes bulkier. The pump power has to be also considered. Larger area and larger mass flow compete here in the opposing directions. So more detailed calculation will be required to optimise the pumping power.

Example A-12.13: *Engine oil is to be cooled from 120°C to 70°C in a double pipe exchanger having an area of 1.4 m². The specific heat of the oil is 2100 J/kg K. Water at 30°C is used to cool the oil and the maximum temperature of water is limited to 90°C. The flow rate of water available is 215.311 kg/hr. The overall heat transfer coefficient is 300 W/m²K. Determine the maximum possible flow rate of oil.*

Solution: The outlet temperature of the hot fluid (70°C) is below the outlet temperature of the cold fluid (90°C). Hence counter flow only is possible

A heat balance will yield.

$$m_c\, C_c\, \Delta T_c = m_h C_h\, \Delta T_h$$

$$215.311 \times 4180 \times (90 - 30) = m_h\, 2100\, (120 - 70)$$

$$\therefore \quad m_h = 514.286 \text{ kg/hr}$$

But the heat flow is to be checked using $Q = UA$ LMTD

$$\textbf{LMTD} = \frac{(120-90)-(70-30)}{\ln\dfrac{120-90}{70-30}} = \mathbf{34.76°C}$$

$$Q = 300 \times 1.4 \times 34.76 = \mathbf{14600\ W}$$

From heat capacity: $\mathbf{Q} = \dfrac{215.311}{3600} \times 4180 \times (90-30) = \mathbf{15{,}000\ W}$

So this flow rate gives a lower heat flow rate and so another direct method is used. There is only a maximum limit on water outlet temperature and so it can be lower. The oil temperature is however fixed. Equating heat flow using LMTD and also heat capacity.

$$\frac{215.311 \times 4180}{3600}(T_{co} - 30) = 300 \times 1.4 \times \frac{(120 - T_{co}) - (60-30)}{\ln\dfrac{120 - T_{co}}{(60-30)}}$$

Simplifying, $0.5952\,(T_{co} - 30) = \dfrac{(120 - T_{co}) - (60 - 30)}{\ln \dfrac{120 - T_{co}}{60 - 30}}$

Solving by trial: $\mathbf{T_{co} = 84.75}$

Oil flow rate: $= \dfrac{215.311}{3600} \times 4180\,(84.75 - 30) = m_{oil} \times 2100\,(120 - 60)$

$\therefore$ $\mathbf{m_{oil}} = 0.10863$ kg/s = **391 kg/hr**

Check: Heat flow using heat capacity and temperature rise.

$$= \frac{215.311}{3600} \times 4180\,(84.75 - 30) = \mathbf{13688\ W}$$

Using $\mathbf{Q} = UA \text{ LMTD} = 300 \times 1.4 \times \dfrac{(120 - 84.75) - (60 - 30)}{\ln \dfrac{120 - 84.75}{60 - 30}}$

$= \mathbf{13673\ W}$**, checks**

The capacity of the unit is 13.688 kW. The water outlet temperature is below the maximum fixed.

Example A-12.14: *An air cooled condenser has a fluid condensing at 50°C with air temperature rising by 8°C, from 25°C. The capacity of the unit is 10 kW. A variable speed fan is used for the air flow and the performance is obtained at 2400 rpm of the fan. If the air mass flow varies directly as the fan speed N and if the overall heat transfer coefficient varies as $N^{0.7}$, determine the heat transfer at a fan speed of 1200 rpm.*

Solution: $Q = UA$ LMTD

$$\textbf{LMTD} = \frac{25 - 17}{\ln \dfrac{25}{17}} = \mathbf{20.74}$$

$$Q = 10{,}000 = UA \text{ LMTD} \quad \therefore \quad UA = 482.08$$

C_{air} is C_{min} as the other fluid is condensing and has infinite heat capacity.

$$C_{min} = \frac{Q}{\Delta T_{min}} = \frac{10{,}000}{8}$$

$\therefore$ $\text{NTU} = UA/C_{min} = 8 \times 482.08/10{,}000 = 3.286$

$\varepsilon = 1 - e^{-\text{NTU}} \quad \therefore \quad \boldsymbol{\varepsilon = 0.32}$, also $8/(50 - 25) = 8/25$ checks

When operating conditions are changed,

$$\mathbf{C_{min2}} = \text{c.m}_2 = c.\ m_1\,(\text{N}_2/\text{N}_1) = \frac{10000}{8} \times 0.5 = \mathbf{625}$$

$$\mathbf{AU_2} = AU_1 \left(\frac{N_2}{N_1}\right)^{0.7} = 482.08 \times (0.5)^{0.7} = \mathbf{296.755}$$

$\therefore$ $$\mathbf{NTU_2} = \frac{AU_2}{C_{min\,2}} = \frac{482.08 \times (0.5)^{0.7}}{625} = \mathbf{0.475}$$

Chapter 12

$$\varepsilon = 0.378 \quad \therefore \quad \varepsilon = \frac{T_{c2} - 25}{50 - 25} \quad \therefore \quad \mathbf{T_{c2} = 34.45°C}$$

$$Q = UA\ \text{LMTD} = 296.755 \times \frac{(50 - 25) - (50 - 34.45)}{\ln \dfrac{50 - 25}{50 - 34.45}}\ \text{W}$$

$$= \mathbf{5906\ W}, \qquad C_{min} = 5000/8,$$

Check
$$Q = \frac{5000}{8} \times (34.45 - 25) = 5906\ \text{W}.$$

Note: NTU and effectiveness increase, but heat transfer decreases due to the decrease in air flow rate, due to reduction in fan speed.

Example A-12.15: *In an evaporator of a refrigerator, the refrigerant evaporates at – 20°C over the tubes. Water flowing inside the tubes enter at 15°C and is cooled to 5°C. The cooling capacity is 5 kW. Determine the cooling rate if the water flow is increased by 25% keeping the inlet temperature to be the same.*

Solution: This is an evaporator where $C_{min}/C_{max} = 0$, so C_{min} is that of water.

$$\mathbf{Q = 5000\ W} = UA\ \text{LMTD} = UA \left(\frac{35 - 25}{\ln \dfrac{35}{25}} \right)$$

$$\therefore \qquad \mathbf{UA = 168.24}$$

$$C_{min} = 5000/(15 - 5) = 500, \text{ After change in flow,}$$

$$\mathbf{NTU} = UA/C_{min} = 168.24/(500 \times 1.25) = \mathbf{0.269}$$

$$\varepsilon = 1 - e^{-NTU} = \mathbf{0.236}, \text{ solving for temperature}$$

$$\therefore \qquad \mathbf{T_{h2} = 6.74°C}$$

$$\mathbf{Q} = UA\ \text{LMTD} = 168.24 \left[\frac{35 - 26.74}{\ln (35/26.74)} \right] = \mathbf{5162.4\ W}$$

Check

$$Q = C_{min} (T_{h1} - T_{h2}) = 625\ (15 - 6.74) = \mathbf{5162.5\ W.}$$

Note: 25% flow increase, increases the capacity only by 3%.

When water flow rate is increased, *h* will increase and *U* will also increase. This is not taken into account in the problem.

Example A-12.16: *In a cross flow heat exchanger gas flow is at a rate of 5 kg/s with a temperature drop of 200°C from 360°C. Water is heated from 40°C to 90°C in the tubes. The air flow is unmixed. Determine the heat transfer rate if the water flow rate is reduced to half the value.*

Solution: The hot gas temperature change of 200°C is higher compared to the 50°C temperature change for water. Hence C_{min} is that of gas as seen from the temperature drop values.

$$C_{min}/C_{max} = 50/200 = 0.25, \ \textbf{capacity} = 5 \times 1005 \times 200/1000 = \mathbf{1005\ kW},$$

$$\varepsilon = (T_{h1} - T_{h2})/(T_{h1} - T_{c1}) = 200/320 = 0.625$$

From chart Fig. 12.12(*e*) for cross flow exchanger with both side fluids unmixed,

$$\text{NTU} = 1.1$$

Under the changed operating conditions, the water flow is halved.

However its heat capacity is still higher than that of the gas. So UA/C_{min} remains unchanged. Hence NTU remains unchanged.

Now $C_{min}/C_{max} = 0.5$. From the chart the value of effectiveness is read as 0.58

$$\therefore \quad \varepsilon = \frac{T_{h1} - T_{h2}}{T_{h1} - T_{c1}} \quad \therefore \quad 0.58 = \frac{360 - T_{h2}}{360 - 40}$$

$$\therefore \quad \mathbf{T_{h2} = 174.4°C}$$

$Q = 5 \times 1005\ (360 - 174.4)$ W $=$ **932.64 kW**, a reduction of 7% only.

Equations available in data book can also be used and in this case trial solution is necessary to solve for NTU.

Water outlet temperature will be high as compared to 90°C *i.e.* 132.8°C may boil.

COMMENT

In this case also change in U is not considered. If necessary data like tube dia etc. are specified a more accurate prediction can be made.

Example A-12.17: ***A shell and tube heat exchanger*** *with one shall pass and two tube passes is used to heat water (flowing in the tubes) at a rate of 10 kg/s from 30°C to 45°C with steam condensing over the tubes at 160°C. If the overall heat transfer coefficient (based on outside area) has a value of 2000 W/m²K, determine the area required. If 20 tubes of 25 mm OD are used, determine the length required.*

Solution: The heat flow is given for a shell and tube exchanger with two tube passes by

$$Q = F\,.\,UA\ \text{LMTD}$$

The value of F is a function of P and R where

$$R = \frac{T_{h1} - T_{h2}}{T_{c2} - T_{c1}} \text{ and } T_{h1} - T_{h2} = 0 \text{ for a condenser.}$$

So $R = 0$, **For R = 0, F = 1 in all cases.**

The problem is meant to stress the idea that for condensers R = 0 ∴ F = 1 and for evaporators P = 0 and F = 1

$$\therefore \quad Q = F\ UA\ \text{LMTD}$$

$$10 \times 4180 \times (45 - 30) = 1 \times 2000 \times A \times \frac{(160 - 30) - (160 - 45)}{\ln \dfrac{160 - 30}{160 - 45}}$$

$$\therefore \quad \mathbf{A = 2.562\ m^2} = N\,\pi\,DL = 20 \times \pi \times 0.025\,L$$

$$\therefore \quad \mathbf{L = 1.631\ m.}$$

Example A-12.18: ***A steam condenser*** *condensing at 70°C has to have a capacity of 100 kW. Water at 20°C is used and the outlet water temperature is limited to 45°C. If the overall heat transfer coefficient has a value of 3100 W/m²K, determine the area required.*

If the inlet water temperature is increased to 30°C determine the increased flow rate of water for maintaining the same water outlet temperature.

Also determine the flow rate for the same amount of heat transfer.

Solution: As the performance is to be evaluated, it is better to use NTU method. Water heat capacity is C_{min}

$$\varepsilon = \frac{45 - 20}{70 - 20} = 0.5$$

NTU is solved for using

$$\varepsilon = 1 - e^{-N} \text{ for } C_{min}/C_{max} = 0 \qquad \text{...(A)}$$

$$\text{NTU} = 0.6931 = \frac{UA}{C_{min}}$$

$$C_{min} = 100{,}000/(45 - 20) = 4000 \text{ W/K}$$

$$\therefore \quad \mathbf{A} = \frac{0.6931 \times 4000}{3100} = \mathbf{0.894\ m^2}$$

New operating conditions:

$$T_{c1} = 30°\text{C},\ T_{c2} = 45°\text{C}$$

$$\varepsilon = (45 - 30)/(70 - 30) = 15/40 = 0.375$$

Using conditions stated by (A)

$$\therefore \quad \mathbf{NTU = 0.47}$$

$$\frac{UA}{C_{min}} = 0.47,\ C_{min} = \frac{UA}{0.47} = \frac{3100 \times 0.894}{0.47}$$

$$= \mathbf{5899\ W/K} \text{ (Compared to 4000)}$$

The flow rate has to be increased by 47.2%

Heat flow decreases by about 11.5% to 88.5 kW (calculate)

Heat transfer rate and the exit temperature cannot be maintained simultaneously. In order to maintain the heat transfer rate.

$$Q = UA \text{ LMTD} = 100{,}000 \text{ W}$$

$$3100 \times 0.894 \times \text{LMTD} = 100{,}000$$

$$\therefore \quad \text{LMTD} = 36.08 \quad \frac{(70 - 30) - (70 - T_{c2})}{\ln \dfrac{(70 - 30)}{(70 - T_{c2})}} = 36.08$$

Solving by trial $\quad T_{c2} = 37.6°\text{C}$

This is within the limits of 45°C

The heat capacity required is

$$100{,}000 = C\,(37.6 - 30) \quad \therefore \quad \mathbf{C = 13158\ W/K}$$

COMMENT

This is compared to 4000 W/K. An increase of more than 300%. Maintaining same heat transfer rate is more expensive compared to maintaining the temperature, when inlet temperature is changed.

Example A-12.19: *A feed water heater uses steam condensing at 120°C to heat water from 30°C to 90°C in a shell and tube heat exchanger at a rate of 5 kg/s. The overall heat transfer coefficient is equal to 2000 W/m²K. Due to usage deposits have formed on the surfaces resulting in a fouling factor of 0.0009 Km²/W. Determine the water exit temperature if flow rates and inlet temperatures are maintained.*

Solution: As it is a condenser, it can be treated as counter flow exchanger. As performance under a different operating condition is to be studied NTU method is used.

$$\varepsilon = \frac{90-30}{120-30} = \frac{60}{90} = \mathbf{0.667} = 1 - e^{-NTU}$$

Solving $\quad \mathbf{NTU = 1.0986}$

$$NTU = UA/C_{min} \quad \therefore \quad \mathbf{A = 11.48\ m^2}$$

Due to fouling, the value of U is changed. C_{min} and area remain unaltered.

$$\frac{1}{U_d} - \frac{1}{U_c} = \text{fouling factor}$$

U_d—overall heat transfer coefficient for dirty surface

U_c—overall heat transfer coefficient for clean surface

$$\frac{1}{U_d} = \frac{1}{2000} + 0.0009 \quad \therefore \quad \mathbf{U_d = 714.29\ W/m^2\ K}$$

$$NTU = UA/C_{min} = 714.29 \times 11.48/5 \times 4180 = 0.3923, \text{ for this value}$$

$$\therefore \quad \varepsilon = 0.325 = \frac{T_{c2} - T_{c1}}{T_{h1} - T_{c1}} = \frac{T_{c2} - 30}{120 - 30}$$

$$\therefore \quad \mathbf{T_{c2} = 59.2°C}$$

Water outlet temperature is reduced from 90 to 59.2°C

Heat flow $= 5 \times 4.180\ (59.2 - 30) =$ **610.44 kW**

as compared to 1254 kW. A reduction more than 50%. This problem can be inverted to determine the fouling factor. This is the reason for use of fouling factor in design.

Example A-12.20: ***A feed water heater having 5.74 m² area** was used to heat water from 30°C at a rate of 2.5 kg/s, using steam condensing at 120°C. The exit temperature was measured as 90°C. Determine the value of overall heat transfer coefficient. After 3 years of operation, for the same flow rates and inlet conditions, the outlet temperature measured only 80°C. Determine the value of fouling resistance.*

Solution: This problem can preferably be solved by NTU method.

As it is a condensing unit $C_c = C_{min}$

$$\varepsilon = \frac{90-30}{120-30} = \frac{60}{90}$$

$$\therefore \quad NTU = -\ln(1 - \varepsilon) = 1.0986$$

$$NTU = UA/C_{min} = U \times 5.74/(2.5 \times 4180)$$

$$\therefore \quad \mathbf{U = U_c = 2000.09\ W/m^2\ K,} \text{ (at clean condition)}$$

Chapter 12

After service the fluid is heated only to 80°C and so,

$$\varepsilon = \frac{80-30}{120-30} = \frac{50}{90} \quad \therefore \quad \text{NTU} = 0.8109$$

$\therefore$ $\qquad$ **U_d = 1476.35 W/m² K, (after deposit formation)**

Fouling factor $= \dfrac{1}{U_d} - \dfrac{1}{U_c}$ = **0.000177 Km²/W**

Example A-12.21: ***A cross flow type air heater*** *with both fluids unmixed has an area of 50 m² and the overall heat transfer coefficient is 45 W/m²K. The flow rate of both hot and cold air are 1 kg/s. Hot air enters at 60°C and cold air enters at 20°C. Determine the exit temperature of the streams. Also determine the exit temperature if (i) the flow rate of both fluids is halved and (ii) the flow rates are doubled. Compare the heat flow rates also.*

Solution: In this case $\quad C_{min} = C_{max} \quad \therefore \quad C_{min}/C_{max} = 1$

$$C_{min} = 1 \times 1005 \text{ W/K}$$

$$\therefore \quad \text{NTU} = UA/C_{min} = 45 \times 50/1005 = 2.24$$

Using the ε–NTU chart for cross flow exchanger with both fluids unmixed, the effectiveness corresponding to NTU = 2.24 is read as 0.64

$$\varepsilon = \frac{T_{h1} - T_{h2}}{T_{h1} - T_{c1}} = \frac{60 - T_{h2}}{60 - 20} = 0.64$$

$\therefore$ $\qquad$ **T_{h2} = 34.4°C, T_{c2} = 45.6°C**

Q = 1 × 1005 × (60 – 34.4) W = **23.73 kW**

When the flow is halved, NTU = 4.48, from chart ε = 0.755

T_{h2} = 29.80°C $\qquad$ T_{c2} = 50.2°C, $\qquad$ Q = 15.18 kW

When the flow is doubled NTU = 1.12, From chart ε = 0.5

T_{h2} = 40°C, $\qquad$ T_{c2} = 40°C, $\qquad$ Q = 40.2 kW.

Note: Higher temperature is achieved by reducing flow rates, but with lower heat flow. Lower temperature is achieved by increasing flow rates, but with higher heat flow.

Example A-12.22: ***In a cross flow type heat exchanger*** *with steam condensing inside tubes at 100°C is used to heat air from 20°C. The air side may be taken as mixed. The effectivenesss of the heat exchanger is found as 0.7. If the area is 20 m² and overall heat transfer coefficient (clean) based on this area is 150 W/m²K, determine the heat transfer rate.*

Also determine the extra area required if a fouling factor of 0.0006 Km²/W is to be allowed for.

Determine the exit temperature when the clean surface is operated with this excess area.

Solution: As steam condensation is used $(C_{min}/C_{max}) = 0$

$$\boldsymbol{\varepsilon = 0.7} = \frac{T_{c2} - 20}{100 - 20} \quad \therefore \quad \mathbf{T_{c2} = 76°C}$$

Using the chart for cross flow heat exchanger the value of NTU is read corresponding to ε = 0.7 and (C_{min}/C_{max}) = 0 as 1.2.

$$\text{NTU} = 1.2 = UA/C_{min}$$

$\therefore$ $\quad\quad C_{min}$ **= 2500 W/K**

$\therefore$ **Heat flow rate** = 2500 × (76 – 20) W = **140 kW**

$$\frac{1}{U_d} - \frac{1}{U_c} = \text{fouling resistance} = 0.0006$$

$\therefore$ $\quad\quad U_d$ **= 137.62 W/m² K**

For the same NTU and effectiveness,

$$\frac{137.62 \times A}{2500} = 1.2 \quad \therefore \quad \mathbf{A = 21.8\ m^2}$$

As the overall heat transfer coefficient is already low, the extra resistance does not add to the area requirement by a large amount.

Under the altered area if original flow rate is maintained.

$$\frac{UA}{C_{mim}} = \frac{150 \times 21.8}{2500} = 1.308$$

From chart $\quad \varepsilon = 0.73 \quad \therefore \quad$ T_{c2} **= 78.4°C**

An increase of 2.4°C.

Example A-12.23: ***Air is to be heated at*** *a rate of 1 kg/s from 30°C to 45°C in a cross flow type of heat exchanger with finned tubes with water at 90°C flowing inside the tubes. The overall heat transfer coefficient is 60 W/m²K referred to an area of 7.5 m². Calculate the flow rate of water and the water exit temperature. Also determine the % change in the flow rate of water if the water inlet temperature is 60°C, 70°C, 80°C and 100°C. Assume the air flow rate and air exit temp. remain unchanged.*

Solution: Both the flow rate and exit temperature of the hot fluid are not known (water)

Using $\quad Q = F.UA \text{ LMTD} = C_c \Delta T_c$

$$1 \times 1005 \times 15 = 60 \times 7.5 \times \text{LMTD} \times F \quad \text{...(A)}$$

First assume $\quad F \approx 1 \quad \therefore \quad \text{LMTD} = 33.5°C$

$$\therefore \quad 33.5 = \frac{(90-45)-(T_{h2}-30)}{\ln \dfrac{90-45}{T_{h2}-30}}$$

Solving by trial, $\quad T_{h2} = 54.2°C$.

For this value, $\quad R = \dfrac{90-54.2}{45-30} = 2.39, P = 0.25, \quad F = 0.948$

Using this value in equation (A), LMTD = 35.34

Solving by trial, $\quad T_{h2} = 57°C, R = 2.2, F = 0.95$

This value is good enough.

Now the flow rate of water is calculated using

$$1 \times 1005 \times 15 = m \times 4180 \times (90 - 57)$$

m = 0.1093 kg/s for water exit temperature of 57°C

The water flow is C_{min}.

When the inlet temperature is changed, it is to be noted that below a certain value of inlet temperature it is not possible to maintain the air outlet temperature even with a very large flow. Take the example of 70°C inlet.

The max value of LMTD in this case is

$$\frac{25-40}{\ln\frac{25}{40}} = 31.91 < 35.34$$

which is lower than the required value for obtaining the heat flow with the given area and overall heat transfer coefficient. So 70°C and 60°C inlet are not compatible.

For 80°C the oulet temperature is about 67°C and the flow rate is 0.494 kg/s. There is an increase of 400%.

For 100°C, the outlet temperature is about 59°C and mass flow is 0.088 kg/s a reduction of about 20%.

Example A-12.24: ***A shell and tube heat exchanger*** *with 4 tube passes is used to cool hot water at 90°C at a rate of 10 kg/s using cold water available at 20°C. With the flow rate of cold water was 5 kg/s the cold water outlet temperature was 70°C.*

Investigate the variation of heat flow rate when the flow rate of cold water is reduced, maintaining the inlet temperature and the hot water flow rate to be the same.

Solution: As performance is to be evaluated, NTU method may be used. In this case $C_c = C_{min}$, $C_{min}/C_{max} = 0.5$

$$\varepsilon = \frac{70-20}{90-20} = \frac{50}{70} = 0.714, \text{ Using the corresponding chart,}$$

NTU is read as 2.42.

$$UA = 2.42 \times C_{min} = 2.42 \times 5 \times 4180 = 50578$$

This value remains constant (assumed)

$$Q = 5 \times 4180\ (70 - 20)\ \text{W} = 1045\ \text{kW}$$

Case (*i*) Flow rate of cold water reduced to, say 4 kg/s

$$C_{min}/C_{max} = 4/10 = 0.4$$

$$\text{NTU} = UA/C_{min} = 50578/4 \times 4180 = 3.025$$

Using the chart $\boldsymbol{\varepsilon = 0.76}$

∴ Cold fluid outlet temperature is determined using

$$0.76 = \frac{T_{c2}-20}{90-20} \quad \therefore \quad \mathbf{T_{c2} = 73.2°C}$$

$$Q = 4 \times 4180 \times (73.2 - 20) = 889504\ \text{W or } \mathbf{889.5\ kW.}$$

Case (*ii*) Using the same procedure, the heat flow and outlet temperatures for 3 and 2.5 kg/s are determined as 78.1°C and 728.6 kW and 80.55°C and 632.75 kW. The values are tabulated below.

Hot water flow rate = 10 kg/s, T_{h1} = 90°C, T_{c1} = 20°C

Cold water flow rate kg/s	*Cold water outlet temp.°C*	*Effectiveness*	*Heat transfer kW*
5	70	0.714	1045.0
4	73.2	0.76	889.5
3	78.1	0.83	728.6
2.5	80.55	0.865	632.75

As the cold water flow rate is decreased, its outlet temperature increases and hence effectiveness increases. But the heat flow decreases.

Example A-12.25: ***For a heat exchanger using condensing steam*** *at T_h heating water at a flow rate with a heat capacity of C_c and water entry temperature of T_{c_1} derive an expression for T_{c_2} in terms of NTU and T_h.*

Solution : Using the above determine the exit temperature of cold water for steam condensing temperatures of 180, 160, 140 and 120°C for a heat exchanger which gave an outlet temperature of 90°C when heating water at a rate of 50 kg/s from 60°C with steam at 200°C.

$$\varepsilon = \frac{T_{c2} - T_{c1}}{T_h - T_{c1}} = 1 - e^{-NTU} \qquad \text{(Capacity ratio = 0)}$$

where $\text{NTU} = UA/C_{min}$

$$\therefore \quad T_{c2} = T_{c1} + (T_h - T_{c1})(1 - e^{-NTU}) \qquad \text{...(A)}$$

For a given exchanger and flow rate all quantities on RHS are known and hence T_{c2} can be calculated.

For the given exchanger, (F = 1, condenser)

$$Q = m_c c_c (T_{c2} - T_{c1}) = UA \times \text{LMTD}$$

$$\therefore \quad UA = \frac{m_c c_c (T_{c2} - T_{c1})}{\text{LMTD}} = \frac{50 \times 4180\,(90 - 60)}{\dfrac{(200 - 60) - (200 - 90)}{\ln \dfrac{200 - 60}{200 - 90}}} = 50403$$

$$\text{NTU} = UA/C_{min}$$

Here C_c is C_{min} (condensing)

$$\therefore \quad \text{NTU} = 50403/(50 \times 4180) = 0.2412$$

This remains constant for the heat exchanger.

$$\therefore \quad 1 - e^{-NTU} = 0.2143, \text{ using equation (A), for}$$

$$T_h = 200°\text{C}, \quad T_{c1} = 60°\text{C} \quad \therefore \quad T_{c2} = 90°\text{C}$$

Check for other values

T_h:	**180**	**160**	**140**	**120**
T_{c2}:	**85.71**	**81.43**	**77.14**	**72.86**

Check for T_h = 120°C

$$Q = UA \text{ LMTD}$$

Also $\qquad Q = m_c c_c (T_{c2} - T_{c1})$

$$\textbf{LMTD} = \frac{60 - 47.14}{\ln \dfrac{60}{47.14}} = \textbf{53.31}$$

$$\boldsymbol{Q} = 50403 \times 53.31 = \textbf{2.687} \times \textbf{10}^6 \textbf{ W, also}$$

$$\boldsymbol{Q} = 50 \times 4180\ (72.86 - 60) = \textbf{2.687} \times \textbf{10}^6 \textbf{ W}$$

so checks.

Example A-12.26: *12 finned tubes of 1.2 m length of ID 10 mm and OD 12 mm with fin length of 6 mm are arranged in a row with a pitch of 26 mm in a heat exchanger. There are 10 row in the unit. Air at 30°C enters the exchanger at a velocity of 12 m/s, and leaves at 70°C. Fin thickness is 0.5 mm and fin pitch is 4 mm. The convective heat transfer coefficient with hot water flowing inside is 950 W/m² K. If the effective LMTD is 22°C determine the heat flow, using kerns method. Assume fin efficiency as 0.85.*

Solution: (Refer Section 12.8). The properties of air at 50°C, (the mean temperature) are

Density = 1.093 kg/m³, $\mu = 19.61 \times 10^{-6}$ Ns/m²

$Pr = 0.698$, $k = 0.02826$ W/m K, Number of fins per/pipe

$= 1.2/0.004 = 300$

Mass flow $\qquad m = (1.2 \times 12 \times 0.026)\ 12 \times 1.093$

$= 4.91$ kg/s

Mass flux $\qquad G = 4.91/(12 \times 0.026 \times 1.2) = 4.91$ kg/m² s.

Area of fin $\qquad AF = \dfrac{\pi}{4} \times 2\ (300)\ (0.024^2 - 0.012^2) = 0.204$ m².

Bare area of pipe, $\qquad AB = \pi \times 0.012\ (1.2 - 300 \times 0.0005) = 0.0396$ m²

Perimeter $\qquad P = (2 \times 2 \times 300 \times 0.006) + 2 \times 1.2 = 9.6$ m

Equivalent diameter $\qquad = 2\ (0.204 + 0.0396)/\pi \times 9.6 = 0.01615$ m.

Reynolds number $\qquad Re = (G\ De/\mu) = 4.91 \times 0.01615/19.61 \times 10^{-6} = 4043.6$

$$\frac{h_o De}{k} . Pr^{-\frac{1}{3}} = \exp\left[\ln 30 + 0.731 \ln\left(\frac{4043.6}{3000}\right)\right] = 37.32$$

$$h_o = 37.32 \times 0.02826 \times 0.698^{1/3}/0.01615$$

$$= 59.92 \text{ W/m}^2 \text{ K.}$$

Fouling factor (from data book) = 0.0003525, m² K/W.

$$h_{oe} = 59.92 \times (1/0.0003525)/(59.92 + (1/0.0003525))$$

$$= 58.68 \text{ W/m}^2 \text{ K.}$$

$$h_i = 950 \text{ W/m}^2 \text{ K, fouling factor} = 0.0001754 \text{ m}^2\text{K/W}^2.$$

$$h_{ie} = 950 \times (1/0.0001754)/950 + (1/0.0001754) = 814.3 \text{ W/m}^2 \text{ K.}$$

$$h = [\{(0.85 \times 0.204) + 0.0396\}\ 58.68/(\pi \times 0.01 \times 1.2)$$

$$= 331.54 \text{ W/m}^2 \text{ K}$$

$$U = \frac{331.54 \times (1/0.0001754)}{331.54 + (1/0.0001754)} = 313.32 \text{ W/m}^2 \text{ K.}$$

$$\text{Total area} = (12 \times 10 \times \pi \times 0.01 \times 1.2) = 4.524 \text{ m}^2$$

$$\mathbf{Q = 313.32 \times 4.524 \times 22 = 31184 \text{ W.}}$$

Example A-12.27: *An air heater uses steam condensing at 110°C to heat air at a rate of 5 kg/ from 20°C to 75°C. It is desired to double the air flow at the same time achieve air outlet temperature of 75°C by increasing the steam condensation temperature. Determine the new condensing temperature.*

Solution: Let N denote NTU for condensers $\varepsilon = 1 - \varepsilon^{-N}$

$$\varepsilon = \frac{75-20}{110-20} = 0.6111 \quad \therefore \quad N = 0.9445$$

As the air flow is doubled C_{min} is doubled. *UA* remains the same.

So the new value of $N = 0.9445/2$ and $\varepsilon = 1 - e^N = 0.3764$

$$\therefore \qquad \varepsilon = 0.3764 = \frac{75-20}{T_h - 20}$$

Solving $\qquad T_h = 166.12°\text{C}$

Check: The heat flow has to be doubled. Also *UA* is the same and so LMTD has to be doubled.

$$\text{Old value of LMTD} = \frac{(110-20)-(110-75)}{\ln \dfrac{110-20}{110-75}} = 58.23°\text{C}$$

$$\text{New value of LMTD} = \frac{(166.12-20)-(166.12-75)}{\ln \dfrac{166.12-20}{166.12-75}} = 116.46°\text{C}$$

so checks.

Note: As the temperature increases the condensing pressure also increases. In this case the pressure increases from about 1.45 to about 7.2 bar. So the tubes may not withstand this pressure.

Example A-12.28: *In an existing heat exchanger of counter flow type dry air is cooled from 70°C to 35°C at a rate of 1.2 kg/s using cold air at 15°C at a rate of 1.5 kg/s. It is desired to cool this stream by another 10°C by increasing the area with the same inlet conditions. Calculate the percentage increase in area.*

Solution: This problem can be worked by LMTD or NTU method. The main assumption is that *U* remains unchanged and effectiveness and C_{min}/C_{max} are known. Using the chart NTU can be determined. The ratio of NTU values is the same as ratio of areas as *U* and C_{min} are the same.

Under the first operating conditions

$$\varepsilon = \frac{70-35}{70-15} = \frac{35}{55} = \mathbf{0.636}, \quad C_{min}/C_{max} = 1.2/1.5 = 0.8$$

From chart $\qquad$ **NTU = 1.5**

Under the altered conditions (with C_{min}/C_{max} being the same)

$$\varepsilon = \frac{70-25}{70-15} = \frac{45}{55} = \mathbf{0.8182}$$

From chart corresponding $\qquad$ **NTU = 3.21**

$$\frac{UA_1}{C_{min}} = 1.5, \ \frac{UA_2}{C_{min}} = 3.21 \qquad \therefore \frac{A_2}{A_1} = \frac{3.21}{1.5} = 2.14$$

Chapter 12

∴ Area increase = 114%.

Note: (*i*) As all the four temperatures can be worked out these values can be checked usin LMTD method

(*ii*) equations can be used to determine NTU, instead of using chart.

As $$N = \frac{1}{C-1} \ln \frac{\varepsilon - 1}{C\varepsilon - 1}$$

These methods may be tried.

REVIEW QUESTIONS

12.1 Define overall heat transfer coefficient. Explain how it comes handy in the analysis of hea exchangers.

12.2 What are the advantages of counter flow arrangement?

12.3 Explain why single tube type of heat exchangers are not in extensive use?

12.4 Explain the function of baffles in shell type heat exchangers.

12.5 What are compact heat exchangers? Indicate where this type is in popular use.

12.6 Explain the terms 'mixing' and 'non-mixing' with reference to cross flow heat exchangers.

12.7 Discuss the effect of fouling on the performance of heat exchangers.

12.8 What is the disadvantage of using arithmetic average of temperature difference in calculatin heat flow in heat exchangers?

12.9 Define LMTD.

12.10 Define capacity ratio, NTU and effectiveness with reference to heat exchangers.

12.11 What is the advantage of NTU method of analysis over LMTD method of analysis?

12.12 Explain how the maximum possible heat transfer arrived at? Under what condition this is possible for achievement?

12.13 What are the special features of condensers and evaporators?

12.14 Under what conditions the equality $\frac{1}{U} = \frac{1}{h_i} + \frac{1}{h_o}$ can be used?

OBJECTIVE QUESTIONS

12.1 Choose the correct statement from each question.

(*i*) The overall heat transfer coefficient is the

(*a*) sum of all resistances (*b*) sum of all conductances

(*c*) sum of the convection coefficients (*d*) resistance due to the wall material.

(*ii*) The range of value of overall heat transfer coefficients fall in the following increasing order

(*a*) boiling or condensation to liquid, liquid to gas, liquid to liquid, gas to gas.

(*b*) liquid to liquid, gas to gas, boiling or condensation to liquid, liquid to gas

(*c*) gas to gas, liquid to gas, liquid to liquid, boiling or condensation to liquid

(*d*) boiling or condensation to liquid, liquid to gas, gas to gas, liquid to liquid.

(*iii*) The decreasing order of effectiveness for a given situation among types of heat exchangers is

(*a*) parallel flow, cross flow, shell and tube, counter flow

(*b*) cross flow, counter flow, shell and tube, parallel flow

(*c*) counter flow, shell and tube, cross flow, parallel flow

(*d*) counter flow, cross flow, shell and tube, parallel flow.

(*iv*) When one of the fluid is condensing the best flow arrangement is
(*a*) counter flow (*b*) parallel flow
(*c*) cross flow (*d*) all are equal.

(*v*) Thermodynamically the type which leads to lower loss in availability is
(*a*) parallel flow (*b*) counter flow
(*c*) cross flow (*d*) shell and tube.

(*vi*) The Net Transfer Unit is (NTU)
(*a*) $U.C_{min}/A$ (*b*) UA/C_{min}
(*c*) AC_{min} (*d*) U/AC_{min}.

(*vii*) Effectiveness of a heat exchanger is
(*a*) actual heat transfer/heat content of hot fluid
(*b*) actual heat transfer/heat content of cold fluid
(*c*) actual heat transfer/heat content of higher heat capacity fluid
(*d*) actual heat transfer/heat transfer when minimum heat capacity fluid goes through the maximum temperature difference in the exchanger.

(*viii*) Cross flow exchangers are popularly used for heat transfer
(*a*) liquid and liquid (*b*) liquid and evaporating fluid
(*c*) condensing fluid and liquid (*d*) gas and gas or liquid and gas.

(*ix*) The minimum heat transfer area for a given situation is for
(*a*) parallel flow (*b*) counter flow
(*c*) cross flow (*d*) shell and tube.

(*x*) Effectiveness is generally represented by (with usual notations)
(*a*) $\frac{C_h}{C_{min}} \frac{T_{h1} - T_{h2}}{T_{c2} - T_{c1}}$ (*b*) $\frac{C_c}{C_{min}} \frac{T_{c2} - T_{c1}}{T_{h1} - T_{h2}}$
(*c*) $\frac{C_h}{C_{min}} \frac{T_{h1} - T_{h2}}{T_{h1} - T_{c1}}$ (*d*) $\frac{C_c}{C_{min}} \frac{T_{h1} - T_{c2}}{T_{h2} - T_{c1}}$.

(*xi*) The flow direction is immaterial in the case of heat exchange from
(*a*) wet or saturated steam to water (*b*) water to gas
(*c*) oil to water (*d*) oil to gas.

Answers

12.1 (*i*) (*b*) (*ii*) (*c*) (*iii*) (*c*) (*iv*) (*d*) (*v*) (*b*) (*vi*) (*b*)
(*vii*) (*d*) (*viii*) (*d*) (*ix*) (*b*) (*x*) (*c*) (*xi*) (*a*).

EXERCISE PROBLEMS

12.1 The inlet and outlet temperatures of the hot and cold fluids in a double pipe heat exchanger are 180°C, 20°C and 140°C and 80°C respectively. Determine the ratio of heat capacities and the LMTD assuming (*i*) counter flow and (*ii*) parallel flow. Also determine the effectiveness.

12.2 For the data of problem 12.1, determine the area ratio for the two flow arrangements. Also determine the Limiting values of outlet temperatures for the two flow arrangements.

12.3 If the heat capacity of the hot fluid for the heat exchanger of problem 12.1 was 6000 W/K. Determine the value of NTU for the two flow arrangements.

Chapter 12

12.4 If the overall heat transfer coefficient was 600 W/m^2 K, for the exchanger of problem 12.1, determine the areas for (*i*) parallel flow (*ii*) counter flow (*iii*) one shell pass and two tube passes type of exchangers (use data of problem 12.3 also).

12.5 In a heat exchanger of double tube arrangement the inner tube is of 25 mm *ID* and 30 mm *OD*. The outer pipe is of 50 mm *ID*. Hot water flows inside at a rate of 1 *l*/s and cold water flows at a rate of 2 *l*/s in the annular space. The inlet temperatures of the hot and cold water are 30°C and 80°C. The tube is 3 m long and is made of material with a conductivity of 40.5 W/mK. Determine the overall heat transfer coefficient for the exchanger assuming that its performance has to be guaranteed for 3 years.

12.6 A cross flow heat exchanger with a tube bank of 10 tube depth uses hot flue gases to heat water flowing inside the tubes. The tubes are of 50 mm *ID* and 58 mm *OD* made of material of conductivity 47 W/mK. The tubes are arranged in a square array of pitch of 95 mm. The hot gases at 280°C approach the bank with a velocity of 10 m/s. The velocity of water inside is 1.2 m/s. The water entry is at 40°C. Assume that water is pressurised and boiling does not occur. Determine the overall heat transfer coefficient.

12.7 A heat exchanger of the counter flow type with inlet temperatures of hot and cold fluids of 200°C and 50°C gave outlet temperature of 100°C for both fluids when the surfaces were clean. After some extended use for the same flow rate and inlet temperatures, the hot fluid outlet was 120°C. Determine the percentage change in the value overall heat transfer coefficient and the heat transfer.

12.8 A clean heat exchanger of counter flow type gave the following inlet and outlet temperatures for the hot and cold fluids. 200°C, 50°C and 100°C for both. If the overall heat transfer coefficient decreased to 80% of the original value due to fouling determine the outlet temperature of the fluids.

12.9 A chemical solution with a specific heat of 3200 J/kg K at 100°C is to be cooled to 60°C at a rate of 30,000 kg/hr using cold water at 30°C with its outlet limited to 50°C. If the overall heat transfer coefficient has a value of 260 W/m^2 K, determine the area required for (*i*) counter flow (*ii*) parallel flow and (*iii*) one shell pass and four tube pass arrangements. Solve the problem using LMTD method and check using NTU method.

12.10 Hot water is used to heat air for comfort conditioning in a heat exchanger of the cross flow type with outside air flow. The flow rate of air is 17000 kg/hr and is to be heated from – 10°C to 22°C. Hot water enters at 62°C and leaves at 30°C.

The overall heat transfer coefficient has a value of 60 W/m^2 K. Determine the area required if the air flow is (*i*) unmixed and (*ii*) mixed. Solve using LMTD method and check using NTU method.

12.11 In a tubular air heater cold air at 2 bar and 20°C flows through the inside of 2.5 cm *ID* pipes entering at a velocity of 6 m/s. 20 pipes of 3 m length are used in the bundle. Hot gases at 90°C flows across the tube bank at 1.33 times the cold air flow rate. The overall heat transfer coefficient based on inside area of tubes has a value of 82 W/m^2 K. Determine the fluid exit temperatures for (*i*) hot gas side unmixed and (*ii*) hot gas side mixed.

12.12 A cross flow heat exchanger of area 4.71 m^2 has a heat transfer rate of 6.6 kW, when heating 0.14 kg/s of air from 20°C using hot gases at 90°C, the overall heat transfer coefficient being 82 W/m^2 K. The gas flow rate is 1.33 times the air flow rate. Determine the area required for maintaining the same heat flow rate when cold air rate and inlet temperatures are maintained but the gas flow rate is changed to (*i*) the same as air flow rate (*ii*) twice the air flow rate. Assume both fluids unmixed, and property values of gas to be equal to that of air.

12.13 Water at the rate of 133.3 kg/min at 25°C is used to cool a chemical flowing at the rate of 166.67 kg/min from 80°C to 50°C, the specific heat of the oil being 2.095 kJ/kgK. Assuming over all heat transfer coefficient as 300 W/m^2 K, calculate the area required for counter flow double tube heat exchanger. [**Ans.** 19.23 m^2]

12.14 Two kg of water is heated in a counter flow double pipe heat exchanger from 25°C to 60°C. Using brine at 140°C. Assuming C_p of brine as 4.31 kJ/kgK and that of water as 4.18 kJ/kgK and the convection coefficient as 0.55 kW/m^2 K determine the heat transfer area required. The flow rate of brine is 3 kg/s. [**Ans.** outlet temp. of brine 117.4°C, 6.642 m^2]

12.15 Water at the rate of 5 kg/s is heated in a two shell pass and multiple tube pass heat exchanger from 20°C to 55°C. Hot oil with C_p = 2.15 kJ/kgK flows in the shell and is cooled from 120°C to 45°C. Heat transfer area is 1.8 m^2. Determine the value of the overall heat transfer coefficient, U. [**Ans.** 13.9 × 10^3 W/m^2 K]

12.16 Water is heated in double pipe heat exchanger at the rate of 3 kg/s from 17°C to 80°C. The heat is provided by steam condensing at 120°C. If h_{fg} = 2203 kJ/kg, and U = 1500 W/m^2 K, determine the heat transfer area. [**Ans.** 7.9 m^2]

12.17 For the heat exchanger specified in problem 12.12, Determine the heat flow rate and exit temperatures if inlet temperature of the fluids are changed to

(*i*) 20°C and 110°C (*ii*) 30°C and 90°C and (*iii*) 30°C and 80°C. Assume both fluids unmixed.

12.18 A counter flow heat exchanger with UA = 1300 W/K is used to heat water from 20°C to a temperature not exceeding 93°C. Using hot air at 260°C at a rate of 1620 kg/hr. Calculate the water flow rate and gas exit temperature.

12.19 A shell and tube heat exchanger with two shell passes and four tube passes cools a chemical with a specific heat of 2385 J/kg K at a rate of 1.25 kg/s from 140°C to 80°C, using water at 35°C. The outlet temperature of the water is 85°C. Determine the area of heat exchanger if U = 800 W/m^2 K. If the unit is operated with flow rates doubled, determine the heat transfer. Also determine the heat transfer if the chemical flow rate is increased by 20%.

12.20 An automobile radiator cools water from 65°C to 40°C at a rate of 0.2 kg/s using air in cross flow with inlet at 23°C. The overall heat transfer coefficient has a value of 73 W/m^{2}K. The air flow rate at this speed when the automobile was travelling at 80 kmph was 2.4 kg/s. Determine the area required. Assume both fluids unmixed. If the mass flow of air is proportional to the speed, what will be the heat flow rate at 40 kmph if the inlet temperatures and water flow rates are the same. Assume that the overall heat transfer coefficient varies as the 0.8 power of speed.

12.21 A comfort air conditioner uses air cooled condenser of the cross flow type with finned tubes. The refrigerant condenses at 40°C, with heat transfer rate of 12 kW. Air at 28°C is used and is heated to 34°C. Determine the area required if the overall heat transfer coefficient has value of 125 W/m^2 K.

During the cooler season air is available at 22°C with the same flow rate maintained by the fan. Determine the condensing temperature if the same heat rate is to be maintained.

12.22 The evaporator of an air conditioner operates at 5°C. The cooling rate required is 10 kW. Air at 26°C is to be cooled to 12°C for the conditioning process. The value of overall heat transfer coefficient is 96 W/m^2 K. Determine the area required. The unit is cross flow type with finned tubes. Assume that condensation does not occur.

During mild weather the load reduces to 6 kW. The air flow rate and inlet temperatures remain unaltered. Determine the evaporation temperature for steady state operation at this load.

12.23 A recuperative air heater having an area of 100 m^2 with an overall heat transfer coefficient of 45 W/m^2 K is used to heat air at a rate of 2.5 kg/s from 40°C. Using gases of same heat capacity available at 220°C. Determine the exit temperatures.

12.24 In a counter flow heat exchanger a cool stream of water at 5°C is heated using a hot stream at 60°C. The flow rate of cold stream is twice that of the hot stream the flow rate of which is 1.3 kg/s. If the heat transfer rate is 150 kW and the overall heat transfer coefficient is 800 W/m^2 K, determine the area required. Also determine the effectiveness. If the flow rate of cold stream is reduced by 12.5% determine the exit temperatures and the heat flow rate.

12.25 In a heat exchanger of the counter flow type the hot fluid is cooled from 260°C to 120°C while the cold fluid is heated from 20°C to 90°C. Determine the effectiveness of the heat exchanger and the heat flow if the hot fluid heat capacity is 10,000 W/K.

Determine the exit temperatures if

(*i*) Both fluid flows are doubled (*ii*) hot fluid flow alone is doubled

(*iii*) fluid flows are halved. (*iv*) cold fluid flow alone is doubled.

12.26 A steam condenser is designed to condense 1 kg/s of steam at 65°C using cooling water at 25°C. The water outlet temperature is to be limited to 40°C. Calculate the area if the overall heat transfer coefficient is 2600 W/m^2 K. The enthalpy of evaporation at 65°C is 2346.2 kJ/kg. If the water temperature is reduced to 18°C calculate the water flow rate for maintaining the same condensation rate.

12.27 In problem 12.22 if the condensation temperature is reduced to 50°C with the same water inlet temperature determine the flow rate required for the same condensation rate. The enthalpy of evaporation is 2382.7 kJ/kg.

12.28 A finned tube air heater is available with the following specifications: Area. 10 m^2. The overall heat transfer coefficient based on this area is 62 W/m^2 K.

Water at a rate of 1 kg/s available at 60°C and air at 2 kg/s to be heated from –10°C. Determine the exit temperature of the fluids and the heat flow.

12.29 For the heat exchanger specified in problem 12.24, determine the exit temperature for the same flow rates for the following water and air inlet temperature combination.

(*i*) 70°C, 0°C, (*ii*) 50°C, – 20°C (*iii*) 40°C, – 30°C. Also calculate the heat flow under these conditions.

12.30 For the heat exchanger specified in problem 12.14, determine the exit temperatures and the heat flow rates for the following flow combinations of hot and cold fluids: per second.

(*i*) 1 kg and 1 kg, (*ii*) 0.5 kg and 2 kg (*iii*) 2 kg and 2 kg and (*iv*) equal heat capacities.

The inlet temperatures are 60°C and –10°C.

12.31 A heat exchanger with 20 m^2 area with an overall heat transfer coefficient of 800 W/m^2 K has hot water at 93°C available at 5 kg/s. Cold water at 25°C is to be heated. Determine the heat transfer for the following flow rates of cold water: 2.5, 3.75, 6.25, and 7.5 kg/s.

Also determine the outlet temperature of cold water in these cases.

12.32 1 kg/s of water of 80°C and 1 kg/s of water at 60°C are available to heat cold water at a rate of 1.33 kg/s from 20°C.

Two exchangers each with area of 2 m^2 are available, with overall heat transfer coefficient of 836 W/m^2 K.

Investigate the following arrangements in terms of heat flow and outlet temperature. (Fig. 12.32).

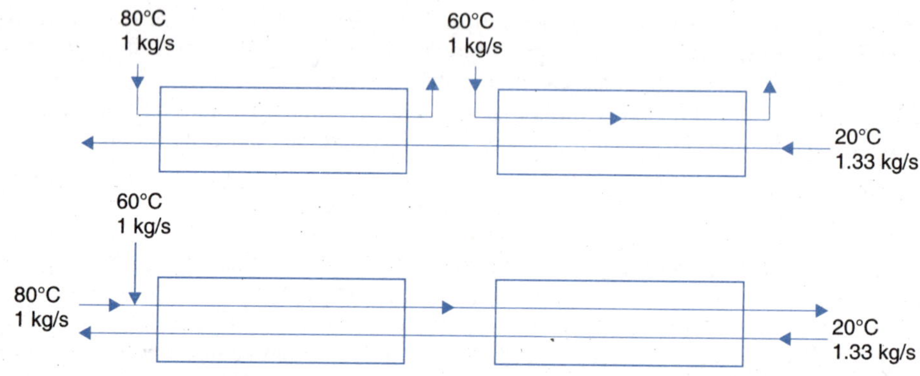

Fig. 12.32

12.33 A heat exchanger cools 1.25 kg/s of a chemical with a specific heat of 2365 J/kg K from 140°C to 80°C, using water which heats up from 35°C to 85°C. The overall heat transfer coefficient has a value of 800 W/m^2 K. If the flow rate of the hot fluid is reduced to 0.5 kg/s. What should be the inlet temperature of water so that its outlet temperature is maintained at 85°C. The water flow rate is not altered. Asssume shell and tube exchanger with 4 tube passes.

2.34 Water is heated at the rate of 4.5 kg/s from 20°C to 70°C in a two shell pass and 12 tube pass exchanger with oil entering at the rate of 10 kg/s at 170°C. Assuming U = 600 W/m^2 K and C_p of oil as 2300 J/kg K determine the area required. [**Ans.** 15 m^2, T_{ho} of oil 129°C]

2.35 Isobutane is condensed in a cross flow exchanger at 75°C, at the rate of 2.7 kg/s. Assume h_{fg} = 255.7 kJ/kg. Air at 21°C is heated to 28°C. Heating surface area is 24.0 m^2. Air is mixed. [**Ans.** 571 W/m^2 K]

12.36 In a cross flow heat exchanger air at the rate of 3 kg/s is heated from 10°C, using water at 95°C at the rate of 1 kg/s. Determine at the limiting condition the exit temperatures. C_p = 4.19 kJ/kg in water and 1.005 kJ/kg for air. [**Ans.** 95°C and 33.8°C]

12.37 Water at the rate of 3 kg/s is heated from 20°C using hot oil at 130°C. The oil flow rate is 3 kg/s. The unit is a two shell pass and multiple tube pass construction. The specific heat of oil is 2.2 kJ/kg K. For water C_p = 4.18 kJ/kg K. Assuming U = 300 W/m^2 K and A = 20 m^2, determine the heat flow. [**Ans.** 385 kW]

12.38 A heat exchanger is to be designed to heat air at a rate of 10 kg/s from 27°C using hot gases at 375°C at the rate of 10 kg/s with an overall heat transfer coefficient of 50 W/m^2 K. If the area is chosen as 200 m^2, determine the heat transfer and exit temperature for the following types.

(*i*) Shell and tube with one shell pass and 4 tube passes

(*ii*) cross flow with both fluids unmixed.

(*iii*) cross flow with one fluid unmixed.

Assume that the property values of the gases are the same as that of air.

12.39 A heat exchanger with an area of 29 m^2 is used to heat water at a rate of 5 kg/s from 30°C to 80°C using steam condensing at 100°C. Determine the overall heat transfer coefficient required to obtain this performance. If the flow rate is changed to 1.5 kg/s with an overall heat transfer coefficient of 80% of the original value and if the water exit temperature is to be 80°C, determine the water inlet temperature.

12.40 Considering data of problem 12.31, determine the condensing temperature which will give the same exit temperature at the altered flow condition, the inlet being at 30°C.

12.41 A double tube heat exchanger has hot and cold air enter at 200°C and 35°C respectively. The flow rate of hot air is 0.75 kg/s and that of the cold air is 1.5 kg/s. Determine the maximum possible heat flow for (*i*) parallel flow and (*ii*) counter flow. For an area of 2.1 m^2 determine the outlet fluid temperatures and heat flow if in both cases. U = 180 W/m^{2}K.

12.42 In an ocean thermal plant 2 kW of electricity is to be generated using a vapour boiler. The plant overall efficiency is 3%. Evaporation takes places at 17°C using water at 27°C. The water outlet is at 19°C. Determine the flow rate required. Determine the area required assuming U = 1200 W/m^2 K. At half load if the efficiency falls to 2% and if the inlet temperature and evaporation temperatures are maintained as at full load, determine the flow rate and water outlet temperature.

12.43 In a recuperative air heater with cross flow arrangement cold air at a rate of 10 kg/s is heated from 30 to 210°C as it flows inside the tubes. Hot gases pass across the tubes and are cooled from 360°C to 160°C. Assuming an overall heat transfer coefficient of 140 W/m^{2}K, determine the area required when (*i*) the gases are mixed and (*ii*) both fluids are unmixed. Assume gas properties to be same as that of air.

12.44 In a water heater using hot combustion gases, the capacity ratio is 0.04 for gas to water, the water flow rate being 12 kg/s. The entry temperature for the gas and water are 400°C and 30°C. If UA = 3850 W/K, determine the exit temperature and the heat flow, if the exchanger is

(*i*) counter flow (*ii*) parallel flow and (*iii*) cross flow with gas flow mixed.

12.45 For the data of problem 12.36 case (*iii*), if the capacity ratio is changed to 0.05, determine the outlet temperature and the heat flow.

12.46 An oil with C = 2100 J/kg K is to be cooled from 82°C to 39°C at rate of 5 kg/s. A cooler having an area of 28 m^2 is available. Water at 27°C is to be used for the cooling process. If the estimated value of overall heat transfer coefficient is 550 W/m^2 K. Determine the water flow rate required for the process. Assume counter flow arrangement.

12.47 In a storage type heat exchanger, the bed is 1 m^2 in area and 2 m long. The specific heat of the solid is 445 J/kg K while air is the fluid used. The surface area per unit length of matrix is 40 m and the mass flow rate of the fluid is 5 kg/s. The mass of solid for m length is 1200 kg. The solid was initially at 200°C and the gas temperature at entry is 600°C. Determine the time required for the matrix at mid point to reach 400°C and also determine the gas temperature at this location at that time. The convective heat transfer coefficient has a value of 82 W/m^2 K.

[**Ans.** 170s, 512°C

12.48 Water is heated at the rate of 0.1 kg/s from 18°C in a two shell pass and multiple tube pass heat exchanger using oil flowing at the rate of 0.2 kg/s and entering at 160°C. Area of exchanger is 2.04 m^2. U = 340 W/m^2 K.

Determine the exit temperature and head flow. Specific heats are: oil 2.2 kJ/kg K, water 4.18 kJ/kg K. [**Ans.** 104.5°C, 77.7°C, 36.2 kW]

12.49 Glycerine at the rate of 0.3 kg/s at 20°C is heated in a counter flow heat exchanger using Ethylene at 60°C at the rate of 0.3 kg/s. Specific heats: Glycerine—2.4 kJ/kg K, Ethylene—2.5 kJ/kg K. Heat transfer area = 5.3 m^2, U = 360 W/m^2 K. Determine the exit temperatures and the heat transfer. [**Ans.** 40.3°C, 40.5°C, 14.63 kW]

Thermal Radiation

13

CHAPTER

13.0 INTRODUCTION

Electromagnetic waves or electromagnetic radiation represent the energy emitted by matter as a result of changes in the electronic configuration of atoms and molecules. Electromagnetic waves transport energy just like other waves and all electromagnetic waves travel at the speed of light in vacuum. Electromagnetic waves are characterised by their frequency ν and wavelength λ. The two properties are related by $\lambda = c/\nu$, where c is the speed of propagation of wave in that medium. The commonly used unit of wave length is micrometer of micron = 10^{-6} m. The frequency of electromagnetic wave depends only on the source and is independent of the medium in which it travels. According to quantum theory electromagnetic radiation is considered as the propagation of a collection of discrete packets of energy called photon or quanta. In this view each photon at frequency ν is considered to have an energy given by $e = h\nu = hc/\lambda$ where h is Planck's constant. As seen in the equation short wavelength radiation possesses larger photon energy. This is the reason why we have to avoid exposure to gamma and X rays, which may cause damage to tissue. Figure 13.1 shows the electromagnetic spectrum and the wavelength of different radiation.

The type of electromagnetic radiation that is pertinent to heat transfer is thermal radiation emitted as a result of energy transition of molecules and atoms and electrons of a substance. Temperature is a measure of strength of the activities at the microscopic level. Radiation emitted by bodies due to their temperature falls in the region of 0.1 to 100 μm and this is the region of thermal radiation which includes visible, infrared and part of ultraviolet radiation.

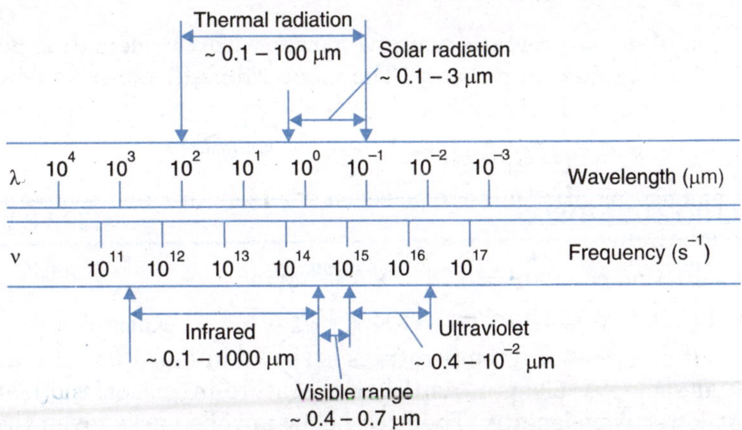

Fig. 13.1. Typical spectrum of electromagnetic radiation due to temperature of a body

Thermal radiation is emitted in the range of wavelengths from 0.1 to 100 micrometers (μm). The lower range is ultraviolet and the upper range is is infrared while the range of 0.4 to 0.7 μm covers the visible region. The total energy radiated at any temperature (emissive power) is found to be distributed all over the wavelengths. The energy radiated at any wavelength is known as monochromatic emissive power. The distribution is known as spectral distribution.

13.1 BLACK BODY

A body at a temperature above absolute zero emits thermal radiation in all directions and over a wide range of wavelengths. The amount of radiation energy emitted from a surface at a given wavelength depends on the material of the body and the condition of its surface as well as the temperature. Therefore different bodies may emit different amounts of radiation per unit surface area even when they are at the same temperature.

Thus it is natural to be curious about the maximum amount of radiation that can be emitted by a surface at a given temperature. Satisfying this curiosity requires the definition of an idealised body called a 'black body' to serve as a standard against which the radiative properties of real surfaces may be compared.

A black body is defined as a perfect emitter and absorber of radiation at all wavelengths and temperatures. No surface can emit more energy than a black body at a specified temperature and wavelength. A black body absorbs all incident radiation regardless of wavelength and direction. Also a black body emits radiation energy uniformly in all directions. Black body is a diffuse emitter that is independent of direction. Real surfaces have directional emissivity.

A body which closely resembles a black body is a large cavity with a small opening. Radiation entering will be absorbed by multiple reflection indicating perfect absorber.

As the surface of the cavity is at constant temperature T. The radiation from the surface also will come out after multiple reflection and thus it will be a diffuse emitter, so the radiation coming out of the opening will be black body radiation.

Such an arrangement therefore can be taken as a black body source for calibration and standardisation.

13.1.1. Black Body Radiation

The radiant energy emission per unit area and unit time from black body over all the wavelengths is defined as its **total emissive power** and is denoted as E_b. The radiant energy emission at any wavelength λ is known as **monochromatic emissive power** and denoted as $E_{b\lambda}$.

The variation of monochromatic emissive power with wavelength is given by Planck's law derived in 1900 using quantum theory. The monochromatic emissive power at any given wavelength and temperature is given by

$$E_{b\lambda}(T) = C_1/[\lambda^5 (e^{C_2/\lambda T} - 1)] \qquad ...(13.1)$$

where $E_{b\lambda}(T)$ is monochromatic emissive power at wavelength λ and temperature T, W/m^2/m

λ—Wavelength, m

C_1—First radiation constant = 3.7415×10^{-16} Wm^2

C_2—Second radiation constant = 1.4388×10^{-2} mK.

This distribution for several temperatures is shown in Fig. 13.2. The distribution has a maximum value at some wavelength. As the temperature increases, the maximum value is found to occur at lower wavelengths. The area below anyone curve given the total emissive power at the corresponding temperature.

The wavelength at which the maximum value of $E_{b\lambda}$ occurs at any given temperature can be determined by differentiating the expression 13.1 with respect to λ and equating it to zero. The result of which is

$$\lambda_{max} T = 2.898 \times 10^{-3} \text{ mK} = \text{constant} \qquad ...[13.2(a)]$$

This relationship is known as Wien's displacement law. It is seen that λ_{max} decreases as T increases. Substituting this value in equation (13.1) we get

$$E_{b\lambda \max T} = 12.87 \times 10^{-6}\, T^5 \text{ W/m}^3 \qquad ...[13.2(b)]$$

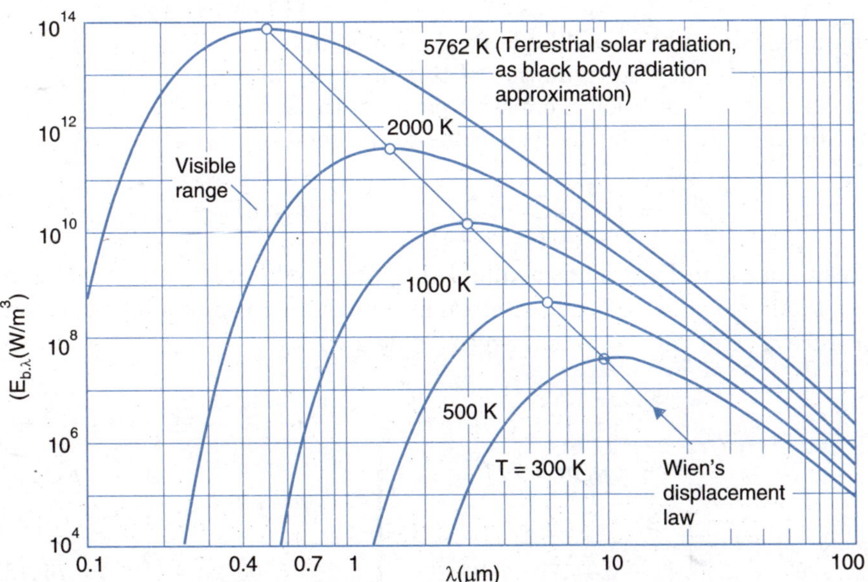

Fig. 13.2. The effects of wavelength and temperature on the monochromatic hemispherical black body emissive power, and the meaning of Wien's displacement law.

Some observations from Fig. 13.2.

1. The emitted radiation is a continuous function of wavelength. At any specified temperature it increases with wavelength reaches a maximum and then decreases with increasing wavelength.
2. At any wavelength the amount of emitted radiation increases with increasing temperature.
3. As the temperature increases, the curve shifts to the left towards lower wavelengths.
4. At the temperature of the sun, the radiation from it falls in the visual wavelength range.

13.1.2. Black Body Emissive Power

The basic relationship known as **Stefan-Boltzmann's equation** for black body emissive power is given by

$$E_b = \sigma T^4 \qquad ...(13.3)$$

where σ is a constant = 5.67 × 10^{-8} W/m^2 K^4 in SI units and E_b is the total emissive power in W/m^2 and T is the temperature in K.

This relationship can be verified using equation (13.1) in the integral given in equation (13.4.), which leads to Stefan-Boltzmann's equation.

$$E_b = \int_0^{\infty} E_{b\lambda}\, d\lambda \qquad ...(13.4)$$

Chapter 13

13.1.3. Another use of the Relationship is the Determination of Radiant Energy Emission in a Wavelength Range λ_1 and λ_2

This is obtained by using equation (13.4) with limits λ_1 and λ_2. This process is shown in Fig. 13.3.

$$E_{b(\lambda_1 - \lambda_2)} = \int_{\lambda_1}^{\lambda_2} E_{b\lambda}\, d\lambda = \int_0^{\lambda_2} E_{b\lambda}\, d\lambda - \int_0^{\lambda_1} E_{b\lambda}\, d\lambda \quad ...(13.5)$$

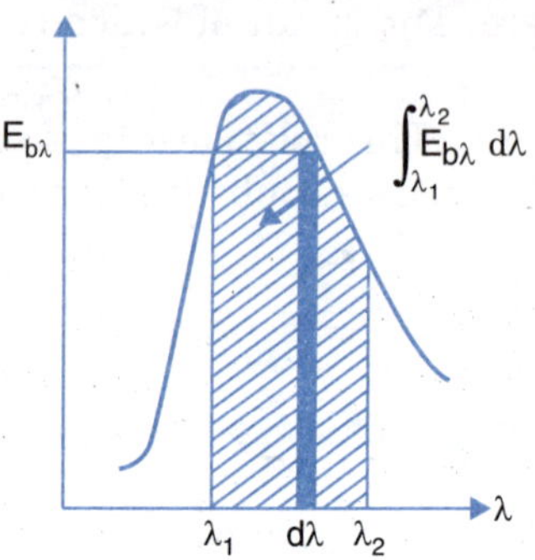

Fig. 13.3. Radiation emission in a given wavelength range

As the integral can be solved by labourious method each time a practical way is the determination of this quantity in a non dimensional form as a fraction of emissive power at the same temperature. In this attempt the analysis leads to the relation.

$$\frac{E_{bT(0-\lambda_1 T)}}{\sigma T^4} = \int_0^{\lambda_1 T} \frac{E_{b\lambda}}{\sigma T^5}\, d(\lambda T) \quad ...(13.6)$$

The result of the integration is available in numerical forms with λT as the independent variable. The tabulation can be used to determine the energy radiated in any band $(\lambda_1 - \lambda_2)$ at any temperature, T, the values are shown in Table 13.1.

Table 13.1. Black body radiation function

λT, μm K	$\frac{E_b(0 \to \lambda T)}{\sigma T^4}$	λT, μm K	$\frac{E_b(0 \to \lambda T)}{\sigma T^4}$
200	0.341796×10^{-26}	6200	0.754187
400	0.186468×10^{-11}	6400	0.769234
600	0.929299×10^{-7}	6600	0.783248
800	0.164351×10^{-4}	6800	0.796180
1000	0.320780×10^{-3}	7000	0.808160
1200	0.213431×10^{-2}	7200	0.819270
1400	0.779084×10^{-2}	7400	0.829580
1600	0.197204×10^{-1}	7600	0.839157
1800	0.393449×10^{-1}	7800	0.848060
2000	0.667347×10^{-1}	8000	0.856344
2200	0.100897	8500	0.874666
2400	0.140268	9000	0.890090
2600	0.183135	9500	0.903147
2800	0.227908	10000	0.914263
3000	0.273252	10500	0.923775
3200	0.318124	11000	0.931956
3400	0.361760	11500	0.939027
3600	0.403633	12000	0.945167

(Contd...)

3800	0.443411	13000	0.955210
4000	0.480907	14000	0.962970
4200	0.516046	15000	0.969056
4400	0.548830	16000	0.973890
4600	0.579316	18000	0.980939
4800	0.607597	20000	0.985683
5000	0.633786	25000	0.992299
5200	0.658011	30000	0.995427
5400	0.680402	40000	0.998057
5600	0.701090	50000	0.999045
5800	0.720203	75000	0.999807
6000	0.737864	100000	1.000000

Example 13.1: ***Determine the maximum value of $E_{b\lambda}$*** *at a temperature of 1400 K. Also determine the value at a wavelength of 4 × 10^{-6} m.*

Solution: Using Wien's law $\lambda_{max} = 2.898 \times 10^{-3}/1400 = \mathbf{2.07 \times 10^{-6}}$ **m**

substituting in equation (13.1)

$$E_{b\lambda} = \frac{3.7415 \times 10^{-16}}{(2.07 \times 10^{-6})^5 \, [e^{(1.4388 \times 10^{-2}/2.898 \times 10^{-3})} - 1]} = \mathbf{6.919 \times 10^{10}\ W/m^2/m}$$

Check: Using equation [13.2(*b*)]

$$E_{b\lambda\,max} = 12.87 \times 10^{-6}\,(1400)^5 = \mathbf{6.922 \times 10^{10}\ W/m^2/m}$$

At 4 micrometer

$$E_{b\lambda} = \frac{3.7415 \times 10^{-16}}{(4 \times 10^{-6})^5 [e^{(1.4388 \times 10^{-2}/1400 \times 4 \times 10^{-6})} - 1]} = \mathbf{3.031 \times 10^{10}\ W/m^2/m.}$$

Example 13.2: ***Determine the radiant energy emission between wavelengths 2 to 5 μm at 1400 K by a black body of area 1 m².***

Solution: $\lambda_1 T = 1400 \times 2 \times 10^{-6} = 2.8 \times 10^{-3}$ mK = 2800 μmK

$\lambda_2 T = 1400 \times 5 \times 10^{-6} = 7 \times 10^{-3}$ mK = 7000 μmK

referring to the Table 9.1 the fractional quantities at these values are read as 0.2281 and 0.8081

i.e.,
$$\frac{E_{b(0-2)}}{E_b} = 0.2279 \text{ and } \frac{E_{b(0-5)}}{E_b} = 0.80816$$

$$\therefore \quad \frac{E_{b(2-5)}}{E_b} = 0.80816 - 0.2279 = 0.58026$$

Radiant energy emission is this range is

$$\mathbf{q} = 0.58026 \times 5.67 \times 10^{-8} \times (1400)^4 = \mathbf{126391\ W/m^2\ or\ 126.391\ kW/m^2}$$

Emissive power at 1400 K is 217.819 kW/m² using (σT^4)

About 60% of the energy is emitted in this wave band at 1400 K.

Example 13.3: ***Window glass is*** *found to transmit radiant energy in the wavelengths of 0.4 to 2.5 μm. Determine the fractions transmitted for a source temperature of (i)* ***5000 K*** *(ii) 300 K.*

Solution: (*i*) **at 5000** *K*, $\lambda_1 T = 5000 \times 0.4 = 2000$ μmK

$$\lambda_2 T = 5000 \times 2.5 = 12500 \text{ μmK}$$

From tables the corresponding fractions are

0.0667347 and 0.9502

∴ **Fraction transmitted = 0.88345**

(*ii*) **at 300 K**

$$\lambda_1 T = 120 \text{ μmK}, \quad \lambda_2 T = 750 \text{ μmK}$$

At these values the fractions are zero as read from tables. Radiant energy will not be transmitted out by the glass at this temperature and the glass is opaque for such radiation.

This is the principle of green house effect. High temperature radiant energy from the sun will be transmitted while low temperature radiation is cut off.

The difference between these two goes to increase the temperature of the space covered by glass. Other materials like CO_2 also transmit selectively. Solar energy is let in by the CO_2 laden atmosphere, while the earths radiation is trapped. This lends to warming up of the atmosphere called as green house effect.

13.2 INTENSITY OF RADIATION

The radiant energy emitted from a surface travels all over the space. This is also called diffuse radiation. When the quantity of energy transferred is to be calculated, the energy emitted in a given direction becomes the basic parameter. **The energy emitted per unit solid angle in a direction is defined as the intensity of radiation in that direction.** Using the definition, a relationship between emissive power and intensity of radiation can be established.

Consider a small black differential area dA_1. Let a hemisphere of radius r with its centre at the centre of dA_1 be placed over this area. All the energy emitted by dA_1 is intercepted by the hemispherical surface and is equal to $E_b \times dA_1$ (Fig. 13.4)

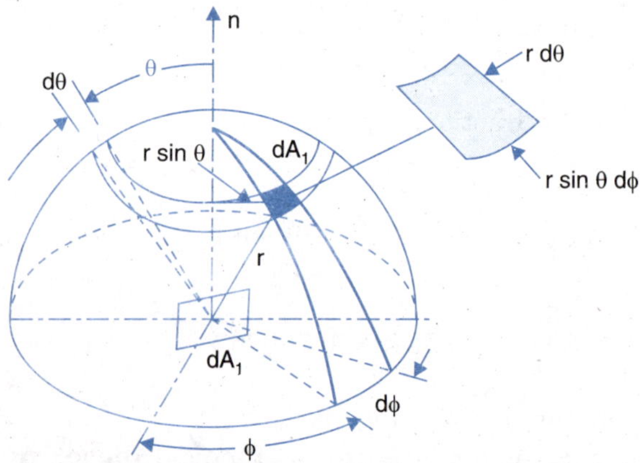

Fig. 13.4. Radiation from a differential are dA_1 to a hemisphere with center at dA_1

Consider the radiation intercepted by small shaded area between angle **ϕ** and **ϕ + dϕ** and **θ** and **θ + $d\theta$.**

The area is $r \sin \theta \,.\, d\phi \,.\, r\, d\theta$

The solid angle $\qquad d\omega = \text{area}/r^2 = \sin \theta \, d\theta \, d\phi$

The projected area of dA_1 in the direction considered is $dA_1 \cos \theta$.

If intensity of radiation is I, then the total energy intercepted by this area is given by $d\omega I \, dA_1 \cos \theta$. Integrating over the hemisphere and dividing by dA_1

$$E = \int_0^{2\pi} \int_0^{\pi/2} I \cos \theta \sin \theta \, d\theta \,.\, d\phi$$

$$= \int_0^{2\pi} I \, d\phi \left[\frac{\sin^2 \theta}{2} \right]_0^{\pi/2} = \int_0^{2\pi} \frac{I}{2} \, d\phi = \pi I$$

$\therefore$ $\qquad \mathbf{E = \pi I}$ ...(13.7)

This is an important relation used in the calculation of radiant heat transfer between surfaces.

13.3 REAL SURFACES

Solids and liquids emit radiation from the surface, with the material involved limited to a small thickness at the surface. Gases however emit radiation over the whole volume. Real surfaces radiate less energy as compared to black surface at the same temperature. The surfaces also do not have a regular spectral distribution.

Some of the surfaces radiate only in limited wave bands. Some samples of real surface monochromatic emission is shown in Fig. 13.5. Curve A represents black body radiation. Curves B, C and D represent different surface characteristics.

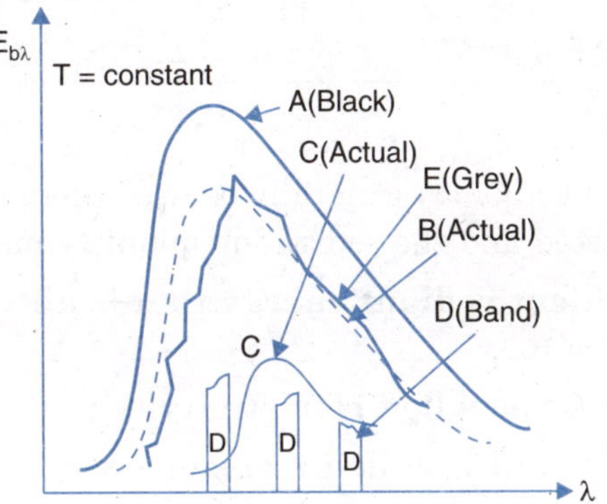

Fig. 13.5. Monochromatic emissive power of different surfaces at temperature T

Curve B represents a more uniform behaviour. Curve D represents what is known as band radiation. The emissive power of these surfaces are obtained by the area below the spectral distribution curve. As it is difficult to deal with such non uniform behaviour, a **fictitious surface called gray body is defined for general use. A gray surface is one whose monochromatic emissive power at all wavelengths is the same fraction of the black body monochromatic emissive power.** This is shown as curve E in Fig. 13.5. The ratio of ordinates of curves A and E at any wavelength is the same. Such a surface does not exist. But real surfaces are approximated to gray surfaces in practical applications. The emissivity of a real surface is not constant. It varies with the temperature of the surface as well as the wavelength and direction of the emitted radiation. There are number of ways in which the emissivity of the surface can be expressed. The basic definition is spectral emissivity defined as the ratio of intensity of radiation emitted by the surface at a specified wavelength and direction to the intensity of radiation of black body at the same temperature, wavelengths and direction.

The ratio E/E_b is known as emissivity (ε) of a surface. Though emissivity of a surface will vary with temperature for general application an average value is used in practice. The emissivity value for various surfaces are listed in hand books and data books. Values for same surfaces is given Table 13.2

$$\varepsilon = E/E_b \qquad ...(13.8)$$

Table 13.2. Hemispherical emissivities of various surfaces

Material		*Temperature*	
	310 K	*530 K*	*800 K*
Aluminium, polished	0.04	0.05	0.08
Aluminium oxidised	0.11	0.12	0.18
Copper polished	0.04	0.05	0.18
Copper oxidised	0.87	0.83	0.77
Iron polished	0.06	0.08	0.13
Cast iron oxidised	0.63	0.66	0.76
Steel rough	0.94	0.97	0.98
Silver polished	0.01	0.02	0.03
Stainless steel 18-8 polished	0.15	0.18	0.22
Stainless steel weathered	0.85	0.85	0.85

13.3.1. Radiation Properties

Radiant energy incident on a body may generally be used up in three ways. Part of the energy may be absorbed, part reflected and the remaining quantity may be transmitted.

The fraction of incident radiant energy that is absorbed by the body is called absorptivity of the surface (α).

The fraction reflected is called reflectivity (ρ);

The fraction transmitted is called transmissivity (τ);

These properties at a particular wavelength are called monochromatic properties *i.e.* α_λ, ρ_λ, and τ_λ). As in the case of emissivity these properties α, ρ, τ will vary with temperature, direction and wavelength. But for general applications these are assumed to be constant for a surface.

In general application the total properties are used. However for special applications monochromatic properties should be used. For example generally it is assumed that $\alpha_\lambda = \alpha$ but it need not be so always and the value of α_λ as measured should be used in specialised applications.

Generally for all surfaces $\alpha + \rho + \tau = 1$ ...[13.9(*a*)]

For opaque surfaces $\alpha + \rho = 1$ ...[13.9(*b*)]

For black surfaces $\alpha = 1$ ...[13.9(*c*)]

Kirchhoff's law: An important relationship between emissivity and absorptivity was postulated by Kirchhoff. The statement of the law is

"The absorptivity of a surface at a temperature and wavelength is equal to the emissivity of the surface at the same temperature and wavelength".

For practical application this statement is simplified as

$$\varepsilon = \alpha \quad ...(13.10)$$

Monochromatic values should be used for critical design situations.

The proof of Kirchhoff law is given in A-13.1 of advanced examples.

Presently surfaces are being developed which do not follow the general statement of $\alpha = \varepsilon$ but $\alpha_{\lambda T} = \varepsilon_{\lambda T}$. This is because emissivity depends on the surface temperature while absoptivity depends on the source temperature. These are called selective surfaces. These surfaces have different values of emissivity at different wavelengths. Particularly the emissivity is high at short wavelengths while it is low at longer wavelengths.

These surfaces are known to have high absorptivity and emissivity for high temperature (short wavelength radiation), but low emissivity for low temperature (longer wavelength) radiation. Such a solar collector surface will absorb a large fraction of incident solar radiation, but will not radiate out much at the lower collector temperature.

Chapter 13

Example 13.4: ***A solar collector surface*** *has an absorptivity of 0.85 for wavelengths upto 3 μm and a value of 0.15 for wavelengths above this value. Determine assuming Kirchhoff's law holds good, the energy absorbed from a source at 5800 K if the flux is 800 W/m². Also determine the energy radiated by the body if its temperature was 350 K.*

Solution. The fraction $\frac{E_{b(0-3)}}{E_b}$ for 5800 × 3 μmK is obtained from tables as 0.9780

∴ **Energy absorbed** = (800 × 0.978 × 0.85) + (0.022 × 0.15 × 800) = **667.68 W**

The fraction $\frac{E_{b(0-3)}}{E_b}$ for 350 K is obtained from tables as 0.00057

∴ **Energy radiated** $= 0.00057 \times 0.85 \times 5.67 \times 10^{-8} \times 350^4$
$+ 0.99943 \times 0.15 \times 5.67 \times 10^{-8} \times 350^4 =$ **128 W**

The loss is about 20%. If the emmissivity was 0.85 throughout, then loss is 723.22 W. This plate temperature of 350 K cannot be maintained at all.

Example 13.5: *In example 13.4, determine the energy absorbed if the source was at 600 K and the flux was 800 W/m².*

Solution: The fraction $\frac{E_{b(0-3)}}{E_b}$ is determined from Tables as 0.0397

corresponding to 600 × 3 = 1800 μmK.

∴ Energy absorbed = 800 × 0.0397 × 0.85 + 800 × (1 – 0.0397) × 0.15 = **142.23 W.**

As most of the energy incident is beyond 3 μm, the energy absorbed is low though the flux is the same.

Note: In addition to the source temperature the flux depends on the geometric location of the source relative to the surface.

Example 13.6: *A surface has emissivities of 0.3 upto 2 μm and 0.85 between 2 and 4 μm and 0.5 beyond this wavelength. Determine the effective emissivity at 4000 K, 2000 K and 400 K.*

Solution: (*i*) **4000 K.** The fraction upto 2 μm, and 4 μm are determined using the tables.

8000 μm K – > 0.856344

16000 μm K – > 0.97389

∴ Effective emissivity is

$$= 0.3 \times 0.856344 - 0.85\ (0.97389 - 0.856344) + (1 - 0.97389) \times 0.5$$

$$= \mathbf{0.36982.}$$

(*ii*) **2000 K**

4000 μm K – > 0.4809

8000 μm K – > 0.8563

Effective emissivity = $0.3 \times 0.4809 + 0.85\ (0.8563 - 0.4809) + 0.5\ (1 - 0.8563) =$ **0.5352**

(*iii*) **400 K**

800 μm K – > zero

1600 μm K – > 0.02

Effective emissivity = $0.02 \times 0.85 + 0.5\ (1 - 0.02) =$ **0.507.**

13.3.2. The Absorptivity and Emissivity for Actual Surfaces

may also very with temperature and direction in addition to wavelength. For better estimates in design it is necessary to use the data available in hand books in chart and table forms for these properties for various surfaces.

Example 13.7: *A directionally emitting surface has an emissivity of 0.6 for angles upto 45° from vertical and 0.2 for the rest of the range. Determine the total emisivity of the surface.*

Solution: The total emissivity can be determined by integrating the local directional emissivities with respect to the angle θ with which the emissivity varies.

(The derivation is given under advanced example A-13.2).

$$\varepsilon = 2 \int_0^{\pi/2} \varepsilon\,(\theta) \cos\theta \sin\theta\, d\theta$$

This is now split up into two integrals.

$$\varepsilon = 2 \int_0^{\pi/4} 0.6 \cos\theta \sin\theta\, d\theta + 2 \int_{\pi/4}^{\pi/2} 0.2 \cos\theta \sin\theta\, d\theta$$

$$= 2 \left[0.6 \frac{\sin^2\theta}{2} \right]_0^{\pi/4} + 2 \left[0.2 \frac{\sin^2\theta}{2} \right]_{\pi/4}^{\pi/2} = 0.4$$

This is also called total hemispherical emissivity.

13.4 RADIATION PROPERTIES OF GASES—ABSORBING, TRANSMITTING AND EMITTING MEDIUM

In a volume like the combustion furnace radiation exchange takes place between gases and the surrounding surfaces. So it becomes necessary to determine the emissivity and absorptivity of gases in a given volume. Unlike in solids and liquids radiation in gases is a volume phenomenon. In solids and liquids the energy emitted inside the material reaches the surface by conduction or convection. At the surface the energy is emitted by radiation.

In the case of gases if the density is low the space between molecules is large and radiant energy emitted inside the volume can reach the surface without being intercepted. The **emissivity and absorptivity of gases** are found to **depend on the linear dimension, geometric shape and the density** (pressure and temperature) **of the volume** in addition to the inherent nature of the gas itself.

The commonly occuring gases like **Oxygen and Nitrogen are transparent** to incident thermal radiation and hence their absorptivity value is zero. By Kirchhoff's law the emissivity is also zero for these cases. Gases like carbon dioxide and water vapour absorb incident thermal radiation. Consequently they also emit thermal radiation. However the emission is not distributed over all values of wavelengths. It is limited to specific wavelength intervals called wave bands.

These aspects make the determination of absorptivity and emissivity values for gases more complicated.

13.4.1. Absorption of Thermal Radiation in Gases

A gas volume is shown in Fig. 13.6.

A thermal radiation beam (monochromatic) with intensity $I_{\lambda 0}$ is incident at face $x = 0$. As it passes through the gas volume partly there is absorption of the beam. At location x, the intensity is $I_{\lambda x}$. As it passes through a differential thickness dx, let the intensity be reduced by dI_λ. Let the absorption coefficient for the gas be a_λ.

Then $$d\,I_\lambda = -\,a_\lambda\,I_{\lambda x}\,dx$$

Integrating (by separating the variables) between limits $x = 0$ and $x = L$

$$\frac{I_{\lambda L}}{I_{\lambda 0}} = e^{-a_\lambda L} \qquad \text{...(13.11)}$$

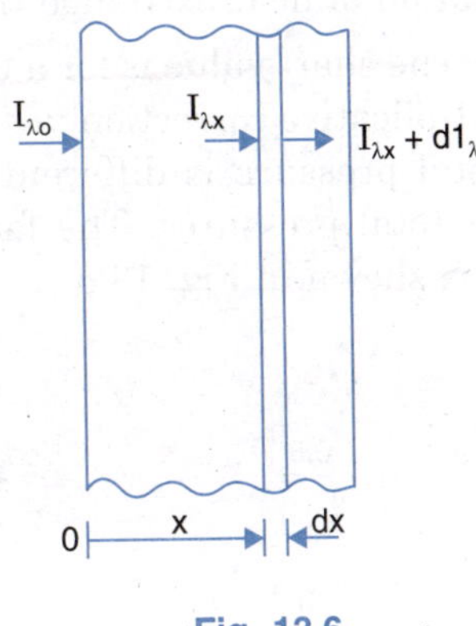

Fig. 13.6

This is known as **Beers law** and this is used in arriving at the monochromatic transmissivity and absorptivity of any gas volume. It should be noted that a_λ is a property of the gas.

Transmissivity τ is the fraction of energy transmitted and is given by equation (13.11).

$$\tau_\lambda = \frac{I_{\lambda L}}{I_{\lambda 0}} = e^{-a_\lambda L}$$

As gases do not reflect any radiation,

$\therefore$ Absorptivity = 1 – Transmissivity

$$\therefore \quad \alpha_\lambda = 1 - \tau_\lambda = 1 - e^{-a_\lambda L} \qquad ...(13.12)$$

Application of Kirchhoff's law gives the relation.

$$\alpha_\lambda = \varepsilon_\lambda$$

The total value of absorptivity or emissivity can be obtained by summing up the absorptivity in the bands in which the gas radiates. For example water vapour has absorptivity values in bands around 1,1.4, 1.8 and 2.75 μm. Similar bands exists for CO_2 at 1.9, 2.7, 4.3 and 15 μm.

13.4.2. Calculation of Absorptivity and Emissivity Values for Water Vapour and Carbon Dioxide

The value of absorptivity is found to depend on the pressure of the gas and the beam length or characteristic length for the volume and the temperature. The emissivity/absorptivity increases with pressure and beam length and decreases with temperature. If water vapour and carbon dioxide are found as a mixture with other gases (particularly air) the partial pressure is to be considered in such cases.

Measured values are available in chart form with emissivity on *y* axis, temperature on *x* axis and *pl* as parameter (*p*—partial pressure, *l*—characteristic length). A schematic chart is shown in Fig. 13.7.

If the pressure, shape and temperature of the gas body are specified, then the emissivity value can be read off from the chart. **The emissivity depends on the gas body temperature while the absorptivity will depend on the radiation source temperature.** Hence in the calculation of heat exchange these values should be calculated at different temperatures.

The chart value is for a total pressure of 1 atm for the mixture of gas/vapour considered. A multiplicative correction factor is necessary to account for the presence of other gases, when the total pressure is different from 1 atm. The correction factor *C* is greater than one for higher total pressures. The factor is less than one for lower partial pressures. A schematic chart is shown in Fig. 13.8.

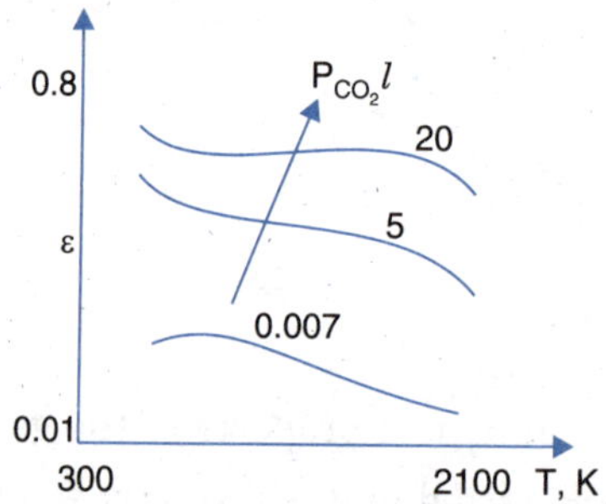

Fig. 13.7. Emissivity of CO_2

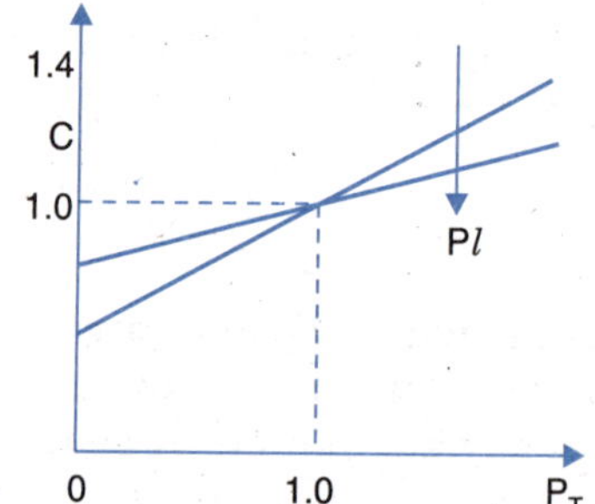

Fig. 13.8. Correction factor for total pressures other than 1 atm (for CO_2)

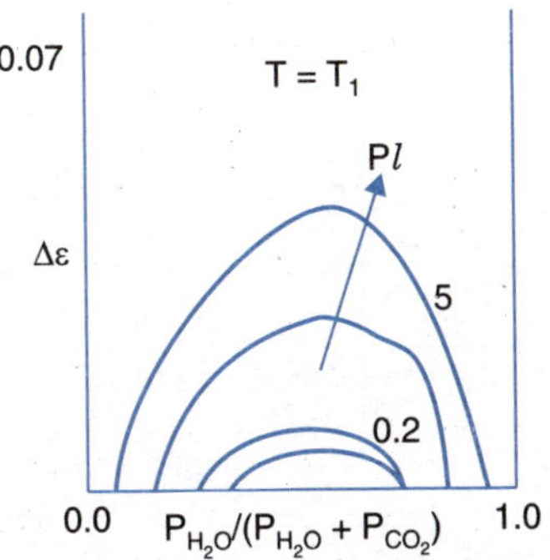

Fig. 13.9. Correction factor for the presence of CO_2 and H_2O in the mixture

Accurate values for the characteristic lengths for various shapes are available in hand books. **Approximate values for all shapes can be obtained by using**

$$l = 3.6 \text{ volume/surface area.} \quad ...(13.13)$$

For water vapour or CO_2 present as a mixture with other non radiating gases, the value of emissivity is obtained by

$$\varepsilon_{CO_2} = \varepsilon_{1\text{ atm}} \cdot C_{CO_2} \quad ...(13.14)$$

For mixtures of both gases in a volume another subtractive factor (Δε) is to be used as the sum of the emissivities calculated using the charts is found to be higher than the actual measured values. This factor is zero when only one of the gas is found in the mixture. The values are of the order of 0.05. A sample chart is shown in Fig. 13.9.

The emissivity when both gases are present is calculated using.

$$\varepsilon = \varepsilon_{CO_2} + \varepsilon_{H_2O} - \Delta\varepsilon \quad ...(13.15)$$

The heat exchange between a black enclosure and the gas is given by

$$Q = \sigma A_s (\varepsilon_g T_g^4 - \alpha_g T_s^4) \quad ...(13.16)$$

Where A_s is the surface area, T_g is the gas temperature and T_s is the surface temperature. ε_g is calculated as indicated in equation (13.15) at T_g. The gas absorptivity is dependent on the source temperature also. The value is calculated as indicated below.

$$\alpha_g = \alpha_{H_2O} + \alpha_{CO_2} - \Delta\varepsilon \quad ...(13.17)$$

$$\alpha_{H_2O} = C_{H_2O} \left(\frac{T_g}{T_s}\right)^{0.45} \varepsilon_{H_2O} \quad ...[13.17(a)]$$

$$\alpha_{CO_2} = C_{CO_2} \left(\frac{T_g}{T_s}\right)^{0.65} \varepsilon_{CO_2}. \quad ...[13.17(b)]$$

Here ε_{H_2O} and ε_{CO_2} should be determined at T_s using a corrected value of *pl* given by $(pl \times T_s/T_g)$.

The charts are shown in Figure (13.10 a, b; 13.11 a, b) and (13.12).

Broken lines indicate extrapolated condition.

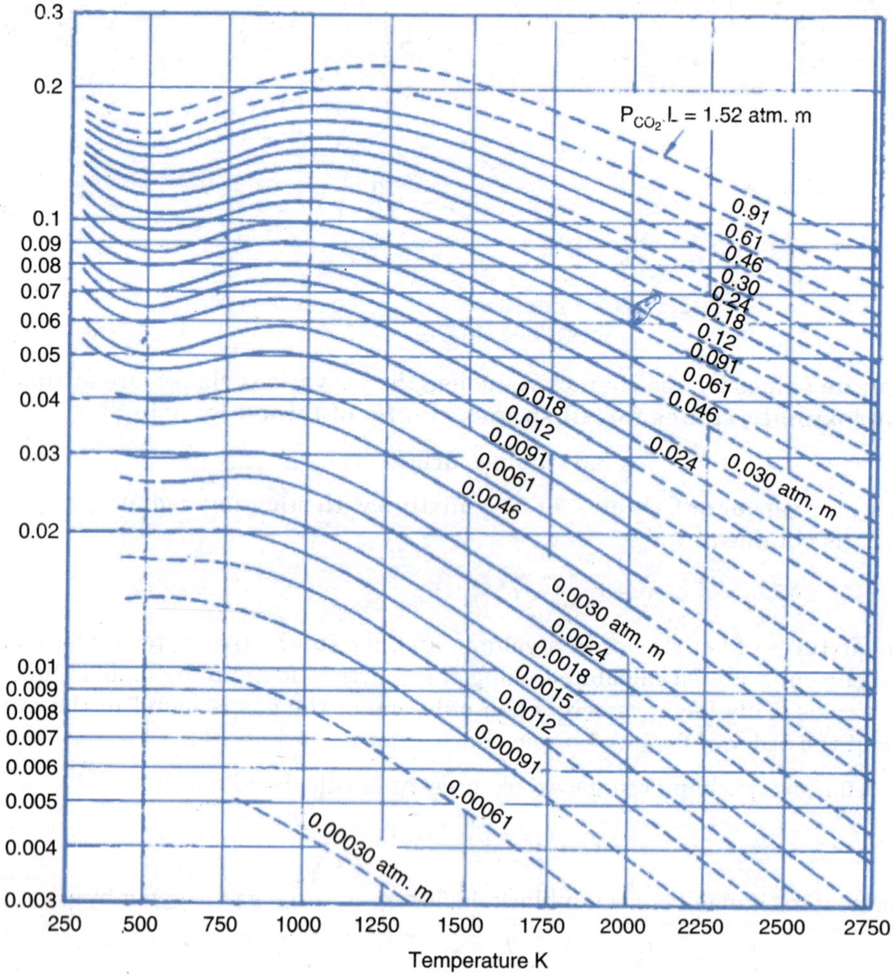

$P_{CO_2}.L$ = 1.52 atm. m
0.91
0.61
0.46
0.30
0.24
0.18
0.12
0.091
0.061
0.046
0.024
0.030 atm. m
0.018
0.012
0.0091
0.0061
0.0046
0.0030 atm. m
0.0024
0.0018
0.0015
0.0012
0.00091
0.00061
0.00030 atm. m
0.3
0.2
0.1
0.09
0.08
0.07
0.06
0.05
0.04
0.03
0.02
0.01
0.009
0.008
0.007
0.006
0.005
0.004
0.003
250 500 750 1000 1250 1500 1750 2000 2250 2500 2750
Temperature K

Fig. 13.10 (a) Emissivity ε_c of carbon dioxide at a total pressure of P_T = 1 atm

P_{CO_2} . L = 0.76 atm.m
0.31
0.152
0 – 0.0061
0.015
0.037
0.076
0.76
0.31
0.152
0.076
0.037
0.015
P_{CO_2}.L = 0 – 0.0061
2.0
1.5
1.0
0.8
0.6
0.5
0.4
0.3
0.05 0.08 0.1 0.2 0.3 0.5 0.8 1.0 2.0 3.0 5.0
P_T (atm)

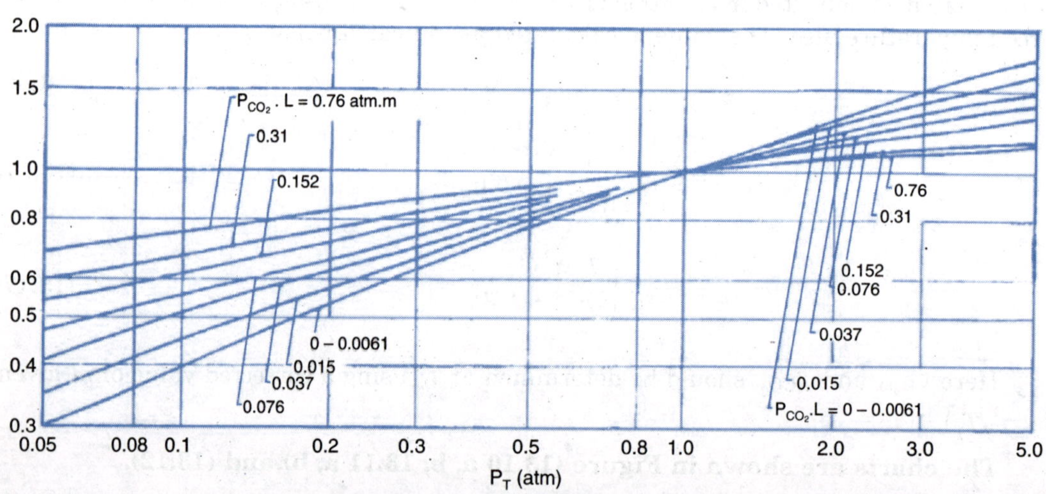

Fig. 13.10 (b) Correction factor C_c for converting the emissivity of CO_2 at 1 atm to emissivity at P_T atm

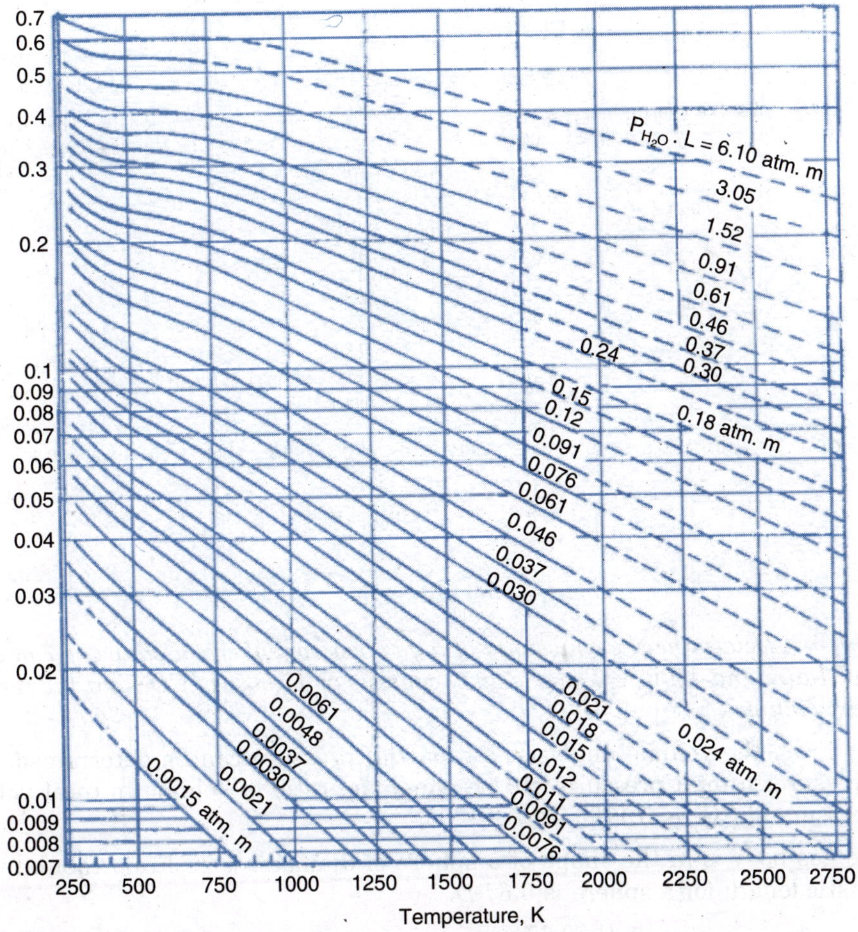

Fig. 13.11 (a) Emissivity ε_w of water vapour at a total pressure of P_T = 1 atm. and corresponding to an idealized case of $P_w \rightarrow 0$

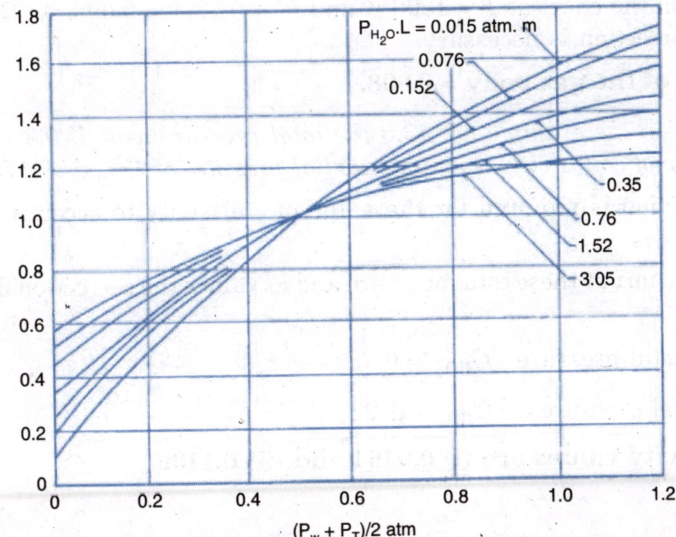

Fig. 13.11 (b) Correction factor C_w for converting emissivity of H_2O to values of P_w and P_T other than 0 and 1 atm. respectively

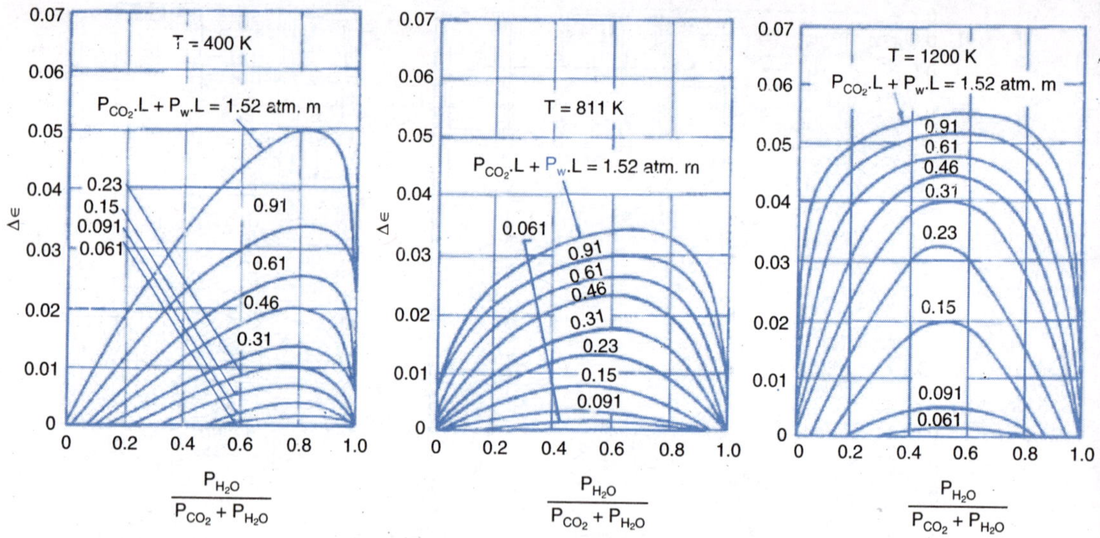

Fig. 13.12. Emissivity correction $\Delta\varepsilon$ for mutual absorption when water vapour and carbon dioxide are present in the gas

Example 13.8: *Determine the emissivity of a gas body of spherical shape of 1 m dia at a total pressure of 1 atm and 1500 K. The gas body consists of 17.96% of CO_2 and the rest oxygen and Nitrogen by volume.*

Solution: From the volume analysis, the partial pressure can be determined. The ratio of partial pressure to total pressure will be equal the ratio of volume to total volume. So the partial pressure of CO_2 is 0.1796 atm.

The gas body is in the shape of a sphere of diameter 1 m. From tabulated values the characteristic length for a sphere is 0.67 *D*.

$$\therefore \qquad l = 0.67 \text{ m}$$

$$\therefore \qquad Pl = 0.1796 \times 0.67 = 0.120$$

Reading from the chart at T = 1500 K and Pl = 0.12, ε = 0.098. As the total pressure is atmospheric, no correction is necessary.

Emissivity of the gas body = 0.098.

Example 13.9 : *If in the example no. 13.8 the total pressure was (i) 0.3 atm and (ii) 3 atm, determine the value of emissivity assuming partial pressure of CO_2 as 0.1796 atm.*

Solution: A correction is required for the value of emissivity to account for pressure other than 1 atm.

Reading the chart at these total pressure and Pl values the correction factors are obtained as

(*i*) 0.3 atm total pressure $\quad C_{CO_2} = 0.78$

(*ii*) 3 atm total pressure $\quad C_{CO_2} = 1.22$

The emissivity values are (i) 0.0764 and (ii) 0.1196.

Example 13.10: *A furnace in the form of a cube of 2 m side has gas in it at 1500 K. The composition of the gas by volume is 16% CO_2, 10% H_2O and the rest are non radiating gases. Determine the emissivity of the gas body. The total pressure is 1 atm.*

Solution: At constant temperature the % partial pressure will have the same value as the % partial volume.

The partial pressure of CO_2 = 0.16 atm

The partial pressure of H_2O = 0.1 atm

The characteristic length = 0.67 × side = 0.67 × 2 = 1.34 m

$$\therefore \quad Pl_{CO_2} = 0.16 \times 1.34 = 0.214$$

$$Pl_{H_2O} = 0.1 \times 1.34 = 0.134$$

The corresponding emissivity values are read from charts entering at 1500 K and reading against the Pl values as

$$\varepsilon_{CO_2} = 0.12 \quad \varepsilon_{H_2O} = 0.105$$

As the total pressure is one atm, no correction is necessary for total pressure. But a correction (deduction) is necessary to account for the combined presence. This is read from the chart at

$P_{H_2O}/(P_{CO_2} + P_{H_2O})$ and $P_c l + P_w l$ values. These are

$$0.1/(0.16 + 0.1) = 0.3846 \text{ and } 0.214 + 0.134 = 0.348.$$

The temperature is 1500 K. The value of $\Delta\varepsilon$ = 0.04 (chart is available for 1200 K only extrapolated)

$$\varepsilon = 0.12 + 0.105 - 0.04 = \mathbf{0.221}.$$

Example 13.11: *In example 13.10, if the total pressure was 3 atm, determine the emissivity of the gas volume.*

Solution. The partial pressures are:

$$P_{CO_2} \rightarrow 0.16 \times 3 = 0.48 \quad P_{H_2O} \rightarrow 0.1 \times 3 = 0.3 \text{ atm}$$

$$Pl_{CO_2} = 0.48 \times 1.34 = 0.6432, \quad Pl_{H_2O} = 0.402$$

The emissivities for the corresponding values are

$$\varepsilon_{CO_2} = 0.17, \quad \varepsilon_{H_2O} = 0.195$$

The correction factors are: for CO_2 at 0.3 and 0.6432, C_{CO_2} = 1.2

For H_2O $(P_{H_2O} + P_T)/2 = (3 + 0.3)/2 = 1.65$

$$Pl = 0.402, \quad C_{H_2O} = 1.41 \text{ (extrapolated)}$$

The value of $\Delta\varepsilon$ is read at $\dfrac{P_{H_2O}}{P_{H_2O} + P_{CO_2}} = 0.3/(0.3 + 0.48) = 0.3846$

and $Pl_{CO_2} + Pl_{H_2O} = 0.6432 + 0.402 = 1.0432, \quad \therefore \quad \Delta\varepsilon = 0.051$

$$\therefore \quad \varepsilon = 0.17 \times 1.2 + 1.41 \times 0.195 - 0.051 = \mathbf{0.428}.$$

Example 13.12: *A furnace of 2 m × 1.5 m × 1.5 m size contains gases at 1500 K while the walls are at 500 K. The gas contains 18% of CO_2 and 12% of water vapour by volume. Determine the heat exchange from the gases to the walls. The total pressure is 2 atm. Assume black surface.*

Solution: The characteristic length = 3.6 volume/surface area

$$= (3.6 \times 1.5 \times 1.5 \times 2)/(2(1.5 \times 1.5 + 2 \times 1.5 \times 2)) = 0.982 \text{ m}$$

The partial pressures are

$$C_{CO_2} = 0.18 \times 2 = 0.36, \qquad \therefore \quad Pl_{CO_2} = 0.354$$

$$P_{H_2O} = 0.12 \times 2 = 0.24 \qquad \therefore \quad Pl_{H_2O} = 0.236$$

The values of emissivities as read from chart for values of 1500 K and *Pl*

$$\varepsilon_{CO_2} = 0.145, \ \varepsilon_{H_2O} = \mathbf{0.155}$$

The correction factors for pressure, as the total pressure is 2 atm are

$$C_{CO_2} = 1.1 \text{ (at 2 and 0.354)}$$

$$C_{H_2O} = 1.42 \text{ (at 1.12 and 0.236)}$$

Correction for the simultaneous presence of the two gases:

$$P_{H_2O}/(P_{CO_2} + P_{H_2O}) = 0.24/(0.36 + 0.24) = 0.4$$

$$Pl_{CO_2} + Pl_{H_2O} = 0.354 + 0.236 = 0.590$$

$$\Delta\varepsilon = 0.047$$

$$\therefore \quad \varepsilon_g = 1.1 \times 0.145 + 1.42 \times 0.155 - 0.047 = \mathbf{0.3326}$$

To determine the absorptivity, the temperature of the wall has to be used. Reading corresponding to 500 K and $P1 = pl \times (T_s/T_g)$ *i.e.*, for $Pl_{CO_2} = 0.118$ and $Pl_{H_2O} = 0.079$

$$\varepsilon_{CO_2} = 0.105, \ \varepsilon_{H_2O} = \mathbf{0.17}$$

The correction factor are:

$$C_{CO_2} = 1.3 \text{ (at 2 atm and 0.118)}$$

$$C_{H_2O} = 1.58 \text{ (at 1.12 atm and 0.079)}$$

The subtractive correction factor is read at

0.4 and (0.118 + 0.079 = 0.197), $\Delta\varepsilon = 0.026$

$$\therefore \quad \alpha_g = 1.3 \times 0.105 \times \left(\frac{1500}{500}\right)^{0.65} + 1.58 \times 0.17 \times \left(\frac{1500}{500}\right)^{0.45} - 0.026$$

$$= 0.2788 + 0.4404 - 0.026 = \mathbf{0.6932}$$

Heat exchange $= \sigma A_s \, [\varepsilon_g T_g^4 - \alpha_g T_s^4]$

$$= 5.67 \times 16.5 \left[0.3326 \times \left(\frac{1500}{100}\right)^4 - 0.6932\left(\frac{500}{100}\right)^4\right]$$

$$= \mathbf{1.5379 \times 10^6 \ W.}$$

13.5 HEAT EXCHANGE BY RADIATION

In the preceding sections the radiation properties and determination of the above were discussed. The important application of the study is heat transfer from one body to another by radiation when a temperature difference exists.

Thermal radiation from a surface travels all over the space in all directions above the surface. Only that portion of radiant energy emitted by a surface which reaches the other body can be absorbed or transmitted or reflected by that body. So it becomes necessary to estimate the fraction of energy emitted by a body that reaches the other body. Out of this portion the energy absorbed by the body can be calculated if the absorptivity of the surface is known. In the case of black body all the incident energy will be absorbed. For opaque grey bodies part will be absorbed and part will be reflected. Part of the reflected energy may again reach the original body which will be disposed off as was the original radiation.

In order to facilitate this study a new concept called shape factor is defined. Several other terms are also used synonymously. **Angle factor, view factor, geometric shape factor** are some of these terms.

Shape factor of surface 1 with respect to surface 2 is defined as fraction of radiant energy emission form surface 1 reaching the surface 2. The fraction is generally denoted by F_{1-2}. Shape factor depends on the geometric positions of the surfaces. If the surface 2 encloses the surface 1 completely then $F_{1-2} = 1$. If surface 2 is not visible from surface 1 then $F_{1-2} = 0$. The shape factor depends on the distance between the surfaces and the areas of the surfaces. Two coaxial circular disks facing each other will have higher value of shape factor if these are moved closer. For the same central distance larger area will give higher value for shape factor.

13.5.1. Expression for Shape Factor

The definition of shape factor was given in the last para. Considering two black surfaces A_1 and A_2 the quantity of radiant energy emission from A_1 reaching A_2 is given by

$$Q_{1-2} = A_1 F_{1-2} \cdot E_{b1} \qquad \text{...[13.18(a)]}$$

This is completely absorbed by surface A_2 as it is black. Similarly the quantity of radiant emission from A_2 reaching A_1 and absorbed by it is given by

$$Q_{2-1} = A_2 F_{2-1} E_{b2} \qquad \text{...[13.18(b)]}$$

When surfaces A_1 and A_2 are at the same temperature, these two quantities should be equal as no net heat transfer is possible.

i.e., $E_{b1} = E_{b2}$ for this condition.

As A_1, A_2, F_{2-1} are not zero

$$\therefore \quad A_1 F_{1-2} = A_2 F_{2-1} \qquad \text{...(13.19)}$$

This is known as reciprocity theorem for shape factors and will be used extensively in many situations.

The net heat exchange

$$Q = Q_{1-2} - Q_{2-1} = A_1 F_{1-2} (E_{b1} - E_{b2}) \qquad \text{...(13.20)}$$

Consider the two surfaces A_1 and A_2. In order to determine the shape factor first consider the differential area dA_1 and dA_2 as shown in Fig. 13.13(*a*). The line joining the centre of these two areas has a length r. The normal to the surfaces make angles θ_1 and θ_2 with this line.

The areas when projected in the direction perpendicular to the line will equal $dA_1 \cos\theta_1$ and $dA_2 \cos\theta_2$. Let the area dA_2 subtend a solid angle $d\omega$ with respect to dA_1. Then the energy emission from dA_1 reaching dA_2 is equal to

$$dq_{dA_1-dA_2} = I_1 \cos\theta_1\, dA_1 \,.\, d\omega$$

$$= I_1 \cos\theta_1\, dA_1 \,.\, \cos\theta_2\, dA_2/r^2 \quad ...[13.21(a)]$$

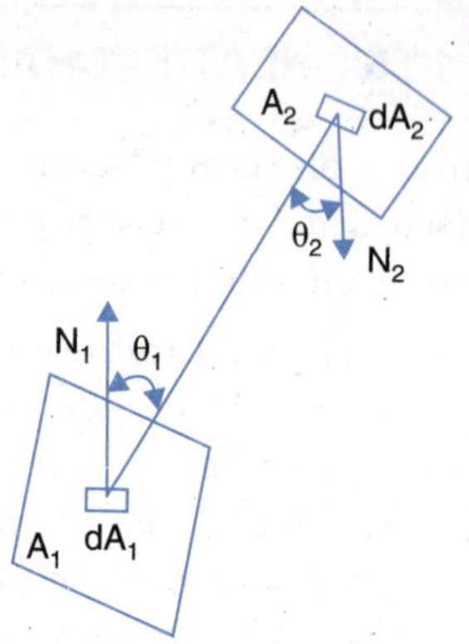

Fig. 13.13 (a) Shape factor determination

Similarly

$$dq_{dA_1-dA_2} = I_2 \cos\theta_2\, dA_2 \,.\, \cos\theta_1\, dA_1/r^2 \quad ...[13.21(b)]$$

The net flow is the difference between these quantities.

Replacing $I = E/\pi$

$$Q_{1-2} = (E_{b1} - E_{b2}) \int_{A_1}\int_{A_2} \frac{\cos\theta_1 \cos\theta_2\, dA_1\, dA_2}{\pi r^2}$$

This is also equal to $(E_{b1} - E_{b2}) A_1 F_{1-2}$

$$\therefore \quad F_{1-2} = \frac{1}{A_1}\int_{A_1}\int_{A_2} \frac{\cos\theta_1 \cos\theta_2\, dA_1\, dA_2}{\pi r^2}$$

The double integral can be evaluated directly only in simple cases. However the values are available for some important configurations in chart form and is determined using the geometric prameters. Perpendicular rectangles or squares with common edge, equal parallel disks or rectangles are the two important configurations for which chart form solutions are available. Solutions are available also for surface element to a large parallel area, small sphere to a parallel plane, for a plane and rows of tubes parallel to it, and unequal coaxial parallel circular disks.

13.5.2. Shape Factor for Special Cases

The shape factor is equal to one for any surface completely enclosed by another surface. The shape factor is also one for infinite parallel planes, from inner cylinder to outer cylinder of long coaxial cylinders and inner to outer concentric spheres. The reciprocity theorem can be used to find the shape factor of outer to inner cylinder or sphere.

In the case of long coaxial cylinder $F_{i-o} = 1$

$$\therefore \quad A_i F_{io} = A_o F_{oi} \quad \therefore \quad F_{o1} = \frac{A_i}{A_o} = \frac{r_i}{r_o} \times 1$$

As the radiation from outer cylinder cannot go out and as only (r_i/r_o) reaches the inner cylinder, the other portion should reach the outer cylinder itself due to the curvature.

So $\quad F_{o-o} = 1 - \dfrac{r_1}{r_o}$

Similarly in the case of concentric spheres, $F_{i-o} = 1$.

$$\therefore \quad F_{oi} = 1 \times \frac{r_i^2}{r_o^2} \quad ...[13.22(a)] \qquad \text{and } F_{o.o} = 1 - \frac{r_i^2}{r_o^2} \quad ...[13.22(b)]$$

The concept of shape factor to itself occurs for all curved surfaces.

For small element dA_1 to a parallel circular disk of radius a refer Fig. 13.13(*b*) at a distance L, the shape factor is $a^2/(a^2 + L^2)$. ...[13.23(*a*)]

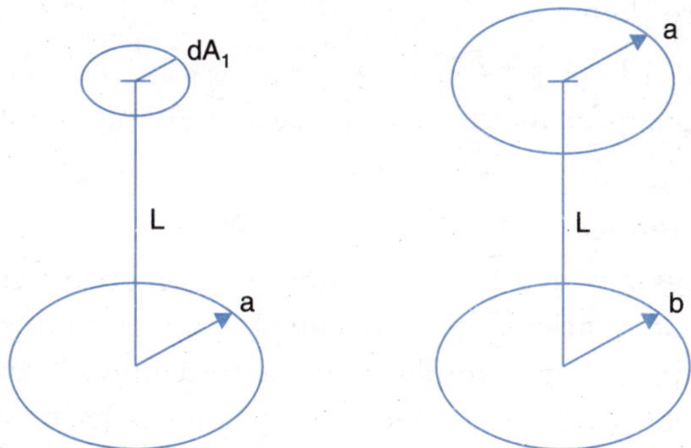

Fig. 13.13 (b)

For two unequal coaxial parallel circular disks of radius a and b at a distance L from each other Ref. Fig. 13.13(*b*) the shape factor from smaller to larger disk of radius b is given by

$$F_{b-a} = \frac{1}{2}\left[S - \left\{S^2 - 4\frac{r_j^2}{r_i^2}\right\}^{0.5}\right] \qquad ...[13.23(b)]$$

where
$$S = \frac{r_j^2 + r_j^2 + L^2}{r_1^2}$$

Crossed string method: The shape factors for surfaces that are two dimensional and infinitely long in one direction with identical cross-sections normal to the infinite direction can be determined using a method called crossed string method due to Hottel and Sarofin.

The configuration is shown in Fig. 13.14.

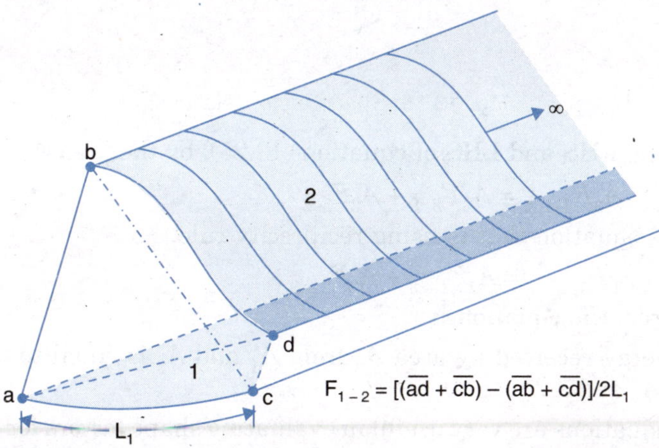

Fig. 13.14. Schematic diagram illustrating the crossed-string method

$$F_{1\text{-}2} = \text{[sum of diagonal lengths} - \text{sum of end lengths]}/2 \times \text{width of plane}$$

$$F_{1\text{-}2} = [(ad + cb) - (ab + cd)]/2L_1 \quad ...(13.24)$$

This method can be used for inclined planes which are longer compared to the distance between them.

The Summation Rule: When several surfaces together form an enclosure the sum of the shape factors from surface 1 to surfaces forming the enclosure will equal to one. Say if 4 surfaces form an enclosure then

$$F_{1-2} + F_{1-3} + F_{1-4} = 1$$

This rule is found helpful in the determination shape factor in certain cases like a closed cylinder.

13.5.3. Shape Factor Relationships

As the shape factor values are available for limited geometric situations only, it becomes necessary to use some basic relationships between shape factors to evaluate the shape factor for other connected geometries. For example shape factor value are available for perpendicular surfaces with a common edge. But shape factor values for perpendicular surfaces which will meet only if extended, may be needed. The shape factor relationship together with the reciprocity theorem are used to evaluate shape of factor value in such situations.

Consider surfaces A_1, A_2 and A_3 shown in Fig. 13.15. The first of such rules is

$$F_{3\text{-}1,\,2} = F_{3\text{-}1} + F_{3\text{-}2} \quad ...(13.25)$$

This is an obvious relation as the energy reaching an area is the sum of energies reaching individual parts of the area. Generally

$$F_{i\text{-}j,\;k,\;l,\;m,\;n.....} = F_{i\text{-}j} + F_{i\text{-}k} + F_{i\text{-}j} + F_{i\text{-}m} + ... \quad ...(13.26)$$

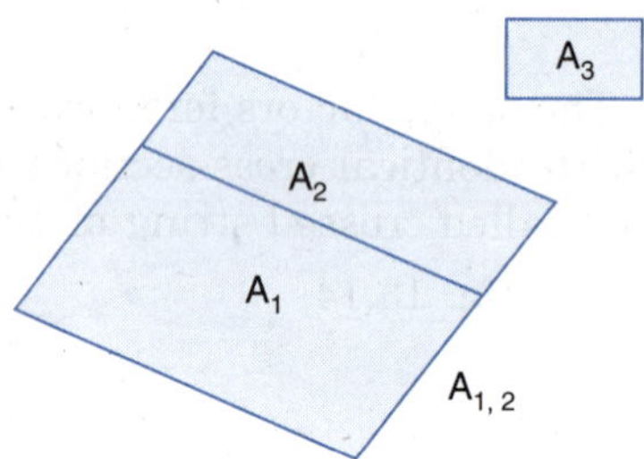

Fig. 13.15. Shape factor relations

Multiplying the RHS and LHS of equation (13.25), by the area A_3

$$A_3F_{3\text{-}1,\,2} = A_3F_{3\text{-}1} + A_3F_{3\text{-}2} \quad ...(13.27)$$

Inverting this equation that is using reciprocity rule,

$$A_{1,\,2-3}\,F_{1,\,2-3} = A_1F_{1-3} + A_2F_{2-3} \quad ...[13.27(a)]$$

Stated in words, the equation is

"The total energy received by area A_3 from A_1 and A_2 is equal to the sum of energies received from area A_1 and A_2"

These there equations are very useful in evaluating shape factors for various geometries.

Various charts for shape factors follow:

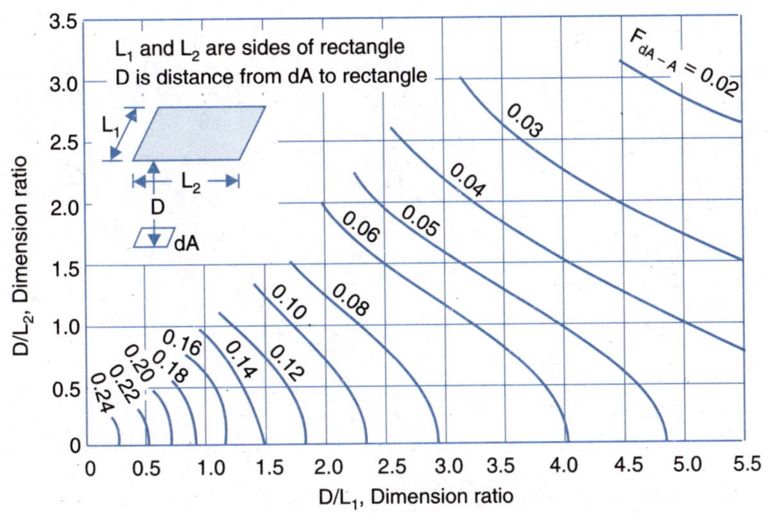

Fig. 13.16. Shape factor for a surface element *dA* and a rectangular surface *A* parallel to it

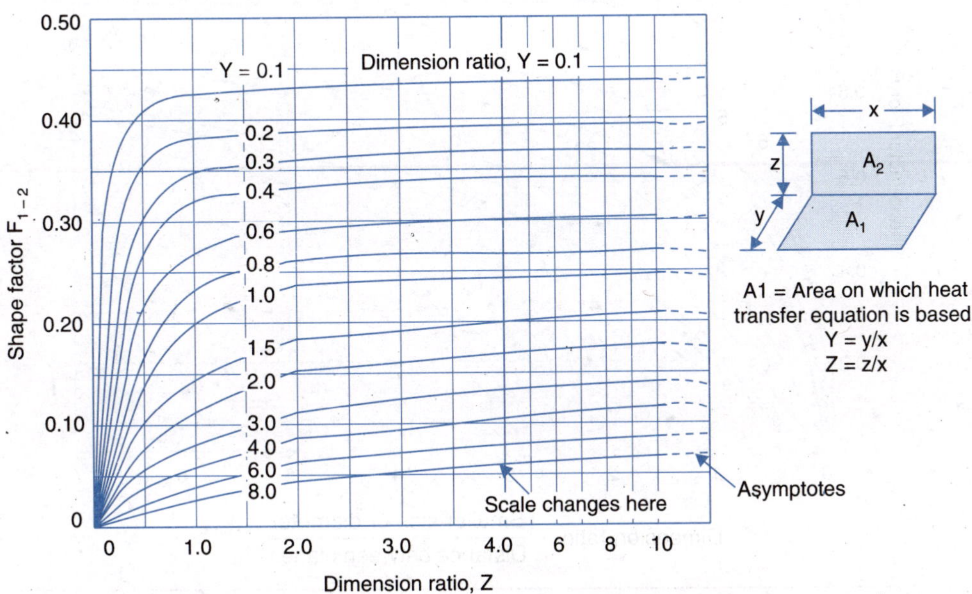

Fig. 13.17. Shape factor for adjacent rectangles in perpendicular planes sharing a common edge

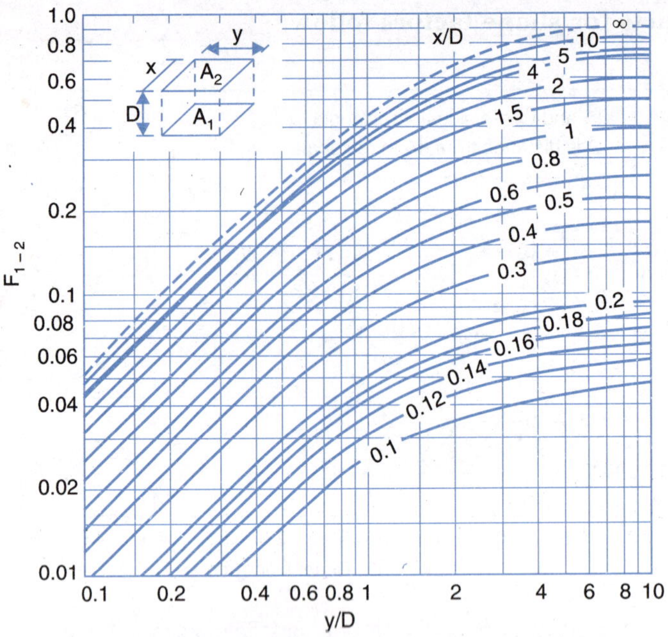

Fig. 13.18. Shape factor for directly opposed rectangles

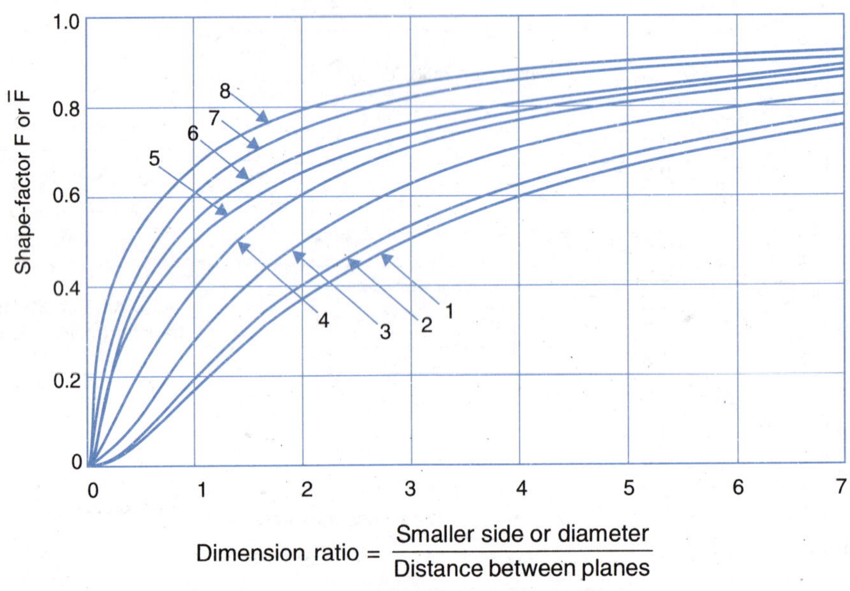

Radiation between parallel planes, directly opposed:

- 1, 2, 3, and 4: Direct radiation between the planes, F
- 5, 6, 7, and 8: Planes connected by nonconducting but reradiating walls, $\overline{F}$
- 1 and 5: Disks
- 3 and 7: 2: 1 Rectangles
- 2 and 6: Squares
- 4 and 8: Long, narrow rectangles

Fig. 13.19. Shape factors for equal and parallel squares, rectangles, and disks

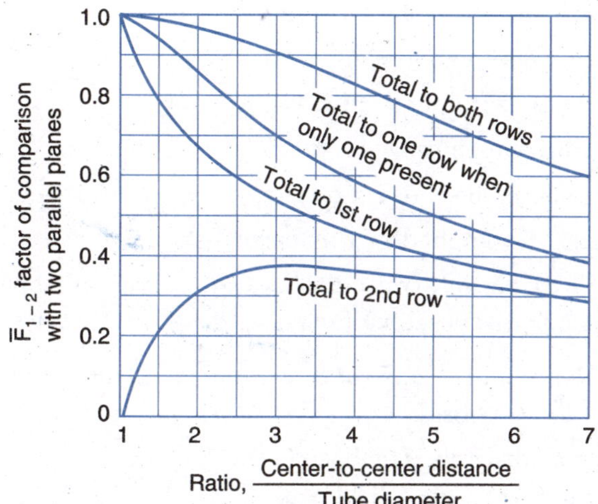

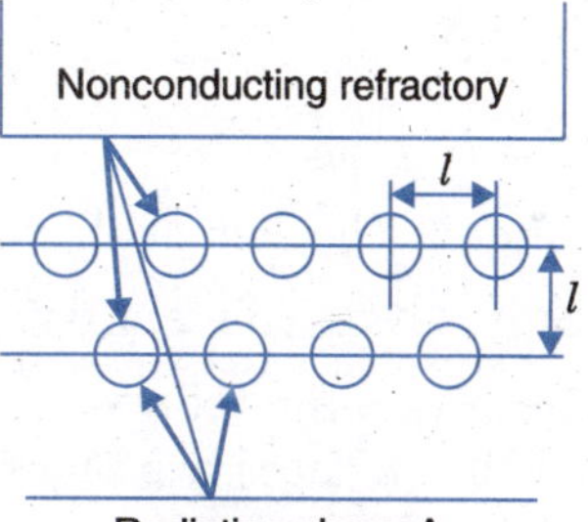

Fig. 13.20. Shape factor for a plane and one or two rows of tubes parallel to it

Example 13.13: *Determine the shape factor from the floor of a furnace of 1 m × 2 m × 1.5 m size to the side surfaces and to the roof.*

Solution: The shape factors **(Also charts and Tables from Data Book are used in all problems).**

F_{1-2}, F_{1-3} and F_{1-4} are to be determined. Refer Fig. Ex. 13.13.

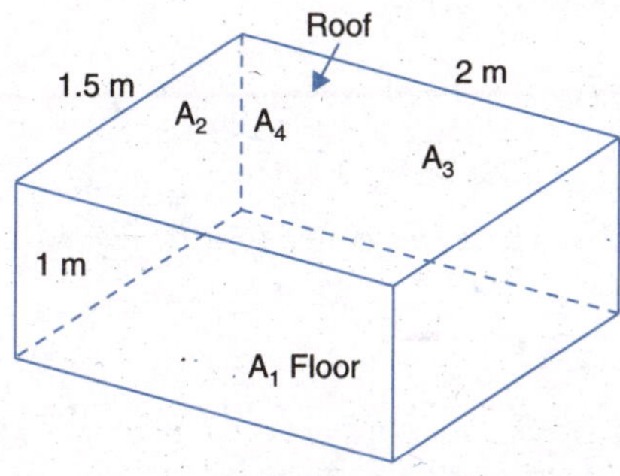

Fig. Ex. 13.13

The parameters for perpendicular surfaces are **(for floor to end wall)**

Z = height of the vertical surface/width = 1/1.5 = 0.67

Y = length of the horizontal surface/width = 2/1.5 = 1.33

As read from the chart the corresponding value of $\mathbf{F_{1-2} = 0.140}$

For F_{1-3} **(Floor to side wall)**

$Z = 1/2.0 = 0.5$

$Y = 1.5/2.0 = 0.75$

The value of F_{1-3} read from chart = **0.180**

For parallel rectangles (**floor to ceiling,** F_{1-4}) the parameters

$$X = \text{longer side/distance between planes} = 2/1 = 2.0$$

$$Y = \text{shorter side/distance between planes} = 1.5/1 = 1.5$$

The corresponding value F_{1-4} = **0.36** (from chart)

Check: The sum of all shape factors should be equal to one $(2 \times 0.14 + 2 \times 0.18 + 0.36) =$

Summation rule and symmetry rules are applied in this case.

Heat exchange between surfaces can be determined if temperatures are specified provided the surfaces are black (by equations discussed so far).

Example 13.14: *Determine the shape factor from the base of a cylinder to the curved surface. Also find the shape factor from curved surface to base and the curved surface to itself.*

Solution: The diameter is 1 m and height is also 1 m. The base (1) is enclosed by the top (3) and curved surface (2) (Fig. Ex. 13.14)

$$\therefore \qquad F_{1-2} + F_{1-3} = 1$$

F_{1-3} can be determined by using the chart for parallel disks. The ratio, diameter/distance between planes = 1.

The corresponding value of shape factor **is 0.17.** Base to curved surface is

$$\therefore \qquad \mathbf{F_{1-2} = 1.0 - 0.17 = 0.83}$$

Fig. Ex. 13.14

Using reciprocity theorem

$$A_1F_{1-2} = A_2F_{2-1}$$

$$\frac{\pi \times 1 \times 1}{4} \times 0.87 = \pi \times 1 \times 1 \times F_{2-1}$$

$$\therefore \qquad \mathbf{F_{2-1} = 0.2175}$$

Considering the curved surface, as concave surface will intercept some radiation from the surface itself.

$$F_{2-1} + F_{2-3} + F_{2-2} = 1$$

As $\qquad \mathbf{F_{2-1}} = F_{2-3}, F_{2-2} = 1 - 2 \times 0.2175 = \mathbf{0.565}$

Concave surfaces intercept part of radiation emitted by themselves. Here it intercepts more than half of the radiation.

Example 13.15: *Determine the shape factor from the surface 1 to surface 3 shown in Fig. Ex. 13.15 (vertical plane and non touching horizontal surface).*

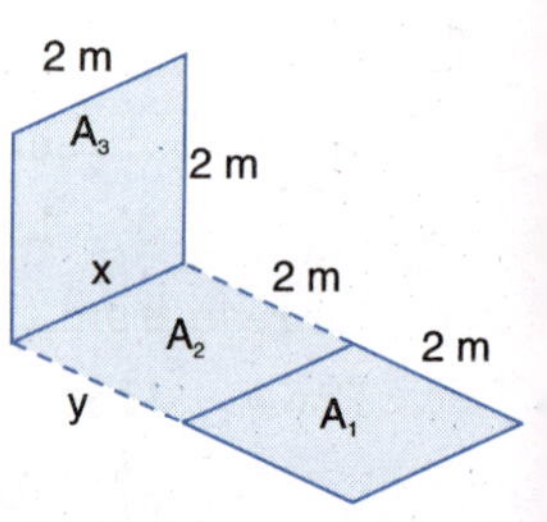

Fig. Ex. 13.15

Solution: Denoting surface in between as 2 and using equation (13.27)

$$(A_1 + A_2)\, F_{1,2-3} = A_1\, F_{1-3} + A_2\, F_{2-3}$$

F_{12-3} and F_{2-3} can be determined

Using the chart Fig. 13.17

For $\qquad F_{1,\,2-3},\ Y = y/x = 4/2 = 2$

$$Z = z/x = 2/2 = 1$$

$\therefore$ $\mathbf{F_{1,\,2\text{–}3} = 0.11643}$

For $F_{2\text{–}3}$, $Y = 2/2 = 1$, $Z = 2/2 = 1$

$\mathbf{F_{2\text{–}3} = 0.20004}$

Substituting

$$(2 \times 2 + 2 \times 2)\ 0.11643 = 2 \times 2F_{1\text{–}3} + 2 \times 2 \times 0.20004$$

$\therefore$ $\mathbf{F_{1\text{–}3} = 0.03282}$

To find $F_{3\text{–}1}$, $A_1 F_{1\text{–}3} = A_3 F_{3\text{–}1}$. In this case, the areas are equal and so $\mathbf{F_{3\text{–}1} = 0.03282.}$

Example 13.16: *Determine the shape factor between the floor and the 45° roof over a long corridor. The width is 2 m and the height on the lower sides is 2 m. (Fig. Ex. 13.16)*

Solution: In this case the crossed string method is to be used. (Approximate)

$$F_{1\text{–}2} = [(ad + bc) - (ab + cd)]/2L$$

$$ad = \sqrt{(2^2 + 4^2)} = 4.472$$

$$bc = \sqrt{(2^2 + 2^2)} = 2.828$$

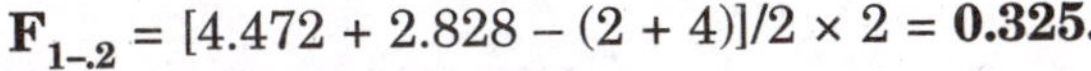

$$\mathbf{F_{1\text{–}2}} = [4.472 + 2.828 - (2 + 4)]/2 \times 2 = \mathbf{0.325.}$$

Fig. Ex. 13.16

13.5.4. Non Absorbing Reradiating Surface

Very often radiating surfaces may be enclosed by surfaces which are well insulated such that very little heat passes through them. The radiant heat received by such surfaces will be reflected back to the surfaces surrounding them. Such surfaces increase the heat transfer between surfaces enclosed by them (Fig. 13.21). For example consider two parallel surfaces. Let the shape factor be $F_{1\text{–}2}$. If these two surfaces are enclosed by four surrounding surfaces which are well insulated, the apparent value of $F_{1\text{–}2}$ will increase. It will not reach a value of one of course because of reradiations to the radiating surface. In the case of two disks 1 m dia and 0.5 m apart, about 37% of radiant emission from surface 1 will reach surface 2. If the surfaces are enclosed by reradiating surface the heat reaching surface 2 increases to 65%. The reradiating surface will be at an intermediate temperature between T_1 and T_2. The geometric shape factor however depends only on the geometric parameters.

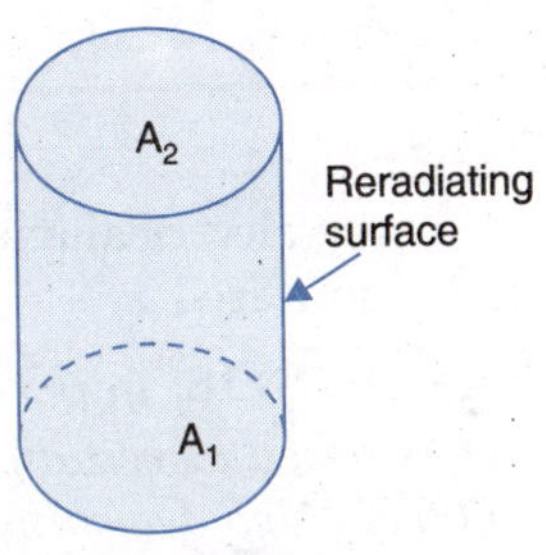

Fig. 13.21

Reradiating surfaces provide an additional parallel path for radiant heat flow, thus increasing the heat transfer.

13.6 RADIANT HEAT EXCHANGE BETWEEN BLACK SURFACES

Equation (13.20) can be used to determine radiant heat exchange between black surfaces.

$$Q_{1-2} = A_1 F_{1-2} (E_{b1} - E_{b2})$$

This can be represented by **electrical analogue** shown in Fig. 13.22.

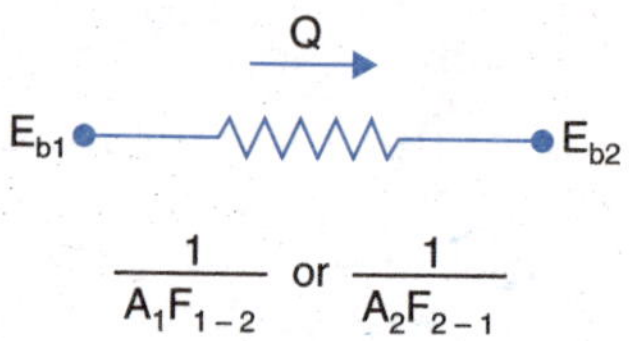

Fig. 13.22. Analogue circuit for radiation

The temperatures and geometric parameters should be specified for solution.

Example 13.17: *Using data of example 13.13 determine the heat exchange between the floor and roof. The shape factor between the floor (2 × 1.5 m²) and roof of the furnace is found as 0.360. The temperatures are specified as 1000 K and 600 K.*

Solution:

$$E_{b2} = 5.67 \times (600/100)^4 = 7348.32 \text{ W/m}^2$$

$$E_{b1} = 5.67 \times \left(\frac{1000}{100}\right)^4 = 56700 \text{ W/m}^2$$

$$\frac{1}{A_1F_{1-2}} = \frac{1}{(2 \times 1.5 \times 0.36)} = 1.08 \text{ m}^{-2}$$

$$Q = \frac{56700 - 7348.32}{1.08} = 45696 \text{ W}$$

The electrical analogy is very useful in solving for situations where reradiating surfaces are encountered.

Example 13.18: *In the case of example 13.13 the side walls are well insulated so that these can be considered as reradiating. Determine the heat exchange if the floor is at 1000 K and the roof at 600 K. Also determine the apparent shape factor.*

Solution: The equivalent circuit is shown in Fig. Ex. 13.18(*a*). The temperature of the reradiating surfaces is not known, but is assumed as T_3. Data is taken from 13.13.

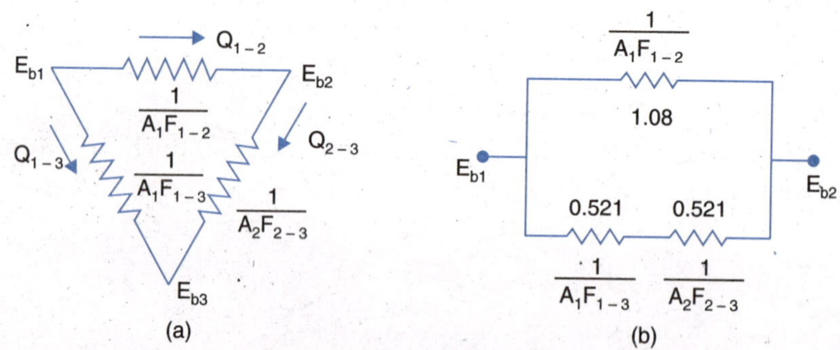

Fig. Ex. 13.18

In addition to the direct flow between 1 and 2 an additional path is now available for the heat flow. Clubbing all the sides as surface 3.

$$F_{1\text{-}2} = 0.360 \quad \therefore \quad F_{1\text{-}3} = 1 - 0.360 = 0.640$$

$$\frac{1}{A_1F_{1\text{-}2}} = 1.08 \text{ m}^{-2}$$

$$\frac{1}{A_2F_{2\text{-}3}} = \frac{1}{A_1F_{1\text{-}3}} = \frac{1}{2 \times 1.5 \times 0.64} = 0.521 \text{ m}^{-2}$$

The circuit can now be represented as in Fig. Ex. 13.18(*b*).

$$\text{Equivalent resistance} = \frac{1}{\dfrac{1}{1.08} + \dfrac{1}{1.42}} = 0.614$$

$$\therefore \qquad Q = \frac{56700 - 7348.32}{0.614} = 80401 \text{ W}$$

This is about twice the heat flow, as compared to direct radiation only.

The apparent shape factor will be:

$$Q = A_1F_{1\text{-}2}(E_{b1} - E_{b2})$$

Solving $\qquad F_{1\text{-}2} = 0.5430$

Note. The total emissive power of the floor surface is 3 × 56700 = 170100 W and that of the roof is 22045 W. The difference between these two and Q goes to heating of the reradiating surfaces.

Example 13.19: *A furnace is in the shape of a cylinder of 1 m dia and 1 m height. The base is at 1000 K while the top is at 600 K. The curved surface is well insulated so that it can be taken as non absorbing reradiating surface (see example 13.14 for shape factors and example 13.18 for emissive power values). Determine the heat exchange between the base and the top with and without the reradiating surface.*

Solution: From example 13.14 the shape factor from base to top is found as 0.17. The shape factor form base or top to the curved surface is 0.83. The equivalent circuit is shown in Fig. Ex. 13.19.

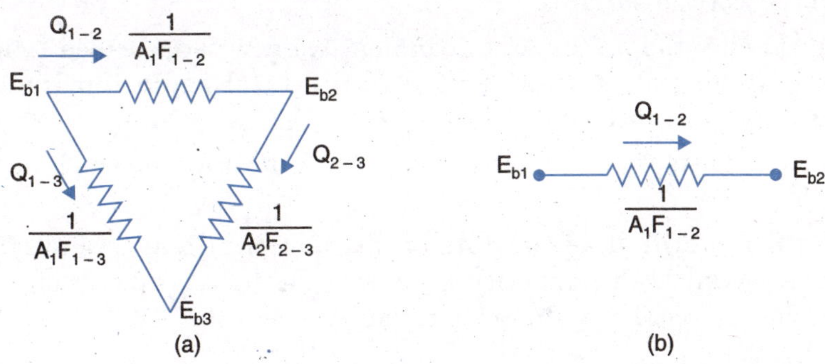

Fig. Ex. 13.19 (a) Circuit with reradiation (b) Circuit without reradiation

$$E_{b1} = 56700 \text{ W/m}^2$$

$$E_{b2} = 7348.32 \text{ W/m}^2$$

$$\frac{1}{A_1F_{1-2}} = \frac{1}{\dfrac{\pi \times 1^2}{4} \times 0.17} = 7.49$$

$$\frac{1}{A_2F_{2-3}} = \frac{1}{A_1F_{1-3}} = \frac{1}{\dfrac{\pi \times 1^2}{4} \times 0.83} = 1.534$$

With reradiation:

The equivalent resistance

$$R = \frac{1}{\dfrac{1}{7.49} + \dfrac{1}{2 \times 1.534}} = 2.1765$$

$$\therefore \qquad \mathbf{Q} = \frac{56700 - 7348.32}{2.1765} = \mathbf{22675\ W}$$

Without reradiation:

$$\mathbf{Q} = \frac{56700 - 7348.32}{7.49} = \mathbf{6589\ W}$$

This is about 30% of the heat flow with reradiation.

The **apparent shape factor** from base to top surface is

$$\mathbf{F_{1\text{-}2}} = \frac{22675}{56700} \times \frac{4}{\pi \times 1 \times 1} = \mathbf{0.5092.} \text{ (Compared to 0.17)}$$

13.7 HEAT EXCHANGE BY RADIATION BETWEEN GRAY SURFACES

In the case of black surfaces all the incident radiant energy is absorbed and hence the determination of heat exchange is rather simple. In the case of gray surfaces there are reflections and rereflections without end.

The calculation of heat exchange involves the summation of the energy absorbed on each incidence on the surface. Additional resistance to heat absorption is introduced by the emissivity/absorptivity of the surface.

In order to simplify the process of calculation two new terms called **"radiosity" and "irradiation"** are introduced. **Irradiation (G) is the total radiation incident upon a surface per unit time and unit area (W/m²).**

This quantity consists of the radiation from other surfaces and the reflected radiation from other surfaces.

Radiosity (J) is defined as the total radiation that leaves a surface per unit time and unit area (W/m²). This quantity consists of the emissive power of the surface and the reflections by the surface. From these definitions we get

$$\therefore \qquad J = \varepsilon E_b + \rho G \qquad \text{...[13.28}(a)\text{]}$$

as

$$\rho = 1 - \alpha = 1 - \varepsilon$$

$$J = \varepsilon E_b + (1 - \varepsilon)G \qquad \text{...[13.28}(b)\text{]}$$

In the calculation of heat transfer between gray surfaces an important assumption is hat radiosity and irradiation are uniform over the surface. Considering a heat balance over he surface, the net energy leaving the surface is the difference between radiosity and rradiation. This can be seen considering the boundary *AA* over the surface (Fig. 13.23). The et energy crossing this surface *AA* is the net energy flow from surface 1 or the net heat ransfer between the surfaces.

$$Q/A_1 = J_1 - G_1$$

ubstituting for *G* from equation (13.28(*b*)) and simplifying

$$Q = \frac{\varepsilon_1 A_1}{1-\varepsilon_1}(E_{b1} - J_1) = \frac{E_{b1} - J_1}{(1-\varepsilon_1)/A_1\varepsilon_1} \text{ W} \qquad \text{...[13.29(a)]}$$

Similarly for surface 2, consider surface *BB*

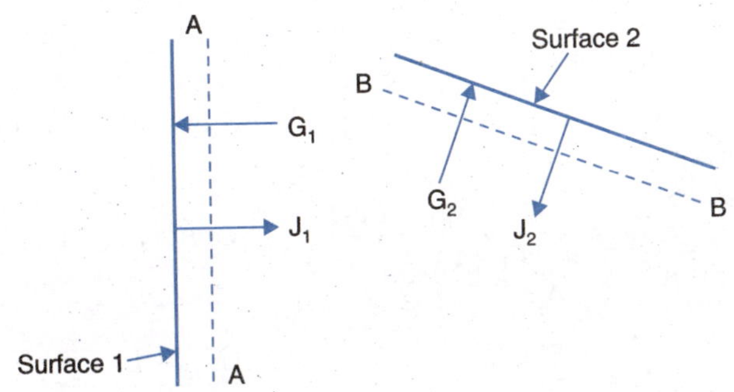

Fig. 13.23. Radiosity-Irradiation concept

$$Q = \frac{J_2 - E_{b2}}{(1-\varepsilon_2)/A_2\varepsilon_2} \text{ W} \qquad \text{...[13.29(b)]}$$

In these equations J_1 and J_2 are unknown.

The energy leaving surface *E*1 and reaching surface 2 is given by $J_1 A_1 F_{1-2}$. Similarly energy leaving the surface 2 and reaching surface 1 is given by $J_2 A_2 F_{2-1}$. So the net interchange between the surfaces *AA* and *BB* equals.

$$Q = J_1 A_1 F_{1-2} - J_2 A_2 F_{2-1},\ Q = \frac{J_1 - J_2}{1/A_1 F_{1-2}} \text{ W} \qquad \text{...[13.29(c)]}$$

Eliminating J_1 and J_2 using equations [13.29(*a*)] and [13.29(*b*)]

$$Q_{1-2} = \frac{E_{b1} - E_{b2}}{\dfrac{1-\varepsilon_1}{A_1\varepsilon_1} + \dfrac{1}{A_1 F_{1-2}} + \dfrac{1-\varepsilon_2}{A_2\varepsilon_2}} \qquad \text{...(13.30)}$$

These equations 13.29(*a*), (*b*) and (*c*) and (13.30) can be represented by an equivalent circuit as in Fig. 13.24.

E_{b1} — $\frac{1-\varepsilon_1}{A_1\varepsilon_1}$ — J_1 — $\frac{1}{A_1F_{1-2}}$ (Q →) — J_2 — $\frac{1-\varepsilon_2}{A_2\varepsilon_2}$ — E_{b2}

or

Fig. 13.24. Equivalent network for radiation between gray surfaces

Equation (13.30) is the most general form which covers radiation heat exchange between any two surfaces whether black or gray. The terms $\frac{1-\varepsilon}{A\varepsilon}$ is known as surface resistance and $1/A_1F_{1-2}$ as space resistance.

If $\varepsilon_1 = \varepsilon_2 = 1$, equation (13.30) reduces to the form already derived for black surfaces.

$$Q = A_1F_{1-2}\,(E_{b1} - E_{b2})$$

If $F_{1-2} = 0$, meaning that the surfaces do not see each other, then

$$Q = \Delta E_b/\infty = 0 \quad \text{obviously.}$$

13.7.1. Special Cases

1. **Infinite parallel planes:** here $F_{1-2} = 1$ and $A_1 = A_2 = 1$

$$\therefore \qquad q = \frac{(E_{b1} - E_{b2})}{\left(\frac{1}{\varepsilon_1} + \frac{1}{\varepsilon_2} - 1\right)} \qquad ...(13.31)$$

2. **Completely enclosed surfaces:** (long coaxial cylinders, concentric spheres etc., from inner to outer surface)

$$Q = \frac{(E_{b1} - E_{b2})}{\frac{1-\varepsilon_1}{A_1\varepsilon_1} + \frac{1}{A_1F_{1-2}} + \frac{1-\varepsilon_2}{A_2\varepsilon_2}}, \text{ as } F_{1-2} = 1$$

$$Q = \frac{A_1(E_{b1} - E_{b2})}{\frac{1}{\varepsilon_1} + \frac{A_1}{A_2}\left(\frac{1-\varepsilon_2}{\varepsilon_2}\right)} = \frac{A_1(E_{b1} - E_{b2})}{\frac{1}{\varepsilon_1} + \frac{A_1}{A_2}\left(\frac{1}{\varepsilon_2} - 1\right)} \qquad ...(13.32)$$

3. **Radiation Shields:** Any surface placed in between two surfaces introduces additional surface resistance reducing heat transfer. This is known as radiation shield and is extensively used in practice in furnace wall construction.

13.7.2. Heat Exchange Among More Than Two Surfaces

It is easy to visualise such exchanges using the resistance network.

Figure 13.25(a) shows such network for three surfaces seeing only each other ($F_{ii} = 0$)

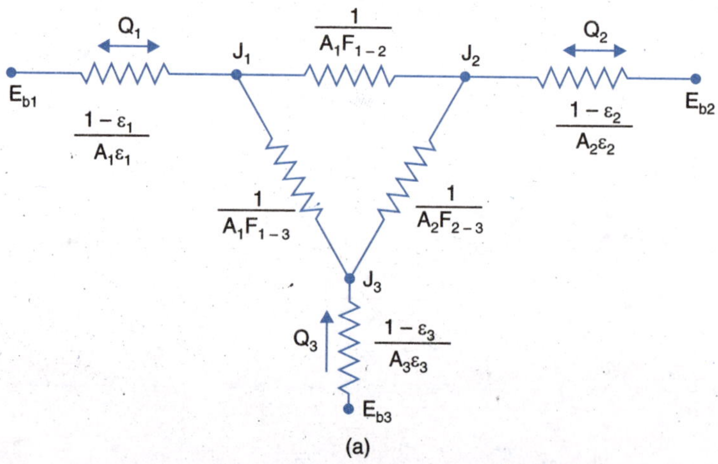

Fig. 13.25. Equivalent network for three gray surfaces which see each other only

These circuits can be solved using electrical network analysis. This concept can be extended to any number of surfaces also.

A special case of such surface is when one of the surfaces becomes a non-absorbing, reradiating one. As there is no net flow into and out of the surface 3, E_{b3} becomes equal to J_3 and the equivalent circuit is shown in Fig. 13.25(*b*).

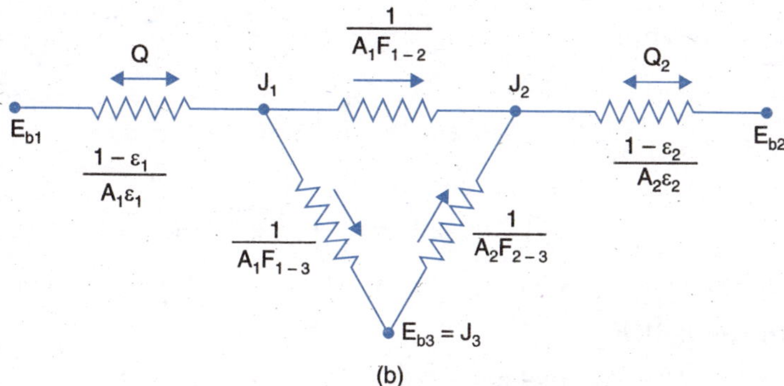

Fig. 13.25. Circuit with reradiation – 3 surface

This circuit can be easily solved for as a simple series parallel circuit (see example 13.18).

Example 13.20: *Consider the data of example 13.13. In a furnace of 2 × 1.5 × 1 m size, floor is at 1000 K and other surfaces are at 600 K. The surface emissivity for the floor is 0.8 and for the other surfaces it is 0.5. Determine the heat exchange by radiation from (i) floor to each of side walls and (ii) floor to roof.*

Solution: The shape factor values are taken from example 13.13 (Also refer to Fig. Ex. 13.13)

In this case heat flow is only from floor to other surfaces

From floor to smaller side walls (end)

$$F_{1-2} = 0.14$$

From floor to longer side walls

$$F_{1-3} = 0.18$$

From floor to ceiling

$$F_{1-4} = 0.360, \text{ check } (0.36 + 2 \times 0.18 + 2 \times 0.14) = 1$$

Heat flow to end walls from floor

$$Q = (E_{b1} - E_{b2}) / \left[\frac{1-\varepsilon_1}{A_1\varepsilon_1} + \frac{1}{A_1F_{1-2}} + \frac{1-\varepsilon_2}{A_2\varepsilon_2} \right]$$

$$E_{b1} = \sigma\, T_1^4 = 5.67 \times \left(\frac{1000}{100}\right)^4 = 56700 \text{ W/m}^2$$

$$E_{b3} = E_{b4} = E_{b2} = \sigma\, T_2^4 = 5.67 \times \left(\frac{600}{100}\right)^4 = 7348.32 \text{ W/m}^2$$

$$(1 - \varepsilon_1)/A_1\varepsilon_1 = (1 - 0.8)/2 \times 1.5 \times 0.8 = 0.08333$$

$$1/A_1F_{1-2} = 1/2 \times 1.5 \times 0.14 = 2.381$$

$$(1 - \varepsilon_2)/A_2\varepsilon_2 = (1 - 0.5)/(1.5 \times 1 \times 0.5) = 0.6667$$

$$\therefore \qquad \mathbf{Q_{1-2}} = \frac{(56700 - 7348.32)}{0.08333 + 2.381 + 0.6667} = \mathbf{15762\ W}$$

This is the heat flow to the two end walls each.

Heat flow to side walls

$$(1 - \varepsilon_1)/A_1\varepsilon_1 = 0.08333$$

$$1/A_1F_{1-3} = 1/2 \times 1.5 \times 0.18 = 1.8519$$

$$(1 - \varepsilon_3)/A_3\varepsilon_3 = (1 - 0.5)/2 \times 1 \times 0.5 = 0.5$$

$$\therefore \qquad \mathbf{Q_{1-3}} = (56700 - 7348.32)/(0.08333 + 1.8519 + 0.5) = \mathbf{20266\ W.}$$

This is to each of the side walls.

Heat flow to the ceiling

$$(1 - \varepsilon_1)/A_1\varepsilon_1 = 0.08333$$

$$1/A_1F_{1-4} = 1/(2 \times 1.5 \times 0.36) = 0.9259$$

$$(1 - \varepsilon_4)/A_4\varepsilon_4 = (1 - 0.5)/2 \times 1.5 \times 0.5 = 0.3333$$

$$\therefore \qquad \mathbf{Q_{1-4}} = (56700 - 7348.32)/(0.08333 + 0.9259 + 0.3333) = \mathbf{36759\ W.}$$

Example 13.21: *Determine the heat transfer from floor to ceiling in example 13.18, if all the side walls act as reradiating walls. Also determine the average temperature of reradiating walls. The equivalent circuit is shown in Fig. Ex. 13.21.*

Solution: Also refer Fig. 13.12

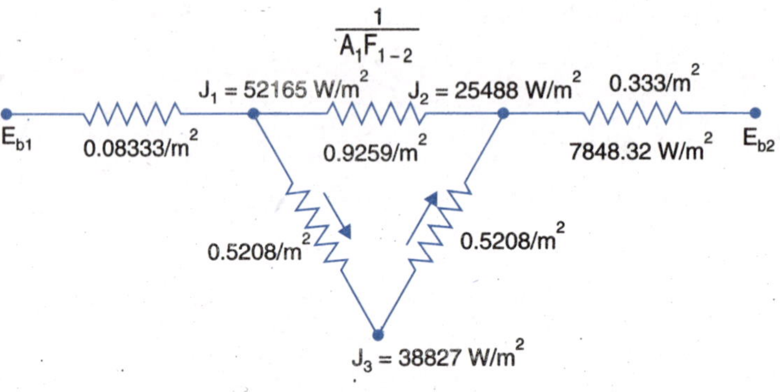

Fig. Ex. 13.21

Using the values of example 13.20

$$(1 - \varepsilon_1)/A_1\varepsilon_1 = 0.08333$$

$$1/A_1F_{1-4} = 1/(2 \times 1.5 \times 0.36) = 0.9259$$

$$(1 - \varepsilon_4)/A_4\varepsilon_4 = (1 - 0.5)/2 \times 1.5 \times 0.5 = 0.3333$$

$$\frac{1}{A_1F_{1-3,2}} = \frac{1}{2 \times 1.5 \times (2 \times 0.14 + 2 \times 0.18)} = 0.5208 = \frac{1}{A_4F_{4-3,2}}$$

The equivalent resistance between J_1 and J_2 is

$$R = \frac{1}{\frac{1}{0.9259} + \frac{1}{2 \times 0.5208}} = 0.4902$$

Total resistance between E_{b1} and E_{b2}

$$\mathbf{R_T} = 0.08333 + 0.4902 + 0.3333 = 0.9068$$

$$\therefore \qquad \mathbf{Q} = \frac{56700 - 7348.32}{0.9068} = \mathbf{54422\ W}$$

as compared to 36759 W. The circuit is shown in Fig. Ex. 13.22

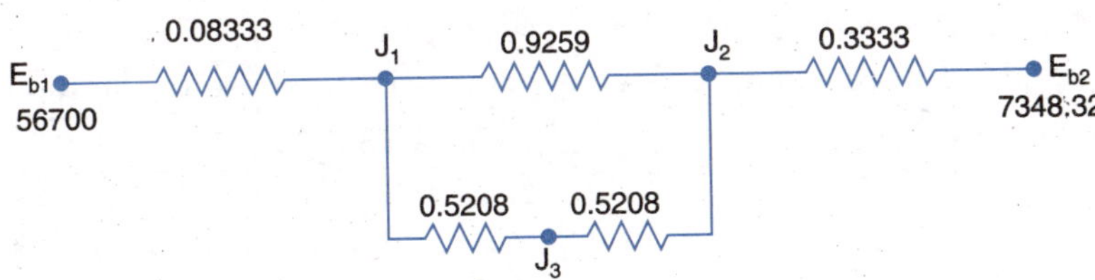

Fig. Ex. 13.22

In order to determine the temperature of reradiating surfaces, J_3 should be determined. This is done by considering the nodes J_1, J_3, J_2 and equating the flow into each node to zero.

Considering node J_1

$$\frac{E_{b1} - J_1}{0.08333} + \frac{J_2 - J_1}{0.9259} + \frac{J_3 - J_1}{0.5208} = 0 \qquad \text{...(A)}$$

For node J_2,

$$\frac{J_1 - J_2}{0.9259} + \frac{J_3 - J_2}{0.5208} + \frac{E_{b2} - J_4}{0.333} = 0 \qquad \text{...(B)}$$

For node J_3

$$\frac{J_1 - J_3}{0.5208} + \frac{J_2 - J_3}{0.5208} = 0 \qquad \text{...(C)}$$

Equation (C) when solved gives

$$J_3 = (J_1 + J_4)/2 \qquad \text{...(D)}$$

Substituting (D) is equation (A)

$$12E_{b1} - 14.04J_1 + 2.04J_2 = 0 \qquad \text{...(E)}$$

Substituting (D) in equation (B)

$$3E_{b4} + 2.04J_1 - 5.04J_2 = 0 \qquad \text{...(F)}$$

Solving for J_1 and J_2 using equation (E) and (F)

$$J_1 = 52165 \text{ W/m}^2, \quad J_2 = 25488 \text{ W/m}^2, \quad J_3 = 38827 \text{ W/m}^2$$

The corresponding temperatures can be determined using

$$\sigma T^4 = J$$

The values are:

Node J_1 $T = 979.4$ K, Node J_2, $T = 818.8$ K

Node J_3 or the **reradiating wall temperature = 909.7 K**

A check for the values:

$$Q = (E_{b1} - J_1)/[(1 - \varepsilon_1)/A_1\varepsilon_1]$$

$$= [56700 - 52165]/0.08333 = 54420 \text{ W}$$

So the values should be correct. Check between J_1 and J_2, J_3 and E_{b2} also.

Example 13.22: *Consider the cylindrical furnace of example 13.19. If emissivities of the base and top surfaces are 0.8 and 0.5, determine the heat exchange.*

Solution: As determined in the example shape factor from the base to the top is 0.17.

$$\therefore \quad Q = \frac{56700 - 7348.32}{\dfrac{(1-0.8)4}{(\pi\times 1\times 1)\times 0.8} + \dfrac{1\times 4}{\pi\times 1\times 1\times 0.17} + \dfrac{(1-0.5)4}{\pi\times 1\times 1\times 0.5}} = \mathbf{5434.5\ W}$$

Compared to 6589 for black surface.

Example 13.23: *Two large parallel planes are at 1000 K and 600 K. Determine the heat exchange per unit area. (i) if surfaces are black (ii) if the hot one has an emissivity of 0.8 and the cooler one 0.5 (iii) if a large plate is inserted between these two, the plate having an emissivity of 0.2.*

Solution: Case (*i*): The equivalent circuit is shown in Fig. Ex. 13.23(*a*)

$$Q = \sigma A\, F_{1-2}\,(T_1^4 - T_2^4). \text{ As } F_{1-2} = 1$$

for large parallel surfaces, considering unit area.

$$\frac{Q}{A} = 5.67 \times 1 \left[\left(\frac{1000}{100}\right)^4 - \left(\frac{600}{100}\right)^4\right] = 49352 \text{ W/m}^2$$

Case (*ii*): The equivalent circuit is shown in Fig. Ex. 13.23(*b*)

$$\frac{Q}{A} = \frac{(E_{b1} - E_{b2})}{\dfrac{1-\varepsilon_1}{A_1\varepsilon_1} + \dfrac{1}{A_1F_{1-2}} + \dfrac{1-\varepsilon_2}{A_2\varepsilon_2}} = \frac{56700 - 7348.32}{\dfrac{0.2}{1\times 0.8} + 1 + \dfrac{0.5}{0.5}} = 21934 \text{ W/m}^2$$

Case (*iii*): The equivalent circuit is shown in Fig. 13.23(*c*)

$$\frac{Q}{A} = \frac{56700 - 7348.32}{\dfrac{1-0.8}{0.8} + 1 + \dfrac{1-0.2}{0.2} + \dfrac{1-0.2}{0.2} + 1 + \dfrac{1-0.5}{0.5}} = 4387 \text{ W/m}^2$$

A considerable heat flow reduction is obtained by shielding. This method finds application in insulation. The shields should have low absorptivity and high reflectivity for economical applications.

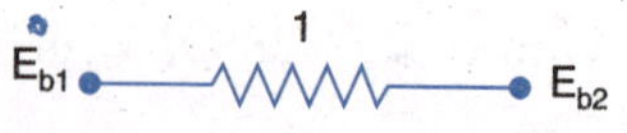

E_{b1} — 1 — E_{b2}

(a)

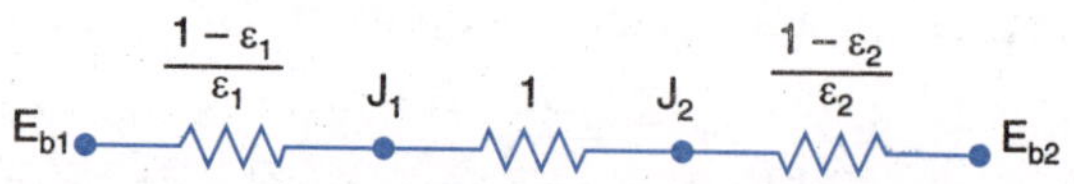

E_{b1} — $\frac{1-\varepsilon_1}{\varepsilon_1}$ — J_1 — 1 — J_2 — $\frac{1-\varepsilon_2}{\varepsilon_2}$ — E_{b2}

(b)

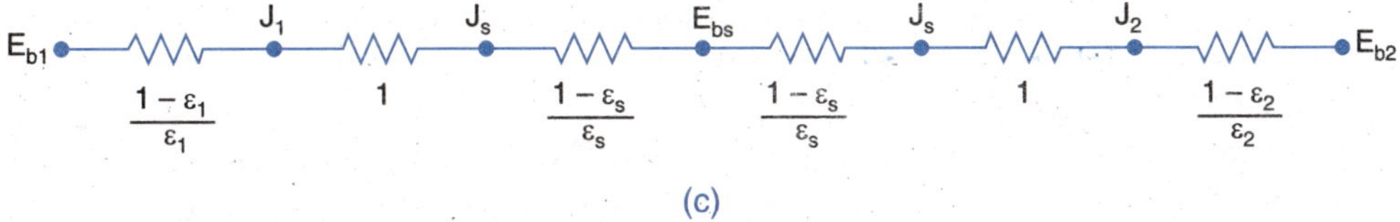

(c)

Fig. Ex. 13.23 (a) Black surface (b) gray surface (c) with shield

Example 13.24: *Two perpendicular planes of each 2 m × 2 m meet at a common edge. The horizontal plane has an emissivity of 0.4 and the vertical plane has an emissivity of 0.6. If these planes are at 400 K and 600 K determine the radiation exchange.*

Solution: Refer example 13.14. The shape factor from the horizontal to the vertical plane is 0.20004 (From chart). The equivalent circuit is shown in several examples.

The heat flow is given by

$$Q = \frac{5.67\left[\left(\frac{600}{100}\right)^4 - \left(\frac{400}{100}\right)^4\right]}{\frac{1-0.4}{4\times0.4} + \frac{1}{4\times0.20004} + \frac{1-0.6}{4\times0.6}} = \mathbf{3292\ W.}$$

13.8 EFFECT OF RADIATION ON MEASUREMENT OF TEMPERATURE BY A BARE THERMOMETER

When a bare thermometer is used to measure the temperature of a flowing gas, errors will be introduced if the container or duct wall temperature is different from that of the gas.

The thermometer when it comes to thermal equilibrium will not be at the gas temperature for this situation. As it will loose heat by radiation to the walls its temperature will lie between the gas temperature and wall temperature. Using the following notations, the energy balance can be written as in equation (13.33).

T_g – gas temperature, K, T_w – wall temperature, K

T_t – thermometer temperature, K, A_t – thermometer surface area (bead or bulb)

$$h\,A_t(T_g - T_t) = A_t\,\sigma\,\varepsilon_t\,[T_t^4 - T_w^4] \qquad ...(13.33)$$

The error in measurement is $(T_g - T_t)$ and this can be reduced by increasing h or by using a shield between the thermometer and the wall, so that the thermometer does not see the wall. It is better to have the shield made of surface with high reflectivity or low emissivity. The shape factor for the thermometer to the walls is unity. If the quantities are specified T_g or T_t can be calculated and the error can be established.

In order to reduce the error a shield is placed over the thermometer. If the area of the shield is A_s and temperature T_s, then two equations for the energy balance of the thermometer and that of the shield can be written as follows.

For the shield

$$2\,A_s\,h_s(T_g - T_s) + A_t\,\sigma\,\varepsilon_t\,[T_t^4 - T_s^4] = A_s\,\varepsilon_s\,\sigma\,(T_s^4 - T_w^4) \qquad ...[13.34(a)]$$

For the thermometer:

$$h_t\,A_t(T_g - T_t) = \sigma\,A_t\,\varepsilon_t\,[T_t^4 - T_s^4] \qquad ...[13.34(b)]$$

In order to reduce error A_s should be large compared to A_t and ε_s should be small compared to ε_t.

Chapter 13

Example 13.25: ***A bare thermometer measuring the temperature*** *of a gas body reads 600 K. The surrounding walls are at 500 K. The thermometer bulb is 3 mm in dia and is spherical, its surface emissivity being 0.7. The convective heat transfer coefficient over the surface is 40 W/m²K. Determine the gas temperature and the error involved.*

Solution: Using equation (13.33),

$$40 \times 4 \times \pi \times (0.0015)^2 (T_g - 600) = 4\pi \times 0.0015^2 \times 5.67 \times 0.7 \left[\left(\frac{600}{100}\right)^4 - \left(\frac{500}{100}\right)^4\right]$$

Solving: $\mathbf{T_g = 666.6\ K}$

An error of 66.6 K is involved. This is rather high. This can be reduced using a shield and also by increasing 'h'. In this case if h = 80, then error will be 33.3 K.

13.9 MULTISURFACE ENCLOSURE

Consider an enclosure of N surfaces with $\alpha_1 \ldots \alpha_N$, $\rho_1 \ldots \rho_N$ and $\varepsilon_1 \ldots \varepsilon_N$ at temperature $T_1 \ldots T_N$ with radiation exchanges between them. Let the shape factors be determinable.

$$F_{1-1}, F_{1-N}, F_{2-1}, \ldots F_{2-N}, F_{N-1} \ldots E_{(N-N)} \text{ etc.}$$

working out the energy balance say for surface A.

Let G_i = irradiation on A_i W/m²

J_i = Radiosity on A_i W/m²

q_i = Net radiation exchange over surface A_i, W/m²

From definitions, and using equations (13.28)(*a*) and (*b*)

$$q_i = J_i - G_i \quad \text{...(13.35)}$$

$$J_i = \varepsilon_i E_{bi} + (1 - \varepsilon_i)G_i \quad \text{...(13.36)}$$

The energy leaving any surface say A_j that reaches surface A_i is given by

$$J_j A_j F_{j-i}$$

Using reciprocity theorem this also equals.

$$J_j A_i F_{i-j}$$

The radiation from all surfaces reaching A_i is given as the sum for j =1 to N *i.e.*,

$$A_i \sum_{j=1}^{N} J_j F_{i-J}$$

This amount divided by the area A_i gives G_i or

$$G_i = \sum_{j=1}^{N} J_j F_{i-J} \quad \text{...(13.37)}$$

Substituting for G_i in (13.36)

$$j_i = \varepsilon_i \cdot E_{bi} + (1 - \varepsilon_i) \sum_{j=1}^{N} J_i F_{i-J} \quad \text{...(13.38)}$$

Rearranging

$$E_{bi} = \frac{1}{\varepsilon_i} J_i - \frac{(1-\varepsilon_i)}{\varepsilon_i} \sum_{j=1}^{N} J_j F_{i-j} \quad \text{...(13.39)}$$

Substituting in the heat flow equation (13.35)

$$q_i = J_i - \sum_{j=1}^{N} J_j F_{i-j} \qquad ...(13.40)$$

by eliminating the summation term (equation 13.38)

$$q_i = \frac{\varepsilon_i}{1-\varepsilon_i} [E_{bi} - J_i] \qquad ...[13.40(a)]$$

These three equations are very useful for solving multienclosure examples.

When all surface temperatures and emissivities are specified, this results in N simultaneous equations with N variables (J_N). These equations can be solved by any of the Matrix methods. When J values are solved the heat flow from each surface can be solved for, using the heat flow equation.

Three special cases are to be recognised.

(*i*) A surface specified as black: In this case, $J_i = E_{bi} = \sigma T_i^4$ and so there will be $N - 1$ equations with $N - 1$ unknowns. The energy balance for the particular area will not yield a separate equation.

$$E_{bi} = \frac{1}{\varepsilon_i} J_i - \frac{1-\varepsilon_i}{\varepsilon_i} \sum_{j=1}^{N} J_j F_{i-j}$$

$$= J_i - \frac{0}{1}\left[\sum_{j=1}^{N} J_j F_{i-j}\right] = J_i$$

So J_i becomes a known quantity to be included in the other equations. (Note: F_{i-i} or F_{j-j} or F_{N-N} will be zero for flat or convex surfaces.)

(*ii*) When net heat flux is specified instead of temperature:

In this case the heat flux equation should be used for the surfaces for which net heat flux is specified.

$$q_i = J_i - \sum_{j=1}^{N} J_j F_{i-j}$$

This provides the relationship for the J_i value, and results in N equations and N unknowns.

The temperature T_i can be determined using

$$\sigma T_i^4 = J_i + \frac{1-\varepsilon_i}{\varepsilon_i} q_i$$

(*iii*) Reradiating surface is encountered: Generally in this case emissivity may not be specified as it is immaterial for the solution. Here also the net heat flux equation has to be used with net heat flux being zero. If surface i is reradiating surface, then

$$q_i = 0 = J_i - \sum_{j=1}^{N} J_j F_{i-j}$$

The equation provides the value for J_i, resulting in N equations and N unknowns. In this case $q_i = 0$, $\sigma T_i^4 = J_i$.

Chapter 13

Example 13.26: *A cylindrical furnace 1 m dia and 1 m long has one end maintained at 1000 K while the other end is at 600 K. The curved surface is maintained at 400 K. The emissivities are 0.25, 0.5 and 0.5. Determine the heat flow by radiation from each surface.*

Solution: This is a 3 surface enclosure. Let the ends be designated as A_1 and A_2 and the curved surface as A_3.

$$A_1 = A_2 = \pi \times 0.5^2 = 0.25\,\pi \text{ m}^2,\ A_3 = \pi \times 1 \times 1 = \pi \text{ m}^2$$

From chart F_{1-2} is determined as 0.18,

(the parameters are $L/R_1 = 2$, $R_2/L = 0.5$)

$$\therefore \quad F_{1-2} = F_{2-1} = 0.18, \qquad F_{1-3} = F_{2-3} = 0.82$$

$$0.25\,\pi\, F_{1-3} = \pi\, F_{3-1} \qquad \therefore \quad F_{3-1} = 0.82/4 = 0.205$$

$$F_{3-2} = 0.205$$

$$F_{1-1} = 0, \quad F_{2-2} = 0, \quad F_{3-3} = 1 - 2 \times 0.205 = 0.59$$

$$E_{b1} = 56700, \qquad E_{b2} = 7348.32, \quad E_{b3} = 1451.52 \text{ W/m}^2$$

The equations derived from (13.39) gives

Surface 1: $$56700 = \frac{1}{0.25} J_1 - \frac{1-0.25}{0.25}\,[J_1 F_{1-1} + J_2 F_{1-2} + J_3 F_{1-3}]$$

Surface 2: $$7348.32 = \frac{1}{0.5} J_2 - \frac{1-0.5}{0.5}\,[J_1 F_{2-1} + J_2 F_{2-2} + J_3 F_{2-3}]$$

Surface 3: $$1451.52 = \frac{1}{0.5} J_3 - \frac{1-0.5}{0.5}\,[J_1 F_{3-1} + J_2 F_{3-2} + J_3 F_{3-3}]$$

Substituting the values of shape factors, these equations yield:

$$56700 = 4\,J_1 - 3(J_2 \times 0.18 + 0.82\,J_3)$$

$$7348.32 = 2J_2 - (0.18\,J_1 + 0.82\,J_3)$$

$$1451.52 = 2J_3 - (0.205\,J_1 + 0.205\,J_2 + 0.59\,J_3)$$

This reduces to

$$\begin{bmatrix} 4 & -0.54 & -2.46 \\ -0.18 & 2 & -0.82 \\ -0.205 & -0.205 & 1.41 \end{bmatrix} \begin{bmatrix} J_1 \\ J_2 \\ J_3 \end{bmatrix} = \begin{bmatrix} 56700 \\ 7348.32 \\ 1451.52 \end{bmatrix}$$

Solving and substituting in equation (13.40)

$$q_1 = 13236 \text{ W/m}^2 \qquad A_1 = 0.25\,\pi \qquad \therefore \quad \mathbf{Q_1 = 10395.6\ W}$$

Similarly $$q_2 = 851.7 \text{ W/m}^2 \qquad A_2 = 0.25\,\pi \qquad \therefore \quad \mathbf{Q_2 = 668.93\ W}$$

$$q_3 = -\,3521.94 \text{ W/m}^2 \qquad A_3 = \pi \qquad \therefore \quad \mathbf{Q_3 = -\,11064.51\ W}$$

Checks are $Q_1 + Q_2 + Q_3 = 0$.

Example 13.27: *In the above example, if the surface 3 acts as reradiating surface, determine the net heat flow from each surface.*

Solution: The first two equations are the same as in the previous case.

$$56700 = 4J_1 - 3(0.18\,J_2 + 0.83\,J_3)$$

$$7348.32 = 2J_2 - (0.18\,J_1 + 0.82\,J_3)$$

The third equation now is

$$q_3 = 0 = J_3 - [J_1 F_{3-1} + J_2 F_{3-2} + J_3 F_{3-3}]$$

Substituting shape factor values,

$$J_3 - 0.205\, J_1 - 0.205\, J_1 - 0.59\, J_3 = 0$$

or

$$0.41\, J_3 - 0.205\, J_1 - 0.205\, J_2 = 0$$

This leads to

$$\begin{bmatrix} 4 & -0.54 & -2.46 \\ -0.18 & 2 & -0.82 \\ -0.205 & -0.205 & 0.41 \end{bmatrix} \begin{bmatrix} J_1 \\ J_2 \\ J_3 \end{bmatrix} = \begin{bmatrix} 56700 \\ 7348.32 \\ 0 \end{bmatrix}$$

Solving: $J_1 = 30705.24,\ J_2 = 16016.32,\ J_3 = 23360.78\ \text{W/m}^2$

The heat flow is calculated using:

$$q_1 = J_1 - [J_1 F_{1-1} + J_2 F_{1-2} + J_3 F_{1-3}]$$
$$q_2 = J_2 - [J_1 F_{2-1} + J_2 F_{2-2} + J_3 F_{2-3}]$$
$$q_3 = J_3 - [J_1 F_{3-1} + J_2 F_{3-2} + J_3 F_{3-3}]$$

Substituting the values and calculating,

$q_1 = 8666.45\ \text{W/m}^2$, $A = 0.25\ \pi$, $\mathbf{Q_1 = 6806.6\ W}$

$q_2 = -\ 8666.45\ \text{W/m}^2$ $\mathbf{Q_2 = -\ 6806.6\ W}$

$\mathbf{Q_3 = 0}$ **checks**

as $\sigma T_3^4 = J_3$ $\mathbf{T_3 = 801.17\ K.}$

13.10 SURFACES SEPARATED BY AN ABSORBING AND TRANSMITTING MEDIUM

Surfaces separated by glass or a gas body are examples for this situation.

The intervening body absorbs part of radiation between the surfaces.

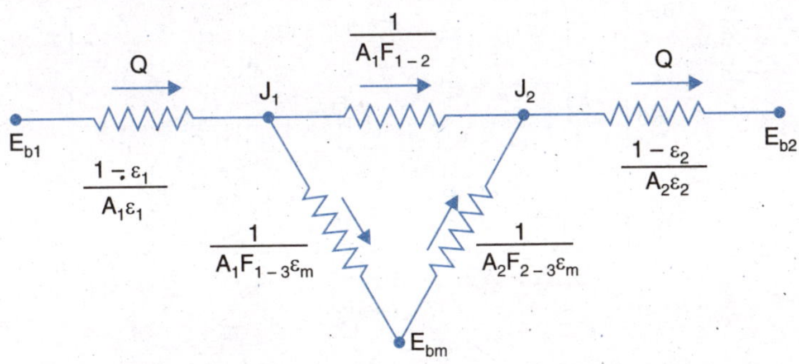

Fig. 13.26. Network for surfaces separated by absorbing and transmitting medium

The space resistance now can be shown to be

$$\frac{1}{A_1 F_{1-2}\,(1 - \varepsilon_m)}$$ where ε_m is the emissivity of the medium.

Similarly the space resistance between the surface and the medium can be shown to be

$$\frac{1}{A_1 F_{1-m} \cdot \varepsilon_m} \text{ and } \frac{1}{A_2 F_{2-m} \cdot \varepsilon_m}$$

The determination of e_m in the case of gases is made complicated because of its dependence on the pressure, temperature and thickness. In the case of glasses the emissivity depends on the wavelength or the source of radiation.

This method however can be used for a first estimate in such situations.

13.11 SOLAR RADIATION AND GREEN HOUSE EFFECT

13.11.1. Solar Spectrum

Solar radiation is, of course, essential to all life on earth. Through the process of photosynthesis, it satisfies our need for food, fibre and fuel. Moreover, through thermal and photovoltaic process, it has the potential to satisfy much of our demand for space heat, process heat, and electricity. The spectral distribution of solar radiation is shown in Fig. 13.27. The distribution approximates that of black body radiation at 5800 K. The radiation is concentrated in the low wavelength region of ($0.2 \le \lambda \le 3$ μm) the thermal spectrum the peak occurring at 0.5 μm.

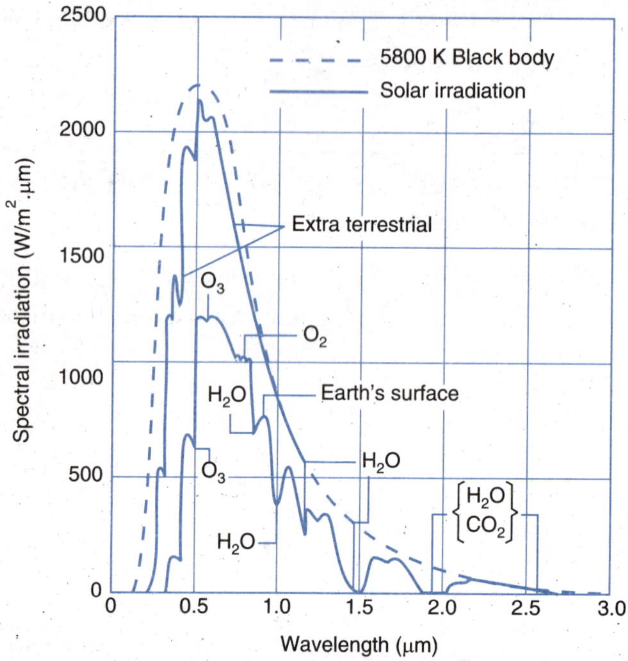

Fig. 13.27. Solar Radiation Spectrum

As solar radiation passes through the earth's atmosphere its magnitude and its spectral and directional distributions experience significant change. The change is due to absorption and scattering of the radiation by atmospheric constituents. The effect of absorption by gases ozone (O_3), water vapour, oxygen and carbon dioxide is shown in the figure by the lower curve. Absorption by ozone is strong in the ultraviolet region, providing considerable attenuation below 0.4 μm and complete attenuation below 0.3 μm. In the visible region there is some absorption by ozone and oxygen and in the near infrared regions absorption is dominated by water vapour. Throughout the solar spectrum there is also absorption by dust and aerosol content of the atmosphere.

13.11.2. The Atmosphere of Earth

While the atmosphere is made up almost entirely of nitrogen and oxygen, other gases and particles existing in very small concentrations determine to a large extent the habitability of earth. These other gases include carbon dioxide (CO_2), nitrous oxide (N_2O), methane (CH_4) and ozone (O_3). The other gases are mostly human made. These are the chlorofluorocarbons (CFCs), hydrofluorocarbons (HCFCs) hydrofluorocarbons (HFCs) etc.

The problem of green house effect enhancement leading to global climate change and atmospheric ozone depletion which increases our exposure to life threatening ultraviolet radiation are linked to changes in these trace gas concentrations.

When the earth was formed some 4.6 billion years ago it probably had an atmosphere made up of helium and compounds of hydrogen such as molecular hydrogen, methane and ammonia. That early atmosphere is thought to have escaped into space after which our current atmosphere slowly began to form.

Through volcanic activity gases such as carbon dioxide, water vapour and various compounds of nitrogen and sulfur were released over time. Molecular oxygen eventually began to form both as a result of photo dissociation of water vapour and by photosynthesis by plants that were evolving under water where life was protected from suns biologically damaging ultraviolet radiation. As photosynthesis gradually increased atmospheric oxygen levels, it is throught that the absorption of incoming ultraviolet radiation by ozone provided the protection necessary for life to begin to emerge onto the land.

Table 13.3 shows the composition by volume of the earths atmosphere as it exists now. The values given are for clean dry air and do not include the relatively small but extremely important amounts of water vapour and particulate matter.

Table 13.3 Composition of clean dry air (fraction by volume in atmosphere, 1994)

Constituent	*Formula*	*Percent by Volume*	*Part per million*
Nitrogen	N_2	78.08	780800
Oxygen	O_2	20.95	209500
Argon	Ar	0.93	9300
Carbon dioxide	CO_2	0.035	358
Neon	Ne	0.0018	18
Helium	He	0.00015	5.2
Methane	CH_4	0.00017	1.7
Krypton	Kr	0.00011	1.1
Nitrous oxide	N_2O	0.00003	0.3
Hydrogen	H_2	0.00005	0.5
Ozone	O_3	0.000004	0.04

While most of the values in the table are essentially unchanging, that is not the case for the principal green house gases carbon dioxide, methane and nitrous oxide which are rising.

It is convenient to think of the atmosphere as being divided into various horizontal, layers, each characterised by the slope of its temperature profile. Starting at the earths surface these layers, are called **troposphere, stratosphere, mesosphere, and thermosphere.**

The troposphere and mesosphere are characterised by decreasing temperature with altitude, while the stratosphere and thermosphere show increasing temperatures. The transition altitude separating these layers are called the tropopause, stratopause and mesopause. Obviously the condition in the true atmosphere of earth vary with time and location, but a useful, idealised temperature profile, known as the US standard atmosphere provides a convenient starting point for atmospheric studies. The profile is shown in Fig. 13.28.

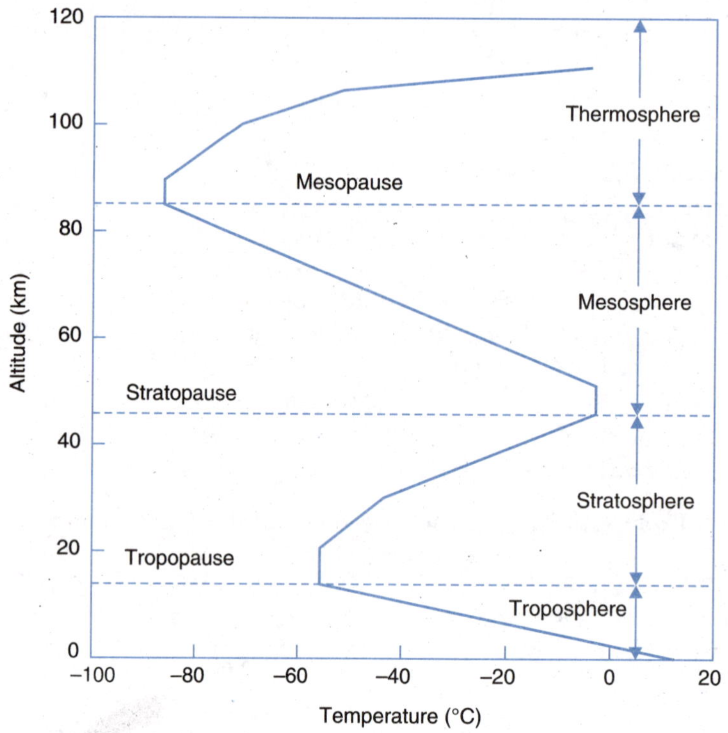

Fig. 13.28. US standard atmosphere showing the four major layers

More than 80% of the mass of the atmosphere and virtually all of water vapour, clouds and precipitation occur in the troposphere. At mid latitudes the troposphere extends up to 10 or 12 km (about the altitude of typical airline flight).

At the poles it may be only 5 or 6 km, while at the equator it is about 18 km. In the troposphere temperature typically decreases at 5 to 7°C per km, which is essentially the wet adiabatic lapse rate corresponding to the rate of change of temperature as water saturated air rises. The troposphere is usually a very turbulent place, due to strong vertical air movements that lead to complete and rapid mixing. This mixing is good for air quality since it rapidly disperses pollutants.

Above the troposphere is the stable layer of very dry air called the stratosphere. Pollutants that find their way into the stratosphere may remain there for many years before they eventually drift back into the troposphere where they can be more easily diluted and ultimately removed by settling or precipitation. In the stratosphere short wavelength ultraviolet energy is absorbed by ozone and oxygen causing the air to be heated. The resulting temperature inversion is what causes the stratosphere to be so stable. The troposphere and stratosphere combined account for about 99.9 per cent of the mass of the atmosphere. Together they extend only about 50 km above the surface of the earth, a distance equal to less than one per cent of the earths radius.

Beyond the stratosphere lies the mesosphere, another layer where air mixes fairly readily and above that the thermosphere. The heating of the thermosphere is due to the absorption of solar energy by atomic oxygen. Within the thermosphere is a relatively dense band of charged particles called the ionosphere. Before satellites the ionosphere was especially important to world wide communication because its ability to reflect the radio waves back to earth.

13.11.3. Global Temperature

The common definition of climate suggests that it is the prevailing or average weather of a place as determined by temperature and other meteorological conditions over a period of years. Average temperature, then is but one measure of climate, and many others including precipitation, winds, glaciation and frequency of extreme events such as typhoons and hurricanes, are also important parameters. Long term variations in average temperature, however, are the single most important attribute of climate change.

Climatologists have used a number of clues to piece together past global temperatures, including evidence gathered from historical documents, tree rings change in ice volume and sea level, fossil pollen analysis, and geological observations related to glacial movements.

Analysis of ice cores taken in Greenland and Antartic has provided a remarkable picture of the climate and composition of the atmosphere over many thousands of years of earth's history. Isotopic analysis of glacial ice provides the temperature record, while analysis of the composition of the air bubbles sealed off in the ice as it was formed gives a corresponding record of the concentration of various atmospheric gases. In one of the most significant scientific studies of past climates an

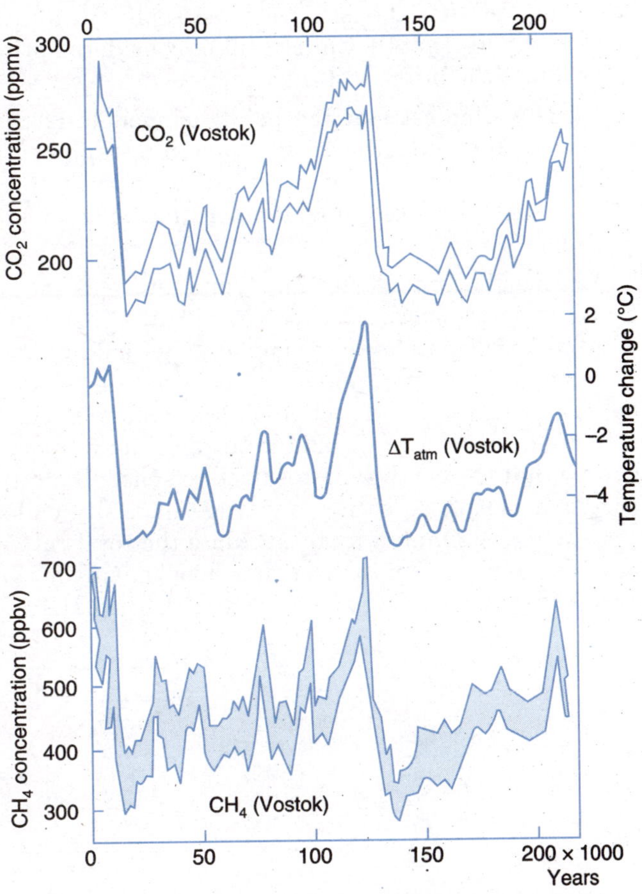

Fig. 13.29. Correlation between the variation of CH_4 and CO_2 with variation of temperature of antartic surface

ice core of 2083 m long was recovered by the sovients at vostok in east Antartica. Careful analysis of hydrogen isotope ratios and gaseous concentrations in the vostok ice core have provided a continuous 160000 year record. A remarkable correlation between atmospheric carbon dioxide and methane concentrations and Antartic surface temperature. This is shown in Fig. 13.29.

In the more recent past, the global mean temperature record is based on actual measurements that have been taken since 1856. During that time, the mean temperature has increased by about 0.3 to 0.6°C, with about 0.2 to 0.3°C of that occurring over the last 40 years.

13.11.4. A Simple Global Temperature Model

Measurement of historical global temperatures show that our planet has maintained its average temperature within a limited range, but within those bounds it seems to be continuously changing. If we hope to predict future impacts of the anthropogenic changes in our environment, we need to develop mathematical models that explain the past. Mathematical modelling of global climate and predicting the impacts of changes in key environmental parameters is an extremely important but difficult task. Such models range from very simple to complex, three dimensional general circulation model (GCMs), which attempt to predict climate on a regional, seasonal and annual basis. The most sophisticated of these models can take weeks to run on super computer, yet they must still be considered primitive. In comparison the model proposed here is just the briefest of introductions.

The simplest starting point for modelling climate begins with models that focus on factors influencing the single variable, temperature. Obviously other factors, such as precipitation patterns, winds and storms, ocean currents, soil moisture sea ice, glacial cover and so fourth are exceedingly important, but they are more difficult to approach with simple models.

Even beginning with just temperature as the single quantity of interest, we could try to find out how temperature varies in all four dimensions, (latitude, longitude, altitude and time), which is the domain of very complex general circulation models. At the other end of complexity scale is a simple zero dimensional model, in which a single average global temperature is obtained that is not a function of location and time. The following is such a model. The basic zero dimensional energy balance model equates solar energy absorbed by the earth with the energy that the earth radiates back to space. Radiation from the sun arrives just outside the earth's atmosphere with an average intensity called solar constant, S, currently equal to 1370 W/m^2. A simple way to calculate the total rate at which energy hits the earth is to note that all the flux passing through a 'hoop' having a radius equal to that of the earth and placed normal to the incoming radiation strikes the earth's surface. This is shown in Fig. 13.30.

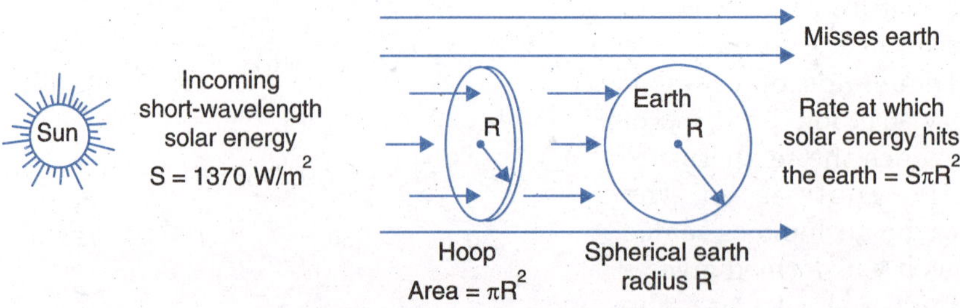

Fig. 13.30. Solar energy incident on earth

Now we can write

Rate at which solar energy strikes the earth = $S\pi R^2$ where S is solar constant and R is the radius of earth, the value of solar constant being 1370 W/m^2.

Some of the incoming solar energy that hits the earth is reflected back into space, and it does not contribute to the heating. The fraction of solar energy that is reflected is called **albedo** and for earth the value of albedo is estimated to be about 31 per cent. What is not reflected is absorbed?

Energy reflected by earth = $S \cdot \pi R^2 \alpha$ ...(13.41)

Energy absorbed by earth = $S\pi R^2 (1 - \alpha)$ ...(13.42)

On the other side of the energy balance equation is the rate at which energy radiates into space. Assuming emissivity as 1, and all over the earth's surface the same temperature prevails,

Energy radiated by earth = $\sigma 4\pi R^2 \cdot T_e^4$

where T_e is the effective earth's black body temperature. Assuming steady conditions equating the energy absorbed and radiated

$$S\pi R^2 (1 - \alpha) = \sigma\, 4\pi R^2 T_e^4$$

Solving for T_e

$$T_e = \left[\frac{S(1-\alpha)}{4\sigma}\right]^{0.25} = \left[\frac{1370\,(1-0.31)}{4 \times 5.67 \times 10^{-8}}\right]^{0.25} = 254 \text{ K or } - 19°\text{C}.$$

The actual value of the earths temperature is 288 K or 15°C. In terms of conditions on earth the estimate of T_e = – 19°C is not sustainable.

There is need find an explanation for why the earth is not that cold. The key factor that makes our model differ so much from reality is that it does not account for the interaction between the atmosphere and the radiation that is emitted from the earth's surface. That is, it does not include green house effect.

13.11.5. The Green House Effect

The surface of the earth is 34°C higher than what is predicted by the model.

The sun can be represented as a black body with temperature 5800 K, so its spectrum peaks at 0.5 μm. The earth at 288 K, has its peak at 10.1 μm.

The solar radiation flux, is the solar constant 1370 W/m^2. All this energy is within the wavelength of 3 μm in the spectrum. The earth's outgoing energy is essentially beyond this wavelength and is mostly in the infrared region. The spectrums are shown in Fig. 13.31.

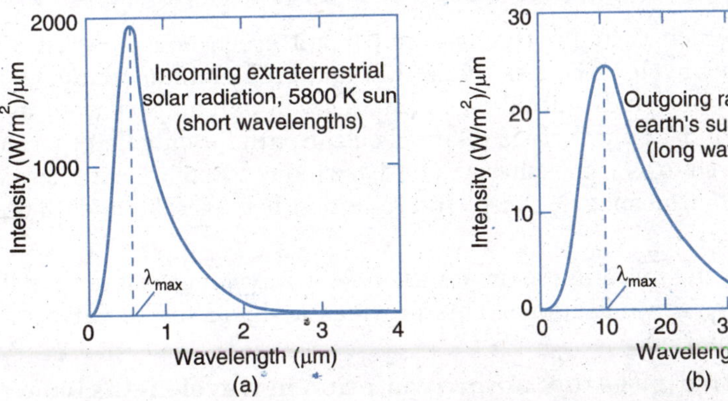

Fig. 13.31 (a) Solar spectrum (b) Earth radiation spectrum

As the radiant energy attempts to pass through the atmosphere, it is affected by various gases and aerosols in air. These atmospheric constituents can either let the radiant energy pass through unaffected, they can scatter the energy by reflection or they can stop it by absorption. The key phenomenon of interest here is the ability of the gases to absorb radiant energy.

Figure 13.32 shows the absorption spectra for the key gases of concern in this respect and their effect on incoming solar radiation and outgoing infrared radiation emitted by the earth's surface. Most of the long wavelength energy radiated by the earth is absorbed by a combination of radiatively active gases, most important of which are water vapour, carbon dioxide, methane, various oxides of nitrogen, molecular oxygen and ozone.

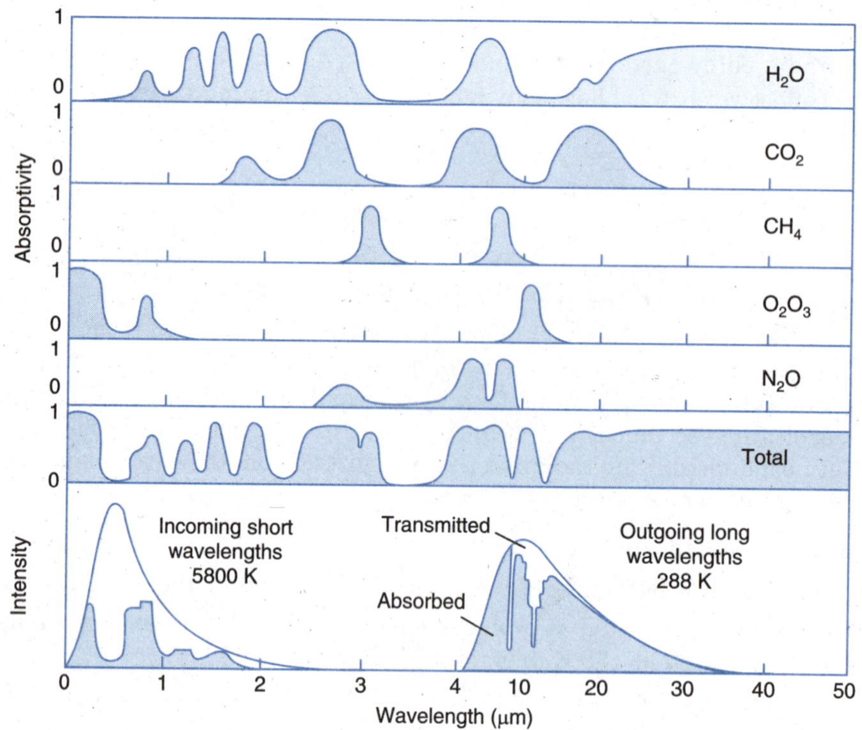

Fig. 13.32. Absorption spectrum of key green house gases. The amount absorbed is shown by shade in the lower figure

Water vapour, which is by far the most important green house gas, strongly absorbs thermal radiation with wavelengths less than 8 μm and greater than 18 μm. Carbon dioxide shows a strong absorption band centred at 15 μm, as well as bands centred at 2.7 μm and 4.3 μm. Between 7 μm and 12 μm there is relatively clear path for outgoing thermal radiation. This wave band is referred to as 'atmospheric radiative window'. Radiation in these wavelengths pass easily through the atmosphere, except the radiation between 9.5 and 10.6 μm which is absorbed by ozone.

Essentially all of the incoming solar radiation with wavelengths less than 0.3 μm ultraviolet is absorbed by oxygen and ozone. This provides a shield to the earth's surface from harmful ultraviolet radiation.

Radiatively active gases that absorb radiation in wavelengths longer than 4 μm are called green house gases. This absorption heats the atmosphere, which in turn, radiates energy back to the earth as well as to the space. The green house gases act as a thermal blanket around the globe, raising the earth's surface temperature beyond the equivalent temperature calculated earlier (– 19°C). The importance of water vapour as a green house gas is quite evident on clear nights, when the earth cools much more rapidly than it does on cloudy nights.

It is also interesting to note that the term green house effect is based on the concept of a conventional green house with the glass roof acting much like the green house gases. The radiation trapping is responsible for the elevated temperature inside green house.

If the earth did not already have a green house effect, its temperature would be –19°C, as predicted by energy balance. One way to quantity the magnitude of the green house effect is to compare the effective temperature T_e with actual surface temperature T_s.

Magnitude of green house effect = $T_s - T_e$.

Thus we can say that the green house effect adds 34°C warming to the surface of the earth.

13.11.6. Global Energy Balance

For the analysis it is convenient to consider the earth, its atmosphere and outer space as three separate regions. The energy flow will be calculated on the basis of W/m^2.

The average incoming solar radiation per unit area of earth is calculated by

$$\frac{\text{Incoming solar radiation}}{\text{Surface area of earth}} = \frac{S\pi R^2}{4\pi R^2} = \frac{S}{4} = \frac{1370}{4} = 342 \text{ W/m}^2.$$

As 31% of the incoming radiation is reflected back to the space,

Energy reflected = 342 × 0.31 = 107 W/m^2.

Of this 107 W/m^2, it is estimated that 77 W/m^2 is reflected by the atmosphere itself with the remaining 30 W/m^2 being reflected by the earth's surface.

Solar energy not reflected is absorbed by the earth's surface and the atmosphere.

$$\text{Energy absorbed} = \frac{S\pi R^2 (1-\alpha)}{4\pi R^2} = \frac{S}{4}(1-a)$$

$$= \frac{1370}{4} \times (1 - 0.31) = 235 \text{ W/m}^2.$$

It is estimated that out of the 235 W/m^2, 67 W/m^2 is absorbed by the atmosphere and the remaining 168 W/m^2 is absorbed by the surface of the earth.

Assuming that the global temperature remains unchanged, the amount of energy received from space should be returned to the space. But of the energy received, 107 W/m^2 is returned to space by reflection from atmosphere and reflection from earth's surface.

The earth and the atmosphere receive (342 – 107) W/m^2 that is 235 W/m^2. Of this 67 W/m^2 are absorbed by the atmosphere. The remaining 168 W/m^2 are absorbed by the surface of the earth.

If global temperature is unchanging with time, then the earth and the atmosphere should return an equal amount of heat to the space. The reflected energy of 107 W/m^2 balances 107 W/m^2 received from space. The remaining 235 W/m^2 should be radiated back into space.

If the earth surface temperature is 254 K as earlier calculated by heat balance, then the earth surface will radiate 235 W/m^2 into space, which will balance the heat from space. But as it is pointed out earlier green house gases would absorb most of this heat and the energy balance will not be realised. Therefore to force enough energy through the atmosphere, the temperature of the earth's surface should be higher than 254 K.

As the surface average temperature has been determined to be 288 K, the earth will radiate more heat to force it through the atmosphere. The energy radiated, assuming the earth's surface as black,

Heat radiated at $288 \text{ K} = 5.67 \times 10^{-8} \times 288^4 = 390 \text{ W/m}^2$.

This amount is radiated out from the surface to the atmosphere. 40 W/m² passes directly to space through the atmospheric radiative window between 7 and 12 μm. The remaining 350 W/m² is received by the atmosphere and absorbed by the green house gases. The atmosphere then radiates 324 W/m² back to the surface. There is also transfer from the earth's surface by convection and evaporation. 78 W/m² is due to evaporation and 24 W/m² is by convection. These transactions are shown in Fig. 13.33.

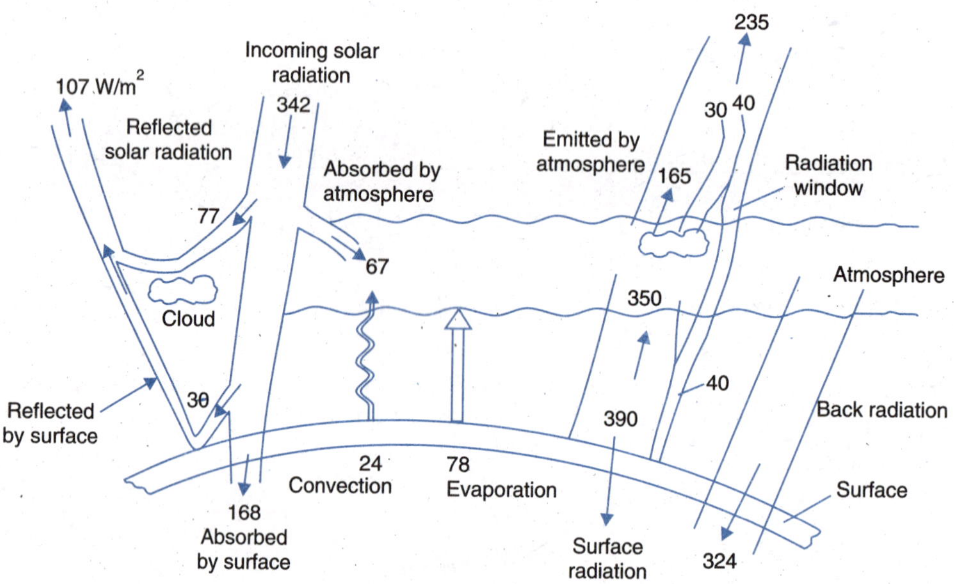

Fig. 13.33. Global average energy flow between space, the atmosphere and the earth's surface. Values given are from IPCC report 1996. Unit used is W/m²

If this model is internally self consistent the rate of heat gain should equal to the rate of energy flow out in each of the three regions. Consider the following for checking the consistency.

	Energy gain = *Energy flow out*	*Check*
Earth Surface	168 + 324 + 30 = 78 + 24 + 30 + 390	yes
Atmosphere	67 + 78 + 24 + 350 = 165 + 30 + 324	yes
Space	107 + 165 + 30 + 40 = 242	yes

13.11.7. Radiative Forcing of Climate Change

The green house effect has been described as a natural phenomenon that is responsible for earth having an average surface temperature of 15°C instead of – 19°C. The radiatively active gases in the atmosphere are responsible for the above. Now due to various sources like industrial development, the fraction of these gases in the atmosphere is increasing and affecting the green house effect, leading the earth into a future of uncertain global climate.

In the model discussed in the previous section total energy balance for each region was seen. If for some reason an additional amount of energy is added to the incoming energy, then the balance will be temporarily upset. Over time, however, the climate system will adjust to the change by either increasing or decreasing the surface temperature of the earth until a balance is once again attained. Mathematically at stable condition, the heat radiated will equal the heat absorbed $Q_{rad} = Q_{abs}$.

In case the heat radiated is changed by ΔQ_{rad}, there will be unbalance and the heat absorbed may also change by ΔQ_{abs} but these quantities will not be equal to start with. So an addition quantity of heat, ΔF will try to gradually change the inbalance and the balance equation will now be

$$Q_{abs} + \Delta Q_{abs} + \Delta F = Q_{rad} + \Delta Q_{rad} \quad \text{or} \quad \Delta F = \Delta Q_{rad} - \Delta Q_{abs}.$$

Because the difference forces changes to obtain balance the quantity ΔF is called radiative forcing.

Determining the radiative forcing associated with changes in green house gas concentration, changes in aerosols, changes in albedo and changes in solar constant has been one of the most important research areas for atmospheric scientists.

Climate Sensitivity Parameter

The main question in climate prediction is how much the surface temperature will change (ΔT_s) for a given change in radiative forcing ΔF. The quantity that links the two is called climate sensitivity parameter, λ, which can be defined by

$$\Delta T_S = \lambda \Delta F = \lambda\,(\Delta Q_{rad} - \Delta Q_{abs})$$

or

$$\lambda = \Delta T_s / (\Delta Q_{rad} - \Delta Q_{abs})$$

or

$$\lambda = \left(\frac{\Delta Q_{rad}}{\Delta T_s} - \frac{\Delta Q_{abs}}{\Delta T_s}\right)^{-1} \quad \text{...(13.43)}$$

Using partial differentials

$$\lambda = \left(\frac{\partial Q_{rad}}{\partial T_s} - \frac{\partial Q_{abs}}{\partial T_s}\right)^{-1} \quad \text{...[13.43(a)]}$$

This equation shows that climate sensitivity factor depends on two components

(1) How much the outgoing radiant energy at the top of the atmosphere changes as the surface temperature changes and

(2) The change in incoming energy that is absorbed as surface temperature changes. The later component (the change in absorbed energy) allows the estimate of change of albedo with temperature.

It is interesting to note that λ is the inverse of change in entropy due to change in energy $\left(\frac{T}{Q}\right)$. Considerable amount of work has gone into trying to determine the appropriate value for the climate sensitivity parameter. One method is based on infrared radiation data

Chapter 13

obtained from satellites. By correlating monthly surface temperature with infrared emissions from the top of the atmosphere, it has been noted that a linear relationship between the two works quite well for moderate temperature changes. The following equation is one such relationship obtained.

$$Q_{rad} = 1.83\, T_s + 209°C \qquad ...(13.44)$$

Example 13.28: *It is estimated that doubling the atmospheric concentration of CO_2 causes a radiative forcing of 4.35 W/m². If the earth's albedo does not change, estimate the climate sensitivity factor λ and use it to estimate the resulting change in the surface temperature of earth needed to balance incoming and out going radiation.*

Solution:

$$\lambda = \left[\frac{\Delta Q_{rad}}{\Delta T_s} - \frac{\Delta Q_{abs}}{\Delta T_s}\right] \qquad ...(A)$$

As albedo is assumed not to change,

$$\frac{\Delta Q_{abs}}{\Delta T_s} = 0,$$

From the equation (13.28)

$$Q_{rad} = 1.83\, T_s + 209,$$

Taking differential,

$$\Delta Q_{rad} = 1.83\, \Delta T_s$$

$$\therefore \qquad \frac{\Delta Q_{rad}}{\Delta T_s} = 1.83$$

Substituting in equation (A),

$$\lambda = (1.83)^{-1} = 0.55°C/(W/m^2)$$

As $\Delta T_s = \lambda \Delta F,$

As ΔF for doubling CO_2 is given as 4.35

$$\Delta \mathbf{T_s} = 0.55 \times 4.35 = \mathbf{2.4°C.}$$

The **International Panel for Climate Change (IPCC)** have estimated that the surface temperature change associated with doubling of CO_2 to lie between 1.5 and 4.5°C, with the best estimate being 2.5°C. As per this estimate λ will have values in the range 0.34 and 1.03 with the best estimate being 0.57.

Example 13.29: *Estimate the radiative forcing if the albedo changes from 0.31 to 0.3 due to glacial melting. If the climate sensitivity factor is in the range 0.34 to 1.03 m²/W, estimate the change in surface temperature.*

Solution: The energy absorbed changes due to change in albedo. The incoming radiation is 342 W/m².

Radiative forcing,
$$\Delta F = \Delta\left[\frac{s}{4}(1-\alpha)\right]$$
$$= 342\,(1 - 0.31) - 342\,(1 - 0.3)$$
$$= 3.42 \text{ W/m}^2.$$
$$\Delta T_s = \lambda \Delta F,$$

For $\lambda = 0.34$, $\Delta T_s = 0.34 \times 3.42 = 1.16°C$

$\lambda = 1.03$, $\Delta T_s = 1.03 \times 3.42 = 3.52°C$

13.11.8. Radiative Forcing Since Industrial Times

The concept of radiative forcing climate change can be applied to the accumulation of green house gases in the atmosphere, changes in aerosols due to natural and industrial development, ozone depletion in the stratosphere, photo chemically produced ozone accumulation in the troposphere and the natural variation in solar intensity reaching the earth's outer atmosphere.

Gases and particulate matter added to the atmosphere can exert both direct and indirect radiative forcing effects. Direct forcing is caused by substances in the atmosphere that have actually been emitted from some source. Indirect forcing are those that occur when these substances go on to cause other atmospheric changes that affect radiative properties of the atmosphere. Aerosols and Halocarbons are examples of this type.

Table 13.4 presents a summary of current estimates of the radiative forcing caused by the direct and indirect effects of green house gases, aerosols and solar radiation.

Table 13.4. Globally averaged radiative forcing due to changes in green house gases, aerosols an solar radiation intensity from 1850 to the present.

	Radiative forcing ***ΔF, W/m^2***	***Confidence level***	***Comments***
Direct Green house gases			
Carbon dioxide	1.56	High	CO_2 increased from 278 to 356 ppm
Methane	0.47	High	CH_4 from 700 to 1714 parts per billion
Halocarbons	0.28	High	*R*11, 12 and 113
Nitrous oxide	0.14	High	NO_2 275 to 311 PPb
Total	2.45	High	Ranges from 2.1 to 2.8
Indirect Green house			
Stratospheric O_3	– 0.1	low	Range – 0.05 to – 0.2
Troposphere O_3	0.4	low	Range, 0.2 to 0.6
Direct Aerosols			
Sulphate, SO_4	– 0.4	low	Range – 0.2 to – 0.8
Fossil fuel, soot	0.1	very low	Range 0.03 to 0.3
Biomass burning	– 0.2	very low	– 0.07 to – 0.4
Indirect Aerosols	0 to – 1.5	very low	causes cloud formation
Solar (Since 1850)	0.3	very low	0.1 to 0.5.

ADVANCED EXAMPLES

Example A-13.1: *State and prove Kirchhoff's law for heat radiation.* *Kirchhoff's law states that the emissivity of a surface is equal to its absoptivity at any given temperature and wavelength. For gray, bodies, this is approximated for all wavelengths. In general application it is also approximated to all temperatures.*

Solution: Rigorously $\varepsilon_{\lambda T} = \alpha_{\lambda T}$, For grey surfaces $\varepsilon_T = \alpha_T$

general application $\varepsilon = \alpha$

Proof: Consider an enclosure at uniform temperature T in which a small body is placed. (Fig. AE-13.1). The walls of the enclosure can be considered as black. After thermal equilibrium is attained, the body temperature will reach that of the walls of the enclosure. The body at this condition has to emit at every wavelength radiation equal to the radiation it absorbs. If G_λ is the irradiation in the space.

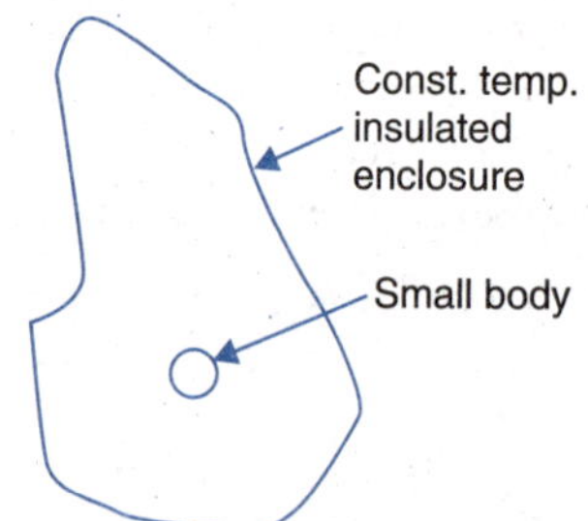

Fig. AE-13.1

$$E_\lambda = \alpha_\lambda G_{b\lambda} \quad \text{...(A)}$$

$$\frac{E_\lambda}{\alpha_\lambda} = G_{b\lambda} \quad \text{...(B)}$$

This equation will hold good for any other body also. As $G_{b\lambda}$ is the same.

$$\therefore \quad \frac{E_{\lambda 1}}{\alpha_{\lambda 1}} = \frac{E_{\lambda 2}}{\alpha_{\lambda 2}} = \frac{E_{\lambda 3}}{\alpha_{\lambda 3}} = G_{b\lambda} \quad \text{...(C)}$$

As α_λ can have a maximum value of unity at which E_λ will equal $E_{b\lambda}$

$$G_{b\lambda} = E_{b\lambda}$$

Using equation (A)

$$\varepsilon_\lambda E_{b\lambda} = \alpha_\lambda G_{b\lambda} \quad \text{...(D)}$$

As this is considered at temperature T,

$$\therefore \quad \boldsymbol{\varepsilon_{\lambda T} = \alpha_{\lambda T}}$$

This is rigourously true. However this is generalised as $\varepsilon = \alpha$.

Example A-13.2: *Show that if emissivity varies with the angle over a surface, the average can be obtained by*

$$\varepsilon = \int_0^{\pi/2} \varepsilon\,(\theta) \cos\theta \sin\theta \, d\theta$$

Solution: Ref. Fig. AE-13.2 the important assumption here is that over the periphery ϕ the emissivity is uniform. The energy radiated from the area dA_1 in the direction (ϕ, θ) over angles ϕ and $d\phi$ is given by $dA_1 \cos\theta\, I\, d\omega$

The sides of the element considered are $r\, d\theta$ and $r \sin\theta\, d\phi$

$$\therefore \quad \text{area} = r^2 \sin\theta \, d\theta \, d\phi$$

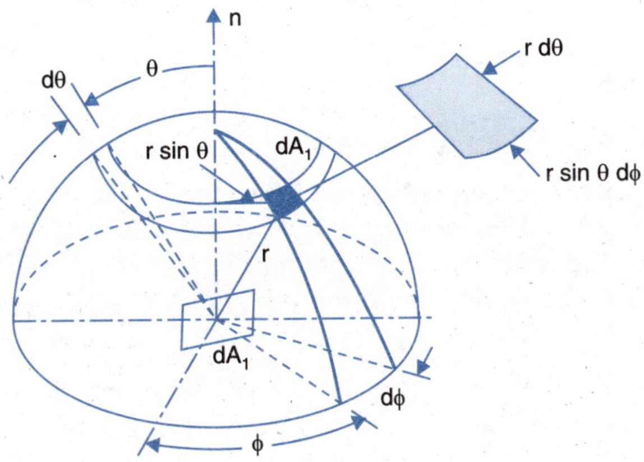

Fig. AE-13.2

The solid angle substended by this area, $d\omega$ = area/r^2

∴ $$d\omega = \sin\theta\, d\theta\, d\phi$$

∴ Energy radiated per unit area in the $d\omega$ direction

$$= I\cos\theta\sin\theta\, d\theta\, d\phi$$

Total over the surface of the hemisphere is

$$= \int_0^{2\pi}\int_0^{\pi/2} I\cos\theta\sin\theta\, d\theta\, d\phi$$

first integrating around the strip,

$$= 2\int_0^{\pi/2} \pi I\cos\theta\sin\theta\, d\theta$$

$$= 2\int_0^{\pi/2} E\cos\theta\sin\theta\, d\theta.$$ Now this can be integrated along θ

For a black surface $$= 2\int_0^{\pi/2} E_b\cos\theta\sin\theta\, d\theta$$

If ε varies with θ, then dividing

$$\varepsilon = \frac{2\int_0^{\pi/2} E(\theta)\cos\theta\sin\theta\, d\theta}{2\int_0^{\pi/2} E_b\cos\theta\sin\theta\, d\theta} = \int_0^{\pi/2} \frac{E(\theta)}{E_b}\cos\theta\sin\theta\, d\theta$$

∴ $$\varepsilon = \int_0^{\pi/2} \varepsilon(\theta)\cos\theta\sin\theta\, d\theta.$$

Example A-13.3: *A surface is radiating diffusely. Determine the fraction of emissive power in the angular region 0° to 45°C from the vertical. (Ref AE-13.2.)*

Solution: Energy radiated in the region 0° to 45° upwards

$$= \int_0^{2\pi}\int_0^{\pi/4} I\sin\theta\cos\theta\, d\theta\, d\phi = 2\pi\int_0^{\pi/4} I\sin\theta\cos\theta\, d\theta$$

$$= 2\pi I \left[\frac{\sin^2 \theta}{2}\right]_0^{\pi/4} = 2\,\pi\, I \left[\frac{0.5 - 0}{2}\right] = 0.5\,\pi\, I = 0.5\, E$$

(For 30° it is 0.25 E and For 60° the value is 0.75 E).

Example A-13.4: ***The energy received from the sun*** *at the earths atmosphere has been measured as 1353 W/m². The diameter of the earth = 1.29 × 10⁷ m. Diameter of the sun = 1.39 × 10⁹ m. Mean distance = 1.5 × 10¹¹ m. Estimate the emissive power of the sun and the surface temperature assuming it to be black. Assuming that the source of energy for the earth is from the sun and earth to be black, estimate the temperature of the earth.*

Solution: Using suffix s to denote sun and e to denote earth

The energy received by earth from the sun considering both as disks

$$= A_s I_s\, \omega = \frac{\pi}{4} D_s^2\, I_s\, \omega = \frac{E_s\, D_s^2}{4}\, \omega$$

$$\omega = A_e/R^2 = \pi\, D_e^{\,2}/4 \times R^2 \text{ as } A_e = \frac{\pi\, D_e^2}{4}$$

$$\therefore \quad \text{Energy received} \quad = A_e\, 1353 \text{ W} = \frac{\pi\, D_e^2}{4}\, 1353 \text{ W}$$

$$\therefore \quad \frac{E_s\, D_s^2}{4} \cdot \frac{\pi\, D_e^2}{4\, R^2} = \frac{\pi\, D_e^2}{4} \cdot 1353 \quad \text{or} \quad E_s = 1353 \cdot \frac{4R^2}{D_s^2}$$

$$D_s = 1.39 \times 10^9 \text{ m}, D_e = 1.29 \times 10^7 \text{ m}, R = 1.5 \times 10^{11} \text{ m}$$

substituting $E_s = 63.025 \times 10^6$ W/m²

$$\sigma\, T_s^{\,4} = E_s = 63.025 \times 10^6 \quad \therefore \quad \mathbf{T_s = 5774\ K}$$

For earth to be in equilibrium.

The earth receives 1353 W/m² normal to the area, but radiates all over the hemispherical surface. So

$$1353 \times \frac{\pi\, D_e^2}{4} = E_e \cdot 2\pi \left(\frac{De}{2}\right)^2$$

$$\therefore \quad E_e = 1353/2 \quad \therefore \quad \mathbf{T_e = 330.5\ K.} \qquad \text{(Assuming no atmosphere)}$$

Example A-13.5: ***The filament of a light bulb*** *is a rectangle of size 5 mm × 2 mm and consumes 60 W. The filament is at 2800 K. Determine the efficiency of the bulb assuming visible radiation in the band 0.4 to 0.7 micrometer.*

Solution: The fraction of energy falling in the range

$$\lambda T = 0.4 \times 2800 \text{ to } 0.7 \times 2800 \text{ is to be estimated}$$

For **λT = 1120** the fraction is **0.001409**

For **λT = 1960** the fraction is **0.061251**

∴ **Fraction in the band is = 0.059842**

Energy in the visible region

$$= 0.059842 \times \sigma \times 2800^4 \times 2 \times 5 \times 2 \times 10^{-6} = \mathbf{4.171\ W}$$

∴ **Efficiency** = 4.171/60 = 0.0695 or **6.95%.**

Example A-13.6: *The temperature of the sun's surface is estimated at 5800 K. Determine the fractions of emissive power falling in the ultraviolet (up to 0.4 μm), visible (0.4 to 0.7 μm) and infrared (above 0.7 μm) regions.*

Solution: λT for the various regions and the corresponding fractions are

$$0.4 \times 5800 = 2320 -> 0.1245088$$

$$0.7 \times 5800 = 4060 -> 0.4914181$$

∴ **Ultraviolet region: 12.45%.**

visible region: 36.69%

Infrared region: 50.85% (by difference)

A seizable fraction is in the visible region.

Example A-13.7: *A filament is in the shape of a cylinder of 1 mm dia and 25 mm length and is in an evacuated space enclosed by transparent medium. The filament is initially maintained at 2900 K by power supply. Determine the time needed for the filament to cool to 1300 K after power supply is cut.*

Density = $19350\ kg/m^3$, *specific heat* = $134\ J/kg\ K$

Emissivity of the filament = 0.45

Solution: The cooling is only by radiation

Energy radiated during a time interval $d\tau$ when the temperature of the filament is T

$$= A_f \sigma \varepsilon T^4 d\tau, A_f = \pi DL, V = \pi D^2L/4$$

This should equal the change in internal energy

$$= -V c \rho\, dT$$

$$\therefore \quad \frac{dT}{T^4} = -\frac{4\sigma\varepsilon}{D c\rho} d\tau$$

Integrating

$$\left[-\frac{T^{-3}}{3}\right]_{2900}^{1300} = \frac{-4 \times 5.67 \times 10^{-8} \times 0.45}{0.001 \times 134 \times 19350} . \tau$$

$$-\frac{1}{3 \times 1300^3} + \frac{1}{3 \times 2900^3} = -3.936 \times 10^{-11} \tau$$

Solving τ = 3.5 seconds. To cool to 300 K, it will take 313s.

Example A-13.8: *Hot gases are produced continuously at 1 atm and 1400 K in a spherical combustion chamber of 1 m dia. The gases contain 16%* CO_2 *and rest being* N_2 *and* O_2*. If the wall is black and is maintained at 500 K, determine the heat flow from the gas to the wall.*

Solution: In this case only carbon dioxide is radiating. The partial pressure of CO_2 = 0.16 atm.

The beam length $= 0.65\ D = 0.65$ m

$$\therefore \quad Pl = 0.104$$

The emissivity value is read from the chart as **0.096.**

No pressure corrections is required in this case as the total pressure is 1 atm. No correction is required for associated gas presence also.

The absorptivity has to be determined corresponding to the wall temperature of 500 K.

$$\alpha = \left(\frac{T_g}{T_s}\right)^{0.65} . \varepsilon Pl'$$

$$Pl' = Pl . T_s/T_g = 0.104 \times 500/1400 = \mathbf{0.0371}$$

The value of ε is read as 0.075

$$\therefore \qquad \alpha = \left(\frac{1400}{500}\right)^{0.65} \times 0.075 = \mathbf{0.1465}$$

$$Q = A\sigma (\varepsilon_g T_g^4 - \alpha_g T_s^4) = 4\pi \times 0.5^2 \times 5.67 [0.096 (14)^4 - 0.1465(5)^4]$$

$$= \mathbf{65.02 \times 10^3 \ W \ or \ 65.02 \ kW.}$$

Example A-13.9: *A gas turbine combustion chamber can be considered as a cylinder of 0.2 m dia and 0.9 m length. The combustion gases are at 8 atm and 1300 K. The partial pressure of CO_2 is 0.8 atm and that of water vapour 0.8 atm. The walls are at 800 K. Determine the heat flux from the gas to the chamber surface.*

Solution: The characteristic length can be determined using long cylinder approximation

$$l = 0.95 D = 0.95 \times 0.2 = 0.190 \text{ m}$$

for both gases $\qquad Pl = 0.8 \times 0.19 = 0.152$ m atm

Emissivity of CO_2: Reading from chart, at 1300 K

$$\varepsilon_{CO_2} = 0.11, C_{CO_2} = 1.4 \text{ (extrapolated)}$$

Emissivity of H_2O

$$\varepsilon_{H_2O} = 0.12, C_{H_2O} = 1.8 \text{ (extrapolated)}$$

The correction factor for the simultaneous presence of the two radiating gases is read as (C denotes CO_2 and w denotes water vapour)

$$P_w/(P_c + P_w) = 0.8/(0.8 + 0.8) = 0.5$$

$$P_w l + P_c l = 0.304 \text{ m atm}$$

$$\Delta\varepsilon = 0.04$$

$$\therefore \qquad \varepsilon = 0.11 \times 1.4 + 0.12 \times 1.8 - 0.04 = \mathbf{0.33}$$

Absorptivity values are determined corresponding to the surface temperature

$$\alpha_w = C_w \left(\frac{T_g}{T_s}\right)^{0.45} . \varepsilon_w Pl$$

$$Pl' = 0.152 \times \frac{800}{1300} = 0.094 \text{ m atm}$$

The value of ε are read at this condition as 0.105 and 0.15 for CO_2 and H_2O

$$\alpha_w = 1.8 \left(\frac{1300}{800}\right)^{0.45} \times 0.15 = 0.336$$

$$\alpha_{CO_2} = 1.4 \left(\frac{1300}{800}\right)^{0.65} \times 0.105 = 0.201$$

(**Note.** The correction factors are, extrapolated values for H_2O at this condition is also 1.8)

$$\Delta\varepsilon = \mathbf{0.04}$$

$\therefore$ $\alpha_g = 0.336 + 0.201 - 0.04 \pm \mathbf{0.497}$

$\therefore$ Heat transfer per unit area:

$$q = \sigma\,[\varepsilon_g T_g^{\,4} - \alpha_g T_s^{\,4}] = 5.67\,[0.33\,(13)^4 - 0.497\,(8)^4]$$

$$= \mathbf{41898 \text{ or } 41.9 \text{ kW/m}^2}.$$

Example A-13.10: *The combustion gases in a Lancashire type of boiler flows through the flue tube of 0.6 m dia and 6 m length. The gas is at 1200 K. The walls are at 500 K. The gases contain 18% CO_2 and 12% H_2O by volume. The total pressure is 1 atm. Assuming the surface to be black, determine the heat flux by radiation from the gases to the wall.*

Solution: Long cylinder configuration can be assumed.

Beam length $= 0.95\,D = 0.570$ m

Partial pressures of CO_2 and H_2O are 0.18 and 0.12 atm

$\therefore$ $Pl_c = 0.1026 \quad Pl_w = 0.0684$

The values of emissivities and correction factors are read from chart:

(c denotes CO_2 and w denotes water vapour)

$$\varepsilon_c = 0.105, \qquad C_c = 1$$

$$\varepsilon_w = 0.085 \qquad C_w = 1.05$$

$$\Delta\varepsilon = 0.02$$

$\therefore$ $\boldsymbol{\varepsilon} = 0.105 + 1.05 \times 0.085 - 0.02 = \mathbf{0.174}$

The absorptivity values are determined using modified Pl values

$$Pl_c' = 0.1025 \times \frac{500}{1200} = 0.043$$

$$Pl_w' = 0.0684 \times \frac{500}{1200} = 0.029$$

$$\alpha_c = C_c \left(\frac{T_g}{T_\omega}\right)^{0.65} .\,\varepsilon_c \qquad \text{reading for } \varepsilon_c \text{ from chart}$$

$$= 1 \times \left(\frac{1200}{500}\right)^{0.65} \times 0.08 = 0.141$$

$$\boldsymbol{\alpha_w} = 1.05 \times \left(\frac{1200}{500}\right)^{0.45} \times 0.098 = \mathbf{0.153}$$

$$\Delta\varepsilon = 0.020$$

$\therefore$ $\boldsymbol{\alpha_g} = 0.141 + 0.153 - 0.02 = \mathbf{0.274}$

$\therefore$ $\mathbf{q} = 5.67\,[0.174\,(12)^4 - 0.274\,(5)^4] = \mathbf{19487 \text{ W/m}^2}.$

Example A-13.11: *A gas flow between two large parallel planes 0.7 m apart is at 1000°C. The plates are at 500°C. The gas is at 3 atm and contains 20% CO_2 and 15% H_2O by volume, the rest being non radiating gases. Calculate the heat exchange with the plates per unit area.*

Solution: $L = 0.7 \times 1.8 = 1.26$ m. $P_{CO_2} = 3 \times 0.2 = 0.6$ atm, $P_{H_2O} = 3 \times 0.15 = 0.45$ atm denoting CO_2 by c and H_2O by w, $P_cL_c = 0.756$, $P_wL_w = 0.567$, From chart $\varepsilon_w = 0.24$, $C_w = 1.4$,

$\varepsilon_c = 0.185$, $C_c = 1.2$, $\Delta\varepsilon = 0.054$, $\boldsymbol{\varepsilon} = \mathbf{0.504}$. Absorptivity is determined corresponding to wall temperature as

$$Pl_c' = \frac{773}{1273} \,.\, 0.756 = 0.459,\ Pl'_w = \frac{773}{1273} \,.\, 0.567 = 0.344$$

Reading from chart

$$\alpha_w = 0.27,\ C_w = 1.45$$

$$\alpha_c = 0.17,\ C_c = 1.13,\ \Delta\varepsilon = 0.028$$

$$\alpha_w = 1.45 \times 0.27 \left(\frac{1273}{773}\right)^{0.45} = 0.49$$

$$\alpha_c = 1.13 \times 0.17 \left(\frac{1273}{773}\right)^{0.65} = 0.266$$

$$\boldsymbol{\alpha_g} = 0.266 + 0.49 - 0.028 = \mathbf{0.728}$$

$$\mathbf{q} = (\varepsilon_g E_{bg} - \alpha_g E_{bw})$$

$$= 5.67\,[0.504\,(12.73)^4 - 0.728\,(7.73)^4]$$

$$= \mathbf{60308\ W/m^2} \text{ or } 60.31\ \text{kW/m}^2.$$

Example A-13.12: *Determine the shape factor F_{1-2} and F_{2-1} for the following cases shown in Fig. AE-13.12. Also find F_{2-2}.*

Solution: (*a*) **Surface 1** is the base of hemisphere. All the radiation from surface 1 reaches surface 2. Hence $F_{1-2} = 1$

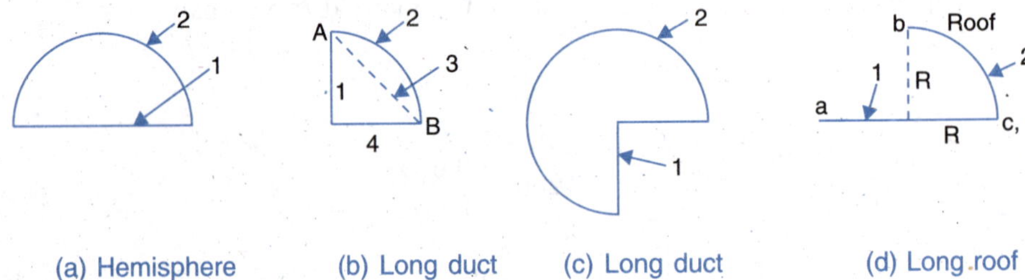

Fig. AE-13.12

Using reciprocity relations, as surface area of hemisphere is $2\pi r^2$

$$A_1F_{1-2} = A_2 F_{2-1},\ \frac{\pi D^2}{4} \times 1 = 2\pi\,(D/2)^2\, F_{2-1}$$

$\therefore$ $\quad \mathbf{F_{2-1} = 0.5}$

Considering surface 2, $F_{2-1} + F_{2-2} = 1 \quad \therefore \quad \mathbf{F_{2-2} = 0.5}$

Half the radiation from the hemisphere surface is intercepted by itself.

(*b*) **Quarter of long cylinder:**

An imaginary surface joining edges *A* and *B* is named surface 3.

$F_{3-2} = 1$ as all radiations from surface 3 reaches surface 2.

By reciprosity rule

$$A_3F_{3-2} = A_2\,F_{2-3}$$

Considering unit length

$$\sqrt{2}\ .\ R\ .\ 1 = \frac{\pi}{2}\ .\ R\ .\ F_{2-3}$$

$$F_{2-3} = 0.9003$$

$$\therefore \qquad F_{2-2} = 1 - 0.9003 = 0.0997$$

Now considering surface 1, and the perpendicular surface,

$$2F_{2-1} + F_{2-2} = 1 \quad \therefore \quad F_{2-1} = 0.4502$$

Using reciprocity theorem,

$$A_1F_{1-2} = A_2\,F_{2-1}$$

Considering, unit length,

$$R.\ F_{1-2} = \frac{\pi}{2}\ .\ R \times 0.4502$$

$$\therefore \qquad F_{1-2} = 0.7070$$

Shape factor to the perpendicular surface from surface 1

$$F_{1-4} = 1 - 0.7070 = \mathbf{0.2930.}$$

This can be checked by crossed string method which gives, (0.2929)

(*c*) **For the long duct considering** the surface 1, (it does not see itself):

So $\qquad F_{1-2} = 1$

By reciprocity rule

$$A_1F_{1-2} = A_2\,F_{2-1}$$

Considering unit length

$$2R \times 1 = 1.5\,\pi\,R\,F_{2-1} \quad \therefore \quad F_{2-1} = 0.4244$$

$$\therefore \qquad \mathbf{F_{2-2}} = 1 - 0.4244 = \mathbf{0.5756}$$

(*d*) **The shape factor is calculated** using crossed string method. (R = 1)

$$F_{1-2} = \frac{(ad + bc) - (ab + cd)}{2L_1} = \frac{(2 + \sqrt{2}) - (\sqrt{2} + 0)}{2 \times 2} = 0.5$$

This can be also checked be extending the roof as a half cylinder. Shape factor for base to full cylinder = 1. To the quarter cylinder shape factor – 0.5.

$$A_1\,F_{1-2} = A_2F_{2-1}$$

$$2 \times 0.5 = \frac{\pi \times 1}{2} \times F_{2-1} \quad \therefore \quad F_{2-1} = 0.6366$$

$\mathbf{F_{2-2}}$ from previous example – > **0.0997**

$$\therefore \qquad F_{2-space} = 0.2637 \rightarrow (1 - 0.6366 - 0.0997)$$

$$\therefore \qquad F_{1-space} = 1 - 0.5 = 0.5.$$

Example A-13.13: *Considering the following long grooves of width 2 R determine the shape factors from the grooves to the outside and also the grooves to themselves. (Fig. AE-13.13).*

Solution: (*a*) **Circular groove:** Considering a fictitious surface to cover the groove and calling it as surface 2.

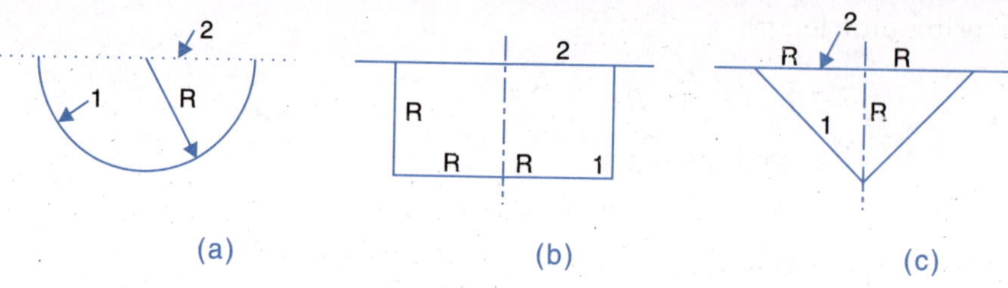

Fig. AE-13.13

$$F_{2-1} = 1 \text{ as all radiation from the cover reaches the groove}$$

By reciprocity rule

$$A_1F_{1-2} = A_2 F_{2-1}$$

(*a*) Considering unit length, **for cylindrical groove**

$$\pi R F_{1-2} = 2 R \times 1$$

$$\therefore \quad \mathbf{F_{1-2} = 2/\pi = 0.6366}$$

$$\mathbf{F_{1-1} = 1 - 0.6366 = 0.3634}$$

In case of hemispherical shape, $F_{2-1} = 1$

$$\pi R^2 \times 1 = 2\pi R^2 \,.\, F_{1-2}, \quad \therefore \quad F_{1-2} = 0.5, F_{1-1} = 0.5$$

(*b*) **Rectangular groove:**

$$F_{2-1} = 1$$

$$A_1 F_{1-2} = A_2 F_{2-1}$$

$$4 R F_{1-2} = 2R \times 1 \qquad \therefore \quad \mathbf{F_{1-2} = 0.5}, \quad \therefore \quad \mathbf{F_{1-1} = 0.5.}$$

(*c*) **Triangular groove:** Considering a fictitious cover – 2,

$$F_{2-1} = 1,$$

$$A_1F_{1-2} = A_2F_{2-1}$$

Considering unit length

$$2\sqrt{2}\, R F_{1-2} = 2 R F_{2-1} \quad \therefore \quad \mathbf{F_{1-2} = 0.7071}, \quad \therefore \quad \mathbf{F_{1-1} = 0.2929}$$

Between the two faces: calling one of the face as 1′ and the other as 1″

$$2F_{2-1'} = 1, \quad \therefore \quad F_{2-1'} = 0.5$$

$$A_2F_{2-1'} = A_{1'} F_{1-2'}$$

$$2R \times 0.5 = \sqrt{2}\, RF_{1'-2}$$

$$2R \times 0.5 = \sqrt{2} R\, F_{1'-2}$$

$$\therefore \quad F_{1'-2} = 1/\sqrt{2} = 0.7071$$

$$\therefore \quad F_{1'-1''} = 1 - 0.7071 = 0.2929$$

Example A-13.14: *In Fig. AE-13.14. determine the shape factor between surfaces 1–2 and also 2 to 1. Also determine the heat flow if T_1 = 1000 K and T_2 = 5000 K.*

Solution:

$$\varepsilon_1 = 0.6, \varepsilon_2 = 0.4$$

$$F_{1-2,\,3} = F_{1-2} + F_{1-3}$$

height of vertical plane/width $= 0.6/1 = 0.6$

length of horizontal plane/width = 1/1 = 1, from chart

$\therefore \quad F_{1-2,3} = 0.16138$

F_{1-3}

height of vertical plane/width = 0.4

length of horizontal plane/width = 1.

$$F_{1-3} = 0.1277$$

$$\therefore \quad \mathbf{F_{1-2}} = 0.16138 - 0.1277 = \mathbf{0.03368}$$

$$A_1 F_{1-2} = A_2 F_{2-1}$$

$$1 \times 1 \times 0.03368 = 1 \times 0.2 \times F_{2-1} \quad \therefore \quad \mathbf{F_{2-1} = 0.1684}$$

$$\mathbf{Q} = \frac{\sigma(T_1^4 - T_2^4)}{\dfrac{1-\varepsilon_1}{A_1\,\varepsilon_1} + \dfrac{1}{A_1\,F_{12}} + \dfrac{1-\varepsilon_2}{A_2\,\varepsilon_2}}$$

$$= \frac{5.67\,(10^4 - 5^4)}{\dfrac{1-0.6}{1\times 0.6} + \dfrac{1}{1\times 0.03368} + \dfrac{1-0.4}{0.2\times 0.4}} = \mathbf{1404.1\ W}$$

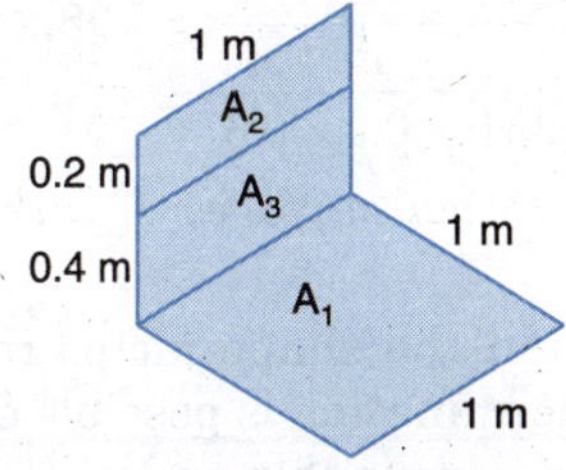

Fig. AE-13.14

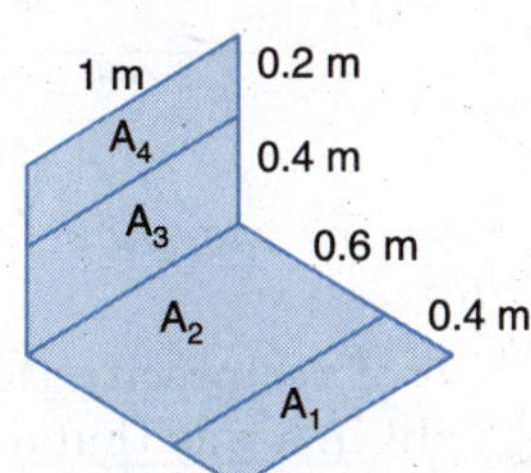

Fig. AE-13.15

Example A-13.15: *In Fig. AE-13.15 determine the shape factor between the surfaces 1–4 and 4–1. Also determine the heat flow if* $\varepsilon_1 = 0.4$ *and* $\varepsilon_4 = 0.6$ *and* $T_1 = 1000$ *K and* $T_4 = 500$ *K.*

Solution:
$$A_{1,2}\,F_{1,2-3,4} = A_1 F_{1-3,4} + A_2\,F_{2-3,4} \quad \text{...(A)}$$

$F_{1,2-3,4}$ and $F_{2-3,4}$ can be evaluated as two perpendicular surfaces and the remaining term is

$$A_1 F_{1-3,4} = A_1 F_{1-3} + A_1 F_{1-4} \quad \text{...(B)}$$

Also $\quad A_1 F_{1-3} = A_3\,F_{3-1}$

As $\quad F_{3-1} = F_{3-2,1} - F_{3-2}$

$F_{3-2,1}$ and F_{3-2} can be evaluated as two perpendicular surfaces, so

(B) reduces to $\quad A_1\,F_{1-3,4} = A_3\,(F_{3-2,1} - F_{3-2}) + A_1\,F_{1-4}$

Substituting all these in (A)

$$A_{1,2}\,F_{1,2-3,4} = A_3\,(F_{3-2,1} - F_{3-2}) + A_1\,F_{1-4} + A_2 F_{2-3,4} \quad \text{...(C)}$$

The only unknown is F_{1-4}

Evaluating the various shape factors:

$F_{1,2-3,4}$ horizontal length to width = 1, vertical length to width = 0.6

$\therefore \quad F_{1,2-3,4} = 0.16138$

$F_{3-2,1}$ Consider 3 as horizontal for use of charts,
horizontal length to width = 0.4
Vertical length to width = 1

$F_{3-2,1} = 0.31924$

F_{3-2} as in $F_{3-2,1}$ horizontal length to width = 0.4
vertical length to width = 0.6

$\therefore$ $F_{3-2} = 0.28809$

$F_{2-3,4}$ horizontal length to width = 0.6
Vertical length to width = 0.6

$\therefore$ $F_{2-3,4} = 0.23147$, substituting in C after rearrangement.

$$F_{1-4} = \frac{1}{A_1}\left[A_{1,2}\,F_{1,2-3,4} + A_3\,F_{3-2} - A_2\,F_{2-3,4} - A_3\,F_{3-2,1}\right]$$

$$= \frac{1}{0.4}\left[1 \times 0.16138 + 0.4 \times 0.28809 - 0.6 \times 0.23147 - 0.4 \times 0.31924\right] = \mathbf{0.0251}$$

$$A_1F_{1-4} = A_4F_{4-1}$$

$\therefore$ $\mathbf{F_{4-1}} = 0.4 \times 0.0251/0.2 = \mathbf{0.0502}$

Heat flow 1 – 4 $$Q = \frac{5.67\,[10^4 - 5^4]}{\dfrac{1-0.4}{0.4\times 0.4} + \dfrac{1}{0.4 \times 0.0251} + \dfrac{1-0.6}{0.2 \times 0.6}} = \mathbf{498.24\ W.}$$

Example A-13.16: *Determine the shape factor between surfaces 1 and 4 in the following Fig. [AE-13.16(a) and (b)].*

Solution: (*a*) By the use of reciprocity theorem and the shape factor relations, the shape factors are step by step split in a way that direct determination is possible except for one of the shape factor, which is then written interms of the determinable ones.

$$A_{1,2}\,F_{1,2-3,4} = A_1F_{1-3,4} + A_2F_{2-3,4} \quad \text{...(A)}$$

$$A_1F_{1-3,4} = A_1F_{1-3} + A_1F_{1-4}$$

$$A_2F_{2-3,4} = A_2F_{2-3} + A_2F_{2-4}$$

$$A_2\,F_{2-3} = A_3F_{3-2}$$

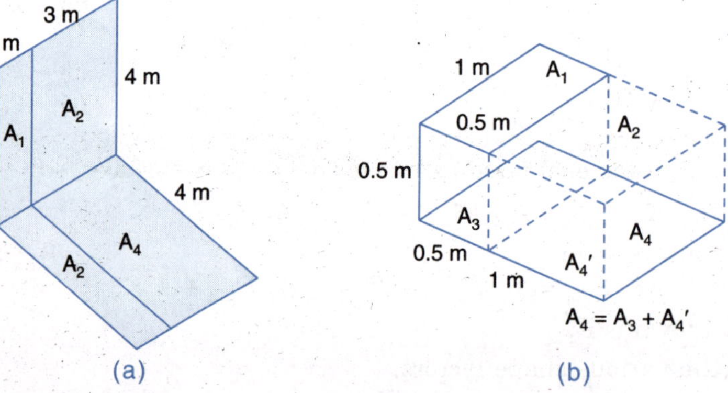

Fig. AE-13.16

As areas A_1 and A_3 are equal and also A_2 and A_4 are equal.

$$A_3F_{3-2} = A_1F_{1-4}$$

Substituting in equation (A),

$$A_{1,2}\,F_{1,2-3,4} = A_1F_{1-3} + A_2\,F_{2-4} + A_1F_{1-4} + A_3F_{3-2}$$
$$= A_1F_{1-3} + A_2F_{2-4} + 2\,A_1F_{1-4} \qquad ...(B)$$

$F_{1,2-3,4}$, F_{1-3}, F_{2-4} can be determined directly from charts as these areas are rectangles with common edge and at right angles.

$F_{1,2-3,4}$: ratios of lengths to common edge are both equal to 1, reading from chart

$\mathbf{F_{1,2-3,4} = 0.2004}$

F_{1-3} Both ratios are 4, $\mathbf{F_{1-3} = 0.10136}$

F_{2-4} The ratios are 4/3 = 1.333 each and $F_{2-4} = 0.182$ (interpolated)

Substituting in equation (B)

$\therefore \quad 4 \times 4 \times 0.2004 = 4 \times 0.10136 + 4 \times 3 \times 0.182 + 2 \times 4\,F_{1-4}$

Solving $\quad \mathbf{F_{1-4} = 0.077}$

(*b*) In this case the shape factor is from the small parallel area to the larger area. Extend it and name the parts as 1, 2, 3, 4′ as in figure, then using similar arguments as in (*a*) equation (B) $A_4 = A_3 + A_{4'}$

$$A_{1,\,2}\,F_{1,2-3,4'} = A_1F_{1-3} + A_2F_{2-4'} + 2\,A_1F_{1-4'}$$

All values except $F_{1-4'}$ can be determined.

$F_{1,2-3,4'}$: longer side to inter plane distance = 1.5/0.5 = 3

shorter side to interplane distance = 1/0.5 = 2

$\therefore \quad \mathbf{F_{1,2-3,4'} = 0.475}$ **(chart)**

F_{1-3} : longer side to inter plane distance = 1/0.5 = 2

shorter side to interplane distance = 0.5/0.5 = 1

$\therefore \quad \mathbf{F_{1-3} = 0.28588}$

$F_{2-4'}$: longer side to inter plane distance = 1/0.5 = 2

shorter side to interplane distance = 1/0.5 = 2

$\mathbf{F_{2-4'} = 0.41525}$

Solving $\quad \mathbf{F_{1-4'} = 0.15431}$

But $\quad \mathbf{F_{1-4}} = F_{1-3} + F_{1-4'} = 0.28588 + 0.15431 = \mathbf{0.44019}$.

Example A-13.17: *A cylindrical electrical heater with heating from the inside curved surface shown in Fig. AE-13.17 radiates to a circular disk coaxial with the cylinder. Determine the shape factor from the internal surface of the cylinder to the disk.*

Solution: Consider the heater base *AA*(2), The shape factor from curved surface to plane *AA* can be first determined. The shape factor between *AA*(2) and the disk can then be determined.

The product will be the shape factor from the curved surface to the disk.

i.e. $\quad F_{1-2} \times F_{2-4} = F_{1-4}$

Considering the base *AA*(2) and top *BB*(3), circles of 1 m dia, displaced by 1 m, the shape factor is obtained from chart as $F_{3-2} = 0.18$

Chapter 13

∴ $F_{3-2} + F_{3-1} = 1$ as surface 2 and 1 enclose 3.

∴ $\mathbf{F_{3-1} = 1 - 0.18 = 0.82}$

$$A_3 F_{3-1} = A_1 F_{1-3} \quad \frac{\pi \times 1 \times 1}{4} \,.\, 0.82 = \pi \times 1 \times 1 \times F_{1-3}$$

Solving $\mathbf{F_{1-3} = F_{1-2} = 0.205}$

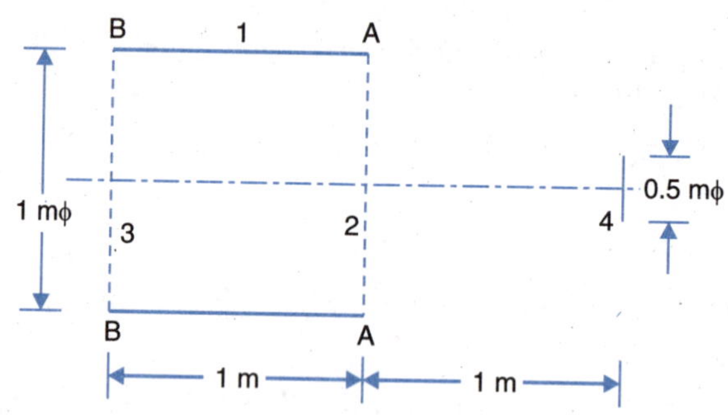

Fig. AE-13.17

To find F_{2-4}: The surfaces 2 and 4 are coaxial unequal disks and using the chart (reference book or data book) with parameters as (*i*) diameter of larger disk to interplane distance = 1 (as L = 1m and the diameter of the larger disc is also 1 m) and (*ii*) interplane distance to diameter of smaller disk = 2 (L/r = 1/0.5 = 2).

$$\mathbf{F_{2-4} = 0.12}$$

∴ $F_{1-4} = 0.205 \times 0.12 = \mathbf{0.0246}$

The heat flow rate can be calculated if other parameters are specified.

Example A-13.18: *A cylindrical shaped oven of section as shown in Fig. AE-13.18 has its curved surface maintained at 600 K. The flat surface is at 300 K. The oven is used for drying. Determine the heat transfer rate per m length of oven. The surfaces may be assumed as black. The oven may be considered as a long one.*

Solution: As the oven is a long one, the heat flow at the ends is negligible $F_{1-2} = 1$ as all the radiation from surface 1 is completely intercepted by surface 2.

$$A_1 F_{1-2} = A_2 F_{2-1}$$

Considering unit length

$$D \times 1 \times 1 = \frac{\pi D}{2} \times 1 \times F_{2-1}$$

∴ $F_{2-1} = 2/\pi = 0.6366.$

$\mathbf{F_{2-2} = 1 - 0.6366 = 0.3634}$

$$\mathbf{q} = A_1 F_{1-2}\, \sigma(E_{b1} - E_{b2}) = \pi \times 0.5 \times 1 \times 0.6366 \times 5.67\,(6^4 - 3^4)$$

$$= \mathbf{6889\ W/m\ length.}$$

Fig. AE-13.18

Example A-13.19: *A small circular disk of 15 mm dia (1) is located as shown in Fig. AE-13.19. Determine the shape factor from the disk to the dome and to the cylinder. Also determine the shape factor for half the length of the cylinder.*

Solution: The shape factor from 1 to 2 and 3 should be equal to one as all the radiation from 1 will be intercepted by these surfaces.

Now considering a fictitious surface 4 forming the base of the hemisphere, the shape factor to this surface from 1 can be calculated.

Using equation [13.23(*a*)] $F_{1-4} = D^2/(D^2 + 4L^2)$ where L is the distance between the planes and D is the diameter of the larger disk,

$$\mathbf{F_{1-4}} = 0.4^2/(0.4^2 + 4 \times 0.4^2) = \mathbf{0.2}$$

$\therefore$ $\mathbf{F_{1-3}} = \mathbf{0.2}$ as the heat intercepted by the base of hemisphere is the quantity reaching the hemisphere.

$$\mathbf{F_{1-2}} = 1 - 0.2 = \mathbf{0.8}$$

(*ii*) Now considering half of the curved surface, introduce a circular surface at the height and call it as 5

$$\mathbf{F_{1-5}} = 0.4^2/(0.4^2 + 4 \times 0.2^2) = \mathbf{0.5}$$

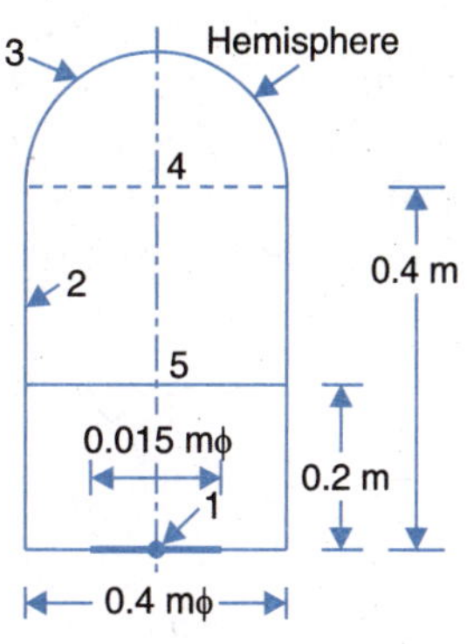

Fig. AE-13.19

$\therefore$ The shape factor **to the bottom half of cylinder** is 1 – 0.5 = 0.5. The shape factor to **the top half** = 0.5 – 0.2 = **0.3.**

(This is because the energy intercepted by the surface 5 is equal to the energy intercepted by top half of cylinder and the hemisphere. The shape factor to the hemisphere was determined as 0.2).

Example A-13.20: *Determine the shape factor for areas 2 to 3 and 4 in the Fig. AE-13.20. Also find F_{1-4} and F_{2-4}.*

Solution: $A_{12}\, F_{1,2-3,4} = A_1\, F_{1-3,4} + A_2\, F_{2-3,4}$ ...(A)

The values of $F_{1,2-3,4}$ and $F_{1-3,4}$ can be determined using charts for equal disks.

For $F_{1,2-3,4}$: $D/L = 0.8/0.2 = 4$

From chart $\mathbf{F_{1,2-3,4} = 0.58}$

$F_{1=3,4}$ is found from $F_{3,4-1}$, which has a value 0.21. (using chart for unequal disks)

By reciprocity principle: $\mathbf{F_{1-3,4} = 0.84}$

Solving equation (A), $\mathbf{F_{2-3,4} = 0.493}$

$$F_{1-3,4} = F_{1-3} + F_{1-4}$$

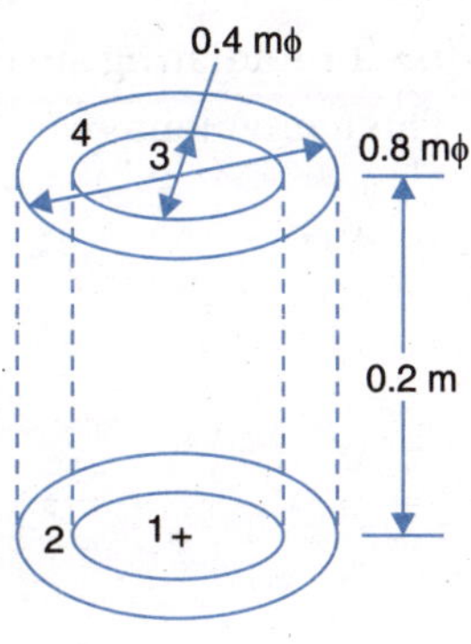

Fig. AE-13.20

F_{1-3} can be determined (equal disks): $D/L = 2$

$$\therefore \quad \mathbf{F_{1-3} = 0.35}$$

$$\therefore \quad \mathbf{F_{1-4}} = 0.85 - 0.35 = \mathbf{0.49}$$

Using reciprocity theorem: $A_1F_{1-4} = A_4F_{4-1}$

$$\frac{\pi \times 0.4^2}{4}\, 0.49 = \frac{\pi}{4}\,(0.8^2 - 0.4^2)\,.\, F_{4-1}$$

$\therefore$ $\mathbf{F_{4-1} = 0.163,\ F_{1-4} = F_{3-2}}$ (by similarity), $\mathbf{F_{2-3} = F_{4-1}}$

Chapter 13

$$\mathbf{F_{2-3,4}} = 0.493 = F_{2-3} + F_{2-4} = \mathbf{0.163 + F_{2-4}}$$

$$\therefore \qquad \mathbf{F_{2-4} = 0.33 = F_{4-2}}.$$

Example A-13.21: *The surface A_1 and A_2 [Fig. AE-13.21(a)] having emissivities of 0.6 and 0.4 are maintained at 800 K and 400 K.*

(i) Determine the heat exchange between the surfaces per unit length considering these are long with the third side open and at 400 K.

(ii) If suface 3 is well insulated, so that the surface is non absorbing determine the heat exchange.

Solution: By symmetry (long duct)

$$F_{1-2} = F_{2-1} = F_{3-2} = F_{2-3} = F_{1-3} = F_{3-1} = 0.5$$

considering surface 1 and 2 the equivalent circuit can be drawn as shown in Fig. AE-13.21(*b*).

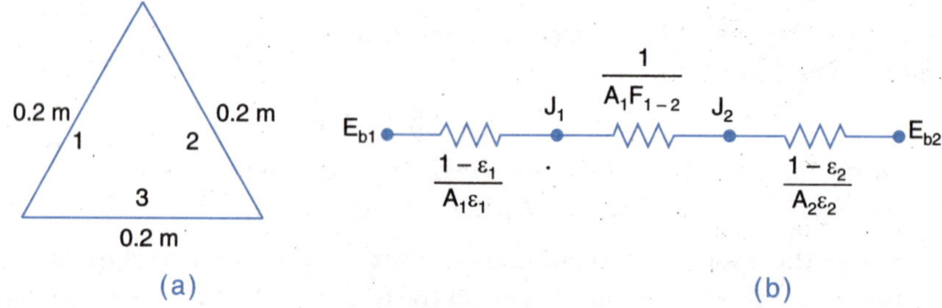

Fig. AE-13.21

$$\mathbf{Q} = \frac{\sigma(T_1^4 - T_2^4)}{\Sigma R} = 5.67\,(8^4 - 4^4)/\left(\frac{1-0.6}{0.6\times0.2\times1} + \frac{1}{0.2\times1\times0.5} + \frac{1-0.4}{0.4\times0.2\times1}\right) = \mathbf{1045\ W}$$

(*ii*) If reradiating surface is added the equivalent circuit is as shown in Fig. AE-13.21(*c*, *d*)

The equivalent resistance is 17.5 (calculate)

$$\therefore \qquad \mathbf{q} = 5.67\,(8^4 - 4^4)/17.5 = \mathbf{1244.4\ W} \text{ for one } m \text{ length.}$$

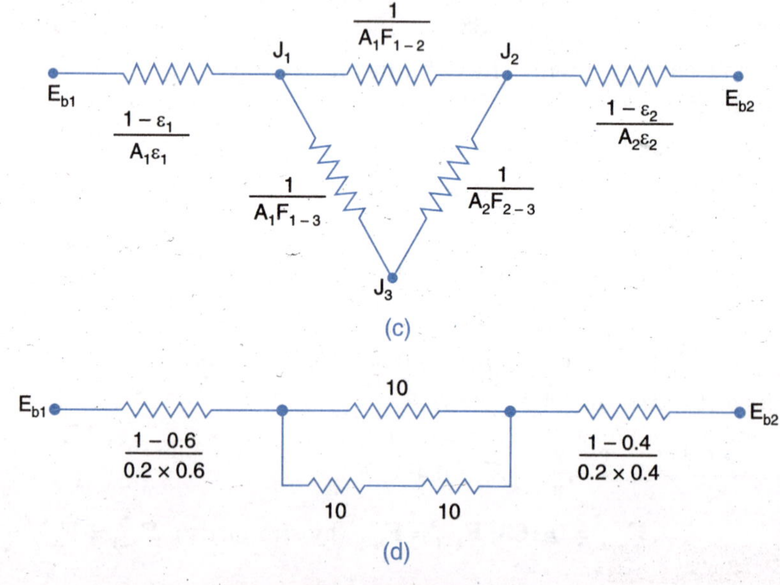

Fig. AE-13.21

Example A-13.22: *In example 13.21 if the surface 3 was open and facing a room at 300 K, determine the net heat flow from surface 1 and 2. Consider the room to be black.*

Solution: The equivalent circuit is shown in Fig. AE-13.22. As the room surface is considered black, there is no surface resistance for E_{b3}. So $J_3 = E_{b3}$

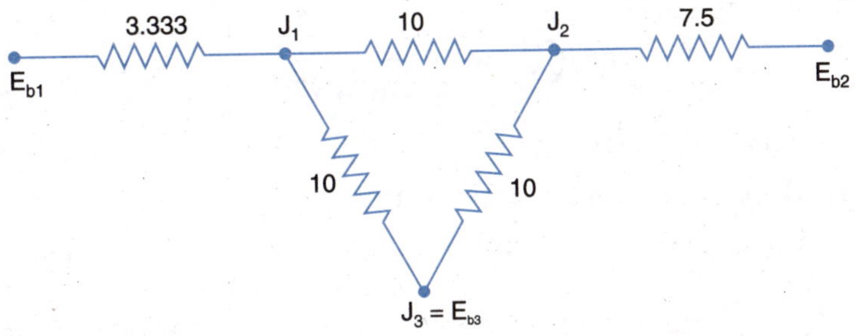

Fig. AE-13.22

The values J_1 and J_2 are solved for by considering net flow into the nodes and using values of E_{b1}, E_{b2} and E_{b3}

Node J_1:

$$\frac{E_{b1} - J_1}{R_1} + \frac{J_2 - J_1}{R_2} + \frac{E_{b3} - J_1}{R_5} = 0$$

$$R_1 = 3.333, R_2 = 10, R_5 = 10, R_3 = 10, E_{b1} = 23224, E_{b3} = 459 = J_3$$

Substituting, we get

$$70132 - 5\,J_1 + J_2 = 0. \quad \text{...(A)}$$

Node J_2:

$$\frac{E_{b2} - J_2}{R_3} + \frac{J_1 - J_2}{R_2} + \frac{E_{b3} - J_2}{R_4} = 0$$

$$R_3 = 7.5, R_2 = R_4 = 10, E_{b2} = 1452$$

Substituting we get

$$2395 + J_1 - 3.33\,J_2 = 0. \quad \text{...(B)}$$

Solving for J_1 and J_2 using equations (A) and (B)

$$\mathbf{J_1 = 15074}\ \text{W/m}^2, \mathbf{J_2 = 5240}\ \text{W/m}^2$$

Net Heat flow from surface 1

$$\mathbf{q_1} = \frac{E_{b1} - J_1}{R_1} = \frac{23224 - 15074}{3.333} = \mathbf{2445\ W}$$

Heat flow from surface 1 to 3 (room)

$$\mathbf{q_{1\text{-}3}} = \frac{J_1 - J_3}{R_5} = \frac{15074 - 459}{10} = \mathbf{1462\ W}$$

Heat flow across J_1, J_2

$$\mathbf{q_{1\text{-}2}} = \frac{15074 - 5240}{10} = \mathbf{983\ W}$$

Check: 1462 + 983 = 2445 W

Net Heat flow to surface 2

$$\mathbf{q_2} = \frac{5240 - 1452}{7.5} = \mathbf{505\ W}$$

Heat flow from J_2 to surface 3

$$\mathbf{q_{2-3}} = \frac{5240 - 459}{10} = \mathbf{478\ W}$$

Check: 505 + 478 = 983 W

Heat received by room = 1462 + 478 = **1940 W.**

Example A-13.23: ***A spherical container*** *with 0.9 m dia stores cryogenic fluid at – 178°C. The surface emissivity is 0.03. It is enclosed by another concentric sphere of 1 m dia with a surface emissivity of 0.04. The interspace is evacuated. The outside surface is at 10°C. Determine the rate of heat leakage.*

Solution: As the interface is evacuated only radiation is to be considered. The resistances are the two surface resistances and the space resistance. Shape factor F_{1-2} = 1 (from inner outer).

$$\mathbf{Q} = 5.67\left[\left(\frac{283}{100}\right)^4 - \left(\frac{95}{100}\right)^4\right] \Bigg/ \left[\frac{1-0.03}{0.03 \times 4\pi \times 0.45^2} + \frac{1}{4\pi \times 0.45^2 \times 1} + \frac{1-0.04}{0.04 \times 4\pi \times 0.5^2}\right] = \mathbf{17.65\ W}$$

If the latent heat of evaporation is about 213 kJ/kg, the evaporation rate will be

$$\frac{17.65}{213000}\ \text{kg/s or}\ \frac{17.65 \times 3600 \times 24}{213000}\ \text{kg/day}\quad \textbf{i.e.}\quad \mathbf{7.16\ kg/day.}$$

Example A-13.24: *A surface at 100 K with emissivity of 0.10 is protected from a radiation flux of 1250 W/m² by a shield with emissivity of 0.05. Determine the percentage cut off and the shield temperature.*

Solution: We have to assume a shape factor of one. As flux is given it is to be taken as radiosity *J*.

The equivalent circuit is shown in Fig. AE-13.24(*b*). Assuming unit area.

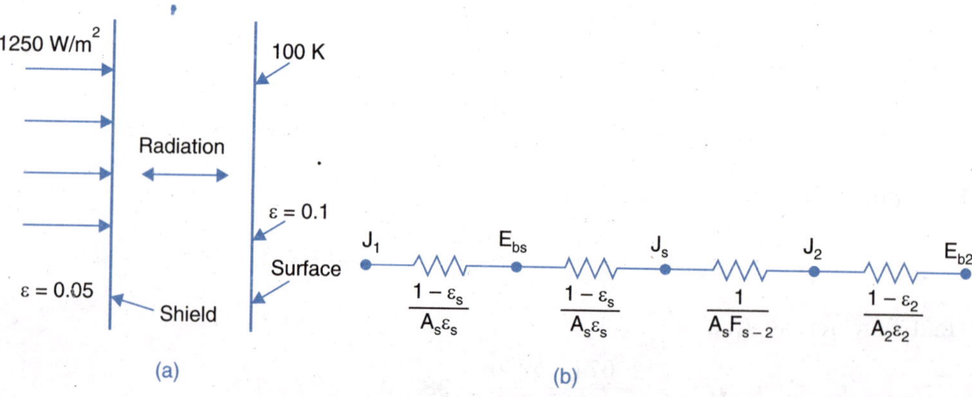

Fig. AE-13.24

$$q = \frac{1250 - 5.67 \times 1^4}{\frac{1-0.05}{1\times 0.05} + \frac{1-0.05}{1\times 0.05} + \frac{1}{1} + \frac{1-0.1}{0.1}} = \mathbf{25.92\ W/m^2}$$

If the **shield was not present** heat flow will be (as only the surface resistance exits = (1 – 0.1)/0.1)

$$= \frac{1250 - 5.67}{9} = \mathbf{138.26\ W/m^2}$$

Percentage cut off = 100(138.26 – 25.92)/138.26 = 81.25%

Shield temperature can be determined using (considering surface resistance)

$$25.92 = \frac{1250 - E_s}{19} \quad \therefore \quad E_s = 757.52$$

$\therefore$ $\mathbf{T_s = 340\ K\ or\ 67°C}$

Check: $Q = \frac{E_s - E_2}{19+1+9} = \frac{757.52 - 5.67}{29}$ = 25.93 W, checks.

Chapter 13

Example A-13.25: *Two large parallel planes with emissivity value of 0.8 are exchanging heat by radiation. It is desired to interpose a radiation shield with emissivity value of ε_s on both sides, so that the heat exchange will be reduced to 1/10 of the original value. Estimate the emissivity requirement.*

Solution: The thermal resistance should be increased to 10 times the original value to achieve the heat reduction.

The thermal resistance without the shield.

$$= \frac{1-0.8}{0.8} + 1 + \frac{1-0.8}{0.8} = 1.5$$

The resistance with shield should be 10 × 1.5 = 15

$$= \frac{1-0.8}{0.8} + 1 + \frac{1-\varepsilon_s}{\varepsilon_s} + \frac{1-\varepsilon_s}{\varepsilon_s} + 1 + \frac{1-0.8}{0.8} = 15$$

Solving: $\mathbf{\varepsilon_x = 0.137931}$

Check: $0.25 + 1 + \frac{1-0.137931}{0.137931} + \frac{1-0.137931}{0.137931} + 1 + 0.25$

= 0.25 + 1 + 6.25 + 6.25 + 1 + 0.25 = **15 checks.**

Example A-13.26: *Two large parallel palnes with emissivity of 0.8 are at 1000 K and 400 K. A shield with one side treated and having an emissivity of 0.05 while the emissivity on the other untreated side was 0.6 was proposed to be used. The designer wanted the low emissivity side to face the hotter plane. During installation by mistake the side with higher emissivity was placed facing the hot side. Investigate the change in performance if any.*

Solution: Considering unit area and using equivalent circuit approach, for the original proposal:

$$Q = \frac{E_{b1} - E_{b2}}{\left[\frac{1-\varepsilon_1}{A_1\,\varepsilon_1} + \frac{1}{A_1\,F_{1S}} + \frac{1-\varepsilon_{s1}}{A_s\,\varepsilon_{s1}} + \frac{1-\varepsilon_{s2}}{A_s\,\varepsilon_{s2}} + \frac{1}{A_s\,F_{s-2}} + \frac{1-\varepsilon_2}{A_2\,\varepsilon_2}\right]}$$

$$A_1 = A_s = A_2 = 1 \text{ and } F_{1-s} = F_{s-2} = 1$$

$$\mathbf{Q} = \frac{5.67\,[10^4 - 4^4]}{\dfrac{1-0.8}{0.8} + 1 + \dfrac{1-0.05}{0.05} + \dfrac{1-0.6}{0.6} + 1 + \dfrac{1-0.8}{0.8}} = \mathbf{2492.41\ W}$$

The shield temperature can be determined using

$$2492.41 = \frac{E_{b1} - E_{bs}}{\dfrac{1-\varepsilon_1}{\varepsilon_1} + \dfrac{1}{F_{1-2}} + \dfrac{1-\varepsilon_s}{\varepsilon_s}} = \frac{5.67\,[10^4 - (T_s/100)^4]}{\dfrac{1-0.8}{0.8} + 1 + \dfrac{1-0.05}{0.05}}$$

$\therefore$ $\mathbf{T_s = 575.7\ K}$

After the wrong installation, the total resistance remains the same as the resistances only shift their location. However the temperature of the shield will be different. It is determined using the relation.

$$2492.41 = \frac{5.67\left(10^4 - \left(\dfrac{T_s}{100}\right)^4\right)}{\dfrac{1-0.8}{0.8} + 1 + \dfrac{1-0.6}{0.6}}$$

$\therefore$ $\mathbf{T_s = 978.2\ K}$

The shield temperature is much higher due to the wrong installation.

Higher resistance should be nearer the hotter surface to reduce shield temperature.

Example A-13.27: ***A cryogenic fluid*** *is carried in a pipe of 10 mm OD at a temperature of 100 K. The pipe is surrounded coaxially by another pipe of OD 13 mm with the space between the pipes evacuated. The other pipe is at 5°C. The emissivity for both surfaces is 0.22. Determine the radiant heat flow for 3 m length. If a shield of emissivity of 0.05 and of dia 11.5 mm is placed between the pipes determine the percentage reduction in heat flow.*

Solution: Considering 3 m length, the pipe areas are:

Inside pipe: $\pi \times 0.01 \times 3 = 0.0942\ m^2$

Shield pipe: $\pi \times 0.0115 \times 3 = 0.1084\ m^2$

Outside pipe: $\pi \times 0.013 \times 3 = 0.1225\ m^2$

The heat flow without shield is calculated using the equivalent circuit method.

$$\mathbf{Q} = \frac{E_{b1} - E_{b2}}{\dfrac{1-\varepsilon_1}{A_1\,\varepsilon_1} + \dfrac{1}{A_1\,F_{1-2}} + \dfrac{1-\varepsilon_2}{A_2\,\varepsilon_2}}$$

$$= \frac{5.67\,(2.78^4 - 1^4)}{\dfrac{1-0.22}{0.0942 \times 0.22} + \dfrac{1}{0.0942 \times 1} + \dfrac{1-0.22}{0.22 \times 0.1225}} = \mathbf{4.315\ W}$$

with shield

$$\mathbf{Q} = \frac{5.67\,(2.78^4 - 1)}{\dfrac{1-0.22}{0.0942\times0.22} + \dfrac{1}{0.0942\times1} + \dfrac{1-0.05}{0.1084\times0.05} + \dfrac{1-0.05}{0.1084\times0.05} + \dfrac{1}{0.1084\times1} + \dfrac{1-0.22}{0.22\times0.1225}}$$

$= \mathbf{0.762\ W.}$

$$\text{Percentage reduction} = \frac{4.315 - 0.762}{4.315} \times 100 = 82.34\%.$$

An air gap and low emissivity filaments are used in furnace wall construction to reduce heat loss.

Example A-13.28: *Show that if surfaces of equal emissivities are used as shields between parallel planes of large size, the heat flow will be reduced to 1/(n + 1) of heat flow without shield, where n is the number of shields.*

Solution: Resistance when no shield is present using unit area, denoting the surface resistance $(1 - \varepsilon)/\varepsilon$ as R_s and the space resistance $1/F_{1-2}$ as R

Resistance $= 2 R_s + R$

When 1 shield is present

Resistance $= 4R_s + 2R = 2(2R_s + R)$ (Draw the circuit and verify)

When 2 shields are present:

Resistance $= 6 R_s + 3R = 3(2R_s + R)$

and so on. So obviously the resistance increases $n + 1$ times and hence the heat flow will become $1/(n + 1)$ time, provided all emissivities are the same.

Chapter 13

Example A-13.29 : *Two circular disks of 1 m dia are placed coaxially, parallely and symmetrically at a distance of 1 m. The disks have an emissivity of 0.6 and are at 1000 K and 500 K. Determine the reduction in radiant heat flow due to the introduction of a shield of equal diameter midway between the two. The shield has an emissivity of 0.1 on both sides. (neglect interactions to the outside space).*

Solution: In this case, the shape factors will change. These are found using charts.

Shape factor between the plates: 0.18, $(D/x = 1)$

Shape factor between plate and shield: 0.39, $(D/x = 2)$

without shield

$$Q = \frac{E_{b1} - E_{b2}}{\dfrac{1-\varepsilon_1}{A_1\,\varepsilon_1} + \dfrac{1}{A_1F_{1-2}} + \dfrac{1-\varepsilon_2}{A_2\,\varepsilon_2}}$$

$$= \frac{5.67\,(10^4 - 5^4)}{\dfrac{(1-0.6)\times 4}{0.6\times\pi\times 1\times 1} + \dfrac{1\times 4}{\pi\times 1\times 1\times 0.18} + \dfrac{(1-0.6)\times 4}{0.6\times\pi\times 1\times 1}} = \mathbf{6060\ W.}$$

with shield

$$Q = \frac{5.67\,(10^4 - 5^4)}{\dfrac{(1-0.6)\,4}{0.6\times\pi\times 1\times 1} + \dfrac{1\times 4}{\pi\times 1\times 1\times 0.39} + \dfrac{(1-0.1)\,4}{0.1\times\pi\times 1\times 1} + \dfrac{(1-0.1)\times 4}{0.1\times\pi\times 1\times 1} + \dfrac{1\times 4}{\pi\times 1\times 1\times 0.39} + \dfrac{(1-0.6)\,4}{0.6\times\pi\times 1\times 1}}$$

$= \mathbf{1707\ W}$

Percentage reduction $= \dfrac{6060 - 1707}{6060} \times 100 = \mathbf{71.83\%}$

Shield temperature $=$ **853.7 K (check)**

When limited size planes are considered, these will also exchange heat through the interspace with surroundings. Hence the assumption.

Example A-13.30: *Show that for two black parallel planes of equal area symmetrically placed, the shape factor with reradiation by enclosure will be $(1 + F_{1-2})/2$ where F_{1-2} is the shape factor without reradiation.*

Solution: Considering the planes as 1 and 2 and the enclosure as 3, $F_{1-2} = 1 - F_{1-3}$ as the radiation from 1 is intercepted completely by plane 2 and the enclosure, 3

Also by symmetry $A_1 F_{1-3} = A_2 F_{2-3}$

Using the equivalent circuit shown in Fig. AE-13.30, as surfaces are black and absorb incident radiation completely.

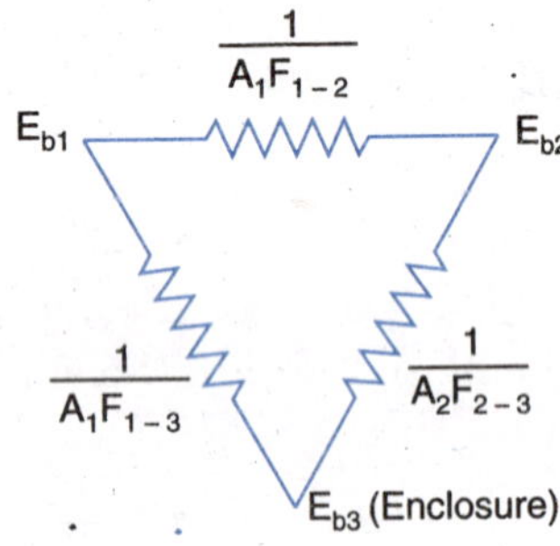

Fig. AE-13.30

Total flow of heat by radiation from surface 1 considering the direct path and path through E_{b3} to E_{b2}.

As the total resistance along the path through E_{b3} is

$$\frac{1}{A_1F_{1-3}} + \frac{1}{A_2F_{2-3}}$$

As $A_1 = A_2$, $\quad F_{1-3} = F_{2-3}$

Total resistance $= \dfrac{2}{A_1F_{1-3}}$.

$\therefore$ Heat flow along their path

$$= \frac{E_{b_1} - E_{b_2}}{2} \cdot A_1F_{1-3}$$

$$Q_{1-2} = (E_{b1} - E_{b2})\, A_1\, F_{1-2} + \frac{(E_{b1} - E_{b2})}{2}\, A_1\, F_{1-3}$$

Denoting the total shape factor as F'_{1-2}

$$Q = A_1\, F'_{1-2}\, (E_{b1} - E_{b2})$$

By similarity $\quad F_{2-3} = F_{1-3}$ and $F_{1-3} = (1 - F_{1-2})$

Substituting,

$$(E_{b1} - E_{b2})\, A_1F_{1-2} + \frac{(E_{b1} - E_{b2})}{2}\, A_1\, (1 - F_{1-2}) = A_1\, F'_{1-2}\, (E_{b1} - E_{b2})$$

Cancelling $(E_{b1} - E_{b2})\, A_1$

$$F'_{1-2} = F_{1-2} + \frac{1-F_{1-2}}{2} = \frac{1+F_{1-2}}{2}, \text{ proved.}$$

Conceptually $$F'_{1-2} = F_{1-2} + \frac{1-F_{1-2}}{2},$$

means that out of the energy from surface 1, F_{1-2} directly reaches surface 2, half of the energy from surface 1 intercepted by surface 3 also reaches surface 2 by reradiation (the other half reaching surface 1 itself).

Example A-13.31: *Show that the temperature of the reradiating enclosure 3 between surface 1 and 2 at T_1 and T_2 is given by*

$$T_3 = \left[\frac{A_1 F_{1-3} T_1^4 + A_2 F_{2-3} T_2^4}{A_1 F_{1-2} + A_2 F_{2-3}}\right]^{0.25}$$

Solution: Using the Fig. AE-13.30, J_3 is solved for by using the node 3

$$A_1 F_{1-3} (E_{b1} - J_3) = (J_3 - E_{b2}) A_2 F_{2-3}$$

Rearranging

$$A_1 F_{1-3} E_{b1} + A_2 F_{2-3} E_{b2} = J_3 (A_1 F_{1-3} + A_2 F_{2-3})$$

or $$A_1 F_{1-3}\, \sigma T_1^4 + A_2 F_{2-3}\, \sigma T_2^4 = \sigma\, T_3^4 (A_1 F_{1-3} + A_2 F_{2-3})$$

Cancelling σ, $$T_3 = \{[A_1 F_{1-3}\, T_1^4 + A_2 F_{2-3} T_2^4]/[A_1 F_{1-2} + A_2 F_{2-3}]\}^{0.25}.$$

Example A-13.32: *A tetrahedron has each face of area 3 m². The surface 4 is insulated and acts a non absorbing reradiating surface. The emissivities of surface 1, 2 and 3 are all equal to 1/3. Calculate the net heat flow from each of the surfaces 1, 2 and 3 if the emissive powers of surfaces 1, 2 and 3 are 12000, 3000 and 750 W/m².*

Solution: The shape factor by similarity will each equal 1/3.

i.e. $$F_{1-2} = F_{2-1} = F_{1-3} = F_{3-1} = F_{2-3} = F_{3-2} = F_{3-4} \ldots = 1/3.$$

The surface 4 acts as reradiating. The equivalent circuit is shown in Fig. AE-13.32.

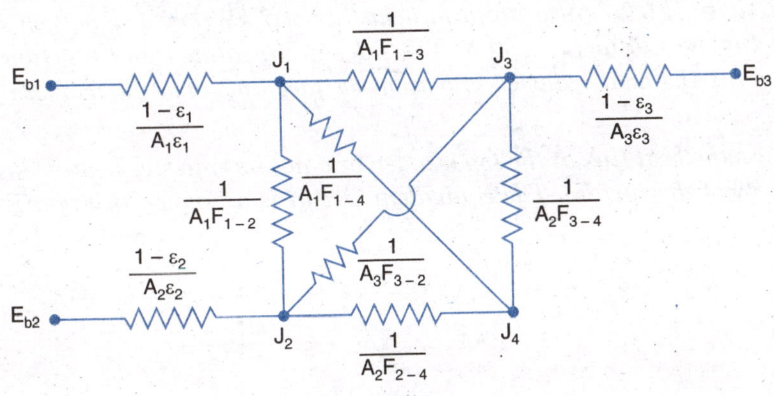

Fig. AE-13.32

The surface resistance are all equal to $\dfrac{1-1/3}{(1/3)\times 3} = 2/3$

The space resistances are all equal to $= \dfrac{1}{3\times 1/3} = 1$

Now considering nodes J_1, J_2, J_3 and J_4 and noting that the algebraic sum of heat flow into the nodes is zero.

For nodes 1, 2, 3, there are four in flows and for node 4 there are three inflows.

Node J_1: $$\frac{3(12000 - J_1)}{2} + \frac{J_3 - J_1}{1} + \frac{J_4 - J_1}{1} + \frac{J_2 - J_1}{1} = 0$$

$$18000 - 4.5\,J_1 + J_2 + J_3 + J_4 = 0$$

Node J_2: $$\frac{3(3000 - J_2)}{2} + \frac{J_1 - J_2}{1} + \frac{J_3 - J_2}{1} + \frac{J_4 - J_2}{1} = 0$$

or $$4500 + J_1 - 4.5\,J_2 + J_3 + J_4 = 0.$$

for Node J_3: $$1125 + J_1 + J_2 - 4.5\,J_3 + J_4 = 0$$

for Node J_4: $$J_1 + J_2 + J_3 - 3J_4 = 0;$$

This set of equations can be solved using any of the methods to solve the matrix

$$\begin{bmatrix} -4.5 & 1 & 1 & 1 \\ 1 & -4.5 & 1 & 1 \\ 1 & 1 & -4.5 & 1 \\ 1 & 1 & 1 & -3 \end{bmatrix} \begin{bmatrix} J_1 \\ J_2 \\ J_3 \\ J_4 \end{bmatrix} = \begin{bmatrix} 18000 \\ 4500 \\ 1125 \\ 0 \end{bmatrix}$$

The answer is $J_1 = 7041$ W/m², $J_2 = 4636$ W/m²

$J_3 = 4023$ W/m², $J_4 = 5250$ W/m²

$$\mathbf{Q_1} = \frac{E_{b1} - J_1}{2/3}, \quad \mathbf{Q_2} = \frac{E_{b2} - J_2}{2/3}$$

$$\mathbf{Q_3} = \frac{E_{b3} - J_3}{2/3}$$ solve and check. **(7363, – 2454, – 4909).**

Example A-13.33: *A cylindrical shaped furnace is 1 m dia and 1 m high. The top surface having an emissivity of 0.7 emit a uniform heat flux of 7 kW/m². The bottom surface with an emissivity of 0.4 is maintained at 350 K. The sides are insulated and function as reradiating surfaces. Determine the heat transfer to bottom surface and also the temperatures of the top and sides.*

In this case the heat flux at the top is given and not its emissive power. So the flux is to be taken as radiosity of the surface. The equivalent circuit is as shown in Fig. AE-13.33.

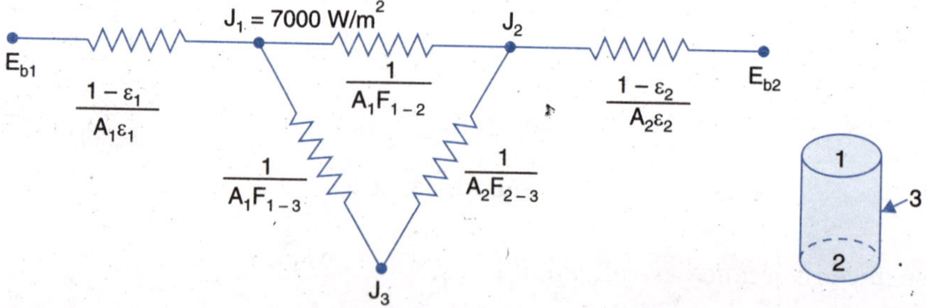

Fig. AE-13.33

Solution: Designating top as 1 bottom as 2 and sides as 3. The shape factor between the base and top

$$F_{1-2} = 0.18 = F_{2-1} \text{ (parallel disks)}$$

$$\therefore \quad F_{1-3} = 1 - 0.18 = 0.82 \quad F_{1-3} = F_{2-3} \text{ as } A_1 = A_2$$

$$\frac{1-\varepsilon_1}{A_1\,\varepsilon_1} = \frac{(1-0.7)4}{\pi \times 1 \times 1 \times 0.7} = 0.5457$$

$$\frac{1-\varepsilon_2}{A_2\,\varepsilon_2} = \frac{(1-0.4)4}{\pi \times 1 \times 1 \times 0.4} = 1.91, \ \frac{1}{A_1F_{1-2}} = \frac{1 \times 4}{\pi \times 1 \times 1 \times 0.18} = 7.074$$

$$\frac{1}{A_1F_{1-3}} = \frac{1 \times 4}{\pi \times 1 \times 1 \times 0.82} = 1.5527 = \frac{1}{A_2F_{2-3}}$$

Considering node J_1

$$\frac{E_{b1} - 7000}{0.5457} + \frac{J_3 - 7000}{1.5527} + \frac{J_2 - 7000}{7.074} = 0$$

$$1.8325\, E_{b1} + 0.644\, J_3 + 0.1414\, J_2 - 18325 = 0 \qquad \text{...(A)}$$

Considering Node J_2 $\quad \dfrac{7000 - J_2}{7.074} + \dfrac{J_3 - J_2}{1.5527} + \dfrac{E_{b2} - J_2}{1.91} = 0$

$$989.54 + 0.644\, J_3 - 1.309\, J_2 + 0.5236 \times 5.67 \times 3.5^4 = 0$$

or

$$0.644\, J_3 - 1.309\, J_2 + 1435.05 = 0 \qquad \text{...(B)}$$

Considering node J_3 $\quad \dfrac{7000 - J_3}{1.5527} + \dfrac{J_2 - J_3}{1.5527} = 0$

$$\therefore \quad 0.644\, J_2 - 1.288\, J_3 + 4508.3 = 0$$

Solving $\quad E_{b1} = 7825 \text{ W/m}^2, \quad \therefore \ \mathbf{T_1 = 609.5\ K}$

$J_2 = 3738 \text{ W/m}^2$

$J_3 = 5389 \text{ W/m}^2 \quad \mathbf{T_3 = 554.7\ K}$

To determine the heat flow:

From surface 1: $\quad \mathbf{Q_1} = \dfrac{E_{b1} - J_1}{(1-\varepsilon_1)/A_1\varepsilon_1} = \dfrac{7825 - 7000}{0.5457} = \mathbf{1512\ W}$

From surface 2 $\quad \mathbf{Q_2} = \dfrac{5369 - 5.67 \times 3.5^4}{1.91} = \mathbf{1512\ W}(J_2 \text{ to } E_{b2})$

Check:

Through $J_1 - J_3$ $\quad \dfrac{7000 - 5369}{1.5527} = 1051 \text{ W}$

Through $J_1 - J_2$ $\quad \dfrac{7000 - 3738}{7.074} = 461 \text{ W}$

The sum of these two = 1512 W. So checks.

Also through $J_3 - J_2$, (5369 – 3738)/1.5527 = 1050.4 so checks.

Note: If flux is specified, then it has to be taken as radiosity.

Chapter 13

Example A-13.34: *A thermocouple is used to measure the temperature of air flowing in a large duct, the walls of which are at 500 K. The thermocouple indicates 400 K.The heat transfer coefficient over the surface of the thermocouple is 40 W/m²K. Determine the true gas temperature if the emissivity of the thermocouple surface is 0.51.*

Solution: Consider the thermocouple to have an area of A. The heat balance for the thermocouple gives.

$$\sigma A\varepsilon\,(T_w^{\,4} - T_1^{\,4}) = hA(T_t - T_a),$$

Where T_w is the wall temperature. T_t is the thermocouple temperature and T_a is the air temperature. Substituting the values.

$$5.67 \times 0.51(5^4 - 4^4) = 40(400 - T_a)$$

Solving $\mathbf{T_a = 373.3\ K}$

Higher wall temperature compared to gas temperature will lead to higher reading of the thermocouple and vice versa.

Example A-13.35: *On clear nights there is radiation from earths surface to the space.On such a night, the water particles on plant leaves radiate to the sky whose temperature may be taken as – 70°C. The water particles receive heat from air surrounding it by convection, the convection heat transfer coefficient having a value of 28 W/m² K. If the water should not freeze, determine the air temperature.*

Solution: For water just to freeze, its temperature has to be 0°C or 273 K. A heat balance, assuming water surface to be black gives,

$$5.67A(2.73^4 - 2.03^4) = 28A(T_a - 273)$$

Solving $\mathbf{T_a = 280.8\ K}$ **or 7.8°C**

Any air temperature lower than this value will cause frost or freezing on the surfaces.

This idea is used to make ice by keeping water in shallow containers during the night exposed to the clear sky, even when air temperature is above 0°C.

Example A-13.36: ***A thermocouple of 3 mm*** *dia with emissivity of 0.6 is used to measure the temperature of gasses in an electrically heated furnace, with heaters fixed on the walls. The thermometer showed 750° C when the gas temperature was 650° C. Estimate the wall temperature if the convection coefficient has a value of 40 W/m² K.*

Note: Absolute units of temperature should be used.

Solution: A heat balance will give

$$0.6 \times 5.67\left[(T_w/100)^4 - \left(\frac{750 + 273}{100}\right)^4\right] = 40[(750 + 273) - (650 + 273)]$$

Solving, $\mathbf{T_w = 1049.4\ K}$ **of 776.4°C**

If the emissivity was 0.05 in the above case what will be the reading shown by the thermocouple.

A heat balance gives

$$0.05 \times 5.67[10.494^4 - (T_t/100)^4] = 40(T_t - 923)$$

This equation has to be solved by trial

$$0.2835[T_t/100]^4 + 40T_t - 40358 = 0$$

or $$(T_t/100)^4 + 141.09\, T_t - 142356.3 = 0$$

Solving T = 951 K or 678°C (an error of 28°C compared to 100°C with higher emissivity).

Example A-13.37: *A flat heater of 1 m dia is covered by a hemisphere of 4 m dia, the surface of which is maintained at 400 K. The emissivity of the surface is 0.8. The heater surface is maintained at 1000 K. The remaining base area is open to surroundings at 300 K. The surroundings may be considered as blak. The emissivity of the heater surface is 0.8. Determine the heat exchange from the heater to the hemisphere and to the surroundings.*

Solution: The model and the equivalent circuit are shown in Fig. AE-13.37. The heater is designated as surface 1, the hemisphere as surface 2 and the remaining base area as surface 3. The shape factor between surface 1 and 3 is zero as they are on the same plane. So J_1 and E_{b3} are not connected. Only the heat reflected and emitted by area 2 passing through area 3 reaches the surroundings. So this surface can be considered for all practical purposes for the determination of heat flow to surroundings.

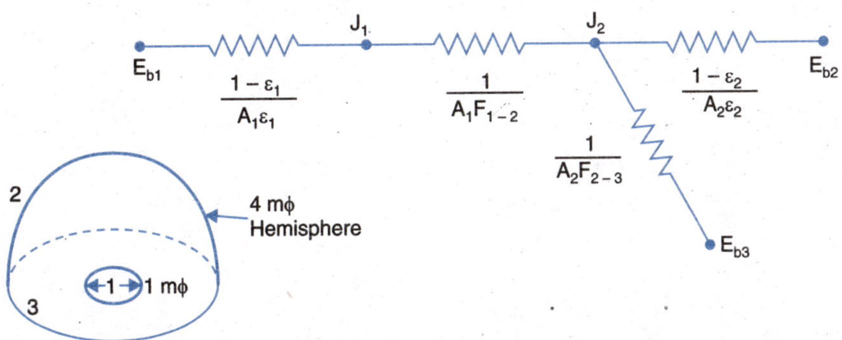

Fig. AE-13.37

Heat flows from heater to the hemisphere, part of which flows from the hemisphere to the surroundings with no direct connection between the heater and surroundings. Part of heat emitted by area 2 also reaches the surroundings. As surrounding is black, surface resistance is zero.

$F_{1-2} = 1$ as all the heat from A_1 is intercepted by the hemisphere surface 2.

$F_{3-2} = 1$ for similar reason

$$A_1F_{1-2} = A_2F_{2-1} \quad \therefore \quad F_{2-1} = \frac{A_1}{A_2}\,.\,1 = \frac{\pi \times 0.5 \times 0.5}{2 \times \pi \times 2 \times 2} = 0.03125$$

$$A_3F_{3-2} = A_2F_{2-3}$$

$$\therefore \quad F_{2-3} = \frac{A_3}{A_2} \times \frac{\pi(2^2 - 0.5^2)}{2\pi \times 2 \times 2} \times 1 = 0.46875$$

$\therefore \quad F_{2-2} = 1 - 0.46875 - 0.03125 = 0.5$ (checks as the shape factor for hemisphere to itself is 0.5).

The resistance are:

$$\frac{1-\varepsilon_1}{A_1\varepsilon_1} = \frac{1-0.8}{\pi \times 0.5 \times 0.5 \times 0.8} = 0.3183$$

$$\frac{1}{A_1F_{1-2}} = \frac{1}{\pi \times 0.5 \times 0.5 \times 1} = 1.273$$

$$\frac{1-\varepsilon_2}{A_2\varepsilon_2} = \frac{1-0.8}{2\pi \times 2 \times 2 \times 0.8} = 9.947 \times 10^{-3}$$

$$\frac{1}{A_2F_{2-3}} = \frac{1}{2\pi \times 2 \times 2 \times 0.46875} = 0.08488$$

$$E_{b1} = 5.67 \times 10^4 = 56700$$

$$E_{b2} = 5.67 \times 4^4 = 1451.52$$

$$E_{b3} = 5.6 \times 3^4 = 459.27$$

Considering node J_1

$$\frac{E_{b1} - J_1}{(1-\varepsilon_1)/A_1\varepsilon_1} = \frac{J_1 - J_2}{1/A_1 F_{1-2}}$$

$$\therefore \quad \frac{56700 - J_1}{0.3183} = \frac{J_1 - J_2}{1.273}$$

Rearranging

$$5J_1 - J_2 + 226800 = 0 \qquad \text{...(A)}$$

Considering node J_2 (Three inflows)

$$\frac{J_2 - E_{b2}}{(1-\varepsilon_2)/A_2\varepsilon_2} + \frac{J_2 - E_{b3}}{1/A_2 F_{2-3}} = \frac{J_1 - J_2}{1/A_1 F_{1-2}}$$

Substituting the values

$$\frac{J_1 - J_2}{1.273} = \frac{J_2 - 1451.52}{9.9472 \times 10^{-3}} + \frac{J_2 - 459.27}{0.08488}$$

Rearranging $J_1 - 144\, J_2 + 192683.61 = 0.$ ...(B)

Multiplying equation (B) by 5 and subtracting and solving,

$$J_1 = \mathbf{45691\ W/m^2},\ J_2 = \mathbf{1655.4\ W/m^2}$$

Heat flow from surface 1 is given by

$$\mathbf{Q_1} = (E_{b1} - J_1)/\{(1-\varepsilon_1)/A_1\varepsilon_1\} = \frac{56700 - 45691}{0.3183} = \mathbf{34586\ W}$$

To surface 2, $$\mathbf{Q_2} = \frac{J_2 - E_{b2}}{(1-\varepsilon_2)/A_2\varepsilon_2} = \frac{1655.4 - 1451.52}{9.9472 \times 10^{-3}} = \mathbf{20494\ W}$$

To outside, $$\mathbf{Q_3} = \frac{J_2 - E_{b3}}{1/A_2 F_{2-3}} = \frac{1655.4 - 459.27}{0.08488} = \mathbf{14092\ W}$$

A check can be made as Q_1 should equal Q_2 and Q_3

34586 = 20494 + 14092 = 34586 checks.

Example A-13.38: *In the case of the above example if surface 3 is insulated and acts as a reradiating surface, determine the heat flow from heater surface to the hemispherical surface.*

In this case, the node J_3 connects with node J_2 and E_{b2}, providing a parallel path between J_2 and E_{b2}.

Solution: Heat flow Q_{1-2} is given by (R_{eq} is for the parallal circuit $J_2 - E_{b2}$)

$$\mathbf{Q_{1\text{-}2}} = \frac{E_{b1} - E_{b2}}{\dfrac{1-\varepsilon_1}{A_1\,\varepsilon_1} + \dfrac{1}{A_1 F_{12}} + R_{eq}}, \; R_{eq} = 9.3966 \times 10^{-3}$$

$$= \frac{56700 - 1451.52}{0.3183 + 1.273 + 9.3966 \times 10^{-3}} = \mathbf{34515\ W.}$$

The heat received by surface 2 increases from 20494 W to 35060 W. The temperature of the reradiating surface can be determined by determining J_2.

$$34515 = \frac{56700 - J_2}{0.3183 + 1.273} \quad \therefore \; J_2 = 1776 \text{ W/m}^2$$

$$\therefore \qquad J_3 = 1614, \quad \therefore \quad T_3 = 411 \text{ K.}$$

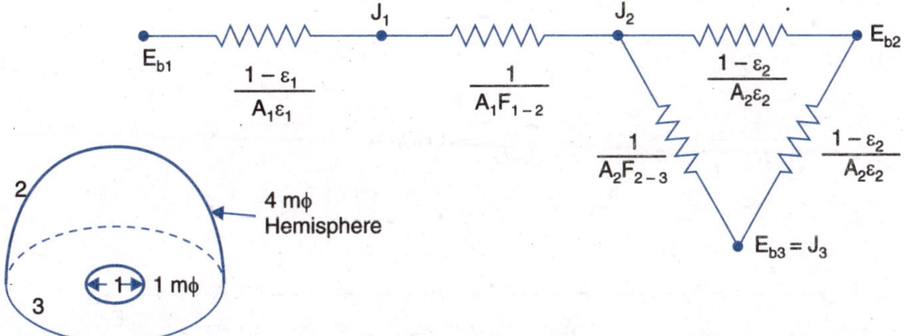

Fig. AE-13.38

Example A-13.39: *Two coaxial cylinders of 0.4 m dia and 1 m dia are 1 m long. The annular top and bottom surfaces are well insulated and act as reradiating surfaces. The inner surface is at 1000 K and has an emissivity of 0.6. The outer surface is maintained at 400 K. The emissivity of the surface is 0.4.*

(i) Determine the heat exchange between the surfaces.

(ii) If the annular base surfaces are open to black surroundings at 300 K, determine the radiant heat exchange.

Solution: These are short cylinders. The shape factor values are determined using Data Book. The outer surface is designated as 2 and the inner as 1. The parameters are r_1/r_2 and L/r_2. The chart provides shape factor from outside cylinder to the inside and to itself.

The parameter values here are 0.4 and 1. Reading from chart

$$F_{2-1} = 0.25, \qquad F_{2-2} = 0.27$$

considering both the annular closing surfaces as 3 (This will not make any difference with total heat flow, see Fig. AE-13.20)

$$F_{2-3} = (1 - 0.25 - 0.27) = 0.48$$

$$A_1 F_{1-2} = A_2 F_{2-1} \quad \therefore \quad F_{1-2} = F_{2-1} \cdot \frac{A_2}{A_1} = 0.25 \times \frac{1}{0.4} = 0.625$$

$$F_{1-3} = 1 - 0.625 = 0.375$$

$$E_{b1} = 56700. \; E_{b2} = 1451.52, \quad E_{b3} = 459.27 \text{ W/m}^2.$$

The equivalent circuit for case 1 is given in Fig. AE-13.39

The resistances are

$$\frac{1-\varepsilon_1}{A_1\,\varepsilon_1} = \frac{1-0.6}{\pi \times 0.4 \times 1 \times 0.6} = 0.5305$$

$$\frac{1-\varepsilon_2}{A_2\,\varepsilon_2} = \frac{1-0.4}{\pi \times 1 \times 1 \times 0.4} = 0.4775$$

$$\frac{1}{A_1 F_{1-2}} = \frac{1}{\pi \times 0.4 \times 1 \times 0.625} = 1.2732$$

$$\frac{1}{A_1 F_{1-3}} = \frac{1}{\pi \times 0.4 \times 1 \times 0.375} = 2.1221$$

$$\frac{1}{A_2 F_{2-3}} = \frac{1}{\pi \times 1 \times 1 \times 0.48} = 0.6631.$$

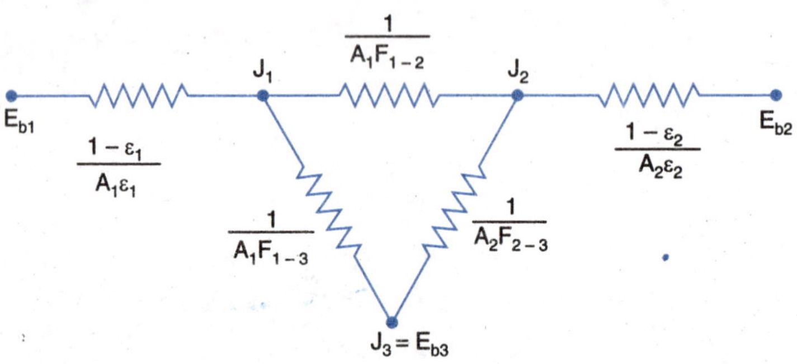

Fig. AE-13.39

Considering node J_1

$$\frac{56700 - J_1}{0.5305} + \frac{J_2 - J_1}{1.2732} + \frac{J_3 - J_1}{2.1221} = 0$$

Simplifying

$$-1.6667\,J_1 + 0.4167\,J_2 + 0.25\,J_3 + 56700 = 0 \qquad \text{...(A)}$$

Considering node J_2

$$\frac{J_1 - J_2}{1.2732} + \frac{1451.52 - J_2}{0.4775} + \frac{J_3 - J_2}{0.6631} = 0$$

Simplifying

$$0.375\, J_1 - 2.095\, J_2 + 0.72\, J_3 + 1451.52 = 0 \quad \text{...(B)}$$

Considering node J_3 $\quad \dfrac{J_1 - J_3}{2.1221} + \dfrac{J_2 - J_3}{0.6631} = 0$

Simplifying $\quad J_1 + 3.2\, J_2 - 4.2\, J_3 = 0 \quad$ ...(C)

Solving: $\quad J_1 = 41123$ W/m^2 $\quad J_2 = 15468$ W/m^2

$J_3 = 21575$ W/m^2

This gives a heat flow about 29360 W from inner to outer surface.

(*ii*) In this case $J_3 = E_{b3} = 459.27$ W/m^2. Forming nodal equations and

Solving $\quad J_1 = 35908.6 \quad J_2 = 7278.23$, $Q_1 = (56700 - 35908.6)/0.5305$

$$\mathbf{Q_1 = 39191\ W,\quad Q_2 = 12203\ W,\quad Q_1 - Q_2 = 26988\ W}$$

Calculating Q_3 by using J_1, J_3 and $J_2 - J_3$,

$Q_3 = 16705 + 10283 = 26988$ W. The values agree.

Example A-13.40: *Determine the shape factors in the following cases between the heater and roof.*

(*i*) *A heater of 15 cm square placed in the centre of the floor of a furnace of 4 m × 4 m size and 2 m height.*

(*ii*) *The heater placed in the corner of the floor*

(*iii*) *The heater placed on one of the corner of one of the vertical walls.*

(*iv*) *A spherical heater of 15 cm dia placed in the centre of the floor.*

(*v*) *The heater placed in one corner.*

Solution: These situations are solved for using charts from databook: (symbols as in charts, Fig. 13.16.)

(*i*) This has to be a sum of four shape factors, dividing the roof into four parts.

$$\frac{D_1}{L_1} = \frac{2}{1} = 2 \quad \therefore \quad \frac{D_2}{L_2} = \frac{2}{1} = 2 \qquad \therefore \text{ chart value} = 0.06$$

$$\therefore \quad \mathbf{F_{1\text{-}2}} = 4 \times 0.06 = \mathbf{0.24}$$

(*ii*) $$\frac{D_1}{L_1} = \frac{2}{4},\ \frac{D_2}{L_2} = \frac{2}{4} = 0.5 \qquad \therefore\ \mathbf{F_{1\text{-}2} = 0.208}$$

(*iii*) $$\frac{b}{a} = \frac{c}{a} = \frac{4}{2} = 0.5 \qquad \therefore\ \mathbf{F_{1\text{-}2} = 0.07}$$

(*iv*) $$\frac{b}{a} = \frac{c}{a} = \frac{1}{2} = 0.5 \qquad \mathbf{F_{1\text{-}2} = 0.018 \times 4 = 0.072}$$

(*v*) $$\frac{b}{a} = \frac{c}{a} = \frac{4}{2} = 2 \qquad \mathbf{F_{1\text{-}2} = 0.07}$$

The closer the plane, and larger the plane, higher will be the value of shape factor.

Example A-13.41: *The direct and diffuse components of solar radiation on surface are 400 and 300 W/m². The direct radiation is incident on the surface at 20° to the normal to the surface. The surface is at 320 K and the effective sky temperature is 260°K. Determine the net radiation heat transfer in the following cases. (a) $\alpha = 0.9$, $\varepsilon = 0.9$ (b) $\alpha = 0.9$, $\varepsilon = 0.1$ and (c) $\alpha = 0.1$, $\varepsilon = 0.9$.*

Solution: The effective solar radiation normal to the surface = 400 × cos 20 + 300 = 676 W/m².

(diffuse radiation is non directional).

$$q = \alpha \times G + \varepsilon\,\sigma\,(T_{sky}^4 - T_s^4)$$

(*a*) $q = 0.9 \times 676 + 0.9 \times 5.67\,[(260/100)^4 - (320/100)^4]$

$= \mathbf{307\ W/m^2}$.

(*b*) $q = 0.9 \times 676 + 0.1 \times 5.67\,[(260/100)^4 - (320/100)^4]$

$= \mathbf{575\ W/m^2}$

(*c*) $q = 0.1 \times 676 + 0.9 \times 5.67\,[(260/100)^4 - (320/100)^4]$

$= \mathbf{-\,234\ W/m^2}$.

These surfaces are called selective surfaces. Case (*b*) type is used for collection of solar radiation so that maximum collection can be achieved.

Example A-13.42: *Two rows of water tubes of 5 cm dia are placed at 10 cm pitch in front of a radiating plane and is backed by insulating wall. The tubes are staggered. Determine the shape factor to (i) direct to row 1 (ii) direct to row 2 (iii) Total including reradiation to row 1 (iv) Total to row 2 (v) Total to both 1 and 2 without reradiation (vi) Total with reradiation.*

Solution: A chart is available which uses the nomenclature as in Fig. AE-13.41 here $P/D = 2$, Refer chart Fig. 13.20.

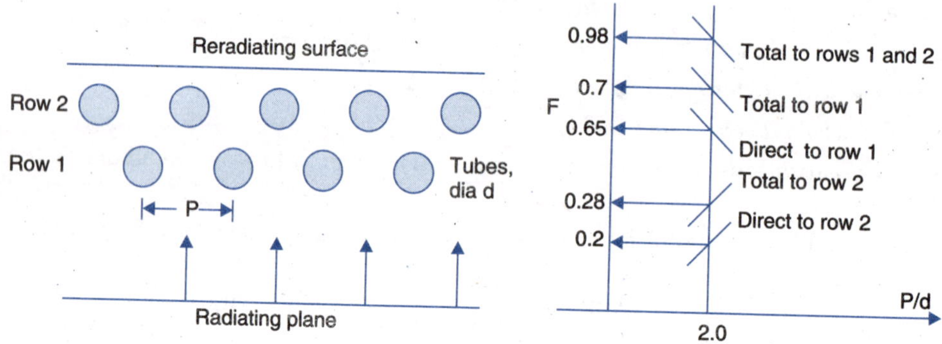

Fig. AE-13.42

(i) Direct to row 1 = 0.65 **(ii) Direct to row 2 = 0.2**

(iii) Total to row 1 = 0.7 **(iv) Total to row 2 = 0.28**

(v) Total to 1 and 2 = 0.98

The heat picked up by first row will be more than double that picked up by row 2.

Example A-13.43: *Determine the shape factor values between **two short coaxial cylinders** of diameters 0.5 m and 1 m of length 1 m and also between the cylinder to each end anuular surface.*

Solution: The outer cylinder is designated as 2 and the inner as 1 and ends as 3 and 3′.

Here $r_2 = 0.5$ m, $r_1 = 0.25$ m $L = 1$ m

The parameters are:

$$r_1/r_2 = 0.5 \qquad L/R_2 = 1/0.5 = 2$$

Reading from chart in data book

$\therefore$ F_{2-1} (outer to inner) = 0.44, $F_{2-2} = 0.32$

$$A_2F_{2-1} = A_1F_{1-2} \qquad \therefore \; F_{1-2} = 0.44 \times \frac{A_2}{A_1} = 0.44 \times 2 = 0.88$$

$$\therefore \qquad F_{1-3'} = F_{1-3} = 0.12/2 = 0.06$$

$$F_{2-3} = F_{2-3'} = (1 - 0.44 - 0.32)/2 = 0.12$$

$$A_1F_{1-3} = A_3F_{3-1} \qquad \therefore \; F_{3-1} = \frac{A_1}{A_3} \times 0.06$$

$$F_{3'-1} = 0.06 \times \pi \times 0.5 \times 1 \times 4 \, \frac{1}{\pi(1^2 - 0.5^2)} = 0.16$$

$$\mathbf{F_{3'-1} = F_{3-1} = 0.16}$$

$$\mathbf{F_{3-2}} = \frac{A_2}{A_3} = 0.12 = \frac{\pi \times 0.51 \times 4}{\pi(1^2 - 0.5^2)} = 0.64$$

$$\mathbf{F_{3-2} = 0.64}$$

$$\therefore \qquad \mathbf{F_{3'-3}} = 1 - 0.16 - 0.64 = \mathbf{0.2}$$

This is the shape factor between annular disks placed coaxially opposite each other. (refer example 13.20 also).

REVIEW QUESTIONS

13.1 State the two important features that characterise radiation.

13.2 Write the equation for photon energy.

13.3 Specify the region in which thermal radiation is concentrated in the electromagnetic spectrum.

13.4 Define monochromatic or spectral properties of radiation and surfaces.

13.5 What is the unit of solid angle? How is solid angle specified?

13.6 Define total emissive power, emissivity, absorptivity transmissivity and reflectivity.

13.7 Define diffuse emitter, radiosity and irradiation.

13.8 Define 'black body'. Explain how black body radiation can be generated artificially?

13.9 State wiens displacement law and give the value for $\lambda_{max} T$.

13.10 Define gray surface.

13.11 Define shape factor. At what condition the shape factor of a surface with another surface will be (*a*) 0 (*b*) 1.

13.12 Explain the application of (*i*) reciprocity relation (*ii*) summation rule (*iii*) the symmetry rule.

13.13 Can a surface have shape factor to itself and if yes what the geometry of such a surface?

13.14 Explain what type of surface is called non absorbing reradiating surface?

13.15 In the case of gray surface define surface resistance and space resistance for radiative transfer of heat.

13.16 What is 'radiation shield'? What purpose does a radiation shield serve?

13.17 What are the factors on which the emissivity of gases depend upon?

Chapter 13

13.18 State Kirchhoff's law.

13.19 Define intensity of radiation. What is its application in heat transfer between surfaces?

13.20 State an application of a surface which has quite different values for its absorptivity and emissivity.

OBJECTIVE QUESTIONS

13.1 Choose the correct statement from each question.

(*i*) The monochromatic emissive power of a black body with increasing wavelength.
- (*a*) decreases
- (*b*) increases
- (*c*) decreases, reaches a minimum and then increases
- (*d*) increases, reaches a maximum and then decreases.

(*ii*) A gray surface is one for which
- (*a*) reflectivity equals emissivity
- (*b*) emissivity equals transmissivity
- (*c*) emissivity is constant
- (*d*) absorptivity equals reflectivity.

(*iii*) As the source temperature increases the wavelength at which the monochromatic emissive power is maximum
- (*a*) decreases continuously
- (*b*) decreases and then increases
- (*c*) increases continuously
- (*d*) increases and then decreases.

(*iv*) The directional emissivity for metallic surfaces is
- (*a*) constant all over the angles from normal
- (*b*) more near the normal and less near the tangential direction
- (*c*) more near tangential direction compared to normal direction
- (*d*) the distribution is affected more by temperature than by direction.

(*v*) For non metallic insulating materials directional emissivity
- (*a*) is constant all over the angles
- (*b*) more at near normal directions than at tangential direction
- (*c*) less at near normal directions compared to tengential directions
- (*d*) can be as case *b* or *c* depending on surface preparation.

(*vi*) Selective surfaces
- (*a*) do not follow Kirchhoff's law
- (*b*) absorb only at definite wavelengths
- (*c*) emit only at definite wave bands
- (*d*) all of these.

(*vii*) Glasses are
- (*a*) opaque for high temperature radiation
- (*b*) opaque for low temperature radiation
- (*c*) transparent at short wavelengths
- (*d*) transparent at long wavelengths
- (*e*) opaque for low temperature radiation transparent at long wavelengths.

(*viii*) For solar collectors the required surface characteristics combination is
- (*a*) high emissivity and low absorptivity
- (*b*) high emissivity and high reflectivity
- (*c*) high reflectivity and high absorptivity
- (*d*) low emissivity and high absorptivity.

(*ix*) Emissivity of gas body of a given composition depends on
- (*a*) shape and temperature
- (*b*) partial pressure and shape
- (*c*) partial pressure and temperature
- (*d*) all of these.

(*x*) For a given shape, partial pressure and temperature the emissivity of

(*a*) O_2 is higher than that of N_2 (*b*) N_2 is higher than that of N_2

(*c*) O_2 is higher than that of CO_2 (*d*) CO_2 is higher than that of O_2.

(*xi*) The combination which will give the highest gas emissivity is

(*a*) low partial pressure, higher temperature and larger thickness

(*b*) higher partial pressure, higher temperature and larger thickness

(*c*) higher partial pressure, lower temperature and larger thickness

(*d*) lower partial pressure, lower temperature and larger thickness.

(*xii*) The value of shape factor will be highest when

(*a*) the surfaces are farther apart (*b*) the surfaces are closer

(*c*) the surfaces are smaller and closer (*d*) the surfaces are larger and closer.

(*xiii*) A radiation shield should have

(*a*) high emissivity (*b*) high absorptivity

(*c*) high reflectivity (*d*) high emissive power.

(*xiv*) Choose the correct statement or statements.

(*a*) Highly reflecting surface is suitable for solar heat collection through flat plates.

(*b*) The emissivity of smooth surface is higher compared to a rough surface of the same material.

(*c*) For a given gas body, the emissivity will decrease with increase in temperature.

(*d*) Snow has high emissivity.

(*xv*) Choose the correct statement or statements.

(*a*) The shape factor of samll enclosed body with respect to the enclosing surface is zero.

(*b*) The shape factor of small enclosed body with respect to the enclosing surface is unity.

(*c*) A small opening from a large enclosure at constant temperature will provide black body radiation.

(*d*) Black paint is an example of black body.

(*xvi*) The reciprocity theorem states

(*a*) $F_{1-2} = F_{2-1}$ (*b*) $A_1 F_{1-2} = A_2 F_{2-1}$

(*c*) $A_2 F_{1-2} = A_1 F_{2-1}$ (*d*) $\varepsilon_1 F_{1-2} = \varepsilon_2 F_{2-1}$.

(*xvii*) Choose the correct statement or statements

(*a*) $F_{1-2,3} = F_{1-2} + F_{1-3}$ (*b*) $A_1 F_{1-2,3} = A_2 F_{2-1} + A_3 F_{3-1}$

(*c*) $F_{1,2-3} = F_{1-3} + F_{2-3}$ (*d*) $A_{1,2} F_{1,2-3,4} = A_1 F_{1,2-3,4} + A_2 F_{1,2-3,4}$.

(*xviii*) Choose the correct statement or statements.

(*a*) Radiosity is another name for emissive power.

(*b*) Radiation intensity is the flux per unit area.

(*c*) Radiation intensity is the radiant energy per unit solid angle.

(*d*) Irradiation is the total radiant energy incident on a surface.

(*xix*) Choose the correct statement or statements.

(*a*) Convex surface have positive value for shape factor with themselves.

(*b*) Concave surface have positive value for shape factor with themselves.

(*c*) Flat surfaces have positive value for shape factor with themselves.

(*d*) Irregular surfaces have possitive value for shape factor with themselves.

(*xx*) If $A_1 = 4$ and $A_2 = 2$ and $F_{1-2} = 0.2$ then

(*a*) $F_{2-1} = 0.2$ (*b*) $F_{2-1} = 0.8$ (*c*) $F_{2-1} = 0.4$ (*d*) $F_{2-1} = 0.1$.

Answers

(*i*) (*d*)	(*ii*) (*c*)	(*iii*) (*a*)	(*iv*) (*c*)	(*v*) (*b*)	(*vi*) (*d*)
(*vii*) (*c*)	(*viii*) (*d*)	(*ix*) (*d*)	(*x*) (*d*)	(*xi*) (*c*)	(*xii*) (*d*)
(*xiii*) (*c*)	(*xiv*) (*c*), (*d*)	(*xv*) (*b*), (*c*)	(*xvi*) (*b*)	(*xvii*) (*a*), (*b*)	(*xviii*) (*c*), (*d*)
(*xix*) (*b*), (*d*)	(*xx*) (*c*).				

EXERCISE PROBLEMS

13.1 Determine the wavelength at which monochromatic emissive power is maximum for the following surface temperatures. Also determine the value of the monochromatic emissive power at these wavelengths.

(*i*) 600 K (*ii*) 2000 K and (*iii*) 5500 K.

13.2 Using Plancks equation derive the Stefan-Boltzmann equation for emissive power of a black surface.

13.3 Determine the fraction of radiant energy emitted up to the wavelength at which monochromatic emissive power is maximum for the following surface temperatures.

(*i*) 600 K (*ii*) 2000 K (*iii*) 5000 K.

13.4 A sphere of 0.2 m diameter at 800 K is suspended in air. Assuming the surface to be black, determine (*a*) The emissive power (*b*) Energy radiated in 300 s and (*c*) The monochromatic emissive power at 3 μm wavelengths. [**Ans.** 23.2×10^3 W/m^2, 876 kJ, 3848 W/m^2 μm]

13.5 A surface is at 2500 K. Assuming the surface to be black determine the fraction of radiant energy emitted in the visible wavelength range of 0.4 μm and 0.76 μm. (visible radiation) Also calculate λ_{max} in this case. [**Ans.** 0.0527135, 1.16 μm]

13.6 A small horizontal surface of 3 cm^2 area at 600 K exchanges radiation with another small surface of 5 cm^2 area at a distance of 75 cm along 55° to the vertical. The angle between the normal to the second surface and the line from the first surface is 40°. Determine the solid angle subtended by the second surface when viewed from the first surface. Also calculate the rate at which the radiation emitted by the first surface that strikes the second surface.

[**Ans.** 6.81×10^{-4} Steradian, 2.74×10^{-4} W]

13.7 The emissivity of a non gray surface is given below.

Wavelength range	*Emissivity*
0–3 μm	0.3
3 μm–7 μm	0.8
7 mμ–λ < ∞	0.1

Determine the average emissivity of the surface and its emissive power if it is at 800 K.

[**Ans.** 0.521, 12100 W/m^2]

13.8 Determine the wavelength up to which the black body radiation at 2000 K will equal.

(*i*) 25% (*ii*) 50% (*iii*) 75% of the emissive power.

13.9 A glass was found to let in 85% of the radiation from the sun at 5800 K. If its transmissivity was 0.9 upto wavelength λ and zero beyond this value determine the value of λ.

13.10 Mylar sheets have a transimissivity of 0.95 upto wavelength of 3 micrometer and zero beyond this value. Determine its total transmissivity for radiation for source at (*i*) 600 K (*ii*) 2000 K and (*iii*) 5800 K.

13.11 Green house effect is nothing but trapping of radiation by letting in radiation of short wavelength and shutting out radiation of long wavelength. A green house has a roof area of 100 m^2 perpendicular to the solar inclination. The material has a transmissivity of 0.9 up to a wavelength of 4 μm and zero beyond. The solar flux has a value of 800 W/m^2. The total wall area is 600 m^2. It the inside is to be maintained at 22°C while the outside is at – 5°C, determine the maximum value of overall heat transfer coefficient for heat flow through the walls. The temperature of solar radiation may be taken as 5000 K.

13.12 A surface is found to have an absorptivity of 0.8 for wavelengths upto 3 μm and 0.2 beyond this value. If the surface is exposed to sunlight with a source temperature of 5000 K, and a flux of 800 W/m^2, determine the rate of heat gain when the surface is at (*i*) 400 K (*ii*) 900 K. Assume only radiant heat flow.

13.13 A 40 W incandescent lamp has a filament area of 1 cm^2 and is at 2800 K. Determine the efficiency of the lamp. Lamp efficiency is defined as the fraction of radiant energy emitted in the wavelength band of 0.4 to 0.7 μm to the power input.

13.14 If in problem 13.9, if the temperature is increased to 3000 K, determine the improvement in the efficiency.

13.15 A surface has an absorptivity of 0.8 in the wavelength range of 0 to 4 μm and 0.2 beyond. Determine the total emissivity of the surface at temperature (*i*) 400 K (*ii*) 900 K (*iii*) 2000 K and (*iv*) 3000 K.

13.16 A surface has an emissivity of 0.5 in the wavelength band of 1.15 to 10 μm and 0.8 above this wavelength. The emissivity is zero below 1.15 μm. Determine the total absorptivity of this surface to radiation from a source at 2000 K.

13.17 A furnace with an emissive power of 3.72×10^5 W/m^2 is used to calibrate a gauge for measuring radiation flux the aperture used is 2 cm in dia. The gauge has an effective area of 1.6×10^{-5} m^2. What should be the distance at which the gauge should be placed so that the flux is 51.8 cm $\times 10^3$ W/m^2?

13.18 An opaque surface having reflectivity values as shown in Fig. 13.18(*a*) receives irradiation as indicated in Fig. 13.18(*b*). Determine the radiant flux absorbed by the surface.

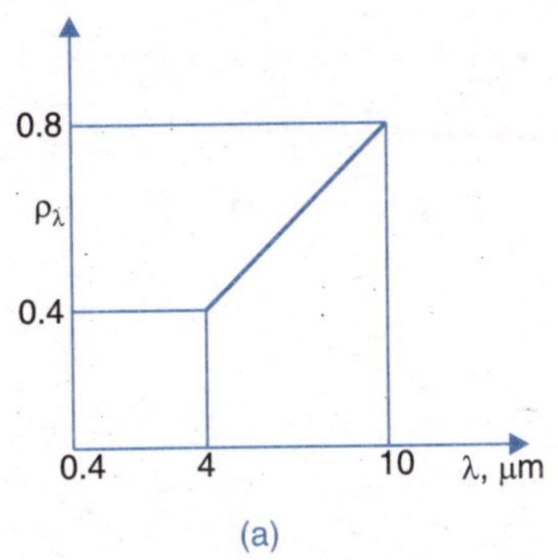

(a)

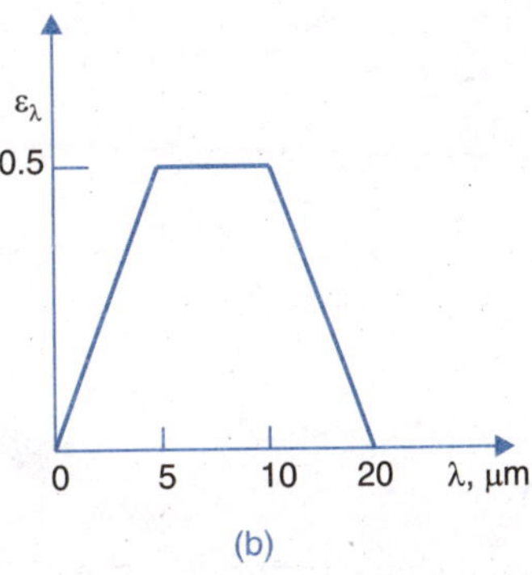

(b)

Fig. 13.18

13.19 A furnace is of cubical shape of side 5 m. The top surface is maintained at 1500 K while the base is maintained at 800 K. The four sides are maintained at 500 K. Determine (*a*) rate of radiation heat transfer between (*a*) the base and side surfaces (*b*) between the base and top and (*c*) the net heat transfer from the base. Assume that surfaces are black.

[**Ans.** 394×10^4 W, -1319×10^3 W, -925 kW)

13.20 A cylindrical furnace of diameter 2 m and height 1 m has its top surface maintained at 700 K. The emissivity of the surface is 0.8. The base is maintained at 500 K and its emissivity is 0.4. The cylindrical surface is black and is maintained at 400 K. At steady condition determine the rate of heat flow from each of the three surfaces.

[**Ans.** From top: 27.6×10^3 W, From base: -2.13×10^3 W,
From cylindrical surface: -25.5×10^3 W]

Chapter 13

13.21 A furnace is in the shape of a long duct of equilateral triangle of side 1 m. One side with emissivity of 0.7 is at 600 K. One side is insulated and reradiating. The third side is black and is at 1000 K. Determine the rate at which heat to be supplied for 1 m length to the side at 1000 K to maintain steady conditions. [**Ans.** 28000 W]

13.22 Determine the total emissivity of two surfaces with the following directional characteristics;

(*i*) Emissivity is 0.3 from normal to 60° and is 0.8 from 60° to 90°

(*ii*) Emissivity is 0.8 from normal to 60° and is 0.3 from 60° to 90°.

13.23 Estimate the equilibrium temperature of a plate in space when exposed solar flux of 1352 W/m^2 (from source at 5800 K) if it has an emissivity of 0.9 upto 3 μm and 0.05 beyond.

13.24 Determine the shape factor F_{1-2} in the following configurations shown in Fig. 13.24.

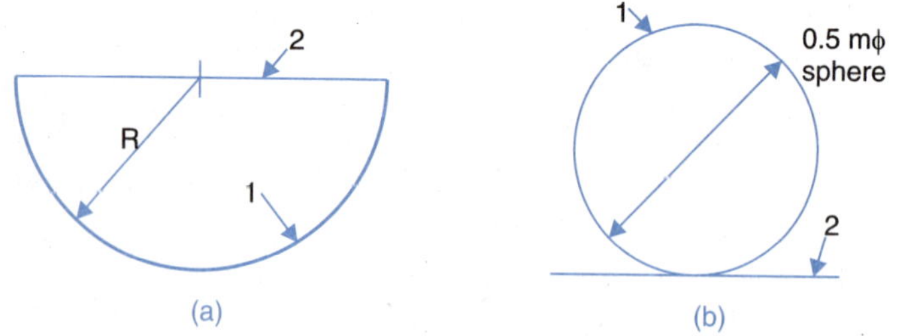

Fig. 13.24

13.25 Determine the shape factor F_{1-2} in the following cases.

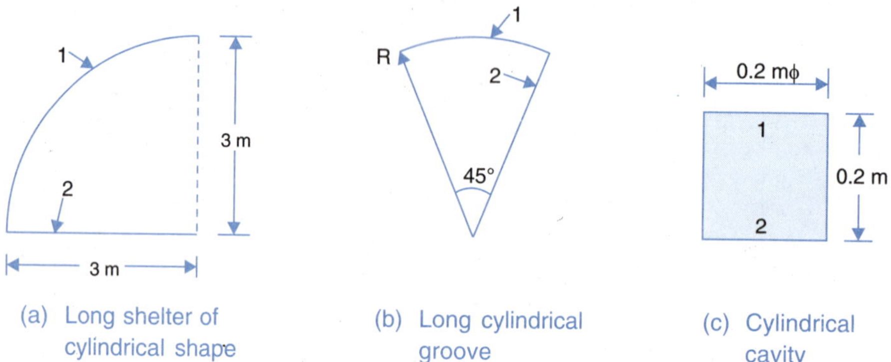

(a) Long shelter of cylindrical shape (b) Long cylindrical groove (c) Cylindrical cavity

13.26 Determine the shape factors F_{1-2} for the following configurations:

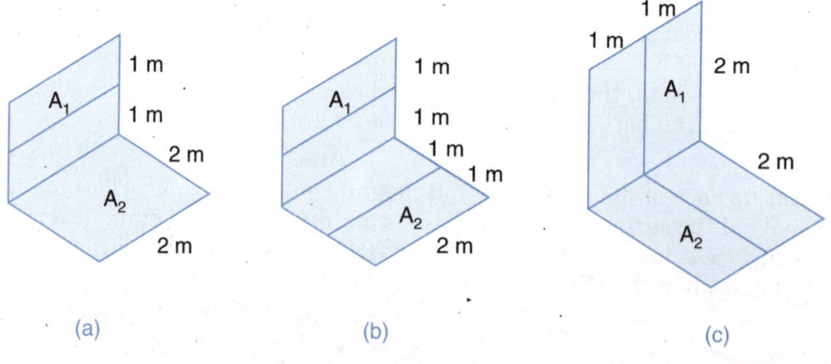

Fig. 13.26

13.27 Determine the shape factor F_{1-2} in the following cases:

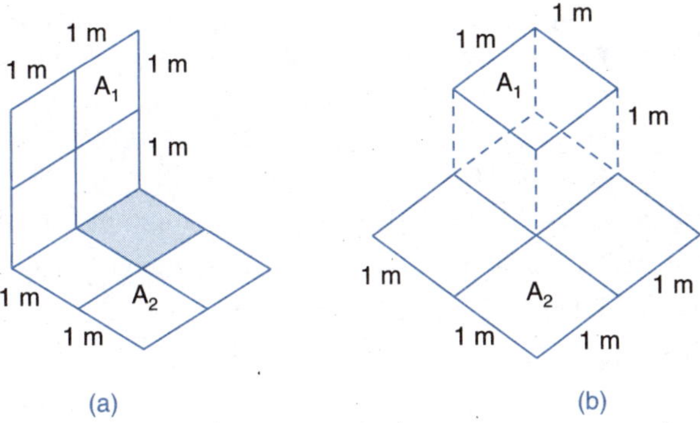

Fig. 13.27

13.28 Determine the shape factor F_{1-2} for the following configurations:

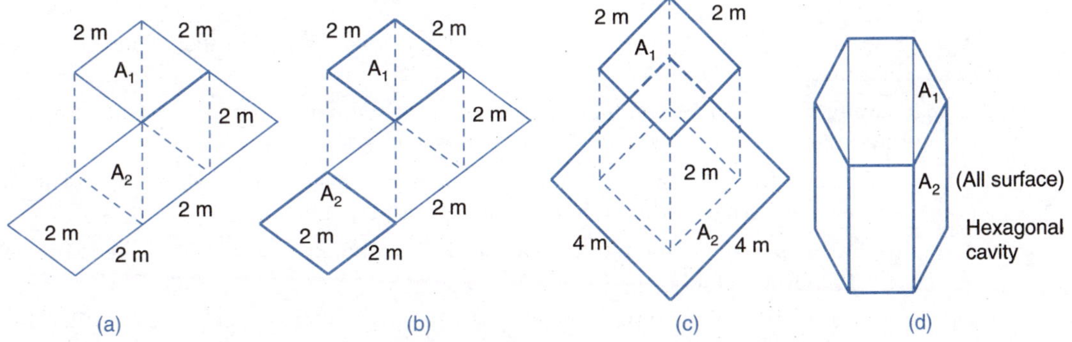

Fig. 13.28

13.29 Determine the emissivity of water vapour in a spherical gas body of 2 m dia when the partial pressure of water vapour is 0.05 atm and the temperature of the mixture is 1000 K. The total pressure is 1 atm.

13.30 Determine the emissivity in the situation of problem 13.29 if the (*i*) total pressure is 1 atm and partial pressure is 0.1 atm (*ii*) if the total pressure is 2.5 atm and partial pressure is 0.05 atm.

13.31 Determine the emissivity of a gas body at 1500 K and 1 atm contained in a cubical furnace of 2 m side if the partial pressure of CO_2 is 0.1 atm in tha gas mixture. Also determine the emissivity if the total pressure is 4 atm while partial pressure is still 0.1 atm.

13.32 An oil heated furnace is of 2 m × 2 m × 3 m size and contains gases at 1600 K. The partial pressure of CO_2 is 0.12 atm and that of H_2O is 0.14 atm. Determine the emissivity of the gas body and also the heat exchange if the surface is at 1000 K. Total pressure = 1 atm.

13.33 A gas turbine combustion chamber is cylindrical and is 0.4 m dia and 0.8 m length. The gas is at 1700 K. The flame tube is at 500 K. The gas is at 6 atm. The partial pressure of CO_2 is 0.12 atm and that of water vapour is 0.1 atm. Determine the radiation exchange between the flame tube and the gas body.

13.34 Pure CO_2 at 1 atm passes through a cylindrical heater of 0.2 m dia and 0.3 m length with surface maintained at 1600 K. The average gas temperature is 800 K. Determine the rate of radiation heat exchange.

13.35 Determine the heat exchange per m^2 of area A_1 between the surface specified below if surface 1 is at 1000 K while surface 2 is at 400 K. All surfaces are black.

(*i*) large parallel planes (*ii*) long coaxial cylinders of 0.5 m and 1 m dia with inner cylinder hotter. (*iii*) concentric spheres of 1 m and 2 m dia with inner surface hotter. (*iv*) parallel planes of 2 m × 2 m size placed symmetrically 1 m apart.

13.36 Determine the heat exchange per m^2 of Area A_1 in the following cases if A_1 is at 1000 K and A_2 is at 400 K, both surfaces being black.

(*i*) Perpendicular planes 2 m × 1 m size with 2 m being the common edge

(*ii*) Perpendicular planes of 2 m × 1 with 1 m being the common edge.

(*iii*) Equal coaxial parallel disks of 1 m dia, placed 1 m apart.

(*iv*) Same as in (*iii*) but placed 0.5 m apart.

(*v*) Concentric circular disks of 1 m dia and 0.5 m dia placed 0.5 m apart with the smaller one as surface A_1.

13.37 Determine the heat exchange between the following surfaces with the smaller surface as surface 1 and at 1000 K while the larger surface is at 400 K, both surfaces being black.

(*i*) A small square of 0.15 m side and a large parallel plane of 2 m square size with its corner along the centre of the small area and 1 m apart.

(*ii*) A small square of 0.15 m side and a large plane of 2 m square kept parallel and centres of both along the same line and 1 m apart.

(*iii*) a small circular disk of 0.15 m dia and large disk of 1 m dia at a distance of 1 m with centres Along the same line and planes parallel to each other.

(*iv*) A small plane of 0.15 m square and a large plane of 2 m square at a distance of 1 m. The small plane is below one corner of the large plane and kept perpendicular to it.

(*v*) A small sphere of 0.1 m dia and a large plane of 1.5 m square, with the sphere kept 1 m below a corner of the large plane.

13.38 Determine the reduction in radiant transfer between two large parallel planes at 1000 K and 500 K when a shield surface with emissivity of 0.1 and 0.6 on either side is placed between them. The emissivity of the surface are 0.6 and 0.8 respectively. Also determine the shield temperature when higher emissivity side of the shield (*i*) faces the hot plane (*ii*) faces the cooler plane.

13.39 Determine the heat exchange between two long coaxial cylinders of diameter 0.5 m and 0.7 m when the outer is at 100 K and inner at 500 K. The surface emissivities are 0.4 and 0.6 for the outer and inner surfaces.

13.40 A spherical container of 1 m dia storing liquid nitrogen at – 120°C is covered by high reflecting coating with reflectivity of 0.96. If it is exposed to surroundings at 20°C, determine the heat flow in due to radiation.

If a concentric spherical surface with equal reflectivity on either side of 0.96 and a diameter of 1.2 m is used over the surface, determine the reduction in the heat flow by radiation, when exposed to the same surroundings.

13.41 Two large parallel planes are at – 10°C and 22°C. The emissivity values are 0.6 and 0.75. If the radiant heat flow is to be reduced by 90% determine the reflectivity of a shield to be placed in between.

13.42 A long cylinder of diameter 0.4 m is at 1000 K and has an emissivity of 0.4. A coaxial cylinder of 1 m dia having surface emissivity of 0.6 maintained at 400°C surrounds the smaller cylinder. It is proposed to use a shield having an emissivity of 0.1. Compare the heat reduction if the shield diameter is (*i*) 0.6 m and (*ii*) 0.8 m based on 1 m^2 of surface used.

13.43 Determine the heat flow between the roof and floor of 4 × 3 m size of a furnace of 4 m × 4 m × 3 m size when the roof is at 1200 K and the floor is maintained at 600 K, with the other surfaces nonabsorbing and reradiating. The surface emissivity of the hotter surface is 0.8 and that of the cooler surface 0.6.

13.44 Two symmetrically placed parallel rectangular surface of 4 m × 3 m size kept at 2 m distance are at 1000 K and 600 K. The emissivities are 0.6 and 0.5. The sides are exposed to black surroundings at 300 K. Determine the heat flow from the surfaces.

13.45 A thin sheet of aluminium is used as a shield between two large parallel surfaces. The emissivity of the shield is 0.1 on both sides. The temperature surface with emissivity of 0.2 is 800 K. The emissivity and temperature of the other surface are 0.7 and 500 K respectively. At steady conditions determine the rate of heat flow between the surfaces without and with the shield per unit area. **[Ans.** 3625 W/m^2, 806 W/m^2]

13.46 A thermo couple used to measure the temperature of hot air flowing in a duct shows 650 K. The duct walls are at 400 K. The emissivity of the thermocouple is 0.6 and convection coefficient over it is 80 W/m^{2}k. Determine the actual temperature of the air. **[Ans.** 715 K]

13.47 A cylindrical furnace of 5 m dia and 5 m high contains combustion gases at 2 atm and 1200 K. The volume analysis shows 5% CO_2 and 8% water vapour, the rest being inert gases. Determine the effective emissivity of the combustion gases. **[Ans.** 0.45]

13.48 For the details of problem 13.47, if the wall temperature is 600 K determine the absorptivity of the gases. Also calculate the rate of heat transfer from gases to walls. **[Ans.** 0.64, 2.79 × 10^4 W]

13.49 Two side walls and the roof of a furnace are maintaind at 600 K and 1200 K respectively, their emissivities being 0.75 and 0.6. The furnace is cubical with sides of 2 m length. The other surfaces are nonabsorbing and reradiating. Determine the heat flow between side walls and the roof.

13.50 If in problem 13.38 the other sides are open to black surroundings at 500 K, determine the heat exchange.

13.51 In a cylindrical furnace of 1.5 m dia and 2 m length the curved surface is maintained at 1400 K, its emissivity being 0.7. The two ends are maintained at 800 K and 600 K. The emissivities are 0.65 and 0.55. Determine the heat flow to each of the surfaces by radiation.

13.52 A furnace is in the shape of a frustum of a cone with base diameter of 2 m and top diameter of 1 m and height 1.5 m. The curved surface having an emissivity of 0.65 is maintained at 1600 K while the base and top are maintained at 700 K and 1200 K, their emissivities being 0.6 and 0.7. Determine the heat flow by radiation from each surface.

13.53 If in problem 13.52, the top surface acts as reradiating surface, with other conditions unchanged. Determine the heat flow between the bottom and curved surface and the temperature of reradiating surface.

13.54 If in problem 13.52, the curved surface acts as reradiating surface. Determine the heat flow and also the temperature of the curved surface.

13.55 A four surface enclosure is in the shape of a tetrahedron of 2 m edge length. The surfaces are at 1400, 1200, 1000 and 800 K respectively with emissivities of 0.8, 0.7, 0.6 and 0.5. Determine the radiant heat flow from each face.

13.56 A thermocouple with an emissivity of 0.8 is used to measure the temperature of air in an enclosure where the walls are at 300 K. The thermometer shows 600 K for a true temperature of 655 K. Determine the value of convection coefficient over the thermocouple surface.

13.57 A thermocouple with an emissivity of 0.9 is used to measure the temperature of air flowing in a duct. The duct walls are at 280 K. The true air temperature is 350 K. Determine the temperature indicated if (*i*) h = 150 W/m^{2}K (*ii*) h = 50 W/m^{2}K and (*iii*) h = 300 W/m^{2}K.

13.58 A shielded thermocouple shows 600 K for hot gases in a duct whose walls are at 400 K. The shield are is large compared to that of the thermocouple. The emissivity of the shield if 0.05. The emissivity of the thermocouple is 0.8. Determine the true temperature of the gas. The convective heat transfer coefficient has a value of 300 W/m^{2}K.

13.59 A large isothermal enclosure is at 2000 K. Determine the maximum monochromatic emissive power at the wavelength at which occurs. Also determine the wavelength below which 10% of emission is concentrated and the wavelength above which 10% of emission is concentrated.

[**Ans.** 9.07×10^5 W/m^2, 1.45 μm, 1.1 μm., 4.69 μm]

13.60 A coaxial radiation shield of emissivity 0.2 and diameter 0.2 m is placed axially between two cylinders of diameter 0.1 m and 0.3 m, with emissivities of 0.7 and 0.4. The inner cylinder is maintained at 700 K while the outer is at 500 K. Determine the heat flow without and with the shield for 1 m length. Assume long cylinders. [**Ans.** 7465 W, 703 W]

13.61 A cube of 0.2 m side with surface temperature of 1000 K is hung in air. Determine the total energy radiated. Also calculate the monochromatic or spectral emissive power at a wavelength of 4 μm. [**Ans.** 1.36×10^4 W, 10.3 kW/m^2 μm]

13.62 The absorptivity of a surface has a value 0.2 in the range 0 to 5 μm and 0.5 for the rest of the spectrum. Determine the average absorptivity for a surface temperature of 5800 K and 300 K.

[**Ans.** 0.203, 0.89]

13.63 600 W/m^2 is incident on a solar collector with surface emissivity of 0.9 and absorptivity 0.87. The plate temperature is 70°C. Air temperature is 25°C convection coefficient is 10 W/m^{2}k. Sky temperature is 15°C. Determine the energy absorbed. [**Ans.** 36.5 W]

13.64 A spherical furnace of 2 m diameter at surface temperature of 600 K contains gases at 1 atm and 1200 K. The gas contains 15% CO_2. Determine the heat flow rate to the wall. [**Ans.** 193.6 kW]

Mass Transfer

14

CHAPTER

14.0 INTRODUCTION

Mass transfer is quite different from flow of fluids like in a pipe. An example of mass transfer is the nice fragrance from a lighted insense stick spreading throughout the room. From the insense stick, the fragrant molecules move or diffuse into the places where such molecules are not present. Another example of sugar diffusing into the water in the glass when the water is not stirred. There are plenty of sugar molecules in the sugar crystals. There are no sugar molecules in the water or there are only a few of them. Here also sugar molecules move from places of higher concentration of molecules towards a lower concentration of such molecules. Two facts emerge from the examples. (1) For mass transfer to take place two regions of different chemical composition should be present. (2) In the mass transfer direction the concentration of the species should decrease. There will be no mass transfer if the concentration of components remain constant throughout the volume.

Mass transfer has several similarities with heat transfer. In both cases a gradient causes the flow. Diffusive mass transfer both under steady and unsteady state is similar to heat transfer in steady and unsteady states. The general differential equations are identical except for the variables. Convection mass transfer is similar to convection heat transfer and the correlations are identical except for the dimensionless numbers and variables. Thus the discussions in the previous chapters on conduction and convection will help in understanding the mass transfer process.

14.1 PROPERTIES OF MIXTURE

As the mass transfer depends on concentration gradients, these are defined in this section.

In a mixture consisting of two or more materials the mass per unit volume of any component is called mass concentration of that component. If there are two components A and B, then the mass concentration of A is

$$\rho_a = \frac{\text{Mass of } A \text{ in the mixture}}{\text{Volume of the mixture}}, \text{ kg/m}^3$$

and concentration of B,

$$\rho_b = \frac{\text{Mass of } B \text{ in the mixture}}{\text{Volume of the mixture}}, \text{ kg/m}^3$$

The total mass concentration is $\rho_a + \rho_b$, which is also the density of the mixture ρ.

Mass concentration ratio is expressed in terms of individual and total densities of the mixture *i.e.*,

Concentration ratio of A, $W_a = \left(\dfrac{\rho_a}{\rho}\right)$, dimensionless.

where ρ_a is the density of A in the mixture and ρ is the density of the mixture.

It is more convenient to express the concentration in terms of the molecular weight of the component as mass transfer is by discrete molecules.

$$\text{Number of moles of a component } A \text{ in a mixture} = \frac{\text{Mass of the component } A \text{ in the mixture}}{\text{Molecular weight of the component}}$$

Mole concentration of component A

$$C_a = \frac{\text{Number of moles of } A \text{ in the mixture}}{\text{Volume of the mixture}} \text{ moles/m}^3.$$

This may vary from point to point if mass transfer of component A is taking place.

$$\text{Molar concentration ratio of } A = \frac{C_a}{C}$$

where C is the total number of moles of all components in the mixture.

In the case of ideal gases, $\rho = P/RT$,

$$\rho_a = P_a/RT,$$

where R is the gas constant, and P_a is the partial pressure of A in the mixture.

This leads to the relation,

Mass concentration ratio = molar concentration ratio = P_a/P.

where P is the total pressure.

The driving force for mass transfer of A is $(d\rho_a/dx)$ or (dC_a/dx).

14.2 DIFFUSION MASS TRANSFER

Diffusion mass transfer occurs without macroscopic mass motion or mixing. A lump of sugar dropped into a cup of tea will dissolve by diffusion even if left unstirred. But it will take a long time for the sugar to reach all of the volume in the cup. However it will diffuse into the volume by and by. Consider a chamber in which two different gases at the same pressure and temperature are kept separated by a thin barrier. When the barrier is removed, the gases will begin to diffuse into each others volume. After some time a steady condition of uniform mixture would be reached. This type of diffusion can occur in solids also. The rate in solids will be extremely slow. Diffusion in these situations occurs at the molecular level and the governing equations are similar to those in heat conduction where energy transfer occurs at the molecular level.

For steady state, the diffusing component should be supplied at one end and removed at the other end, as in the case of heat transfer where heat is supplied at one end and removed at the other end under steady state. In the following sections, the mass transfer at an instant of time is discussed. The basic law governing mass transfer at the molecular diffusion level is known as **Fick's law.** This is similar to the Fourier heat conduction law.

In mass transfer, molal quantities are more convenient to use as compared to mass units, because mass transfer is due to the movement of molecules as discrete quantities. Hence it is convenient to use number of moles, or molar concentration instead of density etc.

14.3 FICK'S LAW OF DIFFUSION

The Fick's law can be stated as

$$\frac{N_a}{A} = -D_{ab}\frac{dC_a}{dx} \quad ...(14.1)$$

where (N_a/A)— number of moles of 'a' diffusing perpendicular to area A, moles/m^2 sec

D_{ab}— diffusion coefficient or mass diffusivity, m^2/s, a into b

C_a — mole concentration of 'a' moles/m^3

x — diffusion direction

The diffusion coefficient is similar to thermal diffusivity, α and momentum diffusivity υ.

Number of moles multiplied by the molecular mass (or more popularly known as molecular weight) will provide the value of mass transfer in kg/m^{2}s.

Equation (14.1) can also be written as

$$\frac{m_a}{A} = -D_{ab}\cdot\frac{d\rho_a}{dx} \quad ...(14.2)$$

but this form is not as popular as the more convenient equation (14.1). The conduction equation similar to this is

$$\frac{Q}{A} = -\left(\frac{k}{\rho c}\right)\cdot\frac{d(\rho cT)}{dx} \quad ...(14.3)$$

$k/\rho c$ is thermal diffusivity α and ρc is the heat capacity (energy density) for unit volume.

The derivation of the general mass diffusion equation is similar to that of the general heat conduction equation with C_a replacing T and D replacing $k/\rho c$.

The general mass diffusion equation for the species A is given by equation (14.4)

$$\frac{\partial^2 C_a}{dx^2} + \frac{\partial^2 C_a}{\partial y^2} + \frac{\partial^2 C_a}{\partial z^2} = \frac{1}{D}\frac{\partial C_a}{\partial \tau} \quad ...(14.4)$$

Generation of mass of the species 'A' by chemical reaction is not considered in the equation. However an additive term N_a/D on the LHS will take care of this similar to heat generation term q/k. In the case of one dimensional steady state, the solution is

$$\frac{N_a}{A} - = -D_{ab}\cdot\frac{dC_a}{dx} \quad ...[14.4(a)]$$

The solutions for this equation are also similar to the solutions of the general conduction equation. However there exist some differences. These are

(*i*) While heat flow is in one direction, the mass of one species may flow opposite to the flow of the other component of the mixture. (here two component mixture is considered).

(*ii*) Even while one component alone diffuses under certain circumstances, a bulk flow has to be generated as otherwise a density gradient will be created spontaneously, which is not possible. For example, when water evaporates into an air body over water surface, an equal quantity of air cannot enter the water phase. The density gradient created is dispersed by some mixture moving away from the surface maintaining a balance. This is termed as bulk flow.

The value of D_{ab} for certain combinations of components are available in literature. It can be proved that $D_{ab} = D_{ba}$. When one molecule of 'A' moves in the x direction, one molecule of 'B' has to move in the opposite direction. Otherwise a macroscopic density gradient will develop, which is not sustainable, (A is area)

$$\frac{N_a}{A} = -D_{ab}\frac{dC_a}{dx}$$

$$\frac{N_b}{A} = -D_{ba}\frac{dC_b}{dx} = -D_{ba}\frac{d(1-C_a)}{dx} = D_{ba}\frac{dC_a}{dx}$$

as $$\frac{N_a}{A} = -\frac{N_b}{A} \text{ and so } D_{ab} = D_{ba}$$

Binary diffusion coefficients for various pairs of species, like air-water, are available in hand books and data books.

The diffusion coefficients are higher in the case of gases by several orders of magnitude compared to liquids. For example, D_{ap} for Nitrogen-water is 2.6×10^{-9}, while Nitrogen-water vapour is 2.5×10^{-5}. This is because of the higher mobility of molecules in gases.

Diffusion coefficients increase with temperature, in the ratio $T^{1.5}$.

14.4 EQUIMOLAL COUNTER DIFFUSION

In two component mixture in a container, if the partial pressure varies with location, the components will diffuse in the opposite directions. As the density is to remain constant, the rate of diffusion of one component should equal the rate of diffusion of the other. This type of diffusion is called equimolar counter diffusion. This is a steady flow process.

The total pressure is constant all through the mixture. Hence the difference in partial pressures will be equal. The Fick's equation when integrated for a larger plane volume of thickness L will give

$$\frac{N_a}{A} = D_{ab}\frac{(C_{a1} - C_{a2})}{L} \quad ...(14.5)$$

$$\frac{N_b}{A} = D_{ba}\frac{(C_{b2} - C_{b1})}{L}$$

as $$\frac{N_b}{A} = -\frac{N_a}{A}, \text{ and } (C_{a1} - C_{a2}) = (C_{b2} - C_{b1}),$$

D_{ab} equals D_{ba}

where C_{a1} and C_{b1} are the mole concentrations at face 1 and C_{a2} and C_{b2} are mole concentrations at face 2 which is at a distance L from the first face.

When applied to gases,

$$\frac{N_a}{A} = \frac{D}{\Re T}\cdot\frac{P_{a1} - P_{a2}}{(x_2 - x_1)} \quad ...(14.6)$$

where P_{a1} and P_{a2} are partial pressures of component 'A' at x_1 and x_2 and $\mathfrak{R}$ is the universal gas constant in J/kg mol *K*. *T* is the temperature in absolute units. The distance should be expressed in metre.

The partial pressure variation and diffusion directions are shown in Fig. 14.1.

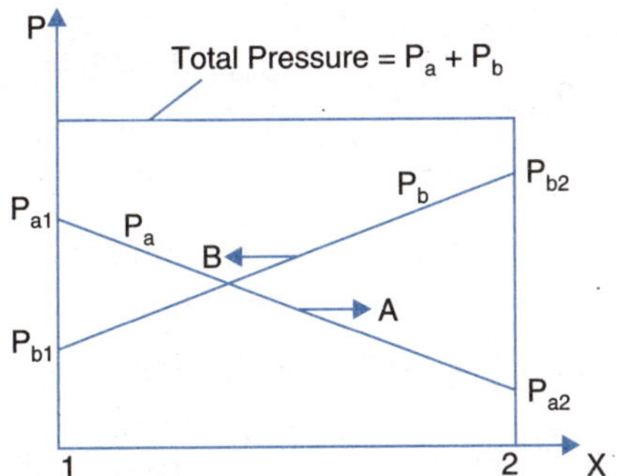

Fig. 14.1. Partial Pressure variation of components in equimolal counter diffusion

Example 14.1: *In order to avoid pressure build up ammonia gas at atmospheric pressure in a pipe is vented to atmosphere through a pipe of 3 mm dia and 20 m length. Determine the mass of ammonia diffusing out and mass of air diffusing in per hour the temperature is 310 K.*

Assume $D = 0.28 \times 10^{-4}$ m²/s, $M = 17$ kg/kg mole

Solution: P_{NH_3} in pipe = 1 atm.

P_{NH_3} at the outlet = 0, As $N_a = m_a/M$

$$m_{NH_3} = \frac{D.A.}{\mathfrak{R}T} \frac{P_{NH_3^1} - P_{NH_3^2}}{L} \times M$$

$$= 0.28 \times 10^{-4} \times \frac{\pi}{4}(0.003)^2 \times \frac{(1.013 \times 10^5 - 0)}{20 \times 310} \times 3600 \times 17/8315$$

$$= \mathbf{23.8 \times 10^{-9}\ kg/hr.}$$

$$m_{air}, N_B = -N_A = -23.8 \times 10^{-9} \times 28.97/17$$

$$= \mathbf{40.56 \times 10^{-9}\ kg/hr.}$$

$$M_{air} = 28.97 \text{ kg/kg mole.}$$

Henry's law: Many processes require absorption of a gas into a liquid. Most gases are weakly soluble in liquids (such as air in water). For such dilute solution the mole fraction of a particular species *i* in a liquid phase at the interphase are observed to be proportional to each other.

Mole fraction in the gas side is proportional to mole fraction in liquid side. For ideal gases, this relation is known as Henry's law.

$$\text{At the interface, mole fraction of } i \text{ on liquid side} = \frac{p_i \text{ gas side}}{H}$$

where *H* is known as Henry's constant and p_i is the partial pressure in the gas side for component *i*.

Example 14.2: *Determine the mole fraction of water vapour at the surface of a lake whose temperature is 18°C and compare it with mole fraction of water in the lake. Atmospheric pressure is 100 kPa. Also determine the fraction of air dissolved in the water at the surface of the lake.*

Solution: Saturation pressure of water vapour at 18°C = **0.02064 bar**

Assuming air and water vapour as ideal gas,

$$\text{mole fraction of water vapour in air} = \frac{P_{\text{vapour}}}{\rho}\ \frac{0.02064}{1} = \mathbf{0.02064}$$

(as 100 kPa = 1 bar)

As water contains negligible amount of air Mole fraction = **1.0**

Henry's constant at this temperature = 62000 bar.

Partial pressure of dry air = 1 – 0.02064 = 0.97936 bar

$$\text{Mole fraction of air on liquid side} = \frac{0.97936}{62000} = \mathbf{15.8 \times 10^{-6}\ bar.}$$

14.5 STATIONARY MEDIA WITH SPECIFIED SURFACE CONCENTRATION

In the diffusion of gas through container walls, gas from inside diffuses to the outside without the metal molecules diffusing into the gas. In these cases the concentration of gas at the surfaces should be known. The solubility of the gas in the surface determines the concentration at the surface.

These cases are similar to conduction through the medium.

In these cases the temperature potential in conduction is replaced by concentration potential $(C_{a1} - C_{a2})$ for component A.

The flow rate can be obtained as in the case of conduction.

$$N_a = (C_{a1} - C_{a2})/R.$$

where R is the **resistance** of diffusion. The resistance in the case **of plane wall is**

$$R_p = \frac{L}{D_{ab}A} \qquad ...(14.7)$$

For **hollow cylindrical** configuration.

$$R_{cyl} = \frac{\ln(r_2 / r_1)}{2\pi D_{ab} L} \qquad ...(14.8)$$

For **hollow sphere,** $R_{sp} = \dfrac{1}{4\pi D_{ab}}\left[\dfrac{1}{r_1} - \dfrac{1}{r_2}\right]$ ...(14.9)

These equations can be derived from the general equation in Cartesian, cylindrical and spherical coordinate systems.

Example 14.3: *Hydrogen stored in a vessel diffuses through the steel wall of 20 mm thickness. The molar concentration at the inner surface is 2 kg mol/m³. At the other surface it is zero. Assuming plane wall condition and* $D_{ab} = 0.26 \times 10^{-12}\ m^2/s$, *determine the rate of hydrogen diffusion.*

Solution:

$$N_a = - D_{ab} \cdot \frac{C_{a2} - C_{a1}}{L} = D_{ab} \cdot \frac{C_{a1} - C_{a2}}{L} = D_{ab} \cdot \frac{C_{a1}}{L}$$

$$= 0.26 \times 10^{-12} \times \frac{2}{0.02} = 2.6 \times 10^{-11} \text{ kg mol/s m}^2.$$

For H_2, molecular weight is 2.

$\therefore$ mass diffused = $2 \times 2.6 \times 10^{-11}$ = $\mathbf{5.2 \times 10^{-11}}$ **kg/m²s.**

Example 14.4: *Hydrogen gas is maintained at 4 bar and 1 bar on the opposite sides of a membrane of 0.5 mm thickness. At this temperature the diffusion coefficient is 8.7×10^{-8} m²/s. The solubility of hydrogen in the material which depends on the pressure is 1.5×10^{-3} m²/s bar.*

Determine the mass diffusion rate of hydrogen through the membrane.

Solution:

$$C_1 = 1.5 \times 10^{-3} \times 4 = 6 \times 10^{-3} \text{ kg mol/m}^3$$

$$C_2 = 1.5 \times 10^{-3} \times 1 = 1.5 \times 10^{-3} \text{ kg mol/m}^3$$

Considering plane wall condition

$$R = \frac{L}{DA} = \frac{0.0005}{8.7 \times 10^{-8} \times 1}$$

$$\therefore \quad \text{Mole flux} = (6 \times 10^{-3} - 1.5 \times 10^{-3}) \Big/ \frac{0.0005}{8.7 \times 10^{-8} \times 1}$$

$$= 7.83 \times 10^{-7} \text{ kg mol/m}^2\text{s}.$$

$$\text{Mass flux} = 2 \times 7.83 \times 10^{-7} \text{ kg/m}^2\text{s} = \mathbf{1.566 \times 10^{-6}} \textbf{ kg/m}^2\textbf{s}.$$

Example 14.5: *Pressurised hydrogen is stored at 358 K in a spherical vessel of 4.8 m OD. The wall is of Nickel and 6 cm thick. The molar concentration of hydrogen at the inner surface of the wall = 0.087 kg mol/m³.*

The concentration at the outer surface is negligible.

Determine the mass rate of diffusion of hydrogen from the vessel. The binary diffusion coefficient for hydrogen in nickel = 1.2×10^{-12} m²/s.

Solution: The flow is through a stationary medium.

For sphere the number of molecules diffusing is

$$N = 4\pi r_1 r_2 D_{ab} \times \frac{C_{a1} - C_{a2}}{r_2 - r_1}$$

$$= 4\pi \times 2.34 \times 2.4 \times 1.2 \times 10^{-12} \times \frac{(0.087 - 0)}{2.4 - 2.34} = \mathbf{1.228 \times 10^{-10}} \textbf{ kg mol/s}$$

$$\text{mass} = N \times M = 1.228 \times 10^{-10} \times 2 = \mathbf{2.456 \times 10^{-10}} \textbf{ kg/s.}$$

Diffusion of gas into metal: The concentration of gas species i in the solid at the interface is proportional to the partial pressure of species i in the gas at the metal surface.

Concentration on solid side = Solubility × P_i gas side.

Solubility also known as permeability is the measure of the ability of the gas to penetrate the solid.

Example 14.6: *Hydrogen gas at 300 kPa and 358 K is in contact with Nickel surface. Determine the molar and mass density of Hydrogen at the surface.*

Solution: Solubility of Hydrogen in Nickel at this temperature

$= 0.00901$ kg mol/m³.

Molar concentration $= 0.00901 \times 3 =$ **0.02703 kg mol/m³**

Mass concentration $= 2 \times 0.2703 =$ **0.05404 kg/m³.**

14.6 DIFFUSION OF ONE COMPONENT INTO A STATIONARY COMPONENT OR UNIDIRECTIONAL DIFFUSION

In this case one of the components diffuses while the other is stationary. For steady conditions the mass diffused should be absorbed continuously at the boundary. In certain cases this is not possible. The popular example is water evaporating into air. In this case, as mentioned earlier, a bulk motion returns the air tending to accumulate at the interface without being absorbed. The diffusion equation for gases can be derived as (with '*a*' as the diffusing medium and P = total pressure) (Refer Advanced example 16 for derivation)

$$\frac{N_a}{A} = \frac{P}{\Re T} \cdot \frac{D}{(x_2 - x_1)} \cdot \ln\left(\frac{P - P_{a2}}{P - P_{a1}}\right) \quad \text{...(14.10)}$$

For liquids (considering '*a*' as diffusing medium)

$$\frac{N_a}{A} = \frac{D \cdot C}{(x_2 - x_1)} \cdot \ln\left(\frac{C - C_{a2}}{C - C_{a1}}\right) \quad \text{...(14.11)}$$

14.7 UNSTEADY DIFFUSION

The equation for one dimensional unsteady diffusion is

$$\frac{\partial^2 C_a}{\partial x^2} = \frac{1}{D} \cdot \frac{\partial C_a}{\partial \tau} \quad \text{...(14.12)}$$

This is similar to heat conduction equation with C_a and D replacing by T and α. All the solutions for various boundary conditions for unsteady conduction are applicable for mass transfer, except that the parameters are different. Heisler charts can also be used, properly recognising the parameters to be used. Refer Chapter 6 for charts.

Conduction parameters	*Equivalent Mass Transfer parameters*
$\frac{T - T_\infty}{T_i - T_\infty}$	$\frac{C_a - C_{as}}{C_{ai} - C_{as}}$
$F_o = \alpha\tau/L^2$	$F_{om} = \frac{D\tau}{L^2}$
$B_i = \frac{hL}{k}$	$B_{im} = \frac{h_m L}{D}$
$\frac{x}{2\sqrt{\alpha\tau}}$	$\frac{x}{2\sqrt{D\tau}}$

Refer Advanced example 24.

Example 14.7: *A steel component is surface hardened by packing it in a carbonaceous material in a high temperature furnace. The surface has a uniform carbon concentration of 0.15% by mass. The diffusion coefficient of carbon in steel at the furnace temperature is 4.8 × 10^{-10} m²/s. The equilibrium concentration of carbon at the interface is 1.2% by mass. Determine the time required for the concentration to reach 1 per cent at 0.5 mm from surface.*

Solution: The problem can be modelled as semi infinite slab. Indicating concentration as C,

$$\frac{C_{A(x,\tau)} - C_{Ai}}{C_{AS} - C_{Ai}} = erf\left(\frac{x}{2\sqrt{D_{AS}\,.\,\tau}}\right)$$

$$\frac{0.01 - 0.0015}{0.012 - 0.0015} = erf\left(\frac{0.0005}{2\sqrt{4.8 \times 10^{-10}\,\tau}}\right) = 0.81$$

From handbook. The value of $\dfrac{0.0005}{2\sqrt{4.8 \times 10^{-10}\tau}}$ is read as 0.17

Solving $0.17 \times 2\sqrt{4.8 \times 10^{-10}}\ \sqrt{\tau} = 0.0005$

$$\tau^{0.5} = 0.0005/0.17 \times 2\sqrt{4.8 \times 10^{-10}} = 67.1$$

$\therefore$ τ = 4505.55, or = **1 hr and 15 min.**

14.8 CONVECTIVE MASS TRANSFER

When a medium deficient in a component, flows over a medium having an abundance of the component, then the component will diffuse into the flowing medium. Diffusion in the opposite direction will occur if the mass concentration levels of the component are interchanged.

In this case a boundary layer develops and at the interface mass transfer occurs by molecular diffusion (In heat flow at the interface, heat transfer is by conduction).

Velocity boundary layer is used to determine wall friction. Thermal boundary layer is used to determine convective heat transfer coefficient. Similarly concentration boundary layer is used to determine convective mass transfer coefficient.

The Fig. 14.2 shows the flow of a mixture of components A and B with a specified constant concentration over a surface rich in component A. A concentration boundary layer develops. The concentration gradient varies from the surface to the free stream. At the surface the mass transfer is by diffusion. Convective mass transfer coefficient h_m is defined by the equation, where h_m has a unit of m/s.

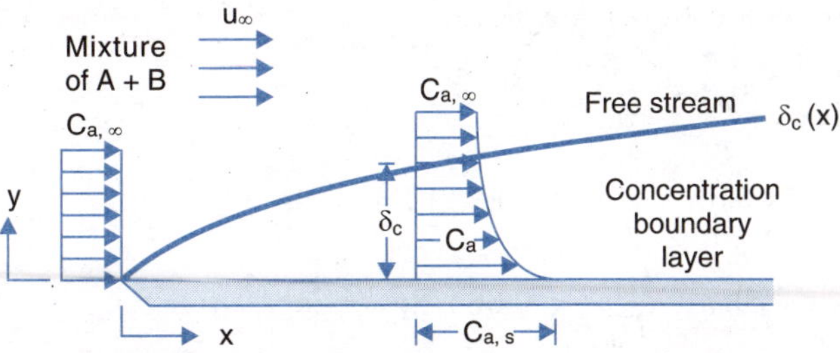

Fig. 14.2. Species concentration boundary layer development on a flat plate

$$\text{Mole flow} = h_m(C_{as} - C_{a\infty})$$

The condition for diffusion at the surface is given by

$$\text{Mole flow} = -D_{ab} \left.\frac{\partial C_a}{\partial y}\right|_{y=0}$$

$$\therefore \qquad h_m = \frac{-D_{ab} \cdot \left.\dfrac{\partial C_a}{\partial y}\right|_{y=0}}{C_{as} - C_{a\infty}} \qquad \text{...(14.13)}$$

In the above case, if mass flow is to be used then

$$h_m = \frac{-D_{ab} \cdot \left.\dfrac{\partial \rho_a}{\partial y}\right|_{y=0}}{\rho_{as} - \rho_{a\infty}} \qquad \text{...(14.14)}$$

Similar to the momentum and energy equation, the mass concentration equation can be obtained as below:

$$u\frac{\partial u}{\partial x} + v\frac{\partial u}{\partial y} = v \,.\, \frac{\partial^2 u}{\partial y^2}$$

$$u\frac{\partial T}{\partial x} + v\frac{\partial T}{\partial y} = \alpha \frac{\partial^2 T}{\partial y^2}$$

$$u\frac{\partial C_a}{\partial x} + v\frac{\partial C_a}{\partial y} = D_{ab} \frac{\partial^2 C_a}{\partial y^2} \qquad \text{...(14.15)}$$

By similarity the solutions for boundary layer thickness for connective mass transfer can be obtained. This is similar to the heat transfer by analogy. In this case, in the place of Prandtl number Schmidt number defined by equation (14.16) is used,

$$\text{Sc} = v/D_{ab} \qquad \text{...(14.16)}$$

Non-dimensionalising the equation leads to the condition as below:

$$\delta_m = f(\text{Re, Sc}) \qquad \text{...(14.17)}$$

$$\text{Sh} = f(\text{Re, Sc}) \qquad \text{...(14.18)}$$

where Sherwood number Sh is defined as

$$\text{Sh} = \frac{h_m x}{D_{ab}}.$$

In the laminar region flow over plate:

$$\delta_{mx} = \frac{5x}{\text{Re}_x^{1/2}} \,.\, \text{Sc}^{-1/3} \qquad \text{...(14.19)}$$

$$\text{Sh}_x = \frac{h_{mx} x}{D_{ab}} = 0.332\ \text{Re}_x^{1/2}\ \text{Sc}^{1/3} \qquad \text{...(14.20)}$$

$$\overline{\text{Sh}}_L = \frac{h_m L}{D_{ab}} = 0.664\ \text{Re}_2^{1/2}\ \text{Sc}^{1/3} \qquad \text{...(14.21)}$$

In the turbulent region $Re > 5 \times 10^5$,

$$\delta_m = \delta_v \quad ...(14.22)$$

$$Sh_x = 0.0296\ Re_x^{0.8}\ Sc^{1/3} \quad ...(14.23)$$

$$\overline{Sh}_L = 0.037\ Re_L^{0.8}\ Sc^{1/3} \quad ...(14.24)$$

For flow through tubes,

In the laminar region, $Re < 2000$

For uniform wall mass concentration,

$$Sh = 3.66 \quad ...(14.25)$$

For uniform wall mass flux

$$Sh = 4.36 \quad ...(14.26)$$

For turbulent region,

$$Sh = 0.023\ Re^{0.83}\ Sc^{1/3} \quad ...(14.27)$$

Significance of dimensionless numbers:

Schmidt Number, $SC = \dfrac{\text{Momentum diffusivity}}{\text{Mass diffusivity}}$

(Similar to Prandtl number in heat transfer)

Sherwood number is similar to Nusselt number

$$= \frac{\text{Concentration gradient at the boundary by diffusion}}{\text{Concentration gradient at the boundary by convection}}$$

Lewis number $= \dfrac{\text{Thermal diffusivity}}{\text{Mass diffusivity}}$

Stanton number for mass transfer = Sherwood number/Re. Sc.

14.9 SIMILARITY BETWEEN HEAT AND MASS TRANSFER

It is possible from similarity between the heat convection equation and mass convection equation to obtain value of h_m. (*i.e.*, called as Lewis number)

$$\frac{h}{h_m} = \rho C_p / Le^{2/3} \quad ...(14.28)$$

where $$Le = \frac{\alpha}{D} \quad ...(14.29)$$

Many of the correlation in heat transfer can be applied to mass transfer under similar condition, by replacing Nusselt number by Sherwood number and Prandtl number by Schmidt number.

Refer advanced examples 24–34 for transient and convective mass transfer.

ADVANCED EXAMPLES

Example A-14.1: *A tank contains a mixture of CO_2 and N_2 in the mole proportions of 0.2 and 0.8 at 1 bar and 290 K. It is connected by a duct of sectional area 0.1 m² to another tank containing a mixture of CO_2 and N_2 in the molal proportion of 0.8 and 0.2. The duct is 0.5 m long. Determine the diffusion of CO_2 and N_2. $D = 0.16 \times 10^{-4}\ m^2/s$.*

Solution: This is a situation of equimolal counter diffusion. The partial pressures have to be used as the diffusing components are gases. The partial pressures are directly proportional to the mole concentration. Denoting CO_2 by a and N_2 by b

$$P_{a1} = 0.2 \text{ bar} \quad P_{a2} = 0.8 \text{ bar}, \quad P_{b1} = 0.8 \text{ bar} \quad \text{and} \quad P_{b2} = 0.2 \text{ bar}$$

The value of gas constant should be obtained in terms of atm pressure or the pressures should be used in N/m² in case the following value is used.

$$\mathfrak{R} = 8315 \text{ J/kg mol K.}$$

$$\frac{N_a}{A} = \frac{D}{\mathfrak{R}T} \cdot \frac{P_{a2} - P_{a1}}{x_2 - x_1}, \quad \frac{N_a}{A} = \frac{0.16 \times 10^{-4}(0.8 - 0.2) \times 10^5}{8315 \times 290 \times 0.5},$$

$$\frac{N_a}{A} = 7.9623 \times 10^{-7} \text{ kg mol/m}^2\text{s, area} = 0.1 \text{ m}^2$$

$$\therefore \quad \mathbf{N_a = 7.9623 \times 10^{-8} \text{ kg mol/s}, m_a = 3.5 \times 10^{-6} \text{ kg/s of } CO_2}$$

As the diffusion is equimolal counter diffusion

$$N_b = 7.9623 \times 10^{-8} \text{ kg mol/s}, m_b = 2.23 \times 10^{-6} \text{ kg/s } N_2.$$

Mass of N_2 diffusing $= 7.9623 \times 10^{-8} \times 28 = 2.23 \times 10^{-6}$ kg/s

Example A-14.2: *A pipe carrying ammonia at 1 bar and 40°C is vented to a large tank containing dry air at 1 bar and 40°C to avoid pressure build up. The pipe is 5 mm in dia and 5 m long. Determine the rate of diffusion of air into the ammonia stream. $D = 0.28 \times 10^{-4}\ m^2/s$.*

Solution: The partial pressure of air at the end of pipe is 1 bar and at the connection with ammonia tank it is zero.

$$N_a = \frac{\pi \times 0.005^2}{4} \times \frac{(10^5 - 0)}{5} \times \frac{0.28 \times 10^{-4}}{8315 \times 313}$$

$$= 4.225 \times 0^{-12} \text{ kg mol/s}$$

or in terms of mass $N_a \times M_a$ gives 1.263×10^{-10} kg/s.

Example A-14.3: *A well is 40 m deep and 9 m dia and the atmospheric temperature is 25°C. The air at the top is having a relative humidity of 50%. Determine the rate of diffusion of water vapour through the well $D = 2.58 \times 10^{-5}\ m^2/s$.*

Solution: This is a situation of diffusion into a stationary medium. The partial pressure is equal to saturation pressure at 25°C = 0.03169 bar. At the top RH = 50%.

$\therefore$ Partial pressure = 0.5 × 0.03169 bar.

If mass is to be calculated use the gas constant for the gas instead of universal gas constant.

$$m_a = A \cdot \frac{DP}{R_a\, T(x_2 - x_1)} \cdot \ln \frac{P - P_{b2}}{P - P_{b1}}$$

$$R_a = \frac{8315}{18} \text{ J/kgK}, T = 298 \text{ K.}, x_2 - x_1 = 40 \text{ m}$$

$$A = \frac{\pi \times 9^2}{4}, P = 10^5 \text{ N/m}^2$$

$$\therefore \quad m_a = \frac{\pi \times 9^2}{4} \times \frac{2.58 \times 10^{-5} \times 18}{8315 \times 298 \times 40} \times 10^5 \cdot \ln \frac{10^5(1 - 0.5 \times 0.03169)}{10^5(1 - 0.03169)}$$

$= \mathbf{4.84 \times 10^{-7}}$ **kg/s** (low because of large depth).

Example A-14.4: *Gaseous hydrogen diffuses through a steel wall of 3 mm thickness. The molar concentration of hydrogen at the interface is 1.5 kg mol / m³ and it is zero on the outer face. Determine the diffusion rate of hydrogen.*

Solution: $D = 0.3 \times 10^{-12}$ m²/s

In this case we can use the one dimensional diffusion equation.

$$N = D \cdot \frac{C_{a1} - C_{a2}}{L} = \frac{0.3 \times 10^{-12} \times 1.5}{0.003}$$

$= \mathbf{1.5 \times 10^{-10}}$ **kg mol/ sm².**

Example A-14.5: *Benzene liquid at 25°C is in a cylindrical glass jar of 5 cm dia at the bottom. Air column is 30 cm above the liquid. The air in the jar is stationary. Sufficient air movement exists at the top to remove the diffused vapour so that the partial pressure of vapour at the top can be assumed as zero. Determine the diffusion rate. The partial pressure at the interface is 0.1 bar.*

Solution: $D = 0.0962 \times 10^{-4}$ m²/s, $R = 8315/78$, as C_6H_6 has a molecular mass of 78.

This is a situation of diffusion into a stationary medium.

$$m_a = \frac{A \cdot D}{R_a T} \cdot \frac{P}{x_2 - x_1} \ln \frac{P - P_{a1}}{P - P_{a2}}$$

$$\therefore \quad m_a = \frac{\pi \times 0.05^2}{4} \times \frac{10^5 \times 0.0962 \times 10^{-4}}{0.3} \times \frac{78}{8315 \times 298} \cdot \ln \frac{1.0}{0.9}$$

$= \mathbf{2.088 \times 10^{-8}}$ **kg/s.**

Example A-14.6: *In a solar pond salt is placed at the bottom of the pond 1.5 m deep. The surface is flushed constantly so that the concentration of salt at the top layer is zero. The salt concentration at the bottom layer is 5 kg mol / m³. Determine the rate at which salt is washed off at the top at steady state conditions per m². D = 1.24 × 10⁻⁹ m² / s.*

Solution: This is also diffusion of one component into a stationary component.

$$N = \frac{D_{ab}}{L} \cdot C \ln \frac{C - C_{a1}}{C - C_{a2}}$$

$C = 1000/18$ kg mol/m³ (This is approximate), $C_{a1} = 0$

$$\therefore \qquad N = \frac{1.24 \times 10^{-9}}{1.5} \times \frac{1000}{18} \cdot \ln \frac{(1000/18)}{\left(\frac{1000}{18}\right) - 5}$$

= **4.33 × 10⁻⁹ mol/m²s,** over a month, 0.0113 kg mol/m^2

mass = 58.5 × 0.0113 = 0.656 kg/month/m^2.

Example A-14.7: *Air at 25°C and 50% RH flows over water surface measuring 12 m × 6 m at a velocity of 2 m/s. Determine the water loss per day considering flow direction is along the 12 m side.* $D_{AB} = 0.26 \times 10^{-4}$ *m²/s, Sc = 0.60,* $\nu = 15.7 \times 10^{-6}$ *m²/s.*

Solution: The density of water vapour at 25°C at saturated condition, from steam tables = 0.023 kg/m^3

$$\text{Re} = \frac{U_\infty L}{\nu} = \frac{2 \times 12}{15.7 \times 10^{-6}} = 1.53 \times 10^6 \quad \therefore \quad \text{flow is turbulent}$$

$$\overline{\text{Sh}} = (0.037\ \text{Re}^{0.8} - 871)\ \text{Sc}^{0.333}$$

$$= \{0.037 \times (1.53 \times 10^6)^{0.8} - 871\}\ 0.6^{0.333} = 2032.$$

The mass transfer coefficient

$$h_m = \text{Sh}\left(\frac{D_{ab}}{L}\right) = 2032 \times \frac{0.26 \times 10^{-4}}{12} = \mathbf{4.1 \times 10^{-3}\ m/s}$$

Mass transfer rate $= Ah_m(\rho_{aw} - \rho_{am})$

$= 12 \times 6 \times 4.4 \times 10^{-3}\ (0.023 - 0.023 \times 0.5) \times 3600 \times 24 = 314.8$ kg/day.

Example A-14.8: *Air at 25°C and 20% RH flows through a pipe of 25 mm ID with a velocity of 5.2 m/s. The inside surface is constantly wetted with water and a thin water film is maintained throughout. Determine the water evaporated per m² surface area.*

Solution: From tables $\quad v = 15.7 \times 10^{-6}$ m²/s, Sc = 0.60

$D_{ab} = 0.26 \times 10^{-4}$ m²/s.

Density of water vapour at saturation = 0.023 kg/m^3

$$\text{Re} = \frac{5.2 \times 0.025}{15.7 \times 10^{-6}} = 8280 \quad \therefore \quad \text{Turbulent}$$

$$\text{Sh} = 0.023\ .\ \text{Re}^{0.83}\ \text{Sc}^{1/3} = 0.023 \times 8280^{0.83} \times 0.6^{1/3} = 34.65$$

$$h_m = \frac{34.65 \times 0.26 \times 10^{-4}}{0.025} = 0.036\ \text{m/s}$$

$\therefore$ mass flow = 0.036 × (0.023 – 0.023 × 0.20) kg/s m^2

= 6.22 × 10^{-4} kg/m²s or 4.886 × 10^{-5} kg/s per m length

(1 m² will equal 1/π × 0.025 = 12.73 m length of pipe).

Example A-14.9: *A thermometer whose bulb is covered by a wetted cloth reads 20°C when dry air is blown over it. Determine the temperature of the air.*

Solution: Heat removed from the air by convection is used for the evaporation of water over the bulb.

The energy balance is obtained as heat transfer by convection = heat used for the evaporation

$$hA(T_a - T_t) = m \cdot h_{fg}$$

where h is the convection coefficient, A is the surface area, T_a is the temperature of air, T_t is the thermometer reading, m is the rate of evaporation and h_{fg} is the enthalpy of evaporation.

The mass evaporated is given by

$$h_m A(C_t - C_a)$$

where h_m is the mass transfer coefficient, C_t is the concentration of water vapour in air over the thermometer surface and C_a is the concentration of water vapour in air.

In this case $C_a = 0$

$$\therefore \quad h(T_a - T_t) = h_m C_t h_{fg}$$

The air over the thermometer is saturated at 20°C

$\therefore$ C_t = 1/57.791 kg/m³ (steam tables) h_{fg} = 2454.1 kJ/kg

As $$(h/h_m) = \rho \cdot c_p \left(\frac{\alpha}{D}\right)^{2/3}$$

$$\frac{\alpha}{D} = \frac{\text{Sc}}{\text{Pr}} = 0.790,\ \rho = 1.205 \text{ kg/m}^3.$$

From property tables for air

$$\alpha = 0.2142 \times 10^{-4},\ D = 2.712 \times 10^{-5}$$

$$\frac{h}{h_m} = 1.205 \times 1005 \times \left(\frac{0.2142 \times 10^{-4}}{2.712 \times 10^{-5}}\right)^{0.666} = 1034.65$$

$$\therefore \quad T_a = T_t + \left(\frac{1}{h/h_m}\right) C_t h_{fg} = 20 + \frac{1}{1034.65} \times \frac{1}{57.791} \times 2.4541 \times 10^6$$

$$= 61°\text{C}.$$

This can be checked using psychrometric chart moving along 20°C wet bulb line going to zero specific humidity by extrapolation.

Example A-14.10: *If air at 30°C flows over a wet bulb thermometer, which reads 22°C, determine the relative humidity of the air.*

Solution: The property values for air are read at 26°C (mean temperature, usual units)

$\rho = 1.181$ $\quad c = 1005,\ \alpha = 0.2228 \times 10^{-4}$ $\quad D_{ab} = 2.583 \times 10^{-5}$ m²/s

At 22°C $\quad C_t = 1/51.447 = 0.01944,$ $\quad h_{fg} = 2449.4 \times 10^3$ J/kg

A heat balance as in example A-14.9 will yield

$$T_a - T_t = \frac{h_m}{h}(C_t - C_a) h_{fg} \quad \text{...(A)}$$

$$\frac{h}{h_m} = \rho c \left(\frac{\alpha}{D}\right)^{2/3} = 1.181 \times 1005 \times \left(\frac{0.2228 \times 10^{-4}}{2.583 \times 10^{-5}}\right)^{0.666} = 1075.5$$

Substituting in equation (A)

$$(30-22) = \frac{1}{1075.5} \times 2.4494 \times 10^6 (0.0194 - C_a) \quad \therefore \quad C_a = 0.01593$$

$$C_t = \text{at saturation at } 30°\text{C} = \frac{1}{32.894} = 0.0304$$

$$\therefore \quad RH = 0.01593/0.0304 = 0.524$$

or $$RH = 52.4\%$$

From Psychrometric chart 50% checks reasonably.

(51.447 and 32.894 are specific volumes of steam at 22°C and 30°C used to calculate C_t).

Example A-14.11: *Determine the value of diffusion coefficient for the given pairs at a total pressure of 101.3 kPa. (i) Hydrogen and air at 0°C, (ii) Water and air at 8°C, and (iii) Oxygen and air at 0°C. (**Note.** Molecular volumes should be specified)*

Solution: Molecular volume V for some gases are tabulated below:

Gas	**V**	***Gas***	**V**
Air	29.9	O_2	7.4
CO_2	34.0	O_2 with S	8.3
H_2	14.3	S	25.6
Nitrogen	15.6	Water	18.8

For this problem the required values are given below:

Gas	***Volume, V***	***Molecular weight, M***
Air	29.9	28.9
Hydrogen	14.3	2.0
Oxygen	7.4	32.0
Water	18.8	18.0

The applicable equation (from references)

$$D_{ab} = \{0.04357 \times T^{15}) \times [(1/M_a) + (1/M_b)]^{0.5} \Big/ \left\{P \times \left(V_a^{1/3} + V_b^{1/3}\right)^2\right\} \text{ m}^2/\text{s}$$

Note. Pressure P in N/m² and temperature T in K.

(*i*) **Hydrogen and air at 0°C or 273 K**

$$D_{ab} = \{(0.04357 \times 273^{1.5}) \times [(1/2) + (1/28.9)]^{0.5}\}/\ \{101.3 \times 10^3 \times (14.3^{1/3} + 29.9^{1/3})^2\} \text{ m}^2/\text{s}$$

$$= \mathbf{46.36 \times 10^{-6} \text{ m}^2/\text{s} \text{ [Data Book value } 54.7 \times 10^{-6}]}$$

(*ii*) **Water and air at 8°C or 281 K**

$$D_{ab} = \{(0.04357 \times 281^{1.5}) \times [(1/18) + (1/28.9)]^{0.5}\}/\ \{101.3 \times 10^3 \times (18.8^{1/3} + 29.9^{1/3})^2\} \text{ m}^2/\text{s}$$

$$= \mathbf{18.32 \times 10^{-6} \text{ m}^2/\text{s} \text{ [Data Book value: } 20.5 \times 10^{-6}]}$$

(*iii*) **Oxygen and air at 0°C or 273 K**

$$D_{ab} = \{(0.04357 \times 273^{1.5}) \times [(1/32) + (1/28.9)]^{0.5}\}/\{101.3 \times 10^3 \times (7.4^{1/3} + 29.9^{1/3})^2\} \text{ m}^2/\text{s}$$

$$= \mathbf{19.5 \times 10^{-6} \text{ m}^2/\text{s} \text{ [Data Book value: } 15.3 \times 10^{-6}]}$$

Example A-14.12: *Estimate the value of diffusivity for ethanol vapour (C_2H_5OH) diffusing into air at 0°C.*

Solution: Diffusivity is determined also in terms of collision function and energy of molecular attraction. In this method the equation to be used for the determination of diffusivity D_{ab} is,

$$D_{ab} = \{10^{-4} (1.084 - 0.249 \times A) \times T^{1.5} \times A\}/\{P \times r^2 \times B\}$$

where $A = \{(1/M_a) + (1/M_b)\}^{0.5}$

B = collision function, $f(kT/\varepsilon_{ab})$, available from charts,

k = Boltzmann constant,

ε_{ab} = energy of molecular attraction = $(\varepsilon_a \times \varepsilon_b)^{0.5}$

r = molecular separation at collision = $(r_a + r_b)/2$, in **nm**,

M = molecular weight, P = total pressure in Pa, (N/m²)

T = temperature in K

The value of r is determined from molecular volume. (k/ε) is determined using boiling temperature.

For the problem in hand T = 273 K, $P = 1.013 \times 10^5$ Pa, M_a = 46,

M_b = 29, A = 0.237, r = 0.416, B = 0.595 (reference)

Note. Obtaining values for r and B for various combination of materials is the crux of the problem.

Substituting the above values

$$D_{ab} = \{10^{-4}(1.084 - 0.249 \times 0.237) \times 273^{1.5} \times 0.237\}/\{1.013 \times 10^5 \times 0.416^2 \times 0.595\}$$

$$= \mathbf{10.5 \times 10^{-6} \text{ m}^2/\text{s}, \text{ (tables value} = 10.2 \times 10^{-6} \text{ m}^2/\text{s}).}$$

Example A-14.13: *Estimate the value of diffusivity of mannitol liquid ($C_6H_{14}O_6$) for diffusion in dilute solution of water at 20°C.*

Solution: The equation applicable for diffusivity of liquid **a** in very dilute solution in solvent **b** is (references)

$$D_{ab} = 117.3 \times 10^{-18} (\phi \times M_b)^{0.5} \times T/\{\mu \times V_a^{0.6}\}$$

where μ = solution viscosity in kg/ms,

ϕ = association factor for solvent, [2.26 for water, 1.9 for methanol, 1.5 for ethanol, 1.0 for benzene, ethyl, ether]

v_a = solute molal volume, values rather scarce to obtain,

The values of parameters for this problem are

v_a = 0.185, M_b = 18, ϕ = 2.26, $\mu = 1.005 \times 10^{-3}$ kg/ms,

Substituting the values

$$D_{ab} = 117.3 \times 10^{-18} \times (2.26 \times 18)^{0.5} \times 293/\{1.005 \times 10^{-3} \times 0.185^{0.6}\}$$

$$= \mathbf{0.601 \times 10^{-9} \text{ m}^2/\text{s}, \text{ (tables value: } 0.56 \times 10^{-9} \text{ m}^2/\text{s})}$$

Chapter 14

Example A-14.14: *Determine the diffusivity of liquid mannitol in dilute water solution at 70°C($C_6H_{14}O_6$).*

Solution: Examining the various parameters in the equation in example A-14.13 only the viscosity is found to be dependent on temperature. So the value of diffusivity at 20°C is used to solve this problem.

$$(D_{ab1} \times \mu_1/T_1) = (D_{ab2} \times \mu_2/T_2)$$

Substituting the values, at 70°C = 0.4061×10^{-3} kg/ms

$$(D_{ab2} \times 0.4061 \times 10^{-3}/343) = (0.601 \times 10^{-9} \times 1.005 \times 10^{-3}/293)$$

$$D_{ab2} = \mathbf{1.741 \times 10^{-9}\ m^2/s, (tables\ value: 1.56 \times 10^{-9}\ m^2/s)}$$

Example A-14.15: *Derive the expression for equimolal counter diffusion of gases a and b.*

Solution: Fick's law applied to one dimensional steady state diffusion gives the mass flux of a as

$$m_a/A = -D_{ab} \times (dC_a/dx) = -D_{ab}(d\rho_a/dx) \quad \text{...(A)}$$

C_a = mass concentration of component a per unit volume in kg/m^3 or the density ρ_a of the particular component, D_{ab} is the diffusion coefficient and A is the area and x is the diffusion direction.

For gases $\rho = P/RT$, R = gas constant

At constant temperature, $d\rho_a = [1(R_a \times T)dP_a$

Substituting in equation (A)

$$(m_a/A) = [-D_{ab}/(R_a \times T)] \times (dP_a/dx)$$

Integrating between sections x_1 and x_2

$$(m_a/A) = [-D_{ab}/(R_a \times T)] \times [(P_{a1} - P_{a2})/(x_2 - x_1)$$

Note. In place of R_a, the ratio universal gas constant/molecular weight of a, may be used.

If number of moles diffusing is needed then $N_a = m_a/M_a$

$$N_a/A = [D_{ab}/(M_a \times R_a \times T) \times (P_{a1} - P_{a2})/(x_2 - x_1)]$$

Note. $N_a = -N_B$ to maintain uniform pressure.

$$M_a \times R_a = \text{universal gas constant} = 8315 \text{ J/kg K.}$$

For liquids, $\dfrac{N_a}{A} = D_{ab} \cdot \dfrac{C_{a_1} - C_{a_2}}{x_2 - x_1}$ mol/m^{2}s.

Example A-14.16: *Derive an expression for the diffusion of one component into a non diffusing (stationary) component.*

Solution: Refer Fig. AE-14.16 : Let w be the diffusing component and a is the non-diffusing (stationary) component.

The diffusion of component a into w is given by

$$m_a = [-D_{aw} \times A/(R_a \times T)] \times (dP_a/dx)$$

But this material diffusing is not absorbed into component **w.** So a bulk flow of the mixture is necessary to balance the diffusion of component **a.** Let the bulk flow velocity be u.

$$\rho_a \times A \times u = -[P_a/(R_a \times T)] \times A \times u$$

Solving for u from the above two equations,

$$u = [D_{aw}/P_a] \times [dP_a/dx]$$

Mass diffusion of component **w**

$$m_{wd} = \{(-D_{wa} \times A)/(R_w \times T)\} \times \{dP_w/dx\} \quad \text{...(A)}$$

Flow of component **w** with the bulk flow

$$\rho_w \times A \times u = \{P_w/(R_w \times T)\} \times A \times u$$

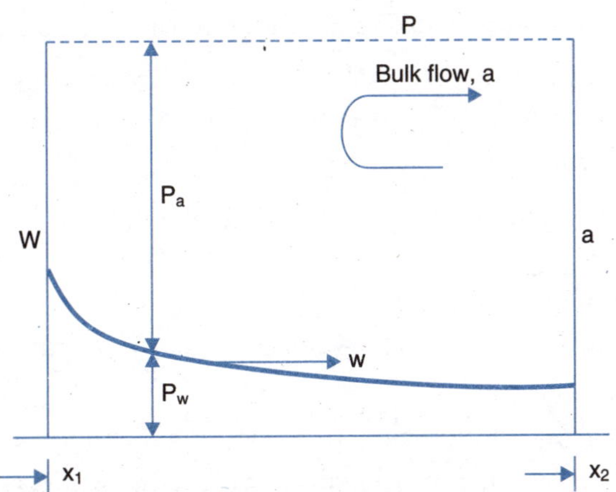

Fig. AE-14.16

Substituting for u, the flow of **w** with bulk flow

$$m_{wb} = \{(P_w \times A \times D_{wa})/(R_w \times T \times P_a)\} \times \{dP_a/dx\} \quad \text{...(B)}$$

Total mass transport of component $m_{w\ \text{total}} = m_{wd} + m_{wb}$

$$m_{w\ \text{total}} = \{(-D_{wa} \times A)/R_w \times T\} \times \{dP_w/dx\} + \{(P_w \times A \times D_{wa})/(R_w \times T \times P_a)\} \times \{dP_a/dx\}$$

As the total pressure $P = P_a + P_w$ is a constant

we get $$dP_a/dx = -dP_w/dx$$

$$m_{\text{total}} = \{(-D_{wa} \times A)/(R_w \times T)\} \times \{1 + P_w/P_a)\} \times (dP_w/dx)$$

$$= \{(-D_{wa} \times A)/(R_w \times T)\} \{P/(P - P_w)\} \times (dP_w/dx)$$

Cross multiplying and integrating (note $D_{wa} = D_{aw}$)

$$\int m\, dx = \{(-D_{wa} \times A \times P)/(R_w \times T)\} \int (dP_w)/(P - P_w)$$

$$m_{\text{total}} = -\frac{D \times A \times P}{R \times T} \times \frac{1}{(x_2 - x_1)} \times \ln\left[\frac{P - P_{w2}}{P - P_{w1}}\right]$$

As $P - P_{w2} = P_{a2}$ and $P - P_{w1} = P_{a1}$

$\ln (P_{a2}/P_{a1})$ can be used in place of $\ln[(P - P_{w2})/(P - P_{w1})]$

For liquids: $$\frac{N_A}{A} = \frac{DC}{x_2 - x_1} . \ln\left[\frac{C - C_{a2}}{C - C_{a1}}\right]$$

where C is the mole concentration.

Chapter 14

Example A-14.17: *Oxygen diffuses through a layer of 3 mm thickness of a mixture of hydrogen and methane in equal volumes. The total pressure is 1 bar and the temperature is 0°C. The partial pressure of oxygen at the two surfaces are 10000 Pa and 5000 Pa. Determine the rate of diffusion of oxygen. The diffusivity of O_2 into H_2 is 69.9×10^{-6} m^2/s and the diffusivity of oxygen into methane is 18.6×10^{-6} m^2/s.*

Solution: The diffusivity into a mixture is obtained by the relation

$$(1/D_{total}) = (C_1/D_1) + (C_2/D_2)$$

where C is the volume fraction or molal fraction.

Substituting the values

$$1/D_{total} = (0.5/69.9 \times 10^{-6}) + (0.5/18.6 \times 10^{-6})$$

$$D_{total} = 29.38 \times 10^{-6} \text{ m}^2/\text{s}$$

The diffusion rate of oxygen is obtained by

$$(m_a/A) = [-D_{ab}/(R_a \times T)] \times [(P_a - P_{a2})/(x_2 - x_1)]$$

Substituting the values

$$(m_a/A) = [(-29.38 \times 10^{-6} \times 32)/(8315 \times 273)] \times [(5000 \times 10000)/(0.003)]$$

$$= \mathbf{0.746 \times 10^{-3} \text{ kg/sm}^2 \text{ or } 23.32 \times 10^{-6} \text{ kg mol/sm}^2}$$

Example A-14.18: *Ethyl alcohol (C_2H_6O) is at the bottom of a glass jar of 30 mm dia at a depth of 80 mm from the top. Air flows at the top and the concentration of alcohol in the air is zero. The total pressure is 1.013 bar and the temperature is 25°C. The partial pressure of alcohol at this temperature is 0.08 bar. Determine the diffusion rate.*

Solution: Diffusion coefficient, $D = 11.9 \times 10^{-6}$ m²/s. Molecular weight = 46, R_b = 8315/46 = 180.76 J/kg K.

This is a situation of one component b (alcohol) diffusing into a stationary component (air).

$$m_b = \{(D \times P \times A)/[(R_b \times T) \times (x_1 - x_2)]\ \{\ln(P - P_{b2})/(P - P_{b1})\}$$

$$P_{b2} = 0, \text{ (Top surface)}, P_{b1} = 0.08 \text{ bar (interface)}$$

$$m_b = \{(11.9 \times 10^{-6} \times 1.013 \times 10^5 \times \pi \times 0.015^2)/$$

$$(180.76 \times 288 \times 0.08)\} \times \{\ln(1.013 - 0.0)/(1.013 - 0.08)]\}$$

$$= \mathbf{16.83 \times 10^{-9} \text{ kg/s or } 60.59 \times 10^{-6} \text{ kg/hr.}}$$

Example A-14.19: *Liquid water diffuses steadily into stationary liquid ethanol layer of 8 mm thickness at 16°C. The concentration of water at the interface is 2 kg mol/m³. At the top of the ethanol layer the concentration is 0.5 kg mol/m³. Determine the rate of diffusion per m². $M = 46$, $D = 0.9 \times 10^{-9}$ m²/s. The total concentration at the interface is 19.6 kg mol/m³.*

Solution: This is a situation where a liquid component (water) diffuses into another stationary liquid component (ethanol).

The equation applicable is (Refer example A-14.16)

$$N_a/A = [(D \times C)/(x_2 - x_1) \times \{\ln[(C - C_{a2})/(C - C_{a1})]\}$$

$$= \{(9 \times 10^{-10} \times 19.6)/0.008\} \times \{\ln(19.6 - 0.5)/(19.6 - 2)]\}$$

$$= 0.18 \times 10^{-6} \text{ kg mol/s m}^2\text{, Multiplying by molecular weight}$$

$$m_a/A = \mathbf{3.24 \times 10^{-6} \text{ kg/sm}^2 \text{ or } 0.03 \text{ kg/m}^2 \text{ hr.}}$$

Example A-14.20: *Acetic acid diffuses into a nondiffusing film of water of 1 mm thickness. The concentration of acid on the opposite sides of the film are 9 and 3% of acid by weight. The diffusivity at this temperature is 0.95×10^{-9} m^2/s. Determine the rate of diffusion of acid. The molecular weight of the acid is 60. The density of the solution with 9% acid is 1012 kg/m^3 and for the 3% solution it is 1003.2 kg/m^3.*

Solution: The equation applicable is

$$N_a/A = [(D \times C)/(x_2 - x_1)] \times \{\ln[(C - C_{a2})/(C - C_{a1})]\}$$

$$C_{a1} = (0.09/60)/\{(0.09/60) + (0.91/18)\} = 0.0288 \text{ (mole fraction)}$$

$$C - C_{a1} = 0.9712.$$

Considering 1 kg of solution the number of moles

$$= (0.09/60) + (0.91/18) = 0.052$$

$$M_{\text{mixture}} = 1/0.052 = 19.21$$

Number of moles per m^3 = 1012/19.21 = 52.7 kg mol/m^3.

$$C_{a2} = (0.03/60)/\{(0.03/60) + (0.97/18)\} = 0.0092$$

$$C - C_{a2} = 0.9908$$

Considering 1 kg of solution the number of moles

$$= (0.03/60) + (0.97/18) = 0.0543$$

$$M_{\text{mixture}} = 1/0.0543 = 18.4,$$

Number of moles per m^3 = 1003.2/18.4 = 54.5 kg mol/m^3.

Average mole concentration, $C = (52.7 + 54.2)/2 = 53.6$

Substituting the values

$$N_a/A = [(0.95 \times 10^{-9} \times 53.6)/(0.001)] \times \{\ln(0.9908/0.9712)\}$$

$$= \mathbf{1.07 \times 10^{-6}\ kg/m^2s.}$$

Example A-14.21: *A steel tank storing hydrogen has a wall thickness of 12 mm. Determine the leakage rate of hydrogen through the wall by diffusion. $D = 0.26 \times 10^{-12}$ m^2/s. At the inner surface the molar concentration of hydrogen is 1.2 kg mol/m^3. At the outside the concentration is 0.2 kg mol/m^3.*

Solution: Hydrogen diffuses through the steel plate. The tank wall is considered as a flat plate.

$$N_a/A = -D(C_{a2} - C_{a1})/(x_2 - x_1) = 0.26 \times 10^{-12}(1.2 - 0.2)/0.012$$

$$= 21.67 \times 10^{-12} \text{ kg mol/sm}^2$$

$$m_a = \mathbf{43.33 \times 10^{-12}\ kg/sm^2 \text{ or } 1.37\ g/m^2\ year.}$$

Example A-14.22: *A spherical tank of 0.18 m radius made of fused silica has a wall thickness of 2.5 mm. It is originally filled with helium at 6 bar gauge and 0°C. Determine the rate of pressure drop with time at this condition due to gas diffusion.*

$D = 0.04 \times 10^{-12}$ m^2/s, the density of gas at the solid surface is given by 18×10^{-9} kg/m^3 Pa. (also termed solubility)

Solution: The wall is approximated as a flat plate.

Gas mass in the container = $m = (PV/RT)$

As the pressure drop due to diffusion is slow, constant temperature can be assumed. As the vessel is rigid constant volume can be assumed. Hence

$$dm/d\tau = (V/RT) \times (dp/d\tau)$$

This rate of change of mass will equal the mass diffusing through the wall of thickness L. As ρ_{a2} is equal to zero.

$$dm/d\tau = A \times D_{ab} \times \rho_{a1}/L$$

where ρ_{a1} is the density of the gas within the solid at the interface

$$(V/RT) \times (dp/d\tau) = (A \times D_{ab} \times \rho_{a1})/L$$

or
$$dp/d\tau = \frac{RT}{V} \cdot \frac{A \times D_{ab}}{L} \cdot \rho_{a1} = R \times T \times \frac{A}{V} \times \frac{D_{ab}}{L} \times \rho_{a1}$$

$$R = 8315/4,\ T = 273 \text{ K},\ L = 0.0025 \text{ m},$$

$$A/V = (4r^2)/\{(4/3)r^2\} = 3/r = 3/0.18$$

$$\rho_{a1} = \text{solubility} \times (P_i - P_o) = 18 \times 10^{-9} \times 6 \times 10^5 \text{ kg/m}^3 = 0.0108$$

Substituting

$$dp/d\tau = \frac{8315}{4} \times 273 \times \frac{3}{0.18} \times \frac{0.04 \times 10^{-12}}{0.0025} \times 0.0108$$

$$= 1.634 \times 10^{-6} \text{ N/m}^2 \text{ per second.}$$

This rate will decrease with time as the pressure drops.

In case the pressure drop over 30 days is to be determined

$$(P_1 - P_2) = 1.634 \times 10^{-6} \times \tau$$

$$= 1.634 \times 10^{-6} \times 30 \times 24 \times 60 \times 60 \ \frac{\text{N}}{\text{m}^2} \cdot \frac{1}{\text{s}} \times \text{s}$$

$$= 4.24 \text{ N/m}^2$$

A very small drop.

Example A-14.23: *Oxygen gas is maintained at a pressure of 4 bar and 1.1 bar on opposite sides of a membrane at 25°C. The plastic membrane is 0.4 mm thick. $D = 0.21 \times 10^{-9}$ m²/s. The solubility of oxygen in the membrane material is 9.984×10^{-7} kg/m³ Pa. Determine the mass flow of oxygen through the membrane.*

Solution: This is a case of diffusion through a stationary medium.

$$m/A = D \times M \times (\rho_{a1} - \rho_{a2})/L$$

$$= 0.21 \times 10^{-9} \times 32 \times 9.984 \times 10^{-7} \times (4 - 1.1) \times 10^5/0.0004$$

$$= \mathbf{4.864 \times 10^{-6} \text{ kg/s.}}$$

This solution is an estimate only as the concentration for diffusion for gas to membrane and membrane to gas are based on different volumes namely unit volume of membrane and unit volume of gas.

Example A-14.24: *A thick layer of water stands still over a slab of sodium chloride salt. The salt diffuses into the water and the concentration of salt at the slab water interface is constant at 380 kg/m³. The water is initially pure. Determine the salt concentration at a distance of 10 mm from the interface after 24 hours. Also determine the amount of salt diffused into the water during this time and the thickness of salt dissolved.*

Solution: $D = 1.2 \times 10^{-9}$ m²/s, density of salt = 2165 kg/m³.

This is a transient diffusion problem. As the salt solution is denser near the salt layer, no bulk flow will occur. The density gradient will maintain a stable layer of water.

The method of solution is similar to that for transient conduction in semi-infinite solid with surface temperature raised and maintained. Let *a* denote salt.

$$\{(\rho_{a(x,\,\tau)} - \rho_{as})/(\rho_{ai} - \rho_{as})\} = erf\{x/2(D \times \tau)^{0.5}\}$$

where $\rho_{a(x,\,\tau)}$ is the required value of salt concentration at x at time τ, ρ_{as} is the concentration at the interface, and ρ_{ai} is the initial concentration. In this case ρ_{ai} is zero and so

$$\rho_{a(x,\,\tau)} = \rho_{as} \times \{1 - erf[x/2(D \times \tau)^{0.5}]\}$$
$$= 380 \times \{1 - erf[0.01/2(1.2 \times 10^{-9} \times 24 \times 3600)^{0.5}]\}$$
$$= 380 \times \{1 - erf(0.491)\} = 380 \times \{1 - 0.52048\}$$
$$= 182.22 \text{ kg/m}^3$$

The diffusion rate at time τ at the interface is given by $\rho_{as} \times \{D/\pi\tau\}^{0.5}$

The total mass m diffused up to time can be obtained by integrating this with respect to time from zero to τ.

$$m = \rho_{as} \times (D/\pi)^{0.5} \int_0^{\tau} \tau^{-0.5}\, d\tau$$
$$= 2 \times \rho_{as}\{D \times \tau/\pi\}^{0.5}, \text{ substituting the values}$$
$$m = 2 \times 380 \times \{1.2 \times 10^{-9} \times 24 \times 3600/\pi\}^{0.5} = 4.366 \text{ kg/m}^2$$

The thickness of salt that will be dissolved can be determined by dividing this quantity by the density of salt.

Thickness dissolved, $L = 4.366/2165 = 0.0020166$ m $= 2.0166$ mm.

Example A-14.25: *Pure water layer of 10 mm thickness stands over a salt slab. Determine the salt concentration at the top after 12 hours. The concentration at the salt-water interface is constant at 380 kg/m³. Diffusion coefficient D = 1.2 × 10⁻⁹ m²/s.*

Solution: This process can be modeled similar to transient conduction in infinite slab with surface temperature raised suddenly and maintained constant at this value, *i.e.*, $h = \infty$.

The parameters are $D\tau/L^2$ for constant surface temperature as $h = \infty$. The layer at 0.01 m from the interface can be considered as the center plane of the slab as there is no mass transfer at this plane.

$$D\tau/L^2 = 1.2 \times 10^{-9} \times 12 \times 3600/0.01^2 = 0.5184$$

Reading from Heisler chart against this value (ref. example 6.10, Fig. 6.12(*b*), Chapter 6)

$$\{\rho_{a(0,\,\tau)} - \rho_{as})/(\rho_{ai} - \rho_{as})\} = 0.36$$

Substituting, $\rho_{ai} = 0.0$ and $\rho_{as} = 380$

$$\rho_{a(0,\,\tau)} = 243.2 \text{ kg/m}^3$$

The concentration will increase with time.

Example A-14.26: *Water flows down on the surface of a vertical plate at a rate of 0.05 kg/s over a width of 1 m. The water film is exposed to pure carbon dioxide. The pressure is 1.013 bar and the temperature is 25°C. Water is essentially CO_2 free initially. Determine the rate of absorption of CO_2. The molal concentration at this condition for CO_2 in water at the surface is 0.0336 kg mol/m³ of solution. D = 1.96 × 10⁻⁹ m²/s, solution density = 998 kg/m³, μ = 0.894 × 10⁻³ kg/ms, G = 0.05 kg/ms, L = 1 m. The notation for convective mass transfer coefficient is h_m.*

Solution: This is a situation where a flowing film of liquid absorbs the surrounding gas. In this case first the film thickness has to be determined. This is done assuming a velocity profile satisfying the boundary conditions. This results in the expression for δ as

$$\delta = \{(3 \times \mu \times G)/(\rho^2 \times g)\}^{1/3} \quad ...(A)$$

Reynolds number is given by $Re = (4 \times G)/\mu$ ...(B)

The average value of convective mass transfer coefficient is

$$h_m = \{(6 \times D_{ab} \times G)/(\pi \times \rho \times \delta \times L)\}^{0.5} \quad ...(C)$$

The average flow velocity in the film is obtained by

$$u = (G/\delta \times \rho) \quad ...(D)$$

The bulk concentration in the film varies along the flow and so an average bulk concentration is determined as the log-mean-difference. Let the bulk mean concentration at entry be C_{a0} and at L be C_{aL} and at the film surface be C_{ai} (at saturation).

Then the log mean concentration can be obtained as

$$\{(C_{ai} - C_a)\}_{\text{log mean}} = \{(C_{ai} - C_{a0}) - (C_{ai} - C_{aL})/ \ln\{C_{ai} - C_{ao})/(C_{ai} - C_{aL})\} \quad ...(E)$$

The total mole diffused up to length L can also be obtained by

$$N_a/A = h_m \times (C_{ai} - C_a)_{\text{log mean}} \quad ...(F)$$

The same can also be obtained by

$$N_a/A = u \times \delta \times (C_{aL} - C_{ao}) \quad ...(H)$$

Using equations (E), (F) and (H), C_{aL} can be solved by trial. Then equations (F) or (H) can be used to determine the mole or mass diffused. For this problem

$$\delta = \{(3 \times \mu \times G)/(\rho^2 \times g)\}^{1/3}$$
$$= \{(3 \times 0.894 \times 10^{-3} \times 0.05)/(998^2 \times 9.81)\}^{1/3}$$
$$= 0.2396 \times 10^{-3} \text{ m.}$$

$$Re = (4 \times 0.05)/0.894 \times 10^{-3} = 203$$

Equation (C) can be used to calculate h_m.

$$h_m = \{(6 \times D_{ab} \times G)/(\pi \times \rho \times \delta \times L)\}^{0.5}$$
$$= \{(6 \times 1.96 \times 10^{-9} \times 0.05)/(\pi \times 998 \times 0.2396 \times 10^{-3} \times 1)\}^{0.5}$$
$$= 27.98 \times 10^{-6} \text{ m/s}$$
$$u = 0.05/(998 \times 0.2396 \times 10^{-3}) = 0.209 \text{ m/s}$$

Using equation (E) and (F)

$$0.209 \times 0.2396 \times 10^{-3} \times (C_{aL} - 0) = (27.98 \times 10^{-6}) \times \{(0.0336 - 0) - (0.0336 - C_{aL})\}/ \ln\{(0.0336 - 0)/(0.0336 - C_{aL})\}$$

Solving by trial $C_{aL} = 0.01438$ kg mol/m^3

Substituting in

$$N_a/A = u \times \delta \times (C_{aL} - C_{a0})$$
$$N_a/A = 0.209 \times 0.2396 \times 10^{-3} \times (0.01438 - 0.0)$$
$$= 0.72 \times 10^{-6} \text{ kg mol/s per } m \text{ width.}$$

Example A-14.27: *Dry air at 30°C flows over a plate also at 30°C with the plate surface constantly covered with a water film. The velocity of flow is 6 m/s. The plate is 1 m long. Determine the average convective mass transfer coefficient and also the mass of water evaporated per second.*

Solution: The properties of air at 30°C are read as $\rho = 1.165$ kg/m^3, kinematic viscosity = 16.0 $\times 10^{-6}$ m^2/s, diffusion coefficient $D = 2.6 \times 10^{-6}$ m^2/s, Sc = v/D = 0.615. The density of water vapour at 30°C = 1/32.9 kg/m^3, the density of water vapour in dry air is zero.

$$\text{Re} = (6 \times 1)/16 \times 10^{-6} = 3.75 \times 10^5 < 5 \times 10^5. \text{ So laminar.}$$

The average value of Sherwood number is given by

$$\text{Sh} = 0.664 \times \text{Re}^{0.5} \times \text{Sc}^{0.333} = 345.86$$

$$\text{Sh} = h_m \times L/D, \text{ so } h_m = 8.99 \times 10^{-3} \text{ m/s.}$$

Mass of water evaporated per second,

$$m = A \times h_m\{\rho_{a1} - \rho_{a2}\} = 1 \times 8.99 \times 10^{-3}\left(\frac{1}{32.9} - 0\right)$$

$$= \mathbf{0.2733 \times 10^{-3} \text{ kg/s or about 1 kg/hr.}}$$

Example A-14.28: *Pure water at 20°C flows over a slab of salt at a velocity of 1 m/s. At the interface the concentration of salt is 380 kg/m^3. Determine over a length of 1 m the average convection coefficient for mass transfer and also the rate of diffusion of salt into the water. Assume turbulent flow from the leading edge. $D = 1.2 \times 10^{-9}$ m^2/s, density of water = 1000 kg/m^3, kinematic viscosity = 1.006×10^{-6} m^2/s, Sc = $1.0006 \times 10^{-6}/1.2 \times 10^{-9}$ = 838.33.*

Solution: Re = $1 \times 1/1.006 \times 10^{-6} = 0.994 \times 10^6$. Flow is turbulent.

The average value of Sherwood number is given by

$$\text{Sh} = 0.0228 \times \text{Re}^{0.8} \times \text{Sc}^{0.333} = 17052 = h_m \times L/D$$

Solving $\quad h_m = 20.46 \times 10^{-6}$ m/s

$$m = h_m \times A \times (\rho_{a1} - \rho_{a2}), \text{ as } \rho_{a2} = 0.0$$

$$= 20.46 \times 10^{-6} \times 1 \times 380 = 7.776 \times 10^{-3} \text{ kg/s m}^2 \text{ or } \mathbf{28 \text{ kg/hr m}^2}.$$

Example A-14.29: *A square plate of side 1 m has one of its sides coated with napthalene and stands vertically in still air at 53°C. Determine diffusion rate. M = 128, $D = 6.11 \times 10^{-6}$ m^2/s, kinematic visocity = 18.8×10^{-6}, Sc = 3.077. The vapour pressure at 53°C is 1.333×10^{-3} bar. R_v = 8315/128 = 64.91 J/kg K, T = 53 + 273 = 326 K.*

Solution: Napthalene diffusing into the air makes the air denser near the surface and creates a downward flow. This problem hence can be treated as a natural convection problem. For mass transfer the equations applicable will be similar to those in natural convection heat transfer. Grashof number for mass transfer is given by

$$\text{Gr}_m = \{g \times (\rho_w - \rho_\infty) \times L^3\}/\{(\rho_\infty \times v^2)\}$$

ρ_w = density of the mixture at the wall surface.

ρ_∞ = density of the mixture well removed from the surface.

Density of the mixture near the surface = density of air + density of vapour

$$= \{P_a/(R_a \times T)\} + \{P_v/(R_v \times T)\}$$

$$P_a = 1.013 \times 10^5 \text{ N/m}^2, P_v = 1.333 \times 10^2 \text{ N/m}^2,$$

$$\rho_v = \{(1.013 - 0.001333) \times 10^5/(326 \times 287)\} + \{0.001333 \times 10^5/(326 \times 64.91)\}$$

$$= 1.0812 + \mathbf{6.299 \times 10^{-3}} = 1.0875744 \text{ kg/m}^3$$

$$\rho_\infty = \{1.013 + 105/(326 \times 287)\} = 1.0827045 \text{ kg/m}^3$$

$$\text{Gr}_m = \{9.81 \times (1.0875744 - 1.827045) \times 1^3\}/\{1.0827045 \times (18.8 \times 10^{-6})^2\}$$

$$= 0.12484 \times 10^9$$

The equation applicable for this range is

$$\text{Sh} = 0.555 \times (\text{Gr}_m \times \text{Sc})^{0.25}$$
$$= 0.555 \times (0.12484 \times 10^9 \times 3.077)^{0.25} = 77.7$$
$$h_m = 77.7 \times 6.11 \times 10^{-6}/1.0 = 0.4747 \times 10^{-3} \text{ m/s}$$
$$m = 0.4747 \times 10^{-3} \times 6.29 \times 10^{-3} = 2.9861 \times 10^{-6} \text{ kg/s m}^2.$$

Example A-14.30: *Derive an expression for the surface temperature of water in evaporative cooling.*

Solution: In evaporative cooling unsaturated air flowing over a wetted surface causes evaporation of water so that the air will become saturated. Part of the sensible heat of air is utilised for the enthalpy of evaporation and the air will get cooled in the process. Energy balance for the process yields.

Heat transferred to the film of water by convection = ethalpy of evaporation of water

$$h \times (T_\infty - T_s) = h_{fg} \times h_m \times (\rho_s - \rho_\infty)$$

where ρ_s and ρ_∞ are the mass concentrations of water vapour at the film surface and in the free steam.

$$T_\infty - T_s = \{(h_{fg} \times h_m)/h\} \times (\rho_s - \rho_\infty), \text{ as } \rho = P/(R \times T),$$
$$T_\infty - T_s = \{(h_{fg} \times h_m)/(h \times R)\} \times \{(P_s/T_s) - (P_\infty/T_\infty)\}$$
$$h_m/h = \rho \times c \times \text{Le}^{0.6667} \text{ and Le} = \alpha/D = \text{Sc/Pr},$$
$$\rho_a \times T = P_a/R_a$$
$$T_\infty - T_s = \{(h_{fg} \times R_a)/(R_w \times P_a \times c \times \text{Le}^{0.6667}) \times \{P_s - P_\infty\}$$

P_s and P_∞ are obtained from the temperature values. A trial solution is necessary when moist air is involved. If dry air is used P_∞ will be zero and direct solution for T_s is obtained.

Example A-14.31: *The outlet of a desert cooler is 28°C and 80% RH. In the inlet air the partial pressure of water vapour is 0.18 bar. Estimate the outside air temperature and the relative humidity.*

Solution: $D = 27.12 \times 10^{-6}$ m²/s, $\alpha = 21.42 \times 10^{-6}$ m²/s.

Saturation pressure of water at 28°C, P_5 = 0.03782 bar.

The equation applicable is

$$T_\infty - T_s = \{(h_{fg} \times R_a)/(R_w \times \rho \times c \times \text{Le}^{0.6667})\} \times \{P_s - P_\infty\}$$
$$R_a/R_w = M_w/M_a = 18/28.9, \; h_{fg} = 2435 \times 10^3 \text{ J/kg}$$
$$Le = \alpha/D = 21.42 \times 10^{-6}/27.12 \times 10^{-6} = 0.79, \; \rho = 1.205 \text{ kg/m}^3,$$
$$T_\infty = 28 + \{18 \times 2435 \times 10^3\} \times \{0.03782 - 0.018\}/$$
$$\{28.9 \times 1005 \times 1.205 \times (0.79^{0.6667})\} = 46°\text{C}$$

The saturation pressure at 46°C is 0.1009 bar. At the inlet

Relative Humidity = 0.018/0.1009 = 0.1784 or 17.84%.

Example A-14.32: *The convective heat transfer coefficient on a body exposed to air flow has a value of 240 W/m² K. If a film of water is maintained on the surface under similar conditions of flow determine the value of convective mass transfer coefficient.*

Solution: $D = 112 \times 10^{-6}$ m²/s, $\alpha = 21.4 \times 10^{-6}$ m²/s, $\rho = 1.205$ kg/m³.

$$(h/h_m) = \rho \times c \times \text{Le}^{0.6667}, \text{ Le} = \alpha/D = 0.191$$
$$h_m = 240/(1.205 \times 1005 \times 0.191^{0.6667}) = 0.5974 \text{ m/s}.$$

Example A-14.33: *A one half scale model tested under heat transfer conditions is used to predict the mass transfer performance of an object the plastic coating on which is to be dried. In the heat transfer test the model was maintained at 100°C with air flow at 0°C.*

The velocity of flow was 120 m/s. The temperature at a point near the model was measured as 80°C.

In the drying process both the object and the air will be at 50°C and flow velocity is to be 60 m/s. In both model testing and drying the air pressure will be 1 bar. The molecular weight of the plastic is 82 and the vapour pressure of the plastic at 50°C is 0.0323 bar. The diffusivity of the vapour in air is 26 × 10^{-6} m^2/s. The freestream does not contain any vapour. At 50°C $\alpha = 25.72 \times 10^{-6}\ m^2/s$. $\rho = 1.093\ kg/m^3$.

(i) Estimate the partial pressure of the vapor and the concentration that can be expected at the corresponding point on the object where the temperature was measured on the model.

(ii) If the convection heat transfer coefficient in the model test was 20 W/m^2 K estimate mass flux in the drying process.

Solution: The mean temperature in both case is 50°C. The product of length and velocity is also the same in both cases. Hence the value of Reynolds number will be same. So the temperature and concentration profiles will be similar. The concentration and partial pressure will be proportional to each other. As the temperature is the same.

$$\{(P_{ay} - P_{as})/(P_{a\infty} - P_{as})\} = \{(T - T_s)/(T_\infty - T_s)\}$$

$$\{(P_{ay} - 0.0323)/(0.0 - 0.0323)\} = \{(80 - 100)/(0.0 - 100)\}$$

Solving, $\mathbf{P_{ay} = 0.02584\ bar.}$

Mass concentration

$$P/(R \times T) = (82 \times 0.02584 \times 10^5)/(8315 \times 323) = 0.07889\ \text{kg.m}^3$$

Mole concentration = 0.07889/82 = 0.9621 × 10^{-3} kg mol/m^3.

$$h/h_m = \rho \times c \times Le^{0.6667} = 1090.63,\ h_m = (20/1090.63) = 0.018338\ \text{m/s}$$

$$m = h_m \times A \times \rho_{as} = 0.018338 \times 0.0323 \times 10^5 \times 82/8315 = \mathbf{0.5841\ kg/m^2.}$$

Example A-14.34: *The partial pressure of diffusing vapour over a surface under steady state of mass transfer was measured and plotted against height above the surface. At the surface the partial pressure was 0.1 bar and in the free stream the partial pressure was 0.02 bar. The tangent to the concentration profile at the surface meets the x-axis at 2.2 mm. Determine the convective mass transfer coefficient D = 28.8 ×* 10^{-6} m^2/s.

Solution: The boundary layer concept leads to the condition that

mass convected = mass diffusion at the interface

$$\text{mass convected} = h_{mx} \times (\rho_{as} - \rho_{a\infty})$$

$$\text{mass diffusing the surface} = -D_{ab}(d\rho_a/dy)_{y=0}$$

$$h_{mx} \times (\rho_{as} - \rho_{a\infty}) = -D_{ab}(d\rho_a/dy)_{y=0}$$

As $\rho = P/(R \times T),\ d\rho = (1/R \times T)dP,$

$$h_{mx} \times (P_{as} - P_{a\infty}) = -D_{ab}(dP_a/dy)_{y=0}$$

$$h_{mx} = -D_{ab}(dP_a/dy)_{y=0}/(P_{as} - P_{a\infty})$$

$$(dP_a/dy)_{y=0} = (0.0 - 0.1)/(0.0022 - 0.0) = -45.45\ \text{bar/m}$$

$$h_m = (-28.8 \times 10^{-3} \times -45.45)/(0.1 - 0.02) = 0.016362\ \text{m/s.}$$

REVIEW QUESTIONS

14.1 Explain the physical mechanism of a diffusion mass transfer.

14.2 Define 'Molar concentration' and 'Mass concentration'. Obtain the relationship between them.

14.3 State 'Fick's law' of diffusion.

14.4 What is stationary or non diffusing medium?

14.5 In heat transfer at the interface of two medium the temperature is the same. Can the concentration of diffusing species will be the same at the interface of two species?

14.6 Define 'Henry's constant'. Explain how it is applied in the determination of concentration of a species at the interface in the two species.

14.7 Explain the equimolar counter diffusion process. Indicate under what condition such processes occur.

14.8 Give an example of a diffusion of a species through a non diffusing species.

14.9 What type of diffusion is similar to heat conduction through a barrier?

14.10 Give an example of transient mass transfer similar to heat transfer in semi infinite body.

14.11 Give an example of transient mass transfer similar to heat transfer in infinite slab.

14.12 Define Lewis number. Explain the application in mass transfer.

14.13 Give an example for mass transfer by forced convection over a surface.

14.14 A slab coated with a liquid is held vertically. What similarity can be used to determine the mass convected?

14.15 What is the unit of mass transfer coefficient?

EXERCISE PROBLEMS

14.1 Two ducts are carrying a mixture of Nitrogen and Ammonia one having ammonia 80% and N_2 20% and the other 80% N_2 and 20% ammonia at the same pressure and temperature. These are connected by a pipe of 10 cm dia and 2 m length with no pressure difference between the ends. The total pressure is 1 bar. Determine the diffusion rates.

14.2 Determine the diffusing rate of benzene standing in a glass jar of 5 cm dia. The air column over it has a height of 60 cm. The air column is stationary. The pressure and temperature are 1 bar and 0°C. The partial pressure of benzene at the interface is 0.01 bar.

14.3 Determine the diffusion rate of water from the floor in still air in a room at 30°C. The air is still upto a height of 2 m and the RH at this height is 30%.

14.4 A thin layer of water is there in a circular pan of 20 cm dia and 10 cm deep. The top is exposed to air at 30°C and 50% RH. Determine the diffusion rate.

14.5 A wet bulb thermometer when exposed to the flow of dry air at 1 bar reads 30°C. Determine the dry bulb temperature.

14.6 Determine the diffusion rate in a lake when wind is blowing at 20 kmph over the surface. The air temperature is 30°C and the relative humidity is 40%. The total pressure is 1 bar. The lake is 1 km wide along the flow.

14.7 Dry air at 20°C and atmospheric pressure flows inside a 100 mm duct, the walls of which are wetted constantly to maintain a film of water. The flow velocity is 5 m/s. Calculate the water concentration at the exit of the pipe if it is 4 m long.

14.8 Estimate the values of diffusivities for the following pairs of gases (*i*) N_2 and CO_2 (*ii*) CO_2 and H_2 (*iii*) H_2O and O_2 (*iv*) Cl_2 and CO_2. The molecular volumes are (*i*) hydrogen-15.6, (*iii*) CO_2-34 (*iv*) H_2-15.6 (*iv*) water-18.8 (*v*) Cl_2-24.6 (Refer problem 14.11).

14.9 Estimate the values of diffusivities for the liquid pairs given below: (*i*) ethanol if dilute solution in water, $M_b = 18$, $\mu = 1.5 = 1.4 \times 10^{-3}$ kg/ms, $v = 0.0592$.

(*ii*) methanol in dilute solution of water, $M_b = 18$, $\mu = 1.9 = 1.4 \times 10^{-3}$ kg/ms, $v = 0.037$. (Refer problem 14.13).

14.10 Hydrogen diffuses through a stagnant gas mixture of nitrogen and oxygen in the ratio 77 : 23 by volume. The total pressure is 1 bar and the temperature is 10°C. The gas film is 0.5 mm thick. The concentration of hydrogen at the surfaces of the gas film are 10% and 5% by volume. Determine the diffusion rate of hydrogen. $D_{h2\text{–}02} = 77.5 \times 10^{-6}$ m²/s, $D_{h2\text{–}n2} = 73.89 \times 10^{-6}$ m²/s (Refer problem 14.17)

14.11 Hydrogen diffuses through a stagnant gas mixture of nitrogen and methane in the ratio 50 : 50 by volume. The total pressure is 1 bar and the temperature is 10°C. The gas film is 1.5 mm thick. The concentration of hydrogen at the surfaces of the gas film are 11% and 3% by volume. Determine the diffusion rate of hydrogen. $D_{h2\text{–}ch4} = 62.5 \times 10^{-6}$ m²/s, $D_{h2\text{–}n2} = 73.89 \times 10^{-6}$ m²/s.

14.12 A cylinder made of napthalene is 600 mm long and its diameter is 75 mm. Pure CO_2 at 100°C flows across it at a velocity of 6 m/s. The total pressure is 1 bar. At this condition the partial pressure of vapour at the interface is 1330 Pa. $D = 5.15 \times 10^{-6}$ m²/s. Determine the diffusion rate.

14.13 A mixture of oxygen and CO_2 in a container is at 27°C. The partial pressure of oxygen is 0.8 bar and that of CO_2 is 1.2 bar. Calculate the molar and mass concentration of each. Also calculate the density of the mixture.

14.14 A spherical steel tank of 120 mm dia has a wall thickness of 2.2 mm and hydrogen gas is stored in the tank at 4.6 bar. Determine the initial rate of mass leakage from the tank caused by the diffusion of hydrogen. Also calculate the rate of pressure drop at this condition. Diffusion coefficient is $D = 0.26 \times 10^{-12}$ m²/s. At the inside surface, the mass concentration of hydrogen is 3 kg/m³ and it is zero at the outside surface.

14.15 The process of diffusion through a membrane is used to separate helium from a gas mixture. The membrane is 0.8 mm thick. The mass concentration of helium at the inner and outer surface of the membrane are 0.08 kg/m³ and 0.02 kg/m³. Diffusion coefficient = 10^{-9} m²/s. Calculate the diffusion rate of helium/m².

14.16 A rubber membrane of 0.6 mm thickness separates two chambers containing oxygen, one at 2.2 bar and the other as 1.1 bar. The system is at 25°C. Determine the diffusion rate of oxygen through the membrane $D = 0.21 \times 10^{9}$ m²/s.

14.17 The humidity level inside a room is such that the water vapour pressure is 0.03 bar. On the outside the air is dry. Plaster board of 10 mm thickness separates the inside from outside. The diffusion coefficient for water vapour into the wall material is about 10^{-9} m²/s. The solubility of water vapour in the wall material is 0.142 kg/m³ for bar of water vapour pressure. Determine the diffusion rate.

14.18 Helium gas at a pressure of 6.2 bar is stored in a fused silica container of spherical shape of 0.21 m diameter, the wall thickness being 2.6 mm. Calculate the mass of helium lost by diffusion. $D = 0.04 \times 10^{-12}$ m²/s. The solubility of helium in the material = 1.8×10^{-3} kg/m³ bar.

14.19 In problem 14.18 if the shape of the container is a long cylinder of the same diameter, determine the loss for 1 m length.

14.20 Two large tanks containing oxygen and nitrogen at 1 bar and 25°C are connected by a pipe of 0.05 m *ID* and 1.2 m long. The partial pressure of oxygen at one end of the tube is 0.2 bar and it is 0.1 bar at the other end. Determine the diffusion rate of O_2 through the tube assuming equimolal counter diffusion. $D = 0.18 \times 10^{-4}$ m²/s.

14.21 The surface of a thick steel plate is packed with carbon and is heated to 1000°C and maintained at this temperature to increase the concentration. The carbon concentration at the surface of the plate is maintained at 30 kg/m³. Originally the carbon concentration in the plate is uniform at 8 kg/m³. Calculate the time required for the concentration to go upto 16 kg/m³ at a depth of 1 mm. $D = 0.26 \times 10^{-12}$ m²/s.

14.22 Atmospheric air at 26°C and 50% RH flows at a velocity of 20 m/s over a porous plate saturated with water also at 26°C. The plate measures 1 m × 1 m. Estimate the amount of water diffusing into the air stream. Use data book for property values.

14.23 The vapour pressure of napthalene at 100°C is 0.02666 bar. Air at 100°C flows over a plate coated with napthalene at a velocity of 6 m/s. The mass diffusivity has a value of 8.258×10^{-6} m²/s. Gas constant for the vapour = 64.91 J/kg K. The free stream air has no trace of napthalene. Determine the diffusion flux.

14.24 Consider the water surface of a swimming pool of square shape of 10 m side as a flat plate. Assuming still air conditions over the water and both water and air to be at 25°C, determine the rate of diffusion of water. RH = 40%.

14.25 If in problem 14.24 air flows parallel to the surface at an average velocity of 4 m/s, determine the rate of diffusion of water.

14.26 The heat transfer correlation for a given convection situation was obtained as

$$\text{Nu} = 0.43\ \text{Re}^{0.8}\ \text{Pr}^{0.4}$$

Dry air at 30°C is passed over a similar object of 1 m characteristic length at 10 m/s. Determine the evaporation rate if at the liquid surface the mass concentration is 0.0077 kg/m³.

14.27 A horizontal cylinder of 90 mm dia and 1 m length made of napthalene is kept in still air at 53°C. Estimate the weight loss in 60 minutes. The partial pressure of napthalene at this temperature is 1.333×10^2 Pa. M = 128, $D = 6.11 \times 10^{-6}$ m²/s, Sc = 3.077.

14.28 A pond of 12 m × 24 m sides is full of water. Air at 25°C and 50% RH flows along the 12 m side at a velocity of 2 m/s. Determine the evaporation loss for hour, $D = 26 \times 10^{-6}$ m²/s, Sc = 0.62.

14.29 Heat transfer studies in flow of air over a smooth body gave the value of convective heat transfer coefficient as 17 W/m² K. Determine the value of convective mass transfer coefficient under similar flow conditions $D = 25.8 \times 10^{-6}$ m²/s, Pr = 0.688, c_p = 1007 J/kgK

$\rho = 0.946_{ky}/\mu^3$, $v = 23.13 \times 10^{-6}$ m².s, $\alpha = 33.64 \times 10^{-6}$ m²/s.

14.30 At a temperature of 25°C, the mass transfer coefficient of a material for air flow was determined as 0.0780 m/s. The diffusion coefficient of the material with air is 0.61×10^{-5} m²/s. Determine the heat transfer coefficient under similar conditions of flow. The kinematic viscosity of air is 2.141×10^{-5} m²/s. Specific heat is 1007 J/kgK. [**Ans.** 215 W/m² K]

14.31 Dry air is blown over the top of a 3 cm ID tube in which water level is maintained at 40 cm from the top. The pressure and temperature at the location are 83.5 kPa and 20°C. After 15 days of experiment, the mass of water evaporated was measured as 1.3 g. Determine the diffusion coefficient under this condition. [**Ans.** 3.06×10^{-5} m²/s]

14.32 Helium is flowing in a pipe at 1 atm and 25°C. In order to avoid pressure build up under transient conditions, the pipe is vented to atmosphere through a pipe of 5 mm ID and 15 m length. The diffusion coefficient of Helium to air is 7.2×10^{-5} m²/s. The molecular mass of air is 29 kg/kg mole. Determine the rate of helium lost to the atmosphere through the tube.

[**Ans.** 1.54×10^{-11} kg/s]

14.33 Pressurised hydrogen is kept in a sphere of 10 cm ID and 12 cm OD. The molar concentration of hydrogen at the inner surface is 0.06 kg mol/m³. The diffusion coefficient is 4.5×10^{-15} m²/s at this temperature. The concentration at the outer surface is negligible. Determine the rate of diffusion. [**Ans.** 1.02×10^{-15} moles/s or 2.04×10^{-15} kg/s]

14.34 On the sides of a plastic membrane 0.3 mm thick hydrogen gas is maintained at 3 bar and 1 bar at a temperature of 25°C. The diffusion coefficient is 8.7×10^{-8} m²/s. The solubility of hydrogen which varies with pressure is 1.5×10^{-3} k mol/m³ bar. Determine the mass diffusion rate.

[**Ans.** 1.74×10^{-6} kg/m²s]

Cooling of Electronic Equipments

15

CHAPTER

15.0 INTRODUCTION

From sophisticated satellites, rockets and aircrafts to simple appliances for everyday use like TV, home computer and cellphone depend on electronic devices for proper functioning. The reliability of operation of such equipments depends very much on the reliability of electronic equipment which control them. All electronic devices generate heat due to the current flow through them. Unless the heat is removed constantly the temperature of the devices will continuously increase leading to failure. Compared to the previous components like vacuum tubes, the heat generated by individual device is very small. But the developments from transistors, integrated circuits and microprocessor have resulted in millions devices being packed into a chip of millimeter size.

Starting in 1960 with 50-1000 components per chip, it increased step by step to more than 10^9 components in a chip of 3 cm × 3 cm size. The trend is shown in Fig. 15.1. The result is the increase of heat output from 0.1 W/cm^2 to 100 W/cm^2, which is comparable in magnitude with those encountered in nuclear reactors and the surface of the sun. Unless the cooling system is properly designed and operated high rate of heat generation will result in high operating temperatures of the equipments resulting in failure of components.

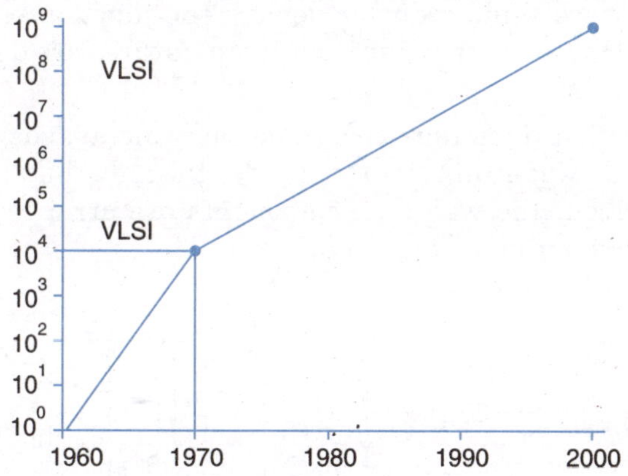

Fig. 15.1. The trend of increase in number of components packed on a chip over the years

The failure rate of electronic equipments increases with temperature almost exponentially. Thermal cycling can cause break at connections. Therefore thermal control has become increasingly important in the design and operation of electronic equipments. The rate of failure of components with temperature is shown in Fig. 15.2.

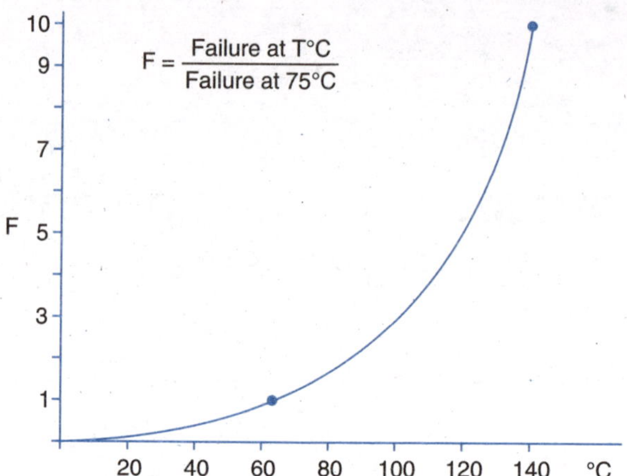

Fig. 15.2. Increase in failure rate of digital devices with temperature

As there are no moving parts in the electronic components there is no wear and tear. Therefore these are inherently reliable and it appears that they can operate safety for many years. This will be the case if they are operated at room temperature.

But electronic components are observed to fail under prolonged use at high temperature, the possible cause being diffusion, chemical reaction and creep of bonding material. A rule of thumb is that failure of electronic components is halved for each 10°C reduction in their operating temperature.

15.1 ELECTRONIC EQUIPMENTS AND COMPONENTS

Heat is generated at the junction where the different regions of semiconductor (such as *p* and *n* type regions) come in contact. The diode is based on a single *p*-*n* junction, while transistor involves two junctions. The junctions being the sites of heat generation are the hottest spots in a component. In silicon based semiconductor devices the junction temperature is limited to 125°C for safe operation. However lower junction temperatures are desirable for extended life and lower maintenance costs.

In a typical application numerous electronic components, some smaller than 1 μm in size, are formed from silicon wafer into a chip. In order to study the various methods of cooling it is necessary to know about the way the components are arranged in an equipment. The process is shown in a line diagram in Fig. 15.3.

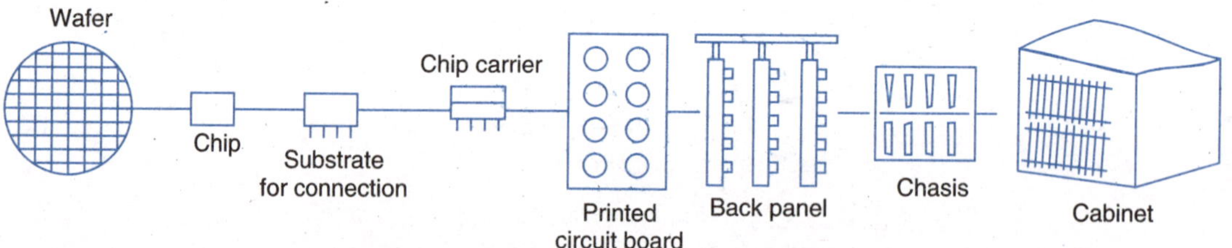

Fig. 15.3. Process of assembly of electronic equipment

The basic or starting point for cooling thus is the chip carrier, the details of which are shown in Fig. 15.4. Chips are housed in a chip carrier or substrate made of ceramic, plastic or glass in order to protect the delicate circuitry from the detrimental effect of the environment.

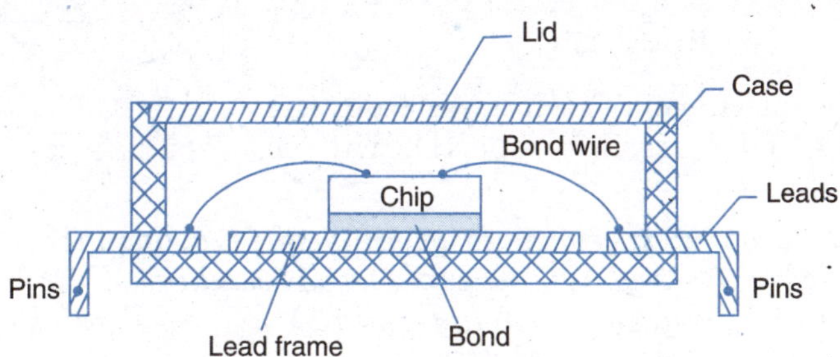

Fig. 15.4. Components of chip carrier

A rugged housing is provided for the safe handling of the chip as well as connectors between the chip and the circuit board. The chip is secured in the carrier by bonding it to the bottom surface. The chip should not be bonded to the plastic casing to avoid distortion. A lead frame made of copper alloy having the same coefficient of expansion as the bond material is used to avoid distortion due to different expansion coefficients.

The design of the chip carrier is the first level in thermal control of electronic equipment, since the transfer of heat from the chip to the chip carrier is the first step in the dissipation of heat generated in the chip. The first resistance to heat flow is called "junction to base resistance". This is due to the conduction/convection/radiation from the chip to the casing. This resistance may be high due to the poor conductivity of the gas filling the enclosure and also the poor conductivity of the plastic material of which the case is generally made. The junction-to-case thermal resistance depends on the geometry, the size of the chip and chip carrier as well as the material properties of the bonding and the case. It varies considerably from about 10°C/W to 100°C/W.

Moisture in the cavity of the chip carrier may cause corrosion of the wiring. Chip carriers are made of materials that will prevent leakage of moisture into the casing by diffusion. The carriers are hermetically scaled in order to prevent entry of moisture through cracks. Materials which on heating may evolve gases are not used for the same cause.

In critical requirements costly ceramic casing is used instead of plastic to have perfect hermeticity. The transistor is formed on a small silicon chip housed in the disc shaped cavity and I/O pins come out from the bottom. The case of the transistor carrier is usually attached directly to a flange to obtain a large surface area for heat dissipation, and to reduce the junction to case thermal resistance.

Hybrid multichip packaging is used where more than one chip is housed with shorter wiring connections. Housing several chips in a single carrier also leads to lower cost and improved reliability.

Example 15.1: *The junction to case resistance of a power transistor is specified by the manufacturer as 10°C/W. The temperature of a power transistor case was measured as 60°C when dissipating 4 W. Determine the junction temperature.*

Solution: The case temperature, T_c, heat dissipation Q and thermal resistance R and known under steady conditions,

$$Q = \frac{\Delta T}{R} = \frac{T_j - T_c}{R}$$

$$4 = (T_j - 60)/10$$

$$\therefore \quad \mathbf{T_j = 100°C.}$$

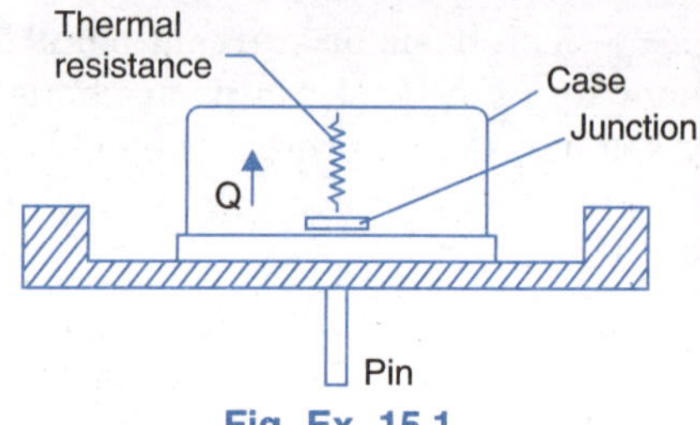

Fig. Ex. 15.1

Determination of thermal resistance from the detailed dimensions and material properties will be discussed in a later section.

Example 15.2: *In a lobaratory experiment to determine the junction to case thermal resistance, the case temperature was measured as 60°C under steady conditions. The junction temperature was measured as 92°C. The voltage across unit was 15 V and the current flow was 0.15 amps. Calculate the junction to case resistance.*

Solution: Refer Fig. Ex. 15.1 for details. Given

The junction temperature = T_j = 92°C,

The case temperature = 60°C.

Power = $VI = 15 \times 0.15 = 2.25$ W

$$Q = \frac{\Delta T}{R}, \quad 2.25 = \frac{92 - 60}{R}$$

$$\therefore \quad \mathbf{R = 14.22°C/W}$$

15.1.1. Printed Circuit Boards (PCBs)

The chip carriers as well several components like ICs, diodes, transistors are mounted on boards where the circuit connecting the components are already etched on copper film. PCB is a properly wired plane board made of polymers and glass epoxy materials on which the components are mounted. Commonly called cards each PCB carries out a specified function like controlling hard disc drive and the cards can be replaced easily during repair. The various connecting wires are taken to a plug and pin type of joint and so the cards are called plug in cards. Commonly the size is 10 cm width and 15 cm length and are few millimeters thick. Copper cladding is added to one or both sides in which the connections are etched. The components can be mounted on one side or both. The power dissipated by a PCB usually ranges from 5 W to 15 W. Electronic systems typically involve several layers of PCBs. The PCBs are usually cooled by direct contact with a fluid such as air flowing between the boards. When the boards are placed in hermetically sealed enclosures they must be cooled by a cold plate (heat exchanger) in contact with the edge of the boards. A diagrammatic view of a PCB is shown in Fig. 15.5.

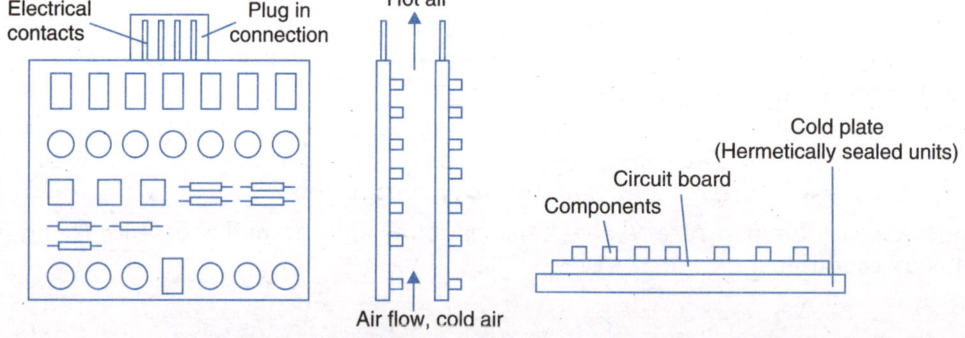

Fig. 15.5. View of PCB with electronic components mounted

The device-to-board thermal resistance of a PCB is usually high, about 20 to 60°C/W, because of the small thickness of the board, and the low thermal conductivity of the board material. A copper cladding on the cold plateside will decrease the resistance for flow of heat.

The components that should be operated at low temperatures and those with higher heat output should be placed near the cooling fluid entry, so that cooler fluid flows around them and chances for failure is reduced.

Generally three types of PCBs are available. These are single sided, double sided and multilayer boards. Single sided PCBs have circuitry lines on one side only. These are suitable for low density electronic devices of 10 to 20 components. Double sided PCBs have circuits on both sides and are suitable for intermediate density devices. Multilayer PCBs contain several layers of circuitry and are suitable for high density devices. Special cooling arrangement may be necessary for this type of boards, as the component temperature will be high in this type. Hermetically sealed boards which are cooled by cold plate are called "conduction cooled". The temperature of the component in this case will depend on the location in the board. The temperature will be higher at the mid section and lower along the sides. High power components should be placed along the sides to reduce failure rate. This is shown in Fig. 15.6.

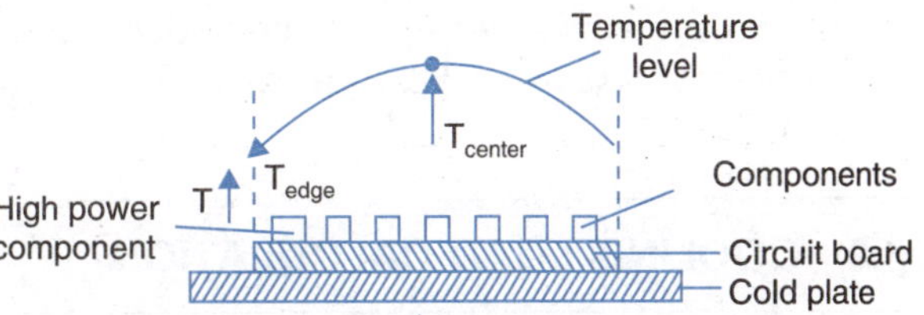

Fig. 15.6. Temperature distribution along cold plate cooled boards

Materials used in the fabrication of PCBs should have the following properties:

1. To prevent electrical breakdown should have effective insulation property.
2. Good heat conductors to conduct away the heat generated.
3. High material strength to withstand forces and maintain dimensional stability.
4. The thermal expansion coefficient should be close to that of copper cladding to prevent cracking during thermal cycling.
5. Should be resistant to moisture absorption.
6. Stability in properties at temperatures encountered.
7. Easy availability and manufacturing should not create problems.
8. Low cost.

No existing material has all the above properties.

The choice will depend on the basic requirement for functioning of the unit.

Glass-Epoxy laminates made of an epoxy or polymide matrix reinforced by several layers of woven glass cloth are commonly used in the production of PCBs. Polymide matrices are more expensive than epoxy but can withstand higher temperatures. Polymer or polymide films are also used without reinforcement for flexible circuits.

The assembled circuit boards are housed in rugged enclosures. In addition the enclosures houses peripheral equipment and connectors. The enclosure should protect the assembly from detrimental effects of the environment and provide cooling mechanism. In small electronic systems like personal computers the enclosure can be simple and inexpensive box made of sheet metal with proper connectors and a small fan. For large system which may have several hundred circuit boards the design and construction of the enclosure will be complex. Particularly thermal design will be challenging.

The enclosure should provide easy access into the system for servicing so that identifying and rectifying faults will be easy and not time consuming.

Long down time is generally unacceptable especially for large systems. So the arrangement of the boards should be such that access is easy. Plug in type boards make it easy to replace defective boards and are commonly used in low power electronic equipments. High power electronic circuits require that the boards be tightly attached to the racks of the cabinet with specially designed brackets. Enclosure should also have provision for switches, indicator lights, message display screen and provision for user interface. The circuit boards are assembled in back panels through edge connectors. The back panel has provision for inter connection and power supply.

The back panels with PCBs assembled in an orderly manner are housed in a cabinet. A wide variety of electronic enclosures are available with varying sizes and made of different materials.

Electronic enclosures have to be sealed if necessary to prevent leakages to inside/outside and to prevent moisture ingression. As it is costly to seal the enclosure, it is done only when it is essential.

15.2 COOLING LOAD CALCULATION

In the assembled condition the cooling requirement of a unit is easily determined by operating it at the desired level and measuring the voltage applied and the current flow.

Cooling load = VI or $I^2 R$ W

At the design stage the load is to be calculated as the sum of heat emission by the components of the designed unit. This is a tedious process. The heat emission from various components should be available from the manufacturers' data or from standard references of technical societies. For margins of safety a percentage of the calculated load should be added before cooling system design is taken up. The duty cycle and frequency of operation of the unit if taken into account may lead to a reduction in the load.

15.3 METHODS OF COOLING

Depending upon the load the cooling method is generally chosen. The methods are:

1. Natural convection cooling.
2. Forced convection cooling with air.
3. Immersion cooling with natural convection.
4. Immersion cooling with boiling.
5. Forced circulation of water.
6. Heat Pipe.

The method adopted, to a great extent, depends on the heat generation by the equipment and the maximum heat transfer capacity of the method. Natural convection can be used where the heat dissipation is low.

For very high loads like in super computers, immersion cooling becomes necessary. Even in forced convection when air is used as the fluid, the heat dissipation capacity is lower than when water is used for cooling. Also direct or indirect cooling depends on other than heat dissipation capacity. The advantages, limitations and the equipment layout will be discussed under each of these topics. The environment and the rigourousness of the duty also dictates the method to be adopted.

Some of the special situations are indicated in the following discussions.

Constant load operating conditions lead to better reliability with well designed systems. Frequent changes in operating conditions can lead to thermal cycling leading to creep and failure of electrical connections. Special care has to be taken in providing the cooling. The cooling load has to be reduced taking into account of the heat storage in the components. In an investigation, when temperature is cycled with 20°C amplitude, the failure rate was found to increase by eight-times the normal rate. Low density electronic systems as in TV or VCR require only natural ventilation which is provided by properly placed slits in the enclosure. Systems for Aircraft have to be completely enclosed. Separate cooling system not affected by environment should be provided. Vibration isolation is a must as aircrafts meet with varying wind and weather conditions. Missiles operating only for short periods need no such arrangement for the electronic system. Long range missiles however need such care in their electronics. Electronic systems for space shuttle is a challenge for designers. Absence of gravity and heat sink only by radiation are the problems involved.

Electronic equipment in ships and submarines is usually housed in rugged cabinets to protect it from vibration and shock.

Communication systems located in remote locations are exposed to extreme environmental conditions. These have to work for long periods, under these conditions with infrequent maintenance. Large communication systems have to be housed in specially built shelters with suitable air conditioning.

In electronic components used in high power microwave equipment, the heat fluxes may be as high as 2000 W/cm^2. Such components can be cooled only by immersion method.

The possible surface heat flux that can be met at different temperature difference by various methods is shown in Fig. 15.7.

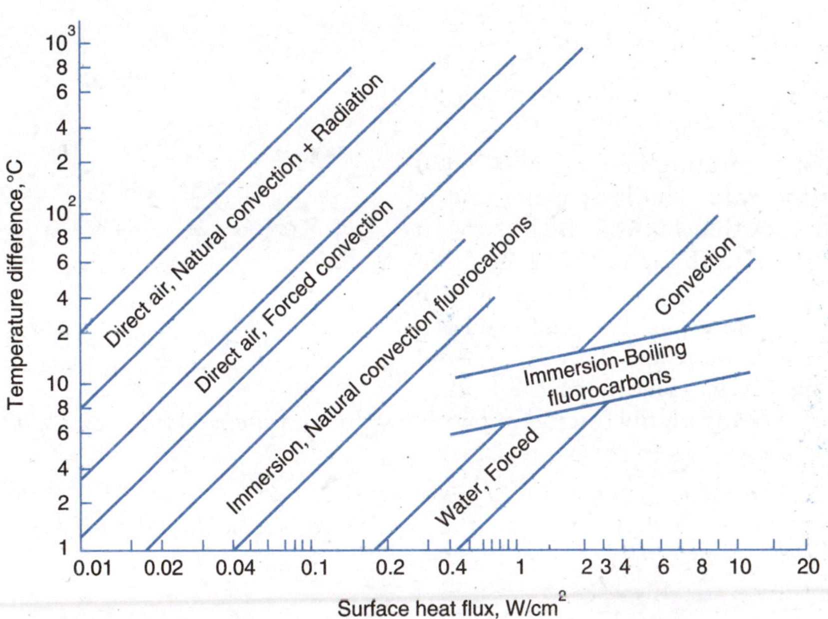

Fig. 15.7. Choice of suitable cooling method for various heat flux and ΔT

The choice of the cooling method can be decided by using the chart if we know the heat flux. The component manufacturers provide the data about the heat dissipation rate and maximum allowable temperature for each of the component. The heat dissipation rate divided by the component area gives the heat flux. For a heat flux of 0.1 W/cm^2 and allowable temperature difference of 60°C natural convection with radiation can be chosen. If direct forced convection is adopted for this flux the temperature rise will be limited to about 15°C. For a heat flux of 1 W/cm^2 forced convection will lead to more than 100°C temperature rise and so the next suitable method, immersion with fluorocarbons may be the choice.

For larger heat fluxes either forced water cooling or immersion with boiling are the choices.

15.4 HEAT TRANSFER PROCESS—CONDUCTION

The heat generated at the component due to the electric current has to be passed finally to the atmosphere through several resistances as heat. The first step is by conduction to the base or the casing. As seen in chapter 2, heat flow equation is

$$Q = -kA\,\Delta T/L \quad \text{or} \quad \Delta T/R \qquad ...(15.1)$$

where $R = L/kA$.

L is the thickness of material in the direction of heat flow, k is the material property thermal conductivity and A is the area perpendicular to flow direction. ΔT is the temperature difference across the thickness. For the same heat flow, higher the resistance, higher will be the temperature drop. But what is required is lower temperature drop. As A and L or fixed k should be high to provide a lower temperature drop. The heat generated at the junction first passes through the chip. The conduction in this case is three dimensional. The resistance for conduction is called constriction resistance and is given as

$$R = \frac{1}{\sqrt{\pi} \times d \times k} \qquad ...(15.2)$$

where d is the diameter of junction and k is the thermal conductivity of the chip material. The heat flow path is shown in Fig. 15.8.

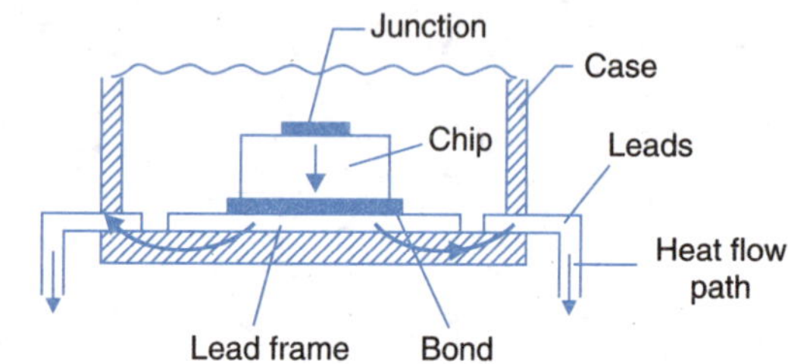

Fig. 15.8. Heat flow path from junction to the leads

The heat generating junction is situated at the top of the chip. The heat generated is conducted through the chip and then through the bond material to the lead frame made of copper. The heat then passes to the plastic case body and then to the leads. The leads and the lead frame are not connected directly to avoid short circuiting. Except for the case all other materials are generally highly conducting with k value in region of 100 – 300 W/mK. The case thickness should be small to reduce resistance.

Example 15.3: *A chip dissipating 0.5 W of power has 14 leads. The material and the dimensions of the device are listed below. If the temperature of the leads is 35°C, determine the temperature at the junction of the chip.*

Material	*Conductivity W/mK*	*Thickness mm*	*Heat transfer surface mm²*
Junction	–	–	0.3 mm dia
Silicon chip	140	0.4	8
Eutectic bond	260	0.025	8
Copper lead frame	380	0.25	8
Plastic case	1	0.2	3
Copper leads (14 leads)	380	5	3

Assuming steady one dimensional heat flow and neglecting radiation and conduction at the top, the resistances are calculated.

The geometry of the set up is shown in Fig. 15.8. The thermal resistance network is shown in Fig. Ex. 15.3.

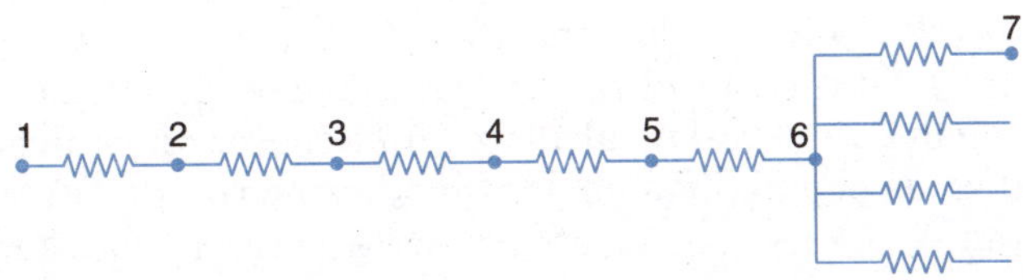

1 Junction, 1–2 constriction resistance, 2–3 chip resistance, 3–4 bond resistance, 4–5 lead frame resistance, 5–6 plastic, from 6–14 leads.

Fig. Ex. 15.3. Thermal resistance network for heat flow from junction to leads

The various resistances are calculated as follows:

1. Constriction resistance

$$= \frac{1}{\sqrt{\pi} \times d \times k} = \frac{1}{\sqrt{\pi} \times 0.0003 \times 140} = 13.433°\text{C/W}.$$

2. Chip resistance $= \left(\frac{L}{kA}\right)_{\text{chip}} = \frac{0.0004}{140 \times 8 \times 10^{-6}} = 0.357°\text{C/W}.$

3. Bond resistance $= (L/kA)_{\text{bond}} = \frac{0.000025}{260 \times 8 \times 10^{-6}} = 0.012°\text{C/W}.$

4. Lead frame resistance $= (L/kA)_{\text{lead frame}}$

$$= \frac{0.00025}{380 \times 8 \times 10^{-6}} = 0.082°\text{C/W}.$$

5. Plastic case $= (L/kA)_{\text{plastic}}$

$$= \frac{0.0002}{1 \times 3 \times 10^{-6}} = 66.667°\text{C/W}.$$

6. Leads $= \dfrac{0.005}{380 \times 3 \times 10^{-6}} = 4.386°C/W.$

The flow from junction to outside is by 14 leads. The equivalent resistance can be taken as that of the total area of cross-section or as 14 parallel circuits. Both methods are found to give the same result. Here the sectional area of 14 leads is taken as 3 mm^2.

Total resistance (in series)

$$= 13.433 + 0.357 + 0.012 + 0.082 + 66.667 + 4.386$$
$$= 84.937°C/W$$

$$Q = \frac{\Delta T}{R} \quad \text{or} \quad \Delta T = QR = 0.5 \times 84.937 = 42.47$$

$$T_j - 35 = 42.47 \quad \therefore \quad \textbf{Junction Temperature = 77.47°C}$$

The highest resistance is at the plastic case (78.5%). Any attempt to reduce the junction temperature should aim at the plastic and find alternate etc.

Analytical determination of the junction to case resistance of various devices is rather complicated and can involve considerable uncertainty. The manufacturers usually determine the value experimentally and list it with their product specifications. The junction to case temperature difference can be accurately calculated from these values and then heat dissipation values. But the actual value of junction temperature depends on the atmospheric temperature also. The actual resistance will be the sum of junction to case resistance and case to atmospheric resistance.

$$R_{actual} = R_{junction\text{-}ambient} + R_{case\ to\ ambient}$$

The actual junction temperature can be calculated by $T_j = T_{ambient} + Q \times R_{actual}$.

A typical chart showing the variation of average thermal resistance with air flow velocity is shown in Fig. 15.9.

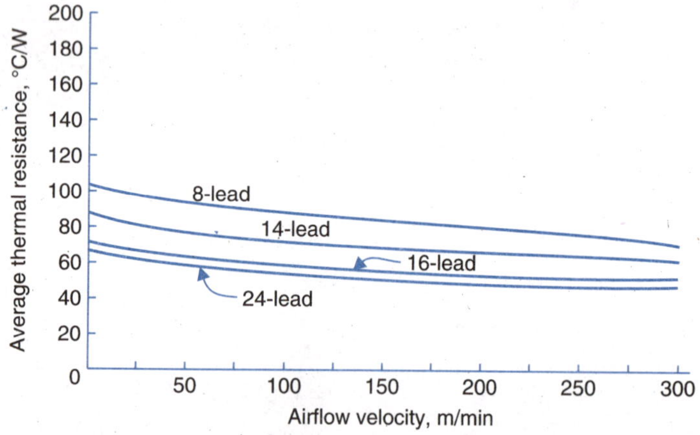

Fig. 15.9. Variation average thermal resistance with airflow velocity for an electronic device

Example 15.4: *A fan blows air at 25°C with a velocity of 150 m/s over a DIP device with 14 leads mounted on a PCB. The variation of resistance is as shown in Fig. 15.9. Determine the junction temperature of the device. If the fan fails what will be the value of junction temperature. The heat dissipation of the device is 1.4 W.*

Solution: Reading the chart at a velocity of 150 m/s for the value of resistance for 14 lead, the value obtained is 65°C/W.

$$\therefore \qquad T_j = 25 + 1.4 \times 65 = 116°\text{C}$$

When the fan fails, the reading at velocity zero and the meeting of the 14 pin line gives R = 85°C/W.

$$\therefore \qquad T_j = 25 + 1.4 \times 85 = 144°\text{C}.$$

The effect of ambient condition has a pronounced effect on junction temperature.

15.4.1. Conduction in Printed Circuit Boards

The next component in the path of heat flow is the printed circuit board on which the chip carriers are mounted. In low heat flow situations air is directly blown over the devices for cooling. The PCB does not play any role in the situation. The other method adopted when PCBs are in sealed enclosures to protect the devices from outside conditions, the PCB is extended at the ends to connect with cooling system. In such situation the PCB conducts the heat towards the end and so the flow resistance of boards have to be calculated. As the PSBs are of electrically insulating materials like glass epoxy, these are poor heat conductors. To get over the problem the boards use aluminium or copper cladding. The copper layer thickness is in the order of 0.036 mm. The heat from the devices is passed on to the cooling medium by both the board and the cladding. As the conductivity of copper is considerably higher than glass epoxy most of the heat flow is along the copper layer. A schematic is shown in Fig. 15.10.

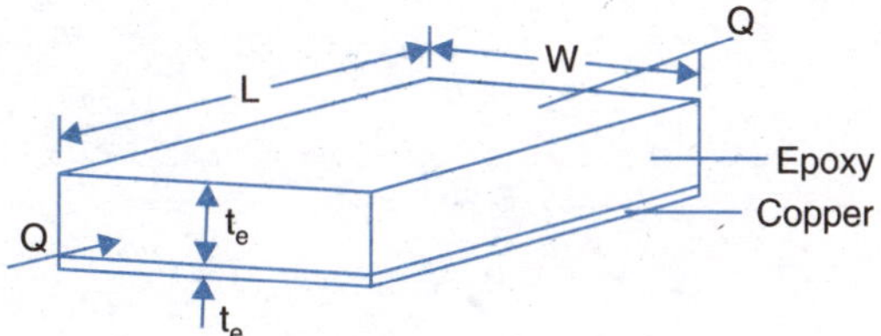

Fig. 15.10. PCB with cladding

Total heat flow = heat flow along epoxy + heat flow along cladding

$$Q_p = Q_e + Q_g = \left(kA\frac{\Delta T}{L}\right)_e + \left(kA\frac{\Delta T}{L}\right)_c$$

$$= \frac{\Delta T}{L}\,[(kA)_e + (kA)_c]$$

As flow area is equal to width W and thickness t,

$$Q_p = \frac{W\Delta T}{L}\,[(kt)_e + (kt_e)_c]$$

The relative magnitude of the heat flow along the two layers depends on the relative magnitude of the product of conductivity and thickness. Numerically if $(kt)_c > 100\,(kt)_c$ then the heat flow along epoxy will be only 1% of the total heat flow and serious error will not occur if the flow along epoxy layer is neglected.

The effective thermal conductivity of the two layers can be calculated as

$$k_{\text{eff}} = \frac{[(kt)_e + (kt)_c]}{t_e + t_c}$$

Heat flow
$$Q = \frac{W\Delta T}{L}\,.\,k_{\text{eff}}\,(t_e + t_c)$$

$$= k_{\text{eff}} \times A_{\text{PCB}} \times \frac{\Delta T}{L}$$

where $A_{\text{PCB}} = W\,(t_e + t_g)$, the area normal to heat flow.

Due to mountings of devices, holes and discontinuities may be there in both epoxy and cladding. The above equations may have to be modified by changing lengths, sections etc.

Example 15.5: *Heat is to be conducted along a copper cladded PCB. The PCB is 15 cm × 15 cm size. Thickness of copper cladding and epoxy are 0.035 and 0.15 mm respectively. Thermal conductivities of epoxy and copper are 0.26 and 380° W/m°C. Neglecting heat conducted along sides and assuming one dimensional flow determine the percentage of heat conducted along cladding. Also calculate the effective conductivity of the PCB.*

Solution: Refer Fig. 15.10. Given

$$L = 0.15 \text{ m}, \quad t_c = 0.035 \text{ mm}, \quad t_e = 0.15 \text{ mm}$$
$$k_c = 380 \text{ W/mK}, \quad k_e = 0.26 \text{ W/mK} \quad W = 0.15 \text{ m}$$

Percentage heat flow through copper layer

$$= \frac{Q_c}{Q_c + Q_e} \times 100 = \frac{(kt)_c \dfrac{W\Delta T}{L}}{[(kt)_c + (kt)_e]\, W \dfrac{\Delta T}{L}}$$

$$= \frac{kt_c}{(kt)_c + (kt)_e} = \frac{380 \times 0.035 \times 10^{-3}}{[(380 \times 0.035) + (0.26 \times 0.15)] \times 10^{-3}}$$

$$= \frac{13.3}{13.3 + 0.039} = \mathbf{0.997 \quad or \quad 99.7\%}$$

Heat flow along epoxy layer is 0.3%

$$k_{\text{eff}} = \frac{(kt)_e + (kt)_c}{t_e + t_c} = \frac{(0.26 \times 0.15) + (380 \times 0.035)}{0.15 + 0.035}$$

$$= \mathbf{72.1 \text{ W/m °C.}}$$

The heat conducted will also be equal to the heat conducted by the thickness of PCB with 72.1 W/m °C conductivity.

Copper cladding on PCB is generally adopted in conduction cooled electronic devices.

15.4.2. The Next Method is Heat Frames

In this case a thicker copper plate is used to pass the heat to the cooling device, termed cold plate which has coolant passing through it or cooled by direct convection. This is suitable for higher heat dissipation as in multilayer PCBs. The schematic is shown in Fig. 15.11. The heat generated first passes through the PCB and then epoxy adhesive and then the heat frame.

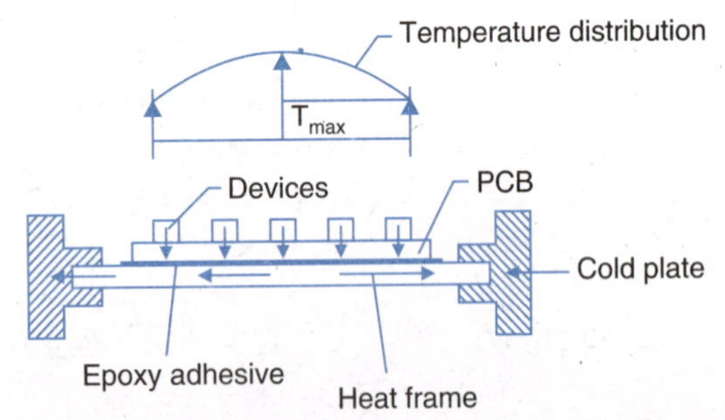

Fig. 15.11. Heat frame method of cooling. Arrows indicate heat flow direction

Heat frame provides a low resistance path from the PCB to the heat sink. Thicker the heat frame lower will be the resistance to heat flow. As heat flow is through the thickness instead of the length, the resistance across the PCB is very much reduced. If the heat sources are evenly distributed and heat is collected at both ends, the temperature variation along flow direction will be symmetric as in Fig. 15.11. The devices at the centre of the PCB will operate at a higher temperature compared to those at the edges. The heat collection can be at one edge only or along two edges as in Fig. 15.11. Heat can also be collected at all four edges. In that case heat flow will be two dimensional.

Example 15.6: *Determine the resistances along the length and across the thickness of a glass epoxy laminate of 0.7 mm thickness and 14 × 14 cm area. Conductivity of the material is 0.2 W/mK.*

Solution: $R = \dfrac{L}{kA}$,

Along length, $L = 0.14$ m, $A = (0.14 \times 0.0007)$ m^2; $k = 0.2$ W/mK

$$\therefore \quad R_l = \frac{0.15}{0.2 \times 0.14 \times 0.0007} = \mathbf{7653°C/W}$$

Across thickness : $L = 0.0007$ m

$A = 0.14 \times 0.14$ m^2

$$\therefore \quad R_t = \frac{0.0007}{0.2 \times 0.14 \times 0.14}$$

$$= \mathbf{0.1785°C/W.}$$

The ratio of temperature drop will be in the ratio of resistances. So flow across the thickness leads to lower temperature drop.

The resistance through the epoxy layer of the PCB can be reduced still by using copper fillings in it, as indicated in Fig. 15.12. Small diameter holes are drilled in the epoxy laminate and filled with copper material. The copper fillings are generally of 1 mm diameter and spacing is few mms. The result is a large reduction in thermal resistance of the board.

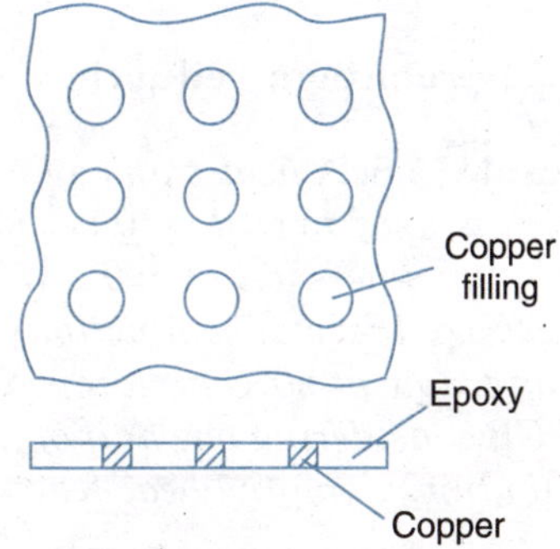

Fig. 15.12. Copper filling in epoxy laminate

Example 15.7: *Refer example 15.6. In the board holes of 1 mm dia are drilled at a spacing of 2 mm and these filled with copper cylinders. Determine the percentage reduction in thermal resistance across the thickness.*

Solution: The board dimensions are:

Area 0.14 × 0.14 m^2, Thickness = 0.7 mm.

$R = 0.1785$°C/W, $k_e = 0.2$ W/mK, $k_c = 380$ W/mK

As spacing is 2 mm, in an area of (2 × 2) mm^2 there is one filling.

$$\therefore \quad \text{Number of fillings} = \frac{\text{Total area in mm}^2}{(2 \times 2)\ \text{mm}^2}$$

$$= \frac{140 \times 140}{2 \times 2} = 4900.$$

Area of copper for heat flow $= (4900 \times \pi \times 0.001^2/4) = 3.84845 \times 10^{-3}$ m^2

Area of epoxy $= (0.14 \times 0.14) - 3.84845 \times 10^3$
$= 19.6 \times 10^{-3} - 3.84845 \times 10^{-3}$
$= 15.7515 \times 10^{-3}$ m^2.

Resistance due to copper fillings

$$= \frac{L}{kA} = \frac{0.0007}{380 \times 3.84845 \times 10^{-3}} = 0.000479°\text{C/W}.$$

Resistance of epoxy layer

$$= \frac{0.0007}{0.2 \times 15.7515 \times 10^{-3}} = 0.2222°\text{C/W}.$$

$$\frac{1}{R_{\text{eff}}} = \frac{1}{R_c} + \frac{1}{R_e} = \frac{1}{0.000479} + \frac{1}{0.2222} = 2093.66$$

$\therefore$ $\mathbf{R_{bound}}$ = 1/2093.66 = **0.000478°C/W**

Without copper fillings **= R = 0.1785°C/W**

With copper fillings **= R = 0.000478°C/W**

Percentage reduction $= \dfrac{0.1785 - 0.000478}{0.1785} \times 100 =$ **99.73%.**

Example15.8: *A heat frame of copper of k = 380 W/mK of size 10 cm by 14 cm and thickness 1.2 mm is used to cool a 10 cm × 12 cm circuit board dissipating 36 W. The epoxy laminate of 0.7 mm thickness with k = 0.2 W/mK is attached to the heat frame. Epoxy adhesive of 0.12 mm thickness of thermal conductivity 1.6 W/mK is used between the laminate and heat frame. The cooling frame at 20°C is clamped at the ends over 5 mm length. Heat is uniformly generated in the PCB considering one half of the PCB board because of symmetry determine the temperature distribution along the heat frame.*

Solution: As heat is generated uniformly each cm length of the PCB will dissipate 3 W. In order to determine the temperature variation along the heat frame, the PCB is divided into 12 strips each dissipating 3 W.

The arrangement is shown in Fig. Ex. 15.8.

Given: $t_e = 0.7$ mm, $t_A = 0.12$ mm, $t_{hp} = 1.2$ mm
$k_e = 0.2$ W/mK, $h_A = 1.6$ W/mK, $k_{hp} = 380$ W/mK

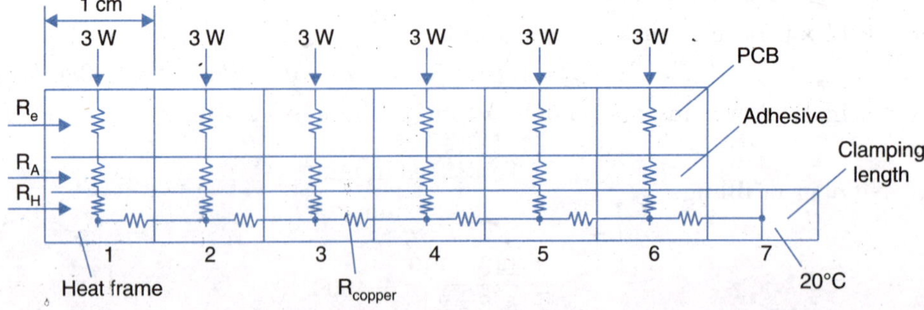

Fig. Ex. 15.8. Schematic of resistance network-one half of heat frame

As the heat input constant over the PCB area, the heat flowing in the heat frame varies continuously. In order to simplify the calculations, the PCB is divided into 6 strips of 1 cm width each dissipating 3 W each. The heat frame extends by one more cm for the clamping purpose. The resistance for heat flow at each strip is made up of three resistances in series. For the heat flow along heat frame there are six resistances in series, upto the cooling connection.

Considering epoxy laminate, the area = 0.1 × 0.01 m^2.

$$L = 0.7 \text{ mm}, \quad k = 0.2 \text{ W/mK}$$

$$\therefore \qquad R_e = \frac{0.0007}{0.2 \times 0.1 \times 0.01} = 3.5\text{°C/W}.$$

For the adhesive $$R_a = \frac{0.00012}{1.6 \times 0.1 \times 0.01} = 0.075\text{°C/W}.$$

For the heat frame R_{hp}, resistance upto half thickness is taken

$$R_{hp} = \frac{0.0006}{380 \times 0.1 \times 0.01} = 0.00158\text{°C/W}.$$

For flow along the Heat frame,

Flow area = 0.1 × 0.0012 m^2, L = 0.01 m

$$\therefore \qquad R_{\text{copper}} = \frac{0.01}{380 \times 0.1 \times 0.0012} = 0.2193\text{°C/W}$$

The temperature at location 1, 2, 3, 4, 6 are calculated using $\Delta T = QR \quad T_7 = 20\text{°C}$

The heat flow along nodes 6 and 7 is 18 W, $\mathbf{T_7 = 20°C}$.

$$T_6 - 20 = 18 \times 0.2193 = 3.95 \qquad \therefore \quad \mathbf{T_6 = 23.95°C}$$

$$T_5 - 23.95 = 15 \times 0.2193 = 3.29 \qquad \therefore \quad \mathbf{T_5 = 27.24°C}$$

$$T_4 - 27.24 = 12 \times 0.2193 = 2.63 \qquad \therefore \quad \mathbf{T_4 = 29.9°C}$$

$$T_3 - 29.9 = 9 \times 0.2193 = 2.0 \qquad \therefore \quad \mathbf{T_3 = 31.9°C}$$

$$T_2 - 31.9 = 6 \times 0.2193 = 1.3 \qquad \therefore \quad \mathbf{T_2 = 33.2°C}$$

$$T_1 - 33.2 = 3 \times 0.2193 = 0.66 \qquad \therefore \quad \mathbf{T_1 = 33.86°C}$$

The PCB temperatures can be found using the same method.

Total resistance from PCB to heat frame

$$= 3.5 + 0.075 + 0.00158 = 3.5766\text{°C/W}$$

The temperature of PCB against node 1

$$T_{\text{PCB1}} - 33.86 = 3 \times 3.5766 = 10.73°$$

This is constant at all nodes.

$$\therefore \qquad \mathbf{T_{PCB1} = 44.59°C}, \quad \mathbf{T_{PCB2} = 43.93°C}, \quad \mathbf{T_{PCB3} = 42.63°C}$$

$$\mathbf{T_{PCB4} = 40.63°C}, \quad \mathbf{T_{PCB5} = 37.97°C}, \quad \mathbf{T_{PCB6} = 34.68°C}$$

As expected the temperature at the central position of the PCB is highest.

A smooth curve through these points will give the actual temperature distribution. In the heat frame arrangement, it is seen that the temperature difference between the frame and the devices at all locations is found to be constant at 10.73°C. A thicker frame will reduce this value still further.

15.4.3. For Conduction Cooling of Two Sided Circuit Boards a Metal Core

in the middle of the PCB can be used. The arrangement is shown in Fig. 15.13.

The metal core receives the heat from the devices through the epoxy laminates. An adhesive layer may also be included to prevent warping as the linear expansion coefficient of laminate is only about half that of metal.

The heat flow is similar to that in the heat frame, only that in this case heat is received from both sides. This may require a larger thickness compared to heat frame.

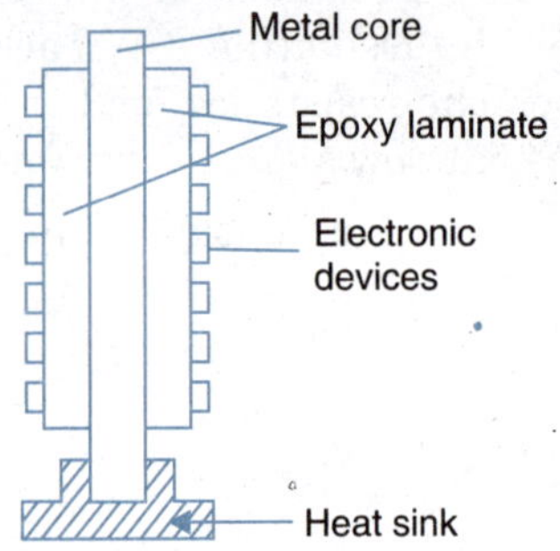

Fig. 15.13. Two sided circuit board with metal core for conduction cooling (sectional view)

15.4.4. The Thermal Conduction Module (TCM)

Due to the circuit density in the chips the heat flux has been steadily increasing from 2 W/cm^2 to more than 20 W/cm^2. Forced air cooling was found inadequate under these circumstances. This led to the development of water cooled thermal conduction module. This differs radically from the previous chip packaging methods.

The TCM houses 100 to 118 logic chips, bonded to a multilayer ceramic substrate of size 9 cm by 9 cm in size with solder balls which provide the electrical connection between the chips and the substrate. Each chip dissipates about 4 W of power. The heat flow path from the chip to the metal casing is through a piston (spring loaded) which is pressed against the back surface of the chip. The tip of the piston is slightly curved to ensure good thermal contact even when there is misalignment. As the contact area is limited, heat is also conducted from the chip to the piston through helium gas filling the space. The heat from the piston is transferred through the helium gas to the module housing and finally to the cooling water.

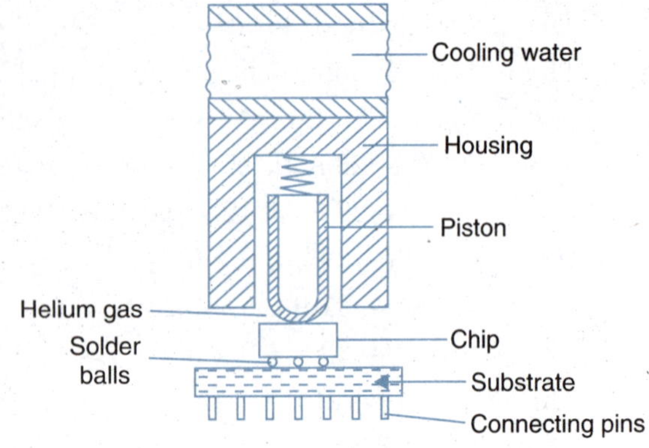

Fig. 15.14. Sectional view of one chip connection to cooling system

A section view of the chip and piston arrangement is shown in Fig. 15.14.

The figure shows only a single chip and the heat flow direction. The following are the thermal resistances.

(1) Chip resistance (2) junction resistance and (3) resistance from the piston upto the cooling water.

The total thermal resistance is around 8°C/W. This will give about 24°C temperature rise above the temperature of the cooling medium.

The chips are packed into a boxlike arrangement with bottom providing the electrical connections and the top providing the cooling water flow path. Air is evacuated and space is filled with helium to provide better thermal conductivity. The compact arrangement reduces the distance between the chips. The signal transmission time is reduced and reliability is increased.

Example15.9: *In a thermal conduction module 108 chips each dissipating 3.6 W of power are packed. Cooling water at 20°C flows through the cooling passage. The thermal resistances are (1) chip resistance = 1.5°C/W, (2) junction to surface of chip = 6.8°C/W, (3) The chip upto the cooling system = 7°C/W. Determine the junction temperature.*

Solution: Refer Fig. 15.14

Under steady conditions,

The total resistance from junction to cooling water

$$= 1.5 + 6.8 + 7 = 15.3°\text{C/W}$$

$$\text{Heat dissipated} = 3.6 \text{ W}$$

$$\therefore \quad T_j - 20 = 3.6 \times 15.3 = 55.08°\text{C}$$

$\therefore$ **Junction Temperature = 75.08°C.**

15.5 AIR COOLING—NATURAL CONVECTION AND RADIATION

Low power electronic systems are conveniently cooled by natural convection and radiation. There are no fans or moving parts involved in free convection, there is no possibility of breakdown and distruption.

In chapter 10 the basics of natural convection and the correlation for determination of convection coefficient has been discussed in detail. So there is no need to repeat them here.

Basically the component or PC board dissipate energy and their temperature is higher than the surroundings. So due to boyant forces the fluid near the part rises upwards giving place to cooler fluid to come in contact with the part. The PC boards should be kept in vertical position for effective cooling by natural convection as shown in Fig. 15.15.

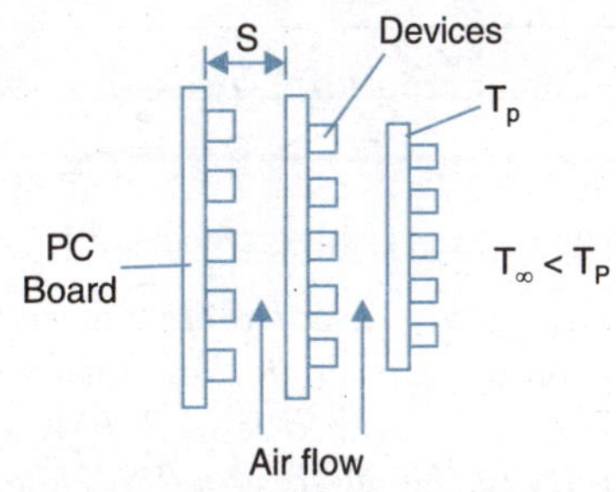

Fig. 15.15. Natural cooling of PCBs

T_∞ is the temperature of the air. T_P is the temperature of the PC board. The spacing S should be sufficient for the air flow, at least about 2 cm.

Boards with heat dissipation around 5 W and flux density of 0.02 W/cm^2 can be conveniently cooled by natural convection.

When PCBs are packed in enclosures like a chassis, sufficient number of vents should be provided both at the bottom and at the top or on the sides near the top. This can been seen in TV or VCR units.

As natural convection is gravity driven, this cannot be used in space application where gravity forces do not exist.

Heat transfer coefficient can be determined by equation of the form below for natural convection.

$$h = K \cdot \left(\frac{\Delta T}{L}\right)^{0.25}$$

where K depends on the geometric shape and positioning of the body, L is the flow length on the body and $\Delta T = T_S - T_\infty$

When radiation is involved, the heat flow is obtained by

$$Q = \varepsilon A \, \sigma \, (T_S^4 - T_\infty^4)$$

ε – emissivity, A – surface area, T_S – surface temperature in kelvin, T_∞ – surrounding surface temperature in kelvin (Refer Chapter 13)

For easy reference the equations for different shapes and orientations are given in Table 15.1.

Table 15.1. Natural Convection Correlation

GEOMETRY	***Vertical cylinder and plate***	***Horizontal cylinder***	***Horizontal plate top hot L = 4 A/P***	***Horizontal plate bottom hot***	***Vertical PCB***	***Small component***	***Sphere***
EQUATION for h, W/m² K	$1.42 \times \left(\frac{\Delta T}{L}\right)^{0.25}$	$1.32 \times \left(\frac{\Delta T}{L}\right)^{0.25}$	$1.32 \times \left(\frac{\Delta T}{L}\right)^{0.25}$	$0.59 \times \left(\frac{\Delta T}{L}\right)^{0.25}$	$2.44 \times \left(\frac{\Delta T}{L}\right)^{0.25}$	$3.53 \times \left(\frac{\Delta T}{L}\right)^{0.25}$	$1.92 \times \left(\frac{\Delta T}{L}\right)^{0.25}$

For surrounding pressures other than 1 atm,

$$h = h_{at} \sqrt{P}. \quad \text{Where P is in atm.}$$

The heat transfer can be increased by providing fins to the surfaces. (Refer Chapter 4).

Example15.10: *A sealed electronic box of 40 cm × 40 cm × 20 cm is placed in room temperature of 30°C on a stand such that natural convection is possible on sides and top only. The heat to be dissipated from the box is 280 W and the surface temperature should not exceed 70°C. The emissivity of the surface is 0.8. Determine whether the condition can be maintained by natural convection.*

Solution: The vertical height is 0.2 m.

Considering the four sides,

Area, $A = 4 \times 0.4 \times 0.2 = 0.32 \text{ m}^2$

$L = 0.2 \text{ m}, \Delta T = 70 - 30 = 40°\text{C}.$

$\therefore$ $h = 1.42 \, (40/0.2)^{0.25} = 5.34 \text{ W/m}^2 \text{ K}$

No correction for pressure is needed.

$Q_{con} = 5.34 \times 0.32 \times 40 = \mathbf{68.352 \text{ W}}$

Horizontal surface: $A = 0.4 \times 0.4 = 0.16 \text{ m}^2$

$L = 4 \times 0.16/1.6 = 0.4 \text{ m} \; \Delta T = 40°\text{C}.$

$$\therefore \quad h = 1.32 \times \left(\frac{40}{0.4}\right)^{0.25} = 4.174 \text{ W/m}^2 \text{ K}$$

$$Q = 4.174 \times 0.16 \times 40 = \mathbf{26.71\ W}$$

Radiation Total area = 0.32 + 0.16 = 0.48 m^2.

$$Q = 0.8 \times 5.67 \times 10^{-8} [(273 + 70)^4 - (273 + 30)^4] = \mathbf{245.5\ W}$$

$$\mathbf{Q}_{total} = 245.5 + 26.71 + 68.352 = \mathbf{340.6\ W > 280\ W}$$

∴ Natural convection can be used.

Example15.11: *A small resistor dissipating 0.25 W is 2 cm long and 0.5 cm in diameter. It is placed horizontally on a PCB undergoing natural convection. Enough gap allows for air flow around the resistor. Neglecting heat flow along connecting wires, determine the surface temperature if surrounding air is at 55°C.*

Solution: Considering the unit as horizontal cylinder, (neglecting end areas)

$$D = 0.005 \text{ m}, \quad A = \pi \times 0.005 \times 0.02 = 314.16 \times 10^{-4} \text{ m}^2$$

From Table 15.1

$$h = 1.32 \times (T_S - 55)^{0.25}/0.005^{0.25} = 4.964\ (T_S - 55)^{0.25}$$

$$Q = hA\ (T_S - 55) = 4.964 \times 314.16 \times 10^{-6} \times (T_S - 55)^{1.25} = 0.25$$

$$\therefore \quad (T_S - 55)^{1.25} = 160.3, \quad \therefore \quad T_S - 55 = 58.07,$$

$$\therefore \quad \mathbf{T_S = 55 + 58.07 = 113.07°C \quad h = 13.7\ W/m^2\ K.}$$

If considered as **vertical cylinder, h** = (1.42/1.32) × 13.7

= **14.73 W/m² K.**

T_S will be 79.3°C.

Example15.12: *A PCB is dissipating 6 W. The size of the PCB is 15 cm × 15 cm. The temperature of the components should not exceed 95°C. The unit operates at a hill station where the atmospheric pressure is 0.65 atm. If natural convection is to be adopted determine the maximum temperature of air. If it is to operate at sea level, will the temperature be lower or higher.*

Solution: In this case, the outside temperature is to be determined. Vertical plate model is to be adopted. $L = 0.15$ m, $A = 0.15 \times 0.15 = 0.0225$ m^2, $T_S = 95°$C. At atmospheric pressure level,

$$h = 1.42 \times (95 - T_\infty)^{0.25}/0.15^{0.25} = 2.282 \times (95 - T_\infty)^{0.25}$$

$$Q = hA\ (95 - T_\infty) = 2.282 \times 0.0225\ (95 - T_\infty)^{1.25}$$

$$6 = 2.282 \times 0.0225\ (95 - T_\infty)^{1.25}$$

$$\therefore \quad (95 - T_\infty)^{1.25} = 116.86 \quad \therefore \quad 95 - T_\infty = 45.09$$

$$\therefore \quad \mathbf{T_\infty = 95 - 45.09 = 49.91°C}$$

In case of pressure is 0.65 atm,

At atmospheric pressure,

$$h = 2.282 \times (95 - 49.91)^{0.25} = 5.913 \text{ W/m}^2 \text{ K}$$

At 0.65 atm,

$$h = 5.915 \times \sqrt{0.65} = 4.77 \text{ W/m}^2 \text{ K}$$

$$6 = 4.77 \times 0.0225\ (95 - T_\infty) \quad \therefore \quad \mathbf{T_\infty = 39.1°C.}$$

At the altitude, the maximum outside temperature can be 39.1°C.

15.6 AIR COOLING: FORCED CONVECTION

The flow velocity and flow volume are much higher and hence the convection coefficient with air in faced convection can be as high as 10 times that in natural convection. So in situations where natural convection is found unsuitable due to the high temperature of devices forced convection is used. The simplest system is arranging the PCBs in an airtight enclosure and using a fan to force air through the PCBs. A simple arrangement is shown in Fig. 15.16.

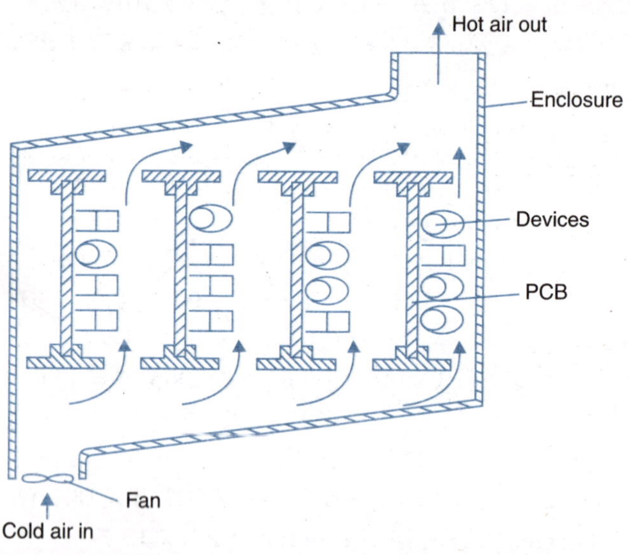

Fig. 15.16. Simple forced air cooling system

Radiation transfer in these cases will be much smaller due to the PCBs facing each other are nearly at the same temperature.

Heat extracted by the air equals the heat dissipated by the components. The heat flow can be calculated using

$$Q = m\, c_p\, (T_{out} - T_{in})$$

Q – heat extracted, W

m – mass flow of air, kg/s

c_p – specific heat of air, J/kg°C

T_{out} – Air outlet temperature

T_{in} – Air inlet temperature

For the same value of Q, if m is increased the temperature difference will be reduced. It is found that for best results, the mass flow should be such that the increase in temperature of the air should be below 10°C. The maximum temperature of air at outlet should not exceed 70°C.

The other route for heat flow calculation is the use of convection coefficient

$$Q = hA_s\, (T_s - T_{air})$$

A_s – heat transfer surface area

T_s – Temperature of the surface

T_{air} – Air temperature

Assuming internal flow through the passage between PCBs shown in Fig. 15.16, hydraulic diameter is calculated as ($4A/P$). This leads to the calculation of Reynold's number and convection coefficient (Refer Chapter 9 on convection–Internal flow correlations)

As the heat input is uniform along the circuit board, the temperature difference between the surface and the fluid is found to be constant after entry length. If air outlet temperature is measured as T_{air}, then the surface temperature of PCB

$$T_s = T_{\text{air}} + \frac{Q}{hA_s}$$

The flow capacity of the fan can be estimated from the total heat dissipation in the unit and the temperature increase in air, which is limited to about 10°C.

$$m = Q/c_p \,.\, \Delta T$$

From the inlet conditions the volume flow can be estimated. The fan characteristics, namely static pressure developed with air flow rate and the system requirement in terms of flow rate and pressure drop should be known to select the fan to suit these conditions. The fan should develop the required static pressure at the flow rate necessary for cooling.

In the case of aircraft systems, as the surrounding pressure is low, the fan should be a variable speed fan to meet the higher pressure ratio requirements.

Another question that arises is whether the fan should force the air into the system by installation at inlet or draw the hot air out by installation at outlet. The volume handled if the fan is at inlet will be lower compared to the fan fixed at the outlet.

Drawing air into the system will also draw dust in keeping the unit above atmospheric pressure is desirable compared to keeping it below atmospheric pressure to avoid dust being drawn in at any small clearance in the enclosure joints, even if filters are installed at inlet.

A set of guidelines for selecting forced convection cooling system for electronic systems is listed below.

1. The inadequacy of natural convection cooling should be first established.

2. Select a fan neither too small or too large. Undersized one will lead to failure of electronic system. Oversizing will lead to higher cost both initial and maintenance.

3. If additional heat from motor is acceptable fan should be installed at inlet with proper filters.

4. Design inlet and outlet passaged to distribute the air evenly.

5. The critical components should be placed near the inlet.

6. Arrange the system such that natural convection will add to forced convection.

In cooling of personal computers, in the old systems, a fan and ventilation slots were used to cool the electronic components. In recent units the CPU is provided with a separate fan and the other units are cooled with a larger fan and ventilation slots.

Hollow core cooling: In cases where air is not allowed to pass over the electronic components, hollow core method is adopted. The schematic is shown in Fig. 15.17. Cooling air flows through the hollow space between the PCBs, collecting heat from the devices mounted on the circuit boards. The flow area is a rectangle of sides equal to width of PCB and the depth of the hollow core.

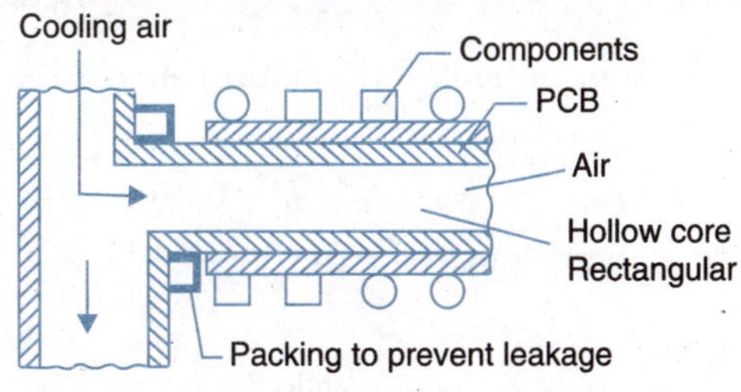

Fig. 15.17. Hollow core method of cooling (sectional view)

Heat generated by the components is conducted through the PCB and a thin layer of epoxy board to reach the cooling air. Heat picked up by air is given by

$$Q = mc_p\,(T_{\text{air out}} - T_{\text{air in}})$$

The heat flow can be also calculated by

$$Q = hA_s\,\Delta T_{\text{mean}}$$

where A_s is the heat transfer area. Convection coefficient h, is to be calculated using the correlations for internal flow. As the temperature difference may vary, ΔT_{mean} is used for property values. In the case of PCBs heat is generated uniformly along the length. So in the hollow core cooling the correlation for constant heat flux should be used.

Example 15.13: *A hollow core PCB is 15 cm high and 20 cm long. The heat dissipation is 50 W. The core gap is 4 mm. Cooling air enters at 20°C at a rate of 1 l/s. Assuming uniform heat generation, determine the temperature of air at exit and the highest temperature at the inner surface of the core.*

Solution: Mass flow of air = *PV/RT*, kg per second

Assuming the pressure as atmospheric

$$= (1.013 \times 10^5 \times 0.1 \times 10^{-3})/(287 \times 293)$$

$$= 1.205 \times 10^{-3} \text{ kg/s}$$

As $$Q = mc_p\,(T_o - T_c)$$

$$50 \text{ J/s} = 1.205 \times 10^{-3} \text{ (kg/s)} \times 1007 \text{ (J/kg °C)} \times (T_o - 20) \text{ °C}$$

Solving $$\mathbf{T_o = 61.2°C.}$$

Determination of convection coefficient.

The properties of air at the average temperature of 40.6°C are extracted from data book.

$$\rho = 1.128 \text{ kg/m}^3, \quad c = 1005 \text{ J/kgK}, \quad k = 0.02756 \text{ W/mK.}$$

$$\text{Prandtl number} = 0.699, \quad \nu = 16.96 \times 10^{-6} \text{ m}^2/\text{s}$$

$$\text{Flow area} = 0.2 \times 0.004 \text{ m}^2 = 0.0008 \text{ m}^2$$

$$\text{Equivalent diameter} = 4A/P = \frac{4 \times 0.0008}{(2 \times 0.2) + 2 \times 0.004}$$

$$De = 0.007843 \text{ m}$$

Velocity, $$V = 0.001/0.0008 = 1.25 \text{ m/s}$$

$$Re = \frac{v\,D}{\nu} = \frac{1.25 \times 0.0008}{16.96 \times 10^{-6}} = 58.962 \text{ laminar}$$

From data book for constant heat flux, for (width/depth) > 8

$$Nu = 6.49, \quad h = Nu\,k/De$$

$$h = 6.49 \times 0.02756/0.007843 = 22.8 \text{ W/m}^2\text{ K}$$

T_{max} occurs at the exit.

$$\therefore \quad \mathbf{T_{max}} = T_{\text{air out}} + \frac{Q}{hA_s}, \quad A_s = (0.2 \times 0.15) \times 2 = 0.06 \text{ m}^2$$

$$= 61.2 + \frac{50}{22.8 \times 0.06} = \mathbf{97.75°C.}$$

Example 15.14: *A transistor with a height of 6 mm and diameter of 5 mm, mounted on a circuit board is shown in Fig. Ex. 15.14. The component is cooled by air flow at a velocity of 1.6 m/s. The temperature of air is 40°C.*

If the surface temperature should not exceed 80°C determine the power the transistor can dissipate.

Solution: The convection coefficient for the cylindrical surface and the flat end surfaces should be determined. Properties of air at (80 + 40)/2 = 60°C.

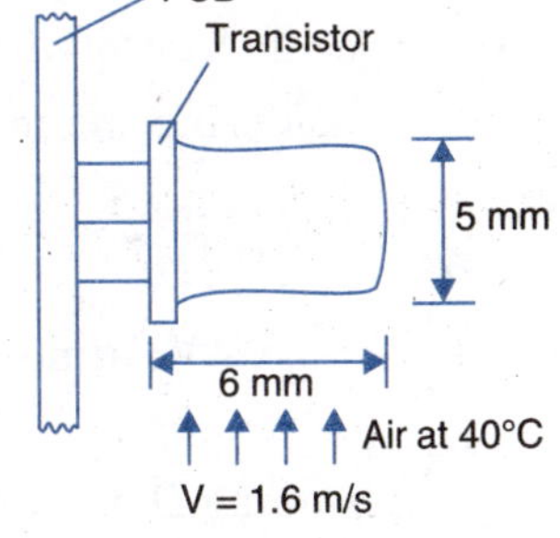

Fig. Ex. 15.14

$$\rho = 1.06 \text{ kg/m}^3, \quad \nu = 18.97 \times 10^{-6} \text{ m}^2\text{/s}$$

$$Pr = 0.696, \quad C_p = 1005 \text{ J/kgK}$$

$$k = 0.02896 \text{ W/mK}$$

(*a*) Cylindrical surface:

$$Re = \frac{VD}{\nu} = \frac{1.6 \times 0.005}{18.97 \times 10^{-6}} = 421.7$$

using Data Book, the generalised form of equation with constants for $Re < 1000$

$$Nu = 0.51\, Re^{0.5}\,.\, Pr^{0.37} \left(\frac{Pr_a}{Pr_w}\right)^{0.25}$$

Pr at 40°C = 0.699 $\quad$ Pr at 80°C = 0.692

$$Nu = 0.51 \times 421.7^{0.5} \times 0.696^{0.37}. \left(\frac{0.694}{0.692}\right)^{0.25} = 9.18$$

$$\mathbf{h} = \frac{Nu \times k}{D} = \frac{9.18 \times 0.02896}{0.005} = \mathbf{53.17\ W/m^2\,K}$$

$$A = \pi DL = \pi \times 0.005 \times 0.006 = 94.25 \times 10^{-6}$$

$$\mathbf{Q_c} = 53.17 \times 94.25 \times 10^{-6}\,(80 - 40) = \mathbf{0.2\ W.}$$

(*b*) Considering the vertical faces:

$$Re = \frac{VL}{\nu} = \frac{1.6 \times 0.005}{18.97 \times 10^{-6}} = 421.7$$

$$Nu = 0.664\, Re^{0.5}\, Pr^{0.333}$$

$$= 0.664 \times 421.7^{0.5} \times 0.696^{0.333} = 12.08$$

$$h = \frac{12.08 \times 0.02896}{0.005} = 69.97 \text{ W/m}^2\text{ K}$$

$$A = 2 \times \pi \times 0.005^2/4 = 39.27 \times 10^{-6}$$

$$Q_v = 69.97 \times 39.27 \times 10^{-6}\,(80 - 40) = \mathbf{0.11\ W}$$

The maximum heat that can be dissipated under these conditions by the transistor is

$$Q_{max} = 0.2 + 0.11 = \mathbf{0.31\ W}$$

If the power is to be increased, air velocity increase is one possibility. The other is lowering the air inlet temperature, generally not easy. The other alternate is using a heat sink.

Example 15.15: *Fan cooling is proposed for a desk top computer, the heat load of which is 80 W. The maximum air outlet temperature for reliable operation is 60°C. The surrounding air is at 30°C and 0.8 bar. The velocity is to be limited 1 m/s in order to reduce noise. Determine the volume handled by the fan. Also calculate the diameter of the fan casing. The fan is to be fitted at the air outlet.*

Solution: $Q = mc_p(T_{out} - T_{in})$

Specific heat of air = 1005 J/kg K,

Considering the maximum outlet temperature so that flow will be minimum,

$$80 = \dot{m} \times 1005 \times (60 - 30)$$

$\therefore$ $m = 0.00265$ kg/s or 0.1592 kg/min.

Volume flow $= V = mRT/P = 0.00265 \times 287 \times 333/0.8 \times 10^5$

$= \mathbf{0.003166\ m^3/s.}$

Assuming D as fan case diameter,

$$(\pi D^2/4)\ \text{V} = 0.003166, \qquad V = 1 \text{ m/s}$$

$$\therefore \quad D^2 \times \left(\frac{\pi}{4} \times V\right) = 0.003166$$

Solving, **D** = 0.0635 m or **6.35 cm.**

If the fan is to be fitted at inlet, the volume handled will be less, but heat load will increase by the fan power.

Example 15.16: *In a system 6 PCBs of 12 cm height and 15 cm length each dissipating 12 W are to be fan cooled. The distance between the PCBs is 5 mm. The available fan power is 12 W. The temperature rise for the air should not exceed 12°C. Determine the flow rate of air. The fan is fixed at the inlet. Also determine the maximum air inlet temperature, if any of the component temperature is not to exceed 60°C.*

Solution: Total amount of heat to be dissipated

$$Q = 12 \times 6 + 12 = 84 \text{ W}, \qquad \Delta T = 12°\text{C}, \qquad c_p = 1005 \text{ J/kgK.}$$

$\therefore$ $m = 84/1005 \times 12 = 0.006965$ kg/s.

Assuming air inlet at 30°C

Volume flow $= (0.006965 \times 287 \times 303)/1.013 \times 10^5 = \mathbf{0.006\ m^3/s.}$

The contribution of fan motor in the load and temperature rise,

$$= 12/12 \times 6 = \frac{1}{6} \quad \text{or} \quad 16.67\% \quad \text{or} \quad 1.67°\text{C out of } 12°\text{C.}$$

To determine the maximum inlet temperature of air, convective heat transfer coefficient has to be determined.

Hydraulic mean diameter = $4A/P$.

$$A = 0.15 \times 0.005 \text{ m}^2, \qquad P = 2\ (0.15 + 0.005) = 0.31$$

$$De = 4\ (0.15 \times 0.005)/0.31 = 0.01 \text{ m}$$

$$\text{Velocity} = 0.006/(0.15 \times 0.005) \times 6 = 1.333 \text{ m/s.}$$

at 30°C $\quad \nu = 16.08 \times 10^{-6}\text{m}^2/\text{s}, \quad k = 0.02588 \text{ W/mK}, \quad Pr = 0.7282$

$$Re = \frac{1.333 \times 0.01}{16.08 \times 10^{-6}} = 829 < 2000.$$

From Data book, $Nu = 6.49$.

$\therefore$ **h** $= 6.49 \times 0.02588/0.01 =$ **16.8 W/m² K**

As the flux is uniform along the flow length, the temperature difference between the fluid and surface will be constant air is given by

$$\Delta T = \frac{Q}{h} = 84/16.8 = 5°\text{C}.$$

As maximum temperature of component is 60°C,

Air outlet maximum $= 60 - 5 = 55°\text{C}$

As air undergoes 12°C rise along the flow,

Air inlet maximum $= 55 - 12 = 43°\text{C}$.

15.7 LIQUID COOLING

Much higher convection coefficients can be achieved with liquid cooling as compared to cooling by gas medium. Leakage, corrosion, weight and pumping power are some of the problems in using liquids for cooling electronic components. Therefore liquid cooling is used only in situations with high power densities that are too high for air cooling.

Liquid cooling systems are classified as direct cooling and indirect cooling systems. Each of them can be classified again as closed loop and open loop systems.

The components are immersed in liquids in the case of direct cooling system. Heat transfer may be by natural or forced convection or boiling depending upon the temperature levels involved and property of the fluid. Only dielectric fluids can be used for direct cooling purposes. Water is excluded for direct cooling systems.

There is no contact between the fluid and the components in indirect cooling system. Heat generated in the components is first transferred to a medium such as cold plate before it is carried away by the liquid. In the open loop system, the liquid is discarded after it has picked up the heat. Most forced air cooling systems are of this type where hot air is let into the atmosphere. In closed loop system the fluid after it gets heated is taken to a heat exchanger and cooled using atmospheric air and recirculated in the system. Open loop systems are not generally used in liquid cooling systems.

In the case of indirect cooling system, the components are mounted on metal plate of highly conducting material such as copper. The metal plate is cooled by circulating cooling fluid through tubes attached to it. The heated liquid is cooled in a heat exchanger and circulated through the system again.

Desirable characteristics of cooling liquids include high thermal conductivity, high specific heat, high surface tension (to reduce leakage) and high dielectric strength, chemical stability and non toxicity.

High conductivity leads to high convection coefficient. High specific heat leads to reduced mass circulation and reduced pump power.

The heat sinks or cold plates of an electronic enclosure is usually cooled by water through channels made for this purpose (as in thermal conduction module) or through tubes attached to the cold plate. For very low temperature requirements refrigerants may be used replacing water.

A liquid cooling arrangement of size 15.2 cm × 18 cm × 2.5 cm in size capable of dissipating upto 2 kW is shown in Fig. 15.18.

The thermal resistance between the case of the devices and the liquid is minimised in this method by directly mounting them over the cooling lines.

The thermal resistances in this system are shown in Fig. 15.18(*a*).

The junction temperature in the case of silicon based devices can go upto 125°C. The junction to case resistance is provided by the manufacturer. The case to liquid thermal resistance can be determined experimentally by measuring the temperatures of the case and the liquid.

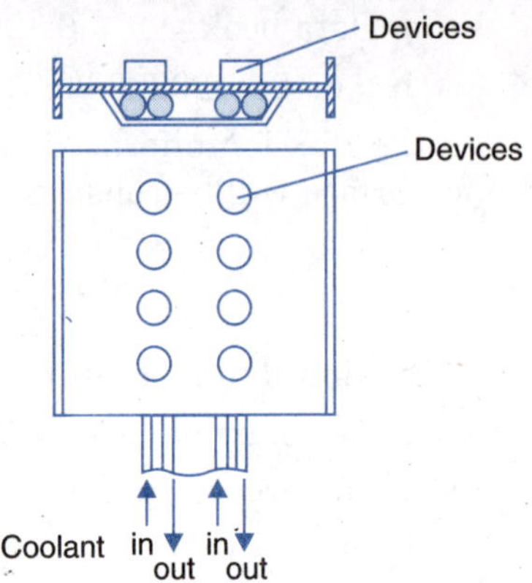

Fig. 15.18. Liquid cooling packages placed on top of the coolant line

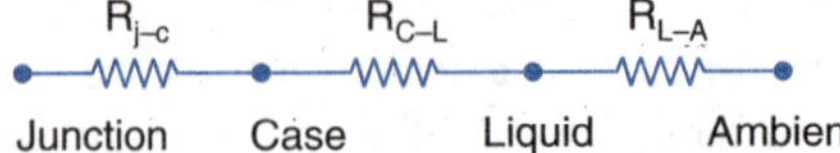

Fig. 15.18(*a*). Thermal resistance network for a liquid cooled electronic device

$$R_{\text{case-liquid}} = \frac{T_{\text{case}} - T_{\text{liquid}}}{Q}$$

$$R_{\text{liquid-air}} = \frac{T_{\text{liquid}} - T_{\text{air}}}{Q}$$

The required mass flow can be determined from the temperatures and total heat to be dissipated. The contact resistance between the plate and devices can be reduced by applying silicon grease and fastening component tightly on the plate.

Example 15.17: *Thirty two power transistors are supported by a cold plate with 16 on each side. The power of each device is 30 W. The cold plate is cooled by water pipes fixed at the edges. Water enters at 30°C and maximum temperature rise is to be limited 4°C. If 10% of heat generated is dissipated by radiation and convection on the top and bottom sides. Determine the amount of water to be circulated per minute. Also calculate the ID of pipe if flow velocity is limited 0.8 m/s. If the thermal resistance from case to water is 0.025°C/W, determine the case temperature.*

Solution: Under steady conditions, the amount of heat to be collected by water

$$Q = 32 \times 30 \times 0.9 = 864 \text{ W}$$

The specific heat of water

$$= 4180 \text{ J/kgK, density} = 1000 \text{ kg/m}^3$$

$$\mathbf{m} = Q/c_p\, \Delta T = 864/(4180 \times 4) = \mathbf{0.05167\ kg/s}$$

or **3.1 kg/min**

Volume of water per second = 0.05167×10^{-3} m³/s

$$\frac{\pi D^2}{4} \times \text{Velocity} = \text{Volume flow.}$$

$$\therefore \quad D^2 = (0.05167 \times 10^{-3} \times 4)/(\pi \times 0.8)$$

∴ $\mathbf{D} = 0.00453$ m or **4.53 mm.**

Case temperature = 30 + 0.025 × 864 = **51.6°C.**

15.8 IMMERSION COOLING

High power electronic components can be cooled effectively by immersing them in a dielectric fluid. The fluid is chosen such that it will boil at the specified maximum temperature of the devices at a reasonable pressure. Boiling has three advantages (1) The convection coefficient will be very high in the order of thousands of W/m^2°C. (2) The temperature remains constant during the process. The fluid does not undergo any cooling or heating. (3) The latent heat has a large value. Previously immersion system was used exclusively for high power radar systems. Now due to increase of packing density many other systems are going in for this system. Super computer CPU is cooled by immersion process.

The simplest type of system is shown in Fig. 15.19.

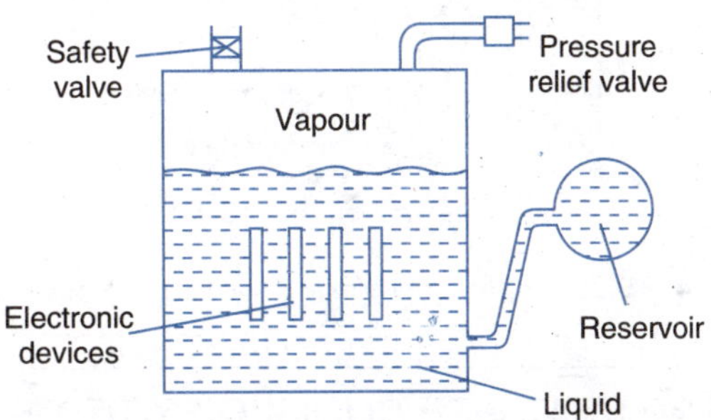

Fig. 15.19. Simple open loop immersion cooling system

As heat is transferred to the liquid, it vapourises. As more vapour is added pressure will increase and pressure relief valve will open to release the vapour.

There are several drawbacks in the use of this system. The vapour released to the atmosphere is objectionable. The cost of liquid is high. An external reservoir makes the system bulky. So this system is limited to applications where a light duty cycle is involved.

Closed loop immersion systems have the advantage that the cooling fluid is not used up by releasing to atmosphere. There are two possible arrangements in closed loop systems. (1) Boiling type (2) Conduction type or all liquid system.

In the boiling type the vapour produced by the heating is condensed by using external coolant like air or water. In the system where vapour is condensed by air, the condenser has to be kept outside the liquid container. In water cooling system, the condenser can be kept inside. In the air cooling there is a possibility of leakages and other gases getting into the system. But it needs rather little attention and power required is lower. In water cooling the heated water has to be cooled externally by a heat exchanger using air. The system thus becomes more complex.

The schematic of the two types are shown in Fig. 15.20 (*a*) and (*b*).

The liquid to liquid cooling systems can reject heat either to air or a cooling fluid like water as in the boiling type. The heat transfer coefficient is one order of magnitude lower because heat transfer is purely by convection and conduction. Allliquid systems are not suitable for high power dissipation systems.

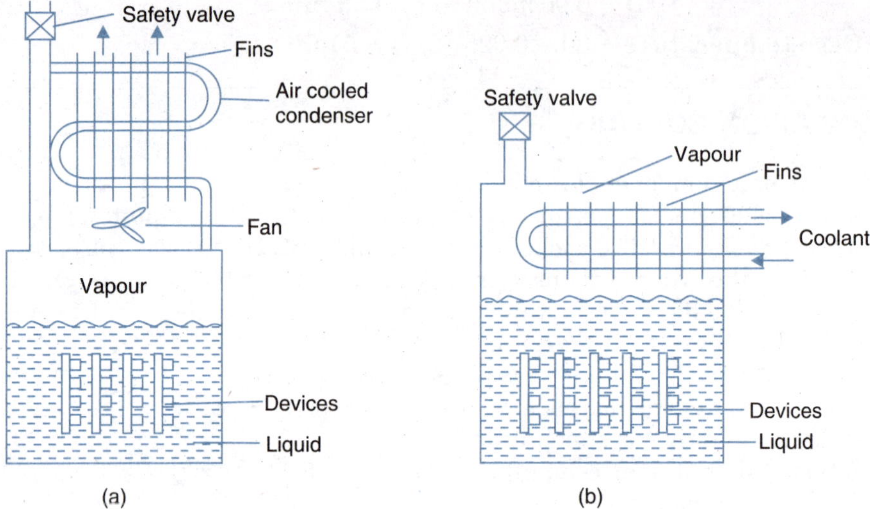

Fig. 15.20. Immersion cooling system of Boiling type (*a*) External condenser (*b*) Internal condenser

15.9 HEAT PIPE COOLING SYSTEMS

Heat pipe is basically heat transporting device. The transporting is by a vapour liquid medium. Detailed working of such devices is explained in Chapter 16. A schematic is given in Fig. 15.21.

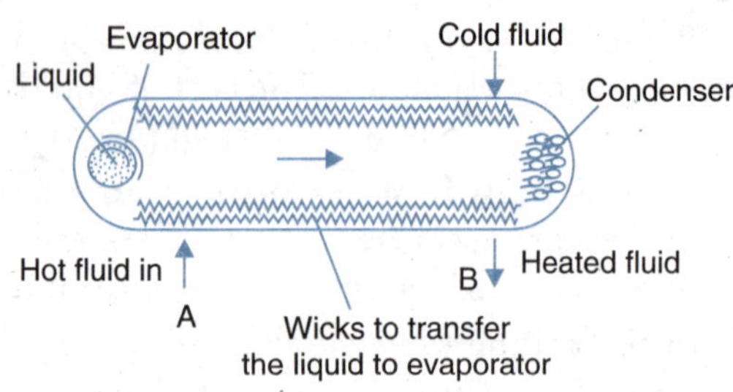

Fig. 15.21. Schematic of heat pipe

An evacuated cylinder is filled with a limited quantity of liquid and sealed. When the liquid is heated it evaporates and moves to the condenser section carrying the latent heat at the saturation temperature. The vapour is condensed at the other end by the fluid which is to be heated. The liquid moves back to the evaporator by capillary action of the wicks. The effect is transporting the heat energy from the hot fluid to the cold fluid at the other end. The temperature inside the system is constant.

The rate of heat transported for a given area is very high, as much as 10 to 20 kW/cm^2.

The operating temperature depends upon the filler fluid. Recent developments have been very rapid and hundreds of types and sizes of heat pipes are in the market from single units to clusters. What makes the use of heat pipes attractive in cooling of electronic equipments? Consider the details of a cold plate. The devices are mounted on the plate and the heat dissipated by the devices is conducted into the cold plate. The cold plate transfers the heat to a fluid by conduction or by a fluid circulated in tubes attached to the cold plate. If heat is transferred by the plate by conduction, the flux is very low. Even with water cooling the flux is not high enough. If the heat pipe evaporator section is attached to the cold plate very high heat flux can be obtained. As water in the other system is cooled by air, the condenser section of the heat pipe will be cooled by air.

Two views of a cold plate cooled by heat pipes are shown in Fig. 15.22. The heat pipes are attached to the lower side while devices are mounted on the top side of the cold plate. This section of the heat pipes works as evaporator. The air cooled portion acts as condenser. The liquid moves back to the evaporator section through the wicks. There are many other types of heat pipes and different ways of attaching them to the devices. Some of the types and method of attachment are described in the later section.

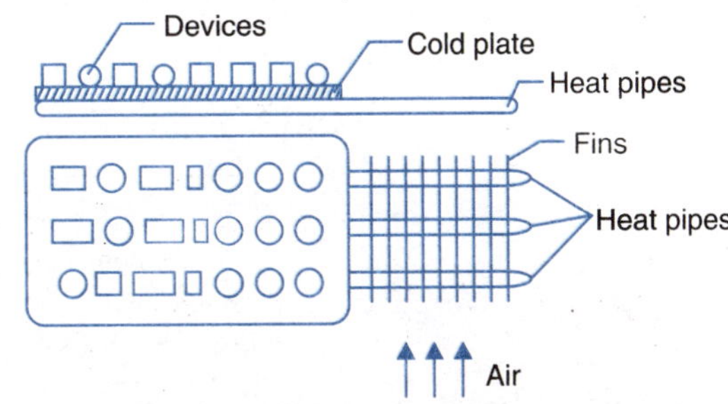

Fig. 15.22. A cold plate cooled by heat pipes

Miniaturisation together with increasing processing speeds decreases the heat transfer surface area at the same time increasing the power. This leads to very high heat fluxes resulting in large temperature rise. To maintain the devices within operating conditions more heat must be removed. The traditional system of forced convection cooling have become inadequate. Use of heat pipes for heat transportation fills the need to a great extent.

15.9.1. Features of the Heat Pipe

The following are some of the functions which can be effectively done by heat pipe for cooling of electronic components:

1. Separation of heat source and sink: In forced circulation cooling or in liquid immersion cooling the components come in contact with fluid. If heat pipe is used there will be no contact between the cooling fluid and components.

2. Temperature flattening: In other cooling system the cooling medium as well as the cold plate temperatures vary with location. In heat pipes the temperature throughout is constant.

3. Transfer of heat to a remote location: Heat pipes are available which are as long as 1 m and are also flexible. So the condenser section can be located away from the heat source.

4. Production of compact heat sink: As the heat conduction per cm^2 is high, the system will be comparatavely small.

5. Air cooling or water cooling of condenser system is possible.

Heat pipes are available in different shapes: These are

(*i*) Tubular
(*ii*) Flat plate
(*iii*) Micro heat pipe arrays
(*iv*) Loop heat pipes
(*v*) Direct contact system

1. The tubular heat pipe is the first type developed and is used extensively for various applications including cooling of electronic devices, Fig. 15.22 illustrates the use.

2. Flat plate heat pipes: This type is well adopted for cooling of electronic components. It is often used for "heat spreading" or temperature flattening. Its use in electronic cooling are: (*i*) multicomponent array temperature flattening (*ii*) multicomponent array cooling (*iii*) can be used as a wall of module or mounting plate. An arrangement is shown in Fig. 15.23.

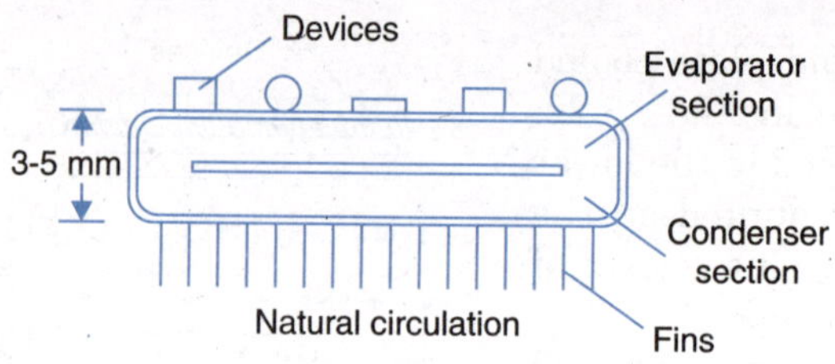

Fig. 15.23. Flat plate heat pipe

3. Embedded heat pipes: In this case heat pipes are embedded in the cold plate instead of placed on them. This arrangement allows the use of high volume low cost heat pipes in heat sink design.

4. Micro heat pipe and arrays: Are used for cooling smaller size high heat flux components. The cross-sections used are triangular, square etc. with the corners as shown in Fig. 15.24 so that condensate collects at the corner and is carried to the evaporator section.

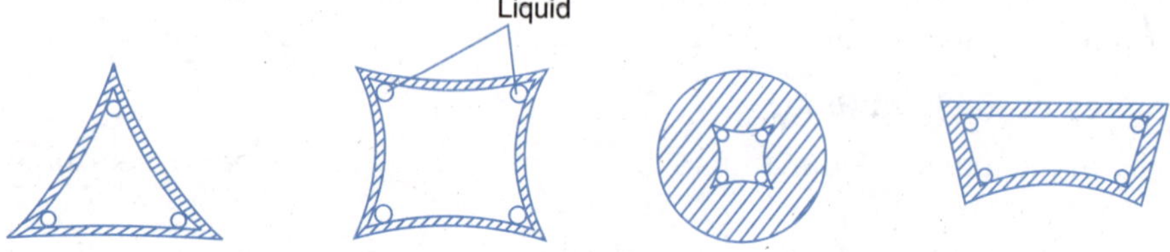

Fig. 15.24. Some sections of micro heat pipes

The internal diameters are in the order of 2-4 mm. No wick is used in this case. Liquid moves down the sharp corners to the evaporator section.

Micro heat pipes are used in laptop computers, note book computers and even cellphones. These can be used for cooling single chips also.

5. Loop heat pipes: The evaporator and condenser sections can be even at 1 m distance. The condenser and evaporator are connected by two capillary tubes one of which conveys the vapour to the condenser and the other returns the liquid to the evaporator. Fig. 15.25 shows an LHPs.

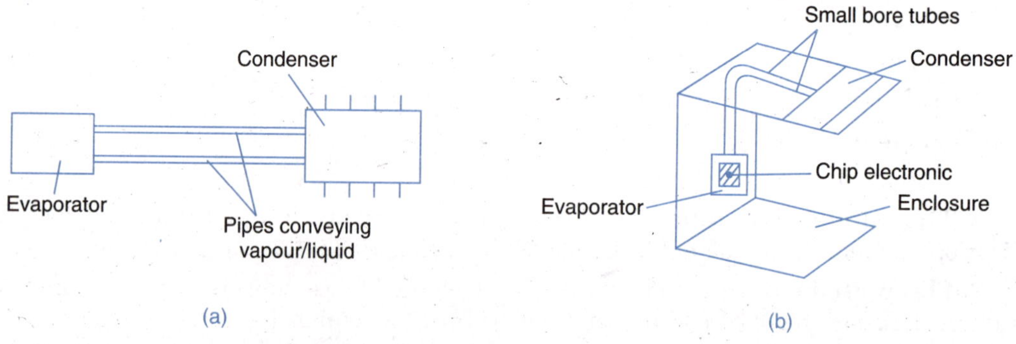

Fig. 15.25. (*a*) Loop heat pipe (*b*) Cooling a chip using LHP

6. Flexible heat pipe: Flexible heat pipes to some extent are similar to loop type heat pipes. In this case the connecting tubes are made flexible and also the distance between the condenser and evaporator can be much longer. The evaporator is fixed at the chip carrier or cold plate and the condenser can be at a distance.

15.10 CONCLUSION

Various methods of cooling of electronic components were discussed. The conduction cooling of chip carriers, use of heat frame natural convection cooling, forced convection cooling with air/water, immersion cooling with boiling and finally use of heat pipes were explained.

Each of the method has its own sphere of application from heat fluxes of 0.01 W/cm^2 to 100 W/cm^2.

Heat pipe appears to be applicable for all ranges and also non intrusive. Already heat pipes are found to be popularly used in laptop computers. Heat pipe manufacturing companies report that they have orders in the range of million pieces per year. There may come some more newer ways of cooling very high power density and low volume and area type devices.

REVIEW QUESTIONS

15.1 Describe with a sketch the growth over the years in the number of components packed on a chip.

15.2 Describe the failure rate of bipolar digital devices with temperature.

15.3 Sketch a chip carrier.

15.4 What is the purpose of a flange attached to the chip carrier?

15.5 What is heat frame? Sketch the temperature variation along the width of a heat frame.

15.6 Describe the various stages involved in the production of an electronic system.

15.7 How is the cooling load of a component or a system determined?

15.8 List the various methods of cooling electronic systems. Indicate the range of heat fluxes for each.

15.9 What is conduction cooling? Describe the heat flow paths from a junction to the leads.

15.10 Sketch the thermal resistance network for a chip carrier with 8 pins leads.

15.11 Discuss the limitation of conduction cooling.

15.12 Sketch the schematic of a heat frame cooling method.

15.13 What is the method adopted to reduce the thermal resistance of epoxy board?

15.14 Describe the arrangement of devices in metal core conduction cooling of boards.

15.15 What is thermal conduction module (TCM)?

15.16 Sketch the schematic cooling of a chip in thermal conduction module.

15.17 What are the advantages and limitations of natural convection cooling of electronic components.

15.18 Sketch the schematic of forced circulation air cooling of an assembly of boards.

15.19 What are the limitations of forced convection air cooling?

15.20 Compare indicating the advantages and limitations of installing cooling air fan at (1) inlet and (2) outlet.

15.21 Explain what is hollow core method of cooling of electronic boards.

15.22 Discuss the advantages of forced convection liquid cooling.

15.23 What is direct cooling method?

15.24 Sketch boiling type of immersion cooling system.

15.25 What are the advantages of heat pipe cooling of electronic devices?

15.26 What is temperature cycling? How does it affect the components and the PCB?

15.27 What is constriction resistance of a chip to heat flow from junction?

EXERCISE PROBLEMS

15.1 The temperature of the case of a power transistor that is dissipating 12 W is measured as 60°C. If the junction temperature should not exceed 120°C, determine the maximum junction to case resistance. [**Ans.** 5°C/W]

15.2 A logic chip in a computer dissipates 6 W of power to an environment at 55°C. The surface area of the chip is 0.32 cm^2. Determine the heat flux on the surface. Also determine the heat dissipated in 8 hours. [**Ans.** 18.8 W/cm^2, 0.048 kWh]

15.3 The material properties and dimensions of components of a DIP with 18 leads are listed. If the chip dissipates 0.8 W and if the temperature of the leads is 50°C, estimate the temperature of the junction of the chip.

Material	*Conductivity W/m°C*	*Thickness mm*	*Heat transfer surface area*
Junction			0.5 mm diameter
Silicon chip	120	0.5	4 mm × 4 mm
Eutectic bond	296	0.05	4 mm × 4 mm
Copper lead frame	386	0.25	4 mm × 4 mm
Plastic separator	1	0.3	10 × 1 mm × 0.25 mm
Copper leads	386	6	18 × 1 mm × 0.25 mm

[**Ans.** 110.1°C]

15.4 Heat is to be conducted along a PCB with copper cladding. The thickness of epoxy layer is 0.5 mm and its conductivity is 0.26 W/mK. The thickness of copper cladding is 0.06 mm and its thermal conductivity is 386 W/mK. Determine the percentage of heat conducted by each layer. Also calculate the effective thermal conductivity of the PCB.

[**Ans.** Epoxy: 0.6%, Copper 99.4%, 41.6 W/m°C]

15.5 A glass epoxy laminate of 15 cm × 18 cm × 1.4 mm of thermal conductivity 0.26 W/m°C is used to transfer heat from chip to the lead frame. In order to reduce the resistance across the thickness copper cylinders of 1 mm diameter with thermal conductivity of 386 W/m°C are used at a center distance of 3 mm. Calculate the thermal resistance of the epoxy board before and after modification. [**Ans.** R = 0.199°C/W, 0.00153°C/W]

15.6 In a thermal conduction module 80 chips of each 4 W of power are packed. Module is cooled by water at 15°C. The thermal resistance between the junction and the chip is 12°C/W. The thermal resistance between the surface of the chip to outer surface of thermal conduction module is 9°C/W. The thermal resistance between the outer surface and water is 7°C/W. Determine the junction temperature of the chip. [**Ans.** 86.8°C]

15.7 A copper plate of 0.5 mm thickness is sandwiched between two epoxy boards of 12 m × 18 cm size and 3 mm thickness. Determine the effective conductivity along the 18 cm length. k = 386 W/m°C for copper, 0.26 W/m°C for epoxy board. Also calculate the % heat flow through copper.

[**Ans.** 29.9 W/m°C, 99.2%]

15.8 A sealed electronic box of 0.5 m × 0.35 m × 0.2 m size is placed on a stand in a room at 30°C. The surface of the box has an emissivity of 0.85. The heat emission is 100 W. The 20 cm side stands vertically. Neglect heat flow at bottom. If the surface temperature should not exceed 65°C can natural convection and radiation meet the need?

[Ans. By natural convection and radiation Q = 200.8 W, So satisfactory]

15.9 A PC board 20 cm wide and 14 cm high is mounted vertically with components on one side. The heat dissipation rate is 7 W. The equipment is working at a location where the surrounding pressure is 70.12 kPa. If the surface temperature should not exceed 90°C, determine the maximum temperature of the environment. [Ans. 52.6°C]

15.10 A circuit board of size 20 cm × 20 cm with 81 square chips mounted on one side is mounted vertically. The surface emissivity on the chip side is 0.65. Surrounding temperature is 25°C. The total heat dissipation rate is 6.48 W. Heat flow on the back side is negligible. Determine the surface temperature of the chips. [Ans. 39.3°C]

15.11 A hollow core PCB is 15 cm high and 20 cm wide. Cooling air flows through the air gap of 2.5 mm, at the rate of 1 L/S. The temperature of cooling air at entry is 30°C. The total heat dissipation is 30 W. Assuming heat generated is uniformly distributed over the two sides determine the (*a*) temperature of the air at exit and (*b*) the highest temperature on the surface of the core. [Ans. 56.4°C, 67.6°C]

15.12 A computer cooled by a fan contains 8 PCBs each dissipating 12 W of power. The height of PCBs is 12 cm and the length is 18 cm. The clearance between the PCBs is 0.3 cm. Cooling air is supplied by a fan of 15 W power is mounted at the inlet. If the rise in air temperature is to be limited to 15°C, determine (*a*) flow rate of air that the fan should deliver (*b*) the fraction of temperature rise due to the heat input of fan and (*c*) the highest allowable air inlet temperature if the temperature of component surface is not to exceed 90°C anywhere in the system.

[Ans. 0.00735 kg/s or 0.00631 m^3/s, 13.5%, 59.8°C]

15.13 An array of power transistors are to be cooled by mounting them on a surface of size 20 cm × 20 cm, by blowing air over the surface at a velocity of 3 m/s. The average temperature of the surface is not to exceed 60°C. The temperature of the air is 30°C. Heat dissipation by transistors is 2 W each. Neglecting heat transfer on the back side, determine the number of transistors that may be mounted. [Ans. 9]

15.14 A cold plate that supports 10 power transistors each dissipating 40 W is to be cooled with water. The water temperature should not increase by more than 4°C and the velocity in the pipe should not exceed 0.5 m/s. Assuming the water cooling is for 75% of the heat dissipated determine the rate of water flow. Also calculate the pipe ID. If the case to liquid thermal resistance is 0.04°C/W and water entry is at 25°C determine the temperature of the device.

[Ans. 1.08 kg/mm, 0.68 cm, 37°C]

15.15 A sealed electronics box has water flowing through channels on two of its sides. The power dissipation is 2000 W. If the water temperature rise should be limited to 3°C, determine the water flow rate. Also calculate the cooling water used up per year of 365 days of 24 hours.

[Ans. 0.1595 kg/s, 5.03 × 10^6 kg/year]

15.16 A logic chip used in a computer dissipates 4 W of power. The heat transfer surface area is 0.3 cm^2. If the surface is to be maintained at 70°C while being cooled by immersion in a dielectric fluid at 20°C, determine the necessary heat transfer coefficient. What type of cooling will be necessary to obtain such value of convection coefficient. [Ans. 2667 W/m^2 °C, Boiling process]

16 CHAPTER

Heat Pipes

16.0 INTRODUCTION

The concept of heat pipe was introduced in 1942 by Gaugler. Only after Grover independently invented the working of heat pipe during early 1960s it started drawing the attention of researchers. During 1960-1980 intensive investigations resulted in rapid development and commercial production and use of heat pipes.

Heat pipes are in extensive use now in cooling of electronic components as well as temperature control. The importance of heat pipes can be seen from the fact that millions of them are being mass produced for cooling of electronic components. The popularity of heat pipe is due to its high conductance for heat as compared with the best heat conducting metals like copper and silver.

The popular standard heat pipe is of cylindrical shape and are mainly used for cooling. Now according to the application heat pipes are being made in different shapes.

16.1 HEAT PIPE

The forerunner to heat pipe is the thermosyphon, a schematic of which is shown in Fig. 16.1. It is a cylindrical tube into which after evacuating the air a small amount of liquid is placed and sealed.

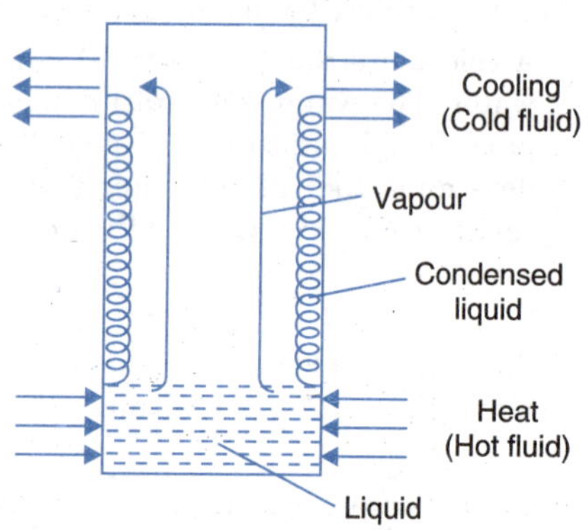

Fig. 16.1. Thermosyphon

When heat is applied to the liquid at the bottom (Evaporator) liquid vapourises and vapours move to the top. At the top (condenser) by cooling, the vapour is condensed and the condensate flows down along the container wall. In this way heat is transferred from the hot fluid to the cold fluid. The high value of conductance is due to the high value of latent heat of the fluid. Also heat is transferred with almost constant tempe-rature. The difference in temperature between the condenser section and the evaporator section is only about two or three degrees. Such a device has been in use for a long time. The main drawback of the thermosyphon is that it is gravity driven. The condensed liquid flows down due to gravity and the unit will work only in the vertical position with condenser at the top. Also it cannot work in zero gravity situation.

The schematic of heat pipe is shown in Fig. 16.2.

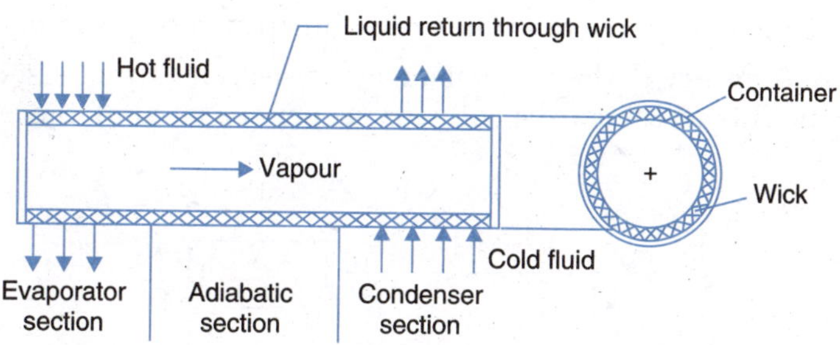

Fig. 16.2. Heat pipe

In the case of the heat pipe the condensed liquid flows by capillary action through the wick to the evaporator. The flow is independent of gravity and the heat pipe will function in any orientation. It can also function under zero gravity conditions. This is the reason why heat pipes are being used in satellites and space shuttles.

The heat transfer capacity of the heat pipe when compared to copper is larger by several hundred times. By the use of different fluids, heat pipes can operate from 4 K through 2500 K.

Helium is used in 4 K range and liquid metal like sodium is used at the higher temperature range.

In between fluids like ammonia, water, halogens and hydrocarbons, and liquid nitrogen and many other fluids are used as per the temperature requirements. Example 16.1 will illustrate the high conductance of heat pipe.

Example 16.1: *A heat pipe with ID 15 mm and length 250 mm using water as the fluid evaporates 0.2 grams per second. The temperature difference between the ends is 2°C. Compare the heat flow rate with that of copper of the same diameter over the same length. The thermal conductivity of copper is 386 W/mK. The latent heat of water is 2258 kJ/kg.*

Solution: Heat flow through 0.2 g of water being evaporated and condensed

$$= \frac{0.2 \times 2258 \times 10^3}{10^3} = \mathbf{451.6\ W.}$$

Heat conducted by the copper rod of same section over the same length

$$= \frac{338 \times 0.15^2 \times \pi}{0.25 \times 4} \times 2 = \mathbf{0.2728\ W.}$$

The ratio is (451.6 / 0.2728) = **1655.4**

So the capacity of the heat pipe is 1655.4 times that of copper of the same sectional area and length.

Important capabilities of heat pipes are listed below:

(*i*) Very high effective thermal conductance.

(*ii*) The ability to act as a flux transformer.

(*iii*) Function as an isothermal surface of low impedance.

(*iv*) Can be made to have variable impedance.

(*v*) Flexibility of mounting.

(*vi*) Heat transfer over long distances possible.

(*vii*) Can function as thermal diode or switch.

(*i*) **The high thermal** conductance has been illustrated by Example 16.1.

(*ii*) **Flux transformer:** If the condenser cross-section is increased as indicated in Fig. 16.3, the flux at condenser will vary as per the circumference of these section.

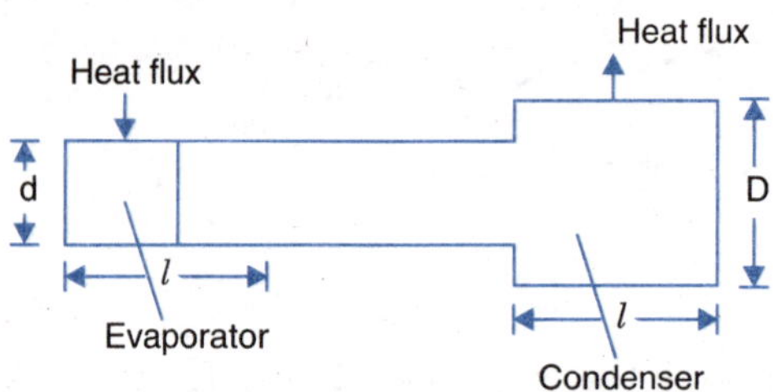

Fig. 16.3. Flux transformer

The heat flux at the smaller diameter is received in the area $\pi d \times l$. The same is transferred over the area πDL.

Thus transforming ratio is $(\pi DL/\pi dl) = (DL/dl)$

Example 16.2: *The diameter of a heat pipe evaporator is 20 mm and heat flux of 10 W/cm² is added over a length of 40 mm. For flux transforming, the condenser section diameter is increased to 40 m and cooling length is also increased to 60 mm. Calculate the heat flux at the condenser.*

Solution: The heat flow at evaporator (using cm unit)

$$= \pi \times 2 \times 4 \times 10 = 80\pi \text{ W.}$$

This heat has to flow out at the condenser.

Condenser area $= \pi \times 4 \times 6 = 24\pi \text{ cm}^2$

$$\therefore \qquad \text{Heat flux} = \frac{80\pi}{24\pi} = \mathbf{3.3\ W/cm^2.}$$

The reverse is also possible. The lower heat flux at the evaporator can be increased to a higher value by having a larger diameter at the evaporator and a smaller diameter at the condenser.

(*iii*) **Isothermalisation:** The condenser surface will tend to operate at uniform temperature. Even if a higher heat load is applied, more vapour will condense and maintain the temperature at the original level.

Heat pipes are also used in chemical reactors, solar heat sources and satellites to equalise the temperature over the whole volume, when the heat generation is non uniform over the volume or heat being received at one side in the case of a satellite.

(*iv*) **Variable thermal impedance:** Ordinarily a heat pipe will operate at steady condition as along as the heat load is constant at the design value. In case the heat load is increased, the evaporator will produce vapour as per the heat load. But the condenser will not be able to condense all the vapour produced. The uncondensed vapour accumulation will increase the pressure in the pipe. So when variable heat load is to be met, the regular heat pipe will not operate steadily. A modification which varies the condenser area internally will only be able to maintain the condition. This is achieved by having a chamber of larger size connected to the condenser with the gas volume much larger than the heat pipe volume. As the heat load changes, due to the change in pressure, the gas volume changes increasing or decreasing the condenser area as per the load change. This leads to constant pressure/temperature operation even when the load changes.

(*v*) **Flexibility of mounting** and heat rejection at a distance.

This is achieved by an arrangement called loop heat pipe (LHP). In this case the condenser and evaporator are connected by two separate tubes, one for the vapour flow and another for liquid flow. These pipes can be very long, can be even several meters. The evaporator is placed on the heat source.

The condenser can be placed at a convenient location removed from the heat source. This type is particularly useful when heat is to be removed from delicate equipments which need to be protected from the surroundings.

(*vi*) **Diode and switch:** Generally a heat pipe can be reversed with the evaporator functioning as condenser and the condenser as evaporator. There are occasions where irrespective of the flow conditions, heat should flow only in one direction. Consider the case where the evaporator side fluid flow instead of being hot becomes cooler, then in the ordinary heat pipe the evaporator will tend to act as condenser and heat flow direction will be reversed. To avoid this, special construction leads to stoppage of heat flow in the reverse direction.

These types of heat pipes will be discussed in detail with suitable schematics, in the section titled "special types of heat pipes."

The cylindrical shape heat pipe used purely for heat transfer has evolved into not only to different shapes but also to different functions.

16.2 COMPONENTS AND MATERIALS OF HEAT PIPES

16.2.1. Introduction

The main components of heat pipes are only three in number. These are

1. The working fluid,
2. The wick and capillary structure, and
3. The container.

As heat pipes are required to work at various temperatures from 4 K to 2300 K, different fluids are required for different temperature bands. For example, Helium is used in the low temperature range of 4 K. Liquid oxygen, Nitrogen are used in the next band. Water, Ammonia and alcohols are used in the 300 to 500 K range. Special materials are required in the 600–900 K range. Above this temperature sodium, lithium and similar materials are used. More than two dozen fluids are in use as working fluid. Similarly, the arrangement and material for wicks also vary as per the application. In order to be compatible with the various fluids more than a dozen materials are in use.

Difficulty always arises in the selection of suitable combination of materials. These conflicts are mainly the material deterioration of one component due to the other component.

16.2.2. Working Fluids

The primary consideration in the choice of the working fluid for an application is the operating vapour pressure and temperature range. Within the chosen temperature range several fluids may be available. A number of other characteristics should be examined before the final selection is made, so that the most acceptable one is identified. The important characteristics to be considered are:

(*i*) Compatibility with wick and container materials

(*ii*) Good thermal stability

(*iii*) Wettability of wick and wall surfaces

(*iv*) Suitable vapour pressure, not too high or too low

(*v*) Large latent heat

(*vi*) Good thermal conductivity

(*vii*) High surface tension

(*viii*) Low viscosity of liquid as well as vapour

(*ix*) Acceptable freezing and pour points.

At any temperature there is also a limitation for the heat flux that can be used. These limitations are due to (*i*) viscosity of the fluid (*ii*) high vapour flow velocity (*iii*) capillary pressure available to support liquid flow back and (*iv*) entrainment of liquid in the vapour flowing to the condenser. A suitable fluid in other respect may not be chosen due to the above restrictions.

A list of fluids and their useful temperature range is given in Table 16.1 The choice of fluid for various temperature range is obvious from the list.

Table 16.1 List of working fluids for Heat pipes

Medium	*Boiling point, °C*	*Useful range, °C*
Helium	– 261	– 271 to – 269
Nitrogen	– 196	– 203 to – 160
Ammonia	– 33	– 60 to 100
Pentane	28	– 20 to 120
Acetone	57	0 to 120
Ethanol	78	0 to 130
Heptane	98	0 to 150
Methanol	64	10 to 130
Flutec PP2	76	10 to 160
Water	100	30 to 200
Toluene	110	50 to 200
Thermax	257	150 to 350
Mercury	361	250 to 650
Caesium	670	450 to 900
Potassium	774	500 to 1000
Sodium	892	600 to 1200
Lithium	1340	1000 to 1800
Silver	2212	1800 to 2300

Compatibility among the three components decides the life of the heat pipe. The production of non condensible gases is one of the problems. For example in the combination of water and steel hydrogen gas is formed during operation. The gases tend to accumulate at the condenser and reduce the capacity of the condenser and after sometime the heat pipe function ceases.

In most cases of failure the cause was identified as gas formation.

Another problem is corrosion and mass diffusion of one material into another. If the wick or container material diffuses into the fluid, the solid is deposited at the condenser. This results in blocking of the pores of wicks and also hot spots weakening the wall.

A considerable amount of compatibility data has been accumulated by long years of research. These may be taken only as a guidance, because the test conditions and the operating conditions may not be identical. So also the manufacturing process. Manufacturers often resort to•test samples of their products for compatibility.

Ammonia is found to be compatible with all metals except copper and alloys containing copper.

Copper is found to be compatible with all fluids except ammonia. Water is found to be compatible only with copper.

Another condition in some cases is that these are compatible only below a certain temperature. For example, Aluminium is compatible with Benzene, *n*-Heptane below 150°C and with ammonia below 60°C. Sodium is compatible with stainless steel but not with titanium. So also potassium. Due to the importance of heat pipe for use in satellites, and nuclear installations, extensive investigations about compatibility have been carried out. For additional information reference can be made to Heat pipes by Dunn.

Thermal stability is another primary requirement for long life. Organic fluids are particularly poor as these degrade under exposure to temperature over long periods and due to temperature excursion. Such reactions lead to gas formation, reduction in surface tension which reduces the capilliary head.

In order to cause flow of liquid by capillary action, the liquid should wet the wick and the container material. The contact angle at the surface should be zero. Mercury is poor in this aspect. Additive may improve wettability, but additives may be left behind at the evaporator as these may not evaporate at the same rate as the fluid.

In the operating range, if the vapour pressure is low the volume flow will be larger and so the flow velocity may be high. This may lead to sonic limit of heat flux. Higher vapour pressure will require higher head for vapour flow and may lead to larger temperature drop between the evaporator and condenser.

Heat is conducted from outside through the wall material and then the liquid soaked wick. The conductivity through wick is made up of the material of the wick and the fluid soaking the wick. The fluid should have a higher conductivity in order to reduce the thermal resistance for heat flow.

High latent heat is desirable because for the same mass flow more heat will be transported. The size of the heat pipe can be smaller for the same heat load. The flow velocity is also lowered due to reduced mass flow, resulting in lower loss of head. Water is high in the list as the latent heat of water is higher compared to all the other liquids. Ammonia is next to water. Hydrocarbons are poor in this aspect.

Lower viscosity will lead to lower pressure drop for vapour flow as well as lower capillary head requirement.

High surface tension will provide higher capillary head which is a primary requirement for the operation of heat pipe.

Instead considering a single property, fluids are evaluated by a combination of properties influencing the working of heat pipes called **Merit number,** it is defined as $\sigma_l L\rho_l/\mu_l$. σ, ρ, μ and L being the surface tension, density viscosity and latent heat of the fluid. The dimension of Merit number is

$$\frac{N}{m}\cdot\frac{J}{kg}\cdot\frac{kg}{m^3}\times\frac{m^2}{Ns} = \frac{J}{sm^2} = W/m^2.$$

Effectively merit number indicates the heat transport capacity of the fluid. The sectional area required for transporting a given thermal load is directly proportional to the merit number.

Example 16.3: *Calculate the Merit number at 60°C for (i) Water (ii) Ammonia and Flutec PP9.*

Solution: The property values of the three fluids at 60°C is tabulated below:

Fluid	*Density, ρ kg/m³*	*Latent heat, L, J/kg*	*Surface tension, σ, N/m*	*Viscosity Ns/m²*
Ammonia	545.2	1026×10^3	1.367×10^{-2}	0.015
Water	983	2359×10^3	6.62×10^{-2}	0.036
Flutec PP9	1891	90.2×10^3	1.24×10^{-2}	0.094

Merit Number

For Ammonia $= \dfrac{1.367\times10^{-2}\times1026\times10^3\times545.2}{0.017}$ = $\mathbf{449.8 \times 10^6\ W/m^2}$

or **44.98×10^3 W/cm²**

For Water $= \dfrac{6.62\times10^{-2}\times2359\times10^3\times983}{0.036}$ = $\mathbf{4.26 \times 10^9\ W/m^2}$

or **426.42 W/cm²**

For Flutec PP9. $= \dfrac{1.24\times10^{-2}\times90.2\times10^3\times1891}{0.094}$ ≐ $\mathbf{22.5 \times 10^6\ W/m^2}$

or **0.225 W/cm².**

Example 16.4: *Determine the evaporation rate per second for a heat pipe of 100 W in the case of Ammonia, Water and Flutec PP9.*

Solution: Ammonia: mass flow per second = $\dfrac{100\ J/s}{1026\times10^3\ J/kg}$ = $\mathbf{97.46 \times 10^{-6}\ kg/s}$

or **0.35 kg/hr.**

Water = **$100/2359 \times 10^3 = 39.4 \times 10^{-6}$ kg/s** or **0.141 kg/hr.**

Flutec PP9 = **$100/90.2 \times 10^3 = 1.11 \times 10^{-3}$ kg/s** or **3.99 kg/hr.**

The Merit number for various fluids is given in Table 16.2

Table 16.2 Merit number of various fluids at the temperature indicated

Material	*Temperature °C*	*Merit number W/m²*
Helium	– 270	170.5×10^3
Nitrogen	– 180	74.8×10^6
Ammonia	60	449.8×10^6
Water	60	4.26×10^9
Pentane	60	134.7×10^6
Heptane	60	123.8×10^6
Acetone	60	364.6×10^6
Flutec PP9	60	22.5×10^6
Flutec PP2	50	17.7×10^6
Methanol	30	377.9×10^6
Ethanol	30	166×10^6
High temperature organic	300	168.6×10^6
Mercury	400	1.63×10^9
Cesium	625	2.1×10^9
Potassium	700	8.0×10^9
Sodium	1000	1.8×10^9
Lithium	1430	64.87×10^{15}

16.2.3. The Wick or Capillary Structure

Wicks differentiate heat pipes from thermosyphons. Wick is a thin layer material with fine pores placed on the inside walls of heat pipes. Wicks provide the capillary head to transport the liquid from condenser to the evaporator. There are two types of wick arrangement namely homogeneous structure and arterial structure. Homogeneous structure is used for normal heat loads. Arterial arrangement is used for high power units. In some cases an arrangement called open channels where longitudinal channels are machined on the inside wall of the heat pipe is also used.

Some forms of wick arrangement is shown in Fig. 16.4.

The important characteristics for an ideal wick for use in heat pipe are:

(*i*) Generate capillary pressure to transport the condensed liquid to the evaporator.

(*ii*) Distribute the liquid around the evaporator section where heat is received.

(*iii*) Permit maximum flow under the head developed.

(*iv*) Should have high thermal conductivity.

(*v*) Compatible with both working fluid and container material.

(*vi*) Should wet the working fluid.

(*vii*) Should be easy to fabricate.

(*viii*) Easily available and cheap.

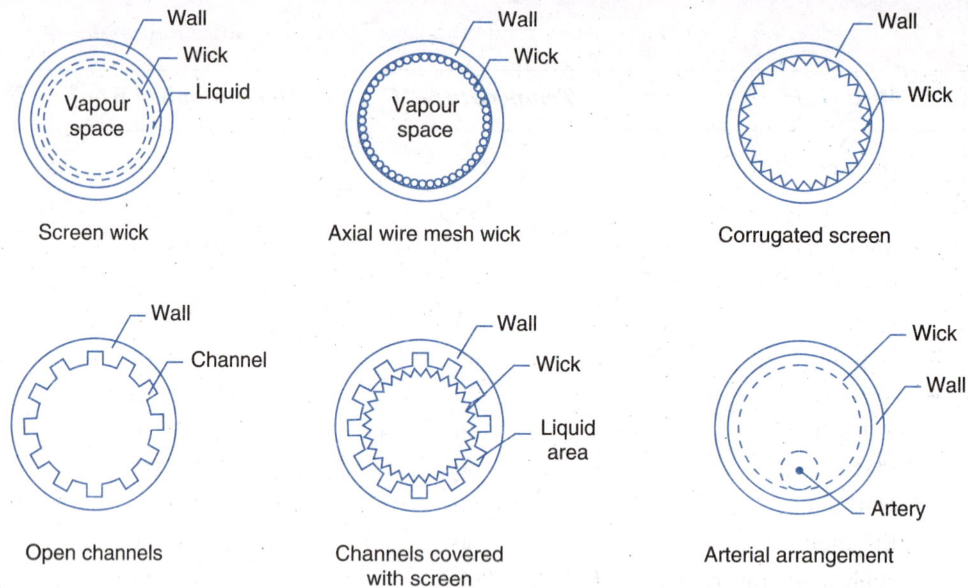

Fig. 16.4. Different wick arrangements

Some of the requirements may conflict with another requirement. For example, for high capillary head the pore size should be small. For maximum liquid flow the pore size should be larger. In such cases an optimum value may be chosen. In the case of high power units, the head as well as flow should be high. So some special arrangements like arterial structure are used.

The wick and the fluid add additional resistance to heat transfer in both condenser and evaporator. It will be desirable to use materials with high conductivity. Unless the wick material wets the working fluid high capillary head cannot be developed.

In the central area of heat pipe the vapour produced flows up to the condenser. Liquid flows along the wick from condenser to evaporator. When high vapour velocities are encountered, the vapour entrains the liquid and carries droplets of liquid back to the condenser. The evaporator will be stared and the capacity reduces drastically. To avoid this, paths are separated such that entrainment is eliminated. This is particularly important for high power units. Arterial wick arrangement is used for this purpose. The liquid is collected into a smaller area. The idea is well illustrated in Fig. 16.5.

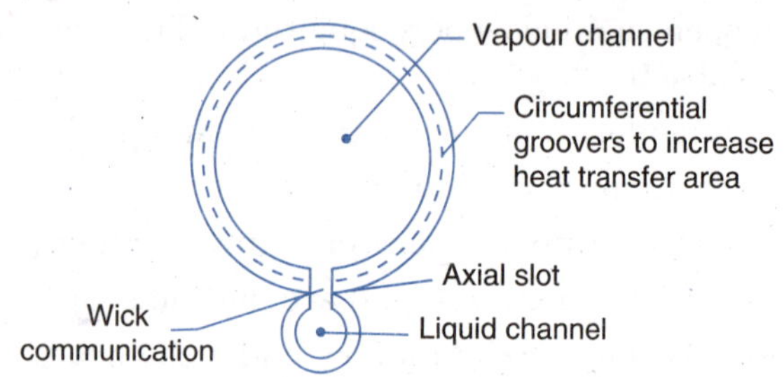

Fig. 16.5. Monogroove heat pipe

The heat pipe consists of two axial channels, one for vapour and one for the liquid flow. A narrow axial slot between these two channels creates a high capillary pressure difference. Circumferential grooves are used for liquid distribution and increased heat transfer, separation of flow paths reduces entrainment. For wicks metal meshes, twills, felt, metal foam, sintered metal powder, metal fibre, and carbon fibre are some of the materials used. By special manufacturing techniques, it is possible to obtain the required permeability, porosity and pore diameter. Table 16.3 gives details about the effect of pore size and capillary head; obtained by various researchers.

Table 16.3 Pore size and capillary head

Material and mesh size	*Capillary head (cm)*	*Pore radius (cm)*	*Porosity (%)*
Monel beads	14.6	0.026	40
30–40	39.5	0.0095	40
70–80	64.6	0.0065	40
100–140	75.0	0.0045	40
Felt metal			
FM 1006	10.0	0.004	–
Nickel fibre 0.01 mm dia	> 40.0	0.001	68.9
Copper powder sintered	156.8	0.0009	52
Nickel–50	4.8	–	62.5
Copper–60	3.0	–	–
Oxidised			
Nickel–120	7.9	0.0019	–
Nickel–200	23.4	0.004	68.9

Thermal resistance of wick structure

The resistance is due to both the solid meshes and the liquid.

For $$k_{\text{eff}} = \left(\frac{\beta - \varepsilon}{\beta + \varepsilon}\right) k_l \quad \text{...(16.1)}$$

where $$\beta = \left[1 + \left(\frac{k_s}{k_l}\right)\right] \Big/ \left[1 - \left(\frac{k_s}{k_l}\right)\right]$$

where ε is the volume fraction of solid in the wick, k_s is conductivity of solid and k_l is conductivity of liquid.

For sintered wicks:

$$k_{\text{eff}} = \left[\frac{2 + (k_l / k_s) - 2\varepsilon \{(1 - (k_l / k_s))\}}{2 + (k_l / k_s) + \varepsilon \{(1 - (k_l / k_s))\}}\right] \quad \text{...(16.2)}$$

For concentric annulus: $k_{\text{eff}} = k_l$. ...(16.3)

16.2.4. The Container

The container isolates the working fluid from the outside environment. It should be leak proof and should be able to withstand the stresses due to pressure difference. If should also offer low resistance to heat transfer with the outside. Some of the factors to be considered when selecting the material for heat pipes are:

1. Compatibility with working fluid, wick and environment.
2. Higher strength
3. Good thermal conductivity
4. Ease of fabrication (Weldability, machinability etc.)
5. Porosity
6. Wettability.

High strength is required for operation at higher pressures and temperatures. Some of the wick materials are:

Aluminium, Brass, Copper, Glass, Nickel (0–100°C), Mild steel and Nickel steel.

Though there are only three components for a heat pipe, it is rather difficult to choose the combination for various applications.

Example 16.5: *In a wick using meshes ammonia is used as working fluid. The conductivity of the liquid at 60°C is 0.255 W/mK. Stainless steel with a conductivity of 48 W/mK is used as wick material. The volume fraction of solid is 52%. Determine the effective conductivity of the wick.*

Solution:

$$k_{eff} = \frac{\beta - \varepsilon}{\beta + \varepsilon} \times k_l.$$

$$\beta = \left[1 + \frac{48}{0.255}\right] \Big/ \left[1 - \frac{48}{0.255}\right] = -1.01068$$

$$\frac{\beta - \varepsilon}{\beta + \varepsilon} = \frac{-1.01068 - 0.52}{-1.01068 + 0.52} = 3.1195$$

$\therefore$ $$\mathbf{k_{eff}} = 3.1195 \times 0.255 = \mathbf{0.7955\ W/mK.}$$

Example 16.6: *A sintered wick of stainless steel of conductivity 48 W/mK is used with ammonia as working fluid. The conductivity of ammonia is 0.255 W/mK. If the volume fraction is 52% determine the effective thermal conductivity of the wick.*

Solution:

$$\mathbf{k_{eff}} = \frac{2 + (k_l / k_s) - 2\varepsilon\,\{1 - (k_l / k_s)\}}{2 + (k_l / k_s) + \varepsilon\,\{1 - (k_l / k_s)\}}$$

$$= \frac{2 + (0.255/48) - 2 \times 0.52\left\{1 - \dfrac{0.255}{48}\right\}}{2 + (0.255/48) + 0.52\,\{1 - (0.255/48)\}} = \mathbf{0.3849\ W/mK.}$$

16.3 SPECIAL TYPES OF HEAT PIPE

A number of different type of heat pipes have been developed for use in different functions. A few of these are discussed in this article.

Variable Conductance Heat Pipe (VCHP)

The equipment to be cooled may not produce a constant heat load. At the same time its temperature has to be maintained constant for proper functioning of the equipment.

The standard heat pipe will not be able to meet this condition of constant temperature under varying heat load. The VCHP is used for such situations. The principle involved in the working of VCHP is automatic variation of condenser area so that the condenser will be able to keept the pressure constant in the heat pipe. As the pressure is constant in the heat pipe, the temperature of the equipment cooled by the heat pipe will remain constant. The working VCHP can be explained using Fig. 16.6.

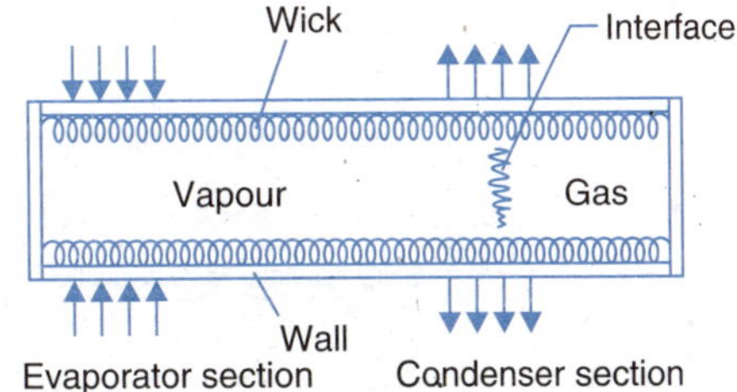

Fig. 16.6. Variable conductance heat pipe

It was observed in some high temperature heat pipes that failed, the reason for the failure was production of gas due to material incompatibility. It was also observed that the gas and the vapour did not mix and the gas has occupied the condenser space leading to failure of the heat pipe. This idea that the gas and vapour do not mix and there is a definite interface is used in the working of VCHP.

A gas is filled to a required amount in the heat pipe in addition to the working fluid. When the pipe starts operation the gas moved to a volume near the condenser. The volume occupied by the gas leaves the designed amount of condenser area for normal operation under design heat load. When the heat load increases, the evaporator will produce more vapour and the condenser will not be able to condense at the same rate. This leads to an increase in pressure which causes the gas to be compressed. The volume of the gas will be smaller and the interface will move outward, providing more area for the condenser. At this condition the condenser will be able to keep the balance. The temperature of operation will remain constant and the equipment temperature will be also constant. If the heat load decreases, the gas interface will move in and the condenser area will be reduced again maintaining the balance and the temperature.

In order to have finer control the volume of the gas filled should be larger compared to the pipe volume. In that case if constant area is used the pipe should be long.

In order to avoid this a reservoir is connected to the pipe as shown in Fig. 16.7 (*a*).

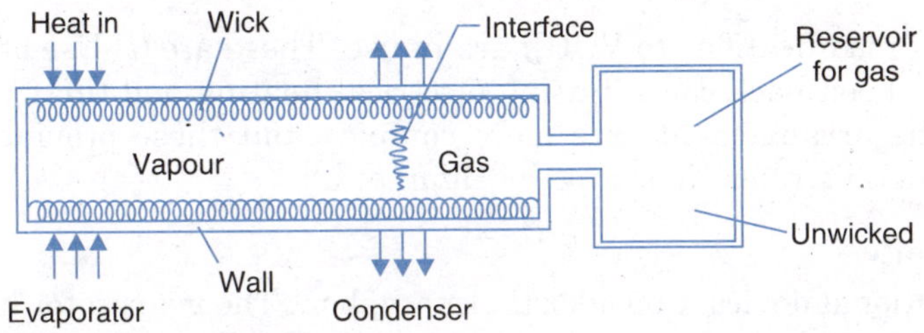

Fig. 16.7 (*a*) Cold reservoir variable conductance heat pipe

As the gas volume is large, the change in volume due to change in pressure will be high. So it is possible to obtain finer control by this arrangement. The pressure increase required for the change of certain area of condenser will be small and that is the reason for finer control.

One drawback of this arrangement is that during prolonged operation, the working fluid tends to diffuse into the reservoir and does not return to the pipe causing failure to meet the load.

To get over this problem the reservoir is also wicked so that any fluid diffusing in will be returned to the operating area. The arrangement is shown in Fig. 16.7 (*b*).

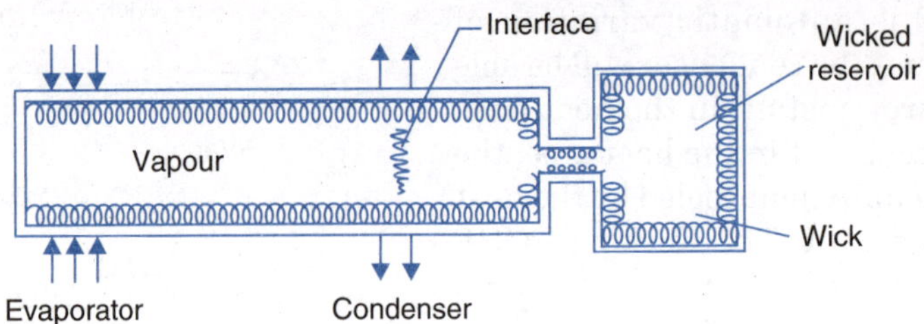

Fig. 16.7 (*b*) Wicked reservoir type of VCHP

Any vapour diffusing into the reservoir will be condensed and returned by the wick to the evaporator.

High power pipes have been successfully operated using this concept.

The cold reservoir type of VCHP is sensitive to the change in temperature of the heat sink. In order to avoid this problem hot reservoir type in which the gas reservoir arranged to be inside the pipe itself. This arrangement is shown in Fig. 16.8.

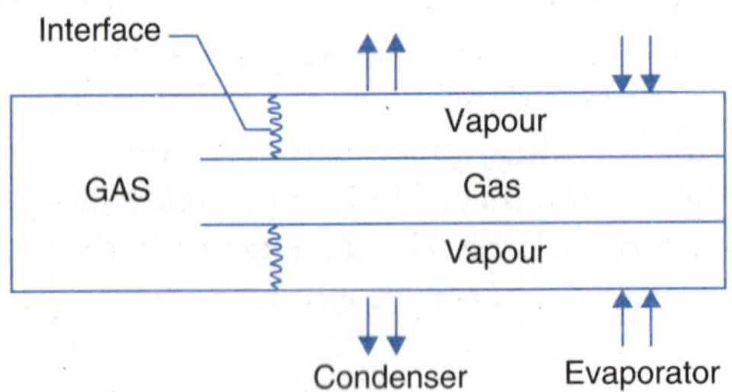

Fig. 16.8. Hot reservoir VCHP

It can be seen that the gas is inside the heat pipe and is not affected by change in sink temperature.

A number of modifications to VCHP are in use. These are (*i*) Use of bellows in place of rigid reservoir. (*ii*) Feed back control using electrical heating and (*iii*) Feed back control by auxiliary fluid. The arrangements are more complex, but these provide precise control of temperature. For details refer Heat Pipe by Dunn et al.

The Thermal Diode

Diode is a unidirectional device. Consider thermosyphon. The evaporator has to be only at the lower level and heat will be transported only upwards. The flow of heat in the other direction is not possible. In that sense, thermosyphon is a thermal diode. It is necessary in some applications like in space and passive cooling of residences using solar energy that a heat pipe should be unidirectional. For example, the occupied space can be cooled by using low temperature at the night times by using thermal diodes. During the day the heat pipe should remain in-operative.

Figure 16.9 (*a*) and (*b*) shows a thermal diode in operation. In this case the reservoir is not connected by a wick with the heat pipe. During normal mode the liquid flow is from condenser to evaporator. If the heat flow direction is reversed, the liquid will get into the reservoir chamber and will not return back and so the operation will cease. This arrangement is called liquid trap diode.

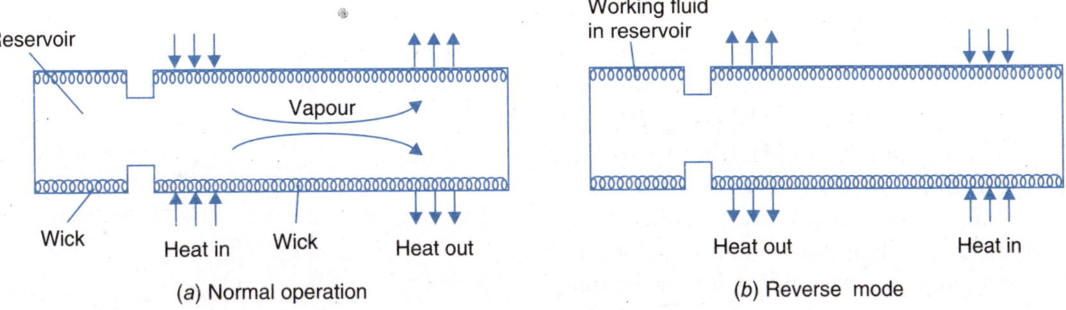

(*a*) Normal operation (*b*) Reverse mode

Fig. 16.9. Thermal diode, liquid trap type

Another type of thermal diode is called "liquid blockage" diode. In this case excess liquid is placed in the reservoir. When operation is reversed all the liquid collects in the condenser and the unit becomes in-operative. The arrangement is shown in Fig. 16.10 (*a*) and (*b*).

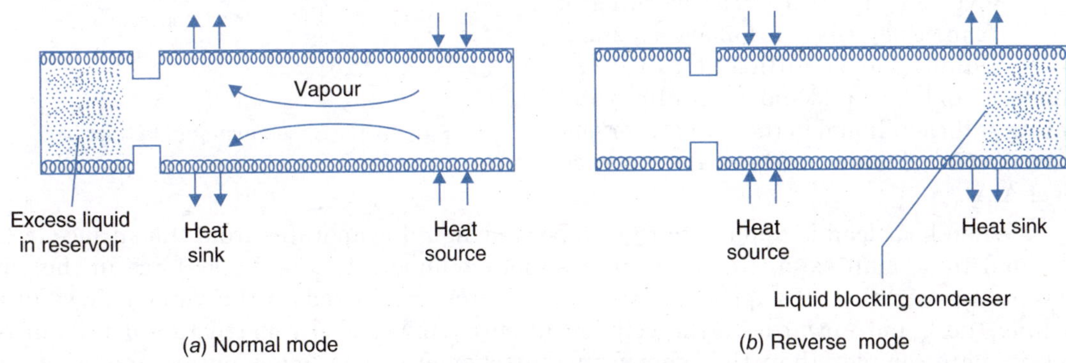

(*a*) Normal mode (*b*) Reverse mode

Fig. 16.10. Liquid blockage diode

Thermal diodes have been in use in space crafts to remove heat to outside and not let in solar heat when the side faces the sun. Thermal diodes are also finding use in renewable energy systems. Diodes are used for passive cooling of space by use of a wall with thermal diode heat pipes fixed in them as shown in Fig. 16.11. During night heat will flow out through the heat pipes. During day the pipes will be in active.

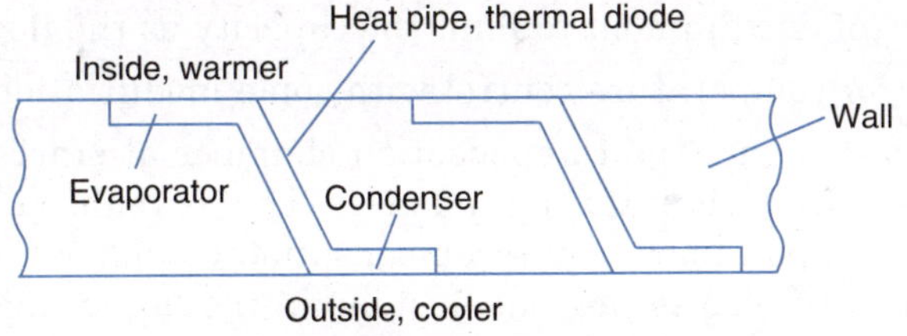

Fig. 16.11. Passive cooling system

During winter solar heat can be collected by a similar arrangement.

Loop Heat Pipes and Capillary Pumped Loops

The operation of a heat pipe depends upon the capillary head within the wick to overcome the pressure drops required for vapour flow and also gravitational head in case evaporator is placed above the condenser. To generate such capillary head the wick should extend the full length of the conventional heat pipe, particularly when heat is to be transported over a distance. In order to obtain a larger capillary head, the pore diameter in the wick should be smaller. This increases the pressure required to transport the liquid. The length of the heat pipe is limited due to these conflicting requirements.

Loop heat pipes (LHP) and capillary pumped heat pipes were developed to overcome the inherent problem of incorporating a long wick. The development of these heat pipes was mainly for use in space applications for higher capacities. The advantages of the two phase loops with small diameter piping systems with no wicks have led to their use for lower capacities also.

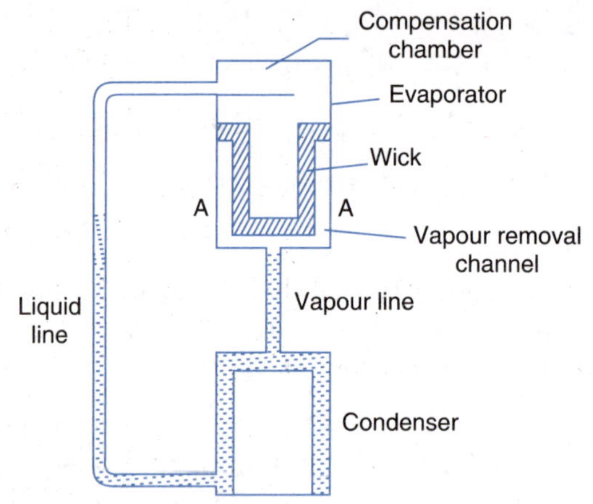

Fig. 16.12. Schematic of loop heat pipe

The schematic of a simple loop heat pipe is given in Fig. 16.12. The working of the LHP can be described as follows. At start-up the liquid load is sufficient to fill the condenser and the liquid and vapour lines and there is sufficient liquid in the evaporator and compensation chamber to saturate the wick (level A-A).

When heat load is applied to the evaporator liquid evaporates from the surface of the wick and to a small extent in the compensation chamber. The capillary force in the wick prevents upward flow of vapour. As the vapour pressure increases the vapour flows down pushing the liquid upwards to the compensation chamber. The construction details of the wick is more complex than that shown in the schematic. The following are the additional advantages of LHP over conventional heat pipe.

(*i*) High heat flux capability

(*ii*) Capability to transport heat over long distances without restriction on the routing of the vapour and liquid lines

(*iii*) Ability to operate over a range of gravitational values

(*iv*) No wick in the liquid and vapour transport lines

(*v*) No possibility of entrainment to limit the capacity as the flows are in separate lines

(*vi*) Can be used for temperature control with some modifications.

For the liquid to flow into the compensation chamber at start up, the pressure in the chamber should be lower than that at the condenser. This is realised by a small temperature difference between the evaporator and the compensation chamber. The liquid flowing out of condenser should be undercooled to prevent flashing in the liquid line.

The ability of LHP to transport heat over tortuous flow paths over a significant distance and its high heat flux capacity is advantageously used in cooling of electronic equipments. In these cases miniaturisation has played a significant role. The capillary pumped loop differs from LHP in that compensation chamber is separate from the evaporator. A schematic of CPL is shown in Fig. 16.13.

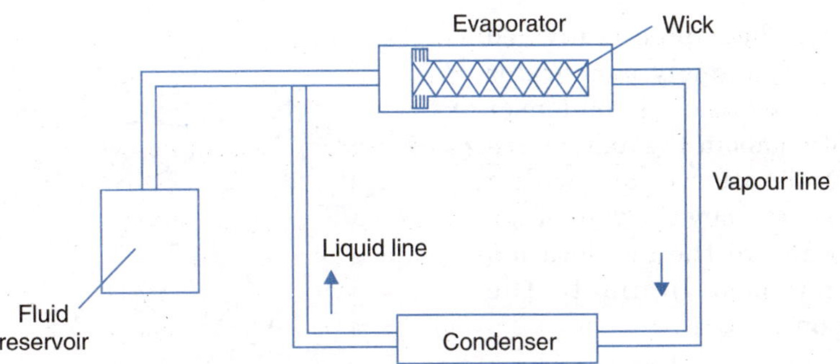

Fig. 16.13. Capillary pumped loop

Before starting of the operation of CPL, the two phase reservoir must be heated up to the required operating temperature. On reaching the operating temperature in the reservoir the internal pressure will rise and the entire loop will be filled with liquid. This operation is called pressure priming.

After priming the CPL is ready for operation. When heat is applied to the capillary evaporator the fluid temperature first rises to the boiling point and then vapour begins to form. After enough evaporation takes place a meniscus is formed at the liquid vapour interface which develops the capillary pressure to cause liquid circulation. Steady state operation is similar to that of LHP.

A set of geometric characteristics of a CPL are given below:

Evaporator: Total length 90 mm, Active length 60 mm.

Diameter: OD 19 mm, ID 12.7 mm

Material: Aluminium alloy 6063 ASTM.

Wick: Polyethylene, mean pore radius 20 μm, porosity 60%.

Vapour line: OD 6.35 mm, ID 4.35 mm, length 200 mm, Stainless steel grade 316 L ASTM.

Condenser: OD 6.35 mm, ID 4.35 mm, length 800 mm, Stainless steel grade 316 L ASTM.

Liquid line: OD 6.35 mm, ID 4.35 mm, length 200 mm, Stainless steel 316 L ASTM.

Two phase reservoir: OD 25.4 mm, length 200 mm, volume 85 cm^3,

Screen mesh, 100-200, Stainless steel 304 L ASTM

Body stainless steel grade 316 L ASTM.

Rotating Heat Pipes

The rotating heat pipe is a two phase thermosyphon in which the condensed liquid is returned to evaporator by means of centrifugal force. The rotating heat pipe consists of a sealed hollow shaft, generally having an internal taper along its axial length, containing a fixed amount of working fluid. A schematics of the unit is shown in Fig. 16.14.

The rotating heat pipe is divided into three sections, namely evaporator region, adiabatic region and condenser region. The rotation about the axis causes a centrifugal acceleration of w^2r with a component $w^2r \sin \alpha$ along the wall of the pipe. The force due to the acceleration moves the condensed liquid to the evaporator section.

Centrifugal forces increase the heat transfer rate in both the regions. As the condensate is removed by these forces the heat transfer coefficient is high in the condenser. The rotating heat pipe has a thermal conductance comparable or higher than that of the capillary type. The applications of rotating heat pipes are discussed in Section 16.4.

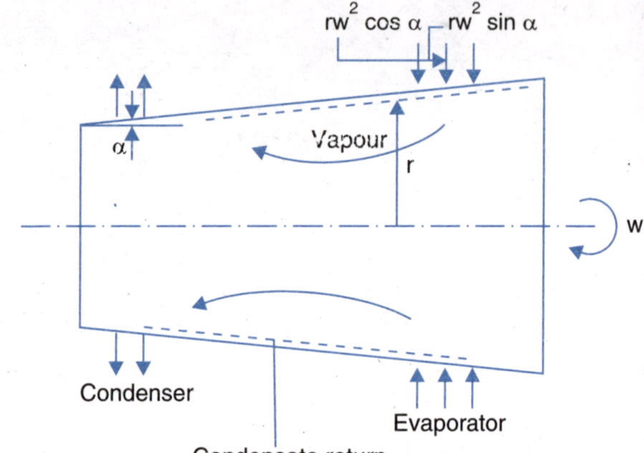

Fig. 16.14. Rotating heat pipe

Micro Heat Pipes

Conventional small diameter wicked heat pipes of circular cross-section have been in use for many years.

Diameters of 2-3 mm permit the use of wicks in them. However when the internal diameter is 1 mm or less wicking is not possible.

Micro heat pipe is defined as one in which $(r_c/r_h) \geq 1$

where r_c is the effective liquid meniscus and r_h is the diameter of the heat pipe. Another condition for defining micro heat pipe is $r_c \leq d_v < l_c$

where d_v is the lesser diameter of the vapour channel cross-section and l_c is defined as

$$l_c = \sqrt{\frac{\sigma}{g(\rho_r - \rho_v)}}$$

Some sections of micro heat pipes are shown in Fig. 16.15. The length of sides is in the range of 0.3 to 1.5 mm and of thickness around 0.1–0.2 mm.

Wicks are not used in micro heat pipes. The liquid is transported by the capillary action of the sharp corners of the pipes.

The heat transport capabilities of micro heat pipes is low. The capacity ranges from 0.25 to 4 W only. The cooling rate of micro heat pipes available in the market is considerably inferior to that offered by forced convection system for applications in cooling of electronic components. The other qualities attributed to micro heat pipes such as isothermalisation, passive operation and ability to control temperatures in environments where loads vary, show the potential for application of micro heat pipes. The thermal management of electronic systems remains a prime candidate for micro heat pipes. Micro heat pipes have also been used for cooling laser diodes of heat capacity of 250 mW. The pipes brought out a temperature reduction of 7°C in the chip. Micro heat pipes as the name indicates are smaller, lighter and of lower cost.

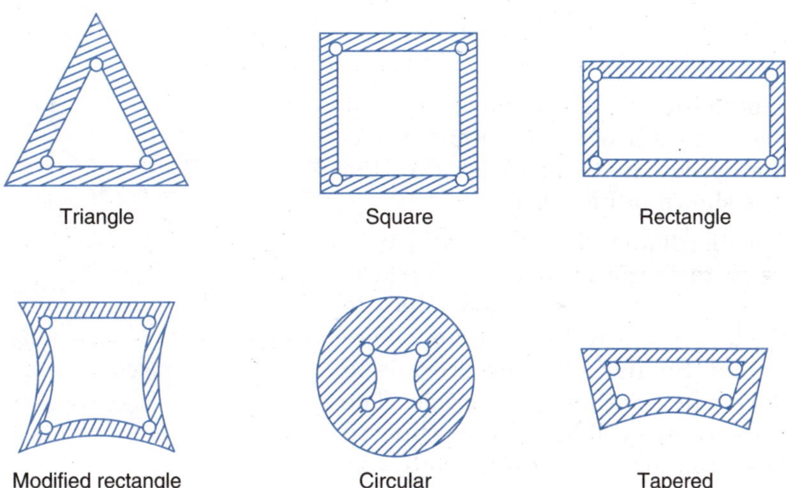

Fig. 16.15. Various sections of Micro Heat Pipes

There are many other types of heat pipes and a few of them are listed below.

(*i*) Thermal switch

(*ii*) Electro kinetics pipes

(*iii*) Electro hydrodynamic pipes

(*iv*) Micro rotating heat pipes

(*v*) The sorption heat pipes

and (*vi*) Magnetic fluid heat pipes.

Readers may refer "Heat Pipes" by Reay and Kew for details of these types.

16.4 APPLICATIONS OF THE HEAT PIPE

In the heat transfer area the application of heat pipe covers the entire range of temperatures starting from 4 K and upto 2000 K. At the lowest level Helium heat pipes are used for target cooling of particle accelerators. At the higher range in the state-of-the-art nuclear reactors.

The broad area of applications as per functions can be as listed below.

(*i*) Separation of heat source from the sink

(*ii*) Isothermalisation or temperature flattening

(*iii*) Heat flux transformation

(*iv*) Temperature control

(*v*) Thermal diodes and switches.

Heat pipes can transport heat over considerable distance with high efficiency due to their higher conductivity. This property is used for the separation of heat source from the sink. In many cases as in cooling of electronic modules it is not desirable to dissipate heat in the near vicinity of the unit. In such cases loop heat pipes can be used to carry and dissipate the heat at a distant location. This is used in thermal control of electronic equipments to a large extent.

Isothermalisation is the process of distributing the heat received or generated, uniformly over the body so that the temperature at all locations in the body will be the same.

For example a satellite in space receives heat on the side exposed to sun while the other side is exposed to cooler space. The temperature over the surface will be as shown in Fig. 16.16.

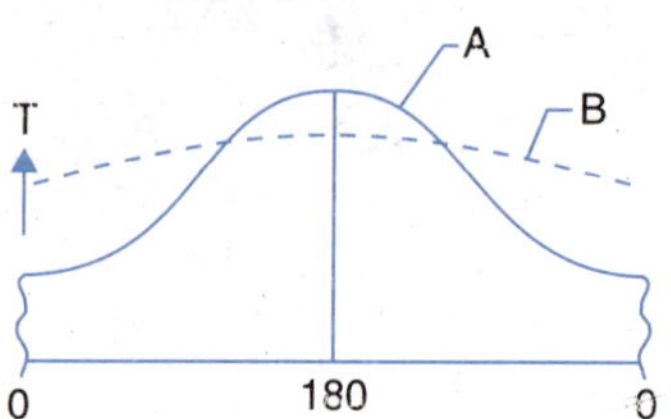

Fig. 16.16. Temperature distribution over the surface of a satellite
A -without isothermalisation
B-with isothermalisation

Without isothermalisation, there will be a large difference in temperature over the surface. This difference is reduced by using heat pipes to transport the heat received to the sections where heat is not received. Similar situation prevails in chemical reactors with exothermic reaction. The temperature at the centre will be higher compared to the sides. This situation is also corrected by use of heat pipes to distribute the heat produced evenly.

With heat pipes low intensity flux can be transformed to high intensity flux for effective utilisation. An example is the use of low heat flux generated by radioactive isotopes to heat thermionic generators. Temperature control is best carried out by variable conductance heat pipe. By mounting the evaporator on the heat source the temperature of the source can be maintained at constant level irrespective of the heat generated.

Though the first major application of VCHP was in space craft it has now become widely accepted in many more applications.

Thermal diode is used when heat flow is desired in one direction only. A case in point is passive heating of house where the low temperature air at night is used to cool the rooms. During the day there should be no heat flow in the other direction. As the criteria for applications vary to a great extent each application must be studied in its own requirements. In all cases it is necessary to ensure reliability, safety and performance. The unit should be installable easily. Investment decision is based on cost effectiveness as for any other equipment. Individual application details are discussed in the following articles.

Heat Pipes in Energy Storage Systems

Energy storage systems operate in a cyclic manner. Heat is delivered to the system during part of the cycle which increases the energy content in the system. There may be some in-active period during which heat loss should be minimum. The next part of the cycle is heat delivery to the load. In phase-change material system, the heat addition will melt the material and extraction will cause solidification. In sensible heat systems, heat is stored by increase in temperature and delivered back by reduction in temperature. In both cases the conductivity of the materials is low and the processes become slow or the extraction equipment should be large.

Heat pipes have been used extensively in a variety of energy storage system. Heat pipes are suited to thermal storage systems, in particular in the role of heat delivery and removal because of their high thermal conductivity and their passive operation. Heat pipes have also been used for heat transfer from the ground for drying of roads. These were among the mass produced heat pipe applications. For maintaining the permafrost near the supporting columns in oil pipe transport, heat pipes are placed with evaporator in the frozen ground and condenser above ground.

A schematic of heat pipe installation is shown in Fig. 16.17(*a*), (*b*).

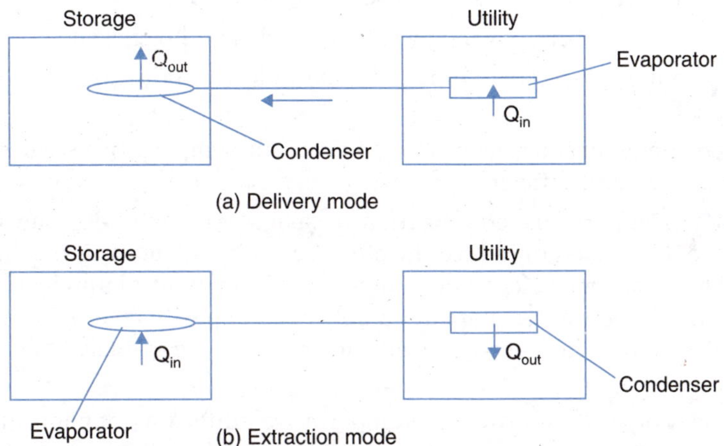

Fig. 16.17. Schematic of heat pipe application in energy storage

The heat pipe is reversible. Also the distance between the storage and the utility can be several meters. Loop heat pipes are suited in this case.

The storage can be the PCM and the utility the conditioned space.

By use of heat pipe both heat delivery into the system and extraction out of the system can progress rapidly.

Due to the separation of the condenser and evaporator heat pipes can be safely used in cooling of nuclear waste storage. The schematic is shown in Fig. 16.18.

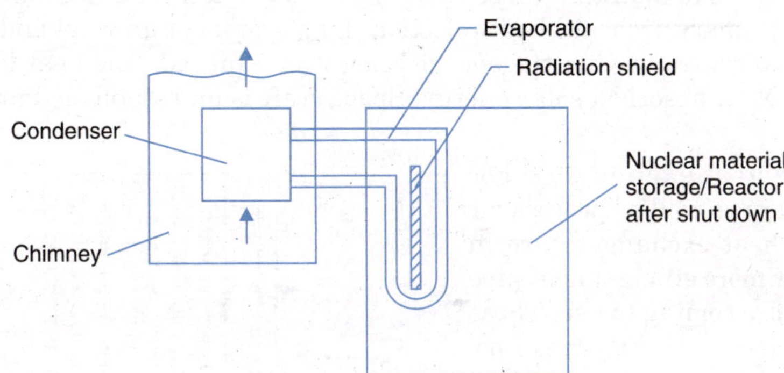

Fig. 16.18. Nuclear reactor cooling

As the main cooling system is shut down when the reactor is shut down, the heat due to stray reaction and the sensible heat of the components is removed by the heat pipe.

Heat Pipes in Chemical Reactors

Heat pipes are used for isothermalisation or temperature flattening. The advantages of using heat pipes is listed below:

1. As the chemical reaction will produce maximum yield when operating at the design temperature, isothermalisation, removing hot spots or cold spots, will maximise yield. This is achieved easily by use of heat pipes.

2. Capability to deal with high heat flux both radial and axial.

3. Heat pipe will give much faster response time compared with other methods.

4. As per the temperature, various working fluids can be chosen.

5. Variable conductance heat pipes will be able to maintain constant temperature irrespective of heat flux.

6. Heat pipes have already been used to even out temperature excursions in reactors, thereby increasing yield and efficiency.

Space craft: Temperature equalisation in space craft, whereby thermal gradients in the structure can be minimised to reduce the effects of external heating such as solar radiation and internal heat generation by electronic components is a potential application for heat pipes. The use of heat pipe connecting two vapour chambers on opposite faces of a satellite is used for reducing the temperature difference between the solar cell array facing the sun and the cold satellite face.

The use of heat pipes to improve the temperature uniformity of nonuniformly irradiated skin structure of satellites has been adopted in some cases. Feasibility studies are being made about a continuous circular heat pipe having many combinations of evaporators and condensers. This was constructed in the form of a toroid with eight heat sources and eight heat sinks. The study indicated that continuous heat pipe properly integrated with a space craft can be a highly effective means for reducing temperature differentials. Temperature equalisation is particularly important to eliminate structural distorsions in space craft.

A full scale thermal model of a space craft was built at General Electric. Heat pipe technology was applied to many of the thermal problems. Earth viewing module of a communication satellite was selected as one potential area for the application of the devices built. The high solar heat loading on the east and west panels of the satellite necessitated the use of super insulation to maintain these surfaces below the temperature limit. The north and south faces were primary areas for heat rejection. By a system of internal and circumferential heat pipes thermal control within the specification was achieved. The heat load was 38 W of internal and 170 W of absorbed solar energy. Space craft is an established area for heat pipe use.

Heat pipe heat exchangers: For waste heat recovery as well as process air heating finned heat exchangers are in common use. Now more efficient heat pipe heat exchangers are coming to use. These consist of heat pipes with external fins. The evaporator section has hot gas flow and the condenser section has flow of gas to be heated. Heat pipe heat exchangers can be in any orientation. It is also possible to adjust the inclination as per the location. A schematic of such unit is shown in Fig. 16.19.

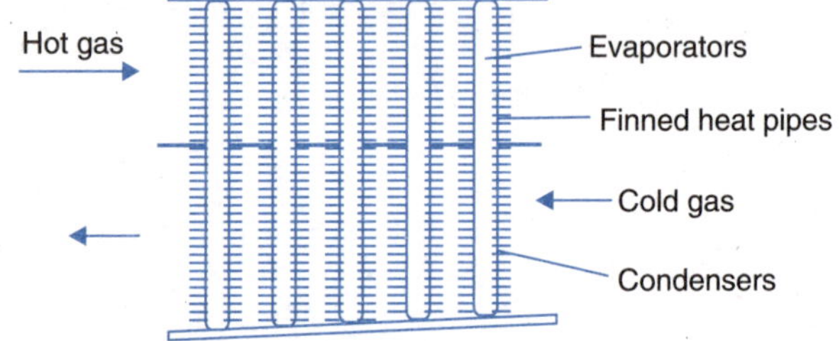

Fig. 16.19. Heat pipe heat exchanger, schematic

Some of the important features of these units are:

1. High reliability

2. No cross contamination

3. Can be made to any size and capacity

4. Fully reversible.

These units can be used for all varieties of heat recovery.

Heat pipes in renewable energy systems: Conventional systems are still more opportunities to exploit heat pipe for energy efficiency. However it is in the area of renewable energy the application of heat pipes is catching up rapidly.

The areas of applications are:

1. Solar energy collection
2. Solar energy concentration
3. Passive heating and cooling systems for buildings
4. Power generation.

The Heat Pipe Turbine:

A schematic of the unit is shown in Fig. 16.20.

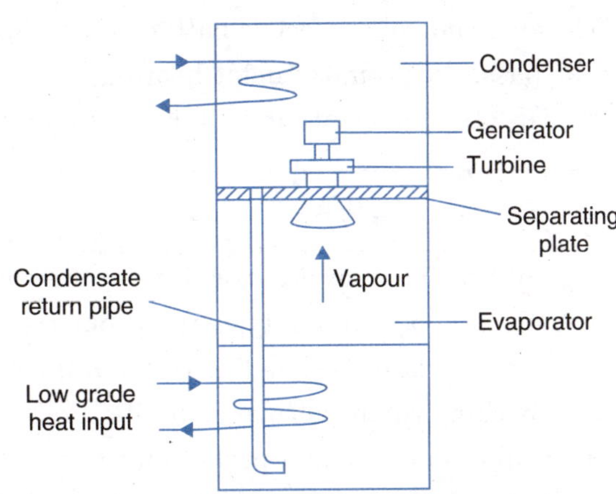

Fig. 16.20. Schematic of thermosyphon turbine

The whole unit is a thermosyphon, consisting of an evaporator section, condenser section and condensate return line, the turbine being in the adiabatic section.

The unit developed in Australia uses geothermal hot water at 55°C. The condensing temperature is 25°C. For a thermal input of 100 kW, the turbine output was 3 kW, the net efficiency being 3%. The ideal efficiency for the temperature range being 9.15%.

A range of commercial manufacturers now market heat pipe based solar water heaters, that use heat pipes to transfer heat from the collectors to the water store or circulation system. Capillary pumped loop in conjunction with solar collectors is found to overcome the limitations of conventional solar collectors.

In order to avoid thawing of permafrost and disturb the oil pipe line support columns, heat pipes have been used in a large way. About 100,000 heat pipes of 5 and 7.5 mm diameter and 9 to 18 m length have been used. The heat pipe is placed in contact with the support columns and heat is removed from the ground and dissipated by radiators on the surface. In this way the temperature of the permafrost remained intact and its temperature was 0.5°C below the normal.

Ground heat is tapped to deice roadways at critical areas by use of heat pipes with the evaporator section placed at depth in the ground and condenser below the road surface.

In the case of a bridge heat pipes were used to maintain the temperature above freezing by using energy from gas.

A list of some more applications is given below:

Heat pipe panels for warming bath room floor (Japan)

Heat pipe cooled dip stick for cooling two wheeler oil

Heat pipes in refrigerated shelves for display

Passive cooling of remote weather station

Cooling of electric motors by rotary heat pipe and cooling of drills

Thermal control of thermoelectric generators

Cooling of gas turbine blades

Cooling soldering iron bit

Cooling bearings of emergency feed water pumps in power stations

Cooling target particles of accelerators

Thermal control of electric batteries

Car passenger compartment heating

Preservation of ice around oil drilling platforms

16.5 DESIGN OF HEAT PIPES

The design of heat pipe for a particular duty involves four broad processes:

(*i*) Selection of suitable type and geometry

(*ii*) Selection of materials like the container, working fluid and wick

(*iii*) Checking the performance limits

(*iv*) Evaluation of actual performance.

As with any design process many of the decisions that must be taken are interrelated and the process is iterative. In the case of heat pipe compatibility may limit the choice of working fluid if the case material and wick material are first specified. If even after the proper selection, performance limits are not satisfied, further choices (which may not be the optimum) may have to be resorted. Two more aspects to be considered in the case of heat pipes are the amount of fluid charge and the start up of the heat pipe.

Decision about the amount of fluid charge is rather tricky. Over charge will reduce the condenser capacity while under charge may lead to drying up of the artery. One way of overcoming the problem is to provide an excess fluid reservoir which will absorb the excess fluid that is not required by the primary wick structure.

In the case of arterial wicks it is necessary to ensure that in the case of an artery becomes depleted of working fluid, it should be able to refill automatically. This is called "Priming".

The maximum diameter of the artery for **priming** is given by

$$d_a = \frac{1}{2}\left[\left(\sqrt{h^2 + \frac{8\sigma_l \cos\theta}{(\rho_l - \rho_v)\, g}}\right) - h\right]$$

where d_a is the maximum diameter of the artery

h is the vertical height to the base of the artery

θ is the contact angle of the fluid

σ_l is the surface tension of the liquid

ρ_l and ρ_v are the densities of the liquid and vapour.

The design procedure can be easily understood by a flow chart shown in Fig. 16.21.

The first detail is the specification, which include the geometry, operating temperature, heat load, orientation and other details like where it is to be used as in space.

The next step is the selection of a set of working fluid suitable for the operating temperature. With reference to the fluids chosen the container material and wick material and type are chosen. The best of the working fluid compatible with these materials is decided upon.

The type of wick capable of providing of the required capillary head is to be selected.

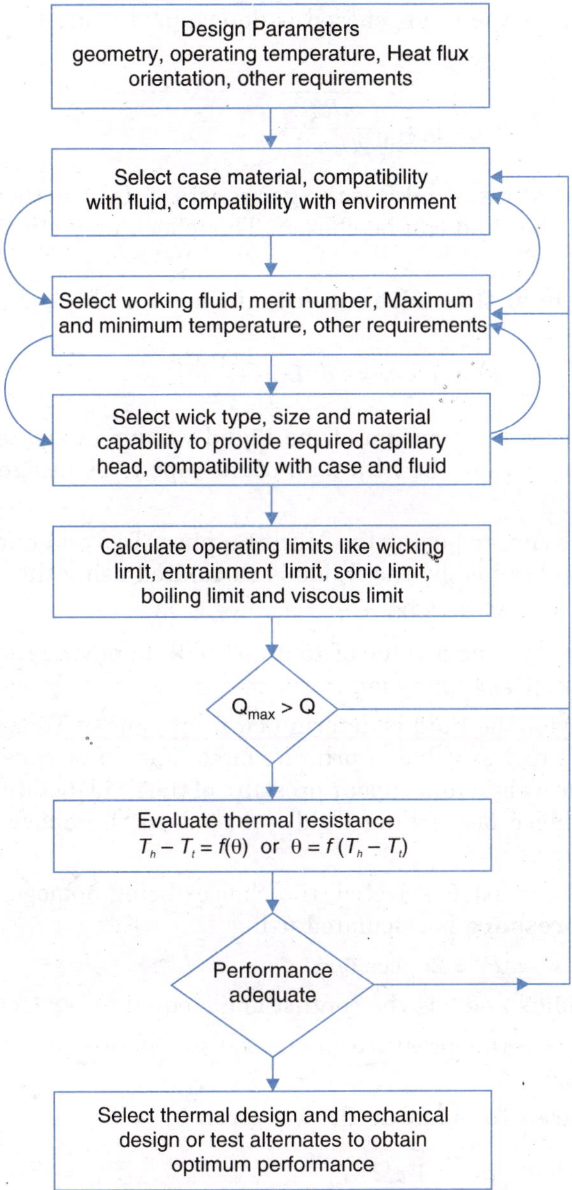

Fig. 16.21. Design procedure for heat pipes

If the requirement is not met, the cycle is repeated till the requirement is satisfied. Next the temperature drop between the condenser and evaporator is checked to be within the maximum allowable value. The detail design is taken up once these conditioned are satisfied.

The specification will give the details of the heat flux, temperature range, maximum dimensions of the heat pipe as well as the condenser and evaporator, where the unit is to work against the 'g' force. Any limitation on weight, temperature drop between evaporator and condenser will also be specified.

Suitable list of working fluids is selected next which can work in the temperature range specified.

The sonic limit for the fluids chosen is determined using the equation for maximum heat flux, given below.

$$\dot{q} = \rho_v L \sqrt{\frac{kRT}{2(k+1)}}$$

where k is the specific heat ratio and R is the gas constant. In case the heat flux is larger than the required heat flux, the fluid can be chosen. This check may eliminate some of the fluids chosen.

The maximum heat flux without entrainment is checked using equation

$$Q = A \times \dot{q} = \pi r^2 L \sqrt{\frac{2\pi \rho_v \sigma_l}{z}}$$

where z is the characteristic dimension of the liquid vapour interface having value of about 0.16. If Q equals or higher than the heat load of the pipe to be designed, the fluids pass this limit.

Wicking limit is checked using the Merit number. To avoid nucleation in the wick, the superheat that may be obtained by the fluids is calculated using the equation

$$\Delta T = (3.06\ \sigma_l\ T_{sat})/(\rho_v \times L \times \delta)$$

where δ is thermal layer, having a value of 15 μm. The fluid having the highest value for ΔT is the most suitable one for the application.

Priming factor for the fluid is determined from plots of Temperature verses Priming factor. The fluid having higher value of priming factor should be considered if it satisfies the other requirements. The weight and vapour pressure of the fluid dictate the container thickness and weight. A final choice of working fluid is based on the above parameters and other specification requirements.

The type of wick for use is selected, the choices being homogeneous wick and arterial wick. **The capillary pressure** is calculated using

$$\Delta P_c = 2\sigma_l \cos\theta / r_c$$

where r_c is the wick radius and θ is the contact angle equal to zero for wetting fluids.

The pipe will work if this pressure equals the sum of pressure drop for vapour flow and gravitational pressure.

Pressure required for vapour flow is given by

$$\Delta P_v = \frac{\mu_l\ Q\ l_{eff}}{\rho_l\ L\ A_w\ K} \quad ...(A)$$

where A_w is the wick area and K is calculated. l_{eff} is the length of pipe – 1/2 (length of evaporator + length of condenser) using equation

$$K = d_w^2\ (1 - \varepsilon)^3 / 66.6\varepsilon^2$$

where d_w is the wick diameter, ε is the volume fraction of solid phase in the wick. The value is substituted in equation (A).

The pressure required for movement against gravity

$$\Delta P_g = \rho_l g\, h$$

where h is the height of movement against gravity using equation

$$\Delta P_c = \Delta P_v + \Delta P_g$$

the wick area A_w is calculated as all quantities except A_w are known in equation (A).

In case A_w is larger than the pipe diameter, then arterial wick is the next choice.

Having chosen arterial wick, **the maximum dimension** of the artery is determined using the condition for priming. The equation is

$$d_a = \frac{1}{2}\left[\sqrt{h^2 + \frac{8\sigma_l \cos\theta}{(\rho_l - \rho_v)\, g}} - h\right]$$

When arterial wick is chosen, a wick is used in the evaporator section to distribute the liquid evenly around the circumference.

The equation for heat flow across the wick is used to determine the thickness of the wick.

$$Q = kA_e\, \Delta T/t$$

where t is the thickness. Q is the heat load, A_e is the evaporator area and ΔT is the allowable temperature drop. The conductivity k is the combined conductivities of the sold material in the wick and the working fluid in the wick.

The maximum value of pressure drop is calculated from the basic equation ΔP_c = sum of other pressure drops. The number of arteries are also decided using these equations.

The design of heat pipe requires a large number of data about properties of the liquid and the wick. These are not readily available and one has to search through the literature and hand books for the same.

REVIEW QUESTIONS

16.1 Explain the working of a thermosyphon.

16.2 What is the main limitation of thermosyphon?

16.3 What is the main difference in the construction of a heat pipe from that of thermosyphon?

16.4 What are the advantages of heat pipe for transport of heat?

16.5 Explain the working of a heat pipe using a schematic.

16.6 List the various functional areas of application of heat pipe.

16.7 Explain the basic reason for the very high thermal conductance of heat pipes.

16.8 List few important types of heat pipes.

16.9 Explain what is meant by flux transformation and indicate how this is achieved.

16.10 Explain what is meant by "isothermalisation" and indicate how this is achieved by variable thermal conductance heat pipe.

16.11 What is the property of working fluid which enables a wick to transport condensed liquid from condenser to evaporator?

16.12 List the main components of a heat pipe.

16.13 List the important properties required for a good working fluid for heat pipes.

16.14 Explain what is meant by "compatibility" of components of a heat pipe.

16.15 What are the limitations in the maximum heat flux that a heat pipe can work with?

16.16 Explain the limitation in flux imposed by viscosity and under what conditions this is applicable.

16.17 Explain how sonic conditions in vapour flow affects the maximum heat flux of a heat pipe. Indicate under what conditions this may occur.

16.18 Explain what is meant by capilliary limitation.

16.19 Explain how the pore size in wicks is related to capillary head developed.

16.20 Explain what is meant by wicking limit.

16.21 Explain the function of priming in heat pipe working.

16.22 Write down the expression for capillary head in heat pipes.

16.23 Explain what is meant by "entraining of liquid". Indicate under what conditions this will occur. How does this limits the maximum heat flux?

16.24 Sketch some forms of wicks used in heat pipes.

16.25 Distinguish between homogeneous wick and arterial wick.

16.26 Define Merit number of working fluids of heat pipes. Indicate the dimension of Merit number. (it is not dimensionless as Reynolds number)

16.27 Write down the equation for the conductivity of a wick saturated with working fluid.

16.28 Explain how gas formation during working affects the functioning of heat pipes.

16.29 List some materials for use in wicks.

16.30 Sketch a channel type of wick arrangement.

16.31 Explain the working of a variable conductance heat pipe.

16.32 Explain the working of a heat pipe to function as thermal diode.

16.33 Explain the working of a looped heat pipe. What are the advantages of this type of heat pipe in transporting heat over a distance?

16.34 Explain the working of a rotary heat pipe. Indicate an application for this type.

16.35 Explain the working of a capillary pumped heat pipe. How does this differ from looped heat pipe?

16.36 For what purpose heat pipe is used in space craft. What type of heat pipe is suited for this application?

16.37 Explain the part played by heat pipe in thermal storage systems.

16.38 Explain the part that can be played by heat pipes renewable energy systems.

16.39 Explain how heat pipe is applied in (*i*) drying of roads, (*ii*) heating roads and (*iii*) maintaining permafrost.

16.40 What are micro heat pipes? Where are these used?

16.41 Sketch some sections of micro heat pipes.

Appendix A

FILL IN THE BLANKS

1. Thermal conductivity of materials (k) generally varies with ___________ .
2. With increase in temperature thermal conductivity of insulators will ___________ .
3. With increase in temperature thermal conductivity of good conductors will ___________.
4. With increase in temperature thermal conductivity of many liquids will ___________ .
5. With increase in temperature thermal conductivity of water and glycerine will ___________ .
6. With increase in temperature thermal conductivity of gases in general will ___________ .
7. Heat conduction in insulators is mainly by ___________ .
8. Heat conduction in good conductors is largely by ___________ .
9. In the SI system, the unit for thermal conductivity is ___________ .
10. The unit for thermal resistance in the SI system is ___________ .
11. The unit for convective heat transfer coefficient in the SI system is ___________ .
12. Convection resistance over a given surface area A is ___________ .
13. Conduction resistance due to a plane wall of thickness L of material with thermal conductivity k and area A is ___________ .
14. For a given amount of heat conducted, higher the value of thermal conductivity ___________ will be the temperature difference.
15. If heat flow across a plane in a given slab should increase the temperature gradient at the plane should ___________ .
16. For a given material and temperature drop, if heat conducted is to be reduced the thickness should be ___________ .
17. In steady conduction for a material whose thermal conductivity increases with increase in temperature, the temperature gradient at the high temperature side will be ___________ .
18. In steady conduction for a material whose thermal conductivity decreases with increase in temperature, the temperature gradient at the low temperature side will be ___________ .

19. The temperature profile in a plane wall under steady conduction will be a straight line if thermal conductivity is ___________ .
20. In the case of a hollow cylinder, the temperature profile along the radius will be ___________ .
21. In the case of a hollow sphere, the temperature profile along the radius will be a ___________ .
22. When conductivity varies linearly with temperature the conductivity at the ___________ can be used in the calculation of heat flow using the general equations.
23. In radial heat conduction in a hollow sphere/cylinder, the temperature gradient at the outer surface will be ___________ compared to the gradient at the inner surface.
24. The pure metal whose thermal conductivity is highest at atmospheric temperatures is ___________.
25. For super insulation, ___________ form will be suitable.
26. Thermal diffusivity of a material is indicative of ___________ ___________ under transient conditions.
27. When multilayers of insulation is used over a sphere, for a given volume, the material with higher thermal conductivity should be applied ___________ the surface.
28. In a conducting slab with uniform heat generation the temperature variation along the thickness is ___________ .
29. In a heat conducting slab of given material, with uniform heat generation the difference in temperature between the center plane and the surface is directly proportional to ___________ and ___________ .
30. In a heat conducting slab of given thickness, with uniform heat generation the difference in temperature between the center plane and surface is ___________ proportional to the thermal conductivity of the material.
31. In a heat conducting sphere with uniform heat generation the difference in temperature between the center and surface is directly proportional to ___________ and ___________ and inversely proportional to ___________ .
32. In heat generation, surface convection also controls the ___________ temperature.
33. For a given heat flow, a fin of rectangular section will require ___________ volume as compared to a fin of triangular section with the same base thickness.
34. With the same cross-section two fins each of length L will dissipate ___________ heat compared to a single fin of length $2L$.
35. To reduce error in temperature measurement of flowing fluids, the thermometer well length should be ___________ .
36. Fins may be modeled as ___________ ___________ system.
37. For given fin volume as the length of the fin is decreased the heat flow will ___________ and then ___________
38. The ratio of heat flow by fin to the heat flow on the bare base areas is called ___________ of the fin.
39. Fin effectiveness is generally ___________ than one.
40. For a given volume tapering fin will dissipate ___________ heat as compared to constant area fin.

41. Along the length of a constant area fin the temperature gradient ___________ .
42. In fins the temperature gradient is highest at the ___________ .
43. For a given fin configuration the heat flow is proportional to ___________ power of h and k.
44. In electronic components Gold is used to improve fin efficiency by reducing ___________ resistance.
45. In transient conduction lumped capacity model can be used if ___________ .
46. For lumped parameter model the surface resistance should be ___________ compared to the internal resistance.
47. For lumped parameter model volume/Area ratio should be ___________ .
48. For lumped parameter model heat capacity of the piece should be ___________ .
49. For a given thermocouple if the response time should be shortened then ___________ should be increased.
50. For lumped parameter model to be used the temperature gradient within the body should be ___________ .
51. In transient conduction for the use of semi infinite solid model, the heat penetration depth should be ___________ over the period considered.
52. For the adoption of semi infinite solid model, the temperature at a depth should not have ___________ in the time considered.
53. A thick concrete slab exposed to flames on one side can be analysed using ___________ model.
54. For the same characteristic dimension, the shape of the object for shorter heating time is ___________ .
55. When Schmidt method is used for temperature calculations in transient conduction the choice of nodal thickness x and time interval $\Delta\tau$ should satisfy the condition.
56. In transient conduction, increasing and maintaining a surface at a temperature is equivalent to Biot number being equal to ___________ .
57. The electrical analogue element for the product of mass and specific heat is ___________.
58. In convection, temperature and velocity gradients vary only in the ___________ .
59. Along the thickness in the boundary layer velocity and temperature gradients ___________ .
60. The thickness of the hydrodynamic boundary layer is defined as the distance from the surface at which velocity ___________ is nearly zero.
61. In laminar flow, the average convection coefficient along the length will ___________ .
62. In laminar flow the velocity at a location with respect to time is ___________ .
63. In turbulent flow the velocity at a point varies randomly about a ___________ velocity.
64. In laminar flow momentum and heat transfer is mainly at the ___________ level.
65. In laminar flow there is no ___________ mixing between layers.
66. In turbulent flow momentum and heat transfer is due to ___________ mixing between layers.
67. If thermal diffusivity equals momentum diffusivity, then the ratio of thermal and velocity boundary layer thickness will be equal to ___________ .

68. The ratio of momentum diffusivity to thermal diffusivity is called ____________ number.
69. Flow transition is generally judged by ____________ number.
70. The thickness of hydrodynamic boundary layer at distance x in laminar flow over a plate is ____________ .
71. The thickness of hydrodynamic boundary layer in flow over plate in turbulent flow is proportional to ____________ .
72. In liquid metal flow over a flate, thermal boundary layer will be ____________ than hydrodynamic boundary layer.
73. In viscous oil flow thermal boundary layer will be ____________ than hydrodynamic boundary layer.
74. In laminar flow over flat plates the convection coefficient will be proportional to the distance raised to the power of ____________ .
75. In turbulent flow over flat plate the convection coefficient will be proportional to the distance raised to the power ____________ .
76. In flow over a flat plate the temperature and velocity gradients above the boundary layer is ____________ .
77. In the case of flow over flat plate the Reynolds number along the length will ____________ continuously.
78. The value of transition Reynolds number in the case of flow over flat plate is ____________.
79. Reynolds number expressed in terms of mass flux G in pipe flow is ____________ .
80. In pipe flow the analogy method of convection analysis relates ____________ number to friction factor.
81. In flow over a bank of tubes effective way to increase heat transfer rate is to ____________ the pitch along the flow direction.
82. In flow over spheres, and cylinders the characteristic length used in the calculation of dimensionless number is ____________ .
83. Hydraulic mean diameter is defined as ____________ .
84. In flow through non circular sections ____________ replaces diameter.
85. In flow through pipes the type of flow is decided by the conditions at ____________ .
86. In fully developed laminar flow through pipes the convection coefficient is ____________.
87. In pipe flow, convection coefficient at entrance region will be ____________ compared to the fully developed region.
88. In pipe flow under constant wall heat flux conditions the convection coefficient will be ____________ as compared to constant wall temperature condition.
89. In fully developed flow in a pipe under constant heat flux condition the temperature difference between the wall and the fluid will be ____________ .
90. In turbulent flow in rough pipes the Stanton number is related to ____________ .
91. In the case of ____________ the flow in pipes can be considered as slug flow.
92. As compared to forced convection the additional force encountered in free convection analysis is ____________ .
93. At constant pressure the value of coefficient of cubical expansion for gases is ____________.
94. The velocity in the case of free convection boundary layer is zero at ____________ .

95. The velocity variation in the natural convection boundary layer can be expressed by ____________ order equation.

96. The temperature variation in free convection in boundary layer can be expressed by ____________ order equation.

97. The dimensionless number which replaces the Reynolds number in the case of free convection is ____________ .

98. The average value of convection coefficient upto length L in free convection is ____________ .

99. Rayleigh number is the product of ____________ number and ____________ number.

100. In free convection the average value of h will be ____________ for short vertical plate as compared to a longer vertical plate.

101. Graetz number is defined as the product of ____________ .

102. Peclet number is the product of ____________ number and ____________ number.

103. In cooling of unattended equipments ____________ convection is more suitable.

104. In some cases where both free convection and forced convection may contribute to heat transfer, the criterion for the determination of predominance of either is the value of ____________ .

105. In boiling excess temperature is the difference between ____________ and ____________ temperatures.

106. In film boiling main mode of heat transfer is ____________ .

107. In boiling surface tension is an important parameter due to ____________ .

108. In nucleate boiling at higher heat flux levels convection coefficient is proportional to the ____________ power of excess temperature.

109. In boiling the excess temperature at which maximum heat flux occurs is about ____________ .

110. The shape of the heating surface ____________ affect boiling heat transfer.

111. The maximum heat flux in boiling of a fluid varies with the pressure ratio ____________.

112. In flow boiling h is maximum at ____________ flow.

113. Maximum heat flux without damage can be sustained only in ____________ boiling.

114. Minimum heat flux occurs in stable ____________ boiling.

115. In film boiling a ____________ exists between surface and liquid.

116. Flow boiling is encountered in ____________ water tubes.

117. In film boiling excess temperature will be ____________ .

118. The important parameter which controls heat flux in boiling is ____________ .

119. Temperature variation is assumed ____________ in condensate film.

120. Velocity in condensate film varies along the film thickness as ____________ of thickness.

121. Average value of convection coefficient in condensation up to distance L is ____________.

122. In film wise condensation the value of h will be ____________ compared to dropwise condensation.

123. The surface should be ____________ for dropwise condensation.

124. As the vapour is in direct contact with more cooler surface the convection coefficient is ____________ in dropwise condensation.

125. The additional force encountered in condensation boundary layer analysis is ___________.

126. In condensation 'h' depends on the ___________ power of conductivity.

127. In condensation 'h' depends on the ___________ power of density.

128. In condensation 'h' depends on the ___________ power of latent heat.

129. Condensation Reynolds number is defined in terms of mass flow G per unit width as ___________ .

130. In condensation, convection coefficient over a single tube is ___________ compared to condensation over a row of tubes.

131. Critical film Reynolds number in condensation is ___________ .

132. In freezing the conduction resistance is due to ___________ .

133. In the regenerator of a gas turbine the ___________ will be nearly equal.

134. The value U_o will be ___________ compared to value U_i in heat exchangers.

135. The product U_oA_o will equal ___________ .

136. During the life of heat exchanger the performance will deteriorate due to ___________.

137. In a shell and tube arrangement, the fluid that can be said to be mixed is the ___________ side fluid.

138. In a shell and tube arrangement, the fluid that can be said to be unmixed is the ___________ side fluid.

139. A heat exchanger in which superheated fluid is cooled, condensed and undercooled, using a cold fluid, the location at which minimum temperature difference occurs is called ___________ .

140. NTU is defined as ___________ .

141. Capacity ratio when cold fluid is having lower heat capacity is defined in terms of terminal temperatures as ___________ .

142. The three main flow arrangement in heat exchangers are ___________ flow.

143. Compact heat exchangers use ___________ tube bundles.

144. Thermodynamically the flow direction that will give minimum loss of available energy is ___________ .

145. The flow direction does not affect the performance in the case of ___________ and ___________ .

146. Fouling resistance is due to ___________ of the heating surface during operation.

147. Shell and tube arrangement is generally adopted because single tube type will be ___________.

148. The heat capacity of the fluid stream is the product of ___________ and ___________ .

149. If heat capacities are equal in a counter flow exchanger the slope of the hot and cold fluid temperature lines at any point will be ___________ .

150. If heat capacities are equal in a parallel flow exchanger the slop of the temperature lines at any point of flow will be ___________ but ___________ .

151. LMTD method of analysis is suitable when ___________ ___________ are known.

152. When inlet flow rates and temperatures for a heat exchanger are specified the easier method of analysis ____________ .
153. For the performance evaluation over the whole operating range of a heat exchanger ____________ ____________ method is more suitable.
154. When heat capacity ratio is zero, effectiveness is equal to ____________ .
155. For a condenser/evaporator the effectiveness is ____________ .
156. If heat capacities are equal for the counter flow type the effectiveness is ____________ .
157. If heat capacities are equal for parallel flow type the effectiveness is ____________ .
158. Effectiveness of a heat exchanger is the ratio of ____________ and ____________ .
159. For a given exchanger if the heat capacity ratio (C_{min}/C_{max}) increases the effectiveness will ____________ .
160. When effectiveness increases the total heat flow ____________ increase.
161. Storage type heat exchanger is also called ____________ heat exchanger.
162. Air preheaters in large thermal plants use ____________ of heat exchangers.
163. The overall heat transfer coefficient will generally be minimum in the case of ____________ exchangers.
164. In a condenser the controlling resistance will be on the ____________ side.
165. In an evaporator the controlling resistance will be on the ____________ side.
166. In a gas to liquid heat exchangers the controlling resistance will be on the ____________ side.
167. In the various flow arrangements the one that may require largest area will be ____________ .
168. The wavelength range of thermal radiation is ____________ .
169. The wavelength range for visual radiation is ____________ .
170. For a black body ____________ is maximum.
171. The radiant energy emitted per unit area and unit wave length called ____________ .
172. The equality of emissivity and absorptivity is postulated by ____________ law.
173. The statement that "the product of the temperature and the wavelength at which the monochromatic emissive power is maximum is a constant" is due to ____________ law.
174. At any temperature as wavelength increases the monochromatic emissive power ____________.
175. The ratio of emissive power of a body to the emissive power of a black body is defined as ____________ .
176. Radiation from a small opening from an isothermal enclosure can be considered as ____________.
177. The relationship between emissive power E and intensity I is ____________ .
178. The radiant energy per unit solid angle in a direction is known as ____________ of radiation.
179. The emissivity is constant in the case of a ____________ .
180. Gases are generally ____________ radiators.
181. Suns radiation is mostly at ____________ wavelengths.
182. Radiation at atmospheric temperature is at ____________ wavelengths.

183. Glasses generally transmit ____________ wavelength radiation.
184. Green house effect is due to ____________ short wavelength radiation and ____________ long wavelength radiation.
185. For real surfaces, the monochromatic emissive power does not vary in a ____________ .
186. The sum of emissivity, and reflectivity will be equal to ____________ in the case of opaque surfaces.
187. For an opaque surface high reflectivity means ____________ emissivity.
188. For selective surface the emissivity is dependent on ____________ and ____________ of radiation.
189. Solar collector surfaces should have high absorptivity for ____________ wavelength radiation and low emissivity at ____________ temperatures.
190. In the case of solids/liquids radiation is a ____________ phenomenon.
191. CO_2 and water vapour are ____________ radiators.
192. In the case of gases radiation is a ____________ phenomenon.
193. Diatomic gases are ____________ for radiation.
194. The emissivity of a gas body depend on ____________, ____________, ____________ .
195. Beers law states that transmissivity of a gas body of length L with transmission coefficient a is ____________
196. Most of the real surfaces have emissivity ____________ with angle from normal.
197. Insulating materials have ____________ emissivity at angles near horizontal.
198. Conducting materials have high emissivity at angles ____________ horizontal.
199. The total radiant energy leaving a gray surface is known as ____________ .
200. The total radiant energy incident on a gray surface is known as ____________ .
201. Radiosity include emissive power and ____________ .
202. Irradiation includes radiation from other surfaces and ____________ .
203. The fraction of radiation emanating from surface 1 incident on surface 2 is called ____________ .
204. Shape factor will ____________ when surfaces are moved closer.
205. Shape factor will ____________ if the areas of the participating planes increase.
206. Shape factor is ____________ of surface properties.
207. Shape factor of body with respect to a fully enclosing body is ____________ .
208. Concave surfaces will have a shape factor with ____________ .
209. The shape factor of a hemisphere to itself is ____________ .
210. Shape factor of a hemispherical surface to its base is ____________ .
211. Shape factor of the base of a cone to the curved surface is ____________ .
212. If n radiation shields of equal emissivity as the parallel large planes are used the heat flow will be reduced ____________ times.
213. Radiation shield should have ____________ reflectivity.
214. Surface resistance is due to the ____________ of the surface.
215. Space resistance between two surfaces is due to the ____________ .
216. Reciprocity theorem for shape factor is ____________ .
217. Well insulated surfaces enclosing surfaces exchanging heat by radiation are called ____________.

218. Reradiating enclosure ___________ the effective shape factor.
219. Mass transfer is due to ___________ gradient.
220. The two types of mass transfer are ___________ and ___________ .
221. Mass transfer at molecular level can be likened to heat ___________ .
222. The unit of mass diffusivity is ___________ .
223. The ratio of mass diffusivity to momentum diffusivity is called ___________ .
224. The two types of molecular diffusion are ___________ and ___________ .
225. For the same concentration gradient, diffusion, into a stationary component will give ___________ rate.
226. Use of kg mole in place of kg is more convenient mass transfer because of ___________.
227. If $D_{ba} <> D_{ab}$ then spontaneous ___________ build up will result.
228. The unit of mass transfer coefficient h_m is ___________.
229. Sherwood number is given by ___________.
230. By similarity h/h_d is proportional to ___________ .
231. Transient mass diffusion problems can be solved similar to ___________ .

TRUE OR FALSE

1. Convection coefficient is a material property.
2. Thermal conductivity is a material property.
3. In good conductors lattice vibration contributes more for heat conduction.
4. Thermal conductivity of water decreases with increase in temperature.
5. For the same amount of heat conduction through a slab, as thickness increases the temperature gradient should increase.
6. Fins for the same flow should be longer if the thermal conductivity of the material is increased.
7. For identical fins of different materials the tip to base temperature difference will be lower if the thermal conductivity is lower.
8. In a hollow cylinder, the temperature variation with radius will be linear.
9. The temperatures gradient at the inner surface will be steeper compared to that at the outer surface in radial heat conduction in a hollow cylinder.
10. Fins are more useful with liquids than with gases.
11. Fins effectiveness is generally greater than one.
12. In three dimensional steady state conduction with uniformly spaced nodes the temperature at a node will be one sixth of the sum of the adjacent nodal temperatures.
13. Lumped capacity model can be used in the analysis of transient heat conduction if Biot number is greater than one.
14. Lumped parameter model can be used if the internal conduction resistance is high compared to the surface convection resistance.
15. To reduce the time constant of a thermocouple, the convection coefficient over its surface should be reduced.
16. To reduce the time constant of a thermocouple its characteristic linear dimension (V/A) should be reduced.

17. A solid of poor conductivity exposed for a short period to surface convection can be analysed as semi infinite solid.
18. A slab will cool the fastest compared to a long cylinder or sphere of the same characteristic dimensions when exposed to the same convection conditions.
19. Higher the value of Biot number slower will be the cooling of a solid.
20. For transient conduction analysis of smaller objects product solution is used.
21. In a slab conducting heat the surface temperatures are 200 and 100°C. The mid plane temperature will be 150°C if k is constant.
22. In a slab of material of variable thermal conductivity, with conductivity increasing with temperature, the surface temperatures are 200°C and 100°C. The mid plane temperature will be greater than 150°C.
23. In a slab material of variable conductivity with conductivity decreasing with temperature the surface temperatures are 200°C and 100°C. The mid plane temperature will be higher than 150°C.
24. In a hollow cylinder with radial conduction the mid plane temperature will be lower than the mean of surface temperatures.
25. In a hollow sphere with radial conduction, the mid plane temperature will be higher than the mean of surface temperatures.
26. With convection on the surface any amount of additional insulation cannot reduce the heat flow through a hollow spherical insulation of the same material to half the original flow rate.
27. In the case of small hollow cylinders or spheres, with outside convection the thermal resistance may decrease by the addition of insulation.
28. Small electronic components may be kept cooler by encasing it in glass like material.
29. If Prandtl number is greater than one, the thermal boundary layer will be thicker compared to hydrodynamic boundary layer.
30. Liquid metal flow in pipes can be approximated to slug flow.
31. The local value of convection coefficient in laminar flow over a flat place will decrease along the length.
32. In flow over a flat plate over length L the, average convection coefficient will be equal to (4/3) h_L.
33. Other conditions remaining the same as viscosity increases the boundary layer thickness will decrease.
34. Momentum and displacement thickness will be more compared to boundary layer thickness.
35. In laminar flow Nusselt is a function of $Re^{0.8}$.
36. In turbulent flow the velocity at point varies about an average value.
37. In turbulent flow in pipes Nusselt is proportional to $Re^{0.8}$.
38. In fully developed flow through a pipe, under laminar flow conditions, average Nusselt number is constant.
39. The hydraulic mean diameter for an annulus is $D_{o2} - D_{12}$.
40. In flow-through tube banks of tubes closer pitch will lead to higher values of h.
41. In free convection, Rayleigh number is similar to Paclet number in forced convection.
42. Gravity force rather than buoyant force plays a more important role in free convection.
43. Grashof number is the ratio between buoyant force and viscous force.

44. Reynolds number is the ratio between viscous force and buoyant force.
45. The value of convection coefficient for the same flow velocity will be lower in the case of water as compared to air.
46. Lower values of kinematic viscosity will lead to higher value of h both in free and forced convection.
47. In pipe flow for similar velocity conditions water will have a higher convection coefficient compared to liquid metal.
48. In cases where both modes of convection may contribute the ratio Gr/Re^2 is a measure of the importance of either mode.
49. As the excess temperature increases, the sustainable heat flux will continuously increase in boiling.
50. The excess temperature range for maximum flux in nucleate pool boiling is about 200°C.
51. In stable film boiling as excess temperature increases sustainable heat flux will increase.
52. In flow boiling mist flow will sustain higher heat flux.
53. In condensation film, linear temperature profile is generally assumed.
54. Dropwise condensation is not sustainable over long periods.
55. Counter flow is always preferable in heat exchanger design.
56. For the same terminal temperatures, LMTD-parallel flow will be higher compared to LMTD-counter flow.
57. NTU method is preferred for the analysis of the complete performance of heat exchangers.
58. For the same NTU, as the capacity ratio increases the effectiveness will decrease.
59. For a given exchanger as the capacity ratio increases the final temperatures will increase.
60. As the capacity ratio in a given exchanger increases, the heat flow will increase.
61. When heat capacities of both fluids are equal, the temperature difference will be constant for parallel flow arrangement.
62. For condensers/evaporators, the flow direction does not affect the heat flow.
63. Capacity ratio is taken as zero for condensers and evaporators.
64. Opaque Gray surfaces have constant reflectivity.
65. Directional emissivity for metals will be lowest at the normal direction.
66. Glasses generally transmit low frequency radiation.
67. Copper dioxide coating can produce selective surface.
68. As temperature increases, the wavelength at which maximum monochromatic emissive power occurs increases.
69. Kirchhoff law states that reflectivity equals absorptivity.
70. As temperature difference increases, radiation resistance will increase.
71. As temperature increases, h_r will increase.
72. Convex surfaces will have shape factor with themselves.
73. Between two surfaces if $F_{1-2} > F_{2-1}$ then $A_1 > A_2$.
74. Shape factor with enclosing surfaces will be one.
75. Gases are truly gray radiators.
76. Radidation from a gas body is a volume phenomenon.
77. Emissivity of a gas body depends on the partial pressure, thickness and temperature.
78. Gases are band radiators.

79. Snow is a very good reflector.
80. Lewis number is used to predict mass transfer rates using heat transfer rates at similar conditions.
81. Schmidt number replaces Nusselt number in convective mass transfer studies.
82. In mass transfer studies the function of Sherwood number is similar to Prandtl number in heat transfer studies.

SHORT QUESTIONS

1. List the basic laws involved in heat transfer studies.
2. List the three modes of heat transfer and differentiate between them.
3. Describe the mechanism of heat transfer by conduction.
4. Describe the mechanism of convection mode of heat transfer.
5. Explain reasons for the involvement of more parameters in the analysis of convection.
6. Explain the importance of thermal conductivity of fluids in convection.
7. Describe the effect of flow velocity and viscosity on convection heat transfer coefficient.
8. Explain the essential conditions for radiation heat exchange.
9. Define steady state conduction giving examples.
10. State the Fourier law of heat conduction.
11. Giving examples explain the use of electrical analogy in heat transfer studies.
12. Define unsteady state conduction giving examples.
13. Explain how contact resistance develops in conduction ?
14. State the reason for the temperature gradient being steeper at the inside compared to the outside in the case of radial heat conduction in a hollow cylinder/sphere.
15. Explain the term "critical thickness of insulation" with reference to insulation of hollow cylinders and spheres with outside convection. Sketch the variation of total resistance against insulation thickness in case of hollow cylinder.
16. Explain the concept of log mean area in the case of heat conduction in hollow cylinders.
17. Draw the equivalent circuit for conduction through a slab under steady state conduction with convection on both surfaces.
18. Sketch the temperature variation along the thickness of a slab under steady conduction when (*i*) thermal conductivity increases with temperature and (*ii*) when thermal conductivity decreases with temperature.
19. Sketch the temperature variation along the radius of a hollow cylinder/sphere under steady radial conduction.
20. Discuss the desirability of tapering the section along the length of a fin exposed to convection.
21. Discuss the conditions for extended surfaces (fins) to be beneficial.
22. Define fin efficiency and explain considering an example.
23. Define fin effectiveness and explain considering an example.
24. State the causes for errors in measurement of temperature of flowing fluids using thermometer well.
25. Define Total fin efficiency and explain considering an example.

26. Sketch qualitatively the temperature variation along the length of fins in the following conditions (*i*) copper fin and (*ii*) steel fin. Assume that similar outside conditions prevail in both cases.
27. Two fins of identical sections and lengths are fixed on a surface for heat transfer enhancement. One is of aluminium and the other is of steel. The tip temperature of which fin will be higher and why ?
28. Explain why for a given volume of material a longer fin may not dissipate as much heat as a shorter fin.
29. Explain why circumferential fins are used in pipes and longitudinal fins are used on motor bodies.
30. For a pin fin which type of shape will be more economical (*i*) constant area (*ii*) conical (*iii*) convex parabolic and (*iv*) concave parabolic. Discuss the reasons.
31. Two rods of same section and length made of material A and B are inserted into a furnace. The temperatures in the rods are found to be equal at lengths L and 1.5 L in materials A and B. Indicate which material has the lower thermal conductivity.
32. Explain how thermal conductivity can be measured using fins.
33. Explain how convection coefficient can be measured using fins.
34. Discuss the effect of conductivity and convection coefficient on the heat dissipation capacity of a fin of a given shape and size.
35. A fin loses heat only by convection. If the same fin is to lose heat only by radiation, will the heat loss (*i*) increase (*ii*) decrease or (*iii*) it cannot be predicted. Discuss.
36. A fin is exposed to a constant heat flux with the base temperature being lower. Sketch the temperature variation along the length.
37. Write down the differential equation for steady two dimensional heat conduction and indicate the method of solving the same.
38. A thin square slab conducting heat along two dimensions has three of its faces at say 400°C and the fourth side at 800°C. Sketch a few equal temperature lines.
39. A thin square slab conducts heat in two directions. Three of its sides are at 100°C and the temperature on the fourth side has a sinusoidal variation with 100°C as minimum. Sketch a few equal temperature lines.
40. List the various methods available for the solution of two dimensional steady conduction problems.
41. Explain the advantages of numerical method in solving two dimensional conduction problems.
42. Describe how a nodal equation can be formed for the temperature at a node in terms of the adjacent nodal temperatures.
43. List the parameters that influence the use of Lumped capacity model in unsteady heat conduction.
44. Explain the significance of Biot number in unsteady conduction.
45. Define "time constant" in the case of thermometer or any other probe used to measure temperature of a flowing gas.
46. Explain the significance of Fourier number in unsteady conduction.
47. Sketch and explain the type of temperature variation with time in the case of a lumped capacity system (*i*) when it cools and (*ii*) when it heats up.
48. Define "semi infinite solid" as used in transient conduction analysis ? Write the differential equation for the problem.

49. Cite some situations where semi infinite solid model can be applied. Give the possible boundary conditions.
50. Explain the effect of thermal diffusivity in transient conduction.
51. In transient conduction sometimes the boundary is specified to be at constant temperature, when transferring heat. Explain how the physical situation can be achieved ?
52. Equal sized spherical shots one of copper and the other of steel are heated in a furnace. Sketch on the same diagram, the variation of temperature with time in these cases.
53. A solid in the shape of a short cylinder is heated in a furnace under convective conditions. The value of Biot number is 0.6. Explain the method of determination of the center temperature.
54. Sketch the temperature at various time periods along the thickness of a slab initially at 100°C suddenly exposed to convection at 800°C on both sides.
55. Sketch the temperature at various time periods along the thickness of a slab initially at 100°C if it suddenly has its surface raised to 800°C on both sides, and maintained at this level.
56. Explain how a cube being heated can be analysed for temperature variation.
57. Differentiate between free and forced convection.
58. Explain the boundary layer concept and indicate its importance.
59. Differentiate between laminar and turbulent flow.
60. State the essential differences in the development of boundary layer in flow over surfaces and flow through ducts.
61. Explain the basic concept used in formulating the equations for the determination of the value of convection coefficient.
62. Explain the significance of Nusselt number.
63. Explain the significance of Reynolds number.
64. Explain the significance of Prandtl number.
65. Explain the significance of "momentum thickness".
66. Explain the significance of "displacement thickness".
67. Explain how the wall temperature gradient at a location in flow over a surface is affected by (*i*) velocity (*ii*) viscosity.
68. Distinguish between eddy diffusivity of heat/momentum and molecular diffusivity.
69. Define and explain the concept of Hydraulic mean diameter. Indicate the application of Hydraulic mean diameter.
70. State the relation between friction coefficient C_f and friction factor f.
71. Define friction coefficient C_f.
72. Define friction factor f.
73. Explain the concept of Bulk mean temperature. Indicate where it is used.
74. Explain the concept of film temperature. Indicate where it is used.
75. Explain what is meant by fully developed flow in pipes.
76. Explain the advantage of the approximate integral method of analysis of boundary layer flow.
77. Explain the use of analogy in heat and mass transfer studies.
78. Explain the advantages and limitations of dimensional analysis method used in convection studies.

79. Give two examples for the use of packed beds in heat transfer situations.
80. State Stefan-Boltzmann law.
81. State Wien's displacement law in heat radiation.
82. State Lambert's cosine law and indicate its uses.
83. State Kirchhoff's law of heat radiation and indicate its uses.
84. State Planck's law of for heat radidation.
85. Distinguish between "total emissive power" and "monochromatic emissive power".
86. Explain the concepts "black body" and "gray body".
87. Explain the concept "intensity of radidation". How does it relate to emissive power ?
88. Define the terms "absorptivity", "reflectivity" and "transmissivity". For a black surface what are the values of each.
89. Explain what is meant by selective coating. Indicate the use of selectively coated surfaces.
90. With an example explain the concept "Band radiators".
91. Explain the concepts "Radiosity" and "Irradiation" and indicate the application of these concepts in the analysis of radiation heat exchange between gray surfaces.
92. Define the explain the concept "shape factor" in radidation heat exchange.
93. Discuss the effect of the following in the value of shape factor (*i*) area of surfaces (*ii*) distance between surfaces (*iii*) enclosing insulated surfaces.
94. Explain what is meant by "green house effect".
95. Describe how an ideal black radiation source can be created.
96. Explain how shields reduce heat transfer by radiation. Indicate the important requirement for shield effectiveness.
97. What is directional emittance ? Describe how directional emittance varies in the case of conducting and insulating surfaces.
98. Describe giving an example the crossed string method of determining shape factor.
99. State reciprocity theorem for shape factors.
100. Explain the concept and write down the expression for "surface resistance" and "space resistance" in case of radiation heat exchange between gray surfaces.
101. Draw the equivalent circuit for radiation heat exchange between gray surfaces.
102. Explain the concept of "nonabsorbing-reradiating surface".
103. Draw the equivalent circuit for heat exchange between two black surfaces connected by a reradiating surface.
104. List the factors affecting the emissivity of a gas body.
105. Explain why glass cover is used in solar collectors of the flat plate type.
106. Explain how error in measurement of temperature using a bare thermometer is introduced due to radiation.
107. Write the expression for the space resistance between surfaces separated by absorbing gas body.
108. Define "excess temperature" and explain its importance in the study of boiling heat transfer.
109. Describe the various regimes of boiling.

110. Differentiate between nucleate boiling and film boiling and indicate in which case maximum heat flux occurs. What is "burnout" in boiling ?
111. Explain the phenomenon of flow boiling and indicate the variation of flux that can be sustained in various regimes in flow boiling.
112. Explain why surface tension becomes important in nucleate boiling.
113. Distinguish between filmwise and dropwise condensation.
114. Describe the assumed variation of temperature in condensate film.
115. Distinguish between recuperative and regenerative heat exchangers.
116. List the classification of heat exchangers based on flow direction.
117. Compare parallel flow and counter flow exchanger in terms of area requirements.
118. Explain how a regenerative (storage type) heat exchanger can be made to work continuously without cycling.
119. Explain why shell and tube arrangement is adopted extensively in heat exchanger construction.
120. Define LMTD, NTU, capacity ratio and effectiveness of heat exchangers.
121. Explain the special advantages of NTU-effectiveness method of analysis of heat exchanger performance.
122. Explain the conditions under which the capacity ratio is taken as zero.
123. State how the ratio of temperature drop is affected by capacity ratio.
124. Explain why the fluid flow direction is not considered when capacity ratio is zero. Give an example.
125. Explain the condition under which the slope of the hot and cold fluid temperature lines will be equal. Indicate the effect of such condition in the case of counter flow exchangers.
126. Distinguish between "flow mixed" and "flow unmixed" in the case of heat exchangers.
127. What is fouling ? What are its effects on heat exchanger performance ?
128. Distinguish between diffusion mass transfer and convective mass transfer.
129. State the dimension for convective mass transfer coefficient. How does it differ from convective heat transfer coefficient?
130. Explain the significance of Schmidt, Sherwood and Lewis numbers in mass transfer analysis.
131. Describe giving examples "equimolal counter diffusion" and **"one component diffusing into a stationary component".**
132. Give an example for simultaneous heat and mass transfer.
133. Write down continuity equation for the boundary layer.
134. List the boundary conditions available for cubic curve fitting of velocity profile for a forced convection boundary layer.
135. List the boundary conditions available for cubic curve fitting of velocity profile in free convection boundary layer.
136. List the initial and boundary conditions in the case of infinite slab of thickness $2L$ exposed on both sides to convection.
137. List the possible boundary conditions in the case of semi infinite slab under transient conduction.
138. Define Radiosity and irradiation.

SHORT PROBLEMS

1. Hot air at 80°C flows over a surface of area 0.2 m^2 at 60°C, the convection coefficient being 25 W/m^2K. The heat flow is __________ .
2. The surface temperatures of a slab conducting heat under steady conditions are 80°C and 60°C. The thermal conductivity of the material of the slab is 25 W/mK. The heat flow rate is 2500 W/m^2. The thickness of the slab is __________ .
3. A slab 0.2 m thick of thermal conductivity 25 W/mK conducting heat under steady conditions at the rate of 2500 W/m^2 has the hotter surface at 100°C. The temperature of the other surface is __________ .
4. A slab of 0.2 m thickness has its surfaces at 120°C and 100°C. The heat conducted at steady conditions is 250 W/m^2. The conductivity of the material is __________ .
5. A slab conducts heat at a steady rate of 2500 W/m^2. The thermal conductivity of the material is 25 W/mK. The temperature gradient in the slab is __________ .
6. The temperature gradient in a slab at steady conduction is – 1000°C/m. The slab conducts heat at 250 W/m^2. The conductivity of the material is __________ .
7. A slab conducts heat at 2500 W/m^2 under steady conditions. The conductivity of the material is 20 W/mK. If the thickness is reduced to half the value, maintaining the heat flow rate, the temperature gradient will be __________ .
8. In a composite slab of two layers the temperature gradient in the first layer of thermal conductivity 50 W/mK is – 50°C/m. If the conductivity of the material of the second layer is 25 W/mK the temperature gradient in the second layer will be __________ .
9. In a composite slab consisting of two layers of equal thickness the temperature drop in the first layer is 40°C. If the conductivity of the material of the second layer is 1/4th of that of the first layer the temperature drop in this layer will be __________ .
10. In a composite slab made of two layers of equal thickness the total temperature drop is 100°C. The conductivity of the first layer material is 1/4th of the conductivity of the second layer material. The temperature drop in the first layer will be __________.
11. A hollow cylindrical insulation of *ID* 0.2 m and *OD* 0.4 m conducts heat radially. If another layer of insulation of the same material of thickness of 0.4 m is added to the heat flow will be changed by the ratio __________ .
12. A hollow spherical insulation of 0.2 m *ID* and 0.4 m *OD* conducts heat under steady conditions. If another layer of insulation of the same material of thickness of 0.4 m is added the heat flow will be reduced to __________ of the original value.
13. A hollow spherical insulation of *ID* 0.2 m and *OD* 0.4 m conducts heat at steady conditions radially. The thickness of additional insulation of the same material needed to reduce the heat flow to 75% of the original value is __________ .
14. To reduce the heat flow through a hollow cylinder to 1/*n*th the value of the original flow the new value of the ratio of the radii should be the __________ power of the original radii ratio.
15. In a hollow spherical insulation of radii r_i and r_o in order to reduce heat flow by fraction *c* of the original flow additional insulation up to radius r_{o2} was added. Then the relation between the new and old radii is __________ .

16. One end of two long rods of same radius of material A and B are in a furnace. The temperature at 10 cm from the furnace in rod A was found equal to the temperature at 20 cm in rod B. The conductivity of material A was 10 W/mK. The conductivity of material B will be ___________ .

17. Two long cylindrical rods A and B of the same material but of different diameters have one of their ends placed in a furnace. The temperature measured at 10 cm from the furnace in rod A was found equal to the temperature at 20 cm in rod B. The ratio of the diameters of the larger to the smaller is ___________ .

18. Two long rods of the same material of diameters 4 cm and 1 cm have one of their ends placed in a furnace. The heat loss from the 4 cm diameter rod will be ___________ times the heat loss from the 1 cm diameter rod.

19. A long rod has one of its ends in a furnace and has reached steady conditions. Suddenly a fan is switched on and the convection coefficient over the surface increases four fold. Under steady conditions the heat loss will increase by ___________ %.

20. A long rod has one of its end in a furnace and has reached steady conditions. The temperature at 16 cm from the end was measured as T°C. Suddenly a fan is switched on and the convection coefficient over the surface increases four fold. Under steady conditions the location at which the temperature will equal T°C will be at ___________ from the end.

21. A long rod of diameter d m has one of its ends in a furnace and has reached steady conditions. The temperature at 10 cm from the end was measured as T°C. If a rod of $4d$ m diameter is in a similar situation the location at which the temperature will equal T°C will be at ___________ from the end.

22. In a heat treatment process a small component of good thermal conductivity is to be cooled. For the same volume if the surface area is doubled the cooling time will be reduced to ___________ the original time.

23. Two sets of spherical pieces are to be cooled in batch process. The diameter of one set is 20 mm and that of the other set is 10 mm. The batch time for the 20 mm shots was 8 min. If other conditions are the same then the batch time for the 10 mm shots will be ___________ .

24. Steel shots of heat capacity 1000 J/K are cooled in a bath, the batch time being 6 min. If similar sized shots of a different material with heat capacity of 1500 J/K are to be cooled under similar conditions the batch time will be ___________ .

25. A thermocouple of volume 1.5×10^{-8} m^3 and area of 9×10^{-6} m^2 with a specific heat of 500 J/kg K and density of 8000 kg/m^3 is exposed to convection at 667 W/m^{2}K. The time constant is ___________ .

26. A slab 16 cm thick generating heat at the rate of 5 MW/m^3 when exposed on both sides to convection has its center temperature 60°C above the surface temperature. If a long cylinder of 8 cm diameter generating heat at the same rate is exposed to convection under similar conditions its center temperature will be ___________ above the surface temperature.

27. The center to surface temperature difference in a heat generating cylindrical rod of 8 cm dia was 30°C. Under similar conditions the center to surface temperature difference in the case of a sphere of 8 cm dia will be ___________ .

28. In a slab 0.2 m thick with a surface temperature of 200°C on both surfaces heat is generated at a rate of 10^6 W/m^3. The conductivity of the material is 50 W/mK. The center temperature will be __________ .

29. In a long solid cylinder the heat generation rate is 10^6 J/m^3. The conductivity is 50 W/mK. The surface temperature is 250°C. The center temperature is 300°C. The radius of the cylinder is __________ .

30. In a solid sphere of 0.2 m dia, heat is generated at the rate of 1.2 × 10^6 W/m^3. The center temperature is 300°C. Conductivity is 50 W/mK. The surface temperature is __________.

31. In a heat generating slab of 0.1 m half thickness the center to surface temperature difference was 100°C. If the half thickness is increased to 0.15 m, the temperature difference will be __________ .

32. A cylindrical rod of 0.1 m radius generating heat has its radius increased to 0.2 m. The center to surface temperature difference is __________ times the original value.

33. A thermocouple has a time constant of 24 seconds, under a certain calibration conditions to improve performance, the convection coefficient is doubled and the surface area is also increased to 1.2 times the value keeping the volume unchanged. The new time constant will be __________ .

34. Inside of a slab under transient conduction three planes 1, 2 and 3 are located at distance of 2 cm intervals. The thermal diffusivity of the material was 1 × 10^{-6} m^2/s. The temperatures at node 1 and 3 at a point of time was 300 and 200°C. After an interval of __________ seconds temperature at node-2 will reach 250°C.

35. Under steady two dimensional conduction the temperatures at nodes $(m + 1, n)$, (m, n), $(m, n + 1)$, $(m, n - 1)$ are respectively 600, 400, 350 and 325. The modes are equal spaced. The temperature at the node $(m - 1, n)$ *i.e.*, $T_{m-1,\, n}$ is __________ .

36. In a flow over a flat plate at a distance of 0.12 m, the Reynolds number is 14,400. The hydro dynamic boundary layer thickness is __________

37. The boundary layer thickness at a point in flow over a plate, is 8 mm. The Reynolds number is 25600. The distance from the leading edge is __________ .

38. The Reynolds number in air flow over a flat plate at 8 m/s was 25000. The kinematic viscosity was 64 × 10^{-6} m^2/s. The distance from the leading edge is __________ .

39. In a flow over a flat plate the distance from the leading edge for the flow to become turbulent is __________ . The free stream velocity is 20 m/s. Kinematic viscosity is 50 × 10^{-6} m^2/s.

40. In a flow over a flat plate the Reynolds number is 25600 and the Prandtl number is 0.834. The average Nusselt number is __________ .

41. The Nusselt number is 100. The location from leading edge is 0.5 m. k = 0.025 W/mK. Convection coefficient is __________ .

42. The temperature gradient at the wall in flow over a flat plate is – 4000°C/m. Conductivity is 0.025 W/mK. Plate temperature is 80°C. Air temperature is 60°C. Convection coefficient is __________ .

43. In flow over a flat plate Reynolds number at a location is 25600. The average friction coefficient is __________ .

44. In flow over a flat plate, in the laminar region, velocity boundary layer thickness is 0.009 m. Pr = 0.729. Thermal boundary layer thickness is __________ .

Appendix A

45. When cubic velocity profile is assumed the hydro dynamic boundary layer thickness in flow over a flat plate is given by __________ .
46. In flow through a pipe the Reynolds number is 1600. The friction factor is __________
47. In laminar flow through a pipe the friction factor is 0.032. The Reynolds number is __________.
48. In fully developed laminar flow of water through a pipe of 0.15 m diameter with uniform wall temperature the thermal conductivity at the condition was 0.655 W/mK. The average convection coefficient is __________ .
49. Water flows at the rate of 0.08 kg/m^{2}s through a 0.1 m diameter pipe the viscosity being 356×10^{-6} kg/ms. The Reynolds number is __________ .
50. In fully developed laminar flow of water through a 0.15 m diameter pipe under constant heat flux conditions the thermal conductivity at the bulk mean temperature was 0.687 W/mK. The value of convection coefficient is __________ .
51. An equilateral triangular duct of side 0.1 m has air flowing through it. The hydraulic mean diameter is __________ .
52. The hydraulic mean diameter of a rectangular duct of 0.12×0.1 m sides is __________
53. In a convection situation both forced and free convections appear to contribute equally The Reynolds number is 4000. The range of value of Grashof number is __________
54. In a flow the convection coefficient has a value of 2600 W/m^{2}K. The thermal conductivity of the material is 0.64 W/mK. The temperature difference is 40°C. The value of temperature gradient at the interface is __________ .
55. In a flow across a tube bundle the tube *OD* is 0.75 m. The tube arrangement is staggered with a pitch of 0.15 m normal to the flow and 0.1 m pitch along the flow. The entrance velocity is 10 m/s. The maximum velocity is __________ .
56. In a heat exchanger the terminal temperatures of the hot fluid are 150 and 80°C. For the cold fluid the terminal temperatures are 30 and 100°C. The flow arrangement is __________ and the value of LMTD is __________ .
57. A clean heat exchanger under test gave a *U* value of 2000 W/m^{2}K. After one year of operation the value of *U* was determined as 1600 W/m^{2}K. The fouling resistance is __________ .
58. The inlet and outlet temperatures of the hot and cold fluids are 200, 40 and 180 and 100°C. The value of the LMTD for (*i*) counter flow and (*ii*) Parallel flow are __________.
59. In a heat exchanger the hot fluid inlet and outlet temperatures are 200 and 180°C. For the cold fluid the terminal temperatures are 40 and 100°C. The capacity ratio is __________ .
60. For a parallel flow heat exchanger the capacity ratio is 1.0 and the NTU value is 2.0. The effectiveness is __________ .
61. For a counter flow heat exchanger the capacity ratio is 1.0 and NTU is 2.0. The effectiveness is __________ .
62. For a condenser the value of NTU is 1.5. The effectiveness is __________ .
63. For a condenser the effectiveness was found as 0.61. The value of NTU is __________ .
64. For a counter flow heat exchanger with equal heat capacities the effectiveness was found as 0.6. The value of NTU is __________ .

65. For an evaporator the effectiveness was 0.5. The value of NTU is ________ .

66. A heat exchanger with 10 m^2 heat transfer area has an overall heat transfer coefficient of 600 W/m^2 K. The minimum heat capacity of the flow is 2100 W/K. The value of NTU is ________ .

67. In an economiser of a large steam generator pressurised water flows at a rate of 20 kg/s. The water temperature increases from 40°C to 160°C. The flue gas temperature drops from 360°C to 190°C. The heat capacity of the gas flow is ________ .

68. In an economiser water gets heated by 120°C while flue gases are cooled by 170°C. The capacity ratio is ________ . If the heat capacity of the gases is 58984 W/K the heat capacity of water is ________ .

69. In a heat exchanger the LMTD was 26.2°C under certain flow conditions. The heat transfer rate was 11000 W. Due to changes in surrounding conditions the inlet temperature alone had changed and the value of LMTD was found as 22.3°C. The heat flow rate at this conditions will be ________ .

70. In an evaporator due to a change in the heat capacity of the hot fluid the effectiveness increased from 0.6 to 0.75. The overall heat transfer coefficient decreased by 20%. The percentage change in the heat capacity is ________ .

71. In a counter flow heat exchanger with equal heat capacities the flow rate changed equally on both sides of flow. This changed the effectiveness from 0.6 to 0.75. If there was no change in the value of overall heat transfer coefficient the percentage change in the flow rates is ________ .

72. In a counter flow heat exchanger with equal heat capacities the effectiveness was 0.75. If both the flow rates were doubled and if this increased the overall heat transfer coefficient by 20% the new effectiveness will be ________ .

73. In the case of parallel flow heat exchangers with equal heat capacities the limiting value of effectiveness is ________ .

74. In a parallel flow heat exchanger the value of NTU is 2.0. The capacity ratio is 0.5. The value of effectiveness is ________ .

75. In a parallel flow heat exchanger the heat capacities are equal. The hot and cold fluid inlet temperatures are 200°C and 100°C. The limiting value of the exit temperature is ________ .

76. In a counter flow heat exchanger with equal heat capacities the inlet temperatures are 100°C and 30°C. The hot fluid exit temperatures was 50°C. The cold fluid exit temperature will be ________ .

77. In a parallel flow heat exchanger the heat capacity of the hot fluid was 0.5 of the cold fluid heat capacity. The entry temperatures are 200°C and 50°C. If the cold fluid leaves at 100°C then the hot fluid will leave at ________ .

78. In a counter flow heat exchanger the hot fluid enters at 200°C and leaves at 100°C. The cold fluid enters at 50°C and leaves at 200°C. The capacity ratio is ________ .

79. In a counter flow heat exchanger the cold fluid enters at 50°C and leaves at 100°C. The capacity ratio is 0.3333. If the hot fluid enters at 200°C and if the hot fluid has the minimum heat capacity it will leave at ________ .

80. The emissive power of a surface is 49000 W/m^2. The intensity of radiation of that surface will be ________ .

81. A black surface is at 1000 K. The fraction of radiation upto a wavelength of 5 m will b ___________ .

82. The fraction of radiation emitted by a black surface upto a wavelength of 10 m wa 0.6337. The temperature of the surface will be ___________ .

83. If the fraction of radiation emitted by a black surface in the wave band 0 to 4 m wa 0.8563 then the surface temperature will be ___________ .

84. At a temperature of 1000 K the monochromatic emissive power will be maximum at th wavelength ___________ .

85. The maximum monochromatic emissive power of a surface was found to occur at a wavelength of 2 μm. The surface temperature should be ___________ .

86. When the surface temperature was increased from 500 K to 1000 K the wavelength a which the monochromatic emissive power is maximum will shift from ________ μm t ________ μm.

87. The shape factor from a completely enclosing surface whose area is 8 m^2 to the enclose surface whose area is 4 m^2 is ___________ .

88. In a cylindrical furnace the shape factor from the base to the top is 0.31. The shape factor from the top to the curved surface will be ___________ .

89. The diameter and the height of a cylindrical furnace are both 2.0 m. The shape facto from the base to the top is 0.18. The shape factor from the curved surface to itself will be ______ .

90. The shape factor from the hemispherical surface to half the area of the enclosing base surface is ___________ .

91. For a right circular cone of slant length equal to the base diameter the shape factor to itself is ___________ .

92. Two long concentric cylinders are of 0.1 m and 0.2 m diameter. The shape factor from the outer to the inner cylinder and to itself are ___________ .

93. A spherical vessel of 1.2 m diameter encloses another concentric spherical vessel of diameter 1.0 m storing cryogenic liquid. The shape factor of the outer vessel to the inner vessel and to itself are ___________ .

94. Two large planes both having an emissivity of 0.5 are parallel to each other. The resistance for radiation heat exchange between them based on 1 m^2 area is ___________ .

95. A radiation shield with emissivity of 0.05 on both sides is placed between two large black parallel black planes. The thermal resistance is ___________ .

96. The shape factor from a surface of 1 m^2 area at 1000 K to another surface is 0.2. The energy radiated by this surface reaching the other surface is ___________ .

97. The emissive power of a surface is 3543.75 W/m^2. The radiosity of the surface is 1451.52 W/m^2. If the emissivity of the surface was 0.5 then the heat flow out of the surface is ___________ .

98. The radiosity of surfaces 1 and 2 are 3543.75 and 1451.52 W/m^2. The shape factor F_{1-2} is 0.5 and the area of the surface 1 is 2.0 m^2. The heat transfer between the surfaces is ___________ .

99. The volume fraction of N_2 and O_2 at surfaces 0.1 m apart are 10% and 90% and 90% and 10% respectively. The diffusion coefficient is 20.6 × 10^{-6} m^2/s. If the total pressure is 1 atm and temperature is 300 K then the diffusion rate of Oxygen and Nitrogen are ___________ .

100. In a flow involving both heat and mass transfer the convection coefficient was 20 W/m^{2}K. Lewis number is 0.85. c_p = 1005 J/kgK. Density is 1.2 kg/m^3. The value of mass transfer coefficient is __________ .

ANSWERS

Fill in The Blanks

1. temperature
2. increase
3. decrease
4. decrease
5. increase
6. increase
7. lattice vibration
8. free electron flow
9. W/mK
10. K/W
11. W/m^{2}K
12. 1/hA
13. L/kA
14. lower
15. increase
16. increased
17. lower
18. lower
19. constant
20. logarithmic
21. 2nd degree curve
22. mean temperature
23. lower
24. Silver
25. powder
26. heat penetration
27. nearer
28. parabolic
29. square of thickness, heat generation rate
30. inversely
31. heat generation rate, square of outer radius, conductivity
32. center
33. more
34. more
35. longer
36. conduction-convection
37. increase, decrease
38. effectiveness
39. greater
40. more
41. increases
42. tip
43. 0.5
44. contact
45. Bi < 0.1
46. higher
47. small
48. low
49. convection coefficient
50. very small
51. small
52. increased
53. semi infinite solid
54. sphere
55. $\Delta x^2/\alpha\Delta\tau = 2$
56. infinite
57. condenser
58. boundary layer
59. decrease
60. gradient
61. decrease
62. constant
63. mean
64. molecular
65. macroscopic
66. macroscopic
67. one
68. Prandtl
69. Reynolds
70. $\left(\dfrac{5x}{\text{Re}^{0.5}}\right)$
71. $\text{Re}_x^{-0.2}$
72. thicker
73. thinner
74. – 0.5
75. – 0.2
76. zero
77. increase
78. 5×10^5
79. $\left(\dfrac{4Gd}{\mu}\right)$
80. Stanton
81. reduce
82. diameter
83. $4A/P$
84. Hydraulic mean diameter
85. entry
86. constant
87. higher
88. higher
89. constant
90. friction factor
91. liquid metals
92. buoyant force
93. $1/T(K)$
94. wall and boundary layer thickness
95. third
96. second
97. Grashof number
98. $\left(\dfrac{4}{3}h_L\right)$
99. Grashof, Prandtl
100. higher

101. Reynolds, Prandtl and D/L **102.** Reynolds, Prandtl **103.** free
104. Gr/Re^2 **105.** surface, saturation **106.** radiation
107. bubble wettability **108.** third **109.** 15°C **110.** does not
111. $P/P_{critical}$ **112.** annular **113.** nucleate **114.** film
115. vapour film **116.** Boiler **117.** high
118. excess temperature **119.** linear **120.** second order
121. {(4/3 h_L} **122.** lower **123.** non wetting **124.** higher
125. gravity force **126.** 0.75 **127.** 0.5 **128.** 0.25
129. $4G/\mu$ **130.** higher **131.** 1800 **132.** ice layer
133. heat capacities **134.** lower **135.** $U_i A_i$ **136.** fouling
137. shell side **138.** tube side **139.** pinch point **140.** UA/C_{min}
141. $T_{h1} - T_{h2} /(T_{c2} - T_{c1})$ **142.** Parallel, Counter, Cross
143. finned **144.** counter flow **145.** condensers, evaporators
146. deterioration **147.** too long **148.** mass flow rate, specific heat
149. the same **150.** equal, of opposite sign
151. all four temperatures **152.** NTU method
153. effectiveness-NTU **154.** $1 - e^{-NTU}$ **155.** $1 - e^{-NTU}$
156. N/(N + 1) + 1 **157.** $1 - e^{-2N} /2$ **158.** actual heat flow/max possible heat flow
159. decrease **160.** need not **161.** regenerative
162. regenerative type **163.** gas to gas **164.** cold/fluid **165.** hot fluid
166. gas **167.** parallel flow **168.** 1 to 100 μm **169.** 0.3 to 0.7 μm
170. emissive power **171.** monochromatic emissive power **172.** Kirchhoff
173. Wien's displacement **174.** increases and then decrease
175. emissivity **176.** black **177.** $E = \Pi I$ **178.** intensity
179. gray surface **180.** band **181.** short, 0-4 μm **182.** longer
183. short **184.** transmitting, cutting off **185.** regular pattern
186. one **187.** low **188.** temperature, wavelength
189. short, low **190.** surface **191.** band **192.** volume
193. transparent **194.** pressure, characteristic length and temperature
195. $1- e^{aL}$ **196.** varying **197.** low **198.** near
199. Radiosity **200.** irradiation **201.** reflected radiation
202. reflected radiation by them **203.** shape factor $F_{1\text{-}2}$ **204.** increase
205. increase **206.** independent **207.** one **208.** themselves
209. 0.5 **210.** 0.5 **211.** one **212.** $1/n + 1$
213. high **214.** emissivity **215.** shape factor **216.** $A_1F_{1-2} = A_2F_{2-1}$
217. Non absorbing reradiating surfaces **218.** improves **219.** concentration
220. diffusion, convective **221.** conduction **222.** m^2/s
223. Schmidt number **224.** equimolal counter diffusion, diffusion into a stationary medium

225. larger   226. diffusion as molecules   227. pressure

228. m/s   229. $h_m x/D$ 230. $Le^{2/3}$

231. Transient conduction problems

True or False

1 False	2. True	3. False	4. False
5. False	6. False	7. False	8. False
9. True	10. False	11. True	12. True
13. False, should be less than 0.1		14. False, should be low	
15. False	16. True	17. True	18. False
19. True	20. True	21. True	22. False
23. True	24. True	25. False	26. True
27. True	28. True	29. False	30. True
31. True	32. False	33. False	34. False
35. False	36. True	37. True	38. True
39. False	40. True	41. True	42. False
43. False	44. False	45. False	46. True
47. False	48. True	49. False	50. False
51. True	52. False	53. True	54. True
55. True	56. False	57. True	58. True
59. False	60. True	61. False	62. True
63. True	64. True	65. False	66. False
67. True	68. False	69. False	70. False
71. True	72. False	73. False	74. True
75. False	76. True	77. True	78. True
79. False	80. True	81. False	82. False

Short Problems

1. 100 W	2. 0.2 m	3. 80°C	4. 2.5 W/mK
5. – 100°C/m	6. 0.25 W/mK	7. – 125°C/m	8. – 100°C/m
9. 160°C	10. 80°C	11. 0.5	12. 2/3
13. 0.2 m	14. *n*th	15. $\{1 - (r_o/r_i\} = c\{1 - (r_{o2}/r_i\}$	
16. 40 W/mK	17. 4	18. 8 times	19. 100%
20. 8 cm	21. 20 cm	22. half	23. 4 min
24. 9 minutes	25. 10s	26. 30°C	27. 20°C
28. 300°C	29. 0.1 m	30. 260°C	31. 225°C
32. 4 times	33. 10 seconds	34. 200 sec	35. 325°C
36. 5 mm	37. 0.256 m	38. 0.2 m	39. 1.25 m
40. 100	41. 5 W/m^2K	42. 5 W/m^2K	43. 8.3125×10^{-3}

44. 10 mm
45. $4.64\, x/\text{Re}^{0.5}$
46. 0.04
47. 2000
48. 16 W/m^2K
49. 2861
50. 20 W/m^2K
51. 0.0577 m
52. 0.109 m
53. 16×10^6
54. – 162500°C/m
55. 20 m/s
56. counter flow, 70°C
57. 1.25×10^{-4} m^2 K/W
58. 118.88, 115.42
59. 0.3333
60. 0.491
61. 0.66666
62. 0.777
63. 0.942
64. 1.5
65. 0.693
66. 2.86
67. 58984 W/K
68. 0.706, 83561 W/K
69. 9363 W
70. – 17.4%
71. – 33.3%
72. 0.375
73. 0.5
74. 0.633
75. 150°C
76. 80°C
77. 100°C
78. 0.66667
79. 50°C
80. 15597 W/str
81. 0.6337 Note : refer table.
82. 500 K Note : Refer table
83. 2000 K
84. 2.8976 μm
85. 1448.2 K
86. 5.7952, 2.8976
87. 0.5
88. 0.69
89. 0.59
90. 0.25
91. 0.5
92. 0.5, 0.5
93. 0.6944, 0.3056
94. 3.0 K/W
95. 40 K/W
96. 11340 W
97. 2092.23 W
98. 2092.23 W
99. 2.11×10^{-10} kg/m^2s, 1.85×10^{-10} kg/m^2s
100. 0.0185 m/s

Appendix B

Property Values of Metals at 20°C

Metal	*Density* ρ *kg/m³*	*Thermal Diffusivity* $\alpha \times 10^6$ *m²/s*	*Specific Heat* c *J/kgK*	*Thermal Conductivity* k *W/mK*
Aluminium, Pure	2707	94.44	896	204.2
Steel 0.5% carbon	7833	14.72	465	53.6
Nickel Steel 20% Ni	7983	5.28	461	19.1
Chrome Steel 20% Cr	7689	6.67	461	22.5
Constantan 60% Cu	8922	6.11	410	22.7
Magnesium, pure	1746	96.94	1013	171.3
Nickel, pure	8906	22.78	444	90.0
Nickel Chrome	8666	4.44	444	17.2
Tungsten Steel 10%W	8313	13.61	419	48.5
Copper, pure	8954	12.22	381	386.0
Bronze	8666	8.61	343	25.9
Brass	8522	33.89	385	110.7
Silver, pure	10524	165.56	235	406.8
Tungsten	19350	62.78	134	162.7
Zinc, pure	7144	41.11	385	112.1
Tin, pure	7304	38.61	226	64.1
Cr Ni steel 15%Cr 10%Ni	7865	5.28	461	19.1

1 W/mK = 0.86 kcal/m hr°C, 1 J/kgK = 238.9 × 10^{-6} kcal/kg°C.

Property Values of Elements at 20°C

Element	*Density* ρ *kg/m³*	*Thermal Diffusivity* $\alpha \times 10^6$ *m²/s*	*Specific Heat* c *J/kgK*	*Thermal Conductivity* k *W/mK*
Berylium	1840	50.97	1675	157
Boron	2500	10.90	1047	28.6
Cadmium	8660	46.67	230	93
Carbon (graphite)	1700	122.22	670	116.3

(Contd.)

Property Values of Elements at 20°C *(Contd.)*

	– 2300			– 174.5
Chromium	7150	218.06	448	69.8
Cobalt	8800	17.64	448	69.8
Gold	19300	127.00	129	317.0
Lithium	5340	40.28	3308	68.6
Molybdenum	10200	54.44	253	140.7
Platinum	21460	24.58	132	69.8
Potassium	870	155.56	737	100.0
Rhodium	12450	48.6	248	150.0
Sodium	975	94.44	1197	109.3
Silicon	2330	93.4	703	153
Thorium	11700	39.17	118	54.0
Uranium	19100	12.70	113	27.4
Vanadium	5900	11.94	496	34.9
Titanium	4540	6.22	532	15.12
Zirconium	6570	12.50	278	22.7

1 W/mK = 0.86 kcal/m hr°C, 1 J/kgK = 238.9×10^{-6} kcal/kg°C

Property Values of Insulating Materials

Material	***Temp-erature*** t **°C**	***Density*** ρ ***kg/m³***	***Thermal Diffusivity*** $a \times 10^6$ ***m²/s***	***Specific Heat*** c ***J/kgK***	***Thermal Conductivity*** $k \times 10^3$ ***W/mK***
Asbestos Fibre	50	470	0.29	816	1105
Asphalt	20	2110	0.16	2093	697.8
Chrome brick	200	3000	0.92	840	2320
Concrete	20	2300	0.49	1130	1279
Cork, plate	30	190	0.12	1884	41.9
Glass	20	2500	0.44	670	744.3
Glass wool	20	200	0.28	670	37.2
Ice	0	920	81.08	2261	2250
Magnesia 85%	100	216	–	–	67.5
Mineral wool	50	200	0.25	921	46.5
Oak, across grain	20	800	0.15	1759	207.0
Porcelain	95	2400	0.40	1089	1035
Quartz, along grain	0	2500 – 2800	3.33	837	7211
Sheet asbestos	30	770	0.20	816	1163
Granite	–	2630	1.37	775	2.79

1 W/mK = 0.86 kcal/m hr°C, 1 J/kgK = 238.9×10^{-6} kcal/kg°C

Property Values of Water in Saturated State

Temperature t °C	Density kg/m³	Kinematic Viscosity $v \times 10^6$ m²/s	Thermal Diffusivity $a \times 10^6$ m²/s	Prandtl Number Pr	Specific Heat c J/kgK	Thermal Conductivity k W/mK
0	1002	1.788	0.1308	13.600	4216	0.5524
20	1000	1.006	0.1431	7.020	4178	0.5978
40	995	0.657	0.1511	4.340	4178	0.6280
60	985	0.478	0.1553	3.020	4183	0.6513
80	974	0.364	0.1636	2.220	4195	0.6687
100	961	0.293	0.1681	1.740	4216	0.6804
120	945	0.247	0.1708	1.446	4250	0.6850
140	928	0.213	0.1725	1.241	4283	0.6838
160	909	0.189	0.1728	1.099	4342	0.6804
180	889	0.173	0.1725	1.044	4417	0.6757
200	867	0.160	0.1701	0.937	4505	0.6652
220	842	0.149	0.1681	0.891	4610	0.6524
240	815	0.143	0.1639	0.871	4756	0.6350
260	786	0.137	0.1578	0.874	4949	0.6106
280	752	0.135	0.1481	0.910	5208	0.5803
300	714	0.135	0.1325	1.019	5728	0.5390

β = (change in density/change in temp.) (1/density)

μ = density × kinematic viscosity

1 W/mK = 0.86 kcal/m hr°C, 1 J/kgK = 238.9×10^{-6} kcal/kg°C

Property Values of Dry Air at One Atm. Pressure

Temperature t °C	*Density* kg/m³	*Coefficient of Viscosity* $\mu \times 10^6$ Ns/m²/s	*Kinematic Viscosity* $v \times 10^6$ m²/s	*Thermal Diffusivity* $\alpha \times 10^6$ m²/s	*Prandtl Number* Pr	*Specific Heat* c J/kgK	*Thermal Conductivity* k W/mK
– 50	1.584	14.61	9.23	12.644	0.728	1013	0.02035
– 40	1.515	15.20	10.04	13.778	0.728	1013	0.02117
– 30	1.453	15.69	10.80	14.917	0.723	1013	0.02198
– 20	1.395	16.18	11.61	16.194	0.716	1009	0.02279
– 10	1.342	16.67	12.43	17.444	0.712	1009	0.02361
0	1.293	17.16	13.28	18.806	0.707	1005	0.02442
10	1.247	17.65	14.16	20.006	0.705	1005	0.02512
20	1.205	18.14	15.06	21.417	0.703	1005	0.02593
30	1.165	18.63	16.00	22.861	0.701	1005	0.02675
40	1.128	19.12	16.96	24.306	0.699	1005	0.02756
50	1.093	19.61	17.95	25.722	0.698	1005	0.02826
60	1.060	20.10	18.97	27.194	0.696	1005	0.02966
70	1.029	20.59	20.02	28.556	0.694	1009	0.03047
80	1.000	21.08	21.09	30.194	0.692	1009	0.03074
90	0.972	21.48	22.10	31.889	0.690	1009	0.03128
100	0.946	21.87	23.13	33.639	0.688	1009	0.03210
120	0.898	22.85	25.45	36.833	0.686	1009	0.03338
140	0.854	23.73	27.80	40.333	0.684	1013	0.03489
160	0.815	24.52	30.09	43.894	0.682	1017	0.03640
180	0.779	25.30	32.49	47.500	0.681	1022	0.03780
200	0.746	25.99	34.85	51.361	0680	1026	0.03931
250	0.674	27.36	40.61	58.500	0.677	1038	0.04268
300	0.615	29.71	48.20	71.556	0.674	1047	0.04605

1 W/mK = 0.86 kcal/mkg°C, 1 J/kgK = 238.9×10^{-6} kcal/kg°C

1 Ns/m² = 0.102 kgf/m², $\beta = \frac{1}{T}$, T in K

Values of Error Function

x	erf(x)	x	erf(x)	x	erf(x)	x	erf(x)	x	erf(x)	x	erf(x)	x	erf(x)
0.00	0.00000	0.35	0.37938	0.70	0.67780	1.05	0.86244	1.40	0.95228	1.75	0.98667	2.20	0.998137
0.01	0.01128	0.36	0.38933	0.71	0.68467	1.06	0.86614	1.41	0.95385	1.76	0.98719	2.22	0.998308
0.02	0.02256	0.37	0.39921	0.72	0.69143	1.07	0.86977	1.42	0.95538	1.77	0.98769	2.24	0.998464
0.03	0.03384	0.38	0.40901	0.73	0.69810	1.08	0.87333	1.43	0.95686	1.78	0.98817	2.26	0.998607
0.04	0.04511	0.39	0.41874	0.74	0.70468	1.09	0.87680	1.44	0.95830	1.79	0.98864	2.28	0.998738
0.05	0.05637	0.40	0.42839	0.75	0.71116	1.10	0.88020	1.45	0.95970	1.80	0.98909	2.30	0.998857
0.06	0.06762	0.41	0.43797	0.76	0.71754	1.11	0.88353	1.46	0.96105	1.81	0.98952	2.32	0.998966
0.07	0.07886	0.42	0.44747	0.77	0.72382	1.12	0.88679	1.47	0.96237	1.82	0.98994	2.34	0.999065
0.08	0.09008	0.43	0.45689	0.78	0.73001	1.13	0.88997	1.48	0.96365	1.83	0.99035	2.36	0.999155
0.09	0.10128	0.44	0.46622	0.79	0.73610	1.14	0.89308	1.49	0.96490	1.84	0.99074	2.38	0.999237
0.10	0.11246	0.45	0.47548	0.80	0.74210	1.15	0.89612	1.50	0.96610	1.85	0.99111	2.40	0.999311
0.11	0.12362	0.46	0.48466	0.81	0.74800	1.16	0.89910	1.51	0.96728	1.86	0.99147	2.42	0.999379
0.12	0.13476	0.47	0.49374	0.82	0.75381	1.17	0.90200	1.52	0.96841	1.87	0.99182	2.44	0.999441
0.13	0.14587	0.48	0.50275	0.83	0.75952	1.18	0.90484	1.53	0.96952	1.88	0.99216	2.46	0.999497
0.14	0.15695	0.49	0.51167	0.84	0.76514	1.19	0.90761	1.54	0.97059	1.89	0.99248	2.48	0.999547
0.15	0.16800	0.50	0.52050	0.85	0.77067	1.20	0.91031	1.55	0.97162	1.90	0.99279	2.50	0.999593
0.16	0.17901	0.51	0.52924	0.86	0.77610	1.21	0.91296	1.56	0.97263	1.91	0.99309	2.55	0.999689
0.17	0.18999	0.52	0.53790	0.87	0.78144	1.22	0.91553	1.57	0.97360	1.92	0.99338	2.60	0.999764
0.18	0.20094	0.53	0.54646	0.88	0.78669	1.23	0.91805	1.58	0.97455	1.93	0.99366	2.65	0.999822
0.19	0.21184	0.54	0.55494	0.89	0.79184	1.24	0.92050	1.59	0.97546	1.94	0.99392	2.70	0.999866
0.20	0.22270	0.55	0.56332	0.90	0.79691	1.25	0.92290	1.60	0.97635	1.95	0.99418	2.75	0.999899
0.21	0.23352	0.56	0.57162	0.91	0.80188	1.26	0.92524	1.61	0.97721	1.96	0.99443	2.80	0.999925
0.22	0.24430	0.57	0.57982	0.92	0.80677	1.27	0.92751	1.62	0.97804	1.97	0.99466	2.85	0.999944
0.23	0.25502	0.58	0.58792	0.93	0.81156	1.28	0.92973	1.63	0.97884	1.98	0.99489	2.90	0.999959
0.24	0.26570	0.59	0.59594	0.94	0.81627	1.29	0.93190	1.64	0.97962	1.99	0.99511	2.95	0.999970
0.25	0.27633	0.60	0.60386	0.95	0.82089	1.30	0.93401	1.65	0.98038	2.00	0.995322	3.00	0.999978
0.26	0.28690	0.61	0.61168	0.96	0.82542	1.31	0.93606	1.66	0.98110	2.02	0.995720	3.20	0.999994
0.27	0.29742	0.62	0.61941	0.97	0.82987	1.32	0.93806	1.67	0.98181	2.04	0.996086	3.40	0.999998
0.28	0.30788	0.63	0.62705	0.98	0.83423	1.33	0.94002	1.68	0.98249	2.06	0.996424	3.60	1.000000
0.29	0.31828	0.64	0.63459	0.99	0.83851	1.34	0.94191	1.69	0.98315	2.08	0.996734		
0.30	0.32863	0.65	0.64203	1.00	0.84270	1.35	0.94376	1.70	0.98379	2.10	0.997020		
0.31	0.33891	0.66	0.64938	1.01	0.84681	1.36	0.94556	1.71	0.98441	2.12	0.997284		
0.32	0.34913	0.67	0.65663	1.02	0.85084	1.37	0.94731	1.72	0.98500	2.14	0.997525		
0.33	0.35928	0.68	0.66378	1.03	0.85478	1.38	0.94902	1.73	0.98558	2.16	0.997747		
0.34	0.36936	0.69	0.67084	1.04	0.85865	1.39	0.95067	1.74	0.98613	2.18	0.997951		

References

Dunn, P.D. and Reay, D.A. (1994). *Heat Pipes,* Pergamon Press.

Eckert, E.R.G. and Drake, R.M. (1959). *Heat and Mass Transfer*, McGraw-Hill.

Gilbert M. Masters, (2004). *Introduction to Environmental Science,* Prentice Hall.

Holman, P.J. (2010, 9th edition). *Heat Transfer*, McGraw-Hill.

Incropera, F.P. and Dewitt, D.P. (2006)., *Introduction to Heat Transfer,* Fourth Edition, John Wiley and Sons.

John, H. Lienhard IV and John H. Lienhard V (2004). *Heat Transfer Text Book.*

Kern, D.Q. (1950). *Process Heat Transfer*, McGraw-Hill.

Kreith, F. and Bohn, M.S. (1997). *Principles of Heat Transfer,* Fifth Edition, PWS Publishing Co. and International Thomson Publishing Co.

Kutateladze, S.S. and Borishankii, V.M. (1966). *A Concise Encyclopedia of Heat Transfer,* Pergamon Press.

McAdams, W.H. (1954). *Heat Transmission,* McGraw-Hill.

Mills, A.F., (1995). *Heat and Mass Transfer*, Richard. D. Irwin Inc.

Mnozisik, (1985). *Heat Transfer–A Basic Approach*

Reay, D.A. and Kew, P.A. (2006). *Heat Pipes,* Elsevier.

Rohsenow, W.M. et al., (1985). *Hand Book of Heat Transfer Applications and Hand Book of Heat Transfer Fundamentals*, McGraw-Hill.

Schneider, P.J. (1957). *Conduction Heat Transfer*, Addison-Wesley Publishing Co.

Spalding, D.B. (1963). *Convective Mass Transfer: An Introduction,* Edward Arnold.

Streeter, V.L. (1983). *Fluid Mechanics*, McGraw-Hill.

Thomas, L.C. (1980). *Fundamentals of Heat Transfer,* Prentice Hill.

Treybal, R.E. (1995). *Mass Transfer Operations*, McGraw-Hill.

W.M. Keys and A.L., *London (3rd ed. 1984), Compact heat Exchangers* McGraw Hill.

Wrangham, D.A. (1961). *The Elements of Heat Flow*, Chatto and Windus.

Yunus, A.Cengel, (2002). *Heat Transfer a Practical Approach.*

Index